Biodegradable Polymers, Blends and Composites

Woodhead Publishing Series in Composites Science and Engineering

Biodegradable Polymers, Blends and Composites

Edited by

Sanjay Mavinkere Rangappa

Jyotishkumar Parameswaranpillai

Suchart Siengchin

M. Ramesh

Woodhead Publishing is an imprint of Elsevier
The Officers' Mess Business Centre, Royston Road, Duxford, CB22 4QH, United Kingdom
50 Hampshire Street, 5th Floor, Cambridge, MA 02139, United States
The Boulevard, Langford Lane, Kidlington, OX5 1GB, United Kingdom

Library of Congress Cataloging-in-Publication Data
A catalog record for this book is available from the Library of Congress

British Library Cataloguing-in-Publication Data
A catalogue record for this book is available from the British Library

ISBN: 978-0-12-823791-5

For information on all Woodhead Publishing publications visit our website at https://www.elsevier.com/books-and-journals

Publisher: Matthew Deans
Acquisitions Editor: Gwen Jones
Editorial Project Manager: Joshua Mearns
Production Project Manager: Swapna Srinivasan
Cover Designer: Victoria Pearson

Typeset by TNQ Technologies

Contents

Contributors

V. Abinaya Department of Mechanical Engineering, PSG Institute of Technology and Applied Research, Coimbatore, Tamil Nadu, India

Radhamanohar Aepuru Materials Engineering Division, Defence Institute of Advanced Technology, Pune, Maharashtra, India; Central Institute of Petrochemicals Engineering and Technology (CIPET): Institute of Plastics Technology (IPT), Kochi, Kerala, India; Departamento de Ingeniería Mecánica, Facultad de Ingeniería, Universidad Tecnológica Metropolitana, Santiago, Chile

Anna Rafaela Cavalcante Braga Department of Chemical Engineering, Universidade Federal de São Paulo (UNIFESP), Diadema, Brazil

Monize Burck Department of Biosciences, Universidade Federal de São Paulo (UNIFESP), Santos, São Paulo, Brazil

Jesús María Frías Celayeta Dublin Institute of Technology, Environmental Sustainability and Health Institute (EHSI), College of Sciences and Health, Dublin, Ireland

Pijush Kanti Chattopadhyay Department of Leather Technology, Government College of Engineering and Leather Technology (Post-Graduate), Maulana Abul Kalam Azad University of Technology, Kolkata, West Bengal, India

Naga Srilatha Cheekuramelli Central Institute of Petrochemicals Engineering and Technology (CIPET): Institute of Plastics Technology (IPT), Kochi, Kerala, India; Materials Engineering Division, Defence Institute of Advanced Technology, Pune, Maharashtra, India; Departamento de Ingeniería Mecánica, Facultad de Ingeniería, Universidad Tecnológica Metropolitana, Santiago, Chile

Kraipat Cheenkachorn Department of Chemical Engineering, Faculty of Engineering, King Mongkut's University of Technology North Bangkok, Bangkok, Bangkok, Thailand

Yu-Shen Cheng Department of Chemical and Materials Engineering, National Yunlin University of Science and Technology, Douliu, Yunlin, Yunlin, Taiwan

Santi Chuetor Department of Chemical Engineering, Faculty of Engineering, King Mongkut's University of Technology North Bangkok, Bangkok, Bangkok, Thailand

Jorge Alberto Vieira Costa College of Chemistry and Food Engineering, Federal University of Rio Grande (FURG), Rio Grande, RS, Brazil

Mousumi Deb Advanced Polymer Laboratory, Department of Polymer Science and Technology, Government College of Engineering and Leather Technology (Post-Graduate), Maulana Abul Kalam Azad University of Technology, Kolkata, West Bengal, India

B.D.S. Deeraj Department of Chemistry, Indian Institute of Space Science and Technology, Kerala, Thiruvananthapuram, India

Mariana Buranelo Egea Goiano Federal Institute of Education, Science and Technology, Rio Verde, Goiás, Brazil

Antônio Gilberto Ferreira Federal University of São Carlos (UFSCar), Nuclear Magnetic Resonance Laboratory, São Carlos, Brazil

Michele Giaconia Department of Biosciences, Universidade Federal de São Paulo (UNIFESP), Santos, São Paulo, Brazil

R.K. Gond Department of Mechanical Engineering, Motilal Nehru National Institute of Technology, Allahabad, Prayagraj, Uttar Pradesh, India

M.K. Gupta Department of Mechanical Engineering, Motilal Nehru National Institute of Technology, Allahabad, Prayagraj, Uttar Pradesh, India

Nishar Hameed Factory of the Future, Swinburne University of Technology, Hawthorn, VIC, Australia

Naman Jain Department of Mechanical Engineering, Meerut Institute of Engineering and Technology, Meerut, Uttar Pradesh, India

Farah Hidayah Jamaludin School of Biomedical Engineering and Health Sciences, Faculty of Engineering, Universiti Teknologi Malaysia, Skudai, Johor, Malaysia

Aswathy Jayakumar School of Biosciences, Mahatma Gandhi University, Kottayam, Kerala, India; Department of Materials and Production Engineering, The Sirindhorn International Thai–German Graduate School of Engineering (TGGS), King Mongkut's University of Technology North Bangkok, Bangkok, Thailand

Jitha S. Jayan Department of Chemistry, School of Arts and Sciences, Amrita Vishwa Vidyapeetham, Amritapuri, Kerala, Kollam, India

Kuruvilla Joseph Department of Chemistry, Indian Institute of Space Science and Technology, Kerala, Thiruvananthapuram, India

Yong Chae Jung Institute of Advanced Composite Materials, Korea Institute of Science and Technology (KIST), Wanju-gun, Jeonbuk, Republic of Korea

Young Nam Kim Institute of Advanced Composite Materials, Korea Institute of Science and Technology (KIST), Wanju-gun, Jeonbuk, Republic of Korea; Department of Chemical and Biomolecular Engineering, Yonsei University, Seodaemun-gu, Seoul, Republic of Korea

A.V. Kiruthika Seethalakshmi Achi College for Women, Karaikudi, Tamil Nadu, India

Sukhila Krishnan Sahrdaya College of Engineering and Technology, Department of Applied Science and Humanities, Kodakara, Thrissur, Kerala, India

Senthilkumar Krishnasamy Department of Materials and Production Engineering, The Sirindhorn International Thai–German Graduate School of Engineering (TGGS), King Mongkut's University of Technology North Bangkok, Bangkok, Thailand

Sudheer Kumar School for Advanced Research in Petrochemicals (SARP), Laboratory for Advanced Research in Polymeric Materials (LARPM), Central Institute of Petrochemicals Engineering & Technology (CIPET), Bhubaneswar, Odisha, India

Suelen Goettems Kuntzler College of Chemistry and Food Engineering, Federal University of Rio Grande (FURG), Rio Grande, RS, Brazil

Maria Carolina Bezerra Di-Medeiros Leal Federal University of São Carlos (UFSCar), Nuclear Magnetic Resonance Laboratory, São Carlos, Brazil

Xiau Yeen Lee Centre for Advanced Materials, Tunku Abdul Rahman University College (TARUC), Setapak, Kuala Lumpur, Malaysia

Ailton Cesar Lemes Department of Biochemical Engineering, Universidade Federal do Rio de Janeiro (UFRJ), Technological Center, School of Chemistry, Ilha do Fundão, Rio de Janeiro, Brazil; Federal University of Rio de Janeiro (UFRJ), School of Chemistry, Department of Biochemical Engineering, Rio de Janeiro, Brazil

Ravi Prakash Magisetty Materials Engineering Division, Defence Institute of Advanced Technology, Pune, Maharashtra, India; Central Institute of Petrochemicals Engineering and Technology (CIPET): Institute of Plastics Technology (IPT), Kochi, Kerala, India; Departamento de Ingeniería Mecánica, Facultad de Ingeniería, Universidad Tecnológica Metropolitana, Santiago, Chile

Abra Mathew Material Sciences and Technology Division, CSIR-National Institute for Interdisciplinary Science and Technology, Thiruvananthapuram, Kerala, India

Sanjay Mavinkere Rangappa Natural Composite Research Group, Department of Materials and Production Engineering, The Sirindhorn International Thai–German Graduate School of Engineering (TGGS), King Mongkut's University of Technology North Bangkok, Bangkok, Thailand

Nyak Syazwani Nyak Mazlan Bioresource and Biorefinery Research Group, Faculty of Science and Technology, Universiti Kebangsaan Malaysia, Bangi, Selangor, Malaysia

C.D. Midhun Dominic Department of Chemistry, Sacred Heart College, Cochin, Kerala, India

K. Madhu Mitha Department of Mechanical Engineering, PSG Institute of Technology and Applied Research, Coimbatore, Tamil Nadu, India

Smita Mohanty School for Advanced Research in Petrochemicals (SARP), Laboratory for Advanced Research in Polymeric Materials (LARPM), Central Institute of Petrochemicals Engineering & Technology (CIPET), Bhubaneswar, Odisha, India

Michele Greque de Morais College of Chemistry and Food Engineering, Federal University of Rio Grande (FURG), Rio Grande, RS, Brazil

Juliana Botelho Moreira College of Chemistry and Food Engineering, Federal University of Rio Grande (FURG), Rio Grande, RS, Brazil

Marhaini Mostapha Bioresource and Biorefinery Research Group, Faculty of Science and Technology, Universiti Kebangsaan Malaysia, Bangi, Selangor, Malaysia

Daniella Carisa Murador Department of Biosciences, Universidade Federal de São Paulo (UNIFESP), Santos, São Paulo, Brazil

M. Muthukrishnan Department of Mechanical Engineering, KIT—Kalaignarkarunanidhi Institute of Technology, Coimbatore, Tamil Nadu, India

Indu C. Nair Department of Biotechnology, Sahodaran Ayyappan Smaraka Sree Narayana Dharma Paripalana Yogam (SAS SNDPYOGAM), Konni College, Pathanamthitta, Kerala, India

Sanjay Kumar Nayak School for Advanced Research in Petrochemicals (SARP), Laboratory for Advanced Research in Polymeric Materials (LARPM), Central Institute of Petrochemicals Engineering & Technology (CIPET), Bhubaneswar, Odisha, India

Josemar Gonçalves de Oliveira Filho São Paulo State University (UNESP), School of Pharmaceutical Sciences, Araraquara, Brazil

Jyotishkumar Parameswaranpillai Department of Bioscience, Mar Athanasios College for Advanced Studies Thiruvalla (MACFAST), Pathanamthitta, Kerala, India

Alessandro Pegoretti University of Trento, Department of Industrial Engineering, Trento, Italy

G.L. Praveen Wimpey Laboratories, Wimpey Building, Al Quoz, Dubai, United Arab Emirates

E.K. Radhakrishnan School of Biosciences, Mahatma Gandhi University, Kottayam, Kerala, India

Sabarish Radoor Department of Mechanical and Process Engineering, The Sirindhorn International Thai—German Graduate School of Engineering (TGGS), King Mongkut's University of Technology North Bangkok, Bangkok, Thailand; Department of Materials and Production Engineering, The Sirindhorn International Thai—German Graduate School of Engineering (TGGS), King Mongkut's University of Technology North Bangkok, Bangkok, Thailand

G. Rajeshkumar Department of Mechanical Engineering, PSG Institute of Technology and Applied Research, Coimbatore, Tamil Nadu, India

L. Rajeshkumar Department of Mechanical Engineering, KPR Institute of Engineering and Technology, Coimbatore, Tamilnadu, India

Abhinay Rajput Material Sciences and Technology Division, CSIR-National Institute for Interdisciplinary Science and Technology, Thiruvananthapuram, Kerala, India

M. Ramesh Department of Mechanical Engineering, KIT−Kalaignarkarunanidhi Institute of Technology, Coimbatore, Tamil Nadu, India

Sergiana dos Passos Ramos Department of Biosciences, Universidade Federal de São Paulo (UNIFESP), Santos, São Paulo, Brazil

Maria Luiza Rezende Ribeiro Federal University of Goiás (UFG), Nuclear Magnetic Resonance Laboratory, Goiânia, Brazil

Gislane Oliveira Ribeiro Federal University of Goiás (UFG), Nuclear Magnetic Resonance Laboratory, Goiânia, Brazil

R. Ronia Richelle Department of Mechanical Engineering, PSG Institute of Technology and Applied Research, Coimbatore, Tamil Nadu, India

Daniele Rigotti University of Trento, Department of Industrial Engineering, Trento, Italy

Haniyeh Rostamzad Fisheries Department, Faculty of Natural Resources, University of Guilan, Sowmeh Sara, Guilan, Iran

Sushanta K. Sahoo Material Sciences and Technology Division, CSIR-National Institute for Interdisciplinary Science and Technology, Thiruvananthapuram, Kerala, India

Kushairi Mohd Salleh Bioresource and Biorefinery Research Group, Faculty of Science and Technology, Universiti Kebangsaan Malaysia, Bangi, Selangor, Malaysia

Sathyaraj Sankarlal Department of Mechanical Engineering, National Institute of Technology, Calicut, Kerala, India

Appukuttan Saritha Department of Chemistry, School of Arts and Sciences, Amrita Vishwa Vidyapeetham, Amritapuri, Kerala, Kollam, India

K. Sekar Department of Mechanical Engineering, National Institute of Technology, Calicut, Kerala, India

S. Arvindh Seshadri Department of Mechanical Engineering, PSG Institute of Technology and Applied Research, Coimbatore, Tamil Nadu, India

Suchart Siengchin Natural Composite Research Group, Department of Materials and Production Engineering, The Sirindhorn International Thai–German Graduate School of Engineering (TGGS), King Mongkut's University of Technology North Bangkok, Bangkok, Thailand

Cleber Klasener da Silva College of Chemistry and Food Engineering, Federal University of Rio Grande (FURG), Rio Grande, RS, Brazil

Harinder Singh Department of Chemical Engineering, Motilal Nehru National Institute of Technology, Allahabad, Prayagraj, Uttar Pradesh, India

Komal Singh Department of Mechanical Engineering, Govind Ballabh Pant University of Agriculture and Technology, Pantnagar, Uttarakhand, India

Vinay Kumar Singh Department of Mechanical Engineering, Govind Ballabh Pant University of Agriculture and Technology, Pantnagar, Uttarakhand, India

Nayan Ranjan Singha Advanced Polymer Laboratory, Department of Polymer Science and Technology, Government College of Engineering and Leather Technology (Post-Graduate), Maulana Abul Kalam Azad University of Technology, Kolkata, West Bengal, India

Malinee Sriariyanun Chemical and Process Engineering, The Sirindhorn International Thai–German Graduate School of Engineering, King Mongkut's University of Technology North Bangkok, Bangkok, Bangkok, Thailand

Aarsha Surendren Central Institute of Petrochemicals Engineering and Technology (CIPET): Institute of Plastics Technology (IPT), Kochi, Kerala, India; Materials Engineering Division, Defence Institute of Advanced Technology, Pune, Maharashtra, India; Departamento de Ingeniería Mecánica, Facultad de Ingeniería, Universidad Tecnológica Metropolitana, Santiago, Chile

Noor Izyan Syazana Mohd Yusoff School of Chemical and Energy Engineering, Faculty of Engineering, Universiti Teknologi Malaysia, Skudai, Johor, Malaysia; Centre for Advanced Composite Materials (CACM), Universiti Teknologi Malaysia, Skudai, Johor, Malaysia; School of Biomedical Engineering and Health Sciences, Faculty of Engineering, Universiti Teknologi Malaysia, Skudai, Johor, Malaysia; Advance Membrane Technology Research Centre (AMTEC), Universiti Teknologi Malaysia, Skudai, Johor, Malaysia

Prapakorn Tantayotai Department of Microbiology, Faculty of Science, Srinakharinwirot University, Bangkok, Bangkok, Thailand

Atthasit Tawai Chemical and Process Engineering, The Sirindhorn International Thai–German Graduate School of Engineering, King Mongkut's University of Technology North Bangkok, Bangkok, Bangkok, Thailand

Weng Hong Tham School of Chemical and Energy Engineering, Faculty of Engineering, Universiti Teknologi Malaysia, Skudai, Johor, Malaysia

Nisa V. Salim Factory of the Future, Swinburne University of Technology, Hawthorn, VIC, Australia

Bruna da Silva Vaz College of Chemistry and Food Engineering, Federal University of Rio Grande (FURG), Rio Grande, RS, Brazil

Akarsh Verma Department of Mechanical Engineering, University of Petroleum and Energy Studies, Dehradun, Uttarakhand, India

C. Vibha Department of Materials and Production Engineering, The Sirindhorn International Thai–German Graduate School of Engineering (TGGS), King Mongkut's University of Technology North Bangkok, Bangkok, Thailand

Mat Uzir Wahit School of Chemical and Energy Engineering, Faculty of Engineering, Universiti Teknologi Malaysia, Skudai, Johor, Malaysia; Centre for Advanced Composite Materials (CACM), Universiti Teknologi Malaysia, Skudai, Johor, Malaysia

Chunhong Wang School of Textile Science and Engineering, Tiangong University, Xiqing District, Tianjin, PR China

Tuck-Whye Wong School of Biomedical Engineering and Health Sciences, Faculty of Engineering, Universiti Teknologi Malaysia, Skudai, Johor, Malaysia; Advance Membrane Technology Research Centre (AMTEC), Universiti Teknologi Malaysia, Skudai, Johor, Malaysia

Nur Amira Zainul Armir Bioresource and Biorefinery Research Group, Faculty of Science and Technology, Universiti Kebangsaan Malaysia, Bangi, Selangor, Malaysia

Sarani Zakaria Bioresource and Biorefinery Research Group, Faculty of Science and Technology, Universiti Kebangsaan Malaysia, Bangi, Selangor, Malaysia

Acknowledgment

The work was supported by Thailand Science Research and Innovation Fund and King Mongkut's University of Technology North Bangkok (KMUTNB), Thailand with contract no. KMUTNB-FF-65-19.

Introduction to biodegradable polymers

1

M. Ramesh [1], *Sanjay Mavinkere Rangappa* [2], *Jyotishkumar Parameswaranpillai* [3] *and Suchart Siengchin* [2]
[1]Department of Mechanical Engineering, KIT—Kalaignarkarunanidhi Institute of Technology, Coimbatore, Tamil Nadu, India; [2]Natural Composite Research Group, Department of Materials and Production Engineering, The Sirindhorn International Thai–German Graduate School of Engineering (TGGS), King Mongkut's University of Technology North Bangkok, Bangkok, Thailand; [3]Department of Bioscience, Mar Athanasios College for Advanced Studies Thiruvalla (MACFAST), Pathanamthitta, Kerala, India

1. Introduction

In developing countries, environmental contamination caused by polymeric materials has reached high levels. Fossil fuel-derived polymers are not biodegradable, and their resistance to microbial degradation causes them to be stored in the environment. Furthermore, oil prices have recently risen dramatically. Evidence like this has fueled research in biodegradable polymers (Mochizuki & Hirami, 1997). These polymers are improved using techniques such as blending and composite forming, resulting in new blends with different properties such as high efficiency, strength, and good processability (Hamad et al., 2014; Nair & Laurencin, 2007; Liu & Zhang, 2011; Ramesh & Rajeshkumar, 2018). Biobased degradable plastics and polymers were first developed during the 1980s. Biodegradable polymers can be used in a variety of forms, from nondegradable to naturally degradable. These naturally degradable polymers can be obtained from renewable sources, while synthetic polymers are made from nonrenewable petroleum-based sources (Jo et al., 1992). Polylactic acid (PLA), polybutylene succinate (PBS), polybutylene succinate adipate (PBSA), polybutylene adipate co-terephthalate (PBAT), polycaprolactone (PCL), and thermoplastic starch (TPS) have all sparked interest in biodegradable polymers (Srithep et al., 2011). Biodegradable polymers have low viscosity, poor resilience, low modulus and thermal sensitivity, and high cost. To overcome these drawbacks, various methodologies such as copolymerization, compositing, and mixing are commonly used. Blending is the most popular among them, owing to its lower cost compared to the preparation of copolymers (Bhatia et al., 2007; Raghavan & Emekalam, 2001; Ramesh, Kumar et al., 2020). In Fig. 1.1, an overview of these polymers is provided based on their origin and production methods (Mochizuki & Hirami, 1997).

Biodegradable Polymers, Blends and Composites. https://doi.org/10.1016/B978-0-12-823791-5.00024-7

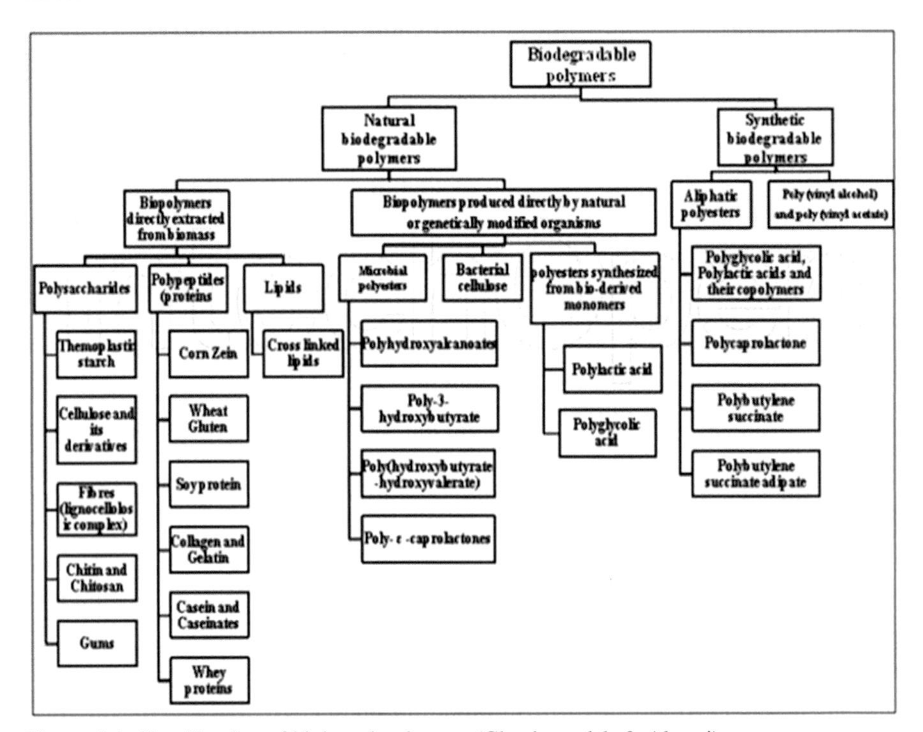

Figure 1.1 Classification of biobased polymers (Ghanbarzadeh & Almasi).

2. Biodegradable polymers

Biodegradation occurs as a result of the reaction of enzymes and/or chemical oxidation in living organisms. This biodegradation happens in two phases. The first phase is the fragmentation of polymers into lower molecular mass organisms by abiotic reactions such as oxidation, photodegradation, or hydrolysis, or biotic reactions such as micro-organism degradation. In the second phase the bioassimilation and mineralization of polymer particles takes place by microorganisms (Jo et al., 1992). For researchers and scientists, reducing the widespread use of petrochemical thermoplastics (polypropylene, polystyrene, polyethylene, polyvinyl chloride, and so on) and thermosets (epoxy, polyester, vinyl ester, phenols, and so on) has become a problem. Researchers and scientists are interested in developing biopolymers to replace petrochemical polymers in this context. The main difference between petrochemical polymers and biopolymers is their structure. Biopolymers have well-defined primary, secondary, and tertiary structures, whereas petrochemical polymers have repeated units called monomers. Starch is a naturally occurring biopolymer, while PLA and polyhydroxybutyrate (PHB) are two more widely used biopolymers.

2.1 Polyhydroxybutyrate

This is a biopolymer with a number of characteristics, including being insoluble in water, and having low moisture absorption and strong ultraviolet resistance. PHB, on the other hand, has lower thermal stability since its glass transition temperature is about 2°C. This also has a poor tolerance for acids and bases. It is suitable for medical use because it is biodegradable. PHB is the most common polymer in the polyhydroxyalkanoates (PHA) family, and it refers to the short-chain length PHA with monomers containing a number of carbon atoms. A biosynthetic method for the production of PHB has been developed due to the fermentation of sugar by the bacterium *Alcaligenes eutrophus*. The PHB homopolymer is highly crystalline in nature, exceedingly brittle, and moderately hydrophobic, much like other PHA homopolymers. As a result, the PHA homopolymers have in vivo degradation with respect to time (Mochizuki & Hirami, 1997).

2.2 Poly-lactic acid

This is a biodegradable polymer made from potato, sugarcane bagasse, maize, and other agricultural fermentation wastes. PLA is a lactic acid cyclic dimer made from D- or L-lactic acid polycondensation or lactide ring opening polymerization. The natural isomer is L-lactide, while the synthetic mixture is D-lactide (Ayala et al., 2009). PLA is a hydrophobic nature polymer due to the availability of $-CH_3$ groups. It is much more resistant to hydrolysis than polyglycolide due to the steric shielding effect of methyl groups. The average glass transition temperature for commercial PLA is 64°C, the elongation at break is around 31%, and the tensile load carrying capacity is about 32 MPa (Ray et al., 2005). It has excellent mechanical properties and is easy to process. However, it has some disadvantages, such as low impact strength, higher water absorption, and high brittleness. It has a wide variety of uses, including pharmaceuticals, packaging, textiles, and household items. PLA can be processed to improve chain mobility and allow the crystallization process. This process is done with oligomeric acid and citrate ester with low-molecular-weight polyethylene glycol (Ray & Bousmina, 2005). The rate of PLA degradation varies depending on the crystallinity index. PLA has a low degradation value when compared to polyglycolide, and so certain copolymers have been examined as bioresorbable orthopedic materials (Ramesh, Rajeshkumar, & Balaji, 2021). The biodegradation nature of PLA also can be improved by a grafting process. During this process L-lactide on chitosan was used to perform ring opening polymerization with a catalyst. The transition temperature and thermal stability of grafted copolymers rise as the grafting percentage rises. As the lactide content rises, the graft polymer's degradation decreases (Zhao et al., 2012).

2.3 Polyglycolide

Polyglycolide is the simple linear aliphatic polyester made by ring opening polymerization of a cyclic lactone and glycolide. It is crystalline in nature, with a crystallinity index of around 50%, and hence is insoluble in several organic solvents. Polyglycolide has a melting temperature of 220–225°C and the range of the glass transition temperature is 35–40°C (Xie et al., 2013). The polyglycolide has good mechanical strength. However, due to its poor solubility and high rate of acid-producing degradation, its biomedical applications are limited. As a result, caprolactone, lactide, or trimethylene carbonate glycolide copolymers have been developed for medical instruments (Nair & Laurencin, 2007; Ramesh & Rajeshkumar, 2021).

2.4 Natural rubber

Natural rubber has been used to prepare biocomposites by reinforcing high-strength natural or synthetic fibers due to their high strength ratio, high water resistance, and high durability. Latex is a product that is typically obtained from a rubber tree. Latex is a milky, sticky colloid that is collected using tapping processes. During tapping, incisions in the bark are made, and the fluid is collected in vessels. The latex is extracted and poured into dry rubber preparation coagulation tanks or ammoniation-sealed airtight containers. Ammoniation keeps the latex in a colloidal state for a long time. It is normally coagulated under formic acid cleaning control or refined into latex concentrate for the manufacture of dipped products. Natural rubber is used mainly in industries, including transportation, pharmaceutical, agricultural, and aerospace (Mochizuki & Hirami, 1997; Jo et al., 1992).

2.5 Starch

Starch, a polymeric carbohydrate, is one of the most widely used biopolymers provided by green plants. It contains a significant number of glucose-bound glycosidic bonds. It is found in a variety of crops, including wheat, corn, and rice. It is a white color powder that is odorless, tasteless, and soluble in water and ethanol. It can be converted to sugar and used to make ethanol for whisky, beer, and biofuels by malting and fermenting it (Jo et al., 1992).

2.6 Polycarbonate

Polycarbonate is made by ring opening trimethylene carbonate that has been catalyzed with diethyl zinc. A flexible polymer with a high molecular weight has been produced, but it performs poorly in mechanical aspects (Raquez et al., 2008). Because of this nature, its applications are restricted, and copolymers are used more frequently. Glycolide and dioxanone were used to make copolymers (Balakrishnan et al., 2010; Reddy et al., 2008). Polypropylene carbonate is made by copolymerizing propylene oxide and CO_2. Polycarbonate has positive qualities such as usability and impact tolerance. Its thermal withstanding nature and biodegradability must be enhanced. A traditional method is to combine it with other polymeric materials (Ramesh, Rajeshkumar, Deepa et al., 2021). Poly[oligo(tetramethylene succinate)-co(tetramethylene carbonate)] a polyester carbonate is marketed by Mitsubishi Gas Chemical Co. The carbonate content of the copolymer can be modified. The melting temperature of this is around 105°C. The addition of carbonate to polyoligotetra-methylene succinate may have caused crystal structure disorder, lowering its melting temperature and making it more susceptible to enzymatic and microbial attacks than polyolefins. It has been shown that this copolyester carbonate has a higher microbial degradability than both of its constituents (Vroman & Tighzert, 2009).

2.7 Soy-based biodegradable polymers

In terms of environmental concerns, polymer disposal has become the most important subject for scientists. As a result, researchers are attempting to create biopolymers from biodegradable and environmentally friendly agricultural materials, such as starches and proteins. Soybeans have numerous advantages, including low cost, ample availability, and suitability for the manufacture of a wide range of chemicals and biodegradable plastics. Other biopolymers include polybutylene succinate and PCL, which are commonly used for fiber reinforcement in biodegradable materials.

2.8 Polycaprolactone

This is a cyclic monomer that is relatively inexpensive. The ring-opening polymerization process of caprolactone in the mixture of a tin octoate catalyst yields a semicrystalline linear polymer (Tan et al., 2011). Polycaprolactone is soluble in different solutions. Its glass transition temperature is about −50°C, and its melting range is around 63°C. It is a semisolid material with a medium- and high-density polyethylene modulus, a low tensile value of around 25 MPa, and a high elongation breakage when used at room temperature. Because of its low transition temperature, polycaprolactone is used as a compatibilizer or a soft block in polyurethane processing. PCL is easily degraded by fungi and enzymes (Nam et al., 2012; Nie et al., 2012). To speed up the degradation process, many copolymers containing lactide or glycolide have been created (Biresaw & Carriere, 2002).

2.9 Polyurethanes

Polyurethane has been widely developed to meet the highly diverse fields of new methodologies such as application coatings, adhesives, fabrics, foams, elastomers, etc. (Middleton & Tipton, 1998). It is a specific polymer substance with an extensive range of physical and chemical characteristics. Polyurethanes are used to make three components: a diisocyanate, a chain extension, and a polyol material. They combine to form a different polymer with alternating hard and soft segments. The soft component is used to make polyols like polyester polyols and polyether polyols. A hard section is used to build the diisocyanate and the extender. The biodegradation process of polyurethanes is influenced by the chemical composition. The deterioration can be modified by selecting an appropriate soft part. Biodegradability is not an issue with polyether-based polyurethanes. Polyurethanes are readily biodegradable if the polyol is a polyester (Maharana et al., 2009).

2.10 Polyanhydrides

Kumar et al. (Saravana Kumar et al., 2017) conducted a study of polyanhydrides and found that they are interesting biodegradable materials due to their hydrolyzable sites in the repeating unit. The rate of degradation is determined by the polymer components. Aromatic polyanhydrides degrade slowly with respect to time, while aliphatic polyanhydrides degrade in a matter of days. Diacyl chloride reaction with coupling agents (Briassoulis, 2004), diacid or diacid ester melt condensation, anhydride polymerization opening loop, interfacial condensation, and diacyl chloride reaction with coupling agents have all been investigated. Aliphatic homopolyanhydrides have restricted applications due to their high crystallinity index and rapid degradation. The rate of degradation of polyanhydride can be corrected by adjusting the hydrophobic and hydrophilic constituents in the polymer. Slower degradation was aided by the polymers' diacid building blocks' increased hydrophobicity. These polymers with hydrophobic aromatic comonomers, such as carboxyphenoxypropane, have been studied widely as biomaterials (Jacobsen & Fritz, 1999). As a wide variety of diacid monomers are available, polyanhydrides with various linkages have been made which include ether, ester, and urethane linkages. Anhydride-amide copolymers have also been developed to enhance the mechanical strengths of polyanhydrides for medical applications (Miller et al., 1977). The other approach is to use acrylic functional groups in the monomeric unit. As a consequence, photo-cross-linkable polyanhydrides are formed. The mechanical properties of these cross-linked polyanhydrides vary depending on the monomeric species.

3. Biodegradable polymer blends

3.1 PLA-based blends

3.1.1 Poly-lactic acid/polystyrene (PLA/PS) blends

PLA/PS blends of interfacial tension were characterized by Biresaw and Carriere (Biresaw & Carriere, 2002; Mohamed et al., 2007a; Ramesh, RajeshKumar, & Bhuvaneshwari, 2021) using various techniques of experiments and simulation. They assessed the thermal behavior and interfacial compatibility between a PLA/PS blend, and it was discovered from the results that a physically formed bonding existed between the unshared electron pair of PLA carbonyl groups and the PS electrons of the aromatic ring that is termed as an n-bond, which was observed by a transition in the carbonyl absorption band at nearly 1760 cm^{-1} of individual PLA to the lower wavelet number in the PLA/PS blend. The consequence of this bond formation was a form of poor interfacial compatibility that resulted in the blend having a higher maximum thermal stability than pure PLA (Hamad et al., 2011; Biresaw & Carriere, 2004). Tensile property tests were used to assess the compatibility of PLA/PS polymer blends, which were compared to other PS/biodegradable systems made using extrusion processes, such as PCL/PS and PBS/PS (Hamad et al., 2010). In contrast to PLA/PS, PCL and PBS blended with PS demonstrated excellent compatibility. The analysis of the rheological characteristics of PLA/PS blends revealed conventional thinning across the shear length of the specimen under normal shear stress rates and it was also observed that as the content of PLA increased, blend viscosity decreased (Li & Shimizu, 2009).

3.1.2 PLA/PE blends

As compared to pure PLA, blending PLA in PE enhances the impact strength, which was even more evident when compatibilizers were used, while tensile properties, such as tensile modulus, Young's modulus, and elongation, were lower in compatible and incompatible materials than in pure PLA. Raghavan and Emekalam prepared PLA/PE blends, and the degradation of blends due to the influence of additives of starch was investigated (Ramesh & Kumar, 2020; Kanzawa & Tokumitsu, 2011; Wang et al., 2012). They discovered that adding starch to the mixture makes it more acid degradable. The mechanical characteristics of PLA/PE blends were studied, and it was discovered that adding the filler materials to be blended increased the Young's modulus, while it lowered the load-bearing capacity and the elongation of the blends (St-Pierre et al., 1997). Many experimenters (Prinos et al., 1998) have manufactured PLA/PE polymer blends and evaluated their mechanical and thermal properties. Weak compatibility was observed and hence the blend's tensile characteristics decreased without an increase in thermal stability.

3.1.3 PLA/ABS blends

ABS's strong mechanical properties, such as impact strength, tensile strength, and tensile modulus, have contributed to the widespread use of novel materials with standalone characteristics in blending technology. Blends of PLA/ABS biopolymers have been also manufactured, made compatible with the aid of styrene/acrylonitrile/glycidyl methacrylate copolymer (SAN-GMA) through the incorporation of ethyltriphenylphosphonium bromide (ETPB) as a catalyst material, and have been characterized by a number of authors (Kaseem et al., 2012a). In the presence of ETPB, SAN-GMA was an important compatibilizer for promoting the reaction between PLA/ABS blends, as demonstrated by a large increase in rubber particle dispersion as well as enhanced impact toughness and Poisson's ratio, with only a slight decrease in tensile modulus and strength as compared with individual PLA.

3.1.4 PLA/PP blends

Experiments have been carried out by a number of experimenters (Ramesh, Deepa et al., 2020; Rosa et al., 2010; Leo et al., 2011; Schlemmer et al., 2007) on manufacturing and analysis of PLA/PP blends for their properties. The morphological and thermal characteristics of PLA/PP blends were only partially compatible at the time of melting. The morphology of the 50% PLA blend revealed ill-defined interfaces similar to those used in compatible PLA/PE systems. PLA/PP blend fibers are also more resistant to hydrolysis than pure PLA, and PP improves PLA's ability to be converted to a dye material.

3.1.5 PLA/PC blends

Some experimenters (Schlemmer et al., 2009; Balaji et al., 2021; Pimentel et al., 2007) have prepared PLA/PC blends with the radical indicator dicumyl peroxide (DCP), and used PBAT to increase the percentage elongation, impact toughness, and tensile characteristics of the above blend. DCP was observed to catalyze the PLA and PBAT reaction, which in turn produced homogeneous dispersion. Meanwhile, PC appeared to be in close proximity to the nuptial blend phase because of its better compatibility, which led to a multiple polymer dispersed PLA/PBAT/PC blend (PBAT-PLA). The newly evolved morphology has assisted in the improvement of the treated material's ductility. Wang et al. (2012) used polybutylene succinate-co-lactate (PBSL) and epoxy (EP) as compatibilizers for twin screw extruder-prepared PLA/PC polymer blends. It was also noted from this study that the influence of the use of a compatabilizer in this combination improved the toughness and heat resistance of PLA/PC biopolymer blends (Oliveira et al., 2010; Kaseem et al., 2012b, 2012c).

3.2 Polycaprolactone blends

3.2.1 PCL/PP blends

Many experimenters (Ratnagifu & Scott, 1998; Tjong & Bei, 1998) have examined the compatibility between the blending elements of PCL and PP polymers manufactured through an extrusion process. It was observed from the blend morphology that PCL and PP precursors were incompatible with each other and the distribution of the dispersoid was coarse within the matrix. PP was also discovered to act as a nucleating agent for PCL. Krucinska et al. (2012) investigated the characteristics of PCL/PP/multi-walled carbon nanotube (MWCNT) composite blends for their thermal and rheological characteristics. Experimental results indicated that when PCL was incorporated with MWCNTs, a small increment in both crystallinity and thermal resistivity occurred, but that incorporating PP into the composite reduced the blends' thermal stability. Similarly with respect to the same composite blends, the impact of MWCNTs on the electrical characteristics of the resulting composites was also discussed by some other authors (Ramesh et al.; Balsamo & Gouveia, 2007). At low levels of MWCNTs, good dispersibility and distribution of MWCNTs in the polymeric matrix was morphologically observed, resulting in the creation of a conductive nature of the MWCNT network.

3.2.2 PCL/PE blends

To prepare the PCL/PE blend and evaluate phase inversion during compounding, Ratnagifu and Scott (Ratnagifu & Scott, 1998) used an internal mixer. The results revealed that the biopolymer blend is not compatible within the composition content range tested. Some experimenters (Ptschke et al., 2011) manufactured PCL/low density polyethylene (LDPE) composites in the presence of maleic anhydride (MA) and compared them with the poly(caprolactone)-block-poly (ethylene glycol). While the latter's mechanical characteristics were superior to those of PCL/LDPE blends, the former was much more compatible between the matrix and the reinforcement.

3.2.3 PLC/PVC blends

Some of the earlier studies on PVC/PCL polymer blends (Mohamed et al., 2007b) investigated the influence of dibasic lead phthalate and dibutyltin dilaurate as thermal stabilizers on PCL phase dispersion within PCL/PVC blends. Following that, the solution rheological behavior of PCL/PVC polymer blends was investigated (David & Homme, 1996). Due to the H-bond between the two-component chains, full compatibility existing between PVC and PCL was observed in the polymer blend. (Bhuvaneswari et al., 2021; Pingping et al., 1998) also confirmed their good compatibility; and a thermal property analysis revealed that the mixture has a single T_g, which was in line with the mechanical properties of the produced mixture, which showed that the mixture break elongation increases as the PCL material is increased.

3.2.4 PCL/PS blends

(Mohamed et al., 2007a) found that PCL/PS blends exhibit better compatibility between them and by comparing the mechanical characteristics of the blends are better than the values of the individual materials (PCL and PS). Mohamed et al. (2007b) also assessed the compatibility within PCL/PS blends. Due to intermolecular interactions between PCL and PS between n-π, the results of differential scanning calorimetry revealed good compatibility between the two components. These blends have better thermal stability than pure PCL; for the PLA/PS blend, the same result was obtained (Bhuvaneswari et al., 2021; Pingping et al., 1998).

3.2.5 PCL/PC blends

Some experiments have been performed on these blends (Chiu & Min, 2000; Don et al., 1996; Ramesh, Maniraj et al., 2021) and the antiplastination characteristics of the blends were studied. Furthermore, Hirotsu et al. (2000a, 2000b) used burial in soil studies to investigate the biodegradable characteristics of PCL/PC polymer blends treated with O_2 and Ar plasma. The findings revealed that when PCL is combined with PC, it becomes less degradable, and that O_2-plasma oxidative therapies increase enzymatic biodegradation, whereas Ar-plasma therapies minimize it. Balsamo et al. (2001) examined the thermal characteristics and biocompatibility of PCL/PC polymer blends and the results indicated that they were compatible in a number of formulations. It is possible to crystallize either one or both components in a blend comprising 40% or more PCL. Some experimenters (Ramesh et al., 2021; Laredo et al., 2005) investigated the isothermal crystallization of a polymer blend containing 10% by volume of PCL and discovered that as the isothermal crystallization of PC progresses, the presence of a rugged nebulous phase in the blend increases in appreciable and considerable amounts (Chiu & Smith, 1984).

3.2.6 PCL/SAN blends

Numerous studies on the preparation of PCL/SAN polymer blends and their applications have been published over the past three decades (Jo et al., 1992; Fernandes et al., 1986; Kressler & Karnmer, 1988). The biocompatibility of PCL/SAN biopolymer blends was tested by numerous researchers. Among them, Ramesh and Rajeshkumar (2021) used testing methods like differential scanning calorimetry and dynamic mechanical analysis to evaluate the blend-compatible behavior of the PCL/SAN composites for different compositions of the composites. The results indicated that AN and PCL materials were compatible when the content of reinforcements were between 8% and 28% by weight. Some other authors (Ayala et al., 2009) evaluated the segmental interaction factors in the PCL/SAN biopolymer blends from research with respect to melting point depression. The influence of AN substance on the rheological characteristics of PCL/SAN polymer composite blends made by solution method was investigated between 0% and 30% by weight of reinforcement (Ray et al., 2005), in which polymer solutions are combined and precipitated. The results indicated that the blend was consistent in the considered variety of formulations, and the viscosity value approached minimum values of 15 wt.% AN.

3.3 Thermoplastic starch (TPS) blends

3.3.1 TPS/PP blends

Many authors (Ray & Bousmina, 2005; Ramesh, Rajeshkumar, & Balaji, 2021) have manufactured glycerol-containing TPS/PP polymer blends as plasticizers with the aid of a single screw extruder and their characteristics have been studied also. The blends exhibited thinning shear behavior while processing in conventional machines, indicating that they were processable. Furthermore, as the glycerol content increased, the influence of lubrication of glycerol between the blend materials and the capillary rheometer die reduced the blend's viscosity. The blend's mechanical properties revealed that as the TPS and glycerol content increased, the Young's modulus increased significantly, while the breaking strain decreased. Zhao et al. (2012) assessed the effect of plasticization of industrial glycerol and biodiesel glycerol used in the preparation of TPS/PP blends and compared them. The findings were almost identical: the blend's tensile strength decreased as the TPS content increased. TPS/PP blends with renewable, biodegradable content and better mechanical characteristics were created using montmorillonite clays (MMT) and cloisite 30B, and it was organically modified with bis-2-hydroxyethyl, methyl, quaternary ammonium, and tallow salts (Nam et al., 2012).

3.3.2 TPS/PE blends

The investigation into the PE and TPS blends characteristics was performed by a number of researchers and the morphological, thermal, and mechanical characteristics of the blend were evaluated (Raquez et al., 2008; Tan et al., 2011; Nie et al., 2012; Hong et al., 2013). In this process, a single TPS screw extruder was coupled with a twin-screw extruder for preparation of the blends, resulting in glycerol plasticized TPS and the blends. The incompatibility between the two blend elements resulted in lower thermal stability as compared to individual PE. The mechanical characteristics showed that the mixture's elongation at break is comparable to PE, but its modulus is lower. Many experimenters (Ramesh et al., 2021; Vroman & Tighzert, 2009; Middleton & Tipton, 1998) have investigated the impact of the ethylene/vinyl acetate (EVA) copolymer over the thermal and mechanical characteristics of TPS/LDPE polymer blends; glycerol was also used as a plasticizer. The results suggested that the mechanical properties and thermal stability of the polymer blend were enhanced with an increase in the EVA content.

3.3.3 TPS/ABS blends

Some authors have discussed the experimental research insights of TPS/ABS blends with respect to their manufacturing and characteristics assessment (Maharana et al., 2009; Saravana Kumar et al., 2017). Glycerol plasticized TPS and ABS materials were blended with the aid of a single screw extruder and their mechanical and rheological characteristics were evaluated. TPS/ABS blends undergo shear thinning when their viscosity decreases with an increase in the rate of shear stress and an increase in the glycerol content, portraying its better material processability using an injection molding method. Breakage stress and ductility were found to be weak mechanical characteristics in various blend formulations, indicating incompatibility between the blend's components (Briassoulis, 2004; Jacobsen & Fritz, 1999; Miller et al., 1977).

3.3.4 TPS/PS blends

Luckachan and Pillai (2006) reported the thermal degradation of glycerol or buriti oil and PS plasticized TPS-containing blends. In contrast to glycerol plasticized TPS buriti oil, it was discovered that the mixture of plasticized TPS buriti oil was more thermally stable. However, research into biodegradability has shown that the former is more biodegradable than the latter (Chandra & Rustgi, 1998; Mochizuki & Hirami, 1997). For the preparation of TPS and PS waste blends, buriti oil was also used (Tokiwa & Suzuki, 1977). A strong TPS dispersion with distinct domains in the PS matrix was observed from the blend morphology, but the thermal stability was lower as compared to individual PS material. Fujimaki (1998) analyzed glycerol-plasticized TPS and MA-treated PS for their interfacial adhesion behavior and the polymer blends were prepared using melt mixing techniques. The influence of TPS volume fraction on degradation of the blend, calculated by weight loss of the specimen, was also evaluated during a soil burial test. Increasing the content of TPS leads to an increase in overall weight loss due to the effect of TPS volume fraction in alleviating the bacterial growth rate at the time of degradation.

3.4 Polybutylene succinate blends (PBS)

PBS is an aliphatic biodegradable polyester that is available commercially as a thermoplastic polymer with biodegradability, thermal and chemical resistance, and melt processability. They are part of the polymer family (alkene dicarboxylate). PBS is a crystalline white thermoplastic with a melting point of 90–120°C (Nair & Laurencin, 2007).

Glycol polycondensation reactions involving ethylene glycol and 1,4-butanediol with aliphatic dicarboxylic acids like succinic and adipic acid have been recorded (Zhu et al., 1991). To improve its characteristics and minimize the manufacturing costs, PBS is reinforced with plant fibers and plant-based fillers. Tao et al. (2009) made PBS/rice straw fiber-reinforced plant-based composite amino coupling agents as catalysts. The results showcased that the mechanical strength was higher for the composites in the presence of an amino group binding agent. The effect of a 5% NaOH solution treatment on the mechanical characteristics of PBS/coir composites has been studied (Pranamuda et al., 1999). The results showed that composite tensile strength and modulus were enhanced when the fiber content increased, while elongation at failure lowered. Kim et al. (2003) studied the kinetics of nonisothermal crystallization of PBS/cotton stalk bast fiber composites and discovered that cotton stalk bast fibers served as both nucleating agents and physical barriers in delaying chain segment transport during crystallization. The microencapsulated ammonium polyphosphate (MCAPP) effect on the thermal characteristics of composites was investigated and compared to magnesium hydroxide and aluminum hydroxide. Melt mixing was used (Nakajima-Kambe et al., 1999) to make PBS/bamboo fiber composites. MCAPP outperformed both the above-described hydroxides in terms of flame-retardant properties. Kumar et al. (2002) investigated the influence of sisal fiber content on the rheological behavior of PBS/sisal composites; the composites experienced shear thinning phases, with viscosity decreasing as shear rate increased. Furthermore, as the fiber content increases, the non-Newtonian composite index (n) decreases, implying that the composite viscosity is robust over a broad range of shear speeds.

3.5 Polybutylene succinate adipate blends

Polybutylene succinate adipate (PBSA) is a biodegradable industrial synthetic polyester with properties similar to linear LDPE, such as high ductility (Tamada & Langer, 1992) and excellent processability with conventional polyolefin equipment. The main disadvantages of this polymer are its lower tensile characteristics and better thermal sensitivity. Different fillers were used to solve these issues in the production of PBSA composites and nanocomposites with higher tensile characteristics and thermal properties without sacrificing the ductile nature of the finished product. Leong et al. (1985) made PBSA/organically modified montmorillonite (OMMT) nanocomposites and investigated the influence of the OMMT atomic structure on the thermal, morphological, and mechanical characteristics of the composites. Tensile load-carrying capacity, modulus, thermal stability, and PBSA break elongation all improved as a result of the research. The use of the chemical structure of OMMT with diols amplified this effect.

3.6 Poly(butylene adipate-co-terephthalate) blends

PBAT, an aliphatic-aromatic polyester material, can deteriorate in the surrounding environment due to the presence of microbial lipases. PBAT possesses outstanding characteristics when used for coatings and film extrusion, as well as a higher elongation at break. Several researches have centered on the manufacture and characterization of

PBAT composites using different raw materials such as OMMT and talc. Ibim et al. (1998) used an extrusion method to make PBAT/talc composites, and MA was grafted to enhance the interfacial adhesion in the PBAT composite. The adhesion at the interface between PBAT-g-MA and talc was enhanced, as observed from SEM images of the composites. The properties of PBAT, such as tensile load, modulus, and break elongation, were all improved. The viscosity of PBAT/talc composites was enhanced as the talc content was increased, according to the composite melt viscosity.

3.7 Starch-based blends

Starch is a fully ecofriendly and biodegradable substance. Furthermore, it is inexpensive. However, starch's use is restricted due to its high sensitivity toward water and lower mechanical characteristics as compared with other petrochemical polymers. Various thermoplastic starch-based biodegradable polymeric blends have been developed and thoroughly researched. Much research has been carried out on the development of starch blends coupled with various artificial biodegradable polymeric materials. There are numerous advantages to using these blends (Chen et al., 1997; Scott & Gilead, 1995). Their material characteristics can be adapted to the requirements of the application area by adjusting the content of the composite. The blending method of the above composites would be inexpensive as compared with the cost of developing new artificial materials. Blends like these need to be further enhanced in terms of their design to make them much more biodegradable than other traditional synthetic plastic materials (Ghanbarzadeh & Almasi).

4. Conclusion

This chapter summarizes recent developments in the preparation of biodegradable polymer blends as well as their properties. The commercial success of these blends is determined by a number of factors, including price, consumer demand, quality, composting infrastructure, and legislation. The literature discussed in this chapter has shown that by developing new manufacturing methods and procedures, these blends with unique properties and low prices can be produced. Biodegradable polymers have benefited from the application of advanced technologies. They offer active packaging technology as well as natural fiber reinforcements. Several studies on the use of nanoclay with biodegradable polymeric materials, specifically starch and aliphatic polyester materials, have recently been published.

References

Ayala, G., Pace, E., Laurienzo, B., et al. (2009). Poly(ε-caprolactone) modified by functional groups: preparation and chemical—physical investigation. *European Polymer Journal, 45*, 3217—3229.

Balaji, D., Ramesh, M., Kannan, T., Deepan, S., Bhuvaneswari, V., & Rajeshkumar, L. (2021). Experimental investigation on mechanical properties of banana/snake grass fiber reinforced hybrid composites. *Materials Today: Proceedings, 42*, 350–355. https://doi.org/10.1016/j.matpr.2020.09.548

Balakrishnan, H., Hassan, A., & Wahit, M. (2010). Mechanical, thermal, and morphological properties of polylactic acid/linear low density polyethylene blends. *Journal of Elastomers and Plastics, 42*, 223–239.

Balsamo, V., & Gouveia, L. (2007). Interplay of fractionated crystallization and morphology in polypropylene/poly(ε-caprolactone) blends. *Journal of Polymer Science Part B: Polymer Physics, 45*, 1365–1379.

Balsamo, V., Calzadilla, N., Mora, G., & Muller, A. (2001). Thermal characterization of polycarbonate/polycaprolactone blends. *Journal of Polymer Science Part B: Polymer Physics, 39*, 771–785.

Bhatia, A., Gupta, R., Bhattacharya, S., & Choi, H. (2007). Combatability of biodegradable poly (lactic acid) (PLA) and poly (butylene succinate) (PBS) blends for packaging application. *Korea-Australia Rheology Journal, 19*, 125–131.

Bhuvaneswari, V., Priyadharshini, M., Deepa, C., Balaji, D., Rajeshkumar, L., & Ramesh, M. (2021). Deep learning for material synthesis and manufacturing systems: a review. *Materials Today: Proceedings*. https://doi.org/10.1016/j.matpr.2020.11.351

Biresaw, G., & Carriere, J. (2002). Interfacial tension of poly(lactic acid)/polystyrene blends. *Journal of Polymer Science Part B: Polymer Physics, 40*, 2248–2258.

Biresaw, G., & Carriere, J. (2004). Compatibility and mechanical properties of blends of polystyrene with biodegradable polyesters. *Composites Part A, 35*, 313–320.

Briassoulis, D. (2004). An overview on the mechanical behavior of biodegradable agricultural films. *Journal of Polymers and the Environment, 12*, 65–81.

Chandra, R., & Rustgi, R. (1998). Biodegradable polymers. *Progress in Polymer Science, 23*, 1273–1335.

Chen, L., Imam, S., Gordon, S., & Greene, R. V. (1997). Starch- polyvinyl alcohol crosslinked film-performance and biodegradation. *Journal of Environmental Polymer Degradation, 5*, 111–117.

Chiu, F., & Min, K. (2000). Miscibility, morphology and tensile properties of vinyl chloride polymer and poly(ε-caprolactone) blends. *Polymer International, 49*, 223–234.

Chiu, S., & Smith, T. G. (1984). Compatibility of poly(ε-caprolactone) (PCL) and poly(styrene-co-acrylonitrile) (SAN) blends. I. Blends containing SAN with 24 wt% acrylonitrile. *Journal of Applied Polymer Science, 29*, 1781–1795.

David, M., & Homme, R. (1996). Influence of thermal stabilizers on diffusion of poly(ε-caprolactone) in poly(vinyl chloride)/poly(ε-caprolactone) blends. *Journal of Applied Polymer Science, 61*, 465–471.

Don, T., Bell, J., & Narkis, J. (1996). Antiplasticization behavior of polycaprolactone/polycarbonate-modified epoxies. *Polymer Engineering & Science, 36*, 2601–2613.

Fernandes, A., Barlow, W., & Paul, D. (1986). Blends containing polymers of epichlorohydrin and ethylene oxide. Part I: Polymethacrylates. *Journal of Applied Polymer Science, 32*, 5481–5508.

Fujimaki, T. (1998). Processability and properties of aliphatic polyesters, “Bionolle”, synthesized by polycondensation reaction. *Polymer Degradation and Stability, 59*, 209–214.

Ghanbarzadeh, B., & Almasi, H. *Biodegradable polymers*. https://doi.org/10.5772/56230

Hamad, K., Kaseem, M., & Deri, F. (2010). Rheological and mechanical properties of poly(lactic acid)/polystyrene polymer blend. *Polymer Bulletin, 65*, 509–519.

Hamad, K., Kaseem, M., & Deri, F. (2011). Effect of recycling on rheological and mechanical properties of poly(lactic acid)/polystyrene polymer blend. *Journal of Materials Science, 46*, 3013–3019.
Hamad, K., Kaseem, M., Ko, Y. G., & Deri, F. (2014). Biodegradable polymer blends and composites: an overview. *Polymer Science – Series A, 56*(6), 812–829.
Hirotsu, T., Ketelaars, A., & Nakayama, K. (2000a). Plasma surface treatment of PCL/PC blend sheets. *Polymer Engineering & Science, 40*, 2324–2331.
Hirotsu, T., Ketelaars, A., & Nakayama, K. (2000b). Biodegradation of poly(ε-caprolactone)–polycarbonate blend sheets. *Polymer Degradation and Stability, 68*, 311–316.
Hong, F., Jie, L., Ping, X., et al. (2013). Effect of fiber morphology on rheological properties of plant fiber reinforced poly(butylene succinate) composites. *Composites Part B, 44*, 193–199.
Ibim, S. E., Uhrich, K. E., Attawia, M., Shastri, V. R., El-Amin, S. F., & Bronson, E. (1998). Preliminary in vivo report on the osteocompatibility of poly(anhydride-co-imides) evaluated in a tibial model. *Journal of Biomedical Materials Research, 43*, 374–379.
Jacobsen, S., & Fritz, H. G. (1999). Plasticizing polylactide – the effect of different plasticizers on the mechanical properties. *Polymer Engineering & Science, 39*, 1303–1310.
Jo, W., Chae, S., & Lee, M. (1992). Rheological properties of poly(ε-caprolactone) and poly(styrene-co-acrylonitrile) blends. *Polymer Bulletin, 29*, 113–118.
Kanzawa, T., & Tokumitsu, K. (2011). Mechanical properties and morphological changes of poly(lactic acid)/polycarbonate/poly(butylene adipate-co-terephthalate) blend through reactive processing. *Journal of Applied Polymer Science, 121*, 2908–2918.
Kaseem, M., Hamad, K., & Deri, F. (2012a). Rheological and mechanical properties of polypropylene/thermoplastic starch blend. *Polymer Bulletin, 68*, 1079–1091.
Kaseem, M., Hamad, K., & Deri, F. (2012b). Rheological and die swell measurements of thermoplastic starch/acrylonitrile- butadiene- styrene blends. *Malaysian Polymer Journal, 7*, 22–27.
Kaseem, M., Hamad, K., & Deri, F. (2012c). Preparation and studying properties of thermoplastic starch/acrylonitrile–butadiene–styrene blend. *International Journal of Plastics Technology, 16*, 39–49.
Kim, B. K., Seo, J. W., & Jeong, H. M. (2003). Morphology and properties of waterborne polyurethane/clay nanocomposites. *European Polymer Journal, 39*, 85–91.
Kressler, J., & Karnmer, W. (1988). Melting point depression in poly(ϵ-caprolactone)/poly(styrene-co-acrylonitrile) blends. *Polymer Bulletin, 19*, 283–288.
Krucinska, I., Surma, B., Chrzanowski, M., Skrzetuska, E., & Puchalski, M. (2012). Application of melt-blown technology for the manufacture of temperature-sensitive nonwoven fabrics composed of polymer blends PP/PCL loaded with multiwall carbon nanotubes. *Journal of Applied Polymer Science, 127*, 869–878.
Kumar, N., Langer, R. S., & Domb, A. J. (2002). Polyanhydrides: an overview. *Advanced Drug Delivery Reviews, 54*, 889–910.
Laredo, E., Grimau, M., Barriola, P., et al. (2005). Effect of isothermal crystallization on the amorphous phase mobility of polycarbonate/poly(ε-caprolactone) blends. *Polymer, 46*, 6532–6542.
Leo, C., Pinotti, C., Goncalves, M., & Velankar, S. (2011). Preparation and characterization of clay nanocomposites of plasticized starch and polypropylene polymer blends. *Journal of Polymers and the Environment, 19*, 689.
Leong, K. W., Brott, B. C., & Langer, R. (1985). Biodegradable polyanhydrides as drug carrier matrices: characterization, degradation and release characteristics. *Journal of Biomedical Materials Research, 19*, 941–955.

Li, Y., & Shimizu, H. (2009). Improvement in toughness of poly(l-lactide) (PLLA) through reactive blending with acrylonitrile–butadiene–styrene copolymer (ABS): morphology and properties. *European Polymer Journal, 45*, 738–746.

Liu, H., & Zhang, J. (2011). Research progress in toughening modification of poly(lactic acid). *Journal of Polymer Science Part B: Polymer Physics, 49*, 1051–1083.

Luckachan, G. E., & Pillai, C. K. S. (2006). Chitosan/oligo L-lactide graft copolymers: effect of hydrophobic side chains on the physico-chemical properties and biodegradability. *Carbohydrate Polymers, 24*, 254–266.

Maharana, T., Mohanty, B., & Negi, Y. S. (2009). Melt-solid polycondensation of lactic acid and its biodegradability. *Progress in Polymer Science, 34*, 99–124.

Middleton, J. C., & Tipton, A. J. (1998). Synthetic biodegradable polymers as medical devices. *Medical Plastics and Biomaterials Magazine, 3*, 30. http://www.devicelink.com/mpb/archive/-98/03/002.html.

Miller, R. A., Brady, J. M., & Cutright, D. E. (1977). Degradation rates of oral resorbableimplants (polylactates and polyglycolates): rate modification with changes in PLA/PGA copolymer ratios. *Journal of Biomedical Materials Research, 11*, 711–719.

Mochizuki, M., & Hirami, M. (1997). Structural effects on biodegradation of aliphatic polyesters. *Polymers for Advanced Technologies, 8*, 203.

Mohamed, A., Gordon, S. H., & Biresaw, G. (2007a). Poly(lactic acid)/polystyrene bioblends characterized by thermogravimetric analysis, differential scanning calorimetry, and photoacoustic infrared spectroscopy. *Journal of Applied Polymer Science, 106*, 1689–1696.

Mohamed, A., Gordon, S. H., & Biresaw, G. (2007b). Polycaprolactone/polystyrene bioblends characterized by thermogravimetry, modulated differential scanning calorimetry and infrared photoacoustic spectroscopy. *Polymer Degradation and Stability, 92*, 1177–1185.

Nair, L. S., & Laurencin, C. T. (2007). Biodegradable polymers as biomaterials. *Progress in Polymer Science, 32*, 762–798.

Nakajima-Kambe, T., Shigeno-Akutsu, Y., Nomura, N., Onuma, F., & Nakarahara, T. (1999). Microbial degradation of polyurethane, polester polyurethanes and polyether polyurethanes. *Applied Microbiology and Biotechnology, 51*, 134–140.

Nam, T. H., Ogihara, S., Tung, N. H., & Kobayashi, S. (2012). Effect of alkali treatment on interfacial and mechanical properties of coir fiber reinforced poly(butylene succinate) biodegradable composites. *Composites, Part B, 42*, 1648–1656.

Nie, S., Liu, X., Dai, G., et al. (2012). Investigation on flame retardancy and thermal degradation of flame retardant poly(butylene succinate)/bamboo fiber biocomposites. *Journal of Applied Polymer Science, 125*, E485–E489.

Oliveira, C., Cunha, F., & Andrade, C. (2010). Evaluation of biodegradability of different blends of polystyrene and starch buried in soil. *Macromolecular Symposia, 290*, 115–120.

Pimentel, T., Duraes, J., Drummond, A., Schlemmer, D., Falcão, R., & Sales, M. J. A. (2007). Preparation and characterization of blends of recycled polystyrene with cassava starch. *Journal of Materials Science, 42*, 7530–7536.

Pingping, Z., Haiyang, Y., & Shiqiang, W. (1998). Viscosity behavior of poly-ε-caprolactone (PCL)/poly(vinyl chloride) (PVC) blends in various solvents. *European Polymer Journal, 34*, 91–94.

Pranamuda, H., Chollakup, R., & Tokiwa, Y. (1999). Degradation of polycarbonate by a polyester degrading strain, *Amycolatopsis* sp. Strain HT-6. *Applied and Environmental Microbiology, 65*, 4220–4222.

Prinos, J., Bikaris, D., Theologidis, S., & Panayiotou, C. (1998). Preparation and characterization of LDPE/starch blends containing ethylene/vinyl acetate copolymer as compatibilizer. *Polymer Engineering & Science, 38*, 954–964.

Ptschke, P., Kobashi, K., Villmow, T., Andres, T., Paivac, M. C., & Covas, J. A. (2011). Liquid sensing properties of melt processed polypropylene/poly(e-caprolactone) blends containing multiwalled carbon nanotubes. *Composites Science and Technology, 71*, 1451–1460.

Raghavan, D., & Emekalam, A. (2001). Characterization of starch/polyethylene and starch/ polyethylene/poly(lactic acid) composites. *Polymer Degradation and Stability, 72*, 509–517.

Ramesh, M., & Kumar, L. R. (2020). Bioadhesives. In R. Inamuddin, M. I. Boddula, Ahamed, & A. M. Asiri (Eds.), *Green adhesives* (pp. 145–167). https://doi.org/10.1002/ 9781119655053

Ramesh, M., & Rajeshkumar, L. (2018). Wood flour filled thermoset composites. Thermoset composites: preparation, properties and applications. *Materials Research Foundations, 38*, 33–65. https://doi.org/10.21741/9781945291876-2

Ramesh, M., & Rajeshkumar, L. (2021). Technological advances in analyzing of soil chemistry. *Applied Soil Chemistry*, 61–78.

Ramesh, M., Kumar, L. R., Khan, A., & Asiri, A. M. (2020). Self-healing polymer composites and its chemistry. *Self-Healing Composite Materials*, 415–427. https://doi.org/10.1016/ b978-0-12-817354-1.00022-3. Elsevier.

Ramesh, M., Deepa, C., Kumar, L. R., Sanjay, M. R., & Siengchin, S. (2020). Life-cycle and environmental impact assessments on processing of plant fibres and its bio-composites: a critical review. *Journal of Industrial Textiles*. https://doi.org/10.1177/1528083720924730

Ramesh, M., Rajeshkumar, L., & Balaji, D. (2021). Aerogels for insulation applications. *Aerogels II: Preparation, Properties and Applications, 98*, 57–76.

Ramesh, M., Rajeshkumar, L., Deepa, C., Tamil Selvan, M., Kushvaha, V., & Asrofi, M. (2021). Impact of silane treatment on characterization of ipomoea staphylina plant fiber reinforced epoxy composites. *Journal of Natural Fibers*. https://doi.org/10.1080/15440478.2021.1902896

Ramesh, M., RajeshKumar, L., & Bhuvaneshwari, V. (2021). Bamboo fiber reinforced composites. In M. Jawaid, S. MavinkereRangappa, & S. Siengchin (Eds.), *Bamboo fiber composites. Composites science and technology*. Singapore: Springer. https://doi.org/ 10.1007/978-981-15-8489-3_1

Ramesh, M., Maniraj, J., & Rajesh Kumar, L. (2021). Biocomposites for energy storage. Bio-based composites: processing, characterization, properties, and applications. In A. Khan, S. M. Rangappa, S. Siengchin, M. Abdullah, & Asiri (Eds.), *Biobased composites: processing, characterization, properties, and applications* (pp. 123–142). Wiley Online Library.

Ramesh, M., Rajeshkumar, L., Balaji, D., & Bhuvaneswari, V. (2021). Green composite using agricultural waste reinforcement. In S. Thomas, & P. Balakrishnan (Eds.), *Green composites. Materials horizons: from nature to nanomaterials* (pp. 21–34). Singapore: Springer. https://doi.org/10.1007/978-981-15-9643-8_2

Ramesh, M., Deepa, C., Tamil Selvan, M., Rajeshkumar, L., Balaji, D., & Bhuvaneswari, V. Mechanical and water absorption properties of Calotropisgigantea plant fibers reinforced polymer composites. *Materials Today: Proceedings*. https://doi.org/10.1016/j.matpr.2020. 11.480

Raquez, J., Nabar, Y., Narayan, R., & Dubois, F. (2008). Novel high-performance talc/poly [(butylene adipate)-co-terephthalate] hybrid materials. *Macromolecular Materials and Engineering, 293*, 310–320.

Ratnagifu, R., & Scott, C. (1998). Phase inversion during compounding with a low melting major component: polycaprolactone/polyethylene blends. *Polymer Engineering & Science, 38*, 1751–1762.

Ray, S., & Bousmina, M. (2005). Poly(butylene sucinate-co-adipate)/montmorillonite nanocomposites: effect of organic modifier miscibility on structure, properties, and viscoelasticity. *Polymer, 46*, 12430–12439.

Ray, S. S., Bousmina, M., & Okamoto, K. (2005). Structure and properties of nanocomposites based on poly(butylene succinate-co-adipate) and organically modified montmorillonite. *Macromolecular Materials and Engineering, 290*, 759–768.

Reddy, N., Nama, D., & Yang, Y. (2008). Polylactic acid/polypropylene polyblend fibers for better resistance to degradation. *Polymer Degradation and Stability, 39*, 233–241.

Rosa, D., Bardi, M., Machado, L., Dias, D. B., Silva, L. G. A., & Kodama, Y. (2010). Starch plasticized with glycerol from biodiesel and polypropylene blends. *Journal of Thermal Analysis and Calorimetry, 102*, 181–186.

Saravana Kumar, A., MaivizhiSelvi, P., & Rajeshkumar, L. (2017). Delamination in drilling of sisal/banana reinforced composites produced by hand lay-up process. In *Applied Mechanics and Materials* (vol. 867, pp. 29–33). Trans Tech Publications.

Schlemmer, D., Oliveira, E., & Sales, M. (2007). Polystyrene/thermoplastic starch blends with different plasticizers. *Journal of Thermal Analysis and Calorimetry, 87*, 635–638.

Schlemmer, D., Sales, M., & Resck, I. (2009). Degradation of different polystyrene/thermoplastic starch blends buried in soil. *Carbohydrate Polymers, 75*, 58–62.

Scott, G., & Gilead, D. (1995). *Degradable polymers: principles and applications* (pp. 247–258). London, UK: Chapman and Hall.

Srithep, Y., Javadi, A., Pilla, S., Turng, L. S., Gong, S., & Peng, J. (2011). Processing and characterization of recycled poly(ethylene terephthalate) blends with chain extenders, thermoplastic elastomer, and/or poly(butylene adipate-co-terephthalate). *Polymer Engineering & Science, 51*(6), 1023–1032.

St-Pierre, N., Favis, B. D., Ramsay, B. A., Ramsay, J. A., & Verhoogt, H. (1997). Processing and characterization of thermoplastic starch/polyethylene blends. *Polymer, 38*, 647–655.

Tamada, J., & Langer, R. (1992). The development of polyanhydrides for drug delivery applications. *Journal of Biomaterials Science, Polymer Edition, 3*, 315–353.

Tan, B., Qu, J., Liu, L., et al. (2011). Non-isothermal crystallization kinetics and dynamic mechanical thermal properties of poly(butylene succinate) composites reinforced with cotton stalk bast fibers. *Thermochimica Acta*, 141–149.

Tao, J., Hu, D., Liu, L., Liu, N., Song, C., & Wang, S. (2009). Thermal properties and degradability of poly (propylene carbonate)/poly (β-hydroxybutyrate-co-β-hydroxyvalerate) (PPC/PHBV) blends. *Polymer Degradation and Stability, 94*, 575–583.

Tjong, S., & Bei, J. (1998). Degradation behavior of poly (caprolactone)–poly(ethylene glycol) block copolymer/low–density polyethylene blends. *Polymer Engineering & Science, 38*, 392–402.

Tokiwa, Y., & Suzuki, T. (1977). Hydrolysis of polyesters by lipases. *Nature, 270*, 76–78.

Vroman, I., & Tighzert, L. (2009). Biodegradable polymers. *Materials, 2*, 307–344.

Wang, Y., Chiao, S., Hung, T., & Yang, S. (2012). Improvement in toughness and heat resistance of poly(lactic acid)/polycarbonate blend through twin-screw blending: influence of compatibilizer type. *Journal of Applied Polymer Science, 125*, E402–E412.

Xie, F., Pollet, E., Halley, P., & Averous, L. (2013). Starch-based nano-biocomposites. *Progress in Polymer Science, 38*, 1590–1628.

Zhao, Y., Qiu, J., Feng, H., & Zhang, M. (2012). The interfacial modification of rice straw fiber reinforced poly(butylene succinate) composites: effect of aminosilane with different alkoxy groups. *Journal of Applied Polymer Science, 125*, 3211–3220.

Zhu, K. J., Hendren, R. W., Jensen, K., & Pitt, C. G. (1991). Synthesis, properties and biodegradation of poly(1.3-trimethylene carbonate). *Macromolecules, 24*, 1736–1740.

Natural rubber-based polymer blends and composites

2

Young Nam Kim [1,2] *and Yong Chae Jung* [1]
[1]Institute of Advanced Composite Materials, Korea Institute of Science and Technology (KIST), Wanju-gun, Jeonbuk, Republic of Korea; [2]Department of Chemical and Biomolecular Engineering, Yonsei University, Seodaemun-gu, Seoul, Republic of Korea

1. Introduction

1.1 What is natural rubber?

Natural rubber (NR) is rubber produced from raw materials of plant origin, and is mainly collected from the rubber tree (*Hevea brasiliensis*), native to the Amazon basin. Raw rubber is manufactured by collecting and drying an emulsion called latex, containing 30%−40% rubber powder, from a rubber tree, and exhibits various properties, depending on the rubber tree.

Research into NR began with Christopher Columbus' second voyage, in 1496, with the first NR balls that were taken from the West Indies to the Iberian Peninsula. Afterward, in 1791, Samuel Peal manufactured a waterproof cloth by coating a fabric with a solution of NR dissolved in turpentine oil, and applied for a patent on this method. This patent contributed to the gradual expansion of the application range of NR. In particular, in 1893, Charles Goodyear and Thomas Hancock devised an independent curing system for NR, by which NR was transformed to a polymer material with improved tensile strength and high resistance to organic solvents. This material modification became the cornerstone of the current global business on NR, dealing with billions of NR products. In 1908, the business of Henry Ford provided an opportunity for explosively increasing the demand for automobile tires using NR. However, Japan's occupation of Southeast Asia during World War II blocked the supply of NR, resulting in a sharp decline in industrial growth.

In the mid-1980s, as the demand for NR latex increased exponentially, and the NR latex industry moved from the West to the East, the total global consumption increased dramatically to 300,000 tons or more. In 2011, the worldwide consumption of NR and synthesized rubber (SR) comprised 42.3% and 57.7%, respectively (30% of NR was consumed by China). According to S. Evans, the former secretary-general of the International Rubber Study Group (IRSG), the expected consumption and production of NR in 2020 were 16.5 and 15.2 million tons, respectively (Chan et al., 2013; Ciesielski, 1999; Hurley, 1981; James & Guth, 1943; Mark et al., 2013; Visakh & Visakh, 2017).

NR is a representative reinforcing agent, and in particular, it has high applicability as an impact modifier in composite materials. At the early stage of research into NR,

Biodegradable Polymers, Blends and Composites. https://doi.org/10.1016/B978-0-12-823791-5.00002-8

latex applications mainly utilized the intrinsic properties of NR. The actual production yield (3000 kg/ha/year) was much less than the theoretical production potential (9000 kg/ha/year), which significantly deviated from the actual industrial potential. This difference is gradually increasing due to the poor working environments in the main production areas, as well as a lack of production manpower (Chan et al., 2013).

To resolve this problem, the Bridgestone Americas Center decoded the major genome sequence of Hevea to increase NR production, and applied the outcome to the development of new production technology and growing methods for NR. However, due to fundamental issues, such as pest attack and mold discoloration of the NR wood, as well as various contaminations occurring in the handling and transportation stages, commercial applications have been restricted, and methods to resolve these problems are still under development.

The major driving force for the recent industrial revival of NR is the discovery of nanomaterials. The application of nanomaterials to conventional polymeric materials has resulted in new physical properties, expansion of the application scope, and enhanced development potential of fusion materials between various materials (Chien et al., 2010; Dong et al., 2014; Kaneto et al., 1999; Yu et al., 2006).

The combination of NR and nanomaterials facilitates the discovery of new materials, and further widens the ripple effect of materials that can meet specific customer requirements. This chapter introduces nanocomposite materials prepared from NR and various nanomaterials and describes their influences on the performance and properties of NR.

2. Natural rubber-based polymer nanocomposites

Recently reported composite materials show differentiated performances compared to conventional materials and are being developed as materials with industrial applications, focusing on high performance, resistance to extreme environment, and eco-friendliness. In the past, materials were designed and developed according to need, while the current and future industries are focused more on the convergence material industry, whose application range is not limited to any one area. Polymer nanocomposite materials are drawing increasing research attention in response to this demand.

Polymer nanocomposite materials have high performance and multifunctionality, which exceed the limitations of existing physical properties, appearance (shape), and synthesis methods of the component nanomaterial. This chapter describes the properties of NR composites added as nanofillers, such as spherical silica particles, multilayered clay, and rod-shaped carbon nanotubes (CNTs).

2.1 Si-based NR nanocomposites

The mechanical/thermal properties of polymer composites are influenced by several factors, including process conditions, filler dispersion, polymer–filler interactions, and filler morphological aspects, and NR composite materials are no exception. Among them, the dispersion state of the filler is essential for determining the overall physical properties of nanocomposite materials, especially inorganic–organic nanocomposites, polymer–clay nanocomposites, metal–clay nanocomposites, etc. prepared by the sol–gel method. Studies to improve the dispersion state of fillers have been reported (Bokobza & Chauvin, 2005; Ikeda & Kameda, 2004; Poompradub et al., 2005).

The sol–gel process can be conducted before or after the cross-linking reaction. Ikeda's group performed a sol–gel reaction before hardening after immersing the unvulcanized NR sheet in tetraethyl orthosilicate (TEOS) solution at room temperature for several hours. The results show that the reference sample NR exhibits high tensile strength and elongation, while the composite material containing silica had enhanced physical properties. In particular, the reinforcing properties of the additives are remarkable at an initial elongation of 200% or more, which can be regarded as a result of the typical improvement of physical properties of composite materials due to the addition of fillers. Furthermore, there was an increase in the initial modulus of elasticity of the composite material synthesized by the simultaneous polymerization method, involving filler addition along with cross-linking of NR rubber, compared to the system fabricated by simple stirring as the filler addition method. This can be interpreted as a result of uniform distribution (variance) of fillers in the matrix (Bokobza & Chauvin, 2005; Ikeda & Kameda, 2004; Murakami et al., 2003; Poompradub et al., 2005).

2.2 Clay particles

In the case of fabricating a nanocomposite using clay, which has a typical plate-like structure, the contact area between the filler and the polymer matrix is relatively larger than that of the spherical filler, resulting in a greater filling effect than that obtained for the conventional composite material. Thus, the clay–polymer nanocomposite has the advantages of reduced dimensional stability, heat resistance, and gas permeability, and exhibits enhanced physical properties in a smaller amount compared to conventional composites. Montmorillonite (MMT) is the most actively applied filler material in the fabrication of polymer–clay nanocomposite materials. MMT is a layered mineral belonging to the smectite family, and has a hydrophilic structure in which cations, such as Ca^{2+}, Mg^{2+}, Na^{+}, and K^{+}, are filled between the layers, and –OH groups are present at the ends of the sheet; therefore, it is difficult to intercalate polymers with lipophilic properties between silicate layers. Thus, layered clay minerals are modified with organic clay, and then used in nanocomposite materials (Christidis & Scott, 1993). As a representative example, the first practical polymer–clay nanocomposite material was the nylon–clay nanocomposite reported by the Toyota Motor Company in 1987, prepared by an exfoliation phenomenon, in which the intercalation of a nylon monomer between silicate layers increased the interlayer distance to approximately 100 Å (Wang & Pinnavaia, 1994).

Joly et al. (2002) fabricated a nanocomposite by dissolving 10 phr of an organomodified clay [dimethyl hydrogenated tallow (2-ethylhexyl) ammonium montmorillonite], an organic solvent (toluene), as a formulation component for synthesizing NR. The XRD results show a diffraction pattern typical of organoclay (OC), as well as a 001 lattice plane at $2\theta = 2.5$ degrees. However, the results suggest that the peaks of the OC lattice included in NR were shifted to a lower diffraction angle. In particular, $d_{001} = 34.0$ Å for pure OC, whereas the interlayer spacing in the composite materials increased to 40.1 Å. This phenomenon can be attributed to the increased interaction between the filler and the polymer upon intercalation of polymer chains between the silicate layers. Moreover, as shown in the TEM images, conventional clay–polymer composites exhibit low dispersion behavior due to coagulation between clays, whereas OC exhibits uniform dispersion of nanofillers in the matrix. In particular, the plate-like structure of clay can be observed partially. These results show the correlation between the dispersion of the nanofiller and the improvement in the physical properties of the polymer nanocomposite (as shown in Figs. 2.1 and 2.2).

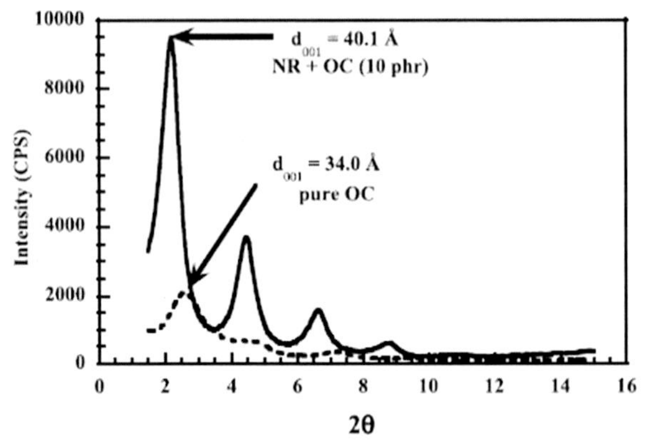

Figure 2.1 X-ray diffraction patterns for pure organoclay (OC) and for the NR composite filled with 10 phr of OC.
Reproduced with permission from Joly, S., Garnaud, G., Ollitrault, R., Bokobza, L., & Mark, J. E. (2002). Organically modified layered silicates as reinforcing fillers for natural rubber. *Chemistry of Materials*, *14*(10), 4202–4208.

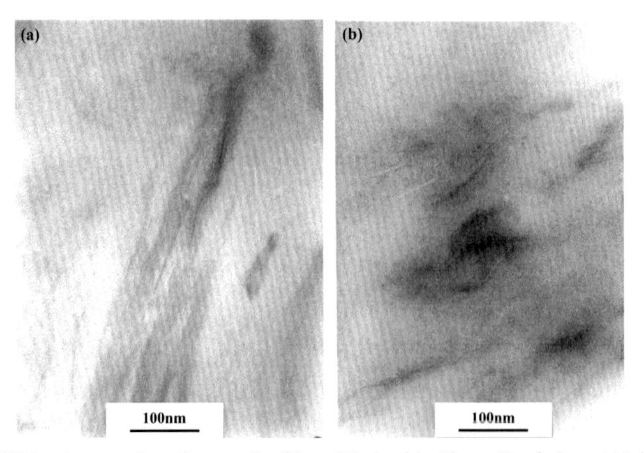

Figure 2.2 TEM micrographs of natural rubber filled with 10 wt.% of clay: (A) is for unmodified clay and (B) is for organomodified clay (Joly et al., 2002).

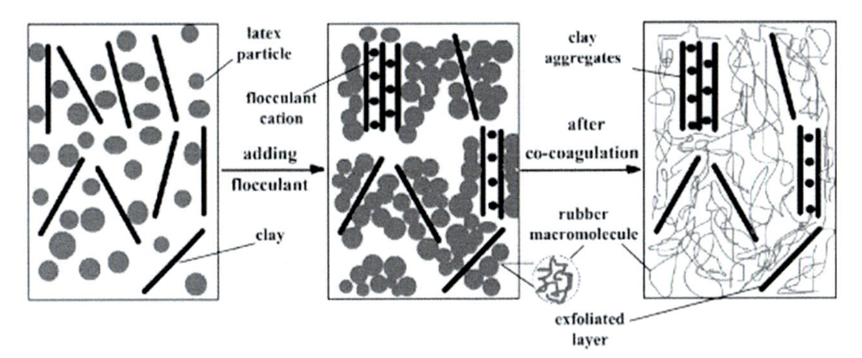

Figure 2.3 Schematic of the mixing and co-coagulating process (Wu et al., 2005).

Wu et al. (2005) further attempted uniform dispersion of the filler in rubber by cocoagulating clay and rubber in an aqueous suspension. Fig. 2.4 illustrates well this dispersion behavior. The filler and the matrix in aqueous suspension allow the rubber to penetrate into the filler by a capillary phenomenon, and subsequent coagulation enables the fabrication of a polymer nanocomposite with a uniform dispersion characteristic (as shown in Fig. 2.3).

2.3 Carbon nanomaterials

NR, which does not undergo reversible deformation, is used in various industries. Because of its mechanical properties of high strength and high elasticity, NR can be utilized for different purposes (Miedzianowska et al., 2019).

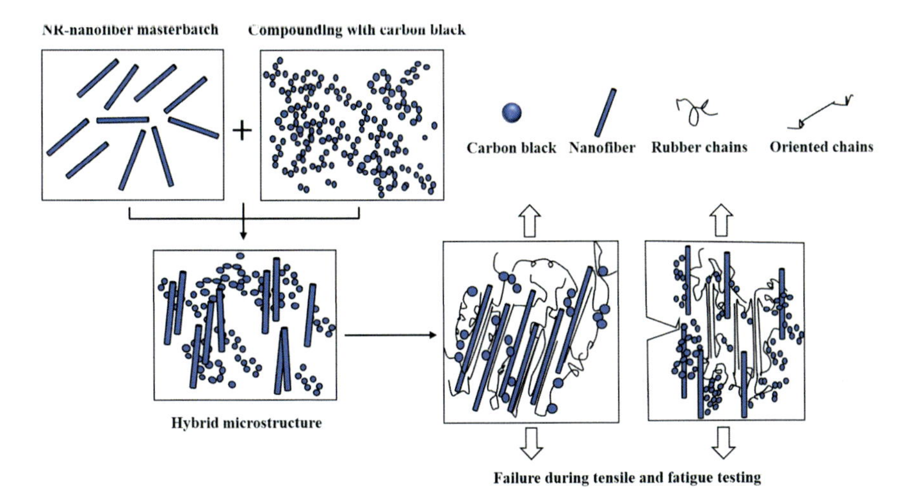

Figure 2.4 Development of hybrid microstructure and its effect on the failure resistance of composites (Parameswaran et al., 2021).

Carbon materials, including carbon nanotubes, carbon black, graphene, and MoS_2, are among the most widely explored nanomaterials (Ajayan et al., 2000; Baughman et al., 2002; Collins et al., 1997; Dresselhaus et al., 1996). Because carbon materials have high elastic modulus, excellent mechanical properties, and chemical stability, they are used in various fields such as for the development of high-strength composite materials, flame retardants, and biofields, as well as the preparation of electrochemical devices, hydrogen storage materials, field emission devices, gas sensors, supercapacitors, and memory devices (Baughman et al., 2002; Berber et al., 2000; Choi et al., 2003; Dzenis, 2004; Subbiah et al., 2005).

Carbon black (CB) is the most widely used reinforced filler in NR formulations, which are frequently used in static-dissipative plastics (Araby et al., 2015; Baughman et al., 2002; Bokobza, 2019; Ponnamma et al., 2014; Sadasivuni et al., 2014). In particular, the use of CB as a filler facilitates controlled conductivity, easy dispersion, improved component properties, and good surface treatment. However, a large amount of CB, that is, 10–30 times higher than that of nanomaterials, is required. Although this high amount enhances the electrical properties, it degrades the mechanical properties, thereby limiting its application. To resolve this problem, the use of nanomaterials as alternatives to CB has been widely investigated.

Zhao and colleagues (Sui et al., 2008) compared the composite formed to conventional NR by mixing CNTs and CB with NR that exhibited considerably higher mechanical strength, which was attributed to the presence of CNTs. In particular, a low tan delta value of the composite using CNTs and CB, determined through dynamic mechanical analysis (DMA), was achieved at 50–60°C, which suggests that the rolling resistance (antislip property) of the composite is high when NR is mixed with the filler (as shown in Fig. 2.4).

Furthermore, a hybrid filler system was prepared by simultaneously adding two or more different carbon-based fillers to the rubber matrix. The hybrid system exhibited higher mechanical properties than the system wherein a single filler was used, because of the physical interaction between the fillers in the rubber matrix. In particular, the composite in which CB and CNT were used showed an increase in the tensile strength value by approximately 110% compared to the one with only CB (Parameswaran et al., 2021). In addition, the tan delta value of the composite was lower than that of the system wherein only CB was used, which could be attributed to the higher rolling resistance of the former. Thus, this NR composite can be used as a tire material. Because abrasion loss is an important property that represents high performance of tires, hybrid composites such as those including CNTs showed a significantly lower abrasion loss value than the system based on only CB.

Das and colleagues illustrated that the simultaneous addition of the two fillers (CB and CNTs) causes an increase in the mechanical properties because when the well-dispersed CNTs and CB are subjected to force, as shown in the Fig. 2.5, CNTs are aligned in the direction of the applied force, facilitating force transmission. In addition to mechanical property enhancement, the piezoelectric properties can be improved by fabricating rubber composites using NR, CB, and CNTs. Although the system using CB alone showed higher sensitivity than the composite with NR + CB + CNTs at the same strain value, the single system exhibited a more stable behavior during the

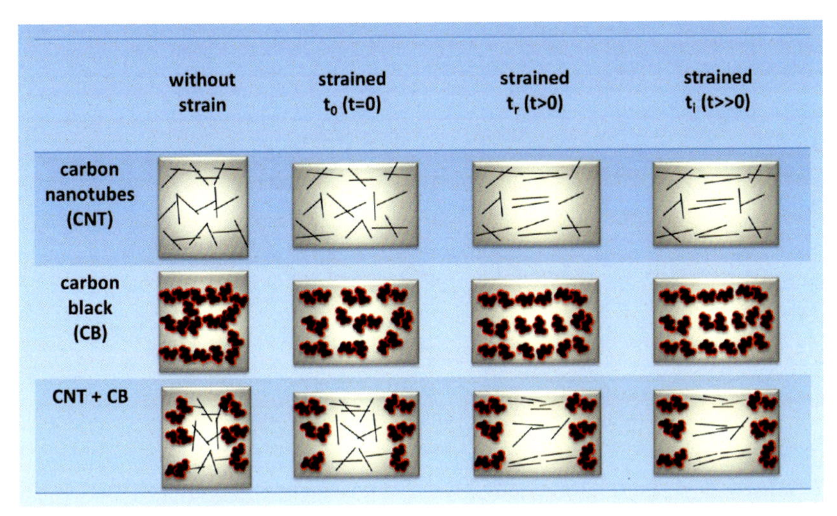

Figure 2.5 Mechanism of time dependence of NR containing CNT, CB, hybrid (CNT/CB) nanocomposites. t0: Breakdown of conductive network immediately upon straining ($t = 0$). tr: Alignment of conductive particles along the tensile direction, as well as reformation of destroyed conductive network ($t > 0$). ti: Slow and gradual reduction in interparticle distances over longer time, leading to slow decay in resistance ($t \gg 0$) (Natarajan et al., 2017).

loading–unloading cycle. This is because when CNTs and CB are mixed in an appropriate amount, the application of strain allows for the alignment of CNTs (as shown in Figs. 2.5 and 2.6) (Natarajan et al., 2017).

CNTs are considered as a one-dimensional cylindrical form of graphene, in which three surrounding carbon atoms form a honeycomb-shaped lattice through sp2 bonding of carbon atoms (Ajayan, 1999; Seidel et al., 2004). The band gap of CNTs is inversely proportional to their diameter. Single-walled CNTs with diameters of 1–2 nm exhibit conducting and semiconducting properties depending on the

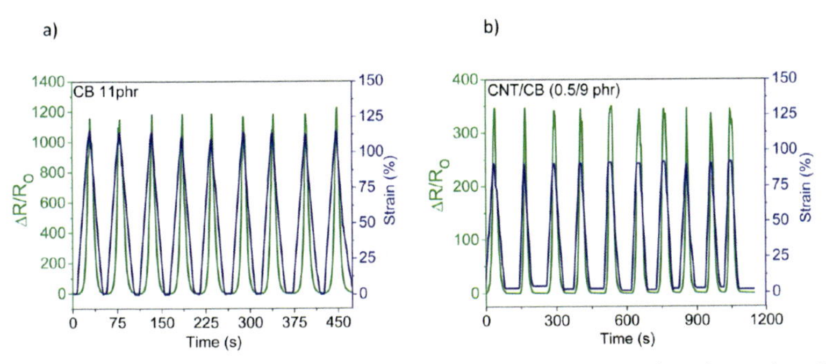

Figure 2.6 Cyclic strain experiment and plots of relative resistance and strain vs. time for (A) CB 11 phr composite and (B) for CNT/CB 0.5/9 phr hybrid composite (Natarajan et al., 2017).

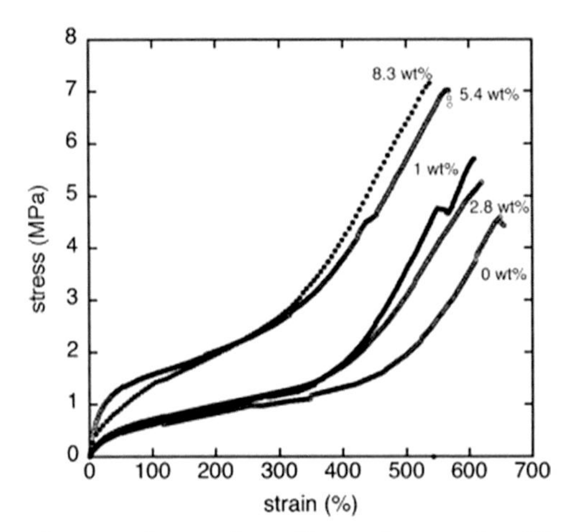

Figure 2.7 Stress–strain curves for pure latex films and composites (Bhattacharyya et al., 2008).

winding direction (Odom et al., 1998; Wilder et al., 1998). As the number of walls of the CNTs increases, the diameter of the outer wall increases to several tens of nanometers and the band gap decreases; thus, multiwalled carbon nanotubes (MWCNTs) exhibit the properties of a conductor (Ouyang et al., 2002; Van Hove, 1953). CNTs are lightweight because of their hollow cylindrical structure and are highly anisotropic owing to their nano-sized diameter and micro-sized length, which impart high electrical and thermal conductivity as well as a higher tensile force than that of iron (Baughman et al., 2002; Endo et al., 2005; Van Hove, 1953).

Significant improvement in mechanical stiffness has been reported with the addition of MWCNTs in NR (Fakhru'l-Razi et al., 2006; Ruoff & Lorents, 1995).

Jean-Paul Salvetata and colleagues (Bhattacharyya et al., 2008) modified the surface of MWCNTs by strong acid treatment to introduce carboxyl functional groups, and added sodium dodecyl sulfate (SDS) into water to make a solution wherein the CNTs were uniformly dispersed. SDS used in this study is an ionic detergent with a strong negative charge, which is widely used for denaturing proteins or as a propellant for CNTs. The overall mechanical properties increased because of the uniform dispersion of the CNTs in the rubber, and the dielectric measurement result at room temperature showed a low transmission threshold (<1 wt.%), which could be attributed to the well-developed network structure between the CNTs uniformly dispersed in the matrix (as shown in Figs. 2.7–2.9).

Typically, the improvement of elongation due to the addition of the nanofiller into the polymer also causes an increase in strain energy, contributing to the improvement of the overall mechanical properties (including tensile strength, elongation, and impact resistance) of nanocomposites. Fig. 2.10 shows that the properties increase to a certain level, temporarily decrease, and then increase again. This phenomenon can be explained by the pull-out phenomenon: when the CNTs are subjected to load application in the tensile direction, the CNTs in the outermost layer are destroyed and the

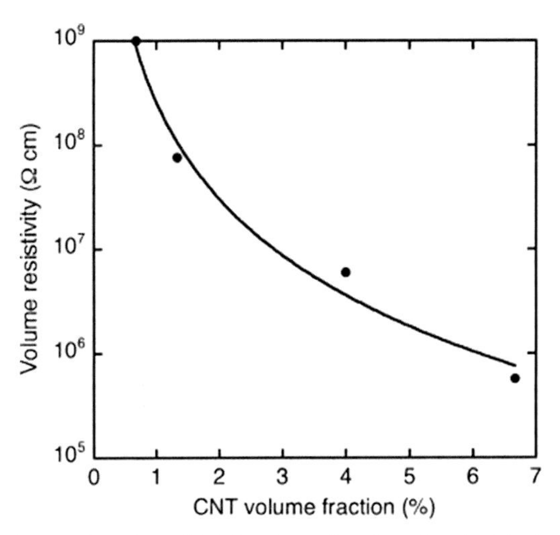

Figure 2.8 Resistivity as a function of CNT volume fraction. An electrical percolation behavior is observed because of CNT network formation (Bhattacharyya et al., 2008).

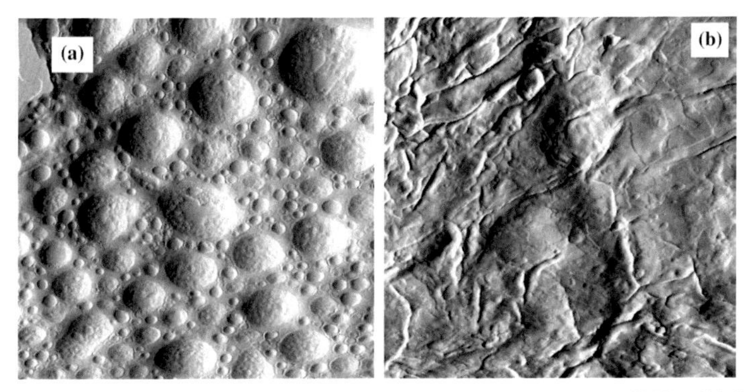

Figure 2.9 AFM image of the surface of a composite with 2.8 wt.% MWCNTs. (A) Before and (B) after coagulation of the latex beads at 60°C. The derivative of the topography signal is shown to highlight the morphology of the film after water evaporation. It was not possible to resolve individual nanotubes at the film surface. No clusters of MWCNTs were observed, showing that macroscopic dispersion was good in agreement with optical microscopy observations (Bhattacharyya et al., 2008).

inner wall layer follows, resulting in the extension of the CNTs. As a result, stretching of the fibers surrounding the CNTs also occurs, which results in an increase in the physical properties of the polyurethane–CNT composite nanowebs (Ajayan et al., 2000; Collins et al., 1997).

Moreover, the comparison between AFM images before and after the coagulation of the composite confirms the uniform dispersion of CNTs (Fig. 2.13). In the images before coagulation, many clusters of irregular size were observed on the surface of the composite, whereas such clusters were not observed after coagulation. In particular,

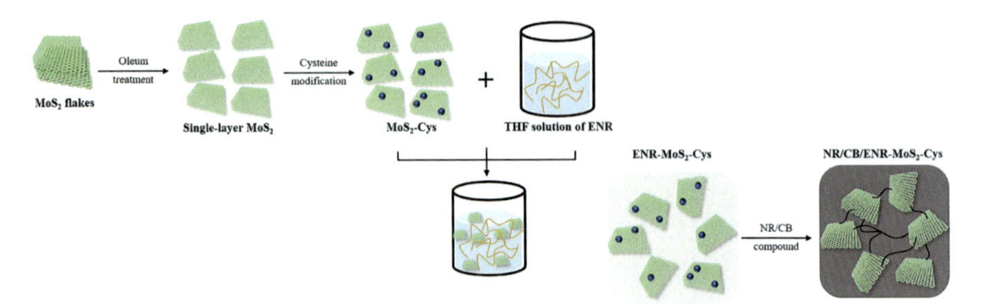

Figure 2.10 Schematic of the entire preparation process of MoS_2-Cys, ENR-MoS_2-Cys, and NR/CB/ENR-MoS_2-Cys (Jiang et al., 2020).

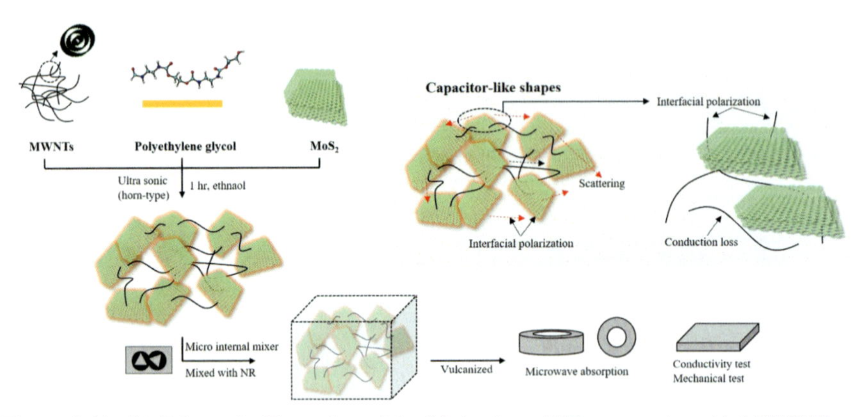

Figure 2.11 (A) Schematic illustration of the fabrication of NR composite with MWCNTs and MoS_2 hybrids. (B) Schematic illustration of possible microwave attenuation mechanisms of the NR/MWCNTs/MoS_2 composite (Geng et al., 2020).

after coagulation, morphological changes such as coagulation and reduced dispersion of CNTs were not identified, which is attributed to the improvement of mechanical properties and electrical properties of the composites (Grunlan et al., 2004, 2006).

2.4 MoS_2

NR is known to exhibit excellent resilience and high tensile strength, while reinforcing fillers are generally used owing to their low wear resistance (Jiang et al., 2020; Spratte et al., 2017). Molybdenum disulfide (MoS_2), a new nanofiller used to effectively respond to this demand, is composed of S−Mo−S triple layers bound by a weak Van der Waals force, which is widely applied in many fields, such as nanoelectronics, sensors, and catalysts.

Studies have shown that MoS_2 has excellent mechanical properties, such as a high elastic modulus, yield stress, and flexural modulus, suggesting that it is highly likely to be an effective reinforcing filler as a type of polymer composite material (Cooper et al., 2013). In addition, MoS_2 has a hierarchical structure similar to graphite, with excellent

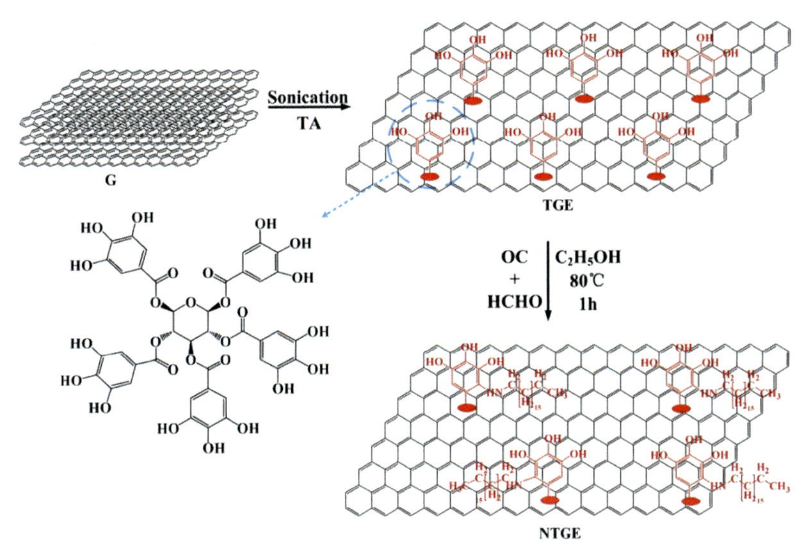

Figure 2.12 Process of liquid-phase exfoliation and mechanism of functionalization of graphene (Zhao et al., 2018).

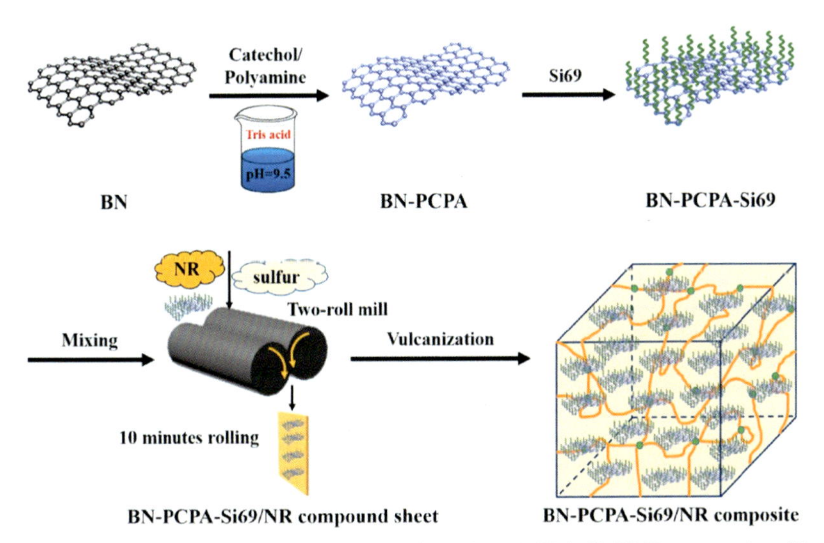

Figure 2.13 Schematic diagram of the preparation of BN-PCPA-Si69/NR composites (Yang et al., 2021).

electronic properties and a high probability of microwave attenuation due to dielectric and magnetic losses by polymorphism. It has been used in various polymers to improve mechanical properties and fire resistance. However, owing to poor dispersion in the matrix and weak interaction with the matrix, this application has not been sufficient for high-performance tires and requires additional application.

The study results on NR/CB/ENR-MoS_2-Cys as an additional application were reported to resolve this problem, where a mixture of epoxide natural rubber (ENR) and cysteine (Cys) modified MoS_2 (MoS_2-Cys) was added into the NR/CB compound (as shown in Fig. 2.10) (Jiang et al., 2020).

The ENR in the MoS_2 nanocomposite showed an excellent interaction between the modified MoS_2-cys and NR, and the MoS_2-cys size of NR/CB/ENR-MoS_2-Cys was smaller than that of NR/CB/MoS_2-Cys. As a result, the ENR effectively increased the dispersion of MoS_2-Cys in the NR matrix (Jiang et al., 2020). In other words, because large-sized MoS_2-Cys may not be properly dispersed in the NR matrix, it could induce stress concentration and severe internal consumption, which can be considered as the cause of the degradation in mechanical properties. In contrast, the size of MoS_2-Cys reduced due to ENR, and it was well dispersed in the CdhsmB region, leading to uniform filler dispersion, which was advantageous in forming a hybrid network in the composite. These advantages resulted in an increase in the complexity of the tensile strength, tensile modulus (100% and 300% strain), and hardness, and further improved the wear resistance and increased cross-linking density. This allowed the control of the dynamic mechanical properties of NR. As a result, NR/CB/ENR-MoS_2-Cys improved the dispersion of the filler and the interaction between the filler and the matrix, and it simultaneously demonstrated the possibility of using MoS_2 for rubber reinforcement.

Electromagnetic pollution has recently deteriorated the performance of electronic equipment and threatened human health (Saini & Arora, 2012). A microwave absorbing material (MA) that absorbs incident electromagnetic waves and converts them into thermal energy has been presented as an effective solution (Arief et al., 2017; Cao et al., 2019; Thomassin et al., 2013). However, owing to the fundamental disadvantages of the narrow effective bandwidth, studies have recently improved the electromagnetic wave attenuation function of polymers by appropriately combining nanopillars (Cao et al., 2012; Geng et al., 2020; Kwon et al., 2014; Maiti et al., 2013).

Yanfang Zhao and Jianhe Liao developed a hybrid NR nanocomposite containing 1D- MWCNT and 2D-MoS_2 to resolve this problem (Geng et al., 2020).

The nanocomposite compounded with NR without mixing the additional filler (MWCNTs and MoS_2) could not properly address the disadvantages of the existing MA. (1) The NR nanocomposite to which MWCNTs were added exhibited a decreased interfacial adhesion between the filler and the matrix as a hole was formed in the matrix owing to the low tensile strength and decreased dispersion of the filler. (2) The MoS_2 could not achieve excellent MA capacity owing to the semiconductive properties of the filler as the matching properties became poor, showing significantly nonuniform dispersion properties. (3) The NR nanocomposite using the mixed filler showed an improved dispersion degree (as shown in Fig. 2.11).

This can be a result of reinforcing free electron transfer by inhibiting the recoagulation of MoS_2, by infiltrating the layers of the keratinized MoS_2 sheet to form a layered structure using the long rod-shaped MWCNTs when manufacturing the nanocomposite. Simultaneously, the surface area of MoS_2 increases, resulting in a widened contact surface area between the MWCNT/MoS_2 and the NR matrix, as well as improved microwave absorption performance. Eventually, this structure had a positive effect on the mechanical, electrical, and electromagnetic properties (Fig. 2.15D) (Geng et al., 2020). In addition, the MA characteristics are superior to those obtained when using a single filler, but inferior to those obtained when using a hybrid (mixed) filler. The RL_{min} of the NR/MWCNT nanocomposite is approximately −5.34 dB at 12.08 GHz, whereas that of the NR/MoS_2 nanocomposite is only −2.24 dB, which is slightly higher than that of NR. However, the hybrid nanocomposite considerably improved the microwave absorption performance.

This characteristic induced a strong interface polarization between the two fillers, or the Maxwell−Wagner effect. In this case, the Maxwell−Wagner polarization model is the simplest for describing the polarization of particles under alternating and direct current electric fields, under the assumption of point dipole when particles and liquids have conductivity as well as a dielectric constant (von Hippel & Morgan, 1955).

This phenomenon mainly occurs in heterogeneous media because charges are accumulated on the filler's interface, and in this case, large dipoles are formed in particles

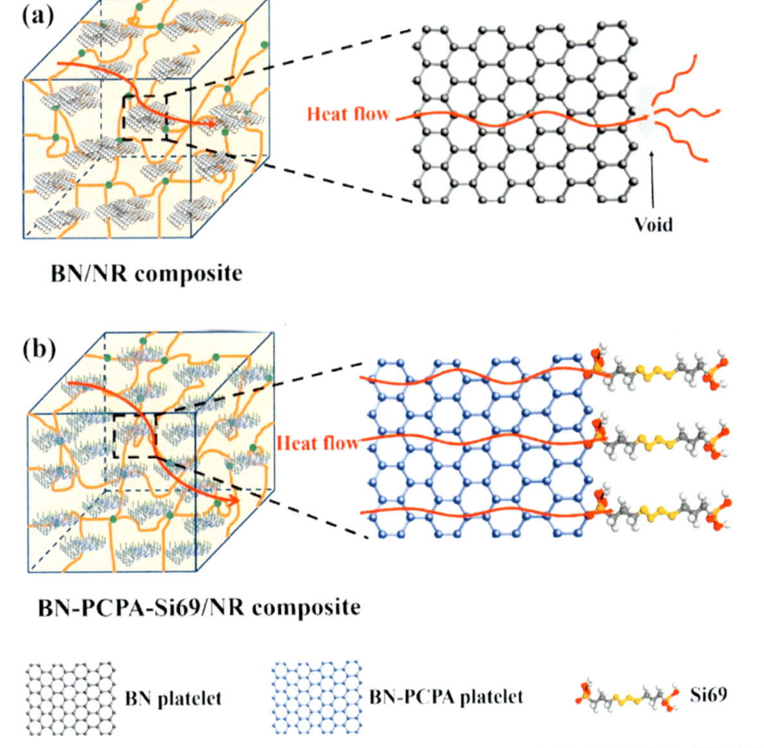

Figure 2.14 Schematic diagram of heat flow of (A) BN/NR and (B) BN-PCPA-Si69/NR composites (Yang et al., 2021).

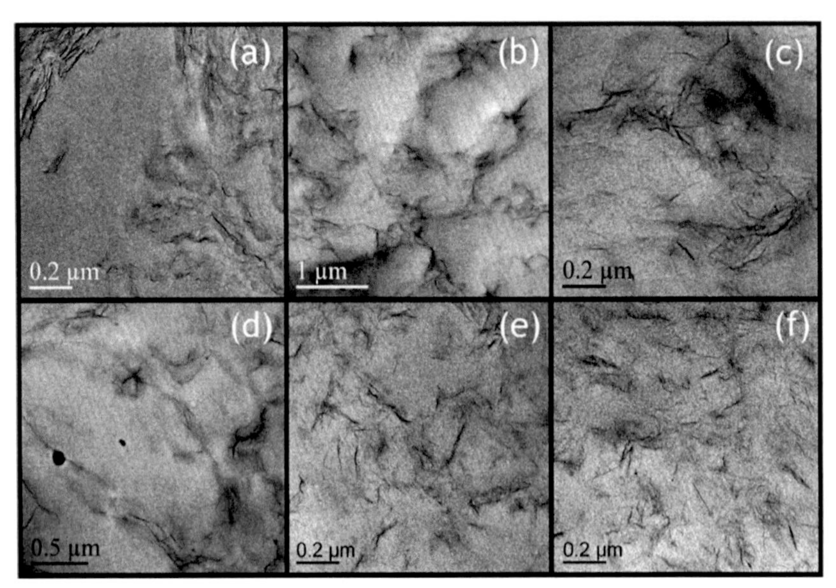

Figure 2.15 TEM micrographs of RG-O/NR nanocomposite sections. (A and B) show the "weblike" dispersion of RG-O platelets in the uncured composites, as obtained directly after latex co-coagulation. (C and D) show the dispersion in the solution-treated samples, while images (E and F) show the morphology of the milled nanocomposites (Potts et al., 2012).

or clusters. Eventually, this study could develop nanocomposite materials with a high absorption efficiency and wide effective absorption bandwidth at a small thickness by mixing fillers with different structures.

2.5 Tannic acid

Tannic acid, which is abundantly present in nature, is a type of polyphenol mainly synthesized by plants. Tannin is used as a flame retardant because it forms graphite during combustion and has the ability to deactivate radicals formed from burning materials.

Zhao et al. (2018) developed functionalized graphene using a new eco-friendly liquid peeling method. They added tannic acid into water as a stabilizer for the dispersion, reduction, and exfoliation of graphene. Tannic acid structurally has many −OH groups and can improve dispersibility in an organic solvent or matrix when combined with nanofillers. For example, when tannic acid is reacted with graphene oxide, reduced graphene is produced. This is because the catechol groups present in the end groups of tannic acid are oxidized to quinone groups, and then, they further bond to the OH groups on the graphene oxide surface or edge via self-polymerization, resulting in reduced graphene coated with TA. This reduced graphene can improve the structural defects and conductivity. In addition, because tannic acid can form graphite during combustion and can deactivate radicals formed in burning materials, it exhibits flame-retardant properties (as shown in Fig. 2.12).

3. Application of NR-based polymer nanocomposites

3.1 Thermal conductivity

Zhang and colleagues used boron nitride (BN) to increase the thermal conductivity of NR and functionalized the surface of BN to further increase the thermal conductivity (as shown in Figs. 2.13 and 2.14) (Yang et al., 2021).

According to the result, BN is functionalized with polyamine and Si69, and BN-PCPA-Si69 is formed wherein polymer chains grow on the surface of BN, and these polymer chains can be cross-linked with NR. Thus, as the heat transfer path is created and the amount of as-prepared BN-PCPA-Si69 is increased, the thermal conductivity increases. The thermal conductivity of functionalized BN-PCPA-Si69 was higher than that of BN, showing that an increase in the filler content to 30 vol% causes an increase in the thermal conductivity from 0.15 to 0.8 W/mk. As shown in Fig. 2.14, the thermal conductivity increases as the phonon moves smoothly. As phonon scattering appears on the interface between BN and NR, the thermal conductivity may decrease. However, the functionalized BN-PCPA-Si69 inhibits phonon scattering at the interface between BN and NR, enabling an increase in the number of phonon movement paths and, thus, higher thermal conductivity.

In the following example, the thermal conductivity of NR was increased upon the addition of R-GO to NR. This study has confirmed that different properties are imparted depending on the fabrication method of rubber composites using R-GO, which includes a two-roll method using the conventional method and a "solution treatment" method using toluene (Potts et al., 2012) (as shown in Figs. 2.15 and 2.16). These two fabrication methods differed in that R-GO had different structures upon filler formation in the rubber matrix. The solution treatment method involved the use of the filler that has a "web-like" structure, whereas the two-roll fabrication method caused homogeneous dispersion after the structure was destroyed.

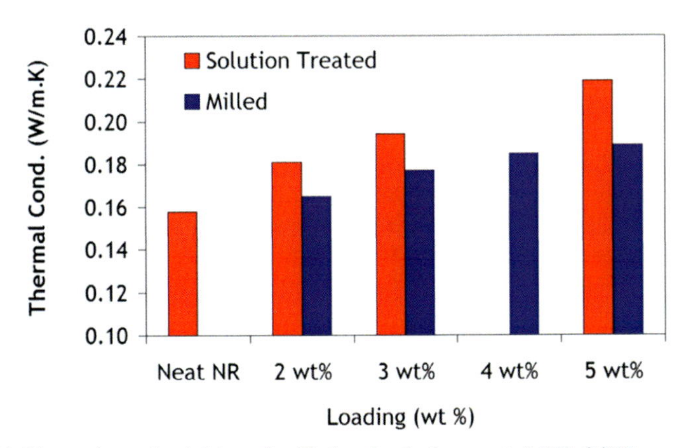

Figure 2.16 Thermal conductivities of milled and solution-treated RG-O/NR nanocomposites at various loadings (Potts et al., 2012).

Because the NR composite fabricated by the solution treatment method is formed upon the addition of the filler having a "web-like" structure, the phonon movement is easier and the thermal conductivity is higher than the composite fabricated by the two-roll method (Fig. 2.16).

4. Conclusion

NR is one of the representative engineering plastics that have been continuously used throughout by numerous industries since its discovery by Christopher Columbus. In this chapter, we have described the preparation of a nanocomposite material by using various additives to overcome the shortcomings and to develop additional functionalities while maintaining the characteristics of the conventional NR. Depending on the purpose, the type of filler to be selected, as well as the application scope, may vary. This chapter will encourage the development of nanocomposite materials.

Acknowledgments

This work was supported by the KIST Institutional Program. The authors would like to thank Mr. Seock Hee Shin and Mr. WonHyeong Seo (KIST-Jeonbuk National University Internship Program, Jeonbuk National University) for report searches and compilation of data.

References

Ajayan, P. M., Schadler, L. S., Giannaris, C., & Rubio, A. (2000). Single-walled carbon nanotube—polymer composites: Strength and weakness. *Advanced Materials, 12*(10), 750—753.

Ajayan, P. M. (1999). Nanotubes from carbon. *Chemical Reviews, 99*(7), 1787—1800.

Araby, S., Meng, Q., Zhang, L., Zaman, I., Majewski, P., & Ma, J. (2015). Elastomeric composites based on carbon nanomaterials. *Nanotechnology, 26*(11), 112001.

Arief, I., Biswas, S., & Bose, S. (2017). Graphene analogues as emerging materials for screening electromagnetic radiations. *Nano-Structures and Nano-Objects, 11*, 94—101.

Baughman, R. H., Zakhidov, A. A., & De Heer, W. A. (2002). Carbon nanotubes-the route toward applications. *Science, 297*(5582), 787—792.

Berber, S., Kwon, Y. K., & Tománek, D. (2000). Unusually high thermal conductivity of carbon nanotubes. *Physical Review Letters, 84*(20), 4613.

Bhattacharyya, S., Sinturel, C., Bahloul, O., Saboungi, M. L., Thomas, S., & Salvetat, J. P. (2008). Improving reinforcement of natural rubber by networking of activated carbon nanotubes. *Carbon, 46*(7), 1037—1045.

Bokobza, L., & Chauvin, J. P. (2005). Reinforcement of natural rubber: Use of in situ generated silicas and nanofibres of sepiolite. *Polymer, 46*(12), 4144—4151.

Bokobza, L. (2019). Natural rubber nanocomposites: A review. *Nanomaterials, 9*(1), 12.

Cao, M. S., Yang, J., Song, W. L., Zhang, D. Q., Wen, B., Jin, H. B., … Yuan, J. (2012). Ferroferric oxide/multiwalled carbon nanotube vs polyaniline/ferroferric oxide/multiwalled carbon nanotube multiheterostructures for highly effective microwave absorption. *ACS Applied Materials and Interfaces, 4*(12), 6949–6956.

Cao, M. S., Cai, Y. Z., He, P., Shu, J. C., Cao, W. Q., & Yuan, J. (2019). 2D MXenes: Electromagnetic property for microwave absorption and electromagnetic interference shielding. *Chemical Engineering Journal, 359*, 1265–1302.

Chan, C. H., Joy, J., Maria, H. J., & Thomas, S. (2013). Natural rubber-based composites and nanocomposites: State of the art, new challenges and opportunities. *Natural Rubber Materials, 2*, 1–33.

Chien, Y. M., Lefevre, F., Shih, I., & Izquierdo, R. (2010). A solution processed top emission OLED with transparent carbon nanotube electrodes. *Nanotechnology, 21*(13), 134020.

Choi, S. W., Jo, S. M., Lee, W. S., & Kim, Y. R. (2003). An electrospun poly (vinylidene fluoride) nanofibrous membrane and its battery applications. *Advanced Materials, 15*(23), 2027–2032.

Christidis, G., & Scott, P. W. (1993). Laboratory evaluation of. *Industrial Minerals*.

Ciesielski, A. (1999). *An introduction to rubber technology*. iSmithers Rapra Publishing.

Collins, P. G., Zettl, A., Bando, H., Thess, A., & Smalley, R. E. (1997). Nanotube nanodevice. *Science, 278*(5335), 100–102.

Cooper, R. C., Lee, C., Marianetti, C. A., Wei, X., Hone, J., & Kysar, J. W. (2013). Nonlinear elastic behavior of two-dimensional molybdenum disulfide. *Physical Review B, 87*(3), 035423.

Dong, P., Zhu, Y., Zhang, J., Hao, F., Wu, J., Lei, S., … Lou, J. (2014). Vertically aligned carbon nanotubes/graphene hybrid electrode as a TCO-and Pt-free flexible cathode for application in solar cells. *Journal of Materials Chemistry A, 2*(48), 20902–20907.

Dresselhaus, M. S., Dresselhaus, G., & Eklund, P. C. (1996). *Science of fullerenes and carbon nanotubes: Their properties and applications*. Elsevier.

Dzenis, Y. (2004). Spinning continuous fibers for nanotechnology. *Science, 304*(5679), 1917–1919.

Endo, M., Muramatsu, H., Hayashi, T., Kim, Y. A., Terrones, M., & Dresselhaus, M. S. (2005). 'Buckypaper'from coaxial nanotubes. *Nature, 433*(7025), 476.

Fakhru'l-Razi, A., Atieh, M. A., Girun, N., Chuah, T. G., El-Sadig, M., & Biak, D. R. A. (2006). Effect of multi-wall carbon nanotubes on the mechanical properties of natural rubber. *Composite Structures, 75*(1–4), 496–500.

Geng, H., Zhao, P., Mei, J., Chen, Y., Yu, R., Zhao, Y., … Liao, J. (2020). Improved microwave absorbing performance of natural rubber composite with multi-walled carbon nanotubes and molybdenum disulfide hybrids. *Polymers for Advanced Technologies, 31*(11), 2752–2762.

Grunlan, J. C., Mehrabi, A. R., Bannon, M. V., & Bahr, J. L. (2004). Water-based single-walled-nanotube-filled polymer composite with an exceptionally low percolation threshold. *Advanced Materials, 16*(2), 150–153.

Grunlan, J. C., Kim, Y. S., Ziaee, S., Wei, X., Abdel-Magid, B., & Tao, K. (2006). Thermal and mechanical behavior of carbon-nanotube-filled latex. *Macromolecular Materials and Engineering, 291*(9), 1035–1043.

Hurley, P. E. (1981). History of natural rubber. *Journal of Macromolecular Science, Chemistry, 15*(7), 1279–1287.

Ikeda, Y., & Kameda, Y. (2004). Preparation of "green" composites by the sol-gel process: In situ silica filled natural rubber. *Journal of Sol-Gel Science and Technology, 31*(1), 137–142.

James, H. M., & Guth, E. (1943). Theory of the elastic properties of rubber. *The Journal of Chemical Physics, 11*(10), 455–481.

Jiang, Y., Wang, J., Wu, J., & Zhang, Y. (2020). Preparation of high-performance natural rubber/carbon black/molybdenum disulfide composite by using the premixture of epoxidized natural rubber and cysteine-modified molybdenum disulfide. *Polymer Bulletin*, 1–18.

Joly, S., Garnaud, G., Ollitrault, R., Bokobza, L., & Mark, J. E. (2002). Organically modified layered silicates as reinforcing fillers for natural rubber. *Chemistry of Materials, 14*(10), 4202–4208.

Kaneto, K., Tsuruta, M., Sakai, G., Cho, W. Y., & Ando, Y. (1999). Electrical conductivities of multi-wall carbon nano tubes. *Synthetic Metals, 103*(1–3), 2543–2546.

Kwon, S., Ma, R., Kim, U., Choi, H. R., & Baik, S. (2014). Flexible electromagnetic interference shields made of silver flakes, carbon nanotubes and nitrile butadiene rubber. *Carbon, 68*, 118–124.

Maiti, S., Shrivastava, N. K., Suin, S., & Khatua, B. B. (2013). Polystyrene/MWCNT/graphite nanoplate nanocomposites: Efficient electromagnetic interference shielding material through graphite nanoplate–MWCNT–graphite nanoplate networking. *ACS Applied Materials and Interfaces, 5*(11), 4712–4724.

Mark, J. E., Erman, B., & Roland, M. (2013). In *The science and technology of rubber*. Academic press.

Miedzianowska, J., Masłowski, M., & Strzelec, K. (2019). Thermoplastic elastomer biocomposites filled with cereal straw fibers obtained with different processing methods—preparation and properties. *Polymers, 11*(4), 641.

Murakami, K., Iio, S., Ikeda, Y., Ito, H., Tosaka, M., & Kohjiya, S. (2003). Effect of silane-coupling agent on natural rubber filled with silica generated in situ. *Journal of Materials Science, 38*(7), 1447–1455.

Natarajan, T. S., Eshwaran, S. B., Stöckelhuber, K. W., Wießner, S., Pötschke, P., Heinrich, G., & Das, A. (2017). Strong strain sensing performance of natural rubber nanocomposites. *ACS Applied Materials and Interfaces, 9*(5), 4860–4872.

Odom, T. W., Huang, J. L., Kim, P., & Lieber, C. M. (1998). Atomic structure and electronic properties of single-walled carbon nanotubes. *Nature, 391*(6662), 62–64.

Ouyang, M., Huang, J. L., & Lieber, C. M. (2002). Scanning tunneling microscopy studies of the one-dimensional electronic properties of single-walled carbon nanotubes. *Annual Review of Physical Chemistry, 53*(1), 201–220.

Parameswaran, S. K., Bhattacharya, S., Mukhopadhyay, R., Naskar, K., & Bhowmick, A. K. (2021). Excavating the unique synergism of nanofibers and carbon black in Natural rubber based tire tread composition. *Journal of Applied Polymer Science, 138*(3), 49682.

Ponnamma, D., Sadasivuni, K. K., Grohens, Y., Guo, Q., & Thomas, S. (2014). Carbon nanotube based elastomer composites–an approach towards multifunctional materials. *Journal of Materials Chemistry C, 2*(40), 8446–8485.

Poompradub, S., Kohjiya, S., & Ikeda, Y. (2005). Natural rubber/in situ silica nanocomposite of a high silica content. *Chemistry Letters, 34*(5), 672–673.

Potts, J. R., Shankar, O., Du, L., & Ruoff, R. S. (2012). Processing–morphology–property relationships and composite theory analysis of reduced graphene oxide/natural rubber nanocomposites. *Macromolecules, 45*(15), 6045–6055.

Ruoff, R. S., & Lorents, D. C. (1995). Mechanical and thermal properties of carbon nanotubes. *Carbon, 33*(7), 925–930.
Sadasivuni, K. K., Ponnamma, D., Thomas, S., & Grohens, Y. (2014). Evolution from graphite to graphene elastomer composites. *Progress in Polymer Science, 39*(4), 749–780.
Saini, P., & Arora, M. (2012). Microwave absorption and EMI shielding behavior of nanocomposites based on intrinsically conducting polymers, graphene and carbon nanotubes. *New Polymers for Special Applications, 3*, 73–112.
Seidel, R., Duesberg, G. S., Unger, E., Graham, A. P., Liebau, M., & Kreupl, F. (2004). Chemical vapor deposition growth of single-walled carbon nanotubes at 600 C and a simple growth model. *The Journal of Physical Chemistry B, 108*(6), 1888–1893.
Spratte, T., Plagge, J., Wunde, M., & Klüppel, M. (2017). Investigation of strain-induced crystallization of carbon black and silica filled natural rubber composites based on mechanical and temperature measurements. *Polymer, 115*, 12–20.
Subbiah, T., Bhat, G. S., Tock, R. W., Parameswaran, S., & Ramkumar, S. S. (2005). Electrospinning of nanofibers. *Journal of Applied Polymer Science, 96*(2), 557–569.
Sui, G., Zhong, W. H., Yang, X. P., Yu, Y. H., & Zhao, S. H. (2008). Preparation and properties of natural rubber composites reinforced with pretreated carbon nanotubes. *Polymers for Advanced Technologies, 19*(11), 1543–1549.
Thomassin, J. M., Jerome, C., Pardoen, T., Bailly, C., Huynen, I., & Detrembleur, C. (2013). Polymer/carbon based composites as electromagnetic interference (EMI) shielding materials. *Materials Science and Engineering: R: Reports, 74*(7), 211–232.
Van Hove, L. (1953). The occurrence of singularities in the elastic frequency distribution of a crystal. *Physical Review, 89*(6), 1189.
Visakh, P. M., & Visakh, P. M. (2017). *Rubber based bionanocomposites*. Springer International PU.
von Hippel, A. R., & Morgan, S. O. (1955). Dielectric materials and applications. *Journal of the Electrochemical Society, 102*(3), 68C.
Wang, M. S., & Pinnavaia, T. J. (1994). Clay-polymer nanocomposites formed from acidic derivatives of montmorillonite and an epoxy resin. *Chemistry of Materials, 6*(4), 468–474.
Wilder, J. W., Venema, L. C., Rinzler, A. G., Smalley, R. E., & Dekker, C. (1998). Electronic structure of atomically resolved carbon nanotubes. *Nature, 391*(6662), 59–62.
Wu, Y. P., Wang, Y. Q., Zhang, H. F., Wang, Y. Z., Yu, D. S., Zhang, L. Q., & Yang, J. (2005). Rubber–pristine clay nanocomposites prepared by co-coagulating rubber latex and clay aqueous suspension. *Composites Science and Technology, 65*(7–8), 1195–1202.
Yang, D., Wei, Q., Yu, L., Ni, Y., & Zhang, L. (2021). Natural rubber composites with enhanced thermal conductivity fabricated via modification of boron nitride by covalent and non-covalent interactions. *Composites Science and Technology, 202*, 108590.
Yu, X., Rajamani, R., Stelson, K. A., & Cui, T. (2006). Carbon nanotube based transparent conductive thin films. *Journal of Nanoscience and Nanotechnology, 6*(7), 1939–1944.
Zhao, Z., Li, L., Shao, X., Liu, X., Zhao, S., Xie, S., & Xin, Z. (2018). Tannic acid-assisted green fabrication of functionalized graphene towards its enhanced compatibility in NR nanocomposite. *Polymer Testing, 70*, 396–402.

Soy protein-based polymer blends and composites

Aswathy Jayakumar[1], *Sabarish Radoor*[2], *E.K. Radhakrishnan*[1], *Indu C. Nair*[3], *Suchart Siengchin*[5] *and Jyotishkumar Parameswaranpillai*[4]

[1]School of Biosciences, Mahatma Gandhi University, Kottayam, Kerala, India; [2]Department of Mechanical and Process Engineering, The Sirindhorn International Thai—German Graduate School of Engineering (TGGS), King Mongkut's University of Technology North Bangkok, Bangkok, Thailand; [3]Department of Biotechnology, Sahodaran Ayyappan Smaraka Sree Narayana Dharma Paripalana Yogam (SAS SNDPYOGAM), Konni College, Pathanamthitta, Kerala, India; [4]Department of Bioscience, Mar Athanasios College for Advanced Studies Thiruvalla (MACFAST), Pathanamthitta, Kerala, India; [5]Natural Composite Research Group, Department of Materials and Production Engineering, The Sirindhorn International Thai—German Graduate School of Engineering (TGGS), King Mongkut's University of Technology North Bangkok, Bangkok, Thailand

1. Introduction

Soybeans are rich sources of fiber, calcium, iron, and vitamin B complex and are a legume with no cholesterol (Lindsay & Claywell, 1998). They are the only vegetable that contains all eight essential amino acids. Soy proteins are globular proteins isolated from soybeans and have received great attention in recent years due to their wide availability, low cost, highly nutritious, biodegradability, and excellent film-forming properties (Montgomery, 2003; Singh et al., 2008). They contains proteins (40%), carbohydrates (30%), oil (20%), moisture (14%), and minerals (Nishinari et al., 2014b). Soy proteins are obtained after the removal of oil and carbohydrates in the form of soy flour, soy protein concentrate, and soy protein isolate. The highly refined or purified form of soy protein obtained from defatted soy flour is soy protein isolate (SPI) which contains 90% protein. Soy protein concentrates are prepared by removing the soluble carbohydrate and are widely used as nutritional ingredients in baked foods and meat (Singh et al., 2008).

The structure of soy protein is composed of conglycinin 7S and glycinin 11S subunits (80% of protein) which contain all eight amino acids. The polypeptide chains of soy protein consist of disulfide and hydrogen bonds and the molecular weight ranges from 300,000 to 600,000 kDa (Sun, 2005). Soy proteins provide a good composition of amino acids and have excellent gelling, emulsifying, and water- and oil-holding capacity (Nishinari et al., 2014a). In the food industries, soy proteins are used for salad dressings, meat analogues, breads, pastas, pet foods, frozen desserts, soups, beverage powders, creamers, cheeses, infant formulas, and breakfast cereals (Singh et al., 2008).

Biodegradable Polymers, Blends and Composites. https://doi.org/10.1016/B978-0-12-823791-5.00012-0

Soy protein-based blends/composites have received great attention due to their excellent thermal, water resistance, and mechanical properties. There are several studies regarding the development of soy protein-based blends and composites. The main factor contributing to the performance of soy protein-based nanocomposite is based on the interaction or interfacial adhesion between soy protein matrices with the nanofillers. The chemical modification is also suggested to improve the miscibility between the nanofiller and the soy protein matrix, as this produces a continuous phase (Huang et al., 2011). This chapter is designed to give an updated overview of soy protein-based blends and composites.

2. Polymer blends and composites

Polymer blends can be defined as the mixture of at least two macromolecular compounds, polymers, or copolymers. Polymer-based blends can be classified in to immiscible or heterogeneous polymer blends, compatible polymer blends, and miscible or homogeneous polymer blends. Mostly, the polymer blends are immiscible in nature while some are miscible. In the immiscible polymer blends, the polymers exist in different phases, whereas single phases are observed in miscible or homogeneous polymer blends due to having the same chemical structure. Compatible polymer blends are also a kind of immiscible polymer blend which have macroscopically uniform physical properties due to the strong interactions between the polymers (Qin, 2016).

Polymer composites are a class of composites having two or more macroscopic components and are low cost in nature (Zagho et al., 2018). It consist of a dispersed phase and a matrix phase with applications in various biomedical, food, and electronic sectors. Generally, biocomposites are a combination of polymers and materials of natural origin (Rudin & Choi, 2013). Based on the type of matrix, the composites are classified into polymer matrix composites, ceramic matrix composites, and metal matrix composites. They are further classified into fibrous, particulate, and laminate composites depending upon the type of reinforcing agents. Electrospinning, melt extrusion, solution mixing, latex technology, and in situ methods are the major techniques used for the preparation of composites (Zagho et al., 2018). Like polymer blends, composites materials are also receiving great attention due to their light weight, good thermal barrier, mechanical performance, water barrier, biodegradable, and low cost nature (Pleşa et al., 2016).

3. Nanofillers as reinforcing agents

Nanofillers can be defined as fillers having a particle size between 1 and 100 nm (Rothon, 2016). They are classified as one dimensional, two dimensional and three dimensional (Fig. 3.1). One-dimensional nanofillers are usually in the form of plates, laminas, or shells, whereas two-dimensional ones include nanotubes and nanofibers. Iso-dimensional nanoparticles are included in the three-dimensional nanofillers. The nanofillers are categorized into organic, inorganic, clays, and carbon nanostructures (Jamróz et al., 2019). The organic nanofillers include natural biopolymers such as chitosan, cellulose, alginate, collagen, and proteins. The most common inorganic

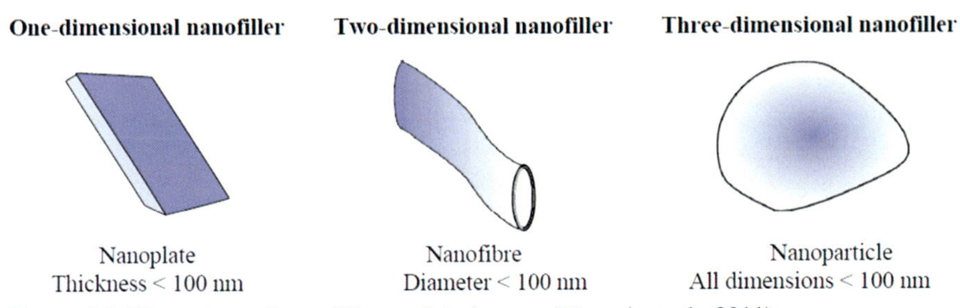

Figure 3.1 Dimensions of nanofillers and their types (Marquis et al., 2011).

nanofillers include various metal and metal oxide nanoparticles such as silver nanoparticles (AgNPs), zinc oxide nanoparticles (ZnONPs), iron oxide nanoparticles (FeONPS), titanium dioxide nanoparticles (TiO_2), gold nanoparticles (AuNPs), copper oxide nanoparticles (CuONPs), graphene oxide nanoparticles (GO NPs), and cerium dioxide nanoparticles (CeO_2 NPs). Carbon-based nanoparticles are categorized into fullerenes, graphene, carbon nanotubes, and nanofibers. Nanofillers are widely used in polymer blends/composites in order to improve the functional properties of the developed material.

4. Fabrication of soy protein-based materials

Soy protein-based materials are generally prepared by solvent-casting and melt-processing methods (Fig. 3.2). The solvent-casting method is the most commonly employed method for the preparation of soy protein-based materials. Among the various solvents, water is most commonly used as the dispersion medium. Here, a soy protein suspension is mixed with a polymer solution or nanoparticle suspension. The higher energy consumption and low production efficiency of this technique restrict its application at an industrial level. However, melt-processing methods are generally preferred for industrial development of soy protein-based materials. This technique generally consists of extrusion, hot-pressing, and injection molding (Tian et al., 2018).

5. Soy protein-based blends and composites

5.1 *Soy protein-based blends and composites for food packaging applications*

Protein-based films have gained great interest in packaging applications due to their excellent oxygen barrier, biodegradability, oil resistance, and nontoxic nature (Wittaya, 2012). However, the low mechanical, water, and thermal barrier properties are major drawback associated with soy proteins. Hence they are widely combined with other biopolymers or nanomaterials for enhancing their barrier properties. There are several

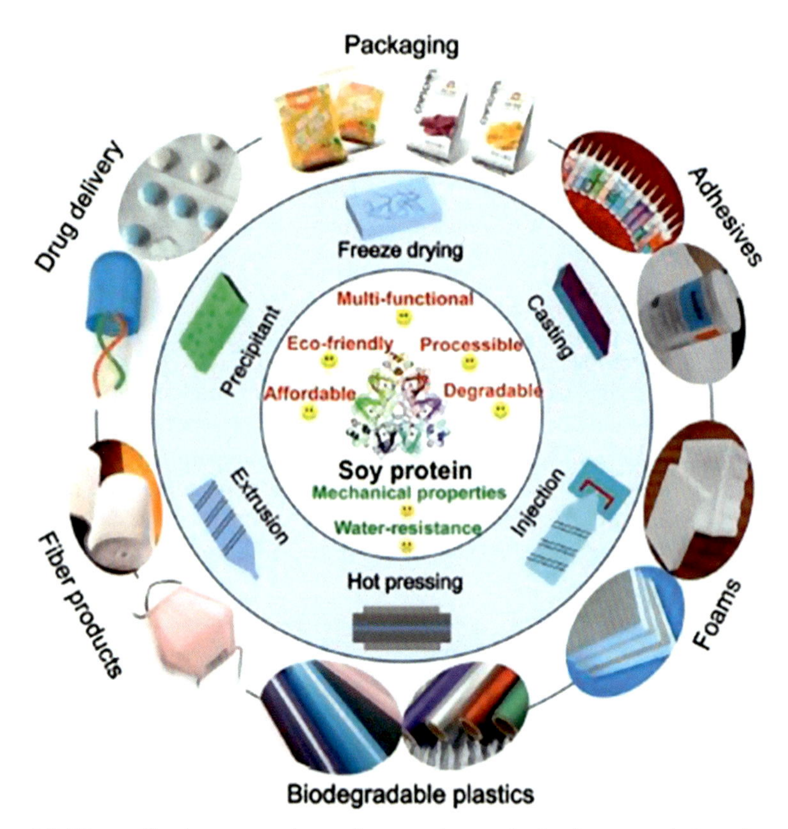

Figure 3.2 The application, properties, and preparation methods of soy protein-based materials (Tian et al., 2018).

studies that suggest the potential food packaging application of soy protein-based materials. Soy protein-based films incorporated with cellulose nanocrystals and zinc oxide nanoparticles have been reported to possess enhanced thermal, mechanical, water, and UV light barrier properties. In addition, the nanocomposite inhibited the growth of foodborne pathogens such as *Escherichia coli* and *Staphylococcus aureus*. After the exposure of these pathogens to zinc oxide nanoparticle-containing films, the bacterial cells were reported to be destroyed and thus resulting in the leakage of intracellular contents (Fig. 3.3). This could be due to the ability of zinc oxide nanoparticles to penetrate the bacterial cell membrane or by electrostatic interaction of zinc oxide nanoparticles with the bacterial surface. The improved properties with excellent antimicrobial activity of the films have shown its potential application in active food packaging (Xiao et al., 2020).

They further studied the effect of soy protein-based films for the preservation of fresh pork. They observed that the zinc oxide nanoparticles/cellulose nanocrystals incorporated nanocomposite and soy protein-ZnONPs films inhibited the growth of spoilage pathogens on the surface of pork. In the case of control films, visible bacterial colonies were reported to appear on the surface of pork after 6 days (Fig. 3.4).

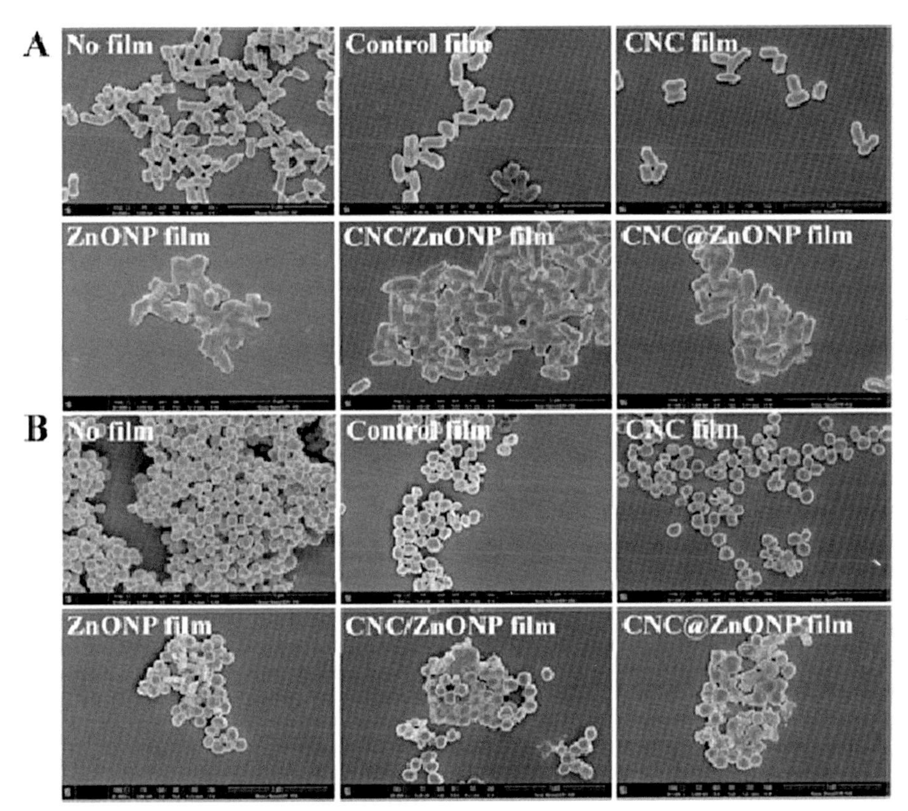

Figure 3.3 Scanning electron micrograph images of nanocomposite-treated food pathogens such as (A) *Escherichia coli* and (B) *Staphylococcus aureus*. (B) *S. aureus* after exposure to the films. [No film represents the untreated control, Control film (soy protein film), CNC film represents cellulose nanocrystals incorporated soy protein film], ZnONP film represents (soy protein film with zinc oxide nanoparticles, CNC/ZnONP: soy protein film with cellulose nanocrystals and zinc oxide nanoparticles, CNC@ZnONP nanohybrid film (soy protein film with cellulose nanocrystals and zinc oxide nanoparticles).
Reproduced with permission from Elsevier, License Number- 4990760554974. From Xiao, Y., Liu, Y., Kang, S., Wang, K., & Xu, H. (2020). Development and evaluation of soy protein isolate-based antibacterial nanocomposite films containing cellulose nanocrystals and zinc oxide nanoparticles. *Food Hydrocolloids*, *106*. 105898. https://doi.org/10.1016/j.foodhyd.2020.105898.

This could be due to the nutrient degradation and microbial proliferation. The improved microbial barrier of the soy protein-based nanocomposites suggested their potential application in preserving the pork.

The development of nanofibrous films by the incorporation of soy protein/gelatin and essential oil from *Zataria multiflora* and *Cinnamon zeylanicum* by electrospinning technique has been analyzed for packaging as well as antimicrobial activity (Raeisi et al., 2020). The fabricated membranes showed good thermal stability along with excellent antimicrobial activity against *Staphylococcus aureus*, *Bacillus cereus*,

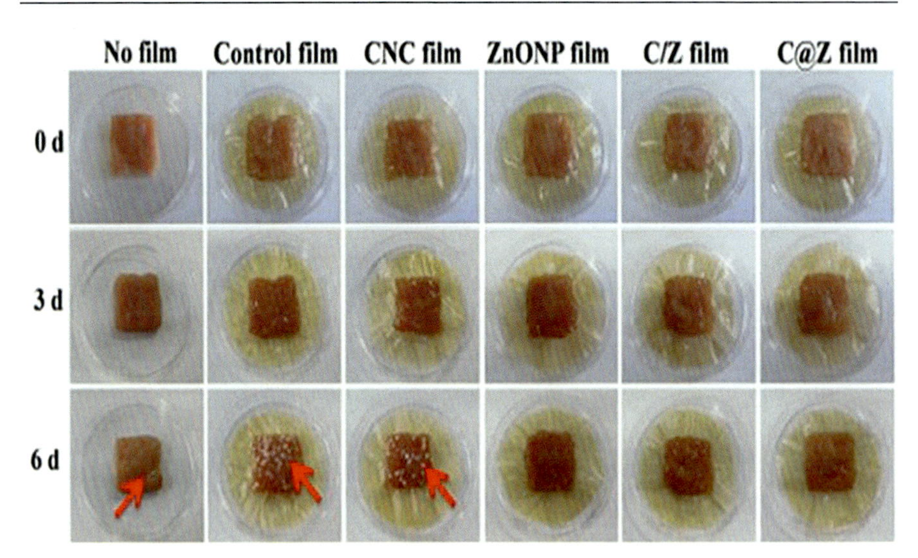

Figure 3.4 (A) Packaging application of soy protein-based nanocomposite films in preserving pork films. Control films showing susceptibility to bacterial infection.
Reproduced with permission from Elsevier, License Number- 4990760554974. From Xiao, Y., Liu, Y., Kang, S., Wang, K., & Xu, H. (2020). Development and evaluation of soy protein isolate-based antibacterial nanocomposite films containing cellulose nanocrystals and zinc oxide nanoparticles. *Food Hydrocolloids*, *106*. 105898. https://doi.org/10.1016/j.foodhyd.2020.105898.

Escherichia coli, *Salmonella typhimurium*, and *Listeria monocytogenes*. This has suggested its potential application as an active food packaging membrane. The development of nanocapsules has been reported for the nanoencapsulation of phenolic extract of grape and apple pomace to chitosan and soy protein. The fabricated nanocapsules enhanced the antioxidant activity. Further, the food preservative effects of the developed nanocapsules were confirmed by testing their efficiency in preserving apple and pineapple juices. Their results revealed the efficiency of nanocapsules as a food-preserving agent (Gaber Ahmed et al., 2020).

In another study, soy protein/organoclay-based nanocomposites developed by solution intercalation method were described for food-packaging applications. The improved thermal stability and oxygen barrier properties of the development films could be due to the loading of organoclay (Swain et al., 2012). The addition of soy protein isolate to egg shell membrane and eugenol-based composite film has been reported to have enhanced mechanical and water barrier properties. By Fourier transform infrared spectroscopy (FT-IR) analysis, the intermolecular interactions such as hydrogen bond formation between soy protein, egg shell membrane, and eugenol were explained. The good compatibility of these additives due to the better dispersion and emulsification paved the way to explore their application in active food packaging (Varshney et al., 2020). The incorporation of aqueous extract of pinhao (*Araucaria angustifolia* (Bertol.) Kuntze) seeds to soy protein isolate has been reported to possess excellent antioxidant activity with application as active packaging (de Souza et al., 2020).

Similarly, the composite based on gamma-aminobutyric acid (GABA)-rich fermented soy protein with different concentration of chitosan (2%, 2.5%, and 3%) developed by solvent casting method has been analyzed for mechanical, antioxidant, and antimicrobial properties. The composite films incorporated with 2.5% of chitosan show enhanced tensile strength, elongation at break, and antioxidant activity. The antimicrobial studies of the same revealed their efficiency against yeast (*Candida albicans*), mold (*Aspergillus niger*), Gram-positive bacteria (*Staphylococcus aureus*, *Bacillus cereus*, and *Listeria innocua*) and Gram-negative bacteria (*Escherichia coli* and *Pseudomonas aeruginosa*). This might be due to the antimicrobial activity of GABA and chitosan. The developed edible films have functional applications in food packaging (Zareie et al., 2020). The composite film based on soy protein/diatomite/thymol complex has been reported to possess antimicrobial activity against *Escherichia coli*. Their study supports its application in active food packaging (Wu et al., 2021).

Protein-based edible films have been receiving great attention in recent years due to their advantages over others (Wittaya, 2012). They have wider applications in packaging and can also be applied as an interface between different layers of components. Soy proteins can be combined with antimicrobial agents/antioxidant agents to prevent microbial attack and the associated oxidation-induced damage. Moisture-induced damage and solute migration in foodstuffs can also be prevented by using these developed systems.

6. Soy protein-based blends and composites for medical applications

The materials based on soy proteins are gaining great interest in biomedical applications due to their biodegradability, biocompatibility, low cost, and low immunogenicity (Tansaz & Boccaccini, 2016). In addition, the nonanimal origin of soy proteins makes them advantageous over others. The similarity of soy proteins with the extracellular matrix (ECM) of tissues makes them suitable for application in biomedical sectors (Tansaz & Boccaccini, 2016). The successful application of biomaterials for wound healing requires better cell adhesion and cell-supporting behavior. An ideal wound dressing material should possess optimum water uptake, biocompatibility, nontoxicity, and good water vapor permeability, along with an antimicrobial barrier. The blending of polymers with natural components is of great significance as this can improve the overall performance of the material.

For example, 2, 2, 6, 6-tetramethylpiperidine-1-oxyl (TEMPO) oxidized cellulose nanofibers grafted with soy protein hydrolysate-based nanocomposite has been analyzed for biomedical applications. The developed scaffold can be used as platforms for the regeneration or repair of hard tissues (Salama et al., 2020). Similarly, electrospun nanofibers based on polyvinyl alcohol/soy protein isolate, sepiolite, and ketoprofen have been reported to have applications in drug delivery. The reinforcement of soy protein isolate and sepiolite resulted in the improved mechanical strength of a nanofibrous mat. The addition of sepiolite has been also reported to result in the controlled release of ketoprofen (Kumar et al., 2010).

The application of biomaterials for tissue regeneration is an emerging field of tissue engineering. The ability of biomaterials to induce tissue regenerations has led to tremendous research in this field. For example, silk fibroin nanofibers/soy protein isolate-based nanofibrous scaffold (SPI/SF) developed by the electrospinning technique has been analyzed for skin tissue regeneration (Varshney et al., 2020). In vitro studies used fibroblast cells and melanocytes as the cell lines for analyzing the skin tissue regeneration potential of the developed scaffold. Surgically induced wounds in rat treated with nanofibrous scaffold (SPI/SF) showed faster healing than that of the sham (control) (Fig. 3.5). The SPI/SF nanofibrous scaffold adhered quickly to the surface of the wound, along with degradation over a period of time (14 days). Their studies suggested that the faster healing of wounds could be due to the ECM-like structure of the nanofibrous scaffold. Histological staining with hematoxylin and eosin of granulation tissue of the nanofibrous scaffold treated and the control shows the infiltration of keratinocytes and collagen deposition in the dermal layer. The control group has been reported to have less reepithelialization and collagen deposition.

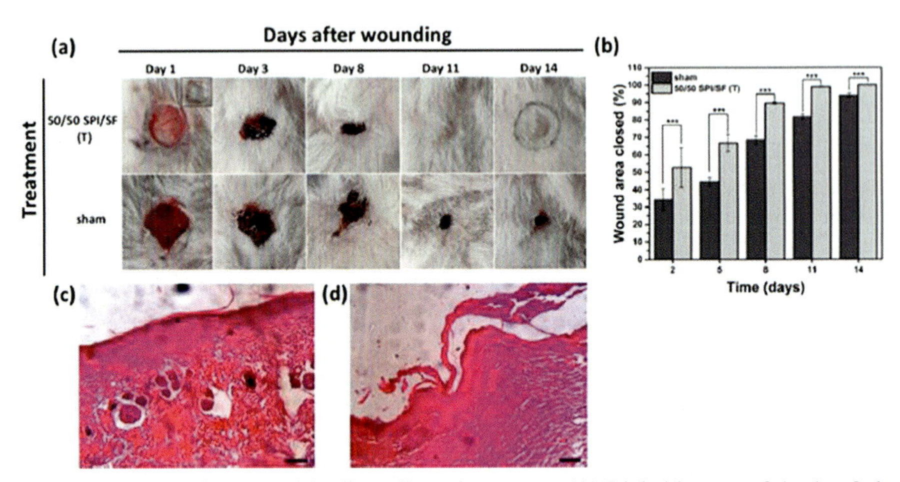

Figure 3.5 The in vivo wound-healing effect of treatment. (A) Digital images of the 1st, 3rd, 8th, 11th, and 14th days of wound closure in rat. (B) Graphical representation of the comparison of wound closure of nanofibrous scaffold treated (SPI/SF) and the control (sham). (C) and (D) Histological staining with hematoxylin and eosin of nanofibrous scaffold treated (SPI/SF) and the control (sham).

Reproduced with permission from Elsevier, License Number-4994671015122. From Varshney, N., Sahi, A. K., Poddar, S., & Mahto, S. K. (2020). Soy protein isolate supplemented silk fibroin nanofibers for skin tissue regeneration: Fabrication and characterization. *International Journal of Biological Macromolecules*, *160*, 112–127. https://doi.org/10.1016/j.ijbiomac.2020.05.090.

Polyvinyl alcohol and soy protein isolate-based electrospun nanofibrous mats (ENMs) and solvent-cast films (CFs) have been analyzed for wound-healing applications (Khabbaz et al., 2019). The ENMs shows improved exudate absorption, mechanical strength, and water vapor transmission rate than CFs. Electrospun nanofibrous mats show better cell adhesion than CFs as shown in the scanning electron microscopic (SEM) images in Fig. 3.6. The porous structure of ENMs might have provided sufficient nutrients that could contribute to the better adhesion of L929 mouse fibroblast cells than CFs.

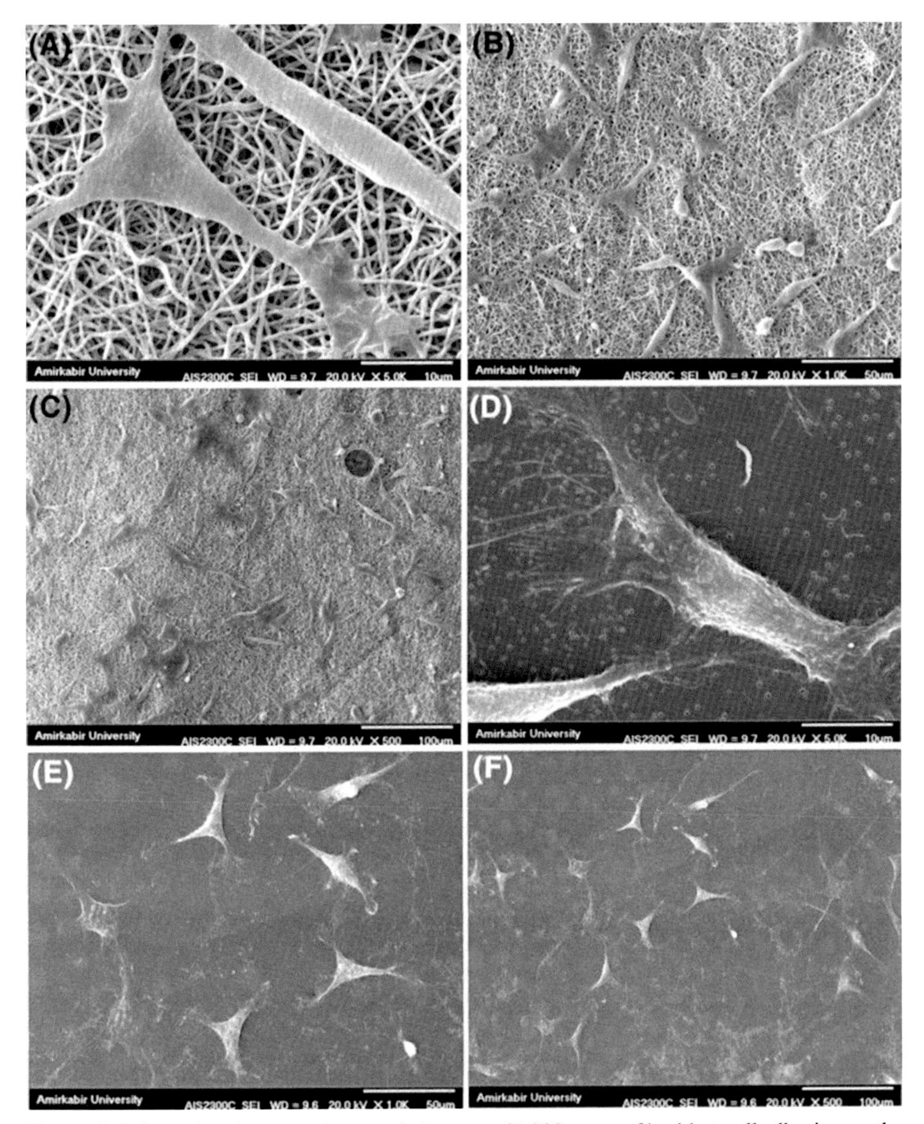

Figure 3.6 Scanning electron microscopic images of L929 mouse fibroblast cell adhesion on the surface of electrospun nanofibrous mats (ENMs) and solvent-cast films (CFs). (A, B, C represent ENMs and D, E, F represent CFs) (Khabbaz et al., 2019).

Electrospun nanofibrous mats based on polyvinyl alcohol and soy protein isolate loaded with ketoprofen have been analyzed for drug-delivery applications (Gutschmidt et al., 2021). They utilized tubular nanoparticles based on sepiolite as the tool for secondary release control. Their result suggests the incorporation of soy protein isolate and sepiolite resulted in the enhanced mechanical strength. Also, the drug-loaded sepiolite has good control of the sustained release of ketoprofen.

7. Soy protein-based blends and composites for horticultural applications

The efficiency of fertilizers can be enhanced by the incorporation of protein-based bioplastic matrices. The controlled release of fertilizers and water retention with good biodegradation potential enhances the growth of plants and is gaining great attention. The water retention potential of soy protein-based matrices due to the hydrophilic nature improves the recycling of water and is of great interest in horticultural applications. Adipic anhydride-plasticized soy protein and poly(lactic acid)-based blends have been also been developed and characterized for horticultural applications. The developed blends possess enhanced thermal stability and biodegradation potential. Their study suggested the incorporation of adipic anhydride-plasticized soy protein has resulted in faster biodegradation potential (Yang et al., 2015). Soy protein-based matrices incorporated with a micronutrient (zinc) have been characterized for horticultural applications (Jiménez-Rosado et al., 2020). The micronutrient-incorporated biomatrices showed enhanced mechanical and water retention ability.

8. Soy protein-based blends and composites as adhesives

Soy protein-based adhesives have low resistance to moisture, high viscosity, and a low level of solid content. The major drawbacks associated with soy protein-based adhesives can be reduced using denaturation techniques or with the modification of cross-linking agents. The denaturation techniques usually rupture the structures of globular proteins that result in the generation of linear polymer. Hence, these linear polymers interact through their functional groups. The cross-linking agents consist of lignin, amino acids, carbohydrates, polyurethanes, resins, and amino acids. The blending of urea formaldehyde resin blended with soy protein concentrate has been studied for its adhesive properties (Bacigalupe et al., 2020). The fabricated adhesive shows enhanced thermal, mechanical, and water resistance. Soy protein/carbon nanotube-based nanocomposites have been reported to possess increased shear strength and water resistance. A nanocomposite adhesive showed an increase in shear strength from 3.48 MPa (without carbon nanotubes) to 6.91 MPa. These enhanced properties of nanocomposites could have applications as adhesives in wood industries (Sadare et al., 2020). Soy protein reinforced with nanochitosan has been reported to

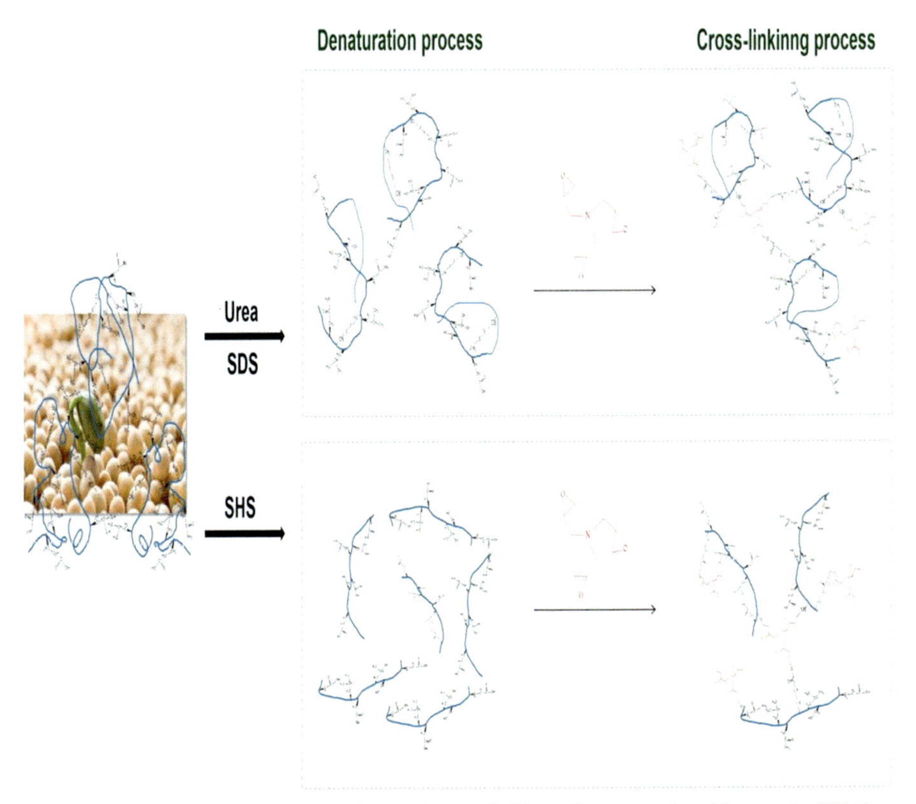

Figure 3.7 Mechanism of denaturation and cross-linking of soy proteins (Yue et al., 2019).

develop biodegradable adhesives. The developed material showed enhanced tensile strength and water resistance, which suits its application in textile industries (Xu et al., 2019). In one study, urea, sodium dodecyl sulfate and sodium hydrogen sulfite were utilized as the denaturing agent for the development of soy protein-based adhesives along with triglycidylamine as a cross-linking agent (CA) (Yue et al., 2019). The denaturation and cross-linking mechanism of soy proteins are described in Fig. 3.7. Their study revealed the increase in wet shear strength of plywood through the use of soy protein isolate-sodium hydrogen sulfate-triglycidylamine (SPI-SHS-CA)-based adhesives. The water resistance properties of the adhesives were also reported to get enhanced.

9. Soy protein-based blends and composites as self-healing agents

A self-healing biodegradable film based on soy protein isolate (SPI), polydopamine (PDA), and polyethyleneimine (PEI) has been reported for functional applications and heavy metal detection (Chang et al., 2021). The developed SPI-PEI-PDA films possess excellent self-healing properties, mechanical strength, are biodegradable and provide a UV barrier with a fluorescent nature.

10. Soy protein-based blends and composites for battery applications

Polymer binders are one of the main components of cathodes, which maintain the integrity of an electrode and ensure sufficient contact between the cathode and current collector (Yuan et al., 2018). The most widely used polymer binder in commercial rechargeable battery is polyvinylidene fluoride (PVDF) (Sung et al., 2020). However, PVDF has low bonding strength and less affinity to anchor the lithium polysulfide. In addition, it is dissolved in a highly toxic, volatile, and flammable organic solvent (N-methyl-2-pyrrolidone) (Li et al., 2021). These drawbacks paved the way for the search for environment-friendly polymers.

Biopolymer-based batteries are gaining great interest due to their sustainable and environment-friendly nature (Lizundia & Kundu, 2020). A soy protein isolate and polyvinyl alcohol-based nanofibrous membrane developed by electrospinning (Fig. 3.8) has been used as the skeleton material in gel polymer electrolyte for the synthesis of lithium ion batteries (Zhu et al., 2016). The affinity between the nanofiber and the electrode, low crystallinity, and porosity of the same resulted in the uptake of a

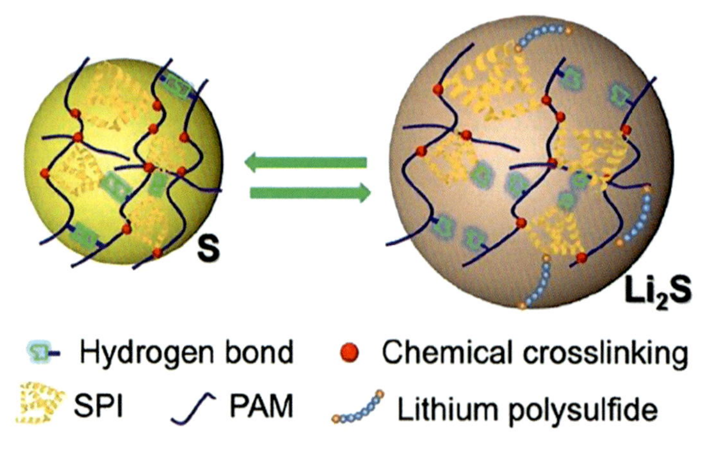

Figure 3.8 Schematic illustration of the charge/discharge process of sulfur anodes using a self-healing SPI-PAM binder.
Reprinted (adapted) with permission from Wang, H., Wang, Y., Zheng, P., Yang, Y., Chen, Y., Cao, Y., Deng, Y., & Wang, C. (2020). Self-healing double-cross-linked supramolecular binders of a polyacrylamide-grafted soy protein isolate for Li–S batteries. *ACS Sustainable Chemistry & Engineering*, *8*(34), 12799–12808. https://doi.org/10.1021/acssuschemeng.0c02477. Copyright (2021) American Chemical Society. https://doi.org/10.1021/acssuscheme,ng.0c02,477.

highly saturated electrolyte. The developed nanofibrous membrane shows excellent electrochemical performance such as improved ionic conductivity, compatibility, and electrochemical stability. The copolymerization of methacrylated soy protein isolate (SPI) and polyacrylamide (PAM) resulted in the development of a self-healing, water-based binder for a sulfur (S) cathode (Wang et al., 2020). The developed SPI-PAM binder possesses good bonding strength, improved electrochemical performance, and self-healing properties (Table 3.1).

Table 3.1 Applications of soy protein-based blends/composites.

Soy protein materials	Properties of developed films	Applications	References
Polyvinyl alcohol/soy protein isolate, sepiolite, and ketoprofen based nanofibers	Enhanced mechanical strength	Controlled drug release/ biomedical applications	Kumar et al. (2010)
Soy protein/ organoclay-based nanocomposites	Enhanced thermal stability and oxygen barrier properties	Food packaging	Swain et al. (2012)
Adipic anhydride-plasticized soy protein and polylactic acid-based blends	Enhanced thermal stability and biodegradation potential	Horticultural applications	Yang et al. (2015)
Soy protein isolate and polyvinyl alcohol-based nanofibrous membrane	Improved ionic conductivity, compatibility and electrochemical stability	Battery applications	Zhu et al. (2016)
Soy protein reinforced with nanochitosan	Enhanced tensile strength and water resistance	Bioadhesives in textile industries	Xu et al. (2019)
Soy protein isolate and montmorillonite	Enhanced mechanical properties due		Guo et al. (2019)
Polyvinyl alcohol and soy protein isolate-based electrospun nanofibrous mats and solvent-cast films	Improved exudate absorption, mechanical strength, water vapor transmission, and cell adhesion	Wound healing	Khabbaz et al. (2019)
Soy protein, cellulose nanocrystals and zinc oxide nanoparticle-based nanocomposites	Enhanced thermal, mechanical, water barrier properties and antimicrobial against *Escherichia coli* and *Staphylococcus aureus*	Active food packaging	Xiao et al. (2020)

Continued

Table 3.1 Applications of soy protein-based blends/composites.—cont'd

Soy protein materials	Properties of developed films	Applications	References
Soy protein isolate/silk fibroin nanofiber-based nanofibrous scaffold	Reepithelialization of wounds in rats	Skin/tissue regeneration	Varshney et al. (2020)
Gamma-aminobutyric acid (GABA)-rich fermented soy protein and chitosan-based composites	Enhanced tensile strength, elongation at break, antioxidant activity and antimicrobial against *Candida albicans*, *Aspergillus niger*, *Staphylococcus aureus*, *Bacillus cereus*, *Listeria innocua*, *Escherichia coli*, and *Pseudomonas aeruginosa*	Active food packaging	Zareie et al. (2020)
Soy protein-zinc matrices	Improved mechanical and water retention	Horticultural applications	Jiménez-Rosado et al. (2020)
Methacrylated soy protein isolate (SPI) and polyacrylamide	Good bonding strength, improved electrochemical performance and self-healing properties	Battery applications	Wang et al. (2020)
Soy protein/carbon nanotube-based nanocomposites	Enhanced shear strength and water resistance	Application in wood industries	Sadare et al. (2020)
Chitosan and soy protein/grape and apple pomace-based nanocapsules	Enhanced antioxidant activity	Food preservation	Gaber Ahmed et al. (2020)
Soy protein/gelatin and essential oil-based nanofibrous film	Enhanced thermal barrier and antimicrobial against *Staphylococcus aureus*, *Bacillus cereus*, *Escherichia coli*, *Salmonella typhimurium*, and *Listeria monocytogenes*	Active food packaging	Raeisi et al. (2020)

Table 3.1 Applications of soy protein-based blends/composites.—cont'd

Soy protein materials	Properties of developed films	Applications	References
Soy protein concentrate/urea formaldehyde resin	Improved thermal, mechanical, and water resistance	Adhesives	Bacigalupe et al. (2020)
Soy protein/carbon nanotube-based nanocomposites	Improved tensile strength and water resistance	Adhesives	Sadare et al. (2020)
Soy protein/diatomite/thymol-based composite film	Antimicrobial against *Escherichia coli*	Active food packaging	Wu et al. (2021)
Polyvinyl alcohol, soy protein isolate ketoprofen, sepiolite-based electrospun nanofibrous mats	Enhanced mechanical strength and controlled release of drug	Drug delivery vehicles	Gutschmidt et al. (2021)

11. Other applications

Bionanocomposites based on soy protein isolate and montmorillonite have been also reported to possess enhanced mechanical properties due to the strong interaction and better dispersion of montmorillonite in the soy protein matrix (Guo et al., 2019). Soy protein and semicrystalline polylactide-based blends developed using a twin-screw extruder have been reported to be characterized for mechanical, thermal, water resistance, and morphological analysis. Their study shows the reduction of water absorption capacity of the soy plastics with the addition of polylactide. The addition of 1−5 phr poly(2-ethyl-2-oxazoline) resulted in enhanced compatibility, tensile strength, and reduced water absorption (Zhang et al., 2006).

12. Conclusion

Soy proteins are globular proteins with excellent film-forming properties with nontoxic and biodegradable nature. Due to the drawbacks such as low water barrier, mechanical, and antimicrobial properties, they are often combined with other polymers or other materials of interest for different applications. Soy protein-based blends and composites are gaining much interest in food packaging, medical, wood industry, self-healing agent, horticulture, and battery applications. The superior thermal, mechanical, water resistance, bonding strength, biodegradability, and other properties makes them more attractive than other polymers.

Acknowledgement

AJ gratefully thanks King Mongkut's University of Technology North Bangkok (KMUTNB), Thailand through the Post-Doctoral Program (Grant No. KMUTNB-64-Post-01).

References

Bacigalupe, A., Molinari, F., Eisenberg, P., & Escobar, M. M. (2020). Adhesive properties of urea-formaldehyde resins blended with soy protein concentrate. *Advanced Composites and Hybrid Materials, 3*(2), 213–221. https://doi.org/10.1007/s42114-020-00151-7

Chang, Z., Zhang, S., Li, F., Wang, Z., Li, J., Xia, C., Yu, Y., Cai, L., & Huang, Z. (2021). Self-healable and biodegradable soy protein-based protective functional film with low cytotoxicity and high mechanical strength. *Chemical Engineering Journal, 404*, 126505. https://doi.org/10.1016/j.cej.2020.126505

de Souza, K. C., Correa, L. G., da Silva, T. B. V., Moreira, T. F. M., de Oliveira, A., Sakanaka, L. S., Dias, M. I., Barros, L., Ferreira, I. C. F. R., Valderrama, P., Leimann, F. V., & Shirai, M. A. (2020). Soy protein isolate films incorporated with Pinhão (*Araucaria angustifolia* (Bertol.) Kuntze) extract for potential use as edible oil active packaging. *Food and Bioprocess Technology, 13*(6), 998–1008. https://doi.org/10.1007/s11947-020-02454-5

Gaber Ahmed, G. H., Fernández-González, A., & Díaz García, M. E. (2020). Nano-encapsulation of grape and apple pomace phenolic extract in chitosan and soy protein via nanoemulsification. *Food Hydrocolloids, 108*, 105806. https://doi.org/10.1016/j.foodhyd.2020.105806

Guo, G., Tian, H., & Wu, Q. (2019). Influence of pH on the structure and properties of soy protein/montmorillonite nanocomposite prepared by aqueous solution intercalating. *Applied Clay Science, 171*, 14–19. https://doi.org/10.1016/j.clay.2019.01.020

Gutschmidt, D., Hazra, R. S., Zhou, X., Xu, X., Sabzi, M., & Jiang, L. (2021). Electrospun, sepiolite-loaded poly(vinyl alcohol)/soy protein isolate nanofibers: preparation, characterization, and their drug release behavior. *International Journal of Pharmaceutics, 594*, 120172. https://doi.org/10.1016/j.ijpharm.2020.120172

Huang, J., Lin, N., Chen, Y., Chang, P. R., & Yu, J. (2011). *Soy protein-based polymer nanocomposites* (pp. 261–282). https://doi.org/10.1093/acprof:oso/9780199581924.003.0011

Jamróz, E., Kulawik, P., & Kopel, P. (2019). The effect of nanofillers on the functional properties of biopolymer-based films: a review. *Polymers, 11*(4), 675. https://doi.org/10.3390/polym11040675

Jiménez-Rosado, M., Perez-Puyana, V., Rubio-Valle, J. F., Guerrero, A., & Romero, A. (2020). Evaluation of superabsorbent capacity of soy protein-based bioplastic matrices with incorporated fertilizer for crops. *Journal of Polymers and the Environment, 28*(10), 2661–2668. https://doi.org/10.1007/s10924-020-01811-x

Khabbaz, B., Solouk, A., & Mirzadeh, H. (2019). Polyvinyl alcohol/soy protein isolate nanofibrous patch for wound-healing applications. *Progress in Biomaterials, 8*(3), 185–196. https://doi.org/10.1007/s40204-019-00120-4

Kumar, P., Sandeep, K. P., Alavi, S., Truong, V. D., & Gorga, R. E. (2010). Preparation and characterization of bio-nanocomposite films based on soy protein isolate and montmorillonite using melt extrusion. *Journal of Food Engineering, 100*(3), 480–489. https://doi.org/10.1016/j.jfoodeng.2010.04.035

Li, M., Zhang, J., Gao, Y., Wang, X., Zhang, Y., & Zhang, S. (2021). A water-soluble, adhesive and 3D cross-linked polyelectrolyte binder for high-performance lithium–sulfur batteries. *Journal of Materials Chemistry A*. https://doi.org/10.1039/d0ta09859k

Lindsay, S. H., & Claywell, L. G. (1998). Considering soy–its estrogenic effects may protect women. *AWHONN Lifelines, 2*(1), 41–44. https://doi.org/10.1111/j.1552-6356.1998.tb00990.x

Lizundia, E., & Kundu, D. (2020). Advances in natural biopolymer-based electrolytes and separators for battery applications. *Advanced Functional Materials, 31*(3), 2005646. https://doi.org/10.1002/adfm.202005646

Marquis, D., Guillaume, E., & Chivas-Joly, C. (2011). *Properties of nanofillers in polymer*. https://doi.org/10.5772/21694

Montgomery, K. S. (2003). Soy protein. *The Journal of Perinatal Education, 12*(3), 42–45. https://doi.org/10.1624/105812403X106946

Nishinari, K., Fang, Y., Guo, S., & Phillips, G. O. (2014a). *Properties and applications of soy proteins* (pp. 28–45). https://doi.org/10.1039/9781782621300-00028

Nishinari, K., Fang, Y., Guo, S., & Phillips, G. O. (2014b). Soy proteins: a review on composition, aggregation and emulsification. *Food Hydrocolloids, 39*, 301–318. https://doi.org/10.1016/j.foodhyd.2014.01.013

Pleşa, I., Noţingher, P., Schlögl, S., Sumereder, C., & Muhr, M. (2016). Properties of polymer composites used in high-voltage applications. *Polymers, 8*(5), 173. https://doi.org/10.3390/polym8050173

Qin, Y. (2016). *Applications of advanced technologies in the development of functional medical textile materials* (pp. 55–70). https://doi.org/10.1016/b978-0-08-100618-4.00005-4

Raeisi, M., Mohammadi, M. A., Coban, O. E., Ramezani, S., Ghorbani, M., Tabibiazar, M., khoshbakht, R., & Noori, S. M. A. (2020). Physicochemical and antibacterial effect of Soy Protein Isolate/Gelatin electrospun nanofibres incorporated with *Zataria multiflora* and *Cinnamon zeylanicum* essential oils. *Journal of Food Measurement and Characterization*. https://doi.org/10.1007/s11694-020-00700-0

Rothon, R. (2016). *Nanofillers*, 1–21. https://doi.org/10.1007/978-3-642-37179-0_78-1

Rudin, A., & Choi, P. (2013). *Biopolymers* (pp. 521–535). https://doi.org/10.1016/b978-0-12-382178-2.00013-4

Sadare, O. O., Daramola, M. O., & Afolabi, A. S. (2020). Synthesis and performance evaluation of nanocomposite soy protein isolate/carbon nanotube (SPI/CNTs) adhesive for wood applications. *International Journal of Adhesion and Adhesives, 100*, 102605. https://doi.org/10.1016/j.ijadhadh.2020.102605

Salama, A., Abou-Zeid, R. E., Cruz-Maya, I., & Guarino, V. (2020). Mineralized nanocomposite scaffolds based on soy protein grafted oxidized cellulose for biomedical applications. *Materials Today: Proceedings*. https://doi.org/10.1016/j.matpr.2019.12.069

Singh, P., Kumar, R., Sabapathy, S. N., & Bawa, A. S. (2008). Functional and edible uses of soy protein products. *Comprehensive Reviews in Food Science and Food Safety, 7*(1), 14–28. https://doi.org/10.1111/j.1541-4337.2007.00025.x

Sun, X. S. (2005). *Thermal and mechanical properties of soy proteins* (pp. 292–326). https://doi.org/10.1016/b978-012763952-9/50010-1

Sung, S. H., Kim, S., Park, J. H., Park, J. D., & Ahn, K. H. (2020). Role of PVDF in rheology and microstructure of NCM cathode slurries for lithium-ion battery. *Materials, 13*(20), 4544. https://doi.org/10.3390/ma13204544

Swain, S. K., Priyadarshini, P. P., & Patra, S. K. (2012). Soy protein/clay bionanocomposites as ideal packaging materials. *Polymer – Plastics Technology and Engineering, 51*(12), 1282–1287. https://doi.org/10.1080/03602559.2012.700542

Tansaz, S., & Boccaccini, A. R. (2016). Biomedical applications of soy protein: a brief overview. *Journal of Biomedical Materials Research Part A, 104*(2), 553–569. https://doi.org/10.1002/jbm.a.35569

Tian, H., Guo, G., Fu, X., Yao, Y., Yuan, L., & Xiang, A. (2018). Fabrication, properties and applications of soy-protein-based materials: a review. *International Journal of Biological Macromolecules, 120*, 475–490. https://doi.org/10.1016/j.ijbiomac.2018.08.110

Varshney, N., Sahi, A. K., Poddar, S., & Mahto, S. K. (2020). Soy protein isolate supplemented silk fibroin nanofibers for skin tissue regeneration: fabrication and characterization. *International Journal of Biological Macromolecules, 160*, 112–127. https://doi.org/10.1016/j.ijbiomac.2020.05.090

Wang, H., Wang, Y., Zheng, P., Yang, Y., Chen, Y., Cao, Y., Deng, Y., & Wang, C. (2020). Self-healing double-cross-linked supramolecular binders of a polyacrylamide-grafted soy protein isolate for Li–S batteries. *ACS Sustainable Chemistry & Engineering, 8*(34), 12799–12808. https://doi.org/10.1021/acssuschemeng.0c02477

Wittaya, T. (2012). *Protein-based edible films: characteristics and improvement of properties.* https://doi.org/10.5772/48167

Wu, H., Lu, J., Xiao, D., Yan, Z., Li, S., Li, T., Wan, X., Zhang, Z., Liu, Y., Shen, G., Li, S., & Luo, Q. (2021). Development and characterization of antimicrobial protein films based on soybean protein isolate incorporating diatomite/thymol complex. *Food Hydrocolloids, 110*, 106138. https://doi.org/10.1016/j.foodhyd.2020.106138

Xiao, Y., Liu, Y., Kang, S., Wang, K., & Xu, H. (2020). Development and evaluation of soy protein isolate-based antibacterial nanocomposite films containing cellulose nanocrystals and zinc oxide nanoparticles. *Food Hydrocolloids, 106*. https://doi.org/10.1016/j.foodhyd.2020.105898, 105898.

Xu, X., Hu, W., Ke, Q., Liu, H., Li, J., & Zhao, Y. (2019). Bio-adhesives from unfolded soy protein reinforced by nano-chitosan for sustainable textile industry. *Textile Research Journal, 90*(9–10), 1094–1101. https://doi.org/10.1177/0040517519886560

Yang, S., Madbouly, S. A., Schrader, J. A., Srinivasan, G., Grewell, D., McCabe, K. G., Kessler, M. R., & Graves, W. R. (2015). Characterization and biodegradation behavior of bio-based poly(lactic acid) and soy protein blends for sustainable horticultural applications. *Green Chemistry, 17*(1), 380–393. https://doi.org/10.1039/c4gc01482k

Yuan, H., Huang, J.-Q., Peng, H.-J., Titirici, M.-M., Xiang, R., Chen, R., Liu, Q., & Zhang, Q. (2018). A review of functional binders in lithium-sulfur batteries. *Advanced Energy Materials, 8*(31), 1802107. https://doi.org/10.1002/aenm.201802107

Yue, L., Meng, Z., Yi, Z., Gao, Q., Mao, A., & Li, J. (2019). Effects of different denaturants on properties and performance of soy protein-based adhesive. *Polymers, 11*(8), 1262. https://doi.org/10.3390/polym11081262

Zagho, M., Hussein, E., & Elzatahry, A. (2018). Recent overviews in functional polymer composites for biomedical applications. *Polymers, 10*(7), 739. https://doi.org/10.3390/polym10070739

Zareie, Z., Tabatabaei Yazdi, F., & Mortazavi, S. A. (2020). Development and characterization of antioxidant and antimicrobial edible films based on chitosan and gamma-aminobutyric acid-rich fermented soy protein. *Carbohydrate Polymers, 244*, 116491. https://doi.org/10.1016/j.carbpol.2020.116491

Zhang, J., Jiang, L., Zhu, L., Jane, J.-l., & Mungara, P. (2006). Morphology and properties of soy protein and polylactide blends. *Biomacromolecules, 7*(5), 1551−1561. https://doi.org/10.1021/bm050888p

Zhu, M., Tan, C., Fang, Q., Gao, L., Sui, G., & Yang, X. (2016). High performance and biodegradable skeleton material based on soy protein isolate for gel polymer electrolyte. *ACS Sustainable Chemistry & Engineering, 4*(9), 4498−4505. https://doi.org/10.1021/acssuschemeng.6b01218

Extraction and properties of cellulose for polymer composites

R.K. Gond[1], *M.K. Gupta*[1], *Harinder Singh*[2], *Sanjay Mavinkere Rangappa*[3] *and Suchart Siengchin*[3]

[1]Department of Mechanical Engineering, Motilal Nehru National Institute of Technology, Allahabad, Prayagraj, Uttar Pradesh, India; [2]Department of Chemical Engineering, Motilal Nehru National Institute of Technology, Allahabad, Prayagraj, Uttar Pradesh, India; [3]Natural Composite Research Group, Department of Materials and Production Engineering, The Sirindhorn International Thai–German Graduate School of Engineering (TGGS), King Mongkut's University of Technology North Bangkok, Bangkok, Thailand

1. Introduction

The use of petroleum-based plastics has led to air, water, and soil pollution and also global warming. There is currently active research to replace petroleum-based plastics with plastics made either from bioresources or by green synthesis. Cellulose is a renewable natural biopolymer and is present in a wide variety of living species including plants, animals, and some bacteria that is suitable for providing such nanoparticles as a reinforcing agent. It is used for load-bearing applications on account of its good stiffness and high degree of crystallinity. It has versatile uses such as in veterinary foods, fibers and clothes, wood and paper, cosmetics, and so on. Cellulose with at least one dimension in the nano-range (less than 100 nm) is called nanocellulose. Nanocellulose is an attractive future material for commercial and medical applications. In this chapter, a detailed discussion about nanocellulose, and its processing, characterization, and applications is presented.

2. Cellulose

Cellulose consists of hundreds—and sometimes even thousands—of carbon, hydrogen, and oxygen atoms. It is one of the most abundant polysaccharides in nature and it is nontoxic, biodegradable, and comes from renewable sources having a long-chain polymer of glucose molecules joined together as shown in Fig. 4.1. It is found in hemp, sugarcane bagasse, wood, rice husk, areca nut, bamboo, cotton, coconut husk, banana, jute, algae, tunicates, and some bacteria (Bettaieb et al., 2015). It is composed of glucose monomers linked together through a β-(1,4) glycosidic bond and possesses properties such as low cost, high strength, low density,

Biodegradable Polymers, Blends and Composites. https://doi.org/10.1016/B978-0-12-823791-5.00011-9

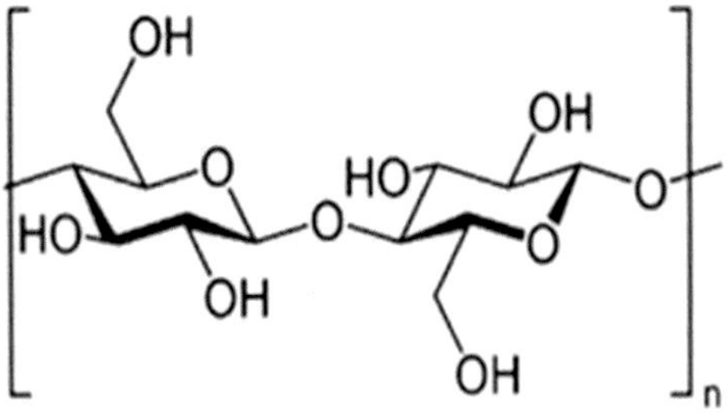

Figure 4.1 Chemical structure of cellulose.

good thermal stability, good biocompatibility, and good mechanical properties (Abraham et al., 2011). The linear and regular structure of hydroxyl groups in the molecule, held together with hydrogen bonds and Van der Waals forces, lead to crystallinity in cellulose. These structures give important mechanical properties to cellulose fibers due to the formation of a hydrogen bond between cellulose. Hydrogen bonding gives load-carrying capacity, insolubility in water, and high resistance to most organic solvents (Anwar et al., 2014; Mautner et al., 2018; Lavoine et al., 2012). The use of cellulose can be improved by treating the raw material with alkali and acid hydrolysis (Hossain et al., 2014; Sari et al., 2017; Asrofi et al., 2017). Alkali and acid hydrolysis breaks the bonding between the hydroxyl groups and results in the improved hydrophilic nature of hemicellulose. Some researchers have suggested that acid hydrolysis can improve crystallinity and reduce its diameter (Chen et al., 2011; Eyholzer et al., 2010). Modification of the surface fibers is commonly carried out to improve the adhesion between the fibers and polymer matrix (Gupta et al., 2018).

3. Nanocellulose

Nanocellulose is cellulose with at least one dimension of cellulose in the nanometer range (1−100 nm) (Ioelovich, 2008). The Technical Association of the Pulp and Paper Industry (TAPPI) has introduced the definitions for nanocellulose and its size (Mariano et al., 2014). The performance of nanocellulose varies with the use of different processing techniques (Abitbol et al., 2016). Nanocelluloses are generally similar in chemical compositions and structure, but differ in morphology, particle size, and crystallinity, which may depend on the various sources of extraction (Sinko et al., 2015; Karimi et al., 2014). Further, nanocelluloses are more transparent and thermally stable than microfibers. Nanocellulose offers high crystallinity, and is more transparent, thermally stable, has good mechanical properties, biocompatibility, biodegradability, optical transparency, high specific surface area, hydrophilicity, and moldability. Based on its physical and mechanical properties nanocellulose is utilized in several fields such as nanocomposite materials, biomedical products, wood adhesives, electronic sensors, batteries, catalytic supports, electroactive polymers, paper and paperboard industry products, food packaging, transparent film application, antimicrobial films, automobile industry applications, painting, cosmetics, etc. (Trache et al., 2020).

3.1 Classification of nanocellulose

Based on the shape and production process, nanocellulose is mainly classified into three categories: cellulose nanocrystals (CNCs), cellulose nanofibrils (CNFs), and bacterial nanocellulose (BNC), as shown in Fig. 4.2. CNCs are also known as cellulose nanocrystalline or cellulose whiskers. CNFs are also called nano/microfibrillated cellulose, and BNC is also called microbial cellulose, electrospun cellulose, or bio-cellulose. The different categories of nanocellulose with high strength and aspect ratio can be prepared using different chemical treatments and acid hydrolysis or in a combination of various processes such as high-pressure homogenization, grinding, chemical or enzymatic treatments, sonication, steam explosion, and cryocrushing. On the other hand, BNCs can be produced using a bacterial process.

3.1.1 Cellulose nanocrystals (CNCs)

CNCs are the mostly commonly used nanocellulose and are obtained with chemical treatments and acid hydrolysis of the amorphous section of fiber surfaces (Trache et al., 2017). Impurities (i.e., wax, hemicellulose, easters group, and lignin) are removed from the surface of fibers using alkaline treatment for different durations. Thereafter, acid treatment using bleaching/H_2SO_4/HCL is done on treated fibers. The amorphous regions of fibers are dissolved in concentrated acid and a high crystalline region is found, as shown in Fig. 4.3. Finally, a short rod-like structure of CNCs with 2–10 nm diameter and 100–500 nm length with high crystallinity is produced. The dimension and crystallinity of CNCs depend on the fiber source and extraction method (Habibi et al., 2010).

3.1.2 Cellulose nanofibrils (CNFs)

The preparation of CNFs is much simpler than the preparation of CNCs. NFCs are generally produced using mechanical processes such as high-pressure homogenizer, grinding, and ultrasonication without using any prechemical treatments (Ifuku et al., 2010; Pääkkö et al., 2007; Khalil et al., 2012). Fig. 4.4 shows a schematic diagram of CNFs which were extracted from cellulose fiber by an externally applied

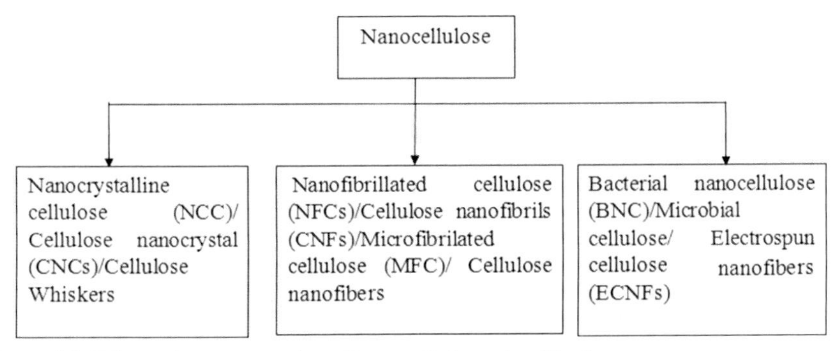

Figure 4.2 Schematic representation of nanocellulose classification.

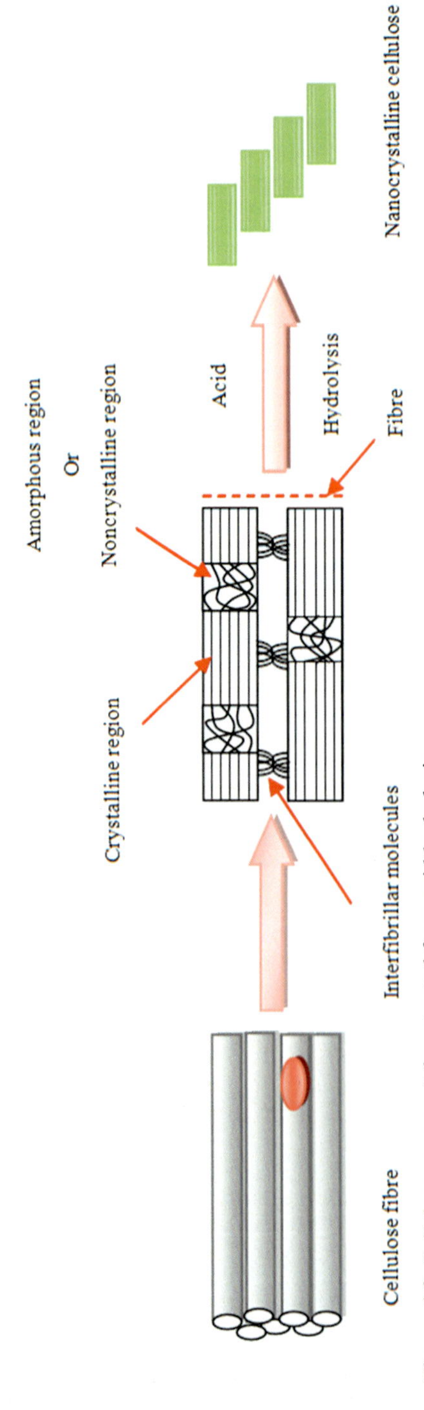

Figure 4.3 Cellulose nanocrystals extracted from acid hydrolysis.

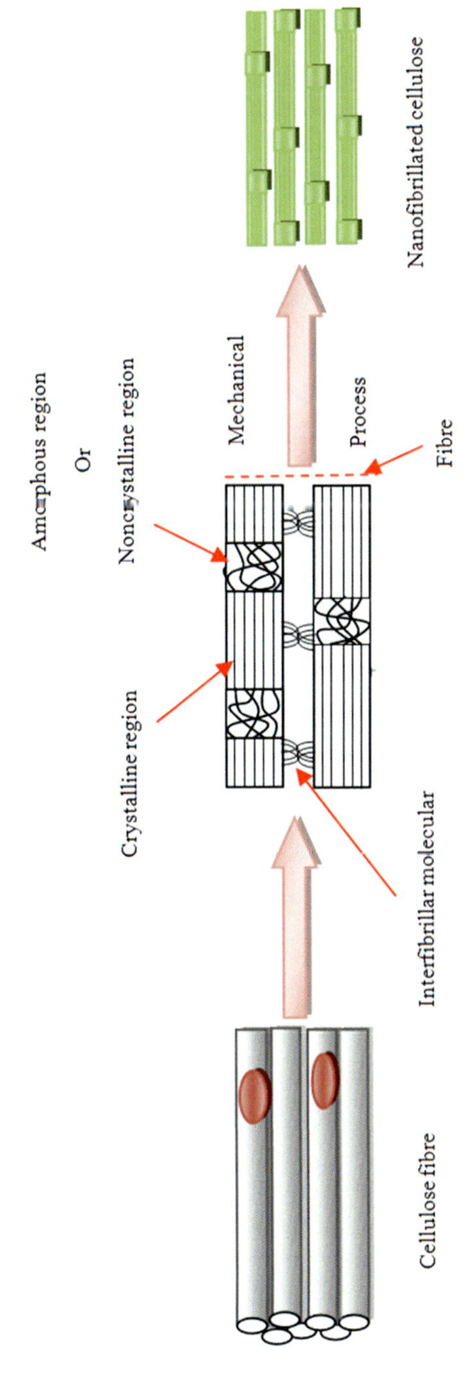

Figure 4.4 Nanofibrillated cellulose extracted from cellulose chain by mechanical process.

mechanical process. CNFs can also be produced using chemical methods and a combination of chemical and mechanical processes also. These chemical treatments are aimed at producing purified cellulose from the fiber. CNFs have long fibril shapes with 20–50 nm diameter and 500–2000 nm length. The extracted CNFs may have a high aspect ratio (L/D) and high surface area as compared to CNCs. The CNFs have a certain limitation in the paper-making application due to their poor compatibility with polymers (hydrophobic) because of the presence of a hydrophilic hydroxyl group in CNFs (Missoum et al., 2013).

3.1.3 Bacterial nanocellulose (BNC)

BNC is an organic compound produced by certain types of bacteria such as: *Acetobacter*, *Rhizobium*, *Agrobacterium*, *Aerobacter*, *Achromobacter*, *Azotobacter*, *Salmonella*, *Escherichia*, and *Sarcina* (Ullah et al., 2016). It does not require additional processing to remove impurities such as lignin, pectin, wax, and hemicellulose (Lin et al., 2013). The properties and extraction process of BNC (bottom-up process) are different from CNCs and CNFs (top-down process). The synthesis of BNC is carried out by biotechnological assembly processes from low-molecular-weight carbon sources, such as the use of uridine diphosphate glucose. It is in the form of ribbons with web-shaped structures and has an average diameter of 20–100 nm and less than 100 micrometers in length with a large surface area. Due to its high production cost and superior structural and physicomechanical features, BNC has been used for various biomedical applications such as tissue engineering, enzyme immobilization, regenerative medicines, drug delivery, and 3D printing. It is also used in industrial wastes such as fruit juices, sugarcane molasses, and agricultural wastes for low-cost production (Kurosumi et al., 2009; Tsouko et al., 2015).

3.2 Isolation processes of nanocellulose

The extraction processes of cellulose from natural fibers include chemical pretreatments, conventional processes, and nonconventional processes, as shown in Fig. 4.5. Several techniques for the isolation of nanocellulose are discussed in the following subsections.

3.2.1 Chemical pretreatments

Chemical pretreatment is the basic step of extraction of nanocellulose before conventional and nonconventional process. It reduces the energy consumption in conversion of nanocellulose by mechanical process (Souza et al., 2010). Energy consumption is the major challenge as reported by researchers. This process increases the surface roughness and also removes a certain amount of lignin, wax, and natural oils that cover the external surface of the fibers. Researchers have reported different types of chemical treatments, such as: enzyme, alkaline (NaOH), acid (sulfuric acid, hydrochloric acid, hydrogen peroxide, hydrogen bromide, nitric acid, and sodium hypochlorite), TEMPO-mediated oxidation, chlorite oxidation, sulfonation, carboxymethylation,

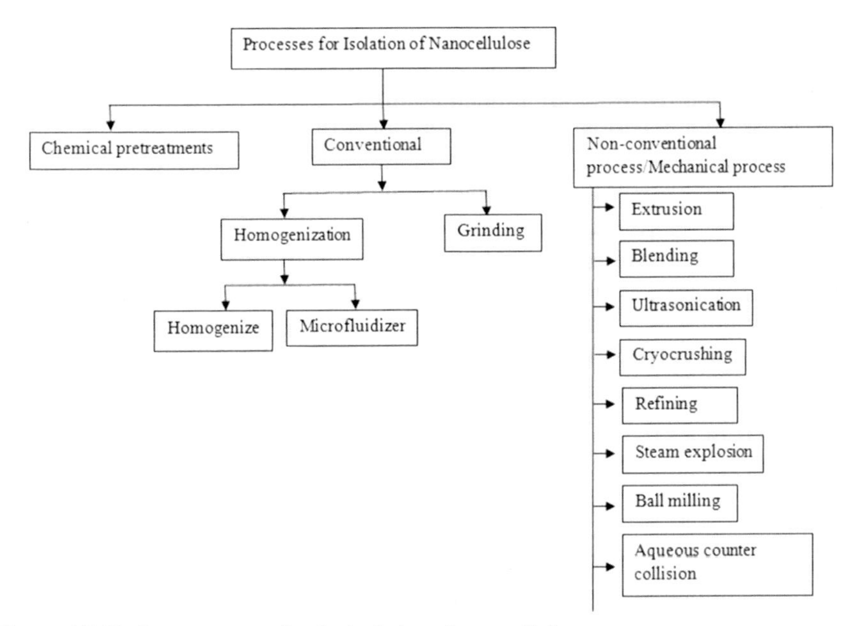

Figure 4.5 Various processes for the isolation of nanocellulose.

quaterization, ionic liquids, and solvent-assisted pretreatment. Alkaline–acid pretreatment is the most common method used to isolate nanocellulose. Most authors have observed that 12–17 wt.% of NaOH can be used for the treatment of fibers for easy hydrolysis (Khalil et al., 2014). To break the linkage between carbohydrate and lignin, a 2 wt.% NaOH solution is used. Sodium hydroxide treatment is an effective method to improve the crystallinity of cellulose fibers and removes the lignin and hemicellulose from wheat straw and soy hull fibers (Alemdar & Sain, 2008). Nanocellulose was prepared by acid hydrolysis followed by a mechanical process. It is the easiest and oldest method for the preparation of nanocellulose. Many strong acids have been used to degrade cellulose fibers. Among them, sulfuric acid (H_2SO_4) and hydrochloric acid (HCl) or combinations of both are the most popular chemical treatments. Researchers have observed that sulfuric acid gives more stable nanofibers than hydrochloric acid.

3.2.2 Conventional method

3.2.2.1 High-pressure homogenization

The high-pressure homogenization (HPH) process is used to refine cellulosic fiber pulp to obtain nanocellulose by a mechanical process. This equipment is used for large-scale production of NFCs in industries, laboratories, and medical applications. Cellulosic fiber pulp is passed between the rotor and stator disk using a small nozzle (100–400 μm) at high pressure (50–2000 MPa) and with repeated cyclic stress. The fiber is subjected to a high-pressure drop with high velocity, impact, and shear forces (Wang et al., 2015). A schematic representation of the high-pressure homogenizer process is illustrated in Fig. 4.6. Using the HPH technique, many researchers

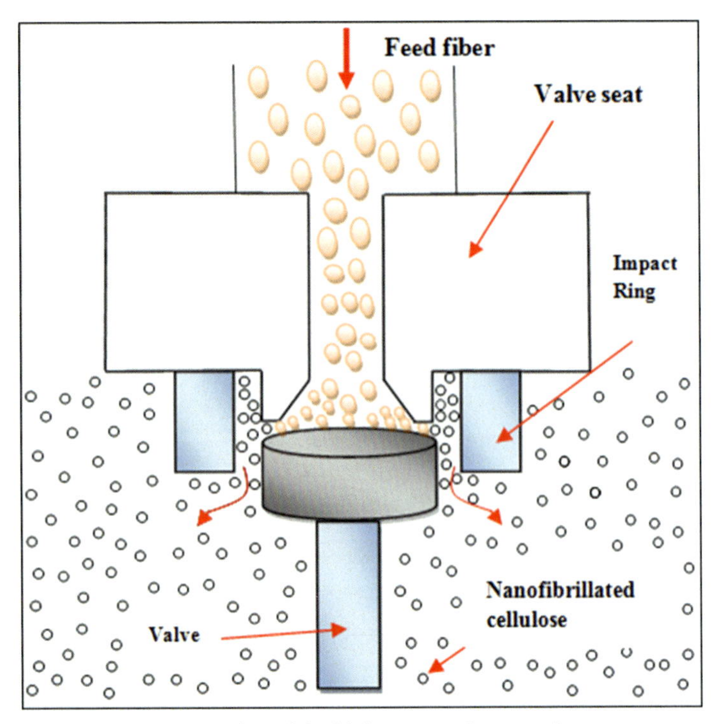

Figure 4.6 Schematic representation of the high-pressure homogenizer process.

have isolated nanofibrillated cellulose from different fibers, such as kenaf bast fiber (Jonoobi et al., 2010), sugar beet (Leitner et al., 2007), prickly pear (Habibi et al., 2009), oil palm leaves (Hussin et al., 2020), oat husk (Qazanfarzadeh & Kadivar, 2016), softwood pulp (Pääkkö et al., 2007), sisal fiber (Siqueira et al., 2010), and sunflower (Chaker et al., 2013). The purity of cellulose fiber was enhanced by researchers using 10,000 revolutions and 8000 psi pressure in the homogenizer, and they obtained NFCs with 100 nm diameters (Turbak et al., 1983; Herrick et al., 1983). Some researchers were able to extract NFCs without chemical treatment or enzymatic hydrolysis using a homogenizing process (Dufresne et al., 2000; Wang et al., 2017).

3.2.2.2 Grinding

Some researchers have attempted also to use a super mass collider grinder (Masuko Sangyo Corporation Limited, Japan) for breaking up cellulose into nano-size fibers. In this process cellulose fiber pulp (5%–10%) passes between two grind stones, where one grind stone is fixed and the other is movable at 1500 rpm. The distance between these grind stones is adjustable to avoiding blocking with the fiber pulp. Some disadvantages of this technique are the maintenance and replacement of grinding stone. Fiber pulp is passed through these grind stones 10–20 times to break the hydrogen and interfibrillar molecules bond. Nanofiber generation from fiber pulp using this method is presented in Fig. 4.7. NFCs in the diameter range of 20–90 nm were

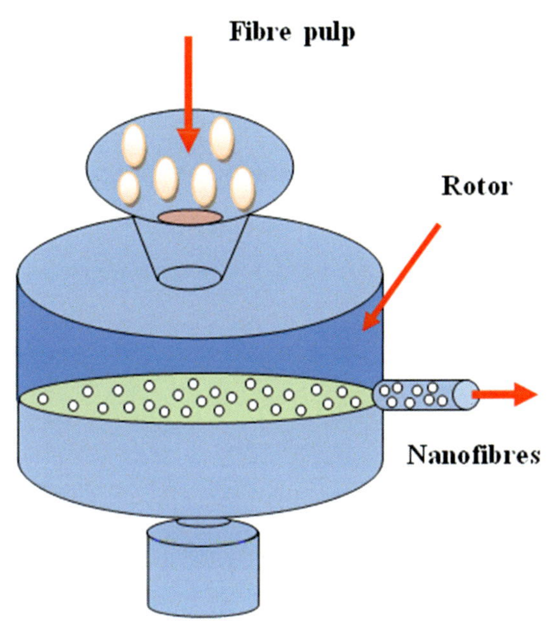

Figure 4.7 Schematic representation of the grinding process.

obtained using this method. Wang and Drzal (2012) isolated NFCs from eucalyptus pulp using a stone grinder. It was observed that the crystallinity of NFCs was dependent on energy consumption, number of passes, and fibrillation time (Iwamoto et al., 2009). CNFs with small and uniform size nanofibers were extracted from rice straw and bagasse through a high shear grinder apparatus using 10−30 passes with small and uniform size nanofibers (Hassan et al., 2012). Iwamoto et al. (2005) also reported that 14 passes were required to obtain 50−100 nm uniform size of nanofibers, and 20−50 nm size were obtained using five passes in a stone grinder at 1500 rpm. The crystallinity of nanofibers decreased with an increase in the number of passes (Iwamoto et al., 2007).

3.2.3 Nonconventional process/mechanical process

3.2.3.1 Ultrasonication

Ultrasonication converts mechanical or electrical energy into high-frequency acoustical energy by a transducer. In this process, high-intensity ultrasonication waves are produce in cavities of cellulose cells to create strong mechanical oscillating power. It is the combination of chemical pretreatments (i.e., sodium chloride and potassium hydroxide) and high-intensity ultrasonication. First, fibers being prepare in cellulose form according to the basic method (Abe & Yano, 2009, 2010). To remove lignin from the cellulose fibers, they are treated with sodium chlorite solution at 70−80°C for 1 h, and again treated with 3−6 wt.% of potassium hydroxide at 80°C for 2 h, with the purpose of this treatment being to remove impurities such as hemicellulose,

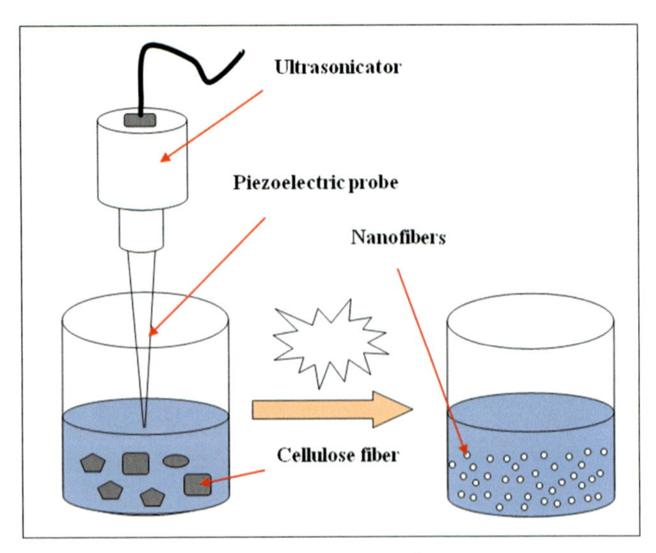

Figure 4.8 Schematic representation of the ultrasonication process.

starch, and pectin. After chemical treatment, a sample is kept in a water-swollen state to prevent a hydrogen bond. A 120-mL solution of purified cellulose fibers with distilled water is prepared and kept in an ultrasonic generator. When molecules in a liquid absorb ultrasonic energy, the formation and expansion of microscopic gas bubbles and shock waves produce nanofibers at high temperature and pressure (Fig. 4.8). Many researchers have reported the synthesis of nanofibers through high-intensity ultrasonication (Frone et al., 2011; Tischer et al., 2010; Qua et al., 2011). Khawas and Deka (2016) obtained nanofibers from banana peel through various chemical treatments followed by high-intensity ultrasonication at various output power levels. It was reported that the thermal stability and crystallinity of nanofibers were increased from 260.81 to 295.33°C when the output power was increased from 0 to 1000 W. Nanocellulose of rice husk and rice straw was obtained from several chemical treatments followed by ultrasonic fibrillation in diameters ranging from 30 to 35 nm (Rezanezhad, Nazanezhad, & Asadpur, 2013).

3.2.3.2 Ball milling

Ball milling is a mechanical process for the production of nanofibers. It is also known as pebble milling or tumbling milling. In this method, a cellulose suspension is placed in a hollow cylinder containing balls (ceramic, zirconia, or metal) which are mounted on a metallic frame (Fig. 4.9). the extraction of nanofibers depends on the diameter of the balls, nature of the balls (hardness), feed rate, rotation speed of the cylinder, and time of the materials in the mill chamber. When the container rotates at high speed, the cellulose cell wall of the fibers breaks down through the high-energy collision between balls. In the ball milling process, the desired particle size is obtained by controlling the operational time, applied energy, size and density of the grinding ball,

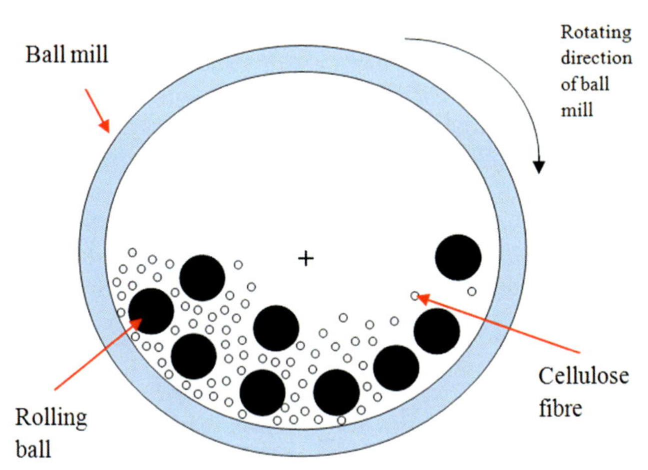

Figure 4.9 Schematic representation of the ball milling process.

moisture content, and carboxylic charges (Piras et al., 2019). Nanofibers with average diameters of 100 nm were obtained from softwood craft pulp using the ball milling process (Zhang et al., 2015). Kekäläinen et al. (2015) reported the production of hardwood kraft pulp nanofibers in the 10–150 nm diameter range using a ball milling process. The major challenge that arises when using this technique is to maintain the homogeneity of nanofibers.

3.2.3.3 Cryocrushing

Cryocrushing is another mechanical method for the isolation of nanofibers by breaking up the cellulose cell wall. The objective of cryocrushing is to form ice crystals within the cellulose cell wall. In this technique, fibers are treated by either freeze drying or suspending in water. First, cellulose fibers are kept in distilled water to absorb water in the cavity. To solidify the water content it is kept in nitrogen liquid and further crushed by high-impact or shear forces applied to the frozen cellulose fibers. After applying these forces, crystals inside the fibers exert pressure on the cell walls of the fibers, causing rupture of the ice crystal which produce nanofibers. Wang and Sain (2007) obtained nanofibers from soybean using a cryocrushing technique and observed diameters in the range 50–100 nm. Bhatnagar and Sain (2005) obtained nanofibers from various sources such as flax, hemp, kraft pulp, and rutabaga fibers with diameters of 5–80 nm using the cryocrushing technique.

3.2.3.4 Steam explosion

Several methods are used to extract nanofibers in which steam explosion is the thermomechanical process to convert nanofibers from plant cell walls by breaking down the structural component of cellulose. It is based on chemical and mechanical treatment (Gupta et al., 2018). In this process, plant fibers are dried and chopped to convert them into powder form, after that they are sieved and washed with distilled water for removal of dust and impurities. Washed fibers are bleached with sodium hydroxide (NaOH) for 2–4 h. They are then treated with sodium hypochlorite (NaOCl) and washed with sodium bisulfate ($NaSO_3$). These bleached fibers are exposed to pressurized steam for a short period of time followed by sudden release of pressure, causing rupture of the cellulose cell wall to produce nanofibers. Using this technique, researchers have extracted nanofibers from different plant sources, such as pineapple leaf (Cherian et al., 2010), banana fiber (Deepa et al., 2011), *Helicteres isora* (Vickers, 2017), and jute fiber (Manhas et al., 2015). Yang et al. (2017) extracted nanofibers from corncobs with diameters of 8 nm using steam explosion pretreatment.

3.2.3.5 Blending

In this technique, nanofibers were obtained by series of chemical treatments and blending using a mechanical high-speed blender. In this process, chopped raw fibers into small size and dipped into 5–10 wt.% sodium hydroxides (NaOH) solution for 4 h at ambient temperature to remove unwanted materials such as hemicellulose and wax. They are further treated with sodium chlorite ($NaClO_2$) and acetic acid (CH_3COOH) at 80°C for 4 h to remove lignin, after that, to break up the cell walls of cellulose and separate microfibrils, hydrochloric acid (HCl) treatment is required. Finally, the treated fibers are washed with distilled water and allowed to dry. The fibers are placed in a household blender at 1500 rpm for 40–60 min. Nanofibers were obtained in rod-like structures from rice straw, 2.7 nm wide and 100–200 nm long, using a high-speed blender (Jiang & Hsieh, 2013). A similar process was used to isolate nanofibers in the diameter range of 15–20 nm (Uetani & Yano, 2011).

3.2.3.6 Extrusion

Twin-screw extrusion uses a combination of a kneading and feeding screw, in which cellulose pulp fiber is passed through several times at higher concentration to obtained nanofibers. The major challenges to this technique are optimizing the screw profile and processing conditions to achieve sufficient shear forces to produce stratification of the fibers. Ho et al. (2015) obtained nanofibrillated cellulose from solid content kraft pulp using this method. Pulp fibers were placed through the extruder up to 14 times at operating room temperature. The same method was used to produce pulp nanofibrils (Suzuki et al., 2013). Rol et al. (2019) optimized the screw profile to produce high-quality nanofibers in one pass. Bleached softwood flour was obtained with a particle size of 200–400 μm. Further, a twin-screw extraction process was used to reduce the diameter of cellulose fibers in the nano-scale range. Good dispersion of nanofibers in the polymer matrix was obtained (Hietala et al., 2014).

3.3 Characterization of nanocellulose

For advanced applications and extensive utilization of nanocellulose, its physical, thermal, mechanical, and morphological properties have been explored. Fig. 4.10 shows

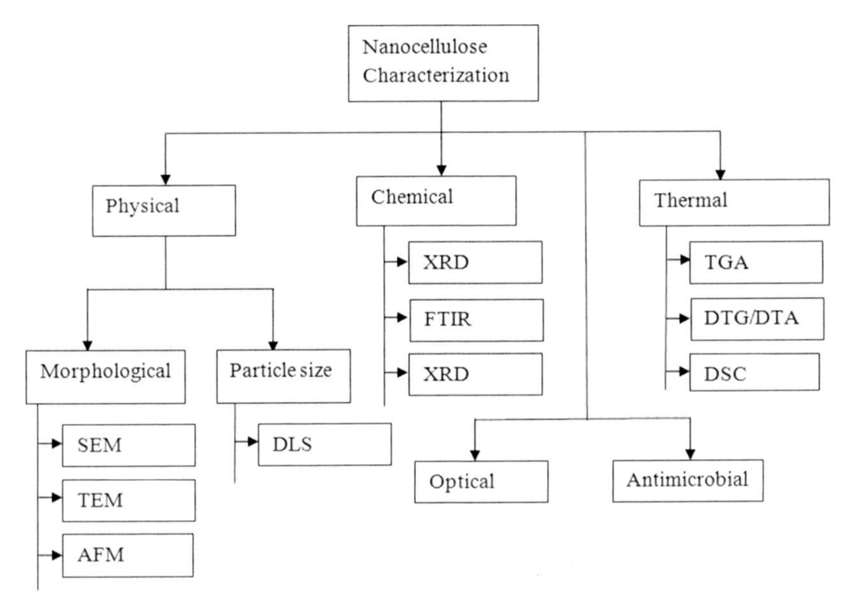

Figure 4.10 Schematic representation of nanocellulose characterization.

different characterization methods carried out on nanocellulose to discover the dimension, shape, aspect ratio, surface roughness, crystalline structure, thermal stability, constituent of the fibers, etc.

3.3.1 Physical properties

3.3.1.1 Dynamic light scattering (DLS)/zeta potential

Dynamic light scattering is also known as photon correlation spectroscopy or quasi-elastic light scattering. DLS is technique for the direct determination of particle size in a solution. It is based on the Brownian motion of a dispersed particle to measure particle size in the range from 0.6 nm to 6 microns. In this process, first nanofibers are dispersed in a solvent. These solvent molecules continuously collide with nanofiber particles based on Brownian motion. Due to these collisions extra energy is generated and transferred to particle molecules. Therefore, smaller particles moving at higher speed are compared to large particles. Particle size can be obtained by measuring the random changes in the intensity of light scattered from particles moving with higher speed. The particle size distribution of sugarcane bagasse nanocellulose is shown in Fig. 4.11. Chandra et al. (2016) obtained dispersion of nanofibers in water with an average diameter below 10 nm from areca nut husk fiber. The same behavior was observed for yerba mate residues (Júnior et al., 2019). Two different types of nanocellulose from usher seed fiber in the nanometric ranges 14–24 nm and 10–20 nm in needle structure were observed (Oun & Rhim, 2016).

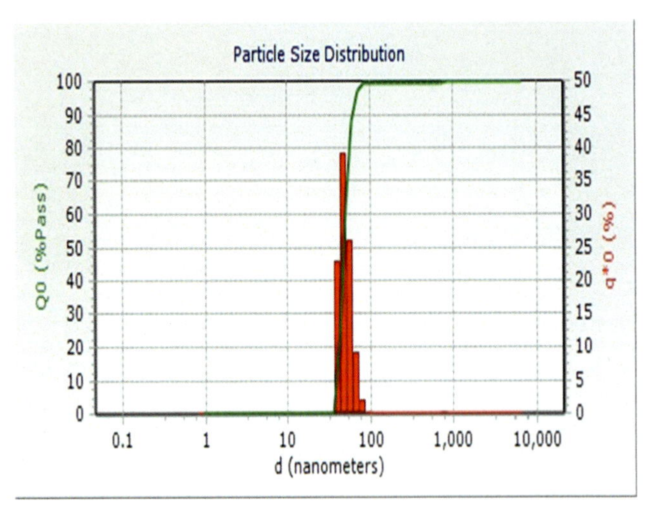

Figure 4.11 Particle size distribution of nanocellulose of sugarcane bagasse.

3.3.1.2 Electron microscopy (EM)

Morphological properties of nanocellulose have been studied by transmission electron microscopy (TEM) and scanning electron microscopy (SEM). TEM is a powerful tool used in nanoscience technology to observe crystal structure (dislocation and grain boundary), growth of layers, compositional information, and shape and size of nanoparticles/nanocellulose. First, nanocellulose is deposited on a carbon-coated grid and allowed to dry at room temperature for 24 h. An electron beam of very short wavelength is emitted from a tungsten filament and irradiated on a thin specimen. To avoid air molecules and scattering of electrons, optical systems are kept in a vacuum. an interaction between the electron beam and specimen surface image is formed and magnified on the fluorescent screen. The resolution of the image is increased by increasing the accelerating voltage of the electron beam.

SEM is a type of electron microscopy that uses a high-intensity electron beam to generate a variety of signals at a fiber surface. These signals give information about the texture, chemical composition, crystalline structure, and orientation of fibers. In SEM analysis, the sample scanning area is approximately 1 cm−5 microns in width and the magnification range 20× to 30,000× with a spatial resolution of 50−1000 nm. Before SEM analysis, a nonconductive sample is sputter coated with gold nanoparticles. A high-intensity electron beam (electron probe) is focused on the sample to produce images. This electron beam is generated by an electron gun and focused with 1 nm size. The morphological structure of nanocellulose is shown in Fig. 4.12.

Johar et al. (2012) obtained nanofibers from rice husk and the size was confirmed by TEM analysis. Most of the nanofibers were observed in the range of 15−20 nm with an aspect ratio 10−15. Morphological changes in rice husk due to chemical treatment are also observed by SEM analysis. TEM micrographs show the variation of nanofiber size

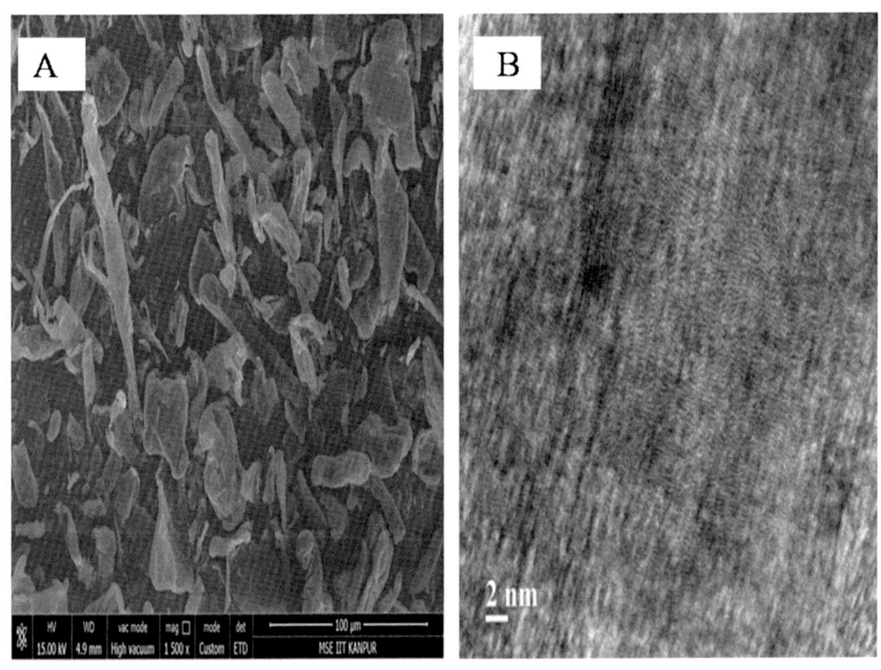

Figure 4.12 Electron microscopy images: (A) FESEM and (B) TEM images of nanocellulose of sugarcane bagasse.

in the range 10–70 nm, of which 64% of fibers were 10–20 nm and with the remainder greater than 50 nm diameter (Li et al., 2014). Reddy and Rhim (2018) studied the surface morphology of extracted cellulose microfibers from onion skin (50–100 μm) and garlic stalk (30–80 μm).

3.3.1.3 Atomic force microscopy (AFM)

Atomic force microscopy is used to analyze the surface characteristics of nanocellulose. It consists of a cantilever sharp tip assembly that directly interacts with and scans the sample surface. The profile of the cantilever tip is in the form of a curve made of silicon or silicon nitride in the nanometer range. This deflection is measured by a sensitive photodetector and generates a quantitative three-dimensional topographic image. AFM can be operated in two modes, i.e., contact mode and noncontact or semicontact mode. In the contact mode, the sample comes directly into contact with the cantilever tip and scans the sample surface. However, in noncontact mode, the cantilever tip generates frequency and force that are exerted on the sample surface due to changes in the amplitude frequency or phase frequency. The noncontact mode is best suited for scanning the nanocellulose surface. Satyamurthy and Vigneshwaran (2013) studied the AFM analysis of two different particles, microcrystalline cellulose and nanocrystalline cellulose, obtained from cotton fibers, with uniform distribution and no aggregation was observed. Similar results have been found in the literature (Herrera et al., 2012).

Table 4.1 Chemical composition present in different fibers.

Fiber	Cellulose	Hemicellulose	Lignin	References
Areca nut husk	34.18	20.83	31.6	Chandra et al. (2016)
Rice husk	35	33	23	Johar et al. (2012)
Kenaf	43.7	34.7	11.5	Kargarzadeh et al. (2012)
Hemp fiber	75.56	10.66	6.61	Wang et al. (2007)
Pineapple leaf	62.5	13.9	15.9	Mahardika et al. (2018)
Garlic straw	41	18	6.3	Kallel et al. (2016)
Soy hull	48.2	24	5.78	Neto et al. (2013)
Mengkuang leaves	37.3	34.4	24	Sheltami et al. (2012)
Wheat straw	43.2	34.1	22	Alemdar and Sain (2008)
Banana waste	43.5	31.7	16.9	Habibi et al. (2008)
Cotton stalk	50.6	28.4	23.1	Habibi et al. (2008)
Jute	60	22.1	15.9	Razera and Frollini (2004)
Cotton lint	90	6	–	Satyanarayana et al. (2007)
Yerba mate	34.85	24.77	25.78	Júnior et al. (2019)
Barley straw	37.7	37.1	15.8	Sun and Sun (2002)
Onion skin	41.1	16.2	38.9	Rhim et al. (2015)
Wheat bran	31.1	34.3	16.3	Xiao et al. (2019)

3.3.2 Chemical properties

3.3.2.1 Chemical composition

The chemical composition of any fiber can be determined according to the Technical Association of Pulp and Paper Industry (TAPPI) method. In this method, cellulose, hemicellulose, lignin, moisture, pectin and ash can be observed with TAPPI standard T222 OS-83. The chemical treatment of the fiber influences the chemical composition. Table 4.1 shows the chemical composition in the form of percentages of cellulose, hemicelluloses, and lignin of several natural fibers.

3.3.2.2 X-ray diffraction (XRD)

X-ray diffraction is a technique to characterize crystalline materials. It is useful for determining the structure, crystal orientation, crystal defect, strain, atomic spacing, and grain size. It is also useful to determine the percentage of crystallinity before and after chemical treatment of natural fibers. In this method, a monochromatic X-ray is generated by a cathode ray tube, this X-ray is filtered and produces a monochromatic electron beam and is allowed to concentrate on a targeted sample. When the electron beam hits the inner cell of the target sample, diffracted rays are produced and satisfy Bragg's law. The diffracted ray of the sample is recorded on an X-ray diffractometer equipped with Cu K_{α} radiation at an operating voltage of 40 kV and current of 35 mA. The diffraction intensities are recorded over the 2θ range from 10–80°C at a scanning rate of 2 degrees per minute. Nishiyama (2009) studied the crystalline structure of cellulose and found diffraction peaks at 22 degrees and 15 degrees. Many researchers have observed that the crystallinity

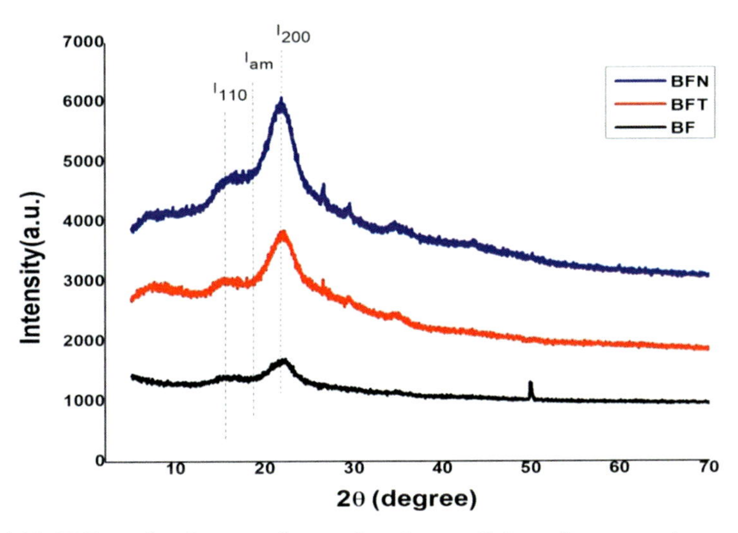

Figure 4.13 XRD graphs of untreated, treated, and nanocellulose of sugarcane bagasse.

of cellulose is affected by chemical treatment. The alkaline treatment performed on natural fibers improves their stiffness and removes unwanted materials (Mwaikambo & Ansell, 2006). Also, the tensile strength and rigidity of cellulose fibers are improved due to the compact molecular structure (Lu et al., 2013). XRD graphs of untreated (BF), treated (BFT), and nanocellulose (BNF) of sugarcane bagasse are shown in Fig. 4.13. Table 4.2 presents the diameter, aspect ratio, and crystallinity of various nanofibers.

The crystallinity index and degree of crystallinity index of fibers are calculated using Segal's method as given in Eqs. (4.1) and (4.2).

$$\text{Crystallinity index}(\%) = \frac{I_{200} - I_{am}}{I_{200}} \times 100 \tag{4.1}$$

$$\text{Degree of crystallinity} = \frac{I_{200}}{I_{200} + I_{am}} \tag{4.2}$$

where I_{200} is the maximum intensity of the diffraction peak ($2\theta = 22.5°$) and I_{am} is the intensity of the amorphous diffraction peak ($2\theta = 18°$). Further, the mean size of the crystalline region is measured using Eq. (4.3).

$$\text{Crystal size } D = \frac{k\lambda}{\beta\cos\theta} \tag{4.3}$$

where k = the dimensionless shape factor which is generally taken as 0.9, λ (1.54Å) is the X-ray wavelength, β is the full width at the half maximum peak, and is the diffraction or Bragg's angle at the peak (200).

Table 4.2 Diameter, aspect ratio, and crystallinity of different fibers.

Nanofibers	Diameter (nm)	Aspect ratio	Crystallinity (%)	References
Areca nut husk	10	120–150	73	Chandra et al. (2016)
Hemp	30–100		71.2	Wang et al. (2007)
Pineapple leaf	40–70	16	69.4	Mahardika et al. (2018)
Rose steams	9.5–28 μm	54–76	56.2	Ventura-Cruz and Tecante (2019)
Garlic straw	6	80	68.8	Kallel et al. (2016)
Rice husk	15–20	10–15	59	Johar et al. (2012)
Grape skin	48.1	10	64.3	Lu and Hsieh (2012)
Spruce bark	28	6.3	84	Le Normand et al. (2014)
Onion skin	20–35	8–10	30	Rhim et al. (2015)
Ramie	10.8	12.4	–	Habibi and Dufresne (2008)
Cotton	5–11	15–42	–	Miller and Donald (2003)
Bagasse	4–10	13	89	Bras et al. (2010)
Soy hull	2.77	44	73.5	Neto et al. (2013)
Pea hull	7–12	34	–	Chen et al. (2011)
Barley	6–14	32	66	Espino et al. (2014)
Mengkuang leaves	5–25	10–20	–	Sheltami et al. (2012)
Wheat bran	17–33	21–32	70.32	Xiao et al. (2019)
Sisal	5–15	40	78	Mondragon et al. (2014)
Maize stalk	3–7	50–64	72.6	Mtibe et al. (2015)
Coconut husk	3–11	16–30	82	Nascimento et al. (2014)
Potato peel	–	41	85	Chen et al. (2012)
Kenaf	9–15	13.2	81	Kargarzadeh et al. (2012)
Tomato peels	1–7	–	80.8	Jiang and Hsieh (2015)
Bamboo	50–100	–	46.08	Liu et al. (2010)
Sugarbeet pulp	10–50	–	62.3	Yang et al. (2018)

3.3.2.3 Fourier-transform infrared spectroscopy (FT-IR)

Fourier-transform infrared spectroscopy is a technique to determine the change in functional groups and structural characteristics of nanocellulose before and after treatment. It is used to record different functional groups and chemical bonds between the wave number range of 5000 to 400 cm^{-1}. In FTIR analysis, a small amount of dry sample is mixed with potassium bromide (KBr) to form a thin layer of sample. Before being subjected to infrared radiation, the entire sample is dried in a hot air oven at 50–60°C to achieve a constant weight. When infrared radiation (10,000 to 100 cm^{-1}) passes through a sample, some of the radiation is absorbed (to convert rotational/vibration energy) and some is transmitted by the sample.

Table 4.3 Functional groups and compounds present in fibers (Morán et al., 2008).

Fibers component	Wave number (cm^{-1})	Functional group	Compounds
Cellulose	4000−2995	OH	Acid, methanol
	2890	H−C−H	Alkyl, aliphatic
	1640	Fibre-OH	Absorbed water
	1270−1232	C−O−C	Aryl-alkyl ether
	1170−1082	C−O−C	Pyranose ring skeletal
	1108	OH	C−OH
Hemicellulose	4000−2995	OH	Acid, methanol
	2890	H−C−H	Alkyl, aliphatic
	1765−1715	C=O	Ketone and carbonyl
	1108	OH	C−OH
Lignin	1000−2995	OH	Acid, methanol
	2890	H−C−H	Alkyl, aliphatic
	1730−1700	−	Aromatic
	1632	C=C	Benzene stretching ring
	1613−1450	C=C	Aromatic skeletal mode
	1430	O−CH_3	Methoxyl O−CH_3
	1270−1232	C−O−C	Aryl-alkyl ether
	1215	C−O	Phenon
	1108	OH	C−OH
	700−900	C−H	Aromatic hydrogen

These signals are recorded by a detector in a given spectrum to generate a different fingerprint peak for different chemical structures. Table 4.3 shows the different functional groups and compounds present at different wave numbers.

3.3.3 Thermal properties

Thermal analysis is the analysis of a change in a property (physical or chemical) of a sample related to a change in temperature. The thermal properties of fibers and polymeric composites are measured using different techniques, such as thermogravimetric analysis (TGA), differential thermal analysis (DTA)/derivative thermogravimetric (DTG), and differential scanning calorimetry (DSC). The thermal decomposition properties of nanocellulose are determined by thermogravimetric analysis as a function of time and temperature at a constant heating rate under a controlled atmosphere. Approximately 3−5 mg of dried sample is placed in a crucible pan and hung on a sensitive microbalance. The sample is heated in a furnace from ambient temperature to 600°C at a heating range from 10 to 20°C/min (maximum heating rate 100°C/min). The whole process is done under a nitrogen atmosphere to remove unwanted gases and avoid thermo-oxidative degradation.

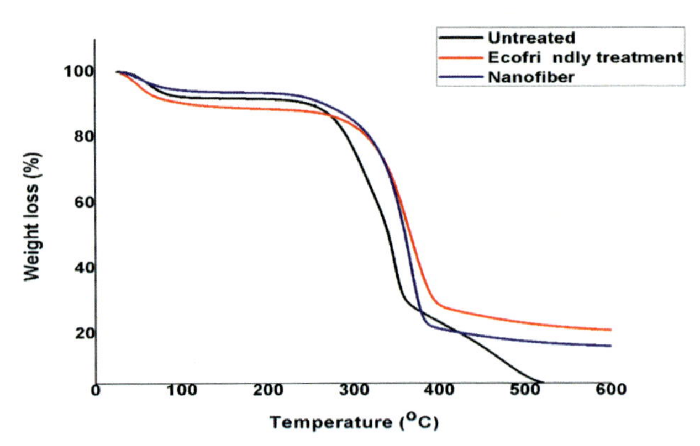

Figure 4.14 TGA graphs of untreated, treated, and nanocellulose of sugarcane bagasse.

A continuous weight loss of fibers with respect to time and temperature is recorded to characterize the humidity, thermal stability, and material purity. On the other hand, exothermic and endothermic processes occur due to changes in temperature or heat in the sample recorded by differential thermal analysis (DTA). Both TGA and DTA analysis give information about the melting point, glass transition temperature, moisture, purity, transformation temperature, and thermal and oxidative stability.

The decomposition of constituents of natural fibers occurs at different temperatures: cellulose from 315 to 400°C, hemicelluloses from 220 to 315°C, and lignin from 200 to 700°C (Yang et al., 2007; Brebu & Vasile, 2010). Fig. 4.14 show the TGA curves of untreated, treated, and nanofibers of sugarcane bagasse. It was observed that minimum weight loss in the temperature range of 200–250°C was offered by nanofibers followed by the treated and untreated fibers (Gond & Gupta, 2020). In another study, kenaf nanocellulose presented a lower thermal stability than raw kenaf (Song et al., 2018). Chandra et al. (2016) reported that an initial weight loss of 35°C up to 120°C and better thermal stability were observed for bleached and homogenized nanocellulose. Another method to measure the thermal properties in the form of glass transition temperature, melting point, crystallization, specific heat, and thermal stability of nanofibers is differential scanning calorimetry (DSC). In this method, the differences in the amount of heat flow rate between the sample chamber and reference chamber are measured as a function of time and temperature. A small sample of 1–15 mg is placed in a temperature-controlled DSC cell. After heating or cooling at a controlled rate, the phase transition of the sample is recorded. The thermal behaviors of *Xanthoceras sorbifolia* husk fiber, cellulose, and nanocellulose were analyzed using DSC and a narrow fusion peak and endothermic peak of cellulose and nanocellulose at 360 and 210°C, respectively, were observed (Ma et al., 2015). Mandal and Chakrabarty (2011) reported a narrow endothermic peak at 253–290°C for nanocellulose of sugarcane bagasse.

4. Applications

It has been reported that nanocellulose is nontoxic, has self-assembly behavior, and is biodegradable and biocompatible, with a high surface area to volume ratio and does not contribute to environmental damage. Owing to the excellent properties and biodegradable nature of nanocellulose, it has been used in various applications such as transparent paper, nanocomposite materials, medical, automotive, electronics, packaging (including food packaging), construction, and wastewater treatment (Klemm et al., 2005). Nanocellulose is highly suitable for biomedical applications. It is being used in several medical areas, including tissue repair, tissue regeneration, implants, drug-delivery systems, hemodialysis membrane, absorbable hemostats, antibacterials, skin disease treatments, and biocatalysts (Lin & Dufresne, 2014; Trache, 2018; Bacakova et al., 2019; Moohan et al., 2020). Further, nanocellulose-based hydrogels have been used in superior drug-delivery systems, contact lenses, sensors, and Tissue engineering and Bioprinted human ear (Dong et al., 2013). Some researchers have prepared aerogels of nanofibers for implantations to repair bone defects in medical applications (Osorio et al., 2019). Besides the above applications, nanocellulose can be used in cosmetics, electronic sensors, biodegradable packaging, and wood adhesives.

5. Conclusion

Cellulose is the most abundant natural polymer obtained from environmentally friendly and biocompatible resources. A cellulose fiber provides different types of nanocellulose, such as cellulose nanocrystals, cellulose nanofibrils, and bacterial nanocellulose. A number of extraction processes have been applied to obtain nanocellulose with different shapes, sizes, and aspect ratios. The properties of the various types of extracted nanocellulose depend on the source and extraction technique. Nanocellulose is used in a number of applications due to its amazing properties such as low weight, biodegradable, high surface area to volume ratio, cost-effectiveness, anisotropic shape, and renewable and better mechanical properties. It has some disadvantages also, such as moisture absorption, low thermal stability, quality variations, and poor compatibility in some cases. Based on the excellent properties of nanocellulose and the excellent performance of its composites, it can be proposed for use in many applications including in the medical, textile, food packaging, cosmetics, and hydrogel industries.

Acknowledgment

The authors would like to thank the Council of Science & Technology, UP, for financial support. The authors are also thankful to Head, Mechanical Engineering Department, Motilal Nehru National Institute of Technology Allahabad-211004, Prayagraj, India.

References

Abe, K., & Yano, H. (2009). Comparison of the characteristics of cellulose microfibril aggregates of wood, rice straw and potato tuber. *Cellulose, 16*(6), 1017–1023.

Abe, K., & Yano, H. (2010). Comparison of the characteristics of cellulose microfibril aggregates isolated from fiber and parenchyma cells of Moso bamboo (*Phyllostachys pubescens*). *Cellulose, 17*(2), 271–277.

Abitbol, T., Rivkin, A., Cao, Y., Nevo, Y., Abraham, E., Ben-Shalom, T., & Shoseyov, O. (2016). Nanocellulose, a tiny fiber with huge applications. *Current Opinion in Biotechnology, 39*, 76–88.

Abraham, E., Deepa, B., Pothan, L. A., Jacob, M., Thomas, S., Cvelbar, U., & Anandjiwala, R. (2011). Extraction of nanocellulose fibrils from lignocellulosic fibres: a novel approach. *Carbohydrate Polymers, 86*(4), 1468–1475.

Alemdar, A., & Sain, M. (2008). Isolation and characterization of nanofibers from agricultural residues—Wheat straw and soy hulls. *Bioresource Technology, 99*(6), 1664–1671.

Anwar, Z., Gulfraz, M., & Irshad, M. (2014). Agro-industrial lignocellulosic biomass a key to unlock the future bio-energy: a brief review. *Journal of Radiation Research and Applied Sciences, 7*(2), 163–173.

Asrofi, M., Abral, H., Kasim, A., & Pratoto, A. (2017). Characterization of the microfibrillated cellulose from water hyacinth pulp after alkali treatment and wet blending. In *IOP Conf. Ser. Mater. Sci. Eng.* (vol. 204, p. 12018).

Bacakova, L., Pajorova, J., Bacakova, M., Skogberg, A., Kallio, P., Kolarova, K., & Svorcik, V. (2019). Versatile application of nanocellulose: from industry to skin tissue engineering and wound healing. *Nanomaterials, 9*(2), 164.

Bettaieb, F., Khiari, R., Dufresne, A., Mhenni, M. F., & Belgacem, M. N. (2015). Mechanical and thermal properties of *Posidonia oceanica* cellulose nanocrystal reinforced polymer. *Carbohydrate Polymers, 123*, 99–104.

Bhatnagar, A., & Sain, M. (2005). Processing of cellulose nanofiber-reinforced composites. *Journal of Reinforced Plastics and Composites, 24*(12), 1259–1268.

Bras, J., Hassan, M. L., Bruzesse, C., Hassan, E. A., El-Wakil, N. A., & Dufresne, A. (2010). Mechanical, barrier, and biodegradability properties of bagasse cellulose whiskers reinforced natural rubber nanocomposites. *Industrial Crops and Products, 32*(3), 627–633.

Brebu, M., & Vasile, C. (2010). Thermal degradation of lignin—a review. *Cellulose Chemistry and Technology, 44*(9), 353.

Chaker, A., Alila, S., Mutjé, P., Vilar, M. R., & Boufi, S. (2013). Key role of the hemicellulose content and the cell morphology on the nanofibrillation effectiveness of cellulose pulps. *Cellulose, 20*(6), 2863–2875.

Chandra, J., George, N., & Narayanankutty, S. K. (2016). Isolation and characterization of cellulose nanofibrils from arecanut husk fibre. *Carbohydrate Polymers, 142*, 158–166.

Chen, W., Yu, H., Liu, Y., Chen, P., Zhang, M., & Hai, Y. (2011). Individualization of cellulose nanofibers from wood using high-intensity ultrasonication combined with chemical pretreatments. *Carbohydrate Polymers, 83*(4), 1804–1811.

Chen, W., Yu, H., Liu, Y., Hai, Y., Zhang, M., & Chen, P. (2011). Isolation and characterization of cellulose nanofibers from four plant cellulose fibers using a chemical-ultrasonic process. *Cellulose, 18*(2), 433–442.

Chen, D., Lawton, D., Thompson, M. R., & Liu, Q. (2012). Biocomposites reinforced with cellulose nanocrystals derived from potato peel waste. *Carbohydrate Polymers, 90*(1), 709–716.

Cherian, B. M., Leão, A. L., De Souza, S. F., Thomas, S., Pothan, L. A., & Kottaisamy, M. (2010). Isolation of nanocellulose from pineapple leaf fibres by steam explosion. *Carbohydrate Polymers, 81*(3), 720–725.

Deepa, B., Abraham, E., Cherian, B. M., Bismarck, A., Blaker, J. J., Pothan, L. A., & Kottaisamy, M. (2011). Structure, morphology and thermal characteristics of banana nano fibers obtained by steam explosion. *Bioresource Technology, 102*(2), 1988–1997.

Dong, H., Snyder, J. F., Williams, K. S., & Andzelm, J. W. (2013). Cation-induced hydrogels of cellulose nanofibrils with tunable moduli. *Biomacromolecules, 14*(9), 3338–3345.

Dufresne, A., Dupeyre, D., & Vignon, M. R. (2000). Cellulose microfibrils from potato tuber cells: processing and characterization of starch–cellulose microfibril composites. *Journal of Applied Polymer Science, 76*(14), 2080–2092.

Espino, E., Cakir, M., Domenek, S., Román-Gutiérrez, A. D., Belgacem, N., & Bras, J. (2014). Isolation and characterization of cellulose nanocrystals from industrial by-products of *Agave tequilana* and barley. *Industrial Crops and Products, 62*, 552–559.

Eyholzer, C., Bordeanu, N., Lopez-Suevos, F., Rentsch, D., Zimmermann, T., & Oksman, K. (2010). Preparation and characterization of water-redispersible nanofibrillated cellulose in powder form. *Cellulose, 17*(1), 19–30.

Frone, A. N., Panaitescu, D. M., Donescu, D., Spataru, C. I., Radovici, C., Trusca, R., & Somoghi, R. (2011). Preparation and characterization of PVA composites with cellulose nanofibers obtained by ultrasonication. *BioResources, 6*(1), 487–512.

Gond, R. K., & Gupta, M. K. (2020). A novel approach for isolation of nanofibers from sugarcane bagasse and its characterization for packaging applications. *Polymer Composites, 41*(12), 5216–5226. https://doi.org/10.1002/pc.25788

Gupta, M. K., Gond, R. K., & Bharti, A. (2018). Effects of treatments on the properties of polyester based hemp composite. *Indian Journal of Fibre and Textile Research (IJFTR), 43*(3), 313–319.

Habibi, Y., & Dufresne, A. (2008). Highly filled bionanocomposites from functionalized polysaccharide nanocrystals. *Biomacromolecules, 9*(7), 1974–1980.

Habibi, Y., El-Zawawy, W. K., Ibrahim, M. M., & Dufresne, A. (2008). Processing and characterization of reinforced polyethylene composites made with lignocellulosic fibers from Egyptian agro-industrial residues. *Composites Science and Technology, 68*(7–8), 1877–1885.

Habibi, Y., Mahrouz, M., & Vignon, M. R. (2009). Microfibrillated cellulose from the peel of prickly pear fruits. *Food Chemistry, 115*(2), 423–429.

Habibi, Y., Lucia, L. A., & Rojas, O. J. (2010). Cellulose nanocrystals: chemistry, self-assembly, and applications. *Chemical Reviews, 110*(6), 3479–3500.

Hassan, M. L., Mathew, A. P., Hassan, E. A., El-Wakil, N. A., & Oksman, K. (2012). Nanofibers from bagasse and rice straw: process optimization and properties. *Wood Science and Technology, 46*(1–3), 193–205.

Herrera, M. A., Mathew, A. P., & Oksman, K. (2012). Comparison of cellulose nanowhiskers extracted from industrial bio-residue and commercial microcrystalline cellulose. *Materials Letters, 71*, 28–31.

Herrick, F. W., Casebier, R. L., Hamilton, J. K., & Sandberg, K. R. (1983). Microfibrillated cellulose: morphology and accessibility. In *J. Appl. Polym. Sci.: Appl. Polym. Symp.; (United States) (Vol. 37, No. CONF-8205234-Vol. 2)*. Shelton, WA: ITT Rayonier Inc.

Hietala, M., Rollo, P., Kekäläinen, K., & Oksman, K. (2014). Extrusion processing of green biocomposites: compounding, fibrillation efficiency, and fiber dispersion. *Journal of Applied Polymer Science, 131*(6).

Ho, T. T. T., Abe, K., Zimmermann, T., & Yano, H. (2015). Nanofibrillation of pulp fibers by twin-screw extrusion. *Cellulose, 22*(1), 421–433.

Hossain, M. K., Karim, M. R., Chowdhury, M. R., Imam, M. A., Hosur, M., Jeelani, S., & Farag, R. (2014). Comparative mechanical and thermal study of chemically treated and untreated single sugarcane fiber bundle. *Industrial Crops and Products, 58*, 78–90.

Hussin, F. N. N. M., Attan, N., & Wahab, R. A. (2020). Extraction and characterization of nanocellulose from raw oil palm leaves (*Elaeis guineensis*). *Arabian Journal for Science and Engineering, 45*(1), 175–186.

Ifuku, S., Nogi, M., Yoshioka, M., Morimoto, M., Yano, H., & Saimoto, H. (2010). Fibrillation of dried chitin into 10–20 nm nanofibers by a simple grinding method under acidic conditions. *Carbohydrate Polymers, 81*(1), 134–139.

Ioelovich, M. (2008). Cellulose as a nanostructured polymer: a short review. *BioResources, 3*(4), 1403–1418.

Iwamoto, S., Nakagaito, A. N., Yano, H., & Nogi, M. (2005). Optically transparent composites reinforced with plant fiber-based nanofibers. *Applied Physics A, 81*(6), 1109–1112.

Iwamoto, S., Nakagaito, A. N., & Yano, H. (2007). Nano-fibrillation of pulp fibers for the processing of transparent nanocomposites. *Applied Physics A, 89*(2), 461–466.

Iwamoto, S., Kai, W., Isogai, A., & Iwata, T. (2009). Elastic modulus of single cellulose microfibrils from tunicate measured by atomic force microscopy. *Biomacromolecules, 10*(9), 2571–2576.

Jiang, F., & Hsieh, Y. L. (2013). Chemically and mechanically isolated nanocellulose and their self-assembled structures. *Carbohydrate Polymers, 95*(1), 32–40.

Jiang, F., & Hsieh, Y. L. (2015). Cellulose nanocrystal isolation from tomato peels and assembled nanofibers. *Carbohydrate Polymers, 122*, 60–68.

Johar, N., Ahmad, I., & Dufresne, A. (2012). Extraction, preparation and characterization of cellulose fibres and nanocrystals from rice husk. *Industrial Crops and Products, 37*(1), 93–99.

Jonoobi, M., Harun, J., Mathew, A. P., Hussein, M. Z. B., & Oksman, K. (2010). Preparation of cellulose nanofibers with hydrophobic surface characteristics. *Cellulose, 17*(2), 299–307.

Júnior, M. A. D., Borsoi, C., Hansen, B., & Catto, A. L. (2019). Evaluation of different methods for extraction of nanocellulose from yerba mate residues. *Carbohydrate Polymers, 218*, 78–86.

Kallel, F., Bettaieb, F., Khiari, R., García, A., Bras, J., & Chaabouni, S. E. (2016). Isolation and structural characterization of cellulose nanocrystals extracted from garlic straw residues. *Industrial Crops and Products, 87*, 287–296.

Kargarzadeh, H., Ahmad, I., Abdullah, I., Dufresne, A., Zainudin, S. Y., & Sheltami, R. M. (2012). Effects of hydrolysis conditions on the morphology, crystallinity, and thermal stability of cellulose nanocrystals extracted from kenaf bast fibers. *Cellulose, 19*(3), 855–866.

Karimi, S., Tahir, P. M., Karimi, A., Dufresne, A., & Abdulkhani, A. (2014). Kenaf bast cellulosic fibers hierarchy: a comprehensive approach from micro to nano. *Carbohydrate Polymers, 101*, 878–885.

Kekäläinen, K., Liimatainen, H., Biale, F., & Niinimäki, J. (2015). Nanofibrillation of TEMPO-oxidized bleached hardwood kraft cellulose at high solids content. *Holzforschung, 69*(9), 1077–1088.

Khalil, H. A., Bhat, A. H., & Yusra, A. I. (2012). Green composites from sustainable cellulose nanofibrils: a review. *Carbohydrate Polymers, 87*(2), 963–979.

Khalil, H. A., Davoudpour, Y., Islam, M. N., Mustapha, A., Sudesh, K., Dungani, R., & Jawaid, M. (2014). Production and modification of nanofibrillated cellulose using various mechanical processes: a review. *Carbohydrate Polymers, 99*, 649–665.

Khawas, P., & Deka, S. C. (2016). Isolation and characterization of cellulose nanofibers from culinary banana peel using high-intensity ultrasonication combined with chemical treatment. *Carbohydrate Polymers, 137*, 608–616.

Klemm, D., Heublein, B., Fink, H. P., & Bohn, A. (2005). Cellulose: fascinating biopolymer and sustainable raw material. *Angewandte Chemie International Edition, 44*(22), 3358–3393.

Kurosumi, A., Sasaki, C., Yamashita, Y., & Nakamura, Y. (2009). Utilization of various fruit juices as carbon source for production of bacterial cellulose by *Acetobacter xylinum* NBRC 13693. *Carbohydrate Polymers, 76*(2), 333–335.

Lavoine, N., Desloges, I., Dufresne, A., & Bras, J. (2012). Microfibrillated cellulose–Its barrier properties and applications in cellulosic materials: a review. *Carbohydrate Polymers, 90*(2), 735–764.

Le Normand, M., Moriana, R., & Ek, M. (2014). Isolation and characterization of cellulose nanocrystals from spruce bark in a biorefinery perspective. *Carbohydrate Polymers, 111*, 979–987.

Leitner, J., Hinterstoisser, B., Wastyn, M., Keckes, J., & Gindl, W. (2007). Sugar beet cellulose nanofibril-reinforced composites. *Cellulose, 14*(5), 419–425.

Li, M., Wang, L. J., Li, D., Cheng, Y. L., & Adhikari, B. (2014). Preparation and characterization of cellulose nanofibers from de-pectinated sugar beet pulp. *Carbohydrate Polymers, 102*, 136–143.

Lin, N., & Dufresne, A. (2014). Nanocellulose in biomedicine: current status and future prospect. *European Polymer Journal, 59*, 302–325.

Lin, S. P., Calvar, I. L., Catchmark, J. M., Liu, J. R., Demirci, A., & Cheng, K. C. (2013). Biosynthesis, production and applications of bacterial cellulose. *Cellulose, 20*(5), 2191–2219.

Liu, D., Zhong, T., Chang, P. R., Li, K., & Wu, Q. (2010). Starch composites reinforced by bamboo cellulosic crystals. *Bioresource Technology, 101*(7), 2529–2536.

Lu, P., & Hsieh, Y. L. (2012). Cellulose isolation and core shell nanostructures of cellulose nanocrystals from chardonnay grape skins. *Carbohydrate Polymers, 87*(4), 2546–2553.

Lu, H., Gui, Y., Zheng, L., & Liu, X. (2013). Morphological, crystalline, thermal and physicochemical properties of cellulose nanocrystals obtained from sweet potato residue. *Food Research International, 50*(1), 121–128.

Ma, N., Liu, D., Liu, Y., & Sui, G. (2015). Extraction and characterization of nanocellulose from *Xanthoceras sorbifolia* husks. *International Journal of Nanoscience and Nanoengineering, 2*(6), 43–50.

Mahardika, M., Abral, H., Kasim, A., Arief, S., & Asrofi, M. (2018). Production of nanocellulose from pineapple leaf fibers via high-shear homogenization and ultrasonication. *Fibers, 6*(2), 28.

Mandal, A., & Chakrabarty, D. (2011). Isolation of nanocellulose from waste sugarcane bagasse (SCB) and its characterization. *Carbohydrate Polymers, 86*(3), 1291–1299.

Manhas, N., Balasubramanian, K., Prajith, P., Rule, P., & Nimje, S. (2015). PCL/PVA nanoencapsulated reinforcing fillers of steam exploded/autoclaved cellulose nanofibrils for tissue engineering applications. *RSC Advances, 5*(31), 23999–24008.

Mariano, M., El Kissi, N., & Dufresne, A. (2014). Cellulose nanocrystals and related nanocomposites: review of some properties and challenges. *Journal of Polymer Science Part B: Polymer Physics, 52*(12), 791–806.

Mautner, A., Hakalahti, M., Rissanen, V., & Tammelin, T. (2018). Crucial interfacial features of nanocellulose materials. In *Nanocellulose and sustainability: production, properties, applications, and case studies*. CRC Press.
Miller, A. F., & Donald, A. M. (2003). Imaging of anisotropic cellulose suspensions using environmental scanning electron microscopy. *Biomacromolecules, 4*(3), 510–517.
Missoum, K., Belgacem, M. N., & Bras, J. (2013). Nanofibrillated cellulose surface modification: a review. *Materials, 6*(5), 1745–1766.
Mondragon, G., Fernandes, S., Retegi, A., Peña, C., Algar, I., Eceiza, A., & Arbelaiz, A. (2014). A common strategy to extracting cellulose nanoentities from different plants. *Industrial Crops and Products, 55*, 140–148.
Moohan, J., Stewart, S. A., Espinosa, E., Rosal, A., Rodríguez, A., Larrañeta, E., & Domínguez-Robles, J. (2020). Cellulose nanofibers and other biopolymers for biomedical applications. A review. *Applied Sciences, 10*(1), 65.
Morán, J. I., Alvarez, V. A., Cyras, V. P., & Vázquez, A. (2008). Extraction of cellulose and preparation of nanocellulose from sisal fibers. *Cellulose, 15*(1), 149–159.
Mtibe, A., Linganiso, L. Z., Mathew, A. P., Oksman, K., John, M. J., & Anandjiwala, R. D. (2015). A comparative study on properties of micro and nanopapers produced from cellulose and cellulose nanofibres. *Carbohydrate Polymers, 118*, 1–8.
Mwaikambo, L. A., & Ansell, M. P. (2006). Mechanical properties of alkali treated plant fibres and their potential as reinforcement materials. I. Hemp fibres. *Journal of Materials Science, 41*(8), 2483–2496.
Nascimento, D. M., Almeida, J. S., Dias, A. F., Figueirêdo, M. C. B., Morais, J. P. S., Feitosa, J. P., & Rosa, M. D. F. (2014). A novel green approach for the preparation of cellulose nanowhiskers from white coir. *Carbohydrate Polymers, 110*, 456–463.
Neto, W. P. F., Silvério, H. A., Dantas, N. O., & Pasquini, D. (2013). Extraction and characterization of cellulose nanocrystals from agro-industrial residue–Soy hulls. *Industrial Crops and Products, 42*, 480–488.
Nishiyama, Y. (2009). Structure and properties of the cellulose microfibril. *Journal of Wood Science, 55*(4), 241–249.
Osorio, D. A., Lee, B. E., Kwiecien, J. M., Wang, X., Shahid, I., Hurley, A. L., & Grandfield, K. (2019). Cross-linked cellulose nanocrystal aerogels as viable bone tissue scaffolds. *Acta Biomaterialia, 87*, 152–165.
Oun, A. A., & Rhim, J. W. (2016). Characterization of nanocelluloses isolated from Ushar (*Calotropis procera*) seed fiber: effect of isolation method. *Materials Letters, 168*, 146–150.
Pääkkö, M., Ankerfors, M., Kosonen, H., Nykänen, A., Ahola, S., Österberg, M., Ruokolainen, J., Laine, J., Larsson, P. T., Ikkala, O., & Lindström, T. (2007). Enzymatic hydrolysis combined with mechanical shearing and high-pressure homogenization for nanoscale cellulose fibrils and strong gels. *Biomacromolecules, 8*(6), 1934–1941.
Piras, C. C., Fernández-Prieto, S., & De Borggraeve, W. M. (2019). Ball milling: a green technology for the preparation and functionalisation of nanocellulose derivatives. *Nanoscale Advances, 1*(3), 937–947.
Qazanfarzadeh, Z., & Kadivar, M. (2016). Properties of whey protein isolate nanocomposite films reinforced with nanocellulose isolated from oat husk. *International Journal of Biological Macromolecules, 91*, 1134–1140.
Qua, E. H., Hornsby, P. R., Sharma, H. S. S., & Lyons, G. (2011). Preparation and characterisation of cellulose nanofibres. *Journal of Materials Science, 46*(18), 6029–6045.
Razera, I. A. T., & Frollini, E. (2004). Composites based on jute fibers and phenolic matrices: properties of fibers and composites. *Journal of Applied Polymer Science, 91*(2), 1077–1085.

Reddy, J. P., & Rhim, J. W. (2018). Extraction and characterization of cellulose microfibers from agricultural wastes of onion and garlic. *Journal of Natural Fibers, 15*(4), 465–473.

Rezanezhad, S., Nazanezhad, N., & Asadpur, G. (2013). Isolation of nanocellulose from rice wast via ultrasonication. *Lignocellulose, 2*(1), 282–291.

Rhim, J. W., Reddy, J. P., & Luo, X. (2015). Isolation of cellulose nanocrystals from onion skin and their utilization for the preparation of agar-based bio-nanocomposites films. *Cellulose, 22*(1), 407–420.

Rol, F., Vergnes, B., El Kissi, N., & Bras, J. (2019). Nanocellulose production by twin-screw extrusion: simulation of the screw profile to increase the productivity. *ACS Sustainable Chemistry & Engineering, 8*(1), 50–59.

Sari, N. H., Wardana, I. N. G., Irawan, Y. S., & Siswanto, E. (2017). The effect of sodium hydroxide on chemical and mechanical properties of corn husk fiber. *Oriental Journal of Chemistry, 33*(6), 3037–3042.

Satyamurthy, P., & Vigneshwaran, N. (2013). A novel process for synthesis of spherical nanocellulose by controlled hydrolysis of microcrystalline cellulose using anaerobic microbial consortium. *Enzyme and Microbial Technology, 52*(1), 20–25.

Satyanarayana, K. G., Guimarães, J. L., & Wypych, F. (2007). Studies on lignocellulosic fibers of Brazil. Part I: source, production, morphology, properties and applications. *Composites Part A: Applied Science and Manufacturing, 38*(7), 1694–1709.

Sheltami, R. M., Abdullah, I., Ahmad, I., Dufresne, A., & Kargarzadeh, H. (2012). Extraction of cellulose nanocrystals from mengkuang leaves (*Pandanus tectorius*). *Carbohydrate Polymers, 88*(2), 772–779.

Sinko, R., Qin, X., & Keten, S. (2015). Interfacial mechanics of cellulose nanocrystals. *MRS Bulletin, 40*(4), 340.

Siqueira, G., Tapin-Lingua, S., Bras, J., da Silva Perez, D., & Dufresne, A. (2010). Morphological investigation of nanoparticles obtained from combined mechanical shearing, and enzymatic and acid hydrolysis of sisal fibers. *Cellulose, 17*(6), 1147–1158.

Song, Y., Jiang, W., Zhang, Y., Wang, H., Zou, F., Yu, K., & Han, G. (2018). A novel process of nanocellulose extraction from kenaf bast. *Materials Research Express, 5*(8), 085032.

Souza, S. F., Leao, A. L., Cai, J. H., Wu, C., Sain, M., & Cherian, B. M. (2010). Nanocellulose from curava fibers and their nanocomposites. *Molecular Crystals and Liquid Crystals, 522*(1), 42–342.

Sun, R. C., & Sun, X. F. (2002). Fractional and structural characterization of hemicelluloses isolated by alkali and alkaline peroxide from barley straw. *Carbohydrate Polymers, 49*(4), 415–423.

Suzuki, K., Okumura, H., Kitagawa, K., Sato, S., Nakagaito, A. N., & Yano, H. (2013). Development of continuous process enabling nanofibrillation of pulp and melt compounding. *Cellulose, 20*(1), 201–210.

Tischer, P. C. F., Sierakowski, M. R., Westfahl, H., Jr., & Tischer, C. A. (2010). Nanostructural reorganization of bacterial cellulose by ultrasonic treatment. *Biomacromolecules, 11*(5), 1217–1224.

Trache, D., Hussin, M. H., Haafiz, M. M., & Thakur, V. K. (2017). Recent progress in cellulose nanocrystals: sources and production. *Nanoscale, 9*(5), 1763–1786.

Trache, D., Tarchoun, A. F., Derradji, M., Hamidon, T. S., Masruchin, N., Brosse, N., & Hussin, M. H. (2020). Nanocellulose: from fundamentals to advanced applications. *Frontiers in Chemistry, 8*.

Trache, D. (2018). Nanocellulose as a promising sustainable material for biomedical applications. *AIMS Materials Science, 5*(2), 201–205.

Tsouko, E., Kourmentza, C., Ladakis, D., Kopsahelis, N., Mandala, I., Papanikolaou, S., & Koutinas, A. (2015). Bacterial cellulose production from industrial waste and by-product streams. *International Journal of Molecular Sciences, 16*(7), 14832–14849.

Turbak, A. F., Snyder, F. W., & Sandberg, K. R. (1983). Microfibrillated cellulose, a new cellulose product: properties, uses, and commercial potential. In *J. Appl. Polym. Sci.: Appl. Polym. Symp.;(United States) (Vol. 37, No. CONF-8205234-Vol. 2)*Shelton, WA: ITT Rayonier Inc.

Uetani, K., & Yano, H. (2011). Nanofibrillation of wood pulp using a high-speed blender. *Biomacromolecules, 12*(2), 348–353.

Ullah, M. W., Ul-Islam, M., Khan, S., Kim, Y., & Park, J. K. (2016). Structural and physico-mechanical characterization of bio-cellulose produced by a cell-free system. *Carbohydrate Polymers, 136*, 908–916.

Ventura-Cruz, S., & Tecante, A. (2019). Extraction and characterization of cellulose nanofibers from Rose stems (*Rosa* spp.). *Carbohydrate Polymers, 220*, 53–59.

Vickers, N. J. (2017). Animal communication: when I'm calling you, will you answer too? *Current Biology, 27*(14), R713–R715.

Wang, T., & Drzal, L. T. (2012). Cellulose-nanofiber-reinforced poly (lactic acid) composites prepared by a water-based approach. *ACS Applied Materials and Interfaces, 4*(10), 5079–5085.

Wang, B., & Sain, M. (2007). Dispersion of soybean stock-based nanofiber in a plastic matrix. *Polymer International, 56*(4), 538–546.

Wang, B., Sain, M., & Oksman, K. (2007). Study of structural morphology of hemp fiber from the micro to the nanoscale. *Applied Composite Materials, 14*(2), 89.

Wang, Y., Wei, X., Li, J., Wang, F., Wang, Q., Chen, J., & Kong, L. (2015). Study on nanocellulose by high pressure homogenization in homogeneous isolation. *Fibers and Polymers, 16*(3), 572–578.

Wang, Y., Wei, X., Li, J., Wang, F., Wang, Q., Zhang, Y., & Kong, L. (2017). Homogeneous isolation of nanocellulose from eucalyptus pulp by high pressure homogenization. *Industrial Crops and Products, 104*, 237–241.

Xiao, Y., Liu, Y., Wang, X., Li, M., Lei, H., & Xu, H. (2019). Cellulose nanocrystals prepared from wheat bran: characterization and cytotoxicity assessment. *International Journal of Biological Macromolecules, 140*, 225–233.

Yang, H., Yan, R., Chen, H., Lee, D. H., & Zheng, C. (2007). Characteristics of hemicellulose, cellulose and lignin pyrolysis. *Fuel, 86*(12–13), 1781–1788.

Yang, W., Cheng, T., Feng, Y., Qu, J., He, H., & Yu, X. (2017). Isolating cellulose nanofibers from steam-explosion pretreated corncobs using mild mechanochemical treatments. *BioResources, 12*(4), 9183–9197.

Yang, W., Feng, Y., He, H., & Yang, Z. (2018). Environmentally-friendly extraction of cellulose nanofibers from steam-explosion pretreated sugar beet pulp. *Materials, 11*(7), 1160.

Zhang, L., Tsuzuki, T., & Wang, X. (2015). Preparation of cellulose nanofiber from softwood pulp by ball milling. *Cellulose, 22*(3), 1729–1741.

Cellulose-based blends and composites

Santi Chuetor[1], *Prapakorn Tantayotai*[2], *Kraipat Cheenkachorn*[1], *Yu-Shen Cheng*[3], *Atthasit Tawai*[4] *and Malinee Sriariyanun*[4]

[1]Department of Chemical Engineering, Faculty of Engineering, King Mongkut's University of Technology North Bangkok, Bangkok, Bangkok, Thailand; [2]Department of Microbiology, Faculty of Science, Srinakharinwirot University, Bangkok, Bangkok, Thailand; [3]Department of Chemical and Materials Engineering, National Yunlin University of Science and Technology, Douliu, Yunlin, Yunlin, Taiwan; [4]Chemical and Process Engineering, The Sirindhorn International Thai–German Graduate School of Engineering, King Mongkut's University of Technology North Bangkok, Bangkok, Bangkok, Thailand

1. Introduction

The current 2030 agenda of the UN is for sustainable development to encourage eradication of poverty and to promote world security in terms of human resources, environment, peace and partnership. To respond to this global agenda, research and development activities have been aimed at implementing the sustainability concept, including bioeconomy, circular economy, and green economy (BCG economy). The BCG economy is the movement of process design in industrialization to utilize wastes or undesired by-products as raw materials for the production of high value-added products with the benefits of minimizing waste disposal and gaining value from the process (Ngammuangtueng et al., 2020). Productions of polymeric materials from fossil resources via petroleum refinery processes are currently considered as conventional commodities for various applications, such as packaging, plastics, paints, medicals, foods, agricultures, etc. However, petroleum-based polymeric materials still provides several inappropriate eco-friendly products. In order to overcome its previous limitations, a biocomposite becomes more interesting in terms of sustainable and eco-friendly products. Biocomposites have been developed in scientific research and implemented at the industrial level and commercialization because of their biodegradability, sustainability, eco-friendly qualities, and societal perception compared to conventional petroleum composites. Moreover, the market shares of biocomposites made from renewable resources grew from 3.5 million tons in 2011 to 12 million tons by 2020 (Aeschelmann & Carus, 2015).

In general, biocomposites are made of a matrix (resin) and reinforced natural fibers, mostly cellulose. Natural fibers are obtained from three main sources: plant (bast, leaf, seed, fruit, grass, and wood), animal (wool, hair, and silk), and mineral (asbestos, glass, mineral wool, ceramic, silicate, and carbon). There were several reports in scientific publications in the early 1800s for the production of utensils and artistic

Biodegradable Polymers, Blends and Composites. https://doi.org/10.1016/B978-0-12-823791-5.00017-X

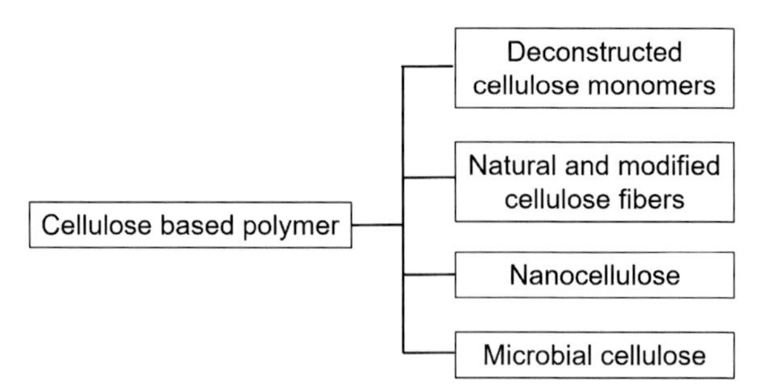

Figure 5.1 Cellulose-based biocomposites produced from various cellulose platforms.

works using composites of rubbers and cottons. Cellulose composite was reported first in 1869, followed by other natural materials, such as polyhydroxybutyrate (PHB) in 1925 and polylactic acid (PLA) in 1932 (Błedzki et al., 2012). Among the large variations of natural fibers, cellulose is one of the highest potential raw materials with wide applications in research and industries due to its characteristics and advantages such as abundancy, degradability, renewability, biocompatibility, and low weight. Annual global production of cellulose was estimated to be approximately 75−150 billion tons (Klemm et al., 2005; Cao et al., 2009). Worldwide, about 3.6 million tons of dissolving cellulose pulps were produced in 2009 (Sundarraj & Ranganathan, 2018). It is estimated that, in 2026, the cellulose market size will increase up to 300 billion USD with a compound annual growth rate (CAGR) of 4.2% between 2019 and 2026 (O'Dea, 2015). This augmentation in the demand and supply of cellulose-derived products is due to multiple routes such as films, or structural materials and their adaptabilities through tailor-made technological developments. The major cellulose-derived products used in current industrial applications are cellulose acetate, cellulose ester, regenerated cellulose, and cellulose ether. To achieve success in the production of different types of cellulose composites, different technologies are selected for the preparation of cellulose materials. With current technologies, cellulose materials can be categorized into four groups: (i) deconstructed cellulose monomers, (ii) natural and modified cellulose fibers, (iii) nanocelluloses, and (iv) microbial celluloses (Shaghaleh et al., 2018) (Fig. 5.1). The final characteristics and properties of the end-products of cellulose composites are typically determined by the properties of the cellulose materials, resulting in variations in the routes of application. This chapter is therefore targeted at present characteristics, research, development, and applications of each category of cellulose composite. The technological developments in each category are demonstrated to foster initiative and ideas for further progress in the cellulose composite industries to meet the goal of sustainable development.

2. Lignocellulose compositions

In nature, cellulose is naturally biosynthesized by various organisms, including plants, algae, and some bacteria. In plants, cellulose is one of the major components of lignocellulose biomass in the plant cell wall. Lignocellulose biomass is a mixture of three biomolecules, including cellulose, hemicellulose, and lignin, which are composed of hydrogen (H), carbon (C), and oxygen (O) atoms in high proportions. A relatively lower proportion of nitrogen (N) atoms in the forms of ubiquitous classes of amines, amides, nitriles, and other trace metal elements are found. Within the plant cell wall, compositions of cellulose, hemicellulose, and lignin vary in the ranges of 40%–60%, 20%–30%, and 15%–20%, respectively, depending on the plant species and other factors, including climate, temperature, growth stage, environment, etc. (Malinee et al., 2017; Wawat & Malinee, 2016). The molecular arrangements of these three components are linked via chemical bonding, in which the chemical and physical properties of lignocellulose biomass are determined (Fig. 5.2).

Cellulose and hemicellulose are classified as polysaccharides and are composed of sugar monomers. Lignin is a large biomolecule composed of monolignol precursors and phenyl derivatives. Cellulose is a linear polymeric chain of repetitions of glucose units (β-D-glucopyranose) linked together with β-1,4-glycosidic bonds (Fig. 5.2). The periodic formula of cellulose is $(C_6H_{10}O_5)_n$, in which n or degree of polymerization varies between hundreds to tens of thousands, and depends on the plant species, plant part, environment, etc. Within each glucose molecule, the C_1 atom is positioned at the reducing end of the closed ring, and reacts with the C_4 atom of another glucose molecule to form a β-1,4-glycosidic bond, with this dimer of glucose called cellobiose. One long chain of polymeric glucose can form intermolecular hydrogen bonds (between the H atom and hydroxyl group) to another glucose chain in a parallel manner at C_2, C_3, and C_6. Additionally, intramolecular hydrogen bonds occur in polymeric cellulose molecules. The numbers of hydrogen bonds and van der Waals forces (these occur proximal to cellulose chain stacking) contribute to the compactness of cellulose fibrils, reflecting the physical and chemical properties, such as stability, inertness, dissolution, gelation, and plasticization (Lindman et al., 2017; Klemm et al., 2005). Multiple cellulose chains are packed together through hydrogen bonding to form natural

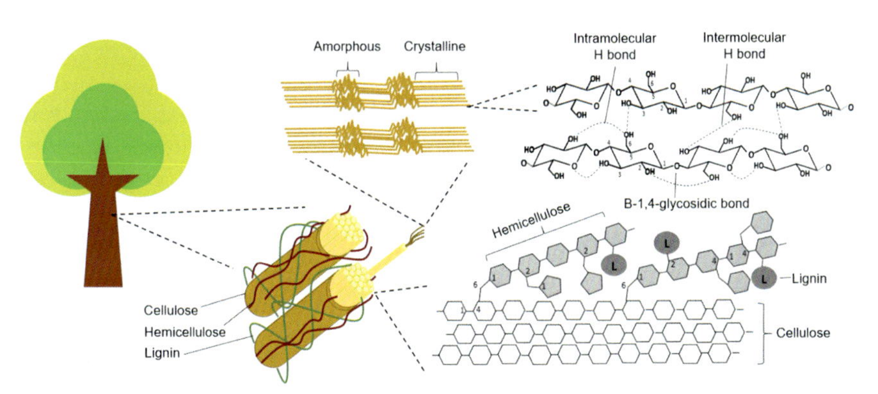

Figure 5.2 Molecular arrangements of lignocellulose composed of cellulose, hemicellulose, and lignin. The cellulose fibril contains amorphous and crystalline regions with different reactivities for chemical and enzymatic conversion. Cellulose fibrils are held together with intra- and intermolecular hydrogen bonds, and each glucose molecule of cellulose is linked to another with a β-1,4-glycosidic bond.

nanofibrils (5–50 nm in diameter). In higher plant species, the arrangements of cellulose structures are classified as crystalline and amorphous (Fig. 5.2). Naturally, cellulose fibrils favor organization in bundles with crystalline and amorphous domains. The crystalline and amorphous structures are interchangeable via different treatments, and the proportion of these structures in cellulose fibrils determines the properties of cellulose composites, such as the glass transition temperature (T_g), appearance, heat resistance, and elasticity (Klemm et al., 2005).

Hemicellulose is composed of hexoses (glucose, mannose, galactose, rhamnose) and pentoses (xylose, arabinose) and forms a heteropolymeric molecule with a linear chain and branch chain. The hemicellulose chain contains 50–200 monomeric units and is cross-linked with cellulose (via ether bond) and lignin (ester bond or ether bond). Due to hemicellulose variants, there are several heteropolymers such as xylan, glucuronoxylan, galactomannan, glucomannan, xyloglucan, and arabinoxylan, in which xylan and arabinoxylan are the most abundant forms of hemicellulose found in plant cell walls (Pauly et al., 2013). Lignin is present in lignocellulose fibrils in a smaller proportion than cellulose and hemicellulose. Lignin is biosynthesized from aromatic monolignol precursors (such as p-coumaryl, coniferyl, and sinapyl alcohol). The lignin molecule is enriched with phenyl compounds, which contain many functional groups. The monolignol units are linked to form macromolecules with covalent bonds to form an amorphous structure. Cellulose, hemicellulose, and lignin link together in a complex via hydrogen bonds and covalent bonds reflecting the stiffness and compactness of lignocellulose fibrils (Ralph et al., 2019) (Fig. 5.2).

3. Cellulose composites

3.1 Deconstructed cellulose monomer platform

Due to the abundance of lignocellulose biomass in nature and to follow the sustainability concept, cellulose has high potential to be harvested from lignocellulose biomass. Currently, parallel with the petrorefining process, a biorefining process was developed around 2010, as a new conceptual transformation process of biomass to produce value-added products from biomass. The biorefining process of lignocellulose biomass is comprised of four main steps: (i) pretreatment, (ii) hydrolysis, (iii) conversion/catalysis, and (iv) fractionation/recovery (Cheng et al., 2020; Malinee & Kanyarat, 2020) (Fig. 5.3). Due to the recalcitrant structure of lignocellulose biomass, the disintegration of cellulose, hemicellulose, and lignin is difficult, therefore a pretreatment process is necessary to disrupt the chemical bonding between these components and, by some pretreatment methods, to solubilize each compound from the complex matrix. Nowadays, pretreatment methods are mainly classified into four groups: physical, chemical, biological, and combined methods. It is suggested that most lignocellulose biomass, primarily, needs to be pretreated by the physical method, such as cutting, grinding, ball milling, and screw press, to reduce the bulky shape to an easier to handle size for further steps (Wawat & Malinee, 2016). Among all pretreatment methods, chemical pretreatment has been demonstrated to be the most often selected option in

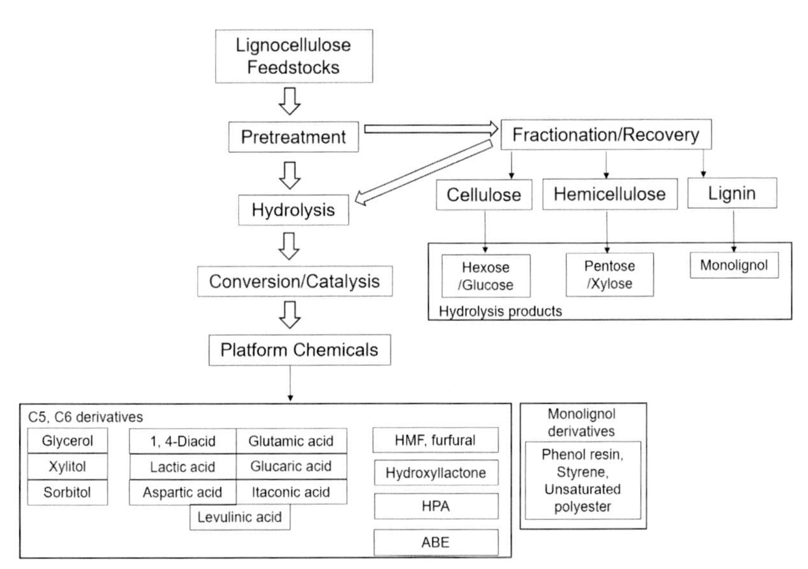

Figure 5.3 Biorefining process of lignocellulosic biomass for the production of platform chemicals as feedstocks of biocomposites.

biomass research, and in industrial applications it was applied in conventional biofuel production processes. This is because it uses conventional chemicals for pretreatment, for example, sulfuric acid and sodium hydroxide, and is a relatively cheap and simple process. However, there remain gaps for further research due to its low efficiency, instrument corrosion, and wastewater treatment. Other chemical pretreatments using milder conditions [e.g., organic acids (Amnuaycheewa et al., 2016; Rattanaporn et al., 2018)] and green solvents [e.g., ionic liquid (Malinee et al., 2017; Cheenkachorn et al., 2016)] have been demonstrated as alternative options for this process.

After lignocellulose biomass is pretreated, the integrations of cellulose, hemicellulose, and lignin are disrupted, making these three components ready for fractionation/separation and/or hydrolysis (Fig. 5.3). As aforementioned, cellulose and hemicellulose are made of sugar monomers and lignin is made of monolignol derivatives. By using a chemical hydrolysis reaction via a thermochemical process and enzymatic hydrolysis with microbial derived cellulase and hemicellulose enzymes, polymeric cellulose and hemicellulose chains are saccharified and release hexose and pentose sugars. Under harsh conditions of pretreatment and hydrolysis, these sugar monomers could be converted to other derivatives, especially furfural and hemifurfural, which are inhibitory compounds of cellulase enzyme and microbial activities in the fermentation process (Phakamas et al., 2019). Lignin is also a polymeric molecule and is depolymerized to monomers or small molecules by thermochemical (Mulat et al., 2018), chemical (Sang et al., 2020), and enzymatic processes (Ayuso-Fernández et al., 2018; Janusz et al., 2017). By different hydrolysis methods for lignin, large variations in lignin hydrolysate products include phenol and its derivatives, guaicol, eugenol, vanillin, syringol and its derivatives, hydrocarbon esters, benzyl alcohol, etc. (Naseem et al., 2016).

After monomers and their derivatives are released from hydrolyzed lignocellulose, these compounds could be converted into different platform chemicals that are subsequently used as raw materials for the synthesis of cellulose-derived biocomposites. Platform chemicals are small molecules, or building blocks, that are used as substrates or intermediates for chemical or biochemical reactions for the synthesis of larger molecules or value-added molecules, for example, hydroxymethylfurfural (HMF) is a substrate to produce more than 100 substances (Cheng et al., 2020; Rachamontree et al., 2020). For C_5 and C_6 sugar-derived platform chemicals, they can be grouped into three main categories, polyols (e.g., glycerol, xylitol, sorbitoln and erythritol), organic acid (e.g., lactic acid, aspartic acid, glutamic acid, itaconic acid, levulinic acid), and unclassified compounds [e.g., HMF, furfural, hydroxypropionic acid (HPA), and acetone-butanol-ethanol (ABE)]. Some other platform chemicals, such as phenol and styrene, can be obtained from the lignin-derived route (Fig. 5.3). However, the method to classify platform chemicals could be done in other ways, for example grouping by numbers of carbon atoms in the molecule, by final application, and by raw materials (Cheng et al., 2020).

Among the large catalog of platform chemicals, furfural ($C_5H_4O_2$), HMF ($C_6H_6O_3$), and 5-methylfurfural ($C_6H_6O_2$) are potential compounds obtained from hydrolysis of C5 and C6 sugars of lignocellulose with mineral acid catalysts and dehydration (Sonsiam et al., 2019). Furfural was demonstrated to be a wide application compound, such as a substitute for fuel additives (such as ethanol, methyl tert-butyl ether, and N-methylaniline) because it has less oxygen content and a high heating value (Rachamontree et al., 2020). One of the commercial processes for the coproduction of furfural and levulinic acid from paper mill sludge, the so-called the Biofine process, is composed of two reactors, a steam hydrolysis and a stirred reactor for levulinic acid with a theoretical yield of 70% (De Jong & Marcotullio, 2010). Another commercial production process, namely the Vedernikovs process, was developed to improve the production yield from 55% to 75% and to reduce losses of raw materials by using concentrated acid and salts as catalysts in hydrolysis and dehydration reactions (Cai et al., 2014).

Another biobased monomer, itaconic acid ($C_5H_6O_4$), is considered to be a potential platform chemical due to its wide range of applications, especially as a substitute for plastic monomers for the production of heteropolymers. Itaconic acid is used in the production of latex, methyl methacrylate, unsaturated polyester resins, and superabsorbent polymers and also as a polymer stabilizer in styrene butadiene rubber. Methyl methacrylate produced from itaconic is subsequently used in acrylic glass for LCD screens and adhesives. Unsaturated polyester resins are used for the production of artificial glass, rubbers, and coating plastics. Itaconic-derived products are applied in construction and decoration industries such as paints, adhesives, translucent elastic fillers, and sealing materials (Sriariyanun et al., 2019). Other applications are related to commodity products, for example, a hardening agent in contact lenses, diapers, napkins, and ionomer cements in dentistry. It is estimated that about 30%–40% of the production cost of itaconic acid is due to the raw materials. Using agricultural waste, such as lignocellulosic biomass, as a feedstock is an interesting source, however there is a constraint in the compatibility of microbial species to utilize a specific type of biomass due to their metabolism. For itaconic acid production, many types of lignocellulosic biomass have been utilized as the nutrition source for culturing microorganisms, mainly fungi, and yields obtained ranged from 19% to 64% conversion, suggesting a potential industrial-scale source (Sriariyanun et al., 2019). Similarly, other platform chemicals derived from lignocellulosic biomass are used as the raw materials for monomers or coadditives for polymerization reactions (Table 5.1). However, several concerns were discussed earlier as challenges to applying platform chemicals in the production of biocomposites, for example, (i) fractionation of cellulose, hemicellulose, and lignin from lignocellulose, (ii) hydrolysis or depolymerization of cellulose, hemicellulose, and lignin, (iii) conversion/catalysis of monomers to active derivatives for polymerization, and (iv) process design for a feasible multistep process (Shaghaleh et al., 2018). As aforementioned, a biorefining process to convert lignocellulosic biomass to platform chemicals is a multistep process, and an earlier step could negatively or positively affect the performance of the next step. In such a scenario, some harsh pretreatment conditions, such as using high temperature combined with strong acid, induced partial degradation of cellulose and hemicellulose and generated furfural and hemifurfural, which are inhibitory compounds of enzyme functions in hydrolysis. Although, pretreatment with organic acid such as acetic acid, is a milder pretreatment condition, the leftover acetic acid residues after pretreatment inhibit the activities of microorganisms in the fermentation process. Therefore, if these inhibitory compounds exist, a separation unit needs to be added to remove these inhibitors before proceeding to the next step, which increases operational cost, capital cost, and the complexity of the process. In the case of chemical hydrolysis and conversion, specific catalysts and optimal operation conditions are needed because there are various monomers and their derivatives released from fractionated lignocellulose biomass, within the same reactor, and these compounds could be converted through multiple routes to different end products. This occurrence leads to the requirement for an extra separation process, and could also reduce the conversion yield of target products due to losses in the form of unwanted by-products. For example, 1,4-butanediol (1,4-BDO), one of the platform chemical derivatives, is obtained from aspartic acid-, itaconic acid- and acetone–butanol–ethanol-based platforms, and can be diversely converted to polyurethane (thermosetting foams), butyrolactone (medical drug, solvent for production of solar cell materials), and tetrahydrofuran (fiber synthesis) (Isikgor & Becer, 2015).

3.2 Natural and modified cellulose fiber platform

The current uses of cellulose composites are mostly in the form of films, and cellulose functions as a filler or matrix in biocomposites. Biocomposite basically is composed of two or more types of materials, one a filler and the other a resin or matrix. Therefore, the properties of biocomposites are dependent on many factors, such as the properties of the raw materials, loading ratio of the filler to resin, orientation of fibers in the composite, and interaction and adhesion of the filler and resin. In the case of cellulose, it can be used as a resin to substitute the use of polyethylene (PE) and polypropylene (PP) and it can also be used as a reinforced fiber, as an alternative option to metal fiber

Table 5.1 Routes of cellulose-derived platform chemicals for the production of polymeric biocomposites.

Platform	Examples of monomeric derivatives	Target polymers and applications	References
Sugar	Glucaro-δ-lactone, methylglucoside, glucuronic acid	Glycopolymers, polyester, polycarbonates, polyurethanes for food packaging, medical devices	Munoz-Guerra et al. (2012)
1,4-Diacid (e.g., succinic acid, fumaric acid)	1,4-BDO, γ-butyrolactone, tetrahydrofuran	Polybutylene terephthalate, poly-4-hydroxybutyrate, polyurethane for yarn, sport wear fabrication, thermoplastic elastomer, compost bag, fishing gear, artificial leather	Minh et al. (2010), Luque et al. (2009)
5-HMF, 2,5-disubstituted furan derivatives	5-Methoxymethylfurfural, furfuryl methacrylate, 2-furyl oxirane, 2,5-bishydroxymethylfuran	Polyethylene-futanoate, vinyl polymer, polyamide, polyurethane, polybenzoimidazole for replacement of PET bottle, filaments for sutures	Buntara et al. (2011), Hong et al. (2006)
3-Hydroxypropyl acrylate (3-HPA)	Acrylic, methacrylic acid, 1,3-propanediol, propiolactone, acrylonitrile, acrolein	Acrylic polymer, polytrimethylene terephthalate, polyester-dimethyl terephthalate for paint, adhesive, thickener in concrete, absorbent, industrial acrylic fibers	Bhatia and Kurian (2008), Slone (2010)
Aspartic acid	Fumaric acid, maleic acid, 2-amino-1,4-butanediol, aspartic anhydride, amino-γ-butyrolactone	Polyaspartic acid, polyaspartamide, poly-β-alanine, polysuccinimide for drug delivery, tissue engineering	Straathof (2014)
Glutamic acid	Benzylglutamate, norvoline, gammaaminobutyric, 1,5-pentanediamine	Polyglutamic acid, polyamino ester, polyamide for fabrication, plasticizer	Werle et al. (2008)
Glucaric acid	1,4-Dioxane-2,5-dione, glucarodilactone, dimethacrylate	Polyesters, glucarodilactone methacrylate, polyglucaramidoamine, polyglucaramides for gene delivery, thermoset polymer, hydrogel	Gallagher et al. (2014)
Itaconic acid	Itaconic anhydride, 2-methyl-1, 4-butanediol	polyitaconic acid, polyester, methacrylate, styrene butadiene polymers for paint, adhesive, translucent glass, LCD screen, diaper, feminine napkins	Werle et al. (2008), Okuda et al. (2012)

Levulinic acid	Diphenolic acid, levulinic ketals, 5-aminolevulinic acid, 2-butanone, adipic acid	Polyarylene ether ketone, polyether ether ketone, methyl methacrylate for plasticizer, PVC plastic, polyurethane foams	Van de Vyver and Román-Leshkov (2013), Foix et al. (2012)
3-hydroxybutyrolactone	Lactone, tetrahydrofuran (THF), epoxide, 3,4-dihydroxybutyric acid	β-Methacryloyloxy-γ-butyrolactones for paint, adhesive, resin	Dhamankar et al. (2014), Lee and Park et al. (2009)
Sorbitol	Isosorbide, 1,4-sorbitan	Heteropolyacid, bis-3,4-dimethylbenzylidene sorbitol, polyester, poly(ethylene-coisosorbide) terephthalate for replacing PET, plasticizer, coating	Gupta et al. (2009), Rose and Palkovits (2012)
Glycerol	Ethylene glycol, propylene glycol, 1,3-propanediol (1,3-PDO), dihydroxyacetone	Polyglycerol methacrylate, polyglycerol caprolactone, polyhydric alcohols, polycarbonates, polyurethanes for plastic, alkyd resin	Behr et al. (2008), Putnam and Zelikin (2010)
Lactic acid	Lactide, phenylpyruvic acid, 2,3-pentanedione, acrylates	Polyester, polycarbonate, polyurethane, polyether polyols, polyalkylene oxalate, polyacetaldehyde-resorcinol for resin, fiber glass, laminate, surface coating, piping, insulation	Rose and Palkovits (2011), Sullivan (2000)
Acetone–butanol–ethanol (ABE)	Vinyl acetate, ethylene, ethylene carbonate, butene, 1,4-BDO, ethyl benzene propionaldehyde, methacryaldehyde	Polyvinyl acetate, ethylenevinyl acetate, cellulose acetate, low-density polyethylene (LLDP), high-density polyethylene (HDPE), PET, PVC, PP, poly(methyl methacrylate) (PMMA) for textile fiber, plastic film, foam rubber, insulator, adhesive, cement coating, desalination membrane, laminate safety glass, packaging	Isikgor and Becer (2015)

Continued

Table 5.1 Continued

Platform	Examples of monomeric derivatives	Target polymers and applications	References
C5 sugar (xylose, furfural, arabinitol)	Xylitol, xylaric acid, furfuryl alcohol, furoic acid, 2-hydroxymethyl-5-vinyl furan,	PET, polybutylene terephthalate, polyurethanes, polyester for resin, adhesive, paper coating, reinforced glass	Rachmontree et al. (2020)
Hydroxyalkanoates	Hydroxyalkanoate	Polyester, poly(3-hydroxybutyrate-co-3-hydroxyvalerate), polyhydroxyvalerate, polyhydroxybutyrate for printing, photography, packaging, plastic, textile	Babu et al. (2013), Chen (2009)
Rubber	Isobutene, isoprene, butadiene, styrene, acrylonitrile	Styrene-butadiene rubber, acrylonitrile-butadiene rubber, polyisoprene for automotive tires	Isikgor and Becer (2015)
Lignin	Styrene, xylene, terephthalic acid, benzene, hydroxybenzoic acid	Nylon, polyaniline, polystyrene polycarbonate, polyether sulfone, PET for bottle, packaging, textile, industrial fiber, coating	Isikgor and Becer (2015)

or glass fiber. Cellulose is selected as a material in composites due to its biocompatibility, high tensile strength, high thermal stability, high flexibility, renewability, and environmental friendliness. Reinforced cellulose fibers could be filled in resins made from petroleum-derived polymers such as PE, PP, PVC, and ethylene vinyl acetate (EVA) (Thakur & Thakur, 2014).

To fractionate cellulose fiber from natural biomass for the production of cellulose composite, the chemical linkages between cellulose, hemicellulose, and lignin are needed to be disrupted in order to enable greater availability of cellulose to the solvents. As aforementioned, individual natural cellulose microfibrils form an intermolecular hydrogen bond to another molecule, and at the same time form an intramolecular hydrogen bond and these bonds determine the compactness of cellulose fibrils (Fig. 5.2). Dissolutions of lignocellulose in solvents occur when the hydrogen bonds are weakened. Several groups of chemical solvents, such as sulfuric acid, ammonia, organic acid, organosolv, and ionic liquid, have been demonstrated by using different types of natural lignocellulose feedstock. Among these chemicals, ionic liquid has been shown to be an efficient solvent for cellulose dissolution due to several advantages including low volatility, recyclability, reusability, low corrosive, low melting point, and low viscosity (Malinee et al., 2017). Most ionic liquids have a liquid phase at room temperature, which makes them suitable to be used as solvents in the reactions. Currently many types of ionic liquids are commercially available, and even homemade synthesis processes are not complicated. Ionic liquids are similar to salts as they are mixtures of cations (such as imidazolium, pyridinium, ammonium, and phosphonium) and anions (such as acetate and chloride). The first type of ionic liquid used for lignocellulosic dissolution, 1-butyl-3-methylimidazolium chloride (BMIM-Cl), was demonstrated in 2002 (Swatloski et al., 2002). After that, many types of ionic liquids were applied to various types of lignocellulose biomass (Table 5.2). After cellulose fibrils are solubilized in ionic liquid, an antisolvent, such as water, methanol, ethanol, acetone, or acetonitrile, is added to induce coagulation of cellulose, leading to precipitation. These antisolvents have polarity properties that induce rearrangements of hydrogen bonds between cellulose molecules to form precipitates. After cellulose recovery, ionic liquids could be recovered by different methods, such as evaporation to remove water residues or electrodialysis, etc. (Hou et al., 2017).

Once the cellulose fibers are extracted from lignocellulose, surface modification is conducted to modify the properties of the cellulose fibers so that they are suitable to be used in biocomposite production. The main purposes of surface modifications are to provide hydrophobicity motifs to cellulose for optimal adhesion and distribution in polymeric resin, and to graft coupling agents to obtain the optimal covalent bond interface between the fiber and the resin, resulting in improvement of the thermophysical properties of biocomposites (Baghaei & Skrifvars, 2020). Additional chemical treatments with alkaline and saline could improve the wettability, tensile strength, and thermal stability of cellulose fibrils. After chemical treatment, other noncellulosic components are removed, leading to morphological changes of fibers. Cellulose can be used as a filler or a resin for biocomposites. As reinforcement fibers, the benefits of these biocomposites are renewable, recyclability, low density, and biocompatibility. Cellulose fibers are applied as fillers to replace the use of glass fibers and mineral

Table 5.2 Cellulose fractionations and regenerations from lignocellulosic biomass using ionic liquids.

Biomass	Ionic liquid	Antisolvents	Fractionation condition	Yield (%w.t)	References
Spruce	[BMIM][Cl]	H_2O	1 h, 120°C, 5% (w/w) loading ratio	Cellulose 94.1%, hemicellulose 85.0%, lignin 87.3.0	Sun et al. (2016)
Willow	[BMIM] [Cl] + Na_2SiO_3	H_2O	1 h, 120°C, 5% (w/w) loading ratio	Glucose 98.6%, cellulose 96.8%, hemicellulose 89.5%, lignin 82.5%	Sun et al. (2016)
Oil palm empty fruit bunch	[BMIM]Ac	H_2O	4 h, 140°C, 10% (w/w) loading ratio	Cellulose 45.51%, hemicellulose 24.49%, lignin 16.01%	Liu et al. (2020)
Cotton linters	[EMIM][OAc]	H_2O	3 h, 60°C, 4% (w/w) loading ratio	Cellulose 15.0%	Stepan et al. (2016)
Pine Kraft pulp	[EMIM][OAc]	H_2O	3 h, 60°C, 10% (w/w) loading ratio	Cellulose 12.0%	Xiaojuan et al. (2017)
Rice straw	[EMIM][OAc]-DMSO (1:1)	Acetone/H_2O	10 min, 80°C, 5% (w/w) loading ratio	Cellulose 28.0%, hemicellulose 60.0%, lignin 36.0%	Mai et al. (2014)
Rice straw	[EMIM][OAc]-DMA (1:1)	Acetone/H_2O	10 min, 80°C, 5% (w/w) loading ratio	Cellulose 28.0%, hemicellulose 20.0%, lignin 11.0%	Mai et al. (2014)
Miscanthus	[EMIM]Ac	H_2O or acetone/H_2O	24 h, 100°C, 30% (w/w) loading ratio	Glucose 98.0%, cellulose 40.0%, hemicellulose 22.0%, lignin 13.3%	Dash and Mohanty (2019)
Miscanthus pulp	[TEA][HSO_4]	H_2O	8 h, 120°C, 10% (w/w) loading ratio	Glucose 77.0%, cellulose 41.0%, lignin 20.0%	Gschwend et al. (2016)
Pine pulp	[TEA][HSO_4]	H_2O	8 h, 120°C, 10% (w/w) loading ratio	Glucose 13.0%, cellulose 38.0%, lignin 5.0%	Gschwend et al. (2016)
Mulberry stem	[Ch][Gly]	H_2O	6 h, 90°C, 10% (w/w) loading ratio	Glucose 74.72%, cellulose 51.36%, hemicellulose 11.66%, lignin 5.2%	Pakdeedachakiat et al. (2019)

fibers for other biodegradable composite matrices, however the mechanical strengths of composites are lower. Since cellulose is naturally degradable, the composite is more vulnerable and sensitive to moisture unless the surfaces of composites are protected by surface modifications and physicochemical treatments. Different approaches have been demonstrated to modify biocomposites to obtain the desired properties, such as increased tensile strength, matrix compatibility, biodegradability, flexibility, thermal stability, adsorption, antibacterial, anti-UV, and electrical conductivity (Table 5.3). A combination of ethyl cellulose, as a filler, and PHA, as a resin, is a biocomposite gaining increasing interest in biomedical and food packaging applications due to its biocompatibility and fully biodegradable properties (Baghaei & Skrifvars, 2020). Cinelli et al. (2019) produced cellulose biocomposites based on poly(3-hydroxybutyrate-3-hydroxyvalerate) (PHB-HV) and 20 wt.% waste wood sawdust fibers, as fillers, using a melt extrusion method. Aiming to create fully degradable properties to further applications in a plant nursery, the biocomposite properties generated by injection molding were monitored for 14 months and the results showed that the tensile strength and elongation slightly decreased, and 78 wt.% of biocomposite mass was lost within 6 months (Cinelli et al., 2019). Meftahi et al. (2019) developed PHB composite films filled with 10 wt.% rice husk flour using triglycidyl isocyanurate (TGIC) as a compatibilizer and dicumyl peroxide (DCP) as an initiator. The obtained composites had transparent appearance, low crystallinity, high thermal stability, and medium performance as barrier for moisture and aroma, which were potentially suitable for food packaging (Meftahi et al., 2019). Rapa et al. (2015) produced biocomposites from mixtures of PHB and PHB-HV with bacterial cellulose and microcrystalline cellulose as fillers by using melt processing. These biomaterials were evaluated for in vitro biocompatibilities in terms of acidity, alkalinity, water extractable quality, and reducing substances and their toxicities were observed using an MTT assay with fibroblast cells and the results showed 100% cell viability (Rapa et al., 2015).

Cellulose and its derivatives can also function as composite matrices or resins. Cellulose esters, cellulose ethers, such as cellulose acetate, cellulose butyrate, and cellulose proprionate are common cellulose derivatives in biocomposites, and are used in film membranes, coatings, and cigarette filters, etc. Numerous works have been conducted on formulations of cellulose resin with various types of fillers by targeting improvements of the mechanical properties of cellulose composites. Soyama and Iji (2017) developed a cellulose-based resin from mixtures of cardanol-bonded cellulose diacetate (CDA), from cashew nut shell, and polyester resins filled with glass fibers. A formulation of this thermoplastic showed an improvement of water resistance and rigidity. The addition of 20–30 wt.% of poly(butylene succinate adipate), polyester resin, and glass fibers increased the flexibility and impact strength of the composite (Soyama & Iji, 2017). Pang et al. (2014), formulated cellulose acetate resin and 30 wt.% of kenaf fiber filler by using a solution casting method and compression molding method. Chemical treatment by soaking with acetone and sodium hydroxide on kenaf fiber was conducted to disintegrate lignocellulose and surface modification. The results showed that the storage modulus of biocomposite improved from 1.23 to 2.76 GPa, but the composites were brittle. The addition of plasticizer, tributyl

Table 5.3 Improvements in cellulose fiber-derived biocomposites for different applications.

Composites	Application	Properties	Condition	References
Cellulose fibers (wood pulp) reinforced polylactic acid (PLA) biocomposites	Biodegradable polymers	Improved tensile modulus	170–180°C, 20 wt.% of wood pulp	Awal et al. (2015)
PLA/EVA-GMA (ethylene-vinyl acetate copolymer modified with glycidyl methacrylate)/cellulose fiber composite	Biodegradable polymers	Improved thermal and mechanical performance	170°C, 30–40 wt.% of cellulose	Pracella et al. (2016)
HEC (hydroxyethyl cellulose)/cellulose (softwood kraft pulp) composite	Packaging applications	Improved mechanical and thermomechanical properties	40°C, 50 wt.% of cellulose	Sirviö et al. (2018)
PVA [poly(vinyl alcohol)]/RC-SP (regenerated cellulose-softwood pulp)	Biomedical applications	Improved mechanical and thermal decomposition properties, and water resistance of the composites	110°C, 5–30 wt.% of RC-SP	Huang et al. (2019)
Cellulose/graphite oxide composite films	Organic conductive films and heat-resistant materials	Improved thermal stability and mechanical properties (tensile strength and Young's modulus)	25°C, 4 wt.% of MCC (microcrystalline cellulose) in NaOH/urea solution	Han et al. (2011)
Bamboo cellulose fibers/epoxy composites	Environment-friendly materials	Improved mechanical properties (tensile strength and elongation at break)	25–60°C, 20 wt.% loading of cellulose fibers	Lu et al. (2013)
Modified starch/carboxymethyl cellulose (CMC) composite films	Biodegradable composites and packaging applications	Improved moisture resistance and physical/mechanical properties	40–75°C, 20 wt.% loading of CMC	Ghanbarzadeh et al. (2010)

Crystalline cellulosic microfibers/polypropylene (PP) composite, crystalline cellulosic microfibers/maleic anhydride grafted polypropylene (MAPP) composite	Thermoplastic composite application	Improved mechanical properties and compatibility of the cellulose fibers/PP matrix	190–200°C, 30 wt.% of cellulose fibers	Qiu et al. (2005)
PLA/microcrystalline cellulose (MCC, oil palm biomass) composite	Packaging, automotive, and medical applications	Improved thermal stability and mechanical properties	60°C, 1–5 wt.% of MCC mixed with PLA solution	Haafiz et al. (2013)
Durable cellulose–sulfur composites (methylpropene-derivatized cellulose, PC)	Portland cement (resistant to degradation in acidic environments)	Improved mechanical properties and remeltable over many cycles	Heating PC with molten S8 at 180°C, 1–20 wt.% of cellulose	Lauer et al. (2019)
Hydroxy propyl methyl-cellulose microcomposite films	Edible films and coatings	Improved organoleptic characteristics of foods	25°C, 1–5 wt.% of MCC	Dogan and McHugh (2007)
Poly(3-hydroxybutyrate-3-hydroxyvalerate) (PHB-HV)/wood sawdust fibers	Biodegradable wood plastic (plant nursery)	Improved mechanical performance and degradation rate	170°C, 20 wt.% waste wood sawdust fibers	Cinelli et al. (2019)
Poly(3-hydroxybutyrate) (PHB)/rice husk flour (RHF) composites	Packaging applications	Transparent appearance, low crystallinity, and high thermal stability	180°C, 10 wt.% rice husk flour	Melendez-Rodriguez et al. (2019)
Plasticized cellulose acetate/kenaf fiber composites	Automotive panels, outdoor furniture, and ceiling panels	Improved flexibility and storage modulus	60°C, 30 wt.% kenaf fiber	Pang et al. (2014)
Wheat gluten/cellulose acetate phthalate (CAP) film	Edible films	Improved permeability to water and oxygen	25°C, 20%–80% of CAP film-forming solution	Fakhouri et al. (2004)

citrate, and moisture decreased the modulus of composites (Pang et al., 2014). Another cellulose derivative, cellulose acetate phthalate, was formulated with wheat gluten to form edible biofilms by using a solution casting method for applications in extending the shelf-life of fruit and vegetables. By varying the thickness and component ratio, the diffusivities of oxygen and moisture, important factors to determine the shelf-life, of biofilm composites were evaluated and it was found that a 1:1 mixture was the formulation with the highest diffusivity. The addition of gluten into the film reduced the tensile strength, and reduced the solubility of this biofilm composite in water and acid (Fakhouri et al., 2004). In medical applications, hydroxyapatite (HA) was in situ synthesized on cellulose fibers using microwave irradiation to create a composite for dental restorations. The results demonstrated that the high-strength biocomposite and high biocompatibility were achieved with 90% cell viability in Alamar Blue assay (Sabir et al., 2020).

Cellulose acetate resin could also be filled with inorganic and organic compounds to form biocomposites with specific properties, such as antimicrobial activities, UV protection, and photocatalysis. Nasouri (2019) produced a lightweight and flexible cellulose acetate resin filled with multiwalled carbon nanotubes impregnated with zinc oxide by using electrospinning methods to function as UV protectant textile composite materials. By variations of thickness, ZnO loading, and carbon nanotube loading, the optimal formulation performed UV protection of UPF at 180.8, suggesting the potential of this composite to be applied in textile commercialization (Nosouri, 2019). The addition of nano-titanium dioxide into cellulose acetate membrane expressed a photocatalysis or photobleaching property for degradation of methyl orange dye using an alkaline destruction method. By activation of UV light, this composite membrane exhibited dye degradation at a rate at 0.01 ppm/min. This immobilized membrane provided the benefit of the possibility of reuse of this composite in a separation process (Rahmawati et al., 2017). Likewise, preparations of cellulose acetate and molybdenum trioxide particles (MoO_3) were conducted and this thermoplastic biocomposite film was able to eradicate both Gram-positive and Gram-negative bacteria, *Staphylococcus aureus*, *Escherichia coli*, and *Pseudomonas aeruginosa* (Shafaei et al., 2017). Additionally, cellulose acetate and polyvinylpyrrolidone (PVP) were formulated as hydrophilic resin materials in nanofiber mats and filled with encapsulated garlic extract using an electrospinning method. Different properties of nanofiber composite mats, including morphology, chemical properties, mechanical properties, and antimicrobial properties were evaluated. The end-product composites expressed hydrophilicity, but when adding glycerol to the mixture, the Young's modulus, tensile strength, and degree of swelling were reduced. The antimicrobial activities of this nanofiber composite mat were observed using an in vitro antibacterial test (Edikresnha et al., 2019).

3.3 Nanocellulose platform

Cellulose nanofibers have been demonstrated for their wide applications in various fields, especially for use as reinforced fibers in composite materials due to their properties, such as low density, high surface area, low manufacturing costs, and degradability. In general, their particles have at least one dimension within the nano-sized scale (1–100 nm). Similar to cellulose fibers, nanocellulose fibers can be extracted from plant cell walls using mechanical or chemical methods. Different types of composites, including thermoset polymer and thermoplastic polymer, could be host resins for nanocellulose reinforced fibers. Currently, several types of nanocellulose are mainly grouped as nanofibrillate cellulose (NFC) or cellulose nanofiber (CNF) and nanocrystalline cellulose (NCC) or cellulose nanocrystal (CNC) (Trache et al., 2020). NFC can be produced by mechanical processes, such as homogenization or sonication, which are high-energy-consuming processes. Additionally, NFC has a hydrophilic property, causing irreversible agglomeration during drying and compounding processes when mixed into hydrophobic resins. Thus, several methods have been developed to promote defibrillation in combination with physical, chemical, and biological treatments. NFC consists of thin cellulose fibers that are naturally fabricated in a three-dimensional network with each cellulose fiber containing both crystalline and amorphous structures. CNCs could be prepared by pretreatment of NFC, such as by using acid hydrolysis to reduce the amorphous structure in NFC and to obtain a high proportion of crystalline structures. Therefore, CNCs appear as rod-like structures of about 100–200 nm in length and 4–25 nm in diameter. Due to the properties of nanocellulose, surface modification is required before mixing in hydrophobic resin to allow adhesion and achieve good distribution of fibers in resins (Bacakova et al., 2020; Trache et al., 2020).

Many studies have been conducted on the production of nanocellulose and surface modifications (Table 5.4). The NFC from oat hulls could be produced by a simple process with a combination of sulfuric acid hydrolysis and ultrasonication. The obtained NFCs had diameters of 70–100 nm and several micrometers in length, which were suitable for application as reinforced fibers in composites (Paschoal et al., 2013). A four-step process with high NFC yield of up to 70% was developed for NFC production from bamboo fiber, including mechanical refining, enzyme treatment,

Table 5.4 Improvements in nanocellulose-derived biocomposites for different applications.

Composites	Properties	Applications	References
SiO_2 nanoparticle/CMC nanocomposite membranes	Improved thermal stability, flame-retardant, and mechanical properties	Flame-retardant composites	Liu et al. (2020)
Hybrid-toughened epoxy/nanocellulose composites	Improved tensile strength and tensile modulus	Biobased material application	Kuo et al. (2017)
Sodium carboxymethyl cellulose (Na-CMC)/ methyl acrylate(TCMC)/ poly(ethylene oxide) (PEO)	Improved mechanical performance and flexibility	Drug-delivery application	Esmaeili and Haseli (2017)

Continued

Table 5.4 Continued

Composites	Properties	Applications	References
Pullulan–nanofibrillated cellulose composite films	Improved thermal and mechanical properties	Polymer applications such as dry food packaging and transparent organic electronics	Trovatti et al. (2012)
Poly(ethylene oxide)-lignin (PEO-L)/ nanocellulose composites	Enhanced adhesion property and shear strength improvement	Cellulose nanofiber film fabrication	Jayaramudu et al. (2019)
Trimethylammonium-modified nanofibrillated cellulose and layered silicates (TMA-NFC/ LS)	Improved vapor permeability and mechanical properties	Packaging applications	Ho et al. (2012)
Glycerol plasticized-starch (GPS)/cellulose nanoparticle (CN) composites	Improved mechanical properties and thermal stability	Polysaccharide composites (medical, agricultural, drug release, and packaging fields)	Chang et al. (2010)
PLA/nanocellulose/ nanoclay composites	Improved thermomechanical resistance and crystallization kinetics	Food packaging applications	Trifol et al. (2016)
Keratin reinforced by surface-functionalized cellulose nanocrystals	Improved mechanical properties	Keratin materials (tissue engineering or drug delivery)	Song et al. (2017)
Electrically conductive nanocellulose/ graphene composites	Enhanced electrical and thermal properties, and improved mechanical properties in high-humidity conditions	Antistatic packages, electromagnetic shielding, and sensors	Dang and Seppälä (2015)
Multilayer zirconium phosphate-reduced graphene oxide (ZrP-RGO) nanoplates/ cellulose nanofiber composites	Improved flame retardation	Flame-retardant materials	Wang et al. (2020)

Table 5.4 Continued

Composites	Properties	Applications	References
Polypyrrole (PPy)/ cellulose nanofiber (CNF) conducting composite films	Improved physical and electrical properties	Electromagnetic shielding applications	Parit et al. (2020)
Lignin/nanocellulose enhanced bio-polyurethane (PU) foams	Improved rigidity	Automotive parts	Faruk et al. (2014)
Cationic cellulose nanocrystals (CNCs)/ anionic alginate biocompatible hydrogel	Biocompatible and sustained drug release	Biomedical application	Lin et al. (2016)
Interpenetrating silica network inside a silylated nanofibrillated cellulose scaffold	Improved thermomechanical properties	Silica aerogel application	Zhao et al. (2015)

carboxymethyl modification, and ultrasonic homogenization (Hu et al., 2017). In brief, bamboo biomass was treated with acidified sodium chloride solution for delignification and treated with potassium hydroxide to remove hemicellulose. A refining step was conducted to increase the surface area for a higher interaction with endoglucanase enzyme hydrolysis. Then hydrolyzed bamboo was impregnated with monochloroacetic acid for carboxymethylation and proceeded to ultrasonication to yield a colloidal form of nanocellulose fibers. The results showed that mechanical refining reduced bamboo crystallinity, and the addition of carboxylate groups improved the efficiency of mechanical disintegration due to repulsion forces between the carboxylate groups (Hu et al., 2017). Later, a two-step process, using 2,2,6,6,-tetramethylpiperidine-1-oxyl radical (TEMPO)-mediated oxidation and high-pressure homogenization (HPH), was conducted. Bamboo kraft pulp was treated by combination of cellulase enzyme and mechanical refining. The carboxyl groups were added by TEMPO-mediated oxidation to promote mechanical disintegration (Yuan et al., 2019). Additionally, several types of enzymes, such as lytic polysaccharide monooxygenases (LPMO), swollenin, and hemicellulases, have been demonstrated to induce amorphogenesis of cellulose and improve the accessibility and reactivity of cellulose. A bleached kraft pulp was delignified and pretreated with sulfuric acid and autoclaving. Then, washed pulp was roughly disintegrated using a mechanical blender and enzymatic hydrolysis by endoglucanase, exoglucanase, xylanase, and LPMO. Finally, the hydrolyzed sample was sonicated to obtained NFC. These enzyme cocktails improved cellulose nanofibrillation by reducing the need for mechanical refining, and helped to stabilize the NFC suspension (Hu et al., 2017).

NFC composites have been demonstrated to be applied in different areas of industries. Yan et al. (2018) produced NFCs as dietary fibers from rice straw using high-density steam flash-explosion and enzymatic hydrolysis with cocktails (xylanase and laccase). The dimensions of rice straw NFCs were 30–200 nm in width with a high water retention capacity (20 g water/g) and swelling capacity (105 mL/g). Additionally, the specific functional group, octenyl succinic anhydride (OSA), was grafted on the NFCs to promote adsorption of bile acid, cholesterol, nitrite ion, and oil/fat ratio in digestive tracts in vitro and in vivo, which could be used in body weight control (Yan et al., 2018). The polymer-grafted nanocellulose/cellulose triacetate (CTA) composite was made of softwood bleached kraft pulp with a highly transparent and hydrophobic surface film. Kraft pulp was first oxidized with TEMPO/NaBr/NaClO and oxidized pulp was mechanical disintegrated in a mixed solvent. Polyethylene glycol-NH_2 was mixed with oxidized fibers to generate interfacial layers. Then, cellulose triacetate (CTA) matrix was additionally mixed with fibers and cast in a Petri dish. Intercalation of PEG molecules within CTA resulted in swelling, and alteration of the interfacial property of composites, which allowed miscibility of oxidized NFC fibers and CTA (Soeta et al., 2020). Zhou et al. (2020) developed an NFC-aerogel composite by mixing an NFC suspension, as a skeleton, with multiwalled carbon nanotubes (MWCNTs) and graphene (GP), to prepare NFC/MWCNTs/GP aerogel (CCGA) using a casting method in a Petri dish and using a freeze-drying method. The electrical conductivity of CCGA was increased by adding the NFC, and this composite had relatively stable electric heating performance at below 150°C, suggesting its potential use as a lightweight composite for electrothermal devices (Zhuo et al., 2020). Wang et al. (2020) developed an NFC composite film containing biocompatible Au nanoclusters (AuNCs) with antibacterial properties for biomedical applications, such as wound healing. The bleached kraft pulp was oxidized by TEMPO/NaBr/NaClO and mechanically disintegrated using a pressurized homogenizer. The carboxyl group was grafted on the oxidized NCF and then AuNCs were mixed in the NFC suspension, and the composite film was cast in a Petri dish. This AuNC–NFC film showed the highest antibacterial activities at 94% when testing with *E. coli* and *S. mutans*. The physical properties of AuNC–NFC films were assessed and showed hydrophilic properties and improvements in thermal stability up to 254–261°C (Chen et al., 2020). Another antibacterial NFC film was made with the addition of chitosan/oregano essential oil for application in food packaging. The NFC composite film was made from bleached bagasse pulp board and showed antibacterial efficiency against *E. coli* and *Listeria monocytogenes* at 99%, suggesting potential to improve food shelf-life (Chen et al., 2020).

Similar to NFC composites, CNC composites have been demonstrated in many research works for various applications to add value to lignocellulose. Prado et al. (2019) isolated CNC from pineapple crown leaf using chemical treatment and sulfuric acid hydrolysis, and this process provided 18–39 nm-diameter fibers with rod-like shape. The mechanical and chemical analyses suggested the crystallinity index and thermal degradation properties of celluloses suitable for reinforced fibers in composites and in liquid media applications (Prado & Spinacé, 2019). Another acid-free process for CNC productions from softwood bleached kraft pulp was conducted by the oxidation method using the TEMPO/NaBr/NaClO system and resulted in increased fiber recovery and carboxylate content. Further ultrasonication was applied to create cavitation force treatment in CNC and the final CNC dimensions ranged between 185–233 nm in length and 3.5–3.6 nm in width with good aqueous dispersion to be used as fillers in polymeric composites (Zhou et al., 2018). To prepare the CNC composite, Xing et al. (2020) hydrolyzed eucalyptus cellulose using sulfuric acid and mechanical treatments by variations in the treatment conditions to study the final forms of CNC products. The CNCs were subsequently mixed with poly(vinyl alcohol) (PVA) to generate transparent biodegradable composite films. Different properties of CNCs prepared under different conditions greatly affected the Young's modulus and the addition of CNCs improved the thermal stability and decrease the melting temperature of composites (Xing et al., 2020).

For application in food packaging, CNCs isolated from wheat bran were produced to increase the value of agricultural waste obtained from the wheat milling industry with concerns about safety for human consumption. Raw wheat bran was washed with ethanol, enzymatic hydrolyzed by α-amylase, and treated with an NaOH solution. The lignin content was removed by bleaching with $NaClO_2$ solution. The isolated cellulose fibrils were subsequently hydrolyzed to reduce the amorphous domains using sulfuric acid hydrolysis and provided a stable aqueous colloidal suspension of CNC solutions. The obtained CNCs showed lower thermal stabilities due to larger surface areas of CNCs allowing larger areas of heat exposure. Additionally, to serve a function in food packaging, CNC's water and oil retention properties and heavy metal adsorption were evaluated and improvement of these properties in CNC composite was observed, which was possibly due to the larger surface area of the nanoparticles. This CNC product was tested for in vitro cytotoxicity with Caco-2 cells, showing that the cell viability was 88.09% when treating with 2000 μg/mL of CNCs, suggesting that lower concentrations of CNCs did not statistically affect the cell viability (Xiao et al., 2019).

Similarly, biobased PLA was used as a resin and mixed with rosin nanoemulsion and CNCs extracted from wood pulp. First, rosin emulsion and CNCs were cast in sheets by the solvent casting method, and this rosin-CNC film was incorporated between two layers of PLA sheets to form multilayer materials in a sandwich system. The rosin–CNC–PLA sheet was then pressed between two metallic plates and two protective nonadhesive paper sheets to create a seven-layer composite sheet. Each component was added to composites with different purposes, for example, rosin nanoparticles exhibited antioxidant property, PLA increased mechanical (Young's modulus) and barrier properties, and metallic plates helped to protect the moisture of meat products. With this idea, these multilayer composite films were suggested for applications in different types of food packaging (Le Gars et al., 2020). Beside food packaging, CNC composites were demonstrated to be synthesized for medical application in the form of a wound-healing film. The composite film of CNC loaded with fucoidan/alginate-based gellan gum was made using photodynamic approaches. The benefits of this composite were evaluated in vitro and in vivo by testing with skin tissue and the contents of collagens and fibroblasts were monitored as indicators of skin tissue recovery. The hydrogel composite induced cell death of cancer cells via mechanisms of released singlet oxygen and reactive oxygen species and it promoted increases in fibroblast and collagen production in gene expression levels. These observations suggested the potential of this CNC-composite hydrogel as a therapeutic agent with medical uses (Shanmugapriya et al., 2020).

3.4 Microbial cellulose platform

In general, nanocellulose is produced based on two strategies, a top-down method and a bottom-up method. As aforementioned, NFCs and CNCs were mostly produced from disintegrations of isolated cellulose fibrils from natural lignocellulose using chemical, physical, or combined methods, classified as a top-down method. For a bottom-up method, nanocellulose is produced from small building block molecules via the metabolism of microorganisms, especially bacteria. Various species of bacteria, both Gram-positive and Gram-negative, have been demonstrated to be able to synthesize nanocellulose via fermentation processes, such as *Gluconacetobacter xylinus*, *Gluconacetobacter hansenii*, *Komagataeibacter medellinensis*, *Sarcina ventriculi*, and *Acetobacter pasteurianus* (Table 5.5) (Choi & Shin, 2020). These bacteria naturally produce cellulose and other polysaccharides as protective agents for themselves from environmental stresses and as cell envelopes. Bacterial cellulose is a nanofibrillated material with a naturally high crystallinity index, polymerization degree, and

Table 5.5 Recent research into the production of microbial cellulose and its various applications.

Species	Production condition	Research synopsis	References
Gluconacetobacter xylinus	Static culturing for 7 days, chemical treatment with 0.1 N NaOH, for 90 min at boiling temperature. Microbial cellulose was immersed in three types of honey	Honey-impregnated NMC for wound dressing. Antibacterial activities against *E. coli* and *S. aureus*. Chemical treatment of microbial cellulose improved absorptivity by 120 times and the product can retain moisture at the wound surface	Meftahi et al. (2020)
Gluconacetobacter xylinus	Shake-flask culturing for 5 days. Microbial cellulose was chemically treated with 1 M NaOH at 90°C for 30 min	*G. xylinus* produced BC on glucose medium. Concentration of glucose higher than 50 g/L resulted in decreased BC production	Adnan et al. (2017)

Table 5.5 Continued

Species	Production condition	Research synopsis	References
Gluconacetobacter hansenii	Cell-free system in static condition at 30°C pH 5.0 for 15 days	Cell-free system gained higher efficiency than cell system at 57.68% yield based on consumed glucose after 15 days. Possibility for further in situ synthesis approach	Ullah et al. (2015)
Gluconacetobacter intermedius	Static culturing condition at 37–48°C for 3 days, and harvested cellulose was washed with 0.1 M K-acetate buffer (pH 4.8)	Using molasse, industrial waste as a main carbon source for cellulose production. Mixtures of cellulose type I and II were produced	Tyagi and Suresh (2016)
Gluconacetobacter oboediens	Static culturing condition at 30°C for 8 days. Cellulose was bleached with 1% calcium hypochlorite	Using crude distillery effluent, and industrial waste as a main carbon source for cellulose production. Nanocelluloses were produced with better thermostability compared to plant cellulose	Jahan et al. (2018)
Komagataeibacter medellinensis	Static culturing condition was conducted for 13 days at 28°C. Cellulose was treated with 5 wt.% KOH for 14 h	Additions of ethanol and acetic acid promoted the production of cellulose, but the thermostability and degree of polymerization were decreased	Molina-Ramírez et al. (2018)
Komagataeibacter europaeus	Static intermittent fed-batch culturing at 30°C for 16 days. Cellulose was treated in boiling 0.5 N NaOH for 1 h and boiling distilled water for 30 min	Using sweet lime pulp waste, an industrial waste, as a main carbon source for cellulose production. Mixtures of cellulose types, Iα and Iβ forms and II, were observed	Dubey et al. (2018)

Continued

Table 5.5 Continued

Species	Production condition	Research synopsis	References
Komagataeibacter xylinus	Static culturing condition at 30°C and shake flask conditions at 30°C, 100 rpm for 12 days. Cellulose was treated with 25 mM NaOH for 4 h at 80°C and pH 4–6	Crystallinity indexes of celluloses obtained from static and shaking culture are comparable in the same level, while the elastic modulus and tensile strength in static culture cellulose were higher than shaking culture	Chen et al. (2019)
Gluconacetobacter entanii	Static culturing condition at 30°C, for 28 days. Cellulose was treated with 0.5 M NaOH. Cellulose samples were modified by mercerization with NaOH/urea (12% and 3%), and treated with dimethyl sulfate to form methylcellulose	Using pecan nutshell, an industrial waste, as a main carbon source for cellulose production. Chemical functionalized cellulose with methylation reduced its crystallinity	Dorame-Miranda et al. (2019)
Acetobacter pasteurianus	Static culturing condition at 30°C for 7 days. Cellulose was treated with 0.5 N NaOH at 80°C for 30 min, and neutralized by 0.5% acetic acid	Using tomato juice, cane molasse, orange pulp, and industrial waste, as a main carbon source for cellulose production	Kumar et al. (2019)
Gluconacetobacter hansenii and *E. coli*	Static culturing condition at 30°C for 5 days. Cellulose was treated with 0.1 M NaOH at 80°C for 1 h. Cellulose was mixed with extracellular polysaccharide (EPS) by coculturing	Coculturing could add EPS in cellulose and help to regulate the bundling process of cellulose microfibrils to get larger bundles. Low loading of EPS increased Young's modulus and stress tolerance to breaking points	Liu and Catchmark (2019)

Table 5.5 Continued

Species	Production condition	Research synopsis	References
Gluconacetobacter xylinus (ATCC 10245)	Static culturing condition at 30°C for 7 days. Cellulose was treated with 4% NaOH for 30 min, and neutralized by 6% acetic acid. Bacterial cellulose was further oxidized to form 2,3-dialdehyde bacterial cellulose DABC. Surface of cellulose was grafted with ethylenediamine and Benzil. Composite was doped with $FeCl_2$ and magnetic $FeCl_3$	The magnetic [Fe_3O_4NP-INS-(DABC-EDA-Bzl)] cellulose composite material was developed by cellulose oxidation treatment, surface modification, grafting, and doping. The composite showed in vitro antibacterial and antifungal activities at a moderate level compared with amphotericin B and its cytotoxicity which was determined in peripheral blood mononucleocyte cells (PBMCs)	Chaabane et al. (2020)

surface area, making it similar to CNCs. Nanofibrillated bacterial cellulose (NFBC) has been demonstrated to contribute to the mechanical strength of composites, including the tensile strength, Young's modulus, and flexibility. NFBC showed good biocompatibility, biodegradable, and environment-friendly properties, which make it suitable for biomedical applications, such as in wound healing and as an organ scaffold (Tayeb et al., 2018).

Synthesized nanofibrillated bacterial cellulose (NFBC) is produced by a fermentation process using different culturing media formulated from variations of carbon and nitrogen sources, including industrial by-products, for example, molasse, whey etc. However, many studies have demonstrated modifications to NFBC properties before using them as fillers in biocomposites. Tajima et al. (2017) produced NFBCs by culturing *Gluconacetobacter intermedius* NEDO-01 in a static culture system supplemented with CMC. The properties of NFBCs were modified by replacing CMC with hydroxyethylcellulose or hydroxypropylcellulose in the culturing media to create amphiphilic NFBCs. The results showed that amphiphilic NFBCs were well dispersed in polar organic solvents and also in poly(methyl methacrylate) (PMMA), therefore supporting the production of transparent NFBC/PMMA composite films with improvement of the tensile strength compared to PMMA films (Tajima et al., 2017). Chitbanyong et al. (2020) produced NFBCs by culturing thermotolerant bacteria,

Komagataeibacter xylinus C30 and *Komagataeibacter oboediens* R37-9, to add as fillers in biocomposite films. NFBCs were oxidized by a TEMPO/NaBr/NaClO system and mechanically treated and cast on films using a heat-drying method. This modified NFBC film showed a transparent appearance with acceptable mechanical properties (Chitbanyong et al., 2020).

Several studies have demonstrated mixtures of nanocellulose, NFC, CNC, and NFBC, as fillers in biocomposites. Sá et al. (2020) produced NFBCs from cashew apple juice by culturing bacteria. The NFBCs were mixed with lignin, as UV-protectants and antioxidant agents, and CNCs extracted from cashew tree pruning fiber. This biocomposite material exhibited enhancement of tensile strength and inhibition of water permeability. With improved properties, this NFBC/CNC/lignin composite had potential for application in food packaging (Sá et al., 2020). Another study produced a biocomposite film for food packaging application by mixing NFBCs and carboxylmethyl chitosan (CC) in a stable emulsion for a 3-month shelf-life. The mechanical evaluation showed that NFBC/CC composite had improved strength and reduced water vapor permeability (Li et al., 2020). In further applications, NFBCs were mixed with MXene ($Ti_3C_2T_x$) to form a composite with electrical conductivity. NFBC/MXene hydrogel composite exhibited mechanical properties, favorable flexibility, good biodegradability, and high water-uptake capacity, and enhanced the proliferation of NIH3T3 cells and improved the wound-healing process. The electro-conductivity was tested because electrical stimulation is another factor to promote wound healing and can be combined with composite materials (Mao et al., 2020). An NFBC-derived composite, produced from *Glucanoacetobacter xylinus*, was blended with polycaprolactone (PCL), gelatin (GEL), bacterial cellulose (BC), and different hydroxyapatites (HAs) to form a PCL/GEL/BC/HA composite scaffold with microporous structures using a 3D printing method for bone tissue engineering. GEL is produced from denatured fibrous collagen and has functions in biocompatibility, regenerative, and therapeutic properties. PCL is a synthetic polymer which promotes mechanical strength, and HA functions as a strong scaffold material for composites. The mechanical strength and physical properties of scaffold composites showed ideal pore size and good uniformity for bone engineering with insignificant effects on cell viability, which is a good biocompatibility sign (Cakmak et al., 2020).

4. Summary

Cellulose biomass is an abundant resource for the production of biocomposites as it offers various potential benefits, especially in terms of biodegradable, biocompatibility, and eco-friendly properties. In this chapter, four routes for cellulose composites have been described including platform monomers, cellulose fibers, nanocellulose, and bacterial cellulose. Each route has significant differences in the production processes, large variations in property modification, and a huge variety of end uses, such as construction, utensils, textiles, environment, automotive, electronic, food, cosmetic, and medical applications. The commercial products of cellulose composites are already available worldwide in different sectors, however research and product development are on-going to create novel composites with wider applications and better properties. The challenging concerns for these developments are up-scaling of the production process, technology transfer to industrial-scale producers, social perceptions of the products, and government policy support, which could possibly accelerate the availability of cellulose composites in the market.

Acknowledgments

The authors would like to thank King Mongkut's University of Technology, North Bangkok (Research University Grant No. KMUTNB-BasicR-64-37) and Srinakarinwirot University (Research University Grant No. 671/2563) for financial support of this work.

References

Adnan, A., Nair, G. R., Lay, M. C., Swan, J. E., Umar, R., & Yusra, A. F. (2017). Influences of saccharides types and initial glucose concentration on microbial cellulose production by G. xylinus. *Journal of Fundamental and Applied Sciences, 9*(2S), 174–181.

Aeschelmann, F., & Carus, M. (2015). Biobased building blocks and polymers in the world: capacities, production, and applications-status quo and trends towards 2020. *Industrial Biotechnology, 11*(3), 154–159. https://doi.org/10.1089/ind.2015.28999.fae

Amnuaycheewa, P., Hengaroonprasan, R., Rattanaporn, K., Kirdponpattara, S., Cheenkachorn, K., & Sriariyanun, M. (2016). Enhancing enzymatic hydrolysis and biogas production from rice straw by pretreatment with organic acids. *Industrial Crops and Products, 87*, 247–254. https://doi.org/10.1016/j.indcrop.2016.04.069

Ayuso-Fernández, I., Ruiz-Dueñas, F. J., & Martínez, A. T. (2018). Evolutionary convergence in lignin-degrading enzymes. *Proceedings of the National Academy of Sciences of the United States of America, 115*(25), 6428–6433. https://doi.org/10.1073/pnas.1802555115

Awal, A., Rana, M., & Sain, M. (2015). Thermorheological and mechanical properties of cellulose reinforced PLA bio-composites. *Mechanics of Materials, 80*, 87–95.

Babu, R. P., O'Connor, K., & Seeram, R. (2013). Current progress on bio-based polymers and their future trends. *Progress in Biomaterials, 2*, 8.

Bacakova, L., Pajorova, J., Tomkova, M., Matejka, R., Broz, A., Stepanovska, J., Prazak, S., Skogberg, A., Siljander, S., & Kallio, P. (2020). Applications of nanocellulose/nanocarbon composites: focus on biotechnology and medicine. *Nanomaterials (Basel), 10*(2), 196. https://doi.org/10.3390/nano10020196

Baghaei, B., & Skrifvars, M. (2020). All-cellulose composites: a review of recent studies on structure, properties and applications. *Molecules, 25*(12), 2836. https://doi.org/10.3390/molecules25122836

Behr, A., Eilting, J., Irawadi, K., Leschinski, J., & Lindner, F. (2008). Improved utilisation of renewable resources: new important derivatives of glycerol. *Green Chemistry, 10*, 13.

Bhatia, S. K., & Kurian, J. V. (2008). Biological characterization of Sorona polymer from corn-derived 1,3-propanediol. *Biotechnology Letters, 30*, 619–623.

Błedzki, A. K., Jaszkiewicz, A., Urbaniak, M., & Stankowska-Walczak, D. (2012). Biocomposites in the past and in the future. *Fibres and Textiles in Eastern Europe, 96*(6 B), 15–22. http://www.fibtex.lodz.pl/file-Fibtex_(6wsz64jlt4bt2aal).pdf-FTEE_96_15.pdf.

Buntara, T., NoelS, P., Phua, H., Melian-Cabrera, I., de Vries, J. G., & Heeres, H. J. (2011). Caprolactam from renewable resources: catalytic conversion of 5-hydroxymethylfurfural into caprolactone. *Angewandte Chemie International Edition, 50*, 7083–7087.

Cai, C. M., Zhang, T., Kumar, R., & Wyman, C. E. (2014). Integrated furfural production as a renewable fuel and chemical platform from lignocellulosic biomass. *Journal of Chemical Technology and Biotechnology, 89*(1), 2–10. https://doi.org/10.1002/jctb.4168

Cakmak, A. M., Semra, U., Ali, S., Oktar, F. N., Mustafa, S., Nazmi, E., Oguzhan, G., & Kalaskar, D. M. (2020). 3D printed polycaprolactone/gelatin/bacterial cellulose/hydroxyapatite composite scaffold for bone tissue engineering. *Polymers*. https://doi.org/10.3390/polym12091962, 1962.

Cao, Y., Wu, J., Zhang, J., Li, H. Q., Zhang, Y., & He, J. S. (2009). Room temperature ionic liquids (RTILs): a new and versatile platform for cellulose processing and derivatization. *Chemical Engineering Journal, 147*, 13.

Chaabane, L., Chahdoura, H., Mehdaoui, R., Snoussi, M., Beyou, E., Lahcini, M., & V Baouab, M. H. (2020). Functionalization of developed bacterial cellulose with magnetite nanoparticles for nanobiotechnology and nanomedicine applications. *Carbohydrate Polymers, 247*, 116707.

Chang, P. R., Jian, R., Zheng, P., Yu, J., & Ma, X. (2010). Preparation and properties of glycerol plasticized-starch (GPS)/cellulose nanoparticle (CN) composites. *Carbohydrate Polymers, 79*(2), 301–305.

Cheenkachorn, K., Douzou, T., Roddecha, S., Tantayotai, P., & Sriariyanun, M. (2016). Enzymatic saccharification of rice straw under influence of recycled ionic liquid pretreatments. In *Energy Procedia* (vol. 100, pp. 160–165). Elsevier Ltd. https://doi.org/10.1016/j.egypro.2016.10.159

Chen, G. Q. (2009). A microbial polyhydroxyalkanoates (PHA) based bio- and materials industry. *Chemical Society Reviews, 38*, 2434–2446.

Chen, G., Wu, G., Chen, L., Wang, W., Hong, F. F., & Jönsson, L. J. (2019). Performance of nanocellulose-producing bacterial strains in static and agitated cultures with different starting pH. *Carbohydrate Polymers, 215*, 280–288.

Chen, S., Wu, M., Wang, C., Yan, S., Lu, P., & Wang, S. (2020). Developed chitosan/oregano essential oil biocomposite packaging film enhanced by cellulose nanofibril. *Polymers (Basel), 12*(8), 1780. https://doi.org/10.3390/polym12081780

Cheng, Y., Mutrakulcharoen, P., Chuetor, S., Cheenkachorn, K., Tantayotai, P., Panakkal, E., & Sriariyanun, M. (2020). Recent situation and progress in biorefining process of lignocellulosic biomass: toward green economy. *Applied Science and Engineering Progress, 13*(4), 299–311. https://doi.org/10.14416/j.asep.2020.08.002

Chitbanyong, K., Pisutpiched, S., Khantayanuwong, S., Theeragool, G., & Puangsin, B. (2020). TEMPO-oxidized cellulose nanofibril film from nano-structured bacterial cellulose derived from the recently developed thermotolerant Komagataeibacter xylinus C30 and Komagataeibacter oboediens R37-9 strains. *International Journal of Biological Macromolecules, S0141–8130*(20), 34484–34486. https://doi.org/10.1016/j.ijbiomac.2020.09.124

Choi, S., & Shin, E. (2020). The nanofication and functionalization of bacterial cellulose and its applications. *Nanomaterials (Basel), 10*(3), 406. https://doi.org/10.3390/nano10030406

Cinelli, P., Seggiani, M., Mallegni, N., Gigante, V., & Lazzeri, A. (2019). Processability and degradability of PHA-based composites in terrestrial environments. *International Journal of Molecular Sciences, 20*(2), 284. https://doi.org/10.3390/ijms20020284

Dang, L. N., & Seppälä, J. (2015). Electrically conductive nanocellulose/graphene composites exhibiting improved mechanical properties in high-moisture condition. *Cellulose, 22*(3), 1799–1812.

Dash, M., & Mohanty, K. (2019). Effect of different ionic liquids and anti-solvents on dissolution and regeneration of Miscanthus towards bioethanol. *Biomass and Bioenergy, 124*, 33–42.

De Jong, W., & Marcotullio, G. (2010). Overview of biorefineries based on co-production of furfural, existing concepts and novel developments. *International Journal of Chemical Reactor Engineering, 8*. https://doi.org/10.2202/1542-6580.2174

Dhamankar, H., Tarasova, Y., Martin, C. H., & Jones Prather, K. L. (2014). Engineering E. coli for the biosynthesis of 3-hydroxy-butyrolactone (3HBL) and 3,4-dihydroxybutyric acid (3,4-DHBA) as value-added chemicals from glucose as a sole carbon source. *Metabolic Engineering, 25*, 72–81.

Dogan, N., & McHugh, T. H. (2007). Effects of microcrystalline cellulose on functional properties of hydroxy propyl methyl cellulose microcomposite films. *Journal of Food Science, 72*(1), E016–E022.

Dórame-Miranda, R. F., Gámez-Meza, N., Medina-Juárez, L.Á., Ezquerra-Brauer, J. M., Ovando-Martínez, M., & Lizardi-Mendoza, J. (2019). Bacterial cellulose production by Gluconacetobacter entanii using pecan nutshell as carbon source and its chemical functionalization. *Carbohydrate Polymers, 207*, 91–99.

Dubey, S., Singh, J., & Singh, R. P. (2018). Biotransformation of sweet lime pulp waste into high-quality nanocellulose with an excellent productivity using Komagataeibacter europaeus SGP37 under static intermittent fed-batch cultivation. *Bioresource Technology, 247*, 73–80.

Edikresnha, D., Suciati, T., Munir, M. M., & Khairurrijal, K. (2019). Polyvinylpyrrolidone/cellulose acetate electrospun composite nanofibres loaded by glycerine and garlic extract with: in vitro antibacterial activity and release behaviour test. *RSC Advances, 9*(45), 26351–26363. https://doi.org/10.1039/c9ra04072b

Esmaeili, A., & Haseli, M. (2017). Optimization, synthesis, and characterization of coaxial electrospun sodium carboxymethyl cellulose-graft-methyl acrylate/poly (ethylene oxide) nanofibers for potential drug-delivery applications. *Carbohydrate Polymers, 173*, 645–653.

Fakhouri, F. M., Tanada-Palmu, P. S., & Grosso, C. R. F. (2004). Characterization of composite biofilms of wheat gluten and cellulose acetate phthalate. *Brazilian Journal of Chemical Engineering, 21*(Issue 2), 261–264. https://doi.org/10.1590/S0104-66322004000200016. Assoc. Brasiliera de Eng. Quimica/Braz. Soc. Chem. Eng.

Faruk, O., Sain, M., Farnood, R., Pan, Y., & Xiao, H. (2014). Development of lignin and nanocellulose enhanced bio PU foams for automotive parts. *Journal of Polymers and the Environment, 22*(3), 279–288.

Foix, D., Ramis, X., Sangermano, M., & Serra, A. (2012). Synthesis of a new hyperbranched-linear-hyperbranched triblock copolymer and its use as a chemical modifier for the cationic photo and thermal curing of epoxy resins. *Journal of Polymer Science Part A Polymer Chemistry, 50*, 1133–1142.

Gallagher, J. J., Hillmyer, M. A., & Reineke, T. M. (2014). Degradable thermosets from sugar-derived dilactones. *Macromolecules, 47*, 498–505.

Ghanbarzadeh, B., Almasi, H., & Entezami, A. A. (2010). Physical properties of edible modified starch/carboxymethyl cellulose films. *Innovative food science & emerging technologies, 11*(4), 697–702.

Gschwend, F. J. V., Brandt, A., Chambon, C. L., Tu, W-. C., Weigand, L., & Hallett, J. P. (2016). Pretreatment of lignocellulosic biomass with low-cost ionic liquids. *JoVE*, 10.3791/54246.

Gupta, V., Singh, S., & Makwana, U. (2009). *WO Patent 2009128087*.

Haafiz, M. M., Hassan, A., Zakaria, Z., Inuwa, I. M., Islam, M. S., & Jawaid, M. (2013). Properties of polylactic acid composites reinforced with oil palm biomass microcrystalline cellulose. *Carbohydrate polymers, 98*(1), 139–145.

Han, D., Yan, L., Chen, W., Li, W., & Bangal, P. R. (2011). Cellulose/graphite oxide composite films with improved mechanical properties over a wide range of temperature. *Carbohydrate Polymers, 83*(2), 966–972.

Ho, T. T., Zimmermann, T., Ohr, S., & Caseri, W. R. (2012). Composites of cationic nanofibrillated cellulose and layered silicates: water vapor barrier and mechanical properties. *ACS applied materials & interfaces, 4*(9), 4832–4840.

Hong, J.-. T., Cho, N.-. S., Yoon, H.-. S., Kim, T.-. H., Koh, M.-. S., & Kim, W.-G. (2006). Biodegradable studies of poly(trimethylenecarbonate-ε-caprolactone)-block-poly(p-dioxanone), poly(dioxanone), and poly(glycolide-ε-caprolactone) (Monocryl®) monofilaments. *Journal of Applied Polymer Science, 102*(1), 737–743.

Hou, Q., Ju, M., Li, W., Liu, L., Chen, Y., Yang, Q., & Zhao, H. (2017). Pretreatment of lignocellulosic biomass with ionic liquids and ionic liquid-based solvent systems. *Molecules, 22*(3). https://doi.org/10.3390/molecules22030490

Hu, Z., Zhai, R., Li, J., Zhang, Y., & Lin, J. (2017). Preparation and characterization of nanofibrillated cellulose from bamboo fiber via ultrasonication assisted by repulsive effect. *International Journal of Polymer Science*. https://doi.org/10.1155/2017/9850814

Huang, B., He, H., Liu, H., Wu, W., Ma, Y., & Zhao, Z. (2019). Mechanically strong, heat-resistant, water-induced shape memory poly (vinyl alcohol)/regenerated cellulose biocomposites via a facile co-precipitation method. *Biomacromolecules, 20*(10), 3969–3979.

Isikgor, F. H., & Becer, C. R. (2015). Lignocellulosic biomass: a sustainable platform for the production of bio-based chemicals and polymers. *Polymer Chemistry, 6*(25), 4497–4559. https://doi.org/10.1039/c5py00263j

Janusz, G., Pawlik, A., Sulej, J., Świderska-Burek, U., Jarosz-Wilkolazka, A., & Paszczyński, A. (2017). Lignin degradation: microorganisms, enzymes involved, genomes analysis and evolution. *FEMS Microbiology Reviews, 41*(6), 941–962. https://doi.org/10.1093/femsre/fux049

Jahan, F., Kumar, V., & Saxena, R. K. (2018). Distillery effluent as a potential medium for bacterial cellulose production: a biopolymer of great commercial importance. *Bioresource Technology, 250*, 922–926.

Jayaramudu, T., Ko, H. U., Kim, H. C., Kim, J. W., Choi, E. S., & Kim, J. (2019). Adhesion properties of poly (ethylene oxide)-lignin blend for nanocellulose composites. *Composites Part B: Engineering, 156*, 43–50.

Klemm, D., Heublein, B., Fink, H. P., & Bohn, A. (2005). Cellulose: fascinating biopolymer and sustainable raw material. *Angewandte Chemie International Edition, 44*(22), 3358–3393. https://doi.org/10.1002/anie.200460587

Kumar, V., Sharma, D. K., Bansal, V., Mehta, D., Sangwan, R. S., & Yadav, S. K. (2019). Efficient and economic process for the production of bacterial cellulose from isolated strain of Acetobacter pasteurianus of RSV-4 bacterium. *Bioresource Technology, 275*, 430–433.

Kuo, P. Y., de Assis Barros, L., Yan, N., Sain, M., Qing, Y., & Wu, Y. (2017). Nanocellulose composites with enhanced interfacial compatibility and mechanical properties using a hybrid-toughened epoxy matrix. *Carbohydrate polymers, 177*, 249–257.

Lauer, M. K., Estrada-Mendoza, T. A., McMillen, C. D., Chumanov, G., Tennyson, A. G., & Smith, R. C. (2019). Durable cellulose–sulfur composites derived from agricultural and petrochemical waste. *Advanced Sustainable Systems, 3*(10), 1900062.

Le Gars, M., Dhuiège, B., Delvart, A., Belgacem, M. N., Missoum, K., & Bras, J. (2020). High-barrier and antioxidant poly(lactic acid)/nanocellulose multilayered materials for packaging. *ACS Omega, 5*(36), 22816–22826. https://doi.org/10.1021/acsomega.0c01955

Lee, S. H., & Park, O. J. (2009). Uses and production of chiral 3-hydroxy-gamma-butyrolactones and structurally related chemicals. *Applied Microbiology and Biotechnology, 84*, 817–828.

Li, Q., Ma, Q., Wu, Y., Li, Y., Li, B., Luo, X., & Liu, S. (2020). Oleogel films through the Pickering effect of bacterial cellulose nanofibrils featuring interfacial network stabilization. *Journal of Agricultural and Food Chemistry, 68*(34), 9150–9157. https://doi.org/10.1021/acs.jafc.0c03214

Lin, N., Geze, A., Wouessidjewe, D., Huang, J., & Dufresne, A. (2016). Biocompatible double-membrane hydrogels from cationic cellulose nanocrystals and anionic alginate as complexing drugs codelivery. *ACS Applied Materials & Interfaces, 8*(11), 6880–6889.

Liu, K., & Catchmark, J. M. (2019). Enhanced mechanical properties of bacterial cellulose nanocomposites produced by co-culturing Gluconacetobacter hansenii and Escherichia coli under static conditions. *Carbohydrate Polymers, 219*, 12–20.

Liu, J., Chen, P., Qin, D., Jia, S., Jia, C., Li, L., & Shao, Z. (2020). Nanocomposites membranes from cellulose nanofibers, SiO2 and carboxymethyl cellulose with improved properties. *Carbohydrate Polymers, 233*, 115818.

Lindman, B., Medronho, B., Alves, L., Costa, C., Edlund, H., & Norgren, M. (2017). The relevance of structural features of cellulose and its interactions to dissolution, regeneration, gelation and plasticization phenomena. *Physical Chemistry Chemical Physics, 19*(35), 23704–23718. https://doi.org/10.1039/c7cp02409f

Lu, T., Jiang, M., Jiang, Z., Hui, D., Wang, Z., & Zhou, Z. (2013). Effect of surface modification of bamboo cellulose fibers on mechanical properties of cellulose/epoxy composites. *Composites Part B: Engineering, 51*, 28–34.

Luque, R., Clark, J. H., Yoshida, K., & Gai, P. L. (2009). Efficient aqueous hydrogenation of biomass platform molecules using supported metal nanoparticles on Starbons. *Chemical Communications*, 5305–5307.

Mai, N. L., Ha, S. H., & Koo, Y. M. (2014). Efficient pretreatment of lignocellulose in ionic liquids/co-solvent for enzymatic hydrolysis enhancement into fermentable sugars. *Process Biochemistry, 49*, 1144–1151.

Malinee, S., & Kanyarat, K. (2020). Trends in lignocellulosic biorefinery for production of value-added biochemicals. *Applied Science and Engineering Progress*. https://doi.org/10.14416/j.asep.2020.02.005

Malinee, S., Ratsarin, A., & Tom, D. (2017). Development of ionic liquid utilization in biorefinery process of lignocellulosic biomass. *The Journal of King Mongkut's University of Technology North Bangkok*. https://doi.org/10.14416/j.ijast.2017.05.002

Mao, L., Hu, S., Gao, Y., Wang, L., Zhao, W., Fu, L., Cheng, H., Xia, L., Xie, S., Ye, W., Shi, Z., & Yang, G. (2020). Biodegradable and electroactive regenerated bacterial cellulose/MXene (Ti_3C_2Tx) composite hydrogel as wound dressing for accelerating skin wound healing under electrical stimulation. *Advanced Healthcare Materials, 9*(19). https://doi.org/10.1002/adhm.202000872

Meftahi, A., Shahriari, H., Khajavi, r, Rahimi, M., Sharifian, Melendez-Rodriguez, B., Torres-Giner, S., Aldureid, Cabedo, L., & Lagaron, J. M. (2019). Reactive melt mixing of poly(3-hydroxybutyrate)/rice husk flour composites with purified biosustainably produced poly(3-hydroxybutyrate-co-3-hydroxyvalerate). *Materials (Basel), 7*(13). Materials Research Express.

Meftahi, A., Shahriari, H. R., Khajavi, R., Rahimi, M. K., & Sharifian, A. (2020). Investigation on nano microbial cellulose/honey composite for medical application. *Materials Research Express, 7*(8), 085003.

Melendez-Rodriguez, B., Torres-Giner, S., Aldureid, A., Cabedo, L., & Lagaron, J. M. (2019). Reactive melt mixing of poly(3-hydroxybutyrate)/rice husk flour composites with purified biosustainably produced poly(3-hydroxybutyrate-co-3-hydroxyvalerate). *Materials (Basel), 12*(13), 2152.

Minh, D. P., Besson, M., Pinel, C., Fuertes, P., & Petitjean, C. (2010). Aqueous-phase hydrogenation of biomass-based succinic acid to 1,4-butanediol over supported bimetallic catalysts. *Topics in Catalysis, 53*, 1270–1273.

Molina-Ramírez, C., Enciso, C., Torres-Taborda, M., Zuluaga, R., Gañán, P., Rojas, O. J., & Castro, C. (2018). Effects of alternative energy sources on bacterial cellulose characteristics produced by komagataeibacter medellinensis. *International Journal of Biological Macromolecules, 117*, 735–741.

Mulat, D. G., Dibdiakova, J., & Horn, S. J. (2018). Microbial biogas production from hydrolysis lignin: insight into lignin structural changes. *Biotechnology for Biofuels, 11*(1). https://doi.org/10.1186/s13068-018-1054-7

Munoz-Guerra, S. (2012). Carbohydrate-based polyamides and polyesters: an overview illustrated with two selected examples. *High Performance Polymers, 24*, 9–23.

Naseem, A., Tabasum, S., Zia, K., Zuber, M., Ali, M., & Noreen, A. (2016). Lignin-derivatives based polymers, blends and composites: a review. *International Journal of Biological Macromolecules, 93*(Pt A), 296–313. https://doi.org/10.1016/j.ijbiomac.2016.08.030

Ngammuangtueng, P., Jakrawatana, N., & Gheewala, S. H. (2020). Nexus resources efficiency assessment and management towards transition to sustainable bioeconomy in Thailand. *Resources, Conservation and Recycling, 160*. https://doi.org/10.1016/j.resconrec.2020.104945

Nosouri, K. (2019). Fabrication of lightweight and flexible cellulose acetate composite nanofibers for high-performance ultra violet protective materials. *Polymer Composites, 40*(8).

O'Dea, N. (2015). *Emerging innovation trends in composites. Paper presented at composites engineering show.*

Okuda, T., Ishimoto, K., Ohara, H., & Kobayashi, S. (2012). Renewable biobased polymeric materials: facile synthesis of itaconic anhydride-based copolymers with poly(l-lactic acid) grafts. *Macromolecules, 45*, 4166–4174.

Pakdeedachakiat, W., Phruksaphithak, N., & Boontawan, A. (2019). Pretreatment of mulberry stem by cholinium-based amino acid ionic liquids for succinic acid fermentation and its application in poly(butylene) succinate production. *Bioresource Technology, 291*, 121873.

Pang, C., Shanks, R., & Daver, F. (2014). *Biocomposites based on cellulose acetate and kenaf fibers: processing and properties* (vol. 1593). AIP Conference Proceedings.

Parit, M., Du, H., Zhang, X., Prather, C., Adams, M., & Jiang, Z. (2020). Polypyrrole and cellulose nanofiber based composite films with improved physical and electrical properties for electromagnetic shielding applications. *Carbohydrate Polymers*, 116304.

Paschoal, Muller, C., Carvalho, G., Tischer, C., & Mali. (2013). Isolation and characterization of nanofibrillated cellulose from oat hulls. *Química Nova, 38*(4), 627–642.

Pauly, M., Gille, S., Liu, L., Mansoori, N., de Souza, A., Schultink, A., & Xiong, G. (2013). Hemicellulose biosynthesis. *Planta, 238*(4), 627–642. https://doi.org/10.1007/s00425-013-1921-1

Phakamas, R., Malinee, S., Surapun, T., & Issaraporn, S. (2019). Optimization of oil production from cassava pulp and sugarcane bagasse using oleaginous yeast. *Oriental Journal of Chemistry*, 668–677. https://doi.org/10.13005/ojc/350222

Pracella, M., Haque, M. M. U., Paci, M., & Alvarez, V. (2016). Property tuning of poly (lactic acid)/cellulose bio-composites through blending with modified ethylene-vinyl acetate copolymer. *Carbohydrate polymers, 137*, 515–524.

Prado, K. S., & Spinacé, M. A. S. (2019). Isolation and characterization of cellulose nanocrystals from pineapple crown waste and their potential uses. *International Journal of Biological Macromolecules, 122*, 410–416. https://doi.org/10.1016/j.ijbiomac.2018.10.187

Putnam, D. A., & Zelikin, A. (2010). *US Patent 7659420.*

Qiu, W., Zhang, F., Endo, T., & Hirotsu, T. (2005). Isocyanate as a compatibilizing agent on the properties of highly crystalline cellulose/polypropylene composites. *Journal of materials science, 40*(14), 3607–3614.

Rachamontree, P., Douzou, T., Cheenkachorn, K., Sriariyanun, M., & Rattanaporn, K. (2020). Furfural: a sustainable platform chemical and fuel. *Applied Science and Engineering Progress, 13*(1), 3–10. https://doi.org/10.14416/j.asep.2020.01.003

Rahmawati, F., Fadillah, I., & Mudjijono, M. (2017). Composite of nano-TiO_2 with cellulose acetate membrane from nata de Coco (Nano-TiO_2/CA(NDC)) for methyl orange degradation. *Journal of Materials and Environmental Science, 8*(1), 287–297. http://www.jmaterenvironsci.com/.

Ralph, J., Lapierre, C., & Boerjan, W. (2019). Lignin structure and its engineering. *Current Opinion in Biotechnology, 56*, 240–249. https://doi.org/10.1016/j.copbio.2019.02.019

Rapa, M., Zaharia, C., Lungu, M., Stanescu, P. O., Stoica, P., Grosu, E., Tatia, R., & Coroiu, V. (2015). Biocompatibility of PHAs biocomposites obtained by melt processing. *Materiale Plastice, 52*(3), 295–300. http://www.revmaterialeplastice.ro/pdf/RAPA%20M.pdf%203%2015.pdf.

Rattanaporn, K., Tantayotai, P., Phusantisampan, T., Pornwongthong, P., & Sriariyanun, M. (2018). Organic acid pretreatment of oil palm trunk: effect on enzymatic saccharification and ethanol production. *Bioprocess and Biosystems Engineering, 41*(4), 467–477. https://doi.org/10.1007/s00449-017-1881-0

Rose, M., & Palkovits, R. (2011). Cellulose-based sustainable polymers: state of the art and future trends. *Macromolecular Rapid Communications, 32*(17), 1299–1311.

Rose, M., & Palkovits, R. (2012). Isosorbide as a renewable platform chemical for versatile applications—quo vadis? *ChemSusChem, 5*, 167–176.

Sá, N. M. S. M., Mattos, A. L. A., Silva, L. M. A., Brito, E. S., Rosa, M. F., & Azeredo, H. M. C. (2020). From cashew byproducts to biodegradable active materials: bacterial cellulose-lignin-cellulose nanocrystal nanocomposite films. *International Journal of Biological Macromolecules, 161*, 1337–1345. https://doi.org/10.1016/j.ijbiomac.2020.07.269

Sabir, M., Ali, A., Siddiqui, U., Muhammad, N., Khan, A. S., Sharif, F., Iqbal, F., Shah, A. T., Rahim, A., & Rehman, I. U. (2020). Synthesis and characterization of cellulose/hydroxyapatite based dental restorative composites. *Journal of Biomaterials Science, Polymer Edition, 31*(14), 1806–1819. https://doi.org/10.1080/09205063.2020.1777827

Sang, Y., Chen, M., Yan, F., Wu, K., Bai, Y., Liu, Q., Chen, H., & Li, Y. (2020). Catalytic depolymerization of enzymatic hydrolysis lignin into monomers over an unsupported nickel catalyst in supercritical ethanol. *Industrial & Engineering Chemistry Research, 59*(16), 7466–7474. https://doi.org/10.1021/acs.iecr.0c00812

Shafaei, S., Dörrstein, J., Guggenbichler, J. P., & Zollfrank, C. (2017). Cellulose acetate-based composites with antimicrobial properties from embedded molybdenum trioxide particles. *Letters in Applied Microbiology, 64*(1), 43–50. https://doi.org/10.1111/lam.12670

Shaghaleh, H., Xu, X., & Wang, S. (2018). Current progress in production of biopolymeric materials based on cellulose, cellulose nanofibers, and cellulose derivatives. *RSC Advances, 8*(2), 825−842. https://doi.org/10.1039/c7ra11157f

Shanmugapriya, K., Kim, H., & Kang, H. W. (2020). Fucoidan-loaded hydrogels facilitates wound healing using photodynamic therapy by in vitro and in vivo evaluation. *Carbohydrate Polymers, 247*. https://doi.org/10.1016/j.carbpol.2020.116624

Sirviö, J. A., Visanko, M., Ukkola, J., & Liimatainen, H. (2018). Effect of plasticizers on the mechanical and thermomechanical properties of cellulose-based biocomposite films. *Industrial Crops and Products, 122*, 513−521.

Slone, R. V. (2010). Acrylic ester polymers. In *Encyclopedia of polymer science and technology*. John Wiley & Sons Inc, 10.1002/0471238961.1921182214152201.a01. pub2.

Soeta, H., Fujisawa, S., Saito, T., & Isogai, A. (2020). Controlling miscibility of the interphase in polymer-grafted nanocellulose/cellulose triacetate nanocomposites. *ACS Omega, 5*(37), 23755−23761. https://doi.org/10.1021/acsomega.0c02772

Song, K., Xu, H., Xie, K., & Yang, Y. (2017). Keratin-based biocomposites reinforced and cross-linked with dual-functional cellulose nanocrystals. *ACS Sustainable Chemistry & Engineering, 5*(7), 5669−5678.

Sonsiam, C., Kaewchada, A., pumrod, S., & Jaree, A. (2019). Synthesis of 5-hydroxymethylfurfural (5-HMF) from fructose over cation exchange resin in a continuous flow reactor. *Chemical Engineering and Processing − Process Intensification, 138*, 65−72. https://doi.org/10.1016/j.cep.2019.03.001

Soyama, M., & Iji, M. (2017). Improving mechanical properties of cardanol-bonded cellulose diacetate composites by adding polyester resins and glass fiber. *Polymer Journal, 49*(6), 503−509. https://doi.org/10.1038/pj.2017.10

Sriariyanun, M., Heitz, J. H., Yasurin, P., Asavasanti, S., & Tantayotai, P. (2019). Itaconic acid: a promising and sustainable platform chemical? *Applied Science and Engineering Progress, 12*(2), 75−82. https://doi.org/10.14416/j.asep.2019.05.002

Stepan, A. M., Monshizadeh, A., Hummel, M., Roselli, A., & Sixta, H. (2016). Cellulose fractionation with IONCELL-P. *Carbohydrate Polymers, 150*, 99−106.

Straathof, A. J. (2014). Transformation of biomass into commodity chemicals using enzymes or cells. *Chemical Reviews, 114*, 1871−1908.

Sundarraj, A. A., & Ranganathan, T. V. (2018). A review on cellulose and its utilization from agro-industrial waste. *Drug Invention Today, 10*(1), 89−94. http://jprsolutions.info/view_journal_new.php?journal_id=2.

Sullivan, C. J. (2000). *Ullmann's encyclopedia of industrial chemistry*. Wiley-VCH. https://doi.org/10.1002/14356007.a22_163

Sun, S., Sun, S., Cao, X., & Sun, R. (2016). The role of pretreatment in improving the enzymatic hydrolysis of lignocellulosic materials. *Bioresource Technology, 199*, 49−58.

Swatloski, R. P., Spear, S. K., Holbrey, J. D., & Rogers, R. D. (2002). Dissolution of cellose with ionic liquids. *Journal of the American Chemical Society, 124*(18), 4974−4975. https://doi.org/10.1021/ja025790m

Tajima, K., Kusumoto, R., Kose, R., Kono, H., Matsushima, T., Isono, T., Yamamoto, T., & Satoh, T. (2017). One-step production of amphiphilic nanofibrillated cellulose using a cellulose-producing bacterium. *Biomacromolecules, 18*(10), 3432−3438. https://doi.org/10.1021/acs.biomac.7b01100

Tayeb, A., Amini, E., Ghasemi, S., & Tajvidi, M. (2018). Cellulose nanomaterials-binding properties and applications: a review. *Molecules, 23*(10), 2684. https://doi.org/10.3390/molecules23102684

Thakur, V., & Thakur, M. (2014). Processing and characterization of natural cellulose fibers/ thermoset polymer composites. *Carbohydrate Polymers, 30*(109), 102–117. https://doi.org/10.1016/j.carbpol.2014.03.039

Trache, D., Tarchoun, A., Derradji, M., Hamidon, T., Masruchin, N., Brosse, N., & Hussin, M. (2020). Nanocellulose: from fundamentals to advanced applications. *Frontiers in Chemistry, 6*(8), 392. https://doi.org/10.3389/fchem.2020.00392

Trifol, J., Plackett, D., Sillard, C., Szabo, P., Bras, J., & Daugaard, A. E. (2016). Hybrid poly (lactic acid)/nanocellulose/nanoclay composites with synergistically enhanced barrier properties and improved thermomechanical resistance. *Polymer International, 65*(8), 988–995.

Trovatti, E., Fernandes, S. C., Rubatat, L., da Silva Perez, D., Freire, C. S., Silvestre, A. J., & Neto, C. P. (2012). Pullulan–nanofibrillated cellulose composite films with improved thermal and mechanical properties. *Composites Science and Technology, 72*(13), 1556–1561.

Tyagi, N., & Suresh, S. (2016). Production of cellulose from sugarcane molasses using Gluconacetobacter intermedius SNT-1: optimization & characterization. *Journal of Cleaner Production, 112*, 71–80.

Ullah, M. W., Ul-Islam, M., Khan, S., Kim, Y., & Park, J. K. (2015). Innovative production of bio-cellulose using a cell-free system derived from a single cell line. *Carbohydrate Polymers, 132*, 286–294.

Van de Vyver, S., & Román-Leshkov, Y. (2013). Emerging catalytic processes for the production of adipic acid. *Catalysis Science & Technology, 3*, 1465.

Wang, D., Peng, H., Wu, Y., Zhang, L., Li, M., Liu, M., Zhu, Y., Tian, A., & Fu, S. (2020). Bioinspired lamellar barriers for significantly improving the flame-retardant properties of nanocellulose composites. *ACS Sustainable Chemistry & Engineering, 8*(11), 4331–4336.

Werle, P., Morawietz, M., Lundmark, S., Sörensen, K., Karvinen, E., & Lehtonen, J. (2008). Alcohols, polyhydric. In *Ullmann's encyclopedia of industrial chemistry*. Wiley-VCH Verlag GmbH & Co. KgaA.

Wawat, R., & Malinee, S. (2016). *Lignocellulosic biomass to biofuel production: integration of chemical and extrusion (screw press) pretreatment*. King Mongkut's University of Technology North Bangkok International Journal of Applied Science and Technology. https://doi.org/10.14416/j.ijast.2016.11.001

Xiao, Y., Liu, Y., Wang, X., Li, M., Lei, H., & Xu, H. (2019). Cellulose nanocrystals prepared from wheat bran: characterization and cytotoxicity assessment. *International Journal of Biological Macromolecules, 140*, 225–233. https://doi.org/10.1016/j.ijbiomac.2019.08.160

Xiaojuan, M., Yunduo, L., Chao, D., Xinxing, L., Shilin, C., Lihui, C., Liulian, H., & Yonghao, N. (2017). Facilitate hemicelluloses separation from chemical pulp in ionic liquid/water by xylanase pretreatment. *Industrial Crops and Products, 109*, 459–463.

Xing, L., Hu, C., Zhang, W., Guan, L., & Gu, J. (2020). Biodegradable cellulose I (II) nanofibrils/poly(vinyl alcohol) composite films with high mechanical properties, improved thermal stability and excellent transparency. *International Journal of Biological Macromolecules, 4*(20), 34080–0.

Yan, J., Hu, J., Yang, R., Zhang, Z., & Zhao, W. (2018). Innovative nanofibrillated cellulose from rice straw as dietary fiber for enhanced health benefits prepared by a green and scale production method. *ACS Sustainable Chemistry & Engineering, 6*(3), 3481–3492. https://doi.org/10.1021/acssuschemeng.7b03765

Yuan, Z., Wei, W., & Wen, Y. (2019). Improving the production of nanofibrillated cellulose from bamboo pulp by the combined cellulase and refining treatment. *Journal of Chemical Technology and Biotechnology, 94*(7), 2178–2186. https://doi.org/10.1002/jctb.5998

Zhao, S., Zhang, Z., Sèbe, G., Wu, R., Rivera Virtudazo, R. V., Tingaut, P., & Koebel, M. M. (2015). Multiscale assembly of superinsulating silica aerogels within silylated nanocellulosic scaffolds: improved mechanical properties promoted by nanoscale chemical compatibilization. *Advanced Functional Materials, 25*(15), 2326–2334.

Zhou, Y., Saito, T., Bergström, L., & Isogai, A. (2018). Acid-free preparation of cellulose nanocrystals by TEMPO oxidation and subsequent cavitation. *Biomacromolecules, 19*(2), 633–639. https://doi.org/10.1021/acs.biomac.7b01730

Zhuo, B., Cao, S., Li, X., Liang, J., Bei, Z., Yang, Y., & Yuan, Q. (2020). A nanofibrillated cellulose-based electrothermal aerogel constructed with carbon nanotubes and graphene. *Molecules, 25*.

Chitin and chitosan-based blends and composites

Nayan Ranjan Singha[1], *Mousumi Deb*[1] *and Pijush Kanti Chattopadhyay*[2]
[1]Advanced Polymer Laboratory, Department of Polymer Science and Technology, Government College of Engineering and Leather Technology (Post-Graduate), Maulana Abul Kalam Azad University of Technology, Kolkata, West Bengal, India; [2]Department of Leather Technology, Government College of Engineering and Leather Technology (Post-Graduate), Maulana Abul Kalam Azad University of Technology, Kolkata, West Bengal, India

1. Introduction

Among the various kinds of biopolymers, polysaccharides are the most frequently available natural macromolecules. Of those polysaccharides, chitin is the second most widely available natural biopolymer next to cellulose. In fact, chitin is a value-added biopolymer extracted commercially from the shells of crustaceans including prawns, crabs, krill, insects, and shrimps. There are many reports published in developing chitin and chitin-based blends/composites suitable for diverse applications in medical and nonmedical sectors. However, chitin inherently suffers from several drawbacks because of the lower amounts of amines. In this regard, the shortcomings of chitin are eventually maneuvered by converting acetamido groups into primary amines via partial deacetylation of chitin in a strong alkaline solution, leading to a chemically modified natural polysaccharide known as chitosan. In contrast to chitin, chitosan has proven to be more useful, as chitosan contains both amine and hydroxyl chelating sites, and these groups can further be modified chemically (Wan Ngah et al., 2004). Though chitosan is a chemically modified form of chitin, the nontoxic, biodegradable, and biocompatible nature of chitin is almost retained in chitosan. Since, chitin is chemically represented as poly(1 → 4)-2-acetamido-2-deoxy-b-D-glucan, the chitosan macromolecules are named as poly(1 → 4)-2-amino-2-deoxy-b-D-glucan (Fig. 6.1).

The functional properties of chitosan-based materials can be improved further by combining them with other components to produce either blends or composites. The main objective of blending or composite making is to produce a new material envisaging variable properties by homogeneous mixing of at least two components. According to the available technical know-how, the key difference between a polymer blend and composite is that a polymer blend is basically a single-phase multicomponent material, whereas a composite is a multiphase multicomponent material (Chattopadhyay, Das, & Chattopadhyay, 2011; Chattopadhyay, Chattopadhyay et al., 2011; Chattopadhyay et al., 2013; Singha, Mahapatra et al., 2018). In case of either chitosan-based blends or composites, one of the components must be chitosan or modified chitosan, and the rest of the component(s) can be polymer(s) or any other material(s).

Biodegradable Polymers, Blends and Composites. https://doi.org/10.1016/B978-0-12-823791-5.00013-2

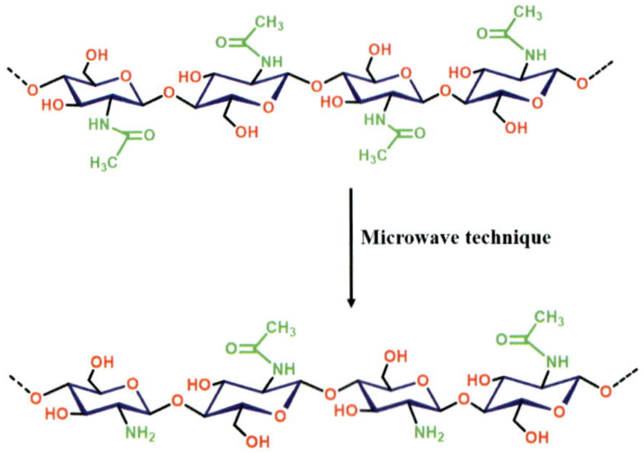

Figure 6.1 Deacetylation of acetamido group of chitin into amino group in chitosan.

Chitosan-based blends and composites have diverse prospective applications in the medical and nonmedical sectors (Tables 6.1–6.3). Nowadays, blends and composites of chitosan have increasingly been utilized in various biomedical fields, such as tissue engineering (Oliveiraa et al., 2006), wound healing (Devi & Dutta, 2017), drug delivery (Constantin et al., 2017), and gene delivery (Guo et al., 2011), along with nonbiomedical sectors including packaging (Hosseini et al., 2016), corrosion prevention (Fayyad et al., 2016), and adsorptive removal of hazardous components from wastewater (Mahapatra et al., 2018; Karmakar et al., 2017). Moreover, such materials can easily be processed into different forms, such as beads (Wan Ngah, Ariff, Hashim, & Hanafiah, 2010; Hameed et al., 2008), membranes (Aliabadi et al., 2014), gels (Mondal et al., 2018), nanofibers (Shalumon et al., 2013), nanofibrils (Liu et al., 2016), nanoparticles (Hosseini et al., 2016), microparticles, scaffolds (Peter, Binulal, Soumya et al., 2010), and sponges (Freag et al., 2018). Accordingly, chitosan can be admixed with various inorganic and organic component(s) to produce different materials suitable for various usages.

Because of the massive applications of so many different kinds of chitosan based blends and composites, it is not possible to cover the entire findings in a single chapter. Therefore, this chapter has been designed to covering an overview of various classes of blends and composites, along with their synthesis, properties, and applications.

2. Structure, properties, and applications of chitosan

2.1 Structure

As stated earlier, chitosan is a copolymer consisting of N-acetyl-2-amino-2-deoxy-D-glucopyranose and 2-amino-2-deoxy-D-glucopyranose in which two types of repeating units are linked by (1 → 4)-glycosidic bonds. Since chitosan is a deacetylated derivative of chitin, several acetyl groups disappear in chitosan depending on the extent of deacetylation. Thus, the number of $-NH_2$ will be greater in chitosan for a higher extent

Table 6.1 Chronological development of some selected chitosan-based blends and composites/nanocomposites in biomedical applications.

Year	Composition	Type#	Application(s)	Advantage(s)	Drawback(s)	References
1991	Chitosan–hydroxyapatite	Composite	Bone substitute material for dental treatment	i. Maximum compressive strength: 22 kg/cm^2	i. In vivo studies have not been done properly	Ito (1991)
1993–96	Bovine atelocollagen and high MW–[a] fully deacetylated chitosan	Blend	Stabilized collagen resisting degradation by collagenase	i. Simple blending procedure	i. Atelocollagen/chitosan association is less stable due to merely ionic interactions	Taravel and Domard (1993, 1995, 1996)
1996	Collagen–chitosan	Blend	For drug delivery	i. Skin-compatible combination is suitable for transdermal delivery of both lipophilic (nifedipine) and hydrophilic (propranolol hydrochloride)	i. In vivo studies have not been performed	Thacharodi and Rao (1996)
2000	Chitosan–alginate	Coacervates (phase separated)	Skin tissue engineering	i. Flexible ii. Porous and hydrophilic films are applicable in packaging, controlled release system, and wound dressing	i. Refinements are necessary in formulation, mechanistic study, and methods of preparation	Yan et al. (2000)
2002	Chitosan–alginate	Complex membrane made of coacervates	Skin tissue engineering by mouse and human fibroblast cells	i. Membrane is flexible, thin, transparent, and aqueous extracts are suitable in wound healing applications ii. Ease of handling, biodegradability, and storage stability	i. When wounds are treated with gauze, gauze fibers adhere to the healing tissues	Wang et al. (2002)
2003	Chitosan–cell scaffolds	Composite	Microporous scaffolds for cartilage tissue engineering	i. Substrate has been attempted for neochondrogenesis in vitro ii. Easy cell attachments iii. Sufficient porosity% (80.7 for frozen sample) and average pore diameter (67 μm) allows cell (diameter = 10 μm) penetration with ease	i. Cellular migration studies are insufficient	Nettles et al. (2002)
2003	poly(ε-caprolactone)–chitosan	Porous matrix	Bone regeneration therapy by scaffold	i. Appreciable cellular affinity and biocompatibility, and elevated bone formation ability ii. Consistently growing osteoblast proliferation	i. Tissue engineering and potential for bone substitution have not been investigated	Im et al. (2003)
2003	Glutaraldehyde cross-linked collagen–chitosan	Blend	Porous scaffolds for skin tissue engineering	i. Improved biostability	Cross-linking mechanism has not been evidenced by spectroscopy	Ma et al. (2003)
2003	Collagen–chitosan	Blend	Matrix for artificial livers	i. Improved biostability via cross-linking with imide type cross-linkers	Only in vitro experimentation has been performed	Wang et al. (2003)
2003	Chitosan–gelatin	Composite	Nerve tissue engineering	i. Soft, elastic complex, better nerve cell affinity than only chitosan, and good biocompatibility	In vivo studies have not been done in humans	Cheng et al. (2003)

Continued

Table 6.1 Continued

Year	Composition	Type#	Application(s)	Advantage(s)	Drawback(s)	References
2004	Chitosan–hydroxyapatite	Nanocomposite	Internal fixation of bone fracture	i. Biodegradable	–	Qiaoling et al. (2004)
2004	Chitosan–collagen–GAG[b]	Composite	Biomaterial for cartilage tissue engineering	i. Improved biostability via cross-linking with imide type cross-linkers ii. Controlled release of TGF-β1 from chitosan microspheres augmented the growth of cartilage	i. In vivo studies have not been done in humans	Lee et al. (2004)
2004	Chitosan–poly-*L*-lysine	Blend	Nerve tissue engineering	i. Soft, elastic complex, better nerve cell affinity than only chitosan, and also good biocompatibility	i. In vivo studies have not been done	Mingyu et al. (2004)
2005	Collagen–chitosan–heparin	Blend	For liver tissue engineering	i. Internal cross-linking between collagen and chitosan via anhydride and ester linkages	i. No external cross-linker has been employed	Wang et al. (2005)
2005	Chitosan–alginate	Blend	Hybrid scaffolds for bone tissue engineering	i. Good tissue compatibility during in vivo attachment of bone marrow to the scaffolds ii. Porosity%: 91.94 ± 0.9 iii. Mechanical properties: Compressive yield strength: 0.46 ± 0.022 MPa and Young's modulus: 8.16 ± 1.57 MPa	i. In vivo study has been done on rats	Li, Ramay et al. (2005)
2005	Chitosan–alginate polyion complex	Hybrid fiber	Hybrid fibers for scaffolds in ligament and tendon tissue engineering	i. Better fibroblast adhesion with high in vitro mechanical strength (tensile strength >200 MPa)	i. In vivo studies on mechanical strength, biocompatibility, and biodegradability have not been done	Majima et al. (2005)
2005	Chitosan–hyaluronan	Composite	Hybrid polymer fibers as a scaffold in ligament tissue engineering	i. Improved mechanical strength and good biological effects on cultured fibroblast	i. In vivo studies have not been done	Funakosi et al. (2005)
2005	Chitosan–hyaluronan	Composite	Hybrid fiber scaffold for ligament tissue engineering	i. In vivo regeneration of tissues ii. Enhanced production of type I collagen iii. Good strength (67.2 ± 2.2 N) and stiffness (18.8 ± 1.2 N/mm) of scaffolds	i. Biodegradability and biocompatibility have not been clearly reported ii. Complex could not be secreted easily from body even after its end of function	Funakoshi et al. (2005)
2005	Chitosan–alginate and chitosan–carrageenan	Blend	Controlled drug release	i. High elastic modulus ($1.8 \times 10^4 \pm 9.5 * 10^2 - 1.3 \times 10^5 \pm 2.9\ 10^4$ Pa) ii. Pore size and diameter (10–60 μm) are suitable for controlled release of drug	i. Fast erosion of the material led to disintegration of tablet due to fast solvent uptake	Tapia et al. (2005)
2005	Chitosan–heparin aminolyzed PET[c]	Composite	Film for surface modification of cardiovascular devices	i. Antiadhesive and antibacterial multilayer film	i. No in vivo studies	Fu et al. (2005)

2006	Chitosan sponges	Composite	Macroporous scaffolds for cartilage tissue engineering	i. Favorable and variable interconnective pore size and diameter (ranging from ≤10 μm to 70–120 μm) improved diffusion through the cell	i. In vivo biomechanical and biological implications have not been studied	Griffon et al. (2006)
2006	Chitosan–poly(lactic acid-glycolic acid)	Composite	Hybrid scaffolds for bone tissue engineering	i. Interconnected porous structure with pore size: 170–200 μm and pore volume: 28%–37% ii. High compressive modulus (10–2000 MPa) and compressive strength (2–180 MPa)	i. Biomolecules have not been introduced in the scaffold, only an in vitro study has been done	Jiang et al. (2006)
2006	Chitosan–GAGs	Composite	Biomaterial for chondrocyte culture	i. Composite having desired combination of chondroitin-4-sulfate and heparin for cartilage tissue engineering has been optimized by fast fractional factorial design considering the effects of individual parameters and interactions between those parameters	i. Results of statistical analysis are not the same as the experimental results	Chen et al. (2006)
2006	Chitosan–hydroxyapatite	Composite bilayered	Scaffold for tissue engineering	i. High elastic modulus (153 ± 12 and 2.9 ± 0.4 MPa)	i. Only in vitro cell culture assay has been done	Oliveiraa et al. (2006)
2007	Chitosan–poly(lactic acid-glycolic acid)	Sintered microsphere matrix	Hybrid scaffolds for bone tissue engineering	i. Good mechanical property with compressive modulus ranging from 622.26 ± 54.53 to 729.63 ± 19.93 MPa	i. Cannot be remodeled by osteoclasts and osteoblasts because of low porosity (porosity% = 19.20 ± 3.05) and pore size (199.62 ± 1.9 μm)	Abdel-Fattah et al. (2007)
2007	Poly(*L*-lactic acid)–chitosan	Composite	For drug delivery	i. Improved mechanical strength (storage moduli: > 7 MPa), porous structure, water uptake capacity (110%), and drug release rate (~45–70% within 2 h)	i. Effectivity as antiinflammatory and antibiotics has not been studied	Prabaharan et al. (2007)
2007	Chitosan–calcium phosphate	Composite	Bone replacement	i. High porosity (66.2%) and good flexural strength (25.2 ± 6.7–10.0 ± 1.4 MPa)	i. No in vitro or in vivo studies have been conducted	Xu et al. (2007)
2007	Chitosan–gelatin	Composite	Tissue engineering scaffold	i. Improved skin regeneration efficacy and enhanced vascularization ii. Fast release rate of drug (71.8% in 14 days)	i. Mechanical property has not been studied in detail	Liu et al. (2007)
2007	Chitosan–chondroitin-6-sulfate–dermatan sulfate	Composite	Scaffolds for cartilage tissue engineering	i. Favorable porosity % (88–91) and pore size (100–200 μm) ii. Improved cell adhesion and cell proliferation	i. In vivo studies have not been performed	Chen et al. (2007)
2007	Phosphorylcholine modified chitosan–hyaluronan	Composite	Biocompatible surface coatings	i. High mechanical strength, water content, viscoelasticity, hydrophilicity, and storage modulus	i. Inferior resistance at low pH owing to ionization of hyaluronan	Kujawa et al. (2007)
2007	*N*-acylated chitosan–iron oxide nanoparticles	Nanocomposite	Bone tissue engineering	i. Stable 3D matrix for controlled fabrication of hydroxyapatite crystals	i. Hydroxyapatite crystal growth mechanism has not been studied	Bhattaraia et al. (2007)

Continued

Table 6.1 Continued

Year	Composition	Type#	Application(s)	Advantage(s)	Drawback(s)	References
2007	Chitosan–hydroxyapatite–silk fibroin composite	Composite	Bone tissue engineering	i. Good chemical interaction and high compressive strength (179.3 ± 4.6 MPa)	i. Neither in vivo nor in vitro studies have been done	Wang and Li (2007)
2008	Chitosan–nano-hydroxyapatite–carboxymethyl cellulose	Nanocomposite	Scaffold for tissue engineering	i. Cross-linker not added ii. Nano-hydroxyapatite acts as reinforcing agent to contribute to biostability	i. Not tested in the human body	Jiang et al. (2008)
2008	Poly(vinyl alcohol)–chitosan bioactive glass	Nanocomposite	Hybrid scaffolds for bone tissue engineering	i. Good hemocompatibility, biodegradability, mechanical property, and low antigenicity ii. Increased hydrolysis grade of PVA shows better mechanical properties suitable for cancellous bone repairing	i. Lower hydrolysis grade of PVA[d] gives relatively poor mechanical properties in hybrid scaffolds	Mansur and Costa (2008)
2008	Chitosan–gelatin	Blend	For nerve tissue engineering	i. Porous cross-linked network with increased wettability and swelling behavior	i. Thermal stability does not improve even after gelatin addition and no in vivo tests	Chiono et al. (2008)
2008	Poly(L-lactic acid)–chitosan	Hybrid scaffolds	Scaffold for bone tissue engineering	i. Biodegradable, biocompatible and osteoconductive	i. In vivo study has not been done in detail	Mano et al. (2008)
2009	rhBMP-2[e] microspheres-loaded chitosan–collagen	Composite	Scaffold for bone tissue engineering	i. Enhanced osseointegration ii. In vivo applications in dog	i. In vivo applications in humans have not been done	Shi et al. (2009)
2009	Chitosan–poly(butylene succinate)	Blend	Scaffold for bone tissue engineering	i. High porosity (avg. porosity %: 44.8 ± 2.1), pore interconnectivity, and mechanical strength (compression modulus: 32.6 ± 12.8 MPa) ii. Suitable for bone regeneration in human bodies	i. Only in vitro studies have been done	Costa-Pinto et al. (2009)
2009	Chitosan–hydroxyapatite	Composite	Scaffold for bone tissue engineering	i. Promising biocompatible material for tissue engineering	i. In vivo studies have not been done	Madhumathi et al. (2009)
2009	Chitosan–polylactic acid–hydroxyapatite	Nanocomposite	For bone tissue engineering	i. High porosity % (85), favorable pore size (100–200 μm with subpores of 2–10 μm), improved elastic modulus, and strength	i. Biocompatibility and biodegradability have not been tested	Cai et al. (2009)
2009	Chitosan–hyaluronic acid	Composite	Extracellular matrix	i. Compressive modulus (28 kPa) suitable for cartilage tissue engineering	i. In vivo studies have not been done	Tan et al. (2009)
2009	Chitosan–silk fibroin	Composite	Scaffolds for liver tissue engineering	i. Good histocompatibility and biodegradability	i. In vivo studies have not been done in humans	She et al. (2009)
2010	Chitosan–galactosylated hyaluronic acid	Scaffolds	Scaffolds for primary hepatocytes culture	i. Enhanced interaction between hepatocyte and scaffold materials in the presence of galactose	i. In vivo studies have not been done	Fan et al. (2010)

2010	Chitosan–poly(lactide-*co*-glycolide)–heparin-recombinant protein	Sintered microsphere scaffolds	Scaffold for bone tissue engineering	i. Both in vitro and in vivo studies have been conducted. ii. Good compressive modulus: 382.1 ± 42.53 MPa and compressive strength: 7.1 ± 1.9 MPa	i. In vivo studies have not been done in humans	Jiang et al. (2010)
2010	Chitosan–glass ceramic	Nanocomposite	Scaffold for bone tissue engineering	i. Improved in vitro biomineralization, controlled swelling, and better cytocompatibility ii. Interconnected pores with diameter 150–300 μm	i. In vivo studies have not been done	Peter, Binulal, Soumya et al. (2010)
2010	Chitosan–gelatin–nano-bioactive glass ceramic	Composite	Scaffold for bone tissue engineering	i. Good bioactivity and better cell attachment	i. Detailed in vivo studies have not been done	Peter, Binulal, Nair et al. (2010)
2010	Chitosan–pectin	Composite	Controlled drug and food delivery	i. Improved mechanical strength, good hydrophilic behavior, and water uptake capabilities	i. Tensile strength is insufficient	Chen et al. (2010)
2010	Poly(ε-caprolactone)–poly(ethylene glycol)–poly(ε-caprolactone) (CEC)–chitosan	Composite	Anticancer agent	i. Enhanced cytotoxicity of the anticancer drug (doxorubicin) toward cancer cell lines	i. Has not been checked in humans. Death of normal cells is still a problem	Li et al. (2010)
2010	Chitosan–silk fibroin	Composite	Nanofibers for wound dressing	i. Improved tensile strength (1.0 ± 0.21–10.3 ± 0.24 MPa), antibacterial property, and biocompatibility	i. In vivo study has not been done	Cai et al. (2010)
2010	Genipin cross-linked chitosan–collagen	Porous scaffolds	Scaffold for cartilage tissue engineering	i. Improved cell viability, biostability, and storage modulus (2.4×10^5–4.6×10^5 Pa)	i. Biocompatibility has been checked only in vitro	Yan et al. (2010)
2010	Polypyrrole–chitosan	Conductive composite	For nerve tissue engineering	i. Enhanced nerve regeneration in scaffolds and cell viability	i. Nerve regeneration has been done by electrical simulation without analyzing the sensitivity	Huang et al. (2010)
2010	Chitosan–hydroxyapatite	Composite	Porous scaffold for drug delivery	i. Sustainable drug release (92%–84% within 72 h) for hybrid scaffolds	i. No in vivo studies have been conducted	Teng et al. (2010)
2011	Chitosan–CNT^f–hydroxyapatite	Nanocomposite	For bone tissue engineering	i. Porous, biodegradable, and biocompatible material of enhanced thermal stability up to 900°C	i. Insufficient water uptake and retention ability, lack of effective bone tissue engineering, only in vitro study has been done	Venkatesan et al. (2011)
2011	Collagen–chitosan	Composite	Skin tissue engineering	i. Fast regeneration and high tensile strength of the repaired tissue [tensile strength of burn wounds treated with petrolatum gauze dressings (MPa): 1.59 ± 0.05 (28 days) 3.22 ± 0.16 (56 days), and 4.35 ± 0.23 (105 days)]	i. Only in vitro study has been executed	Guo et al. (2011)

Continued

Table 6.1 Continued

Year	Composition	Type#	Application(s)	Advantage(s)	Drawback(s)	References
2011	Poly(lactic-*co*-glycolic) acid–chitosan	Nanofiber materials	For drug delivery	i. Increased hydrophilicity and improved drug release behavior	i. No in vivo study has been done	Meng et al. (2011)
2012	Atelocollagen–chitosan	Composite	Skin tissue engineering	i. Enhanced cellular activity, improved tissue repairing, and appreciable biocompatibility	i. In vivo studies on humans have not been done	Judith et al. (2012)
2012	Methoxy poly(ethylene glycol)-*g*-chitosan and curcumin	Composite	Wound healing	i. Noncytotoxic with good antioxidant property (% inhibition of lipid peroxidase: 97.49 ± 0.61) ii. Sustained in vitro release of curcumin, i.e., 44.45% release within 1 week	i. In vivo study has been done in rat only	Li et al. (2012)
2012	Chitosan–peptide	Chitosan film scaffold	Scaffold for bone tissue engineering	i. Improved cell affinity and osteogenic property	i. In vivo study has not been done in humans, and lack of precise control in density of scaffold	Tsai et al. (2012)
2013	Chitosan–hyaluronic acid	Composite	Scaffold for cartilage tissue engineering	i. High cell viability, improved proliferation, and enhanced compressive modulus (10.3 kPa)	i. Only extracellular studies have been done	Park et al. (2013)
2013	Chitosan–gelatin	Composite	Scaffolds for dermal tissue engineering	i. Sufficient pore size (surface pore size: 83.3 ± 11.6 to 178.2 ± 55.5 μm, cross-sectional pore size: 125.3 ± 39.8 to 177.5 ± 31.6 μm), improved modulus, enhanced thermal properties (denaturation temperature: 244–274°C), and high cell proliferation	i. In vivo studies have been done in mice but not in humans	Tseng et al. (2013)
2013	Chitosan hydrogel–nanofibrin	Composite	For skin tissue engineering	i. Improved cell attachment and proliferation ii. Bioactive potential wound dressing material	i. Detailed study of mechanical strength is not available	Sudheesh Kumar et al. (2013)
2013	Chitin–chitosan–gelatin	Composite	Scaffolds for nerve tissue engineering	i. Uniform pore size, improved cell adhesion behavior, and cell viability	i. In vivo studies have not been done	Kuo and Lin (2013)
2013	Chitosan–hyaluronan–nanochondroitin sulfate	Nanocomposite	Composite sponge for wound dressing	i. Improved porosity, biodegradation, cytocompatibility, and good cell adhesion	i. Only in vitro studies have been done	Anisha et al. (2013)
2013	Bioactive glass and chitosan	Nanocomposite	For bone tissue engineering	i. Enhanced mechanical property, bioactivity, and faster biomineralization	i. In vivo studies have not been done	Caridade et al. (2013)
2013	Bioactive glass and hydroxyapatite in PCL^{g}-chitosan	Nanocomposite	Scaffold for bone tissue engineering	i. Higher protein adsorption, better cell spreading, and proliferation	i. Lowered thermal stability despite the presence of nanoparticles	Shalumon et al. (2013)
2014	Chitosan–*g*-poly acrylonitrile–silver	Nanocomposite	Antimicrobial material	i. Antimicrobial performance against *Escherichia coli* and *Staphylococcus aureus* ii. Thermal stability up to 300°C	i. Antimicrobial property has not been studied on other bacteria	Hebeish et al. (2014)

2015	Polycaprolactone nanofibers chitosan–oxidized starch	Nanofibrous composite	Scaffolds for bone regeneration	i. Favorable mean pore size (73.40 ± 28.19 mm), porosity, increased water uptake, improved cell viability, and good compressive modulus: 2.64 ± 0.19 MPa	i. Increasing starch content results decreased compressive strength and modulus	Nourmohammadi et al. (2015)
2016	Chitosan–gelatin	Composite	Scaffold for blood vessel tissue engineering	i. Highly porous structure (porosity: 82% and pore size: 100–230 mm), improved cell adhesion and proliferation, and good tensile strength: 95.81 ± 11 kPa	i. Only in vitro biodegradation study has been done	Badhe et al. (2017)
2016	Chitosan–CNTs–chitosan	Composite foam	Tissue engineering, orthopedics, and nanocarriers	i. High mechanical strength, elasticity, and density	i. Uncontrolled CNT addition imparts decreased mechanical strength	Yan et al. (2016)
2016	Graphene oxide and chitosan	Nanocomposites	Biomaterials	i. Enhanced tensile properties (strength: 41.91–63.48 N/mm^2 and modulus: 1346.07–2032.9 N/mm^2), improved biocompatibility and biodegradability, nontoxicity, good antimicrobial activities, low cost, and low immunogenicity	i. Prospective applications have not been stated clearly	Khan et al. (2016)
2016	Chitosan–silver sulfadiazine	Composite	Sponges as potential wound dressings	i. Excellent porosity (76.9%–78.0%), high swelling, good antibacterial property, and good cell viability	i. Little cytotoxic effect on cell morphology	Shao et al. (2017)
2017	Chitosan–nano-hydroxyapatite	Nanocomposite	For bone tissue engineering	i. Biodegradable, biocompatible, and porous material (porosity%: 72.3 ± 4.04–92.5 ± 0.50) supporting good cell proliferation	i. In vivo study has not been included	Atak et al. (2017)
2017	Chitosan–organomodified rectorite	Composite	Protecting layer for preventing cells from metal toxicities	i. Can efficiently protect the apoptosis of PC12 cells exposed to toxic Pb(II) ions ii. The protective layer is not cytotoxic	i. In vivo applications have not been done	Liu et al. (2017)
2017	Chitosan–bentonite	Nanocomposite	Wound healing	i. Sufficient water vapor transmission rate (1093 ± 20.5–1954 ± 51 g/m/day), porosity (88%), water absorption capacity (1232 ± 14.58–1688 ± 18.52%), folding endurance (145.25 ± 2.21–289.50 ± 0.57), and good antibacterial activity	i. Only in vitro nonenzymatic study has been done	Devi and Dutta (2017)
2017	Chitosan–kaolin	Composite	Hemostasis through quick blood clotting	i. High hemostatic efficacy, kaolin-driven increased water absorption ratio: 1550%–1730%, specific surface area: 28.2–40.7 m^2/g, and pore volume: 52%	i. In vivo study has been done in rat or in human body	Sun et al. (2017)

Continued

Table 6.1 Continued

Year	Composition	Type#	Application(s)	Advantage(s)	Drawback(s)	References
2017	Chitosan–poly(*N*-isopropylacrylamide-*co*-hydroxye thylacrylamide)	Composite	Thermosensitive hydrogel for controlled drug delivery	i. Improved temperature and pH sensitivities, biodegradability, and controlled drug release	i. Only in vitro study has been done	Constantin et al. (2017)
2017	Chitosan–polypyrrole	Composite	Improving surface protection against corrosion in simulated body fluid	i. Improved surface hydrophilicity, reduced contact angle, enhanced protection, good biocompatibility, and bioactivity	i. In vivo study has not been done	Kumara et al. (2017)
2017	Chitosan–starch	Composite	Drug delivery and tissue engineering	i. Improved elastic modulus (28 ± 0. 2 kPa) and high cross-linking density	i. Very low compressive moduli, lower than that of human cartilage tissue, and hence, unsuitable for humans	Baniani et al. (2017)
2017	Chitosan–hyaluronic acid	Nanocomposite	Scaffold sponge for wound healing	i. Improved proper porosity and enhanced wound-healing capacity	i. In vivo study has been done in rats or in human body	Sanad and Abdel-Bar (2017)
2017	Chitosan–hydroxyapatite (Hap[h])	Composite fiber	Scaffold of ligament regeneration	i. High mechanical strength (14.0 ± 1.6–24.0 ± 7.0 MPa) and strain (50.2 ± 8.5–119.0 ± 18.0%)	i. In vivo study for bone treatment has not been done	Takuma et al. (2017)
2017	Chitosan–alginate porous microspheres in *Bletilla striata* polysaccharide	Composite	Hemostatic sponges in surgery	i. Improved porous structure (pore size: 50–200 μm and porosity %: 94.87), higher water absorption capacity, improved biocompatibility, and enhanced cell viability (91.8%–104.2% at 24 h and 101.8%–112.6% at 48 h), and hemostatic property	i. Only in vitro study has been done	Chao et al. (2017)
2017	Chitosan–halloysite nanotubes	Composite	Biocompatible material for drug delivery	i. Maximum drug entrapment efficiency for doxorubicin (45.7%)	i. Slightly cytotoxic toward MC3T3-E1 cells	Huang et al. (2017)
2017	Chitosan–nanocellulose–Ag NP[i]s–curcumin	Nanocomposite	Wound dressing	i. Both Ag NPs and curcumin act as antimicrobial agents	i. Absence of cross-linkers and hence lack long-term durability	Bajpai et al. (2017)
2017	Chitosan–Ag/ZnO	Composite	Dressing for wound healing	i. Enhanced antibacterial activity ii. Hemocompatibility and high porosity (93%)	i. In vivo study has been done on mice	Lu et al. (2017)
2018	Chitosan–poly(L-lactic acid)	Composite	Tunable degradation behavior	i. Improved porosity (89.5 ± 4.6–81.6 ± 7.2%) ii. Compressive strength (2.65 ± 0.32–3.67 ± 0.42 MPa)	i. In vivo study has been done on rat	Guo et al. (2018)
2018	Chitosan–hydroxypropyl methylcellulose	Composite	Sponge for drug delivery	i. The combined matrix can support sustained drug release	i. Cross-linkers have not been used	Freag et al. (2018)

2019	*N*, *O*-carboxymethyl chitosan–oxidized regenerated cellulose	Composite	Biodegradable gauze having antimicrobial functionality and ability to prevent postsurgical peritoneal adhesions	i. Antiadhesion gauze ii. Imide type cross-linking agent enhances the biostability of the gauze material iii. Grafting of modified chitosan and cellulose has been identified by XPS analyses	i. Actual nature of cross-links has not been identified clearly	Cheng et al. (2018)
2019	Chitosan–PVA-PVP[j]	Composite	Wound-healing dressing	i. Improved tensile strength (60.8–79.6 MPa), elongation at break, and good biocompatibility	i. In vivo study has not been done	Rahmani et al. (2020)
2019	Chitosan–amylopectin	Composite	Encapsulation and sustained release of curcumin at the intestine	i. Solubility and stability of curcumin have been increased ii. High encapsulation efficiency = 90.3%	i. Only in vitro cytotoxicity has been evaluated	Liu, Huang et al. (2019)
2019	Hydroxylated lecithin complexed iodine–carboxymethyl chitosan sodium alginate	Composite membrane	Infected burn wound treatment	i. Improved antibacterial property, mechanical properties, water vapor permeability, and pH controllable iodine release ii. Microwave drying of membrane ensures high contents of activated iodine in the composite	i. Wound healing has been tested only in rats	Chen et al. (2018)
2019	Chitosan–silk	Composite	Applications in biomedical devices	i. High tensile strength (97.8 MPa) and Young's modulus (3.5 GPa)	i. Biocompatibility has been tested on rats only	Huang et al. (2020)
2019	Chitosan–hydroxypropyl methylcellulose–mesoporous silica	Composite	Treatment of gastric ulcers by gastroprotective taxifolin delivery	i. Sufficient taxifolin residence time (5 h) and release in the stomach ii. Better efficiency compared to the conventionally used enteric-coated tablets of proton pump inhibitors	i. Not tested in human stomach and humans	Stenger Mouraa et al. (2019)
2019	Chitosan–silica	Composite	Cross-linked chitosan with controllable swelling in drug delivery and wound healing	i. Improved mechanical properties (Young's modulus: 413.21–605.94 MPa) and high thermal stability: exothermic peak at 307°C	i. Genipin polymerization affects the network structure and enlarges the mesh size	Liu, Cai et al. (2019)
2019	Chitosan–biochar–nanosilver	Nanocomposite	Drinking water treatment for killing pathogenic bacteria	i. Very good antimicrobial agent affecting percent wound reduction: 57.8 and 97.2	i. In vivo study on humans has not been done	Hu et al. (2019)
2020	Hydroxybutyl chitosan–diatom–biosilica	Composite	Sponge for hemorrhage control	i. Improved porous structure and good biocompatibility ii. High and fast fluid absorbability (11–16 times that of weight) iii. Good hemostasis effect (clotting time shortened by 70% compared with that of control)	i. Only in vitro study has been done	Zhang et al. (2020)
2020	Chitosan–carrageenan	Composite	Dressing material for wound healing	i. Enhanced surface area ($60.4 \pm 0.2\ m^2/g$), pore volume ($0.06 \pm 0.002\ cm^3/g$), and average pore diameter (4.5 ± 0.1 nm) ii. Combined ability to accelerate hemostatic activity and tissue growth	i. Only in vitro study has been done	Biranje et al. (2020)

Continued

Table 6.1 Continued

Year	Composition	Type#	Application(s)	Advantage(s)	Drawback(s)	References
2020	Chitosan–red marine alga–polysaccharide	Composite	Hydrogels for controlled release of insulin	i. Improved water uptake (80.33 ± 0.33–87.61 ± 0.85) and holding (17.98 ± 0.35–20.92 ± 0.59) capacities ii. Good textural stability (stiffness: 5.36 ± 0.20–13.79 ± 0.79 N/mm and rupture force: 2.39 ± 0.00–14.88 ± 0.67 N) iii. Better sensitivity to pH and ionic strength, and improved thermal stability	i. Only in vitro biodegradability has been tested	Feki et al. (2020)
2020	Chitosan–glucomannan with thrombin-loaded microporous starch particles	Composite	Composite for hemostasis	i. Wound hemorrhage and hemostasis controlling ability, good biocompatibility, and effective degradability ii. Greater hydrophilicity, porous structure, rapid blood adsorption, and cell viability >80%	i. In vivo tests have been done in white rabbits	Shi et al. (2020)

[a] molecular weight
[b] glycosaminoglycan
[c] polyethylene terephthalate
[d] polyvinyl alcohol
[e] recombinant human bone morphogenetic protein-2
[f] carbon nanotube
[g] poly-caprolactone
[h] hydroxyapatite
[i] nanoparticle
[j] polyvinyl pyrrolidine

Table 6.2 Chronological development of some selected chitosan-based blends and composites/nanocomposites in electrical applications, packaging, and corrosion resistance.

Year	Composition	Type#	Application(s)	Advantage(s)	Drawback(s)	References
Electrode/Sensor						
2003	Chitosan–MMT[a]–graphite	Nanocomposite	Sensor for potentiometric determination of several anions	i. Electrodes exhibiting easy surface renewal, ruggedness, and long-term stability ii. Higher selectivity toward monovalent anions iii. Low environmental impact	i. In the absence of graphite, the composite lacks the required electronic conductivity to function as an electrode ii. Not tested for sensing cation	Darder et al. (2003)
2004	Chitosan–glucose oxidase–gold nanoparticles	Nanocomposite	Glucose biosensor applications	i. Rapid response with detection limit of 2.7 μm ii. Good biocompatibility, reproducibility, and storage stability iii. Recovery rate = 94–97%	i. Costly	Luo et al. (2004)
2004	Chitosan–glucose oxidase–ZrO_2[b]	Composite	Glucose biosensor applications	i. Rapid response with detection limit = 1.0×10^{-5} M at 3σ and sensitivity = 0.028 μAm/M ii. Good reproducibility with coefficient of variation = 4.65%	i. Lower stability (45 days)	Yang et al. (2004)
2005	Chitosan–CNT[c]	Nanocomposite	DNA[d] biosensor	i. Highly improved electroactive surface area in the presence of CNT, 0.28 ± 0.03 cm^2 for chitosan–CNT electrodes with detection limit 0.252 nM for fish sperm DNA ii. Improved solubility and stability iii. Recovery rate of DNA = 89–106%	i. The effect of only human serum albumin on the enzyme has been studied	Li, Liu et al. (2005)
2006	Chitosan–sepiolite	Nanocomposites	Bionanocomposites as components of electrochemical devices/fuel cells	i. Low cost and environment-friendly material ii.Enhanced thermal stability and mechanical property [elasticity modulus: 3.0 ± 0.1–6.5 ± 0.1 (nonwashed) and 1.6 ± 0.1–5.1 ± 0.1 (washed)]	i. Not applicable as active phase of potentiometric sensor	Darder et al. (2006)
2007	Chitosan–glucose oxidase on a glassy carbon electrode modified with gold–platinum alloy nanoparticles–MWCNTs[e]	Nanocomposite	Glucose biosensor applications	i. Glucose sensing (sensitivity = 8.53 ± 0.09 μAmM^{-1}) at low potential (0.1 V) with detection limit = 0.2 μM ii. Fast response (<5 s) with good reproducibility, stability, and selectivity	i. Effects of ascorbic acid (AA), uric acid (UA), and acetaminophen (AAP) in glucose sensing need improvement	Kang et al. (2007)

Continued

Table 6.2 Continued

Year	Composition	Type#	Application(s)	Advantage(s)	Drawback(s)	References
2008	Chitosan–SWCNTs[f]–glucose oxidase	Nanocomposite	Glucose biosensor applications	i. Rapid response (<5 s) with LOD[g] = 2.5 μM ii. Improved mechanical stability and film conductivity iii. Excellent catalytic property	i. At acidic and basic pH and at temperature >37°C the sensitivity drops	Qiu et al. (2008)
2009	Substituted polyaniline–chitosan	Composite	Glucose biosensor applications	i. Good thermal stability	i. Only homogeneous samples give high conductivity	Yavuz et al. (2009)
2010	Polypyrrole–chitosan	Composite film	Electrocatalysis, biosensing, and anticorrosive coating	i. Highly stable and extremely electroactive ii. Efficient formation of electroactive protective layer on metal	i. Biocompatibility has not been tested for possible application as biosensor	Yalçınkaya et al. (2010)
2015	Chitosan–carbon (containing catechol)	Nanocomposite	Electrode	i. Composite electrode shows a reversible confined redox behavior by the catechol ii. The electrode catalyzes oxidation of NADH[h] iii. It can bind with Cu^{2+} ions (binding constant 8.7 μM)	i. Aging resistance and durability have not been tested	Jirimali et al. (2015)
2017	Chitosan–N-doped graphene	Nanocomposite	Sensor for determining tartrazine	i. Sensitive with detection limit of 0.036 μmol/L ii. Improved electrical conductivity [electron-transfer resistance = 149.7–335.5 Ω] iii. Good reproducibility	i. Diverse practical applications are to be tested further	An et al. (2017)
Packaging						
2011	Kudzu starch–chitosan	Composite	Edible packaging films	i. High tensile strength (5.46 ± 0.80–13.70 ± 2.52 MPa), maximum elongation (52.47 ± 4.21–56.57 ± 8.81%), and improved flexibility ii. High antibacterial activity (90.91 ± 0.56–99.99 ± 0.00 against *Escherichia coli* and 7.63 ± 1.08–25.11 ± 8.58 against *Staphylococcus aureus*)	i. Antimicrobial resistance should be examined against different Gram-(+ve) and Gram-(−ve) bacteria other than *Escherichia coli* and *Staphylococcus aureus*	Zhong et al. (2011)

2011	Chitosan–graphene oxide	Composite	Packing materials	i. Good thermal stability and mechanical properties (tensile strength: 43.9 ± 1.8 MPa) ii. Biocompatible and water resistant	i. Effectiveness of dry packaging material is lower than that of wet material	Han et al. (2011)
2016	Chitosan–fish gelatin	Composite	Packaging films	i. Bioactive food packaging films with sufficient antibacterial efficacy	i. Further studies should be done by varying different essential oils	Hosseini et al. (2016)
2016	Whey protein–carboxymethylated chitosan	Composite	Edible packaging films	i. Improved transparency and water vapor barrier properties ii. Mechanical property: 300% more tensile strength than other films with improved elongation	i. Surface morphology has not been much improved	Jiang et al. (2016)
2017	Chitosan–soy protein–Cu-nanoclusters	Nanocomposite	Food packaging materials	i. High tensile strength (2.29–5.01 MPa), elongation at break (58.75–197.50 MPa), and enhanced thermal resistance	i. Film properties should be studied by other nanoparticle modifications	Li et al. (2017)
2017	Chitosan–TiO_2[i]	Composite	Efficient antimicrobial activities for food packaging applications	i. Enhanced hydrophilicity, mechanical properties, and efficient antimicrobial activity (99% in the presence of *Escherichia coli*)	i. Poor water resistance due to enhanced hydrophilicity	Zhang et al. (2017)
2018	Rosin modified cellulose nanofiber polylactic acid–chitosan	Composite	Food packaging film	i. Improved mechanical properties compared to unmodified composite owing to better dispersion of nanofibers ii. Excellent antimicrobial performance against *Escherichia coli* and *Bacillus subtilis*	i. Other food pathogens, such as *Salmonella* spp. remain untested ii. Bulky phenanthrene rings of rosin can reduce flexibility of the packaging material	Niu et al. (2018)
2018	Nanocellulose–chitosan–starch Nanocellulose–chitosan–gelatin	Nanocomposite Nanocomposite	Food packaging film Food packaging film	i. Nanocellulose provided high tensile strength ii. Chitosan has been used to enhance antibacterial and antifungal properties	i. Transparency, tensile strength, and elongation to break are lower in starch films	Noorbakhsh-Soltani et al. (2018)
2019	Chito-oligosaccharide–bacterial cellulose Chitosan–bacterial cellulose	Composite Composite	Food packaging and wound dressing Food packaging	i. Better than plant cellulose in terms of purity, crystallinity, tensile strength, and water-holding capacity	i. All the application prospectives have not been investigated	Yin et al. (2019)

Continued

Table 6.2 Continued

Year	Composition	Type#	Application(s)	Advantage(s)	Drawback(s)	References
2019	Chitosan–AgNP[j]	Nanocomposite	Packaging of food, pharmaceutical, and allied products	i. Good tensile strength (51.67 ± 2.34–65.04 ± 1.46 MPa) and elongation at break (25.75 ± 3.51–28.42 ± 2.34%) ii. Good antimicrobial resistance	i. Less effective against Gram-positive bacteria (*Staphylococcus aureus* and *Bacillus subtilis*)	Kadam et al. (2019)
2019	Chitosan–AgNP	Nanocomposite	Food preservation	i. Exhibits strong bacteriostasis against *Escherichia coli*, *Saccharomyces cerevisiae*, and *Penicillum citrinum* ii. Size of the microsphere can be controlled by changing the cross-linker proportion	i. Larger microspheres are relatively less effective in bacteriostasis	Liang et al. (2019)
2020	Chitosan–bacterial cellulose–poly(vinyl alcohol)	Nanocomposite	Food packaging applications	i.Better than plant cellulose in terms of purity, crystallinity, tensile strength, and water-holding capacity	i. All the application prospectives have not been investigated	Ju et al. (2020)
Corrosion protection and other applications						
2016	Oleic acid–grafted chitosan–graphene oxide	Composite	Corrosion protection of carbon steel	i. Corrosion resistance (6 × 103 Ω) has been increased by 100-folds	i. Less conductivity can interrupt ionic transport and electrochemical activity	Fayyad et al. (2016)
2019	Graphene oxide–chitosan	Composite	High-strength screws and more load-bearing materials	i. Good mechanical strength, biocompatibility, and improved bending property (maximum bending strength = 297 MPa)	i. In vitro CCK8 cell assay has been done but in vivo studies have not been performed	Jin et al. (2019)

[a] montmorillonite
[b] zirconium dioxide
[c] carbon nanotube
[d] deoxy ribonucleic acid
[e] multi-walled carbon nanotubes
[f] single-walled carbon nanotubes
[g] limit of detection
[h] nicotinamide adenine dinucleotide (NAD) + hydrogen
[i] titanium dioxide
[j] silver nanoparticles

Table 6.3 Chronological development of some selected chitosan-based blends and composites/nanocomposites in adsorptive removals.

Year	Composition	Type#	Application(s)	Advantage(s)	Drawback(s)	References
Removals of dyes, heavy metal ions, and other hazardous pollutants from wastewater						
2003	Chitosan–ceramic alumina	Composite	Adsorptive removal of Cr(VI)	i. AC[a] = 153.85 mg/g	i. AC decreases at the higher pH	Veera et al. (2003)
2003	Chitosan–perlite	Composite	Adsorptive removal of Cr(VI) well fitted to Langmuir isotherm	i. AC = 153.8 mg/g ii. Absorbed Cr(VI) can be desorbed by alkali treatment followed by reuse	i. AC decreases at the higher pH	Shameem et al. (2003)
2004	Chitosan–activated clay	Composite	Adsorptive removal of methylene blue (λ_{max} = 664.5 nm), reactive dye RR222 (λ_{max} = 502 nm), tannic acid (λ_{max} = 275 nm), and humic acid (λ_{max} = 218.5 nm) fitted to Freundlich equation	i. The ACs of tannic acid, RR222, humic acid, and methylene blue on composite beads are 1490, 1912, 243, and 330 g/kg, respectively	i. Relatively poor ACs for methylene blue and humic acid	Chang and Juang (2004)
2004	Chitosan–PVA[b]	Blend	Adsorptive removal of Cu(II)	i. Maximum monolayer adsorption capacity of Cu(II) on chitosan–PVA beads at pH 6.0 and 500 rpm is 47.85 mg/g	i. Maximum adsorption capacity is not very high ii. Desorption of 87.4% in the presence of EDTA[c]	Wan Ngah et al. (2004)
2006	Chitosan–perlite	Composite	Adsorptive removal of Cd(II) fitted to Langmuir isotherm	i. Maximum AC = 178.6 mg/g at 298 K, pH = 6.0, and Cd(II) concentration of 5000 mg/L	i. The maximum AC should be improved further	Shameem et al. (2006)
2007	Chitosan–MMT[d]	Nanocomposite	Adsorptive removal of Congo red (fitted to pseudosecond-order kinetics and Langmuir isotherm)	i. The AC of chitosan/MMT nanocomposite (i.e., 72.11 mg/g) is higher than the mean value of chitosan (i.e., 96.62 mg/g) and MMT (i.e., 29.52 mg/g) at pH = 7	i. AC decreased with the increasing pH from 7 to 9	Wang and Wang (2007)
2007	Chitosan–sand	Composite	Adsorptive removal of Cu(II) fitted to Langmuir isotherm	i. Cheap adsorbent with maximum AC = 10.87 mg/g	i. Very low AC ii. Binding of sand with chitosan has not been not characterized	Wan et al. (2007)
2008	Chitosan–alginate	Blend	Adsorptive removal of Cu(II) fitted to Langmuir isotherm	i. Maximum adsorption capacity of 67.66 mg/g at pH = 6	i. Only ionic interaction is considered as the driving force for the adsorption	Wan Ngah and Fatinathan (2008)

Continued

Table 6.3 Continued

Year	Composition	Type#	Application(s)	Advantage(s)	Drawback(s)	References
2008	Chitosan–perlite	Composite	Adsorptive removal of Cu(II) fitted to Langmuir isotherm.	i. At 298 K, pH = 4.5, and equilibrium concentration = 812.5 mg/L, the maximum AC of composite bead is 104 mg/g ii. Adsorbent is reusable	i. Controlled addition of EDTA is required to regenerate the adsorbent effectively ii. Excessive EDTA addition is harmful, and the AC deteriorates if deviates from pH = 4.5	Shameem et al. (2008)
2008	Chitosan–oil palm ash	Composite	Adsorptive removal of reactive blue 19 (λ_{max} = 598 nm) fitted to Redlich–Peterson isotherm	i. Maximum AC > 400 mg/g at pH = 6 and 30°C ii. Biodegradable iii. Application of solid waste to remove dyes	i. Recyclability not reported	Hameed et al. (2008)
2008	Chitosan–ceramic alumina	Composite	Adsorptive removal of As(III) and As(V) fitted to Langmuir isotherm	i. Monolayer ACs are 56.5 and 96.46 mg/g for As(III) and As(V), respectively	i. Low AC for As(III)	Veera, Krishnaiah, Jonathan et al. (2008)
2008	Chitosan–ceramic alumina	Composite	Adsorptive removal of Cu(II) and Ni(II) fitted to Langmuir isotherm	i. Monolayer ACs are 86.2 and 78.1 mg/g for Cu(II) and Ni(II), respectively	i. AC should be increased further	Veera, Krishnaiah, Ann, & Edgar (2008)
2008	Chitosan–cotton fiber	Composite	Adsorptive removal of Pb(II), Ni(II), Cd(II), and Cu(II)	i. Better adsorption selectivity for Pb(II) and Cu(II) than those of Ni(II) and Cd(II)	i. Adsorption sites of metal ions have not been determined	Zhang et al. (2008)
2009	Chitosan–cotton fiber	Composite	Adsorptive removal of Hg(II)	i. AC at the optimum pH = 5.0 for SCCH[e] = 0.45 mmol/g and RCCH[f] = 0.47 mmol/g ii. More than 90% recovery of Hg(II) in the presence of K(I), Na(I), Ca(II), Mg(II), Co(II), Mn(II), and Ag(I)	i. Inferior stability of SCCH than RCCH during recycling	Qu, Sun, Fang et al. (2009)
2009	Chitosan–PVC[g]	Composite	Adsorptive removal of Cu(II) and Ni(II)	i. 87.9 mg/g for Cu(II) and 120.5 mg/g for Ni(II)	i. Recyclability aspect and binary adsorption have not been studied	Srinivasa et al. (2009)
2009	Chitosan–PVA	Blend	Adsorptive removal of Cd(II) fitted to Langmuir/Freundlich adsorption isotherm and pseudosecond-order kinetics model	i. Maximum AC of 73.75% at pH = 6	i. Poor desorption by EDTA (only 62.4%)	Kumar et al. (2009)

2009	Chitosan–perlite	Composite	Adsorptive removal of Cu(II), Co(II), and Ni(II) fitted to Lagergren first-order kinetics and Freundlich/Langmuir isotherm	i. Recyclable ii. Order of affinity based on AC = Cu(II) > Co(II) > Ni(II)	i. Adsorption of Ni(II) affects the most in the presence of other two	Kalyani et al. (2009)
2009	Chitosan–magnetite	Nanocomposite	Adsorptive removal of Cr(VI) well fitted to Langmuir isotherm	i. Uptake of Cr(VI) is 69.4 mg/g ii. It can be recycled	i. Magnetic properties have not been investigated	Huang et al. (2009)
2009	Chitosan–magnetite	Nanocomposite	Adsorptive removal of Pb(II), Cu(II), and Cd(II)	i. Can be used as a recyclable tool ii. Concentration of Pb(II) changed from 10 to 0.54 mg/L after 10 min of ultrasound radiation. The removal efficiency of lead is about 94.6% iii. Magnetic property marginally affected after chitosan coating	i. Relatively poor removal efficiencies for Cu(II) and Cd(II) as compared to Pb(II)	Liu et al. (2009)
2009	Chitosan–cotton fiber	Composite	Adsorptive removal of Au(III)	i. 100% selectivity for Au(III) in the presence of Ni(II), Cd(II), Zn(II), Co(II), and Mn(II) at optimum pH = 3.0	i. The nature of bonds in the adsorbents is not characterized	Qu, Sun, Wang et al. (2009)
2009	Chitosan–PU[h]	Composite (foam)	Adsorptive removal of acid violet 48 (λ_{max} = 551 and 591 nm) fitted to pseudo-second-order kinetic model	i. Maximum AC = 30 mg/g ii. Open cell-structure of the foam-type adsorbent helps adsorption and diffusion	i. Intensity at 551 nm has not ignored while determining dye concentration ii. AC substantially decreased above pH = 8.0. iii. Basic dyes as adsorbates have not been examined	Lee et al. (2009)
2010	Chitosan–magnetite	Nanocomposite	Adsorptive removals of Pb(II) and Ni(II) well fitted to Langmuir isotherm	i. Magnetic property marginally affected after chitosan coating ii. Maximum ACs for Pb(II) and Ni(II) at pH = 6 under room temperature are 63.33 and 52.55 mg/g, respectively iii. Heavy metal ion removal from water with the help of an external magnet	i. Relatively poorer removal efficiency for Ni(II) than that of Pb(II)	Tran et al. (2010)

Continued

Table 6.3 Continued

Year	Composition	Type#	Application(s)	Advantage(s)	Drawback(s)	References
2010	Chitosan–bentonite	Chitosan-coated composite beads	Adsorptive removal of tartrazine dye at $\lambda_{max} = 428.4$ nm well fitted to pseudo-second-order kinetics and Langmuir isotherm	i. Maximum AC = 294.1 mg/g at pH = 2.5 and 320 K	i. The absorbent is not suitable for recycling. The reason behind such a strong interaction between dye and adsorbent has not been reported	Wan Ngah, Ariff, & Hanafiah (2010)
2010	Chitosan–bentonite	Chitosan-coated composite beads	Adsorptive removal of malachite green well fitted to pseudo-second-order kinetics and Langmuir isotherm	i. The maximum AC = 435 mg/g at pH = 6	i. Poor adsorption at low pH ii. Not tested for acid dye removals	Wan Ngah, Ariff, Hashim, & Hanafiah (2010)
2010	Chitosan–kaolin–*g*-Fe_2O_3	Nanocomposite	Adsorptive removal of MO[i] at $\lambda_{max} = 464.9$ nm	i. The maximum dye removal is 82.3% at pH = 2.9	i. MO removal decreased significantly at pH > 7.1 ii. Not tested for basic dye removals	Zhu et al. (2010)
2012	Chitosan–Fe(II) complex	Composite	Adsorptive removal of thorium and uranyl ions	i. Recyclable adsorbent with maximum ACs of 312.50 and 666.67 mg/g for thorium and uranyl ions, respectively ii. The material may be regenerated and reused	i. Harmful radiation cannot be stopped from the adsorbed radio cations	Hritcu et al. (2012)
2012	Chitosan–MMT	Nanocomposite	Membrane for adsorptive removal of anionic azo dye bezactiv orange V-3R or remazol brilliant orange 3R ($\lambda_{max} = 493$ nm)	i. The optimum AC of dye at pH = 6 and initial concentration = 280 mg/g	i. Not tested for basic dyes	Nesic et al. (2012)
2013	Chitosan–cloisite15A (hydrogenated tallow modified MMT)-coated PVDF[j]	Nanocomposite	Thin composite membrane for removals of methylene blue and acid orange 7	i. Enhanced adsorptive removal of methylene blue in the presence of cloisite15A ii. Relatively greater intercalation of chitosan with 15A due to the higher interlayer-spacing	i. Inferior removal of acid orange 7 at the acidic pH	Daraei et al. (2013)
	Chitosan–cloisite30B (tallow modified MMT)-coated PVDF		Thin composite membrane for removals of methylene blue and acid orange 7	i. Enhanced adsorptive removal of methylene blue in the presence of cloisite30B ii. Acidic dye (acid orange 7) is well removed by cloisite 30B-chitosan-coated membrane at acidic pH	i. Relatively lesser intercalation of chitosan in cloisite30B due to the lower interlayer spacing	

2014	Chitosan–hydroxyapatite	Fibrous nanocomposite	Membrane for removal of lead, cobalt, and nickel ions fitted to pseudo-second-order and Langmuir models	i. The measured diameter of the composite nanofiber (198 nm) is consistent with the Box–Behnken design predicted diameter (200.6 nm) ii. Reusable up to five adsorption–desorption cycles iii. ACs for Pb(II), Ni(II), and Co(II) are 296.7, 213.8, and 180.2 mg/g, respectively	i. Binary adsorption is not performed, possible coordination of metal ions with the adsorbent has not been examined	Aliabadi et al. (2014)
2014	Chitosan–polyamide-6	Fibrous nanocomposite	Adsorptive removals of solophenyl red 3BL and polar yellow GN	i. 96% and 95% removal for solophenyl red 3BL and polar yellow GN, respectively ii. Marginal errors among the predicted and the experimental ACs	i. Removal efficiency for cationic dye has not been examined ii. Recyclability aspect has not been covered and AC deteriorated at the alkaline pH	Ghani et al. (2014)
2014	Chitosan–cellulose	Nanocomposite membrane	Adsorptive removal of cationic dyes	i. 98%, 84%, and 70% removals of victoria blue 2B, methyl violet 2B, and rhodamine 6G, respectively ii. Roles of ionic interaction and hydrogen bonding in adsorption have been reported iii. Fabricated from renewable resources	i. ACs toward anionic dyes have not been evaluated ii. Recyclability aspect has not been covered	Karim et al. (2014)
2015	Chitosan–cellulose (*Eichhornia crassipes*)-TiO_2[k]	Hybrid composite	Adsorptive removal of reactive black 5 well fitted to Langmuir isotherm and pseudo-second-order	i. The maximum AC of 0.606 mg/g at 25°C within pH = 2.5–3.0 ii. Natural biopolymer composite	i. Not suitable for cationic dye removal ii. Recyclability not reported iii. AC decreased in alkaline pH	El-Zawahry et al. (2015)
2015	Chitosan–polyrhodanine	Nanocomposite	Adsorptive removal of Ni(II) well fitted to Langmuir/Sips isotherm and pseudo-second-order	i. The maximum AC of 67.70 mg/g at pH = 8.0	i. Adsorptive removals of other metal ions have not been tested ii. Recyclability aspect has not been covered iii. Adsorbent–adsorbate bonding has not been studied	Amiri et al. (2015)

Continued

Table 6.3 Continued

Year	Composition	Type#	Application(s)	Advantage(s)	Drawback(s)	References
2015	CNTs–chitosan-PVA[1]	Nanocomposite	Membrane for removal of PAH[m]s	i. Environment-friendly green material	i. Costly ii. Recyclability aspect has not been covered	Bibi et al. (2015)
2015	Triethylenetetramine modified graphene oxide/chitosan	Composite	Adsorptive removal of Cr(VI)	–	–	Ge and Ma (2015)
2016	Chitosan–alunite	Composite	Adsorptive removal of acid red 1 and reactive red 2	i. The maximum ACs of 462.74 and 588.75 mg/g for acid red 1 and reactive red 2, respectively ii. Remarkable recyclability up to 20 cycles	i. Not tested for cationic dyes	Akar et al. (2016)
2016	Chitosan–polymethylmethacrylate	Composite membrane	Adsorptive removal of Cr(VI) fitted to Langmuir model and pseudo-second-order kinetics model	i. Maximum AC of 92.5 mg/g at pH = 3.0 ii. Adsorption of Cr(VI) has been mediated by $-NH_2$ groups iii. Adsorbent can be reused three times	i. Recyclability and AC are insufficient ii. Binding of metal ion in metal-loaded adsorbent has not been studied	Li et al. (2015)
2016	Chitosan–cellulose	Composite	Adsorptive removal of methylene blue (λ_{max} = 663 nm) and anionic dye new coccine (λ_{max} = 508 nm)	i. Maximum ACs at pH = 4.0 for methylene blue (0.067 mmol/g) and new coccine (0.17 mmol/g) at pH = 2.0 ii. Adsorbent can be reused for three times iii. Prepared from renewable resources	i. Maximum AC for cationic dye is low ii. Recyclability and ACs are insufficient	Liu et al. (2016)
2017	Chitosan–cellulose	Composite	Adsorptive removal of Cd(II) well fitted to Langmuir isotherm and pseudo-second-order kinetic model	i. Prepared from renewable resources ii. Maximum AC of 0.995 mg/g at pH = 6 iii. Adsorbent can be regenerated	i. The maximum AC is low.	Lelifajri et al. (2017)
2017	Chitosan-*Lemna gibba*	Composite	Adsorptive removal of boron	i. Prepared from renewable resources ii. Maximum AC of 3.9 mg/g at pH = 7.0 iii. Boron-loaded adsorbent has been characterized	i. Recyclability not studied	Türker and Baran (2017)
2017	Carboxymethyl chitosan–PVDF	Composite membrane	Adsorptive removal of humic acid	i. Fouling resistant due to reduced interfacial roughness in the composite ii. Reusable up to two times or more	i. Not completely made from renewable resources ii. Not studied for foulants other than humic acid	Ekambaram and Doraisamy (2017)

2017.	Chitosan–polymethacrylic acid–halloysite nanotube	Nanocomposite	Adsorptive removals of Pb(II) and Cd(II) best fitted to Freundlich and Langmuir-Freundlich models for single and binary ion adsorption, respectively	i. Maximum ACs = 357.38 and 341.62 mg/g for Pb(II) and Cd(II), respectively, resulting in maximum AC = 699.00 mg/g ii. Recyclable up to five cycles	i. Antagonistic effect while adsorbing from binary mixture of Pb(II) and Cd(II) (i.e., the maximum AC reduces to 622.92 mg/g)	Maity and Ray (2018)
2017	Graphene oxide–chitosan	Composite	Adsorptive removal of safranin O (λ_{max} = 520 nm) well fitted to pseudo-first-order and Redlich Peterson models	i. Maximum adsorption at pH = 6.5	i. Acid dyes not tested, adsorption deteriorates at pH > 8.0 ii. Recyclability not reported	Debnath et al. (2017)
2018	Tetraethyl orthosilicate cross-linked chitosan–polytetrafluoroethylene	Composite membrane	Adsorptive removals of methanol and toluene	i. Restricted membrane swelling and improved separation factors ii. Adequate chemical stability iii. Selective permeability for methanol due to hydrogen bonding of methanol with O–H and $-NH_2$ along with smaller kinetic diameter of methanol than toluene	i. Excessive cross-linking by tetraethyl orthosilicate reduces flux	Moulik et al. (2018)
2018	Chitosan–β-cyclodextrin	Composite	Adsorptive removal of methyl orange	i. Maximum AC of 392 mg/g ii. High selectivity for anionic dye (e.g., methyl orange)	i. Selectivity has been adjudged only by comparing methyl orange with cationic dyes, i.e., methylene blue and rhodamine B	Jiang et al. (2017)
2018	Cu(II) imprinted chitosan–zeolite	Cryocomposites	Selective removal of Cu(II)	i. Highly selective for Cu(II) removal in the presence of competing ions, i.e., Co(II), Ni(II), Zn(II), and Pb(II) ii. Recyclable iii. Higher selectivity arises from the specific imprinting sites created by Cu(II) and the immobilized frozen composite matrix	i. Preparation is complicated	Dinu et al. (2018)
2018	Chitosan–zeolite	Composite	Adsorption and chemical fixation of CO_2^n	i. Conversion of CO_2 gas into cyclic carbonates	i. The adsorbent appears to be for one-time usage	Kumar et al. (2018)

Continued

Table 6.3 Continued

Year	Composition	Type#	Application(s)	Advantage(s)	Drawback(s)	References
2018	Chitosan–waste coffee-grounds–PVA	Hybrid composite	Adsorptive removals of metamizol, acetylsalicylic acid, acetaminophen, and caffeine	i. Utilization of solid waste in separating contaminants of wastewater ii. The used material is reusable, cheap, and environment friendly	i. Recyclability needs further improvement	Lessa et al. (2018)
2018	Polyethylene glycol diglycidyl ether cross-linked chitosan–activated carbon	Composite	Adsorptive removal of Cd(II) well fitted to Langmuir isotherm	i. Maximum AC of 357.14 mg/g ii. Adsorbent is recyclable up to three times	i. Unable to adsorb Cd(II) after fourth regeneration ii. Activated carbon generation needs pyrolysis, which is not environment friendly and consumes energy	Rahmi and Nurfatimah (2018)
2018	Magnetic chitosan–graphene oxide	Composite	Adsorptive removals of phenylurea herbicides (i.e., monuron, linuron, and isoproturon)	i. Maximum ACs of 35.71, 33.33, and 29.41 mg/g for monuron, linuron, and isoproturon, respectively, at room temperature and pH = 5	i. Recyclability not reported	Shah and Tasmia (2018)
2019	$CoFe_2O_4$–chitosan	Magnetic composite	Indigotine blue dye	i. Maximum AC of 380.88 mg/g at 328 K ii. Faster removal in the presence of magnetic field iii. Reused up to four adsorption–desorption cycles	i. Not tested for cationic dyes ii. Recyclability should be improved	dos Santos et al. (2019)
2019	Magadiite–chitosan	Composite	Adsorptive removals of Congo red and methylene blue well fitted to Langmuir isotherm and pseudo-second-order model	i. Maximum ACs are 135.77 and 45.25 mg/g for Congo red and methylene blue, respectively ii. Adsorption mechanism has been studied by characterizing the loaded adsorbent	i. Low AC especially for cationic dye ii. Recyclability not reported	Mokhtar et al. (2020)
2019	Chitosan–polyacrylate–graphene oxide	Composite hydrogel	Simultaneous adsorptive removal of cationic and anionic dyes (i.e., methylene blue and food yellow 3)	i. Polyampholyte hydrogel suitable for both anionic and cationic dye removal ii. Equilibrium ACs are 296.5 ± 31.7 and 280.3 ± 23.9 mg/g for methylene blue and food yellow 3, respectively iii. Recyclable up to five cycles	i. ACs should be improved further	Chang et al. (2019)

2019	Genipin–cross-linked chitosan–graphene oxide–SO_3H	Composite	Adsorptive removal of ibuprofen and tetracycline fitted to pseudo-second-order kinetics and Freundlich isotherm models	i. Maximum ACs for ibuprofen/tetracycline varied from 113.27/473.25 at 298 K to 138.16/556.28 mg/g at 313 K ii. Magnetic field-assisted fast adsorption iii. Microporous structure with ultralarge surface area iv. At least 85% of the maximum AC retained after five adsorption–desorption cycles	i. Relatively inferior performance while adsorbing ibuprofen	Liu, Liu et al. (2019)
2019	Chitosan–Fe-Al-Mn metal oxyhydroxides	Composite	i. Highly efficient fluoride scavenger ii. Well fitted to pseudo-second-order kinetic and Freundlich isotherm models	i. Maximum AC of 55 ± 0.5 mg/g ii. Solid waste (slag) has been used for removing fluoride of effluent	i. Recyclability not reported	Chaydhury et al. (2019)
2019	Chitosan–FeS	Composite	Adsorptive removal of Cr(VI) well fitted tc pseudo-second-order kinetics and Redlich–Peterson isotherm	i. Adsorption-cum-reduction abilities of the composite adsorbent by combining adsorption capacity of chitosan and the reducibility of FeS ii. By adjusting the ratio of chitosan and FeS, the absorption and reduction ability can be tuned	i. Optimization still required to achieve better dual performance of the composite	Zhang et al. (2019)

[a] adsorption capacity
[b] polyvinyl alcohol
[c] ethylene diamine tetra-acetic acid
[d] montmorillonite
[e] chitosan coated cotton fibre
[f] chitosan coated cotton fibre
[g] polyvinyl chloride
[h] polyurethane
[i] methyl orange
[j] poly vinylidene fluoride
[k] titanium dioxide
[l] poly vinyl alcohol
[m] poly aromatic hydrocarbon
[n] carbon dioxide

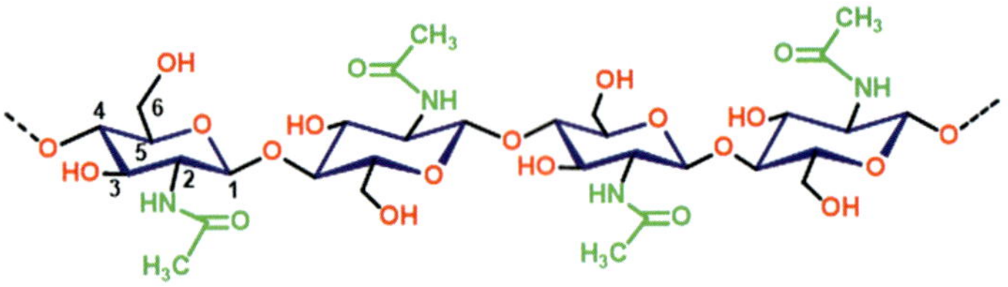

Figure 6.2 Residue of chitosan containing one amino group located in the C2 position and two hydroxyl groups at C3 and C6.

of deacetylation. It can be observed that each glucoside residue of chitosan contains three major reactive sites: one amino group located in C2 and two hydroxyl groups at C3 and C6 (Fig. 6.2). Among the two hydroxyl groups, the hydroxyl group placed at C6 is a primary hydroxyl and hence it is more reactive than that of the secondary hydroxyl group at C3.

2.2 Properties

The main parameters influencing the physical and chemical properties of chitosan are its molecular weight and degree of deacetylation (Shameem et al., 2008). In fact, the degree of deacetylation under a strongly alkaline environment expresses the extent of deacetylation of amides of chitin to amines in chitosan. The choice of chitin and its isolation process also influence the quality of chitosan. In fact, there are three different forms of chitin: α, β, and γ. In this regard, a fully deacetylated chitosan shows maximum crystallinity due to better packing of uniformly arranged macromolecular chains interacting mutually by the maximum extent of hydrogen bonding. Additionally, amino and hydroxyl groups contribute significantly in determining the solubility, biological, and physiochemical properties. Both amine and hydroxyl groups can function as a hydrogen bond donor and acceptor, and thus, chitosan possesses considerable hydrogen bonding at the neutral pH, making it less soluble in water. In addition, hydrogen bonding is responsible for its superior viscosity and film-forming capacity. However, it becomes readily soluble in dilute acidic solutions below pH = 6.0 due to the protonation of amines, as the primary amino groups of chitosan has a pK_a of 6.3. Importantly, the pK_a value of chitosan may vary within 6.0−6.5, and the value is highly dependent on the degree of N-deacetylation. Thus, the solubility and viscosity of chitosan are dependent on the degree of N-deacetylation. The pH-sensitive amino groups of chitosan are responsible for the cationic nature of chitosan in an acidic solution. In fact, the dissolution of chitosan in acetic acid is a reaction between the amine group from chitosan and the acetic acid residue ($NH_3^+Ac^-$) (Nesic et al., 2012). Moreover, both amino and hydroxyl groups are susceptible to physical and chemical modifications. Being a primary aliphatic polyamine, chitosan presents all types of characteristic reactions of amines (Darder et al., 2003). The $-NH_2$ groups are the major metal-chelating sites, and hence, pure chitosan is capable of adsorbing various heavy metal ions. Being a pH-sensitive polyamine, chitosan can effectively function as a polyelectrolyte. In acidic pH, polycationic chitosan has the ability to attach with negative residues at cell surfaces via electrostatic forces, presenting antifungal or antimicrobial activities. Such attachment increases in the presence of more frequently available $-NH_2$ groups. Consequently, cell adhesion and proliferation indirectly depend on the degree of deacetylation. Thus, other biological properties, such as wound healing and osteogenesis enhancement, biodegradation by lysozyme, analgesic, antitumor, hemostatic, hypocholesterolemic, antimicrobial, and antioxidant properties, are affected by the physical properties related to degree of deacetylation, crystallinity, and viscosity of chitosan. Indeed, either physical or chemical modifications of chitosan are executed to improve the pore size, mechanical strength, chemical stability, hydrophilicity, and biocompatibility. The structural integrity, stability, mechanical strength, and chemical resistance of chitosan can be extended through cross-linking by various agents including glutaraldehyde, epichlorohydrin, and ethylene glycol diglycidyl ether (Wan Ngah et al., 2004).

2.3 Applications

The diverse properties of chitosan make it suitable in various biomedical fields including tissue engineering, drug/gene delivery, and wound healing due to its biocompatibility, cytocompatibility, biodegradability, and bioresorbability (Hein et al., 2008). Moreover, its inherent polyelectrolytic nature, chelating ability of the amine group of the macromolecules (Darder et al., 2003), and variable hydrogen bonding ability through O$-$H and $-NH_2$ functionalities are explored and exploited in the fields of dye/metal ion adsorption and food packaging, electrode, and sensing applications.

Indeed, the application prospects of chitosan further enlarge when chitosan is combined with other suitable organic/inorganic components to produce blends and composites showing better overall physical properties and chemical resistance compared to that of pure chitosan. The major drawbacks of pure chitosan including poor mechanical properties, conductivity, and durability can be successfully maneuvered by proper fabrication and structure$-$property$-$performance optimization of blends and composites made from chitosan as a major component. For instance, polyelectrolytes, such as alginate, can be blended with chitosan to overcome the drawbacks of both alginate and chitosan by combining the good characteristics of both polymers. In fact, pure alginate, a polysaccharide biopolymer composed of blocks of (1 $\rightarrow$ 4) linked α-L-gluronic acid and β-D-mannuronic acid, is rich in carboxylate functionalities. Thus, alginate has a very good affinity for metal ions, however, it is not so popular in wastewater remediation due to its mechanical fragility.

Therefore, this chapter deals with the various application prospects and potentials of chitosan-based blends and composites, wherein the inherent positive aspects of chitosan remain unaltered and sometimes improved further while minimizing the negative aspects of the same biopolymer.

3. Classification of chitosan-based blends and composites

In general, all the chitosan-based blends and composites can be classified based on the application possibilities and material composition.

3.1 Classification based on the applications

Based on the application perspectives, chitosan-based blends or composites can be classified according to Chart 6.1.

3.1.1 Applications of chitosan-based blends and composites in tissue engineering

In medical science, often body parts including tissues and organs encounter massive damage and hence repair or replacement of those affected parts is of utmost necessity. To date, the clinical approach or solution to tissue and organ replacement is to follow gold standard therapies, that is, autografts or allografts. However, such an approach suffers from several constraints including limited availability of donor(s), risk of disease transmission, rejection at the graft site due to marginal lack of compatibility, morbidity of the donor as well as the patient, and the enormous cost involved in the entire treatment procedure. In the search for an alternative, the tissue engineering approach has been proven as an effective technique, by which tissues and organs are regenerated with the help of a scaffold matrix with/without cells and biological

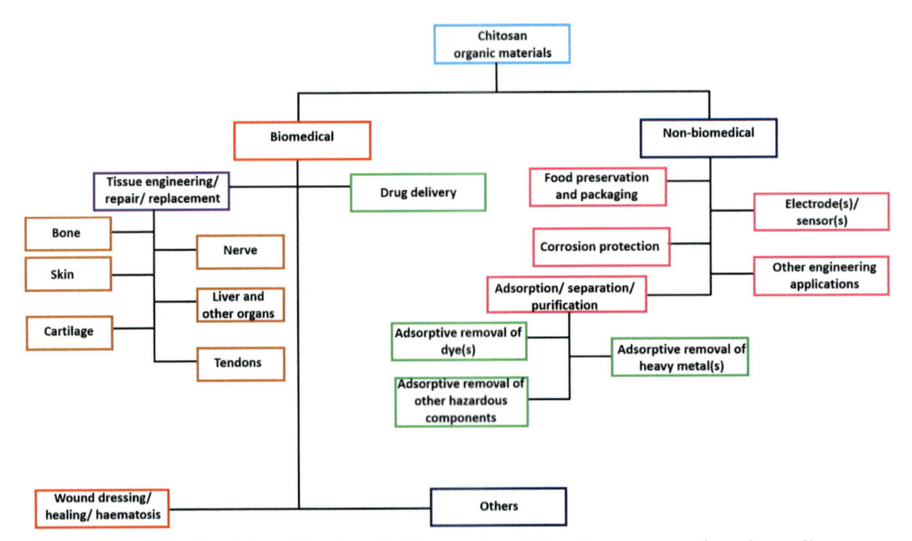

Chart 6.1 Generalized classification of chitosan-based blends or composites depending on the applications.

cues. Indeed, materials based on synthetic/natural polymers, blends, and composites have been used to develop suitable scaffold matrix and other components necessary for the regeneration of tissues and organs. Nevertheless, research is still in progress to discover the best-functioning materials among those diverse materials comprising natural and synthetic polymers and their blends and composites. Among those blends and composites, chitosan-based materials have shown substantial application prospects in the biomedical field (Table 6.1). In fact, pure chitosan matrices possess some inherent advantages including biodegradability, biocompatibility, ease of processing, flexibility, and porosity (Martino et al., 2005). However, the major drawbacks to these materials are their weak mechanical strength and poor stability. Therefore, blends and composites of chitosan have been preferred over pristine chitosan for preparing mechanically strengthened and stable scaffold matrices and other components necessary for the regeneration of tissues and organs. The fabricated materials are tuned by adopting various starting materials and procedures depending on the requirement for replacing tissues and organs, such as bone, ligament, cartilage, tendon, liver, nerve, and skin.

3.1.1.1 Bone tissue engineering

The major objective of bone tissue engineering is to regenerate natural human bone by means of an artificial approach. Bone is an important, high-strength, rigid, highly specialized, and well-organized human body component comprised of dense connective tissues. In regard to its composition, human bone includes organic components, such as collagen, growth factors, and noncollagenous proteins, together with the major inorganic component of carbonated hydroxyapatite. Following the basic criterion of tissue engineering, artificial bones have been regenerated by developing materials suitable for high-performance scaffold materials based on blends and composites of chitosan. Because of its osteoconductive property, chitosan has drawn substantial attention from various research groups. In fact, osteoconduction is broadly defined as the ability to grow bone on the surface of implanted material or scaffold. Successful osteoconduction requires the desired material surface, such as chitosan, to facilitate osteoblast lineage cell adherence, proliferation, and osteogenesis. In this context, an osteogenic approach has been proven to be the most effective methodology, by which bones are constructed by seeding the osteoblast cells onto a suitable matrix composed of both organic and inorganic components.

For preparing the blends of chitosan and other polymeric component(s), the following objectives have been partially or completely fulfilled by various research groups: (1) to strengthen the inherently weaker mechanical properties of chitosan by combining it with suitable natural or synthetic polymers through different fabrication methodologies, such as high-temperature sintering, melt-stretching, and multilayer deposition, (2) to maintain the porous structure of chitosan within the scaffold despite blending with other component(s), (3) to achieve enhanced osteoconductivity via precalcification or through growth factors, (4) to improve stability via restricting degradation of components, and (5) to accelerate bone formation in vivo.

By its nature, pure chitosan is mechanically weak and hence, to enhance its mechanical strength, calcium phosphate (Xu et al., 2007) and calcium carbonate have been incorporated into chitosan to produce bioresorbable composites for orthopedic implants. In the last century, researchers attempted to fabricate artificial bone substitutes for dental treatment by admixing hydroxyapatite with chitosan (Ito, 1991; Qiaoling et al., 2004). In this regard, a few years back, researchers developed a chitosan–alginate hybrid porous scaffold of desired mechanical strength that possessed similar ability to human bones in withstanding sufficient compressive forces (Li, Ramay et al., 2005). Herein, the enhanced mechanical strength of the chitosan–alginate scaffold is related to the strong ionic interactions between $-NH_3^+$ of chitosan and $-COO^-$ of alginate (Fig. 6.3) (Wan Ngah & Fatinathan, 2008). Similar ionic interaction-driven enhanced mechanical strength is also observed when $-COO^-$ bearing poly(lactic acid-co-glycolic acid) is admixed with chitosan (Abdel-Fattah et al., 2007; Jiang et al., 2006). Here, high-strength scaffolds are prepared by a sintering method. It has been observed that the mechanical strength can be increased further at the higher sintering temperature. However, porosity and pore volume are deteriorated when sintering is done at the higher temperature. In addition to the higher mechanical strength and porosity, other essential attributes, such as rapid in vivo bone-forming ability and reduced degradation of the fabricated scaffold, have been achieved by introducing heparin and recombinant human bone morphogenetic protein-2 (Jiang et al., 2010). Herein, heparin and recombinant protein synergistically participated in stimulating a rapid bone formation process through enhanced osteoblastic proliferation and differentiation, connective tissue formation, and mineralization. Inspired by the pivotal role played by collagen in the structural integrity of bones, a group of researchers have developed a chitosan–collagen scaffold loaded with microspheres constituted of recombinant human bone morphogenetic protein-2 and poly(lactic acid-co-glycolic acid) to accelerate the bone formation process (Shi et al., 2009). Likewise, different strategies have been adopted by various research groups to achieve near perfect bone substitutes through scaffold fabrication by combining chitosan with biopolymers, such as polycaprolactones (Im et al., 2003), peptides (Tsai et al., 2012), poly(butylene succinate) (Costa-Pinto et al., 2009), poly(L-lactic acid) (Mano et al., 2008), and other vital ingredients, such as, platelet-derived growth factor (Im et al., 2003) and recombinant proteins, necessary for stimulating rapid bone formation.

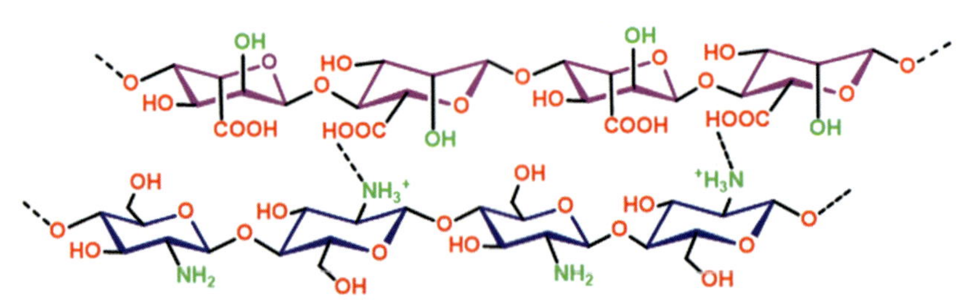

Figure 6.3 Ionic interactions between chitosan and alginate.

Since carbonated hydroxyapatite is the major inorganic component of natural bones, another group of workers have concentrated on developing a mimic of this basic inorganic component. Their developments include bioactive ceramics, such as bioactive glass, hydroxyapatite [$Ca_{10}(PO_4)_6(OH)_2$], and other calcium phosphates, calcium sulfates, nanosilica, zirconia, and titania, which are suitable for inorganic implants. Indeed, some of these inorganic materials have demonstrated osseointegration properties by developing structural and functional connections via direct bond formation to both hard and soft tissues of living bones. Thus, these ceramic materials possess numerous advantages, such as strengthening the polymeric substrates; assisting and influencing the biomineralization, cellular adhesion, proliferation, differentiation, gene expression, vascularization processes; and imparting antibacterial activity. At the same time, these ceramic materials are brittle, and thus, polymer–ceramic composite materials are now used more frequently to exploit the combined advantages of both polymer and ceramics. Thus, hydroxyapatite/nanohydroxyapatite have been used as the ceramic component and chitosan as the biopolymeric ingredient to prepare composite scaffolds for bone tissue engineering applications (Madhumathi et al., 2009). In fact, there are numerous reports on combinations of chitosan with hydroxyapatite and other polymeric or inorganic materials, such as polylactic acid (Cai et al., 2009), silk fibroin (Wang & Li, 2007), carboxymethyl cellulose (Jiang et al., 2008) carbon nanotubes (Venkatesan et al., 2011), gelatin (Zhao et al., 2002), and iron oxide (Bhattaraia et al., 2007), with enhanced mechanical properties. Though hydroxyapatite is the basic inorganic component of bones and teeth, there are several disadvantages to hydroxyapatite. For instance, hydroxyapatite has a slow rate of resorption and is preferable for bone augmentation rather than bone regeneration. Moreover, hydroxyapatite binds directly with hard tissues. To overcome such limitations of hydroxyapatite, some researchers have developed bioactive glass ceramic or bioglass capable of attaching to both hard and soft tissues by forming a hydroxycarbonate apatite layer. Notably, compared to hydroxyapatite, hydroxycarbonate apatite layers make the bioactive glass ceramic or bioglass composition more similar to the carbonated hydroxyapatite component of natural bone. Additionally, bioactive glass ceramics have numerous advantages and variability. These materials have the ability to stimulate angiogenesis and can regulate osteoblast-specific genes and their differentiation without any osteoinductive factors, such as proteins or growth factors. Moreover, the binding abilities of hydroxycarbonate apatite with soft and hard tissues depend upon the rate of formation of the hydroxycarbonate apatite layer. While preparing the class A, that is, osteoproductive/osteoinductive bioactive glass ceramic, the interim hydroxycarbonate apatite layers form within hours and hence are capable of binding with both soft and hard tissues, whereas a much slower forming hydroxycarbonate apatite layer of class B, that is, osteoconductive bioactive glass ceramic, can bind only with hard tissues. Indeed, an osteoproductive approach is better than an osteoconductive approach, as an osteoproductive approach is capable of integrating bone growth within the complex contour of an implant surface in addition to the bone growth at the surface.

Despite several advantages, these bioactive ceramics are not free from defects. Certain limitations, including mechanical weakness and low fracture resistance, confine their applications mostly within the nonload-bearing regions. Again, such shortcomings of pristine bioactive glass ceramic can be maneuvered by replacing them with some better alternatives, such as suitable polymer composite or nanocomposites comprising bioactive glass ceramics as a reinforcing filler (Shalumon et al., 2013; Peter, Binulal, Soumya et al., 2010; Mansur & Costa, 2008; Peter, Binulal, Nair et al., 2010; Caridade et al., 2013). Moreover, researchers have examined the potentials of other inorganic materials, such as silica, titania, and zirconia, in fabricating scaffolds of composites, such as, chitosan-poly(ethylene oxide)/silica nanofiber, hydroxyapatite/chitosan–silica nanocomposite, hydroxyapatite–titania/chitosan–gelatin composite scaffold, and chitin–chitosan/nanozirconia composite scaffold, suitable for augmentation or regeneration of artificial bones. In bone tissue engineering, an osteogenic method for artificial bone construction, i.e., cell seeded constructs, has been established as the most popular method to repair and regenerate natural bone. However, some unavoidable time-bound situations limit the application of this relatively slow osteogenic method. In those emergency cases, other alternatives, such as osteoinductive or osteoconductive constructs, may be attempted for repairing or regenerating native bones.

3.1.1.2 Cartilage tissue engineering

Bones are intimately associated with a resilient, smooth, viscoelastic, relatively less rigid connective tissues known as cartilage. Cartilage is mostly located at the joints, rib cage, ears, nose, bronchial tubes, intervertebral discs, and these structural components function as a protecting layer of the long bone ends, especially at the load-bearing joints, such as knees and hips. Cartilage is composed of two basic components, i.e., specialized cells known as chondrocytes and extracellular matrix. These chondrocytes secrete the extracellular matrix composed of glycosaminoglycans (GAGs), proteoglycans, collagen fibers, and elastin, and the chondrocytes remain embedded in the matrix. The main constraint of cartilage is that it shows very slow turnover of its extracellular matrix and it cannot be repaired. Therefore, unlike other connective tissues including bones, cartilage cannot recover to its original state once it is damaged. Therefore, if damage occurs, self-healing is almost impossible, and hence, the patient requires either conventional surgical intervention or replacement and repair by recently developed artificial cartilage made from cellular scaffolding materials. To avoid difficulties in currently available treatment modalities and surgical intervention, cartilage tissue engineering is a subject of growing interest covering the methodology for creating a completely healed, functional, and scarless tissue. For preparing a successful cartilage tissue-engineered construct, an essential requirement is the ability of chondrocytes to interact, differentiate, and remain functional in the artificial matrix constructed by suitable blends or composites. Among the matrix components, GAGs and hyaluronic acid play active roles in cartilage regeneration and chondrogenesis. Because of the structural similarity of chitosan to GAGs and hyaluronic acid, chitosan-based blends and composites have the potential for constructing an artificial matrix of cartilage. To ensure sufficient growth and interconnectivity of chondrocytes, scientists have strategically selected porous scaffolds so that the chondrocytes penetrate and infiltrate through the pores of the scaffold. However, scaffolds developed initially are microporous and hence the movement, infiltration, and interconnectivity among chondrocytes remain restricted spatially. Because of the lack of sufficient spaces in micropores, the deposited extracellular matrix mostly remains confined at the edges of the scaffold (Nettles et al., 2002). Later, the pore size of chitosan-based scaffolds was increased to the macro-level. These newly employed sponge-type materials with large interconnected pores have facilitated easier movement, distribution, and connectivity among the seeded chondrocytes to develop a relatively homogeneous cell matrix (Griffon et al., 2006). For accelerating the growth of tissue via controlled supply of growth factor, chitosan microsphere-loaded growth factor, i.e., transforming growth factor-β, has been incorporated into chitosan-based scaffold (Lee et al., 2004). Similar to chitosan microspheres, polyelectrolyte complex spheres have been also noted to function as a reservoir of growth factor-accelerating differentiation and activities of chondrocytes. Since collagen is another basic component of natural cartilage matrix, collagen-combined chitosan has been cross-linked with genipin to produce a macroporous scaffold giving longer viability for rabbit chondrocytes. Because of the presence of larger pores, chondrocytes can easily infiltrate and spread through the matrix, however, the same chondrogenic cells are unable to differentiate into chondroblasts (Yan et al., 2010). Some researchers have developed artificial extracellular matrix in the form of injectable hydrogels of chitosan and chitosan-based composites bearing extracellular matrix mimicking polymers, such as chondroitin sulfate and hyaluronic acid (Tan et al., 2009; Park et al., 2013). In this context, to improve the integrity of the artificial extracellular matrix, they have elevated the extent of cross-linking by suitable chemical treatment. Finally, this composition has demonstrated excellent regeneration capacity of cartilage, along with better infiltration of chondrocytes and wound-healing rate. Attempts have been made to improve the cartilage regeneration capacity of those injectable hydrogels by altering the gelation time through variations in the average molecular weight of polymers and initiator concentrations. The stability of the chitosan-based hydrogels has been improved by chemical modification through photo-polymerization under UV light in the presence of a riboflavin initiator. Later, chitosan-based semi-IPN-type hydrogels bearing hyaluronic acid were synthesized as artificial extracellular matrix, which has shown 0% viability of chondrocytes attaining differentiated morphology, leading to enhanced proliferation of auricular chondrocytes in the lacunae of scaffolds. Since natural cartilage is a viscoelastic material, attempts have been made to endow a viscoelastic property in artificial cartilage to act as a shock absorber at load-bearing joints. Recently, researchers have been trying to develop an ideal cartilage-regenerating scaffold equipped with the spatial structure and porosity of a 3D porous construct in combination with the characteristic mechanical strength of a fibrous mesh. Thus, for designing a high-performance artificial cartilage, special structural attributes, such as macroporosity, strength, long-term stability, and viscoelastic property should be incorporated so as to achieve the desired performance parameters, such as viability, differentiation, proliferation, and distribution of chondrocytes/chondroblasts within the scaffold matrix.

3.1.1.3 Ligament and tendon tissue engineering

In general, a ligament is the fibrous connective tissue that connects bones to other bones. Similarly, a tendon is another kind of tough fibrous connective tissue that connects muscle to bones. In regard to the constitution, both tendons and ligaments are made of collagen. The main cellular component of tendons is tenocytes, which are a kind of fibroblast synthesizing the extracellular matrix of tendons. Almost 60%–85% of the extracellular matrix is built of collagen, and the rest is made of noncollagenous components including elastin, proteoglycans, and a very small quantity of inorganic components. In fact, tenocytes synthesize the collagen macromolecules. In the case of ligaments, an essential role is to limit the mobility of articulations or prevent certain movements at the joints. Similar to cartilages, ligaments are viscoelastic. Therefore, an ideal tissue-engineered construct acting as a replacement for a natural tendon or ligament should be composed of collagenous or collagen-like fiber bundles and architecture (Mondal & Chattopadhyay, 2017; Singha, Roy et al., 2019), along with other basic ingredients functioning as self-regeneration catalysts, such as stem cells and growth factors. More importantly, the construct should provide sufficient strength and support comparable to those imparted by native tendon tissue. Thus, attaining high tensile strength is the major criterion for an artificial tendon. Additionally, the artificial matrix should have the desired cell adhesion so that the tenocytes remain attached and embedded in the matrix. It is also desirable that the embedded tenocytes can differentiate, distribute, and proliferate in the surrounding environment provided by the artificial components (Fig. 6.4). In search of such ideal replacements, a group of researchers have observed the superior cell attachment features of a polyelectrolyte complex composed of alginate and chitosan compared to only alginate fibers (Majima et al., 2005). Moreover, hyaluronic acid treatment on chitosan has led to elevated mechanical strength along with retaining the cell attachment features (Funakosi et al., 2005). Studies have also been conducted on the functioning of fibroblast cells embedded in the artificial extracellular matrix composed of chitosan as a major component. The

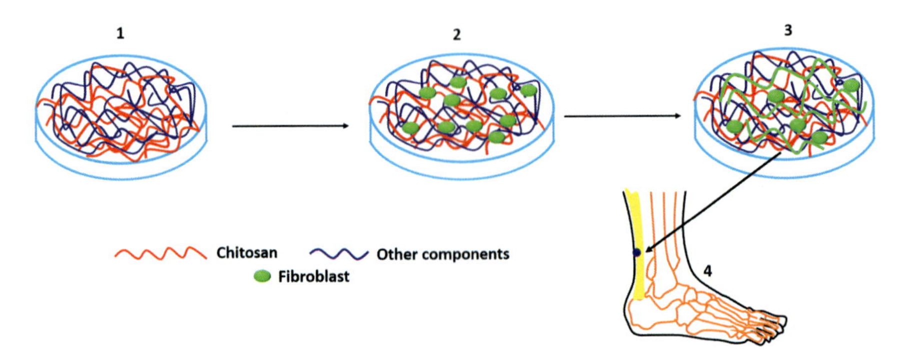

Figure 6.4 Repairing Achilles tendon rupture by tendon tissue engineering: (A) fabrication of artificial construct made of chitosan blend/composite, (B) seeding of fibroblast, (C) differentiation, growth, and proliferation of fibroblast and synthesis of collagen, and (D) implementation at the injury site (Achilles' tendon).

embedded fibroblasts produced the desired level of collagen, leading to improved mechanical strength of the regenerated ligament constituted of collagen and chitosan (Funakoshi et al., 2005). However, faster degradation of chitosan compared to the rate of tissue regeneration is another problem relating to the sustainability of the repaired tendon. Similar problems have been encountered when ligament fibroblasts have been cultured on polycaprolactone and chitosan surfaces, however, collagen synthesis and deposition have been improved markedly due to better expression of transforming growth factor-β1, the gene responsible for wound healing, and collagen deposition (Shao et al., 2010). In addition, chitosan has been employed for the guided growth of tenocytes along with the desired length of aligned collagen fibers in the regenerated tendons. Thus, present research activities are focused on developing near-identical substitutes for tendon and ligament of sufficient strength and durability by producing chitosan-containing collagenous blends or composites as extracellular matrices equipped with essential criteria, such as fibroblast attachment capability; differentiation, guided growth, distribution, and proliferation of the attached fibroblasts; growth factor-mediated adequate synthesis of sufficiently long collagen fibers; and the desired architecture facilitating proper movement at the bone–bone and bone–muscle interfaces.

3.1.1.4 Skin tissue engineering

Skin, the flexible outer covering, has three major functions, being protection, regulation, and sensation. Mammalian skin basically is composed of two layers: epidermis and dermis. Epidermis, the protective barrier over the body's surface, is composed mainly of keratinocytes. These keratinocytes secrete keratin proteins and lipids to form an extracellular matrix giving mechanical strength to the skin. On the other hand, the dermis is the lower layer of epidermis, and this layer is composed of collagen fibrils, microfibrils, and elastic fibers, embedded in hyaluronan and proteoglycans. Dermis imparts strength and elasticity through an extracellular matrix composed mainly of collagenous materials. These collagen fibrils are generated by the fibroblasts residing at the subcutaneous/hypodermal layer lying below the dermis layer. In addition, there are different types of structural components, such as nerve endings, glands, growth factors, etc., distributed in different strata of the skin. Thus, mammalian skin is constituted of a number of layers and sublayers, and these layers have their own compositions and functions. Therefore, skin tissue engineering is a challenging task when trying to provide all of these attributes of natural skin to artificial skin in order to replace or repair damaged natural skin.

Chitosan is widely used as a component for tissue engineering to mediate fabrication of skin substitute materials because of its hemostasis properties and the ability to stimulate collagen synthesis from fibroblasts. Additionally, chitosan has the potential to enhance the wound-healing rate by actuating rapid infiltration of polymorphonuclear cells at the injury site. To improve the stability aspect of chitosan, chitosan–alginate polyelectrolyte complex-based membranes have been fabricated (Yan et al., 2000), which have shown good stability at different pH, and have demonstrated faster healing of incisional wounds in rats (Wang et al., 2002). Since collagen is one of the major components of skin, porous biocompatible scaffolds have been prepared based on the combination of chitosan and collagen. Herein, the porous nature of the scaffold has facilitated infiltration of the surrounding fibroblasts to the bulk of the scaffolds (Ma et al., 2003). Later, collagen was combined with chitosan and silicone to produce microporous composite membranes for treating burns (Guo et al., 2011). The plasmid DNA incorporated within this membrane can express and translate vascular endothelial growth factor 165 necessary for quicker recovery from wounds. Moreover, the viability of some endothelial cells necessary for skin regeneration at the wound site has improved to some extent. In this context, a composite gel developed through the combination of atelocollagen with chitosan has demonstrated better wound healing at excisional wound sites (Judith et al., 2012). This composite gel has shown most of the necessary attributes of an ideal artificial wound-healing gel, such as enhanced cell attachment, proliferation, migration, and collagen deposition. In addition, the composite gel-treated wounds have shown higher levels of platelet-derived growth factor and nerve growth factor (NGF) connected to an elevated concentration of antioxidant and decreased peroxidation of lipid. Thus, collagen–chitosan composite gel has the potential to generate new skin tissues closely matching those of the natural skin. Since gelatin is closely related to the structure of collagen, cross-linked gelatin–chitosan composites have been also employed as scaffolds for regeneration of tissues (Tseng et al., 2013). As skin is composed of epidermis and dermis, another group of workers has developed a combinatorial approach for simultaneous regeneration of both dermal and epidermal layers of the skin (Sudheesh Kumar et al., 2013). To mimic the epidermal structure, they have seeded keratinocytes in a gelatin bioglue, and then the bioglue was combined with chitosan–gelatin scaffolds. Since fibrin plays an important role in natural wound healing, a recent work has demonstrated the fabrication and application of a chitosan–nanofibrin composite hydrogel bandage for wound healing and skin tissue engineering. Thus, the hemostasis, biocompatibility, and fibroblast-stimulating characteristics of chitosan have been exploited by the strategic combination of chitosan with natural skin components including fibrous proteins, DNA, and growth factors to mimic major attributes of the natural skin, leading to accelerated and stable wound healing.

3.1.1.5 Liver tissue engineering

The liver is the largest solid organ in the human body. It carries out more than 500 essential tasks. The major functions of the liver include bile production; absorption and metabolism of bilirubin; supporting blood clotting; metabolism of proteins, fats, and carbohydrates; filtering blood; and immunological functions. The highly specialized tissue of liver consists mostly of hepatocyte cells, and regulates a wide variety of vital biochemical reactions including synthesis and breakdown of small and complex molecules. In the case of partial or complete liver damage due to acute or chronic disorders, liver tissue engineering aims to repair the damaged liver or replace the defective organ with a fully functional artificial liver. In liver tissue engineering, all the technical approaches are based on seeding, survival, and proliferation of the fully developed hepatocytes or stem cell-derived hepatocyte-like cells in a three-dimensional scaffold matrix composed of suitable biomaterials. Obviously, scaffold materials derived from natural resources have advantages over synthetic scaffold materials in terms of physiological bioactivity and biomechanics of natural extracellular matrix. Since GAGs are one of the major components of extracellular matrix surrounding hepatocytes, and chitosan possesses similar structures to GAGs, researchers opted for chitosan as a component of scaffold materials (Li et al., 2003). Thus, cross-linked chitosan—collagen matrix has shown better compatibility toward hepatocytes and blood (Wang et al., 2003). Notably, a heparin—chitosan—collagen ternary composite has shown fair compatibility with blood (Wang et al., 2005). Moreover, the adherence of hepatocytes to the 3D matrix has been noted to improve when alginate/galactosylated chitosan scaffolds have been used as the composite structures (Chung et al., 2002). Hepatocyte attachment has been also observed with self-assembled homogeneous porous composites made from silk fibroin and chitosan (She et al., 2009). Recently, microporous scaffolds composed of chitosan and galactosylated hyaluronic acid have been shown to mimic some natural liver functions, such as albumin secretion, urea synthesis, and ammonia elimination (Fan et al., 2010). However, an ideal replacement for liver tissue is still under investigation, and researchers are continuously working to develop scaffold materials supporting proper biological functioning of the seeded hepatocytes.

3.1.1.6 Nerve tissue engineering

The blends and composites of chitosan are also employed as components in making biomaterials capable of replacing or repairing damaged natural nerve tissues comprised of neurons composed of axons, dendrons, Schwann cells, etc. However, chitosan-based materials have been employed mostly as replacements for tissues in the peripheral nervous system rather than the central nervous system. An artificial construct should possess various attributes including neuron attachment capability, differentiation and growth of embedded neurons, and desired biocompatibility to assist survival of the seeded neurons. In this regard, better cell attachment properties of chitosan-poly(L-lysine)-blended composite materials (Mingyu et al., 2004) and chitosan—gelatin composites (Cheng et al., 2003) are ascribed to an increased hydrophilicity of composite materials due to the presence of hydroxyl and primary amine functionalities, along with the positively charged surface of chitosan. Enhanced cellular adhesion, cell-differentiating ability, and proliferation have been observed when chitosan—gelatin composites have been used as matrices to culture nerve cells (Chiono et al., 2008; Kuo & Lin, 2013). Since nerve tissues have a major function in carrying the nerve impulses from one part to another, conductive polymers have been often employed to produce chitosan-based composites. In this regard, electrical stimulation and conductive polymers have played a substantial role in improving the differentiation, proliferation, and growth of seeded cells, leading to faster regeneration of the nervous tissues. For instance, conductive polypyrrole—chitosan composite membranes have been prepared to electrically stimulate Schwann cells (Huang et al., 2010). In fact, electrical stimulation has enhanced the expression and secretion of neurotrophic factors, and hence, induced the faster growth of cells. However, to date, no satisfactory outcome has been obtained to replace and regenerate tissues of the central nervous system. Moreover, the peripheral nervous system needs ideal constructs acting as true replacements for natural nervous tissues, and conductive aliphatic polymers could be an ideal component replacing the less biocompatible aromatic conductive polymers of the composite constructs used as a component of artificial nerve tissue.

3.1.2 Applications of chitosan-based blends and composites in drug delivery

For a high-performance drug-delivery system, a delivering agent, such as a gel, should fulfill some basic attributes including biodegradability, minimum cytotoxicity, biocompatibility, improved loading efficiency, swelling ability, higher mechanical strength, long-term durability, and resistance to enzymatic degradation, thermostability, thermo-responsive behavior, controlled release, porosity, and permeability. The blends and composites of chitosan are equipped with all of these above-mentioned characteristics, and accordingly, they have already been established as potential drug materials for drug delivery. In fact, the drug release rate of chitosan-based blends and composites can be tuned by varying the relative proportions of the components, degree of cross-linking, and extent of fiber alignments. Chitosan has effectively been combined with other components, such as hydroxyapatite (Teng et al., 2010), alginate (Tapia et al., 2005; Mi et al., 2001), carrageenan (Tapia et al., 2005), starch (Baniani et al., 2017), cellulose (Freag et al., 2018), poly(L-lactic acid) (Prabaharan et al., 2007), poly(lactic-co-glycolic) acid (Meng et al., 2011), collagen (Thacharodi & Rao, 1996), silicate (Park et al., 2001), silica (Liu, Cai et al., 2019) organic rectorite, halloysite nanotubes (Huang et al., 2017), and red algae (Feki et al., 2020), to produce blends and composites for drug-delivery applications. These materials are often cross-linked with different cross-linkers, i.e., glutaraldehyde and cross-linking polymers including poly-carbophil, sodium alginate, gellan gum, to endow mechanical stability in the network. Moreover, the drug-delivery efficiencies of chitosan-based materials for a wide range of drugs, both hydrophobic and hydrophilic, have been tested by various research groups. Some of those drugs include ketoprofen (Prabaharan et al., 2007), fenbufen (Meng et al., 2011), ampicillin, tetracycline hydrochloride (Teng et al., 2010), daptomycin, dexamethasone, betamethasone sodium phosphate, nifedipine (Thacharodi & Rao, 1996), propranolol hydrochloride (Thacharodi & Rao, 1996), berberine (Mi et al., 2001), lidocaine-HCl (Park et al., 2001), sodium salicylate (Park et al., 2001), insulin (Feki et al., 2020), and 4-acetomido phenol (Park et al., 2001). In this regard, molecular size, hydrophobic/hydrophilic character, functionalities, and physical/chemical bonding of the therapeutic agents with the delivering agent can influence the nature of drug delivery. In the case of drug loading by diffusion, larger therapeutics, such as peptides and proteins including insulin, are not readily able to migrate through the small pores of the hydrogel. Therefore, to deliver such large-sized drugs, an effective strategy should be employed to entrap the drug within the gel during gel making. Herein, unwanted cross-linking during gelation should be avoided, as this will reduce the porosity of the carrier, which can badly affect the release of such a bulky drug. To facilitate controlled release at the target, another effective strategy is to control the environmental parameters, such as pH and temperature, surrounding the gel. Therefore, chitosan-based composite gels are made pH and/or temperature sensitive by incorporating suitable polymeric component(s). Moreover, another strategy is the tethering approach by which the loss of the therapeutic reserve and the risk of toxic exposure can be reduced. Accordingly, drugs are often covalently or physically linked to the polymer chains prior to gelation. Research is still in progress to discover chitosan-based ideal nanocomposite-based tunable drug delivering material(s) capable of controlled drug release at specific targets.

3.1.3 Applications of chitosan-based blends and composites in food processing

Modern food industries are engaged in making biopolymer-based edible films capable of facilitating handling properties and achieving an extended shelf-life for food products. Indeed, efficient storage of foods in good condition essentially requires antimicrobial surroundings to protect the food products from possible attack and growth of microorganisms, such as *Escherichia coli* and *Staphylococcus aureus*, responsible for food poisoning. In addition, the protecting materials should be mechanically strong, flexible, durable, and should have the desirable film-forming ability. Because of its good film-forming property, excellent biocompatibility, and certain antimicrobial activity, chitosan is a potential candidate for preparing composites applicable for food packaging and edible film making. The chitosan—starch composites have good antimicrobial properties against *Escherichia coli* and *Staphylococcus aureus* (Zhong et al., 2011). Also, better antimicrobial properties have been achieved when malic acid has been used for dissolution of chitosan. Being a relatively stronger acid, better protonation of amines in chitosan to $-NH_3^+$ has led to improved preventive action of chitosan—starch composites against bacterial growth. Other essential properties including water barrier property, mechanical strength, and flexibility of the chitosan—starch composite films, that have been modulated depending on the type of acid used during dissolution of chitosan prior to preparation of the composite film (Zhong et al., 2011). Another group of researchers has reported the possible application of chitosan—graphene oxide composites as high-strength reliable packaging materials (Yin et al., 2019). Recently, chitosan nanoparticle-filled fish gelatin composite films have exhibited remarkable antimicrobial activities against four different food pathogens, i.e., *Staphylococcus aureus*, *Listeria monocytogenes*, *Salmonella enteritidis*, and *Escherichia coli*, and hence can be utilized as highly effective biodegradable packaging materials (Hosseini et al., 2016). The transglutaminase cross-linking catalyst assisted another protein-based composite comprising whey protein admixed with chitosan which showed better activities, such as improved water barrier properties, transparency, and mechanical properties (Jiang et al., 2016). Researchers have also attempted the reinforcement of chitosan-based films with higher water resistance by adding Cu nanoclusters capped with proteins (Li et al., 2017). Notably, significant antimicrobial activity was observed under visible light, when chitosan—TiO_2 film-based nanocomposite material was used as a packaging film. These films have shown efficient antimicrobial activity against four different strains, i.e., *Escherichia coli*,

Staphylococcus aureus, *Candida albicans*, and *Aspergillus niger* (Zhang et al., 2017). Later, various researchers studied the potential of different organic and inorganic fillers, such as polylactic acid (Niu et al., 2018), cellulose, nanocellulose, starch, gelatin (Noorbakhsh-Soltani et al., 2018), oligosaccharides (Yin et al., 2019), Ag nanoparticles (Kadam et al., 2019; Liang et al., 2019), and polyvinyl alcohol (Ju et al., 2020), to improve the efficiency of chitosan to protect food items from microbial attack.

3.1.4 Applications of chitosan-based blends and composites in corrosion protection

Most steel-made materials suffer from a major unwanted phenomenon called corrosion. The applications of corrosion inhibitors and polymers as protective measures have been adopted to prevent this damage. Steel materials used during acid pickling, oil well acidizing, and heat transfer are more susceptible to corrosion. More importantly, steel used in medical devices should be completely corrosion free. Being a polysaccharide, chitosan has the potential to function as a natural corrosion inhibitor, as its macromolecules can strongly coordinate with metal ions by donating their electron pairs to the empty d orbitals of the metal. In acidic pH, chitosan acts as a polyelectrolyte and hence can bind with metal surfaces, leading to the prevention of corrosion in an acidic medium. In corrosive media, steel surfaces acquire a net positive charge, which leads to the adsorption of anions, such as sulfate and Cl^-, making the steel surface negatively charged. In an acidic medium, chitosan exists in both protonated and neutral forms, and accordingly, the attachment of protonated forms of chitosan on the steel surface by electrostatic interactions is highly feasible, preventing the corrosion process. Compared to pure chitosan, chitosan composites are better corrosion-preventing agents. In this regard, the improved mechanical properties, adhesion, and barrier effect of composites enhance their corrosion-preventive ability. For instance, steel surfaces coated with chitosan–hydroxyapatite composites acquire significant resistance against corrosion (Ashassi-Sorkhabi & Kazempour, 2020). Researchers have also added calcium silicate, Ag, halloysite nanotube, carbon nanotubes, Si, Mg, and cellulose acetate with chitosan–hydroxyapatite combinations to enhance the performance of inhibitors. Moreover, the performance potentials of many corrosion-inhibiting chitosan-based composites and blends have been adjusted by adding metals/nonmetals/oxides, i.e., Zn, ZnO, Cu, Ni, Au, F, SiO_2,TiO_2, graphene oxide (Fayyad et al., 2016; Jin et al., 2019), boron nitride; natural polymers; semisynthetic polymers, i.e., amylose acetate; synthetic polymers, i.e., polypyrrole, poly(vinyl butyral), polyaniline, poly(aniline-anisidine), polyvinyl alcohol (PVA); and ceramics, i.e., bioactive glass as the second/third component(s). As anticipated, a nanocomposite-based coating prevents the corrosion more effectively than a composite having micron-size additives. Thus, nanocomposites, such as chitosan–ZnO nanoparticles, chitosan–TiO_2 nanoparticles, chitosan–Ag nanoparticles, and chitosan–hydroxyapatite nanoparticles, can effectively inhibit corrosion in a more aggressive environment.

3.1.5 Applications of chitosan-based blends and composites in adsorptive removals of dyes, heavy metal ions, and other hazardous species from waste water

Nowadays, various industries regularly release effluents contaminated by several hazardous and toxic substances including dyes, heavy metal ions, polyaromatic hydrocarbons, and radioactive species, imparting a severe threat to the environment and ecosystem. Most organic dyes, categorized under different chemical classes, i.e., azo (Mahapatra et al., 2018), phenothiazine (Singha, Karmakar et al., 2017), phenazine, and triphenyl methanes (Singha, Mahapatra et al., 2017), are highly toxic, carcinogenic, and extremely harmful for aquatic species and plants. These dyes cause several acute health hazards including stomach pain and irritation to the skin, mouth, throat, tongue, and lips and permanent damage to the eyes (Karmakar et al., 2017). Many of these widely applied dyes come under the azo category, and most of those azo dyes decompose into potentially carcinogenic aromatic amines under anaerobic conditions (Chang & Juang, 2004). Recently, industries have relied heavily on the use of reactive dyes, as these dyes strongly attach to the substrate by forming covalent bonds, leading to greater washfastness and wet-rubfastness of the dyed material (Singha, Chattopadhyay et al., 2019). However, industrial effluents containing reactive dyes have limited biodegradability in an aerobic environment (Chang & Juang, 2004). Among the heavy metal ions, Pb(II) inflicts various abnormalities including dysfunction of the brain, liver, kidneys, and bones due to impairment of the central nervous system, reduction in hemoglobin formation, mental retardation, infertility and abnormalities in pregnant women, anemia, headache, dizziness, irritability, and weakness of muscles (Singha, Karmakar et al., 2018; Mitra, Mahapatra, Dutta, Deb et al., 2020). Other heavy metal ions, such as Cu(II) (Mahapatra et al., 2019; Dutta, Mahapatra, Deb, Ghosh et al., 2020), Ni(II), As(III) (Mondal et al., 2020a), As(V), Bi(III) (Mahapatra et al., 2020), Cr(III) (Dutta, Mahapatra, Deb, Mitra et al., 2020), Cr(VI), Hg(II) (Singha, Dutta et al., 2018; Singha, Mahapatra, Karmakar, Mondal et al., 2018), and Cd(II), are more or less harmful if present above certain permissible limits, and inflict diverse damage to humans, the food chain, plants, ecosystem, and the wider environment (Mondal et al., 2018; Karmakar, Mondal, Mahapatra et al., 2019; Shameem et al., 2003). Therefore, considerable attention has been devoted to developing innovative, efficient, economical, and fast treatment methods to remove these toxic species from aqueous systems, of which adsorption using proper biocompatible adsorbents is one of the most widely studied, facile, and accepted techniques for decontaminating polluted wastewater due to its flexible design, potential application prospects, and the easy recyclability of adsorbents. In this regard, various natural and synthetic polymers have been employed to produce highly efficient recyclable adsorbents capable of repeated removals of dyes, heavy metal ions, and other toxic species from wastewater (Mondal et al., 2019; Roy et al., 2019; Singha, Dutta et al., 2019). Some of these adsorbents can be utilized for the separation and purification of vital chemical species from water (Singha, Dutta et al., 2019).

Among those polymeric adsorbents, chitosan-based materials are advantageous, as they contain both amine and hydroxyl functionalities (Karmakar, Mondal, Ghosh et al., 2019), and these groups can be suitably modified by various kinds of fillers, cross-linkers, and other ingredients, depending upon the type and characteristics of the adsorbates and the environmental conditions (Table 6.2). Indeed, the fabrication of both chitosan-based blends and composites for adsorptive removal of dyes, metal ions, and other hazardous species has been done while concentrating mainly on the following aims and objectives:

i. For the preparation of adsorbents, the adsorption capacity should be as high as possible. Accordingly, attempts have been made to produce increasingly hydrophilic materials with higher swelling capacity, as adsorbents are used for the removal of water-soluble dyes from aqueous media. To ensure higher swelling ability, the microstructure of the adsorbents should be flexible enough to facilitate easier conformational rotation of main chains bearing —O— linkages (Mondal et al., 2020b).

ii. The bonding of adsorbent and dye should be reversible via noncovalent bonds, such as ionic and hydrogen bonds, so that the dye-loaded adsorbent can be desorbed, and the adsorbent can be reused. In this regard, the desorbed dye may be reused during a dyeing operation.

iii. In the case of adsorption of metal ions, the nature of bonding is mostly ionic or coordinate, depending on the nature of adsorbent and the metal ions. Most of the transitional metal ions with unfilled d orbitals can frequently attach by coordinate bonds with the available donors of electron pairs, such as O-donors and N-donors (Mitra, Mahapatra, Dutta, Chattopadhyay et al., 2020; Mahapatra et al., 2020). However, such attachment tendency is guided by the coordination number, available vacancies, and soft/hard nature of the ions. For instance, metal ions like Hg(II) preferably bind with N-donors, whereas Pb(II) can coordinate with both O-donors and N-donors, as Pb(II) possesses a variable coordination number ranging from 2 to 10. Notably, with a d^{10} electronic structure, Zn(II) has little tendency to attach by coordinate bonding and hence interacts with adsorbents mostly by ionic bonding. In this regard, chitosan has frequently available O—H and $-NH_2$ functionalities allowing both O-donors and N-donors for coordinate bonding. Additionally, protonation and deprotonation are possible at both O—H and $-NH_2$ functionalities, facilitating ionic bonding. Moreover, both O—H and $-NH_2$ functionalities can function as hydrogen bond donors and acceptors, encouraging variable hydrogen bonding within the loaded and unloaded adsorbents.

iv. The adsorbent should contain a sufficient number of functional groups for holding the maximum number of dye molecules, and the adsorbent should have sufficient structural integrity that the adsorbents remain stable during the repeated adsorption and desorption process. Pure chitosan-based adsorbents cannot offer such structural integrity and hence suitable synthetic aliphatic polymer(s) or inorganic component(s) are incorporated to enhance the overall stability and durability of the adsorbent.

v. The adsorbent should be biodegradable and environment-friendly, and accordingly, as a frequently available natural polysaccharide, chitosan contributes such attributes to adsorbents.

vi. The adsorbent should be pH sensitive so that either dye attachment or detachment during adsorption or desorption can be achieved through protonation/deprotonation of the prevailing functional groups in the adsorbent.

3.1.5.1 Applications of chitosan-based blends and composites in adsorptive removals of dyes

Different types of chitosan-based blends and composites have been employed by several groups of researchers for adsorptive removal of hazardous dyes from wastewater (Table 6.3). All of those blends and composites come under the following types: (1) chitosan-inorganic, (2) chitosan-organic, and (3) chitosan-hybrid. Among the chitosan-inorganic blends and composites, chitosan–clay composites have been reported several times by different researchers. The interactions of clays and organomodified clays, such as montmorillonite (MMT), cloisites, alunite, and bentonite, with the chitosan matrix in enhanced dye absorption and removal at different environmental conditions being examined by spectroscopic and thermal characterization techniques.

The binding site and mechanism of various anionic dyes, i.e., remazol brilliant orange 3R (Nesic et al., 2012) and Congo red (Wang & Wang, 2007), with chitosan–MMT composite absorbents are determined by FTIR. The $-CONH_2$, $-NH_2$, and O–H groups from chitosan and Si–O of MMT are involved in reversible binding with azo dye. The involvement of $-CONH_2$ as the binding site is realized from shifting of amide-I from 1645 to 1629 cm^{-1}. The role of O–H as the binding site is identified from the significant shift of O–H stretching at 3445 cm^{-1} in unloaded membrane to 3430 cm^{-1} in dye-adsorbed membrane (Nesic et al., 2012). Binding of dye with chitosan–MMT composite also has led to a significant shift of Si–O stretching from 1040 and 1118 cm^{-1} to 1034 and 1123 cm^{-1}, respectively (Nesic et al., 2012). In fact, the maximum adsorption has been noted at $pH = 6.0$ (Table 6.3), at which the primary amines of chitosan remain mostly in the protonated form, i.e., $-NH_3^+$, as the pK_a of the primary amines in the chitosan is 6.3. At an acidic pH, the sulfonic groups of the anionic dyes electrostatically interact with $-NH_3^+$. In contrast, in alkaline pH, deprotonation of $-NH_3^+$ to $-NH_2$ results in a poor electrostatic interaction between the dye and adsorbent (Nesic et al., 2012). Moreover, if the pH exceeds 7.0, the increasing number of O–H functionalities can compete with the dye anions, decreasing the anionic dye uptake (Wang & Wang, 2007). Nevertheless, such a reduced extent of dye uptake in the alkaline pH can be ascribed to the nonionic interactions including hydrogen bonding, dipole–dipole, and van der Waals interactions, facilitating the formation of dye stacks at the adsorbent surface. Similarly, cloisite-based chitosan composites are also effective in anionic dye removal (Daraei et al., 2013). It has been observed that the anionic dye removal efficiency of cloisite 30B–chitosan nanocomposite is better than cloisite 15A–chitosan nanocomposite (Daraei et al., 2013). This is because, at acidic pH, the protonated cloisite 30B contains a ternary ammonium and two O–H groups (Fig. 6.5), and hence it should contain more positive charges compared to cloisite 15A bearing no hydrophilic groups. the effective removal of cationic dyes has been achieved by applying chitosan–bentonite composite. The binding of malachite green with the O–H group of chitosan by hydrogen bonding is identified via a small shift in O–H stretching from 3426 cm^{-1} in unloaded adsorbent to 3431 cm^{-1} in a dye-loaded sample (Wan Ngah, Ariff, Hashim, & Hanafiah, 2010). Additionally, N–H + O–H combined stretching at 3616 cm^{-1} of the unloaded composite completely disappeared

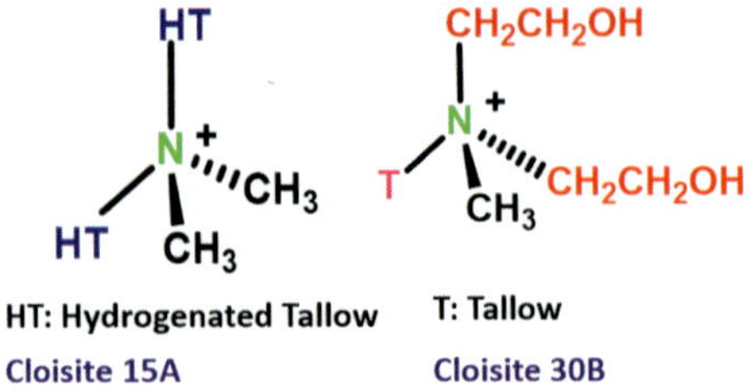

Figure 6.5 Structures of organomodified clays.

in the dye-loaded sample, suggesting massive structural alterations via changes in hydrogen bonding after dye adsorption. Thus, amine and hydroxyl groups of chitosan have been reported as the main functionalities involved in malachite green adsorption (Wan Ngah, Ariff, Hashim, & Hanafiah, 2010). Unlike anionic dyes, relatively poor adsorption at the acidic pH is ascribed to repulsion between the $-NH_3^+$ groups of chitosan and $-N^+(CH_3)_2$ groups of malachite green. Accordingly, enhanced adsorption at alkaline pH is associated with deprotonation of $-NH_3^+$ to $-NH_2$, and the reduced cationic charges actuated a greater attraction between the adsorbent and cationic dye molecules (Wan Ngah, Ariff, Hashim, & Hanafiah, 2010). Active participation of hydrogen bonding is also detected from the marginal shift in O—H stretching of unadsorbed composite from 3426 cm^{-1}—3430 cm^{-1} in tartrazine-loaded chitosan—bentonite composite (Wan Ngah, Ariff, & Hanafiah, 2010). Moreover, protonated amine groups in chitosan—bentonite composite at acidic pH played a substantial role in the adsorption of tartrazine through ionic attraction of $-NH_3^+$ of composite with $R-SO_3^-$ of tartrazine (Fig. 6.6) (Wan Ngah, Ariff, & Hanafiah, 2010).

Accordingly, the maximum adsorption of the dye is registered at pH = 2.5. Such an ionic interaction is identified from the shift of N—H bending from 1646 cm^{-1} in unadsorbed composite to 1638 cm^{-1} in tartrazine-adsorbed composite. However, C—O—C of the composite plays a small role in the adsorbate—adsorbent interactions, as no changes in the frequencies of ether linkages are detected in the dye-loaded sample. However, at pH = 1.0, the adsorption of tartrazine was greatly affected due to protonation of $R-SO_3^-$ of tartrazine (Wan Ngah, Ariff, & Hanafiah, 2010). Since the pK_a of chitosan is around 6.20 (Zhu et al., 2010), the acidic pH mostly favors the adsorption of anionic dyes, whereas the alkaline pH facilitates greater uptake of cationic dyes. However, depending on the shift of pK_a for chitosan-based blends and composites compared to pure chitosan, such pH-dependent adsorption phenomena can fluctuate

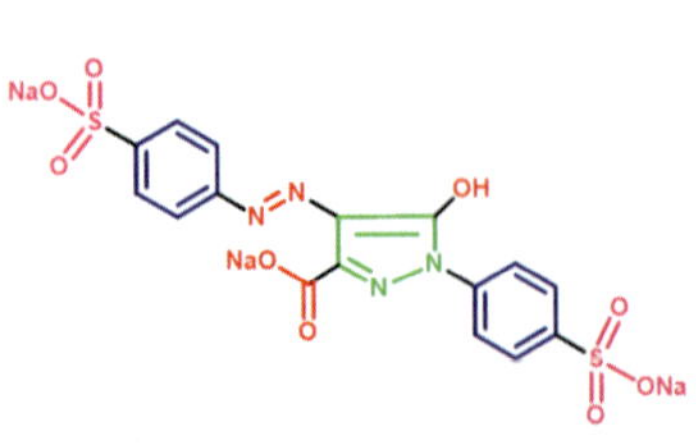

Figure 6.6 Molecular structure of tartrazine [trisodium-5-hydroxy-1-(4-sulfonatophenyl)-4-(4 sulfonatophenylazo)-H-pyrazole-3-carboxylate].

significantly. The adsorption capacity for a particular composite is also dependent on the nature of the adsorbate. For instance, the adsorption capacity of chitosan-activated clay composites for methylene blue is relatively poor, whereas the same adsorbent displayed significantly higher adsorption capacity for reactive dye RR222 (Table 6.3) (Chang & Juang, 2004). Better adsorption of reactive dye RR222 over methylene blue can be associated with stronger binding characteristics of reactive dyes via formation of covalent bond with the $-NH_2$ and O–H functionalities of the chitosan.

3.1.5.2 Applications of chitosan-based blends and composites in adsorptive removals of heavy metal ion(s)

The adsorption of heavy metal ions by chitosan-based blends and composites is mostly mediated by the amine and hydroxyl groups in chitosan (Shameem et al., 2008). Indeed, some chitosan-based blends and composites possess better selectivity for particular metal ion(s). Indeed, such selectivity originates from the better stability of the absorbent–adsorbate adducts, which depends on the structural parameters and binding characteristics of both metal ions and adsorbents. For instance, Schiff base-type chitosan-coated cotton fiber shows better adsorption selectivity for Pb(II)/Cu(II) than those of Ni(II) and Cd(II). Other Schiff base-type adsorbents can be effectively utilized in selective recovery of Hg(II) in the presence of excessive alkali and alkaline earth metal ions, such as K^+, Na^+, Mg(II), and Ca(II), and heavy metals, such as Co(II), Ag(I), and Mn(II), under optimum conditions (Qu, Sun, Fang et al., 2009). Such selectivity arises from the nature of adducts with different stability and structures bearing metal ions and chelating groups, such as O–H, $-NH_2$, and –CH=N– (Zhang et al., 2008). Moreover, the nature of the liquid film diffusion and particle diffusion can contribute significantly to determining the overall adsorption behavior. In addition to selectivity, the recyclability of adsorbents is an important aspect that depends on the stability of the bonds and the overall microstructure. For example, though both RCCH and SCCH fibers are suitable for selective adsorption of Hg(II), RCCH fiber demonstrates better reusability because of the presence of the $-CH_2-NH-$ group in contrast to unsaturated Schiff-base/ –CH=N– in SCCH fiber (Qu, Sun, Fang et al., 2009). In this regard, saturated moiety is less reactive, and hence should offer more stability to the adsorbent. Additionally, such moiety should endow better flexibility to the adsorbent microstructure due to facilitated conformational rotations.

3.1.5.2.1 Cu(II) Of various oxidation states of copper available in aqueous solutions, Cu(II) is the main species of concern. In fact, an excessive concentration of Cu(II) in biological systems is lethal (Wan et al., 2007). During application of chitosan–perlite composites for adsorptive removal of Cu(II), adsorbate attaches to amine groups of chitosan (Shameem et al., 2008), and the maximum adsorption is observed at pH = 4.5 despite the protonation possibility of $-NH_2$ in chitosan. However, such higher adsorption of Cu(II) at pH = 4.5 can be related to the presence of perlite inhibiting the formation of NH_3^+ groups (Shameem et al., 2008). Moreover, some copper may attach as $Cu(OH)^+$, and the conversion of Cu(II) into $Cu(OH)^+$ reduces the extent of positive charges. More importantly, the significant shift in the N1s spectrum after Cu(II) adsorption suggested the formation of coordination complexes of Cu(II) with $-NH_2$ ligands of chitosan (Shameem et al., 2008). Indeed, such coordination of Cu(II) with $-NH_2$ ligands of chitosan is also responsible for the significantly higher extent of Cu(II) adsorption by chitosan–magnetite composites (Liu et al., 2009). In another work, the adsorption of Cu(II) onto chitosan-based composites has been confirmed by the chelation of Cu(II) with $-NH_2$ and O–H groups of the matrix (Kalyani et al., 2009), where chelation has occurred among Cu(II) with deprotonated hydroxyl groups due to the release of H^+ at a higher pH level (Kalyani et al., 2009). Accordingly, the metal uptake is increased by an increase in pH up to 5.0. However, further higher pH values (>5.0) resulted in the formation of metal hydroxide precipitates (Kalyani et al., 2009). The major contributions of $-NH_2$ and O–H groups during adsorption of Cu(II) have also been noted in the case of chitosan–ceramic alumina composites (Veera, Krishnaiah, Ann, & Edgar, 2008).

Cu(II) removal using chitosan-based blends and composites is not only based on the combinations of inorganic components, but also there are several compositions bearing organic components, such as PVA (Wan Ngah et al., 2004) and alginate (Wan Ngah & Fatinathan, 2008). The basic idea behind the incorporation of PVA to chitosan is to increase the number of O–H functionalities, and thus the elevated O–H and $-NH_2$ groups in chitosan–PVA combinations have enhanced adsorption capacity of the blend for Cu(II). Similarly, the main objective behind the addition of alginate is to introduce $-COO^-$ functionalities capable of binding Cu(II) ions through ionic and coordinate bonding (Wan Ngah & Fatinathan, 2008). For chitosan–PVA blends, the maximum adsorption capacity for Cu(II) removal is obtained at $pH = 6.0$. If the pH is either increased or decreased, the adsorption decreases immensely. At $pH > 6.0$, the in situ formed Cu(II) hydroxide precipitates can affect the adsorbing ability of the beads made of chitosan–PVA blends. Such precipitation of Cu(II) hydroxide is observed when Cu(II) adsorption is carried out by Schiff base-type chitosan-coated cotton fiber at high pH ($pH = 7.8$), leading to lowering of the adsorption capacity (Zhang et al., 2008). On the other hand, if $pH < 6.0$, the increasing population of protonated amines, i.e., $-NH_3^+$, can resist adsorption by repelling the incoming Cu(II) ions (Wan Ngah et al., 2004). For the same reason, the drop in Cu(II) adsorption can be noticed when adsorption is conducted by Schiff base-type chitosan-coated cotton fiber at low pH ($pH < 3.0$) (Zhang et al., 2008). In the case of chitosan–alginate combinations, the higher pH has improved adsorption capacity by providing more$-COO^-$ functionalities necessary for binding of Cu(II) (Wan Ngah & Fatinathan, 2008).

3.1.5.2.2 Cr(VI) The adsorption of Cr(VI) is conducted by selecting chromate or dichromate anions as adsorbates. Thus, the maximum adsorption is recorded within a lower pH range at which the population of $-NH_3^+$ is higher. In fact, at the lower pH levels, the amine group of chitosan undergoes protonation, leading to the increased electrostatic attraction among $-NH_3^+$ and sorbate anion. The deprotonation of amino group led to decreased adsorption at the higher pH of the solution (Shameem et al., 2003). Similar pH-dependent adsorption trends are observed in the case of chitosan–alumina (Veera et al., 2003) and chitosan–magnetite composites

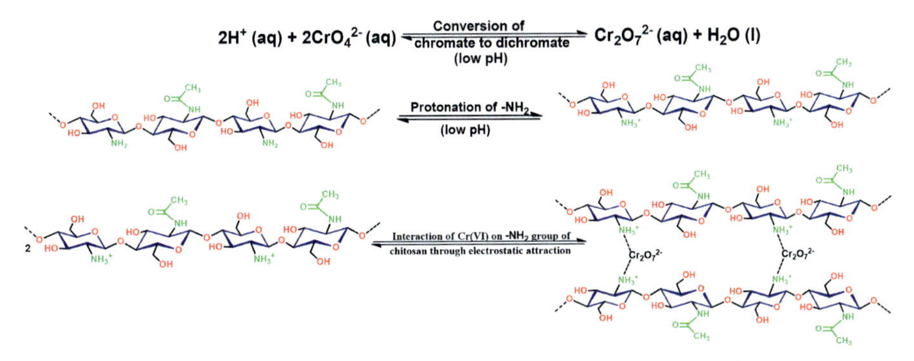

Figure 6.7 pH-dependent adsorption of Cr(VI) on chitosan-alumina composites.

(Fig. 6.7) (Huang et al., 2009). For chitosan–magnetite composites, the adsorption of Cr(VI) is attributed to the formation of N–Cr bonds by the coordination of $-NH_2$ of chitosan with Cr(VI), realized from the significant narrowing of a broad peak at 3380 cm^{-1} (Huang et al., 2009). Moreover, authors have reported the preferable allocation of the adsorbed chromium complexes at the surface rather than the bulk of the chitosan–magnetite composites.

3.1.5.2.3 Cd(II) Long-term exposure to cadmium results in unwanted consequences including damage to the kidneys and liver. The adsorption of Cd(II) into chitosan–perlite composites is associated with the attachment of Cd(II) with $-NH_2$ of chitosan, confirmed by XPS analyses (Fig. 6.8) (Shameem et al., 2006). Interestingly, PZC of Cd(II)-adsorbed chitosan–perlite composite decreases to 6.5 compared to the same of unadsorbed chitosan–perlite composite at 8.5 (Shameem et al., 2006). Here, the reduction in PZC should be ascribed to the scarcity of free basic groups, such as O–H and $-NH_2$,

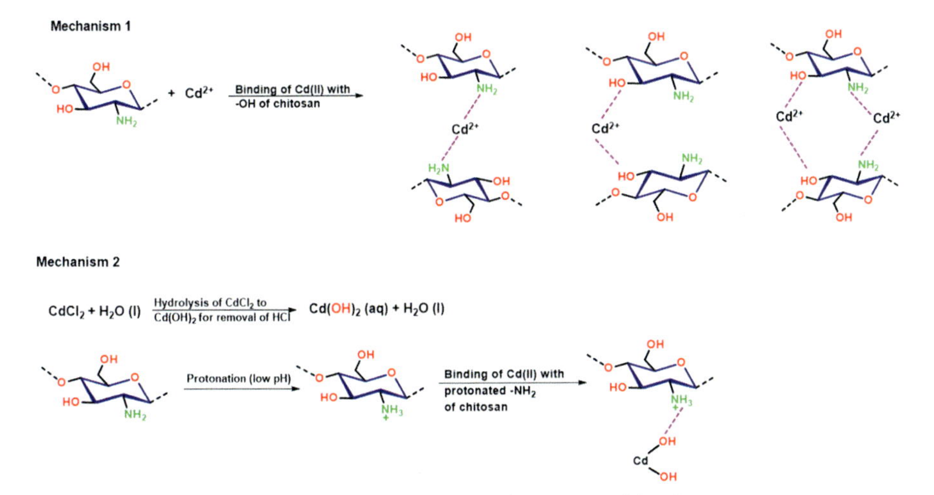

Figure 6.8 Adsorption mechanism of Cd(II) by chitosan-coated beads.

in the Cd(II)-adsorbed chitosan−perlite composite, as O−H and $-NH_2$ may be bonded with Cd(II). In fact, $-NH_2$ is reported to form covalent bonds with Cd(II), confirmed by the larger shift of N1s peak, i.e., 1.78 eV, corresponding to $-NH_2$. In addition, other groups, such as $-CH_2OH$, O−H, or −O−, may act as attachment sites for Cd(II) adsorption. Moreover, the added $CdCl_2$ salt may be hydrolyzed to $Cd(OH)_2$, releasing H^+ ions and lowering the solution pH (Shameem et al., 2006). Thus, the enhanced uptake of Cd(II) by chitosan−perlite composite at the higher pH can be explained by the increased hydrolysis of $CdCl_2$ to $Cd(OH)_2$. In fact, hydrolysis of $CdCl_2$ with increasing pH resulted in the formation of various species, such as $Cd(OH)^+$, $Cd(OH)_2$, and $Cd(OH)_3^-$, of which $Cd(OH)_3^-$ is the predominant species in solutions with a pH between 5−7. However, too great an enhancement of pH results in precipitation of insoluble Cd(II) complexes from the solution (Shameem et al., 2006). Such binding ability of Cd(II) with $-NH_2$ groups can also be the reason behind appreciably higher removal efficiency of chitosan−magnetite composites (Liu et al., 2009). Indeed, binding of Cd(II) ions with $-NH_2$ of chitosan−PVA blends can be realized from the appreciable decrease in the characteristic intensity of the N−H and O−H stretching within 3300−3500 cm^{-1}, substantiated from the lowering of N−H bending from 1644 to 1562 cm^{-1} of an unloaded blend to 1634 and 1553 cm^{-1} in the Cd(II)-loaded adsorbent, respectively (Kumar et al., 2009). Thus, considering $-NH_2$ and O−H as the major adsorption sites of Cd(II) in chitosan−PVA blends, the following chemical reactions are proposed in support of the mechanism explaining adsorption/desorption of Cd(II):

$$R{-}NH_2 + H^+ \rightarrow R{-}NH_3^+ \tag{6.1}$$

$$R{-}NH_2 + Cd^{2+} \rightarrow R{-}NH_2Cd^{2+} \tag{6.2}$$

$$R{-}NH_3^+ + Cd^{2+} \rightarrow R{-}NH_2Cd^{2+} + H^+ \tag{6.3}$$

$$R{-}NH_2Cd^{2+} + H_2O \rightarrow CdOH^+ + R{-}NH_3^+ \tag{6.4}$$

Here, R represents all components other than $-NH_2$ of cross-linked chitosan−PVA blends (Kumar et al., 2009). In fact, the possible coordination of $-NH_2$ with Cd(II) through electron transfer resulted in such alterations in N-donor ligands of Cd(II)-loaded adsorbent (Eqs. 6.2 and 6.3). At the lower pH, $-NH_2$ becomes protonated to $-NH_3^+$ (Eq. 6.1), and hence the absence of a lone pair in$-NH_3^+$ results in complete lack of coordination bonds between Cd(II) and $-NH_3^+$ of adsorbent. Moreover, increased electronic repulsion among $-NH_3^+$ and Cd(II) leads to lower adsorption of Cd(II) in the strongly acidic environment (Kumar et al., 2009).

3.1.5.2.4 Pb(II) As stated earlier, Pb(II) possesses significantly higher coordinating tendency and hence undergo complexation with various biomolecules to obstruct different biochemical pathways, such as interrupting heme biosynthesis, inhibiting several zinc enzymes, and interacting with nucleic acids/tRNA and thereby interrupting the translation of proteins (Liu et al., 2009). Similar to Cu(II) and Cd(II), Pb(II) can coordinate with many$-NH_2$ groups of the amine functionalized magnetite-based

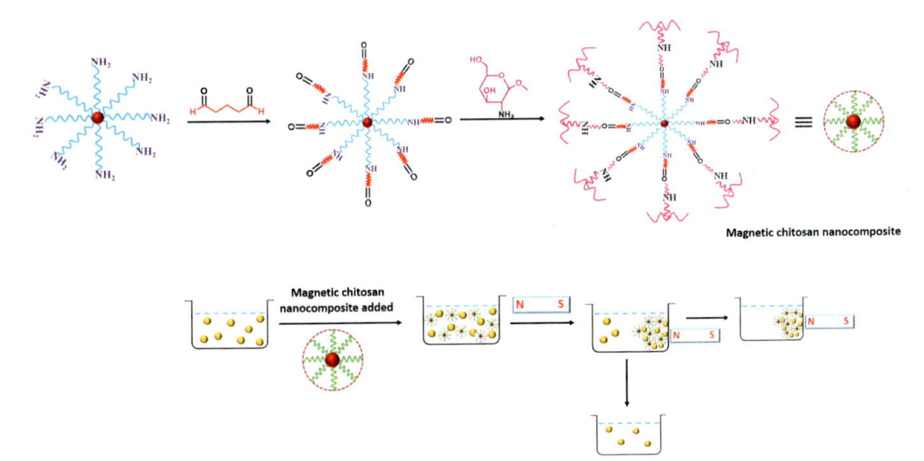

Figure 6.9 Synthesis route of magnetic chitosan nanocomposites and their use as a facile tool for Pb(II) removal with the help of an external magnetic field.

nanocomposites (Fig. 6.9). By varying the initial pH from 2 to 4, the removal efficiency of Pb(II) is increased from 36.8% to 95.3%. However, on increasing the pH further, the adsorption efficiency of Pb(II) was decreased substantially. The adsorption of Pb(II) is contributed by the extent of prevalent nonprotonated chitosan capable of functioning as ligands for complexation with Pb(II). The nature of coordinate bonding of Pb(II) with the nonbonded electron pair of nitrogen in nonprotonated $-NH_2$ can be represented likewise. Thus, at the higher pH, such enhanced deprotonation and complexation can encourage adsorption of Pb(II), whereas increased hydrolysis, polynuclear complex formation, and metal hydroxide precipitates severely affect the absorption capacity of the nanocomposites (Huang et al., 2009). Since Pb(II) is capable of interacting with chitosan mostly through NH_2 and O−H, PVA is blended with chitosan to enrich O−H to achieve better adsorption capacity of the blend compared to that of pure chitosan (Wan Ngah et al., 2004).

3.1.5.2.5 Ni(II) Similar to other heavy metal ions, Ni(II) is also adsorbed by interacting with $-NH_2$ and O−H of chitosan/ceramic alumina composites (Veera, Krishnaiah, Ann, & Edgar, 2008). Thus, chitosan−magnetite composites have presented the maximum adsorption capacity = 52.55 mg/g and 75.07% ion recovery for Ni (II) at a fixed contact time of 120 min and initial concentration = 70 mg/L (Tran et al., 2010).

3.1.5.2.6 As(III) and As(V) The significant removal of arsenic can also be achieved by employing chitosan-based blends and composites as adsorbents. Here, the major aim is to convert highly toxic inorganic arsenic species to stable organic forms through attachment with functionalities of the chitosan. Notably, arsenic-contaminated drinking water imparts the greatest threat to the public health. The intake and bioaccumulation of either pentavalent or trivalent species in the human body can cause cancer in the skin, lungs, liver, kidneys, and bladder. Additionally, the consumption of arsenic imposes acute disturbances in the functioning of cardiovascular and nervous systems, and becomes fatal even at a very low concentration. In 2001, the USEPA adopted 10 ppb as the maximum permissible upper limit of arsenic concentration in drinking water, replacing the old standard of 50 ppb (Veera, Krishnaiah, Jonathan et al., 2008). In this regard, adsorptive removal of arsenic by chitosan–ceramic alumina composite involves coordinate bonding of arsenic with $-NH_2$, O–H, and –CO– of the adsorbent (Veera, Krishnaiah, Jonathan et al., 2008).

3.1.5.2.7 Au(III) Since Au(III) is classified as a soft ion as per the hard soft acid base (HSAB) theory, Au(III) ions should form very strong bonds with functionalities bearing N and S atoms, and hence chitosan and chitosan-based nitrogenous blends and composites have been selected as adsorbents for Au(III) (Qu, Sun, Wang et al., 2009). Adsorption capacities have been studied at different pHs, and the optimum pH for absorption of $AuCl_4^-$ is recorded at 3.0. Such results are related to the ever-increasing population of protonated amines with lowering of pH, leading to a stronger electrostatic attraction between $R-NH_3^+$ and $AuCl_4^-$ (Qu, Sun, Wang et al., 2009). On contrary, at higher pH, hydrolysis of $AuCl_4^-$ into $AuCl_3(OH)^-$ and deprotonation of $R-NH_3^+$ result in decreased electrostatic attractions among Au(III) anions and the positively charged adsorption sites of chitosan-modified cotton fibers (Qu, Sun, Wang et al., 2009).

3.1.5.2.8 Other pollutants In addition to dyes and heavy metal ions, there are several toxic components that directly or indirectly cause several unwanted problems to humans. For instance, the humic substance of soil remains mostly undecomposed by microorganisms. Humic acid, the major component of humic substances, can react with chlorine to produce carcinogenic trihalomethanes (Chang & Juang, 2004). In this regard, chitosan-activated clay composites are employed as adsorbents of humic acid, although the AC is poor for humic acid (Table 6.3). This poor AC could be related to the complicated structure of humic acid contributing to steric hindrance (Chang & Juang, 2004). However, substantially higher AC for tannic acid could be associated with the huge number of O–H and phenolic groups participating in noncovalent interactions with amines, hydroxyls, and other functionalities of the composite.

3.2 Classification based on the material compositions

Chitosan-based blends or composites can broadly be classified in the following way on the basis of material compositions (Chart 6.2).

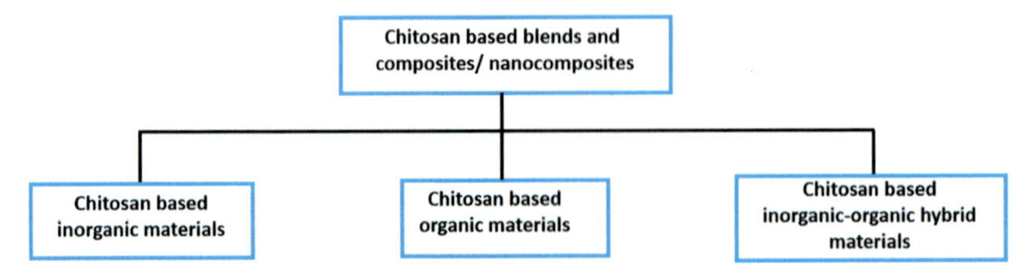

Chart 6.2 Broad classification of chitosan-based blends and composites.

3.2.1 *Chitosan-organic materials*

These are blends and composites of chitosan, wherein inorganic components are absent. Being composed of exclusively organic materials, these materials are more flexible, ductile, and soft. However, they often have inferior thermal stability and mechanical strength. These type of materials can again be subclassified based on the nature of the organic component, i.e., protein, carbohydrate, and synthetic polymer (Chart 6.3).

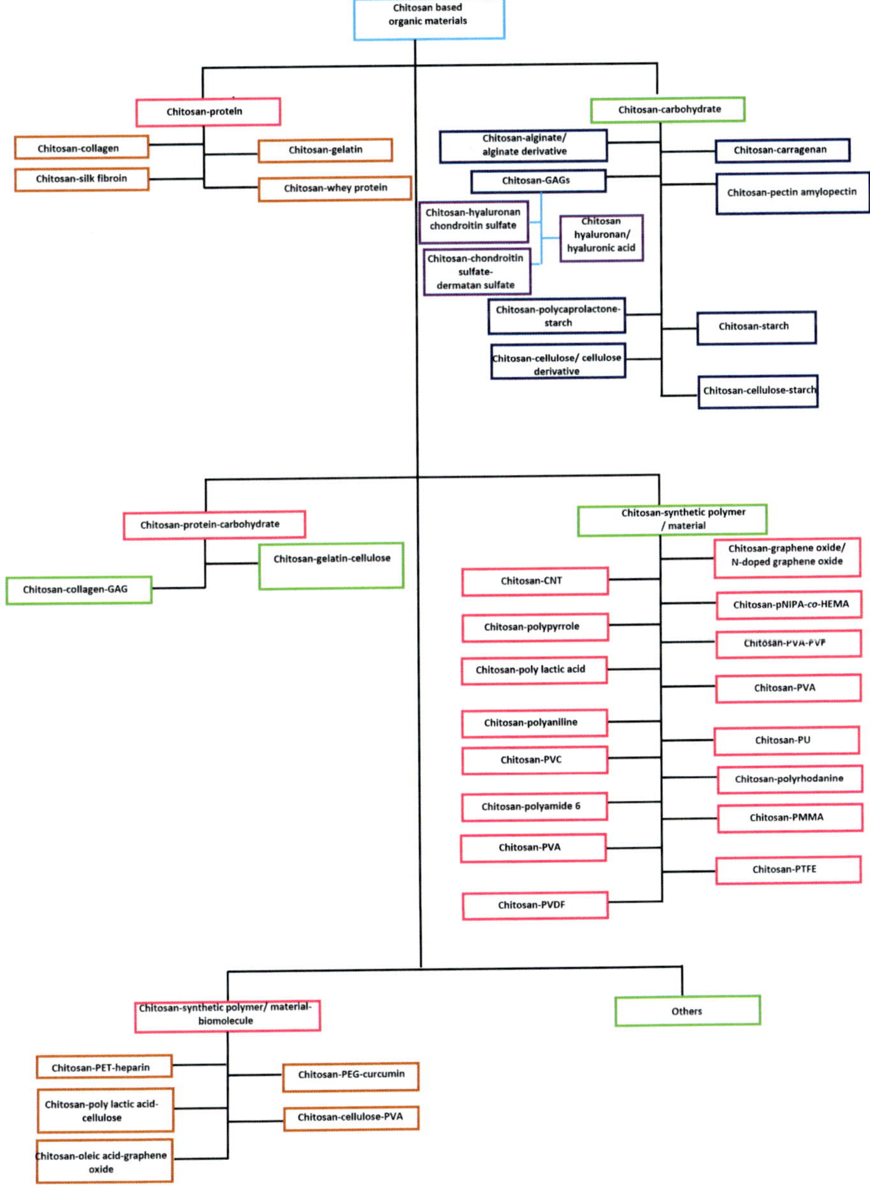

Chart 6.3 Generalized classification of chitosan-organic blends and composites.

Among these chitosan-organic materials, conductive carbonaceous fillers, such as graphenes and carbon nanotubes, and conductive polymers, e.g., polyaniline, are preferably employed in fabricating sensors and biosensors. Indeed, most of the synthetic polymer-containing composites or blends are less suitable for biomedical applications, as these materials have limited biodegradability, biocompatibility, and mechanical strength. However, chitosan blended with poly(vinyl alcohol) possesses good mechanical and chemical properties. Being a nontoxic, water-soluble, biocompatible, and biodegradable synthetic polymer (Singha, Kar, Ray, 2009; Singha, Parya, Ray, 2009; Singha & Ray, 2010; Das et al., 2011), PVA offers good tensile strength, flexibility, and barrier properties to oxygen and aroma. Thus, PVA is applied in the biomedical and biochemical fields. Moreover, chitosan–PVA is used for adsorptive removal of Pb(II) ions from aqueous solutions (Wan Ngah et al., 2004). To further enhance the stability of chitosan–PVA beads, the blends are cross-linked with a suitable cross-linking agent, e.g., glutaraldehyde, which can hold and interconnect both components by covalent bonds (Kumar et al., 2009). Moreover, the complex of alginate and chitosan can be employed for metal ion removal, as alginate salts show high affinity to metal ions (Wan Ngah & Fatinathan, 2008). Notably, the scaffolds made from synthetic material-based composites are rejected within a short time after transplantation due to allergic reaction. In this regard, natural organic polymer-based composites and blends are more suitable for artificial organ or tissue engineering.

3.2.2 Chitosan-inorganic materials

Unlike chitosan-organic materials, these materials are mostly composites containing different types of inorganic components including clays, organoclays, metals, and metal oxides. These materials are naturally thermally more stable and mechanically strong, as inorganic components often act as reinforcing fillers (Singha, Deb et al., 2020; Singha, Chattopadhyay et al., 2020; Samanta et al., 2012; Singha, Karmakar et al., 2019; Chattopadhyay et al., 2010). At the same time, these materials are more suitable to replace and repair hard tissues including bones in the field of tissue engineering. Such materials can be subclassified based on the type of inorganic components present in the composite (Chart 6.4).

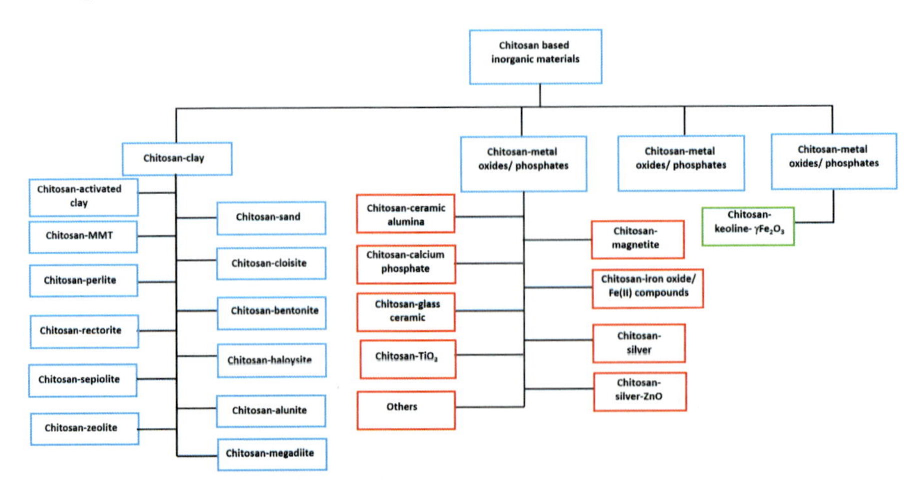

Chart 6.4 Various kinds of chitosan-inorganic materials.

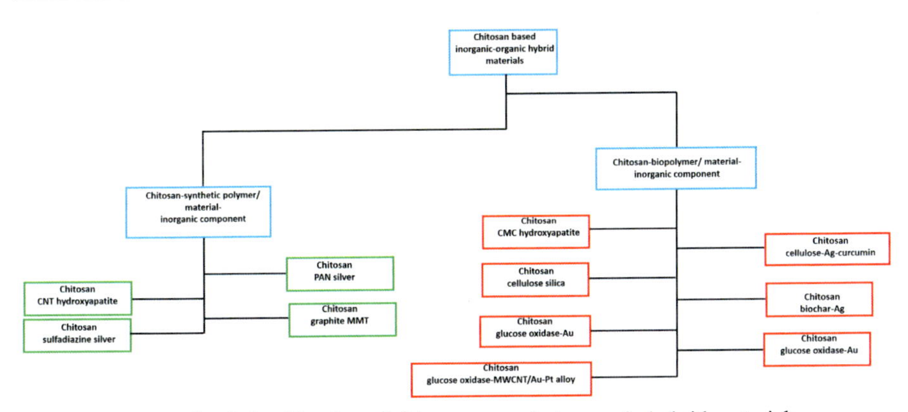

Chart 6.5 Generalized classification of chitosan-organic-inorganic hybrid materials.

Among these inorganic material-based composites, clays and organoclays containing chitosan composites are generally not suitable for tissue engineering applications, as the clay components are usually less biocompatible and not biodegradable. Therefore, the applications of clay containing chitosan composites are confined within some nonmedical applications including adsorptive removals of dyes, heavy metal ions, and other hazardous components from wastewater.

3.2.3 Chitosan-organic-inorganic hybrid materials

Compared to both the previous classes, these materials are more complicated and have been developed in the recent times. These materials have been developed to derive the major advantages of the two previous classes, with different types of these materials now increasingly employed in various sophisticated applications (Chart 6.5).

4. Synthesis of some selected blends and composites of chitosan

Thus, there are different kinds of chitosan-based blends and composites, and each of these composites and blends is shaped into various forms to fulfill varied objectives. In fact, a particular blend/composite is synthesized in different ways depending on its end use. For instance, when a composite or blend is fabricated into a scaffold to be applied in vivo, the electrospinning methodology is used while taking the special precautionary measures to avoid the formation of unwanted impurities. On the other hand, chitosan composites for nonbiomedical applications do not need such stringent precautionary measures. Thus, it is not possible to cover all of the different synthetic procedures employed for manufacturing so many types of chitosan-based blends and composites within the limited space of this chapter. Therefore, some selected methodologies for the synthesis of chitosan-based composites and blends most suitable for adsorptive removal and tissue engineering applications have been assembled.

4.1 Synthesis of some chitosan-inorganic blends and composites

4.1.1 Chitosan—clay composites/nanocomposites

Herein, the basic strategies involve mainly the following: (1) to facilitate dissolution of chitosan by the addition of acetic acid; (2) to functionalize the chitosan and inorganic components by oxalic acid or other agents; (3) to produce stable dispersion by addition of suitable emulsifier(s); (4) to create bondage between chitosan with inorganic components by applying cross-linking agents, such as, epichlorohydrin and glutaraldehyde; and (5) to separate, purify, dry, and reduce the size of as-prepared materials by suitable media, drying technique(s), and mechanical operations.

4.1.1.1 Chitosan—sand composites

For preparing chitosan—sand composites, about 5.0 g chitosan is mixed with 100 g sand and 300 mL 5% HCl. An acidic environment (i.e., pH = 1.5) is necessary to ensure solubilization of chitosan, facilitating the uniform distribution of sand particles. This is followed by 5 h room temperature stirring, and then the mixture is neutralized by gradual dropwise addition of NaOH, leading to the formation of chitosan-coated sand via superficial deposition of chitosan onto the sand surface. Thereafter, these chitosan-coated sand particles are filtered, followed by washing and vacuum drying. Finally, these particles are ground and sieved, and the particles >0.425 mm are finally employed as the adsorbent (Wan et al., 2007).

4.1.1.2 Chitosan-activated clay composites

The first step for the preparation of chitosan-activated clay composites is the removal of proteins from the dried chitin-rich cuttlebone cartilage via 18 h digestion in dilute alkaline environment, i.e., 5 wt.% NaOH, followed by removal of unwanted $CaCO_3$ via mineral acid treatment for 18 h. Thereafter, the as-extracted insoluble chitin is converted to chitosan with the desired degree of deacetylation and molar mass through deacetylation in strongly alkaline conditions, i.e., 50 wt.% NaOH at 90°C for 3 h. Subsequently, the chitosan flakes are washed repeatedly with deionized water to remove impurities followed by drying at 50°C under vacuum. The activation of natural clay of desired particle size is done by refluxing with dilute H_2SO_4 for 2 h at 80°C, followed by air-cooling and filtration of the slurry with a glass fiber. Thereafter, filter cake of the activated natural clay is washed repeatedly with deionized water to remove the excess acid (Chang & Juang, 2004).

4.1.1.3 Chitosan—MMT nanocomposites

For preparing the chitosan—MMT nanocomposites, initially chitosan solution is prepared via dissolution of chitosan in dilute acetic acid at 50°C. Herein, the primary amines of chitosan react with acetic acid to produce $-NH_3^+Ac^-$ (Fig. 6.10) (Nesic et al., 2012). To ensure the greater conversion of $-NH_2$ to $-NH_3^+$, the pH of the chitosan solution is often adjusted to 4.9 with dilute NaOH. Since pK_a of $-NH_2$ is 6.3%, 95% of the $-NH_2$ of chitosan in a chitosan—clay mixture should be protonated

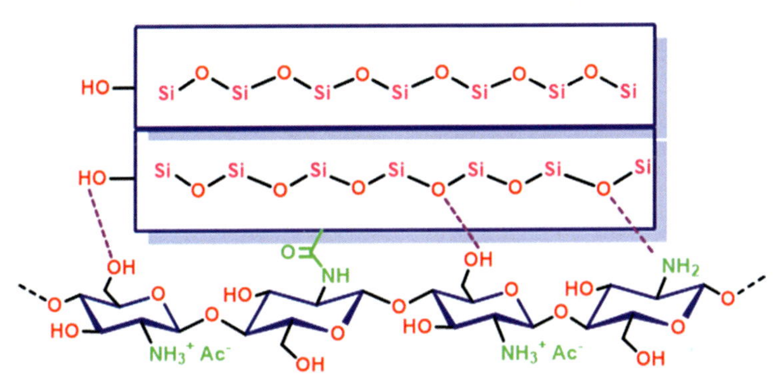

Figure 6.10 Complexation between chitosan and MMT.

to $-NH_3^+$ at pH = 5 (Darder et al., 2003; Wang & Wang, 2007). The MMT component is separately dispersed in dilute acetic acid with constant stirring at 50°C. Later, the chitosan solutions and MMT dispersion are admixed, and the mixture is washed with purified water and cast on Petri dishes through evaporation at room temperature for 24 h followed by size reduction (Nesic et al., 2012; Darder et al., 2003; Wang & Wang, 2007). At this point, being a polycationic biopolymer in acidic media, there is every possibility of chitosan biomacromolecules being intercalated between the layers of Na^+−MMT by cationic exchange processes (Darder et al., 2003). In fact, the negatively charged surface on MMT originates from the isomorphous substitutions of Al(III) for Si(IV) in the tetrahedral layer and Mg(II) for Al(III) in the octahedral layer of MMT (Wan Ngah et al., 2011; Mondal et al., 2010). Amino and two hydroxyl groups of chitosan can form hydrogen bonds with the silicate hydroxylated end groups of clay, leading to a strong interaction between chitosan and clay. Such an interaction results in the shifting of amide of chitosan from 1635 cm^{-1} to 1645 cm^{-1}. In addition, the $-NH_2$ peak shifts from 1540 to 1560 cm^{-1} (Nesic et al., 2012). The chitosan hydroxyl group forms hydrogen bonds with the Si−O−Si groups of silicate multilayer of MMT (Fig. 6.10) (Nesic et al., 2012), resulting in the enhanced thermal stability of chitosan−MMT compared to the individual components. However, if the chitosan concentration is high, the excess chitosan cannot access the MMT surfaces. Accordingly, each of those excess chitosan macromolecules is compelled to interact with one MMT bilayer and another chitosan acetate chain (Fig. 6.11) (Darder et al., 2003). Thus, the layers of MMT move further apart from 1.20 to 2.09 nm to accommodate two chitosan macromolecules within the interspace of the two adjacent MMT layers. In some cases, a portion of the intercalated MMT layers may be exfoliated, leading to a significant decrease in intensity of the prominent peak at $2\theta = 6.94°$ (Wang & Wang, 2007).

4.1.1.4 Chitosan−cloisite nanocomposites

Similar to chitosan−MMT nanocomposites, the desired amount of chitosan is dissolved in a dilute aqueous solution of acetic acid with continuous agitation. However, unlike MMT, cloisite suspensions in acetic acid are not prepared, and the cloisite 15A

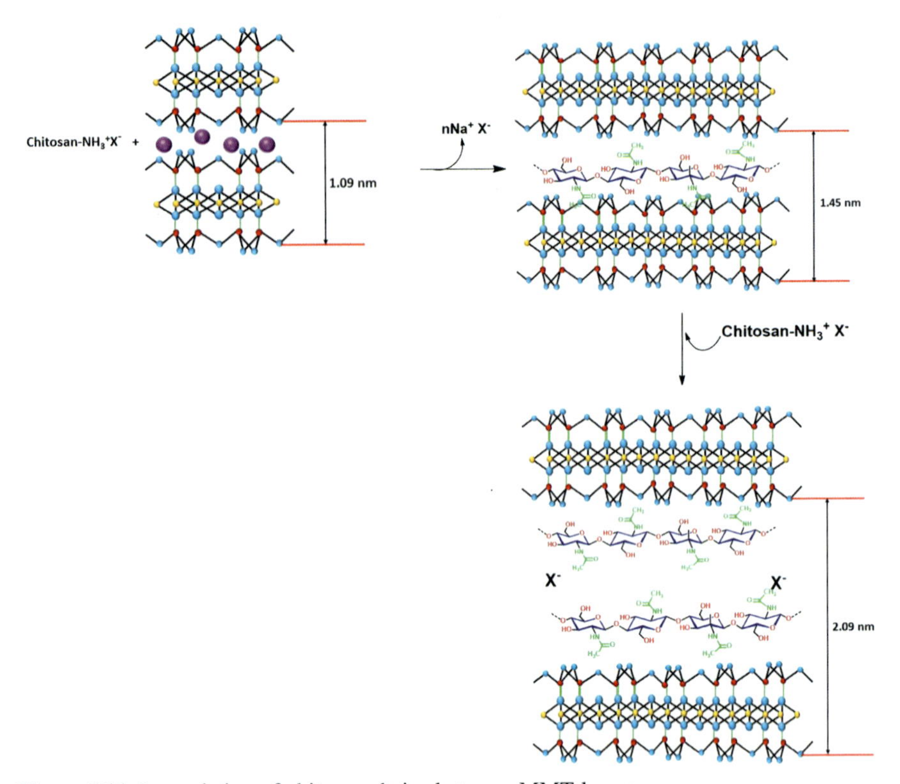

Figure 6.11 Intercalation of chitosan chains between MMT layers.

and 30B nanoparticles are directly added to the chitosan solution with continuous stirring for 48 h followed by final casting of chitosan–cloisite nanocomposites (Daraei et al., 2013). Similar to chitosan–MMT nanocomposites, the penetration of chitosan chains into the silicate galleries results in an increment of the d-spacing of around 0.27 nm compared to that of cloisite 15A (Daraei et al., 2013). However, such penetration of chitosan chains is not good in cloisite 30B, although the initial order of silicate nanolayers is absent in both cloisites. Better penetration and thereby better interactions between silicate layers and chitosan in chitosan–cloisite 15A nanocomposite are substantiated by a substantial shift of protonated amine vibration from 1600 to 1559 cm^{-1} because of ionic interactions among protonated amines in chitosan and negatively charged nanoclay particles. Because of the relatively poor cloisite–chitosan interactions, such shifting is not as prominent in chitosan–cloisite 30B nanocomposite. For preparing the chitosan–cloisite nanocomposite-coated PVDF membrane, moist PVDF membrane is immersed in the nanocomposite solution followed by drying at the ambient temperature for 24 h. Subsequently, to make the chitosan coating insoluble in water and to neutralize the excess acid, the membranes are immersed in 1 M NaOH followed by washing with distilled water and subsequent room temperature drying (Daraei et al., 2013).

4.1.1.5 Chitosan–perlite composites/nanocomposites

Unlike MMT–chitosan composites, for preparing chitosan–perlite composite beads, oxalic acid is used as the medium to treat perlite at room temperature followed by removal of excess oxalic acid by washing with deionized water. Herein, the main objectives of oxalic acid treatment are to remove acid-soluble impurities of perlite as well as to introduce acidic functionalities onto the perlite surface augmenting adhesion of chitosan with perlite (Shameem et al., 2008). Thereafter, the acid-treated perlites are dried at 70°C, and sieved through a 100-mesh sieve. Bead preparation requires viscous chitosan gel, and accordingly, the higher amount of chitosan (30 g) is admixed to oxalic acid solution within 40–50°C. Thereafter, an aqueous dispersion of oxalic acid-treated perlite powder is slowly added to the gel, and highly porous beads are prepared by dropwise addition of perlite–chitosan gel mixture into 0.7 M NaOH. Herein, the concentration of NaOH is critical to ensure the formation of spherical beads through rapid neutralization of oxalic acid with alkali. In this context, a relatively dilute NaOH solution results in the disintegration and deformation of beads. After careful washing and neutralization, the beads are preferably dried using a freeze-drying method to retain the spherical shape (Shameem et al., 2003, 2006; Kalyani et al., 2009). Herein, the chitosan interacts strongly with perlite, an inorganic porous aluminosilicate composed mainly of alumina and silica (Shameem et al., 2003), resulting in more active sites in the nanocomposite. In this regard, the strong interaction of C in chitosan with Si and Al in the perlite is realized from the massive shift of C1s peak from 284.3 eV of chitosan to 283.0 eV in the nanocomposite, along with appreciable widening of all the deconvoluted peaks (Shameem et al., 2008). Thus the functional groups of chitosan, i.e., $-NH_2$ and O–H, may form a complex through cross-linking with a constituent of perlite during the coating process (Fig. 6.12) (Shameem et al., 2006). Such irreversible bonding also resulted in massive changes in pH_{PZC} within 6.0–6.5 of chitosan to 8.5 in nanocomposite.

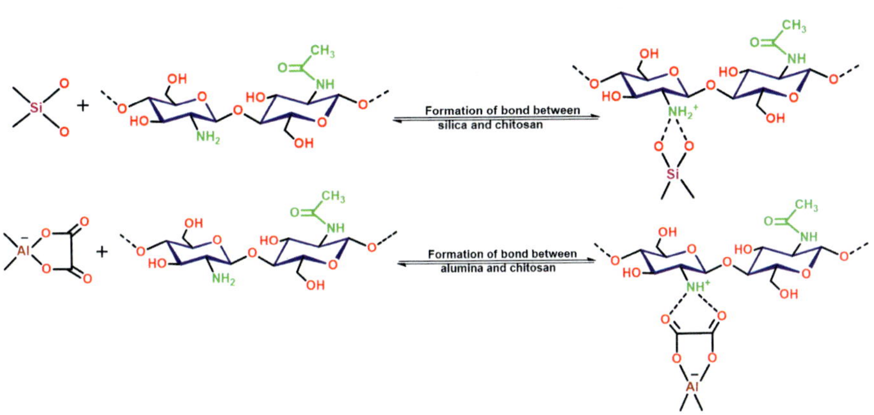

Figure 6.12 Formation of complex through cross-linking of amino/hydroxyl groups of chitosan with constituent of perlite.

4.1.1.6 Chitosan—bentonite composites/nanocomposites

For fabricating chitosan—bentonite composites/nanocomposite beads, both chitosan and bentonite powders are added to dilute acetic acid and soaked overnight. Similar to the chitosan—perlite composite/nanocomposite beads, the mixture of chitosan and bentonite powders is added dropwise to a 0.50 M NaOH solution with continuous stirring at 100 rpm. To achieve stable chitosan coatings on beads, the as-prepared bentonite beads are stirred with 0.01 M epichlorohydrin cross-linker at 50°C followed by rinsing several times thoroughly with hot distilled water and then with cold distilled water. Finally, the cross-linked chitosan-coated bentonite beads are air dried, ground, and sieved to obtain the desired mesoporous adsorbent of sizes between 2 and 50 nm (Wan Ngah, Ariff, & Hanafiah, 2010; Wan Ngah, Ariff, Hashim, & Hanafiah, 2010). Similar to other chitosan—clay composites, a significant drop in the crystalline nature of bentonite is evident from the disappearance of the bentonite-specific peak at $2\theta = 11.6°$ together with lowering of the peak intensity of another bentonite-specific peak at $2\theta = 28.1°$. In addition, such an interaction and cross-linking associated with a drop in crystallinity are understood from the reduced intensity of the chitosan-specific broad peak at $2\theta = 20.3°$ in chitosan—bentonite composite.

4.1.2 *Chitosan—metal oxides/phosphates*

4.1.2.1 Chitosan—ceramic alumina composites

Similar to chitosan—perlite composites/nanocomposites, oxalic acid is used to functionalize the ceramic alumina component of chitosan—ceramic alumina composites. Initially, oxalic acid-treated alumina is washed repeatedly with deionized water to remove excess acids. The main objective behind oxalic acid treatment is to generate a bridge between alumina and chitosan (Fig. 6.13). In fact, one carboxylate group of oxalic acid forms a surface chelate via ester linkage with the alumina, while another carboxylate group forms ionic/electrostatic bond with $-NH_3^+$ groups in chitosan. The oxalic acid can also form hydrogen bonds with O—H, $-CH_2OH$, or$-NH_2$ groups on biopolymer (Veera, Krishnaiah, Ann, & Edgar, 2008). Thereafter, a viscous chitosan gel is prepared by adding 50 g medium-molecular-weight chitosan to 1000 mL 10 wt.% oxalic acid solution with continuous stirring within 40—50°C followed by dilution of gel by water addition. Later, 500 g oxalic acid-treated alumina is added slowly to diluted gel and stirred for about 36 h. The composite material is settled, filtered, washed, dried, and then stored in a desiccator (Veera, Krishnaiah, Ann, & Edgar, 2008; Veera, Krishnaiah, Jonathan et al., 2008; Veera et al., 2003).

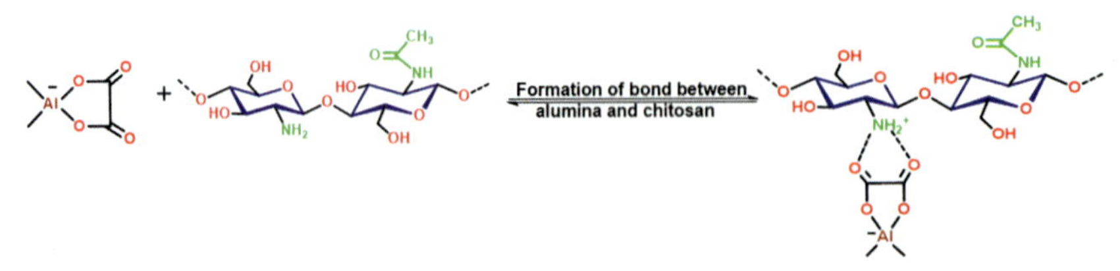

Figure 6.13 Interaction between chitosan and alumina in chitosan—alumina composites.

4.1.2.2 Chitosan–magnetite nanocomposites

The first step in preparing magnetic chitosan nanocomposites normally involves the synthesis of amine-functionalized magnetite nanoparticles by solvothermal reaction between $FeCl_3 \cdot 6H_2O$ and 1,6-hexanediamine in the presence of anhydrous sodium acetate and ethylene glycol. Herein, the mixture is heated at 200°C for 6 h in a Teflon-lined autoclave chamber. Thereafter, the treated nanoparticles are separated using an external magnet. Thus, the magnetic property remains intact despite amine functionalization of magnetite nanoparticles. Such amine functionalization of magnetite nanoparticles can be realized from the peaks at 1629 and 3446 cm^{-1} related to N–H vibrations of free 1,6-hexanediamine and C–H stretching within 2852–2932 cm^{-1} of the alkyl chain (Liu et al., 2009). Thereafter, amine-functionalized magnetite nanoparticles are converted to carbonyl-functionalized magnetite nanoparticles by treating with glutaraldehyde. Herein, the –CHO of glutaraldehyde reacts with $-NH_2$ of amine-functionalized magnetite nanoparticles. Thereafter, the functionalized magnetite nanoparticle-based nanocomposites are fabricated by the addition of these particles to chitosan solution. Finally, chitosan-based functionalized magnetite nanoparticle composites are again separated by an external magnet, indicating marginally unaffected magnetic properties of magnetite nanoparticles within composites (Liu et al., 2009). In this regard, the chitosan-coated magnetite nanoparticle-based nanocomposites show sufficient ferromagnetic behavior. The reported values of magnetic saturations are 80 and 74 emu/g for magnetite and the chitosan–magnetite composite, respectively. The marginal decrease in magnetic saturation may be attributed to the increased mass of glutaraldehyde and chitosan on the surface of the magnetic nanoparticles. Thus, the as-prepared nanocomposites with excellent magnetic property can be separated easily from the solution with the help of an external magnetic force (Fig. 6.14) (Liu et al., 2009). Finally, the materials are dried under vacuum at 40°C (Liu et al., 2009). Sometimes, chitosan–magnetite composites are prepared by chemical coprecipitation of Fe(II) and Fe(III) ions by NaOH in chitosan previously treated with acetic acid. In this case, the chitosan-coated beads presented a lower saturation magnetization of 55 emu/g (Tran et al., 2010). Herein, the interactions between magnetic nanoparticles and chitosan are mediated by the electrostatic interaction between surface-negative charged Fe_3O_4 and positively protonated chitosan, leading to shifting of Fe–O stretching from 610 to 595 cm^{-1} and N–H bending from 1638 to 1681 cm^{-1}. Hence, an external magnet is not used to separate and purify these superparamagnetic composites, and the particles are washed and dried to obtain chitosan–magnetite composite (Tran et al., 2010).

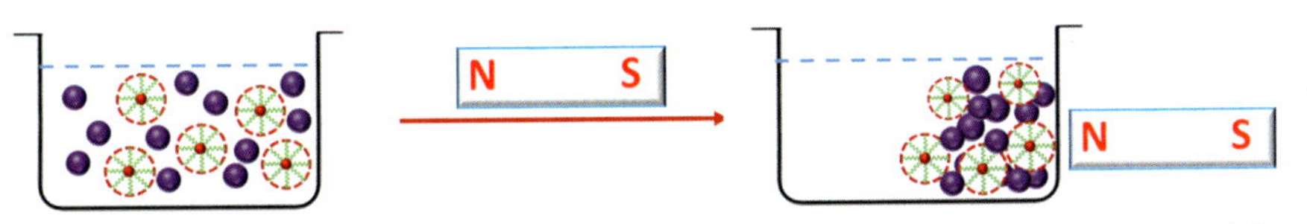

Figure 6.14 Photographs of the magnetic chitosan nanocomposite colloidal solution containing 10 mL/L Pb(II) before and after magnetic separation by an external magnetic field.

4.1.3 Chitosan–clay-metal oxides/phosphates

4.1.3.1 Chitosan–kaolin–*g*-Fe_2O_3 composites

Similar to most chitosan-based inorganic composites, dilute acetic acid solution is used to facilitate the initial dissolution of chitosan. Thereafter, other components, i.e., kaolin and *g*-Fe_2O_3, are added with constant stirring along with emulsifiers including Span-80 to produce a stable dispersion of chitosan–kaolin–$_g$-Fe_2O_3 adduct. Thereafter, glutaraldehyde is used as a cross-linking agent to create stronger bonds among the chitosan and other components. Finally, the pH is adjusted within 9–10 by adding dilute NaOH, and the composite is repeatedly washed with N,N-dimethylformamide, ethanol, and distilled water followed by drying under atmospheric conditions (Zhu et al., 2010). In this ternary composite, the bonding among three components removes the preexisting *g*-Fe_2O_3 peak at 43.36, ascribed to 400 plane along with the disappearance of kaolin-specific peaks at 19.94° and 55.12° (Zhu et al., 2010). Moreover, some new peaks arrive at 14.80°, 18.92°, 33.46°, 41.1° and 49.22°.

4.2 Synthesis of some selected chitosan-organic blends and composites

4.2.1 Chitosan–PVA blend

Followed by initial dissolution of chitosan in acetic acid, a chitosan solution is blended with a dilute aqueous solution of PVA until the attainment of a homogeneous gel blend. Thereafter, the blend is neutralized by adding dilute NaOH, which results in coagulation of chitosan gel into spherical uniform beads of chitosan–PVA, followed by filtration, repeated washing, air-drying, size reduction by grinding, and sieving to achieve chitosan–PVA microbeads of uniform sizes (Wan Ngah et al., 2004).

The aforementioned process is slightly modified for preparing cross-linked chitosan–PVA microbeads. Herein, the first step constitutes of preparing chitosan–PVA microbeads followed by cross-linking in the final step. After making the homogeneous gel blends of chitosan–PVA, it is necessary to reduce the water content of the gels by distillation in the presence of organic solvents, i.e., toluene and chlorobenzene, and surfactant, i.e., tween 80. Herein, the strategy is to produce an azeotropic mixture of water, solvents, and surfactants followed by removal of the mixture through boiling at 90°C. After necessary water removal and cooling, the concentrated glutaraldehyde solution is added as a cross-linker followed by stirring at room temperature for desired cross-linking. The subsequent steps of drying and bead making are very similar to the procedure adopted for making noncross-linked chitosan–PVA microbeads (Fig. 6.15).

4.2.2 Chitosan–alginate blend

After the usual dissolution of chitosan in dilute acetic acid solution, the blended homogeneous gel of chitosan and alginic acid is prepared by mixing for a sufficient period of time. Thereafter, NaOH is added to neutralize the blend and simultaneously to coagulate and transform the chitosan in the form of beads. Finally, the chitosan–alginate beads are filtered and washed to remove excess NaOH followed by air drying, size

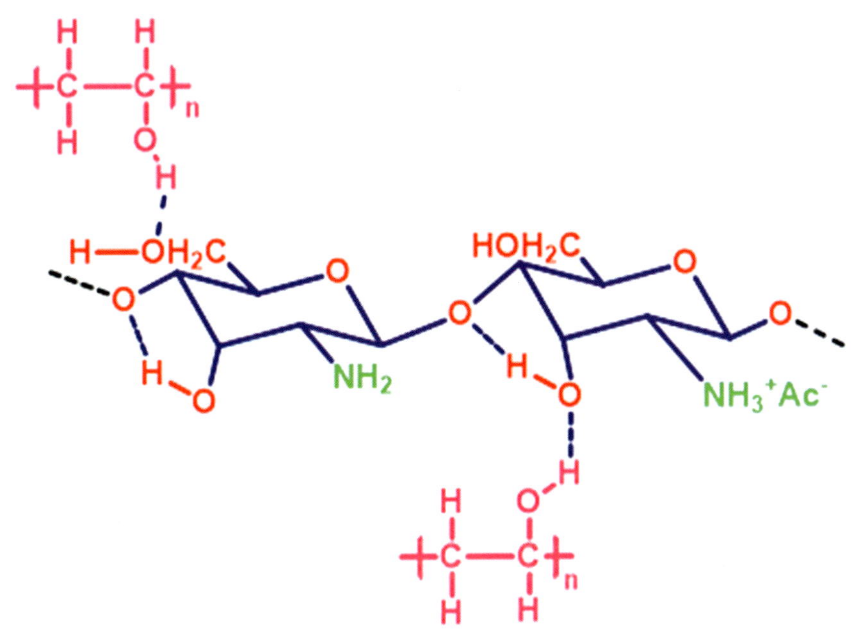

Figure 6.15 Hydrogen bonding interactions of chitosan with PVA in chitosan–PVA blends.

reduction via grinding, and sieving to attain microparticles of the desired size range (Wan Ngah & Fatinathan, 2008). The intimate electrostatic interaction between $-NH_3^+$ of chitosan and $-COO^-$ of alginate is realized from a new peak at around 1420 cm^{-1} along with the complete disappearance of the –COOH peak at 1736 cm^{-1}, suggesting complete dissociation of –COOH to form ion pairs with the protonated amino groups of chitosan through an electrostatic interaction.

4.2.3 Chitosan–cotton fiber composites

As mentioned previously, a selective adsorbent based on Schiff base-type chitosan-coated cotton fiber is prepared using the following steps:

i. Initially, the cotton fibers are functionalized by oxidation using sodium periodate. Indeed, the conversion of cotton to oxidized cotton and thereby transformation of O–H to –CHO can be realized from the arrival of a peak at 1730 cm^{-1} ascribed to C=O stretching of aldehyde. Such a phenomenon is substantiated from the decreased intensity of cotton-specific crystalline peak at 15.5°.
ii. After the termination of reaction by ethylene glycol, the oxidized cotton fibers are filtered, washed to remove the excess oxidizing agent, followed by drying under vacuum at 60°C.
iii. As usual, chitosan is dissolved by adding dilute acetic acid followed by the addition of oxidized cotton fiber fragments in the dispersion of chitosan.
iv. Later, the mixture of chitosan and oxidized cotton fiber is filtered, washed with 0.4% (v/v) acetic acid solution and deionized water, and dried under vacuum at 60°C (Zhang et al., 2008). The particular C=O stretching peak of aldehyde shifts from 1730 to 1716 cm^{-1}

in chitosan-coated cellulose, suggesting the formation of a Schiff base (C=N) through the reaction between the aldehyde group of oxidized cotton and primary amine of chitosan (Zhang et al., 2008). Moreover, an appreciable conversion of aldehyde into Schiff base is reflected with complete destruction of the cotton-specific crystalline peak at 15.5°.

v. Thereafter, C–N bond-linked chitosan-coated cotton fiber is prepared via reduction of Schiff base-type chitosan-coated cotton with $NaBF_4$ (Qu, Sun, Fang et al., 2009). The conversion of Schiff base (C=N) to C–N is realized from the complete disappearance of the Schiff base (C=N) specific peak at 1716 cm^{-1} (Qu, Sun, Fang et al., 2009; Zhang et al., 2008).

4.2.4 Chitosan–cellulose-based blends and composites

Having the highest availability among all the biopolymers, cellulose and cellulose derivatives have been attempted to be combined with chitosan to produce blends and composites suitable for both medical and nonmedical uses. For preparing the composite scaffolds based on carboxymethyl cellulose, hydroxyapatite, and chitosan (Jiang et al., 2008), the required amount of chitosan and carboxymethyl cellulose are admixed to a slurry of nanohydroxyapatite with constant stirring followed by the addition of acetic acid to help dissolution of chitosan. The stirring is continued until the mixture is solidified via the evaporation of solvents. Thereafter, the solidified mixture is freeze-dried at −30°C, neutralized with NaOH to remove excess acetic acid, and washed with deionized water to obtain a porous scaffold. By inserting the carboxymethyl group, i.e., $-CH_2COOH$, in cellulose, the $-COO^-$ of cellulose derivative interacts with $-NH_3^+$ of chitosan by the ionic interaction. Additionally, both conventional and nonconventional hydrogen bonds take part in the intimate interactions among cellulose derivative, chitosan, and nanohydroxyapatite (Jiang et al., 2008). Notably, although no cross-linker is added during fabrication, the necessary biostability of the composite is contributed by the reinforcing effect of nanohydroxyapatite particles. Later, for preparing dressing materials, attempts are made by incorporating cellulose nanocrystals in chitosan films loaded with Ag and curcumin. Here, acetic acid is used for preparing the dispersion of chitosan followed by the addition of cellulose nanocrystals. Curcumin functions as an antimicrobial agent and also the reducing agent for reducing Ag nitrate to Ag nanoparticles, whereas the role of cellulose nanocrystals is to strengthen the matrix (Bajpai et al., 2017). These cellulose nanocrystals envisage enough potential in strengthening chitosan-based composites containing either starch or gelatin (Noorbakhsh-Soltani et al., 2018). Herein, a nanocellulose gel is prepared by dispersing cellulose nanocrystals in water, followed by dispersion of the gel in a chitosan solution in the presence of acetic acid. Thereafter, the mixture is mixed further with either starch/gelatin at 60/40°C, respectively. Finally, the mixture is transformed into the respective films by adopting conventional procedures. Similar to cellulose nanocrystals, rosin-modified cellulose nanofibers are incorporated as reinforcing agent for packaging materials made of polylactic acid and chitosan. Rosin modification of cellulose is conducted to improve the compatibility and dispersion of the reinforcing filler in the polylactic acid matrix (Niu et al., 2018). Rosin modification of cellulose is an esterification reaction in which some O–H groups of cellulose convert to esters by reacting with −COOH of abietic acid, levopimaric acid, and pimaric acid components of rosin, confirmed by the arrival of a new ester carbonyl specific peak at 1730 cm^{-1} coupled with disappearance of the −COOH peak of rosin at 1700 cm^{-1} (Niu et al., 2018). Thereafter, the modified cellulose nanofibers are incorporated into a dispersion of polylactic acid in dichloromethane, and finally, the mixture is cast on a chitosan layer prepared via dissolution in acetic acid, molding, and other conventional steps. A recent publication has described the preparation of composite sponge based on hydroxypropyl methylcellulose and chitosan. The base matrix was prepared by conventional solution mixing in acetic acid followed by gel preparation via neutralization of excess acetic acid (Freag et al., 2018). Recently, a combination based on N,O-carboxymethyl chitosan and regenerated cellulose has been reported in which modifications of chitosan and cellulose were executed to prevent adhesion of wound-healing material with the injured area (Cheng et al., 2018). Herein, the oxidation of cellulose was carried out at 19.5°C by mixing the regenerated cellulose into an oxidizing agent comprised of NO_2 and CCl_4. Thereafter, the N,O-carboxymethyl chitosan was cross-linked with oxidized regenerated cellulose by means of a cross-linking system composed of 1-ethyl-3-(3-dimethylaminopropyl)-carbodiimide (EDC) and N-hydroxyl-succinimide. In fact, such a cross-linking system has been used to bind collagen and chitosan while preparing a biostable scaffold for tissue engineering (Wang et al., 2003). The rest of the steps including freeze-drying and washing were followed as usual. Recently, bacterial cellulose has increasingly been employed to produce high-strength chitosan-based packaging materials with higher purity, crystallinity, tensile strength, and water-holding capacity than those of plant cellulose (Yin et al., 2019; Ju et al., 2020). Cellulose extracted from bacterial species can be admixed with chitosan to obtain materials applicable as wound dressings, artificial skin substrates, edible packing films, and drug-delivery systems.

4.2.5 Chitosan−collagen-based blends and composites

Collagen−chitosan combinations have been under investigation since the second half of the 20th century (Taravel & Domard., 1993, 1995, 1996). Complexes of fully deacetylated chitosan and atelocollagen have been studied by mixing chitosan with atelocollagen obtained from alkaline treatment of calf skin (Taravel & Domard., 1993). The FTIR investigation of the combined film demonstrates purely electrostatic interactions between the carboxylic functions of collagen and ammonium groups of chitosan. Another work described the preparation of a blend in a similar way (Thacharodi & Rao, 1996). Equal volumes of 0.5% (w/v) solutions of chitosan and collagen have been prepared in the presence of 0.5% acetic acid followed by homogeneous mixing. The weakly acidic medium is required to ensure proper dissolution and mixing of both collagen and chitosan chains. In the presence of protons, both chitosan and collagen chains should be cationic and hence minimize the intermolecular coulombic attractions. The mixture is stored at 4°C until complete removal of air bubbles. Thereafter, the blend film is prepared by drying the mixture at 4°C in a Petri dish followed by neutralization by adding dilute NaOH solution. The neutralization is carried out to generate more $-COO^-$, leading to enhanced ionic interactions between $-NH_3^+$ of chitosan and more frequently available $-COO^-$ of collagen. In another work, collagen–chitosan blends showed the formation of esters and anhydrides as cross-linking bridges despite the absence of an external cross-linking agent (Wang et al., 2005). Meanwhile, to improve the stability and longevity of the chitosan–collagen combinations, glutaraldehyde has been used as a cross-linker in preparing scaffolds for tissue engineering (Ma et al., 2003). Herein, the chitosan functioned as a cross-linking bridge. For preparing the scaffold, initially, type I collagen is isolated from fresh bovine tendon by trypsin digestion at 37°C and acetic acid dissolution at 4°C. In fact, enzymatic digestion is conducted at body temperature to exploit the maximum activity of enzyme. Thereafter, for preparing the collagen–chitosan scaffold, the extracted and purified collagen and chitosan are separately dissolved in 0.5 M acetic acid solution to prepare 0.5% (w/v) solutions of each. Subsequently, chitosan solution is added dropwise into the collagen suspension while maintaining a ratio of 9:1 (collagen:chitosan). After the necessary deaeration under vacuum, the collagen–chitosan blend is injected into a mold, frozen in a 70% ethanol bath at −20°C for 1 h, and lyophilization is conducted for 24 h to obtain a porous collagen–chitosan scaffold. The removal of entrapped air is necessary to avoid the formation of air-filled pockets in the scaffold to attain the uninterrupted functioning of the scaffold. The same procedure is adopted by several researchers to prepare collagen–chitosan scaffolds for tissue engineering (Guo et al., 2011; Shi et al., 2009; Judith et al., 2012). The as-prepared scaffold is treated with glutaraldehyde to improve its bio-stability (Ma et al., 2003). The added glutaraldehyde may be involved in cross-linking by reacting with the $-NH_2/O-H$ groups of both collagen and chitosan (Fig. 6.16). In fact, such cross-linking is also evident in collagen during glutaraldehyde tanning of leather. Researchers have reported the cross-linking of collagen and chitosan by a cross-linking system composed of 1-ethyl-3-(3-dimethylaminopropyl)-carbodiimide (EDC) in N-hydroxysuccinimide (NHS) and a 2-morpholinoethane sulfonic acid (MES) buffer system (Wang et al., 2003). In contrast to the previous work, the cross-linking has been characterized by spectrochemical evidences. The preparation methodology is almost the same except for the cross-linking system. Herein, the major strategy is to convert $-NH_2$ of both collagen and chitosan into amides. Thus, as per the FTIR findings, numerous $-NH_2$s are converted into > NH linkages,

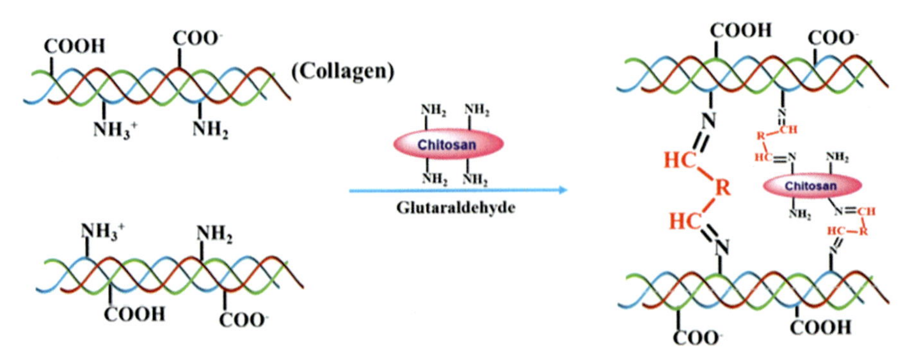

Figure 6.16 Schematic of collagen cross-linked with glutaraldehyde in the presence of chitosan.

suggesting the cross-linking of collagen and chitosan by the imide-type cross-linkers. In addition, the carboxyl groups of collagen react with the amino groups of chitosan and with hydroxyl groups of chitosan resulting in the formation of amide and ester linkages, respectively (Wang et al., 2003). The same cross-linking system and procedure are applied to prepare the collagen–chitosan scaffold matrix for holding chitosan microspheres loaded with growth factor for cartilage tissue engineering (Lee et al., 2004).

4.2.6 Chitosan-based blends and composites bearing graphenes, CNTs, and their derivatives

Graphene obtained from graphite is equipped with diverse properties and application prospective due to its outstanding mechanical properties and conductivity. Graphene oxide obtained by the oxidation of graphene has −COOH and O−H functional groups on the basal planes and at the edges of the sheets. Accordingly, it possesses some unique properties including high specific surface area and surface activity to enhance the mechanical properties and adsorption capacity (Fayyad et al., 2016; Khan et al., 2016; Jin et al., 2019; An et al., 2017; Han et al., 2011; Ge & Ma, 2015; Debnath et al., 2017; Shah & Tasmia, 2018; Chang et al., 2019; Liu, Liu et al., 2019). On the other hand, being a carbonaceous filler with large specific area, highly porous, hollow structure, good antifouling, reinforcing properties, biocompatibility, and biodegradability, CNTs can be incorporated to strengthen chitosan film as well as to enhance the conductivity and porosity (Yan et al., 2016; Li, Liu et al., 2005; Kang et al., 2007; Qiu et al., 2008; Jirimali et al., 2015; Bibi et al., 2015; Maity & Ray, 2018). Thus, both graphene and CNT-filled chitosan composites and related blends can be employed to produce several value-added materials suitable for medical and nonmedical applications (Tables 6.1–6.3).

5. Conclusions

Being the second most abundantly available polysaccharide with some specialized features, such as antimicrobial properties, biocompatibility, biodegradability, availability of ample O−H and $-NH_2$ functionalities, and film-forming ability, chitosan and its derivatives have been increasingly utilized in developing various value-added materials in different forms for both the biomedical and nonbiomedical sectors. However, due to some drawbacks, such as poor mechanical strength, biostability, and flexibility, the application of pure chitosan is limited. Thus, chitosan has been combined with second/third organic/inorganic components to produce binary/ternary blends and composites with elevated mechanical properties, functionalities stability, and porosity. Continuous researches are still in progress to develop high-performance scaffolds for various tissue engineering. Though in vitro and in vivo applications have been conducted in various animals and their body parts, in vivo applications of chitosan-based blends and composites in humans remain very limited. The success rate of transplantation, especially for liver and nervous tissues, has not yet reached the desired level. Chitosan-based blends and composites can be very good materials to manufacture several biomaterials for making artificial organs, limbs, surgical accessories, and equipment. In the nonbiomedical sectors, the development of chitosan-based sensors through a combination of conductive polymeric materials is an interesting field of research. In addition, the development of smart sensors with photophysical properties for real-time monitoring of the status of the ecosystem and environment could be a great challenge to modern researchers.

References

Amiri, A., Ghorbani, M., & Jahangiri, M. (2015). A novel chitosan/polyrhodanine nanocomposite: preparation, characterisation and application for Ni(II) ions removal from aqueous solution. *Journal of Experimental Nanoscience, 10*(18), 1374–1386.

Abdel-Fattah, W. I., Jiang, T., El-Bassyouni, G. E. T., & Laurencin, C. T. (2007). Synthesis, characterization of chitosans and fabrication of sintered chitosan microsphere matrices for bone tissue engineering. *Acta Biomaterialia, 3*, 503–514.

Akar, S. T., Akar, T., & San, E. (2016). Chitosan-alunite composite: an effective dye remover with high sorption, regeneration and application potential. *Carbohydrate Polymers, 143*, 318–326.

Aliabadi, M., Irani, M., Ismaeili, J., & Najafzadeh, S. (2014). Design and evaluation of chitosan/hydroxyapatite composite nanofiber membrane for the removal of heavy metal ions from aqueous solution. *Journal of the Taiwan Institute of Chemical Engineers, 45*, 518–526.

An, Z.-Z., Li, Z., Guo, Y.-Y., Chen, X.-L., Zhang, K.-N., Zhang, D.-X., Xue, Z.-H., Zhou, X.-B., & Lu, X.-Q. (2017). Preparation of chitosan/N-doped graphene natively grown on hierarchical porous carbon nanocomposite as a sensor platform for determination of tartrazine. *Chinese Chemical Letters, 28*, 1492–1498.

Anisha, B. S., Sankar, D., Mohandas, A., Chennazhi, K. P., Nair, S. V., & Jayakumar, R. (2013). Chitosan–hyaluronan/nano chondroitin sulfate ternary composite sponges for medical use. *Carbohydrate Polymers, 92*, 1470–1476.

Ashassi-Sorkhabi, H., & Kazempour, A. (2020). Chitosan, its derivatives and composites with superior potentials for the corrosion protection of steel alloys: a comprehensive review. *Carbohydrate Polymers, 237*, 116110.

Atak, B. H., Buyuk, B., Huysal, M., Isik, S., Senel, M., Metzger, W., & Cetin, G. (2017). Preparation and characterization of amine functional nano-hydroxyapatite/chitosan bionanocomposite for bone tissue engineering applications. *Carbohydrate Polymers, 164*, 200–213.

Badhe, R. V., Bijukumar, D., Chejara, D. R., Mabrouk, M., Choonara, Y. E., Kumar, P., du Toit, L. C., Kondiah, P. P. D., & Pillay, V. (2017). A composite chitosan-gelatin bi-layered, biomimetic macroporousscaffold for blood vessel tissue engineering. *Carbohydrate Polymers, 157*, 1215–1225.

Bajpai, S. K., Ahuja, S., Chand, N., & Bajpai, M. (2017). Nano cellulose dispersed chitosan film with Ag NPs/Curcumin: an in vivo study on Albino Rats for wound dressing. *International Journal of Biological Macromolecules, 104*, 1012–1019.

Baniani, D. D., Bagheri, R., & Solouk, A. (2017). Preparation and characterization of a composite biomaterial including starch micro/nano particles loaded chitosan gel. *Carbohydrate Polymers, 174*, 633–645.

Bhattaraia, S. R., Bahadur, K. C. R., Aryala, S., Khil, M. S., & Kim, H. Y. (2007). N-acylated chitosan stabilized iron oxide nanoparticles as a novel nano-matrix and ceramic modification. *Carbohydrate Polymers, 69*, 467–477.

Bibi, S., Tariq, Y., Hassan, S., Riaz, M., & Nawaz, M. (2015). Chitosan/CNTs green nanocomposite membrane: synthesis, swelling and polyaromatic hydrocarbons removal. *Materials Science and Engineering: C, 46*, 359–365.

Biranje, S. S., Madiwale, P. V., Patankar, K. C., Chhabra, R., Bangde, P., Dandekar, P., & Adivarekar, R. V. (2020). Cytotoxicity and hemostatic activity of chitosan/carrageenan composite wound healing dressing for traumatic hemorrhage. *Carbohydrate Polymers, 239*, 116106.

Cai, X., Tong, H., Shen, X., Chen, W., Yan, J., & Hu, J. (2009). Preparation and characterization of homogeneous chitosan–polylactic acid/hydroxyapatite nanocomposite for bone tissue engineering and evaluation of its mechanical properties. *Journal of Acta Biomaterialia, 5*, 2693–2703.

Cai, Z.-X., Mo, X.-M., Zhang, K.-H., Fan, L.-P., Yin, A.-L., He, C.-L., & Wang, H.-S. (2010). Fabrication of chitosan/silk fibroin composite nanofibers for wound-dressing applications. *International Journal of Molecular Sciences, 11*, 3529–3539.

Caridade, S. G., Merino, E. G., Alves, N. M., Bermudez, V. Z., Boccaccini, A. R., & Mano, J. F. (2013). Chitosan membranes containing micro or nano-size bioactive glass particles: evolution of biomineralization followed by in situ dynamic mechanical analysis. *Journal of the Mechanical Behavior of Biomedical Materials, 20*, 173–183.

Chang, M. Y., & Juang, R. S. (2004). Adsorption of tannic acid, humic acid and dyes from water using the composite of chitosan and activated clay. *Journal of Colloid and Interface Science, 278*, 18–25.

Chang, Z., Chen, Y., Tang, S., Yang, J., Chen, Y., Chen, S., Li, P., & Yang, Z. (2019). Construction of chitosan/polyacrylate/graphene oxide composite physical hydrogel by semi-dissolution/acidification/sol-gel transition method and its simultaneous cationic and anionic dye adsorption properties. *Carbohydrate Polymers, 229*, 115431.

Chao, W., Wenfeng, L., Puwang, L., Sidong, L., Ziming, Y., Zhang, H., Yangyang, L., & Ningjian, A. (2017). Preparation and evaluation of chitosan/alginate porous microspheres/ Bletilla striata polysaccharide composite hemostatic sponges. *Carbohydrate Polymers, 174*, 432–442.

Chattopadhyay, P. K., Basuli, U., & Chattopadhyay, S. (2010). Studies on novel dual filler based epoxidized natural rubber nanocomposite. *Polymer Composites, 31*, 835–846.

Chattopadhyay, P. K., Chattopadhyay, S., Das, N. C., & Bandyopadhyay, P. P. (2011). Impact of carbon black substitution with nanoclay on microstructure and tribological properties of ternary elastomeric composites. *Materials and Design, 32*, 4696–4704.

Chattopadhyay, P. K., Das, N. C., & Chattopadhyay, S. (2011). Influence of interfacial roughness and the hybrid filler microstructures on the properties of ternary elastomeric composites. *Composites Part A: Applied Science and Manufacturing, 42*, 1049–1059.

Chattopadhyay, P. K., Praveen, S., Das, N. C., & Chattopadhyay, S. (2013). Contribution of organomodified clay on hybrid microstructures and properties of epoxidized natural rubber-based nanocomposites. *Polymer Engineering & Science, 53*, 923–930.

Chaydhury, M., Rawat, S., Jain, N., Bhatnagar, A., & Maiti, A. (2019). Chitosan-Fe-Al-Mn metal oxyhydroxides composite as highly efficient fluoride scavenger for aqueous medium. *Carbohydrate Polymers, 216*, 140–148.

Chen, Y.-L., Chen, H.-C., Lee, H.-P., Chan, H.-Y., & Hu, Y.-C. (2006). Rational development of GAG-augmented chitosan membranes by fractional factorial design methodology. *Biomaterials, 27*, 2222–2232.

Chen, Y.-L., Lee, H.-P., Chan, H.-Y., Sung, L.-Y., Chen, H.-C., & Hu, Y.-C. (2007). Composite chondroitin-6-sulfate/dermatan sulfate/chitosan scaffolds for cartilage tissue engineering. *Biomaterials, 28*, 2294–2305.

Chen, P.-H., Kuo, T.-Y., Kuo, J.-Y., Tseng, Y.-P., Wang, D.-M., Lai, J.-Y., & Hsieh, H.-J. (2010). Novel chitosan–pectin composite membranes with enhanced strength, hydrophilicity and controllable disintegration. *Carbohydrate Polymers, 82*, 1236–1242.

Chen, Y., Qiu, H., Dong, M., Cheng, B., Jin, Y., Tong, Z., Li, P., Li, S., & Yang, Z. (2018). Preparation of hydroxylated lecithin complexed iodine/carboxymethyl chitosan/sodium alginate composite membrane by microwave drying and its applications in infected burn wound treatment. *Carbohydrate Polymers, 206*, 432–455.

Cheng, M., Deng, J., Yang, F., Gong, Y., Zhao, N., & Zhang, X. (2003). Study on physical properties and nerve cell affinity of composite films from chitosan and gelatin solutions. *Biomaterials, 24*, 2871–2880.

Cheng, F., Wu, Y., Li, H., Yan, T., Wei, X., Wu, G., He, J., & Huang, Y. (2018). Biodegradable N, O-carboxymethyl chitosan/oxidized regenerated cellulose composite gauze as a barrier for preventing postoperative adhesion. *Carbohydrate Polymers, 207*, 180–190.

Chiono, V., Pulieri, E., Vozzi, G., Ciardelli, G., Ahluwalia, A., & Giusti, P. (2008). Genipin crosslinked chitosan/gelatin blends for biomedical applications. *Journal of Materials Science: Materials in Medicine, 19*, 889–898.

Chung, T. W., Yang, J., Akaike, T., Cho, K. Y., Nah, J. W., Kim, S. I., & Cho, C. S. (2002). Preparation of alginate/galactosylated chitosan scaffold for hepatocyte attachment. *Biomaterials, 23*, 2827–2834.

Constantin, M., Bucatariu, S.-M., Doroftei, F., & Fundueanu, G. (2017). Smart composite materials based on chitosan microspheresembedded in thermosensitive hydrogel for controlled delivery of drugs. *Carbohydrate Polymers, 157*, 493–502.

Costa-Pinto, A. R., Correlo, V. M., Sol, P. C., Bhattacharya, M., Charbord, P., Delorme, B., Reis, R. L., & Neves, N. M. (2009). Osteogenic differentiation of human bone marrow mesenchymal stem cells seeded on melt based chitosan scaffolds for bone tissue engineering applications. *Biomacromolecules, 10*, 2067–2073.

Daraei, P., Madaeni, S. S., Salehi, E., Ghaemi, N., Ghari, H. S., Khadivi, M. A., & Rostami, E. (2013). Novel thin film composite membrane fabricated by mixed matrix nanoclay/chitosan on PVDF microfiltration support: preparation, characterization and performance in dye removal. *Journal of Membrane Science, 436*, 97–108.

Darder, M., Colilla, M., & Ruiz-Hitzky, E. (2003). Biopolymer-clay nanocomposites based on chitosan intercalated in montmorillonite. *Chemistry of Materials, 15*, 3774–3780.

Darder, M., López-Blanco, M., Aranda, P., Aznar, A. J., Bravo, J., & Ruiz-Hitzky, E. (2006). Microfibrous chitosan-sepiolite nanocomposites. *Chemistry of Materials, 18*, 1602–1610.

Das, P., Singha, N. R., Ray, S. K., Kuila, S. B., & Samanta, H. S. (2011). Systematic choice of crosslinker and filler for pervaporation membrane: a case study with dehydration of isopropyl alcohol-water mixtures by polyvinyl alcohol membranes. *Separation and Purification Technology, 81*, 159–173.

Debnath, S., Parashar, K., & Pillay, K. (2017). Ultrasound assisted adsorptive removal of hazardous dye Safranin O from aqueous solution using crosslinked graphene oxide-chitosan (GO-CH) composite and optimization by response surface methodology (RSM) approach. *Carbohydrate Polymers, 175*, 509–517.

Devi, N., & Dutta, J. (2017). Preparation and characterization of chitosan-bentonite nanocomposite films for wound healing application. *International Journal of Biological Macromolecules, 104*, 1897–1904.

Dinu, M. V., Dinu, I. A., Lazar, M. M., & Dragan, E. S. (2018). Chitosan-based ion-imprinted cryo-composites with excellent selectivity for copper ions. *Carbohydrate Polymers, 186*, 140–149.
dos Santos, J. M. N., Pereira, C. R., Pinto, L. A. A., Frantz, T., Lima, É. C., Foletto, E. L., & Dotto, G. L. (2019). Synthesis of a novel $CoFe_2O_4$/chitosan magnetic composite for fast adsorption of indigotine blue dye. *Carbohydrate Polymers, 217*, 6–14.
Dutta, A., Mahapatra, M., Deb, M., Ghosh, N. N., Chattopadhyay, P. K., & Singha, N. R. (2020). Nonconjugated biocompatible macromolecular luminogens for sensing and removals of Fe (III) and Cu (II): DFT studies on selective coordination (s) and on-off sensing macromol. *Rapid Communications*, 2000522.
Dutta, A., Mahapatra, M., Deb, M., Mitra, M., Dutta, S., Chattopadhyay, P. K., Banerjee, S., Sil, P. C., Maiti, D. K., & Singha, N. R. (2020). Fluorescent Terpolymers using two non-emissive monomers for Cr(III) sensors, Removal, and Bio-Imaging. *ACS Biomaterials Science & Engineering, 6*, 1397–1407.
Ekambaram, K., & Doraisamy, M. (2017). Fouling resistant PVDF/carboxymethyl chitosan composite nanofiltration membranes for humic acid removal. *Carbohydrate Polymers, 173*, 431–440.
El-Zawahry, M. M., Abdelghaffar, F., Abdelghaffar, R. A., & Hassabo, A. G. (2015). Equilibrium and kinetic models on the adsorption of reactive black 5 from aqueous solution using Eichhornia crassipes/chitosan composite. *Carbohydrate Polymers, 136*, 507–515.
Fan, J., Shang, Y., Yuan, Y., & Yang, J. (2010). Preparation and characterization of chitosan/galactosylated hyaluronic acid scaffolds for primary hepatocytes culture. *Journal of Materials Science: Materials in Medicine, 21*, 319–327.
Fayyad, E. M., Sadasivuni, K. K., Ponnamma, D., & Al-Maadeed, M. A. (2016). Oleic acid-grafted chitosan/graphene oxide composite coating for corrosion protection of carbon steel. *Carbohydrate Polymers, 151*, 871–878.
Feki, A., Hamdi, M., Jaballi, I., Zghal, S., Nasri, M., & Amara, I. B. (2020). Conception and characterization of a multi-sensitive composite chitosan-red marine alga-polysaccharide hydrogels for insulin controlled-release. *Carbohydrate Polymers, 236*, 116046.
Freag, M. S., Saleh, W. M., & Abdallah, O. Y. (2018). Laminated chitosan-based composite sponges for transmucosal delivery of novel protamine-decorated tripterine phytosomes: ex-vivo mucopenetration and in-vivo pharmacokinetic assessments. *Carbohydrate Polymers, 188*, 108–120.
Fu, J., Ji, J., Yuan, W., & Shen, J. (2005). Construction of anti-adhesive and antibacterial multilayer films via layer-by-layer assembly of heparin and chitosan. *Biomaterials, 26*, 6684–6692.
Funakoshi, T., Majima, T., Iwasaki, N., Suenaga, N., Sawaguchi, N., Shimode, K., Minami, A., Harada, K., & Nishimura, S. (2005). Application of tissue engineering techniques for rotator cuff regeneration using a chitosan-based hyaluronan hybrid fiber scaffold. *The American Journal of Sports Medicine, 33*, 1193–1201.
Funakosi, T., Majima, T., Iwasaki, N., Yamane, S., Masuko, T., Minami, A., Harada, K., Tamura, H., Tokura, S., & Nishimura, S. (2005). Novel chitosan-based hyaluronan hybrid polymer fibers as a scaffold in ligament tissue engineering. *Journal of Biomedical Materials Research Part A, 74*, 338–346.
Ge, H., & Ma, Z. (2015). Microwave preparation of triethylenetetramine modified graphene oxide/chitosan composite for adsorption of Cr(VI). *Carbohydrate Polymers, 131*, 280–287.

Ghani, M., Gharehaghaji, A. A., Arami, M., Takhtkuse, N., & Rezaei, B. (2014). Fabrication of electrospun polyamide-6/chitosan nanofibrous membrane toward anionic dyes removal. *Journal of Nanotechnology*. https://doi.org/10.1155/2014/278418

Griffon, D. J., Sedighi, M. R., Schaeffer, D. V., Eurell, J. A., & Johnson, A. L. (2006). Chitosan scaffolds: interconnective pore size and cartilage engineering. *Acta Biomaterialia, 2*, 313–320.

Guo, R., Xu, S., Ma, L., Huang, A., & Gao, C. (2011). The healing of full-thickness burns treated by using plasmid DNA encoding VEGF-165 activated collagen–chitosan dermal equivalents. *Biomaterials, 32*, 1019–1031.

Guo, Z., Bo, D., He, Y., Luo, X., & Li, H. (2018). Degradation properties of chitosan microspheres/poly(L-lactic acid) composite in vitro and in vivo. *Carbohydrate Polymers, 193*, 1–8.

Hameed, B. H., Hasan, M., & Ahmad, A. L. (2008). Adsorption of reactive dye onto cross-linked chitosan/oil palm ash composite beads. *Chemical Engineering Journal, 136*, 164–172.

Han, D., Yan, L., Chen, W., & Li, W. (2011). Preparation of chitosan/graphene oxide composite film with enhanced mechanical strength in the wet state. *Carbohydrate Polymers, 83*, 653–658.

Hebeish, A. A., Ramadan, M. A., Montaser, A. S., & Faragb, A. M. (2014). Preparation, characterization and antibacterial activity of chitosan-g-poly acrylonitrile/silver nanocomposite. *International Journal of Biological Macromolecules, 68*, 178–184.

Hein, S., Wang, K., Stevens, W. F., & Kjems, J. (2008). Chitosan composites for biomedical applications: status, challenges and perspectives. *Materials Science and Technology, 24*, 1053–1061.

Hosseini, S. F., Rezaei, M., Zandi, M., & Farahmandghavi, F. (2016). Development of bioactive fish gelatin/chitosan nanoparticles composite films with antimicrobial properties. *Food Chemistry, 194*, 1266–1274.

Hritcu, D., Humelnicu, D., Dodi, G., & Ionel Popa, M. (2012). Magnetic chitosan composite particles: evaluation of thorium and uranyl ion adsorption from aqueous solutions. *Carbohydrate Polymers, 87*, 1185–1191.

Hu, Z., Zhang, L., Zhong, L., Zhou, Y., Xue, J., & Li, Y. (2019). Preparation of an antibacterial chitosan-coated biochar-nanosilver composite for drinking water purification. *Carbohydrate Polymers, 219*, 290–297.

Huang, G. L., Zhang, H. Y., Jeffrey, X. S., & Tim, A. G. L. (2009). Adsorption of chromium(VI) from aqueous solutions using cross-linked magnetic chitosan beads. *Industrial & Engineering Chemistry Research, 48*, 2646–2651.

Huang, J., Hu, X., Lu, L., Ye, Z., Zhang, Q., & Luo, Z. (2010). Electrical regulation of Schwann cells using conductive polypyrrole/chitosan polymers. *Journal of Biomedical Materials Research Part A, 93*, 164–174.

Huang, B., Liu, M., & Zhou, C. (2017). Chitosan composite hydrogels reinforced with natural clay nanotubes. *Carbohydrate Polymers, 175*, 689–698.

Huang, J., Qin, J., Zhang, P., Chen, X., You, X., Zhang, F., Zuo, B., & Yao, M. (2020). Facile preparation of a strong chitosan-silk biocomposite film. *Carbohydrate Polymers, 229*, 115515.

Im, S. Y., Cho, S. H., Hwang, J. H., & Lee, S. J. (2003). Growth factor releasing porous poly(ε-caprolactone)–chitosan matrices for enhanced bone regenerative therapy. *Archives of Pharmacal Research, 26*, 76–82.

Ito, M. (1991). In vitro properties of a chitosan-bonded hydroxyapatite bone-filling paste. *Biomaterials, 12*, 41–45.

Jiang, T., Abdel-Fattah, W. I., & Laurencin, C. T. (2006). In vitro evaluation of chitosan/poly(lactic acid–glycolic acid) sintered microsphere scaffolds for bone tissue engineering. *Biomaterials, 27*, 4894–4903.

Jiang, L., Li, Y., Wang, X., Zhang, L., Wen, J., & Gong, M. (2008). Preparation andproperties of nano-hydroxyapatite/chitosan/carboxymethyl cellulose composite scaffold. *Carbohydrate Polymers, 74*, 680–684.

Jiang, T., Nukavarapu, S. P., Deng, M., Jabbarzadeh, E., Kofron, M. D., Doty, S. B., Abdel-Fattah, W. I., & Laurencin, C. T. (2010). Chitosan–Poly(Lactide-Co-Glycolide) microsphere-based scaffolds for bone tissue engineering: in vitro degradation and in vivo bone regeneration studies. *Acta Biomaterialia, 6*, 3457–3470.

Jiang, S.-J., Zhang, X., Ma, Y., Tuo, Y., Qian, F., Fu, W., & Mu, G. (2016). Characterization of whey protein-carboxymethylated chitosan composite films with and without transglutaminase treatment. *Carbohydrate Polymers, 153*, 153–159.

Jiang, Y., Liu, B., Xu, J., Pan, K., Hou, H., Hu, J., & Yang, J. (2017). Cross-linked chitosan/b-cyclodextrin composite for selective removal of methyl orange: adsorption performance and mechanism. *Carbohydrate Polymers, 182*, 106–114.

Jin, X., Li, G., Jiang, H., Zhou, Y., Zheng, W., Bao, X., Wang, Z., & Hu, Q. (2019). High strength graphene oxide/chitosan composite screws with a steel-concrete structure. *Carbohydrate Polymers, 214*, 167–173.

Jirimali, H. D., Saravanakumar, D., & Shin, W. (2015). Preparation of catechol-linked chitosan/carbon nanocomposite modified electrode and its applications. *Bulletin of the Korean Chemical Society, 36*, 1289–1291.

Ju, S., Zhang, F., Duan, J., & Jiang, J. (2020). Characterization of bacterial cellulose composite films incorporated with bulk chitosan and chitosan nanoparticles: a comparative study. *Carbohydrate Polymers, 237*, 116167.

Judith, R., Nithya, M., Rose, C., & Mandal, A. B. (2012). Biopolymer gel matrix as a cellular scaffold for enhanced dermal tissue regeneration. *Biologicals, 40*, 231–239.

Kadam, D., Momin, B., Palamthodi, S., & Lele, S. S. (2019). Physicochemical and functional properties of chitosan-based nano-composite films incorporated with biogenic silver nanoparticles. *Carbohydrate Polymers, 211*, 124–132.

Kalyani, S., Veera, M. B., Siva, K. N., & Krishnaiah, A. (2009). Competitive adsorption of Cu(II), Co(II) and Ni(II) from their binary and tertiary aqueous solutions using chitosan-coated perlite beads as biosorbent. *Journal of Hazardous Materials, 170*, 680–689.

Kang, X., Mai, Z., Zou, X., Cai, P., & Mo, J. (2007). A novel glucose biosensor based on immobilization of glucose oxidase in chitosan on a glassy carbon electrode modified with gold–platinum alloy nanoparticles/multiwall carbon nanotubes. *Analytical Biochemistry, 369*, 71–79.

Karim, Z., Mathew, A. P., Grahn, M., Mouzon, J., & Oksman, K. (2014). Nanoporous membranes with cellulose nanocrystals as functional entity in chitosan: removal of dyes from water. *Carbohydrate Polymers, 112*, 668–676.

Karmakar, M., Mahapatra, M., Dutta, A., Chattopadhyay, P. K., & Singha, N. R. (2017). Fabrication of semisynthetic collagenic materials for mere/synergistic adsorption: a model approach of determining dye allocation by systematic characterization and optimization. *International Journal of Biological Macromolecules, 102*, 438–456.

Karmakar, M., Mondal, H., Ghosh, T., Chattopadhyay, P. K., Maiti, D. K., & Singha, N. R. (2019). Chitosan-grafted tetrapolymer using two monomers: pH-responsive high-performance removals of Cu(II), Cd(II), Pb(II), dichromate, and biphosphate and analyses of adsorbed microstructures. *Environmental Research, 179*, 108839.

Karmakar, M., Mondal, H., Mahapatra, M., Chatterjee, S., Chattopadhyay, P. K., & Singha, N. R. (2019). Pectin-grafted terpolymer superadsorbent via N–H activated strategic protrusion of monomer for removals of Cd(II), Hg(II), ad Pb(II). *Carbohydrate Polymers, 206*, 778–791.

Khan, Y. H., Islam, A., Sarwar, A., Gull, N., Khan, S. M., Munawar, M. A., Zia, S., Sabir, A., Shafiq, M., & Jamil, T. (2016). Novel green nano composites films fabricated by indigenously synthesized graphene oxide and chitosan. *Carbohydrate Polymers, 146*, 131–138.

Kujawa, P., Schmauch, G., Viitala, T., Badia, A., & Winnik, F. M. (2007). Construction of viscoelastic biocompatible films via the layer-by-layer assembly of hyaluronan and phosphorylcholine-modified chitosan. *Biomacromolecules, 8*, 3169–3176.

Kumar, M., Bijay, P. T., & Vinod, K. S. (2009). Crosslinked chitosan/polyvinyl alcohol blend beads for removal and recovery of Cd(II) from wastewater. *Journal of Hazardous Materials, 172*, 1041–1048.

Kumar, S., Prasad, K., Gil, J. M., Sobral, A. J. F. N., & Koh, J. (2018). Mesoporous zeolite-chitosan composite for enhanced capture and catalytic activity in chemical fixation of CO_2. *Carbohydrate Polymers, 198*, 401–406.

Kumara, A. M., Suresh, B., Das, S., Obota, I. B., Adesinae, A. Y., & Ramakrishna, S. (2017). Promising bio-composites of polypyrrole and chitosan: surface protective and in vitro biocompatibility performance on 316L SS implants. *Carbohydrate Polymers, 173*, 121–130.

Kuo, Y. C., & Lin, C. C. (2013). Accelerated nerve regeneration using induced pluripotent stem cells in chitin–chitosan–gelatin scaffolds with inverted colloidal crystal geometry. *Colloids and Surfaces B, 103*, 595–600.

Lee, J. E., Kimb, K. E., Kwonb, I. C., Ahna, H. J., Leea, S.-H., Choa, H., Kima, H. J., Seonga, S. C., & Lee, M. C. (2004). Effects of the controlled-released TGF-b1 from chitosan microspheres on chondrocytes cultured in a collagen/chitosan/glycosaminoglycan scaffold. *Biomaterials, 25*, 4163–4173.

Lee, H. C., Jeong, Y. G., Min, B. G., Lyoo, W. S., & Lee, S. C. (2009). Preparation and acid dye adsorption behavior of polyurethane/chitosan composite foams. *Fibers and Polymers, 10*, 636–642.

Lelifajri, R., Julinawati, J., & Shabrina, S. (2017). Preparation of chitosan composite film reinforced with cellulose isolated from oil palm empty fruit bunch and application in cadmium ions removal from aqueous solutions. *Carbohydrate Polymers, 170*, 226–233.

Lessa, E. F., Nunes, M. L., & Fajardo, A. R. (2018). Chitosan/waste coffee-grounds composite: an efficient and eco-friendly adsorbent for removal of pharmaceutical contaminants from water. *Carbohydrate Polymers, 189*, 257–266.

Li, J., Pan, J., Zhang, L., & Yu, Y. (2003). Culture of hepatocytes on fructose modified chitosan scaffolds. *Biomaterials, 24*, 2317–2322.

Li, J., Liu, Y., Liu, S., & Yao, S. (2005). DNA biosensor based on chitosan film doped with carbon nanotubes. *Analytical Biochemistry, 346*, 107–114.

Li, Z., Ramay, H. R., Hauch, K. D., Xiao, D., & Zhang, M. (2005). Chitosan–alginate hybrid scaffolds for bone tissue engineering. *Biomaterials, 26*, 3919–3928.

Li, X. Y., Kong, X. Y., Shi, S., Wang, X. H., Guo, G., Luo, F., Zhao, X., Wei, Y. Q., & Qian, Z. Y. (2010). Physical, mechanical and biological properties of poly(E-caprolactone)–poly(ethylene glycol)–poly(E-caprolactone) (CEC)/chitosan composite film. *Carbohydrate Polymers, 82*, 904–912.

Li, X., Nan, K., Li, L., Zhang, Z., & Chen, H. (2012). In vivo evaluation of curcumin nanoformulation loaded methoxy poly(ethylene glycol)-graft-chitosan composite film for wound healing application. *Carbohydrate Polymers, 88*, 84–90.

Li, Z., Li, T., An, L., Fu, P., Gao, C., & Zhang, Z. (2015). Highly efficient chromium (VI) adsorption with nanofibrous filter paper prepared through electrospinning chitosan/polymethylmethacrylate composite. *Carbohydrate Polymers, 137*, 119–126.

Li, K., Jin, S., Liu, X., Chen, H., He, J., & Li, J. (2017). Preparation and characterization of chitosan/soy protein isolate nanocomposite film reinforced by Cu nanoclusters. *Polymers, 9*, 247.

Liang, J., Wang, J., Li, S., Xu, L., Wang, R., Chen, R., & Sun, Y. (2019). The size-controllable preparation of chitosan/silver nanoparticle composite microsphere and its antimicrobial performance. *Carbohydrate Polymers, 220*, 22–29.

Liu, H., Fan, H., Cui, Y., Chen, Y., Yao, K., & Goh, J. C. H. (2007). Effects of the controlled-released basic fibroblast growth factor from chitosan-gelatin microspheres on human fibroblasts cultured on a chitosan-gelatin scaffold. *Biomacromolecules, 8*, 1446–1455.

Liu, X. W., Hu, Q. Y., Fang, Z., Zhang, X. J., & Zhang, B. B. (2009). Magnetic chitosan nanocomposites: a useful recyclable tool for heavy metal ion removal. *Langmuir, 25*, 3–8.

Liu, K., Chen, L., Huang, L., & Lai, Y. (2016). Evaluation of ethylenediamine-modified nanofibrillated cellulose/chitosan composites on adsorption of cationic and anionic dyes from aqueous solution. *Carbohydrate Polymers, 151*, 1115–1119.

Liu, X., Huang, R., Zhou, X., Cai, T., Chen, J., Shi, X., Deng, H., & Luo, W. (2017). Presence of nano-sized chitosan-layered silicate composites protects against toxicity induced by lead ions. *Carbohydrate Polymers, 158*, 1–10.

Liu, K., Huang, R.-L., Zha, X., Li, Q.-M., Pan, L.-H., & Luo, J.-P. (2019). Encapsulation and sustained release of curcumin by a composite hydrogel of lotus root amylopectin and chitosan. *Carbohydrate Polymers, 232*, 115810.

Liu, Y., Cai, Z., Sheng, L., Ma, M., Xu, Q., & Jin, Y. (2019). Structure-property of crosslinked chitosan/silica composite films modified by genipin and glutaraldehyde under alkaline conditions. *Carbohydrate Polymers, 215*, 348–357.

Liu, Y., Liu, R., Li, M., Yu, F., & He, C. (2019). Removal of pharmaceuticals by novel magnetic genipin-crosslinked chitosan/graphene oxide-SO_3H composite. *Carbohydrate Polymers, 220*, 141–148.

Lu, Z., Gao, J., He, Q., Wu, J., Liang, D., Yang, H., & Chen, R. (2017). Enhanced antibacterial and wound healing activities of microporous chitosan-Ag/Zno composite dressing. *Carbohydrate Polymers, 156*, 460–469.

Luo, X. L., Xu, J. J., Du, Y., & Chen, H.-Y. (2004). A glucose biosensor based on chitosan–glucose oxidase–gold nanoparticles biocomposite formed by one-step electrodeposition. *Analytical Biochemistry, 334*, 284–289.

Ma, L., Gao, C., Mao, Z., Zhou, J., Shen, J., Hu, X., & Han, C. (2003). Collagen/chitosan porous scaffolds with improved biostability for skin tissue engineering. *Biomaterials, 24*, 4833–4841.

Madhumathi, K., Shalumon, K. T., Divya Rani, V. V., Tamura, H., Furuike, T., Selvamurugan, N., Nair, S. V., & Jayakumar, R. (2009). Wet chemical synthesis of chitosan hydrogel–hydroxyapatite composite membranes fortissue engineering applications. *International Journal of Biological Macromolecules, 45*, 12–15.

Mahapatra, M., Karmakar, M., Dutta, A., Mondal, H., Roy, J. S. D., Chattopadhyay, P. K., & Singha, N. R. (2018). Microstructural analyses of loaded and/or unloaded semisynthetic porous material for understanding of superadsorption and optimization by response surface methodology. *Journal of Environmental Chemical Engineering, 6*, 289–310.

Mahapatra, M., Dutta, A., Roy, J. S. D., Das, U., Banerjee, S., Dey, S., Chattopadhyay, P. K., Maiti, D. K., & Singha, N. R. (2019). Multi C–C/C–N coupled light-emitting aliphatic terpolymers: N–H functionalized fluorophore-monomers and high-performance applications. *Chemistry – A European Journal, 26*, 502–516.

Mahapatra, M., Dutta, A., Roy, J. S. D., Deb, M., Das, U., Banerjee, S., Dey, S., Chattopadhyay, P. K., Maiti, D. K., & Singha, N. R. (2020). Synthesis of biocompatible aliphatic terpolymers via in situ fluorescent monomers for three-in-one applications: polymerization of hydrophobic monomers in water. *Langmuir, 36*, 6178–6187.

Mahapatra, M., Dutta, A., Mitra, M., Karmakar, M., Ghosh, N. N., Chattopadhyay, P. K., & Singha, N. R. (2020). Intrinsically-fluorescent biocompatible terpolymers for detection and removal of Bi(III) and cell-imaging. *ACS Applied Bio Materials, 3*(9), 6155–6166. https://doi.org/10.1021/acsabm.0c00718

Maity, J., & Ray, S. K. (2018). Chitosan based nano composite adsorbent-synthesis, characterization and application for adsorption of binary mixtures of Pb(II) and Cd(II) from water. *Carbohydrate Polymers, 182*, 159–171.

Majima, T., Funakosi, T., Iwasaki, N., Yamane, S. T., Harada, K., Nonaka, S., Minami, A., & Nishimura, S. (2005). Alginate and chitosan polyion complex hybrid fibers for scaffolds in ligament and tendon tissue engineering. *Journal of Orthopaedic Science, 10*, 302–307.

Mano, J. F., Hungerford, G., & Ribelles, J. L. G. (2008). Bioactive poly(L-lactic acid)–chitosan hybrid scaffolds. *Materials Science and Engineering: C, 28*, 1356–1365.

Mansur, H. S., & Costa, H. S. (2008). Nanostructured poly(vinyl alcohol)/bioactiveglass and poly(vinyl alcohol)/chitosan/bioactive glass hybrid scaffolds for biomedical applications. *Chemical Engineering Journal, 137*, 72–83.

Martino, A. D., Sittinger, M., & Risbud, M. V. (2005). Chitosan: a versatile biopolymer for orthopaedic tissue-engineering. *Biomaterials, 26*, 5983–5990.

Meng, Z. X., Zheng, W., Li, L., & Zheng, Y. F. (2011). Fabrication, characterizationand in vitro drug release behavior of electrospun PLGA/chitosan nanofibrous scaffold. *Materials Chemistry and Physics, 125*, 606–611.

Mi, F. L., Tan, Y. C., Liang, H. C., Huang, R. N., & Sung, H. W. (2001). In vitro evaluation of a chitosan membrane cross-linked with genipin. *Journal of Biomaterials Science, Polymer Edition, 12*, 835–850.

Mingyu, C., Kai, G., Jiamou, L., Yandao, G., Nanming, Z., & Xiufang, Z. (2004). Surface modification and characterization of chitosan film blended with poly-L-lysine. *Journal of Biomaterials Applications, 19*, 59–75.

Mitra, M., Mahapatra, M., Dutta, A., Chattopadhyay, P. K., Deb, M., Roy, J. S. D., Roy, C., Banerjee, S., & Singha, N. R. (2020). Light-emitting multifunctional maleic acid-co-2-(N-(hydroxymethyl)acrylamido)succinic acid-co- N-(hydroxymethyl)acrylamide for Fe(III) sensing, removal, and cell imaging. *ACS Omega, 5*, 3333–3345.

Mitra, M., Mahapatra, M., Dutta, A., Deb, M., Dutta, S., Chattopadhyay, P. K., Roy, S., Banerjee, S., Sil, P. C., & Singha, N. R. (2020). Fluorescent guar gum-g-terpolymer via in situ acrylamido-acid fluorophore-monomer in cell imaging, Pb(II) sensor, and security ink. *ACS Applied Bio Materials, 3*, 1995–2006.

Mokhtar, A., Abdelkrim, S., Djelad, A., Sardi, A., Boukoussa, B., Sassi, M., & Bengueddach, A. (2020). Adsorption behavior of cationic and anionic dyes on magadiite-chitosan composite beads. *Carbohydrate Polymers, 229*, 115399.

Mondal, A. K., & Chattopadhyay, P. K. (2017). Influence of the micro-structural factors upon thermal and mechanical properties of various bag leathers. *Bangladesh Journal of Scientific & Industrial Research, 52*, 167–176.

Mondal, M., Chattopadhyay, P. K., Chattopadhyay, S., & Setua, D. K. (2010). Thermal and morphological analysis of thermoplastic polyurethane–clay nanocomposites: comparison of efficacy of dual modified laponite vs. commercial montmorillonites. *Thermochimica Acta, 510*, 185–194.

Mondal, H., Karmakar, M., Dutta, A., Mahapatra, M., Deb, M., Mitra, M., Roy, J. S. D., Roy, C., Chattopadhyay, P. K., & Singha, N. R. (2018). Tetrapolymer network hydrogels via gum ghatti-grafted and N—H/C—H-activated allocation of monomers for composition-dependent superadsorption of metal ions. *ACS Omega, 3*, 10692—10708.

Mondal, H., Karmakar, M., Chattopadhyay, P. K., & Singha, N. R. (2019). Starch-G-tetrapolymer hydrogel via in situ attached monomers for removals of Bi(III) and/or Hg(II) and dye(S): RSM-based optimization. *Carbohydrate Polymers, 213*, 428—440.

Mondal, H., Karmakar, M., Chattopadhyay, P. K., & Singha, N. R. (2020a). Synthesis of pH-responsive sodium alginate-g-tetrapolymers via NC and OC coupled in situ monomers: a reusable optimum hydrogel for removal of plant stressors. *Journal of Molecular Liquids, 319*, 114097.

Mondal, H., Karmakar, M., Chattopadhyay, P. K., & Singha, N. R. (2020b). New property-performance optimization of scalable alginate-g-terpolymer for Ce(IV), Mo(VI), and W(VI) exclusions. *Carbohydrate Polymers*. https://doi.org/10.1016/j.carbpol.2020.116370

Moulik, S., Vani, B., Chandrasekhar, S. S., & Sridhar, S. (2018). Chitosan-polytetrafluoroethylene composite membranes for separation of methanol and toluene by pervaporation. *Carbohydrate Polymers, 193*, 28—38.

Nesic, A. R., Velickovic, S. J., & Antonovic, D. G. (2012). Characterization of chitosan/montmorillonite membranes as adsorbents for bezactiv orange V-3R dye. *Journal of Hazardous Materials, 209—210*, 256—263.

Nettles, D. L., Elder, S. H., & Gilbert, J. A. (2002). Potential use of chitosan as a cell scaffold material for cartilage tissue engineering. *Tissue Engineering, 8*, 1009—1016.

Niu, X., Liu, Y., Song, Y., Han, J., & Pan, H. (2018). Rosin modified cellulose nanofiber as a reinforcing and Co-antimicrobial agents in polylactic acid/chitosan composite film for food packaging. *Carbohydrate Polymers, 183*, 102—109.

Noorbakhsh-Soltani, S. M., Zerafat, M. M., & Sabbaghi, S. A. (2018). Comparative study of gelatin and starch-based nano-composite films modified by nano-cellulose and chitosan for food packaging applications. *Carbohydrate Polymers, 189*, 48—55.

Nourmohammadi, J., Ghaee, A., & Liavali, S. H. (2015). Preparation and characterization of bioactive composite scaffolds from polycaprolactone nanofibers-chitosan-oxidized starch for bone regeneration. *Carbohydrate Polymers, 138*, 172—179.

Oliveiraa, J. M., Rodriguesa, M. T., Silvaa, S. S., Malafayaa, P. B., Gomesa, M. E., Viegasc, C. A., Diasc, I. R., Azevedod, J. T., Manoa, J. F., & Reis, R. L. (2006). Novel hydroxyapatite/chitosan bilayered scaffold for osteochondral tissue-engineering applications: scaffold design and its performance when seeded with goat bone marrow stromal cells. *Biomaterials, 27*, 6123—6137.

Park, S. B., You, J. O., Park, H. Y., Haam, S. J., & Kim, W. S. (2001). A novel pH sensitive membrane from chitosan: preparation and its drug permeation characteristics. *Biomaterials, 22*, 323—330.

Park, H., Choi, B., Hu, J., & Lee, M. (2013). Injectable chitosan hyaluronic acid hydrogels for cartilage tissue engineering. *Acta Biomaterialia, 9*, 4779—4786.

Peter, M., Binulal, N. S., Nair, S. V., Selvamurugan, N., Tamura, H., & Jayakumar, R. (2010). Novel biodegradable chitosan—gelatin/nano-bioactive glass ceramic composite scaffolds for alveolar bone tissue engineering. *Chemical Engineering Journal, 158*, 353—361.

Peter, M., Binulal, N. S., Soumya, S., Nair, S. V., Furuike, T., Tamura, H., & Jayakumar, R. (2010). Nanocomposite scaffolds of bioactive glass ceramic nanoparticles disseminated chitosan matrix for tissue engineering applications. *Carbohydrate Polymers, 79*, 284—289.

Prabaharan, M., Rodriguez-Perez, M. A., de Saja, J. A., & Mano, J. F. (2007). Preparation and characterization of poly(L-lactic acid)—chitosan hybrid scaffolds with drug release capability. *Journal of Biomedical Materials Research B, 81*, 427—434.

Qiaoling, H., Baoqiang, L., Wang, M., & Shen, J. (2004). Preparation and characterization of biodegradable chitosan/hydroxyapatite nanocomposite rods via in situ hybridization: a potential material as internal fixation of bone fracture. *Biomaterials, 25*, 779—785.

Qiu, J. D., Xie, H.-Y., & Liang, R.-P. (2008). Preparation of porous chitosan/carbon nanotubes film modified electrode for biosensor application. *Microchimica Acta, 162*, 57—64.

Qu, R. J., Sun, C. M., Fang, M., Zhang, Y., Ji, C. N., Xu, Q., Wang, C., & Chen, H. (2009). Removal of recovery of Hg(II) from aqueous solution using chitosan-coated cotton fibers. *Journal of Hazardous Materials, 167*, 717—727.

Qu, R. J., Sun, C. M., Wang, M. H., Ji, C. N., Xu, Q., Zhang, Y., Wang, C., Chen, H., & Yin, P. (2009). Adsorption of Au(III) from aqueous solution using cotton fiber/chitosan composite adsorbents. *Hydrometallurgy, 100*, 65—71.

Rahmani, H., Najafi, S. H. M., Ashori, A., Fashapoyeh, M. A., Mohseni, F. A., & Torkaman, S. (2020). Preparation of chitosan-based composites with urethane cross linkage and evaluation of their properties for using as wound healing dressing. *Carbohydrate Polymers, 230*, 115606.

Rahmi, L., & Nurfatimah, R. (2018). Preparation of polyethylene glycol diglycidyl ether (PEDGE) crosslinked chitosan/activated carbon composite film for Cd^{2+} removal. *Carbohydrate Polymers, 199*, 499—505.

Roy, C., Dutta, A., Mahapatra, M., Karmakar, M., Roy, J. S. D., Mitra, M., Chattopadhyay, P. K., & Singha, N. R. (2019). Collagenic waste and rubber based resin-cured biocomposite adsorbent for high-performance removal(s) of Hg (II), safranine, and brilliant cresyl blue: a cost-friendly waste management approach. *Journal of Hazardous Materials, 369*, 199—213.

Samanta, H. S., Singha, N. R., Das, P., & Ray, S. K. (2012). Separation of acid-water mixtures by pervaporation using nano particle filled mixed matrix copolymer membranes. *Journal of Chemical Technology and Biotechnology, 87*, 608—622.

Sanad, R. A.-B., & Abdel-Bar, H. M. (2017). Chitosan—hyaluronic acid composite sponge scaffold enriched with andrographolide loaded lipid nanoparticles for enhanced wound healing. *Carbohydrate Polymers, 173*, 441—450.

Shah, J., & Tasmia, M. R. J. (2018). Magnetic chitosan graphene oxide composite for solid phase extraction of phenylurea herbicides. *Carbohydrate Polymers, 199*, 461—472.

Shalumon, K. T., Sowmya, S., Sathish, D., Chennazhi, K. P., Nair, S. V., & Jayakumar, R. (2013). Effect of incorporation of nanoscale bioactive glass and hydroxyapatite in PCL/chitosan nanofibers for bone and periodontal tissue engineering. *Journal of Biomedical Nanotechnology, 9*, 430—440.

Shameem, H., Abburi, K., Tushar, K. G., Dabir, S. V., Veera, M. B., & Edgar, D. S. (2003). Adsorption of chromium(VI) on chitosan-coated perlite. *Separation Science and Technology, 38*, 3775—3793.

Shameem, H., Abburi, K., Tushar, K. G., Dabir, S. V., Veera, M. B., & Edgar, D. S. (2006). Adsorption of divalent cadmium (Cd(II)) from aqueous solutions onto chitosancoated perlite beads. *Industrial & Engineering Chemistry Research, 45*, 5066—5077.

Shameem, H., Tushar, K. G., Dabir, S. V., & Veera, M. B. (2008). Dispersion of chitosan on perlite for enhancement of copper(II) adsorption capacity. *Journal of Hazardous Materials, 152*, 826—837.

Shao, H. J., Chen, C. S., Lee, Y. T., Wang, J., & Young, T. H. (2010). The phenotypic responses of human anterior cruciate ligament cells cultured on poly(ε-caprolactone) and chitosan. *Journal of Biomedical Materials Research Part A, 93*, 1297–1305.

Shao, W., Wu, J., Wang, S., Huang, M., Liu, X., & Zhang, R. (2017). Construction of silver sulfadiazine loaded chitosan composite sponges as potential wound dressings. *Carbohydrate Polymers, 157*, 1963–1970.

She, Z., Liu, W., & Feng, Q. (2009). Self-assembly model, hepatocytes attachment and inflammatory response for silk fibroin/chitosan scaffolds. *Biomedical Materials, 4*, 0450141–0450148.

Shi, S., Cheng, X., Wang, J., Zhang, W., Peng, L., & Zhang, Y. (2009). rhBMP-2microspheres-Loaded chitosan/collagen scaffold enhanced osseointegration: an experiment in dog. *Journal of Biomaterials Applications, 23*, 331–346.

Shi, Z., Lan, G., Hu, E., Lu, F., Qian, P., Liu, J., Dai, F., & Xie, R. (2020). Puff pastry-like chitosan/konjac glucomannan matrix with thrombin-occupied microporous starch particles as a composite for hemostasis. *Carbohydrate Polymers, 232*, 115814.

Singha, N. R., & Ray, S. K. (2010). Synthesis of chemically modified polyvinyl alcohol membranes for separation of toluene methanol mixtures by pervaporation. *Separation Science and Technology, 45*, 2298–2307.

Singha, N. R., Kar, S., & Ray, S. K. (2009). Synthesis of chemically modified polyvinyl alcohol membranes for dehydration of dioxane by pervaporation. *Separation Science and Technology, 44*, 422–446.

Singha, N. R., Parya, T. K., & Ray, S. K. (2009). Dehydration of 1, 4-dioxane by pervaporation using filled and crosslinked polyvinyl alcohol membrane. *Journal of Membrane Science, 340*, 35–44.

Singha, N. R., Karmakar, M., Mahapatra, M., Mondal, H., Dutta, A., Roy, C., & Chattopadhyay, P. K. (2017). Systematic synthesis of pectin-g-(sodium acrylate-co-N-isopropylacrylamide) interpenetrating polymer network for superadsorption of dyes/M (II): determination of physicochemical changes in loaded hydrogels. *Polymer Chemistry, 8*, 3211–3237.

Singha, N. R., Mahapatra, M., Karmakar, M., Dutta, A., Mondal, H., & Chattopadhyay, P. K. (2017). Synthesis of guar gum-g-(acrylic acid co-acrylamide-co-3-acrylamido propanoic acid) IPN via in situ attachment of acrylamido propanoic acid for analyzing superadsorption mechanism of Pb (II)/Cd (II)/Cu (II)/MB/MV. *Polymer Chemistry, 8*, 6750–6777.

Singha, N. R., Dutta, A., Mahapatra, M., Karmakar, M., Mondal, H., Chattopadhyay, P. K., & Maiti, D. K. (2018). Guar gum-grafted terpolymer hydrogels for ligand-selective individual and synergistic adsorption: effect of comonomer composition. *ACS Omega, 3*, 472–494.

Singha, N. R., Karmakar, M., Mahapatra, M., Mondal, H., Dutta, A., Deb, M., Mitra, M., Roy, C., & Chattopadhyay, P. K. (2018). An in situ approach for the synthesis of a gum ghatti-g-interpenetrating terpolymer network hydrogel for the high-performance adsorption mechanism evaluation of Cd (II), Pb (II), Bi (III) and Sb (III). *Journal of Materials Chemistry, 6*, 8078–8100.

Singha, N. R., Mahapatra, M., Karmakar, M., & Chattopadhyay, P. K. (2018). *Processing, characterization and application of natural rubber based environmentally friendly polymer composites; online*. Cham: Sustainable Polymer Composites and Nanocomposites; Springer, ISBN 978-3-030-05399-4. https://doi.org/10.1007/978-3-030-05399-4_29

Singha, N. R., Mahapatra, M., Karmakar, M., Mondal, H., Dutta, A., Deb, M., Mitra, M., Roy, C., Chattopadhyay, P. K., & Maiti, D. K. (2018). In situ allocation of a monomer in pectin-g-terpolymer hydrogels and effect of comonomer compositions on superadsorption of metal ions/dyes. *ACS Omega, 3*, 4163–4180.

Singha, N. R., Chattopadhyay, P. K., Dutta, A., Mahapatra, M., & Deb, M. (2019). Review on additives-based structure-property alterations in dyeing of collagenic matrices. *Journal of Molecular Liquids, 293*, 111470.

Singha, N. R., Dutta, A., Mahapatra, M., Roy, J. S. D., Mitra, M., Deb, M., & Chattopadhyay, P. K. (2019). In situ attachment of acrylamido sulfonic acid-based monomer in terpolymer hydrogel optimized by response surface methodology for individual and/or simultaneous removal (s) of M (III) and cationic dyes. *ACS Omega, 4*, 1763–1780.

Singha, N. R., Karmakar, M., Chattopadhyay, P. K., Roy, S., Deb, M., Mondal, H., Mahapatra, M., Dutta, A., Mitra, M., & Roy, J. S. D. (2019). Structures, properties and performances-relationships of polymeric membranes for pervaporative desalination. *Membranes, 9*, 58.

Singha, N. R., Roy, C., Mahapatra, M., Dutta, A., Roy, J. S. D., Mitra, M., & Chattopadhyay, P. K. (2019). Scalable synthesis of collagenic-waste and natural rubber-based biocomposite for removals of Hg(II) and dyes: approach for cost-friendly waste management. *ACS Omega, 4*, 421–436.

Singha, N. R., Chattopadhyay, P. K., Karmakar, M., & Mondal, H. (2020). *Lightweight polymer composite structures design and manufacturing techniques*. CRC Press, Taylor & Francis Group, ISBN 9780367199203. https://doi.org/10.1201/9780429244087. Hybrid thermoplastic and thermosetting composites.

Singha, N. R., Deb, M., Mahapatra, M., Mitra, M., & Chattopadhyay, P. K. (2020). *Smart lightweight polymer composites*. CRC Press, Taylor & Francis Group, ISBN 9780367199203. https://doi.org/10.1201/9780429244087. Lightweight Polymer Composite Structures Design and Manufacturing Techniques.

Srinivasa, R. P., Vijaya, Y., Veera, M. B., & Krishnaiah, A. (2009). Adsorptive removal of copper and nickel ions from water using chitosan coated PVC beads. *Bioresource Technology, 100*, 194–199.

Stenger Mouraa, F. C., Periolib, L., Paganob, C., Vivanib, R., Ambrogib, V., Bresolina, T. M., Riccib, M., & Schoubben, A. (2019). Chitosan composite microparticles: a promising gastroadhesive system for taxifolin. *Carbohydrate Polymers, 218*, 343–354.

Sudheesh Kumar, P. T., Raj, N. M., Praveen, G., Chennazhi, K. P., Nair, S. V., & Jayakumar, R. (2013). In vitro and in vivo evaluation of microporous chitosan hydrogel/nanofibrin composite bandage for skin tissue regeneration. *Tissue Engineering Part A, 19*, 380–392.

Sun, X., Tang, Z., Pan, M., Wang, Z., Yang, H., & Liu, H. (2017). Chitosan/kaolin composite porous microspheres with high hemostatic efficacy. *Carbohydrate Polymers, 177*, 135–143.

Türker, O. C., & Baran, T. (2017). Evaluation and application of an innovative method based on various chitosan composites and *Lemna gibba* for boron removal from drinking water. *Carbohydrate Polymers, 166*, 209–218.

Takuma, O., Yuta, N., Toshiisa, K., Tomohiko, Y., Satoshi, H., Ascensão, L. M., Toshiki, M., & Yuki, S. (2017). Preparation of chitosan-hydroxyapatite composite monofiber using coagulation method and their mechanical properties. *Carbohydrate Polymers, 175*, 355–360.

Tan, H., Chu, C. R., Payne, K. A., & Marra, K. G. (2009). Injectable in situ forming biodegradable chitosan–hyaluronic acid based hydrogels for cartilage tissue engineering. *Biomaterials, 30*, 2499–2506.

Tapia, C., Corbalá, V., Costa, E., Gai, M. N., & Yazdani-Pedram, M. (2005). Study of the release mechanism of diltiazem hydrochloride from matrices based on chitosan-alginate and chitosan-carrageenan mixtures. *Biomacromolecules, 6*, 2389–2395.

Taravel, M. N., & Domard, A. (1995). Collagen and its interactions with chitosan II. Influence of the physicochemical characteristics of collagen. *Biomaterials, 16*, 665–671.

Taravel, M. N., & Domard, A. (1996). Collagen and its interactions with chitosan III. Some biological and mechanical properties. *Biomaterials, 17*, 451–455.

Taravel, M. N., & Domard, A. (1993). Relation between the physicochemical characteistics of collagen and its interactions with chitosan. *Biomaterials, 14*, 930–938.

Teng, S. H., Lee, E. J., Wang, P., Jun, S. H., Han, C. M., & Kim, H. E. (2010). Functionally gradient chitosan/hydroxyapatite composite scaffolds for controlled drug release. *Journal of Biomedical Materials Research B, 90*, 275–282.

Thacharodi, D., & Rao, K. P. (1996). Collagen-chitosan composite membranes controlled transdermal delivery of nifedipine and propranolol hydrochloride. *International Journal of Pharmaceutics, 134*, 239–241.

Tran, H. V., Tran, L. D., & Nguyen, T. N. (2010). Preparation of chitosan/magnetite composite beads and their application for removal of Pb(II) and Ni(II) from aqueous solution. *Materials Science and Engineering: C, 30*, 304–310.

Tsai, W. B., Chen, Y. R., Li, W. T., Lai, J. Y., & Liu, H. L. (2012). RGD-conjugated UV-crosslinked chitosan scaffolds inoculated with mesenchymal stem cells for bone tissue engineering. *Carbohydrate Polymers, 89*, 379–387.

Tseng, H. J., Tsou, T. L., Wang, H. J., & Hsu, S. H. (2013). Characterization of chitosan–gelatin scaffolds for dermal tissue engineering. *Journal of Tissue Engineering and Regenerative Medicine, 7*, 20–31.

Veera, M. B., Krishnaiah, A., Jonathan, L. T., & Edgar, D. S. (2003). Removal of hexavalent chromium from wastewater using a new composite chitosan biosorbent. *Environmental Science and Technology, 37*, 4449–4456.

Veera, M. B., Krishnaiah, A., Ann, J. R., & Edgar, D. S. (2008). Removal of copper(II) and nickel(II) ions from aqueous solutions by a composite chitosan biosorbent. *Separation Science and Technology, 43*, 1365–1381.

Veera, M. B., Krishnaiah, A., Jonathan, L. T., Edgar, D. S., & Richard, H. (2008). Removal of arsenic (III) and arsenic (V) from aqueous medium using chitosan-coated biosorbent. *Water Research, 42*, 633–642.

Venkatesan, J., Qian, Z.-J., Ryu, B.-M., A, K., & Nanjundan, Kim, S.-K. (2011). Preparation and characterization of carbon nanotube-grafted-chitosan – natural hydroxyapatite composite for bone tissue engineering. *Carbohydrate Polymers, 83*, 569–577.

Wan, M. W., Kan, C. C., Lin, C. H., Buenda, D. R., & Wu, C. H. (2007). Adsorption of copper (II) by chitosan immobilized on sand. *Chia-Nan Annual Bulletin, 33*, 96–106.

Wan Ngah, W. S., & Fatinathan, S. (2008). Adsorption of Cu(II) ions in aqueous solution using chitosan beads, chitosan-GLA beads and chitosan–alginate beads. *Chemical Engineering Journal, 143*, 62–72.

Wan Ngah, W. S., Kamari, A., & Koay, Y. J. (2004). Equilibrium kinetics studies of adsorption of copper (II) on chitosan and chitosan/PVA beads. *International Journal of Biological Macromolecules, 34*, 155–161.

Wan Ngah, W. S., Ariff, N. F. M., & Hanafiah, M. A. K. M. (2010). Preparation, characterization, and environmental application of crosslinked chitosan-coated bentonite for tartrazine adsorption from aqeous solutions. *Water, Air, and Soil Pollution, 206*, 225–236.

Wan Ngah, W. S., Ariff, N. F. M., Hashim, A., & Hanafiah, M. A. K. M. (2010). Malachite green adsorption onto chitosan coated bentonite beads: isotherms, kinetics and mechanism. *Clean – Soil, Air, Water, 38*, 394–400.

Wan Ngah, W. S., Teong, L. C., & Hanafiah, M. A. K. M. (2011). Adsorption of dyes and heavy metal ions by chitosan composites: a review. *Carbohydrate Polymers, 83*, 1446–1456.

Wang, L., & Li, C. (2007). Preparation and physicochemical properties of A novel hydroxy-apatite/chitosan-silk fibroin composite. *Carbohydrate Polymers, 68*, 740–745.

Wang, L., & Wang, A. (2007). Adsorption characteristics of Congo red onto the chitosan/montmorillonite nanocomposite. *Journal of Hazardous Materials, 147*, 979–985.

Wang, L. S., Khor, E., Wee, A., & Lim, L. Y. (2002). Chitosan-alginate PEC membrane as a wound dressing: assessment of incisional wound healing. *Journal of Biomedical Materials Research, 63*, 610–618.

Wang, X. H., Li, D. P., Wang, W. J., Feng, Q. L., Cui, F. Z., Xu, Y. X., Song, X. H., & van der Werf, M. (2003). Cross-linked collagen/chitosan matrix for artificial livers. *Biomaterials, 24*, 3213–3220.

Wang, X., Yan, Y., Lin, F., Xiong, Z., Wu, R., Zhang, R., & Lu, Q. (2005). Preparation and characterization of o collagen/chitosan/heparin matrix for an implantable bioartificial liver. *Journal of Biomaterials Science, Polymer Edition, 16*, 1063–1080.

Xu, H. H. K., Burguera, E. F., & Carey, L. E. (2007). Strong, macroporous, and in situ-setting calcium phosphate cement-layered structures. *Biomaterials, 28*, 3786–3796.

Yalçınkaya, S., Demetgül, C., Timur, M., & Çolak, N. (2010). Electrochemical synthesis and characterization of polypyrrole/chitosan composite on platinum electrode: its electro-chemical and thermal behaviors. *Carbohydrate Polymers, 79*, 908–913.

Yan, X. L., Khor, E., & Lim, L. Y. (2000). PEC films prepared from chitosan alginate co-acervates. *Chemical and Pharmaceutical Bulletin, 48*, 941–946.

Yan, L. P., Wang, Y. J., Ren, L., Wu, G., Caridade, S. G., Fan, J. B., Wang, L. Y., JiPH, Oliveira, J. M., Oliveira, J. T., Mano, J. F., & Reis, R. L. (2010). Genipin-cross-linked collagen/chitosan biomimetic scaffolds for articular cartilage tissue engineering applica-tions. *Journal of Biomedical Materials Research Part A, 95*, 465–475.

Yan, J., Wu, T., Ding, Z., & Li, X. (2016). Preparation and characterization of carbon nanotubes/chitosan composite foam with enhanced elastic property. *Carbohydrate Polymers, 136*, 1288–1296.

Yang, Y., Yang, H., Yang, M., Liu, Y., Shen, G., & Yu, R. (2004). Amperometric glucose biosensor based on a surface treated nanoporous Zro_2/chitosan composite film as immo-bilization matrix. *Analytica Chimica Acta, 525*, 213–220.

Yavuz, A. G., Uygun, A., & Bhethanabotla, V. R. (2009). Substituted polyaniline/chitosan composites: synthesis and characterization. *Carbohydrate Polymers, 75*, 448–453.

Yin, N., Du, R., Zhao, F., Han, Y., & Zhou, Z. (2019). Characterization of antibacterial bacterial cellulose composite membranes modified with chitosan or chitooligosaccharide. *Carbo-hydrate Polymers, 229*, 115520.

Zhang, G. Y., Qu, R. J., Sun, C. M., Ji, C. N., Chen, H., Wang, C. H., & Niu, Y. H. (2008). Adsorption for metal ions of chitosan coated cotton fiber. *Journal of Applied Polymer Science, 110*, 2321–2327.

Zhang, X., Xiao, G., Wang, Y., Zhao, Y., Su, H., & Tan, T. (2017). Preparation of chitosan-TiO_2 composite film with efficient antimicrobial activities under visible light for food packaging applications. *Carbohydrate Polymers, 169*, 101–107.

Zhang, H., Peng, L., Chen, A., Shang, C., Lei, M., He, K., Luo, S., Shao, J., & Zeng, Q. (2019). Chitosan-stabilized FeS magnetic composites for chromium removal: charac-terization, performance, mechanism, and stability. *Carbohydrate Polymers, 214*, 276–285.

Zhang, K., Li, J., Wang, Y., Mu, Y., Sun, X., Su, C., Dong, Y., Pang, J., Huang, L., Chena, X., & Feng, C. (2020). Hydroxybutyl chitosan/diatom-biosilica composite sponge for hemor-rhage control. *Carbohydrate Polymers, 236*, 116051.

Zhao, F., Yin, Y., Lu, W. W., Leong, C., Zhang, W., Zhang, J., Zhang, M., & Yao, K. (2002). Preparation and histological evaluation of biomimetic three-dimensional hydroxyapatite/chitosan–gelatin network composites caffolds. *Biomaterials, 23*, 3227–3234.

Zhong, Y., Song, X., & Li, Y. (2011). Antimicrobial, physical and mechanical properties of kudzu starch–chitosan composite films as a function of acid solvent types. *Carbohydrate Polymers, 84*, 335–342.

Zhu, H. Y., Jiang, R., & Xiao, L. (2010). Adsorption of an anionic dye by chitosan/kaolin/g-Fe_2O_3 composites. *Applied Clay Science, 48*, 522–526.

Starch-based blends and composites

7

Nayan Ranjan Singha[1] *and Pijush Kanti Chattopadhyay*[2]
[1]Advanced Polymer Laboratory, Department of Polymer Science and Technology, Government College of Engineering and Leather Technology (Post-Graduate), Maulana Abul Kalam Azad University of Technology, Kolkata, West Bengal, India; [2]Department of Leather Technology, Government College of Engineering and Leather Technology (Post-Graduate), Maulana Abul Kalam Azad University of Technology, Kolkata, West Bengal, India

1. Introduction

Modern civilization still relies heavily on nonrenewable petroleum-based resources for manufacturing various polymeric materials fulfilling diverse global demands of the society. Currently, a huge proportion of our daily-use materials are made mostly from petroleum-based polymers. These conventional materials are highly popular and acceptable because of their high specific strength, durability, easy processibility, versatility, and low cost (Peres et al., 2016; Dutta et al., 2020; Mahapatra et al., 2019). In fact, the mechanical properties and thermal stabilities of synthetic polymers are significantly better than those of naturally occurring polymers (Debiagi et al., 2014; Mali et al., 2010; Sionkowska, 2011). However, in addition to being produced from nonrenewable resources, other disadvantages of these petrochemical polymers include increased CO_2 emissions, poor biodegradability, and lack of biocompatibility. Therefore, to overcome the environmental problems arising from petrochemical-based synthetic polymers, the plastic industries and academic communities are searching for new renewable raw materials possessing the capability to replace high-performance synthetic materials derived from nonrenewable resources. In this way, the fossil fuels can be saved, and the greenhouse gas emissions, including carbon dioxide, responsible for global warming can be reduced (Imre & Pukánszky, 2013). Moreover, the unwanted accumulation of conventional plastic materials can be avoided by reducing such nonbiodegradable plastic wastes and the associated environmental pollution. As most of these traditional plastics are resistant to microbial attack, the developed biodegradable materials derived from renewable natural resources are gaining increased attention from research communities (Nafchi et al., 2013; Mondal et al., 2020).

In recent years, among the various renewable materials, starch-based biodegradable products, such as coatings, adhesives, flocculants, superabsorbents, plastics, and rubber biocomposites have been gaining increased attention from researchers engaged in the academic and industrial sectors (Garcia et al., 1998; Imam et al., 2001; Wei et al., 2008; Wu et al., 2003; Pant et al., 2011; Li et al., 2012). Starch, a promising biopolymer, has the ability to replace traditional nonbiodegradable polymers suffering

Biodegradable Polymers, Blends and Composites. https://doi.org/10.1016/B978-0-12-823791-5.00006-5

from poor, difficult, and costly recycling procedures (Kaseem et al., 2012). Starch is one of the most abundantly available biopolymers derived from various natural resources, such as corn, wheat, potato, cassava, rice, and so on. Among these plant-based resources, corn, wheat, and potato are the major contributors, supplying 81.24%, 8.45%, and 5.36% of the total volume of starch produced worldwide, respectively. As per recent statistical data, nearly 48.5 million tons of starch were produced around the world in 2000.

The major objectives behind blending and composite preparation are to combine and extract major advantages of the individual components in either blend or composite. In this regard, such combined materials are prepared to actuate simultaneous property development so as to create high-performance materials (Chattopadhyay et al., 2010). For instance, a synthetic polymer is mechanically stronger than a biodegradable natural polymer, whereas synthetic polymers are prepared often from nonrenewable petroleum resources, and are mostly nonbiodegradable. In this regard, the suitable combination of natural polymer/starch with a synthetic polymer may result in blends and composites presenting high mechanical strength and reasonable biodegradability. Therefore, in recent times, researchers have been trying to develop high-performance green blends and composites constituted of the optimum combination of starch and nonstarch materials. Recently, their research activities have been switching over to the field of self-reinforced composites. By combining a starch or similar polysaccharide-based particulate matter with a continuous starch matrix, scientists have achieved self-reinforced composites with better compatibility, processibility, rheological and mechanical properties, antimicrobial properties, and biodegradability.

Therefore, in this chapter, the discussion is mostly concentrated in three different areas of starch-based blends and composites. The first part deals with the blends and composites of starch and rubber, wherein the conventional modifications of starch and rubber and their roles in improving the overall properties of blends/composites are discussed along with the contributions of compatibilizing and coupling agents. The second part discusses the traditional blends and composites prepared from starch and nonbiodegradable plastic materials, and the final part is associated with the recently developed starch-based biodegradable plastics including self-reinforced composite materials.

2. Chemical structure of starch

Starch is an important member of the homopolysaccharide family. Its basic formula is $(C_6H_{10}O_5)_n$, and the biopolymer consists of a number of glucose units interconnected by two different types of glycosidic linkages, i.e., α-(1 $\rightarrow$ 6) glycosidic and α-(1 $\rightarrow$ 4) glycosidic linkages (Buléon et al., 1998). In general, starch is composed of two major types of microstructure, i.e., amylose and amylopectin (Fig. 7.1). Of these two microstructures, amylose and amylopectin are linear and branched, respectively. In amylose, the glucose units are interconnected by α-(1 $\rightarrow$ 4)

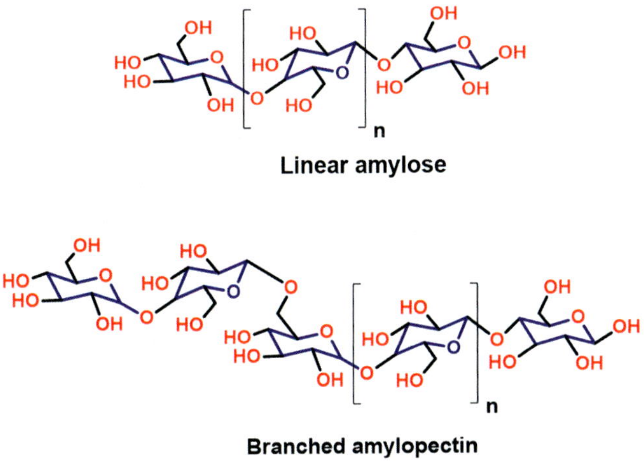

Linear amylose

Branched amylopectin

Figure 7.1 The amylose and amylopectin of starch.

glycosidic linkages, whereas the glucose units in amylopectins are joined by both α-(1 → 6) and α-(1 → 4) glycosidic linkages. Moreover, the average molecular weight of linear amylose, i.e., 20,000–225,000 g/mol, is much lower than that of highly branched amylopectin, i.e., 200,000–1,000,000 g/mol. In fact, the relative proportion of amylose and amylopectin in a starch depends on the botanical origin or species from which the biopolymer has been extracted.

3. Properties of starch

The variable intrinsic properties of starch, such as amylose content, granule shape, particle size, crystallinity, type of crystalline phase, and gelatinization temperature, are dependent on their botanical origins. To understand the structure and intrinsic properties of starch, various characterization techniques including Fourier transform infrared spectroscopy (FTIR), nuclear magnetic resonance (NMR), X-ray diffraction (XRD), and thermogravimetric analyses (TGA) may be utilized. In this regard, the most frequently available corn starch has been often selected as the representative starch sample.

3.1 FTIR analysis

Since starch contains a number of glucose units with multiple O–H groups, the FTIR spectrum of corn starch envisages a wide and strong absorption peak within 3000–3500 cm^{-1} attributed to O–H stretching (Fig. 7.2). The absorption peaks at 2822 and 2928 cm^{-1} are ascribed to the asymmetric stretching of C–H in the $-CH_3$ or $-CH_2-$ group. Peak broadening at 3151 and 2928 cm^{-1} along with a shoulder at 3617 cm^{-1} is related to the coexistence of unconventional C–H···O with the

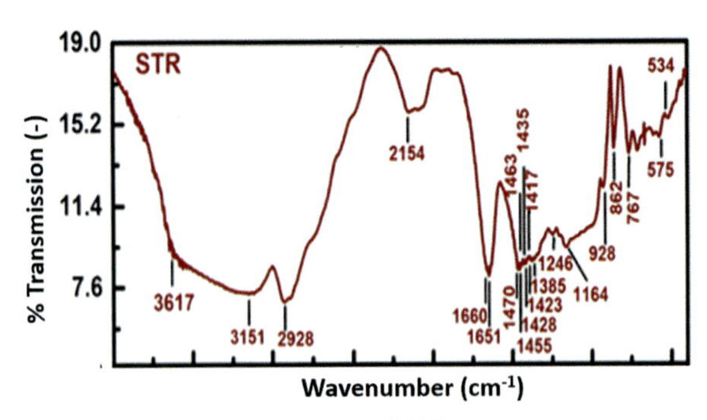

Figure 7.2 FTIR spectra of starch (Singha et al., 2019).

conventional O—H···O type hydrogen bonds (Mitra et al., 2020; Mahapatra, Dutta, Roy et al., 2020). Since starch is mostly populated with methine type C—H, and the hydrogen bond donor strength of methine-type C—H is higher than those of >CH_2 and —CH_3, methine-type C—H of pyranose rings acts as the preferred hydrogen bond donor in C—H···O type hydrogen bonds (Singha et al., 2019). Moreover, the methine-type C—H can be substantiated from peaks at 1435 and 1463 cm^{-1} attributed to the angular deformation of C—H. The peak corresponding to C—O—C is recorded at 1164 cm^{-1}. In addition, peaks at 2154, 1660, and 1651 cm^{-1} are attributed to the vibrations originating from protein impurities often present in the extracted starch.

3.2 NMR analysis

The figures were originally published by Singha et al. (2019) and approved for reuse by ACS. Starch is composed of glucopyranose units connected through α-(1 → 4)- and/or α-(1 → 6)-linkage(s). The ^{1}H NMR of α-(1 → 4)-linkage of starch shows the characteristic peaks of H-1, H-2/H-4, H-3, H-5, and H-6 at 5.39—5.51, 3.57—3.60, 3.93, 3.76, and 3.85 ppm (Fig. 7.3), respectively. Moreover, ^{1}H signals of H-1 and H-2 of α-(1 → 6)-linkage of starch appears within 4.57—5.06 and at 3.36 ppm, respectively. Such structural moieties can more precisely be realized from the ^{13}C NMR of starch. In ^{13}C NMR of starch, peaks at 102.53, 74.42, 75.64, 80.31, 73.83, and 63.74 ppm are assigned to C-1, C-2, C-3, C-4, C-5, and C-6 of α-(1 → 4)-linkage, respectively. Moreover, peaks at 100.15, 71.68, and 60.56 ppm are ascribed to C-1, C-3, and C-6 of the α-(1 → 6)-linkage, respectively.

3.3 XRD analysis T

The XRD diffractogram of corn starch contains five diffraction peaks at θ = 15.1, 17.2, 18.0, 19.9, and 23.0, indicating the presence of A-type semicrystalline polymorphs. In starch, A-type polymorphs are composed of the monoclinic unit cell containing 12 glucose residues at the two left-handed parallel-stranded double helices, wherein four water molecules persist at the interstitial position of the helices (Mondal et al., 2019).

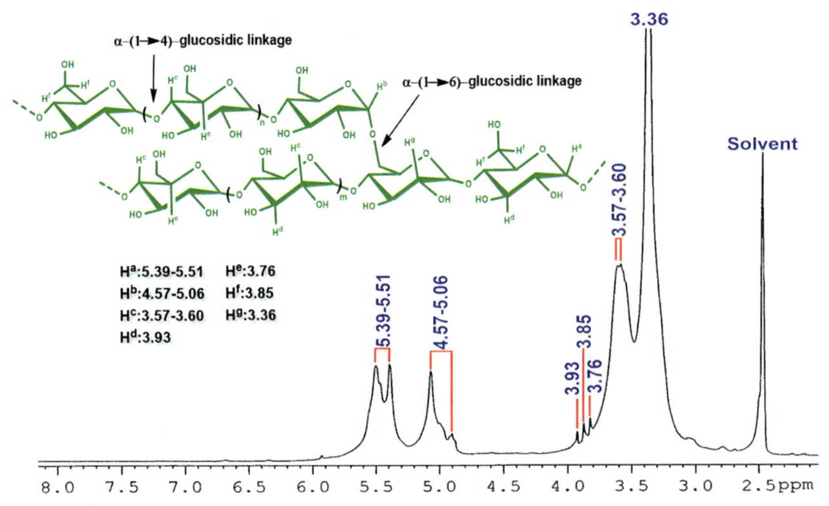

Figure 7.3 ^{1}H NMR spectra of starch (Mondal et al., 2019).
Originally published in Mondal, H., Karmakar, M., Chattopadhyay, P. K., & Singha, N. R. (2019). Starch-g-tetrapolymer hydrogel via in situ attached monomers for removals of Bi(III) and/or Hg(II) and dye(s): RSM-based optimization. *Carbohydrate Polymers 213*, 428–440 and approved for reuse by Elsevier.

3.4 TG analysis

In TG and DTG plots of starch (Fig. 7.4), the initial decomposition stage within 50–150°C is attributed to the removal of moisture, whereas the second stage within 239–353°C shows massive loss of 62 wt.% due to polysaccharide decomposition (Karmakar et al., 2019; Singha, Dutta et al., 2018; Mahapatra et al., 2018; Neto et al., 2016). Such enormous weight loss is also reflected from the corresponding DTG plot envisaging maximum decomposition at 320°C. Finally, the final stage within 353–600°C is assigned to the removal of some compounds originated via decomposition of the starch backbone. Here, almost 17% of the residue is associated with the thermoresistant inorganic impurities of starch.

3.5 Biodegradability

Being a natural polysaccharide, starch can undergo complete biodegradation in the presence of various microorganisms or bacteria, such as *Bacillus amyloliquefaciens*, *Aspergillus oryzae*, *Aspergillus niger*, *Bacillus acidopullulyticus*, *Bacillus licheniformis*, and *Bacillus subtilis* (Jiang et al., 2020). The modes of starch degradation of these microorganisms are different, and accordingly, the end products after enzymatic degradation are different. For instance, *Bacillus licheniformis*, *Bacillus amyloliquefaciens*, and *Aspergillus oryzae* degrade starches by cleaving α-(1 → 4)-glycosidic linkages resulting in the production of α-dextrin, maltose, and oligosaccharides. In contrast, *Aspergillus niger* is involved in cleaving

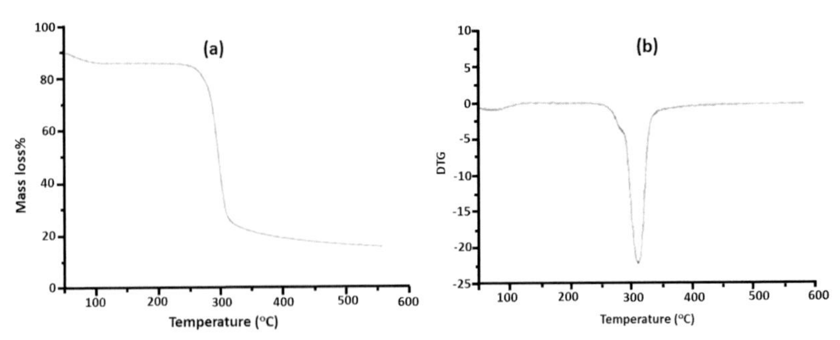

Figure 7.4 (A) TGA and (B) DTG plots of starch.
Originally published by Neto, B.A. de M., Fernandes, B.S., Junior Franco, C.C.M.F., Bonomo, M.R.C.F., de Almeida, P.F., Pontes, K.V. (2016). Thermal-morphological characterisation of starch from peach-palm (Bactris gasipaes Kunth) fruit (Pejibaye), *International Journal of Food Properties 20*, 1007–1015 and approved for reuse by Taylor and Francis Online.

α-(1 $\rightarrow$ 4)- and α-(1 $\rightarrow$ 4)-glycosidic linkages to produce β-glucose. Importantly, these degraded products, such as dextrin, oligosaccharides, and maltose, undergo catabolism to generate carbon dioxide and water. Thereafter, carbon dioxide and water can again be recycled to produce starch via a photosynthesis process. Accordingly, starch is a renewable and biodegradable natural resource.

4. Starch-based blends and composites

4.1 Blends and composites of starch with rubber

Recently, there has been great interest in reinforcing rubber compounds by incorporating starch-based fillers. Earlier, it was observed that the mechanical properties of the rubber composites deteriorate when native starch is incorporated into the rubber matrix (Kiing et al., 2013; Afiq & Azura, 2013; Khalaf & Sadek, 2012). The uniform dispersion of native starch in the rubber matrix is relatively difficult to achieve due to the hydrophilic surface characteristic, large particle size, and high melting point of starch. Moreover, the hydrophobic nature of rubber components makes the dispersion of starch far more difficult. Therefore, the following major strategies can be adopted for producing high-performance rubber composites bearing starch as reinforcing fillers: (1) modification of starch, (2) addition of coupling agent, and (3) modification of rubber matrix.

4.1.1 Modification of starch

Commonly practiced modifications of starch involve several methods, such as gelatinization, plasticization, hydrolysis, and chemical grafting.

4.1.1.1 Gelatinization

The main objective of gelatinization is to break down the crystalline microstructure of starch by rupturing inter/intramolecular hydrogen bonds at a higher temperature. A gelatinization process usually consists of three stages. During the initial stage, mechanical stirring and heat supply are continued to ensure irreversible absorption of water by the added starch granules followed by swelling in water. In the next stage, the intermolecular hydrogen bonds among starch macromolecules become significantly ruptured when the temperature exceeds the gelatinization point. In the final stage, the substantial loss of hydrogen bonds results in easier penetration of water molecules through the starch molecules. Consequently, the starch granules experience massive swelling and destruction of crystallites (Liu et al., 2009). Thus, as result of gelatinization, the branched amylopectin double helices become dissociated (Fig. 7.5), leading to complete disruption and disappearance of the crystalline domains in the structure of starch. Such gelatinization of the starch process is widely applicable in the case of preparing rubber composites by a latex compounding method (Wu et al., 2004). The main objective of this method is to achieve relatively uniform dispersion of starch in the rubber matrix compared to the conventional solid compounding method. Here, the starch granules are converted to starch paste via gelatinization in the aqueous environment at 90°C. Thus, the crystalline microstructure of starch is substantially broken down via massive rupturing of inter/intra-molecular hydrogen bonds at higher temperature followed by penetration of water molecules through the disordered microstructure. After the necessary disintegration, the starch phase is admixed evenly with the rubber latex. Once the desired uniformity in mixing is achieved, the mixture of starch and latex undergo immediate cocoagulation using $CaCl_2$ solution. Here, the added salt, i.e., $CaCl_2$, supplies the necessary Ca^{2+} ions aiding the simultaneous precipitation of both starch and rubber latex. In this way, the particle size of the admixed starch becomes less than 1 μm, leading to the formation of microcomposite and the uniform dispersal of starch. Increasingly smaller particle sizes of starch ensure the formation of a greater starch–rubber interfacial area, resulting in improved mechanical properties of starch–rubber composites prepared by the latex compounding method compared to the conventional solid compounding method. In this context, one group executed comparative studies on the mechanical properties offered by starch–rubber composites prepared by either

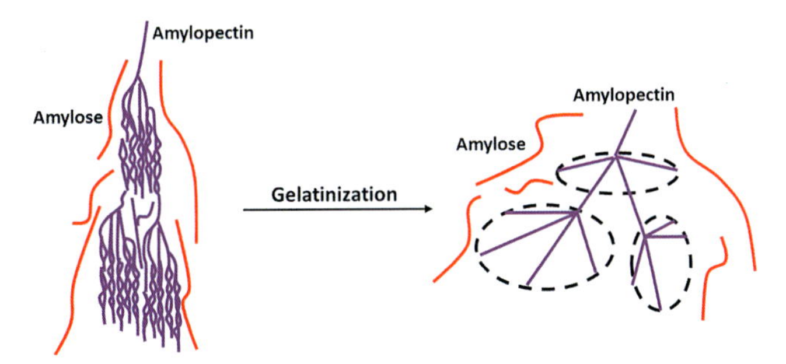

Figure 7.5 Modification of starch by gelatinization.

Table 7.1 Mechanical properties of starch–SBR composites with 10 phr starch.

Materials	Stress at 100%	Tensile strength (MPa)	Elongation at break (%)	Tear strength (kN/m)	Reference
Pure SBR[a]	0.8	2.3	524	12.1	Wu et al. (2004)
Solid compounded SBR composite	0.8	2.6	480	14.0	
Latex compounded SBR composite	1.1	3.6	544	19.9	

[a]styrene butadiene rubber

solid or latex compounding methods (Wu et al., 2004). In the case of starch–rubber composites prepared by the solid compounding method, the addition of 10 phr starch into SBR reduces the elongation at break (Table 7.1). Moreover, the modulus of the composite showed no improvement. On the other hand, the addition of 10 phr starch in latex compounded starch–rubber composites gives rise to an appreciable increase in all parameters, i.e., tensile strength, elongation at break, tearing strength, and modulus. The elongation at break of solid compounded composite deteriorates due to the presence of larger particles of starch restraining the conformational rotation and movements of the polymer chains along the direction of the applied tensile force. The presence of larger particles of starch in solid compounded composite is also evident from SEM photomicrographs (Fig. 7.6A). A large number of starch granules remain scattered on the photomicrograph of the tensile fractured surface, and most of those particles reach as large as 20 μm (Fig. 7.6A). Moreover, there is a large number of voids at the starch granule–rubber interface, indicating poor interfacial compatibility among the hydrophilic starch and hydrophobic SBR matrix. Conversely, the uniform distribution and the smaller dimension of starch particles in a mechanically strengthened latex compounded starch–rubber composite can be manifested in photomicrographs (Fig. 7.6B). Here, the gelatinization of starch causes the massive deformation and enhanced surface area of starch granules, resulting in elevated compatibility among hydrophilic starch and hydrophobic SBR matrix. The gelatinization-driven latex compounding is beneficial in improving the abrasion resistance of the starch–carbon black–SBR composites. However, the compounding of starch deteriorates the tensile strength of carbon black–SBR composites using both solid and latex compounding methods.

4.1.1.2 Plasticization

One of the drawbacks of native starch is its higher melting temperature, i.e., 220–240°C, than its decomposition temperature, i.e., 220°C. Therefore, native starch is highly susceptible to thermal degradation at the time of melting during thermal processing. The higher melting temperature is related to the stronger intermolecular and intramolecular hydrogen bonds among the macromolecules, preventing the translational motion of the chains. Therefore, conventional practice is to reduce the intermolecular and intramolecular hydrogen bonding by the insertion of plasticizing molecules

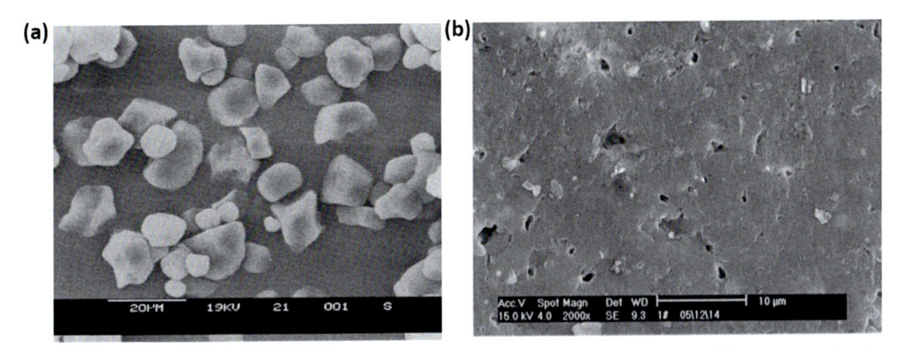

Figure 7.6 SEM micrographs of starch (10 phr)—SBR composites prepared by two methods: (A) solid compounding method and (B) latex compounding method (Wu et al., 2004). Originally published by Wu, Y.P., Qi, Q., Liang, G.H., & Zhang, L.Q. (2006). A strategy to prepare high performance starch/ rubber composites: in situ modification during latex compounding process. *Carbohydrate Polymers 65*, 109—113. and approved for reuse by Elsevier B.V.

that can reduce the cohesive forces among the individual starch macromolecules (Ning et al., 2007). Thus, the melting becomes easier for a plasticized starch compared to a native starch. The plasticized starches of reduced melting temperature and better processibility are commonly known as thermoplastic starch (Fig. 7.7).

To date, various plasticizers, such as water (Van Soest, Benes, & De Wit, 1996; Van Soest, Bezemer et al., 1996; Van Soest & Borger, 1997; Dean et al., 2007; Wang et al., 2008), glycol (Yu et al., 1996; Lourdin et al., 1997; Smits et al., 2001; Da Róz et al., 2006), glycerol (Forssell et al., 1997; Curvelo et al., 2001; Liu et al., 2001; Park et al., 2002; Park et al., 2003; Rodriguez-Gonzalez et al., 2004; Chen & Evans, 2005; Huneault & Li, 2007; Shi et al., 2007; Huang et al., 2004), sorbitol (Wang et al., 2000; Mathew & Dufresne, 2002; Krogars et al., 2003; Teixeira et al., 2009), citric acid (Wang et al., 2009; Wang et al., 2007; Ma et al., 2009; Yu et al., 2005), formamide (Ma & Yu, 2004a, 2004b, 2004d; Ma et al., 2004, 2005), ethanolamide (Ma et al., 2006), urea (Ma & Yu, 2004c), and acetamide (Ma & Yu, 2004c) are effectively applied to plasticize native starch. Among these various plasticizers, polyols, such as glycol, glycerol, and sorbitol, impose a pronounced effect on the

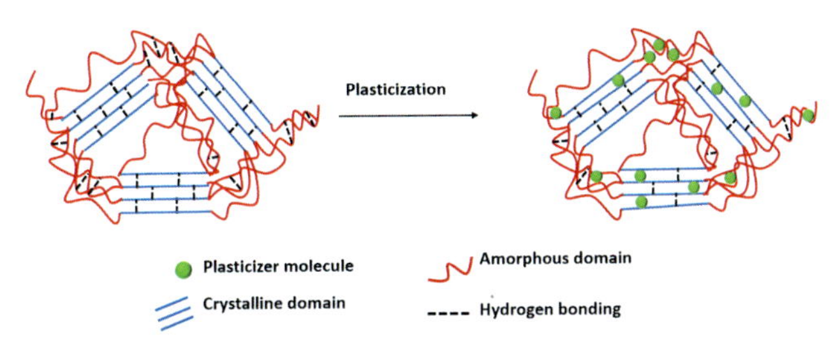

Figure 7.7 Plasticization of starch with plasticizer destroying hydrogen bonds and reducing the melting point.

plasticization of starch. These polyhydric alcohol-based plasticizers are positioned between the adjacent starch macromolecular chains and reduce the interactions among starch molecules. Accordingly, the preexisting strong hydrogen bonds among the starch molecules reduce in the presence of a plasticizer, resulting in a significant drop in the Young's modulus and overall mechanical strength. Thus, the favorable movement of starch macromolecular chains reduces melting viscosity and increases the elongation at break. Moreover, due to the cleavage of preexisting hydrogen bonds, these plasticized starch molecules should contain some free hydroxyl and other hydrophilic functionalities, which can be further involved in binding with moisture. Accordingly, thermoplastic starches are susceptible to moisture when they are stored in high relative humidity. Therefore, the blending of thermoplastic starch with other polymers is necessary to overcome such characteristic inherent drawbacks of thermoplastic starch (Martins & Santana, 2016; Parulekar & Mohanty, 2007). However, lack of compatibility is one of the major problems often encountered during blending of thermoplastic starch with other polymers. The major reasons behind such a lack of compatibility are the mismatch in hydrophilicity/hydrophobicity and differences in chemical structures (Schwach & Averóus, 2004). Thus, for an immiscible blends of two noncompatible polymers, proper compatibilization methods are essential to improve interfacial adhesion between those two noncompatible polymers (Avella et al., 2000). To improve the compatibility, sometimes physical treatment is done by applying compatibilizers and block copolymers. In other cases, chemical treatment is performed to modify the functionalities of the individual components. In this way, the mechanical properties of the blends can be improved via formation of the smaller dispersed phases in the presence of a suitable compatibilizing agent (Imre & Pukánszky, 2013; Martins & Santana, 2016). In this regard, Carvalho et al. (2003) prepared stable thermoplastic starch—NR blends using glycerol as the plasticizer contributing to the plasticization of starch as well as improving the starch—NR interface.

4.1.1.3 Hydrolysis

Hydrolysis is another effective way to produce starch particles in the nano-dimension range. Both acid and alkaline hydrolysis can be used to reduce the particle size of starch. Since native starch is a semicrystalline polymer comprised of amorphous and crystalline phases, both phases essentially possess different dissolving tendencies in water. Being a well-packed domain with stronger intermolecular/intramolecular hydrogen bonds, crystalline domains show greater water resistance and hence are sparingly soluble in water or acidic environments. Conversely, the amorphous domains are less resistant to water and can be dissolved by acid hydrolysis. These almost undissolved small crystalline domains function as starch nanoparticles. Since amorphous and crystalline domains possess different hydrolyzing tendencies, the acid hydrolysis of starch is conducted in two steps (Jayakody & Hoover, 2002). During the first step, amorphous regions undergo hydrolysis, The nature of the hydrolysis step depends on the lipid-complexed amylose chains, particle size, amylose content, and pores on the surface. In the next step, the hydrolysis of crystalline regions is influenced by several factors, such as the mode of distribution of

α-(1 $\rightarrow$ 6)-linkages among amorphous and crystalline regions, amylopectin content, and degree of packing of the double helices within the crystalline domain. Accordingly, the yield and morphology of starch nanocrystals prepared by acid hydrolysis are dependent on the particle size, amylose/amylopectin ratio, crystalline type, and hydrolysis conditions (Jayakody & Hoover, 2002; Wang et al., 2001; Corre et al., 2010; LeCorre et al., 2011; Singn & Ali, 1987). Starch nanocrystals obtained from acidic hydrolysis of native starch are 3D nanoparticles. Because of the very small particle sizes, starch nanocrystals can effectively function as a reinforcing filler in various rubber composites (Angellier et al., 2005a, 2005b). Those rubber composites with fewer voids can improve the gas barrier properties and decrease the water vapor permeability. A substantial reinforcing effect of starch nanocrystals on the nonvulcanized natural rubber matrix results in a massive improvement of storage moduli at 50°C upon continuous addition of starch nanocrystals from 10 to 30 wt.% (Angellier et al., 2005a, 2005b). Moreover, the reinforcing effect of starch nanofillers upon nonvulcanized natural rubber chains can be reflected from the elevated mechanical properties and relatively restricted swelling of nonvulcanized natural rubber nanocomposites (Rajisha et al., 2014). Theoretically, the reinforcing effect of starch nanofillers upon rubber chains is ascribed to the hydrogen bonding among O—H groups of starch nanoparticle clusters and natural rubber chains. In the presence of excessive moisture content, such nonconventional hydrogen bonding among O—H groups of starch nanoparticles and polyisoprene chains of natural rubber is superseded by the stronger conventional hydrogen bonds between moisture and starch moieties, resulting in reduced mechanical strength of the nonvulcanized natural rubber nanocomposites (Angellier et al., 2005a). To throw some light upon the reinforcing mechanism of starch nanocrystals in the nonvulcanized natural rubber matrix, two well-known models, i.e., the Kraus model and the Maier and Goritz model, have been applied to predict the Payne effect (Mele et al., 2011). Herein, adsorption—desorption or slippage of NR chains on the surface of starch nanocrystals influences the nonlinear viscoelastic properties of starch nanocrystals—non-NR (vulcanized) nanocomposites. Additionally, the improved gas barrier properties of nonvulcanized natural rubber nanocomposites have been noted by a continuous decrease in water vapor permeability from 3.41 to 1.88 × 1010 g/(m.s.Pa) with a gradual increase in starch nanocrystals up to 30 wt.%. the effect of the botanical origin of starch on the performances of nonvulcanized NR nanocomposite nanocrystals has been studied (LeCorre et al., 2012). The amylose content of starch nanocrystals is an important parameter determining the reinforcing capacity of starch nanocrystals in starch nanocrystals—NR (nonvulcanized) nanocomposites. The native starch granules with a higher amylose content, used for preparing starch nanocrystals, present a lower water uptake and reinforcing capacity. The influence of the extent of aggregation among starch nanocrystals and the effect of starch nanocrystal loading on the dispersion and electrical properties of nonvulcanized NR nanocomposites has been investigated. The higher extent of aggregation reduces the electrical resistivity, suggesting facilitated electronic conduction by the increasingly higher number of filler—filler interphases within the matrices (Bouthegourd et al., 2011).

4.1.1.4 Chemical reaction

In a blend or composite, the chemically modified starch presents an elevated bonding affinity with other component(s). The modification is often carried out by grafting starch into copolymer of vinyl monomers. Here, the frequently available O—H groups of starch are involved during chemical modifications. Among those O—H groups in glucopyranose units, primary O—H functionalities are inherently more reactive than the secondary —O—H, and accordingly, primary O—H can be more easily involved for grafting into copolymers (Mitra et al., 2019). Sometimes, pendent side chains of the grafted monomers can function as plasticizing agents. Thus, the grafted vinyl polymers can alter the crystallinity (Gao et al., 1994), thermal properties (Wang et al., 2011; Jyothi, 2010), and morphology of starch. In this way, the semisynthetic starches modified with copolymers are widely employed as major ingredients in diverse industrial fields including biodegradable matrix in plastics, reinforcing fillers in rubbers, flooding agents in the oil drilling, packaging films, coatings in wallpaper, adhesives in tapes, superabsorbent polymers in baby diapers, flocculants in water treatments, and biocompatible materials in medical fields. Moreover, being a biodegradable and biocompatible natural polymer, starch-based copolymers can be considered as potential ingredients in bone tissue engineering. For instance, starches modified with either polybutyl acrylate (Liu et al., 2008) or polymethyl methacrylate (Li & Cho, 2013a, 2013b; Li et al., 2013;) have been used as potential fillers in fabricating starch—rubber composites. In addition, encouraging results are obtained when rubbers are filled with chemically modified starches including starch xanthate (Buchanan et al., 1968), resorcinol formaldehyde modified starch (Wu et al., 2006; Tang et al., 2006; Qi et al., 2006), acetylated starch (Jyothi, 2010), and hydroxypropylated starch (Jyothi, 2010). In fact, hydroxypropylation of starch is a strategy to improve the rheological characteristics, processibility, and barrier properties against oxygen and water vapor. Moreover, such chemical treatment elevates the compatibility of starch with hydroxypropyl methylcellulose and rubbers during the preparation of blends. Similarly, starch nanoparticles extracted from maize have been isocyanated and acetylated by means of 1,4-hexamethylene diisocyanate and acetic anhydride, respectively. Such chemically modified starch particles display better compatibility with the natural rubber matrix, resulting in mechanically reinforced rubber matrices, especially in case of composite with isocyanated starch nanoparticles (Valodkar & Thakore, 2011). In this regard, better reinforcing capability of isocyanated starch nanoparticles over acetylated starch nanoparticles in natural rubber composites may be associated with a higher number of available nitrogen/oxygen centers functioning as conventional hydrogen-bonding sites (Mahapatra, Dutta, Mitra et al., 2020). To exploit the reinforcing capability of chemically modified starch, researchers have attempted the replacement of traditionally used carbon black fillers with starch-based fillers (Wu et al., 2009). Simultaneous improvement in the mechanical properties of ternary nitrile rubber-based composites is observed when a suitable combination of polymethyl methacrylate-modified starch and carbon black is used as filler in the nitrile rubber matrix. When the proportions of PMMA-modified starch and carbon black are 25 and 25 phr, respectively, the optimum tensile strength of 18.5 MPa is obtained for ternary composite. This value is significantly higher than the tensile strengths of 11.8 and 13.5 MPa for PMMA-modified starch—NBR and carbon black—NBR binary composites, respectively, containing 50 phr individual filler component (Kim & Cho, 2013). Here, the higher extent of filler—filler and filler—polymer interfaces and their interfacial characteristics can contribute to the greatly elevated mechanical properties of ternary composites (Chattopadhyay et al., 2011).

4.1.2 Addition of coupling agents

The main objective of coupling agent addition is to enhance the interfacial interaction between the inherently hydrophilic starch and hydrophobic rubber matrix within a blend or composite. In a starch—rubber composite, the coupling agent functions as a bridge connecting two mutually incompatible hydrophilic starch and hydrophobic rubber phases and imparts a stronger reinforcing effect. This is a method of compatibilization for stabilizing polymer blends and composites. A coupling agent usually possesses dual functional groups with different reactivities. Of these, one functional group is capable of being attached to the hydrophilic starch by forming covalent and hydrogen bonds, whereas the other functional group reacts with a hydrophobic rubber phase. Thus, numerous coupling agents, such as poly(ethylene vinyl alcohol) (Corvasce et al., 1997), bis(triethoxysilylpropyl) tetrasulfide (Corvasce et al., 1997), glycerol (Carvalho et al., 2003), maleic anhydride (Khalaf & Sadek, 2012), glycidyl methacrylate (Khalaf & Sadek, 2012), *N*-2-(aminoethyl)-3-aminopropyltrimethoxysilane (Wu et al., 2006; Tang et al., 2006; Qi et al., 2006), and 4,4-methylene bis(phenyl isocyanate) (Li & Cho, 2013a), have been employed for preparing various starch—rubber composites. Among the various coupling agents used in preparing PMMA—modified starch—SBR composites, 4,4-methylene bis(phenyl isocyanate) has proven to be the most effective coupling agent (Li & Cho, 2013a). Here, 4,4-methylene bis(phenyl isocyanate) functions as a strong bridge interconnecting the hydroxyl groups of starch and phenyl group of SBR. Isocyanate groups of 4,4-methylene bis(phenyl isocyanate) react with hydroxyl groups of starch to produce strong carbamate bonds, whereas phenyl groups of 4,4-methylene bis(phenyl isocyanate) and SBR can be involved in stacking interactions (Fig. 7.8). On the other hand, silane coupling agents can interact with starch only via hydrogen bonds, which are significantly weaker compared to covalent carbamate bonds. Thus, the cross-linking or bridging ability of 4,4-methylene bis(phenyl isocyanate) is significantly higher than those of silane coupling agents in the case of PMMA—modified starch—SBR composites. Therefore, 4,4-methylene bis(phenyl isocyanate) exhibits superior reinforcement to other coupling agents, resulting in the highest tensile strength and modulus values.

Moreover, the compatibilizing efficiencies of maleic anhydride and glycidyl methacrylate have been compared in producing maize starch—NR vulcanizates (Khalaf & Sadek, 2012). The application of maleic anhydride as a compatibilizer results in a deterioration of the tensile strength of maize starch—NR vulcanizates. On contrast, when glycidyl methacrylate is employed as a compatibilizer, the tensile strength of the composite increases. In the presence of glycidyl methacrylate, the rate of vulcanization is

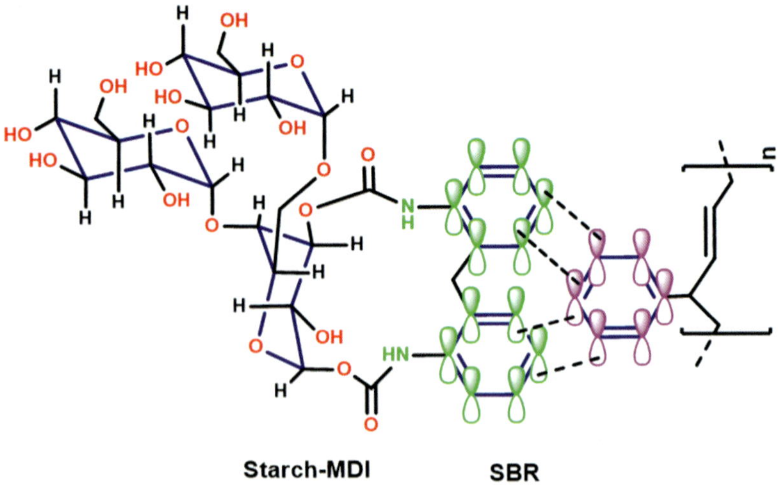

Starch-MDI SBR

Figure 7.8 Stacking interaction of starch-MDI and SBR.

improved, and such an elevated rate of vulcanization is linked with improved tensile strength and overall mechanical properties. Here, the epoxy functionalities of glycidyl methacrylate play an active role in accelerating the rate of vulcanization and improved mechanical properties of maize starch–NR vulcanizates. Wu and colleagues (Wu et al., 2006; Tang et al., 2006; Qi et al., 2006) have reported the simultaneous effect of resorcinol formaldehyde and silane coupling agent on the reinforcement of starch–SBR composites. The significant hydrogen bonding capabilities of resorcinol formaldehyde and silane coupling agent have been exploited to produce cross-links in maize starch–NR vulcanizates.

4.1.3 Modification of rubber matrix

Another important strategy for producing high-performance blends and composites is the modification of rubber matrices. To facilitate the dispersion of starch particles, rubber matrices are often modified by different vinyl monomers, such as poly(dimethylaminoethyl methacrylate) (Rouilly et al., 2004), polymethyl methacrylate (Nakason et al., 2001, 2003), and maleic acid (Garcia et al., 1998) to incorporate polar moieties, such as acrylonitrile, maleic anhydride, and citraconic anhydride, in hydrophobic rubber chains. Thus, the interfacial bonding between the hydrophilic filler, i.e., starch, and the modified rubber with polar functionalities is improved.

However, such a matrix modification strategy is complicated and relatively expensive. Accordingly, researchers are less interested in fabricating starch containing modified rubber matrices. Rouilly et al. (2004) chemically modified natural rubber latex with cationic poly(dimethylaminoethyl methacrylate) via surface grafting. In this way, they achieved better elongation, toughness, and water sensitivity in poly(dimethylaminoethyl methacrylate)-modified natural rubber–starch films. Herein, the tertiary amine moiety functions as a hydrogen bond donor and acceptor, whereas the intermolecular hydrogen bonds and the natural rubber latex facilitate better distribution of starch and interfacial adhesion with the cationic polymer, which is similar to the strategy adopted during the gelatinization process of starch. In other works, natural rubber has been modified by poly(methyl methacrylate) and maleate functionalities (Nakason et al., 2001, 2003, 2005). The as-prepared NR-grafted-poly(methyl methacrylate) and maleated-NR have been compounded individually with cassava starch. However, with the increase in starch portion, both curing time and tensile strength have been noted to decrease. Though electron beam irradiation is one of the effective techniques employed to improve the interfacial adhesion in polymer blends, electron beam irradiation of natural rubber is not suitable for imparting the desired results in starch-filled natural rubber (Senna et al., 2013). In contrast, the mechanical properties are adversely affected due to such irradiation treatment. As a whole, because of the higher cost, processing difficulties, and inferior mechanical properties, the modification of rubber matrix is not a very effective technique for the preparation of high-performance rubber–starch composites.

4.2 Blends and composites of starch with nonbiodegradable plastics

Similar to rubber matrices, plastic components are regularly used to achieve biodegradable blends and composites of starch. Most of these plastic-based blends and composites are composed of polyethylene as the base material. Among the different kinds of polyethylene, low-density polyethylene is the most widely employed plastic, as it possesses the highest amorphous characteristics of all the polyethylenes. Thus, the relatively higher void spaces facilitate an easier dispersion and distribution of starch particles. For the same reason, both linear low-density polyethylene and high-density polyethylene are less studied compared to low-density polyethylene. The better biodegradability of the starch–polyethylene blends and composites has been achieved via consumption of the starch component by the enzymes secreted by microorganisms in soil. Such disappearance of the starch component results in the formation of voids, enhanced porosity, and loss of integrity of plastic matrix, resulting in the ultimate breaking down of polyethylene matrix into smaller particles (Kiatkamjornwong et al., 2001). However, with the increase in starch content, the mechanical property as well as the biodegradability of the polyethylene matrix are enhanced. Accordingly, extensive research is still needed to achieve LDPE-based optimum biodegradable packaging material to create the desirable mechanical strength (Hemati & Garmabi, 2011; Sabetzadeh et al., 2015). As stated earlier, the mechanical properties of LDPE-admixed starch materials are largely dependent on the extent of compatibility between starch and polyethylene. LDPE is naturally mostly hydrophobic, whereas the starch component is highly hydrophilic because of the presence of frequent O–H groups at the surface. Therefore, the uniform distribution of starch granules in the LDPE matrix is obviously difficult. Such difficulty can be circumvented by adding the compatibilizing agents or through the modification of starch and LDPE components. In this regard, a group of researchers has prepared LDPE–starch blends of comparable ductility and moduli as those of virgin LDPE (Rodriguez-Gonzalez et al., 2003), in which the stoichiometric ratio of LDPE and starch is 55:45, and plasticized starch has been used without employing any interfacial modifier. In the case of LLDPE–starch blends, though plasticized starch has been used, the mechanical properties of the blends deteriorate with continuously increasing addition of thermoplastic starch from 10% to 40%. Especially at the higher proportion of thermoplastic starch, the blends clearly envisage a two-phase morphology with unwanted microvoids affecting the mechanical properties (Euaphantasate et al., 2008). In the case of LLDPE–starch blends, the relatively higher crystallinity of LLDPE than that of LDPE may negatively influence the compatibility between LLDPE and plasticized starch. The plasticized starch may not have enough space to achieve intermingling around the starch–LLDPE interface. Sometimes, to improve miscibility and strengthen the mechanical properties, a minute amount of zeolite (1–5 wt.%) is added as a physical compatibilizer and reinforcing filler (Thipmanee & Sane, 2012). Therefore, attaining desirable compatibility between synthetic polymers and thermoplastic starch remains a major challenge for researchers working in the field of starch-based high-strength biodegradable materials. In recent times, researchers have been engaged in developing high-strength blends, wherein different compatibilizers, such as citric acid (Ning et al., 2007; Wang et al., 2007) and maleic anhydride (Sabetzadeh et al., 2015), are added to increase the interfacial adhesion between highly hydrophobic polymer and strongly hydrophilic starch components (Mortazavi et al., 2013). In this regard, the processing method plays an important role in the overall time and cost factors involved during the production of blends. Among all the conventional processing methods, such as extrusion, blown, and injection molding (Ning et al., 2007), extrusion has proven to be more advantageous due to the in situ gelatinization of starch, faster processing, and lower energy consumption (Fishman et al., 2000).

4.3 Blends and composites of starch with biodegradable polymers

In recent times, researchers have been increasingly inclined to the field of environment-friendly biodegradable starch-based blends and composites, as these materials are exclusively composed of renewable components, such as thermoplastic starch and biodegradable polymers (Soykeabkaew et al., 2012). Among these renewable polymers, poly(lactic acid) and poly(hydroxybutyrate) have drawn significant attention because of their aliphatic nature and some favorable characteristics and properties comparable to those of synthetic polymers. However, all of these components, i.e., thermoplastic starch, poly(hydroxybutyrate), and poly(lactic acid), have some limitations including brittleness and much lower elongation values than those of synthetic polymers. Such drawbacks arise from their characteristic chemical structure that lacks the desired flexibility. Notably, despite a synthetic polymer, poly(vinyl) alcohol is a biodegradable polymer. This highly polar and water-soluble polymer is well suited for preparing blends with natural polymers. In fact, poly(vinyl) alcohols are subclassified into different grades of variable polarities depending on the proportion of

O—H and ester groups (Singha et al., 2009; Singha & Ray, 2010). Thus, relatively less polar poly(vinyl) alcohols can be made soluble in some organic solvents (Chiellini et al., 2001). Nevertheless, the biodegradability of poly(vinyl) alcohol is less than that of poly(lactic acid) derived from renewable resources. Therefore, blending of poly(vinyl) alcohol with starch can be an effective way to improve the biodegradation rate and reduce the overall cost (Tang & Alavi, 2011). Interestingly, blending of poly(vinyl) alcohol with starch deteriorates the inherently higher biodegradability of pure starch. The higher biodegradability of starch than that of poly(vinyl) alcohol is substantiated from the observation that the microorganisms initially consume the starch portion followed by degradation of the amorphous phase of poly(vinyl) alcohol (Tudorachi et al., 2000). Herein, the crystalline domain of poly(vinyl) alcohol is well integrated and hence significantly resistant against microbial attack. Indeed, the blending of thermoplastic starch and poly(vinyl) alcohol is facilitated by a high level of compatibility arising from the polar nature of both poly(vinyl) alcohol and starch. However, despite the excellent compatibility of thermoplastic starch with poly(vinyl) alcohol, blends made of thermoplastic starch and poly(vinyl) alcohol suffer from drawbacks, such as inferior water resistance and poor mechanical properties (Xiong et al., 2008). The tensile strength, elongation, and processibility of these blends are appreciably better than those of pure starch. In this regard, although plasticization of starch is an effective way to improve the overall flexibility and mechanical properties of starch—poly(vinyl) alcohol blend, such plasticization of starch should be carried out in a controlled way via limited addition of plasticizer, i.e., glycerol. At the most, 3 mL glycerol can be added to a 100 mL solution containing thermoplastic starch and poly(vinyl) alcohol in 1:1 ratio (Sreekumar et al., 2012). An excessively higher glycerol content causes a sharp drop in the ductility and mechanical and dynamic mechanical properties of the films prepared from a starch—poly(vinyl) alcohol blend. In this regard, comparative efficiencies of glycerol and sorbitol as plasticizers in starch—poly(lactic acid) have been evaluated (Li & Huneault, 2011). Compared to glycerol-plasticized thermoplastic starch—poly(lactic acid) blends, sorbitol plasticized thermoplastic starch—poly(lactic acid) blends exhibit relatively finer dispersion of the thermoplastic starch phase, resulting in much higher tensile strength and modulus than the blends plasticized with glycerol. In addition, the properties of starch—poly(lactic acid) blends are influenced by several variables including the proportion of starch, moisture content, heat treatment, and the quality/quantity of the coupling agents (Ken & Sun, 2003). Moreover, flexibilizers and hydrophobic agents are added to the blend to improve the toughness of the poly(lactic acid) matrix and the hydrophobicity of the blends, respectively (Li et al., 2016). Moreover, blending of thermoplastic starch with poly(hydroxyalcanoates) is effective in reducing the inherent aging tendencies of both thermoplastic starch and poly(hydroxyalcanoates) (Ken & Sun, 2003). Though both thermoplastic starch- and poly(hydroxyalcanoates)-based materials show aging behavior, blending of thermoplastic starch and poly(hydroxyalcanoates) results in almost negligible aging behavior during continuous storage of 30 days at 30°C and 50% relative humidity. For individual thermoplastic starches, such aging behavior and associated loss of mechanical properties are attributed to high moisture uptake by the hydrophilic thermoplastic starch and the continuous time-dependent leaching of the plasticizer from the thermoplastic starch phase. On the other hand, the aging of individual poly(hydroxyalcanoates) is associated with increasingly higher time-dependent crystallization of the amorphous phase of the poly(hydroxyalcanoates) producing small crystallites responsible for the stiffness and brittle nature of the aged material. Due to the combined presence of thermoplastic starch and poly(hydroxyalcanoates), all of the aforementioned phenomena responsible for aging may be restricted significantly. Another group of researchers reported the utility of a surfactant, i.e., Tween 80, in improving the mechanical properties of films made of thermoplastic starch and poly(butylene adipate-*co*-terephthalate), a biodegradable aliphatic-aromatic copolyester. However, Tween 80 addition in thermoplastic starch–poly(butylene adipate-*co*-terephthalate) results in the deterioration of structural integrity and tensile strength compared to the thermoplastic starch–poly(butylene adipate-*co*-terephthalate) films devoid of surfactant (Brandelero et al., 2010). Later, they observed that the addition of soybean oil in Tween 80 compatibilized thermoplastic starch–poly(butylene adipate-*co*-terephthalate) blend elevates the mechanical properties by improving the homogeneity in distribution of the dispersed phases (Brandelero et al., 2012). Likewise, different compatibilizers, such as maleic anhydride (Mohanty & Nayak, 2010), citric acid (Olivato et al., 2012), and tartaric acid (Olivato et al., 2014), have been employed to improve the mechanical and microstructural properties of thermoplastic starch–poly(butylene adipate-*co*-terephthalate) blends. In addition, many researchers have reported the applications of starch in improving the processibility and reducing the manufacturing cost of other materials constituted of natural polymers. In this context, the processibility as well as cost of starch-blended hydrogels can be improved in the presence of starch or modified starch. Moreover, the inherent incompatibility of starch and gelatin can be modulated by altering various factors, such as the processing time, temperature, pH value, and solid concentration. Thus, starch–gelatin blends and other blended films of polysaccharides and proteins prepared through optimization of processing and operating parameters can be effectively utilized as films with improved gas barrier properties compared to the individual components (Arvanitoyannis et al., 1994, 1997). Thus, as per the rule of stoichiometry, a 40:60 starch–gelatin blend is constituted of starch and gelatin as discrete and continuous phases, respectively (Zhang, Liu, Yu, Liu et al., 2013; Zhang, Liu, Yu, Shanks et al., 2013; Zhang, Shanks et al., 2014). Morphological evidence from FTIR spectroscopy substantiates the noticeable chain diffusions on the interface of starch domains, suggesting the existence of starch–protein compatibility in the micro-scale. As a whole, it can be stated that complete separation of starch and gelatin domains does not occur in those starch–gelatin blends. Compatibility of starch with the nonstarch component, such as hydroxypropyl methylcellulose, can also be improved via hydroxypropylation of starch. In this way, blends of hydroxypropyl methylcellulose and hydroxypropyl starch can be endowed with versatile applicability in both scientific and commercial fields (Zhang, Wang, Liu, Zhang et al., 2013; Zhang, Wang, Liu, Yu et al., 2013; Liu et al., 2014; Zhang, Wang et al., 2014). As stated earlier, improved processibility, rheological property, and viscosity modulation can be achieved via hydroxypropylation of starch. More importantly, such modification and blending with another modified polysaccharides, such as hydroxypropyl methylcellulose cellulose, enhances the thermal sensitivity and brings about temperature-dependent noticeable phase transitions and morphological/rheological changes in the as-prepared blends. Thus, such blends of chemically modified polysaccharides can be applied as a model composition for investigating the thermal/cooling gel system. In the commercial field, such blends have already found wide applications in food packaging, medicine capsule making, and cartilage tissue engineering. While studying the effect of the stoichiometric ratio, it has been observed that the viscosity of the blends of hydroxypropyl methylcellulose cellulose and hydroxypropyl starch increases with the increasing proportion of hydroxypropyl starch in the blend at all temperatures. Such a phenomenon is understandable from the fact that pure hydroxypropyl starch possesses higher viscosity compared than individual hydroxypropyl methylcellulose cellulose. This is because of the availability of more O—H groups and fewer esters in pure hydroxypropyl starch compared to a more esterified polysaccharide derivative, such as hydroxypropyl methylcellulose cellulose. Another reason for the enhanced viscosity of hydroxypropyl starch could be the easier involvement of branched chains of starch derivatives in chain entanglement compared to the linear chains of cellulose derivatives.

Similar to the blends of natural components, variable compatibilizers have also been employed to elevate the mechanical properties of fibrous natural polymer composites including those comprising starch and cellulose fiber as basic components. As these natural composites are renewable, biodegradable, abundantly available, and not costly, they have drawn increased attention and appreciation from researchers worldwide, especially in the last 2 decades (Singha, Mahapatra et al., 2018). In addition, these composite materials are usually light-weight and of lower density, with high specific strength and modulus, along with the capability to attenuate high sound. Moreover, these materials are often easily processible, and possess a relatively reactive surface (Nakason et al., 2005). In this regard, different products and associated fabrication techniques have been developed worldwide (Imam et al., 2001). For instance, a fourfold increase in the tensile properties of thermoplastic wheat starch can be seen, if the starch is reinforced by cellulose fibers (Nakason et al., 2001). In this regard, attempts have been made to reinforce starch components by cellulose-based particulate by-products, such as corn/wheat hulls (Corvasce et al., 1997) and apricot/walnut shells (Carvalho et al., 2003), obtained from natural food sources. As anticipated, the addition of these rigid particulate hulls to the starch matrix has resulted in appreciable enhancement of the mechanical strength of the composites compared to pure starch. Such reasonable enhancement originates from the inherent compatibility of both the hydrophilic components exhibiting a polar nature. In addition, the elevated gas barrier properties of hull-reinforced starch composites are ascribed to the rigid particulate matters and reduced void spaces inside the matrix through compaction. In this respect, flat-shaped corn hulls have been proven to be more effective compared to irregular-shaped wheat-hulls in restraining the diffusion of gas molecules through the bulk of composite. It is believed that flat-structured corn hull creates a greater polymer—filler interfacial area compared with the irregular wheat hull, resulting in reduced void spaces restricting movement of the gas molecules. Similarly, the gas permeability of starch–laver composites is reduced significantly compared with that of pure starch matrix (Angellier et al., 2005a). Interestingly, the overall nature of the starch–laver composites is altered appreciably depending on the processing temperature. At a lower temperature, the added laver fibers functioned simply as fibrous reinforcing fillers, resulting in elevated mechanical parameters, such as the modulus and tensile strength, of the starch matrix. However, if processed under hot conditions, a simple composite transforms into a hybrid composite, as the addition of laver fiber into a hot starch suspension results in the release of proteinous components from the laver fiber at the higher temperature followed by subsequent formation of starch protein–laver fiber hybrid composites. Such hybridization results in a further improvement in the mechanical properties compared with simple composites prepared in cold conditions, along with an appreciable decrease in both moisture sensitivity and gas permeability. Invariably, as these components are derived from food sources, the as-prepared composite can be safely used in food packaging and edible film manufacture.

In the recent past, nanoclays have been increasingly employed as reinforcing fillers to achieve strengthened materials capable of replacing nonbiodegradable polymer composites made of polyolefins. By their very nature, unmodified nanoclays are mostly hydrophilic, and one of the major challenges is to exfoliate those stacked nanoclays via delamination. In fact, increased exfoliation generates a greater clay–polymer interfacial area within a clay-filled nanocomposite (Mondal et al., 2010). However, such exfoliation or delamination requires sufficient energy to overcome the preexisting stronger interlaminar forces in clay stacks. In starch-based clay composites, the hydrophilic polysaccharide chains as well as water molecules can exert the necessary force to separate and exfoliate those stacks. Here, water molecules can impart plasticizing action, whereas O–H populated polysaccharide chains can strongly interact with hydrophilic clay functionalities followed by delamination and exfoliation via external mechanical forces during ultrasonication and mixing (Singha, Mahapatra et al., 2018; Mondal et al., 2010; Xie et al., 2013; Chivrac et al., 2008, 2009, 2010; Dean & Yu, 2005; Zhang, Wang, Liu, Zhang et al., 2013; Dean, Do, Petinakis, & Yu, 2008).

Recently, starch has been increasingly employed both as continuous matrix and discrete particulate fillers in the same composite (Lan et al., 2010). These type of composites are categorized as self-reinforced composites or single-polymer composites. Among the starch-based biodegradable composites, these materials are fairly unique because of the minimum difficulty experienced during mixing and compatibilization. Such a self-reinforcing concept imparts relatively improved homogeneity compared to composites comprised of different types of components (Amer & Ganapathiraju, 2001; Hine et al., 2008; Manninen et al., 1992; Suuronen et al., 1992; Gao et al., 2012; Gao et al., 2015). In addition, better compatibility and interfacial characteristics can be achieved among two phases composed of the same materials and functionalities. Accordingly, compared to the composites with different classes of materials, such self-reinforced composites with better homogeneity are better alternatives as biodegradable and biocompatible materials. For instance, a group of researchers has fabricated a self-reinforced composite, wherein both continuous and particulate phases are made of starch as the basic component. Here, hydroxypropylated starch was used as the continuous phase or matrix, whereas cross-linked starch granules were used as particulate matters (Lan et al., 2010). The main objective behind hydroxypropylation is to modify the starch via functionalization, resulting in a decrease in inter/intramolecular hydrogen bonding and the associated melting temperature so as to attain easier processibility (Stading et al., 1998; Dean et al., 2011). Thus, the added fillers can be easily dispersed, as they experience minimum difficulty while being distributed throughout the matrix. In this context, the added particulate fillers are also made of cross-linked starch, and accordingly, better compatibility, thermal resistance, and mechanical strength can be achieved in such self-reinforced starch-based composites. As a result of chemical modification, part of the O—H groups of starch converts into less hydrophilic hydroxypropyl groups, resulting in improved moisture resistance of the composite. Likewise, starch-based matrix reinforced with other polysaccharide-based particulate matters can also be categorized under self-reinforced composites (Senna et al., 2013; Wollerdorfer & Bader, 1998; Dufresne et al., 2000; Ali et al., 2017; Ali, Ali et al., 2019; Chen et al., 2019; El Miri et al., 2015; Li, Qiu et al., 2015; Li, Li et al., 2015; Panaitescu et al., 2015; Ali et al., 2018). Recently, starch and cellulose macrocrystals have been applied as reinforcing fillers for starch-based self-reinforcing edible films (Ali et al., 2018; Liu et al., 2010). Both types of self-reinforced composites appeared to be smooth and transparent. Though the Young's modulus and tensile strength were noted to be improved, the elongation at break decreased slightly. Morphological evidence substantiated good compatibility between starch and two different crystals of polysaccharides constituted of the same monomeric units, i.e., glucopyranose. Compared to starch crystals, cellulose crystals have better thermal stability, and accordingly, cellulose crystals are more capable of improving the mechanical properties of starch matrix-based edible self-reinforced composites. In contrast, starch matrix filled with starch crystals is more efficient in preventing UV degradation. As a whole, both of these edible and biodegradable self-reinforced composite films can be utilized as safe food-packaging materials as all of the components have been derived from food sources. Indeed, nanofibrils of cellulose have enhanced creep resistance and reduced creep recovery rate of starch film (Li, Li et al., 2015). Such composite films reinforced with 7% cellulose nanocrystals achieve comparable strength and stiffness with that of polyolefins (Panaitescu et al., 2015). In addition, starch-based biodegradable composites of elevated antimicrobial properties have been developed by incorporating finely divided pomegranate peel as an antimicrobial agent functioning as a reinforcing filler in the starch matrix (Ali, Chen et al., 2019). The as-developed composite film inhibits the growth of both Gram-positive (*Staphylococcus aureus*) and Gram-negative (*Salmonella* spp.) bacteria. Such a phenomenon can be expected from some previous studies relating to the high antimicrobial potential of pomegranate peel extract (Hayrapetyan et al., 2012; Khalid et al., 2018). The antimicrobial resistance imparted by both pomegranate peel extract and pomegranate peel may originate from the intimate hydrogen bonding of tannin molecules of the pomegranate with the starch macromolecules of the film as well as protein chains present in the bacterial cell wall of both Gram-positive and Gram-negative bacteria. Due to such bonding capability, the fibrous pomegranate peel particles can also function as a reinforcing agent of the starch matrix. More importantly, such antimicrobial materials are potentially used as an edible film and food-grade packaging material, since all the components are derived from food ingredients (Robles et al., 2016).

5. Conclusions

Among the upcoming biodegradable low-cost easily processible polymeric materials derived from renewable biopolymers, starch-based blends and composites are emerging as promising alternatives to the daily usage materials made mostly from petroleum-based nonbiodegradable polymers. To overcome several drawbacks, such as inherent moisture sensitivity, poor compatibility with nonpolar polymeric matrices, lack of homogeneous distribution, and insufficient mechanical properties, several strategies have been proven to be effective, i.e., physicochemical modification of starch, addition of compatibilizers/coupling agents, modification of nonstarch components, i.e., plastics and rubbers, and associated tuning of structures and properties by changing the fabrication procedures. Finally, starch-based nanocomposites and self-reinforced composites/nanocomposites can have the potential to become materials of high demand, that could effectively replace the traditionally used blends and composites made from petroleum-based nonbiodegradable polymers.

6. Future prospects

Very recently, despite different effective approaches, such as fabrication of self-reinforced composites, coatings with reduced moisture sensitivity, and antimicrobial properties, having been adopted to develop increasingly high-performance starch-based blends and composites, several barriers and constraints remain to be resolved for prospective diverse applications. Antimicrobial resistance is still insufficient and needs to be broadened further. The inherent moisture sensitivity needs to be reduced further, and the UV/aging resistance of polysaccharides should be elevated to achieve greater longevity. Moreover, in vivo applications of starch-based blends and composites are still to be diversified. Therefore, nowadays, scientists are engaged in developing some alternative avenues to design biocompatible high-performance but low-cost blends and composites capable of being utilized in drug delivery, scaffold making, and tissue-engineering applications. In the near future, the development of some low-cost versatile green materials could substantially replace the present-day highly popular nonbiodegradable synthetic polymeric materials in various medical and nonmedical applications.

References

Afiq, M. M., & Azura, A. R. (2013). Effect of sago starch loadings on soil decomposition of Natural Rubber Latex (NRL) composite films mechanical properties. *International Biodeterioration & Biodegradation, 85*, 139–149.

Ali, A., Yu, L., Liu, H., Khalid, S., Meng, L., & Chen, L. (2017). Preparation and characterization of starch-based composite films reinforced by corn and wheat hulls. *Journal of Applied Polymer Science, 134*, 45159–45165.

Ali, A., Xie, F., Yu, L., Liu, H., Meng, L., Khalid, S., & Chen, L. (2018). Preparation and characterization of starch-based composite films reinforced by polysaccharide-based crystals. *Composites Part B, 133*, 122–128.

Ali, A., Ali, S., Yu, L., Liu, H., Khalid, S., Hussain, A., Qayum, M. M. N., & Ying, C. (2019). Preparation and characterization of starch-based composite films reinforced by apricot and walnut shells. *Journal of Applied Polymer Science, 136*, 47978.

Ali, A., Chen, Y., Liu, H., Yu, L., Baloch, Z., Khalid, S., Zhu, J., & Chen, L. (2019). Starch-based antimicrobial films functionalized by pomegranate peel. *International Journal of Biological Macromolecules, 129*, 1120–1126.

Amer, M. S., & Ganapathiraju, S. (2001). Effects of processing parameters on axial stiffness of self-reinforced polyethylene composites. *Journal of Applied Polymer Science, 81*, 1136–1141.

Angellier, H., Molina-Boisseau, S., & Dufresne, A. (2005a). Processing and structure properties of waxy-maize starch nanocrystal reinforced natural rubber. *Macromolecules, 38*, 3783–3792.

Angellier, H., Molina-Boisseau, S., & Dufresne, A. (2005b). Mechanical properties of waxy-maize starch nanocrystal reinforced natural rubber. *Macromolecules, 38*, 9161–9170.

Arvanitoyannis, I., Kalichevsky, M., Blanshard, J. M., & Psomiadou, E. (1994). Study of diffusion and permeation of gases in undrawn and uniaxially drawn films made from potato and rice starch conditioned at different relative humidities. *Carbohydrate Polymers, 24*, 1–15.

Arvanitoyannis, I., Psomiadou, E., Nakayama, A., Aiba, S., & Yamamoto, N. (1997). Edible films made from gelatin, soluble starch and polyols Part 3. *Food Chemistry, 60*, 593–604.

Avella, M., Martuscelli, E., & Raimo, M. (2000). Properties of blends and composites based on poly(3-hydroxy)butyrate (PHB) and poly(3-hydroxybutyrate-hydroxyvalerate) (PHBV) copolymers. *Journal of Materials Science, 35*, 523–545.

Bouthegourd, E., Rajisha, K. R., Kalarical, N., Saiter, J. M., & Thomas, S. (2011). Natural rubber latex/potato starch nanocrystal nanocomposites: correlation morphology/electrical properties. *Materials Letters, 65*, 3615–3617.

Brandelero, R. P. H., Yamashita, F., & Grossmann, M. V. E. (2010). The effect of surfactant tween 80 on the hydrophilicity, water vapor permeation, and the mechanical properties of cassava starch and poly(butylene adipate-co-terephthalate) (PBAT) blend films. *Carbohydrate Polymers, 82*, 1102–1109.

Brandelero, R. P. H., Grossmann, M. V. E., & Yamashita, F. (2012). Films of starch and poly(butylene adipate co-terephthalate) added of soybean oil (SO) and tween 80. *Carbohydrate Polymers, 90*, 1452–1460.

Buchanan, R. A., Weislogel, O. E., Russell, C. R., & Rist, C. E. (1968). Starch in rubber. Zinc starch xanthate in latex masterbatching. *Industrial and Engineering Chemistry Product Research and Development, 7*, 155–158.

Buléon, A., Colonna, P., Planchot, V., & Ball, S. (1998). Starch granules: structure and biosynthesis. *International Journal of Biological Macromolecules, 23*, 85–112.

Carvalho, A. J. F., Job, A. E., Alves, N., Curvelo, A. A. S., & Gandini, A. (2003). Thermoplastic starch/natural rubber blends. *Carbohydrate Polymers, 53*, 95–99.

Chattopadhyay, P. K., Basuli, U., & Chattopadhyay, S. (2010). Studies on novel dual filler based epoxidized natural rubber nanocomposite. *Polymer Composites, 31*, 835–846.

Chattopadhyay, P. K., Das, N. C., & Chattopadhyay, S. (2011). Influence of interfacial roughness and the hybrid filler microstructures on the properties of ternary elastomeric composites. *Composites Part A Applied Science and Manufacturing, 42*, 1049–1059.

Chen, B., & Evans, J. R. G. (2005). Thermoplastic starch–clay nanocomposites and their characteristics. *Carbohydrate Polymers, 61*, 455–463.

Chen, Y., Yu, L., Ge, X., Liu, H., Ali, A., Wang, Y., & Chen, L. (2019). Preparation and characterization of edible starch film reinforced by laver. *International Journal of Biological Macromolecules, 129*, 944–951.

Chiellini, E., Cinelli, P., Imam, S. H., & Mao, L. (2001). Composite films based on biorelated agro-industrial waste and poly (vinyl alcohol): preparation and mechanical properties characterization. *Biomacromolecules, 2*, 1029–1037.

Chivrac, F., Gueguen, O., Pollet, E., Ahzi, S., Makradi, A., & Averous, L. (2008). Micromechanical modeling and characterization of the effective properties in starch-based nanobiocomposites. *Acta Biomaterialia, 4*, 1707–1714.

Chivrac, F., Pollet, E., & Averous, L. (2009). Progress in nano-biocomposites based on polysaccharides and nanoclays. *Materials Science and Engineering: R: Reports, 67*, 1–17.

Chivrac, F., Angellier-Coussy, H., Guillard, V., Pollet, E., & Averous, L. (2010). How does water diffuse in starch/montmorillonite nano-biocomposite materials? *Carbohydrate Polymers, 82*, 128–135.

Corre, D. L., Bras, J., & Dufresne, A. (2010). Starch nanoparticles: a review. *Biomacromolecules, 11*, 1139–1153.

Corvasce, F. G., Linster, T. D., & Thielen, G. (1997). *Starch composite reinforced rubber composition and tire with at least one component thereof.* U.S. Patent 5672639.

Curvelo, A. A. S., de Carvalho, A. J. F., & Agnelli, J. A. M. (2001). A thermoplastic starch–cellulosic fibers composites: preliminary results. *Carbohydrate Polymers, 45*, 183–188.

Da Róz, A. L., Carvalho, A. J. F., Gandini, A., & Curvelo, A. A. S. (2006). The effect of plasticizers on thermoplastic starch compositions obtained by melt processing. *Carbohydrate Polymers, 63*, 412–417.

Dean, K., Do, M., Petinakis, E., & Yu, L. (2008). Key interactions in biodegradable thermoplastic starch/PVOH/montmorillonite micro-and nanocomposites. *Composites Science and Technology, 68*, 1453–1462.

Dean, K., Petinakis, E., Goodall, L., Miller, T., Yu, L., & Wright, N. (2011). Nanostabilization of thermally processed high amylose hydroxylpropylated starch films. *Carbohydrate Polymers*, 652–658.

Dean, K., & Yu, L. (2005). Biodegradable protein-nanoparticle composite. In R. Smith (Ed.), *Biodegradable Polymers* (vol. 11, pp. 289–309). Cambridge, U.K: Woodhead.

Dean, K., Yu, L., & Wu, D. Y. (2007). Preparation and characterization of melt-extruded thermoplastic starch/clay nanocomposites. *Composites Science and Technology, 67*, 413–421.

Debiagi, F., Kobayash, R. K. T., Nakazato, G., & Panagio, L. A. (2014). Biodegradable active packaging based on cassava bagasse, polyvinyl alcohol and essential oils. *Industrial Crops and Products, 52*, 664–670.

Dufresne, A., Dupeyre, D., & Vignon, M. R. (2000). Cellulose microfibrils from potato tuber cells: processing and characterization of starch-cellulose microfibril composites. *Journal of Applied Polymer Science, 76*, 2080—2092.
Dutta, A., Mahapatra, M., Deb, M., Mitra, M., Dutta, S., Chattopadhyay, P. K., Banerjee, S., Sil, P. C., Maiti, D. K., & Singha, N. R. (2020). Fluorescent terpolymers using two non-emissive monomers for Cr(III) sensors, removal, and bio-imaging. *ACS Biomaterials Science & Engineering, 6*, 1397—1407.
El Miri, N., Abdelouahdi, K., Barakat, A., Zahouily, M., Fihri, A., Solhy, A., & El Achaby, M. (2015). Bio-nanocomposite films reinforced with cellulose nanocrystals: rheology of film-forming solutions, transparency, water vapor barrier and tensile properties of films. *Carbohydrate Polymers, 129*, 156—167.
Euaphantasate, N., Prachayawasin, P., Uasopon, S., & Methacanon, P. (2008). Moisture sorption characteristic and their relative properties of thermoplastic starch/linear low density polyethylene films for food packaging. *Journal of Materials and Metallurgy, 18*, 103—109.
Fishman, M. L., Coffin, D. R., Konstance, R. P., & Onwulata, C. I. (2000). Extrusion of pectin/starch blends plasticized with glycerol. *Carbohydrate Polymers, 41*, 317—325.
Forssell, P. M., Mikkilä, J. M., Moates, G. K., & Parker, R. (1997). Phase and glass transition behaviour of concentrated barley starch-glycerol-water mixtures, a model for thermoplastic starch. *Carbohydrate Polymers, 34*, 275—282.
Gao, J.-P., Tian, R.-C., Yu, J.-G., & Duan, M.-L. (1994). Graft copolymers of methyl methacrylate onto canna starch using manganic pyrophosphate as an initiator. *Journal of Applied Polymer Science, 53*, 1091—1102.
Gao, C., Yu, L., Liu, H., & Chen, L. (2012). Development of self-reinforced polymer composites. *Progress in Polymer Science, 37*, 767—780.
Gao, C., Meng, L., Yu, L., Simon, G. P., Liu, H., Chen, L., & Petinakis, S. (2015). Preparation and characterization of uniaxial poly(lactic acid)-based self-reinforced composites. *Composites Science and Technology, 117*, 392—397.
Garcia, M. A., Martino, M. N., & Zaritzky, N. E. (1998). Starch-based coatings: effect on refrigerated strawberry (*Fragaria ananassa*) quality. *Journal of the Science of Food and Agriculture, 76*, 411—420.
Hayrapetyan, H., Hazeleger, W. C., & Beumer, R. R. (2012). Inhibition of Listeria monocytogenes by pomegranate (*Punica granatum*) peel extract in meat pate at different temperatures. *Food Control, 23*, 66—72.
Hemati, F., & Garmabi, H. (2011). Compatibilised LDPE/LLDPE/nanoclay nanocomposites: I. structural, mechanical, and thermal properties. *Canadian Journal of Chemical Engineering, 89*, 187—196.
Hine, P. J., Olley, R. H., & Ward, I. M. (2008). The use of interleaved films for optimising the production and properties of hot compacted, self-reinforced polymer composites. *Composites Science and Technology, 68*, 1413—1421.
Huang, M. F., Yu, J. G., & Ma, X. F. (2004). Studies on the properties of montmorillonite-reinforced thermoplastic starch composites. *Polymer, 45*, 7017—7023.
Huneault, M. A., & Li, H. (2007). Morphology and properties of compatibilized polylactide/thermoplastic starch blends. *Polymer, 48*, 270—280.
Imam, S. H., Gordon, S. H., Mao, L., & Chen, L. (2001). Environmentally friendly wood adhesive from a renewable plant polymer: characteristics and optimization. *Polymer Degradation and Stability, 73*, 529—533.
Imre, B., & Pukánszky, B. (2013). Compatibilization in bio-based and biodegradable polymer blends. *European Polymer Journal, 49*, 1215—1233.

Jayakody, L., & Hoover, R. (2002). The effect of lintnerization on cereal starch granules. *Food Research International, 35*, 665–680.

Jiang, T., Duan, Q., Zhu, J., Liu, H., & Yu, L. (2020). Starch-based biodegradable materials: challenges and opportunities. *Advanced Industrial and Engineering Polymer Research, 3*, 8–18.

Jyothi, A. N. (2010). Starch graft copolymers: novel application in industry. *Composite Interfaces, 17*, 165–174.

Karmakar, M., Mondal, H., Ghosh, T., Chattopadhyay, P. K., Maiti, D. K., & Singha, N. R. (2019). Chitosan-grafted tetrapolymer using two monomers: pH-responsive high-performance removals of Cu(II), Cd(II), Pb(II), dichromate, and biphosphate and analyses of adsorbed microstructures. *Environmental Research, 179*, 108839.

Kaseem, M., Hamad, K., & Deri, F. (2012). Thermoplastic starch blends: a review of recent works. *Polymer Science, 54*, 165–176.

Ken, T., & Sun, K. (2003). Melting behavior and crystallization kinetics of starch and poly(lactic acid) composites. *Journal of Applied Polymer Science, 89*, 1203–1210.

Khalaf, A. I., & Sadek, E. M. (2012). Compatibility study in natural rubber and maize starch blends. *Journal of Applied Polymer Science, 125*, 959–967.

Khalid, S., Yu, L., Feng, M., Meng, L., Bai, Y., Ali, A., Liu, H., & Chen, L. (2018). Development and characterization of biodegradable antimicrobial packaging films based on polycaprolactone, starch and pomegranate rind hybrids. *Food Packaging and Shelf Life, 18*, 71–79.

Kiatkamjornwong, S., Thakeow, P., & Sonsuk, M. (2001). Chemical modification of cassava starch for degradable polyethylene sheets. *Polymer Degradation and Stability, 73*, 363–375.

Kiing, S. C., Dzulkefly, K., & Yiu, P. H. (2013). Characterization of biodegradable polymer blends of acetylated and hydroxypropylated sago starch and natural rubber. *Journal of Polymers and the Environment, 21*, 995–1001.

Kim, M. S., & Cho, U. R. (2013). Manufacture and properties of PMMA grafted starch/carbon black/NBR composites. *Polymer, 37*, 764–769.

Krogars, K., Heinamaki, J., Karjalainen, M., Niskanen, A., Leskela, M., & Yliruusi, J. (2003). Enhanced stability of rubbery amylose-rich maize starch films plasticized with a combination of sorbitol and glycerol. *International Journal of Pharmaceutics, 251*, 205–208.

Lan, C., Yu, L., Chen, P., Chen, L., Simon, G., & Zhang, X. (2010). Designing and preparing starch-based self-reinforced composites. *Macromolecular Materials and Engineering, 295*, 1025–1030.

LeCorre, D., Bras, J., & Dufresne, A. (2011). Influence of botanic origin and amylose content on the morphology of starch nanocrystals. *Journal of Nanoparticle Research, 13*, 7193–7208.

LeCorre, D., Bras, J., & Dufresne, A. (2012). Influence of the botanic origin of starch nanocrystals on the morphological and mechanical properties of natural rubber nanocomposites. *Macromolecular Materials and Engineering, 297*, 969–978.

Li, M. C., & Cho, U. R. (2013a). Effectiveness of coupling agents in the poly (methyl methacrylate)-modified starch/styrene-butadiene rubber interfaces. *Materials Letters, 92*, 132–135.

Li, M. C., & Cho, U. R. (2013b). Mechanical performance, water absorption behavior and biodegradability of poly (methyl methacrylate)-modified starch/SBR biocomposites. *Macromolecular Research, 21*, 793–800.

Li, M. C., & Cho, U. R. (2017). Starch in rubber based blends and micro composites. In P. M. Visakh (Ed.), *Rubber Based Bionanocomposites. Advanced Structured Materials* (vol. 56, pp. 109–140). Cham: Springer.

Li, H., & Huneault, M. A. (2011). Comparison of sorbitol and glycerol as plasticizers for thermoplastic starch in TPS/PLA blends. *Journal of Applied Polymer Science, 119*, 2439–2448.

Li, M. C., Lee, J. K., & Cho, U. R. (2012). Synthesis, characterization, and enzymatic degradation of starch-grafted poly(methyl methacrylate) copolymer films. *Journal of Applied Polymer Science, 125*, 405–414.

Li, M. C., Ge, X., & Cho, U. R. (2013). Emulsion grafting vinyl monomers onto starch for reinforcement of styrene-butadiene rubber. *Macromolecular Research, 21*, 519–528.

Li, M., Li, D., Wang, L., & Adhikari, B. (2015). Creep behavior of starch-based nanocomposite films with cellulose nanofibrils. *Carbohydrate Polymers, 117*, 957–963.

Li, X., Qiu, C., Ji, N., Sun, C., Xiong, L., & Sun, Q. (2015). Mechanical, barrier and morphological properties of starch nanocrystals-reinforced pea starch films. *Carbohydrate Polymers, 121*, 155–162.

Li, S., Xia, J., Xu, Y., Yang, X., Mao, W., & Huang, K. (2016). Preparation and characterization of acorn starch/poly(lactic acid) composites modified with functionalized vegetable oil derivates. *Carbohydrate Polymers, 142*, 250–258.

Liu, Z., Yi, X. S., & Feng, Y. (2001). Effects of glycerin and glycerol monostearate on performance of thermoplastic starch. *Journal of Materials Science, 36*, 1809–1815.

Liu, C., Shao, Y., & Jia, D. (2008). Chemically modified starch reinforced natural rubber composites. *Polymer, 49*, 2176–2181.

Liu, H., Xie, F., Yu, L., Chen, L., & Li, L. (2009). Thermal processing of starch-based polymers. *Progress in Polymer Science, 34*, 1348–1368.

Liu, D., Zhong, T., Chang, P. R., Li, K., & Wu, Q. (2010). Starch composites reinforced by bamboo cellulosic crystals. *Bioresource Technology, 101*, 2529–2536.

Liu, X., Wang, Y., Zhang, N., Shanks, R. A., Liu, H., Tong, Z., Chen, L., & Yu, L. (2014). Morphology and phase composition of gelatin-starch blends. *Chinese Journal of Polymer Science, 32*, 108–114.

Lourdin, D., Coignard, L., Bizot, H., & Colonna, P. (1997). Influence of equilibrium relative humidity and plasticizer concentration on the water content and glass transition of starch materials. *Polymer, 38*, 5401–5406.

Ma, X., & Yu, J. (2004a). Formamide as the plasticizer for thermoplastic starch. *Journal of Applied Polymer Science, 93*, 1769–1773.

Ma, X., & Yu, J. (2004b). Studies on the properties of formamide plasticized thermoplastic starch. *Acta Polymerica Sinica, 2*, 240–245.

Ma, X., & Yu, J. (2004c). The effect of plasticizers containing amide groups on the properties of thermoplastic starch. *Starch/starke, 56*, 545–551.

Ma, X., & Yu, J. (2004d). The plastcizers containing amide groups for thermoplastic starch. *Carbohydrate Polymers, 57*, 197–203.

Ma, X., Yu, J., & Jin, F. (2004). Urea and formamide as a mixed plasticizer for thermoplastic starch. *Polymer International, 53*, 1780–1785.

Ma, X., Yu, J., & Kennedy, H. F. (2005). Studies on the properties of natural fibers-reinforced thermoplastic starch composites. *Carbohydrate Polymers, 62*, 19–24.

Ma, X., Yu, J., & Wan, J. (2006). Urea and ethanolamine as a mixed plasticizer for thermoplastic starch. *Carbohydrate Polymers, 64*, 267–273.

Ma, X., Chang, P. R., Yu, J., & Stumborg, M. (2009). Properties of biodegradable citric acid-modified granular starch/thermoplastic pea starch composites. *Carbohydrate Polymers, 75*, 1–8.

Mahapatra, M., Karmakar, M., Dutta, A., Mondal, H., Roy, J. S. D., Chattopadhyay, P. K., & Singha, N. R. (2018). Microstructural analyses of loaded and/or unloaded semisynthetic

porous material for understanding of superadsorption and optimization by response surface methodology. *Journal of Environmental Chemical Engineering, 6*, 289–310.
Mahapatra, M., Dutta, A., Roy, J. S. D., Mitra, M., Mahalanobish, S., Sanfui, M. D. H., Banerjee, S., Chattopadhyay, P. K., Sil, P. C., & Singha, N. R. (2019). Fluorescent terpolymers via in situ allocation of aliphatic fluorophore-monomers: Fe(III)-sensor, high-performance removals, and bio-imaging. *Advanced Healthcare Materials, 8*, 1900980.
Mahapatra, M., Dutta, A., Mitra, M., Karmakar, M., Ghosh, N. N., Chattopadhyay, P. K., & Singha, N. R. (2020). Intrinsically-fluorescent biocompatible terpolymers for detection and removal of Bi(III) and cell-imaging. *ACS Applied Biomaterials, 3*, 6155–6166.
Mahapatra, M., Dutta, A., Roy, J. S. D., Deb, M., Das, U., Banerjee, S., Dey, S., Chattopadhyay, P. K., Maiti, D. K., & Singha, N. R. (2020). Synthesis of biocompatible aliphatic terpolymers via in situ fluorescent monomers for three-in-one applications: polymerization of hydrophobic monomers in water. *Langmuir, 36*, 6178–6187.
Mali, S., Debiagi, F., Grossmann, M. V. F., & Yamashita, F. (2010). Starch, sugarcane bagasse fibre, and polyvinyl alcohol effects on extruded foam properties: a mixture design approach. *Industrial Crops and Products, 32*, 353–359.
Manninen, M. J., Päivärinta, U., Pätiälä, H., Rokkanen, P., Taurio, R., Tamminmäki, M., & Törmälä, P. (1992). Shear strength of cancellous bone after osteotomy fixed with absorbable self-reinforced polyglycolic acid and poly-l-lactic acid rods. *Journal of Materials Science: Materials in Medicine, 3*, 245–251.
Martins, A. B., & Santana, R. M. C. (2016). Effect of carboxylic acids as compatibilizer agent on mechanical properties of thermoplastic starch and polypropylene blends. *Carbohydrate Polymers, 135*, 79–85.
Mathew, A. P., & Dufresne, A. (2002). Morphological investigation of nanocomposites from sorbitol plasticized starch and tunicin whiskers. *Biomacromolecules, 3*, 609–617.
Mele, P., Molina-Boisseau, S., & Dufresne, A. (2011). Reinforcing mechanisms of starch nanocrystals in a nonvulcanized natural rubber matrix. *Biomacromocules, 12*, 1487–1493.
Mitra, M., Mahapatra, M., Dutta, A., Roy, J. S. D., Karmakar, M., Mondal, H., Deb, M., Roy, C., Chattopadhyay, P. K., Bandyopadhyay, A., & Singha, N. R. (2019). Carbohydrate and collagen-based doubly-grafted interpenetrating terpolymer hydrogel via N– H activated in situ allocation of monomer for superadsorption of Pb(II), Hg(II), dyes, vitamin-C, and p-nitrophenol. *Journal of Hazardous Materials, 369*, 746–762.
Mitra, M., Mahapatra, M., Dutta, A., Deb, M., Dutta, S., Chattopadhyay, P. K., Roy, S., Banerjee, S., Sil, P. C., & Singha, N. R. (2020). Fluorescent guar gum-g-terpolymer via in situ acrylamido-acid fluorophore-monomer in cell imaging, Pb(II) sensor, and security ink. *ACS Applied Biomaterials, 3*, 1995–2006.
Mohanty, A. K., & Nayak, S. K. (2010). Starch based biodegradable PBAT nanocomposites: effect of starch modification on mechanical, thermal and morphological and biodegradability behavior. *International Journal of Plastics Technology, 13*, 163–185.
Mondal, M., Chattopadhyay, P. K., Chattopadhyay, S., & Setua, D. K. (2010). Thermal and morphological analysis of thermoplastic polyurethane-clay nanocomposites: comparison of efficacy of dual modified laponite vs. commercial montmorillonites. *Thermochimica Acta, 510*, 185–194.
Mondal, H., Karmakar, M., Chattopadhyay, P. K., & Singha, N. R. (2019). Starch-g-tetrapolymer hydrogel via in situ attached monomers for removals of Bi(III) and/or Hg(II) and dye(s): RSM-based optimization. *Carbohydrate Polymers, 213*, 428–440.
Mondal, H., Karmakar, M., Chattopadhyay, P. K., & Singha, N. R. (2020). New property-performance optimization of scalable alginate-g-terpolymer for Ce(IV), Mo(VI), and W(VI) exclusions. *Carbohydrate Polymers, 245*, 116370.

Mortazavi, S., Ghasemi, I., & Oromiehie, A. (2013). Effect of phase inversion on the physical and mechanical properties of low density polyethylene/thermoplastic starch. *Polymer Testing, 32*, 482–491.

Nafchi, A. M., Moradpour, M., Saeidi, M., & Alias, A. K. (2013). Thermoplastic starches: properties, challenges, and prospects. *Starch Staerke, 61*, 65–72.

Nakason, C., Kaesaman, A., Rungvichaniwat, A., Eardrod, K., & Kiatkamjonwong, S. (2001). Rheological and curing behavior of reactive blending. II. Maleated natural rubber-cassava starch. *Journal of Applied Polymer Science, 81*, 2803–2813.

Nakason, C., Kaesaman, A., Homsin, S., & Kiatkamjonwong, S. (2003). Rheological and curing behavior of reactive blending. I. Natural rubber-g-poly(methyl methacrylate)-cassava starch. *Journal of Applied Polymer Science, 89*, 1453–1463.

Nakason, C., Kaesaman, A., & Eardrod, K. (2005). Cure and mechanical properties of natural rubber-g-poly(methyl methacrylate) cassava starch compounds. *Materials Letters, 59*, 4020–4025.

Neto, B. A.de M., Fernandes, B. S., Junior Franco, C. C. M. F., Bonomo, M. R. C. F., de Almeida, P. F., & Pontes, K. V. (2016). Thermal-morphological characterisation of starch from peach-palm (Bactris gasipaes Kunth) fruit (Pejibaye). *International Journal of Food Properties, 20*, 1007–1015.

Ning, W., Jiugao, Y., Xiaofei, M., & Ying, W. (2007). The influence of citric acid on the properties of thermoplastic starch/linear low-density polyethylene blends. *Carbohydrate Polymers, 67*, 446–453.

Olivato, J. B., Grossmann, M. V. E., Bilck, A. P., & Yamashita, F. (2012). Effect of organic acids as additives on the performance of thermoplastic starch/polyester blown films. *Carbohydrate Polymers, 90*, 156–164.

Olivato, J. B., Müller, C. M. O., Carvalho, G. M., Yamashita, F., & Grossmann, M. V. E. (2014). Physical and structural characterisation of starch/polyester blends with tartaric acid. *Materials Science and Engineering: C, 39*, 35–39.

Panaitescu, D. M., Frone, A. N., Ghiurea, M., & Chiulan, I. (2015). Influence of storage conditions on starch/PVA films containing cellulose nanofibers. *Industrial Crops and Products, 70*, 170–177.

Pant, B. R., Jeon, H. J., & Song, H. H. (2011). Radiation cross-linked carboxymethylated starch and iron removal capacity in aqueous solution. *Macromolecular Research, 19*, 307–312.

Park, H. M., Li, X., Jin, C. Z., Park, C. Y., Cho, W. J., & Ha, C. S. (2002). Preparation and properties of biodegradable thermoplastic starch/clay hybrids. *Macromolecular Materials and Engineering, 287*, 553–558.

Park, H. M., Li, W. K., Park, C. Y., Cho, W. J., & Ha, C. S. (2003). Environmentally friendly polymer hybrids Part I mechanical, thermal, and barrier properties of thermoplastic starch/clay nanocomposites. *Journal of Materials Science, 38*, 909–915.

Parulekar, Y., & Mohanty, A. K. (2007). Extruded biodegradable cast films from polyhydroxyalkanoate and thermoplastic starch blends: fabrication and characterization. *Macromolecular Materials and Engineering, 292*, 1218–1228.

Peres, A. M., Pires, R. R., & Oréfice, R. L. (2016). Evaluation of the effect of reprocessing on the structure and properties of low density polyethylene/thermoplastic starch blends. *Carbohydrate Polymers, 136*, 210–215.

Qi, Q., Wu, Y., Tian, M., Liang, G., Zhang, L., & Jun, M. (2006). Modification of starch for high performance elastomer. *Polymer, 47*, 3896–3903.

Rajisha, K. R., Maria, H. J., Pothan, L. A., Ahmad, Z., & Thomas, S. (2014). Preparation and characterization of potato starch nanocrystal reinforced natural rubber nanocomposites. *International Journal of Biological Macromolecules, 67*, 147–153.

Robles, E., Salaberria, A. M., Herrera, R., Fernandes, S. C., & Labidi, J. (2016). Self-bonded composite films based on cellulose nanofibers and chitin nanocrystals as antifungal materials. *Carbohydrate Polymers, 144*, 41–49.

Rodriguez-Gonzalez, F., Ramsay, D., & Favis, B. (2003). High performance LDPE/thermoplastic starch blends: a sustainable alternative to pure polyethylene. *Polymer, 44*, 1517–1526.

Rodriguez-Gonzalez, F. J., Ramsay, B. A., & Favis, B. D. (2004). Rheological and thermal properties of thermoplastic starch with high glycerol content. *Carbohydrate Polymers, 58*, 139–147.

Rouilly, A., Rigal, L., & Gillbert, R. G. (2004). Synthesis and properties of composites of starch and chemically modified natural rubber. *Polymer, 45*, 7813–7820.

Sabetzadeh, M., Bagheri, R., & Masoomi, M. (2015). Study on ternary low density polyethylene/linear low density polyethylene/thermoplastic starch blend films. *Carbohydrate Polymers, 119*, 126–133.

Schwach, E., & Averóus, L. (2004). Starch-based biodegradable blends: morphology and interface properties. *Polymer International, 53*, 2115–2124.

Senna, M. M., Mohamed, R. M., Shehab-Eldin, A. N., & El-Hamouly, S. (2013). Characterization of electron beam irradiated natural rubber/modified starch composites. *Journal of Industrial and Engineering Chemistry, 18*, 1654–1661.

Shi, R., Zhang, Z., Liu, Q., Han, Y., Zhang, L., Chen, D., & Tian, W. (2007). Characterization of citric acid/glycerol co-plasticized thermoplastic starch prepared by melt blending. *Carbohydrate Polymers, 69*, 748–755.

Singha, N. R., & Ray, S. K. (2010). Synthesis of chemically modified polyvinyl alcohol membranes for separation of toluene methanol mixtures by pervaporation. *Separation Science and Technology, 45*, 2298–2307.

Singha, N. R., Parya, T. K., & Ray, S. K. (2009). Dehydration of 1, 4-dioxane by pervaporation using filled and crosslinked polyvinyl alcohol membrane. *Journal of Membrane Science, 340*, 35–44.

Singha, N. R., Dutta, A., Mahapatra, M., Karmakar, M., Mondal, H., Chattopadhyay, P. K., & Maiti, D. K. (2018). Guar gum-grafted terpolymer hydrogels for ligand-selective individual and synergistic adsorption: effect of comonomer composition. *ACS Omega, 3*, 472–494.

Singha, N. R., Mahapatra, M., Karmakar, M., & Chattopadhyay, P. K. (2018). *Processing, characterization and application of natural rubber based environmentally friendly polymer composites; online*. Cham: Springer, ISBN 978-3-030-05399-4. https://doi.org/10.1007/978-3-030-05399-4_29. Sustainable Polymer Composites and Nanocomposites.

Singha, N. R., Dutta, A., Mahapatra, M., Roy, J. S. D., Mitra, M., Deb, M., & Chattopadhyay, P. K. (2019). In situ attachment of acrylamido sulfonic acid-based monomer in terpolymer hydrogel optimized by response surface methodology for individual and/or simultaneous removal(s) of M(III) and cationic dyes. *ACS Omega, 4*, 1736–1780.

Singn, V., & Ali, S. Z. (1987). Comparative acid modification of various starches. *Starch Staerke, 39*, 402–405.

Sionkowska, A. (2011). Current research on the blends of natural and synthetic polymers as new biomaterials: Review. *Progress in Polymer Science, 36*, 1254–1276.

Smits, A. L. M., Wubbenhorst, M., Kruiskamp, P. H., van Soest, J. J. G., Vliegenthart, J. F. G., & Van Turnhout, J. (2001). Structure evolution in amylopectin/ethylene glycol mixtures by H-bond formation and phase separation studied with dielectric relaxation spectroscopy. *Journal of Physical Chemistry B, 105*, 5630–5636.

Soykeabkaew, N., Laosat, N., Ngaokla, A., Yodsuwan, N., & Tunkasiri, T. (2012). Reinforcing potential of micro- and nano-sized fibers in the starch-based biocomposites. *Composites Science and Technology, 72*, 845–852.

Sreekumar, P. A., Al-Harthi, M. A., & De, S. K. (2012). Effect of glycerol on thermal and mechanical properties of polyvinyl alcohol/starch blends. *Journal of Applied Polymer Science, 123*, 135–142.

Stading, M., Hermansson, A. M., & Gatenholm, P. (1998). Structure, mechanical and barrier properties of amylose and amylopectin films. *Carbohydrate Polymers, 36*, 217–224.

Suuronen, R., Wessman, L., Mero, M., Törmälä, P., Vasenius, J., Partio, E., Vihtonen, K., & Vainionpää, S. (1992). Comparison of shear strength of osteotomies fixed with absorbable self-reinforced poly-l-lactide and metallic screws. *Journal of Materials Science: Materials in Medicine, 3*, 288–292.

Tang, S., & Alavi, S. (2011). Recent advances in starch, polyvinyl alcohol based polymer blends, nanocomposites and their biodegradability. *Carbohydrate Polymers, 85*, 7–16.

Tang, H., Qi, Q., Wu, Y., Liang, G., Zhang, L., & Ma, J. (2006). Reinforcement of elastomer by starch. *Macromolecular Materials and Engineering, 291*, 629–637.

Teixeira, E. D. M., Pasquini, D., Curvelo, A. A. S., Corradini, E., Belgacem, M. N., & Dufresne, A. (2009). Cassava bagasse cellulose nanofibrils reinforced thermoplastic cassava starch. *Carbohydrate Polymers, 78*, 422–431.

Thipmanee, R., & Sane, A. (2012). Effect of zeolite 5A on compatibility and properties of linear low-density polyethylene/thermoplastic starch blend. *Journal of Applied Polymer Science, 126*, E251–E258.

Tudorachi, N., Cascaval, C. N., Rusu, M., & Pruteanu, M. (2000). Testing of polyvinyl alcohol and starch mixtures as biodegradable polymeric materials. *Polymer Testing, 19*, 785–789.

Valodkar, M., & Thakore, S. (2011). Organically modified nanosized starch derivatives as excellent reinforcing agents for bionanocomposites. *Carbohydrate Polymers, 86*, 1244–1251.

Van Soest, J. J. G., & Borger, D. B. (1997). Structure and properties of compression-molded thermoplastic starch materials from normal and high-amylose maize starches. *Journal of Applied Polymer Science, 64*, 631–644.

Van Soest, J. J. G., Benes, K., & De Wit, D. (1996). The influence of starch molecular mass on the properties of extruded thermoplastic starch. *Polymer, 37*, 3543–3552.

Van Soest, J. J. G., Bezemer, R. C., De Wit, D., & Vliegenthart, J. F. G. (1996). Influence of glycerol on the melting of potato starch. *Industrial Crops and Products, 5*, 1–9.

Wang, L., Shogren, R., & Carriere, C. (2000). Preparation and properties of thermoplastic starch–polyester laminate sheets by coextrusion. *Polymer Engineering & Science, 40*, 499–506.

Wang, Y. J., Truong, V. D., & Wang, L. (2001). Structures and physicochemical properties of acid-thinned corn, potato and rice starches. *Starch Staerke, 53*, 570–576.

Wang, N., Yu, J., Chang, P. R., & Ma, X. (2007). Influence of citric acid on the properties of glycerol-plasticized dry starch (DTPS) and DTPS/Poly(lactic acid) blends. *Starch Staerke, 59*, 409–417.

Wang, N., Yu, J., Chang, P. R., & Ma, X. (2008). Influence of formamide and water on the properties of thermoplastic starch/poly(lactic acid) blends. *Carbohydrate Polymers, 71*, 109–118.

Wang, N., Zhang, X., Han, N., & Bai, S. (2009). Effect of citric acid and processing on the performance of thermoplastic starch/montmorillonite nanocomposites. *Carbohydrate Polymers, 76*, 68–73.

Wang, C., Li, X., Chen, J., Fei, G., Wang, H., & Liu, Q. (2011). Synthesis and characterization of polyacrylonitrile pregelled starch graft copolymers using ferrous sulfate-hydrogen peroxide redox initiation system as surface sizing agent. *Journal of Applied Polymer Science, 122*, 2630–2638.

Wei, Y., Cheng, F., & Zheng, H. (2008). Synthesis and flocculating properties of cationic starch derivatives. *Carbohydrate Polymers, 74*, 673–679.

Wollerdorfer, M., & Bader, H. (1998). Influence of natural fibres on the mechanical properties of biodegradable polymers. *Industrial Crops and Products, 8*, 105–112.

Wu, J., Wei, Y., Lin, J., & Lin, S. (2003). Study on starch-graft-acrylamide/mineral powder super absorbent composite. *Polymer, 44*, 6513–6520.

Wu, Y. P., Ji, M. Q., Qi, Q., Wang, Y. Q., & Zhang, L. Q. (2004). Preparation, structure and properties of starch/rubber composites prepared by co-coagulating rubber latex and starch paste. *Macromolecular Rapid Communications, 25*, 565–570.

Wu, Y. P., Qi, Q., Liang, G. H., & Zhang, L. Q. (2006). A strategy to prepare high performance starch/rubber composites: in situ modification during latex compounding process. *Carbohydrate Polymers, 65*, 109–113.

Wu, Y. P., Liang, G. H., & Zhang, L. Q. (2009). Influence of starch on the properties of carbon black filled styrene-butadiene rubber composites. *Journal of Applied Polymer Science, 114*, 2254–2260.

Xie, F., Pollet, E., Halley, P. J., & Averous, L. (2013). Starch-based nano-biocomposites. *Progress in Polymer Science, 38*, 1590–1628.

Xiong, H. G., Tang, S. W., Tang, H. L., & Zou, P. (2008). The structure and properties of a starch-based biodegradable film. *Carbohydrate Polymers, 71*, 263–268.

Yu, J., Gao, J., & Lin, T. (1996). Biodegradable thermoplastic starch. *Journal of Applied Polymer Science, 62*, 1491–1494.

Yu, J., Wang, N., & Ma, X. (2005). The effects of citric acid on the properties of thermoplastic starch plasticized by glycerol. *Starch Staerke, 57*, 494–504.

Zhang, L., Wang, Y., Liu, H., Yu, L., Liu, X., Chen, L., & Zhang, N. (2013). Developing hydroxypropyl methylcellulose/hydroxypropyl starch blends for use as capsule materials. *Carbohydrate Polymers, 98*, 73–79.

Zhang, L., Wang, Y., Liu, H., Zhang, N., Liu, X., Chen, L., & Yu, L. (2013). Development of capsules from natural plant polymers. *Acta Polymerica Sinica, 13*, 1–10.

Zhang, N., Liu, H., Yu, L., Liu, X., Zhang, L., Chen, L., & Shanks, R. (2013). Developing gelatin starch blends for use as capsule materials. *Carbohydrate Polymers, 92*, 455–461.

Zhang, N., Liu, X., Yu, L., Shanks, R., Petinaks, E., & Liu, H. (2013). Phase composition and interface of starch-gelatin blends studied by synchrotron FTIR micro-spectroscopy. *Carbohydrate Polymers, 95*, 649–653.

Zhang, L., Wang, Y., Yu, L., Liu, H., Simon, G., Zhang, N., & Chen, L. (2014). Rheological and gel properties of hydroxypropyl methylcellulose/hydroxypropyl starch blends. *Colloid & Polymer Science, 293*, 229–237.

Zhang, N., Shanks, R., Liu, X., & Yu, L. (2014). Phase composition of starch-gelatin blends studied by FTIR. *Advanced Materials Research, 875–877*, 106–109.

PLA-based blends and composites

B.D.S. Deeraj[1], Jitha S. Jayan[2], Appukuttan Saritha[2] and Kuruvilla Joseph[1]
[1]Department of Chemistry, Indian Institute of Space Science and Technology, Kerala, Thiruvananthapuram, India; [2]Department of Chemistry, School of Arts and Sciences, Amrita Vishwa Vidyapeetham, Amritapuri, Kerala, Kollam, India

1. Introduction

With advancements in material technology, polymers and polymer-based materials have emerged as prominent end use materials because of their inherent properties. The use of synthetic polymers is increasing almost daily, which is alarming as these petroleum-derived polymers are not biodegradable. For this reason, environment-friendly polymers, including degradable polymers and biological polymers, have attracted increasing attention from researchers over the last few decades (Chen & Evans, 2005; Shi et al., 2011; Yi et al., 2020).

The use of synthetic plastics increases the amount of nondegradable material in the land and water and is harmful to nature. Therefore, many researchers are working on alternative polymers that are degradable after use and which exhibit similar or better properties than synthetic polymers. This prevents the concept of nondegradable waste and will not create environmental hazards. Therefore, to enhance their properties these biodegradable polymers are blended with other polymers, incorporated with functional nanoparticles, used as composite matrices, etc. This chapter focuses on one such widely used and studied biodegradable polymer, poly(lactic acid) (PLA). The chapter provides brief details about PLA and focuses more on PLA blends and PLA composites for multifunctional applications.

1.1 Biodegradable polymers

Biodegradable polymers are topics of significant research interest due to their role in sustainable development with minimal global damage. These biodegradable polymers can be classified into agro-based polymers (chitin, starch, etc.) and biodegradable polyesters (PLA, polyhydroxyalkanoates, etc.) (Avérous & Pollet, 2012). In Fig. 8.1, the structures of some biodegradable polymers are presented. These polymers are intended to degrade into byproducts after their specified use. This polymer degradation is generally because of three mechanisms, namely depolymerization, scission, and elimination.

This degradation can be due to many factors, such as heat that leads to thermal degradation. Chemicals lead to hydrolysis or catalytic degradation. Mechanical stress leads to mechanochemical degradation. Visible light or UV light leads to photodegradation. The action of organisms leads to aerobic or anaerobic degradation. Heat and

Biodegradable Polymers, Blends and Composites. https://doi.org/10.1016/B978-0-12-823791-5.00014-4

oxygen lead to thermal oxidation and high-energy beams like X-rays also contribute to degradation.

In the case of polyesters, the aliphatic polymers are mostly studied as biopolymers and these polymers are classified into two groups based on the bonding of monomers (Mochizuki & Hirami, 1997; Vroman & Tighzert, 2009). The first group consists of poly(hydroxy acid)s. These are polyesters synthesized from hydroxy acids (hydroxy-carboxylic acids), or by ring-opening polymerization of cyclic monomers. The second group consists of the poly(alkylene dicarboxylate)s. These are polyesters prepared by polycondensation of diols and dicarboxylic acids. Depending on the polymer backbone, polyester biopolymers are classified as poly(α-hydroxyalkanoic acid)s (PLLA, PDLA, PGA, PLGA, etc.), poly(β-, γ-, δ-hydroxyalkanoate)s (P3HP, P3HB, P4HP, P4HB etc.),Poly(ω-hydroxyalkanoate)s (PCL etc.), Poly(alkylene dicarboxylate)s (PPS, PBS), etc. In Table 8.1, the classification of biopolymers is presented.

1.2 Poly(lactic acid) or polylactide (PLA)

Poly(lactic acid) or polylactide (PLA) is an aliphatic linear poly(α-ester). It is usually obtained by ionic polymerization of lactide, a ring closure of two lactic acid molecules. In another way, PLA can be produced by polycondensation of lactic acid. However, solvent disposal is a major drawback and this process can produce only low-molecular-weight polymers. The structures of PLA and stereoisomeric forms of PLA are presented in Figs. 8.1 and 8.2, respectively. The mechanical properties of these isomeric PLAs are different, as are the degradation times (Auras et al., 2004). The physical properties of PLLA are presented in Table 8.2.

2. PLA-based composites

Conventionally, composites are a combination of two or more individual components having distinct properties to develop a functional material. PLA can be used for the fabrication of composite material both as a reinforcement and a matrix. The incorporation of different fillers in the PLA matrix can enhance its inherent properties and thus ensure the commercial viability of biomedical and packaging products made of PLA (Pluta et al., 2002). The mechanical property of PLA is comparable with other engineering polymers like polypropylene (PP) and polyethylene terephthalate (PET), and hence it finds interesting applications in the automobile industry. Moreover, the biocompatible nature of PLA makes it a potential candidate for pharmaceutical applications. Nowadays, PLA is used to replace short-term packaging materials due to the possibility of the development of temperature-withstanding packing materials. Though biocompatibility is one of the major advantages of PLA, the poor thermomechanical and physical properties encourage its high-end applications. The reduction in stiffness and improvement of mechanical strength are of serious concern and thus researchers are attempting to develop functional PLA-based composites. This will lead to the mass production of green polymers and reduce the use of oil-based polymers in future

Table 8.1 Classification of biopolymers.

Biodegradable				Nonbiodegradable
Bio-based				
Plants	**Microorganisms**	**Animals**	**Fossil-based**	**Bio-based**
Cellulose and derivatives (polysaccharide)	PHAs (e.g., P4HB, PHB, PHBV, PHBH, PHBV)	Chitin (polysaccharide)	Poly (alkylene dicarboxylate)s (e.g., PBA, PBS, PBSA, PBSE, PESE, PESA,PEA, PES, PPF, PPS, PTA, PTMS, PTSE, PTT)	PE (LDPE, HDPE), PP, PVC
Lignin	PHF	Chitosan (polysaccharide)	PGA	PET, PPT
Starch and its derivates (monosaccharide)	Bacterial cellulose	Hyaluronan (polysaccharide)	PCL	PU
Alginate (polysaccharide)	Hyaluronan (polysaccharide)	Casein (protein)	PVOH	PC
Lipids (triglycerides)	Xanthan (polysaccharide)	Whey (protein)	POE	Poly(ether-ester)s
Wheat, corn, pea, potato, soy, potato (protein)	Curdlan (polysaccharide)	Collagen (protein)	Polyanhydrides	Polyamides (PA 11, PA 410, PA 610, PA 1010, PA 1012)
Gums (e.g., *cis*-1, 4- polyisoprene)	Pullulan (polysaccharide)	Albumin (protein)	PPHOS	Polyester amides
Carrageenan	Silk (protein)	Keratin, PFF (protein)		Unsaturated polyesters
PLA (from starch or sugarcane)		Leather (protein)		Epoxy
				Phenolic resins

From Niaounakis, M. (2013). Biopolymers: Reuse, Recycling, and Disposal. *Biopolymers: Reuse, Recycling, and Disposal*pp, 1−413. Elsevier Inc. https://doi.org/10.1016/C2012-0-02583-5.

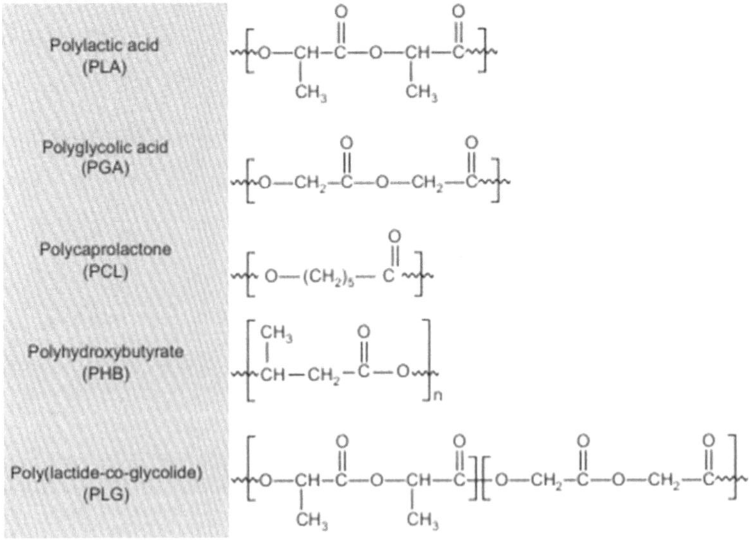

Polylactic acid
(PLA)
Polyglycolic acid
(PGA)
Polycaprolactone
(PCL)
Polyhydroxybutyrate
(PHB)
Poly(lactide-co-glycolide)
(PLG)

Figure 8.1 Structures of some biodegradable polymers.
From Kannan, M. B. (2015). Biodegradable polymeric coatings for surface modification of magnesium-based biomaterials. *Surface Modification of Magnesium and Its Alloys for Biomedical Applications*, *2*, 355–376. Elsevier Inc. https://doi.org/10.1016/B978-1-78242-078-1.00013-X.

isotactic
poly(L-lactide) (PLLA)
isotactic
poly(D-lactide) (PDLA)
heterotactic poly(D,L-lactide)
(heterotactic PDLLA)
syndiotactic poly(D,L-lactide)
(syndiotactic PDLLA)

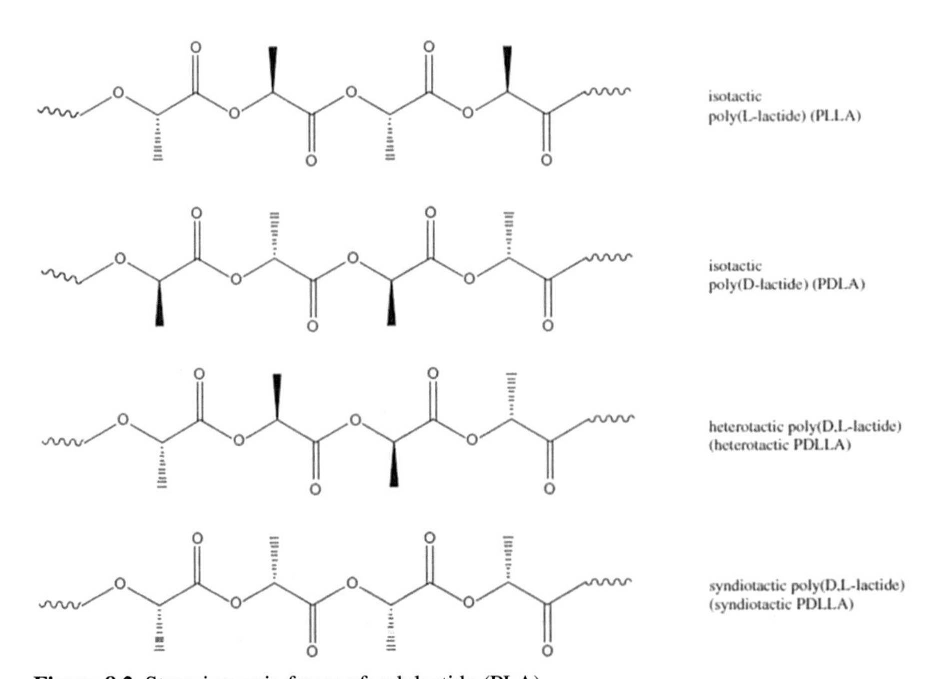

Figure 8.2 Stereoisomeric forms of polylactide (PLA).
From Niaounakis, M. (2013). Biopolymers: Reuse, Recycling, and Disposal. *Biopolymers: Reuse, Recycling, and Disposal*, 1–413. Elsevier Inc. https://doi.org/10.1016/C2012-0-02583-5.

Table 8.2 Physical properties of poly(L-lactic acid).

S. no.	Physical characteristics	Values
1	Glass transition temperature	55–65°C
2	Melting temperature	170–200°C
3	Melt crystallization temperature	90 and 120°C
4	Viscosity	1.24–1.30 g/cm^3
5	Limiting oxygen index	18.5%
6	Elastic modulus	2.7–4.1 GPa
7	Tensile strength	15.5–150 MPa

From Casalini, T., Rossi, F., Castrovinci, A., & Perale, G. (2019). A perspective on polylactic acid-based polymers use for nanoparticles synthesis and applications. *Frontiers in Bioengineering and Biotechnology*, *7*. https://doi.org/10.3389/fbioe.2019.00259.

(Mosanenzadeh et al., 2016). The enhanced availability of PLA and the increase in petroleum cost are driving factors for PLA-based biocomposites. Different kinds of fillers are incorporated in PLA (Mukherjee & Kao, 2011). Nanomaterials, carbonaceous nanomaterials, thermoplastics, synthetic fibers, natural fibers, hybrid fillers, etc. are used as fillers for PLA matrices. Moreover, 2D layered materials like graphene oxide (GO), boron nitride, MoS_2 etc. are used as filler in the PLA matrix. PLA-based composites also find applications in fabricating 3D scaffolds (Serra et al., 2013). Generally, nanoparticles like nanoclay, nanocellulose, and nanometals are used to improve the properties of PLA films like UV-scattering and antibacterial characteristics (Therias et al., 2012). Zinc oxide nanoparticles are used to enhance the properties like availability, nontoxicity, stability, low cost, antimicrobial activity, and ultraviolet absorption capacity. ZnO is considered to be a better component for enhancing the packaging applications (Díez-Pascual & Díez-Vicente, 2014; Shankar et al., 2015). ZnO nanoparticles are capable of improving the UV-light barrier property with reduced transparency and improved barrier properties (Shankar, Rhim, & Won, 2018). Silver–copper (Ag–Cu) nanoparticle (NP)-based PLA composites made by compression molding are capable of making films with enhanced barrier properties and antibacterial action (Ahmed et al., 2018). Ag and lignin NPs are effective for food packaging applications due to the inherent UV screening and antioxidant activity of lignin and the antibacterial properties of AgNPs. Simultaneous incorporation of these nanoparticles enhances the thermal stability, mechanical properties, UV screening effect, and barrier properties (Shankar, Wang, & Rhim, 2018). CuNP-incorporated PLA nanocomposites are effective against Gram-positive and Gram-negative bacteria and the utilization of such nanocomposites has enhanced the food packaging applications of PLA (Longano et al., 2012).

Due to the poor crystallization, PLA generally shows weak mechanical properties and thermal resistance. However, nucleating agents like fibers are efficient in enhancing the crystallization of PLA and have an influence on the thermal properties. Hence many natural and synthetic fibers have been investigated for their good reinforcing effect in PLA. Their studies signify the positive role of synthetic fibers over natural fibers. Glass fiber-incorporated PLA exhibits better thermal resistance due to the presence of silicates, borates, and phosphates, and it is capable of showing enhanced biological activity. Wang et al. (2019) observed that glass fibers (GFs) are capable of enhancing the mechanical performance of PLA by affecting the crystallization. Although, the thermal resistance is feeble, heat treatment is capable of enhancing the thermal stability. An SEM image of GF/PLA composites is depicted in Fig. 8.3 which shows that the incorporation of GF with and without coupling agents results in a different appearance. The absence of a coupling agent presents a smooth surface due to the weak interfacial interaction, whereas the GFs with coupling agents show better covering with PLA due to the better interfacial interaction. Furthermore, this study shows that heat treatment is able to enhance the crystallization and thus thermal and mechanical performance.

Biocompatible polymers like PLA display good applicability in 3D printing applications. Especially, fused filament fabrication is the most used 3D printing method employing PLA. Dominguez-Robles et al. (2019) incorporated lignin in PLA to fabricate fused filaments with excellent antioxidant properties. The enhanced antioxidant properties make them function as a good material for wound-healing applications. The antioxidant characteristics of the lignin nanoparticles are highly useful in imparting antioxidant characteristics to the PLA-based packaging materials, hence incorporation of lignin components is considered as a better choice in packaging applications. The combination of these materials with 3D printable and biocompatible PLA is a good choice for the development of biomedical materials with multiple characteristics. Yu et al. (2019) used an X-ray microscopy technique for the characterization of 3D-printed basalt fiber-reinforced PLA composites. The overall properties of the 3D-printed composites generally depend on the fiber size, shape, orientation, and relative proportions between inter- and inner voids. After 3D printing, inter- and inner filament voids are generated in the composite which are capable of orienting in the printing directions. X-ray microscopic and SEM images are shown in Fig. 8.4. Thus it enhances the mechanical properties much better than that of the composites made using the

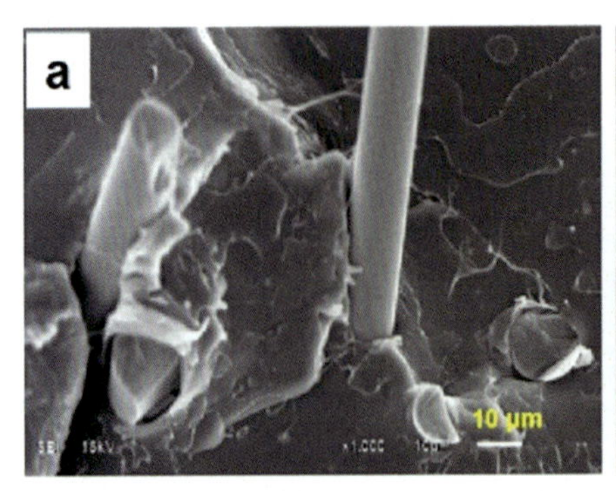

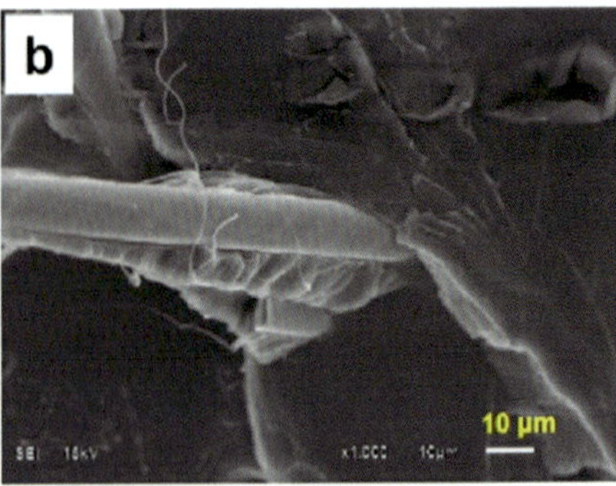

Figure 8.3 The micromorphology of the PLA/GF composites (A) with and (B) without a coupling agent.
From Wang, G., Zhang, D., Li, B., Wan, G., Zhao, G., & Zhang, A. (2019). Strong and thermal-resistance glass fiber-reinforced polylactic acid (PLA) composites enabled by heat treatment. *International Journal of Biological Macromolecules*, *129*, 448–459. https://doi.org/10.1016/j.ijbiomac.2019.02.020.

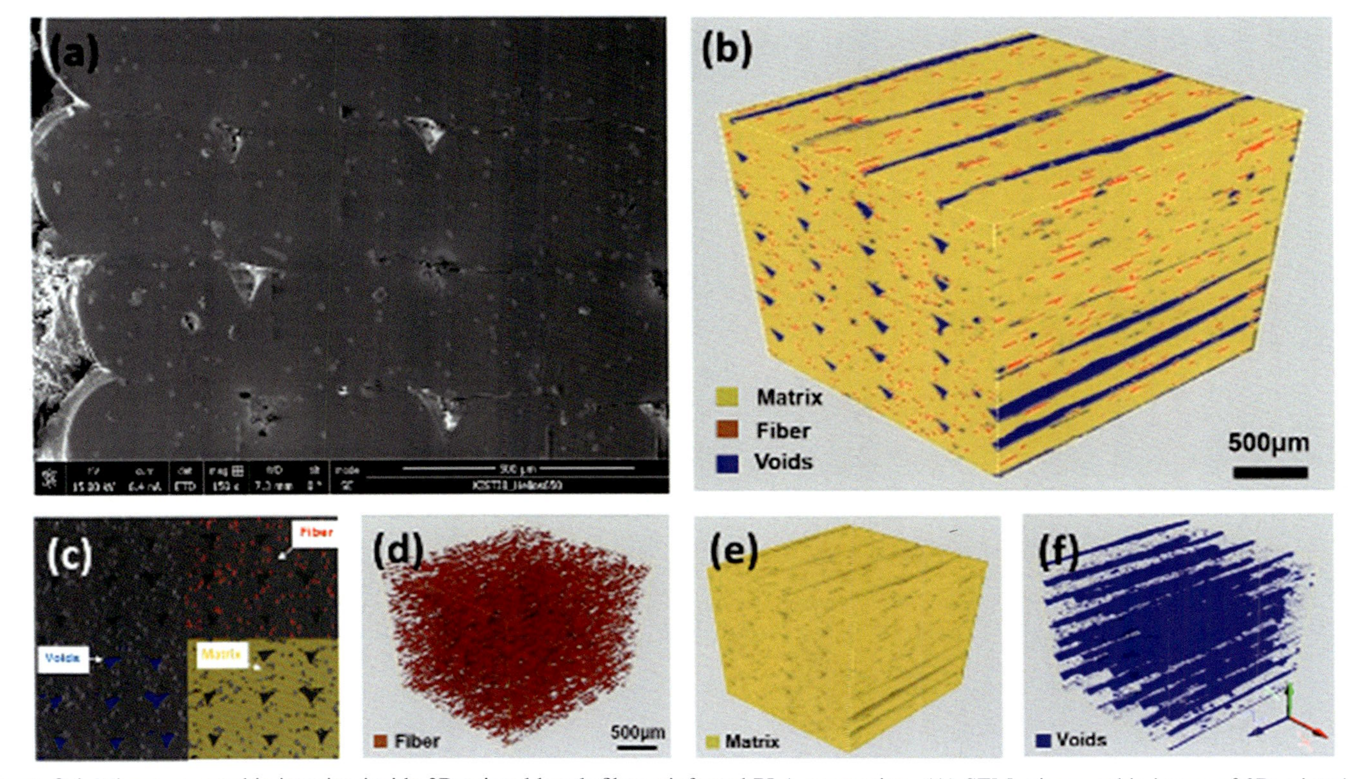

Figure 8.4 Microtomographic imaging inside 3D-printed basalt fiber-reinforced PLA composites. (A) SEM micrographic image of 3D-printed composites, (B) 3D microtomographic image of 3D-printed composites acquired through X-ray microscopy, (C) individual components of 3D-printed composites (D) fiber, (E) matrix, and (F) voids identified by image contrast.
From Yu, S., Hwang, Y. H., Hwang, J. Y., & Hong, S. H. (2019). Analytical study on the 3D-printed structure and mechanical properties of basalt fiber-reinforced PLA composites using X-ray microscopy. *Composites Science and Technology*, *175*, 18–27. https://doi.org/10.1016/j.compscitech.2019.03.005.

normal mold pressing method. Void orientation leads to the fabrication of a lightweight and strong material. By adjusting the printing parameters, the void generation can also be controlled so as to control the interfacial bonding interaction. Tian et al. (2016) carried out 3D printing of carbon fiber (CF)-reinforced PLA composites by controlling the temperature and pressure. Control of both these parameters was found to be effective in determining the mechanical performance of the composites. Hence the 3D printing conditions are good enough for monitoring the properties of the composites. The increase in pressure of the impregnation process is effective in enhancing the bonding between the deposited lines and thus the mechanical strength. The reduction in layer thickness and hatch spacing is capable of increasing the contact pressure and thus the mechanical performance of 3D impregnated composites. The process of 3D printing of CF/PLA composites is schematically demonstrated in Fig. 8.5. Later, the possibility of recycling and remanufacturing of these 3D printed CF/PLA composites was analyzed (Tian et al., 2017). The CF and PLA are recycled from the 3D-printed composites and are used as the resource for further 3D printing process by reversibly applying the original trajectory. The recycled CF and remanufactured composites show better tensile properties. Ferreria et al. (2017) made fused filaments of PLA and PLA + CF, where the CF was able to improve the tensile modulus to a greater extent. However, PLA and CF generally have poor adhesion. In the case of PLA + CF, the short fibers have different orientations than the

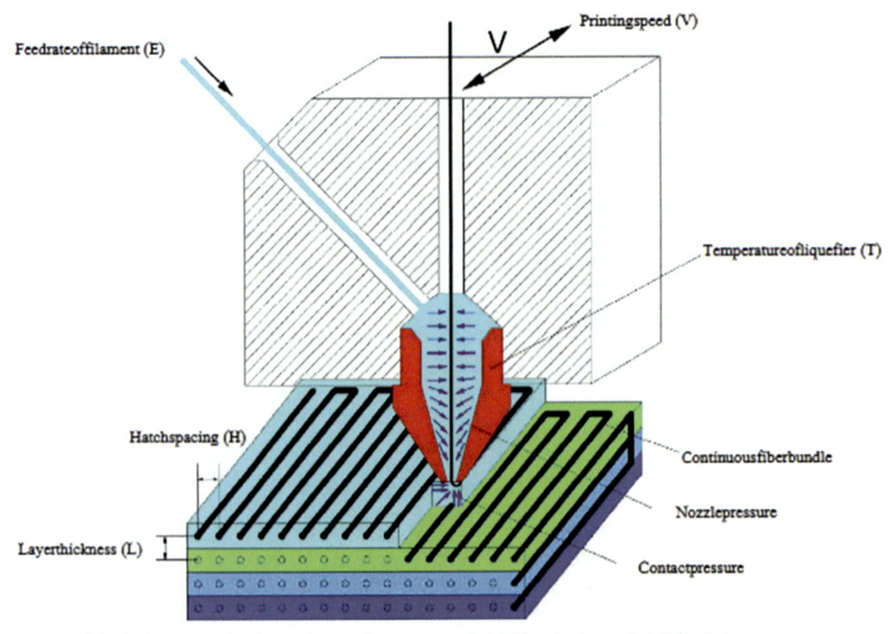

Figure 8.5 Schematic of the process parameters for 3D printing of CFR PLA composite. From Tian, X., Liu, T., Yang, C., Wang, Q., & Li, D. (2016). Interface and performance of 3D printed continuous carbon fiber reinforced PLA composites. *Composites Part A: Applied Science and Manufacturing*, *88*, 198–205. https://doi.org/10.1016/j.compositesa.2016.05.032.

others and they have an important role in determining the properties. 3D printing can be utilized for the fabrication of PLA-based scaffolds by combining PEG and glass particles. This ensures the bioactivity and mechanical integrity of PLA-based scaffolds. Moreover, it also ensures the biodegradability and porous structure of the composite (Serra et al., 2013).

Papon and Haque (2019) utilized the process of additive manufacturing in PLA and carbon fiber (CF)-reinforced PLA, where the CF/PLA filaments showed a 42% improvement in fracture toughness. With the incorporation of CF, the number of intrabead voids was enhanced, which in turn was responsible for the toughness improvement. In the case of 0°, interfacial debonding was observed and the samples showed no contribution to toughness and the outer plane fibers were pulled out from the matrix. Beads at 90° showed delamination between 0° beads which contributed significantly toward toughness. However, the ternary systems are a new class of composite materials with enhanced properties compared with normal composites. CNF and epoxidized soyabean oil (ESO) can be incorporated in PLA to make a ternary composite. The combined effect of both fillers like PLA and ESO in enhancing the tensile strength and ductility were utilized. At higher loadings, the effect of ESO in ductility and toughness was negligible but it enhanced the percolation of CNF, which in turn helped in the stress transfer between fibrils. Excellent mechanical properties were obtained by the combined plasticizing and percolation effect of ESO and CNF. Thus the ternary PLA composite systems are advantageous over the simple PLA/CNF composite (Meng et al., 2018). Different fibers are used in PLA to improve its properties and the details are tabulated in Table 8.3.

PLA has been found to be effective in guided tissue engineering and orthopedic fixation. Moreover, the mechanical strength safeguards its practical use in surgery. Its compatibility and strength make it a potential scaffold material in bone tissue engineering. Hence PLA and its composites are used for the fabrication of biomimetic scaffolds for bone-defect repairing. Liao et al. (2004) made PLA-based composites by combining it with bioactive and biodegradable components like hydroxyapatite (HA) and collagen, which had poor mechanical properties. By combining with PLA they were able to nullify the mechanical properties of these components. These materials were able to mimic the bone-forming components due to its hierarchical nanostructures. Later studies showed that when compared to HA, nanosized HA (n-HA) was able to enhance the porosity as well as the mechanical performance of PLA (Kothapalli et al., 2005). The HA-incorporated PLA composites are able to show a shape memory effect and mechanical strength when 3D printed. HA acts as a nucleation center and thus reduces the molecular mobility. HA-based ceramics are able to prevent crack growth, and HA-incorporated porous PLA composites are capable of showing shape recovery properties. This self-healing is associated with the narrowing of cracks. The higher porosity is a necessary condition for the unhindered diffusion of tissue cells. The composites were able to show 100% recovery, even after two cycles and a slight deformation occurred after the third one. Hence the PLA-based porous scaffolds are efficient candidates for self-fitting tissue engineering (Senatov et al., 2016).

Table 8.3 PLA-based natural fiber composites.

S. no.	System	Filler loading	Advantages	References
1	PLA/flax fiber	30 wt.%	147% improvement in toughness	Oksman et al. (2003)
2	PLA/Kenaf fiber	70 wt.%	Linear increase in flexural strength and elastic modulus and 38% decrease in weight	Ochi (2008)
3	PLA/ microfibrillated cellulose	20 wt.%	72% improvement in storage modulus, in amorphous and crystallized state 22% and 14% improvements in tensile strength are observed	Suryanegara et al. (2009)
4	PLA/chicken feather fiber	5 wt.%	16% improvement in tensile strength without weight change	Cheng et al. (2009)
5	PLA/chopped glass fiber	30 wt.%	152%, 53%, and 244% improvements in flexural modulus, izode impact strength, and storage modulus, respectively	Huda et al. (2006)
	PLA/recycled newspaper cellulose fiber	30 wt.%	63%, 0%, and 188% improvements in flexural modulus, izode impact strength, and storage modulus, respectively	
6	PLA/natural fibers	20 wt.%	44% and 70% reduction in friction coefficient and specific wear rate	Bajpai et al. (2013)
7	PLA/nanocellulose fibers	20 wt.%	50% increase in yield strength due to better fiber matrix interaction, reduction in toughness due to the nucleation on fiber	Kowalczyk et al. (2011)
8	PLA/nanocellulose fibers	1 wt.%	Lowered the coating weight, lowered water vapor transmission rate	Song et al. (2014)
9	PLA/carbon fiber	5 wt.%	42% improvement in fracture toughness	Papon and Haque (2019)
10	PLA/wood fiber	–	Surface-treated mineral additives reduce the content of PLA in fiber–PLA composite system	Ozyhar et al. (2020)
11	PLA/wood fiber	2 5wt.%	Up to 25 wt.% the filaments were able to show enhanced mechanical properties; infill orientation also plays an important role in determining the properties	Kain et al. (2020)

Being a biopolymer, PLA has a sound platform over the oil-based nonbiodegradable polymers. The reinforcing material has a great effect in molding the properties of polymers and, for ensuring the desired application, judicial selection of the filler is a must. Nowadays, exceptional changes are made to nanomaterials to obtaining good functional applications. Recently, hybrid fillers have been developed in order to attain exceptional qualities. The nanofiller hybrid materials made by grafting active functional groups onto carbon nanomaterials exhibit enhanced flame retardancy when incorporated into the PLA matrix. The functionalization, grafting, and hybrid formation are vital in improving the PLA properties. Thus the formation of unique PLA composites is a vibrant research topic (Wen et al., 2020). Boron nitride nanosheets (BNNSs) are analogues to 2D materials like graphene and are capable of tuning the material properties. The overall interaction of BNNS and PLA matrix plays an important role in controlling the overall crystallinity, thermal stability, crystallization rate, and thermal conductivity of PLA. Functionalized BNNSs showed enhanced thermal conductivity and thermal stability over pristine BNNS-incorporated PLA, thus the interaction between filler and matrix is a crucial factor in determining functional properties (Rosely et al., 2019). Biocompatible modifier-grafted biocompatible BNNS-based PLA biocomposites are capable of showing enhanced thermal stability and thus flame retardancy due to the inherent barrier action of BNNSs and phosphorous content in the phytic acid functionalized modifier (Rosely et al., 2020). Hence environment-friendly nanofillers and their functionalization are paving a new path toward the development of multifunctional and biocompatible composites.

3. PLA-based blends

Since PLA has outstanding physical and rheological properties, it can be added to many additives and other polymeric matrices to enhance the properties for high-end applications. Blends can be made by simply mixing PLA with any other additives or polymers, the resulting blend system can be miscible or immiscible depending on the compatibility of the mixing components. The immiscible blends lead to the formation of phase separation and compared to miscible blends the properties of immiscible blends are less effective (NatureWorks, 2007). PLA is a colorless biopolymer having properties similar to those of polystyrene, but high cost, processing difficulties, and brittleness need to be eradicated for better commercial utility. Commercial PLA is a mixture of L and D isomers, having a semicrystalline nature with a melting temperature below 200°C and glass transition range of 55–65°C (Fortelny et al., 2019). The formation of PLA blends is considered as a cost-effective method for improving the performance of epoxy. The blending of PLA offers a better degradation rate, permeability features, thermomechanical properties, and drug release rate (Matta et al., 2014). The selection and composition of components play an important role in tuning the properties of blends. Flexible as well as elastic polymers can be thus blended to PLA to attain better mechanical properties. Grafted rubber, reactive molecules, thermoplastic polymers, biodegradable polymers, petroleum-based polymers, rubber, and thermoplastic elastomers are used for the toughening of PLA, as shown in Fig. 8.6.

Starch is a good choice as it has a lower cost and is used for blending with PLA. The inherent natures of PLA and starch are different; PLA is hydrophobic, whereas starch is hydrophilic in nature. Mixing of starch with PLA avoids water contact, but this results in phase separation and shows a very weak interfacial interaction. Mechanical stress cannot be tolerated by these starch/PLA blends so it shows weak mechanical properties and thus it is more brittle than PLA. However, toughening can be attained by various methods such as adding a plasticizer to PLA, plasticizing starch or PLA, and by softening PLA/starch mixer by adding a third component (Koh et al., 2018).

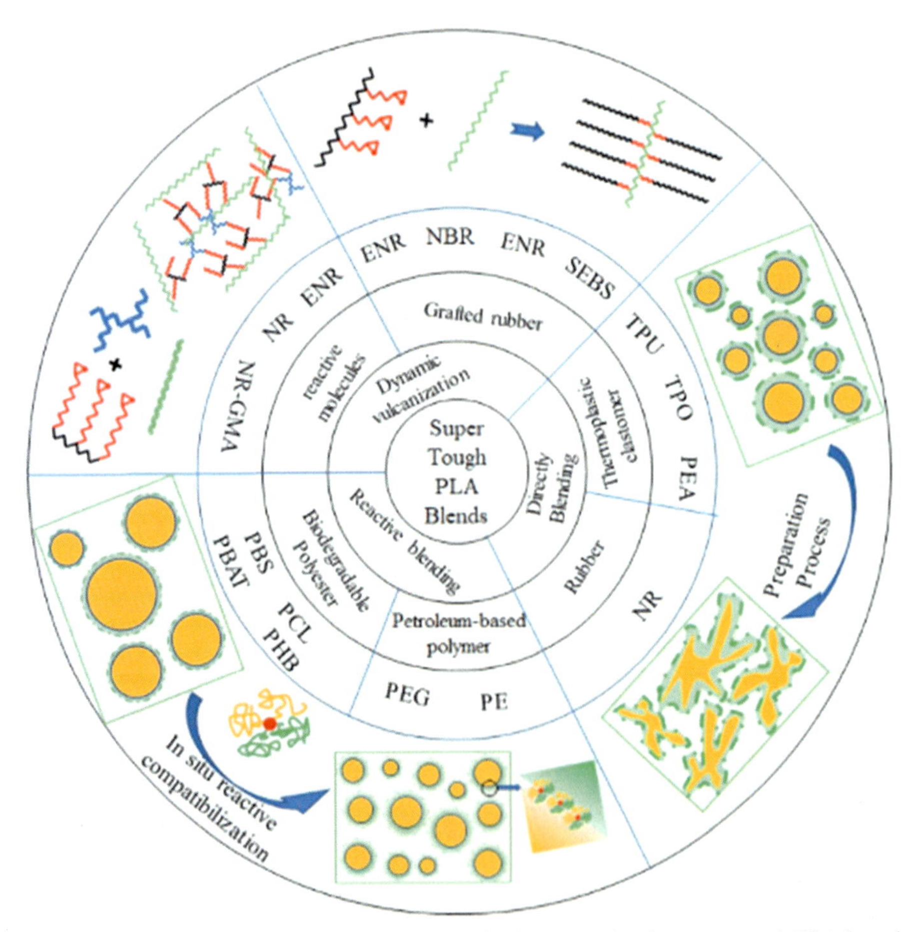

Figure 8.6 The summarized approaches and toughening strategies for super-tough PLA-based blends.
From Zhao, X., Hu, H., Wang, X., Yu, X., Zhou, W., & Peng, S. (2020). Super tough poly(lactic acid) blends: A comprehensive review. *RSC Advances*, *10*(22), 13316–13368. https://doi.org/10.1039/d0ra01801e.

PLA, when blended with copolymer-based polybutylene adipate, results in the ductility of the polymer being improved as it follows the reactive compatibilization method. Reactive compatibilization can be considered as an effective method for improving the ductility of PLA by inducing the polymerization of copolymers during the melt blending process. The possibility of side reactions could enhance the applicability of the blended system in different fields. Therefore, the radical used for initiating the copolymerization will have an effect on the melt behavior of PLA (Coltelli et al., 2010). The blends of PLA with polyhydroxybutyrate (PHB) have excellent flexibility and high strain at break. PLA is miscible with PHB but depending on the molecular weight of both components, the miscibility can be altered. Considering the degradation of PLA, the crystalline part is found to be more resistant against the degradation than the amorphous part. Generally, the addition of plasticizers increases the rate of the crystallization process and affects the biodegradation process. PHB shows enhancement in ductile properties due to the decrease in the glass transition temperature. PHB has been found to be more susceptible to degradation than PLA (Sedničková et al., 2018). Triacetine-plasticized cellulose acetate is a better choice for the improvement of both stiffness and toughness of PLA, so that the extrusion can be considered as an effective method. The adhesion between the phases and pull-out mechanism shown by the fibrils results in better enhancement of the mechanical properties without affecting the elasticity of PLA (Coltelli et al., 2019). In short, PLA can be blended with a variety of polymers such as biodegradable polymers, bio-based biodegradable polymers, bio-based polymers, synthetic/nondegradable thermoplastics, and elastomeric polymers and are capable of forming ternary blends and foams. Table 8.4 outlines the different PLA-based polymer blends.

PLA blends can be generally converted into a variety of structures such as porous scaffolds, films, fibers, and particles and thus they ensure biomedical applications. PLA is generally blended with other polymers to make hybrids so as to improve the thermomechanical properties. The blending ensures the development of certain new properties like shape memory and diverse morphology that can be tuned for biomedical applications. Table 8.5 represents the PLA-blended systems that are used for biomedical applications (Saini et al., 2016).

4. Electrospun PLA fiber composite systems

The application capability of any polymer can be improved effectively when the polymer has a nanofibrous structure. One such technique available to prepare nanofibers from polymers is electrospinning. By carefully optimizing the experimental parameters, we can tune the structure of the resultant fibers (Deeraj et al., 2021; Bhardwaj & Kundu, 2010). Moreover, a variety of polymers can be electrospun into nano- and microfibers for fabricating composites for advanced structural applications (Deeraj et al., 2019, 2020).

Tsuji et al. (2006) reported one of the initial works on how the electrospinning technique was employed to produce nanofibers of stereocomplex PLA. In this work,

Table 8.4 Different types of PLA-based blends.

Type of polymer	PLA/blended system	Properties	References
Bio-based biodegradable	PLA/starch	Improve toughness, ductility	Jacobsen and Fritz (1996)
		Increased water absorption	Ke et al. (2003)
		Increases degree of crystallinity	Ke and Sun (2003)
		Improved elongation	Zhang and Sun (2004)
		Improved tensile strength and elongation at break	Jun (2000)
		Increased toughness and stiffness	Zhang et al. (2013)
		Acetylated starch improves the processability and shows enhanced tensile strength and toughness	Nasseri et al. (2020)
		High strength	Sun et al. (2001)
	PLA/lignin	Increase the thermal stability, reduce the mechanical properties and suppress crystallization	Anwer et al. (2015)
		Loss in mechanical strength	Spiridon et al. (2015)
		Loss in mechanical strength	Xie et al. (2011)
		Loss in mechanical strength	Cicala et al. (2017)
		Alkyl-modified lignin maintains the mechanical strength	Kim et al. (2014)
		Acylated lignin maintains mechanical strength and improves thermal stability	Gordobil et al. (2014)
		Improved Young's modulus and thermal degradation, reduced tensile strength and elongation at break	Li et al. (2019)
		Poor mechanical and thermal properties	Kumar et al. (2019)
		High tensile strength	Guo et al. (2020)
		Transparent and UV protection film, high tensile strength and elongation at break	Kim et al. (2017)
	PLA/cellulose	Highest impact strength and tensile strength	Bax and Müssig (2008)
		Enhanced flexural and impact strength	Dai et al. (2015)
		Improved thermal stability and elongation at break	Claro et al. (2016)

		Improvement in thermal property, better tensile strength, flexural strength, and elasticity	Zhang, Lei et al. (2020)
		Better structural strength and crystallinity	Alashwal et al. (2020)
	PLA/ polyhydroxyalkanoates (PHAs)	Improved strain at break and biodegradability	Han et al. (2012), Weng, Wang et al. (2013), Li et al. (2015)
		Better reinforcing ability and oxygen barrier properties	Arrieta, López, Hernández, and Rayón (2014), Arrieta, López, Rayón, and Jiménez (2014), Arrieta, Castro-López et al. (2014)
		Low thermal stability	Gerard and Budtova (2012)
		Better permeation and fracture toughness	Montes et al. (2020)
		Improves the processability and compatibility	Modi et al. (2012)
		Improves the complex viscosity	Gérard and Budtova (2011)
Biodegradable polymers	PLA/polycaprolactone (PCL)	Improved percentage of elongation, impact toughness, loss factor and decreased strength and modulus	Matta et al. (2014)
		Decreased stiffness, higher toughness	Ostafinska et al. (2015)
		Improved mechanical properties and elongation at break	Gardella et al. (2014)
		Lower elongation at break and stiffness	Urquijo et al. (2015)
		Improved toughness due to addition of lysine triisocyanate	Takayama and Todo (2006)
		High stiffness and toughness	Ostafinska et al. (2017)
		Fastest crystallization behavior	Luyt and Gasmi (2016)
		Increased toughness and reduced tensile strength	Qiu et al. (2020)

Continued

Table 8.4 Different types of PLA-based blends.—cont'd

Type of polymer	PLA/blended system	Properties	References
	PLA/PBAT [poly(butylene adipate-co-terephthalate)]	Immiscible blend system, increased relaxation moduli, and enhanced melt viscosity	Gu et al. (2008)
		After degradation carbon atom content is decreased and oxygen content enhanced, melting temperature shown after degradation increased	Weng, Jin et al. (2013)
		Decreased tensile strength and modulus, increased elongation and toughness, brittle fracture changed to ductile	Jiang et al. (2006)
		Increased plastic deformation, no change in stiffness, increased elongation at break	Gigante, Canesi et al. (2019)
		Increased flexibility without decreasing the mechanical strength and maintenance of the freshness of food items	Wang, Rhim, and Hong (2016)
		Increased compatibility and transition phase morphology tensile properties are enhanced, improved thermal and impact properties	Lu et al. (2017)
	PLA/poly(butylene succinate-co-adipate) (PBSA)	Better mechanical, barrier, and optical properties	Palai et al. (2020)
		Improves ductility and the presence of chain extender enhance both ductility and melt strength	Eslami and Kamal (2013)
		Blends showed shear thinning behavior and high interfacial tension values, storage modulus increased with decreased frequency	Gui et al. (2012)
		Flexible and tough, increased elongation at break, reduction in the thermal stability and modulus	Ojijo et al. (2012)
		Decreased T_g and increased elongation at break, higher crystallinity, and better oxygen barrier properties	Yang et al. (2016)

	PLA/Poly (butylene succinate) (PBS)	Tensile strength and % elongation decreases, reduced brittleness	Bhatia et al. (2007)
		Lower storage modulus, flexible and compatible, increased crystallization, compatibilizer, and plasticizer	Park et al. (2010)
		Improved flexibility and cytocompatibility, improved tear resistance and elongation at break	Gigante, Coltelli et al. (2019)
		Decreased crystallization temperature due to plasticizing nature, higher elongation at break	Ostrowska et al. (2019)
		Low stress, increased elongation at beak, and good mechanical properties	Panichsombat et al. (2019)
	PLA/polyvinyl acetate (PVAc)	Miscible and only one glass transition but after hydroxylation PVAc is phase separated in PLA	Park and Im (2003)
		Single T_g showing miscibility, higher tensile strength, and improved physical properties	Gajria et al. (1996)
		Two polymers interact each other and thus shows reduction in degradation rate	Mahalik and Madras (2006)
	PLA/poly(vinyl alcohol) (PVA)	One T_g showing the miscibility, PVA content increases the thermal stability, higher PVA content reduces tensile strength	Restrepo et al. (2018)
		Decreased moisture absorption and water absorption, PVA reduces the crystallinity of PLA	Wang et al. (2008)
		Better interaction of PVA and PLA makes it miscible, PVA acts as a nucleating agent and improves crystallization	Yeh et al. (2008)
		Phase-separated PLA and PVA, increased contact angle	Tsuji and Muramatsu (2001)
	PLA/ethylene vinyl alcohol (EVOH)	Improved water vapor transmission, homogeneous dispersion	Wu et al. (2016)

Continued

Table 8.4 Different types of PLA-based blends.—cont'd

Type of polymer	PLA/blended system	Properties	References
		Increased elongation at break, improved impact toughness	Zeng et al. (2018)
		Decrease in the intrinsic viscosity, enhanced barrier properties and cold crystallization	Gui et al. (2013)
		Improved water transmission rate, hydrophilic character of EVOH improved melt index of PLA	Gui et al. (2013)
	PLA/poly(propylene carbonate) (PPC)	Compatibility of the polymers enhanced the thermal stability, strong interfacial adhesion	Ma et al. (2006)
		Increased elongation at break, compatibility, good shape memory effect	Qin et al. (2018)
		Increased elongation at break, two T_g showed partial miscibility	Haneef et al. (2019)
		Decreased crystallinity and melting temperature, improved tensile properties	Ploypetchara et al. (2014)
		Improved tensile toughness with maleic anhydride cap-ed PPC, plastic stretching under tensile stress	Zhou et al. (2016)
		Improved thermal stability, increased PCC content leads to shear thinning, improved impact strength	Zou et al. (2016)
	PLA/poly(glycolic acid) (PGA)	Immiscible with each other	You et al. (2006)
		Improved tensile strength and Young's modulus, low creep resistance	Kimble, Bhattacharyya, and Fakirov (2015)
		Less creep resistance due to rapid relaxation of stress	Kimble, Fakirov, and Bhattacharyya (2015)

Bio-based polymers	PLA/polyethylene (PE)	Increase the impact strength, but compatibilizers are needed for improving the interfacial adhesion	Anderson and Hillmyer (2004)
		Interracially localized catalyst enhances the compatibilization	Thurber et al. (2015)
	PLA/polyamide (PA)	Partially miscible, increased crystallinity, strong interfacial adhesion	Feng and Ye 2010
		Bisphenol A used as a compatibilizer, enhanced phase dispersion and interfacial adhesion, showed no break in notched impact	Pai et al. (2013)
		Improved thermal stability by the addition of graphene nanoplatelets into PLA/PA blends	Alhadadi et al. (2019)
		Acrylic melt employed as a compatibilizer in PA/PLA blends, improves processability and engineering applications	Walha et al. (2018)
		Addition of compatibilizer enhances the interfacial adhesion thus showing improved mechanical properties and enhanced thermal stability	Choudhary et al. (2011)
		Pseudoplastic behavior, improved Young's modulus	Hamad et al. (2011b)
		Improved elongation at break and tensile toughness due to compatibilization	Xu et al. (2015)
	PLA/polystyrene(PS)	Shear thinning behavior, viscosity reduced with PLA content	Kaseem and Ko (2017), Hamad et al. (2010)
		Improved moisture resistance, improved biodegradability without compensating useful properties	Biresaw and Carriere (2002)
		Viscosity and mechanical strength are reduced by the processing cycle	Hamad et al. (2011a)

Continued

Table 8.4 Different types of PLA-based blends.—cont'd

Type of polymer	PLA/blended system	Properties	References
	PLA/acrylonitrile-butadiene-styrene (ABS)	Thermodynamically immiscible incorporation of compatibilizer enhances the mechanical properties	Li and Shimizu (2009)
		Epoxidized cardanol-based compatibilizer improves the mechanical strength and elongation at break	Rigoussen et al. (2019)
		Cardanol-grafted ABS shows plasticizing effect and decreases the surface energy	Rigoussen et al. (2017)
		PLA-grafted ABS improves the mechanical strength of PLA, improved impact and tensile strength	Chaikeaw and Srikulkit (2018)
	PLA/polymethylmethacrylate (PMMA)	Miscible with PLA	Cossement et al. (2006)
		Restricted crystallinity in presence of PMMA, improved tensile strength and modulus	Gonzalez-Garzon et al. (2018)
		Amorphous PMMA/semicrystalline PLA shows shape memory effect	Hao et al. (2015)
	PLA/terephthalates — PLA/PET	Immiscible, PLA reduces mechanical properties of PET,	McLauchlin and Ghita (2016)
		but compatibilizers enhance the thermal and mechanical properties	Gere and Czigany (2020)
		Minimum concentration of PET improves crystallization, later reduces the ductility and tensile strength	Topkanlo et al. (2018)
	PLA/terephthalates — PLA/PBT	Along with compatibilizer shows improved mechanical properties like tensile strength, toughness	Chang et al. (2018)
		Highly crystalline and brittle, but reactive blending reduced crystallinity	Irska et al. (2020)

	PLA/PTT	Improved crystallinity	Padee et al. (2013)
		The incorporation of compatibilizer along with PTT enhanced elongation at break, improved toughening	Kultravut et al. (2019)
		Drastic improvement in tensile properties, controlled hydrolytic degradation, improved toughness	Kultravut et al. (2020)
		Improved thermo-degradation, improved % elongation	Tsou et al. (2017)
	PLA/polycarbonate (PC)	High impact strength and heat distortion temperature	Lin et al. (2015)
		The compatibilizer enhances the impact and tensile strength	Wang et al. (2012)
		Excellent heat resistance, good impact strength	Yuryev et al. (2017)
		Improved tensile properties, improved compatibility	Phuong et al. (2014)
Elastomers	PLA/natural rubber(NR)	Molecular weight and compatibilization affect the mechanical properties	Yuan et al. (2014)
		Impact strength 16 times higher than neat PLA	Yuan et al. (2014)
		Epoxidized NR shows reduced thermal resistance, crystallinity and tensile properties than NR-incorporated PLA	Pongtanayut et al. (2013)
		Compatibilizers added to NR enhance the impact strength to a greater extent	Sookprasert and Hinchiranan (2017)
	PLA/SBS	Enhances the crystallization ability and mechanical properties, improved elongation at break, impact resistance	Wu et al. (2015)
		Improved elongation at break and impact strength	Wang, Wei, and Li (2016)
		In presence of compatibilizer shows enhanced dispersion, reduced crystallinity and enhanced thermal stability	Wang et al. (2019)

Continued

Table 8.4 Different types of PLA-based blends.—cont'd

Type of polymer	PLA/blended system	Properties	References
	PLA/styrene ethylene butylene styrene (SEBS)	Improved toughening, impact strength, and reduced thermal stability for SEBS and maleic anhydride-grafted SEBS	Sangeetha et al. (2016)
		36% increase in tensile strength and 17% improved toughness	Nehra et al. (2018a)
		Enhanced shear viscosity and resistance to crack propagation	Nehra et al. (2018b)
PLA/polyurethane(PU)	PLA/TPU(thermoplastic)	Improved toughness using TPU and polyurethane elastomer prepolymer, improved interfacial interaction	Zhang, Kang et al. (2020)
		Submicron structural formation leads to an improvement in mechanical properties	Imre et al. (2013)
		Improved interfacial compatibility and adhesion lead to improved tensile strength and impact strength	Zhao et al. (2017)
	PLA/polyurethane (TPU) (thermoset)	Improved flexibility, decreased T_g, improved impact strength	Zeng et al. (2011)
		Improved complex viscosity with the addition of PU, increased elastic character, improved elongation at break	Feng et al. (2013)
		Improved interfacial interaction, compatibility, and hence obtained super-toughened materials	Lu et al. (2014)

Table 8.5 Biomedical applications of different PLA-based blends.

Polymer	Technique	Application	Properties affected	References
Polyhydroxybutyrate (PHB)	Melt extrusion	Packaging	Enhanced ductility, elongation at break Barrier properties	Armentano et al. (2015)
Thermoplastic polyurethane	Melt extrusion	Actuators Tissue engineering	Shape memory, enhanced tensile strength and elongation at break	Song et al. (2015a), Mi et al. (2013)
Thermoplastic polyurethane	Foaming	Tissue engineering	Shape memory, porous morphology, higher crystallinity	Song et al. (2015b)
Polycaprolactone (PCL)	Electrospinning	Tissue engineering, grafts	Increased elasticity, fiber Orientation, decreased hydrophilicity	Sankaran et al. (2014)
Polycaprolactone (PCL)	Solvent cast film	Nerve conduits	PLA induced more decreased crystallinity and increased degradation than PCL, introduced surface porosity for cell attachment	Sun et al. (2010)
Polyglycolic acid (PGA)	Electrospinning	Soft tissue engineering	Similar mechanical properties to structural materials of soft tissues	Ramdhanie et al. (2006)
Poly(ethylene glycol) (PEG)	Nanoparticles/ microparticles	Drug delivery	Increased hydrophilicity, tuned degradation	Mainardes et al. (2010)
Poly(ethylene glycol) (PEG)	Foaming	Scaffolds	Decreased viscoelasticity, increased crystallinity and hydrophilicity	Chen et al. (2015)
Hydroxy-apatite	Melt blending/ compression molding	Bone grafts, wound healing, tissue engineering	PLA induced higher mechanical strength and elongation at break than HA alone	Wu and Liao (2005)
Lignin	Melt extrusion/ compressed film	Food packaging	Introduction of antioxidant and barrier properties by lignin	Domenek et al. (2013)

Reproduced with permission from Elsevier, Saini, P., Arora, M., & Kumar, M. N. V. R. (2016). Poly(lactic acid) blends in biomedical applications. *Advanced Drug Delivery Reviews*, *107*, 47–59. https://doi.org/10.1016/j.addr.2016.06.014.

Table 8.6 Properties of a PLLA and PLLA/PDLA blend (1:1) at different voltages.

Specimen	Applied voltage for spinning (kV)	Diameter (nm)	Glass transition temperature (°C)	Cold crystallization temperature (°C)
PLLA	0		45	82
	−12	1700−2700	55	83.5
	−25	1000−1800	58.1	83.7
Blend	0		49.8	82.9
	−12	830−1400	54.4	85.6
	−25	400−970	59.4	83.5

From Tsuji, H., Nakano, M., Hashimoto, M., Takashima, K., Katsura, S., & Mizuno, A. (2006). Electrospinning of poly(lactic acid) stereocomplex nanofibers. *Biomacromolecules*, 7(12), 3316−3320. https://doi.org/10.1021/bm060786e.

stereocomplex nanofibers of PLLA/PDLA (1:1) were prepared using chloroform as the solvent for electrospinning. In Table 8.6, properties of a PLLA and PLLA/PDLA blend (1:1) at different voltages are presented. The electrospun nanofibers with diameter ranges of 830−1400 nm and 400−970 nm were obtained at voltages of −12 and −25 kV, respectively. The SEM images of the fibers are presented in Fig. 8.7. The wide-angle X-ray-scattering studies revealed that with an increasing value of voltage from 0 to 25 kV, the crystallinity decreased from 5% to 1%, whereas the crystallinity of stereocomplex crystallites increased from 16% to 20%. The obtained results indicate that the electrospinning process is an efficient method to prepare nanofibers with a small amount of homocrystallites, and the orientation caused by the high voltage during the process improves the formation of stereocomplex crystallites and decreases the homocrystallite formation.

Rashedi et al. (2020) successfully prepared core−shell electrospun fibers of poly(-lactic acid)−gelatin (PLA-GT) by employing a coaxial electrospinning method with a new solvent combination. They employed dimethylformamide (DMF) as a solvent for PLA and employed concentrated acetic acid for gelatin. The confirmation of these core−shell structures is done by attenuated total reflectance Fourier transform spectroscopy (ATR-FTIR) and transmission electron microscopy (TEM). The morphology of these prepared spun nanofiber mats was observed by scanning electron microscopy (SEM) and, from the micrographs, it is clear that fibers have a bead-free ribbon shape for core−shell PLA-gelatin nanofibers with a mean fiber diameter of 347 ± 88 nm. From the differential scanning calorimetry (DSC) compositional analysis, the PLA composition in the prepared spun cross-shell fibers was low (7.8%), but the presence of these PLA was able to enhance the mechanical properties. Further, by using glutaraldehyde, the shell of the fibers was cross-linked. This cross-linking treatment improved the strength of nanofibers and preserved the nanofibrous and porous structure. The in vitro degradation test of these mats show that these fibrous structures retain the cross-linking even after 11 days. They concluded that these core−shell fibers have great potential in biomedical applications. TEM micrographs of prepared fibers are illustrated in Fig. 8.8.

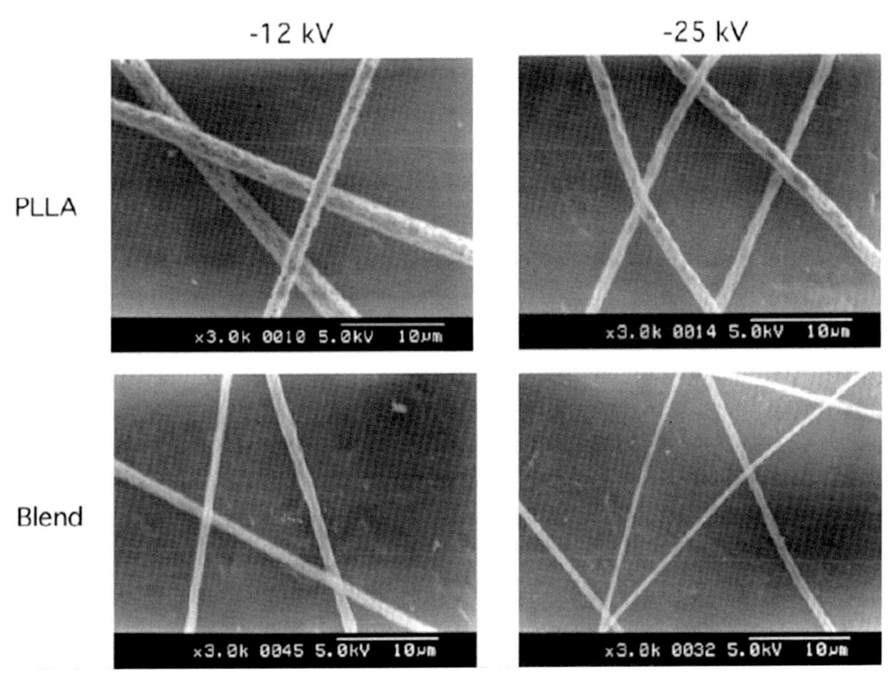

Figure 8.7 SEM micrographs of PLLA and PLLA blends.
From Tsuji, H., Nakano, M., Hashimoto, M., Takashima, K., Katsura, S., & Mizuno, A. (2006). Electrospinning of poly(lactic acid) stereocomplex nanofibers. *Biomacromolecules*, *7*(12), 3316–3320. https://doi.org/10.1021/bm060786e.

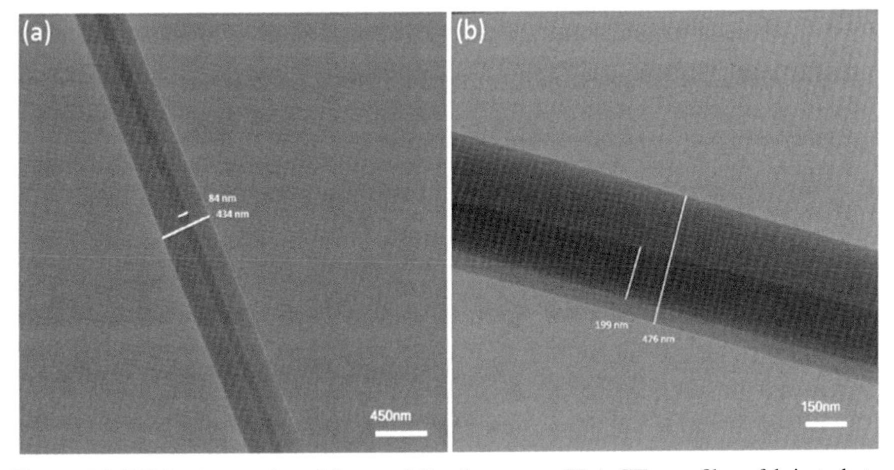

Figure 8.8 TEM micrographs of the coaxially electrospun PLA-GT nanofibers fabricated at core and shell flow rates of 0.4 mL/h and 1.3 mL/h, showing (A) concentric and (B) eccentric core–shell structures of nanofibers with sharp boundaries.
From Rashedi, S., Afshar, S., Rostami, A., Ghazalian, M., & Nazockdast, H. (2020). Co-electrospun poly (lactic acid)/gelatin nanofibrous scaffold prepared by a new solvent system: morphological, mechanical and in vitro degradability properties. *International Journal of Polymeric Materials and Polymeric Biomaterials*, 1–9.

Touny et al. (2010) prepared halloysite nanotube (HNT)-reinforced polylactic acid (PLA) electrospun fibers for biomedical applications. They investigated process parameters such as solvent type, concentration, feed rate, and HNT loading that influence the electrospun fibers and optimum quantities were reported. The prepared electrospun PLA/HNT composite fiber mats were tested using X-—ray diffraction techniques and scanning electron microscopy. The results display that the replacement of dimethylformamide by chloroform improved the electrospinnability as the solution electric conductivity and viscosity were improved. After the addition of HNT to the polymer, nanosized fibers were obtained and HNT loadings had a profound effect on the nanofiber morphology. Uniform fibers were obtained when the employed feed rates were between 1 and 4 mL/h.

Cheng et al. (2020) prepared a nanocomposite fiber system, where silver nanoparticles (Ag-NPs) were embedded uniformly in polylactic acid and gelatin fibers using a coelectrospinning technique and the fibers employed to improve the osseointegration and antiinfection ability. They optimized the composition of polylactic acid/gelatin to be 90/10 (mass ratio) and silver content to be 7%. It was observed that Ag nanoparticles played a role as heterogeneous nucleation centers for the surface growth of calcium phosphate. It is also seen that bone marrow-derived mesenchymal stem cells are attached to the surface of fibers and the existence of osteoblasts is confirmed by positive fluorescence staining. The vascular endothelial cells also had adherence on the fiber surface, displaying good angiogenic properties. These composite fibers have a >97% antibacterial rate against *Monilia albicans*, *Escherichia coli*, *Staphylococcus aureus*, and *Pseudomonas aeruginosa*, showcasing the high antibacterial performance. They concluded that these multifunctional biomacromolecule composite fibers are useful for advanced applications.

Threepopnatkul et al. (2010) studied the influence of polyethylene glycol (PEG) on the mechanical and antibacterial properties of polylactic acid (PLA) spun mats. They prepared spun PLA/PEG mats along with gentamicin sulfate (GS) by an electrospinning process. Polyethylene glycol was mixed with PLA at different w/w% and dissolved in a cosolvent system of dichloromethane and N,N-dimethylformamide (at 70:30 w/w). The quantity of GS in the polymer solution was fixed at 0.1 w.%. The test results signified that the increased PEG amount deteriorated the tensile modulus, while there was an increase in the tensile strength of PLA/PEG mats. The strain at break of these electrospun PLA/PEG mats was observed to increase with PEG loading. By employing an agar diffusion test, investigations into antibacterial *Staphylococcus aureus* (*S. aureus*) of electrospun PLA- and PLA/PEG-containing GS fiber mats were performed. It was found that GS can provide resistance or susceptibility to *S. aureus* on electrospun PLA and PLA/PEG mats.

Valente et al. (2016) experimented on developing polymer scaffolds for medically approved sterility methods. In their work, PLA fiber membranes were prepared by electrospinning with aligned and random fiber alignment and sterilized under ethylene oxide (EO), UV, and γ-radiation. They observed that both UV light and γ-radiation had no effect on fiber alignment or morphology, but the electrospun mats treated with ethylene oxide had fiber orientation loss. Fiber morphology was observed to change from cylindrical type to ribbon-type profiles, and an increase in polymer crystallinity of up to 28% was observed. UV light and γ-radiation methods were observed to be cause disturbance to morphology, without noteworthy changes in the mechanical and thermal properties, but a slight improvement in wettability was observed. In vitro studies indicate that both UV and γ-radiation treatments of PLA membranes allow the adhesion and proliferation of MG 63 osteoblastic cells in a close interaction with the fiber meshes and with a growth pattern that is highly sensitive to the underlying random or aligned fiber orientation. These results suggest γ-radiation-sterilized PLA membranes are potential candidates for clinical applications.

Liu and colleagues (Wenqiang et al., 2018) mixed PLA solutions with cellulose nanowhiskers and successfully prepared polylactic acid (PLA)/cellulose nanowhisker composite nanofibers by electrospinning. The microscopic observations using TEM reveal the homogeneous distribution of CNWs within the PLA nanofibers. The Fourier transform infrared spectroscopy (FTIR) spectra of composite nanofibers reveal characteristic hydroxyl groups of CNWs. The results also prove that hydrophilic CNWs improved the water absorption ability of nanofibers. With the increase in CNW content, the initial cold crystallization temperature decreases, denoting the nucleating agent role of CNWs. The degree in crystallinity is observed to be enhanced from 6.0% for pure electrospun PLA fibers to 14.1% and 21.6% for 5CNWs/PLA and 10CNWs/PLA composite fibers, respectively. They concluded that the loading of CNWs into the PLA matrix is expected to offer functionalities to prepared composite fibers in the area of tissue engineering.

Zhang, Ye et al. (2020) fabricated a silver (I) metal-organic framework Ag_2[HBTC][im] (1) and polylactic acid composite electrospun mat (1-PLA) with good antimicrobial ability. The resultant 1-PLA shows antibiosis potency against *Pseudomonas aeruginosa* (*P. aeruginosa*), *Escherichia coli* (*E. coli*), *Staphylococcus aureus* (*S. aureus*), and *Mycobacterium smegmatis* (*M. smegmatis*). In vivo studies indicate that the 1-PLA electrospun mat can remarkably increase the healing rate of wounds. This work demonstrated that the fibrous mat is promising for application in antiinfective therapy.

Peng et al. (2020) developed polylactic acid (PLA) fiber membranes loaded with hyperbranched PLA-modified cellulose nanocrystals (H-PLA-CNCs) by an electrospinning technique. The H-PLA-CNCs and the fiber membranes were characterized by Fourier transform infrared spectroscopy (FTIR), dynamic mechanical analysis (DMA), scanning electron microscopy (SEM), differential scanning calorimetry (DSC), and thermogravimetric analysis (TGA). The test results proved that the cellulose nanocrystals (CNCs) could be successfully improved by the hyperbranched PLA, because of CNCs/matrix interfacial adhesion. Both the shape memory and mechanical properties of PLA can be increased by adding H-PLA-CNCs. Notably, when 7 wt.% of H-PLA-CNCs was added, the tensile property and strain of composite membranes were 15.56 MPa and 25%, respectively. The increase in tensile strength and strain were 228% and 72.4% higher than those of virgin PLA. The shape recovery rate of the PLA/H-PLA-CNCs (5 wt.%) fiber membrane was 93%, which is higher than neat ones. They concluded that this study helps in the development of shape memory membranes with good mechanical properties.

Yan et al. (2020) prepared electrospun PLA membranes with silver nanoparticles (AgNPs) and cellulose nanofibrils (CNFs) for cell proliferation and antimicrobial application. In their work, AgNPs (below 0.1%) were successfully attached to the surface of PLA membranes by CNFs, which improved the membrane hydrophilicity and antibacterial ability. The in vitro studies indicated that the membranes had good biocompatibility with ocular epithelial cells. The antibacterial and antifungal activities were observed to be good. The prepared PLA membranes with CNFs and AgNPs are promising materials to promote cell proliferation and destroy pathogens, displaying potential in wound-healing applications. The schematic representation of this system is presented in Fig. 8.9.

Bi et al. (2020) compared the wound-healing ability of PLA membranes and PLA/PVA/SA membranes using in vitro and in vivo methods. They made PLA and PLA/PVA/SA membranes by an electrospinning process. In vitro studies indicated that

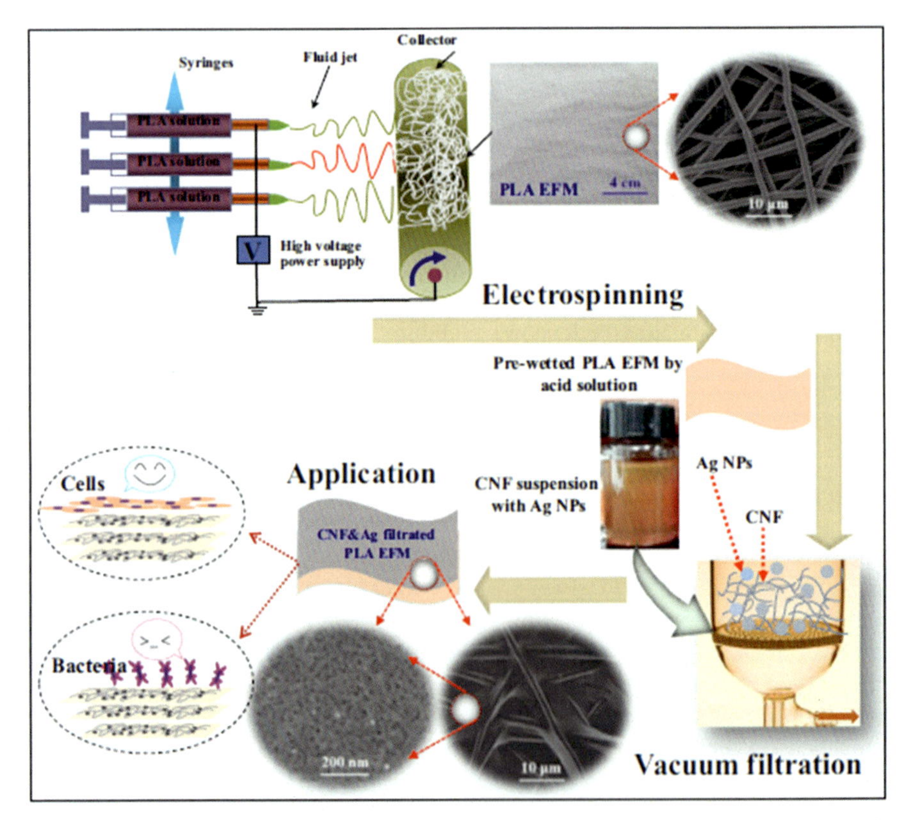

Figure 8.9 A schematic showing the fabrication of PLA EFMs with CNFs and Ag NPs used for cell proliferation and antibacterial application.
From Yan, D., Yao, Q., Yu, F., Chen, L., Zhang, S., Sun, H., Lin, J., & Fu, Y. (2020). Surface modified electrospun poly(lactic acid) fibrous scaffold with cellulose nanofibrils and Ag nanoparticles for ocular cell proliferation and antimicrobial application. *Materials Science and Engineering C, 111*. https://doi.org/10.1016/j.msec.2020.110767.

PLA and PLA/PVA/SA membranes could support the growth of rat fibroblasts (L929). The rat fibroblasts showed better adhesion on PLA/PVA/SA membranes than on PLA membranes. In vivo experiments showed that both membranes had good effects on all properties compared to the control group. The PLA membrane displayed higher collagen deposition than the PLA/PVA/SA membrane. Both fiber membranes showed similar angiogenesis and PLA/PVA/SA dressings, with notably decreased inflammatory responses during early wound healing. They concluded that this work illustrates the use of fiber membranes as potential wound-healing dressings.

Fan et al. (2020) prepared a poly(L-lactic acid) (PLLA) scaffold by loading tetrabutylammonium bromide (TBAB) to the PLLA. The influence of TBAB on the morphology and mechanical properties of PLLA scaffolds were studied. The results indicate that the incorporation of TBAB improved both the crystallinity and mechanical performance of PLLA. The infrared spectroscopy reveals that the chain orientation of the PLLA fibers increases with the TBAB amount and helps in the formation of the α/α' crystal phase structure, which is the major contributor to the significant improvement of the mechanical performance of the scaffold.

Ashraf et al. (2020) prepared polyaniline nanoparticles (PANI NPs) in emeraldine base (EB) form. These PANI NPs were incorporated into polylactic acid nanofibers by using an electrospinning technique. The incorporation of nanoparticles in PLA resulted in a decrease in the fiber diameter and the surface of fibers appeared to be smooth. There is an improvement in the mechanical and electrical properties of prepared PLA/PANI fibers and wettability tests showed that prepared fibers were less hydrophobic. The effect of irradiation on the physical properties of the nanofibers was studied by using gamma rays at 25 kGy and irradiated and nonirradiated fibers were compared. It was concluded that gamma irradiation samples have better properties such as tensile strength, wettability, electrical conductivity, and tensile strength because of the improved interactions between nanoparticles and PLA.

Lopresti et al. (2020) studied the reinforcing capability of both micrometric and nanometric hydroxyapatite particles in n electrospun PLA scaffold using aligned and random orientation of fibers. The incorporation or presence of particles were investigated and confirmed with the help of Fourier transform infrared spectroscopy. The impact of hydroxyapatite particles on the crystallinity was tested by differential scanning calorimetry. Finally, cell culture studies with preosteoblastic cells were done and compared with those for polylactic acid scaffolds. From the test results, we can conclude that hydroxyapatite/polylactic acid composites are promising materials for bone tissue regeneration. In Fig. 8.10, the variation of dimensionless parameters with respect to filler concentration is provided. A list of some electrospun fibers and applications is presented in Table 8.7.

5. Conclusion and future perspectives

In this chapter, a brief introduction to biodegradable polymers was presented along with a general idea for developing PLA-based functional composites. Owing to their

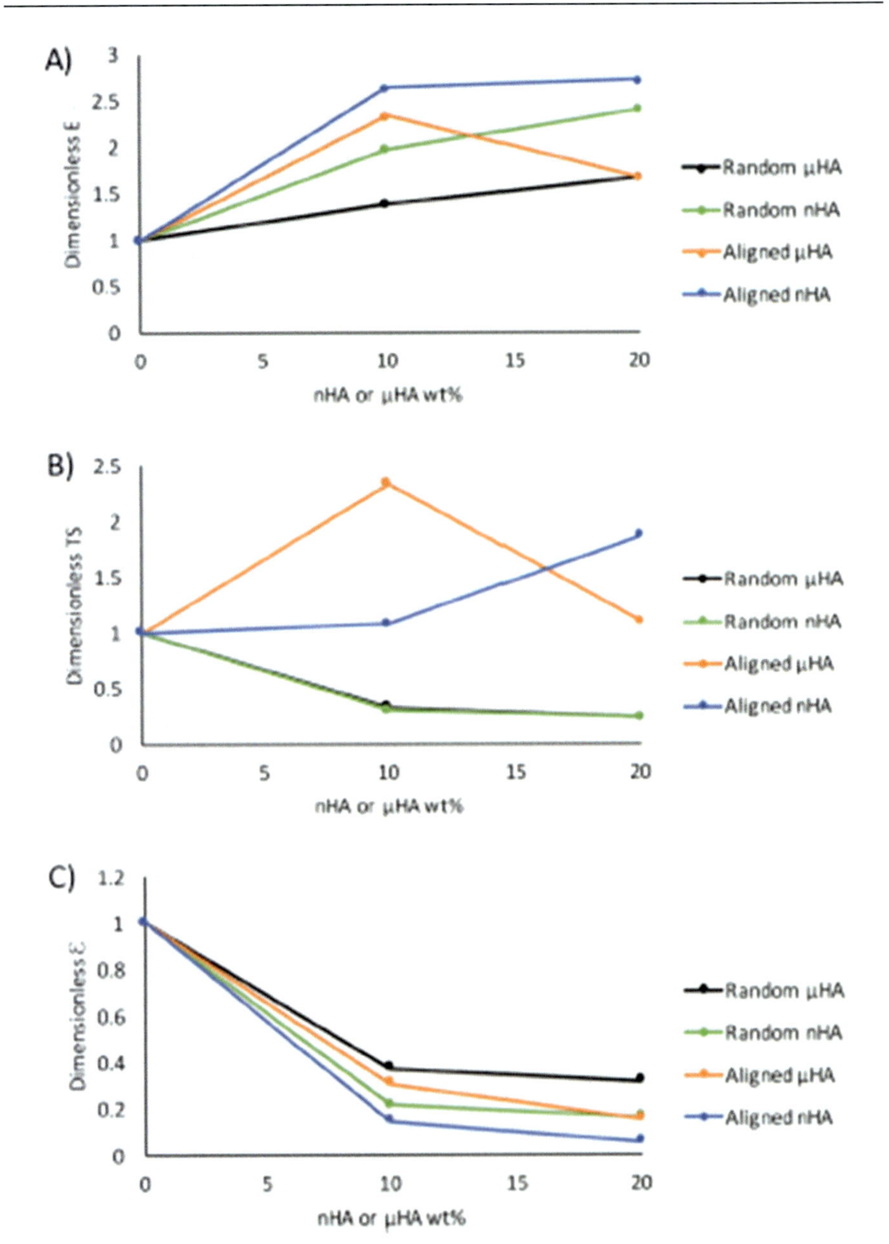

Figure 8.10 Dimensionless (A) elastic modulus, (B) tensile strength, and (C) elongation of electrospun PLA and nanocomposite as a function of filler concentration.
From Lopresti, F., Carfì Pavia, F., Vitrano, I., Kersaudy-Kerhoas, M., Brucato, V., & La Carrubba, V. (2020). Effect of hydroxyapatite concentration and size on morpho-mechanical properties of PLA-based randomly oriented and aligned electrospun nanofibrous mats. *Journal of the Mechanical Behavior of Biomedical Materials, 101*. https://doi.org/10.1016/j.jmbbm.2019.103449.

Table 8.7 List of some electrospun PLA fibers and applications.

S. no.	Material	Solvent	Application	References
1	Poly(lactic acid)	–	Structure–property	Gómez-Pachón et al. (2014)
2	Poly(lactic acid)	Chloroform/acetone (3/1)	Substrates for biosensors	Li et al. (2006)
3	Poly-D,L-lactide fibers	Chloroform and acetone	Structure	Gu and Ren (2005)
4	Poly(lactic acid)/polyethylene glycol	DCM:DMF	Antibacterial activity	Threepopnatkul et al. (2010)
5	Poly L-lactide	DCM/DMF	Structure	Yang et al. (2004)
6	poly(lactic acid)	Different solvents	Structure	Casasola et al. (2014)
7	Polylactic acid and graphene oxide		Structure–property	Davoodi et al. (2018)
8	Poly L-lactide	Hexafluoroisopropanol	Stem cells	Prabhakaran et al. (2011)
9	Poly L-lactide	Hexafluoroisopropanol	Tissue response	Ishii et al. (2008)
10	Poly(lactic acid) with multiwalled carbon nanotubes	Different systems	Structure–property	McCullen et al. (2007)
11	Carbon nanotubes/poly(lactic acid)	Dichloromethane/ dimethylformamide	Structure–property	Magiera et al. (2018)
12	Curcumin-incorporated poly(lactic acid) composite fibers	–	Drug-eluting stent	Chen et al. (2010)
13	Poly(lactic acid)	Dichloromethane	Scaffold in drug delivery	Santos et al. (2013)
14	Poly(lactic acid)/poly(glycidyl methacrylate)/fluorescein	Methylene chloride/N,N-dimethylformamide	Fluorescence	Wei et al. (2013)

biodegradable nature, these materials have been studied on a large scale. PLA can be used for the fabrication of both blends and composites with advanced functional applications, more interestingly it can be utilized as a filler/reinforcing agent as well as a matrix material. The ability of PLA to form micro- or nanofibrous structures has attracted many researchers and active research in this area could lead to the fabrication of functional composite materials. However, research in the development of PLA nanofibers and their composites is still in its infancy and has a long way to go. The outstanding biomedical and structural applications of PLA nanofiber-based composites along with their biodegradable nature could contribute greatly toward the development of sustainable materials. PLA and PLA-based materials are used for medical implants in various forms taking advantage of their degradability. PLA-based materials are also used as packaging materials for food, as cups, disposable tableware, tea bags, etc. PLA-based conductive materials are used for microcoils, 3D printing, etc. Electrospun PLA materials are used for biomedical and drug-delivery applications. These electrospun nonwovens are used in fluorescence, as drug-eluting agents, and as tissue scaffolds. Thus it is anticipated that there is a vast prospect in this excellent material and a wide possibility for PLA materials to capture a sustainable market in the near future.

References

Ahmed, J., Arfat, Y. A., Bher, A., Mulla, M., Jacob, H., & Auras, R. (2018). Active chicken meat packaging based on polylactide films and bimetallic Ag–Cu nanoparticles and essential oil. *Journal of Food Science, 83*(5), 1299–1310. https://doi.org/10.1111/1750-3841.14121

Alashwal, B. Y., Bala, M. S., Gupta, A., Sharma, S., & Mishra, P. (2020). Improved properties of keratin-based bioplastic film blended with microcrystalline cellulose: a comparative analysis. *Journal of King Saud University Science, 32*(1), 853–857.

Thermal stability of melt-blended poly (lactic acid)(PLA)/polyamide 66 (PA66)/graphene nanoplatelets (GnP). In Alhadadi, W., Almaqtari, A., Hafidzah, F., Bijarimi, M., Desa, M., Merzah, H., et al. (Eds.), *IOP conference series: materials science and engineering*, (2019). IOP Publishing.

Anderson, K. S., & Hillmyer, M. A. (2004). The influence of block copolymer microstructure on the toughness of compatibilized polylactide/polyethylene blends. *Polymer, 45*(26), 8809–8823.

Anwer, M. A., Naguib, H. E., Celzard, A., & Fierro, V. (2015). Comparison of the thermal, dynamic mechanical and morphological properties of PLA-Lignin & PLA-Tannin particulate green composites. *Composites Part B: Engineering, 82*, 92–99.

Armentano, I., Fortunati, E., Burgos, N., Dominici, F., Luzi, F., Fiori, S., et al. (2015). *Processing and characterization of plasticized PLA/PHB blends for biodegradable multiphase systems*.

Arrieta, M. P., Castro-López, M. M., Rayón, E., Barral-Losada, L. F., López-Vilariño, J. M., López, J., et al. (2014). Plasticized poly (lactic acid)–poly (hydroxybutyrate)(PLA–PHB) blends incorporated with catechin intended for active food-packaging applications. *Journal of Agricultural and Food Chemistry, 62*(41), 10170–10180.

Arrieta, M. P., López, J., Hernández, A., & Rayón, E. (2014). Ternary PLA—PHB—limonene blends intended for biodegradable food packaging applications. *European Polymer Journal, 50*, 255—270.

Arrieta, M. P., López, J., Rayón, E., & Jiménez, A. (2014). Disintegrability under composting conditions of plasticized PLA—PHB blends. *Polymer Degradation and Stability, 108*, 307—318.

Ashraf, S. S., Frounchi, M., & Dadbin, S. (2020). Gamma irradiated electro-conductive polylactic acid/polyaniline nanofibers. *Synthetic Metals, 259*. https://doi.org/10.1016/j.synthmet.2019.116204

Auras, R., Harte, B., & Selke, S. (2004). An overview of polylactides as packaging materials. *Macromolecular Bioscience, 4*(9), 835—864. https://doi.org/10.1002/mabi.200400043

Avérous, L., & Pollet, E. (2012). Biodegradable polymers. *Green Energy and Technology, 50*, 13—39. https://doi.org/10.1007/978-1-4471-4108-2_2

Bajpai, P. K., Singh, I., & Madaan, J. (2013). Tribological behavior of natural fiber reinforced PLA composites. *Wear, 297*(1—2), 829—840.

Bax, B., & Müssig, J. (2008). Impact and tensile properties of PLA/Cordenka and PLA/flax composites. *Composites Science and Technology, 68*(7—8), 1601—1607.

Bhardwaj, N., & Kundu, S. C. (2010). Electrospinning: a fascinating fiber fabrication technique. *Biotechnology Advances, 28*(3), 325 347. https://doi.org/10.1016/j.biotechadv.2010.01.004

Bhatia, A., Gupta, R. K., Bhattacharya, S. N., & Choi, H. (2007). Compatibility of biodegradable poly (lactic acid)(PLA) and poly (butylene succinate)(PBS) blends for packaging application. *Korea-Australia Rheology Journal, 19*(3), 125—131.

Bi, H., Feng, T., Li, B., & Han, Y. (2020). In vitro and in vivo comparison study of electrospun PLA and PLA/PVA/SA fiber membranes for wound healing. *Polymers, 12*(4). https://doi.org/10.3390/POLYM12040839

Biresaw, G., & Carriere, C. (2002). Interfacial tension of poly (lactic acid)/polystyrene blends. *Journal of Polymer Science Part B: Polymer Physics, 40*(19), 2248—2258.

Casasola, R., Thomas, N. L., Trybala, A., & Georgiadou, S. (2014). Electrospun poly lactic acid (PLA) fibres: effect of different solvent systems on fibre morphology and diameter. *Polymer, 55*(18), 4728—4737.

Chaikeaw, C., & Srikulkit, K. (2018). Preparation and properties of poly (lactic acid)/PLA-g-ABS blends. *Fibers and Polymers, 19*(10), 2016—2022.

Chang, B. P., Mohanty, A. K., & Misra, M. (2018). Tuning the compatibility to achieve toughened biobased poly (lactic acid)/poly (butylene terephthalate) blends. *RSC Advances, 8*(49), 27709—27724.

Chen, B., & Evans, J. R. G. (2005). Thermoplastic starch-clay nanocomposites and their characteristics. *Carbohydrate Polymers, 61*(4), 455—463. https://doi.org/10.1016/j.carbpol.2005.06.020

Chen, Y., Lin, J., Fei, Y., Wang, H., & Gao, W. (2010). Preparation and characterization of electrospinning PLA/curcumin composite membranes. *Fibers and Polymers, 11*(8), 1128—1131.

Chen, B. Y., Jing, X., Mi, H. Y., Zhao, H., Zhang, W. H., Peng, X. F., et al. (2015). Fabrication of polylactic acid/polyethylene glycol (PLA/PEG) porous scaffold by supercritical CO_2 foaming and particle leaching. *Polymer Engineering and Science, 55*(6), 1339—1348.

Cheng, S., Lau, K-t, Liu, T., Zhao, Y., Lam, P.-M., & Yin, Y. (2009). Mechanical and thermal properties of chicken feather fiber/PLA green composites. *Composites Part B: Engineering, 40*(7), 650—654.

Cheng, X., Wei, Q., Ma, Y., Shi, R., Chen, T., Wang, Y., Ma, C., & Lu, Y. (2020). Antibacterial and osteoinductive biomacromolecules composite electrospun fiber.

International Journal of Biological Macromolecules, 143, 958–967. https://doi.org/10.1016/j.ijbiomac.2019.09.156

Choudhary, P., Mohanty, S., Nayak, S. K., & Unnikrishnan, L. (2011). Poly (L-lactide)/polypropylene blends: evaluation of mechanical, thermal, and morphological characteristics. *Journal of Applied Polymer Science, 121*(6), 3223–3237.

Cicala, G., Saccullo, G., Blanco, I., Samal, S., Battiato, S., Dattilo, S., et al. (2017). Polylactide/lignin blends. *Journal of Thermal Analysis and Calorimetry, 130*(1), 515–524.

Claro, P., Neto, A., Bibbo, A., Mattoso, L., Bastos, M., & Marconcini, J. (2016). Biodegradable blends with potential use in packaging: a comparison of PLA/chitosan and PLA/cellulose acetate films. *Journal of Polymers and the Environment, 24*(4), 363–371.

Coltelli, M. B., Bronco, S., & Chinea, C. (2010). The effect of free radical reactions on structure and properties of poly(lactic acid) (PLA) based blends. *Polymer Degradation and Stability, 95*(3), 332–341. https://doi.org/10.1016/j.polymdegradstab.2009.11.015

Coltelli, M. B., Mallegni, N., Rizzo, S., Cinelli, P., & Lazzeri, A. (2019). Improved impact properties in poly(lactic acid) (PLA) blends containing cellulose acetate (CA) prepared by reactive extrusion. *Materials, 12*(2). https://doi.org/10.3390/ma12020270

Cossement, D., Gouttebaron, R., Cornet, V., Viville, P., Hecq, M., & Lazzaroni, R. (2006). PLA-PMMA blends: a study by XPS and ToF-SIMS. *Applied Surface Science, 252*(19), 6636–6639.

Dai, X., Xiong, Z., Ma, S., Li, C., Wang, J., Na, H., et al. (2015). Fabricating highly reactive bio-based compatibilizers of epoxidized citric acid to improve the flexural properties of polylactide/microcrystalline cellulose blends. *Industrial & Engineering Chemistry Research, 54*(15), 3806–3812.

Davoodi, A. H., Mazinani, S., Sharif, F., & Ranaei-Siadat, S. O. (2018). GO nanosheets localization by morphological study on PLA-GO electrospun nanocomposite nanofibers. *Journal of Polymer Research, 25*(9), 204.

Deeraj, B. D. S., Saritha, A., & Joseph, K. (2019). Electrospun styrene-butadiene copolymer fibers as potential reinforcement in epoxy composites: modeling of rheological and visco elastic data. *Composites Part B: Engineering, 160*, 384–393. https://doi.org/10.1016/j.compositesb.2018.12.102

Deeraj, B. D. S., Harikrishnan, R., Jayan, J. S., Saritha, A., & Joseph, K. (2020). Enhanced visco-elastic and rheological behavior of epoxy composites reinforced with polyimide nanofiber. *Nano-Structures and Nano-Objects, 21*. https://doi.org/10.1016/j.nanoso.2019.100421

Deeraj, B. D. S., Jayan, J. S., Saritha, A., & Joseph, K. (2021). Electrospun biopolymer-based hybrid composites. In *Hybrid Natural Fiber Composites* (pp. 225–252). Woodhead Publishing.

Díez-Pascual, A. M., & Díez-Vicente, A. L. (2014). ZnO-reinforced poly(3-hydroxybutyrate-co-3-hydroxyvalerate) bionanocomposites with antimicrobial function for food packaging. *ACS Applied Materials and Interfaces, 6*(12), 9822–9834. https://doi.org/10.1021/am502261e

Domenek, S., Louaifi, A., Guinault, A., & Baumberger, S. (2013). Potential of lignins as antioxidant additive in active biodegradable packaging materials. *Journal of Polymers and the Environment, 21*(3), 692–701.

Domínguez-Robles, J., Martin, N. K., Fong, M. L., Stewart, S. A., Irwin, N. J., Rial-Hermida, M. I., Donnelly, R. F., & Larrañeta, E. (2019). Antioxidant PLA composites containing lignin for 3D printing applications: a potential material for healthcare applications. *Pharmaceutics, 11*(4). https://doi.org/10.3390/pharmaceutics11040165

Eslami, H., & Kamal, M. R. (2013). Effect of a chain extender on the rheological and mechanical properties of biodegradable poly (lactic acid)/poly [(butylene succinate)-co-adipate] blends. *Journal of Applied Polymer Science, 129*(5), 2418−2428.

Fan, C. J., Sun, T. S., Luo, J. Q., Liu, H. Y., & Zhou, X. D. (2020). Effect of introduction of tetrabutylammonium bromide on properties of poly (L-lactic acid) tubular scaffold prepared by electrospinning. *Micro and Nano Letters, 15*(5), 277−282. https://doi.org/10.1049/mnl.2019.0544

Feng, F., & Ye, L. (2010). Structure and property of polylactide/polyamide blends. *Journal of Macromolecular Science®, Part B: Physics, 49*(6), 1117−1127.

Feng, L., Bian, X., Cui, Y., Chen, Z., Li, G., & Chen, X. (2013). Flexibility improvement of poly (l-lactide) by reactive blending with poly (ether urethane) containing poly (ethylene glycol) blocks. *Macromolecular Chemistry and Physics, 214*(7), 824−834.

Ferreira, R. T. L., Amatte, I. C., Dutra, T. A., & Bürger, D. (2017). Experimental characterization and micrography of 3D printed PLA and PLA reinforced with short carbon fibers. *Composites Part B: Engineering, 124*, 88−100. https://doi.org/10.1016/j.compositesb.2017.05.013

Fortelny, I., Ujčić, A., Fambri, L., & Slouf, M. (2019). Phase structure, compatibility and toughness of PLA/PCL blends: a review. *Frontiers in Materials, 6.*

Gajria, A. M., Dave, V., Gross, R. A., & McCarthy, S. P. (1996). Miscibility and biodegradability of blends of poly (lactic acid) and poly (vinyl acetate). *Polymer, 37*(3), 437−444.

Gardella, L., Calabrese, M., & Monticelli, O. (2014). PLA maleation: an easy and effective method to modify the properties of PLA/PCL immiscible blends. *Colloid & Polymer Science, 292*(9), 2391−2398.

Gérard, T., & Budtova, T. (Eds.). (2011). *Preparation and characterization of polyhydroxyalkanoates (PHA) and polylactide (PLA) blends.*

Gerard, T., & Budtova, T. (2012). Morphology and molten-state rheology of polylactide and polyhydroxyalkanoate blends. *European Polymer Journal, 48*(6), 1110−1117.

Gere, D., & Czigany, T. (2020). Future trends of plastic bottle recycling: compatibilization of PET and PLA. *Polymer Testing, 81*, 106160.

Gigante, V., Canesi, I., Cinelli, P., Coltelli, M. B., & Lazzeri, A. (2019). Rubber toughening of polylactic acid (PLA) with Poly (butylene adipate-co-terephthalate)(PBAT): mechanical properties, fracture mechanics and analysis of ductile-to-brittle behavior while varying temperature and test speed. *European Polymer Journal, 115*, 125−137.

Gigante, V., Coltelli, M.-B., Vannozzi, A., Panariello, L., Fusco, A., Trombi, L., et al. (2019). Flat die extruded biocompatible poly (lactic acid)(PLA)/poly (butylene succinate)(PBS) based films. *Polymers, 11*(11), 1857.

Structure of poly (lactic-acid) PLA nanofibers scaffolds prepared by electrospinning. In Gómez-Pachón, E., Vera-Graziano, R., & Campos, R. M. (Eds.), *IOP conference series: materials science and engineering*, (2014). IOP Publishing.

Gonzalez-Garzon, M., Shahbikian, S., & Huneault, M. A. (2018). Properties and phase structure of melt-processed PLA/PMMA blends. *Journal of Polymer Research, 25*(2), 1−13.

Gordobil, O., Egüés, I., Llano-Ponte, R., & Labidi, J. (2014). Physicochemical properties of PLA lignin blends. *Polymer Degradation and Stability, 108*, 330−338.

Gu, S. Y., & Ren, J. (2005). Process optimization and empirical modeling for electrospun poly (D, L-lactide) fibers using response surface methodology. *Macromolecular Materials and Engineering, 290*(11), 1097−1105.

Gu, S.-Y., Zhang, K., Ren, J., & Zhan, H. (2008). Melt rheology of polylactide/poly (butylene adipate-co-terephthalate) blends. *Carbohydrate Polymers, 74*(1), 79−85.

Gui, Z-y, Wang, H-r, Gao, Y., Lu, C., & Cheng, S-j (2012). Morphology and melt rheology of biodegradable poly (lactic acid)/poly (butylene succinate adipate) blends: effect of blend compositions. *Iranian Polymer Journal (English Edition), 21*(2), 81–89.

Gui, Z., Zhang, W., Lu, C., & Cheng, S. (2013). Improving the barrier properties of poly (lactic acid) by blending with poly (ethylene-co-vinyl alcohol). *Journal of Macromolecular Science, Part B., 52*(5), 685–700.

Guo, J., Chen, X., Wang, J., He, Y., Xie, H., & Zheng, Q. (2020). The influence of compatibility on the structure and properties of PLA/lignin biocomposites by chemical modification. *Polymers, 12*(1), 56.

Hamad, K., Kaseem, M., & Deri, F. (2010). Rheological and mechanical properties of poly (lactic acid)/polystyrene polymer blend. *Polymer Bulletin, 65*(5), 509–519.

Hamad, K., Kaseem, M., & Deri, F. (2011a). Effect of recycling on rheological and mechanical properties of poly (lactic acid)/polystyrene polymer blend. *Journal of Materials Science, 46*(9), 3013–3019.

Hamad, K., Kaseem, M., & Deri, F. (2011b). Rheological and mechanical characterization of poly (lactic acid)/polypropylene polymer blends. *Journal of Polymer Research, 18*(6), 1799–1806.

Han, L., Han, C., Zhang, H., Chen, S., & Dong, L. (2012). Morphology and properties of biodegradable and biosourced polylactide blends with poly (3-hydroxybutyrate-co-4-hydroxybutyrate). *Polymer Composites, 33*(6), 850–859.

Haneef, I., Buys, Y., Shaffiar, N., Shaharuddin, S., & Khairusshima, M. N. (2019). Miscibility, mechanical, and thermal properties of polylactic acid/polypropylene carbonate (PLA/PPC) blends prepared by melt-mixing method. *Materials Today: Proceedings, 17*, 534–542.

Hao, X., Kaschta, J., Liu, X., Pan, Y., & Schubert, D. W. (2015). Entanglement network formed in miscible PLA/PMMA blends and its role in rheological and thermo-mechanical properties of the blends. *Polymer, 80*, 38–45.

Huda, M. S., Drzal, L. T., Mohanty, A. K., & Misra, M. (2006). Chopped glass and recycled newspaper as reinforcement fibers in injection molded poly (lactic acid)(PLA) composites: a comparative study. *Composites Science and Technology, 66*(11–12), 1813–1824.

Imre, B., Bedő, D., Domján, A., Schön, P., Vancso, G. J., & Pukánszky, B. (2013). Structure, properties and interfacial interactions in poly (lactic acid)/polyurethane blends prepared by reactive processing. *European Polymer Journal, 49*(10), 3104–3113.

Irska, I., Paszkiewicz, S., Gorący, K., Linares, A., Ezquerra, T., Jedrzejewski, R., et al. (2020). Poly (butylene terephthalate)/polylactic acid based copolyesters and blends: miscibility-structure-property relationship. *Express Polymer Letters, 14*(1), 26–47.

Ishii, D., Ying, T. H., Mahara, A., Murakami, S., Yamaoka, T., Lee, W-k, et al. (2008). In vivo tissue response and degradation behavior of PLLA and stereocomplexed PLA nanofibers. *Biomacromolecules, 10*(2), 237–242.

Jacobsen, S., & Fritz, H. (1996). Filling of poly (lactic acid) with native starch. *Polymer Engineering & Science, 36*(22), 2799–2804.

Jiang, L., Wolcott, M. P., & Zhang, J. (2006). Study of biodegradable polylactide/poly (butylene adipate-co-terephthalate) blends. *Biomacromolecules, 7*(1), 199–207.

Jun, C. L. (2000). Reactive blending of biodegradable polymers: PLA and starch. *Journal of Polymers and the Environment, 8*(1), 33–37.

Kain, S., Ecker, J., Haider, A., Musso, M., & Petutschnigg, A. (2020). Effects of the infill pattern on mechanical properties of fused layer modeling (FLM) 3D printed wood/polylactic acid (PLA) composites. *European Journal of Wood and Wood Products, 78*(1), 65–74.

Kaseem, M., & Ko, Y. G. (2017). Melt flow behavior and processability of polylactic acid/polystyrene (PLA/PS) polymer blends. *Journal of Polymers and the Environment, 25*(4), 994–998.

Ke, T., & Sun, X. (2003). Melting behavior and crystallization kinetics of starch and poly (lactic acid) composites. *Journal of Applied Polymer Science, 89*(5), 1203–1210.

Ke, T., Sun, S. X., & Seib, P. (2003). Blending of poly (lactic acid) and starches containing varying amylose content. *Journal of Applied Polymer Science, 89*(13), 3639–3646.

Kim, S., Oh, S., Lee, J., Ahn, N., Roh, H., Cho, J., et al. (2014). Effect of alkyl-chain-modified lignin in the PLA matrix. *Fibers and Polymers, 15*(12), 2458–2465.

Kim, Y., Suhr, J., Seo, H.-W., Sun, H., Kim, S., Park, I.-K., et al. (2017). All biomass and UV protective composite composed of compatibilized lignin and poly (lactic-acid). *Scientific Reports, 7*, 43596.

Kimble, L., Bhattacharyya, D., & Fakirov, S. (2015). Biodegradable microfibrillar polymer-polymer composites from poly (L-lactic acid)/poly (glycolic acid). *Express Polymer Letters, 9*(3).

Poly (L-lactic acid)/poly (glycolic acid) microfibrillar polymer-polymer composites: Preparation and viscoelastic properties. In Kimble, L., Fakirov, S., & Bhattacharyya, D. (Eds.), *AIP conference proceedings*, (2015). AIP Publishing LLC.

Koh, J. J., Zhang, X., & He, C. (2018). Fully biodegradable Poly(lactic acid)/Starch blends: a review of toughening strategies. *International Journal of Biological Macromolecules, 109*, 99–113. https://doi.org/10.1016/j.ijbiomac.2017.12.048

Kothapalli, C. R., Shaw, M. T., & Wei, M. (2005). Biodegradable HA-PLA 3-D porous scaffolds: effect of nano-sized filler content on scaffold properties. *Acta Biomaterialia, 1*(6), 653–662. https://doi.org/10.1016/j.actbio.2005.06.005

Kowalczyk, M., Piorkowska, E., Kulpinski, P., & Pracella, M. (2011). Mechanical and thermal properties of PLA composites with cellulose nanofibers and standard size fibers. *Composites Part A: Applied Science and Manufacturing, 42*(10), 1509–1514.

Kultravut, K., Kuboyama, K., & Ougizawa, T. (2019). Effect of blending procedure on tensile and degradation properties of toughened biodegradable poly (lactic acid) blend with poly (trimethylene terephthalate) and reactive compatibilizer. *Macromolecular Materials and Engineering, 304*(11), 1900323.

Kultravut, K., Kuboyama, K., & Ougizawa, T. (2020). Annealing effect on tensile property and hydrolytic degradation of biodegradable poly (lactic acid) reactive blend with poly (trimethylene terephthalate) by two-step blending procedure. *Polymer Degradation and Stability, 179*, 109228.

Kumar, A., Tumu, V. R., Chowdhury, S. R., & SVS, R. R. (2019). A green physical approach to compatibilize a bio-based poly (lactic acid)/lignin blend for better mechanical, thermal and degradation properties. *International Journal of Biological Macromolecules, 121*, 588–600.

Li, Y., & Shimizu, H. (2009). Improvement in toughness of poly (l-lactide)(PLLA) through reactive blending with acrylonitrile–butadiene–styrene copolymer (ABS): morphology and properties. *European Polymer Journal, 45*(3), 738–746.

Li, D., Frey, M. W., & Baeumner, A. J. (2006). Electrospun polylactic acid nanofiber membranes as substrates for biosensor assemblies. *Journal of Membrane Science, 279*(1–2), 354–363.

Li, H., Lu, X., Yang, H., & Hu, J. (2015). Non-isothermal crystallization of P (3HB-co-4HB)/PLA blends. *Journal of Thermal Analysis and Calorimetry, 122*(2), 817–829.

Li, X., Hegyesi, N., Zhang, Y., Mao, Z., Feng, X., Wang, B., et al. (2019). Poly (lactic acid)/lignin blends prepared with the Pickering emulsion template method. *European Polymer Journal, 110*, 378–384.

Liao, S. S., Cui, F. Z., Zhang, W., & Feng, Q. L. (2004). Hierarchically biomimetic bone scaffold materials: nano-HA/collagen/PLA composite. *Journal of Biomedical Materials Research Part B: Applied Biomaterials, 69*(2), 158–165. https://doi.org/10.1002/jbm.b.20035

Lin, L., Deng, C., Lin, G.-P., & Wang, Y.-Z. (2015). Super toughened and high heat-resistant poly (lactic acid)(PLA)-based blends by enhancing interfacial bonding and PLA phase crystallization. *Industrial & Engineering Chemistry Research, 54*(21), 5643–5655.

Longano, D., Ditaranto, N., Cioffi, N., Di Niso, F., Sibillano, T., Ancona, A., Conte, A., Del Nobile, M. A., Sabbatini, L., & Torsi, L. (2012). Analytical characterization of laser-generated copper nanoparticles for antibacterial composite food packaging. *Analytical and Bioanalytical Chemistry, 403*(4), 1179–1186. https://doi.org/10.1007/s00216-011-5689-5

Lopresti, F., Carfì Pavia, F., Vitrano, I., Kersaudy-Kerhoas, M., Brucato, V., & La Carrubba, V. (2020). Effect of hydroxyapatite concentration and size on morpho-mechanical properties of PLA-based randomly oriented and aligned electrospun nanofibrous mats. *Journal of the Mechanical Behavior of Biomedical Materials, 101*. https://doi.org/10.1016/j.jmbbm.2019.103449

Lu, X., Wei, X., Huang, J., Yang, L., Zhang, G., He, G., et al. (2014). Supertoughened poly (lactic acid)/polyurethane blend material by in situ reactive interfacial compatibilization via dynamic vulcanization. *Industrial & Engineering Chemistry Research, 53*(44), 17386–17393.

Lu, X., Zhao, J., Yang, X., & Xiao, P. (2017). Morphology and properties of biodegradable poly (lactic acid)/poly (butylene adipate-co-terephthalate) blends with different viscosity ratio. *Polymer Testing, 60*, 58–67.

Luyt, A., & Gasmi, S. (2016). Influence of blending and blend morphology on the thermal properties and crystallization behaviour of PLA and PCL in PLA/PCL blends. *Journal of Materials Science, 51*(9), 4670–4681.

Ma, X., Yu, J., & Wang, N. (2006). Compatibility characterization of poly (lactic acid)/poly (propylene carbonate) blends. *Journal of Polymer Science Part B: Polymer Physics, 44*(1), 94–101.

Magiera, A., Markowski, J., Pilch, J., & Blazewicz, S. (2018). Degradation behavior of electrospun PLA and PLA/CNT nanofibres in aqueous environment. *Journal of Nanomaterials, 2018*.

Mahalik, J., & Madras, G. (2006). Enzymatic degradation of poly (D, L-lactide) and its blends with poly (vinyl acetate). *Journal of Applied Polymer Science, 101*(1), 675–680.

Mainardes, R. M., Khalil, N. M., & Gremião, M. P. D. (2010). Intranasal delivery of zidovudine by PLA and PLA–PEG blend nanoparticles. *International Journal of Pharmaceutics, 395*(1–2), 266–271.

Matta, A. K., Umamaheswara, R. R., Suman, K. N. S., & Rambabu, V. (2014). Preparation and characterization of biodegradable PLA/PCL polymeric blends. *Procedia Materials Science*, 1266–1270. https://doi.org/10.1016/j.mspro.2014.07.201

Matta, A., Rao, R. U., Suman, K., & Rambabu, V. (2014). Preparation and characterization of biodegradable PLA/PCL polymeric blends. *Procedia Materials Science, 6*, 1266–1270.

McCullen, S. D., Stano, K. L., Stevens, D. R., Roberts, W. A., Monteiro-Riviere, N. A., Clarke, L. I., et al. (2007). Development, optimization, and characterization of electrospun poly (lactic acid) nanofibers containing multi-walled carbon nanotubes. *Journal of Applied Polymer Science, 105*(3), 1668–1678.

McLauchlin, A. R., & Ghita, O. R. (2016). Studies on the thermal and mechanical behavior of PLA-PET blends. *Journal of Applied Polymer Science, 133*(43).

Meng, X., Bocharova, V., Tekinalp, H., Cheng, S., Kisliuk, A., Sokolov, A. P., Kunc, V., Peter, W. H., & Ozcan, S. (2018). Toughening of nanocelluose/PLA composites via bio-epoxy interaction: mechanistic study. *Materials and Design, 139*, 188–197. https://doi.org/10.1016/j.matdes.2017.11.012

Mi, H.-Y., Salick, M. R., Jing, X., Jacques, B. R., Crone, W. C., Peng, X.-F., et al. (2013). Characterization of thermoplastic polyurethane/polylactic acid (TPU/PLA) tissue engineering scaffolds fabricated by microcellular injection molding. *Materials Science and Engineering: C, 33*(8), 4767–4776.

Mochizuki, M., & Hirami, M. (1997). Structural effects on the biodegradation of aliphatic polyesters. *Polymers for Advanced Technologies, 8*(4), 203–209. https://doi.org/10.1002/(SICI)1099-1581(199704)8:4<203::AID-PAT627>3.0.CO;2-3

Modi, S., Koelling, K., & Vodovotz, Y. (2012). Miscibility of poly (3-hydroxybutyrate-co-3-hydroxyvalerate) with high molecular weight poly (lactic acid) blends determined by thermal analysis. *Journal of Applied Polymer Science, 124*(4), 3074–3081.

Montes, M. I., Cyras, V., Manfredi, L., Pettarín, V., & Fasce, L. (2020). Fracture evaluation of plasticized polylactic acid/poly (3-HYDROXYBUTYRATE) blends for commodities replacement in packaging applications. *Polymer Testing, 84*, 106375.

Mosanenzadeh, S. G., Khalid, S., Cui, Y., & Naguib, H. E. (2016). High thermally conductive PLA based composites with tailored hybrid network of hexagonal boron nitride and graphene nanoplatelets. *Polymer Composites, 37*(7), 2196–2205. https://doi.org/10.1002/pc.23398

Mukherjee, T., & Kao, N. (2011). PLA based biopolymer reinforced with natural fibre: a review. *Journal of Polymers and the Environment, 19*(3), 714–725. https://doi.org/10.1007/s10924-011-0320-6

Nasseri, R., Ngunjiri, R., Moresoli, C., Yu, A., Yuan, Z., & Xu, C. C. (2020). Poly (lactic acid)/acetylated starch blends: effect of starch acetylation on the material properties. *Carbohydrate Polymers, 229*, 115453.

NatureWorks, L. (2007). *Technology focus report: blends of PLA with other thermoplastics. USA*.

Nehra, R., Maiti, S., & Jacob, J. (2018b). Effect of thermoplastic elastomer on melt rheological and fracture behavior of poly (lactic acid). *Polymer Plastics Technology and Engineering, 57*(12), 1254–1264.

Nehra, R., Maiti, S., & Jacob, J. (2018a). Poly (lactic acid)/(styrene-ethylene-butylene-styrene)-g-maleic anhydride copolymer/sepiolite nanocomposites: investigation of thermo-mechanical and morphological properties. *Polymers for Advanced Technologies, 29*(1), 234–243.

Ochi, S. (2008). Mechanical properties of kenaf fibers and kenaf/PLA composites. *Mechanics of Materials, 40*(4–5), 446–452.

Ojijo, V., Sinha Ray, S., & Sadiku, R. (2012). Role of specific interfacial area in controlling properties of immiscible blends of biodegradable polylactide and poly [(butylene succinate)-co-adipate]. *ACS Applied Materials & Interfaces, 4*(12), 6690–6701.

Oksman, K., Skrifvars, M., & Selin, J.-F. (2003). Natural fibres as reinforcement in polylactic acid (PLA) composites. *Composites Science and Technology, 63*(9), 1317–1324.

Ostafinska, A., Fortelny, I., Nevoralova, M., Hodan, J., Kredatusova, J., & Slouf, M. (2015). Synergistic effects in mechanical properties of PLA/PCL blends with optimized composition, processing, and morphology. *RSC Advances, 5*(120), 98971–98982.

Ostafinska, A., Fortelný, I., Hodan, J., Krejčíková, S., Nevoralová, M., Kredatusová, J., et al. (2017). Strong synergistic effects in PLA/PCL blends: impact of PLA matrix viscosity. *Journal of the Mechanical Behavior of Biomedical Materials, 69*, 229–241.

Ostrowska, J., Sadurski, W., Paluch, M., Tyński, P., & Bogusz, J. (2019). The effect of poly(butylene succinate) content on the structure and thermal and mechanical properties of its blends with polylactide. *Polymer International, 68*(7), 1271–1279.

Ozyhar, T., Baradel, F., & Zoppe, J. (2020). Effect of functional mineral additive on processability and material properties of wood-fiber reinforced poly (lactic acid)(PLA) composites. *Composites Part A: Applied Science and Manufacturing, 132*, 105827.

Padee, S., Thumsorn, S., On, J. W., Surin, P., Apawet, C., Chaichalermwong, T., et al. (2013). Preparation of poly (lactic acid) and poly (trimethylene terephthalate) blend fibers for textile application. *Energy Procedia, 34*, 534–541.

Pai, F. C., Lai, S. M., & Chu, H. H. (2013). Characterization and properties of reactive poly (lactic acid)/polyamide 610 biomass blends. *Journal of Applied Polymer Science, 130*(4), 2563–2571.

Palai, B., Mohanty, S., & Nayak, S. K. (2020). Synergistic effect of polylactic acid (PLA) and Poly (butylene succinate-co-adipate)(PBSA) based sustainable, reactive, super toughened eco-composite blown films for flexible packaging applications. *Polymer Testing, 83*, 106130.

Biodegradable fibers from poly (lactic acid)/poly (butylene succinate) blends. In Panichsombat, K., Panbangpong, W., Poompiew, N., & Potiyaraj, P. (Eds.), *IOP conference series: materials science and engineering*, (2019). IOP Publishing.

Papon, E. A., & Haque, A. (2019). Fracture toughness of additively manufactured carbon fiber reinforced composites. *Additive Manufacturing, 26*, 41–52. https://doi.org/10.1016/j.addma.2018.12.010

Park, J. W., & Im, S. S. (2003). Miscibility and morphology in blends of poly (L-lactic acid) and poly (vinyl acetate-co-vinyl alcohol). *Polymer, 44*(15), 4341–4354.

Park, S. B., Hwang, S. Y., Moon, C. W., Im, S. S., & Yoo, E. S. (2010). Plasticizer effect of novel PBS ionomer in PLA/PBS ionomer blends. *Macromolecular Research, 18*(5), 463–471.

Peng, Q., Cheng, J., Lu, S., & Li, Y. (2020). Electrospun hyperbranched polylactic acid–modified cellulose nanocrystals/polylactic acid for shape memory membranes with high mechanical properties. *Polymers for Advanced Technologies, 31*(1), 15–24. https://doi.org/10.1002/pat.4743

Phuong, V. T., Coltelli, M.-B., Cinelli, P., Cifelli, M., Verstichel, S., & Lazzeri, A. (2014). Compatibilization and property enhancement of poly (lactic acid)/polycarbonate blends through triacetin-mediated interchange reactions in the melt. *Polymer, 55*(17), 4498–4513.

Ploypetchara, N., Suppakul, P., Atong, D., & Pechyen, C. (2014). Blend of polypropylene/poly (lactic acid) for medical packaging application: physicochemical, thermal, mechanical, and barrier properties. *Energy Procedia, 56*, 201–210.

Pluta, M., Galeski, A., Alexandre, M., Paul, M. A., & Dubois, P. (2002). Polylactide/montmorillonite nanocomposites and microcomposites prepared by melt blending: structure and some physical properties. *Journal of Applied Polymer Science, 86*(6), 1497–1506. https://doi.org/10.1002/app.11309

Pongtanayut, K., Thongpin, C., & Santawitee, O. (2013). The effect of rubber on morphology, thermal properties and mechanical properties of PLA/NR and PLA/ENR blends. *Energy Procedia, 34*, 888–897.

Prabhakaran, M. P., Ghasemi-Mobarakeh, L., Jin, G., & Ramakrishna, S. (2011). Electrospun conducting polymer nanofibers and electrical stimulation of nerve stem cells. *Journal of Bioscience and Bioengineering, 112*(5), 501–507.

Qin, S.-X., Yu, C.-X., Chen, X.-Y., Zhou, H.-P., & Zhao, L.-F. (2018). Fully biodegradable poly (lactic acid)/poly (propylene carbonate) shape memory materials with low recovery

temperature based on in situ compatibilization by dicumyl peroxide. *Chinese Journal of Polymer Science, 36*(6), 783–790.
Qiu, H., Hou, K., Zhou, J., Liu, W., Wen, J., & Gu, Q. (2020). Preparation of biodegradable PLA/PCL composite filaments: effect of PLA content on strength. *MS&E., 770*(1), 012059.
Ramdhanie, L. I., Aubuchon, S. R., Boland, E. D., Knapp, D. C., Barnes, C. P., Simpson, D. G., et al. (2006). Thermal and mechanical characterization of electrospun blends of poly (lactic acid) and poly (glycolic acid). *Polymer Journal, 38*(11), 1137–1145.
Rashedi, S., Afshar, S., Rostami, A., Ghazalian, M., & Nazockdast, H. (2020). Co-electrospun poly (lactic acid)/gelatin nanofibrous scaffold prepared by a new solvent system: morphological, mechanical and in vitro degradability properties. *International Journal of Polymeric Materials and Polymeric Biomaterials, 1–9.*
Restrepo, I., Medina, C., Meruane, V., Akbari-Fakhrabadi, A., Flores, P., & Rodríguez-Llamazares, S. (2018). The effect of molecular weight and hydrolysis degree of poly (vinyl alcohol)(PVA) on the thermal and mechanical properties of poly (lactic acid)/PVA blends. *Polímeros, 28*(2), 169–177.
Rigoussen, A., Verge, P., Raquez, J.-M., Habibi, Y., & Dubois, P. (2017). In-depth investigation on the effect and role of cardanol in the compatibilization of PLA/ABS immiscible blends by reactive extrusion. *European Polymer Journal, 93*, 272–283.
Rigoussen, A., Raquez, J.-M., Dubois, P., & Verge, P. (2019). A dual approach to compatibilize PLA/ABS immiscible blends with epoxidized cardanol derivatives. *European Polymer Journal, 114*, 118–126.
Rosely, C. V. S., Shaiju, P., & Gowd, E. B. (2019). Poly(L-lactic acid)/boron nitride nanocomposites: influence of boron nitride functionalization on the properties of poly(L-lactic acid). *Journal of Physical Chemistry B, 123*(40), 8599–8609. https://doi.org/10.1021/acs.jpcb.9b07743
Rosely, C. V. S., Joseph, A. M., Leuteritz, A., & Gowd, E. B. (2020). Phytic acid modified boron nitride nanosheets as sustainable multifunctional nanofillers for enhanced properties of poly(l -lactide). *ACS Sustainable Chemistry and Engineering, 8*(4), 1868–1878. https://doi.org/10.1021/acssuschemeng.9b06158
Saini, P., Arora, M., & Kumar, M. N. V. R. (2016). Poly(lactic acid) blends in biomedical applications. *Advanced Drug Delivery Reviews, 107*, 47–59. https://doi.org/10.1016/j.addr.2016.06.014
Sangeetha, V., Varghese, T., & Nayak, S. (2016). Toughening of polylactic acid using styrene ethylene butylene styrene: mechanical, thermal, and morphological studies. *Polymer Engineering & Science, 56*(6), 669–675.
Sankaran, K. K., Krishnan, U. M., & Sethuraman, S. (2014). Axially aligned 3D nanofibrous grafts of PLA–PCL for small diameter cardiovascular applications. *Journal of Biomaterials Science, 25*(16), 1791–1812. Polymer Edition.
Santos, L. G., Oliveira, D. C., Santos, M. S., Neves, L. M. G., de Gaspi, F. O., Mendonça, F. A., et al. (2013). Electrospun membranes of poly (lactic acid)(PLA) used as scaffold in drug delivery of extract of Sedum dendroideum. *Journal of Nanoscience and Nanotechnology, 13*(7), 4694–4702.
Sedničková, M., Pekařová, S., Kucharczyk, P., Bočkaj, J., Janigová, I., Kleinová, A., Jochec-Mošková, D., Omaníková, L., Perďochová, D., Koutný, M., Sedlařík, V., Alexy, P., & Chodák, I. (2018). Changes of physical properties of PLA-based blends during early stage of biodegradation in compost. *International Journal of Biological Macromolecules, 113*, 434–442. https://doi.org/10.1016/j.ijbiomac.2018.02.078
Senatov, F. S., Niaza, K. V., Zadorozhnyy, M. Y., Maksimkin, A. V., Kaloshkin, S. D., & Estrin, Y. Z. (2016). Mechanical properties and shape memory effect of 3D-printed PLA-

based porous scaffolds. *Journal of the Mechanical Behavior of Biomedical Materials, 57*, 139–148. https://doi.org/10.1016/j.jmbbm.2015.11.036

Serra, T., Planell, J. A., & Navarro, M. (2013). High-resolution PLA-based composite scaffolds via 3-D printing technology. *Acta Biomaterialia, 9*(3), 5521–5530. https://doi.org/10.1016/j.actbio.2012.10.041

Shankar, S., Teng, X., Li, G., & Rhim, J. W. (2015). Preparation, characterization, and antimicrobial activity of gelatin/ZnO nanocomposite films. *Food Hydrocolloids, 45*, 264–271. https://doi.org/10.1016/j.foodhyd.2014.12.001

Shankar, S., Rhim, J. W., & Won, K. (2018). Preparation of poly(lactide)/lignin/silver nanoparticles composite films with UV light barrier and antibacterial properties. *International Journal of Biological Macromolecules, 107*, 1724–1731. https://doi.org/10.1016/j.ijbiomac.2017.10.038

Shankar, S., Wang, L. F., & Rhim, J. W. (2018). Incorporation of zinc oxide nanoparticles improved the mechanical, water vapor barrier, UV-light barrier, and antibacterial properties of PLA-based nanocomposite films. *Materials Science and Engineering: C, 93*, 289–298. https://doi.org/10.1016/j.msec.2018.08.002

Shi, Q., Chen, C., Gao, L., Jiao, L., Xu, H., & Guo, W. (2011). Physical and degradation properties of binary or ternary blends composed of poly (lactic acid), thermoplastic starch and GMA grafted POE. *Polymer Degradation and Stability, 96*(1), 175–182. https://doi.org/10.1016/j.polymdegradstab.2010.10.002

Song, Z., Xiao, H., & Zhao, Y. (2014). Hydrophobic-modified nano-cellulose fiber/PLA biodegradable composites for lowering water vapor transmission rate (WVTR) of paper. *Carbohydrate Polymers, 111*, 442–448.

Song, J. J., Chang, H. H., & Naguib, H. E. (2015a). Biocompatible shape memory polymer actuators with high force capabilities. *European Polymer Journal, 67*, 186–198.

Song, J. J., Chang, H. H., & Naguib, H. E. (2015b). Design and characterization of biocompatible shape memory polymer (SMP) blend foams with a dynamic porous structure. *Polymer, 56*, 82–92.

Sookprasert, P., & Hinchiranan, N. (2017). Morphology, mechanical and thermal properties of poly (lactic acid)(PLA)/natural rubber (NR) blends compatibilized by NR-graft-PLA. *Journal of Materials Research, 32*(4), 708–800.

Spiridon, I., Leluk, K., Resmerita, A. M., & Darie, R. N. (2015). Evaluation of PLA–lignin bioplastics properties before and after accelerated weathering. *Composites Part B: Engineering, 69*, 342–349.

Sun, M., Kingham, P. J., Reid, A. J., Armstrong, S. J., Terenghi, G., & Downes, S. (2010). In vitro and in vivo testing of novel ultrathin PCL and PCL/PLA blend films as peripheral nerve conduit. *Journal of Biomedical Materials Research Part A: An Official Journal of The Society for Biomaterials, The Japanese Society for Biomaterials, and The Australian Society for Biomaterials and the Korean Society for Biomaterials, 93*(4), 1470–1481.

Sun, X. S., Seib, P., & Wang, H. (2001). *High strength plastic from reactive blending of starch and polylactic acids*. Google Patents.

Suryanegara, L., Nakagaito, A. N., & Yano, H. (2009). The effect of crystallization of PLA on the thermal and mechanical properties of microfibrillated cellulose-reinforced PLA composites. *Composites Science and Technology, 69*(7–8), 1187–1192.

Takayama, T., & Todo, M. (2006). Improvement of impact fracture properties of PLA/PCL polymer blend due to LTI addition. *Journal of Materials Science, 41*(15), 4989–4992.

Therias, S., Larché, J. F., Bussière, P. O., Gardette, J. L., Murariu, M., & Dubois, P. (2012). Photochemical behavior of polylactide/ZnO nanocomposite films. *Biomacromolecules, 13*(10), 3283–3291. https://doi.org/10.1021/bm301071w

Threepopnatkul, P., Vichitchote, K., Saewong, S., Tangsupa-Anan, T., Kulsetthanchalee, C., & Suttiruengwong. (2010). Mechanical and antibacterial properties of electrospun PLA/PEG mats. *Journal of Metals, Materials and Minerals, 20*(3), 185−187.

Thurber, C. M., Xu, Y., Myers, J. C., Lodge, T. P., & Macosko, C. W. (2015). Accelerating reactive compatibilization of PE/PLA blends by an interfacially localized catalyst. *ACS Macro Letters, 4*(1), 30−33.

Tian, X., Liu, T., Yang, C., Wang, Q., & Li, D. (2016). Interface and performance of 3D printed continuous carbon fiber reinforced PLA composites. *Composites Part A: Applied Science and Manufacturing, 88*, 198−205. https://doi.org/10.1016/j.compositesa.2016.05.032

Tian, X., Liu, T., Wang, Q., Dilmurat, A., Li, D., & Ziegmann, G. (2017). Recycling and remanufacturing of 3D printed continuous carbon fiber reinforced PLA composites. *Journal of Cleaner Production, 142*, 1609−1618. https://doi.org/10.1016/j.jclepro.2016.11.139

Topkanlo, H. A., Ahmadi, Z., & Taromi, F. A. (2018). An in-depth study on crystallization kinetics of PET/PLA blends. *Iranian Polymer Journal, 27*(1), 13−22.

Touny, A. H., Lawrence, J. G., Jones, A. D., & Bhaduri, S. B. (2010). Effect of electrospinning parameters on the characterization of PLA/HNT nanocomposite fibers. *Journal of Materials Research, 25*(5), 857−865. https://doi.org/10.1557/jmr.2010.0122

Tsou, C.-Y., Wu, C.-L., Tseng, Y.-C., Chiu, S.-H., Suen, M.-C., Hung, W., et al. (2017). Isothermal crystallization kinetics effect on the tensile properties of PLA/PTT polymer composites. *Strength of Materials, 49*(1), 171−179.

Tsuji, H., & Muramatsu, H. (2001). Blends of aliphatic polyesters. IV. Morphology, swelling behavior, and surface and bulk properties of blends from hydrophobic poly (L-lactide) and hydrophilic poly (vinyl alcohol). *Journal of Applied Polymer Science, 81*(9), 2151−2160.

Tsuji, H., Nakano, M., Hashimoto, M., Takashima, K., Katsura, S., & Mizuno, A. (2006). Electrospinning of poly(lactic acid) stereocomplex nanofibers. *Biomacromolecules, 7*(12), 3316−3320. https://doi.org/10.1021/bm060786e

Urquijo, J., Guerrica-Echevarría, G., & Eguiazábal, J. I. (2015). Melt processed PLA/PCL blends: effect of processing method on phase structure, morphology, and mechanical properties. *Journal of Applied Polymer Science, 132*(41).

Valente, T. A. M., Silva, D. M., Gomes, P. S., Fernandes, M. H., Santos, J. D., & Sencadas, V. (2016). Effect of sterilization methods on electrospun poly(lactic acid) (PLA) fiber alignment for biomedical applications. *ACS Applied Materials and Interfaces, 8*(5), 3241−3249. https://doi.org/10.1021/acsami.5b10869

Vroman, I., & Tighzert, L. (2009). Biodegradable polymers. *Materials, 2*(2), 307−344. https://doi.org/10.3390/ma2020307

Walha, F., Lamnawar, K., Maazouz, A., & Jaziri, M. (2018). Biosourced blends based on poly (lactic acid) and polyamide 11: structure−properties relationships and enhancement of film blowing processability. *Advances in Polymer Technology, 37*(6), 2061−2074.

Wang, H., Sheng, M., Zhai, L., & Li, Y. (2008). Study on hydrophilicity and degradability of polyvinyl alcohol/polylactic acid blend film. *Sheng wu yi xue gong cheng xue za zhi= Journal of Biomedical Engineering= Shengwu yixue gongchengxue zazhi., 25*(1), 139−142.

Wang, Y., Chiao, S., Hung, T. F., & Yang, S. Y. (2012). Improvement in toughness and heat resistance of poly (lactic acid)/polycarbonate blend through twin-screw blending: influence of compatibilizer type. *Journal of Applied Polymer Science, 125*(S2), E402−E412.

Wang, L.-F., Rhim, J.-W., & Hong, S.-I. (2016). Preparation of poly (lactide)/poly (butylene adipate-co-terephthalate) blend films using a solvent casting method and their food packaging application. *LWT-Food Science and Technology, 68*, 454−461.

Wang, Y., Wei, Z., & Li, Y. (2016). Highly toughened polylactide/epoxidized poly (styrene-b-butadiene-b-styrene) blends with excellent tensile performance. *European Polymer Journal, 85*, 92–104.

Wang, G., Zhang, D., Li, B., Wan, G., Zhao, G., & Zhang, A. (2019). Strong and thermal-resistance glass fiber-reinforced polylactic acid (PLA) composites enabled by heat treatment. *International Journal of Biological Macromolecules, 129*, 448–459. https://doi.org/10.1016/j.ijbiomac.2019.02.020

Wang, B., Tu, Z., Wu, C., Hu, T., Wang, X., Long, S., et al. (2019). Effect of poly (styrene-ran-methyl acrylate) inclusion on the compatibility of polylactide/polystyrene-b-polybutadiene-b-polystyrene blends characterized by morphological, thermal, rheological, and mechanical measurements. *Polymers, 11*(5), 846.

Wei, J., Yang, S., Wang, L., Wang, C.-F., Chen, L., & Chen, S. (2013). Electrospun fluorescein-embedded nanofibers towards fingerprint recognition and luminescent patterns. *RSC Advances, 3*(42), 19403–19408.

Wen, X., Liu, Z., Li, Z., Zhang, J., Wang, D. Y., Szymańska, K., Chen, X., Mijowska, E., & Tang, T. (2020). Constructing multifunctional nanofiller with reactive interface in PLA/CB-g-DOPO composites for simultaneously improving flame retardancy, electrical conductivity and mechanical properties. *Composites Science and Technology, 188*. https://doi.org/10.1016/j.compscitech.2019.107988

Weng, Y.-X., Jin, Y.-J., Meng, Q.-Y., Wang, L., Zhang, M., & Wang, Y.-Z. (2013). Biodegradation behavior of poly (butylene adipate-co-terephthalate)(PBAT), poly (lactic acid)(-PLA), and their blend under soil conditions. *Polymer Testing, 32*(5), 918–926.

Weng, Y.-X., Wang, L., Zhang, M., Wang, X.-L., & Wang, Y.-Z. (2013). Biodegradation behavior of P (3HB, 4HB)/PLA blends in real soil environments. *Polymer Testing, 32*(1), 60–70.

Wenqiang, L., Yu, D., Dongyan, L., Yuxia, B., & Xiuzhen, L. (2018). Polylactic acid (PLA)/Cellulose nanowhiskers (CNWs) composite nanofibers: microstructural and properties analysis. *Journal of Composites Science, 4*. https://doi.org/10.3390/jcs2010004

Wu, C.-S., & Liao, H.-T. (2005). A new biodegradable blends prepared from polylactide and hyaluronic acid. *Polymer, 46*(23), 10017–10026.

Wu, C.-P., Wang, C.-C., & Chen, C.-Y. (2015). Enhancing the PLA crystallization rate and mechanical properties by melt blending with poly (styrene-butadiene-styrene) copolymer. *Polymer-Plastics Technology and Engineering, 54*(10), 1043–1050.

Wu, J.-H., Wu, C.-P., Kuo, M., & Tsai, Y. (2016). Characterization and properties of reactive poly (lactic acid)/ethylene–vinyl alcohol copolymer blends with chain-extender. *Journal of Polymers and the Environment, 24*(2), 129–138.

Xie, Y.-C., Yu, D.-M., Zhang, N.-N., & Liang, H.-L. (2011). Enhanced mechanical and dielectric properties of PU networks with hyperbranched structures. *Polymer-Plastics Technology and Engineering, 50*(2), 168–172.

Xu, Y., Loi, J., Delgado, P., Topolkaraev, V., McEneany, R. J., Macosko, C. W., et al. (2015). Reactive compatibilization of polylactide/polypropylene blends. *Industrial & Engineering Chemistry Research, 54*(23), 6108–6114.

Yan, D., Yao, Q., Yu, F., Chen, L., Zhang, S., Sun, H., Lin, J., & Fu, Y. (2020). Surface modified electrospun poly(lactic acid) fibrous scaffold with cellulose nanofibrils and Ag nanoparticles for ocular cell proliferation and antimicrobial application. *Materials Science and Engineering: C, 111*. https://doi.org/10.1016/j.msec.2020.110767

Yang, F., Xu, C., Kotaki, M., Wang, S., & Ramakrishna, S. (2004). Characterization of neural stem cells on electrospun poly (L-lactic acid) nanofibrous scaffold. *Journal of Biomaterials Science, Polymer Edition, 15*(12), 1483–1497.

Yang, X., Xu, H., Odelius, K., & Hakkarainen, M. (2016). Poly (lactide)-g-poly (butylene succinate-co-adipate) with high crystallization capacity and migration resistance. *Materials, 9*(5), 313.
Yeh, J.-T., Yang, M.-C., Wu, C.-J., Wu, X., & Wu, C.-S. (2008). Study on the crystallization kinetic and characterization of poly (lactic acid) and poly (vinyl alcohol) blends. *Polymer-Plastics Technology and Engineering, 47*(12), 1289–1296.
Yi, Z., Yang, J., Liu, X., Mao, L., Cui, L., & Liu, Y. (2020). Enhanced mechanical properties of poly(lactic acid) composites with ultrathin nanosheets of MXene modified by stearic acid. *Journal of Applied Polymer Science, 137*(17). https://doi.org/10.1002/app.48621
You, Y., Youk, J. H., Lee, S. W., Min, B.-M., Lee, S. J., & Park, W. H. (2006). Preparation of porous ultrafine PGA fibers via selective dissolution of electrospun PGA/PLA blend fibers. *Materials Letters, 60*(6), 757–760.
Yu, S., Hwang, Y. H., Hwang, J. Y., & Hong, S. H. (2019). Analytical study on the 3D-printed structure and mechanical properties of basalt fiber-reinforced PLA composites using X-ray microscopy. *Composites Science and Technology, 175*, 18–27. https://doi.org/10.1016/j.compscitech.2019.03.005
Yuan, D., Chen, K., Xu, C., Chen, Z., & Chen, Y. (2014). Crosslinked bicontinuous biobased PLA/NR blends via dynamic vulcanization using different curing systems. *Carbohydrate Polymers, 113*, 438–445.
Yuryev, Y., Mohanty, A. K., & Misra, M. (2017). Novel biocomposites from biobased PC/PLA blend matrix system for durable applications. *Composites Part B: Engineering, 130*, 158–166.
Zeng, J.-B., Li, Y.-D., He, Y.-S., Li, S.-L., & Wang, Y.-Z. (2011). Improving flexibility of poly (L-lactide) by blending with poly (L-lactic acid) based poly (ester-urethane): morphology, mechanical properties, and crystallization behaviors. *Industrial & Engineering Chemistry Research, 50*(10), 6124–6131.
Zeng, Q., Feng, Y., Wang, R., & Ma, P. (2018). Fracture behavior of highly toughened poly (lactic acid)/ethylene-co-vinyl acetate blends. *E-Polymers, 18*(2), 153–162.
Zhang, J. F., & Sun, X. (2004). Mechanical properties and crystallization behavior of poly (lactic acid) blended with dendritic hyperbranched polymer. *Polymer International, 53*(6), 716–722.
Zhang, S., Feng, X., Zhu, S., Huan, Q., Han, K., Ma, Y., et al. (2013). Novel toughening mechanism for polylactic acid (PLA)/starch blends with layer-like microstructure via pressure-induced flow (PIF) processing. *Materials Letters, 98*, 238–241.
Zhang, S., Ye, J., Sun, Y., Kang, J., Liu, J., Wang, Y., Li, Y., Zhang, L., & Ning, G. (2020). Electrospun fibrous mat based on silver (I) metal-organic frameworks-polylactic acid for bacterial killing and antibiotic-free wound dressing. *Chemical Engineering Journal, 390*. https://doi.org/10.1016/j.cej.2020.124523
Zhang, H.-C., Kang, B.-H., Chen, L.-S., & Lu, X. (2020). Enhancing toughness of poly (lactic acid)/Thermoplastic polyurethane blends via increasing interface compatibility by polyurethane elastomer prepolymer and its toughening mechanism. *Polymer Testing, 87*, 106521.
Zhang, Q., Lei, H., Cai, H., Han, X., Lin, X., Qian, M., et al. (2020). Improvement on the properties of microcrystalline cellulose/polylactic acid composites by using activated biochar. *Journal of Cleaner Production, 252*, 119898.
Zhao, X., Ding, Z., Lin, Q., Peng, S., & Fang, P. (2017). Toughening of polylactide via in situ formation of polyurethane crosslinked elastomer during reactive blending. *Journal of Applied Polymer Science, 134*(2).
Zhou, L., Zhao, G., & Jiang, W. (2016). Effects of catalytic transesterification and composition on the toughness of poly (lactic acid)/poly (propylene carbonate) blends. *Industrial & Engineering Chemistry Research, 55*(19), 5565–5573.
Zou, W., Chen, R., Zhang, G., Zhang, H., & Qu, J. (2016). Mechanical, thermal and rheological properties and morphology of poly (lactic acid)/poly (propylene carbonate) blends prepared by vane extruder. *Polymers for Advanced Technologies, 27*(11), 1430–1437.

PHBV based blends and composites

A.V. Kiruthika
Seethalakshmi Achi College for Women, Karaikudi, Tamil Nadu, India

1. Introduction

Over the last two decades biopolymers have received increased attraction because of growing environmental awareness and they have led a new way to manage nonbiodegradable waste, especially in the packaging industry. Polyhydroxyalkanoates (PHA) are one of the thermoplastic polymers that are similar to petro-based products like polypropylene (PP). Poly(3-hydroxybutyrate-co-3-hydroxyvalerate) (PHBV) is a member of the PHA aliphatic polyester family with randomly distributed polymer 3-hydroxybutyrate (PHB) and 3-hydroxyvalerate(HV). PHBV is the most commonly used biodegradable biopolymer, which can be synthesized from a microorganism or genetically modified bacteria. PHBV is hydrophobic, compostable, nontoxic, renewable, and biocompatible, giving better options than oil-based polymers.

PHBV is a rigid polymer with a low melting temperature (lower than PHB) and it can be easily dissolved in chlorinated solvents (Avella et al., 2000). It has low thermal stability, high water vapor resistance, and considerably greater crystallinity than the other biodegradable aliphatic polyesters like polylactic acid (PLA), polycaprolactone (PCL), etc. Manufacturing of PHBV can be done by extrusion, molding, and electrospinning on conventional plastic processing instruments (Mekonnen et al., 2013). PHBV exhibits a tensile strength of 22−34 MPa, which is similar to PP, and its water permeability is 9−10 g/($m^2 \times 24$ h), which is similar to that of polyvinyl chloride (PVC) (Pilla, 2011). In order to improve the mechanical properties, Daiane et al. (2014) used natural plasticizers (cotton seed oil) and nucleation agent (Licowax) on PHBV. The improvements in impact resistance (58%) and elongation (46%), and reduction in elastic modulus (35%) and crystallinity (18%) were noted in PHBV.

Bourbonnais and Marchessault (2010) prepared PHBV coated paperboard packages for the dry products, dairy items, and beverages, whereas PHB coated boards were used for frozen meals packaging (Kuusipalo, 2000). However, the commercial use of PHBV is limited due to its brittleness and it could be achieved by chemical modifications and blending with flexible polymers such as PBS and PBAT. Highly expensive, too brittle, and narrow processing temperature windows are the major obstacles to the widespread utilization of PHBV. For this reason, the use of biopolymer is only economically feasible for certain applications. The literature available on PHBV has proved that the effects of various fillers, plasticizers, and surface modification provide a new pathway to utilize this polymer on a large scale.

Biodegradable Polymers, Blends and Composites. https://doi.org/10.1016/B978-0-12-823791-5.00008-9

1.1 Structure of PHBV

PHBV production, with the trademark name Biopol, was first commercialized in 1990 by Imperial Chemical Industries (ICI). At present, PHBV production is around 2000 tons/year but it is expected to increase to about 50,000 tons annually (Wenjian et al., 2018).

The structure of PHBV (Fig. 9.1) could be employed by a carbon type supplemented into the medium (Larissa et al., 2017). 3HB and 3HV are the monomers joined together by ester bonds, with the backbone of the polymer being made up of carbon and oxygen atoms. The characteristics of PHBV are highly dependent upon the introduction of the monomer units in the polymeric chains. 3HB provided stiffness, whereas 3HV provided flexibility. Enhancements of the melting point, water permeability, T_g, and tensile strength (TS) were noted with an increase in the monomer ratio (Jacquel et al., 2008).

1.2 Sources of PHBV

Bacterial fermentation is the primary source for the production of PHBV polymers. This involves a two-step process: (a) fermentation process—the bottleneck of the production stage where the microorganisms are fed into reactors (batch or fed batch) with butyric acid or fructose and then the available sugar is metabolized and PHB stored as a power source in the inner cells; (b) extraction process—polymer stored in the microorganism inner cells is detached and cleansed with solvents when the final product is reached. PHBV, the copolymer of PHB, has also produced with fermentation, with the only difference being in the use of propionic acid together with glucose as a carbon source (Daiane et al., 2014). In order to introduce the 3HV content, odd-carbon number substances were used, which were derived from *Haloferax mediterranei* and *Halomonas campisalis*. This species generated PHBV without the inclusion of any specific substrates. There are various substrates available to induce 3HV, these are propionic acid, propanol, valeric acid, pentanol, sodium propionate, valerate, heptanoic acid, levulinic acid, and olive and sunflower oils. Compared to Gram-negative bacteria (which create immunogenic lipopolysaccharides), Gram-positive bacteria such as *Nocardia*, *Rhodococcus*, *Bacillus*, and *Corynebacterium* generate PHBV. For the production of PHBV, *Nocardia* and *Rhodococcus* could assemble in nature but they are not feasible from an economic point of view (Wenjian et al., 2018). The introduction of an excessive amount of threonine and cyanocobalamin may generate PHBV in *Bacillus* whereas in *Corynebacterium* some propionate is needed in the medium. Nathalie (2012) enhanced the 3HV component in bioplastic PHBV production by *Cupriavidus necator* which was initially done by inoculation growth and then by fermentation with the carbon sources (such as levulinic acid and sodium propionate).

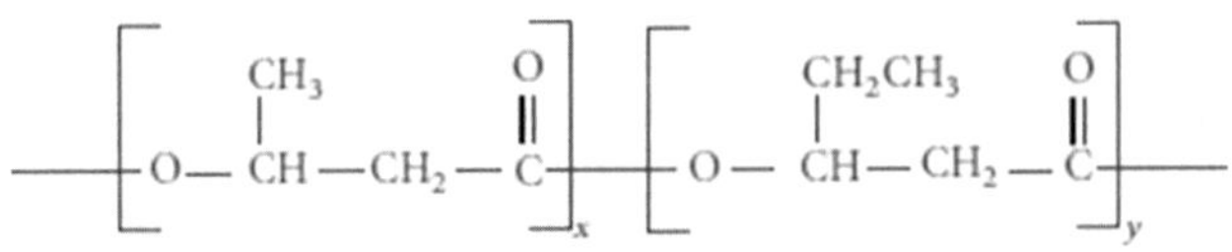

$$-\left[O-\underset{\displaystyle |}{\overset{\displaystyle CH_3}{}}\!\!CH-CH_2-\overset{\displaystyle O}{\overset{\|}{C}}\right]_x-\left[O-\overset{\displaystyle CH_2CH_3}{\overset{|}{CH}}-CH_2-\overset{\displaystyle O}{\overset{\|}{C}}\right]_y-$$

Figure 9.1 Structure of PHBV.

1.3 Properties of PHBV

The physical, mechanical, and thermal properties of PHBV are greatly influenced by the 3HV content in the determined copolymer. Hence it is more important to choose PHBV with 3HV content depending upon its requirements. The thermal characteristics were analyzed for PHBV with different HV monomers of 4.6, 9.5, and 20.7 mol%. The HV fraction had an intensive effect on the properties of PHBV, that is, crystallinity was decreased, when the HV content was increased. This leads to an enhancement in the rate of degradation and workability. It was noted that there is no decrement in the molecular weight of polymer specimens from the degradation mechanism. Weight loss was observed for the enzymatic degradation at the polymer surface (Longan et al., 2012).

Michael et al. (2017) evaluated the contact angle (for polar and nonpolar solvents) and surface tension of PLA, PP, and PHBV polymers. These three polymers have hydrophobic properties, in the case of polar solvent. For nonpolar solvents, such as diiodomethane, PHBV had greater wettability followed by PLA and PP. Due to the presence of intermolecular attraction in the polymer chain structure, the surface tension value was approximately 10 m Nm^{-1}, which was higher for PHBV and PLA. Using the Hoy, Hoftyzer-Van Krevelen, and Hansen methods, the solubility parameters (δ) were determined for cellulose acetate, PHBV, and triethyl citrate. The results showed that PHBV and the other polymers are partially miscible, but their blends are miscible (Kjeld et al., 2020). Peter et al. (2020) measured the mechanical properties of neat PHBV. A greater modulus of 3.3 GPa, TS of 41 MPa, high crystallinity, but lower elongation at break of 3.6% and impact strength of 21 J/m were noted. The molecular weight (MW) may vary, based on the different kinds of microorganisms used and the extraction procedure adopted. PHBV from *Pseudomonas pulida* have low MW and that from *Methylocystis* sp. has a high MW of $1.5-1.8 \times 10^6$ kDa.

1.4 Modification of PHBV

Several studies have been reported to enhance the brittleness of PHBV, they are (a) modification with dextran grafting and plasma radiations, (b) blends with other polymers (like NR, PLA, PBAT, PBS), (c) effects of plasticizers to increase the ductility and impact strength, and (d) the influence of fillers on PHBV to improve the stiffness, spherulite size, and crystallinity.

PHBV is an outstanding polymer in the biomedical field, as artificial skin scaffolds in tissue engineering were proved to be by Yongjing et al. (2020). Photochemical surface modification was given to PHBV/PEO (polyethylene oxide)/AZ (azidobenzoic acid) gelatin electrospinning mats. Increases in fiber diameter (600 nm), porosity (88%), and WVP ($230/g^{-2}$ day^{-1}), and a decrease in mechanical properties (4.2 MPa) were noticed for PHBV/PEO/AZ mats. Compared to virgin PHBV and PHBV/PEO mats, the PHBV/PEO/AZ mats showed superior in vitro degradation and proliferation properties.

Pilon and Kelly (2016) described the modification of PHBV on thermal and mechanical properties by a one-step reactive blending process with linalool (a monoterpene derivative). An increase in TS and decrease in T_m were noticed. Modified PHBV has less brittle behavior and crystallization. Hence linalool, a free radical cross-linking agent, played an important role in the modification of PHBV.

The surface modification of superparamagnetic iron oxide nanoparticles (SPIONs) with lauric acid and oleic acid at different proportions on PHBV microspheres via a solid-in-oil-in-water (S/O/W) emulsion solvent extraction process has been reported (Maizlinda et al., 2018). The result revealed that the iron content (155.7 μg mg^{-1} sample), loading (22.9%), and encapsulation efficiency (80.1%) were higher for PHBV/$SPIONs^{lauric}$ at a 2.5:1 ratio. With the same ratio, magnetic susceptibility (7×10^{-6} mg^{-1}) was higher for PHBV/$SPIONs^{oleic}$. The PHBV and PHBV microspheres had a particle size of <1 μm. The latter were nontoxic with mouse embryotic fibroblast cells (MEF). The authors recommended that the prepared PHBV microspheres have the capability of being utilized in biomedical applications.

The modification of PHBV fibrous scaffolds with dextran grafting was studied to enhance the proliferation of bone marrow-derived mesenchymal stem cells in tissue-engineering applications (Peng et al., 2016). After modification with dextran the water contact angle value was decreased from 117.3 ± 4.30 degrees to 56.03 ± 1.13 degrees. The results indicated that the modified PHBV had superior hydrophilicity properties, due to the presence of hydroxyl groups in dextran. The fiber diameters of unmodified and modified PHBV were 1.07 and 1.03 μm, respectively.

PHBV nanofibrous mats were surface modified with plasma radiation (O_2 gas) and laminin protein (Mohammadali et al., 2015). Using energy-dispersive X-ray analysis, the carbon (67.2%) and oxygen contents (27.2%) were measured and the values were higher for PHBV scaffolds. Similarly, the contact angle Θ (105) and UTS (6.37 MPa) were higher for PHBV scaffolds and laminin- coated PHBV scaffolds, respectively.

2. PHBV based blends

The poor mechanical properties, low thermal stability, and arduous processing methods are several factors limiting the use of PHBV in a wide range of applications. Hence blending of PHBV with any other polymeric materials results in improved properties for each of the polymers. That means that the existence of one polymer might have the effect of increasing the overall performance of the other polymers.

The interfacial bonding between the NR and PHBV was upgraded with the influence of peroxide treatments. The blends were enhanced via a melt blending method. The incorporation of PHBV into NR improved the thermal stability and melt strength, also the flexibility and toughness of the blends were increased by 59% and 20%, respectively (Xiaoying et al., 2019). The influence of a chain extender (styrene-acrylic-co-glycidyl methacrylate) and a plasticizer [triethyl citrate (TEC)] on the biodegradable PHBV polymers and plasticized cellulose acetate (PCA) blends were prepared by an injection molding process. From this experiment, it was observed that the percent of crystallinity (84.53%) and density (1.799 gcm^{-3}) were higher for a 70:30 blend ratio of PHBV and PCA. Compared to pristine PHBV and PCA, the mechanical properties of all ratios of blends were decreased. However, the notched impact strength was enhanced by 110% (Kjeld et al., 2020).

Mechanical recycling effect (up to six repeated cycles) on the characteristics of PHBV/PLA (50/50 wt.%) blends were studied (Zembouai et al., 2014). Using a single-screw extruder process, the blending was done, and then it was granulated. After that, the blended pellets were dried and injection molded. It was noticed that a large reduction in molecular weight and viscosity of PHBV took place, because of its degradation caused by six repeated cycles. The improvement in chain mobility associated with PHBV with respect to processing cycles could be evidenced by DSC. Moreover, a considerable reduction in the thermomechanical properties of PHBV due to the existence of PLA (a sensitive polymer) was also observed.

The same PLA/PHBV blends with various ratios from 10/90 to 90/10 w/w were used by another research group (Gurmeet et al., 2018). Using a melt mixing process, the specimen was prepared and its properties like FESEM, XRD, tensile test, PALS (positron annihilation lifetime spectroscopy), and thermal conductivity were studied. For virgin PLA (0.2 Wm^{-1} k^{-1}) and virgin PHBV (0.23 Wm^{-1} k^{-1}) respectively, the thermal conductivity was low, due to the amorphous nature of materials, whereas for PLA/PHBV blends it was increased with increasing PHBV content. Similarly, the mean size and relative concentration of free volume was increased with increasing PHBV content. The occurrence of open- volume cavities with 1.02 ns diameters was noted for blends using PALS. Further, the TS and YM were decreased with increasing content of PHBV.

Three-dimensional printing filaments were used to prepare the PLA/PHBV/PBAT blends with various compositions and its characteristics studied by Yang et al. (2019). From the thermal analysis, it was noticed that the T_g, T_m, and ΔH_m of PLA were decreased with increasing PHBV. A small increment in T_{cc} and decrement in ΔH_{cc} were noticed with a further increase of PHBV. The melting temperature of PLA and PHBV decreased considerably due to the interaction of various molecular segments and the PBAT retarded the motion of segments. For the blends, the melt flow index (MFI) was greater with increased PHBV content. The motion of molecular chains of PLA was inhibited by the PHBV and PBAT phases, which slow down the flow ratio. MFI is higher for both phases than PLA. The motion of molecular chains of PHBV improved the MFI of blends with increasing PHBV content.

Melt blending of PHBV/PBAT and PHBV/PBSeBT (70/30 wt.%) was carried out by Cunha et al. (2016) using a twin-screw extruder process. The experimental results revealed that the MFI and T_m were higher for PHBV + PBSeBT than the pristine polymers (PHBV, PBAT, PBSeBT). The combination of PHBV and PBAT provided better results in tear resistance, sealability, and WVP ($1.0-4.2 \times 10^{-11}$ g $m^{-2}s^{-1}$ Pa^{-1}), which were the essential properties for food-packaging applications. The blends of pristine polymers exhibited comparable mechanical performance that were achieved mostly by PHBV resins. PHBV exhibited isotropic properties with YM of 2.3–3.2 GPa and elongation at break of 1.2%–1.6%.

Solution casting processes were utilized to prepare PPC [poly(propylene carbonate)]/PHBV blends at various compositions of 100/0, 80/20, 60/40, 40/60, 20/80, and 0/100. The thermal stability of PPC was developed with the inclusion of PHBV content. Compared to PPC, PHBV was biodegraded more rapidly in the soil suspension cultivation because of the microorganism and these were responsible for PHBV degradation. Conversely, PPC was degraded more rapidly by in vitro degradation, due to the chemical hydrolysis and hydrolytic chain scissions. The authors also suggested that if the third element could be included such as PL or OMMT, to produce a PPC/PHBV blend, then it would be considered as a more useful material in the packaging industries (Jian et al., 2009).

Blended PHBV and polyethylene oxide (PEO) fibrous mats with various proportions (100/0, 80/20, 70/30, 50/50, 0/100% wt/wt) were prepared by electrospinning techniques. The average fiber diameter was higher for PHBV (2.8 μm), which was comparatively lower in PEO (0.24 μm). The obtained result was related to the thermal conductivity result, 0.059 μs cm^{-1} and 0.004 μs cm^{-1} for PHBV and PEO, respectively. The blends of PHBV/PEO increased the thermal conductivity and also viscosity with increasing PEO. At the same time, the fiber size was decreased with respect to increasing PEO. The mechanical properties were higher for PHBV (TM-80MPa, UTS-1.8 MPa, strain at failure 30%) and lower for PEO (4 MPa, 0.2 MPa, 15%). The mechanical performance of PHBV/PEO blends was in between that of neat PHBV and neat PEO (Alessandra et al., 2013).

The compatibility of PHBV/PBS blends was enhanced by in situ compatibilization with the help of dicumyl peroxide (DCP). After reactive compatibilization, an increment in interfacial adhesion and decrease in particle sizes of PBS was observed by the morphological studies (Ma et al., 2011). The TS and elongation at break of the blend were improved. The unnotched Izod impact toughness values varied from 10 to 50 kJ m^{-2} for the blends of PHB/PBS (at 70:30).

Blends from PCL and PHBV were prepared using fused deposition modeling. The two polymers were blended at various proportions of 100:0, 75:25, 50:50, and 25:75, then fabricated to 3D scaffolds. The low-pressure O_2 treatment was given to scaffolds, which is attributed to the increase in surface hydrophilicity. From the experimental observations, the average surface roughness values (98.18 nm), O/C atomic ratios (47.04%), and compressive strength (15.93 MPa) were higher for plasma-treated blends with weight ratios of 25:75. Hence the authors proposed that increases in the roughness and hydrophilic character of the blends enhanced the cellular functions, resulting in the scaffolds being promising constructs for cartilage tissue engineering applications (Kosorn et al., 2017).

Xueyan et al. (2017) performed an experiment to prepare blends of PHBV/PBS with PEG as a plasticizing agent. The different weight ratios of the blends were 70/0, 60/10, 50/20, 40/30, 30/40, 20/50, 10/60, and 0/70, and PEG as 30%. From this study, the TS (20.4 MPa) was higher for PHBV/PBS blends at the 70:0 ratio. Compared to other ratios of blends, the T_c (84.57°C) and T_m (172.11°C) were also higher for these blends. The inclusion of a plasticizer decreased the T_c and T_m of blends. There is no effect on degradability of blends with the addition of PEG and it was also verified that the *Fusarium* sp. FS1301 had higher degradability in the blends than *P. mendocian* DSO4-T.

3. Fabrication of PHBV composites

Numerous processes are available for the manufacture of composite specimens. Fabrication plays an important role in deciding the mechanical behavior of composite materials. An emulsion freezing/freeze-drying method was used for the preparation of PHBV/PLLA composite scaffolds with the influence of a biocompatible polymer such as nanohydroxyapatite (HA). The prepared scaffolds have a promising application for bone tissue engineering (Naznin and Min, 2012).

Tao et al. (2018) used a three-step process to manufacture the composites. The first process was to ball mill the ternary cellulose fiber to access the filler, next, the incorporation of PHBV and PLA is done by extrusion and then with injection molding, and the final process is true compounding of the cellulose/PHBV/PLA (Fig. 9.2).The other fabrication method, that is the hot press and salt particulate leaching technique, was used to prepare the PHBV/HV composites (Jin-Young et al., 2012).

An overview of various scaffold fabrication techniques used, their characteristics, limitations, and applications was discussed by Vahideh et al. (2017). The other researchers used solvent casting techniques to manufacture the HA/PHBV nanocomposite films. The addition of HA nanoparticles improved the mechanical strength of films and the adsorption of human fibrinogen. Also, these films have been used as an effective biomedical material (Yuan et al., 2011).

Nanobiocomposites from PHBV/cellulose nanofibrils (CNFs) were fabricated by a solution casting and electrospinning process. In this study, the inclusion of CNFs did not affect the film transparency and a considerable increment in the rate of crystallization was noted for solution casting composites. Conversely, a significant increase in process raising time and improved quality of mats (Fig. 9.3) were observed for the electrospinning process (Kelly et al., 2017).

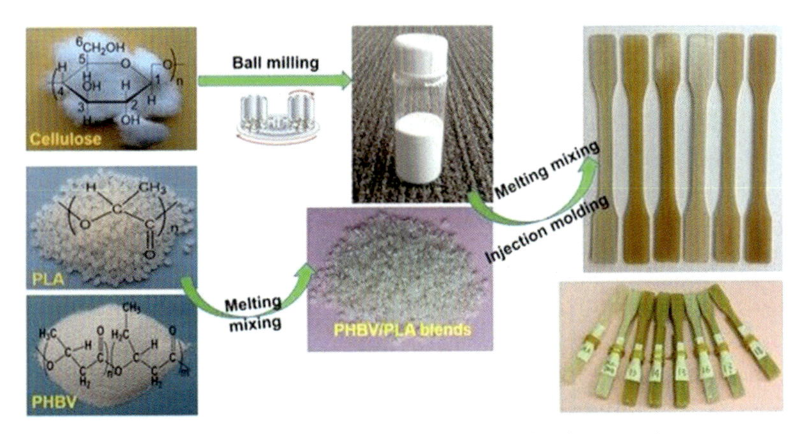

Figure 9.2 Fabrication of ternary cellulose-modified PHBV/PLA composites.

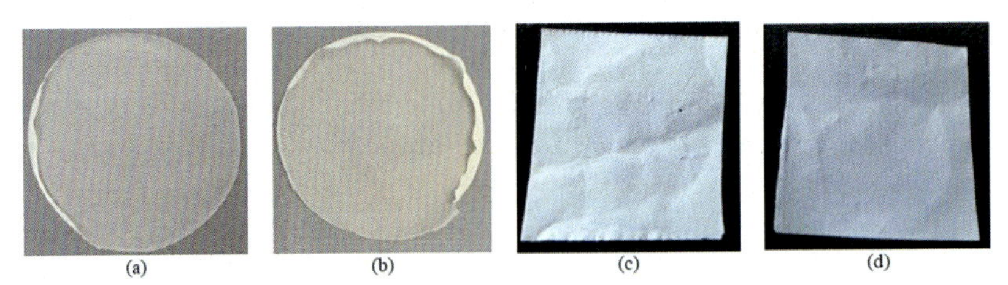

Figure 9.3 Films obtained by solution casting (A, B), mats obtained by electrospinning (C, D).

3.1 PHBV based composites

In recent years, the PHBV could be used as a matrix in the production of composites with plant fibers (Jiratti et al., 2016; Modi et al., 2016; Rapeephun et al., 2014). Various fabrication methods and properties of PHBV based plant fiber composites reported by various researchers are given in Table 9.1.

Yong et al. (2019) prepared novel composites using luffa fiber(LF)/PHBV by hot press forming techniques. In this work, 5% NaOH and 5% H_2O_2 treated LF/PHBV composites had superior flexural strength than untreated ones. The highest FS was achieved at 60% LF (Fig. 9.4). Increasing the content of LF (0%–90%) increased the WA. After immersion in water, the FS was decreased for treated LF composites. The good interfacial bonding between LF and PHBV was also observed from the SEM analysis. Kunyu et al. (2014) investigated the influence of DCP (as a compatibilizing agent) on the blends of PHBV, PBAT, and ENR reinforced with *Miscanthus* fiber (MF) composites. The inclusion of MF to the matrix resulted in a considerable increase in mechanical properties. The results also indicated that 20% MF enhanced the impact strength of 240 Jm^{-1} with stiffness. Furthermore, these composites have balanced heat deflection temperature, MFI, and density, which determined the applications of these materials in automotive sectors.

Biodegradable composite films were prepared with PHBV reinforced *Ceiba pentandra* (CP) fiber (5%, 10%, 15%, 20%) using a casting process. In this study, the addition of CP fibers improved the mechanical, thermal stability, and crystallinity percentage of the composites compared with the pristine PHBV. An antimicrobial test was performed with two food pathogens; *E. coli* did not show any activity but the other pathogen, *S. aureus*, showed antibacterial activity with increasing fiber content. Finally, the composite films were used as a fruit-packaging material where the freshness was maintained compared to unpacked fruits even after 7 days of packing (Sandhya et al., 2020).

PHBV based novel composites consisting of SF/clay particles were developed using a solution casting process. The influence of silane coupling agent at 3 vol% and 5 vol% on the characterization of composites was studied. From the experimental results, the TM had the largest value of 1501 MPa at 5 wt.% of sisal fiber. The addition of clay particles in the matrix resulted in a significant increment in dynamic mechanical properties. WA studies showed that PHBV/SF composites had the highest WA, whereas PHBV/silane-treated SF had the lowest, because silane enhanced the

Table 9.1 Properties of PHBV reinforced with plant fiber composites.

Fiber type	Properties of composites	Fabrication process	References
Wheat straw	WVPR, 11 to 110 g^{-2} day^{-1}; strain at break, 1.15%; stress at break, 27.2 MPa; YM, 3.13 GPa	Twin screw extrusion	Berthet et al. (2015)
Coconut fiber, CF (for 5 wt.% CFs)	TM, 4093 MPa; WVP, 3.96×10^{15}kg mm^{-2} Pa^{-1} s^{-1}; LP, 1.48×10^{14} kg mm^{-2} Pa^{-1} s^{-1}; CI, 58.2%; residual mass, 5.8%	Twin-screw extrusion followed by compression molding	Torrer et al. (2018)
Switch grass	IS, 53 Jm^{-1} [for PHBV/PBAT/SG (75:25)]; MFI, 13 g/10 min [PHBV/PBAT/SG (80:20)]	Melt-mixing technique, i.e., extrusion followed by injection molding	Vidhya et al. (2013)
Wheat straw/olive mill	WVTR, 32.2 g^{-2} day^{-1}; WVP, 5.2×10^{-11}g s^{-1} Pa^{-1} m^{-1}; CI from WAXD, 46.8%; CI from DSC, 63.8% (20 wt.% of wheat straw)	Twin-screw extruder	Marie et al. (2015)
Maple wood fiber	TM, 2.73 GPa; FM, 3.44 GPa (at 40 wt.% of WF); IS, 30 Jm^{-1} (at 30 wt.% of WF)	Extrusion followed by injection molding	Sanjeev and Mohanty (2007)
Jute	TM, 7.0 GPa; TS, 35.2 MPa; elongation at break, 0.8%	Injection molding	Bledzki and Jaszkiewicz (2010)
Bamboo	TS, 39 MPa; TM, 4.6 GPa; FS, 57.2 MPa; FM, 3.9 GPa	Melt compounding and injection molding	Long et al. (2008)
PHB/sisal	FS, 35.3 MPa; FM, 2898 MPa; TS, 18.2 MPa; MOE, 958 MPa; CI, 44.82%; ΔH, 58.98 (Jg^{-1})	Compression molding	Meire et al. (2016)
Sisal	Elastic modulus, 3.8 GPa (at 30 wt.% of SF); CI, 64.9%; TS, 28.6 MPa; elongation at break, 1.2% (at 10 wt.% of SF)	Compression molding	Alberto et al. (2019)
Coir	CI, 57%; average cell size, 41 μm; cell density, 1.37 (number/cm^3)	Injection molding	Alireza et al. (2010)
Sisal	Elastic modulus, 3.84 GPa; TS, 20.29 MPa; elongation at break, 0.73%	Compression molding	Cristina et al. (2017)
Sisal	TS, 21 MPa; TM, 1690 MPa; CI, 76%; ΔH, 134 kJ mol^{-1} (30 wt.% of SF)	Compression molding	Thossak et al. (2018)
PHBV/PLA/switch grass (30 wt.%)	TS, 54 MPa; FS, 52 MPa; residual mass, 8.9%; diffusion coefficient, 0.89×10^8 $cm^2 s^{-1}$	Injection molding	Malaya et al. (2012)
Sugarcane bagasse	Tensile stress, 7.6 MPa; YM, 753.7 MPa; flexural stress, 15.2 MPa; FM, 2344 MPa	Injection molding	Franscisco et al. (2012)

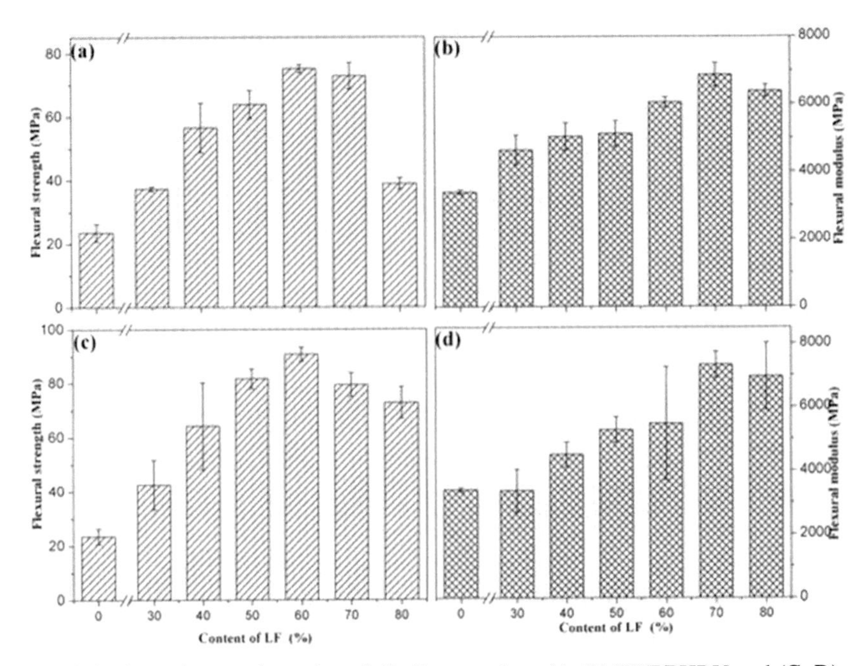

Figure 9.4 Flexural strengths and moduli of composites: (A, B) ULF/PHBV and (C, D) TLF/PHBV.

crystallinity and the interfacial bonding between the reinforcement and the matrix used. Moreover, nucleating density was also increased with increasing SF content (Jiratti et al., 2016).

In the case of hot compression techniques, sisal fiber (0.25 mm short and 5 mm long) reinforced PHBV composites were prepared and the results showed that an increase in TM was noted at 20 wt.% of SF. Then the hardness was also increased for the short SF composites and silane-treated SF composites. A small decrement in TS and IS, and a significant increment in water resistance, were noted for PHBV/SF composites with the inclusion of nanoclay particles (Rapeephun et al., 2014).

Different concentrations of three different fibers (common reed, reed canary grass, and water celery) reinforced with PHBV were fabricated using extrusion compounding followed by injection molding techniques with the influence of fiber weights (2%, 5%, 10%). An increase in modulus was noticed for 2% common reed and 5% canary grass fibers. Compared with the other fibers, water celery/PHBV composites had good compatibility between the fiber and matrix. Reed canary fiber (10%)/PHBV had the greatest complex viscosity due to the higher degradation temperature, which provided better thermal stability. Hence these properties opened the narrow processing window of these PHBV matrices to maintain the mechanical properties (Modi et al., 2016).

3.2 Effects of plasticizers on PHBV based blends/composites/ films/mats

PHBV, a biobased polymer, could act as a better replacement for manmade polymers, however its application is constrained due to its brittle behavior. Keeping this in mind, Hasanul et al. (2020), used various types of plasticizer (sorbitol, ethylene glycol, formamide) and different nanoclays (hectorite, montmorillonite, kaolinite) to reduce the brittle behavior of PHBV. The starch-based NC were prepared by facile and solution casting techniques. A significant increase in the mechanical and thermal properties of nanocomposite films with the influence of nanoclays was observed. At the same time, the WA behavior was decreased with the inclusion of plasticizers and there was a further decrease with the inclusion of various nanoclays. This starch-based clay NC has potential applications in drug-delivery systems, sensors, and is a suitable alternate for plastic products.

Plasticizers include dioctylphthalate (DOP), epoxidized soybean oil (ESO), and triethyl citrate (TEC) on the PHBV/organo-modified montmorillonite (OMt). Using melt processing, the composites were prepared. From the experimental observations, it was noted that TEC has been used as a biodegradable plasticizer (instead of DOP), primarily with the combination of OMt. Among the three plasticizers, ESO does not have an effective thermal result or mechanical characteristics of PHBV. Compared to all composites, PHBV reinforced with TEC nanocomposites has promising application as disposable packaging. The impact strength of PHBV/TEC OMt has the highest value of 29.1 Jm^{-1} (Matheus et al., 2018).

The influence of plasticizers including polyethylene glycol (PEG 200, 1000, 4000), lauric acid, and stearic acid (SA) on PHBV films has been studied (Raquel et al., 2016). The results reported that except for SA, the rest of the plasticizers improved the WVP, solubility, and water sorption capacity. A small reduction in film stiffness was also noted. Thus, PEG 1000 could act as an effective plasticizer for PHBV films, however for packaging purposes, a greater ratio would be required to adapt the PHBV mechanical properties.

3.3 Effects of various fillers on PHBV based composites

Stanislaw et al. (2019) evaluated the incorporation of various fillers like nanocellulose, walnut shell flour, eggshell flour, and tuff (at 15 wt.%) biorenewable composite reinforced with PHBV. The lowest IS in PHBV/walnut shell flour composites and higher thermal stability for PHBV/nanocellulose composites were obtained. An increase in the stiffness of composites was observed with the incorporation of fillers. The authors recommended that compared with petro-based products, these materials could be utilized for the manufacture of disposable products in airline packaging, such as containers for vegetables and sandwiches. This is attributed to the reduced environmental effects of a short biodegradation period in water, CO_2, soil, and marine conditions.

Paddy straw powder (PSP) filled composites were prepared from PHBV matrix with the effectiveness of filler content (0 wt.%, 5 wt.%, 10 wt.%,15 wt.%, 20 wt.%) and alkali treatment of PSP. It was mentioned that the alkali-treated composites had greater TS and modulus of elasticity with the addition of PSP, whereas they were lower for untreated composites. Improved filler matrix adhesion was observed for the chemically treated PSP, which was evidenced via SEM analysis (Yaacob et al., 2017).

A simple solution casting technique was utilized to prepare the PHBV biocomposites, where spent coffee bean powder (SCBP) was used as a waste filler as a reinforcement for the composites. The SCBP had greater thermal stability than the PHBV and its composites. A decrease in crystallinity affected the tensile properties with increasing SCBP content, that is, TS and modulus decreased with increasing SCBP. Finally, the PHBV/SCBP composites at 15 wt.% filler had the optimum parameters and the possibility of replacing manmade packaging materials (Senthil et al., 2019).

PHBV/cellulose nanocrystals (CNCs), a green filler with different CNC contents of 1%, 5%, and 10%, were used for the manufacture of composites (Hou-Yong et al., 2011). From the non-isothermal crystallization kinetics, the parameters like n (Avrami exponent), k (crystallinity rate constant), k_c (corrected crystallinity constant), and $t_{0.5}^{-1}/\text{min}^{-1}$ were increased with increasing filler content, whereas this was reversed for $t_{0.5}$/min (half-life crystallization time). The results from a contact angle analyzer indicated the hydrophilic nature of nanocomposite films. With the addition of filler content from 0% to 10%, the spherulite size was decreased from 58.3 to 2.2 μm, and water contact angle was from 60.1 to 32.5 degrees. The introduction of filler into the matrix could reduce the disadvantages of PHBV, which stretched the applications in the biomedical field.

3.4 Biodegradation mechanisms of PHBV based blends/composites

Degradation denotes an alteration in the mechanical and chemical properties of PHBV with effectiveness of such factors as soil, heat, and light. A soil biodegradation test was performed for the PHBV/peach palm particles (PPP) biocomposites using the soil burial test. The experiment was visually monitored and the results indicated that the prepared composites degraded faster because of the introduction of PPP content. The creation of holes with increasing PPP allowed water and microorganisms into the matrix, causing improved degradation of PHBV (Batista et al., 2010).

Biodegradation of PHBV film under a marine environment (solid, liquid, solid–liquid) was studied by Morgan et al. (2015). In a foreshore sand environment (solid state), the PHBV is fully biodegradable after 600 days. In the case of seawater inoculum (liquid) the biodegradation kinetics are faster. Depending on the sand, under seawater conditions (solid–liquid) the bacterial populations present in the seawater encouraged PHBV biodegradation after a short delay period. A great loss of mass is noted, when the PHBV film was submerged in seawater. After 180 days, a considerable reduction in thickness and surface erosion were observed in PHBV films.

According to NF ISO 14855, the biodegradation kinetics of two samples, PHBV and PBS, were measured under laboratory-scale composting conditions. For this work, mature green compost was taken and the humidity maintained with distilled water. The result explained that the films were completely analyzed into CO_2 serving its final assimilation by microorganisms. After 70–90 days of incubation, the rate of mineralization was 100% for both of the films (Salomez et al., 2019).

PHBV biodegradation studies were evaluated by Longan et al. (2012) and the films kept in an enzymatic environment (lipase medium) for up to 49 days. From the result it was noted that the rate of degradation of PHBV film was raised with decreasing PHBV crystallinity. The film's surface roughness and weight loss were noticed under the lipase medium.

The effectiveness of nZnO on PLA and PHBV degradation has been illustrated (Alojz et al., 2018). SEC chromatography indicated that degradation occurred for both polymeric composites. The incorporation of nZnO considerably increased the rate of PLA polymer degradation, whereas it was negligible in PHBV. The result was also confirmed by FTIR and NMR spectroscopy. The authors finally concluded that the nZnO increased the degradation process of PLA more than by PHBV.

Hydrothermal aging characteristics of PHBV reinforced with olive husk flour (OHF) in an acid medium (CH_3–COOH) have been studied (Hamour et al., 2020). A compatibilizer maleic anhydride and benzoyl peroxide was used for the preparation of composites. The specimens were submerged in water and monitored periodically for the studies. From the DSC results, a considerable drop in T_m, hence decreasing its T_m due to chain cuts, was observed for PHBV. Compared to plain PHBV, the degradation rate of the prepared composites, after the aging period, was increased, with the inclusion of an untreated filler. This is due to its greater crystallinity. At the same time, the thermal stability was low, which was analyzed by TGA.

Biodegradation tests of PHBV composites reinforced with wood fibers (WF, 7.5%) or basalt fibers (BF, 15%) were conducted in saline solution (40°C). It was noted from the result that the coefficient of thermal expansion (CTE) (59.923×10^{-6}/°C) was higher for PHBV/WF composite (at 7.5 wt.%). In the case of other BF composites, the CTE was decreased. A considerable decrement (10%) in the strength of specimens was noticed after the 2 weeks of biodegradation in silane solution (Fig. 9.5). These composites with natural fibers as a reinforcement could be proved as a material for longer life materials (Karolina and Stanisław, 2019).

PHBV and PHBV/TiO_2 nanocomposites underwent accelerated weathering after 500, 1000, and 2000 h. In this work, the SEM and AFM images showed the formation of cracks on the surface of PHBV. The crack size and surface roughness values increased with increasing accelerated weathering. However, in PHBV/TiO_2 samples, there were no cracks and negligible changes in roughness were observed. It was proposed that the degradation mechanism was slowed down with the influence of TiO_2 (Fig. 9.6). Also, the melt flow index (MFI; g/10 min) was increased with increasing accelerated weathering for PHBV and PHBV/TiO_2 composite, whereas elongation at break and strength at break were decreased for the same (Ana et al., 2020).

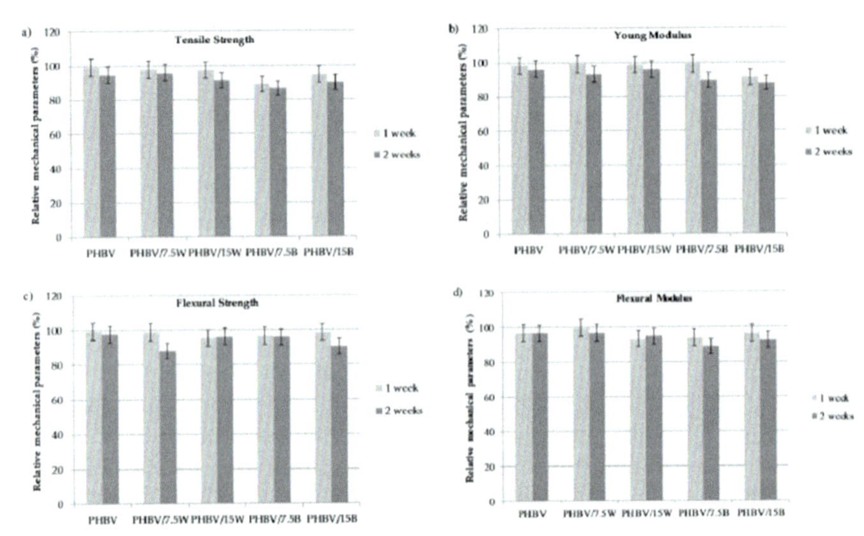

Figure 9.5 Strength of specimens after 2 weeks of biodegradation.

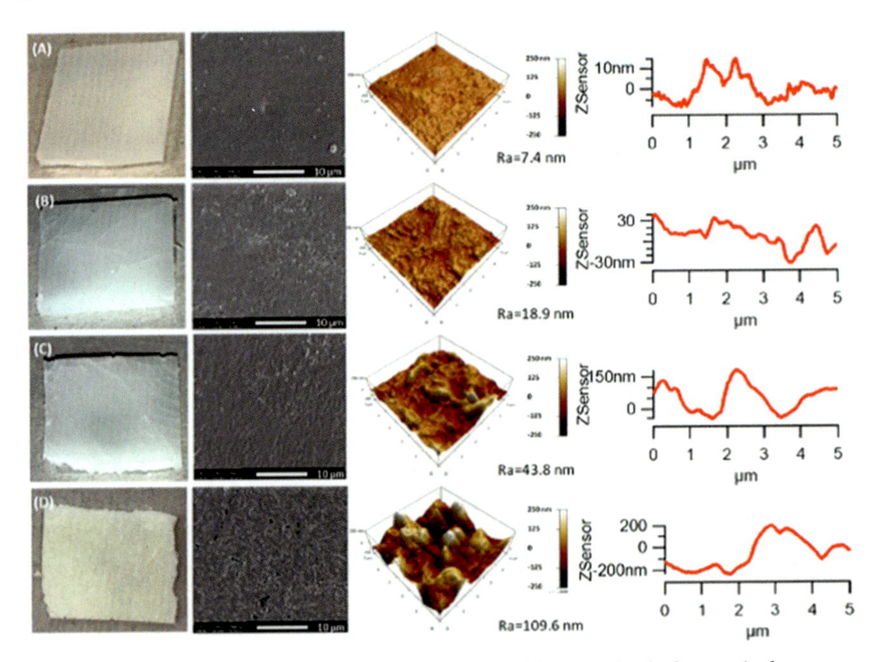

Figure 9.6 SEM images of (i) PHBV and (ii) PHBV/TiO_2 samples before and after accelerated weathering.

The biodegradability of alginic acid-treated flax fiber/PHBV composites (88%) was higher than that of untreated flax/PHBV composites (86%) and plain PHBV (63.2%). This is because both flax and alginic acid are biodegradable materials, that is, alginic acid is a water-soluble polysaccharide and flax fiber has greater moisture absorption.

Hence the biocomposites under compost conditions have a limited duration of survival and their useful organic nutrients could be used for enriching the soil at the end of composting (Sudhakar et al., 2019).

In another research study, samples from PHBV/flax, PHBV/PBAT/flax, and PHBV/ENR/flax were kept in a natural soil burial environment (for 112 days). After this period, weight losses of 6%, 9%, and 17%, respectively, were noticed for these composites (Zain et al., 2019).

4. PHBV based nanocomposites

The influence of nanocrystalline cellulose (NCC) obtained from oil palm empty fruit bunch fiber (OPEFBF at 0.25 wt.%, 0.50 wt.%, 0.75 wt.%, 1.00 wt.%, 1.50 wt.%, 2.00 wt.%) in PLA/PHBV (70/30) bionanocomposites was developed by a solution casting method. TEM and AFM micrographs evaluated the average diameter of treated (alkali cum bleached) NCC as 63 and 8.75 nm, respectively. Similarly the crystallinity index (71.69%) was higher for 0.25 wt.% NCC-derived composites than the other fiber weight composites and virgin blends. The incorporation of 0.25 wt.% NCC was the optimum parameter which enhanced the mechanical, morphological, and barrier properties. In the same way, an increase in NCC wt.% indicated the negative outcomes on nanocomposites, because of the aggregation of NCC in blends and constraints in the movement of the chain. This work also recommended that the prepared nanocomposites are a highly influential material for thermoplastic-based products (Dasan et al., 2016).

A comparative study of organically modified montmorillonite cloisite and unmodified halloysite nanoparticles of PHBV NCs was undertaken by Carli et al. (2011). A considerable reduction in strength and an increase in brittleness of the NC were observed with the inclusion of cloisite. In contrast, the unmodified halloysite NC had an enhanced impact strength with stiffness and is a potential material for the improvement of matrix characteristics without any modification of the composites.

Compression molding and the cast film extrusion method were used for the preparation of PHBV/PBAT nanocomposites (Akilesh et al., 2020). The effect of organically modified nanoclay (0.6 wt.%, 1.2 wt.%, 1.8 wt.%) was to act as a reinforcement with PBAT as a toughening agent to improve the barrier and mechanical properties. The analysis of TS results showed that the compression molded samples were found to be 46.1%, 53.1%, and 19.7% with 0.6 wt.%, 1.2 wt.%, and 1.8 wt.% nanoclay, whereas cast extrusion had 77.5%, 75.1%, and 34.5% improvements of the same formulations. The reverse was the case for the Young's modulus outcomes. Similarly, compared to compression molding, the oxygen (~79%) and water vapor (~70%) barrier properties of NC reflected good dispersion, interaction of PHBV/PBAT, and the size and shape of the blends in the cast extrusion samples. The uniform dispersion of nanoclay in the composites has been proved by rheological properties. From these studies, it was confirmed that nanoclay played a major role in increasing the barrier, thermal, and mechanical characteristics of polymers.

A three-step process was performed to develop the graphite nanosheets (GNS) via (a) intercalation by chemical exfoliation, (b) expansion by thermal treatment, and (c) physical treatment by ultrasonic exfoliation. The influence of GNS (0.25%, 0.50%,1.00%) on the properties of PHBV was studied by Larissa et al. (2017). Through XRD, the crystal size (35.52 nm) was identified for PHBV/GNS (0.50 wt.%). The strong peak at $2\Theta = 26$ degrees in the planes and d-spacing in the PHBV/GNS NCs were noted. Similar diffraction peaks with a clear indication of PHBV's crystal structure remain unchanged, although the existence of GNS in the PHBV was noticed (Fig. 9.7). The obtained results confirmed that the casting process used better techniques to fabricate PHBV/GNS nanocomposites with good dispersion and distribution of nanosheets into the PHBV polymers.

The synergistic effect of CNC (1 wt.%, 2 wt.%, 3 wt.%) with graphene oxide nanohybrids (0.5 wt.%, 0.6 wt.%, 0.7 wt.%) obtained by chemical grafting (CNC-GO, covalent bonds) and physical blending (CNC/GNO noncovalent bonds) on the properties of PHBV nanocomposites was determined. Hydrogen bond fractions (F_{H-CO}, 0.27), thermal degradation parameters (T_0, 243.7°C; T_{max}, 272.1°C), and antibacterial rate(*S. aureus*, 99.9%; *E. coli*, 100.0%) were higher for PHBV NCs (1 wt.% CNC-GO covalent bond). The same trend was followed for the crystallinity index (62.7%), melt crystallization temperature (81.1°C), and k (17.12×10^{-2}). Hence these NCs showed good robust barrier properties, better antibacterial rate, and lower migration levels, which led to high-performance food-packaging materials (Fang et al., 2019).

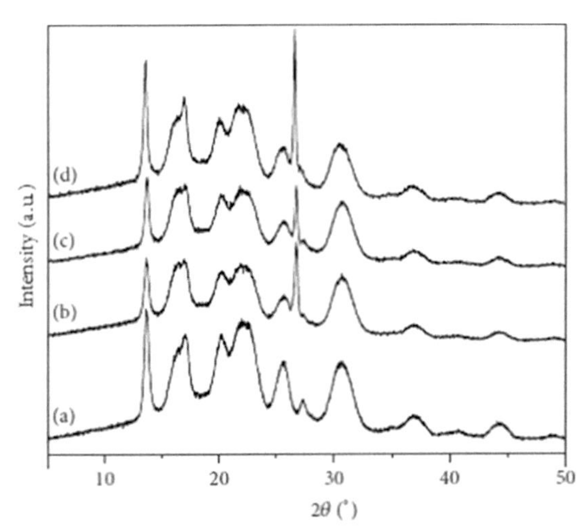

Figure 9.7 XRD of neat PHBV (A) and PHBV with different concentrations of GNS:PHBV/ 0.25 wt.% GNS (B), PHBV/0.50 wt.% GNS (C), and PHBV/1.00 wt.% GNS (D).

5. PHBV applications

A wide variety of highly favorable applications have been established for the utilization of PHBV. Many research works have been carried out in the development of biocomposites where reinforcement was changed or even an entirely new one added. This naturally derived polymer, PHBV, has great effective applications in specialty packaging, reinforced materials in composites, and biomedical applications like drug-delivery carriers, tissue engineering, scaffolds, bone repair, healing of diabetic wounds, and the manufacture of orthopedic devices. Generally, the properties such as low crystallinity and low thermal stability caused a decrease in the mechanical behavior and barriers that may constrain the utilization of PHBV in food packaging applications. In order to strengthen the possibilities for PHBV as packaging, many scientists and industrialists utilized organic and inorganic reinforcements to enhance the PHBV matrix.

5.1 PHBV in packaging applications

Using a solution casting process, bionanocomposites were prepared with ZnO nanoparticles (nucleating agent) reinforced PHBV polyesters (Ana and Angel, 2014). ZnO increased the T_c and mechanical properties, and decreased the WA, WVP, and oxygen permeability of composites. PHBV/ZnO films displayed antimicrobial functions against human pathogenic bacteria and have greater benefits in this regard as they are ideal for packaging material as well as for disposable goods.

Hybrids of rGO-ZnO (reduced graphene oxide) blended with glycerol plasticized PHBV composites were prepared by a melt extrusion method and the characteristics studied. The results predicted that the inclusion of ZnO nanoparticles improved the thermal and optical properties of the prepared nanocomposites. Due to the surface antibacterial activity of the fillers, the PHBV/hybrid composites have wider applications as active packaging (Rodrigo et al., 2018).

In another interesting research work, Mina et al. (2019) fabricated ZnO-silver (Ag) nanocomposites, using *Thymus vulgaris* leaf extract as a stabilizing agent. Increased mechanical and antimicrobial characteristics were observed for these composites. The NCs could be used for the sensory assessment of chicken breast refrigerated for a period of 15 days, and it also demonstrated a high degree of antimicrobial action, which provided better opportunities for the substitution of conventional petro-based polymers, currently used for packaging of poultry products.

The incorporation of PHBV matrix reinforced with CNC (2 wt.%, 4 wt.%, 6 wt.%) nanocomposites was fabricated by solution casting techniques. With the optimum parameters of 4 wt.% of CNC, NC revealed superior barrier properties against WVP and oxygen permeability. Further, these biodegradable NC could be a better option for synthetic plastic packaging materials (Sara et al., 2017). PHBV reinforced with NR (15%) was prepared by a melt blending process with balanced mechanical properties, indicating favorable products for food-packaging applications. According to the sealability and WVP, it was a promising bioalternative to nonbiodegradable plastics for the production of microwave food packaging.

5.2 PHBV in biomedical applications

PHBV and its copolymers have been extensively used in biomedical applications. Both natural and synthetic polymers could be used for the application of artificial dressings, especially in wound healing and surgical sutures. Zine and Sinha (2017) designed nanofibrous PHBV/collagen/GO scaffolds for wound coverage. The addition of GO (an antibacterial agent) enhanced the WA, porosity, and mechanical properties, and decreased the hydrophilic nature of nanofibers. Moreover, the incorporation of collagen with PHBV improved the cell attachment with increasing hydrophilicity of scaffolds. The authors concluded that these scaffolds constitute potential wound coverage products in medical applications.

In joint arthroplasty, bacterial infection is a major issue, and so, to overcome this, Zhi-Cai et al. (2010) carried out an in vitro assessment of the antibacterial activity and cytocompatibility of PHBV monomers with Ag (5−13 nm) nanoparticles. From the experimental analysis it was noted that at 1 wt.% of Ag, the PHBV/Ag scaffolds exhibited better assessment of antibacterial and cytocompatibility and could be used in joint arthroplasty.

A comparative study was carried out by Chin-San (2019), in which 3D printing filaments were used for the preparation of PHBV/fish scale (FS) and PHBV-g-MA/FS (grafted maleic anhydride-g-MA) composites. Researchers found that PHBV/FS and PHBV-g-MA/FS had higher water resistance and mechanical properties. Conversely, human foreskin fibroblast (FB) proliferation was higher for PHBV/FS composites. Here, the fish scale improved both the antioxidant and antimicrobial activities of these composites, indicating the potential of using these composites in biomedical applications.

Natalia et al. (2020) prepared a new biocomposite hydrogel for a dual drug-delivery carrier. PHBV microparticles with κ-carrageenan/locust bean gum composites were used to release ketoprofen and mupirocin. The composite hydrogel had biocompatibility with fibroblast cells and was considered as a drug-delivery carrier of poorly water-soluble drugs, especially wound-healing applications.

Another interesting comparative study of the biocompatibility and osteoinductivity between single-walled carbon nanotubes (SWCNTs)/PHBV and multiwalled carbon nanotubes (MWCNTs)/PHBV was determined by Pan et al. (2018). Using thermal injection molding techniques, the composites were prepared and implanted in rat femoral bone defects. The results indicated that both CNTs at 2.4% enhanced the mechanical and thermal characteristics. The authors also proposed that MWCNTs/PHBV showed superior biocompatibility and osteoinductivity both in vitro and in vivo.

Coelectrospinning techniques were employed for the 3D fibrous scaffolds with diatom shells (DS). The spinning process has the possibility of combining the hydrophobic PHBV/PCL fibers and hydrophilic pullulan (PUL) fibers. In this study, DS were included in a fibrous scaffold to improve the osteogenic properties. From this experiment, it was noted that DS-loaded, coelectronspinning scaffolds are potential candidates for bone tissue engineering (Ali et al., 2019).

Novel composite hydrogels from chitosan PHBV with chondroitin sulfate (CS) nanoparticles have been developed (Nair et al., 2015). This composite can withstand varying stresses, and has WA and viscoelastic behavior which is similar to native tissue that did not alter in PBS for 2 weeks. From the results it was clear that this composite hydrogels offered a greater ability for nucleus pulposus tissue engineering.

Almeida et al. (2019) prepared nanocomposites from PHBV, nanodiamond (nD, 1.5 wt.%) and nanohydroxyapatite (nHA, 5 wt.%) filled with vancomycin (VC). Rotary evaporator and spray-dryer methods were employed for the preparation of NCs. The percentage of crystallinity and dispersion of nanoparticles were higher for NC prepared with spray-dryer method. It was observed that the inclusion of nanoparticles enhanced the PHBV flexural elastic modulus by 34%, corresponding to that of human bone. In vitro arrays exhibited better adhesion and growth of cells, which indicate noncytotoxic and noncytostatic properties. The results also illustrated that these materials could be explored as bone defect filling and infection treatment materials.

6. Conclusions

This chapter discusses some of the recent works related to PHBV based blends and composites. Based on the previous works, (1) PHBV, an aliphatic polyesters, is one of the most extensively used polymers by researchers and polymer engineers, (2) two common methods were employed to obtain PHBV, they are bacterial fermentation and extraction processes, (3) PHBV is derived from microorganisms and have attracted attention as green plastics or bioplastics, (4) various surface modifications like polyethylene oxide (PEO), SPIONs, plasma radiation, and dextran grafting on PHBV were discussed, (5) PHBV has been compounded with various blends (PPC, PBAT, PLA, PBSeBT, PEO, PBS), (6) different fabrication methods (such as compression molding, injection molding, screw extrusion, solvent casting) have been utilized by researchers to manufacture PHBV based composites, blends, and nanocomposites, (7) various fillers, plant fibers, and plasticizers have been incorporated with PHBV to reduce the brittle characteristics of PHBV, (8) the surface roughness and weight loss for the biodegradation studies of PHBV, and (9) at the end of the chapter, various applications of PHBV based composites/mats/fillers in food packaging and in biomedical fields were described.

List of abbreviations

FM	Flexural modulus
FS	Flexural strength
IS	Impact strength

MFI	Melt flow index
MoE	Modulus of elasticity
PBAT	Poly(butylene adipate-co-terephtalate)
PBS	Poly(butylene succinate)
PHBV	Poly(3-hydroxybutyrate-co-3-hydroxyvalerate)
PLA	Polylactic acid
PP	Polypropylene
T_g	Glass transition temperature
TM	Tensile modulus
T_m	Melting temperature
TS	Tensile strength
WVTR	Water vapor transmission rate
YM	Young's modulus
ΔH	Activation energy

References

Akhilesh, K. P., Feng, W., Manjusri, M., & Amar, K. M. (2020). Reactive extrusion of sustainable PHBV/PBAT-based nanocomposite films with organically modified nanoclay for packaging applications: Compression moulding vs. cast film extrusion. *Composites Part B: Engineering, 198*, 108141.

Alberto, L., Cristina, M., Barbara, B., Rodolfo, B., & Elisabetta, A. (2019). Evaluation of the mechanical and thermal properties decay of PHBV/sisal and PLA/sisal biocomposites at different recycle steps. *Polymers, 11*, 1477.

Alessandra, B., Manuela, C., & Ilaria, C. (2013). Electrospun PHBV/PEO co-solution blends: Microstructure, thermal and mechanical properties. *Materials Science and Engineering: C, 33*, 1067–1077.

Ali, D. D., Deniz, A., Ayten, K., Aysen, T., & Dilek, K. (2019). Diatom shell incorporated PHBV/PCL-pullulan co-electrospun scaffold for bone tissue engineering. *Materials Science and Engineering: C, 100*, 735–746.

Alireza, J., Yottha, S., Srikanth, P., Jungjoo, L., Shaoqin, G., & Lih-Sheng, T. (2010). Processing and characterization of solid and microcellular PHBV/coir fiber composites. *Materials Science and Engineering: C, 30*, 749–757.

Almeida, N. G. R., Barcelos, M. V., Ribeiro, M. E. A., Folly, M. M., & Rodríguez, R. J. S. (2019). Formulation and characterization of a novel PHBV nanocomposite for bone defect filling and infection treatment. *Materials Science & Engineering C-Materials for Biological Applications, 104*, 110004.

Alojz, A., Andrej, K., & Ema, Z. (2018). Degradation of PLA/ZnO and PHBV/ZnO composites prepared by melt processing. *Arabian Journal of Chemistry, 11*, 343–352.

Ana, A., Anton, P., Omar, A., Mohammad, K. H., Peter, K., & Adriaan, S. L. (2020). Accelerated weathering effects on poly(3-hydroxybutyrate-co-3-hydroxyvalerate) (PHBV) and PHBV/TiO_2Nanocomposites. *Polymers, 12*, 1743.

Ana, M. D., & Angel, L. D. (2014). ZnO-reinforced poly(3-hydroxybutyrate-co-3-hydroxyvalerate) bionanocomposites with antimicrobial function for food packaging. *ACS Applied Materials and Interfaces, 6*, 9822–9834.

Avella, M., Martuscelli, E., & Raimo, M. (2000). Properties of blends and composites based on poly(3-hydroxy)butyrate(PHB) and poly(3-hydroxybutyrate-hydroxyvalerate) (PHBV) copolymers. *Journal of Materials Science, 35*, 523–545.

Batista, K. C., Silva, D. A. K., Coelho, L. A. F., Pezzin, S. H., & Pezzin, A. P. T. (2010). Soil biodegradation of PHBV/peach palm particles biocomposites. *Journal of Polymers and the Environment, 18*, 346–354.

Berthet, M.-A., Angellier-Coussy, H., Chea, V., Guillard, V., Gastaldi, E., & Gontard, N. (2015). Sustainable food packaging: Valorising wheat straw fibres for tuning PHBV-based composites properties. *Composites Part A: Applied Science and Manufacturing, 72*, 139–147.

Bledzki, A. K., & Jaszkiewicz, A. (2010). Mechanical performance of biocomposites based on PLA and PHBV reinforced with natural fibres - a comparative study to PP. *Composites Science and Technology, 70*, 1687.

Bourbonnais, R., & Marchessault, R. H. (2010). Application of polyhydroxyalkanoate granules for sizing of paper. *Biomacromolecules, 11*, 989–993.

Carli, L. N., Janaina, S. C., & Raquel, S. M. (2011). PHBV nanocomposites based on organomodified montmorillonite and halloysite: The effect of clay type on the morphology and thermal and mechanical properties. *Composites Part A, 42*, 1601–1608.

Chin-San, W. (2019). Comparative assessment of the interface between poly(3-hydroxybutyrate-co-3-hydroxyvalerate) and fish scales in composites: Preparation, characterization, and applications. *Materials Science and Engineering: C, 104*, 109878.

Cristina, M., Jose, D. B., Barbara, B., Elisabetta, A., Alberto, L., Marco, C., & Amparo, R. (2017). Mechanical and thermal performance of PLA and PHBV based biopolymers as potential alternatives to PET. *Chemical Engineering Transactions, 57*, 1417–1422.

Cunha, M., Fernandes, B., Covas, J. A., Vicente, A. A., & Hilliou, L. (2016). Film blowing of PHBV blends and PHBV-based multilayers for the production of biodegradable packages. *Journal of Applied Polymer Science, 133*, 42165.

Daiane, G. B., Wagner, M. P., Carla, D., & José, A. M. A. (2014). Natural additives for poly (hydroxybutyrate-CO-hydroxyvalerate) - PHBV: Effect on mechanical properties and biodegradation. *Materials Research, 17*, 1145–1156.

Dasan, Y. K., Bhata, A. H., & Ahmad, F. (2016). Polymer blend of PLA/PHBV based bionanocomposites reinforced with nanocrystalline cellulose for potential application as packaging material. *Carbohydrate Polymers, 157*, 1323–1332.

Fang, L., Hou-Yong, Y., Yan-Yan, W., Ying, Z., Heng, Z., Ju-Ming, Y., Somia, Y., Hussain, A., & Kam, C. T. (2019). Natural biodegradable poly(3-hydroxybutyrate-co-3- hydroxyvalerate) nanocomposites with multifunctional cellulose nanocrystals/graphene oxide hybrids for high-performance food packaging. *Journal of Agricultural and Food Chemistry, 67*, 10954–10967.

Francisco, A. C., Lucia, H. I., Ana, P. L., Sandra, G. M., & Nelson, D. (2012). Processing and characterization of composites of poly(3-hydroxybutyrate-co-hydroxyvalerate) and lignin from sugar cane bagasse. *Journal of Composite Materials, 46*, 417–425.

Gurmeet, S. K., Ilham, A., & Adriaan, S. L. (2018). Morphology and property changes in PLA/PHBV blends as function of blend composition. *Journal of Polymer Research, 25*, 196.

Hamour, N., Boukerrou, A., & Beaugrand, J. (2020). Influence of hydrothermal ageing of PHBV/olive husk flour composite in acid medium. *Materials Today: Proceedings, 36*, 54–60.

Hasanul, B. M., Islam, Z., Abu Bin Hasan Susan, M., & Bin Imran, A. (2020). Effects of plasticizers and clays on the physical, chemical, mechanical, thermal, and morphological properties of potato starch-based nanocomposite films. *ACS Omega, 5*, 17543–17552.

Hou-yong, Y., Zong-yi, Q., & Zhe, Z. (2011). Cellulose nanocrystals as green fillers to improve crystallization and hydrophilic property of poly(3-hydroxybutyrate-co-3-hydroxyvalerate). *Progress in Natural Science: Materials International, 21*, 478–484.

Jacquel, N., Lo, C. W., Wei, Y. H., & Wu, H. S. (2008). Isolation and purification of bacterial poly(3-hydroxyalkanoates. *Biochemical Engineering Journal, 39*, 15–27.

Jian, T., Cunjiang, S., Mingfeng, C., Dan, H., Li, L., Na, L., & Shufang, W. (2009). Thermal properties and degradability of poly(propylene carbonate)/poly(β-hydroxybutyrate-co-β-hydroxyvalerate) (PPC/PHBV) blends. *Polymer Degradation and Stability, 94*, 575–583.

Jin-Young, B., Zhi-Cai, X., Giseop, K., Keun-Byoung, Y., Soo-Young, P., Lee, S. P., & Inn-Kyu, K. (2012). Fabrication and characterization of collagen-immobilized porous PHBV/HA nanocomposite scaffolds for bone tissue engineering. *Journal of Nanomaterials, 2012*. https://doi.org/10.1155/2012/171804. Article ID 171804, 11p.

Jiratti, T., Pawinee, B., Rapeephun, D., & Suchart, S. (2016). Characterization of poly(-hydroxybutyrate-co-hydroxyvalerate)/sisal fiber/clay bio-composites prepared by casting technique. *Periodica Polytechnica - Mechanical Engineering, 60*, 103–112.

Yuan, J., Qixin, Z., & Jian, C. (2011). Fabrication, characterization and adsorptive of human fibrinogen of HA/PHBV nano composite films. *Advanced Materials Research, 160–162*, 1325–1330.

Karolina, M., & Stanisław, K. (2019). Mechanical and hydrothermal aging behaviour of polyhydroxybutyrate-Co-valerate (PHBV) composites reinforced by natural fibres. *Molecules, 24*, 3538.

Kelly, C. C. B., Cioffi, M. O. H., & Herman, J. C. (2017). PHBV/cellulose nanofibrils composites obtained by solution casting and electrospinning process. *Matèria. Revista Internacional d'Art, 22*, e11837.

Kjeld, W. M., Akhilesh, K. P., Manjusri, M., & Amar, K. M. (2020). Sustainable PHBV/cellulose acetate blends: Effect of a chain extender and a plasticizer. *ACS Omega, 5*, 14221–14231.

Kosorn, W., Sakulsumbat, M., Uppanan, P., Kaewkong, P., Chantaweroad, S., Jitsaard, J., Sitthiseripratip, K., & Janvikul, W. (2017). PCL/PHBV blended three dimensional scaffolds fabricated by fused deposition modeling and responses of chondrocytes to the scaffolds. *Journal of Biomedical Materials Research Part B: Applied Biomaterials, 105B*, 1141–1150.

Kunyu, Z., Manjusri, M., & Amar, K. M. (2014). Toughened sustainable green composites from poly(3hydroxybutyrate-co-3-hydroxyvalerate) based ternary blends and Miscanthus biofiber. *ACS Sustainable Chemistry and Engineering, 2*, 2345–2354.

Kuusipalo, J. (2000). PHBV in extrusion coating of paper and paperboard: Part I: Study of functional properties: Study of functional properties. *Journal of Polymers and the Environment, 8*, 39–47.

Larissa, S. M., Thaís, L. A. M., Joao, P. B. M., Fábio, R. P., Ana, P. L., & Mirabel, C. R. (2017). Effect of graphite nanosheets on properties of poly(3-hydroxybutyrate-co-3-hydroxyvalerate). *International Journal of Polymer Science, 2017*. Article ID 9316761, 9 p.

Long, J., Jijun, H., Jun, Q., Feng, C., Jinwen, Z., Michael, P. W., & Yawei, Z. (2008). Study of poly(3-hydroxybutyrate-co-3-hydroxyvalerate) (PHBV)/Bamboo pulp fiber composites: Effects of nucleation agent and compatibilizer. *Journal of Polymers and the Environment, 16*, 83–93.

Longan, S., Qiang, F., Yan, H. Z., Xian, Z. W., Dai-Di, F., & Ho, N. C. (2012). Thermal properties and biodegradability studies of poly(3-hydroxybutyrate-co-3-hydroxyvalerate). *Journal of Polymers and the Environment, 20*, 23–28.

Ma, P., Hristova, B. D. G., Lemstra, P. J., Zhang, Y., & Wang, S. (2011). Toughening of PHBV/PBS and PHB/PBS blends via in situ compatibilization using dicumyl peroxide as a free-radical grafting initiator. *Macromolecular Materials and Engineering, 297*, 402–410.

Maizlinda, I. I., Jan, Z., Rainer, D., Judith, A. R., Harald, U., Christoph, A., & Aldo, R. B. (2018). Surface modification of SPIONs in PHBV microspheres for biomedical applications. *Scientific Reports, 8*, 7286.

Malaya, R. N., Manjusri, M., & Amar, K. M. (2012). Performance evaluation of biofibers and their hybrids as reinforcements in bioplastic composites. *Macromolecular Materials and Engineering*. https://doi.org/10.1002/mame.201200112

Marie-Alix, B., Hélène, A., Diogo, M., Loic, H., Andreas, S., Antonio, V., & Nathalie, G. (2015). Exploring the potential of using lignocellulosic fibres derived from three food by-products as constituents of biocomposites for food packaging. *Industrial Crops and Products, 69*, 110–122.

Matheus, D. S., Suelen, D. F. B., Tales, S. D., Raquel, S. M., Marcelo, G., Janaina, S. C., & Larissa, N. C. (2018). Comparison of the effect of plasticizers on PHBV and organoclay based biodegradable polymer nanocomposites. *Journal of Polymers and the Environment, 26*, 2290–2299.

Meire, N. H., Andressa, B. D., Virgínia, A. S. M., & Jane, M. F. P. (2016). Polyhydroxybutyrate composites with random mats of sisal and coconut fibers. *Materials Research*. https://doi.org/10.1590/1980-5373-MR-2016-0254

Mekonnen, T., Paolo, M., Hamdy, K., & David, B. (2013). Progress in bio-based plastics and plasticizing modifications. *Journal of Materials Chemistry A, 1*, 13379–13398.

Michael, R. S., Amar, K. M., & Manjusri, M. (2017). Miscibility and performance evaluation of biocomposites made from polypropylene/poly(lactic acid)/Poly(hydroxybutyrate-cohydroxyvalerate) with a sustainable biocarbon filler. *ACS Omega, 2*, 6446–6454.

Mina, Z., Keerthiraj, N., Shaista, I., Abdo, H., Sanjay, M., & Kullaiah, B. (2019). Smart fortified PHBV-CS biopolymer with ZnO–Ag nanocomposites for enhanced shelf life of food packaging. *ACS Applied Materials and Interfaces, 11*, 48309–48320.

Modi, S., Cornish, K., Koelling, K., & Vodovotz, Y. (2016). Mechanical and rheological properties of PhbvBioplastic composites engineered with invasive plant fibers. *Transactions of the ASABE, 59*, 1883–1891.

Mohammadali, S., Esmaeil, B., Mostafa, S., Hesam, H., & Hayedeh, R. (2015). Surface modification of PHBV nanofibrous mat by laminin protein and its cellular study. *International Journal of Polymeric Materials and Polymeric Biomaterials, 64*, 149–154.

Morgan, D., Guy, C., Antoine Le, D., Peter, D., & Stephane, B. (2015). Natural degradation and biodegradation of poly(3-hydroxybutyrate-co-3-hydroxyvalerate) in liquid and solid marine environments. *Journal of Polymer Environment, 23*, 493–505.

Nair, M. B., Baranwal, G., Vijayan, P., Keyan, K. S., & Jayakumar, R. (2015). Composite hydrogel of chitosan-poly(hydroxybutyrate-co-valerate) with chondroitin sulfate nanoparticles for nucleus pulposus tissue engineering. *Colloids Surface B Biointerfaces, 136*, 84–92.

Natalia, P., Saddys, R., Yousof, F., Rebeca, B., Luis, B., Sandra, F., & Francisca, L. (2020). Poly(hydroxybutyrate-co-hydroxyvalerate) microparticles embedded in κ-carrageenan/locust bean gum hydrogel as a dual drug delivery carrier. *International Journal of Biological Macromolecules, 146*, 110–118.

Nathalie, B. (2012). Enhancing the 3-hydroxyvalerate component in bioplastic PHBV production by Cupriavidus necator. *Biotechnology, 7*, 304–309.

Naznin, S., & Min, W. (2012). PHBV/PLLA-based composite scaffolds fabricated using an emulsion freezing/freeze-drying technique for bone tissue engineering: Surface modification and in vitro biological evaluation. *Biofabrication, 4*, 015003.

Pan, W., Xiao, X., Li, J., Deng, S., Shan, Q., Yue, Y., Tian, Y., Nabar, N. R., Wang, M., & Hao, L. (2018). The comparison of biocompatibility and osteoinductivity between multi-walled and single-walled carbon nanotube/PHBV composites. *Journal of Materials Science: Materials in Medicine, 29*, 189.

Peng, Z., Hairong, L., Yongsheng, L., Jiying, H., & Yao, D. (2016). Surface dextran modified electrospun poly(3-hydroxybutyrate-co-3-hydroxyvalerate)(PHBV) fibrous scaffold promotes the proliferation of bone marrow-derived mesenchymal stem cells. *Materials Letters, 179*, 109–113.

Peter, Z., Feng, W., Manjusri, M., & Amar, K. M. (2020). Toughening of biodegradable poly(3-hydroxybutyrate-co-3hydroxyvalerate)/poly(ε-caprolactone) blends by in situ reactive compatibilization. *ACS Omega, 5*, 14900–14910.

Pilla, S. (2011). *Handbook of bioplastics and biocomposites engineering applications* (vol. 81). New Jersey: John Wiley & Sons.

Pilon, L., & Kelly, C. (2016). Modification of poly(3-hydroxybutyrate-co-3-hydroxyvalerate) properties by reactive blending with a monoterpene derivative. *Journal of Applied Polymer Science, 133*, 42588.

Rapeephun, D., Jiratti, T., Pawinee, B., & Suchart, S. (2014). Sisal natural fiber/clay-reinforced poly(hydroxybutyrate-cohydroxyvalerate) hybrid composites. *Journal of Thermoplastic Composite Materials*, 1–17.

Raquel, R., Alberto, J., Maria, V., & Amparo, C. (2016). Effect of plasticizers on thermal and physical properties of compression-moulded poly[(3-hydroxybutyrate)-co-(3-hydroxyvalerate)] films. *Polymer Testing, 56*, 45e53.

Rodrigo, F. G., Eduardo, M. D. A., Vânia, M. F. P., & Cristina, T. A. (2018). Extruded hybrids based on poly(3-hydroxybutyrate-co-3-hydroxyvalerate) and reduced graphene oxide composite for active food packaging. *Food Packaging and Shelf Life, 16*, 77–85.

Salomez, M., Matthieu, G., Fabre, P., Touchaleaume, F., & Cesar, G. (2019). A comparative study of degradation mechanisms of PBSA and PHBV under laboratory scale composting conditions. *Polymer Degradation and Stability, 167*, 102–113.

Sandhya, A. V., Harikrishnan, P., Sanjay, M. R., Suchart, S., & Jyotishkumar, P. (2020). Novel biodegradable polymer films based on poly(3-hydroxybutyrate-co-3hydroxyvalerate) and *Ceiba pentandra* natural fibers for packaging applications. *Food Packaging and Shelf Life, 25*, 100538.

Sanjeev, S., & Mohanty, A. K. (2007). Wood fiber reinforced bacterial bioplastic composites: Fabrication and performance evaluation. *Composites Science and Technology, 67*, 1753–1763.

Sara, M., Belén, M., Maite, R., Luis, B., & Rebeca, B. (2017). Morphology, thermal and barrier properties of biodegradable films of poly (3-hydroxybutyrate-co-3-hydroxyvalerate) containing cellulose nanocrystals. *Composites Part A: Applied Science and Manufacturing, 93*, 41–48.

Senthil, M. K. T., Krittirash, Y., Rajini, N., Suchart, S., Nadir, A., & Varada, R. (2019). Mechanical and thermal properties of spent coffee bean filler/poly(3-hydroxybutyrate-co-3-hydroxyvalerate) biocomposites: Effect of recycling. *Process Safety and Environmental Protection, 124*, 187–195.

Stanislaw, K., Karolina, M., & Paulina, J. (2019). Novel biorenewable composites based on poly (3-hydroxybutyrateco-3-hydroxyvalerate) with natural fillers. *Journal of Polymers and the Environment, 27*, 803–815.

Sudhakar, M., Osei, O., Boobalan, T., Angelin, S. T. R., Satheesh, M. R., Saravanan, S., Jothi, B. M., & Arun, A. (2019). Thermal-chemical and biodegradation behaviour of alginic acid treated flax fibres/poly(hydroxybutyrate-co-valerate) PHBV green composites in compost medium. *Biocatalysis and Agricultural Biotechnology, 22*, 101394.

Tao, Q., Jinwu, W., & Michael, P. W. (2018). Facile fabrication of 100% bio-based and degradable ternary cellulose/PHBV/PLA composites. *Materials, 11*, 330.

Thorsak, K., Raminatun, M., Emma, S., Monica, E. K., & Sigbritt, K. (2018). Enhancement of mechanical, thermal and antibacterial properties of sisal/polyhydroxybutyrate-co-valerate biodegradable composite. *Journal of Metals, Materials and Minerals, 28*(1), 52–61.

Torres-Giner, S., Hilliou, L., Melendez-Rodriguez, B., Figueroa-Lopez, K. J., Madalen, D., Cabedo, L., Covas, J. A., Vicente, A. A., & Lagaron, J. M. (2018). Melt processability, characterization, and antibacterial activity of compression-molded green composite sheets made of poly(3-hydroxybutyrate-co-3-hydroxyvalerate) reinforced with coconut fibers impregnated with oregano essential oil. *Food Packaging and Shelf Life, 17*, 39–49.

Vahideh, R. H., Soodabeh, D., Ali, R., & Roya, S. (2017). Design and fabrication of porous biodegradable scaffolds: A strategy for tissue engineering. *Journal of Biomaterials Science, Polymer, 28*, 1797–1825.

Vidhya, N., Manjusri, M., & Amar, K. M. (2013). New engineered biocomposites from poly(3-hydroxybutyrate-co-3-hydroxyvalerate) (PHBV)/poly(butylene adipate-co-terephthalate) (PBAT) blends and switchgrass: Fabrication and performance evaluation. *Industrial Crops and Products, 42*, 461–468.

Wenjian, M., Jianli, W., Ye, L., Lianghong, Y., & Xiaoyuan, W. (2018). Poly(3-hydroxybutyrate-co-3-hydroxyvalerate) co-produced with l-isoleucine in *Corynebacterium glutamicum* WM001. *Microbial Cell Factories, 17*, 93.

Xiaoying, Z., Kuihao, J., Koelling, K., Katrina, C., & Yael, V. (2019). Optimal mechanical properties of biodegradable natural rubber-toughened PHBV bioplastics intended for food packaging applications. *Food Packaging and Shelf Life, 21*, 100348.

Xueyan, H., Huifang, L., Zhaoying, G., Tingting, S., Zhanyong, W., & Lei, Y. (2017). Blending modification of PHBV/PBS/PEG and its biodegradation. *Polymer-Plastics Technology and Engineering, 56*, 1128–1135.

Yaacob, N. D., Hanafi, I., & Sam, S. T. (2017). Paddy straw powder filled PHBV biocomposites: The effects of filler loading and surface modification. *AIP Conference Proceedings, 1865*, 040023-1–040023-7.

Yang, M., Hu, J., Xiong, N., Xu, B., Weng, Y., & Liu, Y. (2019). Preparation and properties of PLA/PHBV/PBAT blends 3D printing filament. *Materials Research Express, 6*.

Yong, G., Li, W., Yuxia, C., Panpan, L., & Tong, C. (2019). Properties of Luffa fiber reinforced PHBV biodegradable composites. *Polymers, 11*, 1765.

Yongjing, X., Hongwei, L., Chenkai, S., Liming, Z., Hai, Y., & Wengang, L. (2020). Preparation and characterization of PHBV/PEO/AZ-Gelatin electrospun mats with photochemical surface modification: The role of AZ-gelatin. *Macromolecular Materials and Engineering, 305*, 2000344.

Zain, Z., Damia, M., & Alan, C. (2019). Soil biodegradation of unidirectional polyhydroxybutyrate-Co-valerate (PHBV) biocomposites toughened with polybutylene-adipate-Co-terephthalate (PBAT) and epoxidized natural rubber (ENR). *Frontier in Materials, 6*. https://doi.org/10.3389/fmats.2019.00275

Zembouai, I., Bruzaud, S., Kaci, M., Aida, B., Corre, Y., & Yves, G. (2014). Mechanical recycling of poly(3-hydroxybutyrate-co-3-hydroxyvalerate)/polylactide based blends. *Journal of Polymers and the Environment, 22*, 449–459.

Zhi-Cai, Z., Won-Pyo, C., Jin-Young, B., Moon-Jeong, C., Yongsoo, J., & Inn-Kyu, K. (2010). In vitro assessment of antibacterial activity and cytocompatibility of silver-containing PHBV nanofibrous scaffolds for tissue engineering. *Biomacromolecules, 11*, 1248–1253.
Zine, R., & Sinha, M. (2017). Nanofibrouspoly(3-hydroxybutyrate-co-3-hydroxyvalerate)/collagen/graphene oxide scaffolds for wound coverage. *Materials Science and Engineering C-Materials for Biological Applications, 80*, 129–134.

PVA-based blends and composites

10

Akarsh Verma[1], *Naman Jain*[2], *Komal Singh*[3], *Vinay Kumar Singh*[3], *Sanjay Mavinkere Rangappa*[4] and *Suchart Siengchin*[4]
[1]Department of Mechanical Engineering, University of Petroleum and Energy Studies, Dehradun, Uttarakhand, India; [2]Department of Mechanical Engineering, Meerut Institute of Engineering and Technology, Meerut, Uttar Pradesh, India; [3]Department of Mechanical Engineering, Govind Ballabh Pant University of Agriculture and Technology, Pantnagar, Uttarakhand, India; [4]Natural Composite Research Group, Department of Materials and Production Engineering, The Sirindhorn International Thai—German Graduate School of Engineering (TGGS), King Mongkut's University of Technology North Bangkok, Bangkok, Thailand

1. Introduction

Polyvinyl alcohol (PVA) is a thermoplastic, nontoxic, whitish, tasteless, and semi-crystalline polymer (Abdullah et al., 2017; Saini et al., 2017; Ye et al., 2014). It is soluble in water with a backbone of carbon atoms, and under anaerobic and aerobic conditions PVA is completely biodegradable (Marušincová et al., 2013; Li et al., 2012). The discovery of PVA occurred in 1924 with the hydrolysis of polyvinyl acetate. The degree of hydrolysis is completely dependent upon the synthesis method employed for the production of PVA, and on that basis it is classified as partial hydrolysis and full hydrolysis. PVA has versatile physical and chemical properties such as film forming, viscosity, emulsifying, moderate tensile strength, dispersing powder, flexibility, dielectric strength, optical properties, and excellent charge-storage ability (Saini et al., 2017). In the market it is available in many grades depending upon the degree of hydrolysis and viscosity (Marin et al., 2014; Aslam et al., 2017). Due to its versatile properties, it has applications in a number of industries such as packaging, coating, paper sizing, emulsion polymerization, and the production of polyvinyl butyral (Shimao, 2001; Kawai & Hu, 2009). Due to its film-forming properties, it has wide applications in the packaging and coating industries. Some of its limitations can be overcome by the blending, reinforcing, and cross-linking of PVA. PVA also acts as an immobilization carrier, due to which it is used in biocatalysts (El-Naas et al., 2013; Surkatti & El-Naas, 2014). Another application is in the field of antibacterial films where it is blended with different antibacterial agents (Balasubramanian, 2014). PVA-based hydrogels are another very practice application of PVA by cross-linking with different chemicals such as the glutaraldehyde, boric acid, etc. In vinyl polymers, PVA is the only one which is biodegradable,

Biodegradable Polymers, Blends and Composites. https://doi.org/10.1016/B978-0-12-823791-5.00010-7

that is, it is utilized by microorganisms (Suzuki et al., 1973; Kim et al., 2003). Despite its applications in different fields, PVA is used as a matrix but it still needs some modification by altering the hydroxyl groups. The most commonly used modification reactions are etherification, carbamation, esterification, acetalization, etc. (Awada & Daneault, 2015).

2. Chemistry of PVA

The physical, chemical, and mechanical properties of PVA are governed by the presence of hydroxyl (−OH) present at every other carbon atom (DeMerlis & Schoneker, 2003). Just like many other polymers, PVA is not obtained by the polymerization of vinyl alcohol (monomer), as it is unstable in nature. The raw material used for the synthesis of PVA is vinyl acetate. The reaction involves partial replacement of the acetate group of vinyl acetate with a hydroxyl group in the presence of an alkali. The precipitate received after the partial hydrolysis is known as polyvinyl alcohol. The degree of hydrolysis depends upon the saponification reaction and is one of the important parameters on which the various physical and chemical properties of PVA depend. On the basis of hydrolysis, PVA can be divided into two categories (1) fully hydrolyzed and (2) partially hydrolyzed. The melting point, pH, refractive index solubility, adhesiveness, and flexibility are some of the major properties which depend upon the degree of hydrolysis and molecular weight of PVA (Kirk-Othmer). The specific gravity of commercially available PVA is about 0.6−0.7 g/cc, whereas with a decrease in the degree of hydrolysis, the true specific gravity (ratio of density of PVA to the density of water at the same temperature) of PVA also decreases. A similar type of behavior is also shown by PVA in the case of reflective index. The crystallinity of PVA mainly depends upon the degree of hydrolysis, and crystallinity increases with an increase in the degree of hydrolysis. Another external factor that influences the degree of crystallinity is the heat treatment process and, in fact, PVA at a temperature of 200°C with any degree of polymerization has a crystallinity of about 0.50−0.55. The most common technique for measuring the crystallinity is wide-angle X-ray. This is measured as the ratio of area under the crystalline region to the total area of scattering by waveform separation by the least square method (Hindeleh & Johnson, 1972). However, for PVA, difficulties occur in the waveform separation of amorphous and crystalline scatterings. Another method is differential scanning calorimetry (DSC), which measures the melting enthalpy (ΔH) which is proportional to the crystallinity. The melting point of PVA increases with an increase in the degree of hydrolysis and is obtained from the endothermic peak in the DSC. The increment in melting point with an increase in the degree of hydrolysis is almost linear when the residual acetate group is randomly distributed. However, there is no change in the melting point with an increase in the acetate groups when distributed blockwise.

Another property, solubility in water, also depends on the degree of hydrolysis. Due to the presence of the hydroxyl group in their domain, inter- and intramolecular hydrogen bonding with water molecules take place (Jain et al., 2017, 2018; Jain, Ali et al., 2019; Jain, Verma, & Singh, 2019). The acetate functional group present in partial-hydrolysis PVA is hydrophobic in nature. Sometimes, the high solubility of PVA acts as a limitation in the case of PVA-based composite applications, and to overcome these limitations blending and cross-linking are usually done. Jain et al. (2018) fabricated PVA-based blended films with starch and protein to overcome the solubility limitation of PVA. The solubility of PVA decreases to some extent due to the formation of hydrogen bonding between the hydroxyl groups of PVA and the hydroxyl groups of the glucose unit, and hydroxyl groups of PVA and amino groups in protein. This results in the reduction of hydroxyl groups of PVA resulting in hydrogen bonding with the water molecules. Jain, Verma, and Singh (2019) fabricated PVA-based cross-linked films with hydrochloric acid (HCl) to remove the hydroxyl groups in PVA through the dehydration and esterification reaction. In both these research works, PVA was used in the same way. A blending phenomenon results in a decrease in solubility, whereas the cross-linking overcomes this limitation and only the water absorption takes place. PVA is soluble in low concentrations of alcoholic solution (with water). Some organic solvents can also be used to dissolve PVA, such as glycols (ethylene glycol, glycerin, and diethylene glycol), amides, and amines.

3. History of PVA

Polyvinyl alcohol, commonly known as PVA or PVOH, is a biodegradable, synthetic thermoplastic polymer with numerous hydroxyl groups that can interact with other substances. PVA was first synthesized in 1924 by Herrmann (German Nobel chemist majorly working in the field of macromolecules) and Haehnel. However, the mass production of PVA through industrial manufacturing occurred 2 years after its discovery in Germany. PVA was not directly obtained from the polymerization of vinyl alcohol as compared to the other polymers. The polymerized form of vinyl alcohol was obtained by polymerizing the vinylic (polyvinyl acetate) compounds and was then treated with an alkaline solution (ethanol). Herrmann and Haehnel fabricated PVA fibers by the hydrolysis of polyvinyl acetate in the presence of anhydrous sodium methylate. In 1912, Klatte was the first to synthesize vinyl acetate, the ester of vinyl alcohol. It was obtained as a by-product during the fabrication of ethylidene diacetate, when the liquid acetic acid and acetylene were created (Klatte, 1913). During the research of Herrmann and Haehnel in 1924, one day they had added alkali to the alcoholic solution of clear polyvinyl acetate to obtain polymeric ester as per their expectation. However, this saponification reaction of monomeric vinyl acetate results in an impure resinous precipitation. After their examination, they discovered an ivory white-colored substance known as polyvinyl alcohol (Haehnel & Herrmann, 1924). They both independently studied the substance and, in 1926, reported the transformation of PVA into polyvinyl acetate through saponification and esterification (Staudinger & Arbeitserinnerunger, 1961). Other than Germany, the production of PVA started also in France, the United States, and the United Kingdom. PVA had its first application in rayon textiles as a warp sizing material. Other than that, PVA was also used as a stabilizer and emulsifier in emulsion polymerization.

The first patent was applied for in 1931 by WO Herrmann and colleagues on PVA fiber. They employed a dry and wet spinning method for the synthesis of PVA fiber and proposed the use of that fiber for surgical threads to replace catgut and silk (Herrmann & Haehnel, 1932). In industry, the raw material for the production of PVA is vinyl acetate, which is obtained from either ethylene or acetylene. Most use acetylene-based raw materials. In the PVA industry, vinyl acetate is first synthesized by passing acetic acid and acetylene vapor through a layer of the catalyst containing the zinc acetate. On the other hand, in the ethylene process, a gaseous mixture of oxygen acetic acid and ethylene is passed through the catalyst layer containing palladium. After that, polymerization of polyvinyl acetate through heating in the presence of methanol is performed. Only about 70% of monomer is converted into polyvinyl acetate and the residual monomer is removed. Finally, to obtain PVA, a small amount of alkali is added resulting in methanolysis and then the PVA is precipitated. The remaining residual after removal of the PVA contains methyl acetate, methanol, and a small amount of sodium acetate. Through the acidic catalyst the already-mentioned residuals are hydrolyzed to obtain the methanol and acetic acid, which can be again used in the manufacturing process.

PVA was first fabricated in Germany, but later most research into it was conducted in Japan. Japan had started research into the natural fibers in 1939. First was the production of PVA which was started in 1950 in Japan by the Kuraray company. Kuraray was the first company to manufacture polyvinyl alcohol fiber under the name of KURALON. In 1948, the PVA fiber was known by the name of "vinylon," but later it was named "vinal" in the United States. In 1938, Japan initiated research into PVA as a textile fiber. With the increase in demand for synthetic fibers, the production of PVA increased. With this increase in the demand, the manufacturing process to synthesize PVA fibers also was improved, resulting in a reduction in the fiber cost. On the other hand, by the end of 1955, the quality of the PVA fiber was also improved as there was a 30%–40% increment in the tensile strength and 70%–80% knot tensile strength. In addition to Japan, PVA industries were also established in other countries; for example, in 1953, a plant was established in the Democratic People's Republic of Korea. The project was led by S. Lee from Kyoto University who was researching PVA fiber. The People's Republic of China also established a plant in Beijing in 1965 which used acetylene as the raw material and employed the wet spinning method.

4. PVA-based blended/cross-linked composites

Significant research works have been aimed at the growing concerns about the environmental impact from nondegradable polymers with the hope of finding a solution in order to reduce the amount of plastic waste. Environmental concerns, utilization of natural resources in more efficient ways, and efforts the toward development of biodegradable plastics are some of the major challenges being faced by synthetic polymer composites (Jawaid & Abdul Khalil, 2011; Xu et al., 2005). Of these synthetic polymers, polyvinyl alcohol has emerged as one of the prominent matrix materials for composites fabrication due to its low manufacturing cost, moderate tensile strength, biodegradable nature, insolubility to most organic solvents, etc. However, despite its many advantages, it also has some disadvantages, including its solubility in water. To overcome these disadvantages and improve the physical and mechanical properties of PVA blending, reinforcing and cross-linking have been done by different researchers over the past few years. Some researchers also used different additives such as acids, glycerol, urea, formamide, etc. usually, additives are used as a plasticizer. Depending upon the application of PVA composites, different additives are employed, such as to increase the elongation percentage, antibacterial activities, improve water resistance, etc. Additives used with PVA polymers are classified as plasticizers, for example, glycerol, sorbitol; fillers, for example, nanoparticles, fibers; cross-linking agents, for example, glutaraldehyde, citric acid, etc.; and natural polymers, such as starch, chitosan, etc.

5. PVA-based blended films

One of the techniques used to improve the PVA-based composites is blending them with other polymers. Over the past few years, researchers have been working to overcome the disadvantages of PVA. One of the major advantages of PVA is its biodegradability, however the decomposition rate remains comparatively slow. PVA shows good miscibility and compatibility with other polymers by the formation of hydrogen bonds due to the presence of hydroxyl groups (Fukushima & Camino, 2016; Rahmat et al., 2009; Gupta et al., 2012). Blending results in unusual properties that may be completely different from the parent polymer. Natural polymers such as starch, cellulose, chitosan, and lignocelluloses show good compatibility and miscibility with PVA. Tian et al. (2017) used the melt processing technique for the fabrication of PVA/starch blended films. The major focus of their study was to investigate the structure and properties of the blends. The findings illustrated that due to the presence of the hydroxyl functional groups in PVA and starch, compatibility between the components was good. These findings were supported by the FTIR results. However, on the other hand, with an increase in starch content the crystallinity of the PVA films was decreased. Starch acted as a nucleating agent for the PVA that resulted in an increase in the crystallization temperature and also destruction of the strong interaction between the PVA chains that would result in a decrease in the crystalline structure. Wu et al. (2017) proposed a ternary blend of PVA/starch/citric acid for food-packaging applications. Blended films show good antibacterial activities on foodborne pathogenic bacteria such as *E. coli* and *Listeria monocytogenes*. The antibacterial activity of blended films was due to the presence of citric acid which combines with the cell membrane of bacteria to break down the synthesis system. Fabricated films were also used for the packaging of figs and their freshness was tested. Blended films prevented the formation of condensed water; moreover, there was no deterioration of the quality of the figs. In some of the applications, high flexibility and elongation percentage of the PVA films required moderate tensile strength also. For this purpose, plasticizers are blended with PVA (Sothornvit & Krochta, 2001). Some of the compatible polymers which are used as a plasticizer for PVA are glycerol, urea, polyethylene glycol, sorbitol, ascorbic acid, citric acid, tartaric acid, etc. (Ismail & Zaaba, 2014). These plasticizers are also used with PVA-blended biofilms. The mechanical properties of PVA-blended films mainly depend upon the number of functional groups (mostly the hydroxyl and carboxyl chemical groups) present in the plasticizer molecular structure. Succinic acid has a structure of H-0, C-2, which means no hydroxyl and two carboxyl chemical groups, as compared to the glycerol, which has a structure of H-3, C-0. The elongation percentage of glycerol-blended films was higher as compared with the succinic acid-blended films (Yoon et al., 2006). On the other hand, if both groups are present then the tensile strength and elongation percentage of the films increases, such as with malic acid (H-1, C-2) (Yoon et al., 2006a; Yun et al., 2006). The glass transition temperature of PVA films is also affected by the plasticizers, which makes the films more rubber-like (Jayasekara et al., 2003; Sreedhar et al., 2006; Yoon et al., 2006b). Aydin and Ilberg studied the effect of different plasticizers such as 1,4-butanediol, 1,2,6,-hexanetriol, mannitol, xylito, and pentaerythritol on the mechanical and glass transition temperature of PVA/starch-blended films (Aydin & Ilberg, 2016). The results show that, with an increase in the plasticizer content, the glass transient temperature of the PVA/starch-blended films decreases. In PVA-based blended films two types of hydrogen bonding take place: first between the PVA and starch; and second between the functional group of the plasticizer with PVA and the starch functional group. The second type of hydrogen bonding may disturb the hydrogen bonding between the PVA and starch. Due to the low molecular size of plasticizers, they penetrate through the matrix and reduce the crystallinity of blended films (lower glass transition temperature) with an improvement in their flexibility (Raj et al., 2004; Forssell et al., 1997; Chin & Te, 2008).

An antibacterial film for food packaging is also one of the important applications of PVA-blended films. For antimicrobial activity, chitosan, which is a natural biodegradable polymer, is used as the blended material in conjunction with PVA. Other than its antimicrobial activity, chitosan also has low oxygen permeability, which can help in the preservation of food, and it is also hydrophilic in nature, nontoxic, and has good film formability. Chitosan restricts microbial growth through the ionic interactions with negatively charged cell walls, and for the past few years many researchers have worked on PVA films modified with chitosan biopolymer to enhance the antibacterial growth of films. Some works also have shown an additional antibacterial agent to further improve upon the antibacterial activity of blended films. Olewnik-Kruszkowska et al. (2019) used poly-hexamethylene guanidine (PHMG) to further improve the antibacterial properties of PVA/chitosan blended films. To study the bacterial growth, *S. aureus* and *E. coli* were utilized to observe the antibacterial performance of the blended films. Neat PVA does not affect the growth of bacteria, whereas the PVA/chitosan show some reduction in the proliferation of both bacteria, but no such inhibition zone was observed. On the other hand, films containing PHMG show a proper inhibition zone as compared to the neat PVA and PVA/chitosan blend.

6. PVA-based composites

PVA composites have attracted enormous attraction in view of their satisfactory performance, properties, and biodegradable nature. Their performance in many applications such as the consumer, biomedical, and agriculture markets is well defined and promising. PVA composites have potential for use in numerous other applications. PVA has been widely used in the preparation of blends and composites with several natural, renewable polymers like chitosan, nanocellulose, starch, or lignocellulosic fillers. It is important that composite manufacturing technology is developed to achieve ecological sustainability while striving to meet consumer needs. The harmful effects of our disposable consumer lifestyle on our environment are well known and well documented; and as a result, governments are increasingly introducing legislation to control the effects on the environment of the materials used in the manufacture of many everyday products. Composite materials should be able to be recycled, reused, reprocessed, or biodegradable, to minimize their impact on the ecosystem. At the same time, the supply of materials should be sustainable and renewable. PVA is one of the most promising examples of biodegradable matrix polymers used in mulch films. These polymers have great potential as biodegradable matrices in environmental-friendly composites, in comparison to carbon fibers composites or other nonbiodegradable, recyclable fillers. PVA is widely used in agricultural mulch film or biodegradable packaging. For many innovative and environmentally conscious manufacturers, a composite consisting of PVA, a biopolymer, with natural fibers, that will further improve the PVA biodegradability and physical properties, is the choice of eco-sustainable materials.

PVA is a hydrophilic polymer which can be used as a matrix because of its good mechanical properties, low cost, processability, etc. Because of the high interfacial adhesion (Kuljanin et al., 2006) of PVA with the reinforcing material (fibers, particles, and flakes) it can be used for the fabrication of composites which are mechanically stronger and tougher, and have the potential for use in several applications. Because of the better interfacial adhesion between the PVA matrix and fibers, fiber-reinforced PVA composites were mechanically stronger and tougher than the PVA film, which results in the wide application of PVA for composite materials. PVA-based materials are used in the pharmaceutical and biomedical fields as drug carriers, and are also applied in tissue engineering science. PVA has various applications in the food industries as a binding and coating agent also. It is a film-coating agent, especially in applications where moisture barrier/protection properties are required. As a component of tablet-coating formulations intended for products including food supplement tablets, PVA protects the active ingredients from moisture, oxygen, and other environmental components, while simultaneously masking the taste and odor. The viscosity of PVA allows for the application of PVA coating agents to tablets, capsules, and other forms to which film coatings are typically applied at relatively high solids contents. PVA hydrogels have certain advantages which make them ideal candidates for biomaterials. The advantages of PVA hydrogels are that they are nontoxic, noncarcinogenic, and bioadhesive in nature. PVA also shows a high degree of swelling in water (or biological fluids) and a rubbery and elastic nature, and therefore closely simulates natural tissues and can be readily accepted into the body. PVA gels have been used for contact lenses, the lining for artificial hearts, and drug-delivery applications. PVA hydrogels have been used for various biomedical applications. Many researchers have been working on PVA-based composites recently.

In 2008, Lu et al. prepared PVA-based composites reinforced with microfibrillated cellulose (MFC). A solution of 5 wt.% PVA with water was stirred at 95°C for 2 h, and varying contents of microfibrillated cellulose (1, 5, 10, and 15 wt.%) were added after cooling at room temperature. It was observed that the tensile strength and Young's modulus were enhanced with an increase in the MFC content up to 10 wt.%, but they decrease at a relatively higher MFC content. The Young's modulus of composite film was increased by 40% at a 10 wt.% MFC content. In 2008, Roohani et al. prepared a nanocomposite based on PVA which was reinforced with cellulose whiskers (0, 3, 6, 9, and 12 wt.% for PVA) prepared from cotton linter. The tensile strength of nanocomposites was studied under different relative humidities and it was found that there was an increase in the tensile modulus of nanocomposite films when the whisker content increased for each relative humidity condition. For a given filler content, there was a significant decrease in the tensile modulus with an increase in the humidity. In 2012, Qiu and Netravali fabricated PVA-based biodegradable composite reinforced with microfibrillated cellulose at different contents (5%, 10%, 15%, 20%, 30%, 40%, and 50% by weight). Herein, glyoxal was used as a cross-linking agent. Tensile test results showed that the tensile strength of neat PVA was 34.1 MPa, which increased to 89.9 MPa with the increasing content of MFC up to 40%. In 2013, Baheti et al. reinforced small particles of jute in the PVA domain. Jute fibers were processed into microscale particles in the shape of nanofibrillar cellulose (NFC) using a high-energy planetary ball milling method. It was noted that the PVA film glass transition temperature (T_g) increased from 84.36°C to 95.22°C after adding 5% jute particles without affecting the PVA crystallinity and melting temperature (T_m). Dynamic mechanical analysis (DMA) of the film revealed 5% jute particles with an increase in the value of storage modules 14×10^8 Pa compared to the 9×10^8 Pa of neat film. In 2013, Xu et al. prepared the PVA/cellulose nanocrystals/silver nanoparticle nanocomposite. In 100 mL distilled water, 10% PVA solution was mixed with a specific amount of CNs/AgNPs and CNs. The tensile results showed that the maximum tensile strength obtained for the PVA/CNs/AgNPs was 81.21 MPa, and was 57.02 MPa for the neat PVA. This shows an increase in the tensile properties of PVA after reinforcement of the cellulose nanocrystals or silver nanoparticles.

In 2015, Ching et al. reported the thermomechanical and morphological properties of the nanocellulose and nanosilica reinforced polyvinyl alcohol nanocomposites.

To prepare the films, 10 wt.% PVA solution was stirred at 90°C for approximately 2 h, and after cooling to room temperature nanocrystals cellulose (1−7 wt.%) and nanosilica (0.5%) were added. Mechanical testing of PVA-based film was done to examine its mechanical behavior. It was found that the tensile strength and tensile modulus were optimized at 3 wt.% of nanocellulose and 0.5 wt.% of nanosilica composition. The tensile and modulus properties achieved for PVA composites were up to 5 wt.% of nanocellulose. The tensile strength was improved due to the crystalline nature of nanosilica particles. A study has shown a significant improvement in the thermomechanical properties of nanomaterial-reinforced PVA composites with 3 wt.% of nanocellulose and 0.5 wt.% of nanosilica, and good optical properties due to the effective dispersion and polymer−filler interaction. In 2016, Ye et al. prepared PVA-based film reinforced with eucalyptus lignosulfonate calcium (HLS) particles. Different ratios of HLS to PVA (0/100, 5/95, 25/75, 35/65, 50/50, and 60/40 wt./wt.) were dissolved in 100 mL distilled water stirred at 95°C for 3 h and then dried at 70°C in a vacuum oven. Pure PVA has a tensile strength of 48.4 MPa, elongation at break 220.7%, and Young's modulus 707.9 MPa. With the addition of HLS (35 wt.%) the tensile strength increased to 124.15%, whereas the Young's modulus increased to 57.19%. In 2016, Kashyap et al. developed graphene oxide (GO)-reinforced PVA nanocomposite. The results showed that the tensile strength of neat PVA film was 25.3 ± 3 MPa, whereas PVA-GO had 63 ± 5 MPa and the Young's modulus of the PVA film was 2.32 ± 0.3 MPa while PVA-GO had 5.82 ± 0.6 MPa. In 2016, Choo et al. prepared bionanocomposite films from chitosan (CS) and polyvinyl alcohol (PVA) blended with 2,2,6,6-tetramethylpiperidine-1-oxyl radical (TEMPO) oxidized cellulose fiber (TOCN) at different contents of 0, 0.5, 1.0, and 1.5 wt.% through a solution casting method. FTIR showed the compatible and miscible owing hydrogen bonding interaction between PVA and CS. It was found that at lower filler loading, TOCNs were homogeneously distributed and began to agglomerate at 1 wt.% of TOCNs. The tensile profile showed that the tensile strength of PVA/CS composite films was higher than the films without filler reinforcement. However, at low loading the flexibility of PVA/CS composite films was reduced. A mild change was found in thermal stability. The FTIR and XRD analyses illustrated the strong interaction between PVA/CS polymer matrix and TCONs, which resulted in better dispersion of nanofillers into the polymer matrix. In 2018, Sarwar et al. prepared PVA nanocomposite films reinforced with nanocellulose and Ag nanoparticles for antimicrobial food packaging. The mechanical and thermal properties were enhanced by the reinforcement of Ag nanoparticles and nanocellulose. Various attributes such as the mechanical properties, water vapor transmission rate, and moisture retention capability were examined. The tensile strength of neat PVA was $5.23 + 0.27$ MPa. With the addition of 8 wt.% nanocellulose the highest stress value was obtained at around $12.32 + 0.61$ MPa. The highest value of elongation at break of $518.10 + 0.25$ was obtained 12 wt.%. There was a further increase in the mechanical properties with the addition Ag nanoparticles in the PVA/NC film.

7. PVA-based cross-linked composites

Polyvinyl alcohol is a biodegradable, nontoxic polymer used in food packaging due to its high oxygen and carbon dioxide barrier properties. However, owing to its water solubility, it is often coupled in multilayer structures with other polymers. The cross-linking of PVA would be an alternative method of avoiding the use of several polymers. Cross-linking is a process of bonding (usually chemically) polymer chains with each other to improve their properties. Cross-linking is performed with chemical reagents with multifunctional groups having good reactivity with the functional groups of the polymer that needs to be modified. Cross-linking is performed with chemical reagents with multifunctional groups having good reactivity with the functional groups of the polymer that needs to be modified. Depending upon the nature of the polymer, different techniques may be used for cross-linking. Cross-linking may occur through polymerization of monomers having more than two functionalities (by condensation) or by covalent bonding between a polymeric chain through irradiation, sulfur vulcanization, or chemical reactions by adding different chemicals in conjunction with heating and, sometimes, pressure. In all cases, the chemical structure of the polymer is altered through the cross-linking process. Adding cross-links between polymer chains affects the physical properties of the polymer, depending upon the degree of cross-linking and the presence or absence of crystallinity. As a result of cross-linking, the polymer will become less viscous and less elastic. The viscosity of the polymer will decrease and, in order for polymers to flow, the chains must move past each other, which cross-linking prevents. As a result of the restriction in flow there is an improvement in the creep behavior. Cross-linking results in insolubility as the chains are tied together by strong covalent bonds. Cross-linking increases the molecular mass of a polymer. Cross-linked polymers are important because they are mechanically strong and resistant to heat, wear, and attack by solvents.

For polymers such as PVA, dialdehydes like glutaraldehyde, glyoxal, and polycarboxylic acids are some common cross-linkers used for PVA. Maleic acid, boric acid, malic acid, fumaric acid, and citric acid can also be used for the cross-linking of PVA. The cross-linking of PVA reduces its hydrophilic nature and it is not completely soluble in water after cross-linking has been done. The effect of cross-linking can be examined by FT-IR spectroscopy. Many researchers have studied the effects of cross-linking on the properties of PVA.

In 2009, Zhou et al. prepared and characterized surface cross-linked thermoplastic starch (TPS)/PVA by applying ultraviolet (UV) irradiation. Cross-linking was done by soaking the blend films in an aqueous solution of sodium benzoate. The most effective surface photo cross-linking in TPS/PVA films was found at 0.75% of aqueous sodium benzoate solution for 30 s. The surface hydrophilic property of TPS/PVA films was reduced by surface photo cross-linking and the water resistance of the films increased. Increases in tensile strength and Young's modulus but a decrease in elongation at break of TPS/PVA films were observed due to the surface photo cross-linking modification. PVA is used in preservation, and hence dehydration of ethylene glycol was studied by Hyder and Chen (2009). They prepared chitosan–polyvinyl alcohol

(CS-PVA) blend membrane was cross-linked with trimesoyl chloride (TMC)/hexane. A 5 wt.% solution of PVA with DI water and 1 wt.% solution with acetic acid was mixed in ratios of CS−PVA: 25−75, 50−50, 70−30, 75−25, and 80−20, and cross-linking was done by dipping the membrane into 0.5 wt./vol.% TMC/hexane. It was found that the tensile strength of cross-linked CS-PVA films was higher in films which had a higher percentage of CS and the maximum value obtained for CS−PVA (80−20) ratio was 74.5 ± 2.7. In 2009, Figueiredo et al. studied the effects of polyvinyl alcohol cross-linked films under mild conditions. Cross-linking of PVA with glutaraldehyde was done in the absence of an organic solvent and an acid catalyst in order to improve the water resistance of the hydrophilic polymer. Glutaraldehyde had been selected as the cross-linker as it favors an intermolecular reaction with PVA and is capable of binding with protein nonspecifically. The film's oxygen permeability was decreased 2.7-fold, indicating that the polymer cross-linking had been achieved effectively. For a GA/PVA mass ratio of 0.01, the highest cross-linking density was observed. A further increase in GA content causes PVA branching instead of cross-linking, showing greater water solubility due to a loss of packing density. In 2010, Wang and Hsieh studied the cross-linking of polyvinyl alcohol (PVA) fibrous membrane with polyethylene glycol (PEG), diacyl chloride, and glutaraldehyde. It was stated that an electrospun aqueous solution of PVA cross-linked with GA and PEG diacyl chloride was found to be effective in fabricating water-stable hydrophilic PVA fibrous membrane. There was an enhancement in crystalline structure in PVA fibrous membrane cross-linked with GA/ethanol, whereas there was a reduction in the crystalline structure in PVA fibers cross-linked with GA in an aqueous solution of sodium sulfate (Na_2SO_4). Complete loss of the crystalline structure was found in PEG diacyl chloride cross-linked in toluene/pyridine. A reaction with a lower and shorter extent of PEG produced hydrophobic PVA fibrous membrane with a minimum change in interfiber porous structure, among all cross-linking agents. In 2014, Wang et al. investigated the properties of polyvinyl alcohol/xylan composite with citric acid as a cross linking agent or a new plasticizer. Composite PVA/xylan films were prepared by a solution casting method in the presence of citric acid. By altering the PVA/xylan weight ratio and content of CA, enhancements in tensile strength and elongation properties could be accomplished. CA resulted in elevated tensile strength but low elongation at break as a cross-linking agent. The CA content had a greater impact on the moisture permeability; however, the PVA/xylan weight ratio had little effect on the moisture permeability. These findings showed clearly that the CA content in film formulations had important impacts on mechanical properties, degree of swelling, and permeability of moisture. In 2014, Birck et al. fabricated films of PVOH (polyvinyl alcohol) and HPβCD (hydroxypropyl-β-cyclodextrin) cross-linked with CTR (citric acid) in 80/10/10 weight ratios, using a solvent casting method followed by heat treatment. Different film features (chemical structure, glass transition temperature, crystallinity, and tensile properties) have been researched in relation to their formulation and cross-linking timing. Films with elevated CTR content (40 wt.%) are too fragile for food implementation applications, whereas films with a low CTR content (10 wt.%) are transparent and demonstrate ductile behavior.

In 2015, Chen et al. prepared a PVA composite reinforced with graphene oxide (GO) and used boric acid as a cross-linking agent by an aqueous solution method. An improvement in mechanical properties and thermal stability was reported. Tensile strength was increased from 23.3 to 67.7 MPa by adding 5 wt.% of boric acid, while with the addition of 0.2 wt.% GO alone the tensile strength was improved from 23.3 to 53.4 MPa. It was found that by adding both 0.2 wt.% GO and 5 wt.% of boric acid, the enhancement in tensile properties was 65−85.5 MPa. In 2016, Jahan et al. studied the effect of ionic cross-linkage with KNO_3 on chitosan−PVA blend film to enhance the mechanical strength and percentage elongation. The result showed that the tensile strength of the CS-PVA (50/50) blend was increased by 298.75% and percentage of elongation from 7.16% to 67% after cross-linking with KNO_3 (0.5 g). In 2017, Jose and Al-Harthi prepared polyvinyl alcohol/starch/graphene nanocomposites by solution mixing and casting method with the addition of citric acid as cross-linking agent and glycerol as the plasticizer. The interaction between the PVA and/or starch hydroxyl groups and oxygen-containing groups on the graphene sheet was enhanced by the cross-linking reaction. Adding starch to the PVA matrix tends to decrease PVA's tensile strength. The tensile strength and modulus increased markedly with the addition of graphene. It was found that in the polymer matrix 0.5 wt.% of graphene gave the highest characteristics, tensile strength increased from 6.7 to 10 MPa, and Young's modulus from 24.2 to 53.3 MPa. The increase in elongation-at-break (from 27.5% to 57.2%) also verified that the composite toughness was enhanced as a result of the enhanced interaction between the components. In 2017, Sonker et al. studied the PVA cross-linked films using aliphatic (suberic) and aromatic (terephthalic) dicarboxylic acids. The results showed that suberic acid is a better cross-linking agent than terephthalic acid. The maximum strength in the cross-linked samples is 32.5 MPa for PVA cross-linked suberic acid than the neat PVA (22.6 MPa). Cross-linked film using terephthalic acid (35% w/w) with a curing time of 8 h showed at least 5.4% water intake relative to the smooth PVA, which is easily dissolved in water. DTGA showed that the decomposition temperature of the cross-linked PVA film was 345°C, while that of a neat PVA was 315°C. FTIR confirms cross-link ester bond formation in film. There have also been very significant works carried out by authors in the field of bio- and hybrid composites (Verma, Singh, Verma, & Sharma, 2016; Verma, Parashar, & Packirisamy, 2018; Verma, Gaur, & Singh, 2017; Verma, Parashar, & Packirisamy, 2019; Verma, Negi, & Singh, 2019; Verma & Singh, 2019; Verma et al., 2018a, 2018b; Verma & Singh, 2016; Verma, Budiyal et al., 2019; Verma, Joshi et al., 2018; Verma, Singh et al., 2019; Verma, Kumar, & Parashar, 2019; Chaurasia et al., 2019; Verma, Singh, & Arif, 2016; Verma, Baurai et al., 2020; Rastogi et al., 2020; Verma, Parashar et al., 2020; Singh et al., 2020; Verma, Jain et al., 2020; Bharath et al., 2021; Marichelvam et al., 2021; Bharath et al., 2020).

Acknowledgment

Financial and academic support from the University of Petroleum and Energy Studies, Dehradun, India, is highly appreciated by the authors.

References

Abdullah, Z. W., Dong, Y., Davies, I. J., & Barbhuiya, S. (2017). PVA, PVA blends, and their nanocomposites for biodegradable packaging application. *Polymer – Plastics Technology and Engineering, 56*(12), 1307–1344.

Aslam, M., Kalyar, M. A., & Raza, Z. A. (2017). Fabrication of reduced graphene oxide nanosheets doped PVA composite films for tailoring their opto-mechanical properties. *Applied Physics A, 123*(6), 424.

Awada, H., & Daneault, C. (2015). Chemical modification of poly (vinyl alcohol) in water. *Applied Sciences, 5*(4), 840–850.

Aydin, A. A., & Ilberg, V. (2016). Effect of different polyol-based plasticizers on thermal properties of polyvinyl alcohol: starch blends. *Carbohydrate Polymers, 136*, 441–448. https://doi.org/10.1016/j.carbpol.2015.08.093

Baheti, V., Militky, J., & Marsalkova, M. (2013). Mechanical properties of poly lactic acid composite films reinforced with wet milled jute nanofibers. *Polymer Composites, 34*(12), 2133–2141.

Balasubramanian, K. (2014). Antibacterial application of polyvinylalcohol-nanogold composite membranes. *Colloids and Surfaces A: Physicochemical and Engineering Aspects, 455*, 174–178.

Bharath, K. N., Madhu, P., Gowda, T. G., Verma, A., Sanjay, M. R., & Siengchin, S. (2020). A novel approach for development of printed circuit board from biofiber based composites. *Polymer Composites, 41*(11), 4550–4558.

Bharath, K. N., Madhu, P., Gowda, T. Y., Verma, A., Sanjay, M. R., & Siengchin, S. (2021). Mechanical and chemical properties evaluation of sheep wool fiber–reinforced vinylester and polyester composites. *Materials Performance and Characterization, 10*(1), 99–109.

Birck, C., Degoutin, S., Tabary, N., Miri, V., & Bacquet, M. (2014). New crosslinked cast films based on poly (vinyl alcohol): preparation and physico-chemical properties. *Express Polymer Letters, 8*(12).

Chaurasia, A., Verma, A., Parashar, A., & Mulik, R. S. (2019). Experimental and computational studies to analyze the effect of h-BN nanosheets on mechanical behavior of h-BN/polyethylene nanocomposites. *Journal of Physical Chemistry C, 123*(32), 20059–20070.

Chen, J., Li, Y., Zhang, Y., & Zhu, Y. (2015). Preparation and characterization of graphene oxide reinforced PVA film with boric acid as crosslinker. *Journal of Applied Polymer Science, 132*(22).

Chin, A. L., & Te, H. K. (2008). Shear and elongational flow properties of thermoplastic polyvinyl alcohol melts with different plasticizer contents and degrees of polymerization. *Journal of Materials Processing Technology, 200*, 331–338.

Ching, Y. C., Rahman, A., Ching, K. Y., Sukiman, N. L., & Cheng, H. C. (2015). Preparation and characterization of polyvinyl alcohol-based composite reinforced with nanocellulose and nanosilica. *BioResources, 10*(2), 3364–3377.

Choo, K., Ching, Y., Chuah, C., Julai, S., & Liou, N. S. (2016). Preparation and characterization of polyvinyl alcohol-chitosan composite films reinforced with cellulose nanofiber. *Materials, 9*(8), 644.

DeMerlis, C. C., & Schoneker, D. R. (2003). Review of the oral toxicity of polyvinyl alcohol (PVA). *Food and Chemical Toxicology, 41*(3), 319–326.

El-Naas, M. H., Mourad, A. H. I., & Surkatti, R. (2013). Evaluation of the characteristics of polyvinyl alcohol (PVA) as matrices for the immobilization of Pseudomonas putida. *International Biodeterioration and Biodegradation, 85*, 413–420.

Figueiredo, K. C., Alves, T. L., & Borges, C. P. (2009). Poly (vinyl alcohol) films cross linked by glutaraldehyde under mild conditions. *Journal of Applied Polymer Science, 111*(6), 3074–3080.
Forssell, P. M., Mikkila, J. M., Moates, G. K., & Parker, R. (1997). Phase and glass transition behaviour of concentrated barley starch-glycerol-water mixtures, a model for thermoplastic starch. *Carbohydrate Polymers, 34*, 275–282.
Fukushima, K., & Camino, G. (2016). Polymer nanocomposites biodegradation. In A. Dasari, & K. Fukushim (Eds.), *Functional and physical properties of polymer nanocomposites* (pp. 57–91). Chichester: Wiley.
Gupta, B., Agarwal, R., & Alam, M. S. (2012). Preparation and characterization of polyvinyl alcohol-polyethylene oxide-carboxymethyl cellulose blend membranes. *Journal of Applied Polymer Science, 127*, 1301–1308.
Haehnel, W., & Herrmann, W. O. (1924). *German Pat. 450286 to consort. F. Elektrochem. Inc. G.m.b.H.*
Herrmann, W. O., & Haehnel, W. (1932). *German Pat. 685048 (1931) to consort. F. Elektrochem. Ind. G.m.b.H.; Brit. Pat. 386161.*
Hindeleh, A. M., & Johnson, D. J. (1972). Peak resolution and X-ray crystallinity determination in heat-treated cellulose triacetate. *Polymer, 13*(1), 27–32.
Hyder, M. N., & Chen, P. (2009). Pervaporation dehydration of ethylene glycol with chitosan–poly(vinyl alcohol) blend membranes: effect of CS–PVA blending ratios. *Journal of Membrane Science, 340*, 171–180.
Ismail, H., & Zaaba, N. F. (2014). Effect of unmodified and modified sago starch on properties of (sago starch)/silica/PVA plastic films. *Journal of Vinyl and Additive Technology, 20*(3), 185–192. https://doi.org/10.1002/vnl.21344
Jahan, F., Mathad, R. D., & Farheen, S. (2016). Effect of mechanical strength on chitosan-pva blend through ionic crosslinking for food packaging application. *Materials Today: Proceedings, 3*, 3689–3696.
Jain, N., Singh, V. K., & Chauhan, S. (2017). A review on mechanical and water absorption properties of polyvinyl alcohol based composites/films. *Journal of the Mechanical Behavior of Materials, 26*(5–6), 213–222. https://doi.org/10.1515/jmbm-2017-0027
Jain, N., Singh, V. K., & Chauhan, S. (2018). Dynamic and creep analysis of polyvinyl alcohol based films blended with starch and protein. *Journal of Polymer Engineering, 39*(1), 26–35. https://doi.org/10.1515/polyeng-2018-0032
Jain, N., Ali, S., Singh, V. K., Singh, K., Nitesh, & Chohan, S. (2019). Creep and dynamic mechanical behaviour of Cross-linked polyvinyl alcohol reinforced with cotton fiber laminate composites. *Journal of Polymer Engineering, 39*(4), 326–335. https://doi.org/10.1515/polyeng-2018-0286
Jain, N., Verma, A., & Singh, V. K. (2019). Dynamic mechanical analysis and creep-recovery behaviour of polyvinyl alcohol based cross-linked biocomposite reinforced with basalt fiber. *Materials Research Express, 6*(10), 105373.
Jawaid, M., & Abdul Khalil, H. P. S. (2011). Cellulosic/synthetic fibre reinforced polymer hybrid composites: a review. *Carbohydrate Polymers, 86*(1), 1–18.
Jayasekara, R., Harding, I., Bowater, I., Christie, G. B. Y., & Lonergan, G. T. J. (2003). Biodegradation by composting of surface modified starch and PVA blended films. *Journal of Polymers and the Environment, 11*, 49–56.
Jose, J., & Al-Harthi, M. A. (2017). Citric acid cross linking of poly (vinyl alcohol)/starch/graphene nanocomposites for superior properties. *Iranian Polymer Journal, 26*(8), 579–587.

Kashyap, S., Pratihar, S. K., & Behera, S. K. (2016). Strong and ductile graphene oxide reinforced PVA nanocomposites. *Journal of Alloys and Compounds*, 684254–684260.

Kawai, F., & Hu, X. (2009). Biochemistry of microbial polyvinyl alcohol degradation. *Applied Microbiology and Biotechnology, 84*(2), 227.

Kim, B. C., Sohn, C. K., Lim, S. K., Lee, J. W., & Park, W. (2003). Degradation of polyvinyl alcohol by *Sphingomonas* sp. SA3 and its symbiote. *Journal of Industrial Microbiology and Biotechnology, 30*(1), 70–74.

Kirk-Othmer, *Encyclopedia of chemical technology, index to Volumes 1 – 26* (5th ed.)

Klatte, F. (1913). *German Pat. 271381 (1912) to chem. Fabrik Griesheim Elektron; U.S. Pat. 1084581.*

Kuljanin, J., Omor, M. I. C., Djokovic, V., & Nedeljkovic, J. M. (2006). Synthesis and characterization of nanocomposite of polyvinyl alcohol and lead sulfide nanoparticles. *Materials Chemistry and Physics, 95*, 67–71.

Li, M., Zhang, D. X., Du, G. C., & Chen, J. (2012). Enhancement of PVA-degrading enzyme production by the application of pH control strategy. *Journal of Microbiology and Biotechnology, 22*(2), 220–225.

Lu, J., Wang, T., & Drzal, L. T. (2008). Preparation and properties of microfibrillated cellulose polyvinyl alcohol composite materials. *Composites Part A: Applied Science and Manufacturing, 39*(5), 738–746.

Marichelvam, M. K., Manimaran, P., Verma, A., Sanjay, M. R., Siengchin, S., Kandakodeeswaran, K., & Geetha, M. (2021). A novel palm sheath and sugarcane bagasse fiber based hybrid composites for automotive applications: an experimental approach. *Polymer Composites, 42*(1), 512–521.

Marin, E., Rojas, J., & Ciro, Y. (2014). A review of polyvinyl alcohol derivatives: promising materials for pharmaceutical and biomedical applications. *African Journal of Pharmacy and Pharmacology, 8*(24), 674–684.

Marušincová, H., Husárová, L., Růžička, J., Ingr, M., Navrátil, V., Buňková, L., & Koutny, M. (2013). Polyvinyl alcohol biodegradation under denitrifying conditions. *International Biodeterioration and Biodegradation, 84*, 21–28.

Olewnik-Kruszkowska, E., Gierszewska, M., Jakubowska, E., Tarach, I., Sedlarik, V., & Pummerova, M. (2019). Antibacterial films based on PVA and PVA–chitosan modified with poly(hexamethylene guanidine). *Polymers, 11*, 2093.

Qiu, K., & Netravali, A. N. (2012). Fabrication and characterization of biodegradable composites based on microfibrillated cellulose and polyvinyl alcohol. *Composites Science and Technology, 72*(13), 1588–1594.

Rahmat, A. R., Rahman, W. A., Sin, L. T., & Yussuf, A. (2009). Approaches to improve compatibility of starch filled polymer system: a review. *Materials Science and Engineering: C, 29*, 2370–2377.

Raj, B., Sidddaramaiah, S., & Somashekar, R. (2004). Structure-property relation in polyvinyl alcohol/starch composites. *Journal of Applied Polymer Science, 91*, 630–635.

Rastogi, S., Verma, A., & Singh, V. K. (2020). Experimental response of nonwoven waste cellulose fabric–reinforced epoxy composites for high toughness and coating applications. *Materials Performance and Characterization, 9*(1), 151–172.

Roohani, M., Habibi, Y., Belgacem, N. M., Ebrahim, G., Karimi, A. N., & Dufresne, A. (2008). Cellulose whiskers reinforced polyvinyl alcohol copolymers nanocomposites. *European Polymer Journal, 44*(8), 2489–2498.

Saini, I., Sharma, A., Dhiman, R., Aggarwal, S., Ram, S., & Sharma, P. K. (2017). Grafted SiC nanocrystals: for enhanced optical, electrical and mechanical properties of polyvinyl alcohol. *Journal of Alloys and Compounds, 714*, 172–180.

Sarwar, M. S., Niazi, M. B. K., Jahan, Z., Ahmad, T., & Hussain, A. (2018). Preparation and characterization of PVA/nanocellulose/Ag nanocomposite films for antimicrobial food packaging. *Carbohydrate Polymers, 184*, 453–464.

Shimao, M. (2001). Biodegradation of plastics. *Current Opinion in Biotechnology, 12*(3), 242–247.

Singh, K., Jain, N., Verma, A., Singh, V. K., & Chauhan, S. (2020). Functionalized graphite–reinforced cross-linked poly (vinyl alcohol) nanocomposites for vibration isolator application: morphology, mechanical, and thermal assessment. *Materials Performance and Characterization, 9*(1), 215–230.

Sonker, A. K., Rathore, K., Nagarale, R. K., & Verma, V. (2017). Crosslinking of polyvinyl alcohol (PVA) and effect of cross linker shape (aliphatic and aromatic) thereof. *Journal of Polymers and the Environment, 26*(5), 1782–1794.

Sothornvit, R., & Krochta, J. M. (2001). Plasticizer effect on mechanical properties of beta-lactoglobulin films. *Journal of Food Engineering, 50*(3), 149–155. https://doi.org/10.1016/S0260-8774(00)00237-5

Sreedhar, B., Chattopadhyay, D. K., Karunakar, M. S. H., & Sastry, A. R. K. (2006). Thermal and surface characterization of plasticized starch polyvinyl alcohol blends cross linked with epichlorohydrin. *Journal of Applied Polymer Science, 101*, 25–34, 415.

Staudinger, H., & Arbeitserinnerunger, D. (1961). *Alfred Hutig Verlag G.m.b.H., Heidelberg* (p. 196).

Surkatti, R., & El-Naas, M. H. (2014). Biological treatment of wastewater contaminated with p-cresol using Pseudomonas putida immobilized in polyvinyl alcohol (PVA) gel. *Journal of Water Process Engineering, 1*, 84–90.

Suzuki, T., Ichihara, Y., Yamada, M., & Tonomura, K. (1973). Some characteristics of Pseudomonas 0–3 which utilizes polyvinyl alcohol. *Agricultural and Biological Chemistry, 37*(4), 747–756.

Tian, H., Yan, J., Rajulu, A. V., Xiang, A., & Luo, X. (2017). Fabrication and properties of polyvinyl alcohol/starch blend films: effect of composition and humidity. *International Journal of Biological Macromolecules, 96*, 518–523.

Verma, A., & Singh, V. K. (2016). Experimental investigations on thermal properties of coconut shell particles in DAP solution for use in green composite applications. *Journal of Materials Science and Engineering, 5*(3), 1–5.

Verma, A., & Singh, V. K. (2019). Mechanical, microstructural and thermal characterization of epoxy-based human hair–reinforced composites. *Journal of Testing and Evaluation, 47*(2), 1193–1215.

Verma, A., Singh, V. K., & Arif, M. (2016). Study of flame retardant and mechanical properties of coconut shell particles filled composite. *Research Review: Journal of Materials Science, 4*(3), 1–5.

Verma, A., Singh, V. K., Verma, S. K., & Sharma, A. (2016). Human hair: a biodegradable composite fiber–a review. *International Journal of Waste Resources, 6*(206), 2.

Verma, A., Gaur, A., & Singh, V. K. (2017). Mechanical properties and microstructure of starch and sisal fiber biocomposite modified with epoxy resin. *Materials Performance and Characterization, 6*(1), 500–520.

Verma, A., Joshi, K., Gaur, A., & Singh, V. K. (2018). Starch-jute fiber hybrid biocomposite modified with an epoxy resin coating: fabrication and experimental characterization. *Journal of the Mechanical Behavior of Materials, 27*(5–6).

Verma, A., Negi, P., & Singh, V. K. (2018a). Experimental investigation of chicken feather fiber and crumb rubber reformed epoxy resin hybrid composite: mechanical and microstructural characterization. *Journal of the Mechanical Behavior of Materials, 27*(3–4).

Verma, A., Negi, P., & Singh, V. K. (2018b). Physical and thermal characterization of chicken feather fiber and crumb rubber reformed epoxy resin hybrid composite. *Advances in Civil Engineering Materials, 7*(1), 538–557.

Verma, A., Parashar, A., & Packirisamy, M. (2018). Atomistic modeling of graphene/hexagonal boron nitride polymer nanocomposites: a review. *Wiley Interdisciplinary Reviews: Computational Molecular Science, 8*(3), e1346.

Verma, A., Budiyal, L., Sanjay, M. R., & Siengchin, S. (2019). Processing and characterization analysis of pyrolyzed oil rubber (from waste tires)-epoxy polymer blend composite for lightweight structures and coatings applications. *Polymer Engineering and Science, 59*(10), 2041–2051.

Verma, A., Kumar, R., & Parashar, A. (2019). Enhanced thermal transport across a bi-crystalline graphene–polymer interface: an atomistic approach. *Physical Chemistry Chemical Physics, 21*(11), 6229–6237.

Verma, A., Negi, P., & Singh, V. K. (2019). Experimental analysis on carbon residuum transformed epoxy resin: chicken feather fiber hybrid composite. *Polymer Composites, 40*(7), 2690–2699.

Verma, A., Parashar, A., & Packirisamy, M. (2019). Effect of grain boundaries on the interfacial behaviour of graphene-polyethylene nanocomposite. *Applied Surface Science, 470*, 1085–1092.

Verma, A., Singh, C., Singh, V. K., & Jain, N. (2019). Fabrication and characterization of chitosan-coated sisal fiber–Phytagel modified soy protein-based green composite. *Journal of Composite Materials, 53*(18), 2481–2504.

Verma, A., Baurai, K., Sanjay, M. R., & Siengchin, S. (2020). Mechanical, microstructural, and thermal characterization insights of pyrolyzed carbon black from waste tires reinforced epoxy nanocomposites for coating application. *Polymer Composites, 41*(1), 338–349.

Verma, A., Jain, N., Parashar, A., Singh, V. K., Sanjay, M. R., & Siengchin, S. (2020). Design and modeling of lightweight polymer composite structures. In *Lightweight polymer composite structures* (pp. 193–224). CRC Press.

Verma, A., Parashar, A., Jain, N., Singh, V. K., Rangappa, S. M., & Siengchin, S. (2020). Surface modification techniques for the preparation of different novel biofibers for composites. In *Biofibers and biopolymers for biocomposites* (pp. 1–34). Cham: Springer.

Wang, Y., & Hsieh, Y. L. (2010). Crosslinking of polyvinyl alcohol (PVA) fibrous membranes with glutaraldehyde and PEG diacylchloride. *Journal of Applied Polymer Science, 116*(6), 3249–3255.

Wang, S., Ren, J., Li, W., Sun, R., & Liu, S. (2014). Properties of polyvinyl alcohol/xylan composite films with citric acid. *Carbohydrate Polymers, 103*, 94–99.

Wu, Z., Wu, J., Peng, T., Li, Y., Lin, D., Xing, B., Li, C., Yang, Y., Yang, L., Zhang, L., Ma, R., Wu, W., Lv, X., Dai, J., & Han, G. (2017). Preparation and application of starch/polyvinyl alcohol/citric acid ternary blend antimicrobial functional food packaging films. *Polymers, 9*, 102. https://doi.org/10.3390/polym9030102

Xu, Y. X., Kim, K. M., Hanna, M. A., & Nag, D. (2005). Chitosan-starch composite film: preparation and characterization. *Industrial Crops and Products, 21*(2), 185–192.

Xu, X., Yang, Y. Q., Xing, Y. Y., Yang, J. F., & Wang, S. F. (2013). Properties of novel polyvinyl alcohol/cellulose nanocrystals/silver nanoparticles blend membranes. *Carbohydrate Polymers, 98*(2), 1573–1577.

Ye, M., Mohanty, P., & Ghosh, G. (2014). Morphology and properties of poly vinyl alcohol (PVA) scaffolds: impact of process variables. *Materials Science and Engineering: C, 42*, 289–294.

Ye, D. Z., Jiang, L., Hu, X. Q., Zhang, M. H., & Zhang, X. (2016). Lignosulfonate as reinforcement in polyvinyl alcohol film: mechanical properties and interaction analysis. *International Journal of Biological Macromolecules, 83*, 209–215.

Yoon, S. D., Chough, S. H., & Park, H. R. (2006a). Effects of additives with different functional groups on the physical properties of starch/PVA blend film. *Journal of Applied Polymer Science, 100*(5), 3733–3740. https://doi.org/10.1002/app.23303

Yoon, S. D., Chough, S. H., & Park, H. R. (2006b). Properties of starch-based blend films using citric acid as additive. *Journal of Applied Polymer Science, 100*, 2554–2560.

Yun, Y. H., Na, Y. H., & Yoon, S. D. (2006). Mechanical properties with the functional group of additives for starch/PVA blend film. *Journal of Polymers and the Environment, 14*(1), 71–78. https://doi.org/10.1007/s10924-005-8709-8

Zhou, J., Ma, Y., Ren, L., Tong, J., Liu, Z., & Xie, L. (2009). Preparation and characterization of surface cross linked TPS/PVA blend films. *Carbohydrate Polymers, 76*(4), 632–638.

PBAT-based blends and composites

Sudheer Kumar[1], *Sukhila Krishnan*[2], *Smita Mohanty*[1] *and Sanjay Kumar Nayak*[1]
[1]School for Advanced Research in Petrochemicals (SARP), Laboratory for Advanced Research in Polymeric Materials (LARPM), Central Institute of Petrochemicals Engineering & Technology (CIPET), Bhubaneswar, Odisha, India; [2]Sahrdaya College of Engineering and Technology, Department of Applied Science and Humanities, Kodakara, Thrissur, Kerala, India

1. Introduction

Despite several benefits, petroleum-based conventional nonbiodegradable plastics have a serious drawback in terms of their environmental effects, nondegradable waste, higher energy consumption, and health risks, etc. Thus, there is a need to find an alternate eco-friendly biodegradable polymer to be employed mostly in the packaging and agricultural film industries. Biodegradable polymers are greener alternatives to petro-based nonbiodegradable plastic materials, with frequently specified advantages such as being completely biodegradable, with a lower carbon footprint and eco-friendly nature (Bastioli, 2020; Kale, 2011; Klemm et al., 2005; Martins et al., 2009; Nair & Laurencin, 2007; Chandra & Rustgi, 1998; Thakur et al., 2016; Vieira et al., 2011). The design of an integrated blends and composites system of eco-friendly greener packaging and agriculture film materials and development process is of great significance to the world, because of the gradual increase in packaging and agriculture film-related consumption. The global biodegradable plastics market is projected to reach $6.0 billion by 2026 at a compound annual growth rate of 21.3% for applications such as horticulture, packaging, textiles, consumer goods, and agriculture sectors (Gross & Kalra, 2002; Guan et al., 2014; Khan et al., 2020; Sander, 2019; Touchaleaume et al., 2016; Van De Velde & Kiekens, 2002).

Biodegradable polymers have an immense diversity of benefits for environmental protection; based on their nonhazardous characteristics, they can be categorized into two most important types, natural polymers and synthetic polymers. The polymers procured mostly from sustainable resources are an innovative generation of materials able to extensively alleviate the environmental effects while attaining specific technical demands and also being completely biodegradable. Also, natural polymer-based materials are an appropriate alternative to conventional plastic materials when recycling of synthetic polymers is not cost-effective or technically impossible (Bastioli, 2020; Kale, 2011; Klemm et al., 2005; Martins et al., 2009; Nair & Laurencin, 2007; Chandra & Rustgi, 1998; Thakur et al., 2016; Vieira et al., 2011).

Biodegradable Polymers, Blends and Composites. https://doi.org/10.1016/B978-0-12-823791-5.00018-1

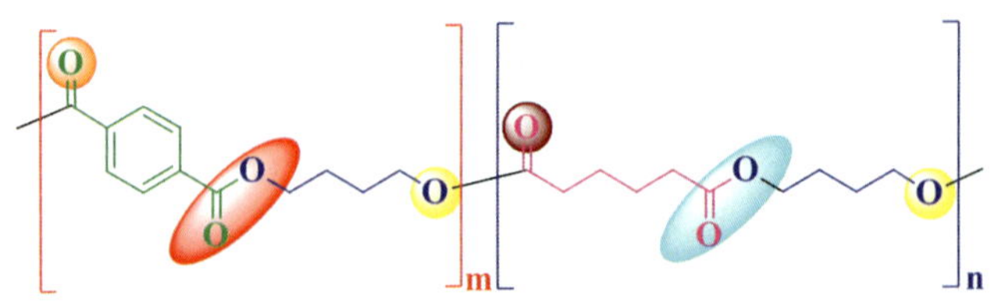

Figure 11.1 Chemical structure of poly(butylene adipate-co-terephthalate) (PBAT).

Among all the commercially available biodegradable polymers, PBAT is fossil resource based biodegradable, synthetic random copolyester, with high elongation, and extreme flexibility which is synthesized by the polycondensation of 1,4-butanediol (BDO), adipic acid (AA), and terephthalic acid (PTA). Fig. 11.1 shows the chemical structure of PBAT (Ferreira et al., 2019; Rodrigues et al., 2016). However, PBAT suffers from higher manufacturing cost, and poor mechanical and thermal properties which limit its commercial applications in certain areas compared to other traditional nonbiodegradable polymers (Nunes et al., 2019; Pappu et al., 2016).

Researchers have addressed that problem by blending PBAT with other more cost-effective biodegradable polymers and the addition of organic/inorganic fillers, compatibilizers, and plasticizers to generate cost-effective blends and composites to expand the applications and improve the properties of the PBAT matrix while maintaining its biodegradability (Beber et al., 2018; Freitas et al., 2017; Jiang et al., 2009; Zhang et al., 2009). PBAT is a random copolymer that possesses both flexible and tough parts. These characteristics make it suitable for blending with other biodegradable polymers and impart greater strength and elastic modulus. The accurate composition of polymer and filler can provide better characteristics for particular applications (Chinsirikul et al., 2015; Nofar et al., 2019; Tiimob et al., 2018). However, PBAT is a better alternative to nonbiodegradable conventional polymers, for example, polyethylene (PE), polypropylene (PP), and polystyrene materials which could resolve the single-use plastic pollution problem (Hamaide et al., 2014).

On the other hand, commercially existing biodegradable polymers show similar properties to nonbiodegradable polymers, for example, poly(lactic acid) (PLA), polyhydroxybutyrate (PHB), polycaprolactone (PCL), poly(butylenes succinate) (PBS), poly(butylene succinate-co-adipate) (PBSA), and PBAT, which are presently utilized in the polymer industry in a variety of applications, for example, packaging, biomedical, agriculture, and food containers products, and industrial composting (Marinho et al., 2017; Mohanty et al., 2018).

Therefore, it is confirmed from the above discussion that PBAT-based blends and composites are in great demand for the current scenario to decrease the surplus utilization of petroleum-based polyolefin single-use plastics products and to formulate new environmental friendly materials. The following section summarizes the preparation of PBAT blends and composites with higher mechanical, thermal, and biodegradation properties.

2. PBAT-based blends and composites

The renewable resource-based biodegradable polymer materials have many great possibilities to replace commercial petroleum-based plastics in different applications because of their lower environmental impact. Of all the biodegradable polymers, PBAT is widely used for the production of blown film due to its high flexibility, high elongation (%), biodegradability, and biocompatibility, with better hydrophilic and processing characteristics. Nevertheless, PBAT exhibits high cost and low toughness which limit its applications. For that reason, the development of PBAT blends and composites with other biodegradable polymers or reinforcing fillers to enhance the properties and expand its scope in various areas is a more economically feasible and better alternative to synthetic nonbiodegradable materials.

2.1 PBAT/TPS-based blends and composites

2.1.1 PBAT/TPS-based blends

Dammak et al. (2020) developed new bioplastic materials after blending PBAT with plasticized thermoplastic starch (TPS) and incorporating suitable compatibilizers (e.g., maleic anhydride) for packaging applications. Further, they studied the influence of maleated PBAT (PBATg-MA) and maleic anhydride (MA), on the mechanical, morphological, and biodegradability properties of blend systems with various (40, 50, and 60 wt.%) TPS contents. The mechanical properties of the PBAT/TPS blend system are depicted in Fig. 11.2A–C with different wt.% TPS with and without MA and PBATg-MA compatibilizers. All the specimens revealed three regions in the tensile–strain plot such as elastic, plastic deformation, and strain hardening actions as illustrated in Fig. 11.2A. The elastic region demonstrated linear stretching with recoverable deformation, and the second region exhibited plastic deformation following the neck formation for particular specimens. However, the last region displayed a strain hardening character with regular improvement of the tensile stress before breakage of the specimen.

Fig. 11.2B–C expresses the comparative properties for the tensile strength and elongation (%) of the various components. The virgin PBAT and TPS revealed that the tensile strength, elongation at break at 52.6 MPa increased by 750% and 35%, respectively. This confirmed that PBAT is a highly flexible soft polymer, with sufficient strength as compared to TPS, which has poor mechanical performance. Moreover, by reducing the content of TPS from 60 wt.% to 40wt.% with and without PBATg-MA, the tensile strength and elongation at break of the materials were significantly enhanced.

Conversely, with the addition of MA, the tensile strength and elongation at break indicated comparable values to the pure TPS, remaining approximately constant at various TPS (%). Finally, the 50% TPS-based blend system achieved 12.8 MPa and 205% tensile strength and elongation, respectively, without any additive, with the further addition of PBATg-MA compatibilizers the values were greatly improved to 15.1 MPa and 614%. However, after the incorporation of a 2% MA blends system

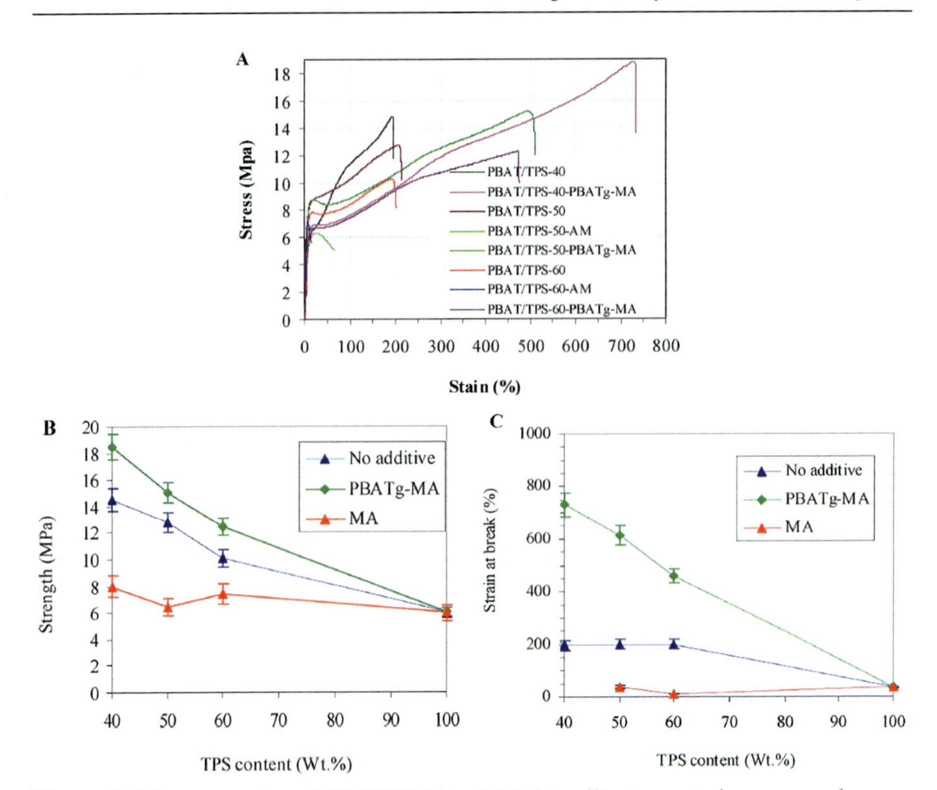

Figure 11.2 Representation of PBAT/TPS blends' (A) tensile stress—strain curves and (B) tensile strength, and (C) strain at break with various TPS content loading.
Adapted from: blends of PBAT with plasticized starch for packaging applications: Mechanical properties, rheological behaviour and biodegradability. (2020). *Industrial Crops and Products*, *144*, 112061. https://doi.org/10.1016/j.indcrop.2019.112061.

these values suddenly decreased to 6.5% and 36%, respectively, due to the poor compatibilization effect of MA. As a result, PBATg-MA acts as a better compatibilizer for the enhancement of the mechanical performance of the PBAT/TPS blend systems to improve the interfacial adhesion among the introduced MA onto PBAT and TPS.

Moisture absorption is one of the main shortcomings that restrict the application of starch-based materials. One of the main objectives of blending TPS with hydrophobic polyester is to reduce the water absorption of the film to meet the requirements for maintaining acceptable levels of mechanical properties. Fig. 11.3A and B confirms that the moisture absorption of the PBAT/TPS blends system with various contents of TPS at 50% relative humidity (RH). The MA was enhanced abruptly in the first 20 h after which it moderately declined until a saturation point was reached after 48 h depending on the TPS content.

The neat PBAT and TPS exhibited very diverse water uptake properties at 50% RH, achieving about less than 1% and 10%, respectively. Furthermore, the moisture absorption equilibrium enhanced after an increased amount of TPS (40%, 50%, and

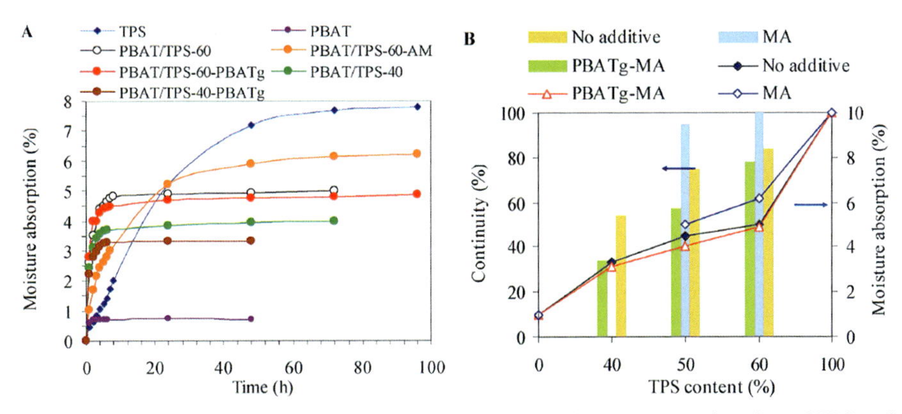

Figure 11.3 Moisture uptake properties of PBAT/TPS blend systems as a function of TPS and compatibilizers.
Adapted from Blends of PBAT with plasticized starch for packaging applications: Mechanical properties, rheological behavior and biodegradability. (2020).

60%) without any compatibilizers achieved 3.3, 4.3, and 4.8 respectively. Nevertheless, the moisture absorption did not alter extensively the existence of PBATg-MA compatibilizers. After the addition of a higher amount of MA, the moisture absorption was attained at a higher level with 40% and 50% TPS content of 5.0% and 6.3%, respectively. This is attributed to the superior continuity in TPS in the case of MA employed as a compatibilizer. Further, in the presence of PBATg-MA, the TPS phase enhanced the dispersion of the PBAT matrix and was completely encapsulated by the polyester, particularly at less than 50% TPS content. For that reason, the water absorption of the PBAT/TPS blend system declines.

In the case of a biodegradation study, mineralization of carbon at various formulations and standard cellulose is illustrated in Fig. 11.4 as per ISO 14855-2 standard. In the curve PBAT, TPS and cellulose displayed a degree of biodegradation of more than 90% after 90 days, confirming that all materials are completely biodegradable but there are various rates of CO_2 production, dependent on the compatibilizers employed. However, all systems showed different rates of CO_2 production; this is attributed to the various compositions of the blends and the addition of compatibilizers. For instance, the PBAT/TPS blend system (TPS content 40%–60%) revealed a similar biodegradation rate, that is, 82%–87% after 90 days without any compatibilizers. Further, with the addition of MA, the degradation curve was similar to that for cellulose and PBAT. Conversely, in the presence of PBATg-MA, the biodegradation rate is reduced, achieving between 70% and 75% after 80 days due to the insertion of the PBATg-MA coupling agent.

Wei et al. (2015) prepared a PBAT/TPS blends system to decrease the cost and improve the mechanical properties of the blend systems with the addition of traditional compatibilizers such as Joncryl-ADR-4368, containing synthesized styrene–maleic anhydride–glycidyl methacrylate (SMG) reactive compatibilizers and modified PBAT. As a result, the mechanical properties were mostly influenced by the TPS

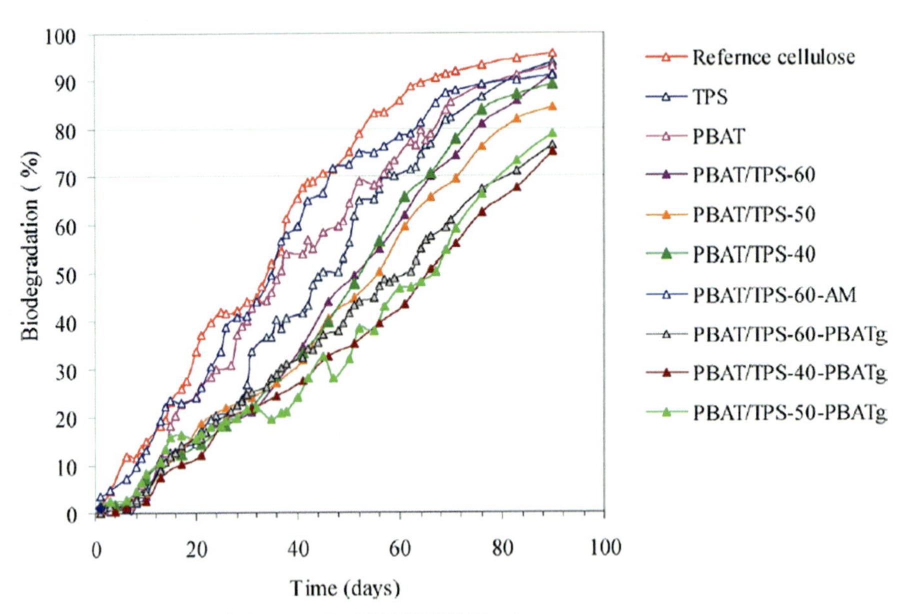

Figure 11.4 Biodegradation test for PBAT/TPS blend systems.
Adapted from Blends of PBAT with plasticized starch for packaging applications: Mechanical properties, rheological behavior and biodegradability. (2020). https://doi.org/10.1016/j.indcrop.2019.112061.

content and dispersion. The PBAT/TPS (70/30) blend systems exhibited higher mechanical properties due to the better dispersion of TPS particles in the PBAT matrix, as depicted in Fig. 11.5.

The traditional and synthesized SMG compatibilizer enhanced the compatibility and dispersion but does not improve the tensile strength, as shown in Fig. 11.6. Further, a higher molecular weight modified PBAT-(M-PBAT) improved the mechanical performance of the PBAT (40)_M-PBAT (20)_TPS (40) blend system (Fig. 11.7). This was attributed to the good dispersion of TPS. Therefore, good mechanical performance was found that is suitable for different applications, such as packaging and agriculture mulching films.

2.1.2 *PBAT/TPS-based composites*

da Silva et al. (2019) developed a PBAT/TPS blend film with the addition of eucalyptus cellulose nanowhiskers (CNW) with various wt.% (0–3) in the presence of different compatibilizer agents. In particular, stearic acid (SA), glycerol (G), and citric acid (CA) have been investigated. The PBAT/TPS_CNW demonstrated suitable properties for different applications. CNW enhanced the interaction between the TPS and PBAT matrix. The rate of water vapor permeability and permeability exhibited no significant effect after the addition of CNW, which confirmed that the presence of TPS had a greater impact on the blend system, supporting the formation of pores produced

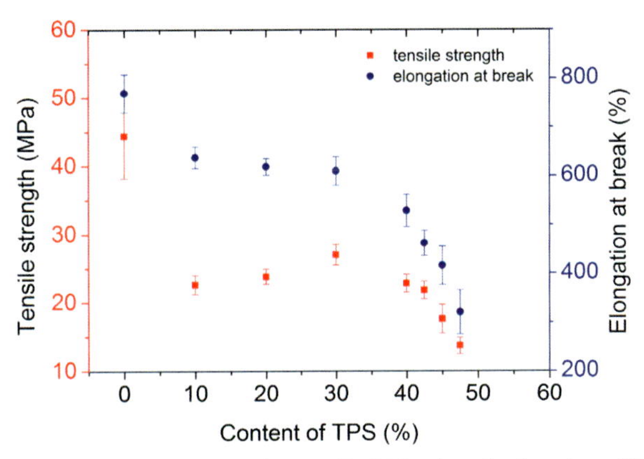

Figure 11.5 Tensile and elongation (%) of PBAT/TPS blend on the function of TPS contents (%). Adapted from Morphology and mechanical properties of poly(butylene adipate-co-terephthalate)/ potato starch blends in the presence of synthesized reactive compatibilizer or modified poly(butylene adipate-co-terephthalate). (2015). *Carbohydrate Polymers, 123*, 282. https://doi.org/10.1016/j.carbpol.2015.01.058.

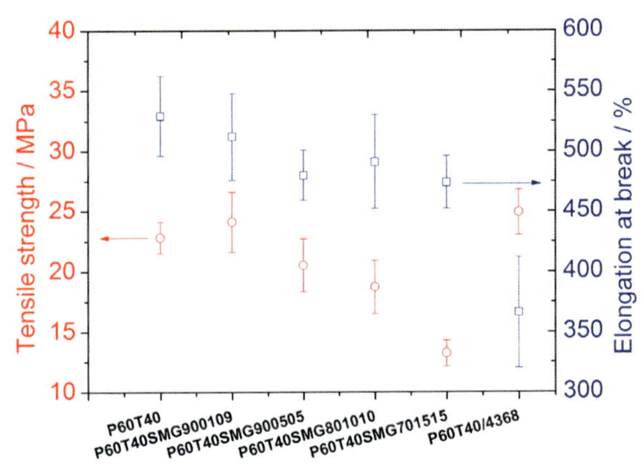

Figure 11.6 Mechanical properties of PBAT/TPS with various compatibilizers. Adapted from Morphology and mechanical properties of poly(butylene adipate-co-terephthalate)/potato starch blends in the presence of synthesized reactive compatibilizer or modified poly(butylene adipate-co-terephthalate). (2015). *Carbohydrate Polymers, 123*, 282. https://doi.org/10.1016/j.carbpol.2015.01.058.

by the combination of TPS and CNW in the presence of CA. The developed film exhibited antimicrobial characteristics and was suitable for vegetable packaging.

Further, cellulose nanocrystals (CNC) extracted from munguba fibers modified with octadecyl isocyanate were also reported by Pinheiro et al. (2019). These modified cellulose nanocrystals (MCNC) were used for the development of a biodegradable PBAT nanocomposite with 3, 5, and 7 wt.% of MCNC. As a result, the incorporation of MCNC improved the modulus of the elasticity up to 7 wt.% as compared to the pure

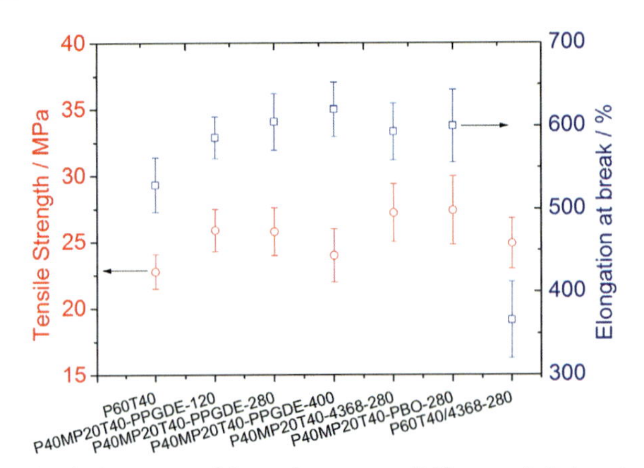

Figure 11.7 Mechanical property of the various compatibilizers and chain extender (PPGDE), (120 shows the rotation speed of the twin-screw extruder).
Adapted from Morphology and mechanical properties of poly(butylene adipate-co-terephthalate)/potato starch blends in the presence of synthesized reactive compatibilizer or modified poly(butylene adipate-co-terephthalate). (2015). *Carbohydrate Polymers*, 123, 282. https://doi.org/10.1016/j.carbpol.2015.01.058.

PBAT (Fig. 11.8). This is attributed to the MCNC which dissipates external energy and absorbs it via the MCNC particles. Conversely, the tensile stress of PBAT is reduced by the addition of 5 wt.% MCNC, leading to brittleness through the agglomeration of the MCNC particles. The elongation at break exhibits a great deal of deviation, as depicted in Fig. 11.8. The thermal stability of the CNC is enhanced after modification, due to the elimination of sulfate groups from the CNC and hence improves the thermal stability of the specimens. The prepared nanocomposite is a potential alternative to the commercial polymer for future applications because of its biodegradable nature.

da Silva et al. (2017) investigated the addition of starch nanoparticles (SNPs) formed via ultrasound in blends of PBAT and TPS. Films with 1%, 2%, 3%, 4%, and 5% w/w of SNPs were prepared by extrusion. In the absence of any chemical reagent, the SNPs were synthesized in water. In this method, SNPs smaller than 100 nm with an amorphous nature, lower thermal stability, and gelatinization temperature in contrast to cassava starch were formed. The relative crystallinity of films reduces as the concentration of SNPs increases. With the incorporation of 1% SNPs, the blends showed 36% and 35% enhancements in Young's modulus and elongation (%), respectively. Meanwhile, a 53% decrease in water vapor permeability and extensive reduction in the water absorption of these films at this concentration, indicates the preference for low concentrations of SNPs in a polymeric matrix. During the extrusion process, blending of starch to the PBAT matrix increases its crystallinity, which is associated with amylose recrystallization (Santos et al., 2014). PBAT/TPS films show a high degree of crystallinity in contrast to PBAT/TPS/SNP films duc to the amorphous character of the nanoparticles. Water vapor permeability of the films increases abruptly

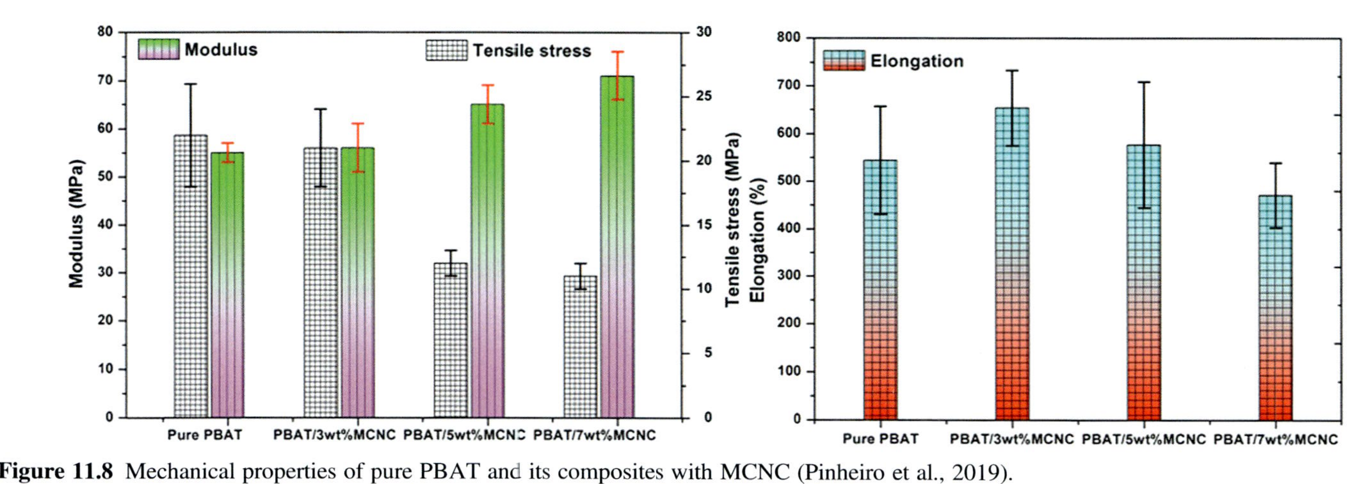

Figure 11.8 Mechanical properties of pure PBAT and its composites with MCNC (Pinheiro et al., 2019).
Adapted from Pinheiro, I. F., Ferreira, F. V., Alves, G. F., Rodolfo, A., Morales, A. R., & Mei, L. H. I. (2019). Biodegradable PBAT-based nanocomposites reinforced with functionalized cellulose nanocrystals from *Pseudobombax munguba*: Rheological, thermal, mechanical and biodegradability properties. *Journal of Polymers and the Environment*, 27(4), 757−766. https://doi.org/10.1007/s10924-019-01389-z.

in the presence of TPS in the PBAT matrix which is attributed to the hydrophilic behavior of the starch. The results revealed that with a high SNP content a strong hydrogen bond network can be formed within SNPs and starch, and hence, the stress distribution is reduced, which reduced the elongation of the films.

Liu et al. (2020) studied the PBAT/TPS composite which reduces the cost but the mechanical properties of the prepared PBAT/TPS composites are also greatly decreased. Thereby, the PBAT/TPS composites with high mechanical properties were blended by a simple melt extrusion method by adding reinforcing and compatibilizing agents. The reinforced and double-compatibilized composite tensile strength improved to 50%, and its elongation at break was raised to about 18%. As compared to a virgin PBAT matrix, the PBAT/TPS blend showed a much lower tensile strength of 11.6 MPa and a reduced elongation at break of 1264%, which was attributed to the weak interaction between PBAT and TPS. In the presence of nano-SiO_2 (0.3 phr) PBAT/TPS blend exhibits a high tensile strength of 15.2 MPa, although with slightly less elongation at break of 1138%. By incorporating maleic anhydride (MA), the blend has similar tensile strength as the PBAT/TPS/SiO_2 blend, while the PBAT/TPS/SiO_2/MA demonstrates improved elongation at break and better impact strength than the virgin PBAT. The toughness of the PBAT/TPS/SiO_2/MA composite improved with the incorporation of MA, which was attributed to the good compatibility between PBAT and TPS in the presence of MA. Thus, to improve the compatibility and mechanical performance of the PBAT/TPS composite, a double-compatibilizer (MA and ethylene-co-vinyl alcohol) was incorporated. By bonding within ethylene-co-vinyl alcohol, PBAT and TPS improve the compatibility between PBAT and TPS, as hydrophilic TPS is inserted into the continuous phase of the PBAT matrix. As a result, the mechanical properties of the composites show great enhancement with better thermal stability. Incorporation of the additives into the composite melt shows pseudoplastic behavior, with simple processability. PBAT-TPS-based composites prepared by the melt extrusion process exhibit better properties and low cost with various potential applications.

2.2 PBAT/PLA-based blends and composites

2.2.1 PBAT/PLA-based blends

PLA is an aliphatic polyester exhibiting several physical properties provided it an appropriate substitute commodity polymer. PLA also demonstrated high strength, better processability with important gas barrier properties, and was used in various applications from medical to packaging, etc. Zhao et al. (2020) prepared PBAT and stereocomplex polylactide (sc-PLA) blends by a melt blending process by incorporating 10–30 wt.% sc-PLA. Fig. 11.9 reveals the accurate mechanism of sc-PLA formation, as well as its dispersion in the PBAT matrix after the melt blending process.

Fig. 11.10 demonstrates the tensile strength and elongation (%) of a pristine PBAT and PBAT/sc-PLA blend system. The pristine PBAT exhibits very low modulus and yield stress, but high ductility. On the other hand, all the blend system displays increased modulus but with reduced elongations as compared to the pristine PBAT,

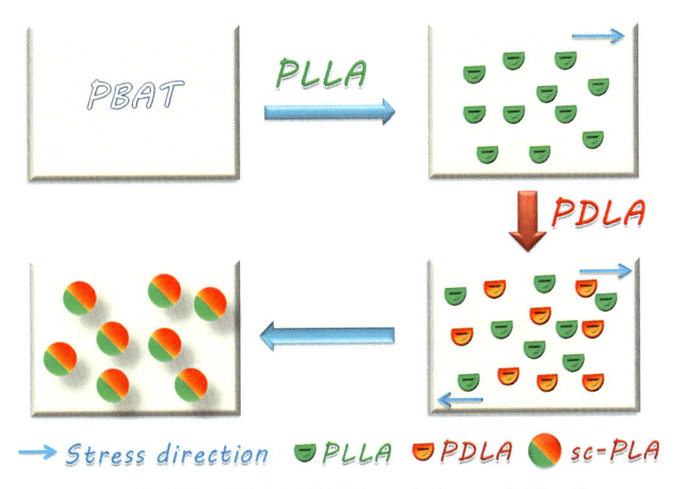

Figure 11.9 Formation of solid sc-PLA in PBAT matrix in a melt blending process.

as shown in Fig. 11.10. The enhancement in the modulus was attributed to the chain movement of the PBAT via spreading of the sc-PLA network structure. Furthermore, the sc-PL content enhances by up to 20 wt.% the modulus of the blend which improved significantly. This is due to the formation of the spherical sc-PLA network in the blend system. Hence, the elongation (%) also decreased significantly with the formation of the sc-PLA network. As a result, after the inclusion of sc-PLA into the PBAT matrix, significant improvement was achieved.

The development of the biodegradable PBAT/PLA polymer blend is employed in different ionic liquid-based tetraalkyl phosphonium cations such as trihexyl(tetradecyl) phosphonium bis-2,4,4-(trimethylpentyl)phosphinate (IL-TMP), trihexyl(tetradecyl)phosphonium chloride (IL-Cl), (trihexyl)tetradecylphosphonium 2-ethylhexanoate (IL-EHT), trihexyl(tetradecyl)phosphonium bistriimide (IL-TFSI), and (trihexyl)tetradecylphosphoniumbis(2-ethylhexyl)phosphate (IL-EHP) as compatibilizers was reported by Lins et al. (2015) and the mechanical and thermal properties of the PBAT/PLA blend system were studied also. The mechanical properties of pristine PLA, PBAT, and PBAT/PLA blends with and without ILs were analyzed. The incorporation of PLA in PBAT improves the rigidity of PBAT/PLA blends in contrast to the pristine PBAT, while the PBAT enhances the strain at break of the PBAT/PLA blend in contrast to the PLA brittleness. The addition of ILs as compatibilizer agents in a blend system promotes stiffness and ability to prolong great deformation, in particular for IL-TMP, IL-EHP, and IL-EHT. The incorporation of a 1 wt.% of different ILSs in a PBAT/PLA matrix reduced the size of the dispersed PLA areas to a submicron scale. Nevertheless, the ILS position in an interfacial area about the dispersed PLA phase acts as a compatibilizer and enhances the distribution of PLA in the PBAT matrix (Fig. 11.11). As a result, improved mechanical and thermal performances are achieved due to the chemical nature of the counteranions. Furthermore, phosphate and hexanoate anions play the main role in the formation of PLA fibrils which promote a notable improvement in Young's modulus of the PBAT/PLA blend systems.

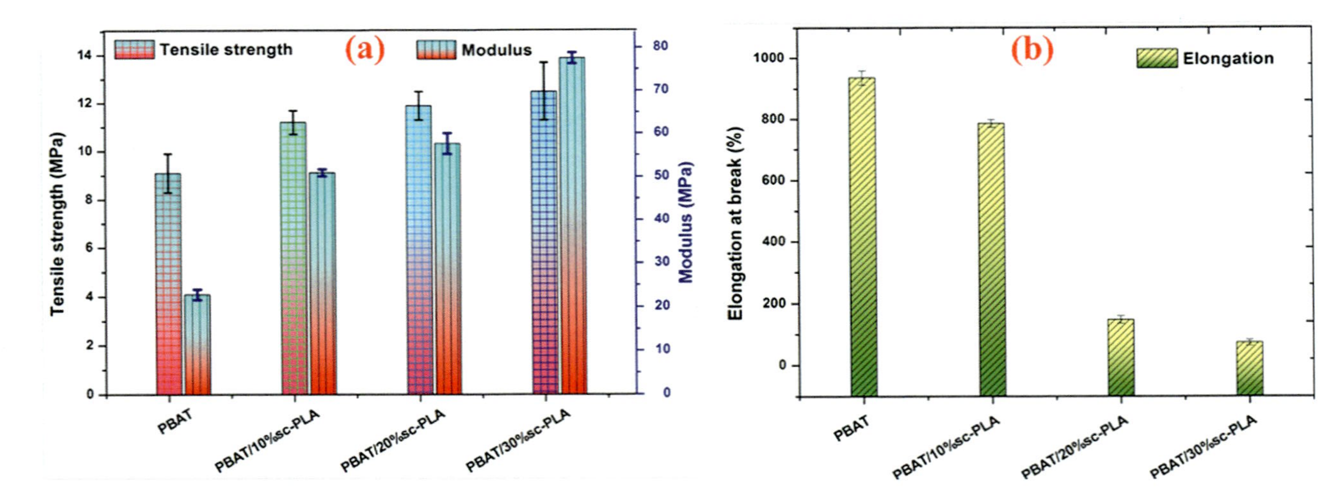

Figure 11.10 Tensile strength and elongation (%) of pristine PBAT and sc-PLA blend systems (Zhao et al., 2020).
Adapted from Pinheiro, I. F., Ferreira, F. V., Alves, G. F., Rodolfo, A., Morales, A. R., & Mei, L. H. I. (2019). Biodegradable PBAT-based nanocomposites reinforced with functionalized cellulose nanocrystals from *Pseudobombax munguba*: Rheological, thermal, mechanical and biodegradability properties. *Journal of Polymers and the Environment*, *27*(4), 757–766. https://doi.org/10.1007/s10924-019-01389-z.

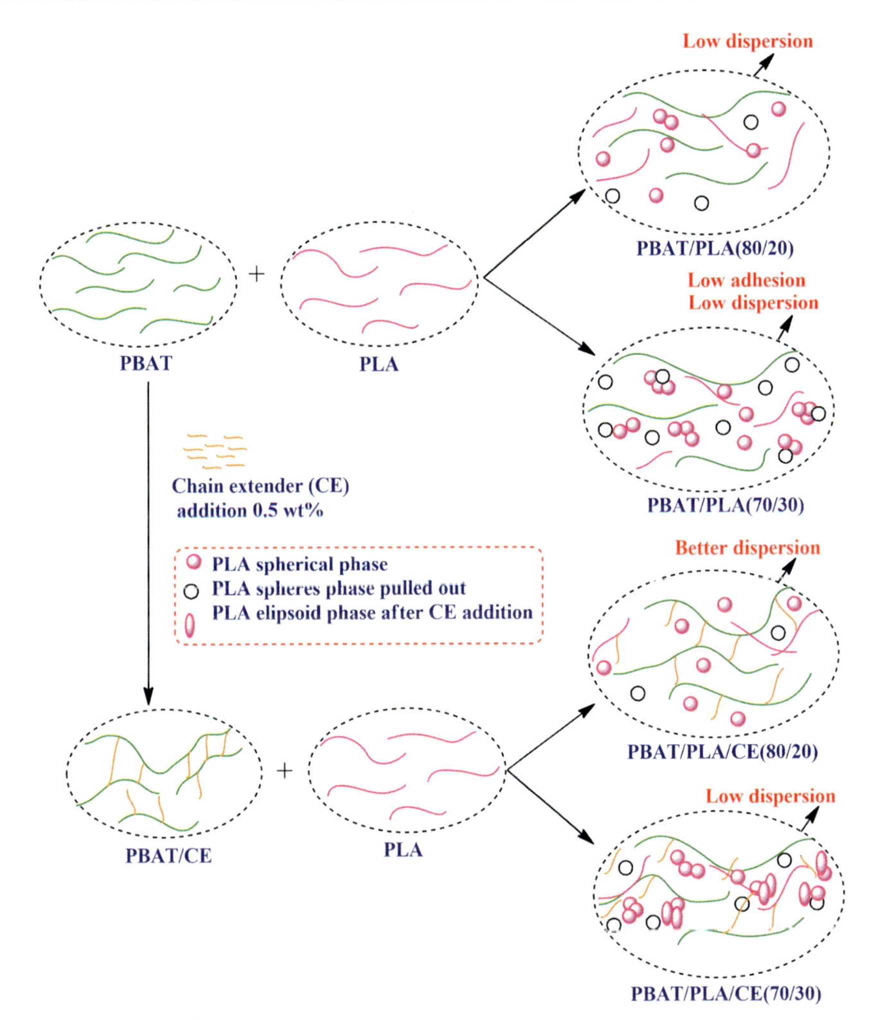

Figure 11.11 Schematic representation of fractured pristine PBAT, PLA, and its blends in the presence and absence of EC.

Conversely, IL-Cl, IL-TFSI, and IL-EHP stimulate the formation of PLA particles joined by an enhancement of strain at break without any decrease in stiffness. However, the ILs show great potential to be a better alternative to commercial compatibilizers used in very small amounts and to prepare high-performance polymer blends.

Increasing exploratory development of biodegradable products with the addition of the new material Joncryl ADR chain extender into the PBAT/PLA blends was explored by De et al. (Nunes et al., 2019; Pappu et al., 2016). The incorporation of a 0.5 wt.% chain extender (CE) into the pristine PBAT increased elongation (%) from 1269.3% to 1848.7%, but reduced the tensile strength from 25.1 to 19.9 MPa. The improvement in the elongation was attributed to the CE performance due to the interaction with the PBAT chain end. However, in the case of virgin PLA, elongation and tensile strength were both improved. After the inclusion of CE into the PBAT/PLA (80/20) blend, both elongation and elastic modulus increase. Similarly, PBAT/PLA/CE(80/20) also demonstrated a small improvement in tensile strength owing to the better compatibility and interfacial adhesion among the PBAT and PLA phases. Therefore, tensile and strain properties were relocated from the PBAT matrix to the PLA areas with the incorporation of CE.

Further, in the case of PBAT/PLA (70/30), the inclusion of CE exhibits a tiny decline in the modulus of elasticity which is related to the structural alteration in the PLA particles in the mixture from a spherical to an ellipsoid phase. They act as a stress concentrator depending on the PLA portion present in the blend, as represented in Fig. 11.11. However, the used CE (0.5 wt.%) was not sufficient to enhance the adhesion among the phases of the blend for the dispersion of PLA particles. The thermal properties of the pristine PBAT, PLA, and its blend did not show any notable changes.

2.2.2 PBAT/PLA-based composites

Increasing the nonbiodegradable petroleum-based polymer in the environment encouraged researchers to develop biodegradable, eco-friendly, and cost-effective PBAT materials reinforced with bio-based nanochitin of various wt.% (0.5, 1, 2, 4, 8), as suggested by Meng et al. (2020). The tensile strength of PBAT was improved significantly by the inclusion of 0.5 wt.% nanochitin. The tensile strength of the PBAT/nanochitin was reduced particularly after the introduction of a nanochitin content of 1, 2, 4, and 8 wt.%. The results confirmed that the incorporation of even a small content of nanochitin (0.5 wt.%) increased the stiffness of the PBAT matrix and for that reason significantly improves its tensile strength. Moreover, the better distribution of nanochitin throughout the PBAT matrix enhanced the interfacial adhesion among the nanochitin and PBAT chains. Further, the thermal degradation of neat PBAT and its composites was similar when the chitin content was lower than 4%, whereas the char % also was enhanced with the chitin content, because chitin acted as an additional carbon source for char formation. The incorporation of the nanochitin obstructed heat conduction in the PBAT matrix, as a result changing the carbonization kinetic of PBAT.

Rocha et al. (2018) introduced an economical PBAT/PLA film for agricultural applications by the addition of 10–20 wt.% $CaCO_3$ with improved mechanical properties. Flexible films show enhanced compatibility with good tensile strength, Young's modulus, and maximum strain, as shown in Fig. 11.12. Poor water absorption and degradation in the simulated soil are revealed due to the highly orientated amorphous arrangements. The calcium carbonate catalytic effect during ester bond depolymerization helps in free radical and reactive end-group formation through polyester decomposition owing to the presence of a metal ion which finally leads to low degradation (Sirisinha & Somboon, 2012; Teamsinsungvon et al., 2013). This low degradation reveals that the prepared films exhibit potential applications for developing flexible mulch films, mainly for the Muridori plantation system for which long-term plantations are essential.

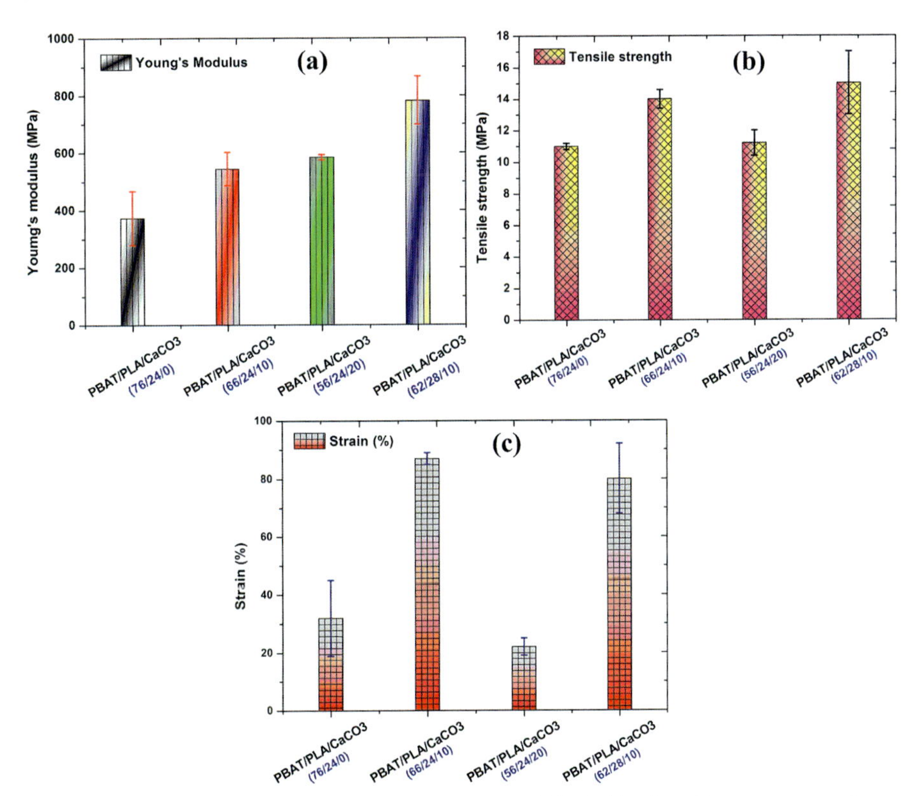

Figure 11.12 Mechanical properties of a pristine PBAT/PLA blend and its nanocomposites (Rocha, Carvalho, Oliveira, & Rosa, 2018).
Adapted from: A new approach for flexible PBAT/PLA/$CaCO_3$ films into agriculture. (2018). *Applied Polymer Science*, *13*(35), 46660. https://doi.org/10.1002/app.46660.

Yao et al. (2020) prepared hydroxyapatite (HA)/PLA and HA/PBAT/PLA composite films by the incorporation of synthesized nanohydroxyapatite and PBAT as a filler. The tensile strength, elongation (%), and modulus of the HA/PLA composites were improved via incorporation of 10 wt.% PBAT as compared to the 1 wt.% HA. The inclusion of HA and PBAT played a significant role in enhancing the crystallinity of the composite films. Further, the water vapor and oxygen permeability indicate its immense prospects in green and fresh packaging applications.

Blending natural and synthetic polymers has become a new basis for the creation of biomaterials. Nakayama et al. (2018) studied the blending of silk fibroin powder and the biodegradable PBAT/PLA system. SEM studies revealed the weak dispersion of the silk powder agglomerates that are caused by the better hydrogen interactions among the silk powder chains in the PBAT/PLA matrix. With increased silk powder fractions, the mechanical performance reduces as the silk powder agglomerates but, still, a ternary blend with 10 wt.% silk powder shows improved impact strength and tensile modulus of 108 J/m and 1.2 GPa, respectively. The mechanical behavior of ternary blends indicates that the tensile modulus and strength, and flexural modulus and strength attain a maximum value at 0.3 phr Joncryl and then reduce at 0.5 phr Joncryl. The same trend was also revealed by a PBAT/PLA binary matrix indicating that Joncryl improved compatibility between the PBAT/PLA matrix in preference to the silk powder and PBAT/PLA interface. Blends with Joncryl showed a reduced storage modulus. Additionally, rheological studies indicate the lower viscosity of the PBAT/PLA/silk powder blends, because of the weak hydrogen bonds between the silk chains, which was attributed to the reaction between the Joncryl epoxy groups. Due to this the processability of this biomaterial is enhanced with improved interfaced polymer blends.

2.3 PBAT/lignin-based blends and composites

Lignin is the most abundant renewable and biodegradable natural polymer with a cross-linked and complex phenolic structure. Lignin and its derivatives are attractive to researchers because of its easy availability on a large scale, with lignosulfonate being the main product of the paper industry. Nevertheless, lignin is a low-cost, nontoxic, high-carbon content biodegradable with antioxidant and antibacterial activities. Lignosulfonates contain several functional groups such as phenolic hydroxyl, methoxy, carbonyl, hydroxyl, and carboxyl groups, as depicted in Fig. 11.13, which enhances the interaction and compatibility within the polymer, as well as the PBAT matrix. Further, lignosulfonate fillers also improve the mechanical properties of materials (Stewart, 2008; Tavares et al., 2018; Xiao et al., 2001; Xing et al., 2017).

Yang and Zhong (2020) first developed maleic anhydride-modified lignin sulfonate nanoparticles (MLSs) and added them to the PBAT matrix at 5–30 wt.% to prepare PBAT/MLS composites via melt blending and studied the mechanical and rheological properties. As a result, the tensile strength and elongation (%) of PBAT/MLS composites improved by 20% and 40%, respectively, after the addition of 5 wt.% MLS in contrast to the neat PBAT, owing to the better dispersion

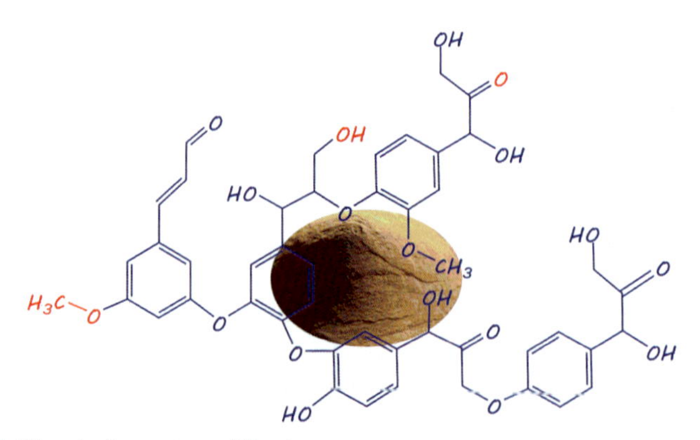

Figure 11.13 Chemical structure of lignin.

of MLSs into the PBAT matrix. However, the increment of filler % reduces the tensile strength, and is associated with the weak dispersion of MLSs and PBAT matrix. The tensile and flexural modulus indicated continuous enhancement with an increasing amount of MLSs, owing to the high modulus of lignin. Finally, the results indicated that the addition of MLSs can extensively improve the tensile strength, elongation (%), and tensile and flexural modulus of PBAT composites. It is confirmed that MLSs have great potential to be used as a filler in PBAT and other polymers. The rheological analysis also indicated that the inclusion of MLSs reduced the viscosity and improved the processing ability.

Biodegradable PBAT/lignin composite film with a high content of lignin (60 wt.%) was prepared via melt extrusion PBAT as reported by Xiong et al. (2020) and reduced the overall cost to ~36% as compared to the pristine PBAT film. They studied the effect of neat lignin (L), methylated lignin (ML), and MA-g-PBAT compatibilizer on the mechanical and thermal properties of PBAT, as shown in Fig. 11.14. The neat PBAT film exhibited higher tensile strength and elongation (%) of 23.70 MPa and 816%, respectively. The addition of lignin into the PBAT matrix reduced the overall tensile strength. Also, the addition of 60% lignin into the PBAT decreased the tensile and elongation properties from 23.70 to 14.41 MPa and 816.49% to 378.94%, respectively. Further enhancement of the lignin to 50 wt.% to 60 wt.%, reduced the tensile and elongation properties owing to the agglomeration of lignin particles, weak interfacial interactions, and reduction of PBAT matrix.

Moreover, the inclusion of 60 wt.% ML into the PBAT matrix increased the tensile strength and elongation (%) properties to 19.48 MPa and 545.08% respectively, compared to the PBAT/lignin composites. This is attributed to the size reduction of the lignin particles in PBAT/ML composites and increased flexibility of lignin because of the interaction of the strong hydrogen bond among −OH groups. Similarly, PBAT/MA-g-PBAT showed better tensile performance than

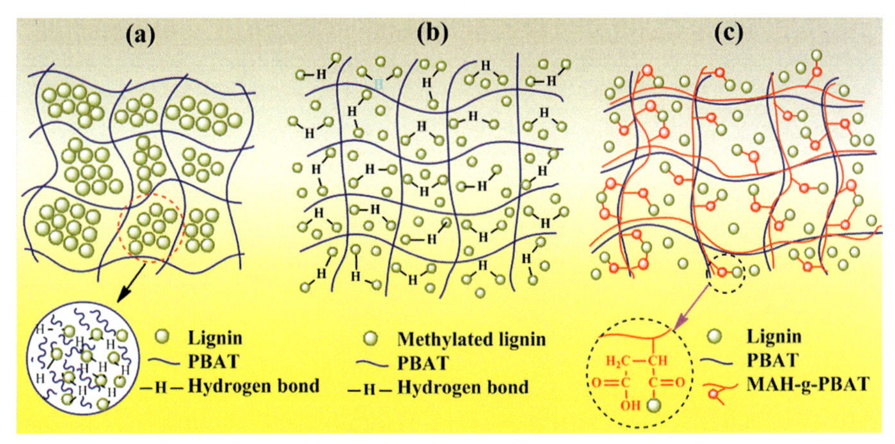

Figure 11.14 Representation of PBAT/lignin composites film with (A) neat lignin, (B) ML, and (C) neat lignin with compatibilizer.

PBAT/lignin composites. Because of the strong interfacial interactions there was better stress transfer from the PBAT matrix to lignin fillers and also the mobility of the PBAT was enhanced and as a result it exhibited better mechanical properties. The thermal degradation of the PBAT/MA-g-PBAT slightly decreased to 10–19°C with the addition of lignin. However, PBAT/ML composites exhibited slightly higher thermal stability compared to PBAT/lignin and PBAT/MA-g-PBAT composites with the same filler content.

Bauli and others (Bastioli, 2020; Kale, 2011; Klemm et al., 2005; Martins et al., 2009; Nair & Laurencin, 2007; Chandra & Rustgi, 1998; Thakur et al., 2016; Vieira et al., 2011) prepared cellulose nanostructure (CNS)-based PBAT biodegradable polymeric composite film with the addition of 1–5 wt.% CNS fillers via solvent casting in the presence of nonionic surfactant for surface modification. Incorporation of 1 and 3 wt.% CNS fillers into the PBAT/CNS pretreated sample enhanced the modulus to about 11% and 12% whereas, tensile strength, to about 12% and 5%, respectively. However, by increasing the filler to 5 wt.%, the modulus and tensile strength declined to 10% and 33%, respectively. These results confirmed that the incorporation of filler up to 3 wt.% enhanced the properties but a further increment in the filler wt.% reduced its properties due to the agglomeration. Fig. 11.15 displays the expected interaction between the filler and PBAT matrix in the composite film. In the case of surfactant-modified specimens, the behavior did not alter this. Thermal properties of the composite film sample were enhanced to 10°C with the addition of 3% filler for all the sample. On the other hand, incorporation of surfactants into the matrix supported a reduction in the degradation rate of the composite film, making PBAT a versatile material for employment in packaging applications.

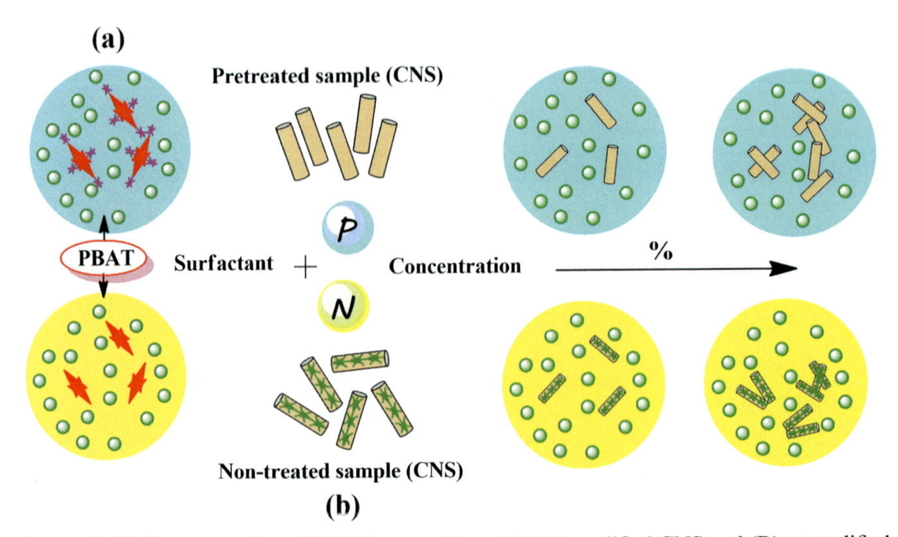

Figure 11.15 Representation of PBAT composites with (A) modified CNS and (B) unmodified CNS with 1–5 wt.% N samples.

2.4 PBAT and other polymer blends and composites

Zhang and others (Beber et al., 2018; Freitas et al., 2017; Jiang et al., 2009; Zhang et al., 2009) prepared poly(ethylene 2, 5-furancarboxylate) (PEF)-based PBAT polymer blends by loading 1−50 wt.% PEF as a filler and studied the mechanical and thermal properties. With the addition of 1%−7% PEF biodegradable rigid organic filler, the thermal stability ($T_{d5\%}$ ˃ 371°C) and mechanical properties improved for the PBAT/PEF blend, demonstrating high elongation (%) due to the reduced crystallinity of PBAT. With a further increase in the PEF content from 20% to 40%, the impact toughness of the PBAT/PEF blends were significantly enhanced, which was attributed to the strong interfacial adhesion. The 40% PEF-added PBAT blend exhibited 91.0 ± 8.7 kJ/m^2 impact toughness. The developed biobased PBAT/PEF blend has great potential for various applications.

Moustafa, Guizani, Dupont et al. (2017) developed biodegradable PBAT/coffee grounds (CG)-based blend materials with or without polyethylene glycol (PEG) as plasticizers. The lignocellulose CGs were utilized as a reinforcing agent for the PBAT matrix due to the reduced cost, huge availability, and ecofriendly, nontoxicity, and biodegradable nature. The blend can be used in various applications such as packaging, compositing, and biomedical, in contrast to the virgin PBAT which is costly and for that reason its application is restricted in various fields. However, CG powder and PBAT both shows hydrophilic and hydrophobic natures, respectively, so they are not compatible and, for this reason, it is very complicated to obtained satisfactory properties of PBAT/CG biocomposite blends. However, to achieve good compatibility and better interfacial bonding among the PBAT matrix and CG filler, a PEG plasticizer was added which forms a bonding between the −OH group of PEF and the −OH group of the CG filler, resulting in a new ether group (−O−), as shown in Fig. 11.16.

The plasticized PBAT/PEG/CG composite blend exhibited good tensile properties compared to unplasticized PBAT/CG composite blend, while the elastic modulus was reduced with the inclusion of a PEG plasticizer. The thermal stability of the PBAT/CG composite blend system reveals the TGA curve that shifts toward higher temperatures as PEG was included in the PBAT matrix, confirming the

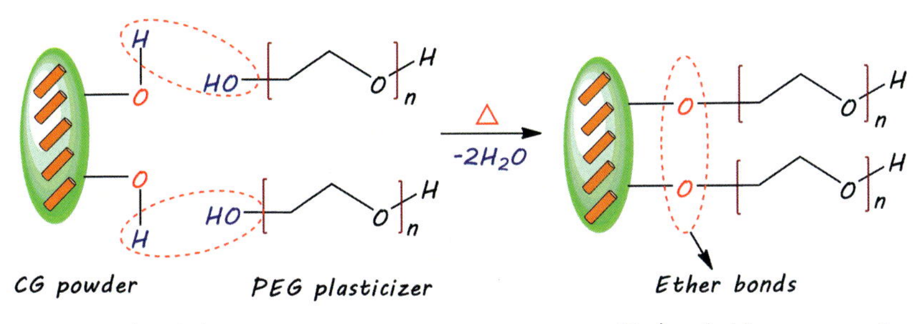

Figure 11.16 The formation of the ether bond (−O−) among the CG filler and PEG plasticizer in the PBAT matrix.

interaction between the filler and PEG. Further, Moustafa, Guizani, and Dufresne (2017) prepared new PBAT/CG composites without compatibilizer with enhanced mechanical and thermal properties.

Wu et al. (2018) synthesized acetylated maltodextrin (AMDs) with a high degree of substitution (DS), exhibiting higher thermal stability than commercial maltodextrin (MD), prepared with a novel PBAT/AMD and PBTA/MD blend system. The tensile strength and elongation (%) of PBAT/AMD blends were extensively enhanced compared to the PBAT/MD blend system, showing better interfacial adhesion between PBAT and AMD. Therefore, the PBAT/AMD blend revealed improved performance in a very cost-effective manner.

Fabrication of a biodegradable PBAT/PHB blend composite film with the addition of a natural cost-effective filler (1−3 wt.%), such as babassu, was reported by Hoffmann et al. As a result of the 3 wt.% babassu, the thermal stability was slightly enhanced. The tensile strength of the PBAT/PHB (50/50) blend revealed higher mechanical properties appropriate for packaging and mulch film applications. Bbabassu is an eco-friendly and cost-effective filler which maintains the overall properties of the blend system (Bauli et al., 2019).

3. Application of PBAT-based blends and composites

Recently, there has been an increasing awareness of the design of novel biodegradable polymeric materials to reduce the environmental pollution problems and replaced them with petroleum-based nonbiodegradable plastic materials in packaging fields, for instance, polyethylene (PE), polypropylene (PP), and polystyrene, etc. (Jiao et al., 2020), because these wastes cannot be degraded. On the other hand, biodegradable aliphatic−aromatic copolyesters have been used widely for the development of biodegradable blend and composite polymeric materials and also for various applications such as garbage bags, shopping bags, etc. (Yi et al., 2020), however here only the recent market applications, such as packaging and mulching film, are discussed, as shown in Fig. 11.17.

Figure 11.17 PBAT blend and composite applications.

3.1 Packaging

Traditional plastic packaging is commonly employed in several customer goods and garbage collection applications owing to its better characteristics and low cost when compared to other packaging materials. Over the last few years, approximately 14 million tons of traditional plastics packaging waste were produced every year, and only 1.6 million tons were reclaimed by recycling and the rest went to landfill. However, to reduce the traditional plastic packaging, another composting recovery biodegradation process was retained (Reichert, Bugnicourt, & Maria-Beatrice, 2020).

Currently, many researchers and academics are developing biodegradable compostable PBAT-based materials that have been commercialized. These packaging materials achieved great interest in several areas due to their exclusive performance compared to traditional plastic materials (Brusseau et al., 2019; Waterer, 2010; Youssef et al., 2018, 2019). A very few eminent companies have developed compostable PBAT-based materials, for instance, BASF, KINGFA Novamont, and BIOTECH, etc. Kingfa is the world leader in the field of biodegradable plastics and has developed many compostable materials based on PBAT, PLA, and starch. On the other hand, starch–PBAT-based blends have been extensively employed in China in high-quality supermarkets.

3.2 Agriculture (mulch film)

Recently, farmers have developed an extreme dependence on the use of traditional plastic mulch film in modern agriculture technology. For crop production, plastic mulch films have various advantages, such as weed and insect control, soil and air temperature maintenance, less evaporation, reduced soil erosion, and inhibition of soil splashing on fruits or vegetables, as shown in Fig. 11.18. Overall, to the contribute

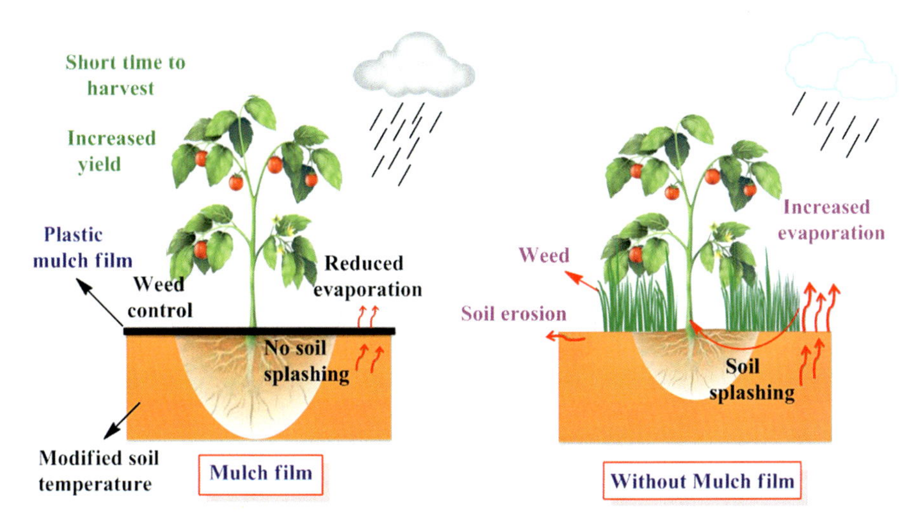

Figure 11.18 Schematic depiction of a plastic mulch film used in an agriculture cropping system with and without the mulch film.

to low usage of pesticides, timely planting in spring, conservation of water, and also improving the crop productivity and quality (Kasirajan & Ngouajio, 2012; Sintim & Flury, 2017; Steinmetz et al., 2016). However, the overall agriculture film market was predicted to reach 7.5 million tons by 2021 (Kwiecien et al., 2018). Most countries use PE mulch film and it is very difficult to recover it from agriculture fields completely, which reduces the soil productivity by blocking water penetration, obstructing root growth, hindering soil—gas exchange, and altering the soil microbial colonies. For that reason, it is planned to replace traditional PE mulch film with biodegradable mulch film because these films are degradable by soil microorganisms. PBAT-based blend and composite film has an insignificantly impacted by water, high temperatures, and UV radiation for the period of its useful lifetime and is completely biodegradable. In the current scenario, food requirements are increasing with the growing global population, thereby the usage of a plastic mulch film in agriculture has great prospects to enhance food supply. Instead of the presently used polyethylene-based mulches, biodegradable plastic mulches are a potential substitute, however severe testing is required to confirm their eco-friendly usage.

Kwiecie et al. (2018) developed new biodegradable and bioactive PBAT/PLA mulch films in the presence of 2-methyl-4-chlorophenoxyacetic acid (MCPA) conjugated with poly(3-hydroxybutyrate-co-3-hydroxyvalerate) (PHBV) at 5%, 10%, and 15% respectively, which was used as an herbicide to target broadleaf weed species. Fig. 11.19 demonstrates the effectiveness of the bioactive mulch film on weed prevention and the growth of the fava bean plant in a greenhouse experiment. Fig. 11.19 illustrates that the weed species could easily grow when no mulch was applied, and the mulch film did not contain MCPA-PHBA. On the other hand, no weed growth

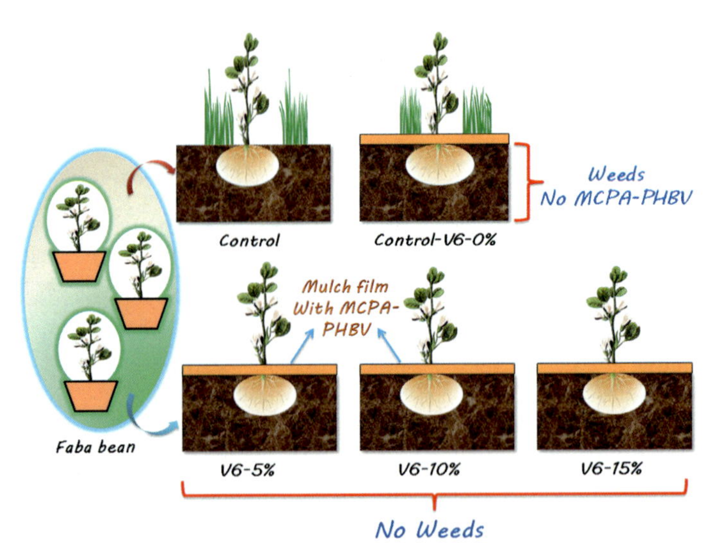

Figure 11.19 Schematic representation of the effect of bioactive MCPA-PHBV-incorporated PBAT/PLA films at different wt.% (0, 5, 10, 15) on the inhibition of weed growth on faba bean during a greenhouse experiment.

was seen after the incorporation of 5 wt.% MCPA-PHBV into the PBAT/PLA mulch film. This confirmed that the addition of MCPA-PHBV into the mulch film controls broadleaf weed growth in all concentration ratios. However, the developed bioactive and biodegradable mulch film is a better candidate for future agriculture applications, which increased the growth of crops and reduced water consumption.

4. Conclusion

PBAT blends and composites from biodegradable polymers have emerged as versatile materials compared to unmodified PBAT and have been widely used in packaging and mulch film applications. On the other hand, the developed composite materials exhibited unique properties after blending with low-cost polymers and the addition of nanofillers using different techniques, such as melt blending and twin-screw extrusion, etc. to improve the dispersion of filler within the PBAT matrix. The filler enhances the interfacial interaction between the filler and PBAT matrix to improve the properties of the PBAT blends and composites. Therefore, the developed PBAT blends and composites possess exclusive biodegradability, eco-friendliness, cost-effectiveness, and better mechanical and thermal properties, and overall they reduce the dependency on nonbiodegradable polymeric materials for various applications. PBAT blends and composites have exhibited great potential for uses in new industrial applications.

References

Bastioli, C. (2020). *Handbook of biodegradable polymers*.

Beber, V. C., de Barros, S., Banea, M. D., Brede, M., de Carvalho, L. H., Hoffmann, R., Costa, A. R. M., Bezerra, E. B., Silva, I. D. S., Haag, K., Koschek, K., & Wellen, R. M. R. (2018). Effect of Babassu natural filler on PBAT/PHB biodegradable blends: an investigation of thermal, mechanical, and morphological behavior. *Materials, 11*(5). https://doi.org/10.3390/ma11050820

Brusseau, M. L., Pepper, I. L., & Gerba, C. (2019). *Environmental and pollution science*.

Lins, C. L., Sébastien, L., Jannick, D.-R., & Jean-François, G. (2015). Phosphonium ionic liquids as new compatibilizing agents of biopolymer blends composed of poly(butylene-adipate-co-terephtalate)/poly(lactic acid) (PBAT/PLA). *RSC Advances*, 59082−59092. https://doi.org/10.1039/C5RA10241C

Chinsirikul, W., Rojsatean, J., Hararak, B., Kerddonfag, N., Aontee, A., Jaieau, K., Kumsang, P., & Sripethdee, C. (2015). Flexible and tough poly(lactic acid) films for packaging applications: property and processability improvement by effective reactive blending. In *Packaging Technology and Science (Vol. 28, Issue 8, pp. 741−759)*John Wiley and Sons Ltd. https://doi.org/10.1002/pts.2141

da Silva, J. B. A., Santana, J. S., de Almeida Lucas, A., Passador, F. R., de Sousa Costa, L. A., Pereira, F. V., & Druzian, J. I. (2019). PBAT/TPS-nanowhiskers blends preparation and application as food packaging. *Journal of Applied Polymer Science, 136*(26). https://doi.org/10.1002/app.47699

da Silva, N. M. C., Correia, P. R. C., Druzian, J. I., Fakhouri, F. M., Fialho, R. L. L., & De Albuquerque, E. C. M. C. (2017). PBAT/TPS composite films reinforced with starch nanoparticles produced by ultrasound. *International Journal of Polymer Science*. https://doi.org/10.1155/2017/4308261

Dammak, M., Fourati, Y., Tarrés, Q., Delgado-Aguilar, M., Mutjé, P., & Boufi, S. (2020). Blends of PBAT with plasticized starch for packaging applications: mechanical properties, rheological behaviour and biodegradability. *Industrial Crops and Products, 144*. https://doi.org/10.1016/j.indcrop.2019.112061

Ferreira, F. V., Cividanes, L. S., Gouveia, R. F., & Lona, L. M. F. (2019). An overview on properties and applications of poly(butylene adipate-co-terephthalate)—PBAT based composites. *Polymer Engineering and Science, 59*(2), E7—E15. https://doi.org/10.1002/pen.24770

Freitas, A. L. P. D. L., Tonini Filho, L. R., Calvão, P. S., & Souza, A. M. C. D. (2017). Effect of montmorillonite and chain extender on rheological, morphological and biodegradation behavior of PLA/PBAT blends. *Polymer Testing, 62*, 189—195. https://doi.org/10.1016/j.polymertesting.2017.06.030

Gross, R. A., & Kalra, B. (2002). Biodegradable polymers for the environment. *Science, 297*(5582), 803—807. https://doi.org/10.1126/science.297.5582.803

Guan, D., Su, X., Zhang, Q., Peters, G. P., Liu, Z., Lei, Y., & He, K. (2014). The socioeconomic drivers of China's primary PM2. 5 emissions. *Environmental Research Letters, 9*(2).

Hamaide, T., Deterre, R., & Feller, J. F. (2014). Environmental impact of polymers. In *Environmental impact of polymers* (pp. 1—368). Wiley Blackwell. https://doi.org/10.1002/9781118827116, 9781848216211.

Jiang, L., Liu, B., & Zhang, J. (2009). Properties of poly(lactic acid)/poly(butylene adipate-co-terephthalate)/nanoparticle ternary composites. *Industrial and Engineering Chemistry Research, 48*(16), 7594—7602. https://doi.org/10.1021/ie900576f

Jiao, J., Zeng, X., & Huang, X. (2020). An overview on synthesis, properties and applications of poly(butylene-adipate-co-terephthalate)—PBAT. *Advanced Industrial and Engineering Polymer Research*, 19—26. https://doi.org/10.1016/j.aiepr.2020.01.001

Kale, G. (2011). Overview of biodegradable packaging, methods, and current trends. In *Formulating, packaging, and marketing of natural cosmetic products* (pp. 411—419). John Wiley and Sons. https://doi.org/10.1002/9781118056806.ch21

Kasirajan, S., & Ngouajio, M. (2012). Polyethylene and biodegradable mulches for agricultural applications: a review. *Agronomy for Sustainable Development, 32*(2), 501—529. https://doi.org/10.1007/s13593-011-0068-3

Khan, H., Kaur, S., Baldwin, T. C., Radecka, I., Jiang, G., Bretz, I., Duale, K., Adamus, G., & Kowalczuk, M. (2020). Effective control against broadleaf weed species provided by biodegradable PBAT/PLA mulch film embedded with the herbicide 2-methyl-4-chlorophenoxyacetic acid (MCPA). *ACS Sustainable Chemistry and Engineering, 8*(13), 5360—5370. https://doi.org/10.1021/acssuschemeng.0c00991

Klemm, D., Heublein, B., Fink, H. P., & Bohn, A. (2005). Cellulose: Fascinating biopolymer and sustainable raw material. *Angewandte Chemie International Edition, 44*(22), 3358—3393. https://doi.org/10.1002/anie.200460587

Kwiecien, I., Adamus, G., Jiang, G., Radecka, I., Baldwin, T. C., Khan, H. R., Johnston, B., Pennetta, V., Hill, D., Bretz, I., & Kowalczuk, M. (2018). Biodegradable PBAT/PLA blend with bioactive MCPA-PHBV conjugate suppresses weed growth. *Biomacromolecules, 19*(2), 511—520. https://doi.org/10.1021/acs.biomac.7b01636

Liu, W., Liu, S., Wang, Z., Liu, J., Dai, B., Chen, Y., & Zeng, G. (2020). Preparation and characterization of compatibilized composites of poly(butylene adipate-co-terephthalate)

and thermoplastic starch by two-stage extrusion. *European Polymer Journal, 122*. https://doi.org/10.1016/j.eurpolymj.2019.109369

Marinho, V. A. D., Pereira, C. A. B., Vitorino, M. B. C., Silva, A. S., Carvalho, L. H., & Canedo, E. L. (2017). Degradation and recovery in poly(butylene adipate-co-terephthalate)/thermoplastic starch blends. *Polymer Testing, 58*, 166–172. https://doi.org/10.1016/j.polymertesting.2016.12.028

Martins, I. M. G., Magina, S. P., Oliveira, L., Freire, C. S. R., Silvestre, A. J. D., Neto, C. P., & Gandini, A. (2009). New biocomposites based on thermoplastic starch and bacterial cellulose. *Composites Science and Technology, 69*(13), 2163–2168. https://doi.org/10.1016/j.compscitech.2009.05.012

Meng, D., Xie, J., Waterhouse, G. I., Zhang, K., Zhao, Q., Wang, S., & Pan, Y. (2020). Biodegradable Poly (butylene adipate-co-terephthalate) composites reinforced with bio-based nanochitin: preparation enhanced mechanical and thermal properties. *Journal of Applied Polymer Science*, 12.

Mohanty, A. K., Vivekanandhan, S., Pin, J. M., & Misra, M. (2018). Composites from renewable and sustainable resources: challenges and innovations. *Science, 362*(6414), 536–542. https://doi.org/10.1126/science.aat9072

Moustafa, H., Guizani, C., & Dufresne, A. (2017). Sustainable biodegradable coffee grounds filler and its effect on the hydrophobicity, mechanical and thermal properties of biodegradable PBAT composites. *Journal of Applied Polymer Science, 134*(8). https://doi.org/10.1002/app.44498

Moustafa, H., Guizani, C., Dupont, C., Martin, V., Jeguirim, M., & Dufresne, A. (2017). Utilization of torrefied coffee grounds as reinforcing agent to produce high-quality biodegradable PBAT composites for food packaging applications. *ACS Sustainable Chemistry and Engineering, 5*(2), 1906–1916. https://doi.org/10.1021/acssuschemeng.6b02633

Nair, L. S., & Laurencin, C. T. (2007). Biodegradable polymers as biomaterials. *Progress in Polymer Science, 32*(8–9), 762–798. https://doi.org/10.1016/j.progpolymsci.2007.05.017

Nakayama, D., Wu, F., Mohanty, A. K., Hirai, S., & Misra, M. (2018). Biodegradable composites developed from PBAT/PLA binary blends and silk powder: compatibilization and performance evaluation. *ACS Omega, 3*(10), 12412–12421. https://doi.org/10.1021/acsomega.8b00823

Nofar, M., Oguz, H., & Ovalı, D. (2019). Effects of the matrix crystallinity, dispersed phase, and processing type on the morphological, thermal, and mechanical properties of polylactide-based binary blends with poly[(butylene adipate)-co-terephthalate] and poly[(butylene succinate)-co-adipate]. *Journal of Applied Polymer Science, 136*(23). https://doi.org/10.1002/app.47636

Nunes, E. D. C., Souza, & Rosa, D. D. S. (2019). Use of a chain extender as a dispersing agent of the $CaCO_3$ into PBAT matrix. *Journal of Composite Materials, 54*(10), 1373–1382.

Pappu, A., Saxena, M., Thakur, V. K., Sharma, A., & Haque, R. (2016). Facile extraction, processing and characterization of biorenewable sisal fibers for multifunctional applications. *Journal of Macromolecular Science, Part A: Pure and Applied Chemistry, 53*(7), 424–432. https://doi.org/10.1080/10601325.2016.1176443

Pinheiro, I. F., Ferreira, F. V., Alves, G. F., Rodolfo, A., Morales, A. R., & Mei, L. H. I. (2019). Biodegradable PBAT-based nanocomposites reinforced with functionalized cellulose nanocrystals from *Pseudobombax munguba*: rheological, thermal, mechanical and biodegradability properties. *Journal of Polymers and the Environment, 27*(4), 757–766. https://doi.org/10.1007/s10924-019-01389-z

Bauli, C. R., Rocha, D. B., Santos, R., & dos, D. (2019). Composite films of ecofriendly lignocellulosic nanostructures in biodegradable polymeric matrix. *SN Applied Sciences*. https://doi.org/10.1007/s42452-019-0765-0

Chandra, R., & Rustgi, R. (1998). Biodegradable polymers. *Progress in Polymer Science*, 1273–1335. https://doi.org/10.1016/s0079-6700(97)00039-7

Reichert, Corina L., Bugnicourt, E., Maria-Beatrice, Coltelli, et al. (2020). Bio-based packaging: Materials, modifications, industrial applications and sustainability. *Polymers, 12*(7), 1558. https://doi.org/10.3390/polym12071558

Rocha, D. B., Carvalho, S.de, Oliveira, & Rosa, dos S. (2018). A new approach for flexible PBAT/PLA/$CaCO_3$ films into agriculture. *Journal of Applied Polymer Science, 135*(35).

Rodrigues, B. V. M., Silva, A. S., Melo, G. F. S., Vasconscellos, L. M. R., Marciano, F. R., & Lobo, A. O. (2016). Influence of low contents of superhydrophilic MWCNT on the properties and cell viability of electrospun poly (butylene adipate-co-terephthalate) fibers. *Materials Science and Engineering: C, 59*, 782–791. https://doi.org/10.1016/j.msec.2015.10.075

Sander, M. (2019). Biodegradation of polymeric mulch films in agricultural soils: concepts, knowledge gaps, and future research directions. *Environmental Science and Technology, 53*(5), 2304–2315. https://doi.org/10.1021/acs.est.8b05208

Santos, R. A. L., Muller, C. M. O., Grossmann, M. V. E., Mali, S., & Yamashita, F. (2014). Starch/poly (butylene adipate-co-terephthalate)/montmorillonite films produced by blow extrusion. *Quimica Nova, 37*(6), 937–942. https://doi.org/10.5935/0100-4042.20140170

Sintim, H. Y., & Flury, M. (2017). Is biodegradable plastic mulch the solution to agriculture's plastic problem? *Environmental Science and Technology, 51*(3), 1068–1069. https://doi.org/10.1021/acs.est.6b06042

Sirisinha, K., & Somboon, W. (2012). Melt characteristics, mechanical, and thermal properties of blown film from modified blends of poly(butylene adipate-co-terephthalate) and poly(lactide). *Journal of Applied Polymer Science, 124*(6), 4986–4992. https://doi.org/10.1002/app.35604

Steinmetz, Z., Wollmann, C., Schaefer, M., Buchmann, C., David, J., Tröger, J., Muñoz, K., Frör, O., & Schaumann, G. E. (2016). Plastic mulching in agriculture. Trading short-term agronomic benefits for long-term soil degradation? *The Science of the Total Environment, 550*, 690–705. https://doi.org/10.1016/j.scitotenv.2016.01.153

Stewart, D. (2008). Lignin as a base material for materials applications: chemistry, application and economics. *Industrial Crops and Products, 27*(2), 202–207. https://doi.org/10.1016/j.indcrop.2007.07.008

Tavares, L. B., Ito, N. M., Salvadori, M. C., dos Santos, D. J., & Rosa, D. S. (2018). PBAT/kraft lignin blend in flexible laminated food packaging: peeling resistance and thermal degradability. *Polymer Testing, 67*, 169–176. https://doi.org/10.1016/j.polymertesting.2018.03.004

Teamsinsungvon, A., Ruksakulpiwat, Y., & Jarukumjorn, K. (2013). Preparation and characterization of poly(lactic acid)/Poly(butylene adipate-co-terepthalate) blends and their composite. *Polymer – Plastics Technology and Engineering, 52*(13), 1362–1367. https://doi.org/10.1080/03602559.2013.820746

Thakur, M. K., Thakur, V. K., Gupta, R. K., & Pappu, A. (2016). Synthesis and applications of biodegradable soy based graft copolymers: a review. *ACS Sustainable Chemistry and Engineering, 4*(1), 1–17. https://doi.org/10.1021/acssuschemeng.5b01327

Tiimob, B. J., Rangari, V. K., Mwinyelle, G., Abdela, W., Evans, P. G., Abbott, N., Samuel, T., & Jeelani, S. (2018). Tough aliphatic-aromatic copolyester and chicken egg white flexible biopolymer blend with bacteriostatic effects. *Food Packaging and Shelf Life, 15*, 9–16. https://doi.org/10.1016/j.fpsl.2018.01.001

Touchaleaume, F., Martin-Closas, L., Angellier-Coussy, H., Chevillard, A., Cesar, G., Gontard, N., & Gastaldi, E. (2016). Performance and environmental impact of biodegradable polymers as agricultural mulching films. *Chemosphere, 144*, 433–439. https://doi.org/10.1016/j.chemosphere.2015.09.006

Van De Velde, K., & Kiekens, P. (2002). Biopolymers: overview of several properties and consequences on their applications. *Polymer Testing, 21*(4), 433–442. https://doi.org/10.1016/S0142-9418(01)00107-6

Vieira, M. G. A., Da Silva, M. A., Dos Santos, L. O., & Beppu, M. M. (2011). Natural-based plasticizers and biopolymer films: a review. *European Polymer Journal, 47*(3), 254–263. https://doi.org/10.1016/j.eurpolymj.2010.12.011

Waterer, D. (2010). Evaluation of biodegradable mulches for production of warm-season vegetable crops. *Canadian Journal of Plant Science, 90*(5), 737–743. https://doi.org/10.4141/CJPS10031

Wei, D., Wang, H., Xiao, H., Zheng, A., & Yang, Y. (2015). Morphology and mechanical properties of poly(butylene adipate-co-terephthalate)/potato starch blends in the presence of synthesized reactive compatibilizer or modified poly(butylene adipate-co-terephthalate). *Carbohydrate Polymers, 123*, 275–282. https://doi.org/10.1016/j.carbpol.2015.01.058

Wu, D., Tan, Y., Han, L., Zhang, H., & Dong, L. (2018). Preparation and characterization of acetylated maltodextrin and its blend with poly(butylene adipate-co-terephthalate). *Carbohydrate Polymers, 181*, 701–709. https://doi.org/10.1016/j.carbpol.2017.11.092

Xiao, B., Sun, X. F., & Sun, R. (2001). Chemical modification of lignins with succinic anhydride in aqueous systems. *Polymer Degradation and Stability, 71*(2), 223–231. https://doi.org/10.1016/S0141-3910(00)00133-6

Xing, Q., Ruch, D., Dubois, P., Wu, L., & Wang, W. J. (2017a). Biodegradable and high-performance poly(butylene adipate-co-terephthalate)-Lignin UV-blocking films. *ACS Sustainable Chemistry and Engineering, 5*(11), 10342–10351. https://doi.org/10.1021/acssuschemeng.7b02370

Xiong, S. J., Pang, B., Zhou, S. J., Li, M. K., Yang, S., Wang, Y. Y., Shi, Q., Wang, S. F., Yuan, T. Q., & Sun, R. C. (2020). Economically competitive biodegradable PBAT/lignin composites: effect of lignin methylation and compatibilizer. *ACS Sustainable Chemistry and Engineering, 8*(13), 5338–5346. https://doi.org/10.1021/acssuschemeng.0c00789

Yang, X., & Zhong, S. (2020). Properties of maleic anhydride-modified lignin nanoparticles/polybutylene adipate-co-terephthalate composites. *Journal of Applied Polymer Science, 137*(35). https://doi.org/10.1002/app.49025

Yao, Q., Song, Z., Li, J., & Zhang, L. (2020). Micromorphology, mechanical, crystallization and permeability properties analysis of HA/PBAT/PLA (HA, hydroxyapatite; PBAT, poly(-butylene adipate-co-butylene terephthalate); PLA, polylactide) degradability packaging films. *Polymer International, 69*(3), 301–307. https://doi.org/10.1002/pi.5953

Yi, T., Qi, M., Mo, Q., Huang, L., Zhao, H., Liu, D., & Liu, Y. (2020). Ecofriendly preparation and characterization of a cassava starch/polybutylene adipate terephthalate film. *Processes, 8*(3).

Youssef, A. M., Assem, F. M., Abdel-Aziz, M. E., Elaaser, M., Ibrahim, O. A., Mahmoud, M., & Abd El-Salam, M. H. (2019). Development of bionanocomposite materials and its use in coating of Ras cheese. *Food Chemistry, 270*, 467–475. https://doi.org/10.1016/j.foodchem.2018.07.114

Youssef, A. M., El-Sayed, S. M., El-Sayed, H. S., Salama, H. H., Assem, F. M., & Abd El-Salam, M. H. (2018). Novel bionanocomposite materials used for packaging skimmed milk acid coagulated cheese (Karish). *International Journal of Biological Macromolecules, 115*, 1002–1011. https://doi.org/10.1016/j.ijbiomac.2018.04.165

Zhang, N., Wang, Q., Ren, J., & Wang, L. (2009). Preparation and properties of biodegradable poly(lactic acid)/poly(butylene adipate-co-terephthalate) blend with glycidyl methacrylate as reactive processing agent. *Journal of Materials Science, 44*(1), 250–256. https://doi.org/10.1007/s10853-008-3049-4

Zhao, H., Liu, H., Liu, Y., & Yang, Y. (2020). Blends of poly (butylene adipate-co- terephthalate) (PBAT) and stereocomplex polylactide with improved rheological and mechanical properties. *RSC Advances, 10*(18), 10482–10490.

Hydrogel- and aerogel-based composites

12

Biodegradable hydrogel and aerogel polymer blend-based composites

Kushairi Mohd Salleh [1], *Nur Amira Zainul Armir* [1], *Nyak Syazwani Nyak Mazlan* [1], *Marhaini Mostapha* [1], *Chunhong Wang* [2] *and Sarani Zakaria* [1]
[1]Bioresource and Biorefinery Research Group, Faculty of Science and Technology, Universiti Kebangsaan Malaysia, Bangi, Selangor, Malaysia; [2]School of Textile Science and Engineering, Tiangong University, Xiqing District, Tianjin, PR China

1. Introduction

A polymer blend is defined as an intimate mixture of two polymers with no covalent bonds (Asano, 2017). The main objective of a multicomponent polymeric system, or polymer blend, is to have a full set of the desired properties at the lowest cost. The goal of combining two or more polymers is to achieve a combination of each polymer's desirable properties without compromising the intended final properties. Polymer blends seldom form a homogeneous mixture but rather exhibit macro- or microphase separation (Martuscelli et al., 1979). Often known as a heterogeneous system, these phase separations offer a more robust performance than a homogeneous system. It even appears that many useful blending polymers are attributable to the properties obtained from the two-phase system (Vasile & Kulshreshhta, 2003). There are certain inherent benefits to incompatibility between polymers that produce a new polymer mixture with superior properties. The benefits of polymer blends lie in the synergistic interaction between polymers, even though polymers that are usually mutually immiscible are prone to have weak polymer−polymer interfaces, therefore resulting in low mechanical properties. It is important to note that the miscibility is the main criterion for blending. Miscible polymer blending improves processability and cost efficiency, and promises enhanced material properties. Blended polymers are justified as "compatible" or miscible only if they are blended sufficiently and a particular purpose is satisfied (Asano, 2017). Polymer pairs can be qualitatively deemed incompatible, semicompatible, or compatible after mixing together, based on whether two different or immiscible phases exist, if partial mixing of the two polymers occurs at the molecular level, or whether a single thermodynamically stable phase is created (Asano, 2017). Blends of immiscible polymers usually involve graft, block copolymerization, and interpenetration of two networks. To ensure the miscibility of immiscible polymers, chemical modifications via ionic or donor−acceptor interactions are considered. Modification is a way to improve the properties and increase

Biodegradable Polymers, Blends and Composites. https://doi.org/10.1016/B978-0-12-823791-5.00019-3

compatibilization to achieve the required properties for specific applications (Imre & Pukánszky, 2013). Further research is needed into the relationships of miscibility–structure–property of polymer blends.

Biodegradation is a natural process that breaks down substances into the essential elemental components with a naturally occurring biodegrading microorganism. Biodegradation processes are characterized by certain pathways, namely abiotic involvement, biodeterioration, biofragmentation, and assimilation (Lucas et al., 2008). Initially, biodegradation involves an enzyme attack on the amorphous regions, then the crystalline regions. Long polymer chains are reduced to monomers, and aerobic or anaerobic degradation occurs to complete the degradation chain. When degradation is complete, it is known as mineralization. If decomposition is due to other than microorganisms—for example, degraded by light—the material cannot be considered biodegradable. Biodegradation is typically measured as the amount of CO_2 evolved, or the amount of carbon that has disappeared over time (Kjeldsen et al., 2019). This results in a gradual decrease in the molecular weight of the polymer (Banerjee et al., 2014). A long time for a substance to degrade reveals that the biological impact was ineffectual. The progressiveness of biodegradation is primarily dependent on the chemical structure, crystallinity, weight, and distribution of the molecules. A paradigm change from biostable products to biodegradable ones that are compliant with the environment has been created by increasing environmental awareness to ensure a healthy, eco-friendly atmosphere for agriculture and medical products. Nowadays, the ever-increasing use of nondegradable materials has led to biodegradability becoming an integral aspect of products. An additional benefit of utilizing biodegradable products is that they serve as fertilizers and soil conditioners upon disintegration and composting, improving crop production while capitalizing on recycling by natural resources in an environmentally sustainable and cleaner atmosphere (Makhijani et al., 2015). Medical and agricultural biodegradable materials typically are made from hybridized natural and biodegradable synthetic polymers. The development of novel biodegradable polymer blends for agricultural and medical uses is a relatively new science.

Natural polymers, which may be plant- or animal-based materials, are biodegradable. They are environmentally sustainable and natural, making them the most attractive materials for various applications, but they are not entirely economical. They are mainly carbohydrates and proteins functioning as structural support in plants and animals. For plant-based polymers, typically obtained from biomass, they are practically produced from three different kinds of constituent: polysaccharides (i.e., cellulose, hemicellulose, starch, chitin, and chitosan), proteins (i.e., wheat gluten, soy, and collagen/gelatin), and specialty polymers (i.e., lignin and natural rubber) (Sam et al., 2013). Thanks to their useful physical and chemical properties, such as emulsifying strength, viscoelasticity, polyelectrolyte conformation, adherence, biocompatibility, and stabilizing power, these macromolecules are gaining renewed interest as biopolymers, and they are typically exploited in a wide variety of applications, especially in biodegradable products (Makhijani et al., 2015). Their biocompatibility and minimal inflammatory response are gaining uses as reliable alternative materials, especially in medical applications. The focus on polymers is derived primarily from natural bulk forms, such as cellulose, hemicellulose, and lignin. Cellulose, along with hemicellulose and lignin, is the key structural polysaccharide in plants. Their abundance and versatility in the medical and agriculture sectors have proved to be useful. Further recognition was gained when they were blended with biodegradable synthetic polymer. Biodegradable synthetic and natural polymer combinations are capable of creating a new class of materials with enhanced mechanical and thermal properties and biocompatibility (Sionkowska, 2011).

Biodegradable synthetic polymers are the most commonly employed scaffold components in tissue engineering, including various polyesters (Kim et al., 2013). They are typically easy to process, and are constructed by connecting monomers via direct covalent bonding. The malleability during synthesis, production, and modification enables them to form virtually any shape, size, and functionality (Kim et al., 2013). Synthetic polymers possess identical physicochemical and mechanical properties to biological tissues. They are sterilizable, can be easily processed, and their surface can be modified. They are also a suitable material as a polymeric matrix for plant growth; for example, aliphatic-aromatic poly(butylene adipate-co-butylene terephthalate) (PBAT) and blends containing it have shown promise as a biodegradable polymer in agricultural soil (Agarwal, 2020). The most commonly used biodegradable synthetic polymers are polyvinyl alcohol (PVA), poly(glycolic acid) (PGA), poly(lactic acid) (PLA), and their copolymers poly(lactide-co-glycolide), polyanhydride, poly(propylene fumarate), polycaprolactone (PCL), polyethylene glycol (PEG), and polyurethane (Kundu et al., 2013). Synthetic polymers are a desirable substance class because of the potential to customize material properties for particular applications such as molecular weight modification or chemistry to influence biodegradability and stability (Vyas et al., 2017). The limitations of synthetic polymers in medicine are that they have no specific cell attachment binding motifs and are usually hydrophobic. Similar to natural polymers, degradation causes interference with their physical stability, making them ineffective for particular applications, and there can even be cytotoxic by-products of degradation (Vyas et al., 2017). The degradation process for biodegradable synthetic polymers usually involves simple hydrolysis and often harsh conditions for hydrolysis, for example, polyolefins [polyethylene (PE) and polypropylene] and polyethylene terephthalate (PET) are not considered biodegradable (Agarwal, 2020). The major contributor to the variations in the degradability of a substance is the individual monomer functional group. Some chemical groups and bonds are more easily degraded by biological agents (Kjeldsen et al., 2019). These vary from simple carbon chains [e.g., polypropylene (PP)] to more complex sidechains [e.g., poly(methyl methacrylate) (PMMA)].

Both natural and synthetic biodegradable polymers are synergistically mutually miscible to some extent. The value of polymer-blending technologies was demonstrated by large arrays of commodity manufacturing using them. They often are known as bioartificial and biosynthetic polymeric materials (Sionkowska, 2011). For example, if natural polymers have low structural resilience and are unable to maintain their native structure from high-temperature exposure, the properties could be improved by blending with a synthetic biodegradable polymer. The interaction between them is symbiotically riveting. In this field, the development of research activities has not yet been thoroughly explored and characterized. From materials selection to process design, end-products and their intended properties focus on research advancement. Biodegradable polymer blends, especially complex materials like gels, biomacromolecules, and micelles are often referred to as "soft materials" The focus here is on hydrogels and aerogels, which can be synthetically built, naturally built, or a combination of both. To obtain hydrogels and aerogels, processing of two or more polymers is generally involved at a liquid state where they can interact at the molecular level.

Hydrogel is a three-dimensional (3D) polymeric network of physically or chemically cross-linked hydrophilic polymers that have the capacity to absorb water and preserve a large fraction of water within their framework (Salleh et al., 2018, 2019). The features of the hydrogel depend on the network configuration and the response to the external environment. Due to hydrophilic functional groups such as—OH, —COOH, and C—O—C attached to the polymer backbone, they are hydrophilic, and the hydrophilicity of hydrogel is often impaired by the capillary effect and the osmotic pressure difference. Hydrogel is able to imbibe water from 10 to 100,000 times the actual skeletal weight of the cryogel (Salleh et al., 2019). They also resistant to dissolution due to cross-link density between the network chains (Pal et al., 2009). The higher the cross-link density, the more the equilibrium swelling will be decreased, the hydrophobicity is consequently improved, and the stretchability of the polymer network reduces accordingly (Pal et al., 2009; Salleh et al., 2020). Due to their swelling and deswelling properties that enable water movement while retaining solid-like mechanical properties, they are ideal for mobilizing scaffolds for molecular and nano-scale species (Sirousazar et al., 2014). As a consequence of these properties, hydrogels can be commonly employed in several areas, such as drug delivery and tissue engineering in medicine and nutrient transfer in agriculture. Hydrogel has been shown to mimic the extracellular matrix (ECM) and supply surrounding cells with structural and biochemical support. They also have proven to be capable as a matrix for watering plants (Elbarbary & Ghobashy, 2017).

Aerogel is an extremely porous ultralight solid material derived by a supercritical drying method from sol—gel materials in which the liquid portion has been substituted by gas to leave intact solid micro- or nanostructures without substantial collapsing of the gel structure (Salleh et al., 2020; Zuo et al., 2015). It was first introduced by Steven Kistler in 1931 (Kistler, 1931). While they are still considered to be evolving, they are not seen as entirely new materials. Aerogels' textural characteristics, typically nanostructured solid phase, mesoporosity, low density, etc., and their chemical versatility enable organic—inorganic materials to be hybridized (Rigacci et al., 2017). They may be tuned into different sizes and forms, from beads to monoliths of discs, and graded according to their numerous microstructures (microporous, mesoporous, mixed porous) or by describing their composition (Stergar & Maver, 2016). Though many aerogel products, either organic, inorganic, or hybrid, have proven to be biocompatible, their ability has been used only in nonmedical fields of application. Nonetheless, countless experiments have been attempted and proven to be workable for biomedicine applications. The high surface area and porosity of aerogel plus biodegradability and biocompatibility render them efficient nanoparticle immobilizing scaffolds to fulfill multiple functions (Zuo et al., 2015). For example, two separate methods may be used to integrate drugs: (1) the in situ method is focused on incorporating the drug during the sol−gel phase and (2) the ex situ approach includes the addition of drugs through various postprocessing means (Stergar & Maver, 2016). Aerogel is a successful candidate in agriculture as a promising fast nutrient absorber material that avoids nutrient leaching and cleanses environmental contaminants at the same time.

Biodegradable hydrogel and aerogel polymer blends from natural and synthetic polymers for medical and agriculture products are covered in this chapter. Due to their specific mechanisms and properties, both hydrogels and aerogels have been widely studied in many applications, especially in medicine and agriculture. With polymer blends, enhancement of their properties makes them more desirable and reliable for many applications.

2. Lignocellulosic-based hydrogel and aerogel

Hydrogel and aerogel from lignocellulosic-based materials such as cellulose, hemicellulose, and lignin have immense potential. The properties are enhanced when they are blended with another natural polymer and/or biodegradable synthetic polymer. The development of hydrogel and aerogel from blending biodegradable materials is a significant area for commercialization—combining sustainability with performance. Both have proven to work well in medical and agricultural applications, although recent discoveries have revealed that some restrictions are ingrained in them. The scarcity of lignocellulosic-based material properties in hydrogel and aerogel for medical and agricultural applications has pressed scientists to investigate and discover better products. The production of hydrogel and aerogel from lignocellulosic materials concentrates on two key blending principles in this section: (1) lignocellulosic−natural polymers and (2) lignocellulosic−synthetic polymers. Blending systems are then classified based on the number of homopolymers: (1) binary, (2) ternary, and (3) multicomponent polymer blends.

2.1 Cellulose blends

Cellulose is a polysaccharide first mentioned in 1838 by Anselme Payen (Payen, 1838). It was used as a precursor for chemical modifications and polymer mixing long before the cellulose polymeric structure was understood and defined. Cellulose is the world's most abundant renewable polymer and is sourced from wood, hemp, linen, cotton, etc., with limited applications due to its insoluble and infusible properties. Factors that influence the solubility of cellulose are due to the crystallized fibril structure, the existence of inter- and intramolecular hydrogel bonding, high molecular weight (Mw), and degree of polymerization (DP) and hydrophobic interactions (Labafzadeh, 2015; Salleh et al., 2021). These factors must be tackled prior to the dissolving process. The insolubility makes them less desirable materials when modifications and dissolving processes are not completely resolved. Modifications of cellulose improve the functionalities with different functional groups attached to their polymer backbones and changes in their Mw and DP. Chemical modification of cellulose entails derivatization reactions in derivatizing solvents that mainly target cellulose substitutions of −OH groups at C6, C2, and C3, resulting in significantly altered functional properties in several applications that resolve native cellulose limitations (Salleh et al., 2018). Chemically modified cellulose, also known as a cellulose derivative, can be easily dissolved in water and other organic solvents, making them suitable for the blending process. It is important to note that dissolving the cellulose is the most important component for the blending process in producing regenerated products. Therefore, gel and solution blending involves a molecular-level blending process. Some of the most well-known cellulose derivatives are carboxymethyl cellulose, hydroxyethyl cellulose, methylcellulose, etc. Polymer blending of cellulose is also made possible without pretreatments and modifications using a special nonderivatizing solvent such as aqueous alkaline/urea and ionic liquid solvent systems. However, these systems have their limits, especially aqueous alkaline/urea solvent, as they required specific Mw and DP to be fully effective. Thermal pretreatment or simple acid hydrolysis on native cellulose have been proven to significantly decrease the Mw and DP of cellulose and thus increase specialty solvents' effectiveness. Cellulose dissolution's primary objective is to enable it to be used and blended with different polymer types and regenerated into the designated products and applications.

Blending cellulose with other polymers is to alter and improve potential polymer properties to produce enhanced materials with intermediate characteristics. Regenerated cellulose-based hydrogel and aerogel lack specific functionalities as cellulose is usually used as a supporting scaffold but not to provide a particular property. However, this applies only to native cellulose and not entirely to its derivatives. Regenerated cellulose hydrogel and aerogel in medical and agricultural applications have been attempted. Aerogel derived from cellulose is also called aerocellulose (Innerlohinger et al., 2006). Since cellulose has been proven to be biocompatible, it is suitable for long-term implantable materials with minimal reactivity with biological systems. They are also confirmed as nontoxic and nonimmunogenic and have no undesirable local or systematic effects on the biological host (Vyas et al., 2017). These criteria make them suitable for medical and agricultural applications. In the polymer blending system, cellulose usually serves to support the skeletal structure that improves the mechanical properties and provides a reaction site for the cross-linking process (Salleh et al., 2018). These features render them useful in blending systems as many probable polymers and materials can be physically cross-linked, allowing targeted applications to become feasible with promising effectivity. They can also provide the cells with suitable anchoring sites, contributing to improved cell adhesion, proliferation, and differentiation (Maharjan et al., 2021). Table 12.1 lists recent publications on cellulose blends regenerated as hydrogels and aerogels for medical and agricultural applications.

From Table 12.1 it can be seen that hydrogels and aerogels from cellulose blended with natural and biodegradable synthetic polymers enhance structural integrity and provide a reaction site for the embedment of functional materials for medical and

Table 12.1 Cellulose blends with natural and synthetic biodegradable polymers, their classification, and applications.

Cellulose–natural polymer	Classification	Applications	References
Hydrogel			
Cellulose nanofiber/chitosan	Binary	Hydrogel scaffolds for bone tissue engineering with improved osteogenic differentiation	Maharjan et al. (2021)
Cellulose nanofibrils/gelatin	Binary	High mechanical strength hydrogel with unique beads-on-a-string for artificial articular cartilage	Liu et al. (2020)
Nano-CMC-alginate/chitosan	Ternary	Hydrogel beads for controlled drug delivery with excellent pH sensitivity	Jeddi and Mahkam (2019)
Cationic cellulose nanocrystals/anionic alginate	Binary	Double-membrane hydrogels for complexing drugs codelivery	Lin et al. (2016)
Aerogel			
Chitosan/CMC/graphene oxide	Ternary	pH-controlled drug-delivery aerogel as the carrier for an effective chemotherapeutic agent in the treatment of cancers	Wang et al. (2017)
Bacterial cellulose/gelatin	Binary	Surface-modified microporous aerogel showed complete skin regeneration within 2 weeks	Khan et al. (2018)
Peptide/cellulose	Binary	Nanocellulosic aerogel for potential use in chronic wound sensor/dressing design platforms	Edwards et al. (2018)

Continued

Table 12.1 Continued

Cellulose–synthetic polymer	Classification	Applications	References
Hydrogel			
Cellulose/gelatin/AgNPs	Ternary	High-porosity injectable hydrogel with improved wound-healing ability	Gou et al. (2020)
Methylcellulose/PVA	Binary	Carrier for doxycycline hyclate (DOX-h) drug with controlled-release behavior that is sensitive to pH changes	El-Naggar et al. (2017)
Polyacrylamide/methyl cellulose/montmorillonite	Ternary	Hydrogel as a slow-release fertilizer with the capability to be loaded with a considerable amount of fertilizer	Bortolin et al. (2013)
PVP/CMC	Binary	Physically cross-linked hydrogel as controlled-release fertilizer	Elbarbary and Ghobashy (2017)
Aerogel			
Bacterial cellulose/AgNP/polyaniline	Ternary	The fabricated aerogel is a promising tool to provide an active site to spread live cells	Hosseini et al. (2020)
PVA/cellulose nanofibril	Binary	Highly porous aerogel successfully facilitates cell attachment, differentiation, and proliferation	Zhang et al. (2017)

agricultural applications. Physically and chemically cross-linked polymers, apart from covalent bonding, have been proven to play an immense role as a bridge to instill intermediate properties of blended polymers. Though a heterogeneous and mutually immiscible mixture is unlikely favored, a two- or more-phase system synergistically improves when intermolecular interactions induce interactions between polymers. As long as thermodynamic changes to the polymer state in a cellulose blend are controlled, the intended properties are attainable. Cellulose has been proven to be sustained in soil for certain periods before it fully degrades and is mineralized. This degradability makes it suitable to be used as a slow-release fertilizer for plants. In medical and agricultural applications, polymer blending between cellulose–natural and cellulose–biodegradable synthetic polymers for hydrogel and aerogel regeneration has been proven to be practical. However, the effectiveness of polymer blending still bounds to the homogeneity, miscibility, and synergistic effect between polymers. Even though many proposed researches have been carried out, technology deployments and commercialization have still failed due to product usefulness continually being overshadowed by a lack of cost-effectiveness. However, this has not stopped researchers from developing ideas and solutions to render cellulose and its derivatives as a reliable, sustainable, and renewable material for many high-end applications, especially in medical and agricultural developments.

2.2 Hemicellulose blends

The plant cell wall is comprised of two main building blocks, known as the primary and secondary cell walls. In the first part of lignocellulosic materials, hemicellulose and pectin structurally form the polysaccharide matrix that encloses the crystalline structure of cellulose microfibrils in the primary cell wall. Known as the cellulose-binding polysaccharides and cross-linked by pectin, they result in the formation of a strong and resilient network (Gibson, 2012). During the cell wall expansion, the secondary cell wall is formed, and the cellulose microfibrils, as well as the hemicellulose, are more structurally organized in the secondary cell wall (Loix et al., 2017). There is very limited understanding of hemicellulose biosynthesis. Nevertheless, the term hemicellulose is still conveniently used, although a number of ideas regarding the structure of hemicellulose and its name have been suggested. In this regard, hemicellulose is known to be a pentose-rich heterogeneous polysaccharide consisting of xylans, mannans, xyloglucans, and β-glucans, with new findings stating that galacturonic acid is also present (Scheller & Ulvskov, 2010). Hemicellulose is largely underutilized at a value-added scale because it usually remains in the fiber for the pulp and paper industry as a strength provider and degrades into low-molecular-weight components, as reported in the Kraft pulping industry (Farhat et al., 2017).

Increasing interest in hemicellulose-based hydrogels and aerogels has been noted, although limited studies are reported related to their abundant availability, antiinflammatory effect, and anticancer, antioxidant, and nonhazardous properties. Depending on the applications, the hydrophilic nature of hemicellulose means that it exhibits poor mechanical strength properties and is immiscible when blended with other polymers, alongside being a low-molecular-weight polymer. This weakness has limited its full application in bioproducts. In contrast, incorporating a carboxylic group into hemicellulose makes it fit to develop absorbent and adsorbent-functioning materials (Salam et al., 2011). As for chemical modification, aldehyde groups are introduced into hemicellulose, in a process known as oxidation, enabling hemicellulose to form supermolecular materials with silver nanoparticles. The pendant of the hydroxyl group possessed in the hemicellulose structure allows it to undergo chemical modifications. However, most synthetic polymers are hydrophobic, and, as mentioned before, the blending of hemicellulose and other synthetic polymers will potentially be less compatible. To such a degree, correlating with the research and development paradigm to increase the polymer blending process, hemicellulose-based products are being pursued by chemically hydrophobizing its structure by incorporating a hydrophobic backbone (Farhat et al., 2017). A few common hydrophobization methods have been used to treat hemicellulose, ranging from redox reaction, esterification, and etherification (Junli et al., 2012).

Hemicellulose hydrophobization offers a broad area of versatility in hemicellulose applications. It improves the fabricated materials to acquire diversified new properties such as water resistance, thermal stability, solubility in organic solvents, as well as functioning as thermoplastics (Farhat, 2018).

On the subject of hemicellulose-based aerogel, it is recognized as an excellent material for adsorbent applications such as removing heavy metal ions from water with excellent adsorption capacity; however, very limited studies have looked at hemicellulose-based aerogel for direct agricultural and medical purposes. It is also considered a fascinating material for electrical conductivity and would greatly help with medical applications such as cell—cell interactions. In addition, hemicellulose usage as the precursor of electrically conductive hydrogels has shown a greater swelling ratio and flexibility to enhance the conductivity magnitude. The use of dyes in the cosmetic, textile, printing, and dyeing industries can endanger the environment and living organisms upon release into wastewater. Thus, this issue has been addressed using several approaches, including adsorption, biodegradation, oxidation, etc. (Gan et al., 2020; Sun et al., 2015). The versatility of hemicellulose with chemical modification such as carboxylic acid introduction enables the formation of high water absorption and adsorption materials. For instance, to combat this problem, xylan-based hydrogels are being developed via physical cross-linking mechanisms. Carboxymethyl-xylan/poly N-isopropyl acrylamide hydrogel is formed through a semiinterpenetrating (IPN) network and a high swelling behavior with stability in response to environmental stimuli that are regarded as an intelligent material for drug delivery (Sun et al., 2018). Next, a thermo-gelling xylan/chitosan composite hydrogel was formed via ammonium hydrogen phosphate that acts as a salt for the gelation process (Bush et al., 2016). The multifaceted xylan used as the hydrogel precursor from hemicellulose eases the drug delivery when subjected to the blending interaction between the free complex xylan's glucuronic acid side chains and amino groups of chitosan. An interconnected macroporous hydrogel of dextran and PEG is also formed through radical polymerization of the dextran derivatives scaffold and polymerizable group via liquid—liquid phase separation caused by immiscible blending between PEG and dextran that resulted in an interconnected macroporous hydrogel that is important for tissue engineering using tissue adhesion (Ullah & Chen, 2020).

2.3 Lignin blends

The exploitation of hydrogels and aerogels is not limited to cellulose and hemicellulose; in fact, current interest is shifting toward lignin-based materials. Lignin is abundantly available and makes up approximately 25%—30% of the biomass in softwood and hardwood. It is highlighted in the research and development of biomass technology in making full use of biomass waste (Cesarino et al., 2012). Since the early years, lignin has been abundantly derived from the pulp and paper industry and ethanol production from lignocellulosic biomass. In terms of its structure, it comprises cross-linked hydroxylated and methoxylated phenylpropane units in a random manner (Larrañeta et al., 2018). It exists as a hydrophobic and amorphous polymer with three aromatic alcohols known as monolignols that are comprised of *p*-coumaryl, coniferyl, and sinapyl, where their aromatic constituents are *p*-hydroxyphenyl (H), guaiacyl (G), and syringyl (S) moieties (Espinoza-acosta et al., 2016). Lignin is also regarded as a promising phenolic polymer because of its high porosity and compressibility, and strong mechanical performance. Also, the variety of functional groups that exist in lignin allow extensive chemical modifications to take place.

There is limited literature on the application of lignin on biobased materials; however, it has shown promising developments in recent years toward the agriculture and medical sector, as shown in Table 12.2. It is widely incorporated into various products such as dispersants, flocculants, and hydrogels due to its low toxicity, high-strength, and low production cost (Zerpa et al., 2018). Lignin is a vital component of plant cell walls in terms of giving strength and protection from biochemical stresses via an enzymatic degradation inhibition mechanism. Additionally, the advantage of using lignin as a natural primary supplier of soil organic matter makes it suitable for agriculture purposes, while also exhibiting antibacterial properties that could be useful in the medical field, acting as an antioxidant and anticarcinogenic. Its application as an energy-storage device has been highlighted because lignin's economic and ecologic viability act as an alternative to nonrenewable fossil fuel resources. Accordingly, the low mechanical property of lignin-based hydrogel is overcome through the double cross-link approach, which refers to alkaline and acidic cross-linking steps, and by cross-linking lignin as a single material that shows poor mechanical strength and low swelling capacity because of its aromatic rings that make the fabricated products rigid and inflexible, which eventually causes the products to become brittle (Shen et al., 2015). The flexibility of lignin is imposed by the presence of hydroxyl groups and aromatic rings enable it to be physically blended with a wide range of polymers. The physical blend technique improves the material mechanical properties. For example, lignin/PVA/chitosan hydrogel is formed by incorporating lignin into the PVA/chitosan hydrogel. It improves the thermal stability and strength that is required in drug delivery and wound dressing (Ingtipi & Moholkar, 2019). Also, lignin is sufficiently adaptable to be employed in the electrospinning method in fabricating tissue scaffolds with PCL as well as enhancing the porosity due to its hydrophilic nature (Duarah et al., 2020). In addition, lignin/PEGMA hydrogel is also developed via physical cross-linking using an ultrasonicating technique and 800% water retention self-healing hydrogel is fabricated with less unreacted material waste (Kai et al., 2015).

3. Synthetic-based hydrogel and aerogel

Due to limited physical-chemicals portrayed by natural-based polymer hydrogels and aerogels, synthetic-based polymers are in demand today. The research into biodegradable synthetic polymers is increasing, despite the limitations that they might cause, with successful modifications that are widely employed nowadays. Extensive and continuous research is being executed to modify synthetic polymers to make them biodegradable and biocompatible, or to produce satisfying applications in blending

Table 12.2 Lignin blends with a natural and synthetic biodegradable polymer, their classification, and applications.

Lignin−natural polymer	Classification	Application	References
Hydrogel			
Lignin/cellulose	Binary	High thermal resistance and excellent stability hydrogel beads that can be used in immobilization technology such as in the medical, enzymology, and bioelectronic devices fields	Park et al. (2015)
Aerogel			
Lignin/arabinoxylan	Binary	High porosity and mechanical strength aerogel for tissue scaffold engineering	Berglund et al. (2018)
Lignin−synthetic polymer			
Hydrogel			
Lignin/PVA/chitosan	Ternary	Improves the thermal stability of hydrogel for drug-delivery and wound-dressing materials	Ingtipi and Moholkar (2019)
Lignin/PAA/PVA	Ternary	Provides excellent swelling behavior, biodegradability, and reusability in agriculture and horticulture equipment subjected to a water-release system	Tanan et al. (2018)
Lignin/PCL	Binary	Scaffolds technology as biomaterials	Duarah et al. (2020)
Lignin/PVA	Binary	High swelling capacity, high porosity hydrogel, good thermal and mechanical properties for various fields such as self-healing technology	Morales et al. (2019)
Lignin/grafted PEGMA	Binary	Increases properties of self-healing hydrogel in term of physicochemical and biocompatible properties	Kai et al. (2015)

technology that do not cause environmental sustainability issues. Furthermore, the synthetic polymers have controllable molecular weight, linkages, structure, and chemical composition subject to the modifications made (Tan & Marra, 2010). Apart from producing a biodegradable synthetic polymer, the novel methods used to prepare these synthetic polymers in forming a polymeric network are also essential in determining their functional properties. For instance, the cross-linking methods applied to the biodegradable synthetic polymeric network are prominently determined by mechanical properties, including flexibility and degradation characteristics, in various applications. In the medicals applications field, a hydrogel with self-healing ability, also known as 'reversible' hydrogel produced via physical crosslinking, is notably used for drug delivery systems. However, until today, the physically crosslinked hydrogel is still being improvised through combination with biodegradable synthetic polymers as it is cost-effective and safe to be used in vitro and in vivo. From an agriculture perspective, biodegradable materials are being pursued to achieve sustainable farming, such as hydrogels for soilless systems, while aerogels are used to absorb heavy metal ions and reduce irrigation water salinity. Blending of numerous biodegradable synthetic polymers is gaining wide research interest with the aim of achieving a sustainable future farming.

As an example, the photo cross-linking method is applied to synthetic biodegradable polymers in order to form hydrogels or aerogels and to enhance their properties in forming a polymeric network (van Bochove & Grijpma, 2019). The mechanisms employed in such a method include radical chain steps when the photo-initiators dissociate, react with the double bond in the macromere, and form a robust carbon—carbon bond. The availability of promising biodegradable synthetic polymers such as PLA, PEG, poly(propylene fumarate) (PPF), poly(L-glutamic acid) (PLGA), and others has opened up a number of applications for hydrogel and aerogel fabrications. These biodegradable synthetic polymer-based hydrogels and aerogels show improved characteristics upon the aforementioned cross-linking method in regard to their swelling behavior, porosity, and long degradation time, but they have maintained strength (Hong et al., 2011).

Natural polymers provide an excellent range of applications, especially in medical and agricultural fields, because they are safe and sustainable. However, in some cases, these advantageous properties of natural-blend applications can sometimes be considered as some sort of failure. For instance, from a tissue engineering perspective, natural polymers' hydrophilic nature will cause the hydrogels to absorb a massive amount of water and lead to destruction in the polymeric network density (Zhang et al., 2018). This phenomenon is regarded as a disadvantage to the tissue engineering field because it drives the lack of mechanical strength and support. Thus, researchers seek alternatives that mimic natural polymers, with acceptable and satisfactory characteristics in mechanical strength, degradation mechanisms, porosity, and biocompatibility to widen the prospects for the applications.

3.1 Polyurethane blends

Polyurethane (PU) has received great attention due to its versatility in fabricating tissue engineering products, controlled drug delivery, regenerative medicine, textiles, and wound-healing technology. It is synthesized from a diisocyanate, chain extender, and polyol and consists of alternating soft and hard segments (Vroman & Tighzert, 2009). It is also flexible, lightweight, highly environment-compatible, and possesses hydrophobic properties (Chang et al., 2014). However, some limitations have been reported in reference to PU properties, such as low bioactivity and poor healing mechanisms. Also, most are nonbiodegradable; however, with the burgeoning and intensifying research to enable PU to become a biodegradable material, an improvement by modifying it chemically by applying a cross-linking method with the compound of interest—hydrolyzable moieties—into its polymer backbone has been made (Amran et al., 2019; Ates et al., 2014). For example, polyols with hydrolyzable bonds are used as the starting materials to produce biodegradable and biocompatible PU.

The use of PU as a biomaterial has gained a lot of interest from a tissue engineering viewpoint in developing biological membranes after surgery is performed to prevent cell adhesion between tissues (Xiao et al., 2019). Self-healing hydrogels respond to the stimuli through dynamic bonds preceding their formation, such as hydrophobic association, hydrogen bonding, imine bonds, disulfide bonds, and host−guest interaction trending interests (Zhang et al., 2018). Antibacterial self-healing hydrogels are commonly made by incorporating water-soluble chitosan, for example, glycol chitosan and N-carboxyethyl chitosan, which have been reported as having problems in in vivo decomposition and in vitro cross-linking. In this respect, water-based PU nanoparticles have been utilized as a biodegradable cross-linker through the Schiff base method to form a physical blend with a chitosan-self-healing hydrogel, exhibiting excellent characteristics in swelling degree and full recovery upon shape damage (Lin & hui Hsu, 2020). PU properties have been regarded as tunable physicochemical properties because they can be transformed into sheets, foams, and sponges with excellent added characteristics such as high absorbency, good barrier properties, permeability to oxygen, and acting as a thermal stimulator. These properties have made PU suitably employed in wound healing, coating, foams, adhesive technology, and the agricultural sector. PU in coating technology is utilized as a coating material in chemical protective textiles production due to its absorbency while simultaneously being hydrophobic.

A variety of PU−hydrogel and PU−aerogel applications have been available in research and also at a commercial scale. For instance, PU−hydrogel is being adopted in bioprinting technology due to its excellent water capacity, outstanding biocompatibility, elasticity, and low toxicity to print three-dimensional network-functional living tissues (Hsiao & Hsu, 2018). Also, polycaprolactone (PCL) is added into PU to enhance the biodegradability of PU as PCL can fabricate the thermodynamically miscible products with other polymers (Wen et al., 2020). In addition, recent photo-printing technology has expanded its state-of-the-art system by employing PU−hydrogel based on its structure maintenance and enhancing the rapid printing process. Moreover, PU−hydrogel provides an excellent bioprinting layout through combination with other naturally occurring polymers such as gelatin via hydrophobic and hydrophilic gelation interaction to form the bioink that is employed in 3D-printing tissue engineering (Hsieh & Hsu, 2019). Apart from bioprinting utilization, PU−PEG hydrogel that is also formed through a balance between hydrophobic and hydrophilic interaction has been trending in recent years due to the promising thermo-responsive property reflected by the adjustable chemical structure on soft and hard segments, thus developing into excellent biocompatibility and mechanical properties (Wen et al., 2020).

3.2 Poly(lactic acid) blends

PLA belongs to the polyester group, which is formed from the monomer of lactic acid (LA) (Cheng et al., 2009). PLA is easily absorbed by animals and humans, hence it is extensively used in medicines. The degradation of polymers in animals and humans is processed via nonenzymatic hydrolysis (Doble & Kumar, 2005). There are several enzymes involved in the degradation process, including proteinase K, pronase, and bromelain. PLA is readily degraded in compost to carbon dioxide and achieves 90% degradation in 90 days (Doble & Kumar, 2005). PLA can be derived from various polymer sources, for example, corn starch, sugarcane, cassava roots, potatoes, etc. The chiral nature of LA results in the formation of two types of enantiomerically pure homopolymers, which are poly(L-lactic acid) (PLLA) and poly(D-lactic acid) (PDLA) (Mulchandani et al., 2018; Tabasum et al., 2019). These homopolymers have different degradation rates, where PLLA prolonged biodegradation and has superior biocompatibility compared with PDLA. Studies also have reported that PDLA has a lower strength than PLLA (Jones et al., 2015). Therefore, PLA requires a longer time to reach its hydrolysis half-life. Some other important factors affecting the degradation of polymers are their molecular weight and crystallinity. Crystalline PLA is more resistant to degradation compared to amorphous PLA. Table 12.3 is a list of publications on PLA blends.

Biobased hydrogels are very important in pharmaceutical and medical field applications. They should be harmless to human tissues when used in vivo, and their degradation products should be nontoxic. As listed in Table 12.3, PLA can be blended with natural (alginate, chitosan, cellulose, etc.) and synthetic (PEG, PVA, etc.) polymers to be used in medical and agricultural applications. Several physical cross-linking techniques, such as cross-linking by ionic interaction, hydrogen bonding, and stereocomplex formation, have been used to make PLA-based hydrogel. As for chemical cross-linking, the methods are cross-linking via free radical polymerization, photopolymerization, and graft-polymerization. Note that the cross-linking of these polymers does not involve any covalent bonding in the process.

Physical cross-linking of PLA through hydrogen bonding formed hydrogels in vitro. This happens due to the interaction between the polymer chain and compatible geometries. This type of hydrogel is injectable and resembles the ECM polymer (Li et al., 2016). Meanwhile, the stereocomplexes formation of hydrogels also has gained the attention of researchers. This stereo-selective association between polymers having configurations that join and form new composites with new physical properties is compared to parent polymers. A strong association between L-lactides and D-lactide leads to a stereocomplex formation with different crystal morphologies having a high melting temperature of 230°C, higher mechanical properties, and stability

Table 12.3 PLA blends with a natural and synthetic biodegradable polymer, their classification, and applications.

PLA−natural polymer	Classification	Application	References
Hydrogel			
PLA/chitosan	Binary	Controlled dual drug-delivery system	Pircher et al. (2014)
Aerogel			
Bacterial cellulose/PLA	Binary	Temporary scaffolds for the creation of porous aerogels with a morphology resembling the guiding host network	Pircher et al. (2014)
PLA−synthetic polymer			
Hydrogel			
PLA/ poly(ethylene oxide)	Binary	Drug-delivery system	Yamamoto et al. (2016)
PLA/PEG	Binary	Sol−gel systems producing strengthened gel are versatile as injectable scaffolds in tissue engineering	Hsu et al. (2015)

compared to PLA alone (Slager & Domb, 2003). Mao and coworkers investigated the stereocomplexed driven hydrogel based on diblock polymers of PLA−PEG and the triblock of PLA−PEG−PLA. The physical cross-linking occurs via connecting bridges between diblock and triblock polymers through stereocomplexation of PLLA and PDLA (Mao et al., 2015). Free-radical cross-linking is preferable for the formation of hydrogel in larger quantities. In some studies, dextrin was cross-linked with the PLA chain through free radical polymerization using potassium sulfate as the initiator (Cheng et al., 2009). However, free radical and monomer productions from producing hydrogel utilizing this process are harmful and inappropriate for medical applications. This has become a significant problem in the industry because it is difficult to remove from the system (Munim & Raza, 2019).

Pircher and colleagues successfully reinforced bacterial cellulose aerogel with biocompatible PLA to improve the bacterial cellulose properties, which are known to be fragile, ultra-lightweight, open porous, and transversally isotropic (Pircher et al., 2014). PLA had enhanced mechanical strength resistance toward compressive stress at the far-reaching end of the open-porous bacterial cellulose morphology. This composite was proposed to be used as a temporary scaffold to create porous aerogels with a morphology resembling the guiding host network (Pircher et al., 2014). In 2019, Calcagnile and colleagues formed a PLA/cellulose hydrogel at high temperatures that can retain and release fertilizer solution into the soil in a controlled manner (Calcagnile et al., 2019). PLA's primary function in this blend was to delay the solution released by the hydrogels (Calcagnile et al., 2019).

3.3 Polyvinyl alcohol blends

PVA is a creamy or whitish granule that is tasteless, odorless, nontoxic, biocompatible, and thermostable. PVA is well known for its remarkable optical properties, sizable dielectric strength, and excellent charge storage capability (Saini et al., 2017). All of these properties can be modified by mixing with another polymer in the polymer-blending process. The PVA grade in the market depends on different characteristics such as melting point, viscosity, pH, refractive index, and bandgap (Demerlis & Schoneker, 2003). Common polymers blended with PVA are chitosan, PEG, carrageenan, and cellulose. PVA is one of the few biodegradable synthetic polymers known to be mineralized by microorganisms. PVA is wholly degraded and utilized by a bacterial strain of *Pseudomonas* 0–3, as the sole source of carbon and energy (Eghbalifam et al., 2020). However, PVA-degrading microorganisms are not ubiquitous within the environment. Table 12.4 lists some of the more recent studies on PVA blends.

The physical cross-linking methods do not require any chemical cross-linker, and the freezing/thawing technique is one of them. For example, Nho and Park successfully formed a PVA/PVP–chitosan polymer blend using a freezing/thawing method (Nho & Park, 2002; Park & Nho, 2003). The benefit of this process is that the hydrogel

Table 12.4 PVA blends with a natural and synthetic biodegradable polymer, their classification, and applications.

PVA–natural polymer	Classification	Application	References
Hydrogel			
PVA/cellulose	Binary	Tissue engineering such as scaffolds and drug-delivery system	Ciolacu et al. (2019)
PVA–synthetic polymer			
Hydrogel			
PVA/polysaccharide/carrageenan	Ternary	Wound dressing	Chowdhury et al. (2006)
PVA/PVP-chitosan	Ternary	Wound dressing	Nho and Parl (2002)
Chitosan/PVA/PEG	Ternary	Nonirritating wound-dressing materials	Nacer Khodja et al. (2013)

will not produce harmful residues or materials. This process works by freezing the hydrogel fluid for 2–20 h at −5 to −20°C and thawing it for 5–24 h (Kamoun et al., 2013; Sung et al., 2010). Some of the advantages of using this method are that hydrogel personifies bioadhesive characteristics, having a higher swelling degree and rubbery elastic nature. The produced hydrogels also displayed higher mechanical strength and good biocompatibility. The produced hydrogel's quality can be further enhanced by adding poly(acrylic acid) (PAA) into the system. PVA hydrogels are often used to make implants, artificial organs, contact lenses, drug-delivery devices, and wound dressings (Kokabi et al., 2007). While PVA hydrogel has shown considerable versatility in many applications, hydrophilicity can often lead to hydrogel's poor physical stability. Therefore, the PVA hydrogels can be insolubilized by copolymerization, grafting, cross-linking, and blending to enhance this property (Abd El-Mohdy & Ghanem, 2009).

Biomass-based aerogels could not work as efficiently as intended, thus blending with biodegradable synthetic polymers such as PVA offers more favorable properties. The realization that biomass must be hybrids with PVA improves the structural networking to support the aerogel structure. Chowdhury and coworkers modified PVA hydrogel to form polymer hydrogel by radiation cross-linking (Chowdhury et al., 2006). This hydrogel can be used in medical applications. Optimization of the concentration of ingredients for preparing the PVA with polysaccharide (flour) hydrogel is crucial to forming a good-quality hydrogel. The concentration of polysaccharide used also plays a major role in the blended polymer. The increasing amount of polysaccharides caused a reduction in gel fraction and decreased the cross-linking density of the PVA. Radiation can induce a chemical reaction to modify polymer at low temperatures (Nho & Park, 2002). As for wound-dressing applications, the hydrogel can be prepared using a freezing–thawing process with irradiation, which subsequently affected the properties of the hydrogel. For example, the gel fraction of hydrogel increases when using the freeze–thawing method with radiation. On the other hand, chitosan addition in the blending improves the PVA/PVP–chitosan hydrogel (Nho & Park, 2002).

Recently, Ciolacu formed a PVA/cellulose hydrogel using a physical cross-linking method to avoid any toxic residues from being trapped in the hydrogel (Ciolacu, 2019). Their research compared the properties of PVA/cellulose hydrogel formed using chemical and physical cross-linking. By using epichlorohydrin (ECH) as a cross-linker, the swelling ratio of hydrogel increases, but the hydrogen bonding formed is weak. The presence of ECH will result in the formation of covalent bonding between the polymers. Hence, it cannot be classified as polymer blending. However, hydrogel formed by repeated freeze–thawing cycles (physical cross-linking) produces hydrogels with a denser structure between the PVA and cellulose and improved its mechanical strength (Ciolacu, 2019). Air-drying and cryo-gelation via repeated freeze–thaw cycles are some of the physical cross-linking methods that are much safer and biocompatible, without adding other chemicals (Otsuka & Atsushi, 2009). However, this method has a disadvantage as it cannot control the amount of cross-linking that occurs. PVA and its composites are considered biomaterials interested in forming a 3D network, closely matching the human tissues.

To form a suitable hydrogel for medical applications, a cryogenic gelation method of PVA creates a porous structure suitable for fluid absorption in its surroundings. With good compatibility when blended with other synthetic and natural materials, including poly(vinyl pyrrolidone) (PVP), gelatin, alginates, collagen, and composites with calcium phosphates, the PVA blend is reliable for scaffolds for bone tissue engineering (Nkhwa et al., 2014). The preparation of hydrogel blends of tertiary components of PVA/gelatin/PVP via physical cross-linking of air–dry and freeze–thaw methods showed that the degree of cross-linking was influenced by different concentrations of PVA that yielded a superior hydrogel with considerable mechanical strength for medical applications.

3.4 Polyethylene glycol blends

PEG is a hydrophilic polymer of ethylene oxide, which falls under the polyether class and is usually used in pharmaceuticals, cosmetics, lubricants, ink, and surfactants. It has a linear or branched structure containing dissymmetric or asymmetric hydroxyl ions as tail groups. Hydroxyl groups can be replaced by functional groups such as acetylene, acrylate, amine azide, carbonyl, methoxyl, thiol, and vinyl chloride (Zhu, 2010). The properties of PEG include being nonimmunogenic, biocompatible, and flexible. These properties make PEG suitable to be used as a wound-dressing material. The stability of polymeric materials greatly depends on the applied conditions, and essential factors include water content, pH, temperature, exposure to UV light, presence of enzymes, etc. (Ulbricht et al., 2014). The mechanical properties of PEG can be enhanced by blending PEG with chitosan and PLGA (Hasan & Abdel-Raouf, 2019).

Regarding PEG degradability, PEG is known to be a nondegradable polymer. Therefore, many studies have been conducted to make PEG biodegradable. Ulbricht reported that PEG, polypeptoids, and poly(2-oxazoline)s are degradable by oxidative degradation under biologically relevant conditions (Ulbricht et al., 2014). They investigated the effect of transition metal-catalyzed generation of reactive oxygen species (ROS) leading to pronounced time- and concentration-dependent degradation of all polymers. *Flavobacterium* sp. and *Pseudomonas* sp. together mineralized PEG completely under aerobic conditions (Doble & Kumar, 2005). During PEG degradation, their molecules are reduced to one glycol unit after each oxidation cycle. *Pelobacter venetianus* was found to degrade PEG and ethylene glycol under anaerobic conditions (Doble & Kumar, 2005). For PEG with a higher molecular weight of 4000–20,000, it was degraded by *Sphingomonas macrogoltabidus* and *S. terrae* (Doble & Kumar, 2005). These properties make PEG and its blends suitable to be used in agricultural applications because it can degrade naturally. Table 12.5 presents a list of publications on PEG blends hydrogel and aerogel.

PEG hydrogel has gained much attention for tissue engineering applications. For example, Timothy D. Sargeant and colleagues managed to form an in situ formed collagen–PEG hydrogel via the photopolymerization process (Sargeant et al., 2012). They developed a unique injectable hydrogel system composed of collagen and multiarmed PEG. In 2011, Zhang et al. reported that PCLA degrades slower

Table 12.5 PEG blends with a natural and synthetic biodegradable polymer, their classification, and applications.

PEG–natural polymer	Classification	Applications	References
Hydrogel			
Collagen–PEG	Binary	Tissue regeneration	Sargeant et al. (2012)
PEG–synthetic polymer			
Hydrogel			
PEG/PLA	Binary	Drug delivery, encapsulation	Burdick et al. (2002), Burdick and Anseth (2002)
PAA/PEG	Binary	Controlled release fertilizer	Tyliszczak et al. (2010)
PCLA/PEG/PCLA	Ternary	Postoperative peritoneal adhesion	Zhang et al. (2011)

from the blends, hence reducing the accumulation of acid in the system (Zhang et al., 2011). These properties make PCLA/PEG polymer blends useful in the medical field. PEG hydrogels have a porous structure that can provide a moist and humid microenvironment that facilitates exudate absorption from wounded areas (Masood et al., 2019). There are several experiments that have been performed on aerogel PEG. The study was aimed at improving the durability, thermal conductivity, and latent heat diffusion of the PEG aerogel type. Jia Shen and coworkers also managed to form a new shape-stabilized phase-change material based on phosphorylated PVA/graphene aerogel/PEG as a double-network support material (X. Shen et al., 2015). Graphene aerogel (GA) is an oxygen-containing functional group. Therefore, this functional group aids the formation of hydrogen bonding between phosphorylated PVA and PVA with PEG. This study was conducted to overcome the PEG phase of changing property. These blended polymers can be used as a potential injectable hydrogel in the medical field.

4. Applications

4.1 Medicine

In medical applications, alternative products for drug delivery, tissue engineering, bone scaffolds engineering, and wound dressing with specific properties including antimicrobial, controllable biodegradability rate, durable with high mechanical strength, and promising physical stability are always favored. Nonetheless, one material cannot deliver all these requirements. Thus, polymer blending comes to into effect. Polymers are extensively used in the formulation of pharmaceutical and healthcare products. They are commonly used in most major dosage forms, including tablets, films, capsules, semisolids, suspensions, gels, and transdermal patches (Nyamweya, 2021). While the synthesis of new polymers to acquire preferred functionalities is feasible, constraints in extensive safety testing for new materials are often a limiting barrier in the employment of new medical products. Contemplating this, polymer blends represent a gratifying alternative means by which to address drug formulation and delivery. Medicine applications require a biocompatible polymer that is harmless, nontoxic, and has no long-term or side effects on the body. Referring to compatibility, polymer blends must be compatible not only with each other but also with the human body. For instance, biocompatible polymers with biodegradable property are often blended with a biocompatible polymer with slow degradation properties. The intermediate property of the resulting product can be further altered by not sacrificing the intended property. Al though polymer blend technology can address this discrepancy, the absence of a comprehensive study remains. Subsequently there have been calls for continuous study of polymer blending for medical applications. In tissue engineering technology, the developments halted at biobased materials, however these are susceptible to bacterial attack. Although chitosan is seen to have significant antimicrobial properties, it may cause subsequent side effects of constipation, nausea, etc. Thus, additional materials with nano-size antibacterial properties (i.e., silver nanoparticles) with no adverse characteristics are required; also, the polymer—polymer precursor must be able to hold, chelate, or bind the additional materials to avoid seepage.

Apart from typical formulation ratios between polymer—polymer in polymer-blending technology, treatment, modification, and formation are the areas often disregarded in blending technology. Thus, numerous researches have endeavored to improve the blended polymers by converging on the above-mentioned. For instance, bilayer hydrogel beads showed improved swelling behavior and drug-loading capacity, preventing release in the gastrointestinal tract (Karzar Jeddi & Mahkam, 2019). The polymer blends of cationic cellulose nanocrystals/anionic alginate were used to prepare double-membrane hydrogels for complexing drugs codelivery, due to the outer layer for rapid drug release, followed by prolonged slow release from the inner-layer hydrogel (enabling two different drugs to be carried) with synergistic release potential in medical applications (Lin & hui Hsu, 2020). Modified hemicellulose and chitosan with the addition of citrate reported forming soft, highly porous, and durable aerogel that is able to absorb saline solutions at up to 100 g per gram material and 80 g water per gram material due to its hygroscopic behavior, which makes it suitable as a biomaterial for medical applications (Salam et al., 2011). Nonetheless, all the outlined factors are still bound to the potential of the polymer—polymer property itself. A combination of natural polymers of cellulose nanofiber/chitosan hydrogel yielded unique porous morphology, increased biomineralization, improved osteogenic differentiation, and increased compressive strength compared to pure chitosan hydrogel. Temperature-responsive aerogel blends of methylcellulose/N-isopropylacrylamide are highly porous, with abundant functional groups that are able to load 5-fluorouracil (FU) cancer drugs via hydrogen bond and physical embedment. A photoresponsive gelation hydrogel synthesized by a combination of a cyclic polymer of PLA/chitosan, PLA/polyethylene oxide(PEO), and PLA-PEO-PLA blends was reported as influencing the "topological conversion" of a polymer that is applicable in drug-delivery systems (Yamamoto et al., 2016). Meanwhile, cellulose nanofibers/gelatin blended 3D scaffold biocomposite aerogels have improved mechanical properties (high compressive strength of 61.35 kPa), high surface area, and slow in vivo degradation with high-porosity scaffolds >90% (7−135 μm) which are suitable for cell attachments and proliferation (Mirtaghavi et al., 2020). Hence, in polymer blending technology, to form biodegradable hydrogel and aerogel composites for medical applications, of critical importance is the rational polymer selection and processing technique that is required to serve the designated purpose of a medical appliance.

4.2 Agriculture

The conventional method of agriculture is unable to entirely meet the global food demand. Polymers have fostered sustainable agriculture by playing prominent roles in growing media, shelters, mulches, and monitoring and fertilizing systems. A wide range of polymers that are both natural and synthetic has made sustainable future farming achievable. For healthy and efficient agricultural practice, polymer blending technology is required. The common method of release of agrochemical-based materials such as direct spraying on crops is considered hazardous and impractical. The widespread use of herbicides also causes drainage problems and uncontrolled environmental pollution. The seepage of nutrients or chemicals from the substrate is detrimental and causes insufficient nutrient supply to plants. If this is not appropriately managed, adverse effects will trigger an irrecoverable hazard. Therefore, an assortment of delivery methods, substrate, or media that can effectively carry herbal pesticides or nutrients should be highlighted for more dynamic release and control of these substances. Thus, biodegradable polymer-blending products such as hydrogels and aerogels are tailored to address this problem.

Hydrogel fabricated from lignocellulose sources, such as biomass, or any biodegradable synthetic polymer, is currently gaining attention due to its outstanding properties, by which it can absorb substantial herbicide substances or be chelated with the required nutrients for plants. Numerous hydroxyl groups that are possessed by lignocellulosic materials have made them suitable for modification and blending with other synthetic and natural polymers—hemicellulose has more hydroxyl groups than cellulose, and lignin is at an advantage as it has aromatic rings. However, some studies reported the some of the lignocellulosic materials formed a heterogeneous blend with other synthetic polymers and limited homogeneity subjected to the type of polymers introduced. Thus, various techniques have been employed in polymer blending to increase the homogeneity and this could be considered for most compatible polymers. For instance, polymer blend hydrogels such as polyacrylamide/methylcellulose/montmorillonite showed a synergistic effect as a nutrient carrier of nitrogenated fertilizer and urea (Bortolin et al., 2013). The addition of 50%

montmorillonite on the hydrogel resulted in controlled nutrient liberation at different pH ranges and the best urea desorption almost 200 times slower than for pure urea. Superabsorbent hydrogel with irradiation of carboxymethyl cellulose/acrylic acid blends enhanced the water retention of sand and soil for seed germination in agriculture (Chandra Sutradhar et al., 2015). The combination of PLA/cellulose presented fully degradable superabsorbent hydrogel, especially when designed as water and fertilizer controlled-release materials for agriculture and horticulture applications. This hydrogel material reveals a significantly different water-retention capability on two soil (red or white), particularly on sandy red soil for Mediterranean cultivars, supporting plant growth and increasing fruit yield (Calcagnile et al., 2019). Hydrogel blends via IPN lignin/PAA/PVA networks showed an increase in swelling capacity, good water retention, salt, and pH sensitivity, which are suitable for agricultural applications as fertilizer and water-absorption hydrogel with retention capacity and reswelling ability (Tanan et al., 2018). Cellulose blend hydrogel was suitable for leafy vegetable growth. It served as a slow-release process, with germination, biodegradation, water retention, and other qualities that are important to establish a new growth media for a free-watering system.

PVA is widely used to prepare polymer blends and composites with several other natural, renewable polymers such as chitosan cellulose starch or lignocellulose. Semi-IPN polymer networks (semi-IPNs) composed of cross-linked and linear polymers can improve the polymer composite performance (Teodorescu et al., 2018). For instance, semi-IPNs with improved structure and environment-friendly properties were synthesized by free radical and graft copolymerization of wheat straw and poly(potassium acrylate) in the presence of PVA. This polymer blend can be used in agriculture as a water-retention material by improving the water retentivity into the soil. Hence, it will support plant growth and quality. Other than that, PLA can be obtained from agriculture biomass wastes such as cellulose, starch, etc. It is composable, disappearing entirely in less than 1 month, and it is mostly used in biodegradable and biobased products in short-term applications (Tabasum et al., 2019). A fully degradable superabsorbent composite material was produced by introducing PLA into a cellulose hydrogel (Calcagnile et al., 2019). This material is used in agricultural applications, especially in the case of water shortages. PLA had aided superabsorbent hydrogels by retaining water and releasing fertilizer from the composite in a controlled manner (Calcagnile et al., 2019).

5. Future prospects

There are different opinions and multifaceted concepts to interpreting sustainability; however, all these ideas must be based on protecting what we have in the present while not compromising the future. Apropos to this statement, the environmental aspect is one of the pillars of sustainability, alongside social and economic factors. In the past decades, inventions of biodegradable materials have been brought to the fore and continue to grow until this day. The usage of nonbiodegradable materials such as plastics in food-packaging films, household materials, shopping bags, etc., creates waste disposal issues. These wastes have overburdened our landfills and cannot be sustained in the long term amidst the continued environmental damage. The process of recycling and disposing of nonbiodegradable materials does not guarantee a positive impact on the environment. The accumulation of polymer waste and its "disposition" in landfills also has shifted the burden on the future generation to solve this global damage and to actively avoid compromising the future.

As per their definition, biodegradable polymers are able to decompose into several components: carbon dioxide, inorganic compounds, water, and biomasses, as well as not causing toxicity or damage to the environment (Rydz et al., 2018). Numerous natural and biodegradable polymers have undergone inventions and research resulting in unique properties applicable to the applications of interests. The biodegradable polymeric materials have sealed great interest in extensive prospects for the medical and agricultural fields (van Bochove & Grijpma, 2019). Biodegradable natural, synthetic, and semisynthetic polymer are utilized in fabricating hydrogels and aerogels. Studies have shown that multiple polymeric materials are better at conveying the intended applications rather than using single materials. Thus, various polymer blend techniques have been employed to mix two or more polymers via physical cross-linking mechanisms ranging from natural and synthetic polymers such as hydrophobic and hydrophilic interaction and temperature applied, and semi-IPN for miscible and immiscible blending, respectively. Physically cross-linking hydrogel and aerogel is considered to be the "green" method substantially impacting agriculture and medical areas that involve human and animal consumption. Research is centered on continuous improvement. Studies have shown that natural polymer products sometimes result in variability in material properties due to their compositional heterogeneity and limited purity, despite being abundantly available (Germershaus et al., 2015). Thus, the new research direction aims toward the semisynthetic polymers, where these are natural polymers that have undergone some chemical modification such as modified gelatin and chitosan to improve their physicochemical properties. In an attempt to meet the properties required by particular polymer blends for maximum performance, in situ and ex situ compatibilization strategies are being employed to increase the miscibility, as immiscibility is regarded as one of the greatest challenges in polymer blends technology (Muthuraj et al., 2017). Also, biodegradable tunable thermal-responsive thermogel blends are attracting a great deal of attention in medical polymer blend technology as they are more convenient to be used in vivo and in vitro for excellent cell adhesion (Lee et al., 2020).

However, from a holistic point of view, the cost of biodegradable materials is expensive due to their processing costs and so only a few are marketable. The price of these products is not competitive and they not equally affordable. The future outlook regarding this subject is the "synthetically inspired" biomaterials that can pull together the great and unique properties of natural polymers in each area by managing the fundamental aspects of polymer blends in terms of compatibility and miscibility. Despite the challenges that have been highlighted concerning their price, properties, and reproducibility, it is believed that the pertinent optimization and modification of the fabricated products would offer a great range of possible applications and allow better control of the products. Extensive and detailed research is still necessary to pursue the goal to make these products universally affordable, promoting biodegradable polymers in the global market, and maintaining the stability of functionalities and properties of the fabricated products.

6. Conclusion

From the literature, we can conclude that biodegradable hydrogel and aerogel polymer blend composites have tremendous potential for many high-end applications. The natural- and synthetic-based biodegradable polymers will always contribute high value to the industry, especially currently, as people have grown increasingly aware of the importance of biodegradable materials and the subsequent products. To form hydrogels and aerogels using biodegradable natural- and synthetic-based materials, molecular-level blends in the solution state are required. The formation of hydrogels and aerogels frequently involves the regeneration process of the dissolved polymer. However, it was found that literature concerning aerogels both in agriculture and medicine is deficient compared to that for hydrogels. This is generally due to regenerated natural-based aerogel having a soft texture and not being durable enough for agriculture and medicine applications. Nonetheless, the fundamentals of blended polymer regenerated for hydrogels and aerogels are similar—via noncovalent interactions of either physical or chemical interactions. The interactions of polymer blends must be intimate and thermodynamically compatible. Although the solution state promotes miscibility of polymer−polymer blends, immiscibility still occurs. Phase separation of the immiscible polymer−polymer requires physical or chemical noncovalent interactions to instigate hydrogel and aerogel formation. Nonetheless, each polymer material discussed in this chapter has its advantages and is beneficial in certain polymer-blending systems. Polymer blending intensifies the properties of the materials where the intended properties can be altered accordingly. The blended polymer products will have intermediate properties and meet the qualities of the proposed application. For instance, biodegradable hydrogel and aerogel polymer blends are often used in the medical and agricultural industries. The applications are vastly extending from tissue engineering, biosensors, implant materials, as well as controlled-release fertilizers for plants.

Acknowledgments

The authors thank the Ministry of Higher Education LRGS/1/2019/UKM-UKM/5/1 and Universiti Kebangsaan Malaysia MI-006-2020 for the financial support and facilities provided.

References

Abd El-Mohdy, H. L., & Ghanem, S. (2009). Biodegradability, antimicrobial activity and properties of PVA/PVP hydrogels prepared by $γ$-irradiation. *Journal of Polymer Research, 16*(1), 1–10. https://doi.org/10.1007/s10965-008-9196-0

Agarwal, S. (2020). Biodegradable polymers: present opportunities and challenges in providing a microplastic-free environment. *Macromolecular Chemistry and Physics, 221*(6), 1–7. https://doi.org/10.1002/macp.202000017

Amran, U. A., Zakaria, S., Chia, C. H., Roslan, R., Jaafar, S. N. S., & Salleh, K. M. (2019). Polyols and rigid polyurethane foams derived from liquefied lignocellulosic and cellulosic biomass. *Cellulose, 26*(5), 3231–3246. https://doi.org/10.1007/s10570-019-02271-w. Available from:.

Asano, A. (2017). Polymer blends and composites. In *Modern magnetic resonance* (pp. 1–15). https://doi.org/10.1007/978-3-319-28275-6_57-1

Ates, B., Koytepe, S., Karaaslan, M. G., Balcioglu, S., & Gulgen, S. (2014). Biodegradable non-aromatic adhesive polyurethanes based on disaccharides for medical applications. *International Journal of Adhesion and Adhesives, 49*, 90–96. https://doi.org/10.1016/j.ijadhadh.2013.12.012

Banerjee, A., Chatterjee, K., & Madras, G. (2014). Enzymatic degradation of polymers: a brief review. *Materials Science and Technology, 30*(5), 567–573. https://doi.org/10.1179/1743284713Y.0000000503

Berglund, L., Forsberg, F., Jonoobi, M., & Oksman, K. (2018). Promoted hydrogel formation of lignin-containing arabinoxylan aerogel using cellulose nano fibers as a functional biomaterial. *RSC Advances, 8*, 38219–38228.

Bortolin, A., Aouada, F. A., Mattoso, L. H. C., & Ribeiro, C. (2013). Nanocomposite PAAm/methyl cellulose/montmorillonite hydrogel: evidence of synergistic effects for the slow release of fertilizers. *Journal of Agricultural and Food Chemistry, 61*, 7431–7439. https://doi.org/10.1021/jf401273n

Burdick, J. A., & Anseth, K. S. (2002). Photoencapsulation of osteoblasts in injectable RGD-modified PEG hydrogels for bone tissue engineering. *Biomaterials, 23*(22), 4315–4323.

Burdick, J. A., Mason, M. N., Hinman, A. D., Thorne, K., & Anseth, K. S. (2002). Delivery of osteoinductive growth factors from degradable PEG hydrogels influences osteoblast differentiation and mineralization. *Journal of Controlled Release, 83*(1), 53–63.

Bush, J. R., Liang, H., Dickinson, M., & Botchwey, E. A. (2016). Xylan hemicellulose improves chitosan hydrogel for bone tissue regeneration. *Polymers for Advanced Technologies, 27*(8), 1050–1055. https://doi.org/10.1002/pat.3767

Calcagnile, P., Sibillano, T., Giannini, C., Sannino, A., & Demitri, C. (2019). Biodegradable poly (lactic acid)/cellulose-based superabsorbent hydrogel composite material as water and fertilizer reservoir in agricultural applications. *Journal of Applied Polymer Science, 47546*, 1–9. https://doi.org/10.1002/app.47546

Cesarino, I., Araújo, P., Pereira, A., Júnior, D., & Mazzafera, P. (2012). An overview of lignin metabolism and its effect on biomass recalcitrance. *Brazilian Journal of Botany, 35*(4), 303–311.

Chandra Sutradhar, S., Khan, M. M. R., Rahman, M. M., & Chandra Dafadar, N. (2015). The synthesis of superabsorbent polymers from a carboxymethylcellulose/acrylic acid blend using gamma radiation and its application in agriculture. *Journal of Physical Science, 26*(2), 23–39. https://www.researchgate.net/publication/281551156.

Chang, K., Wang, Y., & Peng, K. (2014). Preparation of silica aerogel/polyurethane composites for the application of thermal insulation. *Journal of Polymer Research, 21*. https://doi.org/10.1007/s10965-013-0338-7, 338–338.

Cheng, Y., Deng, S., Chen, P., & Ruan, R. (2009). Polylactic acid (PLA) synthesis and modifications: a review. *Frontiers of Chemistry in China, 4*(3), 259–264. https://doi.org/10.1007/s11458-009-0092-x

Chowdhury, M. N. K., Alam, A. K. M. M., Dafader, N. C., Haque, M. E., Akhtar, F., Ahmed, M. U., et al. (2006). Radiation processed hydrogel of poly (vinyl alcohol) with biodegradable polysaccharides. *Bio-Medical Materials and Engineering, 16*(3), 223–228.

Ciolacu, D. E. (2019). *Structure-property relationships in cellulose-based hydrogels*.

Demerlis, C. C., & Schoneker, D. R. (2003). Review of the oral toxicity of polyvinyl alcohol (PVA) – PDF free download. *Food and Chemical Toxicology, 41*(3), 319–326.

Doble, M., & Kumar, A. (2005). Degradation of polymers. In *Biotreatment of industrial Effluents* (pp. 101–110).

Duarah, P., Haldar, D., & Purkait, M. K. (2020). Technological advancement in the synthesis and applications of lignin-based nanoparticles derived from agro-industrial waste residues: a review. *International Journal of Biological Macromolecules*. https://doi.org/10.1016/j.ijbiomac.2020.09.076. Available from:.

Edwards, J. V., Fontenot, K. R., & Liebner, F. (2018). Peptide-cellulose conjugates on cotton-based materials have protease sensor/sequestrant activity. *Sensors, 18*(7), 234.

Eghbalifam, N., Shojaosadati, S. A., Hashemi-Najafabadi, S., & Khorasani, A. C. (2020). Synthesis and characterization of antimicrobial wound dressing material based on silver nanoparticles loaded gum Arabic nanofibers. *International Journal of Biological Macromolecules, 155*, 119–130. https://doi.org/10.1016/j.ijbiomac.2020.03.194. Available from:.

Elbarbary, A. M., & Ghobashy, M. M. (2017). Controlled release fertilizers using superabsorbent hydrogel prepared by gamma radiation. *Radiochimica Acta, 105*(10), 865–876. https://doi.org/10.1515/ract-2016-2679

El-Naggar, A. W. M., Senna, M. M., Mostafa, T. A., & Helal, R. H. (2017). Radiation synthesis and drug delivery properties of interpenetrating networks (IPNs) based on poly(-vinyl alcohol)/methylcellulose blend hydrogels. *International Journal of Biological Macromolecules, 102*, 1045–1051. https://doi.org/10.1016/j.ijbiomac.2017.04.084. Available from:.

Espinoza-acosta, J. L., Torres-chávez, P. I., & Ramírez-wong, B. (2016). Antioxidant, antimicrobial, and antimutagenic properties of technical lignins and their applications. *BioResources, 11*(2), 5452–5481. Available from: https://ojs.cnr.ncsu.edu/index.php/BioRes/article/view/BioRes_11_2_Acosta_Review_Technical_Lignins_Applications_/4463.

Farhat, W., Venditti, R. A., Hubbe, M., Taha, M., Becquart, F., & Ayoub, A. (2017). A review of water-resistant hemicellulose-based materials: processing and applications. *ChemSusChem, 10*(2), 305–323. https://doi.org/10.1002/cssc.201601047

Farhat, W. (2018). *Investigation of hemicellulose biomaterial approaches: the extraction and modification of hemicellulose and its use in value- added applications* (pp. 1–215). Jean Monnet University.

Gan, S., Zakaria, S., Salleh, K. M., Anuar, N. I. S., Moosavi, S., & Chen, R. S. (2020). An improved physico-mechanical performance of macropores membrane made from synthesized cellulose carbamate. *International Journal of Biological Macromolecules, 158*, 552–561. https://doi.org/10.1016/j.ijbiomac.2020.04.166

Germershaus, O., Luhmann, T., Rybak, J. C., Ritzer, J., & Meinel, L. (2015). Application of natural and semi-synthetic polymers for the delivery of sensitive drugs. *International Materials Review, 60*(2), 101–130. https://doi.org/10.1179/1743280414Y.0000000045

Gibson, L. J. (2012). The hierarchical structure and mechanics of plant materials. *Journal of The Royal Society Interface, 9*(76), 2766–2794. https://doi.org/10.1098/rsif.2012.0341

Gou, L., Xiang, M., & Ni, X. (2020). Development of wound therapy in nursing care of infants by using injectable gelatin-cellulose composite hydrogel incorporated with silver nanoparticles. *Materials Letters, 277*, 128340. https://doi.org/10.1016/j.matlet.2020.128340. Available from:.

Hasan, A. M. A., & Abdel-Raouf, M. E.-S. (2019). *Cellulose-based superabsorbent hydrogels* (pp. 245–267). https://doi.org/10.1007/978-3-319-77830-3_11

Hong, Y., Huber, A., Takanari, K., Amoroso, N. J., Hashizume, R., Badylak, S. F., et al. (2011). Mechanical properties and in vivo behavior of a biodegradable synthetic polymer microfiber-extracellular matrix hydrogel biohybrid scaffold. *Biomaterials, 32*(13), 3387–3394. https://doi.org/10.1016/j.biomaterials.2011.01.025. Available from:.

Hosseini, H., Zirakjou, A., Goodarzi, V., & Mohammad, S. (2020). Lightweight aerogels based on bacterial cellulose/silver nanoparticles/polyaniline with tuning morphology of polyaniline and application in soft tissue engineering. *International Journal of Biological Macromolecules, 152*, 57–67. https://doi.org/10.1016/j.ijbiomac.2020.02.095. Available from:.

Hsiao, S. H., & Hsu, S. H. (2018). Synthesis and characterization of dual stimuli-sensitive biodegradable polyurethane soft hydrogels for 3D cell-laden bioprinting. *ACS Applied Materials and Interfaces, 10*(35), 29273–29287. https://doi.org/10.1021/acsami.8b08362

Hsieh, C., & Hsu, S. (2019). Double-network polyurethane-gelatin hydrogel with tunable modulus for high-resolution 3D bioprinting. *Applied Materials and Interfaces, 11*, 32746–32757. https://doi.org/10.1021/acsami.9b10784

Hsu, Y. I., Masutani, K., Yamaoka, T., & Kimura, Y. (2015). Strengthening of hydrogels made from enantiomeric block copolymers of polylactide (PLA) and poly(ethylene glycol) (PEG) by the chain extending Diels-Alder reaction at the hydrophilic PEG terminals. *Polymer (Guildf), 67*, 157–166. https://doi.org/10.1016/j.polymer.2015.04.026. Available from:.

Imre, B., & Pukánszky, B. (2013). Compatibilization in bio-based and biodegradable polymer blends. *European Polymer Journal, 49*(6), 1215–1233. https://doi.org/10.1016/j.eurpolymj.2013.01.019

Ingtipi, K., & Moholkar, V. S. (2019). Sonochemically synthesized lignin nanoparticles and its application in the development of nanocomposite hydrogel. *Materials Today: Proceedings, 17*, 362–370. https://doi.org/10.1016/j.matpr.2019.06.443. Available from:.

Innerlohinger, J., Weber, H. K., & Kraft, G. (2006). Aerocellulose: aerogels and aerogel-like materials made from cellulose. *Macromolecular Symposia, 244*, 126–135. https://doi.org/10.1002/masy.200651212

Jeddi, M. K., & Mahkam, M. (2019). Magnetic nano carboxymethyl cellulose-alginate/chitosan hydrogel beads as biodegradable devices for controlled drug delivery. *International Journal of Biological Macromolecules, 135*, 829–838. https://doi.org/10.1016/j.ijbiomac.2019.05.210. Available from:.

Jones, K., Ramakrishnan, G., Uchimiya, M., & Orlov, A. (2015). New applications of X-ray tomography in pyrolysis of biomass: Biochar imaging. *Energy and Fuels, 29*(3), 1628–1634. https://doi.org/10.1021/ef5027604

Junli, R., Xinwen, P., Linxin, Z., Feng, P., & Runcang, S. (2012). Novel hydrophobic hemicelluloses: synthesis and characteristic. *Carbohydrate Polymers, 89*(1), 152–157. https://doi.org/10.1016/j.carbpol.2012.02.064. Available from:.

Kai, D., Low, Z. W., Liow, S. S., Abdul Karim, A., Ye, H., Jin, G., et al. (2015). Development of lignin supramolecular hydrogels with mechanically responsive and self-healing properties. *ACS Sustainable Chemistry and Engineering, 3*(9), 2160–2169.

Kamoun, E. A., Kenawy, E. R. S., Tamer, T. M., El-Meligy, M. A., & Mohy Eldin, M. S. (2013). Poly (vinyl alcohol)-alginate physically crosslinked hydrogel membranes for wound dressing applications: characterization and bio-evaluation. *Arabian Journal of Chemistry, 8*(1), 38–47. https://doi.org/10.1016/j.arabjc.2013.12.003

Khan, S., Ul-islam, M., Ikram, M., Ul, S., Wajid, M., Israr, M., et al. (2018). Preparation and structural characterization of surface modified microporous bacterial cellulose scaffolds: a potential material for skin regeneration applications in vitro and in vivo. *International Journal of Biological Macromolecules, 117*, 1200–1210. https://doi.org/10.1016/j.ijbiomac.2018.06.044. Available from:.

Kim, N. J., Lee, S. J., & Atala, A. (2013). Biomedical nanomaterials in tissue engineering. In *Nanomaterials in tissue engineering* (pp. 1–26). Woodhead Publishing Limited. https://doi.org/10.1533/9780857097231.1

Kistler, S. S. (1931). Coherent expanded aerogels and jellies. *Nature, 127*(3211). https://doi.org/10.1038/127741a0, 741.

Kjeldsen, A., Price, M., Lilley, C., Guzniczak, E., & Archer, I. (2019). A review of standards for biodegradable plastics with support from. *Industrial Biotechnology Innovation Centre, 28*.

Kokabi, M., Sirousazar, M., & Hassan, Z. M. (2007). PVA-clay nanocomposite hydrogels for wound dressing. *European Polymer Journal, 43*(3), 773–781. https://doi.org/10.1016/j.eurpolymj.2006.11.030

Kundu, J., Pati, F., Jeong, Y. H., & Cho, D. (2013). Chapter 2. Biomaterials for biofabrication of 3D tissue scaffolds. In *Biofabrication* (First Edit, pp. 23–46). Elsevier Inc. https://doi.org/10.1016/B978-1-4557-2852-7.00002-0

Labafzadeh, S. R. (2015). *Cellulose-based materials* (p. 74). University of Helsinki. Available from: http://hdl.handle.net/10138/153410.

Larrañeta, E., Imízcoz, M., Toh, J. X., Irwin, N. J., Ripolin, A., Perminova, A., Domínguez-Robles, J., Rodríguez, A., & Donnelly, R. F. (2018). Synthesis and characterization of lignin hydrogels for potential applications as drug eluting antimicrobial coatings for medical materials. *ACS Sustainable Chemistry and Engineering, 6*(7), 9037–9046. https://doi.org/10.1021/acssuschemeng.8b01371

Lee, E. J., Kang, E., Kang, S. W., & Huh, K. M. (January 2020). Thermo-irreversible glycol chitosan/hyaluronic acid blend hydrogel for injectable tissue engineering. *Carbohydrate Polymers, 244*. https://doi.org/10.1016/j.carbpol.2020.116432, 116432.

Li, Y., Zhu, L., Fan, Y., Li, Y., Cheng, L., Liu, W., Li, X., & Fan, X. (2016). Formation and controlled drug release using a three-component supramolecular hydrogel for anti-*Schistosoma japonicum* cercariae. *Nanomaterials, 6*(3). https://doi.org/10.3390/nano6030046

Lin, T. W., & hui Hsu, S. (2020). Self-healing hydrogels and cryogels from biodegradable polyurethane nanoparticle crosslinked chitosan. *Advanced Science, 7*(3). https://doi.org/10.1002/advs.201901388, 1901388–1901388.

Lin, N., Gèze, A., Wouessidjewe, D., Huang, J., & Dufresne, A. (2016). Biocompatible double-membrane hydrogels from cationic cellulose nanocrystals and anionic alginate as complexing drugs co-delivery. *ACS Applied Materials and Interfaces, 8*(11), 6880–6889. https://doi.org/10.1021/acsami.6b00555

Liu, Q., Liu, J., Qin, S., Pei, Y., Zheng, X., & Tang, K. (2020). High mechanical strength gelatin composite hydrogels reinforced by cellulose nanofibrils with unique beads-on-a-string morphology. *International Journal of Biological Macromolecules, 164*, 1776–1784. https://doi.org/10.1016/j.ijbiomac.2020.08.044. Available from:.

Loix, C., Huybrechts, M., Vangronsveld, J., Gielen, M., Keunen, E., & Cuypers, A. (2017). Reciprocal interactions between cadmium-induced cell wall responses and oxidative stress in plants. *Frontiers in Plant Science, 8*, 1–19. https://doi.org/10.3389/fpls.2017.01867

Lucas, N., Bienaime, C., Belloy, C., Queneudec, M., Silvestre, F., & Nava-saucedo, J. (2008). Polymer biodegradation: mechanisms and estimation techniques. *Chemosphere, 73*, 429–442. https://doi.org/10.1016/j.chemosphere.2008.06.064

Maharjan, B., Park, J., Kaliannagounder, V. K., Awasthi, G. P., Joshi, M. K., Park, C. H., & Kim, C. S. (2021). Regenerated cellulose nanofiber reinforced chitosan hydrogel scaffolds for bone tissue engineering. *Carbohydrate Polymers, 251*(August 2020), 117023. https://doi.org/10.1016/j.carbpol.2020.117023

Makhijani, K., Kumar, R., & Sharma, S. K. (2015). Biodegradability of blended polymers: a comparison of various properties. *Critical Reviews in Environmental Science and Technology, 45*(16), 1801–1825. https://doi.org/10.1080/10643389.2014.970682

Mao, H., Pan, P., Shan, G., & Bao, Y. (2015). In situ formation and gelation mechanism of thermoresponsive stereocomplexed hydrogels upon mixing diblock and triblock poly(lactic acid)/poly(ethylene glycol) copolymers. *Journal of Physical Chemistry B, 119*(21), 6471–6480. https://doi.org/10.1021/acs.jpcb.5b03610

Martuscelli, E., Palumbo, R., & Kryszewski, M. (1979). *Polymer blends processing, morphology, and properties*. New York and Landon: Plenum Press.

Masood, N., Ahmed, R., Tariq, M., Ahmed, Z., Masoud, M. S., Ali, I., et al. (January 2019). Silver nanoparticle impregnated chitosan-PEG hydrogel enhances wound healing in diabetes induced rabbits. *International Journal of Pharmaceutics, 559*, 23–36. https://doi.org/10.1016/j.ijpharm.2019.01.019. Available from:.

Mirtaghavi, A., Luo, J., & Muthuraj, R. (2020). Recent advances in porous 3D cellulose aerogels for tissue engineering applications: a review. *Journal of Composites Science, 4*, 152. https://doi.org/10.3390/jcs4040152

Morales, A., Labidi, J., & Gullón, P. (2019). Assessment of green approaches for the synthesis of physically crosslinked lignin hydrogels. *Journal of Industrial and Engineering Chemistry*. https://doi.org/10.1016/j.jiec.2019.09.037. Available from.

Mulchandani, N., Gupta, A., & Katiyar, V. (2018). *Polylactic acid based hydrogels and its renewable characters: tissue engineering applications* (pp. 1–24). https://doi.org/10.1007/978-3-319-76573-0_51-1

Munim, S. A., & Raza, Z. A. (2019). Poly(lactic acid) based hydrogels: formation, characteristics and biomedical applications. *Journal of Porous Materials, 26*(3), 881–901. https://doi.org/10.1007/s10934-018-0687-z. Available from:.

Muthuraj, R., Misra, M., & Mohanty, A. K. (2017). Biodegradable compatibilized polymer blends for packaging applications: a literature review. *Journal of Applied Polymer Science, 45726*, 1–35.

Nacer Khodja, A., Mahlous, M., Tahtat, D., Benamer, S., Larbi Youcef, S., Chader, H., et al. (2013). Evaluation of healing activity of PVA/chitosan hydrogels on deep second degree burn: pharmacological and toxicological tests. *Burns, 39*(1), 98–104. https://doi.org/10.1016/j.burns.2012.05.021. Available from:.

Nho, Y. C., & Park, K. R. (2002). Preparation and properties of PVA/PVP hydrogels containing chitosan by radiation. *Journal of Applied Polymer Science, 85*(8), 1787–1794. https://doi.org/10.1002/app.10812

Nkhwa, S., Lauriaga, K. F., Kemal, E., & Deb, S. (2014). Poly(vinyl alcohol): physical approaches to designing biomaterials for biomedical applications. *Conference Papers in Science, 2014*, 1–7. https://doi.org/10.1155/2014/403472

Nyamweya, N. N. (2021). Applications of polymer blends in drug delivery. *Future Journal of Pharmaceutical Sciences, 7*(1). https://doi.org/10.1186/s43094-020-00167-2

Otsuka, E., & Atsushi, S. (2009). A simple method to obtain a swollen PVA gel crosslinked by hydrogen bonds. *Journal of Applied Polymer Science, 114*, 10–16. https://doi.org/10.1002/app

Pal, K., Banthia, A., & Majumdar, D. (2009). Polymeric hydrogels: characterization and biomedical applications. *Designed Monomers and Polymers, 12*, 197–220. https://doi.org/10.1163/156855509X436030

Park, K. R., & Nho, Y. C. (2003). *Synthesis of PVA/PVP hydrogels having two-layer by radiation and their physical properties* (vol. 67, pp. 361–365). https://doi.org/10.1016/S0969-806X(03)00067-7

Park, S., Kim, S. H., Kim, J. H., Yu, H., Kim, H. J., Hyungsup, Y., et al. (2015). Application of cellulose/lignin hydrogel beads as novel supports for immobilizing lipase. *Journal of Molecular Catalysis B: Enzymatic.* https://doi.org/10.1016/j.molcatb.2015.05.014. Available from:.

Payen, M. (1838). Mémoire sur la composition du tissu propre des plantes et du ligneux. *Comptes-rendus l'académie des Sci., 7*, 1052–1057.

Pircher, N., Veigel, S., Aigner, N., Nedelec, J. M., Rosenau, T., & Liebner, F. (2014). Reinforcement of bacterial cellulose aerogels with biocompatible polymers. *Carbohydrate Polymers, 111*, 505–513. https://doi.org/10.1016/j.carbpol.2014.04.029. Available from:.

Rigacci, A., Budtova, T., & Smirnova, I. (2017). Aerogels: a fascinating class of materials with a wide potential of application fields. *Journal of Sol-Gel Science and Technology, 84*(3), 375–376. https://doi.org/10.1007/s10971-017-4538-1

Rydz, J., Musioł, M., Zawidlak-w, B., & Sikorska, W. (2018). Polymers for food packaging applications. In A. M. Grumezescu, & A. M. Holban (Eds.), *Biopolymers for food design* (pp. 431–467). Netherlands: Elsevier. https://doi.org/10.1016/B978-0-12-811449-0/00014-1

Saini, I., Sharma, A., Dhiman, R., Aggarwal, S., Ram, S., & Sharma, P. K. (2017). Grafted SiC nanocrystals: for enhanced optical, electrical and mechanical properties of polyvinyl alcohol. *Journal of Alloys and Compounds, 714*, 172–180. https://doi.org/10.1016/j.jallcom.2017.04.183. Available from:.

Salam, A., Venditti, R. A., Pawlak, J. J., & El-tahlawy, K. (2011). Crosslinked hemicellulose citrate – chitosan aerogel foams. *Carbohydrate Polymers, 84*(4), 1221–1229. https://doi.org/10.1016/j.carbpol.2011.01.008. Available from:.

Salleh, K. M., Zakaria, S., Sajab, M. S., Gan, S., Chia, C. H., Jaafar, S. N., et al. (2018). Chemically crosslinked hydrogel and its driving force towards superabsorbent behaviour. *International Journal of Biological Macromolecules, 118*, 1422–1430. https://doi.org/10.1016/j.ijbiomac.2018.06.159. Available from:.

Salleh, K. M., Zakaria, S., Sajab, M. S., Gan, S., & Kaco, H. (2019). Superabsorbent hydrogel from oil palm empty fruit bunch cellulose and sodium carboxymethylcellulose. *International Journal of Biological Macromolecules, 131*, 50–59. https://doi.org/10.1016/j.ijbiomac.2019.03.028. Available from:.

Salleh, K. M., Zakaria, S., Gan, S., Baharin, K. W., Ibrahim, N. A., & Zamzamin, R. (2020). Interconnected macropores cryogel with nano-thin crosslinked network regenerated cellulose. *International Journal of Biological Macromolecules, 148*, 11–19. https://doi.org/10.1016/j.ijbiomac.2019.12.240. Available from:.

Salleh, K. M., Armir, N. A. Z., Mazlan, N. S. N., Wang, C., & Zakaria, S. (2021). Cellulose and its derivatives in textiles: primitive application to current trend. In *Fundamentals of natural fibres and textiles* (pp. 33–63). Woodhead Publishing.

Sam, S. T., Nuradibah, M. A., Ismail, H., Noriman, N. Z., & Ragunathan, S. (2013). Recent advances in polyolefins/natural polymer blends used for packaging application. *Polyermer-Plastics Technology and Engineering, 53*, 631–644. https://doi.org/10.1080/03602559.2013.866247

Sargeant, T. D., Desai, A. P., Banerjee, S., Agawu, A., & Stopek, J. B. (2012). An in situ forming collagen-PEG hydrogel for tissue regeneration. *Acta Biomaterialia, 8*(1), 124–132. https://doi.org/10.1016/j.actbio.2011.07.028. Available from:.

Scheller, H. V., & Ulvskov, P. (2010). Hemicelluloses. *Annual Review of Plant Biology, 61*, 263–289. https://doi.org/10.1146/annurev-arplant-042809-112315

Shen, X., Shamshina, J. L., Berton, P., Gurau, G., & Rogers, R. D. (2015). Hydrogels based on cellulose and chitin: fabrication, properties, and applications. *Green Chemistry, 18*(1), 53–75. https://doi.org/10.1039/c5gc02396c

Sionkowska, A. (2011). Current research on the blends of natural and synthetic polymers as new biomaterials: review. *Progress in Polymer Science, 36*(9), 1254–1276. https://doi.org/10.1016/j.progpolymsci.2011.05.003

Sirousazar, M., Forough, M., Farhadi, K., Shaabani, Y., & Molaei, R. (2014). Hydrogels: properties, preparation, characterization and biomedical, applications in tissue engineering, drug, delivery and wound care. *Advanced Healthcare Materials*, 295–357. https://doi.org/10.1002/9781118774205.ch9, 9781118773.

Slager, J., & Domb, A. J. (2003). Biopolymer stereocomplexes. *Advanced Drug Delivery Reviews, 55*(4), 549–583. https://doi.org/10.1016/S0169-409X(03)00042-5

Stergar, J., & Maver, U. (2016). Review of aerogel-based materials in biomedical applications. *Journal of Sol-Gel Science and Technology, 77*(3), 738–752. https://doi.org/10.1007/s10971-016-3968-5

Sun, X., Gan, Z., Jing, Z., Wang, H., Wang, D., & Jin, Y. (2015). Adsorption of Methylene Blue on hemicellulose-based stimuli-responsive porous hydrogel. *Journal of Applied Polymer Science, 132*(10). https://doi.org/10.1002/app.41606, 41606.

Sun, X., Zeng, Q., Wang, H., & Hao, Y. (2018). Preparation and swelling behavior of pH/temperature responsive semi-IPN hydrogel based on carboxymethyl xylan and poly (N -isopropyl acrylamide). *Cellulose*. https://doi.org/10.1007/s10570-018-2180-x. Available from:.

Sung, J. H., Hwang, M. R., Kim, J. O., Lee, J. H., Kim, Y. I., Kim, J. H., et al. (2010). Gel characterisation and in vivo evaluation of minocycline-loaded wound dressing with enhanced wound healing using polyvinyl alcohol and chitosan. *International Journal of Pharmaceutics, 392*(1–2), 232–240. https://doi.org/10.1016/j.ijpharm.2010.03.024. Available from:.

Tabasum, S., Younas, M., Zaeem, M. A., Majeed, I., Majeed, M., Noreen, A., et al. (2019). A review on blending of corn starch with natural and synthetic polymers, and inorganic nanoparticles with mathematical modeling. *International Journal of Biological Macromolecules, 122*, 969–996. https://doi.org/10.1016/j.ijbiomac.2018.11.064. Available from:.

Tan, H., & Marra, K. G. (2010). Injectable, biodegradable hydrogels for tissue engineering applications. *Materials, 3*, 1746–1767. https://doi.org/10.3390/ma3031746

Tanan, W., Panichpakdee, J., & Saengsuwan, S. (2018). Novel biodegradable hydrogel based on natural polymers: synthesis, characterization, swelling/reswelling and biodegradability. *European Polymer Journal*. https://doi.org/10.1016/j.eurpolymj.2018.10.033. Available from.

Teodorescu, M., Bercea, M., & Morariu, S. (2018). Biomaterials of poly(vinyl alcohol) and natural polymers. *Polymer Reviews, 58*(2), 247−287. https://doi.org/10.1080/15583724.2017.1403928

Tyliszczak, B., Polaczek, J., Pielichowski, J., & Pielichowski, K. (2010). Synthesis of control release KH_2PO_4-based fertilizers with PAA matrix modified by PEG. *Molecular Crystals and Liquid Crystals, 523*(October 2014), 297/[869]−303/[875].

Ulbricht, J., Jordan, R., & Luxenhofer, R. (2014). On the biodegradability of polyethylene glycol, polypeptoids and poly(2-oxazoline)s. *Biomaterials, 35*(17), 4848−4861. https://doi.org/10.1016/j.biomaterials.2014.02.029

Ullah, S., & Chen, X. (2020). *Fabrication, applications and challenges of natural biomaterials in tissue engineering* (p. 20). https://doi.org/10.1016/j.apmt.2020.100656

van Bochove, B., & Grijpma, D. W. (2019). Photo-crosslinked synthetic biodegradable polymer networks for biomedical applications. *Journal of Biomaterials Science, Polymer Edition, 30*(2), 77−106. https://doi.org/10.1080/09205063.2018.1553105. Available from:.

Vasile, C., & Kulshreshhta, A. K. (2003). *Handbook of polymer blends and composites Volume 4B. Shawbury*. Shrewsbury, Shropshire: Rapra Technology Limited. https://doi.org/10.1080/10426919008953291

Vroman, I., & Tighzert, L. (2009). Biodegradable polymers. *Materials, 2*(2), 307−344. https://doi.org/10.3390/ma2020307

Vyas, C., Poologasundarampillai, G., Hoyland, J., & Bartolo, P. (2017). 3D printing of biocomposites for osteochondral tissue engineering. In *Biomedical composites* (Second Edi, pp. 261−302). Elsevier Ltd. https://doi.org/10.1016/B978-0-08-100752-5.00013-5

Wang, R., Shou, D., Lv, O., Kong, Y., Deng, L., & Shen, J. (2017). pH-Controlled drug delivery with hybrid aerogel of chitosan, carboxymethyl cellulose and graphene oxide as the carrier. *International Journal of Biological Macromolecules, 103*, 248−253. https://doi.org/10.1016/j.ijbiomac.2017.05.064. Available from:.

Wen, J., Jia, Z., Zhang, X., Pan, M., Yuan, J., & Zhu, L. (2020). Tough, thermo-responsive, biodegradable and fast self-healing polyurethane hydrogel based on microdomain-closed dynamic bonds design. *Materials Today Communications, 25*, 101569. https://doi.org/10.1016/j.mtcomm.2020.101569. Available from:.

Xiao, K., Wang, Z., Wu, Y., Lin, W., He, Y., Zhan, J., Luo, F., Li, Z., Li, J., Tan, H., & Fu, Q. (2019). Biodegradable, anti-adhesive and tough polyurethane hydrogels crosslinked by triol crosslinkers. *Journal of Biomedical Materials Research Part A, 107*(10), 2205−2221. https://doi.org/10.1002/jbm.a.36730

Yamamoto, T., Inoue, K., & Tezuka, Y. (2016). Hydrogel formation by the "topological conversion" of cyclic PLA-PEO block copolymers. *Polymer Journal, 48*(4), 391−398. https://doi.org/10.1038/pj.2015.134

Zerpa, A., Pakzad, L., & Fatehi, P. (2018). Hardwood kraft lignin-based hydrogels: production and performance. *ACS Omega, 3*(7), 8233−8242. https://doi.org/10.1021/acsomega.8b01176

Zhang, Z., Ni, J., Chen, L., Yu, L., Xu, J., & Ding, J. (2011). Biodegradable and thermoreversible PCLA-PEG-PCLA hydrogel as a barrier for prevention of post-operative adhesion. *Biomaterials, 32*(21), 4725−4736. https://doi.org/10.1016/j.biomaterials.2011.03.046

Zhang, C., Zhai, T., & Turng, L.-S. (2017). Aerogel microspheres based on cellulose nanofibrils as potential cell culture scaffolds. *Cellulose, 24*, 2791−2799.

Zhang, K., Wu, J., Zhang, W., Yan, S., Ding, J., Chen, X., Cui, L., & Yin, J. (2018). In situ formation of hydrophobic clusters to enhance mechanical performance of biodegradable poly(l-glutamic acid)/poly(e-caprolactone) hydrogel towards meniscus tissue engineering. *Journal of Materials Chemistry B, 6*(47). https://doi.org/10.1039/c8tb01453a, 7822.

Zhu, J. (2010). Bioactive modification of poly(ethylene glycol) hydrogels for tissue engineering. *Biomaterials, 31*(17), 4639–4656. https://doi.org/10.1016/j.biomaterials.2010.02.044. Available from:.

Zuo, L., Zhang, Y., Zhang, L., Miao, Y. E., Fan, W., & Liu, T. (2015). Polymer/carbon-based hybrid aerogels: preparation, properties and applications. *Materials, 8*(10), 6806–6848. https://doi.org/10.3390/ma8105343

Further reading

Bai, T., Zhu, B., Liu, H., Wang, Y., Song, G., Liu, C., et al. (2020). Biodegradable poly(lactic acid) nanocomposite reinforced and toughened by carbon nanotubes/clay hybrids. *International Journal of Biological Macromolecules, 151*, 628–634. https://doi.org/10.1016/j.ijbiomac.2020.02.209

Polyhydroxybutyrate (PHB)-based blends and composites

13

Juliana Botelho Moreira, Suelen Goettems Kuntzler, Bruna da Silva Vaz, Cleber Klasener da Silva, Jorge Alberto Vieira Costa and Michele Greque de Morais
College of Chemistry and Food Engineering, Federal University of Rio Grande (FURG), Rio Grande, RS, Brazil

1. Introduction

The use of plastic materials in the domestic and industrial sectors in excess has been causing several concerns concerning improper disposal and environmental contamination. Conventional plastics take decades to fully decompose and produce toxins during the degradation process (Li & Wilkins, 2020; Sirohi, Pandey, Tarafdar et al., 2020a). In this sense, there is a need to advance research in the production of biodegradable polymers from materials that can be ecofriendly and eliminated from the biosphere to replace petrochemical-based plastics (Choi et al., 2020; Sirohi, Pandey, Tarafdar et al., 2020a).

Polyhydroxybutyrate (PHB) is a short-chain biogenic polymer, belonging to the group of polyhydroxyalkanoates (PHAs). PHB is an alternative material to conventional polymers and its physical properties are similar to those of polypropylene. PHB deterioration can occur through the action of enzymes and organisms, such as bacteria, yeasts, and fungi (Sirohi, Pandey, Gaur et al., 2020b). PHB can come from chemical or biological synthesis from renewable and sustainable sources. The naturally synthesized polymer acts as an energy-storage compound intracellularly and is present in microorganisms such as bacteria, cyanobacteria, and/or microalgae (Balaji et al., 2013; Price et al., 2020). PHB can be produced by combining various substrates under different growing conditions, including aerobic/anaerobic, temperature, pH, nutrients, and submerged/solid-state fermentation. Industrially, PHB is produced by fermentation with heterotrophic bacteria in aerobic conditions (Levett et al., 2016). Microalgae and/or cyanobacteria also have polymer production capacity and could be a promising alternative, as they use carbon dioxide from the atmosphere as a carbon source and convert it directly into PHB (Singh et al., 2017).

The use of PHB has great environmental and economic potential, due to the advancement of research to develop new products in several sectors. The market was forecast to grow by $18.66 million during the 2020–24 period (Technavio, 2020). PHB can be applied to compose composite materials or blends, due to the mechanical and thermal properties, biodegradation, and biocompatibility (Moura et al., 2019; Price et al., 2020). In addition, this polymer has been explored in the areas of

Biodegradable Polymers, Blends and Composites. https://doi.org/10.1016/B978-0-12-823791-5.00007-7

bone tissue regeneration (Esposti et al., 2019), food packaging with antimicrobial action (Kuntzler et al., 2018), agriculture in the development of fertilizers (Arrieta et al., 2020), in the environmental sector for wastewater treatment (Heitmann et al., 2016), and biorefinery (Prieto et al., 2017).

This chapter approaches the physical, chemical, and biological characteristics of PHB and the main methods of producing the polymer through chemical and biological syntheses. Moreover, strategies for using PHB in composites and blends, as well as the applications of these materials in different areas, are explored.

2. General properties of PHB

PHB is a homopolymer belonging to the family of PHAs (Bugnicourt et al., 2014). PHB is characterized by having a methyl (CH_3) as an alkyl substitution group. This chemical characteristic is related to the material's hydrophobicity. PHB is a short-chain polymer, composed of 3-hydroxybutyrate monomers that have four to five carbon atoms. This monomer contains β-hydroxybutyric acid, which has an alcoholic group (−OH) and a carboxylic acid (−COOH). In addition, the chemical structure of PHB has a chromophoric carbonyl group (Santos et al., 2017).

The molecular weight of PHB is dependent on the producing organism, growing conditions, and extraction and purification methods (Bugnicourt et al., 2014; Sudesh et al., 2000). PHB is a nontoxic biopolymer, relatively resistant to hydrolytic degradation, and is soluble in chloroform and other chlorinated hydrocarbons. PHB has good ultraviolet resistance, however, it has low resistance to acids and bases. Moreover, PHB is a fragile material due to recrystallization with aging at room temperature. Therefore, the use of plasticizers and nucleating agents is important to reduce the crystallization process and improve product flexibility and elongation (Bugnicourt et al., 2014).

PHB is similar to polypropylene (PP) in terms of molecular structure. Both polymers have monomers of similar sizes with a single methyl side pendant group. As a result, these polymers have comparable physical properties, such as water resistance, melting temperature, and tensile strength. However, PHB monomers are linked by ester bonds, which allow biodegradation (Price et al., 2020) when exposed to biologically active environments (soils, seawater or freshwater, aerobic and anaerobic compost, activated sludge, landfill) (Santos et al., 2017). In addition, PHB sinks in water, which facilitates its biodegradation in sediments (Bugnicourt et al., 2014). On the other hand, PP and most petrochemical-based plastics are inaccessible to microorganism enzymes and biodegradation due to their carbon−carbon bonds (Price et al., 2020). PHB thermally decomposes at temperatures slightly above its melting point. Thus, its exposure to a temperature close to 180°C can induce severe degradation accompanied by the production of crotonic acid and oligomers (Bugnicourt et al., 2014).

Another important property of PHB is biocompatibility, which makes it suitable for medical applications since it is not rejected by the body and assimilation generates nontoxic residues. Biocompatibility is associated with the fact that this biopolymer is present in the human bloodstream, in the form of low-molecular-weight PHB.

Furthermore, PHB's biocompatibility is also related to its degradation product (3-hydroxybutyric acid), which is a common metabolite in living beings (Doyle et al., 1996). Thus, due to its outstanding features of biodegradability and biocompatibility, PHB has potential for applications in several sectors, such as pharmacological, environmental, packaging, veterinary, industrial, agricultural, and biomedical (Santos et al., 2017).

3. Chemical synthesis

The synthetic production of PHAs has been explored in studies that seek to synthesize these polymers through specific catalysts. Among the techniques explored for the chemical synthesis of PHAs, PHB retrosynthesis and ring-opening polymerization (ROP) stand out (Singh et al., 2017; Winnacker, 2019). In PHB retrosynthesis, the process is based on the union of propylene oxide (PO) and carbon monoxide structures (Fig. 13.1). Carbon monoxide comes from petroleum gas, fuel oil, coal, and biomass, while PO can be obtained from propylene from crude oil (Winnacker, 2019).

The formation of polymers via ring-opening reactions of cyclic compounds is one of the most promising methods for obtaining synthetic PHB (Fig. 13.2). The formation of the polymeric polymer chain occurs through successive additions of the open structures of racemic β-butyrolactone (BL) (Reichardt & Rieger, 2011). The PHB produced by this technique is an economically beneficial alternative to organic production. However, the source material and the catalyst are used to influence the thermomechanical behavior and biodegradability of synthetic PHB (Ravve, 2012). In this sense, research is dedicated to the development of suitable catalysts for ROP of BL. Several catalytic systems have been described for BL polymerization. However, only a few allow control of polymerization in terms of the microstructure and molecular weight of PHB (Singh et al., 2017; Vagin et al., 2015).

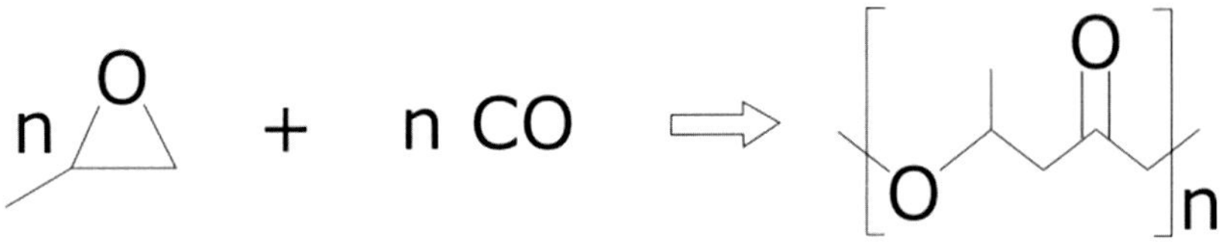

$$n\,\text{PO} + n\,\text{CO} \Rightarrow [\,\text{O–CH(CH}_3\text{)–CH}_2\text{–C(=O)}\,]_n$$

Figure 13.1 PHB retrosynthesis.

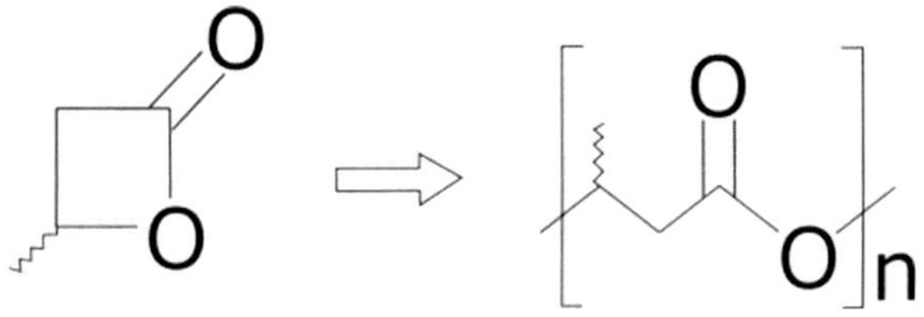

$$\beta\text{-BL} \Rightarrow [\,\text{CH(CH}_3\text{)–CH}_2\text{–C(=O)–O}\,]_n$$

Figure 13.2 ROP of racemic-β-butyrolactone to PHB.

The main challenge to chemical synthesis is to produce PHB with different degrees of tacticity to expand the polymer applications. Atactic PHB has a low melting temperature, and as a result, it is in the form of oil. Isotactic PHB in solid form has a high melting point and crystallinity (Singh et al., 2017). Structures other than butyrolactone can also be used in the ROP. Tang and Chen (2018) produced P(3HB) using racemic cyclic diode (rac-DL) derived from biological succinate. With rac-DL stereoselective catalysts, the rac-DL ROP took place under environmental conditions and rapidly produced P3HB with similar isotacticity to biological, high melting temperature (Tm = 171°C), and high molecular weight (Mn = 1.54x 10^5 g/mol). According to Winnacker (2019), most of the catalysts used contain metals that can be toxic. Thus, for commercialization of the polymer, it is necessary to remove these compounds. Therefore, the choice of catalyst must consider the economic feasibility studies.

4. Biological synthesis

PHB is a biodegradable polymer that can be biologically synthesized and has properties similar to conventional plastics. The synthesis can be carried out by several bacteria, as well as by cyanobacteria and/or microalgae. PHB is stored in these organisms intracellularly as an energy reserve under conditions with limited nutrients and excess carbon (Hiremath et al., 2015). Although microbial PHB is promising as a biodegradable polymer, its production costs may limit its application on an industrial scale. However, low-cost carbon sources and nutritional supplements can improve competitiveness in the market due to considerable cost savings. In addition, the production parameters of this biopolymer can be optimized to increase the biomass concentration and/or PHB yields (Anjum et al., 2016; Hiremath et al., 2015).

4.1 Bacteria

The occurrence of PHAs in bacteria was observed in 1888 by Beijerincka. However, its composition has not been elucidated. In 1926, the PHB biopolymer was the first PHA to be identified by the French researcher Maurice Lemoigne, who obtained poly-3-hydroxybutyric acid [P(3HB)] from the bacterium *Bacillus megaterium* (Lemoigne, 1926). Since then, bacterial PHB has been studied and characterized by the accumulation of up to 80% by dry cell weight (Keshavarz & Roy, 2010; Yu & Stahl, 2008). About 300 species of bacteria have been reported as capable of producing and accumulating large amounts of the biopolymer. In this sense, Gram-positive and Gram-negative bacteria are included, with the emphasis on wild and recombinant strains *of Ralstonia eutropha*, *Azotobacter* sp., *Bacillus* sp., *Pseudomonas* sp., *Halomonas* sp., *Aeromonas* sp., *Methylobacterium* sp., *Thermus thermophilus*, *Hydrogenophaga pseudoflava*, *Saccharophagus degradans*, *Comamonas* sp., *Alcaligenes latus*, *Rhodobacter sphaeroides*, *Zobellella denitrificans*, cyanobacteria, *Chromobacterium* sp., *Erwinia* sp., and *Escherichia coli* (Anjum et al., 2016).

PHB is a lipid reserve material naturally accumulated by bacteria in intracellular granules in its cytoplasm. This accumulation is related to the defense mechanism of the microorganism when exposed to nutrient stress conditions. Usually, it occurs in an environment containing excess carbon sources and limitation of nutrients, such as nitrogen, phosphorus, or oxygen (Rathore, 2014). PHB is synthesized by bacteria via route I (Fig. 13.3), better known among the biosynthetic pathways of PHAs. In this way, the bacteria produce acetyl-coenzyme-A (acetyl-CoA) which is converted to PHB by three enzymes. In the first step, the enzyme 3-ketothiolase (PhaA) combines two molecules of acetyl-CoA to form acetoacetyl-CoA. Acetoacetyl-CoA reductase (PhaB) acts to reduce acetoacetyl-CoA to form 3-hydroxybutyryl-CoA. In the last stage of synthesis, the enzyme PHB synthase (PhaC) catalyzes the polymerization of three hydroxybutyryl-CoA in P(3HB) (Anjum et al., 2016). During normal bacterial growth, 3-ketothiolase is inactivated by the free coenzyme-A that leaves the Krebs cycle. However, when the entry of acetyl-CoA in the Krebs cycle is restricted (during nutrient limitation), excess acetyl-CoA triggers PHB biosynthesis (Ratledge & Kristiansen, 2001).

During the bacterial synthesis of PHAs, substrates are metabolized in the cell by different metabolic pathways. Three metabolic pathways are most studied, according to the carbon substrate. Thus, the composition of the monomer is related to the carbon source used. For example, if sugars (glucose or fructose) are processed via I, the PHB homopolymer is obtained. However, copolymers are produced when fatty acids or

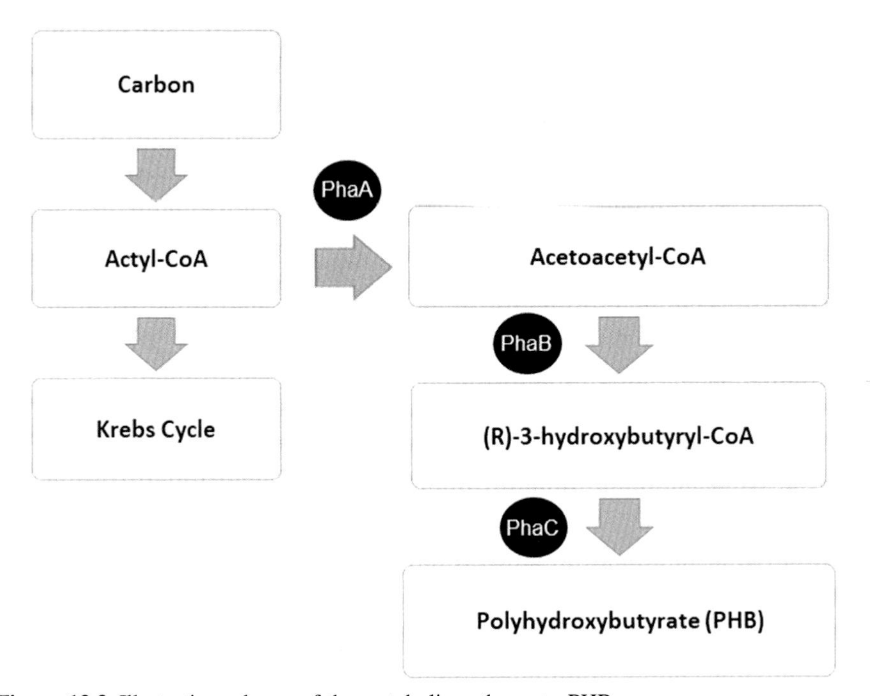

Figure 13.3 Illustrative scheme of the metabolic pathway to PHB.

sugars are metabolized via route II, III, or other routes (Aldor & Keasling, 2003; Steinbüchel & Lütke-Eversloh, 2003; Verlinden, 2007). Thus, when bacteria use mixed substrates, they can produce copolymers, such as poly(3-hydroxybutyrate-co-3-hydroxyvalerate) [P(3HB-co-3HV)] or poly(3-hydroxybutyrate-co-4-hydroxybutyrate) [P(3HB-co-4HB)] (PHAs with 3–5 carbon atoms) and poly(3-hydroxyhexanoate-co-3-hydroxyoctanoate) [P(3HHx-co-3HO)] (PHAs with 6–14 atoms of carbon) (Mozejko-Ciesielska & Kiewisz, 2016).

4.2 Microalgae

Microalgae and/or cyanobacteria are considered sustainable and renewable alternatives for the production of biopolymer. These microorganisms have a photoautotrophic nature, with high photosynthetic efficiency and potential for CO_2 fixation; minimum nutritional requirements for biomass generation such as lower carbon requirement (compared to heterotrophic bacteria); rapid growth rate; use of wastewater and nonarable land; and the production of PHB from microalgae and/or its secondary metabolites (Rahman & Miller, 2017; Singh et al., 2017). In addition, cyanobacteria can produce PHB through genetic modification of strains through transformation with genes involved in the metabolic pathway of *R. eutropha* (Miyake et al., 2000).

In 1966, the production of PHAs in cyanobacteria was first reported by Prof. Carr at the University of Warwick (United Kingdom), obtained with the cultivation of *Chlorogloea fritschii* with supplementation of the organic compound acetate (Carr, 1966). Over the years, PHA accumulation has been observed by supplementation with organic compounds, such as acetate, glucose, citrate, maltose, valerate, and propionate (Singh et al., 2017). In 1971, there was the first report on the photoautotrophic accumulation of biopolymers in cyanobacteria (Rippka et al., 1971). Several species of cyanobacteria have been reported to be capable of accumulating PHB, such as *Chlorogloea fritschii*, *Spirulina* spp., *Spirulina platensis*, *Aphanothece* spp., *Gloeothece* spp., *Synechococcus* spp., *Synechocystis* sp., *Gloeocapsa* sp., and *Phormidium* sp. (Balaji et al., 2013).

Among the heterotrophic and photoautotrophic prokaryotes, there are similarities regarding the biopolymer biosynthesis (Wang et al., 2013). As in bacteria, the biosynthetic pathway involves three sequential reactions mediated by the enzymes 3-ketothiolase, acetoacetyl-CoA reductase, and PHB synthase. In the first stage, a condensation reaction occurs, where the two acetyl-CoA molecules act as precursors for the synthesis of acetoacetyl-CoA with the enzyme β-ketothiolase. After, acetoacetyl-CoA is converted to R-3-hydroxybutyryl-CoA by the enzyme acetoacetyl-CoA reductase, involving NADPH molecules. Finally, the PHB synthase enzyme catalyzes the incorporation of the monomeric unit (R-3-hydroxybutyryl-CoA) in the PHB polymer chain, through an ester bond (Singh et al., 2017). During microorganism growth, 3-ketothiolase is inhibited by free coenzyme A, leaving the Krebs cycle. However, when nutrients are limited, acetyl-CoA is prevented from entering the Krebs cycle, and its excess is directed to PHB biosynthesis (Ratledge & Kristiansen, 2001).

In addition, as cyanobacteria fix CO_2 through the Calvin–Benson cycle, carbon is accumulated in the form of intracellular polymers, such as glycogen and PHB. The glycogen biosynthetic pathway may have some connections with the pathway for the production of PHA in cyanobacteria, at the level of 3-phosphoglycerate (3PG). 3PG is synthesized as a result of CO_2 assimilation by Rubisco through the carboxylation of ribulose-1,5-bisphosphate (RuBP). Thus, the biosynthetic pathway of glycogen that competes for 3PG should be blocked to improve PHB synthesis in cyanobacteria (Singh & Mallick, 2017; Singh et al., 2017; Wu et al., 2002).

4.3 Production factors

PHB is produced as an intracellular carbon reserve and energy under conditions containing excess carbon source and limitations of essential nutrients, such as nitrogen and/or phosphorus. Organisms respond to nutrient limitations by diverting their metabolic pathways. Macromolecule degradation of essential nutrients results in the accumulation of carbon-rich compounds such as PHAs (Coelho et al., 2015).

The PHB production process is usually carried out in two stages. In single batch-fed fermentations, there is not enough biomass accumulation. Thus, consequently, there is lower biopolymer production (Katırcıoğlu et al., 2003). The first phase involves ideal nutritional conditions with the necessary nutrients (carbon source, nitrogen source, oxygen) to provide a high biomass concentration. In the second phase, the nutritional stress, caused by the reduced supply of specific nutrients (nitrogen and phosphorus), induces PHB production. This limitation of nutrients leads to reduced cell growth and PHB accumulation in the form of intracellular granules (Carpine et al., 2020).

The composition and/or concentration of PHB is influenced by the strain, type, and concentration of substrates used, and environmental growth conditions. For example, copolymers of PHB poly(3-hydroxybutyrate-co-3-hydroxyvalerate) (PHBV) or poly(3-hydroxybutyrate-4-hydroxybutyrate) (PHB4B) are formed by bacteria when mixed substrates, such as the mixture of glucose and valerate, are used (Rodgers & Wu, 2010; Yan et al., 2005). The cyanobacterium *Nostoc muscorum* has been accumulated up to 71% of PHB when grown in a medium supplemented with glucose, acetate, valerate, and nitrogen and phosphate limitation (Bhati and Mallick, 2012). The bacterium *Aulosira fertilissima* produced 77% of PHB in a culture medium with acetate and under phosphate limitation (Samantaray & Mallick, 2012), and 66% copolymer P(3HB-co-3HV) with fructose and valerate under phosphorus deficiency (Samantaray & Mallick, 2014).

Sustainable cultivations using wastes as substrates and natural environmental conditions to synthesis contribute to economically viable PHB production to replace chemical PHB production (Costa et al., 2019). PHB production by bacteria requires organic molecules that are more expensive than inorganic compounds used for microalgal cultivation (Jin & Melis, 2003). Thus, microalgae and/or cyanobacteria stand out as low-cost PHB-producing organisms. These microorganisms have a photoautotrophic nature, where they use light energy (solar) and CO_2 as the main sources of energy. Moreover, microalgae have high productivity, can be grown in wastewater due to the ability to use inorganic substances, are tolerant to climate change, and can be grown in areas that are not suitable for agriculture (Costa et al., 2019).

Growth conditions can also be optimized for the adaptation of biopolymer structures through genetic modifications (Chen et al., 2016). Genetic engineering experiments have been reported to increase PHB production in microorganisms. Wu et al. (2016) changed the binary fission in bacteria to multiple fission by excluding genes related to minC and minD fission, which resulted in cell division in more than two daughter cells at the same time. In addition, genes related to the cell division process (*ftsQ*, *ftsL*, *ftsW*, *ftsN*, and *ftsZ*) and the cell shape control gene (*mreB*) were overexpressed in *E. coli* JM109 ΔminCD to improve cell growth and PHA production. Thus, changing the pattern of cell morphology and division helps to accelerate cell growth and PHB accumulation. With this, it was possible to obtain more than 80% polymer accumulation, compared to its binary fission control (Chen & Jiang, 2017; Wu et al., 2016).

Jiang and Chen (2016) reported that genetic manipulations promote the enlargement of bacterial forms (rods for fibers or small spheres or large spheres) for a higher accumulation of intracellular granules. Furthermore, easy rupture of the cell wall also allows more space for storing PHB granules. This disruption can be achieved by excluding the genes responsible for the synthesis of the cell wall. Jiang et al. (2015) modified the *E. coli* strain JW3175 to obtain better PHB yield by excluding the *mtg-A* gene, which increased the biopolymer production about 1.6-fold compared to the control strain. In other studies, there was a more than 100% increase in PHB accumulation in *E. coli* overexpressing the *mreB* gene, through the inducible expression of the FtsZ inhibitory SulA protein (Jiang & Chen, 2016; Wang et al., 2014).

Random mutagenesis has also been used to obtain cyanobacterial strains with higher PHB productivity, by introducing an exogenous biosynthetic pathway into the cell (Kamravamanesh et al., 2019). Thus, cyanobacteria can accumulate significant amounts of PHB through the modification of strains by genetic engineering and using genes involved in the PHB pathway of *R. eutropha*. The synthesis of PHB from the heterologous expression of the *R. eutropha* gene in *Synechocystis* sp. PCC 6803 increased the PHB content from 7% to 11% (Carpine et al., 2020; Miyake et al., 2000; Sudesh et al., 2002). Hempbel et al. (2011) reported that after introducing the PHB bacterial pathway in the microalgal species, it was possible to obtain 10.6% PHB by *Phaeodactylum tricornutum*. Kamravamanesh et al. (2019) showed that a mutated strain of *Synechocystis* sp. PCC 6714 produced up to 37% PHB.

5. Steps involved in PHB production

The process of obtaining PHB involves six steps: biomass production, biomass harvesting, pretreatment, extraction, recovery, and drying (Fig. 13.4). Biomass harvesting is performed by techniques such as filtration, centrifugation, flotation, or precipitation. The pretreatment step can combine two or more methods to facilitate the extraction and recovery of the microbial PHB. Freeze-drying and thermal drying, grinding, chemical,

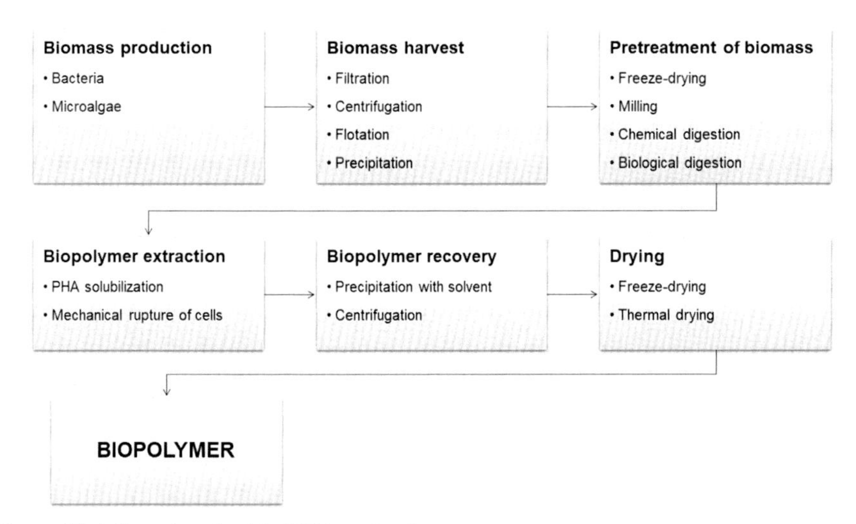

Figure 13.4 Steps involved in PHB production.

and biochemical pretreatments are the most explored (Tang & Chen, 2018). The PHB extraction stage precedes the recovery process. The extraction can occur by direct solubilization of the polymers. Extraction methods generally employ solvent, which permeates through the cell wall and solubilizes the polymer. The solvents more used for PHB extraction are chloroform, dichloroethane, propylene carbonate, ethylene carbonate, esters, and ketones. Moreover, PHB extraction can be performed by the combined action of cell disruption techniques followed by solubilization and separation (Koller, 2017). After extraction, the solubilized polymer is precipitated with a more polar solvent, for example, water or ethanol. Finally, the polymer is dried by drying in air-circulation greenhouses (Posada et al., 2011).

Pretreatments can be physical, chemical, or biological. These methods weaken the cell structure that protects and surrounds the PHB granules. The most commonly applied techniques are traditional drying and freeze-drying. These techniques are usually the first stage of pretreatment (Tang & Chen, 2018). The use of sodium chloride alters the osmotic pressure, favoring the extraction process. Sodium hypochlorite can also be used to digest part of the microbial cell structure, releasing PHB into the extraction medium (Samorì et al., 2015). This extraction medium has the advantage of degrading compounds that can pigment the extracted polymers, such as chlorophylls and phycocyanin (Silva et al., 2018). Physical methods such as glass spheres and the application of ultrasound waves cause the cell wall to rupture and release the intracellular material to the external environment (Leong et al., 2017).

The application of solvents in the extraction process of PHAs is the main way of obtaining these biopolymers, mainly on a laboratory scale (Meixner et al., 2018). Among the solvents most used in the extraction are acetone, chlorinated hydrocarbons such as chloroform and 1,2-dichloromethane, or some cyclic carbonates such as ethylene carbonate and 1,2-propylene carbonate (Jiang et al., 2006). Solvent extraction has advantages over other methods, such as obtaining PHB with a high degree of purity and high molecular weight (Meereboer et al., 2020).

In recent years, studies into the application of extraction techniques with less potential to harm the environment have intensified. Some techniques use wet biomass; thus, they represent a reduction in the number of steps involved in the obtaining process of polymer and cost reduction. The addition of chemicals under a suspension containing PHB-loaded microbial cells disrupts the cell wall, releasing biopolymers and other materials into the medium. The three most explored chemicals in this step are sodium hydroxide, sodium hypochlorite, and sodium dodecyl sulfate (Tang & Chen, 2018).

Purified enzymes or crude extracts can be used to break the cell wall of microorganisms. These have the advantages of low energy requirements, use water as a reaction and recovery solvent, and low capital investment. However, the cost of producing enzymes is often high as they are purified, which can make the process unfeasible (Rodriguez-Perez et al., 2018). Proteases are the main enzymes used. However, other enzymes can also be used, individually or in combination (Mannina et al., 2020). Studies to reduce enzyme costs are necessary to facilitate the application of these techniques on an industrial scale. The use of immobilized enzymes or integration of enzyme production as part of the process of obtaining PHAs are alternatives to reduce production costs (Rodriguez-Perez et al., 2018).

The biological disruption of microbial cells is technological and economical, with reduced costs for obtaining PHA. In this technique, biological agents are used to forcing the release of PHB from bacterial cells. Viral particles break down bacterial cells using the lytic cycle of viruses to release PHB granules. When the lytic cycle is complete, the virus leaves the host cell breaking the cell wall. This break also enables the release of the PHB particles, allowing their recovery (Steinbuchel et al., 2013). The addition of natural bacterial predators can also be exploited to disrupt microbial cells and release PHA (Gonzalez et al., 2020; Ong et al., 2018). Despite the advantages of biological disruption as it reduces the use of chemicals and solvents, the process requires more time than other recovery processes and requires biomass pretreatment (Tang & Chen, 2018).

6. PHB blends

Polymeric blends have been used to improve the physical and mechanical properties of new materials. In addition, blends are an effective and economical approach to changing aspects related to the disadvantages of original components, varying the composition of the blends and the preparation factors. Recently, the choice of materials has been directed to the concept of sustainability and ecology (Yeo et al., 2018).

As PHB is relatively fragile, expensive, and sensitive to thermal degradation, the use of blends is advantageous to improve its general properties (Santos et al., 2017). Due to their biocompatibility and total biodegradability, PHB and starch are promising biopolymers for the manufacture of plastic items and various biomaterials. Moreover, to overcome PHB limitations, a PHB blend with starch has been studied, as well as the inherent properties of starch (Yeo et al., 2018). Zhang and Tomas (2010) studied a

PHB blend with two types of corn starch. Starch 1 contained 70% amylose and starch 2 consisted of 72% amylopectin. PHB/starch blends were prepared in a 70:30 ratio, by melting. The authors observed that the starch granules act as a filler and nucleating agent. These properties caused a reduction in the size of PHB spherulites. Furthermore, the study found that enhanced hydrogen bonds between PHB and starch I were responsible for improvements in thermal, rheological, and mechanical properties.

The chemical modifications of the starch were also investigated to evaluate the effects caused on the thermomechanical properties of PHB. The blends of natural starch and derivatives with PHB contributed to the reduction of melting (T_m) and transition (T_g) temperatures compared to pure PHB. On the other hand, the PHB/modified starch blends showed low T_m and T_g values, which was associated with internal plasticization (Innocentini-Mei et al., 2003). Starch modified with polyvinyl acetate has also been investigated to improve the thermal and mechanical properties of PHB (Don et al., 2010).

Lignocellulosic materials are widely distributed and available in abundance in nature. With this, they have become promising resources for applications in polymer blends. In this context, Angelini et al. (2014) used a residue rich in lignin (obtained from the production process of bioethanol) and a commercial alkaline lignin for the preparation of blends with PHB. The authors found that the residue did not affect the thermal stability of the PHB. On the other hand, commercial lignin had a strong pro-degrading effect. In addition, the study considered the residue as a heterogeneous nucleating agent capable of controlling the physical aging of PHB. Angelini et al. (2016) investigated the rheological properties in PHB and PHB-based blends with lignocellulosic biomass derivatives (lignin and holocellulose). According to this study, the PHB/lignin blend had higher modules and the viscoelastic properties were dependent on the lignin content. Moreover, with the observation of the fracture surface of PHB/lignin in SEM, a smoother and more homogeneous aspect was found. In contrast to lignin, the addition of holocellulose promoted PHB melt crystallization during rapid cooling. The authors associated this response with the high melt viscosity of the PHB/lignin blend, which hindered chain mobility. The results obtained by both studies demonstrated the environmentally correct and economic viability of converting an agro-industrial by-product into a biomaterial that can serve different fields of application (Angelini et al., 2014; Angelini et al., 2016).

In addition to the addition of starch and lignocellulosic materials, polyesters derived from renewable raw materials such as polylactic acid (PLA) can also be used for the preparation of blends with PHB to improve the mechanical properties (Arrieta et al., 2020). D'Amico et al. (2016) developed a blend of PHB/PLA with tributyrin by fusion. The study found the immiscibility between PHB and PLA from the thermal analysis when finding two T_g. Compared to pure PHB, the elongation at break for the blend prepared with 30% PHB/70% PLA was increased by approximately 190%. The images obtained by SEM corroborated this result by revealing detachment between the interfacial surfaces. Thus, the authors concluded that for blends rich in PLA, a certain amount of plasticized PHB can significantly improve elongation at the break without loss of tensile strength.

Polycaprolactone (PCL) is a polymer with a linear structure, which can be used as a biodegradable plasticizer to improve ductility and elongation in material breakage. In this sense, Garcia-Garcia et al. (2016) produced PHB/PCL blends by melt compounding in a twin screw corotating extruder and injection molded, in the PHB:PCL proportions 0:100, 25:75, 50:50, 75:25, 100:0. The authors found that the PCL acts as an impact modifier, increasing the flexibility and ductility of the PHB/PCL blend. The tensile strength and modulus of elasticity decreased with increasing PCL content. However, the flexural strength and the flexural modulus reach the highest values for the blend prepared with 75% PHB/25% PCL. Nishida et al. (2020) observed that when preparing a PHB/PCL blend with a PCL concentration greater than that of PHB, the deformation at rupture was improved, maintaining high maximum stress. The 50% PHB/50% PCL blend adopted a bicontinuous morphology with major hemispheric defects on the fracture surfaces after a tensile test.

Flexible polyvinyl alcohol (PVA) is a polymer that has also been used in blends to improve the physical properties of PHB. Ol'khov et al. (2015) evaluated the morphology of films developed from PHB/PVA blends with different proportions (PHB:PVA, 100: 0, 90:10, 80:20, 70:30, 50:50, and 0:100%) and found that, with a 20%–30% PHB content, the mechanical characteristics and water diffusion coefficients changed significantly. In addition, this study reported that despite the limited PHB concentration, the PHB/PVA blend is interesting as a biodegradable coating, as well as for applications in the packaging, agricultural, and biomedical industries.

Elastomers and polyolefin-based thermoplastics can be applied as a second component to toughen PHB (Bartczak & Grala, 2016). Polypropylene is one of the most commonly used polymers of petrochemical origin due to its physical properties being similar to those of PHB. In this sense, blends of PHB and polypropylene (PP) showed better mechanical properties than pure PHB. With the PHB/PP blend, there was a tendency for less crystallinity and rigidity of the matrix with an increase in the percentage of PP. Furthermore, both pure PHB and blend 90% PHB/10% PP showed traces of degradation during the accelerated degradation tests. The PHB/PP blend stood out with greater degradation in alkaline soil since an alkaline medium favors the PHB hydrolytic degradation mechanism (Pachekoski et al., 2009). Although elastomers contribute to obtaining a toughened PHB, these nonbiodegradable polymers can contribute to a reduction in stiffness and tensile strength (Bartczak & Grala, 2016). Therefore, the proportions of each component must be further investigated.

7. PHB-based composites

Composite systems are manufactured with several multilayer materials containing fibers, films, or particles, which improve the mechanical, thermal, and barrier properties of individual compounds (Anjum et al., 2020; Rodríguez-Tobías et al., 2019). Composites based on PHB have interesting characteristics related to biodegradation and biocompatibility properties, ecotoxicity, and nontoxicity (Isola et al., 2017; Yeo et al., 2018). Another important factor in the production of composites is the addition

Table 13.1 PHB-based composites containing nanostructures.

Type of material	Complementary or plasticizer material	Type of nanostructure	References
PHB film	Glyceryl tributyrate and poly [di(ethylene glycol) adipate]	Cellulose nanoparticles	Seoane et al. (2019)
	Polyvinyl acetate and acetyltributyl citrate	Nano-clay based on montmorillonite	El-Hadi (2014)
	PCL	Cellulose nanocrystals	Garcia-Garcia et al. (2018)
	Poly(3-hydroxyhexanoate-co-3-hydroxyoctanoate)	Cellulose nanofibers	Panaitescu et al. (2018)
	–	Cadmium sulfide nanoparticles	Riaz et al. (2020)
PHB nanofibers	PLA	Cellulose nanocrystals	Arrieta et al. (2016)
	Calcium alginate	Carbon nanotubes	Zhijiang, Cong, Ping et al. (2018a)
PHB/PLA blends	–	Cellulose nanofibrils	Aydemir and Gardner (2020)
	–	Cellulose nanocrystals	Frone et al. (2020)
	Epoxidized canola oil	Cellulose nanocrystals	Lopera-Valle et al. (2019)
PHB/ chitosan scaffold	–	Bioglass nanoparticles	Foroughi et al. (2017)
PHB scaffold	–	Carboxyl multi-walled carbon nanotubes	Zhijiang, Cong, Jie et al. (2018b)

of nanostructures in the matrix (Table 13.1). Nanoscale structures provide the interaction between layers and assist in the physical and thermal properties of materials (Anjum et al., 2020).

Seoane et al. (2018) produced a multilayer composite containing paperboard reinforced with PHB and cellulose nanocrystals. Paperboard has a low water vapor barrier property, which causes limitations in applications (Rovera et al., 2020). Therefore, in the study by Seoane et al. (2018), PHB was added to increase the mechanical properties and decrease water vapor permeability. According to the authors, the cellulose nanocrystals promoted interfacial interaction, guaranteeing the adhesion between the hydrophilic paperboard and the hydrophobic PHB layer. Therefore, the addition of nanocrystals also increased the mechanical and barrier properties of the composite coated with PHB. In this way, PHB is promising as a replacement for nonrenewable polymers to coat hydrophilic materials and, together with nanocrystals, provide several possibilities for applying the formed composite.

Composites of biopolymers and natural fibers have been investigated by researchers searching to produce biodegradable eco-materials (Aigbodion & Atuanya, 2016). Melo et al. (2012) produced a composite containing 10% fiber from the leaves of the carnauba palm tree, reinforced with PHB. The authors found a 45% increase in mechanical properties compared to the composite produced with untreated fibers. In the study by Moura et al. (2019), coconut fibers and PHB were used to produce the composite. In this case, the addition of 10% and 20% of fibers increased the elastic modulus and the material stiffness compared with the matrix composed only of PHB.

Another study that introduced natural fibers to the polymeric matrix to obtain biodegradable compost from renewable resources was developed by Battegazzore et al. (2018). The authors added cotton fibers to the matrices composed of PHB and poly(lactic acid) through an epoxy coupling agent. The composite showed greater adhesion between the fibers and the dies due to the epoxy. The highest Young's modulus, tensile strength, and strain at maximum stress were obtained with the PHB matrix. Therefore, the use of natural fibers improves the mechanical properties of composites and assists in the viability of waste recovery and environmental protection.

PHB is also used to form composites with conductive and thermal properties. One of the approaches to producing thermal material is the incorporation of inorganic compounds. Hybrid system biopolymers and inorganic compounds provide covalent bonds or physical interactions between inorganic filler and biopolymer matrix, which results in composites with thermal stability (Otsuka & Chujo, 2010; Silva et al., 2021).

Silva et al. (2021) developed a composite with PHB film and the addition of zirconium oxide or zirconium hydroxide. Regardless of the inorganic fraction, the composites showed an increase in thermal stability about the PHB film. In addition, the incorporation of zirconium hydroxide decreased the degree of crystallinity of the PHB. In this case, there was a strong interaction between the phases, resulting in smaller crystals and a more homogeneous system. In a study by El-hadi et al. (2018), composites formed of PHB, polyaniline, soot (carbon black), and plasticizer showed semiconductive properties. The composite presented an increase in the dielectric constant, a dielectric loss factor, and provided thermal stability in the PHB matrix. Therefore, PHB composites formed with inorganic materials promote an increase in thermal, physical, and conductive properties.

8. Innovative applications and products

The technological advance of biobased research is related to new product development and opportunities that meet environmental and economic requirements (Adeleye et al., 2020). PHB biopolymer has been explored in several applications such as biomedicine for bone tissue regeneration (Esposti et al., 2019), food packaging with antimicrobial action (Kuntzler et al., 2018), agriculture in the development of fertilizers (Arrieta et al., 2020), environmental sector for wastewater treatment (Heitmann et al., 2016), and biorefinery systems (Prieto et al., 2017).

Tissue engineering is one of the biomedical applications that have the capacity for cell regeneration through scaffolds. PHB has potential use in the scaffold coating that stimulates bone growth due to biocompatibility with osteoblasts, epithelial cells, and chondrocytes (Holmes, 1985; Hutmacher, 2000). Parvizifard and Karbasi (2020) evaluated the application of the PHB nanocomposite, chitosan, and carbon nanotubes as a coating for filling the deficiencies in scaffolds composed of bioglass and titania. In this study, the scaffolds coated with the nanocomposite had a smaller contact angle (32.17°) with water. This response suggests the possibility of improving cell adhesion, migration, and spreading. On the seventh day of the cell culture test, coated scaffolds showed a cell-covered surface and increased cell viability compared to uncoated scaffolds.

A matrix for drug release in the human organism is another potential biomedical application of the PHB biopolymer. Aguilar-Rabiela et al. (2020) produced PHB microparticles loaded with 5%, 10%, and 15% curcumin. The microparticles were porous and had an encapsulation efficiency of approximately 98%. The results of curcumin release showed biphasic kinetic behavior considering all drug concentrations. The initial burst release occurred during the first 7 h. After 10 h, the performance was slow and gradual, resulting in stabilization of the curcumin concentration.

The quest to reduce the use of conventional petroleum-based polymers by the packaging industry makes biopolymers promising to minimize adverse environmental impacts. In this context, Manikandan et al. (2020a) developed PHB nanocomposites and graphene nanoplatforms (NPsGr). This study investigated the barrier properties, estimated food shelf-life (chips and dairy products), and biodegradability. The addition of 0.7% NPsGr reduced the oxygen and water vapor permeability of the PHB film. Moreover, the addition of 0.7% NPsGr increased the shelf-life to 245 and 26 days for chips and dairy products, respectively. The nanocomposite biodegradation analysis indicated that, after 30 days, all samples were 100% degraded.

Rech et al. (2020) produced antimicrobial and biodegradable packaging with eugenol in the PHB matrix. The films containing 40% of the bioactive compound showed action against *Salmonella* sp., *Staphylococcus aureus*, *Escherichia coli*, and *Aspergillus niger*. Biodegradation of the PHB films was evaluated in different types of soils (sandy, agricultural, and landfill). PHB films with eugenol exposed in agricultural soil presented a faster degradation rate than other types of soil. The degradation occurs due to the diversity of populations of bacteria and fungi, acidity, and phosphorus availability.

Biopolymers can also be used in agriculture to coat fertilizers and agrochemicals. These biomaterials have low toxicity, high degradability, and controlled release capacity of the active compound (Grillo et al., 2012; Volova et al., 2016). Thus, Souza et al. (2019) developed a composite of PHB and sodium montmorillonite clay to release the nutrients KNO_3 and NPK (NH_4^+, P_2O_5, K_2O). This study found that the addition of clay controlled the release of cationic nutrients while accelerating the release of anionics.

The use of composites for the release of fertilizer seeks to protect nutrients from adverse environmental conditions. Moreover, it increases the useful life and ensures the promoters of plant health (Azeem et al., 2014; Ghormade et al., 2011).

Souza et al. (2018) also evaluated the release of nutrients KNO_3 and NPK using a composite of PHB, starch, glycerol, and sodium montmorillonite clay. The authors produced each composite containing the KNO_3 and NPK fertilizers individually. Thus, the results showed that KNO_3 was completely released after 60 h, with 60% NPK being released.

Industrial waste fluids containing organic pollutants are harmful to the ecosystem, causing environmental pollution and having adverse effects on human health (Mekonnen et al., 2016). Therefore, researchers have studied various polymeric supports to catalyze pollutants, as they have high stability in the aqueous environment, are low cost, and inert (Yew et al., 2006). A PHB film was produced by Heitmann et al. (2020) as a catalyst via photoactivation to remove methylene blue dye from aqueous solutions. The authors found that the material can be used for effluent treatment of up to five cycles with complete degradation of pollutants.

Innovations in the production process are necessary to reduce economic and environmental impacts. In this sense, Manikandan et al. (2020b) proposed a new biorefinery model for PHB production. In this system, lignin from carob pods was used as a self-sufficient feedstock for upstream to downstream in the PHB production process. In addition, lignin acted as a precursor to cell disruption for the extraction of PHB produced by bacterial fermentation. The extracted biopolymer was tested as a film and showed good mechanical properties and antibacterial activity, demonstrating suitability in food packaging applications. Thus, the biorefinery model proposed by this study was appropriate for the sustainable production of PHB.

9. Conclusions

Research on new components and ecofriendly products to reduce impacts on the ecosystem is expanding to minimize dependence on compounds derived from fossil fuels. For this reason, it stands out in the use of biodegradable polymers such as PHB for the development of blends and composites, as it has properties similar to conventional polymers. PHB is a biopolymer, which can be biologically synthesized by bacteria, cyanobacteria, and/or microalgae. The synthesis of biopolymers by microalgae is considered a sustainable, economic, and renewable alternative due to the high photosynthetic efficiency of these microorganisms and their ability to use industrial wastes (liquid, solid, and gaseous) as a carbon source for cultivation. PHB-based composites and blends are innovative products with the ability to combine qualities of individual components, resulting in materials with improved properties for application in various industrial sectors. In this way, advances in research can sustainably create new products, which positively affect the thermal, surface, and mechanical properties of materials composed of PHB, as well as in reducing environmental impacts.

Acknowledgment

This research was developed within the scope of the Capes-PrInt Program (Process # 88887.310848/2018-00).

References

Adeleye, A. T., Odoh, C. K., Enudi, O. C., Banjoko, O. O., Osiboye, O. O., Odediran, E. T., & Louis, H. (2020). Sustainable synthesis and applications of polyhydroxyalkanoates (PHAs) from biomass. *Process Biochemistry, 96*, 174–193. https://doi.org/10.1016/j.procbio.2020.05.032

Aguilar-Rabiela, A. E., Hernández-Cooper, E. M., Otero, J. A., & Vergara-Porras, B. (2020). Modeling the release of curcumin from microparticles of poly(hydroxybutyrate) [PHB]. *International Journal of Biological Macromolecules, 144*, 47–52. https://doi.org/10.1016/j.ijbiomac.2019.11.242

Aigbodion, V. S., & Atuanya, C. U. (2016). Improving the properties of epoxy/melon shell biocomposites: effect weight percentage and form of melon shell particles. *Polymer Bulletin, 73*, 3305. https://doi.org/10.1007/s00289-016-1657-8

Aldor, I. S., & Keasling, J. D. (2003). Process design for microbial plastic factories: metabolic engineering of polyhydroxyalkanoates. *Current Opinion in Biotechnology, 14*, 475–483. https://doi.org/10.1016/j.copbio.2003.09.002

Angelini, S., Cerruti, P., Immirzi, B., Santagata, G., Scarinzi, G., & Malinconico, M. (2014). From biowaste to bioresource: effect of a lignocellulosic filler on the properties of poly (3-hydroxybutyrate). *International Journal of Biological Macromolecules, 71*, 163–173. https://doi.org/10.1016/j.ijbiomac.2014.07.038

Angelini, S., Cerruti, P., Immirzi, B., Scarinzi, G., & Malinconico, M. (2016). Acid-insoluble lignin and holocellulose from a lignocellulosic biowaste: bio-fillers in poly (3-hydroxybutyrate). *European Polymer Journal, 76*, 63–76. https://doi.org/10.1016/j.eurpolymj.2016.01.024

Anjum, A., Zuber, M., Zia, K. M., Noreen, A., Anjum, M. N., & Tabasum, S. (2016). Microbial production of polyhydroxyalkanoates (PHAs) and its copolymers: a review of recent advancements. *International Journal of Biological Macromolecules, 89*, 161–174. https://doi.org/10.1016/j.ijbiomac.2016.04.069

Anjum, M. N., Malik, S. A., Bilal, C. H., Rashid, U., Nasif, M., & Zia, K. M. (2020). Polyhydroxyalkanoates-based bionanocomposites. In *Bionanocomposites* (pp. 321–333). Elsevier. https://doi.org/10.1016/B978-0-12-816751-9.00013-1

Arrieta, M. P., López, J., López, D., Kenny, J. M., & Peponi, L. (2016). Biodegradable electrospun bionanocomposite fibers based on plasticized PLA–PHB blends reinforced with cellulose nanocrystals. *Industrial Crops and Products, 93*, 290–301. https://doi.org/10.1016/j.indcrop.2015.12.058

Arrieta, M. P., Perdiguero, M., Fiori, S., Kenny, J. M., & Peponi, L. (2020). Biodegradable electrospun PLA-PHB fibers plasticized with oligomeric lactic acid. *Polymer Degradation and Stability, 179*, 109226. https://doi.org/10.1016/j.polymdegradstab.2020.109226

Aydemir, D., & Gardner, D. J. (2020). Biopolymer blends of polyhydroxybutyrate and polylactic acid reinforced with cellulose nanofibrils. *Carbohydrate Polymers, 250*, 116867. https://doi.org/10.1016/j.carbpol.2020.116867

Azeem, B., Kushaari, K., Man, Z. B., Basit, A., & Thanh, T. H. (2014). Review on materials & methods to produce controlled release coated urea fertilizer. *Journal of Controlled Release, 181*, 11–21. https://doi.org/10.1016/j.jconrel.2014.02.020

Balaji, S., Gopi, K., & Muthuvelan, B. (2013). A review on production of poly β hydroxybutyrates from cyanobacteria for the production of bio plastics. *Algal Research, 2*(3), 278–285. https://doi.org/10.1016/j.algal.2013.03.002

Bartczak, Z., & Grala, M. (2016). Toughening of semicrystalline and amorphous polylactide with atactic poly(hydroxy butyrate). *Polymer-Plastics Technology and Engineering, 56*, 29–43. https://doi.org/10.1080/03602559.2016.1185664

Battegazzore, D., Frache, A., Abt, T., & Maspoch, M. L. (2018). Epoxy coupling agent for PLA and PHB copolymer-based cotton fabric bio-composites. *Composites Part B: Engineering, 148*, 188–197. https://doi.org/10.1016/j.compositesb.2018.04.055

Bhati, R., & Mallick, N. (2012). Production and characterization of poly (3-hydroxybutyrate-co-3-hydroxyvalerate) co-polymer by a N_2-fixing cyanobacterium, *Nostoc muscorum* agardh. *Journal of Chemical Technology and Biotechnology, 87*, 505–512. https://doi.org/10.1002/jctb.2737

Bugnicourt, E., Cinelli, P., Lazzeri, A., & Alvarez, V. (2014). Polyhydroxyalkanoate (PHA): review of synthesis, characteristics, processing and potential applications in packaging. *Express Polymer Letters, 8*, 791–808. https://doi.org/10.3144/expresspolymlett.2014.82

Carpine, R., Olivieri, G., Hellingwerf, K. J., Pollio, A., & Marzocchella, A. (2020). Industrial production of polyhydroxybutyrate from CO_2: can cyanobacteria meet this challenge? *Processes, 8*, 323. https://doi.org/10.3390/pr8030323

Carr, N. G. (1966). The occurrence of poly-β-hydroxybutyrate in the blue-green alga, *Chlorogloea fritschii. Biochimica et Biophysica Acta – Biophysics including Photosynthesis, 120*, 308–310. https://doi.org/10.1016/0926-6585(66)90353-0

Chen, G.-Q., & Jiang, X.-R. (2017). Engineering bacteria for enhanced polyhydroxyalkanoates (PHA) Biosynthesis. *Synthetic and Systems Biotechnology, 2*, 192–197. https://doi.org/10.1016/j.synbio.2017.09.001

Chen, G.-Q., Jiang, X.-R., & Guo, Y. (2016). Synthetic biology of microbes synthesizing polyhydroxyalkanoates (PHA). *Synthetic and Systems Biotechnology, 1*, 236–242. https://doi.org/10.1016/j.synbio.2016.09.006

Choi, S. Y., Rhie, M. N., Kim, H. T., Joo, J. C., Cho, I. J., Son, J., Jo, S. Y., Sohn, Y. J., Baritugo, K. A., Pyo, J., Lee, Y., Lee, S. Y., & Park, S. J. (2020). Metabolic engineering for the synthesis of polyesters: a 100-year journey from polyhydroxyalkanoates to non-natural microbial polyesters. *Metabolic Engineering, 58*, 47–81. https://doi.org/10.1016/j.ymben.2019.05.009

Coelho, V. C., Silva, C. K., Terra, A. L. M., Costa, J. A. V., & Morais, M. G. (2015). Polyhydroxybutyrate production by Spirulina sp. LEB 18 grown under different nutrient concentrations. *African Journal of Microbiology Research, 9*, 1586–1594. https://doi.org/10.5897/AJMR2015.7530

Costa, S. S., Miranda, A. L., Morais, M. G., Costa, J. A. V., & Druzian, J. I. (2019). Microalgae as source of polyhydroxyalkanoates (PHAs) — a review. *International Journal of Biological Macromolecules, 131*, 536–547. https://doi.org/10.1016/j.ijbiomac.2019.03.099

D'Amico, D. A., Iglesias Montes, M. L., Manfredi, L. B., & Cyras, V. P. (2016). Fully bio-based and biodegradable polylactic acid/poly(3-hydroxybutirate) blends: use of a common plasticizer as performance improvement strategy. *Polymer Testing, 49*, 22–28. https://doi.org/10.1016/j.polymertesting.2015.11.004

Don, T., Chung, C., Lai, S., & Chiu, H. (2010). Preparation and properties of blends from poly(3-hydroxybutyrate) with poly (vinyl acetate)-modified starch. *Polymer Engineering and Science, 50*, 709–718. https://doi.org/10.1002/pen.21575

Doyle, V., Pearson, R., Lee, D., Wolowacz, S., & Taggart, S. Mc (1996). An investigation of the growth of human dermal fibroblasts on poly-L-lactic acid in vitro. *Journal of Materials Science, 67*, 381–385. https://doi.org/10.1007/BF00154554

El-Hadi, A. M. (2014). Investigation of the effect of nano-clay type on the non-isothermal crystallization kinetics and morphology of poly(3(R)-hydroxybutyrate) PHB/clay nanocomposites. *Polymer Bulletin, 71*, 1449–1470. https://doi.org/10.1007/s00289-014-1135-0

El-hadi, A. M., Al-Jabri, F. Y., & Altaf, W. J. (2018). Higher dielectric properties of semi-conducting biopolymer composites of poly(3-hydroxy butyrate) (PHB) with polyaniline (PANI), carbon black, and plasticizer. *Polymer Bulletin, 75*, 1681−1699. https://doi.org/10.1007/s00289-017-2118-8

Esposti, M. D., Chiellini, F., Bondioli, F., Morselli, D., & Fabbri, P. (2019). Highly porous PHB-based bioactive scaffolds for bone tissue engineering by in situ synthesis of hydroxyapatite. *Materials Science and Engineering: C, 100*, 286−296. https://doi.org/10.1016/j.msec.2019.03.014

Foroughi, M. R., Karbasi, S., Khoroushi, M., & Khademi, A. A. (2017). Polyhydroxybutyrate/chitosan/bioglass nanocomposite as a novel electrospun scaffold: fabrication and characterization. *Journal of Porous Materials, 24*, 1447−1460. https://doi.org/10.1007/s10934-017-0385-2

Frone, A. N., Batalu, D., Chiulan, I., Oprea, M., Gabor, A. R., Nicolae, C. A., Raditoiu, V., Trusca, R., & Panaitescu, D. M. (2020). Morpho-structural, thermal and mechanical properties of PLA/PHB/cellulose biodegradable nanocomposites obtained by compression molding, extrusion, and 3D printing. *Nanomaterials, 10*, 1−18. https://doi.org/10.3390/nano10010051

Garcia-Garcia, D., Ferri, J. M., Boronat, T., Lopez-Martinez, J., & Balart, R. (2016). Processing and characterization of binary poly(hydroxybutyrate) (PHB) and poly(caprolactone) (PCL) blends with improved impact properties. *Polymer Bulletin, 73*, 3333−3350. https://doi.org/10.1007/s00289-016-1659-6

Garcia-Garcia, D., Lopez-Martinez, J., Balart, R., Strömberg, E., & Moriana, R. (2018). Reinforcing capability of cellulose nanocrystals obtained from pine cones in a biodegradable poly(3-hydroxybutyrate)/poly(ε-caprolactone) (PHB/PCL) thermoplastic blend. *European Polymer Journal, 104*, 10−18. https://doi.org/10.1016/j.eurpolymj.2018.04.036

Ghormade, V., Deshpande, M. V., & Paknikar, K. M. (2011). Perspectives for nano-biotechnology enabled protection and nutrition of plants. *Biotechnology Advances, 29*, 792−803. https://doi.org/10.1016/j.biotechadv.2011.06.007

Gonzalez, K., Navia, R., Liu, S., & Cea, M. (2020). Biological approaches in polyhydroxyalkanoates recovery. *Current Microbiology, 78*, 1−10. https://doi.org/10.1007/s00284-020-02263-1

Grillo, R., Santos, N. Z. P., Maruyama, C. R., Rosa, A. H., Lima, R., & Fraceto, L. F. (2012). Poly(εcaprolactone) nanocapsules as carrier systems for herbicides: physico-chemical characterization and genotoxicity evaluation. *Journal of Hazardous Materials*, 231−232. https://doi.org/10.1016/j.jhazmat.2012.06.019, 1−9.

Heitmann, A. P., Patrício, P. S. O., Coura, I. R., Pedroso, E. F., Souza, P. P., Mansur, H. S., Mansur, A., & Oliveira, L. C. A. (2016). Nanostructured niobium oxyhydroxide dispersed Poly(3-hydroxybutyrate) (PHB) films: highly efficient photocatalysts for degradation methylene blue dye. *Applied Catalysis B: Environmental, 189*, 141−150. https://doi.org/10.1016/j.apcatb.2016.02.031

Heitmann, A. P., Rocha, I. C., Souza, P. P., Oliveira, L. C. A., & Patrício, P. S. O. (2020). Photoactivation of a biodegradable polymer (PHB): generation of radicals for pollutants oxidation. *Catalysis Today, 344*, 171−175. https://doi.org/10.1016/j.cattod.2018.12.024

Hempel, F., Bozarth, A. S., Lindenkamp, N., Klingl, A., Zauner, S., Linne, U., Steinbüchel, A., & Maier, U. G. (2011). Microalgae as bioreactors for bioplastic production. *Microbial Cell Factories, 10*, 81. https://doi.org/10.1186/1475-2859-10-81

Hiremath, L., Kumar, S. N., Ravishankar, H. N., Angadi, S., & Sukanya, P. (2015). Design, screening and microbial synthesis of bio-polymers of poly-hydroxy-butyrate (PHB) from low cost carbon sources. *International Journal of Advanced Research, 3*(2), 420−425.

Holmes, P. A. (1985). Applications of PHB-a microbially produced biodegradable thermoplastic. *Physics in Technology, 16*(1), 32–36.

Hutmacher, D. W. (2000). Scaffolds in tissue engineering bone and cartilage. In *The biomaterials: silver jubilee compendium* (pp. 175–189). Woodhead Publishing Limited, Elsevier. https://doi.org/10.1016/s0142-9612(00)00121-6

Innocentini-Mei, L. H., Bartoli, J. R., & Baltieri, R. C. (2003). Mechanical and thermal properties of poly (3-hydroxybutyrate) blends with starch and starch derivatives. *Macromolecular Symposia, 197*, 77–88. https://doi.org/10.1002/masy.200350708

Isola, C., Sieverding, H. L., Raghunathan, R., Sibi, M. P., Webster, D. C., Sivaguru, J., & Stone, J. J. (2017). Life cycle assessment of photodegradable polymeric material derived from renewable bioresources. *Journal of Cleaner Production, 142*(4), 2935–2944. https://doi.org/10.1016/j.jclepro.2016.10.177

Jiang, X. R., & Chen, G. Q. (2016). Morphology engineering of bacteria for bio-production. *Biotechnology Advances, 34*, 435–440. https://doi.org/10.1016/j.biotechadv.2015.12.007

Jiang, X. R., Wang, H., Shen, R., & Chen, G. Q. (2015). Engineering the bacterial shapes for enhanced inclusion bodies accumulation. *Metabolic Engineering, 29*, 227–237. https://doi.org/10.1016/j.ymben.2015.03.017

Jiang, X., Ramsay, J. A., & Ramsay, B. A. (2006). Acetone extraction of mcl-PHA from *Pseudomonas putida* KT2440. *Journal of Microbiological Methods, 67*(2), 212–219. https://doi.org/10.1016/j.mimet.2006.03.015

Jin, E., & Melis, A. (2003). Microalgal biotechnology: carotenoid production by the green algae Dunaliella salina. *Biotechnology and Bioprocess Engineering, 8*, 331–337. https://doi.org/10.1007/BF02949276

Kamravamanesh, D., Kiesenhofer, D., Fluch, S., Lackner, M., & Herwig, C. (2019). Scale-up challenges and requirement of technology-transfer for cyanobacterial poly (3-hydroxybutyrate) production in industrial scale. *International Journal of Biobased Plastics, 1*(1), 60–71. https://doi.org/10.1080/24759651.2019.1688604

Katırcıoğlu, H., Aslim, B., Yüksekdað, Z. N., Mercan, N., & Beyatli, Y. (2003). Production of poly-beta-hydroxybutyrate (PHB) and differentiation of putative Bacillus mutant strains by SDS-PAGE of total cell protein. *African Journal of Biotechnology, 2*, 147–149. https://doi.org/10.5897/AJB2003.000-1029

Keshavarz, T., & Roy, I. (2010). Polyhydroxyalkanoates: bioplastics with a green agenda. *Current Opinion in Microbiology, 13*(3), 321–326. https://doi.org/10.1016/j.mib.2010.02.006

Koller, M. (2017). Advances in polyhydroxyalkanoate (PHA) production. *Bioengineering, 4*(4), 88. https://doi.org/10.3390/bioengineering4040088

Kuntzler, S. G., Almeida, A. C. A., Costa, J. A. V., & Morais, M. G. (2018). Polyhydroxybutyrate and phenolic compounds microalgae electrospun nanofibers: a novel nanomaterial with antibacterial activity. *International Journal of Biological Macromolecules, 113*, 1008–1014. https://doi.org/10.1016/j.ijbiomac.2018.03.002

Lemoigne, M. (1926). Produit de déshydratation et de polymérisation de l'acide b-oxybutyrique. *Bulletin de la Société de Chimie Biologique, 8*, 770–782.

Leong, Y. K., Lan, J. C.-W., Loh, H.-S., Ling, T. C., Ooi, C. W., & Show, P. L. (2017). Cloud-point extraction of green-polymers from *Cupriavidus necator* lysate using thermoseparating-based aqueous two-phase extraction. *Journal of Bioscience and Bioengineering, 123*(3), 370–375. https://doi.org/10.1016/j.jbiosc.2016.09.007

Levett, I., Birkett, G., Davies, N., Bell, A., Langford, A., Laycock, B., Lant, P., & Pratt, S. (2016). Techno-economic assessment of poly-3-hydroxybutyrate (PHB) production from methane – the case for thermophilic bioprocessing. *Journal of Environmental Chemical Engineering, 4*(4), 3724–3733. https://doi.org/10.1016/j.jece.2016.07.033

Li, M., & Wilkins, M. R. (2020). Recent advances in polyhydroxyalkanoate production: feedstocks, strains and process developments. *International Journal of Biological Macromolecules, 156*, 691−703. https://doi.org/10.1016/j.ijbiomac.2020.04.082

Lopera-Valle, A., Caputo, J. V., Leão, R., Sauvageau, D., Luz, S. M., & Elias, A. (2019). Influence of epoxidized canola oil (eCO) and cellulose nanocrystals (CNCs) on the mechanical and thermal properties of polyhydroxybutyrate (PHB)—poly(lactic acid) (PLA) blends. *Polymers, 11*(6), 1−18. https://doi.org/10.3390/polym11060933

Manikandan, N. A., Pakshirajan, K., & Pugazhenthi, G. (2020b). A closed-loop biorefinery approach for polyhydroxybutyrate (PHB) production using sugars from carob pods as the sole raw material and downstream processing using the co-product lignin. *Bioresource Technology, 307*, 123247. https://doi.org/10.1016/j.biortech.2020.123247

Manikandan, N. A., Pakshirajan, K., & Pugazhenthi, G. (2020a). Preparation and characterization of environmentally safe and highly biodegradable microbial polyhydroxybutyrate (PHB) based graphene nanocomposites for potential food packaging applications. *International Journal of Biological Macromolecules, 154*, 866−877. https://doi.org/10.1016/j.ijbiomac.2020.03.084

Mannina, G., Presti, D., Montiel-Jarillo, G., Carrera, J., & Suárez-Ojeda, M. E. (2020). Recovery of polyhydroxyalkanoates (PHAs) from wastewater: a review. *Bioresource Technology, 297*, 122478. https://doi.org/10.1016/j.biortech.2019.122478

Meereboer, K. W., Misra, M., & Mohanty, A. K. (2020). Review of recent advances in the biodegradability of polyhydroxyalkanoate (PHA) bioplastics and their composites. *Green Chemistry, 22*(17), 5519−5558. https://doi.org/10.1039/D0GC01647K

Meixner, K., Kovalcik, A., Sykacek, E., Gruber-Brunhumer, M., Zeilinger, W., Markl, K., Haas, C., Fritz, I., Mundigler, N., Stelzer, F., Neureiter, M., Fuchs, W., & Drosg, B. (2018). Cyanobacteria Biorefinery — production of poly(3-hydroxybutyrate) with *Synechocystis salina* and utilisation of residual biomass. *Journal of Biotechnology, 265*, 46−53. https://doi.org/10.1016/j.jbiotec.2017.10.020

Mekonnen, T., Mussone, P., & Bressler, D. (2016). Valorization of rendering industry wastes and co-products for industrial chemicals, materials and energy. *Critical Reviews in Biotechnology, 36*(1), 120−131. https://doi.org/10.3109/07388551.2014.928812

Melo, J. D. D., Carvalho, L. F. M., Medeiros, A. M., Souto, C. R. O., & Paskocimas, C. A. (2012). A biodegradable composite material based on polyhydroxybutyrate (PHB) and carnauba fibers. *Composites Part B: Engineering, 43*(7), 2827−2835. https://doi.org/10.1016/j.compositesb.2012.04.046

Miyake, M., Takase, K., Narato, M., Khatipov, E., Schnackenberg, J., Shirai, M., Kurane, R., & Asada, Y. (2000). Polyhydroxybutyrate production from carbon dioxide by cyanobacteria. *Applied Biochemistry and Biotechnology, 84*, 991−1002. https://doi.org/10.1385/ABAB:84-86:1-9:991

Moura, A. S., Demori, R., Leão, R. M., Frankenberg, C. L. C., & Santana, R. M. C. (2019). The influence of the coconut fiber treated as reinforcement in PHB (polyhydroxybutyrate) composites. *Materials Today Communications, 18*, 191−198. https://doi.org/10.1016/j.mtcomm.2018.12.006

Możejko-Ciesielska, J., & Kiewisz, R. (2016). Bacterial polyhydroxyalkanoates: still fabulous? *Microbiological Research, 192*, 271−282. https://doi.org/10.1016/j.micres.2016.07.010

Nishida, M., Yasuda, K., & Nishida, M. (2020). Correlative analysis between morphology and mechanical properties of poly-3-hydroxybutyrate (PHB) blended with polycarprolactone (PCL) using solid-state NMR. *Polymer Testing, 91*, 106780. https://doi.org/10.1016/j.polymertesting.2020.106780

Ol'Khov, A. A., Iordanskii, A. L., & Danko, T. P. (2015). Morphology of poly(3-hydroxybutyrate)-polyvinyl alcohol extrusion films. *Journal of Polymer Engineering, 35*, 765–771. https://doi.org/10.1515/polyeng-2014-0202

Ong, S. Y., Zainab-L, I., Pyary, S., & Sudesh, K. (2018). A novel biological recovery approach for PHA employing selective digestion of bacterial biomass in animals. *Applied Microbiology and Biotechnology, 102*(5), 2117–2127. https://doi.org/10.1007/s00253-018-8788-9

Otsuka, T., & Chujo, Y. (2010). Poly(methyl methacrylate) (PMMA)-based hybrid materials with reactive zirconium oxide nanocrystals. *Polymer Journal, 42*, 58–65. https://doi.org/10.1038/pj.2009.309

Pachekoski, W. M., Agnelli, J. A. M., & Belem, L. P. (2009). Thermal, mechanical and morphological properties of poly (hydroxybutyrate) and polypropylene blends after processing. *Materials Research, 12*, 159–164. https://doi.org/10.1590/S1516-14392009000200008

Panaitescu, D. M., Frone, A. N., Chiulan, I., Nicolae, C. A., Trusca, R., Ghiurea, M., Gabor, A. R., Mihailescu, M., Casarica, A., & Lupescu, I. (2018). Role of bacterial cellulose and poly (3-hydroxyhexanoate-co-3-hydroxyoctanoate) in poly (3-hydroxybutyrate) blends and composites. *Cellulose, 25*, 5569–5591. https://doi.org/10.1007/s10570-018-1980-3

Parvizifard, M., & Karbasi, S. (2020). Physical, mechanical and biological performance of PHB-Chitosan/MWCNTs nanocomposite coating deposited on bioglass based scaffold: potential application in bone tissue engineering. *International Journal of Biological Macromolecules, 152*, 645–662. https://doi.org/10.1016/j.ijbiomac.2020.02.266

Posada, J. A., Naranjo, J. M., López, J. A., Higuita, J. C., & Cardona, C. A. (2011). Design and analysis of poly-3-hydroxybutyrate production processes from crude glycerol. *Process Biochemistry, 46*(1), 310–317. https://doi.org/10.1016/j.procbio.2010.09.003

Price, S., Kuzhiumparambil, U., Pernice, M., & Ralph, P. J. (2020). Cyanobacterial polyhydroxybutyrate for sustainable bioplastic production: critical review and perspectives. *Journal of Environmental Chemical Engineering, 8*(4), 104007. https://doi.org/10.1016/j.jece.2020.104007

Prieto, C. V. G., Ramos, F. D., Estrada, V., Villar, M. A., & Diaz, M. S. (2017). Optimization of an integrated algae-based biorefinery for the production of biodiesel, astaxanthin and PHB. *Energy, 139*, 1159–1172. https://doi.org/10.1016/j.energy.2017.08.036

Rahman, A., & Miller, C. D. (2017). Microalgae as a source of bioplastics. In *Algal green chemistry* (pp. 121–138). https://doi.org/10.1016/B978-0-444-63784-0.00006-0

Rathore, P. (2014). Bioprospects of PHB: a review. In *International journal of emerging trends in science and technology* (pp. 529–532).

Ratledge, C., & Kristiansen, B. (2001). *Basic biotechnology*. Cambridge: Cambridge University Press (584 pp).

Ravve, A. (2012). *Principles of polymer chemistry*. New York, NY: Springer New York. https://doi.org/10.1007/978-1-4614-2212-9

Rech, C. R., Brabes, K. C. S., Silva, B. E. B., Bittencourt, P. R. S., Koschevic, M. T., Silveira, T. F. S., Martines, M. A. U., Caon, T., & Martelli, S. M. (2020). Biodegradation of polyhydroxybutyrate films incorporated with eugenol in different soil types. *Case Studies in Chemical and Environmental Engineering, 2*, 100014. https://doi.org/10.1016/j.cscee.2020.100014

Reichardt, R., & Rieger, B. (2011). Poly(3-hydroxybutyrate) from carbon monoxide. *Advances in Polymer Science, 245*, 49–90. https://doi.org/10.1007/12_2011_127

Riaz, S., Raza, Z. A., & Majeed, M. I. (2020). Preparation of cadmium sulfide nanoparticles and mediation thereof across poly(hydroxybutyrate) nanocomposite. *Polymer Bulletin, 77*, 775–791. https://doi.org/10.1007/s00289-019-02775-2

Rippka, R., Neilson, A., Kunisawa, R., & Cohen-Bazire, G. (1971). Nitrogen fixation by unicellular blue-green algae. *Archiv für Mikrobiologie, 76*, 341–348. https://doi.org/10.1007/BF00408530

Rodgers, M., & Wu, G. (2010). Production of polyhydroxybutyrate by activated sludge performing enhanced biological phosphorus removal. *Bioresource Technology, 101*, 1049–1053. https://doi.org/10.1016/j.biortech.2009.08.107

Rodriguez-Perez, S., Serrano, A., Pantión, A. A., & Alonso-Fariñas, B. (2018). Challenges of scaling-up PHA production from waste streams. A review. *Journal of Environmental Management, 205*, 215–230. https://doi.org/10.1016/j.jenvman.2017.09.083

Rodríguez-Tobías, H., Morales, G., & Grande, D. (2019). Comprehensive review on electrospinning techniques as versatile approaches toward antimicrobial biopolymeric composite fibers. *Materials Science and Engineering: C, 101*, 306–322. https://doi.org/10.1016/j.msec.2019.03.099

Rovera, C., Türe, H., Hedenqvist, M. S., & Farris, S. (2020). Water vapor barrier properties of wheat gluten/silica hybrid coatings on paperboard for food packaging applications. *Food Packaging and Shelf Life, 26*, 100561. https://doi.org/10.1016/j.fpsl.2020.100561

Samantaray, S., & Mallick, N. (2012). Production and characterization of poly-β-hydroxybutyrate (PHB) polymer from Aulosira fertilissima. *Journal of Applied Phycology, 24*, 803–814. https://doi.org/10.1007/s10811-011-9699-7

Samantaray, S., & Mallick, N. (2014). Production of poly(3-hydroxybutyrate-co-3-hydroxyvalerate) co-polymer by the diazotrophic cyanobacterium *Aulosira fertilissima* CCC 444. *Journal of Applied Phycology, 26*, 237–245. https://doi.org/10.1007/s10811-013-0073-9

Samorì, C., Abbondanzi, F., Galletti, P., Giorgini, L., Mazzocchetti, L., Torri, C., & Tagliavini, E. (2015). Extraction of polyhydroxyalkanoates from mixed microbial cultures: impact on polymer quality and recovery. *Bioresource Technology, 189*, 195–202. https://doi.org/10.1016/j.biortech.2015.03.062

Santos, A. J., Valentina, L. V. O. D., Schulz, A. A. H., & Duarte, M. A. T. (2017). From obtaining to degradation of PHB: material properties. Part I. *Ingeniería y Ciencia, 13*, 269–298. https://doi.org/10.17230/ingciencia.13.26.10

Seoane, I. T., Cerrutti, P., Vazquez, A., Cyras, V. P., & Manfredi, L. B. (2019). Ternary nanocomposites based on plasticized poly(3-hydroxybutyrate) and nanocellulose. *Polymer Bulletin, 76*, 967–988. https://doi.org/10.1007/s00289-018-2421-z

Seoane, I. T., Manfredi, L. B., & Cyras, V. P. (2018). Bilayer biocomposites based on coated cellulose paperboard with films of polyhydroxybutyrate/cellulose nanocrystals. *Cellulose, 25*, 2419–2434. https://doi.org/10.1007/s10570-018-1729-z

Silva, C. K., Costa, J. A. V., & Morais, M. G. (2018). Polyhydroxybutyrate (PHB) synthesis by Spirulina sp. LEB 18 using biopolymer extraction waste. *Applied Biochemistry and Biotechnology, 185*(3). https://doi.org/10.1007/s12010-017-2687-x

Silva, D. C. P., Menezes, L. R., Silva, P. S. R. C., & Tavares, M. I. B. (2021). Evaluation of thermal properties of zirconium–PHB composites. *Journal of Thermal Analysis and Calorimetry, 143*, 165–172. https://doi.org/10.1007/s10973-019-09106-7

Singh, A. K., & Mallick, N. (2017). Advances in cyanobacterial polyhydroxyalkanoates production. *FEMS Microbiology Letters, 364*(20). https://doi.org/10.1093/femsle/fnx189

Singh, A. K., Sharma, L., Mallick, N., & Mala, J. (2017). Progress and challenges in producing polyhydroxyalkanoate biopolymers from cyanobacteria. *Journal of Applied Phycology, 29*(3), 1213–1232. https://doi.org/10.1007/s10811-016-1006-1

Sirohi, R., Pandey, J. P., Gaur, V. K., Gnansounou, E., & Sindhu, R. (2020b). Critical overview of biomass feedstocks as sustainable substrates for the production of polyhydroxybutyrate (PHB). *Bioresource Technology, 311*, 123536. https://doi.org/10.1016/j.biortech.2020.123536

Sirohi, R., Pandey, J. P., Tarafdar, A., Sindhu, R., Parameswaran, B., & Pandey, A. (2020a). Applications of poly-3-hydroxybutyrate based composite in advanced applications of polysaccharides and their composites. *Materials Research Foundations, 68*, 45–59. https://doi.org/10.21741/9781644900659-2

Souza, J. L., Campos, A., França, D., & Faez, R. (2019). PHB and montmorillonite clay composites as KNO_3 and NPK support for a controlled release. *Journal of Polymers and the Environment, 27*, 2089–2097. https://doi.org/10.1007/s10924-019-01498-9

Souza, J. L., Chiaregato, C. G., & Faez, R. (2018). Green composite based on PHB and montmorillonite for KNO_3 and NPK delivery system. *Journal of Polymers and the Environment, 26*, 670–679. https://doi.org/10.1007/s10924-017-0979-4

Steinbüchel, A., & Lütke-Eversloh, T. (2003). Metabolic engineering and pathway construction for biotechnological production of relevant polyhydroxyalkanoates in microorganisms. *Biochemical Engineering Journal, 16*(2), 81–96. https://doi.org/10.1016/S1369-703X(03)00036-6

Steinbuchel, A., Madkour, M. H., Heinrich, D., Alghamdi, A. M., & Shabbaj, I. I. (2013). PHA recovery from biomass. *Biomacromolecules, 14*(9), 2963–2972. https://doi.org/10.1021/bm4010244

Sudesh, K., Abe, H., & Doi, Y. (2000). Synthesis, structure and properties of polyhydroxyalkanoates: biological polyesters. *Progress in Polymer Science, 25*, 1503–1555. https://doi.org/10.1016/S0079-6700(00)00035-6

Sudesh, K., Taguchi, K., & Doi, Y. (2002). Effect of increased PHA synthase activity on polyhydroxyalkanoates biosynthesis in *Synechocystis* sp. PCC 6803. *International Journal of Biological Macromolecules, 30*, 97–104. https://doi.org/10.1016/s0141-8130(02)00010-7

Tang, X., & Chen, E. Y. X. (2018). Chemical synthesis of perfectly isotactic and high melting bacterial poly(3-hydroxybutyrate) from bio-sourced racemic cyclic diolide. *Nature Communications, 9*(1), 1–11. https://doi.org/10.1038/s41467-018-04734-3

Technavio. (2020). *Polyhydroxyalkanoate market, size, growth, trends*. Industry Analysis & Forecast, 120 pages.

Vagin, S., Winnacker, M., Kronast, A., Altenbuchner, P. T., Deglmann, P., Sinkel, C., Loos, R., & Rieger, B. (2015). New insights into the ring-opening polymerization of β-butyrolactone catalyzed by chromium(III) salphen complexes. *ChemCatChem, 7*(23), 3963–3971. https://doi.org/10.1002/cctc.201500717

Verlinden, R. A. J., Hill, D. J., Kenward, M. A., Williams, C. D., & Radecka, I. (2007). Bacterial synthesis of biodegradable polyhydroxyalkanoates. *Journal of Applied Microbiology, 102*, 1437–1449. https://doi.org/10.1111/j.1365-2672.2007.03335.x

Volova, T. G., Prudnikova, S. V., & Boyandin, A. N. (2016). Biodegradable poly-3-hydroxybutyrate as a fertiliser carrier. *Journal of the Science of Food and Agriculture, 96*, 4183–4193. https://doi.org/10.1002/jsfa.7621

Wang, B., Pugh, S., Nielsen, D. R., Zhang, W., & Meldrum, D. R. (2013). Engineering cyanobacteria for photosynthetic production of 3-hydroxybutyrate directly from CO_2. *Metabolic Engineering, 16*, 68–77. https://doi.org/10.1016/j.ymben.2013.01.001

Wang, Y., Wu, H., Jiang, X., & Chen, G. Q. (2014). Engineering *Escherichia coli* for enhanced production of poly(3-hydroxybutyrate-co-4-hydroxybutyrate) in larger cellular space. *Metabolic Engineering, 25*, 183–193. https://doi.org/10.1016/j.ymben.2014.07.010

Winnacker, M. (2019). Polyhydroxyalkanoates: recent advances in their synthesis and applications. *European Journal of Lipid Science and Technology, 121*(11), 1–9. https://doi.org/10.1002/ejlt.201900101

Wu, G., Boa, T., Shen, Z., & Wu, Q. (2002). Sodium acetate stimulates PHB biosynthesis in *Synechocystis* sp. PCC, 6803. *Tsinghua Science and Technology, 7*, 435–438.

Wu, H., Fan, Z., Jiang, X., Chen, J., & Chen, G. Q. (2016). Enhanced production of polyhydroxybutyrate by multiple dividing *E. coli*. *Microbial Cell Factories, 15*, 128. https://doi.org/10.1186/s12934-016-0531-6

Yan, Q., Sun, Y., Ruan, L. F., Chen, J., & Yu, P. H. F. (2005). Biosynthesis of short-chain-length-polyhydroxyalkanoates during the dual-nutrient-limited zone by *Ralstonia eutropha*. *World Journal of Microbiology and Biotechnology, 21*, 17–21. https://doi.org/10.1007/s11274-004-0877-5

Yeo, J. C. C., Muiruri, J. K., Thitsartarn, W., Li, Z., & He, C. (2018). Recent advances in the development of biodegradable PHB-based toughening materials: approaches, advantages and applications. *Materials Science and Engineering: C, 92*, 1092–1116. https://doi.org/10.1016/j.msec.2017.11.006

Yew, S. P., Tang, H. Y., & Sudesh, K. (2006). Photocatalytic activity and biodegradation of polyhydroxybutyrate films containing titanium dioxide. *Polymer Degradation and Stability, 91*(8), 1800–1807. https://doi.org/10.1016/j.polymdegradstab.2005.11.011

Yu, J., & Stahl, H. (2008). Microbial utilization and biopolyester synthesis of bagasse hydrolysates. *Bioresource Technology, 99*, 8042–8048. https://doi.org/10.1016/j.biortech.2008.03.071

Zhang, M., & Thomas, N. L. (2010). Preparation and properties of polyhydroxybutyrate blended with different types of starch. *Journal of Applied Polymer Science, 116*, 688–694. https://doi.org/10.1002/app.30991

Zhijiang, C., Cong, Z., Jie, G., Qing, Z., & Kongyin, Z. (2018b). Electrospun carboxyl multi-walled carbon nanotubes grafted polyhydroxybutyrate composite nanofibers membrane scaffolds: preparation, characterization and cytocompatibility. *Materials Science and Engineering: C, 82*, 29–40. https://doi.org/10.1016/j.msec.2017.08.005

Zhijiang, C., Cong, Z., Ping, X., Jie, G., & Kongyin, Z. (2018a). Calcium alginate-coated electrospun polyhydroxybutyrate/carbon nanotubes composite nanofibers as nanofiltration membrane for dye removal. *Journal of Materials Science, 53*, 14801–14820. https://doi.org/10.1007/s10853-018-2607-7

Active and intelligent biodegradable films and polymers

Haniyeh Rostamzad
Fisheries Department, Faculty of Natural Resources, University of Guilan, Sowmeh Sara, Guilan, Iran

1. Introduction

Today, an adequate food supply has become one of the main concerns of researchers, so that to address this issue, many efforts in various fields, including increasing production, increasing productivity at the time of harvesting, improving storage methods, and developing methods to protect food from destructive agents such as fungi and bacteria. The use of suitable polymers and the packaging process are of special importance to protect foods from external harmful agents and ultimately ensure their long-term storage. A polymer is a substance included in large molecules made up of small repeating units called monomers. The number of repeating units in a large molecule is the degree of polymerization. Polymers that consist of only one type of repeating unit are called homopolymers and those that consist of two repeating units are called copolymers. The term terpolymer is sometimes used for products resulting from the polymerization of three monomers. Products made up of more than three monomers are called heteropolymers. The word "polymer" is derived from the two Greek words "poly" meaning many and "mer" meaning part, torn, or piece. Polymers are divided into two categories, natural polymers and synthetic polymers, and are classified in different ways. Polymers are divided into two categories in terms of heat resistance: thermoplastics and thermostats. Thermoplastics are polymers that melt when heated and solidify when cooled. Thermostats, on the other hand, are polymers that do not melt when heated but decompose irreversibly at very high temperatures. These polymers have viscoelastic properties and the source of this phenomenon is the bonding of chains.

2. Active packaging

Active packaging is a type of packaging that, in addition to having the main inhibitory properties of conventional packaging (such as inhibitory properties against gas and water vapor and mechanical stresses), improves the safety, shelf-life, or sensory properties of the food by changing packaging conditions while at the same time, the quality

Biodegradable Polymers, Blends and Composites. https://doi.org/10.1016/B978-0-12-823791-5.00023-5

of the food is maintained (Domínguez et al., 2018; Khumkomgool et al., 2020; Mousavi Khaneghah et al., 2018). Many common methods of food processing and storage are not particularly suitable for fresh foods (such as fresh meat) and, in addition to reducing the nutritional value of the product, they can have adverse effects on its final quality. The use of active packaging is a new method to store this type of food and, in recent years, extensive research has been carried out to make it economical and practical. Active packaging technology includes interactions between the food, packaging material (or coating), and the gaseous atmosphere inside the package, which must be able to increase the shelf-life of the product while maintaining its quality and safety. Active packaging can have many roles that are not present in common packaging. These roles include: (1) antimicrobial activity, (2) removing oxygen, moisture, and/or ethylene (scavenger feature), (3) releasing flavor enhancers or ethanol, and (4) immersing the product in an antimicrobial solution or spraying an antimicrobial solution on the product (Carina et al., 2021; Domínguez et al., 2018). Active packaging is a type of packaging that is the result of the interaction between the components of the packaging with its contents (such as food) or the atmosphere inside the packaging, and which maintains the quality and freshness of the product. In active packaging, the release of antimicrobials from the polymer matrix to the surface of the food is done slowly, over a long period of time, resulting in a high concentration of antimicrobials on the surface of the product for an extended period of time. Antimicrobials increase the shelf-life of food products by slowing the growth rate and prolonging the delayed phase of microorganisms or inactivating and destroying microbes (Mousavi Khaneghah et al., 2018; Varghese et al., 2020). The history of the use of oral antimicrobial films and coatings and synthetic films containing antimicrobial compounds goes back a long way, when materials containing chemical preservatives and organic acids were used as sausage coatings. These natural wrappers were mostly tubes of edible protein stuffed with sausage paste. Various methods such as salting, smoking, and acid treatments were used to improve the shelf-life of these natural wrappers. In addition to covering the sausages, the meat pieces and carcasses were coated with gels containing antimicrobial compounds to prevent the growth of spoilage bacteria and pathogens. However, most of these coatings were made from nonprotein gums such as calcium alginate. Antimicrobial compounds added to calcium alginate coatings include nisin and lactic/acetic acid. In the following, the types of antimicrobial substances that can be used in the production of these films are mentioned. Many factors must be considered in the design of an antimicrobial or antioxidant package. Most of these factors depend on the chemical nature of the active ingredient used in the packaging, the type of food, and the target microorganism, as discussed below. Research into films and coatings containing antimicrobial compounds began in 1980, with a focus on the use of sorbic acid and potassium sorbate, with films and coatings of carbohydrates and fats as sorbate-containing films also being studied. These coatings included methylcellulose, hydroxypropyl methylcellulose, starch, fatty acids, and carnauba wax. Since 1990, various antimicrobial films of chitosan and organic acids have been produced.

3. Active ingredients used in active packaging

In nonfood packaging systems, many kinds of preservatives such as organic acids and their salts, fungicides, bacteriocins, antibiotics, enzymes, alcohols, thiols, antioxidants, metals, and disinfectant gases can be used. However, in the case of food films and coatings, one of the most important parameters of the modern food packaging industry is choosing the type of antimicrobial agent, biodegradability, and antimicrobial activity of that substance. The features of an antimicrobial compound used in active packaging should include: approved by regulatory agencies and allowed to come into contact with food, an affordable price, affects a wide range of microorganisms, is effective against microorganisms in low concentrations, and does not have a negative effect on the sensory properties of food.

Active materials that enter the polymer directly. These can be classified into the groups described below.

3.1 *Organic acids and their salts*

These acids include propionic acid, benzoic acid, sorbic acid, acetic acid, lactic acid, and malic acid. Most of these acids are present in plants or fermented products. However, they are mainly chemically synthesized and placed in the category of chemical preservatives.

3.2 *Enzymes*

Lysozyme and glucose oxidase are the most common enzymes used in research related to antimicrobial packaging. Lysozyme is an enzyme that breaks down the glycosidic bonds in the cell wall of Gram-positive bacteria. Therefore, it can potentially affect a wide range of Gram-positive bacteria, including lactobacilli, micrococci, and bacilli. However, the presence of a lipid outer membrane on the cell wall peptidoglycans of Gram-negative bacteria limits the antimicrobial activity of this enzyme against Gram-negative bacteria. Lysozyme can be isolated from a variety of sources such as mammalian tears, milk, serum, and poultry eggs; however, chicken egg lysozyme is the most important and major source that is used for food applications. In addition to the two enzymes mentioned, it is possible to use other enzymes, such as lactoperoxidase and lactoferrin, which should be further studied.

3.3 *Bacteriocins*

These are compounds that are produced by microbes during fermentation and are mostly peptides. The most common bacteriocins used in active packaging are nisin, pyocyanin, and lactin.

3.4 Antioxidants

There are substances that slow down the oxidation process in materials. Various antioxidants have been used in food packaging and have had a beneficial effect on increasing the shelf-life of foods, especially oxidation-sensitive foods (such as seafood). Numerous studies have been performed on the shelf-life of food due through the use of different antioxidants in packaging polymers. Rostamzad and colleagues conducted several studies on the effects of using thyme essential oil, ginger extract (Rostamzad et al., 2019), and licorice extract (Rostamzad and Eshagh, 2020) in biodegradable films. In addition, they examined the shelf-life of fish fillets packaged with films and concluded that the extracts used had a significant effect on the oxidation process of packaged fish fillets.

3.5 Inorganic gases

These include sulfur dioxide, chlorine dioxide, carbon dioxide, ozone, quinitol, and allyl isothiocyanates. These gases can penetrate the entire mass of the food and do not need to be in direct contact with the food. It is worth noting that the use of gases should be accompanied by thorough studies on their permeability and reactivity with the packaging film.

3.6 Metals

The most important metal with antimicrobial activity is silver. Copper also has antimicrobial and antiviral properties, but due to its toxicity, direct contact with food is not safe and it is also a strong catalyst for the oxidation of fats and vitamin C. Unlike copper ions, silver ions are not easily released and are not toxic with contact. Therefore, the type of additive and the properties of the polymer are important in this regard. In general, the quality of the film is often reduced by adding active ingredients to the film or coating constituents because the film or coating structure becomes more heterogeneous. Therefore, there must be compatibility between the active substance and the polymer in terms of polarity. In addition, the composition of the film or food coating should not affect the organoleptic properties of the food. In general, to minimize the effects of the film or coating on food taste, it is more appropriate to use tasteless films and food coatings.

4. Active films

The use of edible gelatin films in food packaging was first introduced in 1895. In 1930, high-melting-point paraffin was used to prevent reducting the moisture. In the late 1950s, an oil emulsion in water and carbonaceous wax was also used to coat fresh fruit and vegetables. The initial idea of using edible coatings was taken from natural coatings that are on the surface of fruits and vegetables. Bae et al. (2009) studied the effect of transglutaminase on the properties of composite films made of gelatin and nanoclay.

They concluded that the addition of microbial transglutaminase increased the molecular weight of the fish gelatin solution. The tensile strength of the films decreased, while the percentage of resistance increased until they were torn (Bae et al., 2009). Piers et al. (2011) used different concentrations of thyme oil as antioxidants in a film made from fish protein powder. The results showed that the addition of thyme oil reduces the film thickness and permeability to vapor in the produced films. According to that study, a concentration of 0.25% thyme oil had better antioxidant properties than the control film. Rostamzad et al. (2016) evaluated a biofilm from fish protein with nanoclay and transglutaminase in increasing the shelf-life of fresh rainbow trout fillets and concluded that the application of nanotechnology enhances the mechanical and physical properties and thermal stability of the fish protein film (Rostamzad et al., 2016). Song et al. (2012) investigated the composite film properties of gelatin and barley crust protein in combination with grapefruit kernel extract and its effect on salmon preservation. In that study, the tensile strength and tear resistance of films decreased with increasing barley crust protein concentration. Although with the addition of gelatin, the tensile strength of the films increased, the tensile strength of the films decreased (Song et al.,2012). Piers et al. investigated the physical, chemical, antioxidant, and antibacterial properties of films prepared with vegetable oils (spikenard, coriander, tarragon, and thyme). In that study, the addition of oils led to reduced water vapor permeability and increased solubility of the films in water (Pires et al., 2013). Rostamzad et al. (2016) produced and evaluated an active nanocomposite based on fish nanoparticle myofibril protein in combination with thyme extract. In the mentioned study, the shelf-life and quality of fish fillets packed with films were investigated and it was concluded that by adding nanoclay alone or in combination with thyme extract, the shelf-life of fish fillets increased. Rostamzad et al. examined the effect of using chitosan coating and film in combination with ginger extract (Rostamzad et al., 2019) and licorice extract (Rostamzad and Eshagh, 2020) and concluded that the extracts of these plants significantly increased the shelf-life of preserved fish fillets in a refrigerator and the packaged fillets were edible for up to 12 days. However, the fillets packed with chitosan film alone were of good quality for only 8 days. Therefore, the use of the mentioned active films is recommended to maintain the quality and increase the shelf-life of fish fillets and other meat foods. Lessani, Rostamzad, & Zakipour Rahimabadi (2020) developed and evaluated an active nanocomposite based on carrageenan–nanoclay in combination with a microencapsulation clove extract with maltodextrin. The results of the mentioned research indicated that the addition of nanoclay up to 3% resulted in better functional properties in the film. Also, the addition of a microencapsulation clove extract improved the biofilm properties.

5. Intelligent packaging

Intelligent packaging is considered to be the next generation of packaging systems. Intelligent packaging is a type of packaging in the food industry that shows the condition of packaged food or the environment around food. In a broader sense, intelligent food packaging involves communication of the packaging system by monitoring changes in the internal and external environments and establishing the conditions of packaged food. Intelligent food packaging does not act directly to extend the life of the food, but provides information on food quality to stakeholders in food supply chains. For example, an intelligent packaging system can show the consumer whether the food is fresh or expired, thereby informing the consumer or producer, and by providing quality indicators, provides information about food storage status, packaging, or the indoor and outdoor environment. These packaging systems are able to perform intelligent functions such as detection, recording, monitoring, and communication to facilitate decision-making, and alert to potential problems. This type of packaging analyzes the system, and processes and presents information. Intelligent packaging is also used to protect the product against the adverse effects of external environmental conditions such as heat, light, the presence or absence of moisture, pressure, microorganisms, emissions, etc. It also provides the consumer with ease of use and time saving, convenience, and the production of products in different sizes and shapes (Bhargava et al., 2020; Ghoshal et al., 2018; Gregor-Svetec et al., 2018; Kalpana et al., 2019; Lloyd et al., 2019; Soltani Firouz et al., 2021; Yang et al., 2021).

6. Types of intelligent packaging

6.1 Oxygen-absorbent packaging systems

Currently, the active packaging technology used for most foods is the oxygen-inhibitor system. The presence of oxygen in packaging intensifies the oxidation of food. Oxygen promotes the growth of aerobic microbes, which can cause bad odors, discoloration, and nutritional damage, and reduce the shelf-life of food. As a result, it is advisable to control the level of oxygen in food packages.

6.2 Antibacterial packaging

Antibacterial packaging in the food industry is an active type of packaging that aims to reduce, inhibit, or delay the growth of microorganisms that may be present in packaged foods. Natural antimicrobials such as plant extracts can be used to produce these packages (in some sources, active packaging is classified as intelligent packaging).

6.3 Controller packaging of moisture

One of the causes of food spoilage is the presence of food moisture. In intelligent food packaging, the purpose of the moisture regulator is to reduce the activity of water to prevent microbial growth in the product. Moisture-absorbing pads are used to control food moisture.

6.4 *Ethylene-absorbent packaging*

Ethylene is a natural plant growth hormone. This hormone increases the respiration rate of fruits and vegetables. Ethylene ripens fruits and softens vegetables. The presence of ethylene causes yellowing of vegetables, lettuce rot, and reduces the shelf-life of many fruits and vegetables. Activated carbon combined with packaging in the presence of various metal catalysts can remove ethylene.

7. Other types of intelligent packaging

Packaging can adjust the entry and exit of gases into a food package, including sulfur dioxide-releasing packaging, ethanol- and water vapor-releasing packaging, and sensitive markers in packaging to highlight microbial and biochemical changes of foods (Bhargava et al., 2020; Cerqueira et al., 2018; Enescu et al., 2019; Ghoshal et al., 2018; Gregor-Svetec et al.,2018).

8. Indicators used in intelligent food packaging

8.1 *One- time and temperature indicators (TTIs)*

This index shows the temperatures associated with the overall reaction from the product history continuously. These indicators, as a single unit with a specific unit of measurement, can display the entire temperature—time history as the average temperature during distribution. A major method of the working of a temperature—time index is based on how they work, which is classified into mechanical, chemical, enzymatic, microbial, polymeric, and electronic.

8.2 *Integrity indicator*

The most common of these are leak indexes for detecting holes in a packed container. The most common indexes of gas are oxygen (O_2) and carbon dioxide (CO_2). The simplest indexes of integrity are time indicators that provide information about the period of a leak starting in a product.

8.3 *Freshness indicator*

These indicators monitor the quality of food products during storage and transportation. The loss of quality may be due to exposure to destructive conditions or the expiration of a useful life. These indicators show sufficient information about the quality of the product due to the microbial growth of chemical changes. For example, hydrogen sulfide indexes can be used to determine the quality of meat products. The hydrogen sulfide released through the meat matrix is related to the color of myoglobin, which is considered as a quality feature for meat products. Other indicators of novelty detection are based on other microbial metabolites such as ethanol, diethyl, and carbon dioxide.

8.4 Radio frequency detection (RFID)

RFID is a wireless data collection technology that uses electromagnetic waves (EWs) to transmit between a transmitter and/or a receiver. Intelligent radio frequency packaging provides integrity control, originality, antitheft protection, anticounterfeiting, quality measurement, and traceability (Fuertes et al., 2016).

9. Types of sensors used in packaging

9.1 Chemical sensors

Chemical sensors are sensory receptors that detect specific chemical stimuli in the environment. The use of chemical sensors is one of the most advanced methods in analytical chemistry that makes it possible to quantitatively measure different species instantly. Available electronic and optical technology has expanded these tools. Scientists from different sciences such as biochemistry, biology, electronics, and various branches of chemistry and physics have been involved in the design of chemical sensors. Chemical sensors include a sensor layer in which electrical signals are generated by the interaction of a chemical species with the layer. This signal is then amplified and processed. Therefore, the operation of chemical sensors includes two main steps: detection and amplification. In general, the device that performs this process is called a chemical sensor. The main part of a chemical or biological sensor is its sensor element. The sensor element is in contact with a detector. This element is responsible for identifying and linking with the species in a complex specimen. The detector then converts the chemical signals generated by the bonding of the sensor element to the target species into a measurable output signal. This device collects information about the chemical composition of its operating environment and transmits it to the processor as an optical or electrical signal. An objective example of these sensors in nature is the human nose, where a nerve signal is generated by the collision of material molecules with nerve cells and then amplified and sent to the brain (Gregor-Svetec et al., 2018; Lloyd et al., 2019; Ramos et al., 2012). Ideally, a chemical sensor is in direct contact with the sample and provides good results with high accuracy and selectivity in a short time. In clinical chemistry, the monitoring and detection of specific gases such as oxygen and carbon monoxide, etc. are used to monitoring the causes of diseases such as diabetes. Sensors are also used to determine the amount of environmental pollutants. One of the most important features of chemical sensors is that they can be made in very small sizes. This allows different species to be measured, even in the cells of living organisms. Chemical sensors are divided into four categories based on the converter used to convert the chemical change into a processable signal: thermal sensors, mass sensors, electrochemical sensors (potentiometry, amperometry, conductivity), and optical sensors. Two types of thermal detector are used in the construction of heat sensors. Among these detectors, the thermistor is the most common, and is more often used due to its low price, wide availability, stability, and high sensitivity. Pyroelectrics are another type of converter used in thermal sensors that have high sensitivity as thermal sensors. They can be used to track the heat absorbed by a gas layer. Another type of thermal sensor is a microbiosensor made of silicon chips that are more sensitive than conventional thermistor sensors. Applications of thermal sensors include: cholesterol determination, measuring of the catalytic properties of stabilized cells, control of biological processes, and the measurement of water in food.

9.2 Mass sensors

Measuring mass change, like measuring the heat generated by a reaction, can be used as a suitable criterion for chemical sensors. This property can be used for reactions that result in a change in net mass due to the release of a selective catalytic reactant. These sensors have two important features, first, they can be used in the liquid phase, and second, they are used for immunoassay applications due to their use in the gas phase and the selectivity in this phase.

The oldest and largest group of chemical sensors is electrochemical sensors. The response generated by these sensors is due to the interaction between chemistry and electricity. Currently, many of these sensors are commercially made and available on the market, and many are in development.

Chemical optical sensors are one of the newest chemical sensors. There are several reasons for paying close attention to these sensors. They have many applications for remote control of processes, which makes their use safer. They can be made in small sizes and even placed on the tip of an optical fiber. Optical-chemical sensors, like electrochemical sensors, use extensive spectroscopic knowledge that can be easily converted to remote sensors. Optical sensors are used in a variety of ways such as measuring absorption, fluorescence, and luminescence over a wide range of wavelengths. Chemical optical sensors are divided into several categories based on their application, such as: chemical safe optical sensors, pH sensors, gas optical sensors, humidity sensors, ion optical sensors, and sensors used in petroleum chemistry.

Synthetic sensors are nanosensors produced by attaching specific particles to the ends of carbon nanotubes and calculating the vibrational frequency in the presence or absence of particles. These nanosensors are often used to detect and control chemical reactions by nanoparticles.

With the advent of sensors and their introduction to the field of decomposition of biochemical species, another solution was provided to control the treatment of patients: biosensors. These sensors, which work electrochemically and still have many applications, are commonly used in extracorporeal measurements due to their relatively large dimensions, and in measuring the amount of fat and skin moisture, body sweating rate, etc. With the advancements in science and technology, biomicrosensors have been created. These tools are used in biochemical reactions performed on the electrode surface and recording the amount of current or following potential changes, and these factors are related to the concentration of the desired species. Biosensors rely on biological components such as antibodies. Enzymes, receptors, or whole cells can be used as sensors. An example of the work done at this level is the measurement of epinephrine by the adrenal cells.

10. Materials used in the manufacture of sensors

10.1 Classic materials

The basis of solid state sensors is the electrical response to the chemical environment. For example, electrical properties change at the liquid or gas phases. These changes are used to identify chemical species. Although chemical silicon sensors such as the field effect transistor (FET) have been developed, the price and technology required to manufacture them, as well as problems such as reproducibility, stability, sensitivity, and selectivity, have led to their use remaining limited. Metal oxide semiconductor sensors, such as compact powders and SnO_2 thin films, also have a catalytic activation effect.

10.2 Polymers

These materials have many advantages for use in the manufacture of sensors, including the following. Polymers can be deposited on different types of substrates. Some chemical species can be chemically deposited on polymer substrates to reduce the amount of reactant leakage into the sample solution. There is a variety of polymers in terms of structural properties such as having side chains or being charge or neutral, etc. that causes the appropriate physical and chemical properties for the sensor. It is possible to dissolve or disperse the chemical detector evenly in the polymer contexture. The mechanical stability of polymers makes it possible to use the sensor for a long time and create high stability. The polymer structure in some cases improves properties such as selectivity and sensitivity. Polymers are relatively inexpensive materials and their manufacturing techniques are simple and do not require special conditions. Polymers and materials suitable for making sensors must have a number of special properties, which include: detector solubility in polymer contexture; appropriate life span; no crystallization, migration, or reorientation of species; having the effects of aging and depreciation; stability to chemicals such as acids, bases, and oxidants; transparency to light; biocompatibility (nontoxic substances); and no inherent color or luminescence (in use as a background). A polymer fabric is used to make such sensors. A compound with optical properties is stabilized as a sensitizer inside the polymer substrate (mass sensor) or on its surface (surface sensor). Therefore, the polymer only acts as a reactant retainer or a solid substrate. It should be noted that, in these sensors, the polymer membrane used has other functions in addition to holding the detector in a specific and fixed place. The permeability properties of the membrane can repel other species or ions that are considered to be interfering. For example, hydrophobic membranes are used to repel nonvolatile substances in sensors used to measure the concentration of gases in water.

10.3 pH-monitoring sensors

Over the past decade, pH sensors based on colorimetric, optical, electromagnetic, and electrochemical detection have been used for continuous monitoring of food pH. Some of these devices involve advanced technologies, which makes production more expensive and, in some cases, impractical. One simple and cost-effective tool for pH monitoring is the use of detector colors, especially colors that are absorbed due to changes in pH along the visible wavelength. Although most synthetic pH-sensing dyes have a potentially toxic effect, the use of natural dyes can be beneficial because of their safety properties, and the change in pH can be easily observed due to the change in natural colors because they show significant changes in a wide range of pH values. One of the major challenges to using dyes as a colorimetric indicator in wound dressing is how to minimize or prevent dye washing. Polymeric matrices, especially those based on hydrogels, may provide color immobilization. Hydrogels are used in many biomedical and pharmaceutical activities, especially in wound-healing applications. Typically, hydrogel films are transparent, making them suitable matrices for the development of color pH sensors. In addition, polysaccharide-based hydrogel films may exhibit some ideal properties for wound dressings, including wound care, conformational contact with the skin, biocompatibility, and biosecurity. PH-sensitive hydrogels are divided into two types, neutral and ionic. Ionic hydrogels consist of polymeric networks containing positively charged parts, while neutral hydrogels contain positively charged and negatively charged polymer chain components. Swelling of ionic hydrogels depends on these two issues along with ionic interactions between the charged parts of the polymer network and the free ions. The presence of functional groups that have the ability to ionize, such as carboxylic acid and sulfonic acid or ammonium salt groups, increase the hydrophilicity of the polymer, which enables an increase in water absorption. In the following, some of the research that has been done in this field is reviewed. Kamali Sabeti, Rostamzad, & Babakhani (2019) produced and evaluated an intelligent biodegradable film based on carrageenan for packaging fish fillets. According to the tests performed, the results obtained showed that the film of carrageenan containing cabbage extract was initially colorless and with the passage of time and spoilage progress in packaged fillets TVB-N and pH factors exceeded their limits and the smoky film color darkened, which could inform consumers of the quality and freshness of the packaged fish. Therefore, according to the results of that research, carrageenan film containing cabbage extract can be used as an indicator of spoilage of fish and other meat foods of high economic value (Kamali Sabeti, Rostamzad, & Babakhani, 2019). Ebrahimi Tirtashi et al. (2019) developed a cellulose/chitosan detector with carrot anthocyanins that was sensitive to pH for intelligent food packaging. Based on these results, the highest value for the parameter (L) was observed at a pH of 2, which indicates a high intensity red color and greater transparency at this pH. In regard to parameter a (green-red), with increasing pH a decreasing trend emerged, and a similar trend was observed for parameter b, with the values indicating the tendency of the film to turn blue, purple, and yellow (Ebrahimi Tirtashi et al., 2019). Kuswandi et al. (2012) conducted a study on a new label of food packaging to discover the quality of the product, which found that these labels were sensitive to bacterial growth patterns and showed real-time spoilage of the product (Kuswandi et al., 2012). Yoshida et al. (2014) produced a film based on chitosan and anthocyanin composition. The results of this study showed that the mechanical properties, such as tensile strength in the film containing anthocyanin composition, improved compared to the blank film. These compounds were sensitive to pH changes, and in alkaline, acidic, and neutral media they turned purple, pink, and green-blue, respectively. The use of these compounds could replace the control packaging (blank) and create a safe and quality packaging (Yoshida et al., 2014). Wu et al. (2014) prepared a film of gelatin–chitosan with oregano essential oil and investigated its effects on silver carp skin. The results showed that oregano essential oil had the best antimicrobial activity compared to cinnamon and fennel essential oil. Adding oregano essential oil to the gelatin–chitosan film improved the antimicrobial properties and prevented the absorption of light and the passage of water vapor, with the best outcome being obtained at the 4% oregano value, and this extract can increase the shelf-life of fish meat (Wu et al., 2014). Silva et al. (2015) produced a film of chitosan/corn starch with red cabbage extract which was described as an intelligent film. The red cabbage extract was added to the film to change its color with a change in the pH during spoilage. The film was tested on fish and the results showed that the film was sensitive to changes in pH, and the film color also changed with a change in the pH, and it could be used as an intelligent packaging in the food industry (Silva-Pereira et al., 2015). In addition, anthocyanin extract of sour tea has a significant effect on the microstructure of the composite film also, and has good compatibility with polyvinyl alcohol/starch matrix and has the best color stability. Therefore, this film can be used in research into the freshness of pork meat. Sour tea anthocyanins could be a suitable initial material for intelligent indicator films and a sign of freshness of meat products, with the use of natural anthocyanins of the sour tea plant in the structure of intelligent indicator films being well compatible with different substrates of film formation. Other researchers conducted research on the production of intelligent wound dressings based on carrageenan/arabic gum containing anthocyanin cranberry extract. For this, they placed films containing cranberry extract close to *Staphylococcus* and *Pseudomonas* bacteria. The results of that study showed that the use of substances (with anthocyanin plant extracts) that are sensitive to pH changes in the production of wound dressings allows the consumer to be informed of the onset of wound infection (by changing the color of the wound dressing) because bacteria from the wound infection agents change the color of the pH of the environment (Zepon et al., 2019) Hashemi et al. (2021) developed and evaluated an intelligent wound dressing based on chitosan–gum arabic in combination with anthocyanins. Their results showed that the wound dressings made in the vicinity of *Pseudomonas aeruginosa* and *Staphylococcus aureus* (which are important in the infection of superficial wounds), change color and inform of possible wound infection. This color change is due to a change in the pH of the environment created by the activity of bacteria (Hashemi et al., 2021).

11. Nanotechnology in intelligent packaging

The nanosensor interacts with internal factors (food components) and environmental factors. As a result of this interaction, the nanosensor generates a response (visual signal, electrical signal) that is related to the condition of the food product. The information generated is not only useful for communicating with consumers, but also gives them information about the safety and quality of products. Nanosensors can also be used by manufacturers in decision-support systems to determine the time and necessary actions throughout the whole production process. These scientific advances make it possible to produce a new generation of nanosensors with food packaging applications. Nanosensors have great potential for accelerating the identification and quantification of pathogens, decayed materials, and allergenic proteins. Therefore, these nanodevices can have a significant impact on food security. In general, nanosensors are placed on food packaging to monitor the internal and external conditions of products and to accurately identify various food contaminants. Nanosensors are the physical, chemical, or biological sensors at the nano size that detect changes at the nanoscale and transmit them with high accuracy and sensitivity. With the advancement of science in the world and the advent of electronic equipment and the great changes that have taken place in recent decades and during the 20th century, the need to build more accurate, smaller, and more capable sensors emerged. High-sensitivity sensors are used today that are sensitive to small amounts of gas, heat, or radiation. Increasing the sensitivity, utility, and accuracy of these sensors requires the discovery of new materials and tools. Nanosensors are nanometer-sized sensors that, due to their nanometer size, are so accurate and responsive that they react even to the presence of several atoms of the relevant gas (Gallocchio et al., 2015; Mihindukulasuriya and Lim, 2014; Villena de Francisco & García-Estepa, 2018; Wahab et al., 2021).

12. Characteristics and properties of nanosensors

The output signal should be proportional to the type and amount of the target species. They act very specifically to the desired species, have high resolution and selectivity, high repeatability and accuracy, and a high response speed (up to milliseconds). There is however a lack of response to environmental disturbances such as temperature, ionic strength of the environment, etc.

12.1 Food industry uses

Food industry uses include (1) control of flavor, (2) increasing the bioavailability of bioactive compounds, and (3) detection of harmful substances in foods.

13. Classification of nanosensors

13.1 Carbon-based nanotubes

In general, the use of nanotubes in sensors can be divided into two categories: carbon nanotubes as chemical sensors and carbon nanotubes as mechanical sensors:

13.2 Polymer nanoparticles

These are derived based on biocompatible and biodegradable polymers from natural and artificial resources.

13.3 Liposomes

Liposomes are two concentric lipid layers. These nanocarriers are made of a surfactant-encapsulated aqueous nucleus that can be natural or synthetic phospholipids.

13.4 Dendrimers

These are a macromolecular of single scattered that is composed of molecules of duplicate branches around the inner nucleus. Typically, dendrimers are symmetric about the core, and often adopt a spherical three-dimensional morphology.

13.5 Nanoemulsions

These are composed of droplets with a size range of 10–100 nm and can be classified into two types of nanoemulsions, oil-in-water (O/W) and water-in-oil (W/O), based on their spatial structure.

13.6 Hydrogel nanoparticles

These are three-dimensional polymer networks that can absorb large amounts of water or biological fluids. The water absorption ability of hydrogels depends on the hydrophilic group.

13.7 Nanotools

The use of these sensors makes it possible to detect very small amounts of chemical contamination or viruses and bacteria in agricultural and food systems.

References

Bae, H. J., Darby, D. O., Kimmel, R. M., Park, H. J., & Whiteside, W. S. (2009). Effects of transglutaminase-induced cross-linking on properties of fish gelatin–nanoclay composite film. *Food Chemistry, 114*(1), 180–189.

Bhargava, N., Sharanagat, V. S., Mor, R. S., & Kumar, K. (2020). Active and intelligent biodegradable packaging films using food and food waste-derived bioactive compounds: A review. *Trends in Food Science and Technology, 105*, 385–401.

Carina, D., Sharma, S., Jaiswal, A. K., & Jaiswal, S. (2021). Seaweeds polysaccharides in active food packaging: A review of recent progress. *Trends in Food Science and Technology, 110*, 559–572.

Cerqueira, M.Â. P. R., Nurmi, M., & Gregor-Svetec, D. (2018). *Intelligent packaging: for interactive communication, consumer convenience, improved management and higher food safety*. OTGO: Faculty of Natural Sciences and Engineering.

Domínguez, R., Barba, F. J., Gómez, B., Putnik, P., Bursać Kovačević, D., Pateiro, M., Santos, E. M., & Lorenzo, J. M. (2018). Active packaging films with natural antioxidants to be used in meat industry: A review. *Food Research International, 113*, 93–101.

Ebrahimi Tirtashi, F., Moradi, M., Tajik, H., Forough, M., Ezati, P., & Kuswandi, B. (2019). Cellulose/chitosan pH-responsive indicator incorporated with carrot anthocyanins for intelligent food packaging. *International Journal of Biological Macromolecules, 136*, 920–926.

Enescu, D., Cerqueira, M. A., Fucinos, P., & Pastrana, L. M. (2019). Recent advances and challenges on applications of nanotechnology in food packaging. A literature review. *Food and Chemical Toxicology, 134*, 110814.

Fuertes, G., Soto, I., Vargas, M., Valencia, A., Sabattin, J., & Carrasco, R. (2016). Nanosensors for a monitoring system in intelligent and active packaging. *Journal of Sensors, 2016*, 7980476.

Gallocchio, F., Belluco, S., & Ricci, A. (2015). Nanotechnology and food: Brief overview of the current scenario. *Procedia Food Science, 5*, 85–88.

Ghoshal, G. (2018). Chapter 10 - Recent trends in active, smart, and intelligent packaging for food products. In A. M. Grumezescu, & A. M. Holban (Eds.), *Food Packaging and Preservation* (pp. 343–374). Academic Press.

Gregor-Svetec, D. (2018). Chapter 8 – Intelligent packaging. In M.Â. P. R. Cerqueira, J. M. Lagaron, L. M. Pastrana Castro, & A. A. M. de Oliveira Soares Vicente (Eds.), *Nanomaterials for Food Packaging* (pp. 203–247). Elsevier.

Kalpana, S., Priyadarshini, S. R., Maria Leena, M., Moses, J. A., & Anandharamakrishnan, C. (2019). Intelligent packaging: Trends and applications in food systems. *Trends in Food Science and Technology, 93*, 145–157.

Kamali Sabeti, N., Rostamzad, H., & Babakhani, A. (2019). Production and evaluation of smart biodegradable film based on carrageenan for fish fillet packaging. *Journal of Fisheries (Iranian Journal of Natural Resources), 72*(1), 85–95.

Khumkomgool, A., Saneluksana, T., & Harnkarnsujarit, N. (2020). Active meat packaging from thermoplastic cassava starch containing sappan and cinnamon herbal extracts via LLDPE blown-film extrusion. *Food Packaging and Shelf Life, 26*, 100557.

Kuswandi, B., Jayus, Restyana, A., Abdullah, A., Heng, L. Y., & Ahmad, M. (2012). A novel colorimetric food package label for fish spoilage based on polyaniline film. *Food Control, 25*(1), 184–189.

Lessani, S., Rostamzad, H., & Zakipour Rahimabadi, E. (2020). Strengthening the physical and mechanical properties of biodegradable film of carrageenans using nanoclay. *Innovation in Food Science and Technology, 12*(3), 137–149. https://doi.org/10.30495/JFST.2020.674393.

Lloyd, K., Mirosa, M., & Birch, J. (2019). Active and intelligent packaging. In L. Melton, F. Shahidi, & P. Varelis (Eds.), *Encyclopedia of food chemistry* (pp. 177–182). Oxford: Academic Press.

Mihindukulasuriya, S. D. F., & Lim, L. T. (2014). Nanotechnology development in food packaging: A review. *Trends in Food Science and Technology, 40*(2), 149–167.

Mousavi Khaneghah, A., Hashemi, S. M. B., & Limbo, S. (2018). Antimicrobial agents and packaging systems in antimicrobial active food packaging: An overview of approaches and interactions. *Food and Bioproducts Processing, 111*, 1–19.

Pires, C., Ramos, C., Teixeira, B., Batista, I., Nunes, M. L., & Marques, A. (2013). Hake proteins edible films incorporated with essential oils: Physical, mechanical, antioxidant and antibacterial properties. *Food Hydrocolloids, 30*(1), 224–231.

Pires, C., Ramos, C., Teixeira, G., Batista, I., Mendes, R., Nunes, L., & Marques, A. (2011). Characterization of biodegradable films prepared with hake proteins and thyme oil. *Journal of Food Engineering, 105*(3), 422–428.

Ramos, M., Jiménez, A., Peltzer, M., & Garrigós, M. C. (2012). Characterization and antimicrobial activity studies of polypropylene films with carvacrol and thymol for active packaging. *Journal of Food Engineering, 109*(3), 513–519.

Rostamzad, H., Abbasi Mesrdashti, R., Akbari Nargesi, E., & Fakouri, Z. (2019). Shelf life of refrigerated silver carp, *Hypophthalmichthys molitrix*, fillets treated with chitosan film and coating incorporated with ginger extract. *Caspian Journal of Environmental Sciences, 17*(2), 143–153.

Rostamzad, H., Paighambari, S. Y., Shabanpour, B., Ojagh, S. M., & Mousavi, S. M. (2016). Improvement of fish protein film with nanoclay and transglutaminase for food packaging. *Food Packaging and Shelf Life, 7*, 1–7.

Rostamzad, H. Z. R., & Eshagh. (2020). Production and evaluation of chitosan film incorporated licorice extract for fish packaging. *Innovation in Food Science and Technology, 12*(3), 79–94.

Silva-Pereira, M. C., Teixeira, J. A., Pereira-Júnior, V. A., & Stefani, R. (2015). Chitosan/corn starch blend films with extract from *Brassica oleraceae* (red cabbage) as a visual indicator of fish deterioration. *Lebensmittel-Wissenschaft und -Technologie (Food Science and Technology), 61*(1), 258–262.

Soltani Firouz, M., Mohi-Alden, K., & Omid, M. (2021). A critical review on intelligent and active packaging in the food industry: Research and development. *Food Research International, 141*, 110113.

Song, H., Shin, Y., & Song, K. (2012). Preparation of a barley bran protein–gelatin composite film containing grapefruit seed extract and its application in salmon packaging. *Journal of Food Engineering, 113*, 541–547.

Varghese, S. A., Siengchin, S., & Parameswaranpillai, J. (2020). Essential oils as antimicrobial agents in biopolymer-based food packaging: A comprehensive review. *Food Bioscience, 38*, 100785.

Villena de Francisco, E., & García-Estepa, R. M. (2018). Nanotechnology in the agrofood industry. *Journal of Food Engineering, 238*, 1–11.

Wahab, A., Rahim, A. A., Hassan, S., Egbuna, C., Manzoor, M. F., Okere, K. J., & Walag, A. M. P. (2021). Chapter 10 - Application of nanotechnology in the packaging of edible materials. In C. Egbuna, A. P. Mishra, & M. R. Goyal (Eds.), *Preparation of Phytopharmaceuticals for the Management of disorders* (pp. 215–225). Academic Press.

Wu, J., Ge, S., Liu, H., Wang, S., Chen, S., Wang, J., Li, J., & Zhang, Q. (2014). Properties and antimicrobial activity of silver carp (*Hypophthalmichthys molitrix*) skin gelatin-chitosan films incorporated with oregano essential oil for fish preservation. *Food Packaging and Shelf Life, 2*(1), 7–16.

Yang, J., Shen, M., Luo, Y., Wu, T., Chen, X., Wang, Y., & Xie, J. (2021). Advanced applications of chitosan-based hydrogels: From biosensors to intelligent food packaging system. *Trends in Food Science and Technology, 110*, 822–832.

Yoshida, C. M. P., Maciel, V. B. V., Mendonça, M. E. D., & Franco, T. T. (2014). Chitosan biobased and intelligent films: Monitoring pH variations. *Lebensmittel-Wissenschaft und Technologie (Food Science and Technology), 55*(1), 83–89.

Zepon, K. M., Martins, M. M., Marques, M. S., Heckler, J. M., Dal Pont Morisso, F., Moreira, M. G., Ziulkoski, A. L., & Kanis, L. A. (2019). Smart wound dressing based on κ-carrageenan/locust bean gum/cranberry extract for monitoring bacterial infections. *Carbohydrate Polymers, 206*, 362–370.

Biodegradable biosourced epoxy thermosets, blends, and composites

Abra Mathew[1], *Sathyaraj Sankarlal*[2], *Abhinay Rajput*[1], *K. Sekar*[2] *and Sushanta K. Sahoo*[1]

[1]Material Sciences and Technology Division, CSIR-National Institute for Interdisciplinary Science and Technology, Thiruvananthapuram, Kerala, India; [2]Department of Mechanical Engineering, National Institute of Technology, Calicut, Kerala, India

1. Introduction

The excessive use of fossil resources for the preparation of prepolymers and polymers has raised key concerns among environmental policy makers and researchers around the world because of its depletion and the generation of severe hazardous concerns. Polymer industries are considerably affected by the expensive petro-resources and stringent disposal regulations (de Espinosa & Meier, 2011; Galià et al., 2010). Epoxy networks, a major class of the thermoset polymer industry, are commonly used in coatings, adhesives, laminates, electrical castings, etc. due to their excellent mechanical properties, high cross-link density, higher thermal stability, good chemical resistance, and low curing shrinkage, with the most widely used epoxies being diglycidyl ether-based bisphenol A (DGEBA), bisphenol F (DGEBF), novalac epoxy, etc. (Baroncini et al., 2016). However, these classic petro-based monomers are hazardous and endocrine disruptors, which can mimic body hormones and induce serious health disorders, especially in the very young children (Supanchaiyamat et al., 2012). Further, these cross-linked epoxy thermosets are not recyclable or degradable, which results in huge amounts of land fill and environmental pollution caused by the post-consumption plastic waste. Thus, there is growing attention worldwide to develop eco-friendly epoxy monomers from sustainable green and renewable sources, which can degrada safely in the environment (Ding et al., 2015; Kadam et al., 2015; Pawar et al., 2016; Sahoo et al., 2018c).

In this context, biodegradable bio-sourced epoxy polymers have been receiving greater attention as they are renewable, environment friendly, and nontoxic in nature. Plant oils, lignin, and its derivatives, eugenol, cardanol, saccharides, rosins, catechins, tannins, etc. are some of the natural sources used to prepare biobased and biodegradable epoxies (Atta et al., 2017; Cho et al., 2013; Deng et al., 2015; El-Ghazawy et al., 2015; Fache et al., 2015; Hernandez et al., 2016; Hu et al., 2014; Jaillet et al., 2014; Jung et al., 2017; Kanehashi et al., 2013; Li et al., 2018; Lorenzini et al., 2015; Ma et al., 2017; Miao et al., 2017; Pas & Torr, 20171; Sahoo et al., 2018c;

Biodegradable Polymers, Blends and Composites. https://doi.org/10.1016/B978-0-12-823791-5.00020-X

Shibata & Ohkita, 2017; Voirin et al., 2014; Wang et al., 2017; Xin et al., 2016; Zhao & Abu-Omar, 2017). Epoxidized triglycerides are prominent resources to make bio-epoxies and modified bio-epoxies through several chemical functionalization methods. Oils like soybean, corn, sunflower, castor, linseed, cotton seed, etc. are widely used in the chemical and petrochemical industries as they are inexpensive and widely available (Altuna et al., 2013; Baroncini et al., 2016; Ding et al., 2015; Kadam et al., 2015; Lu & Larock, 2009; Pawar et al., 2016; Sahoo et al., 2018c; Supanchaiyamat et al., 2012). However, cured functionalized plant oils, devoid of aromatic moieties, exhibit inferior properties to petro-based polymers. Another commendable epoxy feedstock that has received major attention in recent years is lignin and its derivatives, as it is the second most abundant polymer from biomass (Hernandez et al., 2016; Jung et al., 2017; Pas & Torr, 2017; Wang et al., 2017; Xin et al., 2016; Zhao & Abu-Omar, 2017). The large number of propylphenols in the lignin structure makes it an excellent raw material for preparing epoxies and derivatives. Cardanol, a component of cashew nut shell liquid (CNSL), is another potential resource for polymer products. This can be used to produce novalac epoxy resins with characteristics at a par with petro-based epoxies, when cured in combination with other biobased components (Voirin et al., 2014). Thus, such natural resources help researchers to adopt them as green raw materials in producing polymers which are more green, industrially viable, and environment friendly.

In this chapter, the preparation of bioderived epoxy resins, biobased thermoset blends, and bio-epoxy composites with partially or complete degradable potential is discussed. This includes their chemical modifications, polymerization techniques, processes, and biodegradation properties.

2. Application prospective: adhesives, coatings, and composites

Epoxy thermosets and composites find their major applications in the automotive, construction and building, adhesives, and coating industries, because of their strong adhesion to metal and non-metal surfaces, thermal stability, low shrinkage, good flexibility, and excellent stability in solvents. Biobased epoxies have not been widely accepted in the industrial field till now due to their inferior properties to petro-based epoxy products. However, epoxy blends with other biosourced reactive components can have enhanced processability and toughening properties comparable to unmodified resins (Baroncini et al., 2016; de Espinosa & Meier, 2011; Lu & Larock, 2009; Supanchaiyamat et al., 2012). In the global market, some key companies such as Entropy resin, Change Climate, pond ApS, Sicomin, ABT, COOE, ALPAS, Spolchemie, Nagase ChemteX Corporation, etc. have commercialized bio-epoxy resins from different plant oils and other renewable resources for coating, electronics, adhesive, and composite applications. Similarly, Ford, Toyota, Dieffenbacher, BASF, and Mercedes Benz companies have developed biobased composites using soy resin and different natural fibers for different interior and exterior automotive parts. However, after their use, discarded automotive thermoset components massively pollute the environment, as they are non-degradable. Similarly, printed circuit boards (PCBs) in electronic products consisting of epoxy laminates and coating do not degrade because of the strong cross-linking. Therefore, more efforts are required to synthesize novel biobased epoxies and composites with biodegradable potential.

3. Bio-sourced epoxy resins: chemical modifications and synthesis

Today, the production of epoxy with added properties and performance faces a barrier due to the unsustainability of the petroleum-driven materials (de Espinosa & Meier, 2011). Biosourced epoxy resins have drawn huge attention not only from academic researchers also from some prominent industries. The presence of unsaturation and other unique functional groups such as hydroxyl, carboxyl, and alkoxy groups makes renewable resources suitable for the development of novel cross-linked polymers. These renewable resources can be chemically modified, as shown in Fig. 15.1 to make prepolymers and polymers that are more industrially adaptable.

3.1 Plant oil-based epoxy polymers

The plant oils are a major source for synthesizing biobased thermosets and blends. They have triglycerides backbone whose compositions are dependent on the kind of plant and growing environments. About three-fourths of worldwide plant oil

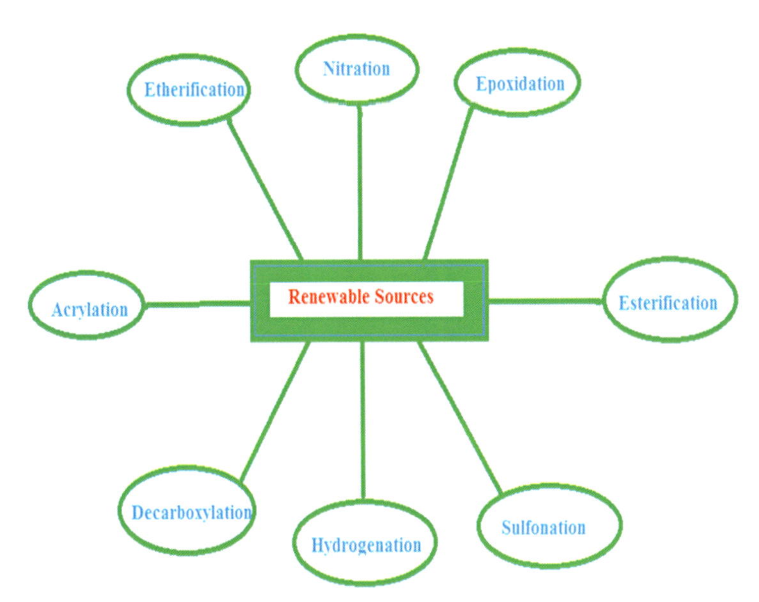

Figure 15.1 Chemical modifications of renewable resources.

manufacturing is for food and the remainder is used for adhesives, coatings, inks, and plasticizers, etc. (de Espinosa & Meier, 2011). The loads of C=C bonds in plant oils which act as sites for polymerization to occur makes them excellent for thermoset resins (Baroncini et al., 2016; de Espinosa & Meier, 2011; Galià et al., 2010). Also, the inherent functional groups present in triglycerides, like epoxide or hydroxyl, can be cross-linked with the help of various polymerization techniques (Lu & Larock, 2009). The shortcoming of the inferior reactivity of triglycerides present in plant oils can be overcome by including effortlessly polymerizable functional groups by chemical modifications. The common compositions of some major plant oils are listed in Table 15.1. Vegetable/plant oils in chemically unaltered form cannot be used as a matrix or for a blend, as the unsaturated C=C bonds are not sufficient to provide stiffness and strength while undergoing polymerization. Therefore, the plant oils can be polymerized to one of the following forms (Lu & Larock, 2009):

- Directly polymerizing the C=C already there in the fatty acid
- Converting the C=C bonds into functional groups so that they can be easily polymerized by reactions like epoxidation, transesterification, acrylation and maleation
- Conversion of the oils into monoglycerides or diglycerides which can act as base monomers for polymer synthesis.

In the last decade, many approaches have been adopted to prepare completely biosourced epoxies from plant oils that have been compared with petro-sourced epoxy (Manthey et al., 2011). ELO-based bio-epoxy was prepared by Supanchaiyamat et al., which was cross-linked with biobased diacid cross-linker in the presence of different catalysts (Supanchaiyamat et al., 2012). Therein, a higher tensile strength is noticed in the presence of dimethylaminopyridine (DMAP) catalyst and a much greater Young's modulus is achieved using an imidazole catalyst. Similarly, self-healing bio-epoxy films have been developed from epoxidized soybean oil cured with citric acid (CA)-based aqueous solution in the absence of any catalyst (Altuna et al., 2013). Ding et al. investigated the thermal and mechanical properties of ELO-based epoxies based on biorenewable dicarboxylic acid-based curing agents with DMAP as the catalyst (Ding et al., 2015). Recently, biobased epoxy has been prepared from epoxidized sucrose soyate with adequate mechanical and thermal properties cured with CA and DL-malic acid. Although the properties are acceptable for specific applications, the biodegradability of these materials was not studied to certify them as biodegradable epoxies (Kadam et al., 2015).

Kadam et al. prepared biodegradable bioepoxy from karanja oil curing with citric acid and tartaric acid. The mechanical properties and thermal stability achieved were found to be higher when cured with CA (Pawar et al., 2016). Similarly, Sahoo et al. cross-linked epoxy, epoxidized castor oil (ECO), and epoxidized linseed oil (ELO) with citric acid as the cross-linker and found that the storage modulus and elongation at break for bio-epoxies were reasonable for flexible elastomeric applications (Sahoo et al., 2018c). The ductile behavior of the epoxidized oil was found to be very high with respect to epoxy due to the existence of a long aliphatic chain which reduces its stiffness and increases its flexibility. Pin et al. developed bio-epoxy from linseed oil with excellent thermal stability (340 °C), higher Tg (>100 °C) and good

Table 15.1 Availability, composition, and physical properties of various plant oils (Lu & Larock, 2009; Supanchaiyamat et al., 2012).

Plant oil	Average yearly production (mega tonnes)	Type of fatty acid						Iodine value (mg/100 g)	Kinematic viscosity at 40 C (mm^2/s)
		Oleic	Palmitic	Stearic	Linoleic	Linolenic	Double bonds/ triglyceride		
Soybean oil	26.52	23.4	11.0	4.0	53.3	7.8	4.6	117–143	29
Palm	23.53	40.5	42.8	4.2	10.1	–	1.7	44–58	39.4
Rapeseed	15.29	56	4	4.2	26	10	3.8	94–120	35
Sunflower	15.29	37.2	5.2	2.7	53.8	1.0	4.7	110–143	36
Cottonseed	4.49	18.6	21.6	2.6	54.4	0.7	3.9	90–119	34
Olive	2.52	71.1	13.7	2.5	10.0	0.6	2.8	75–94	–
Linseed	0.83	19.1	5.5	3.5	15.3	56.6	6.6	168–204	26–29
Canola	15.28	60.9	4.1	4.2	21.0	8.8	3.9	11–126	–
Fish	1.13	11–25	10–22	–	–	–	–	104–110	–
Sesame	0.76	41	9	6	43	1.0	3.9	103–116	36
Castor	0.56	5.0	1.5	0.5	4.0	0.5	2.7	82–88	251

impact strength. Methyl hexahydrophthalic anhydride (MHHPA) and Benzophenone-tetracarboxylic dianhydride (BTDA) were used as anahydride based crosslinkers to achieve toughened material with desired thermal properties (Pin et al., 2015).

3.2 Lignin-based epoxy resins

Ligno-cellulosic biomass is an important natural resource that has the potential to replace the fossil fuels. It is mainly composed of cellulose, hemicellulose, and lignin. Among them, lignin is an aromatic hetereopolymer cross-linked by C—O and C—C bonds, where ether linkages are more predominant. Lignin accounts for 15%—30% of the total biomass. It is the second most abundant natural polymer and comprises about 30% of the organic carbon in the biosphere. Lignin derivatives, such as vanillin, that are industrially available have been explored as they are commercially available for production. Thermosets made of biobased epoxy from vanillin-derived oligomers were developed by Fache et al. Methoxyhydroquinone is synthesized from vanillin and then epoxidized, oligomerized, and cross-linked to form an epoxy thermoset along with the diamine agent IPDA, as shown in Fig. 15.2 (Fache et al., 2015).

In 2017, Shibata et al. documented a novel bio-epoxy resin (DGEDVCP) which is biobased and was prepared from the glycidylation reaction of 2,5-bis(4-hydroxy-3-methoxybenzylidene) cyclopentanone (DVCP), which is the crossed-aldol condensation product of the reaction between vanillin and cyclopentanone, as shown in Fig. 15.3 (Shibata & Ohkita, 2017).

In 2018, Li et al. reported a successful method for the synthesis of new high-yield silicone-bridged difunctional epoxy monomers which are synthesized from the natural compound eugenol through incorporation into their molecular backbones—phenyl siloxane and methyl siloxane connectors of various chain length dimensions—as shown in Fig. 15.4. The viscosity of these was much lower than that of industrial DGEBA epoxy, and is ideal for composites and prepregs. They exhibited strong

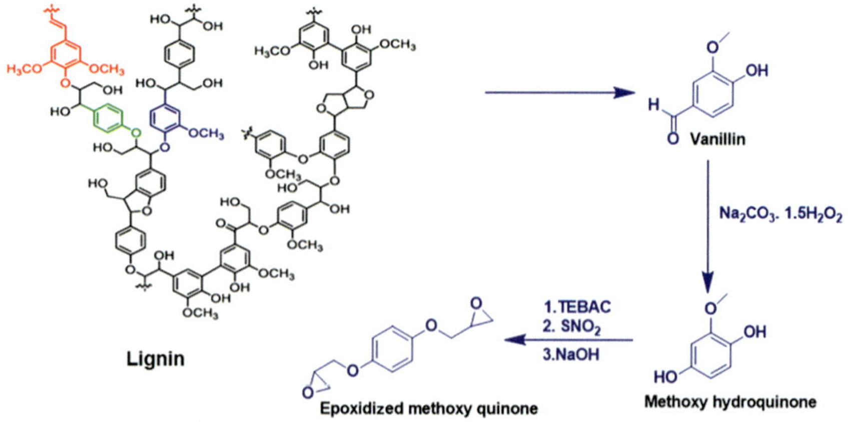

Figure 15.2 Synthesis of bio-epoxy from lignin-based derivative: Vanillin (Fache et al., 2015).

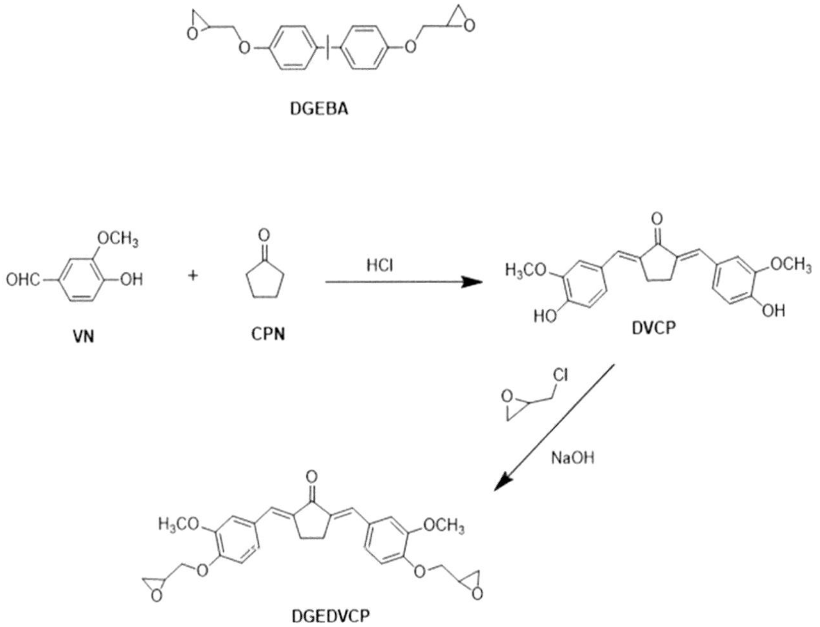

Figure 15.3 Structure of 2,5-bis(4-hydroxy-3-methoxybenzylidene) cyclopentanone (DGEBA) and synthesis of the bio-epoxy resin DGEDVCP (Shibata & Ohkita, 2017).

properties after healing, which were superior to those of DGEBA. All these advantages are due to the low polar property, greater dissociation strength, helical structure, and high molar volume of their siloxane-contained segments (Li et al., 2018).

An epoxy resin with a large biomass and high performance was recorded by Miao et al., which is important for sustainable development. The specific bis(2-methoxy-4-(oxiran-2-ylmethyl)phenyl)furan-2,5-dicarboxylate (EUFU-EP) structure was synthesized from biosourced components, 2,5-furan dicarboxylic acid and eugenol, as shown in Fig. 15.4, which had a very high biomass content. Furthermore, by using methyl hexahydrophthalic anhydride (MHHPA) and 2-ethyl-4-methylimidazole as the curing agent and the curing accelerator, respectively, EUFU-EP was prepared as shown in Fig. 15.5 (Miao et al., 2017).

Zhao et al. prepared lignin-incorporated epoxy prepolymers in liquid phase using phenolated organosolv lignin, which was condensed with salicyl alcohol (SA) to form biobased polyphenols (PL-SA). Through the glycidylation process, these polyphenols were epoxidized to give liquid epoxy prepolymers. The properties were found to be closer to or higher than those for DGEBA-epoxy (Zhao & Abu-Omar, 2017).

Jung et al. epoxidized fractionated kraft lignin using epichlorohydrin to form lignin-epoxy resin. Kraft lignin-derived epoxy resin is used to form biopolyester through an esterification process. The thermal properties like thermal stability (T_5) were found to be 257.1°C, which revealed that the epoxidized kraft lignin can replace BPA in the epoxy. However, the mechanical properties have not been reported (Jung et al., 2017).

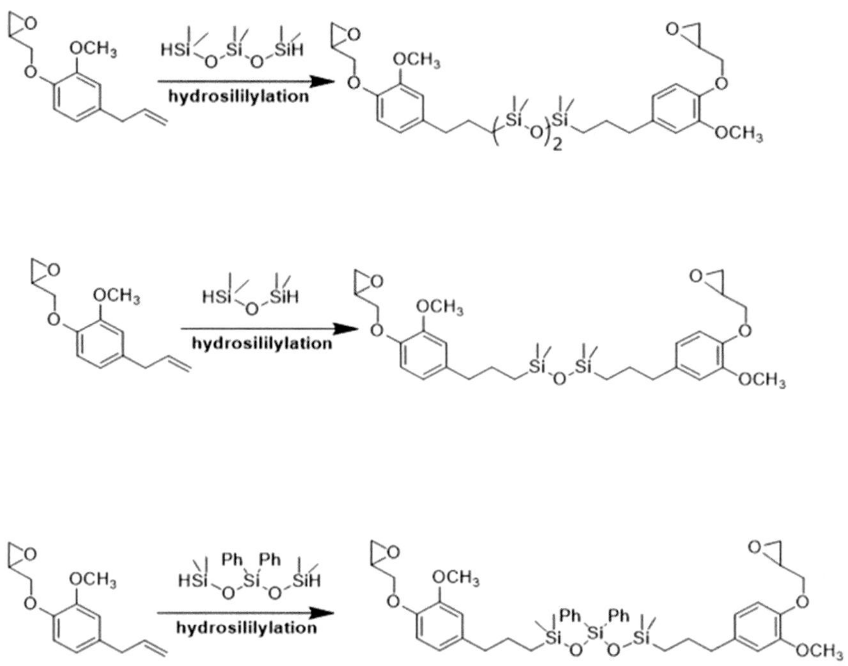

Figure 15.4 Synthesis of new silicone-bridged difunctional epoxy monomers from eugenol (Li et al., 2018).

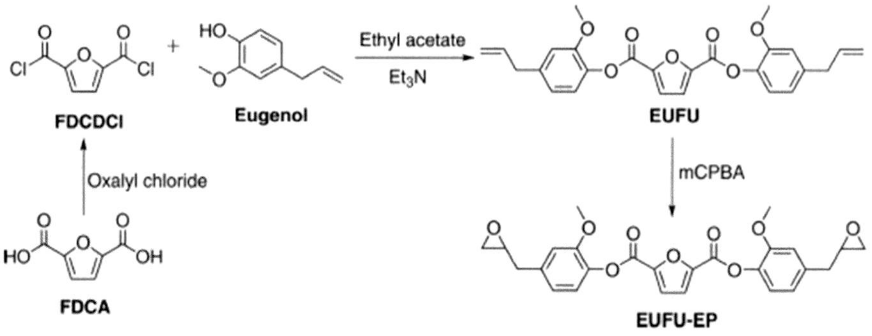

Figure 15.5 Synthesis of bio-epoxy EUFU-EP from eugenol (Miao et al., 2017).

Pas et al. synthesized the biobased prepolymer from native softwood lignin through hydrogenolysis and by reacting this lignin-based polyol with epichlorohydrin through a glycidylation process. The bioresin was blended with BADGE or GDGE, up to 25%–75% with an enhancement of 52% increase in flexural modulus and 38% increment was observed in the strength of cured epoxy resins (Pas & Torr, 20171). Wang et al. developed two novel biobased epoxy resins from the vanillin through a one-pot reaction containing Schiff base formation and phosphorus–hydrogen addition, subsequently reacting with epichlorohydrin. The cured biobased epoxies presented excellent flame retardancy with a UL-94 V0 rating and high LOI of 31.4% and 32.8%. Biobased epoxies showed higher T_g (214°C) and mechanical strength and modulus (80.3 MPa of 2709 MPa) compared with DGEBA-epoxy (Wang et al., 2017). Xin et al. synthesized the lignin-based epoxy resin by reacting epichlorohydrin (ECH) and partially depolymerized lignin (PDL) at 117°C and cured with the tung oil-derived anhydride (Methyl ester-MA) curing agent. The cured system demonstrated greater mechanical features than bisphenol A-based epoxy and the dynamic mechanical properties of cured PDL-epoxy were found to be higher than those of cured bisphenol-epoxy (Xin et al., 2016). Hernadez et al. reported the synthesis of biobased bisguaiacol (BG) through electrophilic aromatic condensation of vanillyl alcohol, a lignin derivative, and guaiacol. Three biobased aromatic diglycidyl ethers (DGEBG, DGEVA, DGEGD) were prepared and cured with cycloaliphatic diamine (PACM). The thermal and thermo-mechanical properties of biothermosets were found to be comparable with those of commercial epoxy (Hernandez et al., 2016).

3.3 Cardanol-based epoxy polymers

Cashew nut shell liquid (CNSL) is a reddish brown viscous liquid that is not edible and is commonly collected from cashew nut shells. Cardanol, cardol, anacardic acid, and 2-methyl cardol are the primary components of natural CNSL (Voirin et al., 2014). All of these components are meta-substituted phenolic compounds that include almost all of the phenol reactions and can therefore be used as an excellent phenol substitute. Cardanol is considered to be a potential feedstock that could replace petro-based polymer products, with its unique molecular structure, commercial availability, and inexpensive nature. It is widely used for the production of materials such as polyurethane, epoxies, vinyl esters, phenolics, cross-linked resins, benzoxazine, and photochromic molecules. Cardanol is a phenolic compound which, in its meta position, has a long unsaturated side chain of 15 carbon atoms (Fig. 15.6). Some of the major chemical modifications performed on cardanol are epoxidation, esterification, hydrogenation, nitration, etherification, sulfonation, and decarboxylation.

There are many works on prepolymers based on cardanol and CNSL. The photopolymerization of cardanol/CNSL-based prepolymers with various photoinitiators and photosensitizers has been studied (Kanehashi et al., 2013). The thermal stability of these UV-curable prepolymers was up to 250–300°C, suggesting potential for their application in heat-resistant materials. The phenolic hydroxyl group of epoxidized cardanol was converted into epoxide by reacting with epichlorohydrin in presence of $ZnCl_2$ at temperature of 95 °C. Two of the significant epoxy resins are the mono/difunctional glycidal and polyepoxidized cardanol resins, as shown in Fig. 15.6. Many research works have been carried out on epoxidized cardanol and new types of bio-epoxies for binders and marine coatings (Atta et al., 2017; Jaillet et al., 2014). Cardolite Corporation has already commercialized different cardanol-based epoxies globally for industrial applications.

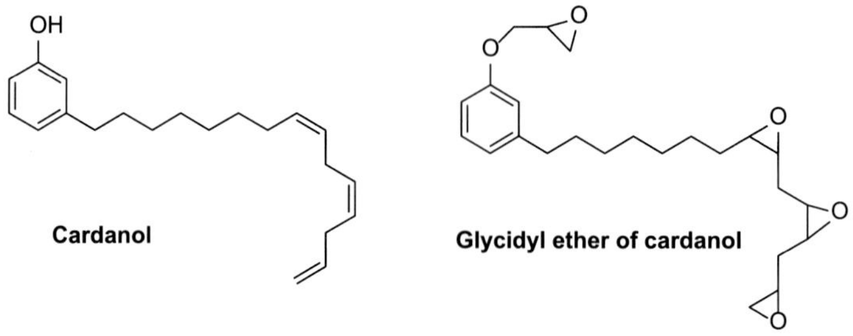

Cardanol

Glycidyl ether of cardanol

Figure 15.6 Chemical structure of cardanol and glycidyl ether of epoxidized cardanol.

3.4 Other renewable-sourced bio-epoxy resin

Isosorbide is prepared initially from starch, involving the hydrogenation of glucose and the dehydration of sorbitol. To produce desirable epoxy resins, isosorbides and furans extracted from sugars are used. In the presence of diaryl iodonium chloride, a new biodegradable UV-cured epoxy network from DGEADS and PHA-diepoxy by the process of photoinitiated cationic ring-opening polymerization. PHA diepoxy, bis-epoxidized terminated PHA oligomers, were first prepared in three stages: (1) microwave-assisted alcoholysis, (2) allyl isocyanate condensation, and (3) alkene terminal group epoxidation. Moreover, isosorbide diglycidyl ether (DGEDAS) was prepared by epoxidation of the allylic derivative (Lorenzini et al., 2015). Biobased epoxy copolymerized networks containing DGEDAS and PHA are shown in Fig. 15.7.

The preparation of monofuran diepoxide (2,5-bis[(2-oxiranylmethoxy)methyl]-furan (BOF)) and a bis-furan diepoxide, using 2,5-bis(hydroxymethyl)furan which are furan-based epoxy monomers, has recently been reported by Cho et al., as shown in Fig. 15.8 (Cho et al., 2013). These epoxies were cationically photocured and their applications were investigated for their adhesive property. Monofuran epoxy (2,5-bis [(2 oxiranylmethoxy)methyl]-furan (BOF) was crosslinked with the alicyclic and aromatic curing agents as 4,4 methylene biscyclohexanamine (PACM) and diethyl toluene diamine (EPIKURE W), respectively. Promising glass transition temperature values (71 and 88°C, respectively) and a storage moduli (3.5 GPa) at room temperature were achieved. These can be attributed to the restricted rotation of the furan ring in the polymer network due to the presence of 2,5-furandicarboxylic groups adjacent to the furan ring in the diglycidyl ester (Cho et al., 2013; Deng et al., 2015; El-Ghazawy et al., 2015; Hu et al., 2014).

Hu et al. synthesized novel epoxy resin containing both naphthalene moiety and cycloaliphatic group by chemical bonding in the same molecule with naphthol and limonene as the starting products, as shown in Fig. 15.9 (Hu et al., 2014). The rosin derivative was prepared by dehydrocarboxylation of isomerized abietic acid in forming dipimaryl ketone which was coupled with maleic anhydride catalyzed by a Diels–Alder reaction. Then dipimaryl ketone was epoxidized to obtain the corresponding tetra glycidyl ester. Rosin acid-based tetra-functional epoxy was found to have comparable viscoelastic properties and a higher glass transition temperature (El-Ghazawy et al., 2015).

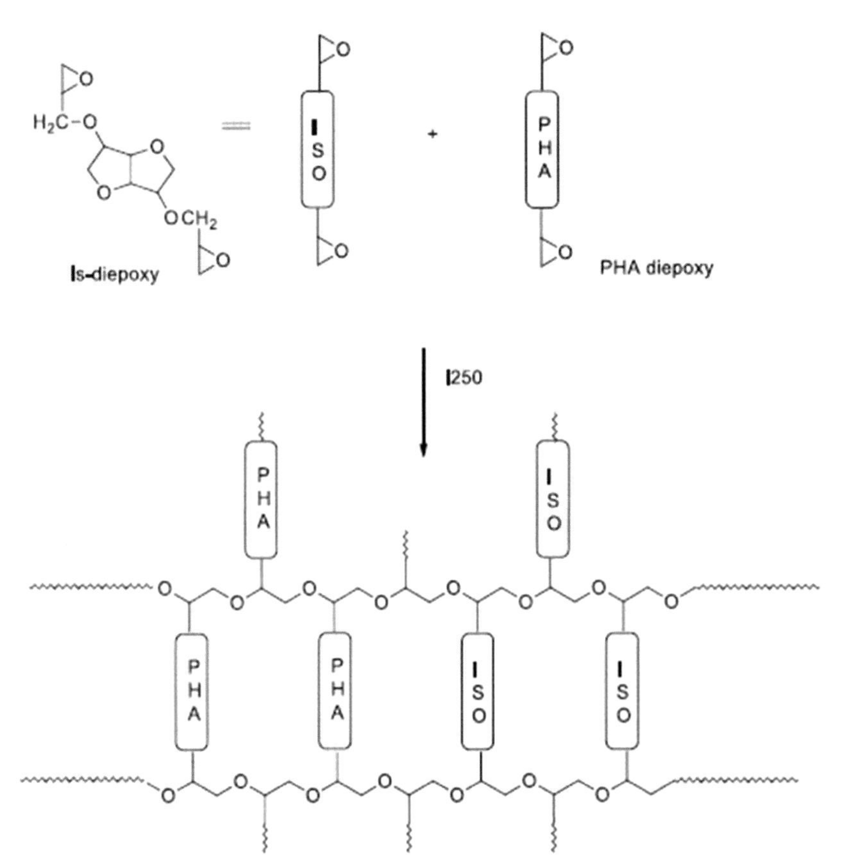

Figure 15.7 Synthesis of bio-epoxy resins based on network obtained by photo-initiated cationic copolymerization of PHA diepoxy and Is-diepoxy, DGEDAS (Lorenzini et al., 2015). Reprinted with permission from Lorenzini, C., Versace, D.L., Renard, E., & Langlois, V. (2015). Renewable epoxy networks by photoinitiated copolymerization of poly (3-hydroxyalkanoate) s and isosorbide derivatives. *Reactive and Functional Polymers*, *93*, 95—100.

3.5 Biodegradation studies

Kadam et al. studied the biodegradability of CA- and TA-cured epoxidized karanja oil-based bio-epoxies using a bacterial consortium, in which it was noticed that bioepoxy-CA got degraded up to 82% in 69 days, while the bioepoxy-TA was degraded by 95% in 259 days, as depicted in Fig. 15.10 (Kadam et al., 2015). The bacteria attack on the polymer chain led to visible changes such as cracks and the formation of a larger hole, as observed in a morphological study. Similarly, Sahoo et al. noticed complete degradation of the CA-cured ELO- and ECO-based epoxies in 2 wt.% NaOH alkali solution. This might have occurred due to the reaction of OH groups with an ester-bonded acid-cured epoxy system (Ding et al., 2015). Ma et al. published a biobased vitrimer produced from isosorbide-sourced epoxy and aromatic diamines, which has disulfide

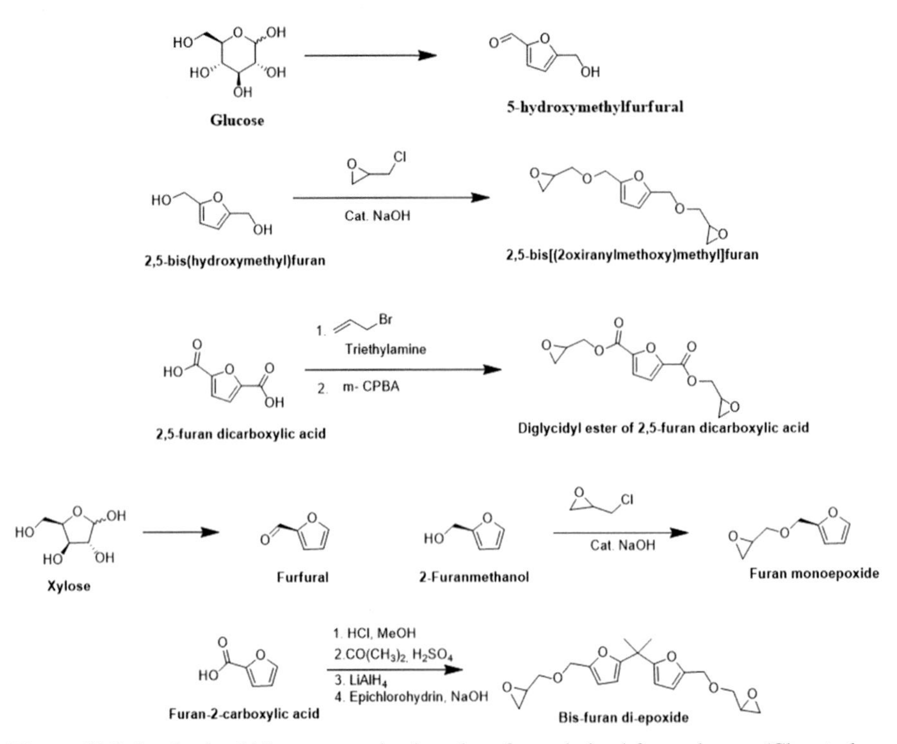

Figure 15.8 Synthesis of bio-epoxy resins based on furan derived from glucose (Cho et al., 2013; Deng et al., 2015; El-Ghazawy et al., 2015; Hu et al., 2014).

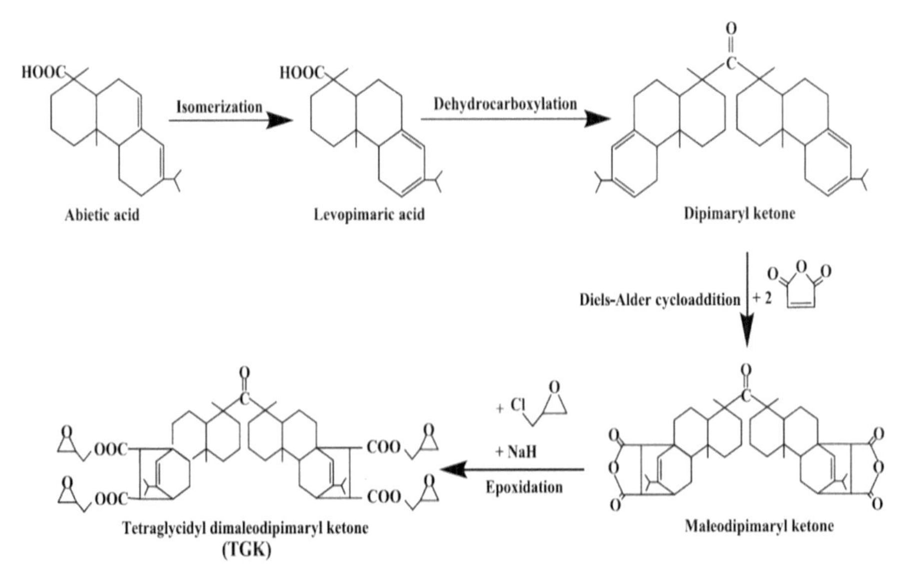

Figure 15.9 Synthesis of rosin-based tetra-functional bio-epoxy (El-Ghazawy et al., 2015). Reprinted with permission from El-Ghazawy, R. A., El-Saeed, A. M., Al-Shafey, H. I., Abdul-Raheim, A. R. M., & El-Sockary, M. A. (2015). Rosin based epoxy coating: Synthesis, identification and characterization. *European Polymer Journal*, *69*, 403.

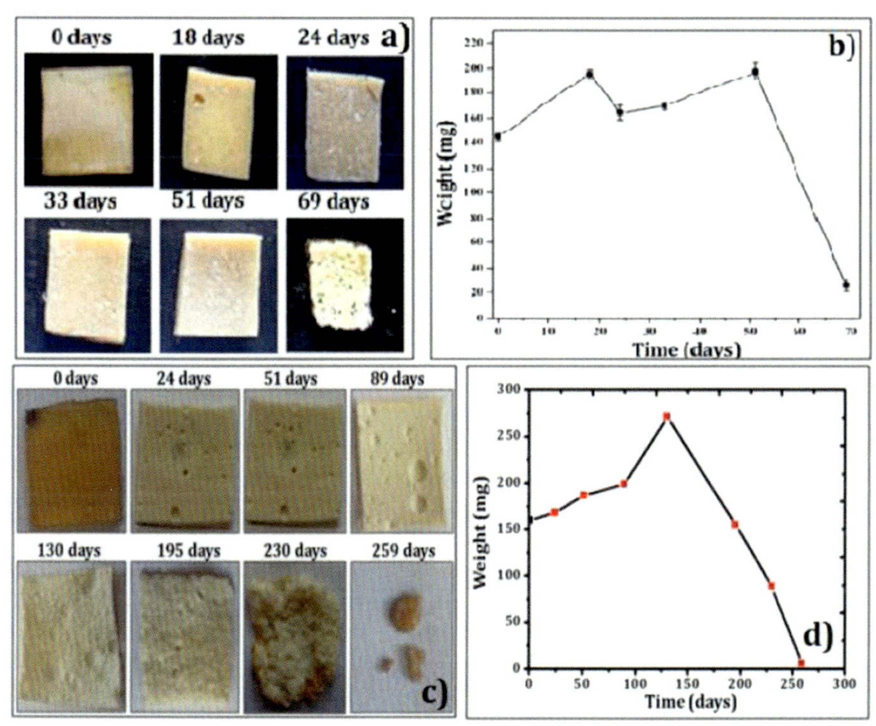

Figure 15.10 Biodegradation vs time and Weight change vs time of (A, B) Bioepoxy-CA and (C,D) Bioepoxy-TA (Kadam et al., 2015).
Reprinted with permission from Kadam, A., Pawar, M., Yemul, O., Thamke, V., & Kodam K. (2015). Biodegradable biobased epoxy resin from karanja oil. *Polymers (United Kingdom), 72*, 82–92.

linkages. In contrast to such a type of epoxy networks cured by conventional curing agents, the resulting complex showed comparable properties. The degradability of MDS-EPO was also demonstrated by dipping it in 5 wt.% sodium hydroxide (NaOH) solution due to the decrease in disulfide linkages in an alkaline atmosphere. It is anticipated that this could provide an alternative approach from both economic and green perspectives to material design (Ma et al., 2017). To enhance the biodegradation potential of the material, PHA was introduced into the isosorbide-based epoxy network. The samples were found to be thermally and hydrolytically degradable. The degradation behavior was studied at a temperature of 37°C and pH of 7.4 in the presence of a commercial lipase AK. These degradable materials paved the new way for developing biodegradable and environment-friendly bio-epoxies derived from a renewable resource (Lorenzini et al., 2015). Chow et al. studied the soil-burial biodegradability of ESO biothermosets by burying the ESO specimens in compost soils for 8 months. It was noted that the ESO biothermosets with lower thickness and lower cross-link density experienced a significant weight loss and could be degraded under these environmental conditions. From the 16S rDNA sequencing technique, it was found that *Comamonas* sp., *Bacillus* sp., *Streptomyces* sp., and *Acinetobacter* sp. are potential soil microbes to biodegrade ESO biothermosets (Chow et al., 2014).

As the methyl esters are low-molecular-weight compounds, they can undergo degradation easily and hence the bioepoxy blends have the possibility of undergoing degradation if kept in compost soil (Sahoo et al., 2018b). In a work by Salam et al., it was seen that the biodegradability of the blend increased when the percentage of epoxidized oil content increased beyond 40% due to the over-plasticizing effect, leading to a lower cross-linking density owing to the presence of long fatty acid chains in the cross-linking points. The hardener also plays a major role in the cross-linking density, which in turn affects the biodegradability (Salam & Dong, 2019).

4. Bio-epoxy thermoset blends: value-added matrix

Despite having excellent mechanical and thermal stability, synthetic epoxy resins have serious drawbacks in terms of brittleness or poor crack resistance, which limit their wider exploitation in structural applications. On the other hand, pure bioresins lack adequate mechanical and thermophysical properties to be useful for structural applications. In order to achieve a stiffness–toughness balance, vegetable oil-based bioresins have been reported as successful modifiers or reactive diluents to overcome the inherent brittleness of cross-linked networks (Cadu et al., 2019). Further, petroleum-derived resins are blended with plant oil-based resins as a reactive diluent to ensure good processability, reduced viscosity, improved wettability, and enhanced thermophysical properties. To be able to act as a successful modifier the reactive diluent must be compatible and have low viscosity, and must also have a functional group that can react with the other components during the curing process. The functional groups must be of the right amount and reactivity degree to prevent in-can storage problems, retardation in curing, and over plasticizing effect of non-reactive components in the final product (Majid et al., 2018). Simultaneously, reduction and control of the viscosity of epoxy resins have been of key interest to enhance processibility in the liquid-molding technologies like resin transfer molding, pultrusion, etc. The rheological behavior of resin plays a vital role in determining the processibility and end-use property of a thermoset material. The toughening mechanism of bioresin blends is shown in Fig. 15.11.

Further, viscosity of the resin plays a key part in the properties of the end-product of a polymer. The commercially available DGEBA-based epoxy has a viscosity in the range of 9000–12,000 cps, which makes it quite difficult for proper mixing with hardener and for proper dispersion of fillers into the matrix. Most of the bioresins have lower viscosity then epoxy and hence could help in reducing the viscosity of the resin when blended to act as diluents. The thermal properties play a major role in the case of polymer applications. The glass transition temperature of bioresins is comparatively very low with respect to epoxy resin due to the low cross-linking density resulting from flexible aliphatic chains. Hence, to achieve a balance between toughness, stiffness, and thermo-mechanical properties, bio-epoxy blends act as a value addition to the matrix.

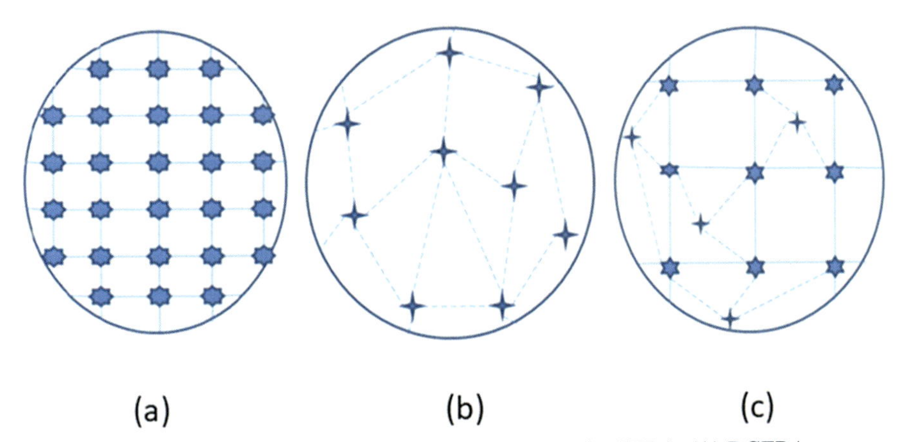

Figure 15.11 Diagrammatic representation for toughening of DGEBA: (A) DGEBA, (B) bioresin, and (C) blends.

4.1 Plant oil-based epoxy blends

Plan oil-based bioresins are better alternatives, which can substitute petro-based resins partly due to functionalization with different groups, abundant availability, low cost, and wide possibilities for chemical transformation. Furthermore, if the reactive diluent is biorenewable it would have a greater positive environmental impact since it would not only reduce the VOC content but also make the material partially biobased and biodegradable.

The structure of common plant oil and castor oil with unique functionality is presented in Figs. 15.12 and 15.13 respectively, for better understanding of the feasibility.

4.1.1 Soyabean oil-based blends

Soybean oil is one of the cheapest and most plentiful natural resources existing in the planet. It is extensively used as an ideal substitute for petro-based compounds for the synthesis of biobased epoxy resins. However, due to the long aliphatic chains, as is the case for most plant-based bioresins, polymerized soybean oil-based polymers do not yield satisfactory mechanical and thermo-physical properties. However, the modifications like epoxidation, esterification, maleation, and acrylation of soybean oil have shown better properties than the unmodified counterpart. The incorporation of ESO in parent epoxy enhances the toughness and at the same time overcomes the brittleness, thereby acting as a diluent. Many researchers have shown that the mechanical and thermo-physical properties have reduced on blending a higher amount of ESO with DGEBA due to surplus plasticization caused by long aliphatic chains and low cross-link density (Sahoo et al., 2015a). Further modification, such as transesterification, maleation, etc. reduces the viscosity and enhances the reactivity, which results in superior cross-linked networks with better properties at the same bio-content. The synthesis scheme of ESO and transesterified ESO is shown in Fig. 15.14. The works carried out on different oil-based blends cured by different cross-linkers with the

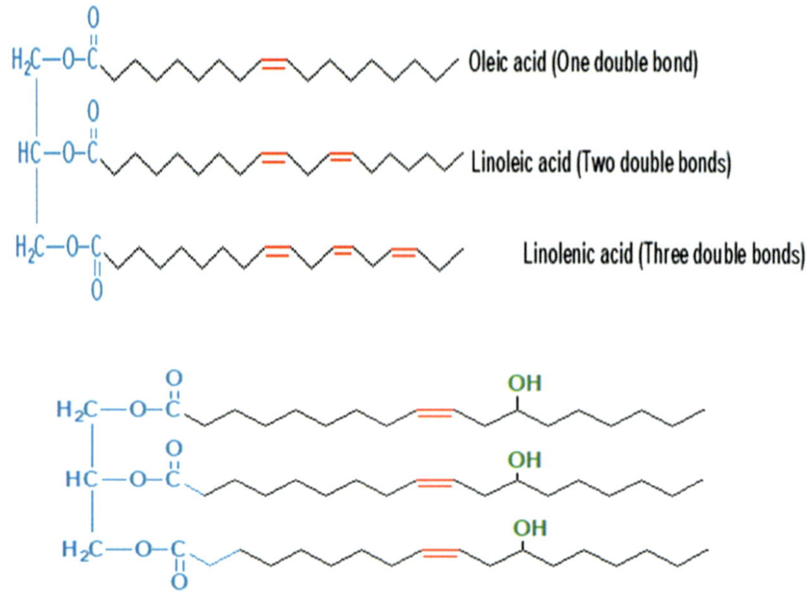

Oleic acid (One double bond)

Linoleic acid (Two double bonds)

Linolenic acid (Three double bonds)

Figure 15.12 Common chemical structure of triglycerides of plant oil.

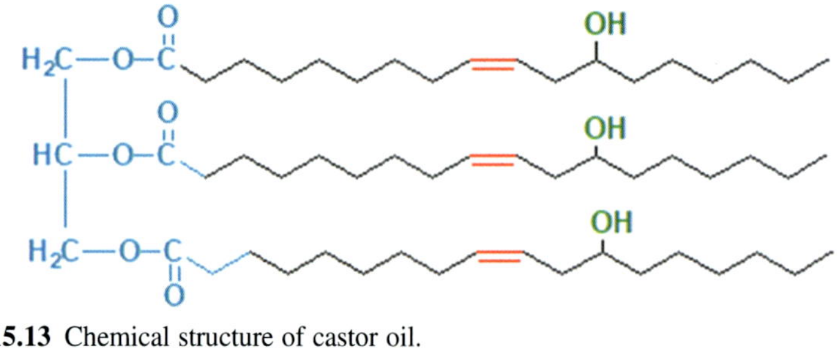

Figure 15.13 Chemical structure of castor oil.

achieved properties are presented in Table 15.2. To yield tougher thermosetting polymers, ESO can be polymerized with maleate half-esters of resols that are oil-soluble, p-nonyl phenol (p-NP) and p-tertiary butyl phenol (p-TBP) (Cavusoglu & Cayli, 2015). Maleated hydroxylated soybean oil (MHSO) and maleated acrylated epoxidized soybean oil (MAESO) were synthesized and blended with styrene for SMC applications by Lu and Wool (Lu & Wool, 2007). Maleic anahydride was used to modify acrylated ESO (AESO) to form maleinized acrylated ESO (MAESO). The flexural properties of these blends showed higher values, which revealed their suitability for industrial applications.

4.1.2 Linseed oil-based epoxy blends

Among the nonedible oils, linseed oil is one of the most used, with an oil iodine value of 170–200 mg/100 g for epoxidation as it has a high number of unsaturated double

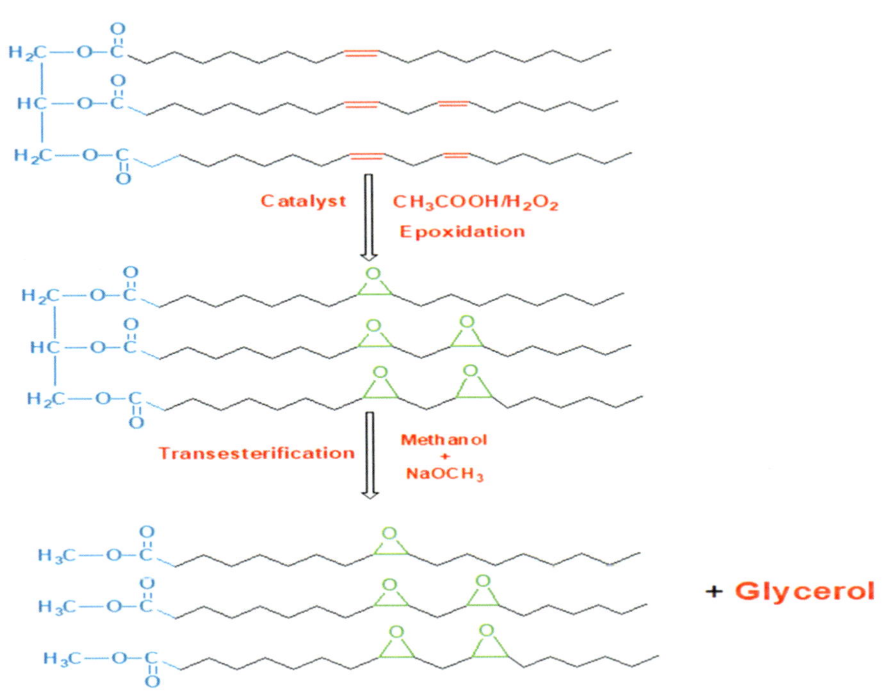

Figure 15.14 Synthesis of ESO and EMR (epoxidized methyl soyate) from SO (Sahoo et al. 2015b).

bonds with nearly 56% linolenic fatty acid (3 C=C) and as a result epoxidized linseed oil (ELO) has a high epoxide content and is more reactive toward polymerization. It has 6.6 double bonds per triglyceride chain. ELO has yielded a high cross-linked network suitable for industrial applications when cured with dicarboxylic acids or anhydrides owing to its higher oxirane content (Ding et al., 2015; Sahoo et al., 2018c). Miyagawa et al. prepared biobased epoxy material with ELO cross-linked with a methyltetra-hydrophthalic anhydride (MTHPA) curing agent (Miyagawa et al., 2004) and poly(oxypropylene) triamine (POPTA) curing agent (Miyagawa et al., 2004). DMA study revealed that the storage modulus and T_g were slightly decreased for anhydride-cured epoxy blends and significantly reduced for an amine-cured epoxy system with as increase in the ELO content. The peak factor of the amine-cured epoxy system remained almost constant irrespective of the ELO loading, confirming the homogeneity of the epoxy network. In contrast, the peak factor of the anhydride-cured epoxy system increased with an increase in the ELO content. The izod impact strength of the amine-cured epoxy blend was determined to be higher than for the anhydride-cured epoxy blend. Sahoo et al. blended ELO with DGEBA at up to 30% and cured with biobased phenalkamine hardener, resulting in a reduction in stiffness and improvement in toughness due to the surplus bioresin undergoing macro-phase separation, thereby incorporating ductility (Sahoo et al., 2018a).

Table 15.2 List of study of plant oil-based epoxy blends with different curing agents.

S. no.	Resin	Curing agent	Properties	References
1.	Epoxidized soybean oil/ epoxy	TETA (triethylenetetramine)	**1.** Maximum impact strength and fracture toughness of 28.2 J/m and 2.07 MPa $m^{1/2}$ respectively at 20 phr ESO **2.** Break at elongation increased significantly to 3.9% for 30 phr	(Sahoo et al., 2015a)
2.	Epoxidized soybean oil/ epoxy	MHHPA with 2-methyl imidazole (2-MI) as the catalyst	**1.** Tensile properties at 20 phr ESO > virgin epoxy **2.** Impact strength, critical stress intensity factor, critical strain energy release rate were maximum at 20 phr ESO	(Fache et al., 2015)
3.	Epoxidized soybean oil/ epoxy	Methyltetrahydrophthalic anhydride (MTHPA) with 1-methyl imidazole (1-MI) as a catalyst	**1.** Impact properties increased by 38% at 40 phr ESO compared to pure epoxy **2.** Compression properties were reduced with an increase in ESO content	(Altuna et al., 2011)
4.	Epoxy methyl soyate/ epoxy and epoxidized soybean oil/epoxy	TETA	**1.** 10% EMS/epoxy showed a tensile strength of 45.29 MPa and almost the same modulus as that of epoxy **2.** EMS/epoxy blends showed better properties then ESO counterparts **3.** Flexural strength of 10% EMS/epoxy showed the same as that of epoxy **4.** Impact strength was maximum for 30% EMS/epoxy	(Sahoo et al., 2015b)
5.	Partially ESO/epoxy and fully ESO/epoxy	TETA	pESO displayed superior tensile strength and elongation at break than the system with fESO at ESO 30 wt.%	(Lim & Kim, 2021)

6.	Epoxidized soybean oil/ epoxy	Alicyclic acid anhydride	Maximum tensile, flexural, and impact strength were seen at 10% ESO	(Lee, 2018)
7.	Epoxidized allyl soyate (EAS)/epoxy	Aliphatic amine and phthalic anhydride	**1.** Anhydride-cured EAS blends showed higher initial storage modulus **2.** The amine-cured resins showed higher tan delta values **3.** Cross-linking for anhydride-cured samples was more than for amine-cured ones	(Shabeer et al., 2005)
8.	Epoxidized linseed oil/ epoxy	Phenalkamine	**1.** The impact strength and elongation at break increased by 43% and 285% for 30 phr ELO **2.** Tensile strength and modulus decreased by 170% and 95% for 30 phr ELO	(Sahoo et al., 2018a)
9.	Epoxidized linseed oil/ epoxy epikote 240	Diethylenetriamine (DETA)	**1.** The 10% ELO blend exhibited improved tensile and flexural properties than epoxy **2.** Impact strength increased by 12% for 30% ELO and was maximum	(Nguyen, 2020)
10.	Epoxidized castor oil (ECO)/Epoxy	TETA	**1.** Tensile properties decreased with an increase in ECO content **2.** Flexural property increased by 5% at 20 phr ECO **3.** Maximum impact strength was seen at 20 phr ECO	(Sudha et al., 2017)

Continued

Table 15.2 Continued

S. no.	Resin	Curing agent	Properties	References
11.	**i)** Castor oil glycidyl ether (COGE)/ epoxy **ii)** Hydrogenated castor oil glycidyl ether (HCOGE)/ epoxy **iii)** Epoxidized castor oil glycidyl ether (ECOGE)/epoxy	Diethylenetriamine (DETA)	**1.** All the three bioresin blends showed maximum tensile property at 5% **2.** Break at elongation and impact strength increased with an increase in bioresin for all bioresins **3.** Highest tensile strength was shown by 5% ECOGE blend (81.6 MPa) **4.** 20% HCOGE blend displayed the maximum impact strength of all	(Fu et al., 2020)
12.	**i)** Epoxidized castor oil (ECO)/epoxy **ii)** Epoxy methyl ricinoleate	Phenalkamine	**1.** The tensile properties decreased with addition of ECO and EMR **2.** EMR blends showed better mechanical properties than the ECO counterparts **3.** Highest impact strength was seen for 30% EMR, which was 315% more than epoxy	(Sahoo et al., 2018b)
13.	Epoxidized palm oil/ epoxy	Methylhexahydrophthalic anhydride (MHHPA)	**1.** Fracture toughness increased with increase in EPO content	(Tan & Chow, 2010)
14.	Epoxidized palm oil/ epoxy	m-Xylylene diamine	**1.** Two methods for blends were investigated **2.** The premixing of EPO with hardener gave inferior mechanical properties to direct mixing **3.** T_g values also reduced in the case of premixing	(Sarwono et al., 2012)

4.1.3 Castor oil-based epoxy blends

Castor oil embodies unsaturated functionality and ricinoleic acid (12-hydroxy-9-*cis*-octadecenoic acid), the lone vegetable oil with hydroxyl groups with a hydroxyl value of around 156–160 (Fu et al., 2020; Hernandez et al., 2017). Unlike other plant oils, castor oil has a hydroxyl group which can be glycidated by reacting with epichlorohydrin to overcome the problem of low reactivity of alicyclic epoxy groups, also improving the toughness of epoxy resins. To overcome the oversoftening property, a group of researchers developed castor oil glycidyl ether (COGE), epoxidized castor oil glycidyl ether (ECOGE), and hydrogenated castor oil glycidyl ether (HCOGE) and blended it with epoxy. The increase in tensile strengths of these blends at small amounts was due to suitable long flexible chains that reduced the steric hindrance and led to an increase in the cross-linking density. The toughness of the entire three blends increased as the percentage of bioresin increased due to plasticization. Similarly, acrylated epoxidized castor oil (AECO) resin was blended with epoxy to achieve higher mechanical strength and thermal properties due to the existence of pendant acrylic double bonds that can take part in the free radical reaction (Akesson et al., 2009; Behera & Banthia, 2008). Similar works on modified castor oil-based epoxy blends have been reported with greater toughness and moderate mechanical properties, that are adequate for structural applications (Fache et al., 2015; Fang et al., 2020; Hu et al., 2015).

Apart from the above-mentioned plant oil blends, epoxy blends of crambe oil, palm oil, and rapeseed oil have been reported in the last decade as a low-viscous toughened resin system (Deng et al., 2013; Kumar et al., 2017, 2018; Ma et al., 2013; Okabe et al., 2009).

4.2 Other biosourced epoxy blends

Researchers around the world have attempted to replace the aromatic group of DGEBA with biobased aromatic resins such as vanillin, eugenol, magnolol, etc. as the polymers from plant/vegetable oil, showed inferior mechanical and thermomechanical properties. Similarly, the cardanol based epoxy blends show reasonable properties for coating applications (Darroman et al., 2015) and epoxidized cardanol-formaldehyde was used to toughen DGEBA resin with reduced mechanical properties (Natarajan and Murugavel, 2016). Therefore, biobased aromatic structures could be introduced into epoxy by blending and subsequent crosslinking. A great deal of work has been done on biobased aromatic resins like methacrylated vanillin (Zhang et al., 2015), diglycidyl ether of magnolol (DGEM) (Qi et al., 2020), diglycidyl ether of vanillic acid (Stanzione et al., 2012), diglycidyl ether of vanillyl alcohol (Fache et al., 2015), epoxidized divanillin (Fang et al., 2020), eugenol based, furan based (Hu et al., 2014, 2015), lignin based (Hirose et al., 2001) and rosin based (Deng et al., 2013). Most of these display brittle behavior, which hinders their usage in structural applications and also very limited works have been reported in the area of aromatic bioresin blends. Zhang et al. (2015) synthesized and copolymerized methacrylated vanillin (MV) and acrylated epoxidized soybean oil in different proportions.

Copolymers resulted in an increase in the storage modulus, glass transition temperature, Young's modulus, and tensile strength with an increase in methacrylated vanillin due to the rising quantity of rigid aromatic rings in the polymer chains, and thereby the methacrylated vanillin acted as hard phase and acrylated epoxidized soybean oil acted as soft phase. However, the thermal stability and elongation at break of these blends decreased with increasing MV content. The fully biobased copolymer or blend properties could be tailored by varying the content of either aromatic or aliphatic bioresin.

Itaconic acid (IA) is a nontoxic, easily biodegradable crystalline unsaturated diprotic compound which is prepared by distillation of citric acid. A total of 80,000 tons per year of itaconic acid are produced globally (Ma et al., 2013; Okabe et al., 2009). It consists of dual carboxylic groups, one carboxylic group that is conjugated to the methylene group without a flexible bond or rigid ring. IA can be used for blends in resin, plastic, adhesive, and binder applications. Trifunctional epoxy resin (TEIA) can be synthesized by reacting IA with meta-chloroperoxybenzoic acid followed by cooling the mixture at room temperature then washing with 10 wt.% of Na_2SO_3 solution and distilled water (Kumar et al., 2017). The increase in TEIA content increase the impact strength, tensile strength, and modulus due to inclusion of the flexible segment of TEIA into brittle natured epoxy. This TIA enhanced the internal stress and superior interfacial bonding with the pure epoxy as result of its end functionality and better cross-linking density (Kumar et al., 2018).

5. Bio-epoxy thermoset composites

5.1 Natural fiber-based bio-epoxy green composites

As per the advancements in the fields of science and technology, biocomposites have gained increased attention from the research community because of their economic and environmental benefits. The viability of biocomposites for commercial applications has become tangible only due to the huge abundance of biobased polymers as matrix and natural fibers as reinforcement. Since improper disposal of vast amounts of nonbiodegradable fiber wastes creates serious environmental hazards, these plant-derived fibers, because of their biodegradability, could be excellent alternatives for conventional synthetic fibers. They are noncorrosive, nontoxic, biocompatible, and eco-friendly with high biological abundance, low-cost production, and high energy efficiency. They also minimize concerns about an increasing carbon footprint due to the use of synthetic composites. Bio-epoxies derived from green natural resources are mostly from plants, soyabean oil, linseed oil, castor oil, Karanja oil, cardanol, natural resin, terpenes, rosin, lignin, and derivatives (Baroncini et al., 2016).

Pure bioresin or its blends with high loading of bioresin are not suitable for structural applications on account of their inadequate stiffness and strength. Thus, in recent years, natural fiber-reinforced plant oil-based thermoset composites have drawn growing attention from academia and industries because of their higher stiffness and strength, light weight, sustainability, and eco-friendliness. Several authors have used natural fibers including jute, cellulose, bamboo, kenaf, hemp, kapok, flax, sisal, ramie, vine, ratten, treated-palm tree fiber, date palm fiber, luffa, coir fiber, etc., to enhance the performance of bio-epoxy polymers in recent years (Väisänen et al., 2017). Some of the recent works on biobased epoxy composites including epoxidized oils and other bioresins are presented in Table 15.3.

Minerals spewed from volcanic eruptions can be used as natural fibers. Espana et al. researched a green composite based on basalt fibers and epoxy based on vegetable resource by vacuum-assisted resin transfer molding (España et al., 2013). The tensile modulus of untreated basalt fibers is limited to 7 GPa but treatment with a silane coupling agent increases the cross-linking density, which results in a two- to threefold increase in the tensile modulus as silane acts as a bridge between epoxy and basalt fibers, providing optimum wetting properties to the basalt fibers (España et al., 2013). Flax fibers are extensively used in the clothing and garment industry, but they also have potential applications in moderate- to high-strength biocomposites. Marrot et al. researched a biocomposite based on flax fibers and partially biobased epoxy based on linseed oil for green composite fabrication (Marrot et al., 2014). A tensile modulus of 28 GPa was observed for 47% volume fraction of flax fibers, which illustrated strong sensitivity of high interfacial shear strength with respect to the in-plane shear strength of the composite (Marrot et al., 2014). In the same year Luca Di Landro researched another bio composite based on hemp fibers. With a tensile modulus of 5.8 GPa, it has satisfactory damping properties which are indirectly related to the improved shear modulus and stiffness of the composite. The observed results show promising applications in secondary structures and insulating panels (Di Landro & Janszen, 2014). Scalici investigated a green composite based on plasma-treated leaf of the giant reed *Arundo donax* L. as fibers and biobased epoxy resin (Scalici et al., 2016). The treated lignocellulosic natural fibers provided better adhesion at the interface of resin and fiber, which resulted in enhanced modulus and strength. The addition of plasma treatment improved its interfacial adhesion properties by activation of functional groups. It has a tensile modulus of 18 GPa, which is comparable to that observed in lignocellulosic fibers composite (Scalici et al., 2016). In the same year, Scarponi et al. investigated a biobased epoxy composite using hemp natural fibers. The composite was subjected to a low-velocity impact test. It has a flexural modulus of 5.33 GPa at 20 J impact energy, which decreases by 50% for 20 J impact energy, which is a similar test standard as is used for traditional synthetic epoxy (nonecofriendly) (Scarponi et al., 2016). Based on further experiments, hemp fiber-biobased epoxy composites could be the potential candidates for the next generation of semistructural materials. Recent works stated in Table 15.3 revealed that scaling up with higher mechanical properties can be achievable with a mixture of two or more natural fibers and blends of biobased epoxies. Further, the additional pretreatment process upon natural fibers can tailor specific functional groups to achieve the desired mechanical properties in biobased epoxy composites and paved way to be the next generation high performance structural materials. However, the biodegradability of these composites has not been explored under different environmental conditions.

Table 15.3 Most recent research into natural fiber-reinforced biobased epoxy composites.

S. no.	Biobased epoxy	Natural fiber reinforcement	Properties	Highlights	References
1.	Epoxidized soyabean oil (30%), DGEBA—base component, epoxy oxirane (6.5%) + triethylenetetramine (TETA) (hardener)	Woven sisal fibers	Tensile strength 80–85 MPa	Increase in tensile strength by NaOH and silane treatment	(Rajkumar et al., 2017)
2.	Epoxidized soyabean oil (20%), DGEBA—base component	Short sisal fibers (15% volume fraction) + TETA hardener	Tensile strength 80–85 MPa, tensile modulus 1.3–2.5 GPa, impact strength 17.1–41.6 J/m	ESO enhances ductility nature of modified epoxy matrix increases elongation, impact strength, tensile modulus	(Sahoo et al., 2017)
3.	Araldite AW106 epoxy resin (biobased) + HV953 hardener	Jute, sisal, and banana fibers (40%) volume fraction	Tensile strength (288–432 MPa), flexural strength (75–89 MPa), impact strength (300–450 kJ/m^2)	Twofold tensile strength, flexural strength increment by NaOH treatment of fibers	(Singh et al., 2017)
4.	Epoxy resin (CHS-EPOXY G520) ~ 30% biobased + hardener ~100% biobased	Unidirectional flax fibers fabric	Tensile strength (288–432 MPa), tensile modulus 15 GPa), bond strength (3.5–4 MPa)	Accelerated aging conditions cause slight decrease in mechanical properties	(Benzarti et al., 2018)
5.	Super Sap CPM/CPL (resin/ hardener) ~ 31% biobased	Biotex flax unidirectional mats (23%) volume fraction	Tensile strength (45–75 MPa), tensile modulus (3–5.5 GPa), shear strength (4.5–7 MPa)	Acetic anhydride treatment showed improvements with moisture resistance and tensile strength	(Loong & Cree, 2018)

6.	Epoxamite ~ biobased (resin/ hardener)	Kenaf fibers	Increasing crack length of 47–100 mm for 3, 6, 9 J impact energy	Slightly lower impact damage resistance relative to glass/epoxy	(Majid et al., 2018)
7.	Epoxidized neem seed oil (25%), diglycidyl ether of bisphenol-A, + triethylenetetramine (hardener)	Kenaf fibers + sea urchin particles (40% volume fraction)	Tensile strength (118–135 MPa), flexural strength (178–193 MPa), impact strength (4.23–4.75 J), shear strength (27–28 MPa)	Surface-treated kenaf fibers with sea urchin particles brought 20% in mechanical properties	(Prakash & Viswanthan, 2019)
8.	DGEBA epoxy resin (SR 8500) + amine hardener (SZ 8525)	Pure unidirectional flax fibers (47% volume composition)	Tensile strength (275–325 MPa), tensile modulus (27–30 GPa)	Influence of hygrothermal aging brings a slight decrease in tensile properties	(Cadu et al., 2019)
9.	GreenPoxy-56/SD GP 505 V2, with about 56% biobased	Hemp fiber woven fabric (60% volume fraction)	Bending stiffness (23–26 GPa), torsional stiffness (12–11 GPa)	High epoxy retention capacity but lower bending and torsional stiffness	(Maino et al., 2019)
10.	Greenepoxy (InfuGreen) 810/SD 8822, mixing ratio of 100:31, from Sicomin, Chateauneuf (38% biobased)	Noncrimp unidirectional flax fiber fabric (40% volume fraction)	Ultimate tensile strength (240–300 MPa), tensile modulus (25–28 GPa), 96 h creep resistance at 60% UTS	Pretreatment on flax fiber using a green chemical FA was applied aiming to reduce the creep deformation of FFCs	(Jia & Fiedler, 2020)
11.	Epoxidized vegetable oil (surf clear) ~ (40% biobased)+ SD O hardener, Sicomin resin Greenpoxy 56) ~ (56% biobased) + Sicominhardner SZ 8525	Ixtel, henequen, jute (varying volume fraction 15%–29%)	Ultimate tensile strength (35–75 MPa), tensile modulus (3–8 GPa), fleuxral modulus (3.5–7 GPa)	Bioresin's mechanical performance was highly favored with ixtle fiber, jute biolaminates present higher stiffness	(Torres-Arellano et al., 2020)

Continued

Table 15.3 Continued

S. no.	Biobased epoxy	Natural fiber reinforcement	Properties	Highlights	References
12.	Biobased epoxidized oligo-isosorbide glycidyl ethers + isophorone diamine (IPD) hardener	Scutched textile flax (Aramis variety) (60% volume fraction)	Ultimate tensile strength (965–1338 MPa), tensile modulus (3–8 GPa), stiffness (270–340 MPa)	Feasibility to synthesize biobased epoxy prepolymers with high reactivity toward amine hardener, quasi fully biobased composites	(Musa et al., 2020)
13.	"SR GreenPoxy 56"—(56% biobased) + SD surf clear hardener "Sicomin"	Kenaf and sisal fiber layers (volume fraction 18% approximately)	Tensile strength (28–40 MPa), tensile modulus (0.9–1.3 GPa), impact strength (6–8 kJ/mm^2)	Addition of kenaf/sisal fibers dampens the mechanical properties degradation due to accelerated weathering, water absorption also contributed	(Yorseng et al., 2020)
14.	Greenpoxy 55 resin (56% biobased), biobased epoxy resin system Super Sap 100/1000 (37% biobased)	Hemp fibers, plain weaves (35%–45% volume fraction)	Tensile strength (40–54 MPa), tensile modulus (3.7–6.4 GPa), flexural modulus (3.59–5.84 GPa)	Resistance to deformation of composites made with bioresins has a minimum reduction of 20% by natural fiber addition, allowing the substitution of traditional resins based on biomaterials	(Colomer-Romero et al., 2020)
15.	Greenpoxy 56 resin (56% biobased) + SD surf clear hardener	Water hyacinth fibers	Tensile strength (40–65 MPa), tensile modulus (1.6–2.2 GPa), flexural modulus (3.06–3.3 GPa), impact strength (3.83–4.42 GPa)	Silane-treated WHFs reinforced biobased epoxy 0° has highest tensile strength value (65.3 MPa) and raw WHFs reinforced biobased epoxy composite 90° orientation has lowest (25.7 MPa)	(Sumrith et al., 2020)

16.	Surf Clear EVO (biobased content 40%) + SD EVO fast hardener	Three-ply yarn of kenaf fibers, organic GrO particles	Tensile strength (70 MPa), tensile modulus (6 GPa), viscoelastic storage modulus (3915.2 MPa)	Presence of functional groups enhances interface interaction between the fillers and polymer chains. GrO particles enhance mechanical and viscoelastic properties of composite	(Franco-Urquiza & Rentería-Rodríguez, 2020)
17.	Araldite LY 556 epoxy (biobased) + Ardur HY 951 (hardener)	Caryota and sisal fibers (total 40% volume fraction, but varying amounts)	Tensile strength (22.2−38.2 MPa), tensile modulus (4.2−6.44 GPa), flexural modulus (2.14−3.4 GPa), impact strength (88−97 J)	15% caryota and 25% sisal fibers has best set of mechanical properties in composite	(Atmakuri et al., 2021)
18.	Prime 27 epoxy resin (biobased), benzoxazine resin (biobased)	Plain weave flax fabric	Normalized flexural modulus and flexural strength (70%−100%) of initial values during hydrothermal aging	Due to hydrophobic nature of benzoxazine resin, diffusion coefficient lower than flax epoxy composites. Benzoxazine-based composites have higher mechanical properties than glass epoxy resin composites	(Pisupati et al., 2021)

6. Biodegradability prospective

Biodegradable substances are environment-friendly materials which do not have a negative impact on nature. Biodegradability is the ultimate aim behind the preparation of all biobased thermosetting resins. Biodegradation is the breakdown of matter by the activity of microorganisms into several by-products such as carbon dioxide, water, ammonia, etc. The microorganisms cleave a specific chemical bond and eventually leads to the production of lower molecular weight products. It is reported that microorganisms are genetically modified in order to produce enzymes that will cleave the desired chemical bonds in existing polymers. In order to determine the long-term environmental effects of biobased materials, biodegradability studies are therefore critical. A number of factors make biobased polymers and composites appealing to be eco-friendly, including their potential for biodegradation, petroleum demand conservation, accessibility, low toxicity, economic efficiency, and low carbon footprint. The thermoset class of polymers, which is very much useful in structural applications, lacks biodegradability. Bio-epoxies are one of the emerging biodegradable resins made from vegetable oils, saccharides, tannins, cardanol, terpene, rosin, lignin, etc. Sustainability also relies on a decrease in the impacts on workplace and public health as well as health and safety throughout their life cycles. Biobased epoxy and its natural fiber composite materials are important candidates for sustainability as they are natural sourced via CO_2 sequestration, minimize greenhouse emissions, reduce the cost of raw materials by utilization of biomaa, and generate opportunities for growth and jobs in agriculture sector.

7. Conclusion

In this chapter, the preparation methods and applications prospective of biosourced epoxies, blends, and composites are explored through an extensive literature survey. Biobased epoxy resin can be used in a variety of applications, such as resins in the food packaging, coatings, adhesives, laminates, electronic materials, etc. Starting with completely biobased epoxies from different renewable resources through different chemical modifications and polymerization methods, toughened biobased blends and natural fibers bio-epoxy composites are briefly described with biodegradation potential. Further, the properties are compared with petroleum-based epoxy to reveal their suitability for industrial applications. The incorporation of less viscous oil-based bioresins not only enhanced the processibility and toughness, but also ensured better packing of fibers through proper impregnation. Based on recent studies, it is concluded that the combination of bioresin and natural fibers enhances the properties with a stiffness–toughness balance with adequate viscoelastic behavior. Higher toughness and damping ability of the biobased epoxy blends makes them a promising material for structural applications. Although many noteworthy results have been reported in the field of biobased epoxy composites in the last decade, the study of biodegradation of composites remains unexplored and needs to be addressed in future with a circular economy approach.

References

Akesson, D., Skrifvars, M., & Walkenstrom, P. (2009). Preparation of thermoset composites from natural fibres and acrylate modified soybean oil resins. *Journal of Applied Polymer Science, 114*, 2502.

Altuna, F. I., Esposito, L. H., Ruseckaite, R. A., & Stefani, P. M. (2011). Thermal and mechanical properties of anhydride-cured epoxy resins with different contents of biobased epoxidized soybean oil. *Journal of Applied Polymer Science, 120*, 789–798.

Altuna, F. I., Pettarin, V., & Williams, R. J. J. (2013). Self-healable polymer networks based on the cross-linking of epoxidised soybean oil by an aqueous citric acid solution. *Green Chemistry, 15*(12), 3360–3366.

Atmakuri, A., Palevicius, A., Kolli, L., Vilkauskas, A., & Janusas, G. (2021). Development and analysis of mechanical properties of Caryota and sisal natural fibers reinforced epoxy hybrid composites. *Polymers, 13*(6), 864.

Atta, A. M., Al-Hodan, H. A., Abdel Hameed, R. S., & Ezzat, A. O. (2017). Preparation of green cardanol-based epoxy and hardener as primer coatings for petroleum and gas steel in marine environment. *Progress in Organic Coatings, 111*, 283–293.

Baroncini, E. A., Kumar Yadav, S., Palmese, G. R., & Stanzione, J. F., III (2016). Recent advances in bio-based epoxy resins and bio-based epoxy curing agents. *Journal of Applied Polymer Science, 133*(45). https://doi.org/10.1002/app.44103

Behera, D., & Banthia, A. K. (2008). Synthesis, characterization, and kinetics study of thermal decomposition of epoxidized soybean oil acrylate. *Journal of Applied Polymer Science, 109*, 2583.

Benzarti, K., Chlela, R., Zombré, W., Quiertant, M., & Curtil, L. (2018). Durability of flax/bio-based epoxy composites intended for structural strengthening. In *, Vol. 199. MATEC Web of Conferences* (p. 07014). EDP Sciences.

Cadu, T., Van Schoors, L., Sicot, O., Moscardelli, S., Divet, L., & Fontaine, S. (2019). Cyclic hygrothermal ageing of flax fibers' bundles and unidirectional flax/epoxy composite. Are bio-based reinforced composites so sensitive? *Industrial Crops and Products, 141*, 111730.

Cavusoglu, J., & Cayli, G. (2015). Polymerization reactions of epoxidized soybean oil and maleate esters of oil-soluble resoles. *Journal of Applied Polymer Science, 132*(7), 41457.

Cho, J. K., Lee, J. S., Jeong, J., Kim, B., Kim, B., Kim, S., Shin, S., Kim, H. J., & Lee, S. H. (2013). Synthesis of carbohydrate biomass - based furanic compounds bearing epoxide end group(s) and evaluation of their feasibility as adhesives. *Journal of Adhesion Science and Technology, 27*, 2127.

Chow, W. S., Tan, S. G., Ahmad, Z., Chia, K. H., Lau, N. S., & Sudesh, K. (2014). Biodegradability of epoxidized soybean oil based thermosets in compost soil environment. *Journal of Polymers and the Environment, 22*(1), 140–147.

Colomer-Romero, V., Rogiest, D., García-Manrique, J. A., & Crespo, J. E. (2020). Comparison of mechanical properties of hemp-fibre biocomposites fabricated with biobased and regular epoxy resins. *Materials, 13*(24), 5720.

Darroman, E., Durand, N., Boutevin, B., & Caillol, S. (2015). New cardanol/sucrose epoxy blends for biobased coatings. *Progress in Organic Coatings, 83*, 47–54.

de Espinosa, L. M., & Meier, M. A. (2011). Plant oils: The perfect renewable resource for polymer science?! *European Polymer Journal, 47*(5), 837–852.

Deng, L., Ha, C., Sun, C., Zhou, B., Yu, J., Shen, M., & Mo, J. (2013). Properties of bio-based epoxy resins from rosin with different flexible chains. *Industrial and Engineering Chemistry Research, 52*(37), 13233–13240.

Deng, J., Liu, X. Q., Li, C., Jiang, Y. H., & Zhu. (2015). Synthesis and properties of a bio-based epoxy resin from 2,5-furandicarboxylic acid (FDCA). *Journal of RSC Advances, 5*, 15930.

Di Landro, L., & Janszen, G. (2014). Composites with hemp reinforcement and bio-based epoxy matrix. *Composites Part B: Engineering, 67*, 220–226.

Ding, C., Shuttleworth, P. S., Makin, S., Clark, J. H., & Matharu, A. S. (2015). New insights into the curing of epoxidized linseed oil with dicarboxylic acids. *Green Chemistry, 17*(7), 4000–4008.

Ding, C., Shuttleworth, P. S., Makin, S., Clark, J. H., & Matharu, A. S. (2015). New insights into the curing of epoxidized linseed oil with dicarboxylic acids. *Green Chemistry, 17*, 4000–4008.

El-Ghazawy, R. A., El-Saeed, A. M., Al-Shafey, H. I., Abdul-Raheim, A. R. M., & El-Sockary, M. A. (2015). Rosin based epoxy coating: Synthesis, identification and characterization. *European Polymer Journal, 69*, 403.

España, J. M., Samper, M. D., Fages, E., Sánchez-Nácher, L., & Balart, R. (2013). Investigation of the effect of different silane coupling agents on mechanical performance of basalt fiber composite laminates with biobased epoxy matrices. *Polymer Composites, 34*(3), 376–381.

Fache, M., Auvergne, R., Boutevin, B., & Caillol, S. (2015). New vanillin-derived diepoxy monomers for the synthesis of biobased thermosets. *European Polymer Journal, 67*, 527–538.

Fache, M., Viola, A., Auvergne, R., Boutevin, B., & Caillol, S. (2015). Biobased epoxy thermosets from vanillin-derived oligomers. *European Polymer Journal, 68*, 526.

Fang, Z., Nikafshar, S., Hegg, E. L., & Nejad, M. (2020). Biobased Divanillin as a precursor for formulating biobased epoxy resin. *ACS Sustainable Chemistry and Engineering, 8*(24), 9095–9103.

Franco-Urquiza, E. A., & Rentería-Rodríguez, A. V. (2020). Effect of nanoparticles on the mechanical properties of kenaf fiber-reinforced bio-based epoxy resin. *Textile Research Journal*, 0040517520980459.

Fu, Q., Tan, J., Han, C., Zhang, X., Fu, B., Wang, F., & Zhu, X. (2020). Synthesis and curing properties of castor oil-based triglycidyl ether epoxy resin. *Polymers for Advanced Technologies, 31*(11), 2552–2560.

Galià, M., de Espinosa, L. M., Ronda, J. C., Lligadas, G., & Cádiz, V. (2010). Vegetable oil-based thermosetting polymers. *European Journal of Lipid Science and Technology, 112*(1), 87–96.

Hernandez, E. D., Bassett, A. W., Sadler, J. M., La Scala, J. J., & Stanzione, J. F., III (2016). Synthesis and characterization of bio–based epoxy resins derived from vanillyl alcohol. *ACS Sustainable Chemistry and Engineering, 4*(8), 4328–4339.

Hernandez, N. P. L., Bonon, A. J., Bahu, J. O., Barbosa, M. I. R., Wolf Maciel, M. R., & Filho, R. M. (2017). Epoxy monomers obtained from castor oil using a toxicity-free catalytic system. *Journal of Molecular Catalysis A: Chemical, 426*, 550–556.

Hirose, S., Kobayashi, M., Kimura, H., Hatakeyama, H., & Kennedy, J.F. (2001). England: Woodhead Publishing, Chapter 9, 73–78.

Hu, F. S., La Scala, J. J., Sadler, J. M., & Palmese, G. R. (2014). Synthesis and characterization of thermosetting furan-based epoxy systems. *Macromolecules, 47*, 3332.

Hu, F., Yadav, S. K., La Scala, J. J., Sadler, J. M., & Palmese, G. R. (2015). Preparation and characterization of fully furan-based renewable thermosetting epoxy-amine systems. *Macromolecular Chemistry and Physics, 216*, 1441–1446.

Jaillet, F., Darroman, E., Ratsimihety, A., Auvergne, R., Boutevin, B., & Caillol, S. (2014). New biobased epoxy materials from cardanol. *European Journal of Lipid Science and Technology, 116*(1), 63–73.

Jia, Y., & Fiedler, B. (2020). Tensile creep behaviour of unidirectional flax fibre reinforced bio-based epoxy composites. *Composites Communications, 18*, 5–12.

Jung, J. Y., Park, C.-H., & Lee, E. Y. (2017). Epoxidation of methanol-soluble kraft lignin for lignin-derived epoxy resin and its usage in the preparation of biopolyester. *Journal of Wood Chemistry and Technology, 37*(6), 433–442.

Kadam, A., Pawar, M., Yemul, O., Thamke, V., & Kodam, K. (2015). Biodegradable biobased epoxy resin from karanja oil. *Polymers (United Kingdom), 72*, 82–92.

Kanehashi, S., Yokoyama, K., Masuda, R., Kidesaki, T., Nagai, K., & Miyakoshi, T. (2013). Preparation and characterization of cardanol-based epoxy resin for coating at room temperature curing. *Journal of Applied Polymer Science, 130*(4), 2468–2478.

Kumar, S., Samal, S. K., Mohanty, S., & Nayak, S. K. (2017). Synthesis and characterization of itaconic-based epoxy resins. *Polymers for Advanced Technologies, 29*(1), 160–170.

Kumar, S., Samal, S. K., Mohanty, S., & Nayak, S. K. (2018). Bio-based tri-functional epoxy resin (TEIA) blend cured with anhydride (MHHPA) based cross-linker: Termal, mechanical and morphological characterization. *Journal of Macromolecular Science, Part A: Pure and Applied Chemistry, 55*(6), 496–506.

Lee, S. (2018). Properties of epoxidized soybean oil modified epoxy resins 2: Thermal, mechanical, and morphological properties. *Polymer (Korea), 42*(3), 498–503.

Li, C., Fan, H., Aziz, T., Bittencourt, C., Wu, L., Wang, D. Y., & Dubois, P. (2018). Biobased epoxy resin with low electrical permissivity and flame retardancy: From environmental friendly high-throughput synthesis to properties. *ACS Sustainable Chemistry and Engineering, 6*(7), 8856–8867.

Lim, S. J., & Kim, D. S. (2021). Effect of functionality and content of epoxidized soybean oil on the physical properties of a modified diglycidyl ether of bisphenol A resin system. *138*(20), e50441.

Loong, M. L., & Cree, D. (2018). Enhancement of mechanical properties of bio-resin epoxy/flax fiber composites using acetic anhydride. *Journal of Polymers and the Environment, 26*(1), 224–234.

Lorenzini, C., Versace, D. L., Renard, E., & Langlois, V. (2015). Renewable epoxy networks by photoinitiated copolymerization of poly (3-hydroxyalkanoate) s and isosorbide derivatives. *Reactive and Functional Polymers, 93*, 95–100.

Lu, Y., & Larock, R. C. (2009). Novel polymeric materials from vegetable oils and vinyl monomers: preparation, properties, and applications. *ChemSusChem, 2*(2), 136–147.

Lu, J., & Wool, R. P. (2007). Sheet molding compound resins from soybean oil: Thickening behavior and mechanical properties. *Polymer Engineering and Science, 47*(9), 1469–1479.

Ma, S., Liu, X., Jiang, Y., Tang, Z., & Zhu, J. (2013). Bio-based epoxy resin from itaconic acid and its thermosets cured with anhydride and comonomers. *Green Chemistry, 15*, 245–254.

Ma, Z., Wang, Y., Zhu, J., Yu, J., & Hu, Z. (2017). Bio-based epoxy vitrimers: Reprocessibility, controllable shape memory, and degradability. *Journal of Polymer Science Part A: Polymer Chemistry, 55*(10), 1790–1799.

Maino, A., Janszen, G., & Di Landro, L. (2019). Glass/epoxy and hemp/bio based epoxy composites: Manufacturing and structural performances. *Polymer Composites, 40*(S1), E723–E731.

Majid, D. L., Jamal, Q. M., & Manan, N. H. (2018). Low-velocity impact performance of glass fiber, kenaf fiber, and hybrid glass/kenaf fiber reinforced epoxy composite laminates. *BioResources, 13*(4), 8839–8852.

Manthey, N. W., Cardona, F., Aravinthan, T., & Cooney, T. (2011). Cure kinetics of an epoxidized hemp oil based bioresin system. *Journal of Applied Polymer Science, 122*(1), 444–451.

Marrot, L., Bourmaud, A., Bono, P., & Baley, C. (2014). Multi-scale study of the adhesion between flax fibers and biobased thermoset matrices. *Materials and Design, 62*, 47–56 (1980–2015).

Miao, J. T., Yuan, L., Guan, Q., Liang, G., & Gu, A. (2017). Biobased heat resistant epoxy resin with extremely high biomass content from 2, 5-furandicarboxylic acid and eugenol. *ACS Sustainable Chemistry and Engineering, 5*(8), 7003–7011.

Miyagawa, H., Mohanty, A. K., Misra, M., & Drzal, L. T. (2004). Thermo-physical and impact properties of epoxy containing epoxidized linseed oil. *Macromolecular Materials and Engineering, 289*, 636–641.

Miyagawa, H., Mohanty, A. K., Misra, M., & Drzal, L. T. (2004). Thermo-physical and impact properties of epoxy containing epoxidized linseed oil. *Macromolecular Materials and Engineering, 289*, 629–635.

Musa, C., Kervoelen, A., Danjou, P. E., Bourmaud, A., & Delattre, F. (2020). Bio-based unidirectional composite made of flax fibre and isosorbide-based epoxy resin. *Materials Letters, 258*, 126818.

Natarajan, M., & Murugavel, S. C. (2016). Cure kinetics of bio-based epoxy resin developed from epoxidized cardanol–formaldehyde and diglycidyl ether of bisphenol–A networks. *Journal of Thermal Analysis and Calorimetry, 125*, 387–396.

Nguyen, T. A. (2020). Mechanical and flame-retardant properties of nanocomposite based on epoxy resin combined with epoxidized linseed oil, which has the presence of nanoclay and MWCNTs. *Journal of Chemistry*. https://doi.org/10.1155/2020/2353827

Okabe, M., Lies, D., Kanamasa, S., & Park, E. Y. (2009). Biotechnological production of itaconic acid and its biosynthesis in Aspergillus terreus. *Applied Microbiology and Biotechnology, 84*, 597–606.

Pas, D. J. V., & Torr, K. M. (2017). Biobased epoxy resins from deconstructed native softwood lignin. *Biomacromolecules, 8*, 2640–2648.

Pawar, M., Kadam, A., Yemul, O., Thamke, V., & Kodam, K. (2016). Biodegradable bioepoxy resins based on epoxidized natural oil (cottonseed & algae) cured with citric and tartaric acids through solution polymerization: a renewable approach. *Industrial Crops and Products, 89*, 434–447.

Pin, J. M., Sbirrazzuoli, N., & Mija, A. (2015). From epoxidized linseed oil to bioresin: an overall approach of epoxy/anhydride cross-linking. *ChemSusChem, 8*(7), 1232–1243.

Pisupati, A., Bonnaud, L., Deléglise-Lagardère, M., & Park, C. H. (2021). Influence of environmental conditions on the mechanical properties of flax fiber reinforced thermoset composites. *Applied Composite Materials*, 1–7.

Prakash, V. A., & Viswanthan, R. (2019). Fabrication and characterization of echinoidea spike particles and kenaf natural fibre-reinforced *Azadirachta-indica* blended epoxy multi-hybrid bio composite. *Composites Part A: Applied Science and Manufacturing, 118*, 317–326.

Qi, Y., Weng, Z., Zhang, K., Wang, J., Zhang, S., Liu, C., & Jian, X. (2020). Magnolol-based bio-epoxy resin with acceptable glass transition temperature. *Processability and Flame Retardancy Chemical Engineering Journal, 387*, 124115.

Rajkumar, S., Tjong, J., Nayak, S. K., & Sain, M. (2017). Permeability and mechanical property correlation of bio based epoxy reinforced with unidirectional sisal fiber mat through vacuum infusion molding technique. *Polymer Composites, 38*(10), 2192–2200.

Sahoo, S. K., Mohanty, S., & Nayak, S. K. (2015a). Synthesis and characterization of bio-based epoxy blends from renewable resource based epoxidized soybean oil as reactive diluents. *Chinese Journal of Polymer Science, 33*, 137–152.

Sahoo, S. K., Mohanty, S., & Nayak, S. K. (2015b). Toughened bio-based epoxy blend network modified with transesterified epoxidized soybean oil: synthesis and characterization. *RSC Advances, 5*(18), 13674–13691.

Sahoo, S. K., Mohanty, S., & Nayak, S. K. (2017). Mechanical, thermal, and interfacial characterization of randomly oriented short sisal fibers reinforced epoxy composite modified with epoxidized soybean oil. *Journal of Natural Fibers, 14*(3), 357–367.

Sahoo, S. K., Khandelwal, V., & Manik, G. (2018a). Development of toughened bio-based epoxy with epoxidized linseed oil as reactive diluent and cured with bio-renewablecrosslinker. *Polymers for Advanced Technologies*, 1–10.

Sahoo, S. K., Khandelwal, V., & Manik, G. (2018b). Renewable approach to synthesize highly toughened bioepoxy from Castor oil derivative–epoxy methyl ricinoleate and cured with biorenewable phenalkamine. *Industrial and Engineering Chemistry Research, 57*(33), 11323–11334.

Sahoo, S. K., Khandelwal, V., & Manik, G. (2018c). Development of completely bio-based epoxy networks derived from epoxidized linseed and castor oil cured with citric acid. *Polymers for Advanced Technologies, 29*(7), 2080–2090.

Salam, H., & Dong, Y. (2019). Property evaluation and material characterization of soybean oil modified bioepoxy/clay nanocomposites for environmental sustainability. *Materials Today Sustainability, 5*, 100012.

Sarwono, A., Man, Z., & Bustam, M. A. (2012). Blending of epoxidised palm oil with epoxy resin: the effect on morphology, thermal and mechanical properties. *Journal of Polymers and the Environment, 20*, 540–549.

Scalici, T., Fiore, V., & Valenza, A. (2016). Effect of plasma treatment on the properties of *Arundo donax L.* leaf fibres and its bio-based epoxy composites: a preliminary study. *Composites Part B: Engineering, 94*, 167–175.

Scarponi, C., Sarasini, F., Tirillò, J., Lampani, L., Valente, T., & Gaudenzi, P. (2016). Low-velocity impact behaviour of hemp fibre reinforced bio-based epoxy laminates. *Composites Part B: Engineering, 91*, 162–168.

Shabeer, A., Garg, A., Sundararaman, S., Chandrashekhara, K., Flanigan, V., & Kapila, S. (2005). Dynamic mechanical characterization of a soy based epoxy resin system. *Journal of Applied Polymer Science, 98*, 1772–1780.

Shibata, M., & Ohkita, T. (2017). Fully biobased epoxy resin systems composed of a vanillin-derived epoxy resin and renewable phenolic hardeners. *European Polymer Journal, 92*, 165–173.

Singh, J. I., Dhawan, V., Singh, S., & Jangid, K. (2017). Study of effect of surface treatment on mechanical properties of natural fiber reinforced composites. *Materials Today: Proceedings, 4*(2), 2793–2799.

Stanzione, J. F., Sadler, J. M., La Scala, J. J., Reno, K. H., & Wool, R. P. (2012). Vanillin-based resin for use in composite applications. *Green Chemistry, 14*(8), 2346–2352.

Sudha, G. S., Kalita, H., Mohanty, S., & Nayak, S. K. (2017). Biobased epoxy blends from epoxidized Castor oil: effect on mechanical, thermal, and morphological properties. *Macromolecular Research, 25*(5), 420–430.

Sumrith, N., Techawinyutham, L., Sanjay, M. R., Dangtungee, R., & Siengchin, S. (2020). Characterization of alkaline and silane treated fibers of 'water hyacinth plants' and reinforcement of 'water hyacinth fibers' with bioepoxy to develop fully biobased sustainable ecofriendly composites. *Journal of Polymers and the Environment, 28*(10), 2749–2760.

Supanchaiyamat, N., Shuttleworth, P. S., Hunt, A. J., Clark, J. H., & Matharu, A. S. (2012). Thermosetting resin based on epoxidised linseed oil and bio-derived crosslinker. *Green Chemistry, 14*(6), 1759–1765.

Tan, S. G., & Chow, W. S. (2010). Thermal properties, fracture toughness and water absorption of epoxy-palm oil blends. *Polymer-Plastics Technology and Engineering, 49*, 900–907.

Torres-Arellano, M., Renteria-Rodríguez, V., & Franco-Urquiza, E. (2020). Mechanical properties of natural-fiber-reinforced biobased epoxy resins manufactured by resin infusion process. *Polymers, 12*, 2841.

Väisänen, T., Das, O., & Tomppo, L. (2017). A review on new bio-based constituents for natural fiber-polymer composites. *Journal of Cleaner Production, 149*, 582–596.

Voirin, C., Caillol, S., Sadavarte, N. V., Tawade, B. V., Boutevin, B., & Wadgaonkar, P. P. (2014). Functionalization of cardanol: towards biobased polymers and additives. *Polymer Chemistry, 5*(9), 3142–3162.

Wang, S., Ma, S., Xu, C., Liu, Y., Dai, J., Wang, Z., Liu, X., Chen, J., Shen, X., Wei, J., & Zhu, J. (2017). Vanillin-derived high-performance flame retardant epoxy resins:facile synthesis and properties. *Macromolecules, 50*, 1892–1901.

Xin, J., Li, M., Li, R., Wolcott, M. P., & Zhang, J. (2016). Green epoxy resin system based on lignin and tung oil and its application in epoxy asphalt. *ACS Sustainable Chemistry and Engineering, 4*, 2754–2761.

Yorseng, K., Rangappa, S. M., Pulikkalparambil, H., Siengchin, S., & Parameswaranpillai, J. (2020). Accelerated weathering studies of kenaf/sisal fiber fabric reinforced fully biobased hybrid bioepoxy composites for semi-structural applications: Morphology, thermomechanical, water absorption behavior and surface hydrophobicity. *Construction and Building Materials, 235*, 117464.

Zhang, C., Yan, M., Cochran, E. W., & Kessler, M. R. (2015). Biorenewable polymers based on acrylated epoxidized soybean oil and methacrylated vanillin. *Materials Today Communications, 5*, 18–22.

Zhao, S., & Abu-Omar, M. M. (2017). Synthesis of renewable thermoset polymers through successive lignin modification using lignin–derived phenols. *ACS Sustainable Chemistry and Engineering, 5*(6), 5059–5066.

Analysis and characterization of starches from alternative sources

Maria Carolina Bezerra Di-Medeiros Leal[1], *Gislane Oliveira Ribeiro*[2], *Maria Luiza Rezende Ribeiro*[2], *Antônio Gilberto Ferreira*[1], *Anna Rafaela Cavalcante Braga*[3], *Mariana Buranelo Egea*[4] *and Ailton Cesar Lemes*[5]

[1]Federal University of São Carlos (UFSCar), Nuclear Magnetic Resonance Laboratory, São Carlos, Brazil; [2]Federal University of Goiás (UFG), Nuclear Magnetic Resonance Laboratory, Goiânia, Brazil; [3]Department of Chemical Engineering, Universidade Federal de São Paulo (UNIFESP), Diadema, Brazil; [4]Goiano Federal Institute of Education, Science and Technology, Rio Verde, Goiás, Brazil; [5]Federal University of Rio de Janeiro (UFRJ), School of Chemistry, Department of Biochemical Engineering, Rio de Janeiro, Brazil

1. Introduction

Starch is a natural, renewable, and biodegradable polysaccharide (Di-Medeiros et al., 2014; Hedayati et al., 2016), which is used in food and nonfood products (Vanier et al., 2017). Large amounts of starch from different cereal, tuber, and root crops are obtained in different ways and applied to meet the needs around the world (Copeland et al., 2009). This extraction yields enormous pressure on the few official commercialized sources, demanding a continuous search for new economically viable sources (Builders et al., 2010). In addition, global projections for the food sectors reveal the problem of scarcity of food sources due to the burgeoning population growth (Fukase & Martin, 2020). In this regard, starch from unconventional sources may substitute the official ones for commercial applications.

Starches are used in various sectors of industry in a wide range of products besides food applications. This kind of biopolymer covers the pharmaceutical, cosmetic, paper, and mineralogical industries among others. Therefore, starches with different functional properties are interesting to ensure its suitability for a diverse range of applications. New sources of starch have been investigated in a variety of cereals, seeds, and tubers. In addition to alternative sources, industrial waste rich in starch is a way of obtaining the material. This type of obtention brings important aspects of sustainability and environmental preservation, thus adding value to the by-products.

The applicability of starch matrices depends on their characteristics. For example, the variation of starch isoforms is important for the production of noble materials such as the synthesis of biocomposites with improved thermal, mechanical, and barrier properties (Abral et al., 2019; Forouzandehdel et al., 2020; Lopez-Rubio et al., 2008; Tak et al., 2019), production of nanoparticles (Chen et al., 2019; Mohammad

Biodegradable Polymers, Blends and Composites. https://doi.org/10.1016/B978-0-12-823791-5.00025-9

Amini & Razavi, 2016; Xiao et al., 2020), functional foods like those rich in resistant starch (Hao et al., 2018; Mao et al., 2018; Tsuiki et al., 2016), and the development of new products, among others.

The choice of a certain starch for proper application is made considering its physicochemical properties and availability (Pascoal et al., 2013). Starch—despite its apparently simple chemical composition—has diverse functions and variable structures, which may change according to its origin (Liu et al., 2016). The structure of starch granules depends upon the atomic arrangement in the polysaccharide molecules according to the distribution of intra- and intermolecular forces (Correia et al., 2012). In general, starch has variability in granule morphology, amylose concentration, amylopectin architecture, and in the arrangement of the crystalline and amorphous region of the granules (Liu et al., 2016).

These technological, physical-chemical, and structural characteristics influence the functional characteristics of starch such as water solubility and absorption index, gelatinization, retrogradation, viscous properties, and susceptibility to enzymatic digestion (Liu et al., 2016). In this sense, a thorough investigation of the properties of starch is essential. This must be carried out through different highly regarded methods, and with highly reliable results. Examples include imaging techniques using polarized light microscopy, scanning and transmission electron microscopy (STEM), spectroscope methods such as Fourier transform infrared spectroscopy, solution- and solid-state nuclear magnetic resonance, X-ray diffraction, and thermal and viscoamylograph analysis.

2. Molecular characterization

2.1 Microscopic techniques

2.1.1 Electron and polarized light microscopy

Optical and electronic microscopy are valuable tools used for the characterization of starch and food products containing it. Much of what is known about the granular structure of starch such as the shape, size, and growth rings was determined by microscopic means (Bertolini, 2009; Thomas & Atwell, 1999). Among the most used microscopic techniques are polarized light microscopy and scanning and transmission electron microscopy. These are complementary tools for studying the morphology of the granules which helps with their characterization and later choice of industrial application (Copeland et al., 2009).

The study of morphology is important for the characterization of powder products (Zabot et al., 2019), and for the understanding of their structural basis, as this basis interferes in their physical-chemical properties. Knowledge of the properties of starch, of the changes during its processing, and of its interactions with other substances allows one to control the quality of the product and to design starch systems (Xiao et al., 2020).

Starch is stored in the form of granules and these granules are different in structure, shape, and size depending on the plant source (Xiao et al., 2020). Starch granules present a hierarchical structure that can be easily observed by electron microscopy and polarized light microscopy. Multiple concentric shells called growth rings of increasing diameter extend from the hilum (the center of growth) toward the surface of the granules (Copeland et al., 2009).

Through a polarized light microscope, it is possible to easily identify the shape, size, and position of the hilum in starches (Jay-Lin et al., 1994; Wang et al., 2020). The technique consists, basically, in the dispersion of the starch in water and subsequent fixation in a slide suitable for a microscope (Bertolini, 2009; Thomas & Atwell, 1999). Subsequently, the starch samples are illuminated by a polarized light. The transmitted polarized light is blocked by a perpendicular polarizer, which allows depolarized light to be detected. The procedure is generally applied to observe anisotropic objects (Xiao et al., 2020).

A light microscope equipped with a polarizer can be used to quantify the loss of birefringence of a starch sample, which is closely associated with the gelatinization phenomenon. This determination is possible based on the principle that highly ordered structures produce intense scattering patterns under polarized light (Bertolini, 2009; Thomas & Atwell, 1999).

The formation of the image is based on the ability of the polarized light to interact with the polarizable bonds present in the ordered molecules in a direction-sensitive manner. The disturbances in the polarized light waves in aligned molecules cause a phase delay among the sample beams. This delay allows interfering changes in the image plane amplitude. Thus, the formation of the image is based not only on the diffraction and interference principles, but also on the existence of molecular arrangements. The degree of order found in the materials varies from almost perfect crystals to vaguely ordered associations of asymmetric molecules or molecular assemblies. Under the polarization microscope, such structures usually appear bright against a dark background (Murphy, 2001). A characteristic dark cross—Maltese cross—is seen centered on the hilum when the starch granules are observed under the microscope under polarized light. This observation is characteristic of the crystalline substances. Their refractive index varies depending on the direction in which the light beam travels through them (this phenomenon is known as birefringence) (Tester et al., 2004), which leads granules to be considered as distorted spherical crystals. The intensity of birefringence depends greatly on the shape and orientation of the granules in relation to the light beam (Buléon et al., 1998). The crossed diffraction pattern, known as the "Maltese cross," is created by the rotation of polarized light through a crystalline or highly ordered region, such as those found in starch granules.

Despite having less resolving power in relation to electron microscopy, optical microscopy presents (1) low cost of the equipment, (2) easy sample preparation, and (3) easy micrographic acquisition. With the development of confocal scanning laser microscopy which obtains high-resolution 3D images at different depths, the use of optical microscopy has allowed advances in structural research and evaluation of the behavior of amylaceous structures during gelatinization (Bertolini, 2009).

Scanning electron microscopy (SEM), in turn, allows the shape and characteristics of the surface of the starch granules to be visualized in three dimensions. The sample requires preparation for the experiment. It consists of coating the material with a thin layer of a reflective metal, and then irradiating it with an electron beam. The electrons then are reflected back to a sensor, allowing the surface characteristics of the starch granules to be seen in great detail (Bertolini, 2009; Thomas & Atwell, 1999). SEM has a depth of focus hundreds of times greater than optical microscopy, with a much more efficient order of resolution and magnification (Jay-Lin et al., 1994).

When starch granules are visualized by scanning electron microscopy, interesting features become apparent. Shapes, sizes, indentations, and marks on the surface are clearly visible in some granules. Surface packaging characteristics can also be visualized (Bertolini, 2009; Thomas & Atwell, 1999). There are several types of starch granule sizes and shapes which can be correlated with the various types of biological sources from which starches are isolated.

A comparative study of the starch granule morphology of a wide variety of plant species consisting of roots and tubers, grains, corn, peas and beans, fruits, and nuts has been conducted. The studied starches present a wide variety of morphologies. At the same time, a set of common and specific characteristics for each particular biological source was noted. The authors observed that root and tuber starches have relatively large structures with smooth surfaces. Some of these starches have unusual morphologies, occurring as relatively flat plates or lenticular-shaped discs. Some grain starches also have disc shapes, although they tend to be much thicker and almost round. Grain starches are known for their bimodal size distribution. Corn starches are characterized mainly as irregular-shaped polygons having edges with varying degrees of sharpness. As the amount of amylose increases in corn starches, the granule loses its polygonal shape, and its surface becomes smoother. Fruit and nut starches are varied (Jay-Lin et al., 1994).

The size, shape, and appearance of granules of native and modified pea starch was observed through SEM (Jay-Lin et al., 1994; Wang et al., 2020). The granules varied in size between 3 and 40 μm. Most of the granules were oval, although spherical, round, elliptical, and irregularly shaped granules were also found. Other findings such as wrinkles on the surfaces of the granules could be investigated. In addition to the natural morphological aspects, this technique allows the evaluation of chemical and physical interventions such as those observed in this study, in which the starch granules showed cracks and imperfections after acid modification. Through the 4000× magnification, alternating growth rings and their thickness, where the amorphous regions are regularly arranged, were also observed. The thickness of the semicrystalline growth rings was about 200 nm. After acid hydrolysis, rupture of the growth rings was observed.

The types and content of synthetic enzymes that work in the starch biosynthesis are included among the forms of genetic control that cause individual characteristics of starch granules. Biosynthesis occurs in the amyloplast organelle in which the membrane structure and physical characteristics can give a particular shape and morphology to the individual types of starch granules. This, in turn, can have an effect on the arrangement and association of amylose and amylopectin molecules within the granule, thus also having an effect on the granule morphology (Jay-Lin et al., 1994).

In addition to the characteristics presented so far, other observations can be made through transmission electron microscopy (TEM). This technique is capable of displaying enlarged images in the range of $10^3-10^6\times$. The instrument can be used to produce electron diffraction patterns which are useful for analyzing the properties of a crystalline material (Egerton, 2005).

The internal arrangement of the starch components in concentric rings can be highlighted through TEM. In addition, this technique has been used to obtain information about the changes in the starch (Villar et al., 2017). Wang et. al. observed the structure of starch nanocrystals produced by acid hydrolysis (Wang et al., 2008). After acid hydrolysis of the pea starch granules, aggregates of nanometer-sized fragments were observed in the TEM images. It was also observed that the amorphous and crystalline phases coexist in the granule fragments after the acid hydrolysis.

2.2 Spectroscopic techniques

2.2.1 Fourier transform infrared spectroscopy (FT-IR)

The infrared spectroscopy technique with Fourier transform can be used in several applications. It provides fundamental information about the structures of organic molecules, revealing the types of functional groups present.

By absorbing radiation in the infrared region of approximately 400–800 nm, the molecules are excited to reach a state of greater energy. The radiation in this energy range includes vibrational frequencies of stretching and deformation of the connections. The absorbed radiation frequencies are equivalent to the natural vibrational frequencies of the molecule, and the absorbed energy serves to increase the range of motion of the connections. The characteristic absorptions of the molecule's functional groups are observed only in small regions of the vibrational spectrum in the infrared wavelength. A small absorption range can be defined for each type of connection. Outside this range, absorptions are usually due to some other type of connection (Pavia et al., 2010). Thus, these vibrational transitions generate the fundamental bands in the infrared spectrum.

The FT-IR technique is widely used for the characterization of starches. Through this technique it is possible to identify the main functional groups present in these polymers, as well as compounds that may be associated with the structure of the granule. According to Sevenou and Hill, polysaccharides, such as starch, absorb in the region of $1200-800\ cm^{-1}$, that is, in the wavelength of $8-12\ \mu m$, this band being known as the "fingerprint," therefore, the printing region contains the polysaccharides (Sevenou et al., 2002).

The results of FT-IR when combined with sampling methods with the introduction of light, such as attenuated total reflectance (ATR) followed by spectrum deconvolution procedures, expand the fields of application of spectral information in the infrared, which is useful in studies of starch structure. ATR is often considered a surface method. It is a technique in which the radiation beam at the wavelength of the infrared region penetrates the first micrometers of the sample (~2 μm). This penetration depth is generally less than the average size of the starch granules, thus, ATR-FTIR is an analytical method that can acquire information about the external region of the starch granules, making this type of study relevant for understanding the relationship between the starch granules and the surrounding environment (Sevenou et al., 2002).

The change in the molecular order, that is, in the crystallinity of the starch, can be observed in the FT-IR spectra. Changes in the shape and intensity of some bands are demonstrated in the spectra of amorphous starches. The 1047, 1022, and 861 cm^{-1} bands are sensitive to changes in crystallinity. It was observed that the 1022 cm^{-1} band increases from the decrease in crystallinity and has been shown to be characteristic of hot starch gel systems. The wide band at 1047 cm^{-1} in the starch spectrum is related to the crystallinity of the starch because it increases with the molecular order of the starch. The 1047 cm^{-1} band is probably composed of two overlapping bands at 1040 and 1053 cm^{-1}, and neither is seen on hot starch gels. During the starch retrogradation, the cm^{-1} 1040 cm^{-1} band develops within a few hours, while the absorbance at 1053 cm^{-1} increases for longer periods. The infrared bands at 1077 and 994 cm^{-1}, however, do not show significant changes with the decrease in crystallinity (van Soest et al., 1995).

The change in the degree of crystallinity of the starch can also be determined by using the intensity of the most characteristic bands of crystalline and amorphous starch that are in range 1040–1020, and for the ratio between them. The intensity of the bands and the reason were determined on which the peaks in the FTIR's spectrum 1047/1022 were signaled and quantified or by using the intensity of the characteristic band of crystalline starch at 1047 cm^{-1}, expressed as the ratio 1047/1035 (Lin et al., 2016; Wang et al., 2016). Instead of the areas, the heights of the bands measured from the baseline are obtained because the bands at 1047 and 1022 cm^{-1} show an increasing asymmetric shape with a change in crystallinity. This is due to the overlap of both bands and interference with the band at 994 cm^{-1} (Sevenou et al., 2002; van Soest et al., 1995).

The water content influences the crystallinity of the starch, and the humidity variation can be seen in the FT-IR spectra. The bands 1047, 1022, and 994 cm^{-1} change proportion with the variation in the amount of water, and the ratio 1047/1035 is shown to be very sensitive to the water content. Thus, water has a great influence on the elongation and flexion of the C–O and C–C bonds, and on the flexion of C–O–H (van Soest et al., 1995).

In order to determine the change in crystallinity expressed by the ratio 1047/1022, it is important to have some analytical care during the experiment and data processing. For instance, the experiment must be quantitative, the spectral window and the number of scans must be constant, the baseline must be flat, and the signals must be deconvolved.

FT-IR data can be combined with the results of X-ray diffraction and differential calorimetric analysis (DSC) to assess the crystallinity change in the starch granules. Also, the results of NMR nuclear magnetic resonance spectroscopy enable the structure of a molecule to be fully determined.

The structures and properties of phytoglycogen—an amorphous analogue of amylopectin—and waxy corn starch were subjected to carboxymethylation (Liu et al., 2016). FT-IR analysis was used to characterize both native and modified starch. In order to assess the formation of carboxymethylated starches with different degrees of substitution, peaks at 2930 cm^{-1} and 1640 cm^{-1} were analyzed, being indicative of asymmetric CH_2 elongation vibration, and of O—H bonds referring to the absorption of water molecules, respectively. For carboxymethylated starch, a strong band was found around 3430 cm^{-1}. In the starch fingerprint region, five characteristic peaks were found between 800 and 1200 cm^{-1}, which were attributed to the C—O bonds. The technique was also used to confirm the formation of carboxymethylated starches, once in comparison with native starch, with typical absorption peaks around 1601 and 1327 cm^{-1}.

2.2.2 Nuclear magnetic resonance (NMR)

The NMR technique consists of the modulation of active nuclear spins for the resonance phenomenon (NMR—active nucleus) through magnetic fields and radiofrequency pulses under specific thermodynamic conditions. The analytical signals obtained in NMR come from free induction decay (FID) with the applied Fourier transform aiming at converting the spectrum from the time domain to the frequency domain. The typical graphical result of an NMR spectrum is a histogram of intensity versus frequency in one dimension. In two dimensions, a contour map with information on homonuclear or heteronuclear correlations is observed (Simpson, 2017).

From the NMR spectra, it is possible to obtain information about the chemical displacement which allows the recognition of patterns of chemical groups, and their respective functions. It is also possible to measure the intensity of the signals, which reflects the abundance of the detected spins and the area of the signals/peaks, which can be correlated with the concentration of the species of interest in the studied matrix. In addition, one can observe the nuclear couplings via chemical connection called scalar couplings, and the dipolar couplings (NOE) through the spatial interactions of the spins, which inform the multiplicity of the signals (Simpson, 2017). This entire arsenal of information allows for qualitative and quantitative characterization regarding their structures, conformations, and molecular dynamics. The NMR technique is versatile in terms of the state of aggregation of the material to be studied. The matrix of interest can be analyzed in solution, semisolid or colloidal, and in solid state, representing multifunctionality for starch analysis.

The signals of the NMR spectrum in solution are different from those presented in the solid-state experiments. According to Nokab and Van der Wel this is because, in the liquid state, small soluble molecules undergo rapid thermal isotropic movements, which average all nuclear magnetic interactions dependent on orientation (Nokab & Wel, 2020). Therefore, isotropic components are the only remaining detectable interactions, resulting in a high-resolution solution NMR spectrum with excellent signal-to-noise ratio. However, in the solid state, the molecules are in an immobilized state where they have restricted movements and are too large to fall quickly. This lack of sufficiently rapid isotropic mobility reveals in the studies of solid-state NMR the presence of different types of nuclear and internuclear interactions dependent on orientation (for example, anisotropic and dipolar interactions) normally hidden in the liquid-state NMR of small molecules. In the absence of line-narrowing techniques, the NMR spectra of most solids are broad and weak, limiting the accessible insights by this technique in such conditions (Nokab & Wel, 2020).

NMR spectroscopy is a powerful method for investigating natural and synthetic macromolecule structures, and it has been widely used in the field of food science in recent years. It is a user-friendly, effective tool for quality control, which assists in the designation of the origin of products (geographical designation), and in the development of new products (Zhu, 2017). It can be used in starch systems to study the structure, composition, and interactions, and to investigate the crystalline nature and type of packaging of the granule. It is also possible to determine through this method the amylose content and the degree of phosphorylation in the starch chains. It can also be used to understand the structural changes that can occur in the granule due to physical, chemical, and enzymatic changes, and the retrogradation process.

The structural information of the starch molecule is fundamental to better understanding the intermolecular interactions. The application of one-dimensional and two-dimensional spectroscopy—in combination—improved the accuracy of structural information through HMBC, HSQC, and Cosy analysis (Chen et al., 2019; Mohammad Amini & Razavi, 2016; Xiao et al., 2020). Laignel, Bliard used the ^{1}H and Cosy experiments to evaluate the substitution pattern in modified starches and to locate the point of attachment of the substituting groups (Laignel et al., 1997). The high-resolution solid-state NMR is an excellent technique to elucidate structure change in a highly cross-linked polymeric system, especially when other techniques do not provide detailed information about starch structure. This NMR technique has also been applied in the investigation of the thermal decomposition process, and polymer carbonization in many systems. Zhang and Golding investigated the chemical structure and thermal behavior of four samples of native and modified starches after being heated to 600°C by solid-state NMR spectroscopy through a ^{13}C CP/MAS experiment. They evaluated the mechanisms of thermal condensation and dehydration in the formation and degradation of compounds during the heating of starch. The results of this type of study are useful for understanding the applications of starch materials as packaging or bonding materials for use at high temperatures (Zhang et al., 2002).

The molecular order in the starch granule is made up of two types of amylopectin side-chain helices: propellers that are packaged in regular matrices forming crystallinity, and propellers that are not regularly packaged or packaged within short range. The ^{13}C NMR experiment in solid state provides information on both molecular organization (crystalline nature) and noncrystalline (but rigid) chains (Atichokudomchai et al., 2004).

The signal corresponding to anomeric carbon (C-1) with a chemical displacement of approximately 100 ppm contains information of a crystalline nature. Its multiplicity corresponds to the type of packaging in the starch granules (Lin et al., 2016). In type A starches, maltotriose is the repetition unit, and the double axis generates the double helix, so A-type starches show a signal with a triplet-like split in the anomeric carbon (C-1) region. In B-type starches, maltose is the repetition unit, and the axis of the triple screw generates the double helix, so the C-1 peak in B-type starch spectra presents a signal with the characteristic of a doublet (Atichokudomchai et al., 2004). C-type starches generally have a C-1 triplet if an A-type crystalline structure is predominant in the sample or a C-1 doublet if a B-type crystalline structure is prevalent. These differences are attributed to the specific arrangement of the crystals in the starch granules (Zhu, 2017).

The signals in the ^{13}C NMR CP/MAS spectrum correspond to the carbon atoms present in the starch structure in hexopyranose. The signal at 90–110 ppm corresponds to anomeric carbon (C-1), and the signal at 58–65 ppm is attributed to carbon C-6. The overlapping signals around 68–78 ppm are associated with C-2, C-3, and C-5, and the signal at 80–84 ppm with carbon C-4 (Lin et al., 2016).

The region of greatest intensity in the anomeric carbon range is characteristic of double helices, and the shoulders of lesser intensity observed in the C-1 region in the range of 93–99 ppm are developed from unordered material. Broad signals in the C-4 region also reveal amorphous domains. C-4 signals are absent in crystal spectra (Atichokudomchai et al., 2004). In the study conducted by Atichokudomchai and Varavinit, the signs of C-1 and C-4 decreased with acid hydrolysis and almost disappeared in hydrolyzed starches for long periods (Atichokudomchai et al., 2004).

The degree of crystallinity of the starch is commonly determined through X-ray diffractograms, by the determination of the crystalline and the amorphous area (Buléon et al., 1998). However, the relative crystallinity of the starch can also be calculated through the proportion of the peak areas (doublet or triplet) of the anomeric carbon in relation to the total area of the C-1 signal in the ^{13}C CP/MAS NMR spectrum, having its result expressed as a percentage (Atichokudomchai et al., 2004).

The ^{13}C NMR technique in solid state can be used successfully in the study of the short-range order in starches (Ya Bogracheva et al., 2001). It is also possible to calculate the content of double helices in the starch structure using Eqs. (16.1) and (16.2). Before doing so, spectra need to have their peaks properly adjusted using appropriate software.

$$\%\ of\ amorphous\ part\ of\ native\ starch = \frac{native\ starch\ PPA}{amorphous\ starch\ PPA} \times 100 \quad (16.1)$$

$$\%\ of\ double\ propellers = 100 - amorphous\ part\ (\%) \quad (16.2)$$

where, PPA is defined as the ratio of the adjustment area of C-4 peak in relation to the total area of the spectrum.

The NMR technique is complementary to other analytical techniques, such as X-ray diffraction (XRD) for the characterization of starch. The combination of these techniques allows one to infer the pattern and degree of crystallinity of starches.

The structure and physicochemical properties of native and resistant starch in Chinese yam were evaluated (Zou et al., 2020). One of the used techniques was ^{13}C NMR with cross-polarization and rotation in a magic angle (CP/MAS). The analysis showed that both native and resistant starches of the expansion and dormant stage had triple peaks in the region 99.28–103.72 ppm, therefore indicating that they were a type A crystalline structure. In relation to starch from the expansion stage, the native and resistant starch from the dormant stage presented lower proportions of amorphous regions. In addition, the dormant stage samples revealed higher proportions of double helices. These results were supported by the crystallinity characteristics measured by XRD. The crystalline region of the starch presented mainly in its composition double helix ordered in the lateral clusters of amylopectin, and by the single helix of the amylose–fat complex. The amorphous region presented dispersed amylose molecules, and branched regions of amylopectin molecules. An increase in crystallinity was observed from the increase in the level of double helix in starches in the dormant stage. The results found using the NMR technique are useful for the subsequent industrial application of Chinese yam starch in food products.

2.3 X-ray diffraction (XRD)

X-ray diffraction has great applicability and importance in revealing internal structures in the order 10^{-8} cm. X-rays are electromagnetic radiation that has the same nature as visible light, but with a much shorter wavelength in the range 0.5–2.5 Å. Within the full electromagnetic spectrum, X-rays are in the region between gamma and ultraviolet rays and are produced when electrically charged particles are decelerated rapidly. Radiation is produced when electrons at high speed originating from the heated filament hit the target. X-rays are produced at the point of impact and are scattered in all directions, like a shower of randomly distributed photons (Cullity, 1978).

This technique is widely used to characterize starches because it allows to distinguish the pattern and degree of crystallinity in the starch granules. Factors such as chain length, presence of water, and density and type of packaging in the starch granule can influence the crystallinity pattern (Lima et al., 2012). According to Buléon and Colonna, the starch granules exhibit A-, B-, C-, and V-type crystallinity patterns. The C-type pattern has been demonstrated to be a mixture of A- and B-type diagrams, and the V-type crystalline characteristic is formed by the complexation of amylose with lipids such as fatty acids and monoglycerides (Buléon et al., 1998). According to Ferreira and Araujo, amylose occurs mainly in the amorphous state and presents the ability to form a complex with endogenous lipids during the biosynthesis process and adopts a single helical conformation. This single helix structure can stack in parallel in crystals, thus forming a type V diffraction pattern (Shirlyanne et al., 2019).

The type of packaging in the granules is closely related to the length of the amylopectin chains. Short chains exhibit A-type crystallinity, while long chains show B-type, and intermediate-length chains are associated with C-type crystallinity. Another difference between A- and B-type starches is that the A-type adopts a closed arrangement with molecules of water between each double helical structure, while the B-type is more open, with more water molecules, and is located in a central cavity surrounded by six double helices. These factors are related to the biopolymer's three-dimensional structure (Cheetham & Tao, 1998). It is important to understand the packaging inside starch granules because the type of polymorph greatly affects the physicochemical properties of the biopolymer, and consequently, its applicability (Zhu, 2018).

In an X-ray diffractogram, the A-type crystallinity pattern can be evidenced by the presence of peaks at approximately 15 degrees and 23 degrees, and an ill-resolved doublet around 17 degrees and 18 degrees. In diffractograms of starches with B-type crystallinity, a more intense peak appears around 17 degrees. Small peaks appear at 15 degrees, 20 degrees, 22 degrees, and 24 degrees, in addition to a characteristic peak at 5.6 degrees. C-type starch can be classified as CA-type (closest to type A), C-type, and CB-type (closest to type B) according to the proportion of A- and B-types polymorphs. The C-type standard presents singlets around 17 degrees and 23 degrees, and small peaks at approximately 5.6 degrees and 15 degrees. CA- and CB-type patterns are similar to C-type, but there is a signal split around 18 degrees, and a strong singlet at 23 degrees for CA-type. For CB-type there are two peaks with splits such as a shoulder in about 22 degrees and 24 degrees (Cai et al., 2014). The XRD of V-type starch shows peaks in the 2θ diffraction angles at approximately 5.8 degrees, 12.3 degrees, and 20.4 degrees (Shirlyanne et al., 2019). The presence of peaks at 12.6 degrees, 13.2 degrees, 19.4 degrees, and 20.6 degrees has been described (Lima et al., 2012).

The amylose/amylopectin ratio within the starch granule modifies the width and intensity of the signals shown in the X-ray diffractogram. With an increase in amylose content, the peak at 15 degrees 2θ becomes progressively weaker and broader (Cheetham & Tao, 1998) while the peaks at 17 degrees and 18 degrees, typical of A-type starch, is merged into a singlet, and the sign at 22degrees changes the width and spread into a doublet. In general, all diffraction peaks decrease except for the 20 degrees signal. These changes are indicative of crystallinity transition.

Another factor that modifies the signals in the X-ray diffractogram is the moisture content present in the granules. The crystallinity of starch must be determined with well-defined hydration, since the degree of crystallinity depends strongly on the amount of water present in the granules (Buléon et al., 1998).

The degree of crystallinity can be defined as the percentage of the area of the crystalline region of the diffractogram in relation to the sum of the amorphous and crystalline areas and, therefore, the total area (Lopez-Rubio et al., 2008). This measure reflects the repeatability pattern or habit of the amylopectin fraction. The starch crystallinity is measured by determining the amorphous area, and the crystalline area using the Gaussian curve plotted on the X-ray diffractogram (Fig. 16.1).

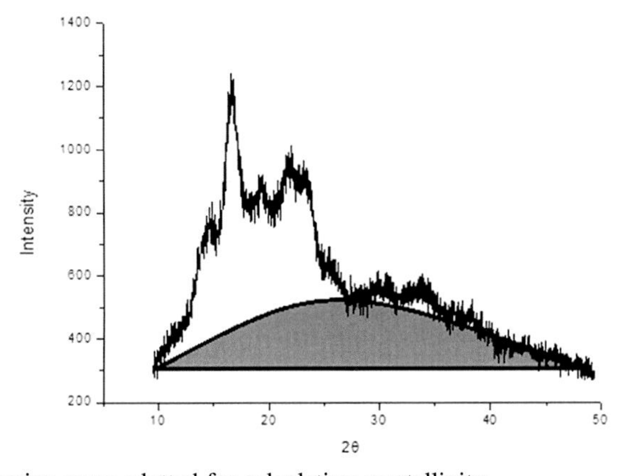

Figure 16.1 Gaussian curve plotted for calculating crystallinity.

The degree of crystallinity is estimated quantitatively, and is expressed as a percentage according to Eq. (16.3):

$$Dc\ (\%) = \frac{Ac}{Ac + Aa} \times 100 \tag{16.3}$$

where, Dc is the degree of crystallinity; Ac is the crystalline area in the X-ray diffractogram; and Aa is the amorphous area on the X-ray diffractogram.

The starch's crystallinity is due, in large part, to the amylopectin chains, which form a double helical structure packed laterally in a crystalline network. Parts of the amylose are observed in the amorphous structure, and part of it is present as a helical complex with lipids. Amylose is intertwined with amylopectin to maintain the integrity of the granules (He & Wei, 2020).

The degree of crystallinity is very important for the study of the hydrodynamic characteristics of this molecule, as the crystalline structure will give this polymer less solubility. The crystalline nature of starch probably depends on genetic control and the climatic conditions during plant growth. The length of the chains involved in the crystalline phase and the branching pattern in the amylopectin molecules influence both the crystallinity and the crystalline type. In addition, among the factors that contribute to crystallinity are the temperature and hydration conditions during plant growth (Buléon et al., 1998). These statements, together with the evidence presented by Almeida and Batista, allow us to infer the importance of crystallographic analysis for botanical monitoring, in addition to the physical-chemical evaluation of starch (Almeida et al., 2019).

XRD experiments and the determination of starch crystallinity can be applied in the quantitative estimate of the extent of gelatinization. Starch crystallinity is known to decrease with gelatinization. A completely gelatinized starch has an amorphous structure (Teixeira et al., 2015). A study conducted in China with flour and starch evaluated the physical, functional, paste, and thermal properties from three potato cultivars, namely Xisen 6, Hongmei, and Heimeiren, in addition to sweet potato cultivars Yu 15, Qin 7, and 10-6-5. X-ray diffraction was used to investigate the type of crystal, and the crystallinity of starches and flours from different cultivars (Wang et al., 2020). The diffraction peaks showed A-type crystals. The reflection intensities of the flours were lower than those of their respective starches, possibly due to the effects of other components of the flours such as proteins, lipids, and fibers. It was also evident that the potato starch granules were larger than the sweet potato granules. A 5.6 degrees signal was found in the analysis of potato starches and flours. Another research found the same peak in fermented cassava starch, and it was suggested that it may determine the formation of nanocrystals (Alonso-Gomez et al., 2016). Thus, this may be a crystalline nanomaterial with future industrial applications.

Studies involving starch commonly investigate the properties and characteristics of retrogradation. Retrogradation is a phenomenon that involves the interaction between amylose and amylopectin chains to allow helical aggregation, and an increase in the crystalline order after the gelatinization process. The extent of the breakdown of the starch granule is strongly influenced by the structural arrangement of the chains in the amorphous and crystalline regions of the granule. Therefore, the retrograde property of a starch gel is influenced by its structural characteristics that direct the interactions between the starch chains during gel storage with a great influence on the acceptance, quality, and shelf-life of starch-containing foods, highlighting the importance of studies into the structural characterization of starches (Cozzolino et al., 2013; Wang et al., 2015).

2.4 Thermal properties (DSC and TG)

The term thermal analysis is applied to any technique that involves measurement of the specific properties of a material while the temperature is controlled (changed or maintained) and monitored (Cheng et al., 2000). This thermoanalytical technique is interdisciplinary, useful in several scientific and technological sectors, among which are natural sciences (chemistry, metallurgy, and ceramics); geosciences (geology, mineralogy, and oceanography); biological sciences (botany, agronomy, ecology, technology in chemistry and food technology), and forensic sciences (legal and criminalistic science). The techniques of TG (thermogravimetry) and DSC (differential scanning calorimetry) have been widely disseminated and used (Ionashiro, 2005).

Thermogravimetric analysis (TG) is carried out in thermal scales, with instruments that allow the continuous weighing of a sample as a function of temperature, that is, the extent to which it is heated or cooled. The mass variation curves (in general loss of mass) as a function of temperature allow conclusions about the thermal stability of the sample, the composition and stability of the intermediate compounds, and the composition of the residue (Ionashiro, 2005). Structural changes occur when a sample is subjected to high temperatures. These changes can be evaluated through the alterations caused in the sample by heating, usually associated with the loss of masses related to thermolabile chemical groups. Thermogravimetric analysis can also define the temperature range in which the sample decomposition processes occur.

Through the TG curves, the thermal stability of the materials can be evaluated. In starch thermograms, a typical pattern of three stages of degradation can be observed. The first mass decay corresponds to the loss of volatile components (Cuenca et al., 2020), and mainly to the evaporation of water adsorbed to the starch granules (Di-Medeiros et al., 2014). This loss is usually associated with the initial moisture content in starches, and provides useful information for practical applications. In general, a loss of mass of 8%–12% is observed at temperatures below 130°C at this stage (Zabot et al., 2019). After the temperature range in which the mass loss of volatile compounds occurs, a plateau with virtually no loss of mass is observed in the thermogram. Then, a second mass decay is noticed, which corresponds to the process of amylose degradation, and breaking of the amylopectin chains (Elmi Sharlina et al., 2017). This is the main stage of decomposition, with greater mass loss. This sharp interval of decline in the curve indicates the presence of large amounts of compounds with very similar thermal properties, which is characteristic of homopolysaccharides such as starches. This step consists of rapid dehydration and decomposition of the hydroxyl groups of the glucose residues to form water molecules (Di-Medeiros et al., 2014). The third mass decay is the formation of inert carbonaceous residues with mass stabilization (Londoño-Restrepo et al., 2014). According to Sharlina and Yaacob, thermally stable samples do not decompose below 190°C (Elmi Sharlina et al., 2017).

TG experiments must be carefully monitored because some factors can influence the appearance of the curves. Examples are instrumentals (oven heating ratio, oven atmosphere, sample holder, and oven geometry), and factors related to the sample characteristics (particle size, amount of sample, solubility of the gases released in the sample itself, reaction heat, sample compaction, sample nature, sample thermal conductivity) that influence the results (Ionashiro, 2005).

Care must be taken when preparing the sample and carrying out the experiment in order to minimize these adverse factors. The amount of sample must be observed because it influences the results and, in general, it can be said that the curve obtained with 1 mg of sample will be different from the curve obtained with 100 mg of sample. Commonly, the larger the sample mass, the higher the initial thermal decomposition temperature, as well as the final temperature, but with an exception if the decomposition reaction is exothermic. In order to detect the presence of intermediate compounds, it is better to use small amounts of samples instead of large ones (Ionashiro, 2005).

Other limiting factors include the size of the particles in the sample, thermal conductivity, and packaging (compaction). It can be generally stated that a decrease in particle size causes a decrease in temperatures at which the decomposition reaction begins and ends. The thermal conductivity of the sample depends on its density. In turn, this density depends on the size of the particles, and on the compaction to which it was subjected. In addition, the sample density may vary as the reaction proceeds due to the processes of fusion, conversion into different substances, sintering, and swelling, that occur with the sample (Ionashiro, 2005). In thermogravimetric studies, it is advisable to experimentally standardize either the sample preparation through granulometric control, drying, and weighing or heating ramp conditions.

The applicability of sorghum starch as a superabsorbent biopolymer was investigated and one of the characteristics evaluated was the thermal behavior (Teli & Mallick, 2018). The thermal stability of a superabsorbent polymer under high temperature is a very useful and important physical property. Thus, the sorghum starch in natura, sorghum starch thermograms, and the chemically modified ones were subjected to TG experiments. Unmodified starch lost 10% of its mass at 133°C, corresponding to the loss of moisture present in it. After that, up to 300°C, the unmodified starch showed good thermal stability with a total of 20% mass loss attributed to carbon skeleton degradation. However, in addition to this temperature, it showed a marked and considerable weight loss within the range of 300–335°C, with a loss of carbon skeleton and inorganic material contents. As the temperature increased, the rate of weight loss was again reduced, and a total of 90.71% weight loss occurred at 600°C. The modified starch showed relatively better thermal stability compared to the unmodified one. No marked inflection was observed in the thermogram of the modified starch, and gradual thermal decomposition was observed throughout the temperature range. A total of 76.1% of weight loss was observed until the temperature reached 600°C. This clearly indicates that the modified starch sample presented better thermal stability. This may be due to the formation of the cross-linked three-dimensional structure, and to the chemical modification of the product after grafting and cross-linking with vinyl monomers.

Differential scanning calorimetry (DSC) is another analytical approach that assesses the thermal behavior of materials. This is a technique that measures the difference in supplied energy to a substance, and a reference material as a function of temperature, while the substance and the reference material are subjected to controlled temperature programming. In this technique, the sample and reference material are isothermally maintained by the application of electrical energy when they are heated or cooled to a linear ratio. The obtained curve is the recording of the heat flow dH/dt in mcal s^{-1} as a function of temperature (Ionashiro, 2005).

The process of disorder in the starch granules caused by heating of the material in the presence of water can be studied through the endothermic peak in the DSC curve, through the gelatinization temperature [(T_o, initial temperature), (T_p, peak temperature), (T_c, final temperature)], gelatinization temperature range (T_c−T_o), and gelatinization enthalpy (ΔH) (Fig. 16.2).

The initial gelatinization temperature (T_o) indicates the temperature at which the structural disorder begins in the starch granules. This is where the weakest fractions in the crystalline region melt, and high T_o values indicate that the crystalline region has a higher degree of stability and/or lower quantity of structural defects (Ratnaningsih et al., 2016).

The gelatinization temperature range (T_c−T_o) represents the thermal stability range. Accordingly, Ratnaningsih and Suparmo stated that ΔT is the variation in the organization of the crystalline structure in the starch granules. The study presents that low values of ΔT indicate that the crystalline region has a higher degree of perfection, with better alignment of the double helices within the crystalline coverslip (Ratnaningsih et al., 2016).

The gelatinization enthalpy reflects the necessary energy for the double helix to unfold with a consequent loss of crystallinity (Mbougueng et al., 2012; Tianyu et al., 2020). It is related to the characteristics of the starch granule such as amylose content and lipid content in granules (Singh et al., 2003). The study by Lopez-Rubio et al. concluded that the enthalpy associated with gelatinization is due to the melting of imperfect crystals, with potential contributions from the packaging of the crystals and the fusion enthalpy of the helices (Lopez-Rubio et al., 2008).

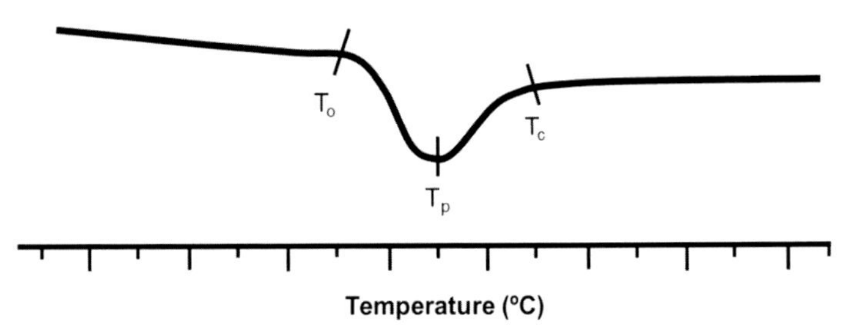

Figure 16.2 DSC curve demonstrating the temperature, range, and enthalpy of gelatinization, where: T_o is the gelatinization temperature; T_p is the peak temperature; T_c is the final temperature; and ΔH is the gelatinization enthalpy variation.

Through the DSC curve, the heat effects associated with physical or chemical changes in the sample can be monitored, such as phase transitions (melting, boiling, sublimation, freezing, inversion of crystalline structures) or dehydration, dissociation, decomposition, oxide-reduction, etc., capable of causing variations in heat. In general, phase transitions, dehydrations, reductions, and certain decomposition reactions produce endothermic effects, whereas crystallizations, oxidations, and some decomposition reactions produce exothermic effects. These techniques also allow the study of transitions involving entropy variations (second-order transitions), the most common of which are glass transitions that certain polymers can undergo (Ionashiro, 2005).

The study of the thermal decomposition of starches occurs through the calorimetric analytical techniques, as previously mentioned, such as DSC and TGA. The swelling, degradation, and retrogradation characteristics of starch granules during processing and retrogradation during storage largely determine the texture and stability of high-moisture starch-based foods.

DSC has been widely used to study the phenomena of gelatinization, glass transition temperature, and starch crystallization. The knowledge of the transition endothermic phenomena from an ordered to a disordered phase is important and fundamental to understanding the behavior of the materials.

The study of the thermal behavior of starches is highly complex, and can vary with different measurement conditions. The thermal behavior of starches is much more complex than that of conventional thermoplastics. This is because the physicochemical changes that occur during the heating of starch or starch products may involve gelatinization, melting, glass transition, crystallization, alteration of the crystal structure, expansion of volume, molecular degradation, and water movement. All of these thermal behaviors depend on the moisture content, and the water contained in the starch granules is not stable during heating. In addition, the thermal conductivity of starch is very low once its apparent density is low (Yu & Christie, 2001).

The moisture content of the system strongly influences the gelatinization process. When the gelatinization process is investigated with DSC, only one endothermic curve is observed when the starch has high water content (>65%). However, as the amount of water decreases (35%–65%), the endothermic curve develops a shoulder at higher temperatures. At a sufficiently low water level (<35%), only one endothermic curve is observed at high temperature (De Bondt et al., 2020).

During the heating process of the starch, phase transition changes are observed such as fusion, evaporation, and sublimation, as well as chemical condensation, decomposition, and finally carbonization at very high temperatures. In this sense, thermal analysis is useful for monitoring the mass loss and the endothermic or exothermic nature of any physical-chemical changes involved in thermal processes, providing important information about thermal decomposition. For instance, examination of changes in the chemical structure that occur at elevated temperatures, and the thermal reaction pathways of starch which are important. The results play a fundamental role in understanding the thermal behavior and physical properties of the materials, and they are critical to the understanding of the applications of starch materials as packaging or bonding materials for use at high temperatures (Zhang et al., 2002).

Thermal properties, among others, were analyzed in order to evaluate the rheological and physico-chemical properties of *Solanum lycocarpum* starch (Di-Medeiros et al., 2014). In the thermogravimetric analysis, a typical pattern of weight loss in three stages was noted. The first refers to the vaporization of volatile compounds with a weight loss of 15.1%. In the second stage, a weight loss of 62.4% occurred in the interval between 302.1 and 328.3°C. The marked loss of weight in a small temperature range indicates the presence of a large amount of compounds with very similar thermal properties, which is a characteristic of homopolysaccharides such as starches. In the last stage there was an accumulation of carbon. The heat-treated samples were also examined by DSC to assess the extent of the retrogradation process. The presence of an endothermic peak at 65°C, in the thermogram of the sample treated at 50°C, was observed. This peak changed to 68°C when the sample treated at 50°C had been stored at 4°C for 7 days. These peaks indicate the presence of intact or partially gelatinized granules. This observation is in accordance with the data obtained from scanning electron microscopy and X-ray. The hydrothermal treatment at 65°C was sufficient to completely break the starch granule, as evidenced by the absence of an endothermic peak in the thermogram. The recovery of the structural order of *S. lycocarpum* starch after heat treatment at 65°C was not sufficient to result in an endothermic peak in the DSC curve. These analyzes are interesting to evaluate the characteristics of *Solanum lycocarpum* starch for inclusion in food.

2.4.1 Pasting properties (RVA)

Heating the starch granules above the gelatinization temperature results in irreversible changes with a modification of the viscosity. The pasting properties of starch and water suspension are measured during the heating and cooling cycle with a constant shear force, using a rapid visco-analyzer (RVA) (Delcour & Hoseney, 2010). These properties depend on the physical and chemical characteristics of the starch granules such as average granule size, granulometric distribution, amylose/amylopectin ratio, and mineral content (Singh et al., 2003).

The starch granule behavior during the viscosity analysis is: when the native starch is heated in the presence of sufficient moisture, the granules absorb water and swell. The absorption of water by the granules starts predominantly in the amorphous regions, and due to the disturbing stress, the water molecules are leached out from the amorphous into the crystalline regions (Copeland et al., 2009).

The increase in viscosity in starch suspensions occurs mainly due to the products leached out by the starch granules under shear that form a network, allowing the development of gel (Rincón-Londoño et al., 2016). Under continuous heating, the viscosity of the paste increases to its maximum which corresponds to the point with the largest number of swollen starch granules, however it remains intact. Subsequently, a reduction in viscosity caused by the rupture of the granules and dispersion of the starch in the aqueous phase is observed. The rate and extent of swelling and cracking depend on the type and amount of starch, the temperature gradient, the shear force, and the presence of lipids and proteins in the starch composition (Copeland et al., 2009).

The starch paste profile displayed through the RVA curves commonly presents important characteristics such as maximum viscosity, paste temperature, drop in viscosity, cold paste viscosity, and hot paste stability. In addition, it may indicate the tendency to retrograde due to the reassociation of leachate compounds (Rincón-Londoño et al., 2016).

The pasting temperature is the temperature where gelatinization of the first starch granules occurs, and corresponds to the point where the curve formation in the graph begins. According to Copeland and Blazek, the amylose content of starch influences the gelatinization temperature (Copeland et al., 2009).

Peak viscosity, or maximum viscosity, is the highest viscosity reached by the starch–water system during prolonged heating up to a temperature of 95°C (Karim et al., 2000), and it reflects the ability of the granules to swell freely before of their physical breakdown (Copeland et al., 2009). According to Ratnaningsih and Suparmo, the peak height is influenced by the friction between the swollen granules, amylose content, and relative crystallinity (Ratnaningsih et al., 2016).

The minimum viscosity is the lowest value found on the RVA curve. It is observed when the temperature remains constant at 95°C. This is the phase in which broken granules and dispersed starch molecules in the aqueous phase are found (Copeland et al., 2009).

After the heating period to 95°C, controlled cooling to 50°C occurs. This cooling gives rise to a rapid increase in viscosity, which is known as retrogradation. This phenomenon is caused by a decrease in the energy of the system that allows more hydrogen bonds, and entanglement between the starch chains (Delcour & Hoseney, 2010). The tendency to retrograde indicates the ability of the gelatinized starch to recrystillize during the cooling period, and it reflects the interaction between the amylose chains leached out during the cooling cycle, and the presence of intact and fragmented granules embedded in amylose (Ratnaningsih et al., 2016). It is measured by the difference between the final viscosity and the minimum viscosity.

The starch concentration, temperature, and shear applied during the steps prior to the cooling phase determine the structure of the starch suspension. The concentration of starch used in the analysis can drastically affect the RVA curve. Very low concentrations of starch are insufficient to generate a signal in the equipment that makes this type of analysis. In a very concentrated regime, the granules cannot swell to their full extent, thus resulting in granule stiffness that drastically affects the system rheology (Delcour & Hoseney, 2010).

As the temperature reduction continues, the viscosity of the starch continues to increase until reaching a final viscosity. The cooling cycle is finished when the system temperature is at 50°C. Another important parameter that can be measured on the RVA curve is the resistance to breaking or breaking viscosity, which illustrates the stability of the paste after cooking (Karim et al., 2000), and it corresponds to the difference between the maximum viscosity reached, and minimum viscosity during the maintenance of the temperature at 95°C.

Each starch exhibits a unique viscosity behavior with the changes in temperature, concentration, and shear rate. These results provide important information for application in food processing (Alimi & Workneh, 2018).

Pasting and rheological properties are the most important characteristics to be studied in starches and their derivatives. This great importance is given because many foods that contain starch in their composition undergo high-temperature processes, which cause major changes in the structure and morphology of the granules, resulting in modifications of the final product (Rincón-Londoño et al., 2016).

Understanding the steps that occur during the gelatinization and retrogradation of a specific starch are fundamental steps to better predicting the functional properties of the processed starch from the knowledge of the structure of native granules (Copeland et al., 2009). Starches used in the human diet are generally heated in the presence of water under shear, and then cooled down. During the heat treatment, the starch granules are gelatinized, losing their crystallinity and structural organization. Upon cooling, the disaggregated starch molecules form a gel, and then they gradually reassociate in semicrystalline aggregates that differ in the form of native granules. These foods contain substantial amounts of retrograded starch.

In a study by Zabot and Silva, the physical-chemical, morphological, thermal, and pasting properties of the native starch of degreased and depigmented annatto seeds were performed. In the evaluation of the pasting property, the annatto starch showed an average peak of 6660 cP and a pasting temperature of 64.7°C. When compared to commercial corn starch, the annatto seed starch showed higher peak and final viscosity but lower pasting temperature. The pasting temperature of annatto starch was approximately 8°C lower than that of corn starch. From the analysis performed, it was observed that annatto starch can be used in preparations that require low heating, such as sensitive foods (Zabot et al., 2019).

3. Conclusion

Currently, great emphasis is being given to the use of alternative sources of starch, promoting the potential of these raw materials, in addition to adding value to sources not exploited. Alternative sources can result in starches with different properties and characteristics, with characterization by methods that are able to provide accurate results and sufficient information for its application being extremely important. Techniques of polarized light microscopy, scanning and transmission electron microscopy, and spectroscope methods such as Fourier transform infrared spectroscopy, solution- and solid-state nuclear magnetic resonance, X-ray diffraction, thermal and viscoamylograph analysis standing out for this purpose.

References

Abral, H., Hartono, A., Hafizulhaq, F., Handayani, D., Sugiarti, E., & Pradipta, O. (2019). Characterization of PVA/cassava starch biocomposites fabricated with and without sonication using bacterial cellulose fiber loadings. *Carbohydrate Polymers, 206*, 593–601. https://doi.org/10.1016/j.carbpol.2018.11.054

Alimi, B. A., & Workneh, T. S. (2018). Structural and physicochemical properties of heat moisture treated and citric acid modified acha and iburu starches. *Food Hydrocolloids, 81*, 449–455. https://doi.org/10.1016/j.foodhyd.2018.03.027

Almeida, V. O., Batista, K. A., Di-Medeiros, M. C. B., Moraes, M. G., & Fernandes, K. F. (2019). Effect of drought stress on the morphological and physicochemical properties of starches from *Trimezia* juncifolia. *Carbohydrate Polymers, 212*, 304–311. https://doi.org/10.1016/j.carbpol.2019.02.015

Alonso-Gomez, L., Niño-López, A. M., Romero-Garzón, A. M., Pineda-Gomez, P., del Real-Lopez, A., & Rodriguez-Garcia, M. E. (2016). Physicochemical transformation of cassava starch during fermentation for production of sour starch in Colombia. *Starch Staerke, 68*(11–12), 1139–1147. https://doi.org/10.1002/star.201600059

Atichokudomchai, N., Varavinit, S., & Chinachoti, P. (2004). A study of ordered structure in acid-modified tapioca starch by 13C CP/MAS solid-state NMR. *Carbohydrate Polymers, 58*(4), 383–389. https://doi.org/10.1016/j.carbpol.2004.07.017

Bertolini, A. (2009). *Starches: characterization, properties, and applications.*

Builders, P. F., Nnurum, A., Mbah, C. C., Attama, A. A., & Manek, R. (2010). The physicochemical and binder properties of starch from Persea americana Miller (Lauraceae). *Starch Staerke, 62*(6), 309–320. https://doi.org/10.1002/star.200900222

Buléon, A., Colonna, P., Planchot, V., & Ball, S. (1998). Starch granules: structure and biosynthesis. *International Journal of Biological Macromolecules, 23*(2), 85–112. https://doi.org/10.1016/S0141-8130(98)00040-3

Cai, J., Cai, C., Man, J., Zhou, W., & Wei, C. (2014). Structural and functional properties of C-type starches. *Carbohydrate Polymers, 101*(1), 289–300. https://doi.org/10.1016/j.carbpol.2013.09.058

Cheetham, N. W. H., & Tao, L. (1998). Variation in crystalline type with amylose content in maize starch granules: an X-ray powder diffraction study. *Carbohydrate Polymers, 36*(4), 277–284. https://doi.org/10.1016/S0144-8617(98)00007-1

Chen, L., Xiong, Z., Xiong, H., & Din, Z.u. (2020). Investigating the structure and self-assembly behavior of starch-g-VAc in starch-based adhesive by combining NMR analysis and multi-scale simulation. *Carbohydrate Polymers, 246*. https://doi.org/10.1016/j.carbpol.2020.116655

Chen, Y., Hao, Y., Ting, K., Li, Q., & Gao, Q. (2019). Preparation and emulsification properties of dialdehyde starch nanoparticles. *Food Chemistry, 286*, 467–474. https://doi.org/10.1016/j.foodchem.2019.01.188

Cheng, S. Z. D., Li, C. Y., Calhoun, B. H., Zhu, L., & Zhou, W. W. (2000). Thermal analysis: the next two decades. *Thermochimica Acta, 355*(1–2), 59–68. https://doi.org/10.1016/S0040-6031(00)00437-8

Copeland, L., Blazek, J., Salman, H., & Tang, M. C. (2009). Form and functionality of starch. *Food Hydrocolloids, 23*(6), 1527–1534. https://doi.org/10.1016/j.foodhyd.2008.09.016

Correia, P., Cruz-Lopes, L., & Beirão-da-Costa, L. (2012). Morphology and structure of chestnut starch isolated by alkali and enzymatic methods. *Food Hydrocolloids, 28*(2), 313–319. https://doi.org/10.1016/j.foodhyd.2011.12.013

Cozzolino, D., Roumeliotis, S., & Eglinton, J. (2013). Relationships between starch pasting properties, free fatty acids and amylose content in barley. *Food Research International, 51*(2), 444–449. https://doi.org/10.1016/j.foodres.2013.01.030

Cuenca, P., Ferrero, S., & Albani, O. (2020). Preparation and characterization of cassava starch acetate with high substitution degree. *Food Hydrocolloids, 100*. https://doi.org/10.1016/j.foodhyd.2019.105430

Cullity, B. (1978). *Element of X-ray diffraction*. Addison-Wesley Publishing Company.

De Bondt, Y., Liberloo, I., Roye, C., Goos, P., & Courtin, C. M. (2020). The impact of wheat (*Triticum aestivum* L.) bran on wheat starch gelatinization: a differential scanning calorimetry study. *Carbohydrate Polymers, 241*. https://doi.org/10.1016/j.carbpol.2020.116262

Delcour, J. A., & Hoseney, R. C. (2010). Principles of cereal science and technology authors provide insight into the current state of cereal processing. *Cereal Foods World, 55*(1), 21–22. https://doi.org/10.1094/CFW-55-1-0021

Di-Medeiros, M. C. B., Pascoal, A. M., Batista, K. A., Bassinello, P. Z., Lião, L. M., Leles, M. I. G., & Fernandes, K. F. (2014). Rheological and biochemical properties of *Solanum lycocarpum* starch. *Carbohydrate Polymers, 104*(1), 66–72. https://doi.org/10.1016/j.carbpol.2014.01.023

Egerton, R. F. (2005). The transmission electron microscope. In R. F. Egerton (Ed.), *Physical principles of electron microscopy* (pp. 57–92). Springer. https://doi.org/10.1007/0-387-26016-1_3

Elmi Sharlina, M. S., Yaacob, W. A., Lazim, A. M., Fazry, S., Lim, S. J., Abdullah, S., Noordin, A., & Kumaran, M. (2017). Physicochemical properties of starch from *Dioscorea* pyrifolia tubers. *Food Chemistry, 220*, 225–232. https://doi.org/10.1016/j.foodchem.2016.09.196

Forouzandehdel, S., Forouzandehdel, S., & Rezghi Rami, M. (2020). Synthesis of a novel magnetic starch-alginic acid-based biomaterial for drug delivery. *Carbohydrate Research, 487*. https://doi.org/10.1016/j.carres.2019.107889

Fukase, E., & Martin, W. (2020). Economic growth, convergence, and world food demand and supply. *World Development, 132*. https://doi.org/10.1016/j.worlddev.2020.104954

Hao, H., Li, Q., Bao, W., Wu, Y., & Ouyang, J. (2018). Relationship between physicochemical characteristics and in vitro digestibility of chestnut (*Castanea mollissima*) starch. *Food Hydrocolloids, 84*, 193–199. https://doi.org/10.1016/j.foodhyd.2018.05.031

He, W., & Wei, C. (2020). A critical review on structural properties and formation mechanism of heterogeneous starch granules in cereal endosperm lacking starch branching enzyme. *Food Hydrocolloids, 100*.

Hedayati, S., Shahidi, F., Koocheki, A., Farahnaky, A., & Majzoobi, M. (2016). Physical properties of pregelatinized and granular cold water swelling maize starches at different pH values. *International Journal of Biological Macromolecules, 91*, 730–735. https://doi.org/10.1016/j.ijbiomac.2016.06.020

Ionashiro, M. (2005). *Fundamentos de termogravimetria e análise térmica diferencial/calorimetria exploratória diferencial, 80*. São Paulo: Giz.

Jay-Lin, J., Tunyawat, K., Sharon, L., Henzy, Z., & F., R. J. (1994). Anthology of starch granule morphology by scanning electron microscopy. *Starch*, 121–129. https://doi.org/10.1002/star.19940460402

Karim, A. A., Norziah, M. H., & Seow, C. C. (2000). Methods for the study of starch retrogradation. *Food Chemistry, 71*(1), 9–36. https://doi.org/10.1016/S0308-8146(00)00130-8

Laignel, B., Bliard, C., Massiot, G., & Nuzillard, J. M. (1997). Proton NMR spectroscopy assignment of D-glucose residues in highly acetylated starch. *Carbohydrate Research, 298*(4), 251–260. https://doi.org/10.1016/S0008-6215(96)00314-X

Lima, B. N. B., Cabral, T. B., Neto, R. P. C., Tavares, M. I. B., & Pierucci, A. P. T. (2012). Estudo do amido de farinhas comerciais comestíveis. *Polimeros, 22*(5), 486–490. https://doi.org/10.1590/S0104-14282012005000062

Lin, L., Guo, D., Zhao, L., Zhang, X., Wang, J., Zhang, F., & Wei, C. (2016). Comparative structure of starches from high-amylose maize inbred lines and their hybrids. *Food Hydrocolloids, 52*, 19–28. https://doi.org/10.1016/j.foodhyd.2015.06.008

Liu, J., Wang, X., Wen, F., Zhang, S., Shen, R., Jiang, W., Kan, J., & Jin, C. (2016). Morphology, structural and physicochemical properties of starch from the root of *Cynanchum auriculatum* Royle ex Wight. *International Journal of Biological Macromolecules, 93*, 107–116. https://doi.org/10.1016/j.ijbiomac.2016.08.063

Liu, Y., Lu, K., Hu, X., Jin, Z., & Miao, M. (2020). Structure, properties and potential applications of phytoglycogen and waxy starch subjected to carboxymethylation. *Carbohydrate Polymers, 234*. https://doi.org/10.1016/j.carbpol.2020.115908

Londoño-Restrepo, S. M., Rincón-Londoño, N., Contreras-Padilla, M., Acosta-Osorio, A. A., Bello-Pérez, L. A., Lucas-Aguirre, J. C., Quintero, V. D., Pineda-Gómez, P., del Real-López, A., & Rodríguez-García, M. E. (2014). Physicochemical, morphological, and rheological characterization of Xanthosoma robustum Lego-like starch. *International Journal of Biological Macromolecules, 65*, 222–228. https://doi.org/10.1016/j.ijbiomac.2014.01.035

Lopez-Rubio, A., Flanagan, B. M., Gilbert, E. P., & Gidley, M. J. (2008). A novel approach for calculating starch crystallinity and its correlation with double helix content: a combined XRD and NMR study. *Biopolymers, 89*(9), 761–768. https://doi.org/10.1002/bip.21005

Mao, X., Lu, J., Huang, H., Gao, X., Zheng, H., Chen, Y., Li, X., & Gao, W. (2018). Four types of winged yam (*Dioscorea alata* L.) resistant starches and their effects on ethanol-induced gastric injury in vivo. *Food Hydrocolloids, 85*, 21–29. https://doi.org/10.1016/j.foodhyd.2018.06.036

Mbougueng, P. D., Tenin, D., Scher, J., & Tchiégang, C. (2012). Influence of acetylation on physicochemical, functional and thermal properties of potato and cassava starches. *Journal of Food Engineering, 108*(2), 320–326. https://doi.org/10.1016/j.jfoodeng.2011.08.006

Mohammad Amini, A., & Razavi, S. M. A. (2016). A fast and efficient approach to prepare starch nanocrystals from normal corn starch. *Food Hydrocolloids, 57*, 132–138. https://doi.org/10.1016/j.foodhyd.2016.01.022

Murphy, D. B. (2001). *Fundamentals of light microscopy and electronic imaging*.

Nokab, E., & Wel. (2020). Use of solid-state NMR spectroscopy for investigating polysaccharide-based hydrogels: a review. *Carbohydrate Polymers, 240*.

Pascoal, A. M., Di-Medeiros, M. C. B., Batista, K. A., Leles, M. I. G., Lião, L. M., & Fernandes, K. F. (2013). Extraction and chemical characterization of starch from *S. lycocarpum* fruits. *Carbohydrate Polymers, 98*(2), 1304–1310. https://doi.org/10.1016/j.carbpol.2013.08.009

Pavia, Lampman, G., Kriz, G., & Vyvyan, J. R. (2010). *Introdução à Espectroscopia* (2nd ed.). Cengage Learning.

Ratnaningsih, N., Suparmo, Harmayani, E., & Marsono, Y. (2016). Composition, microstructure, and physicochemical properties of starches from Indonesian cowpea (*Vigna unguiculata*) varieties. *International Food Research Journal, 23*(5), 2041–2049. http://www.ifrj.upm.edu.my/23%20(05)%202016/(27).pdf.

Rincón-Londoño, N., Vega-Rojas, L. J., Contreras-Padilla, M., Acosta-Osorio, A. A., & Rodríguez-García, M. E. (2016). Analysis of the pasting profile in corn starch: structural, morphological, and thermal transformations, Part I. *International Journal of Biological Macromolecules, 91*, 106–114. https://doi.org/10.1016/j.ijbiomac.2016.05.070

Sevenou, O., Hill, S. E., Farhat, I. A., & Mitchell, J. R. (2002). Organisation of the external region of the starch granule as determined by infrared spectroscopy. *International Journal of Biological Macromolecules, 31*(1–3), 79–85. https://doi.org/10.1016/S0141-8130(02)00067-3

Shirlyanne, F., Thais, A., Natalia, S., Layanne, R., M., L. H., Matheus, P., Gilmar, T., & Paula, R. A. (2019). Physicochemical, morphological and antioxidant properties of spray-dried mango kernel starch. *Journal of Agriculture and Food Research, 100012*. https://doi.org/10.1016/j.jafr.2019.100012

Simpson, J. H. (2017). Chapter 1 - introduction to the methods of nuclear magnetic resonance. In J. H. Simpson (Ed.), *NMR case studies* (pp. 1–9). Elsevier.

Singh, N., Singh, J., Kaur, L., Sodhi, N. S., & Gill, B. S. (2003). Morphological, thermal and rheological properties of starches from different botanical sources. *Food Chemistry, 81*(2), 219–231. https://doi.org/10.1016/S0308-8146(02)00416-8

Tak, H. Y., Yun, Y. H., Lee, C. M., & Yoon, S. D. (2019). Sulindac imprinted mungbean starch/PVA biomaterial films as a transdermal drug delivery patch. *Carbohydrate Polymers, 208*, 261–268. https://doi.org/10.1016/j.carbpol.2018.12.076

Teixeira, A. S., Navarro, A. S., Molina-García, A. D., Martino, M., & Deladino, L. (2015). Corn starch systems as carriers for yerba mate (*Ilex paraguariensis*) antioxidants: effect of mineral addition. *Food and Bioproducts Processing, 94*, 39–49. https://doi.org/10.1016/j.fbp.2015.01.002

Teli, M. D., & Mallick, A. (2018). Application of sorghum starch for preparing superabsorbent. *Journal of Polymers and the Environment, 26*(4), 1581–1591. https://doi.org/10.1007/s10924-017-1057-7

Tester, R. F., Karkalas, J., & Qi, X. (2004). Starch - composition, fine structure and architecture. *Journal of Cereal Science, 39*(2), 151–165. https://doi.org/10.1016/j.jcs.2003.12.001

Thomas, D. J., & Atwell, S. S. (1999). *Practical guides for the food industry, 94*. St. Paul, MN: Eagan Press.

Tianyu, J., Qingfei, D., Jian, Z., Hongsheng, L., & Long, Y. (2020). Starch-based biodegradable materials: challenges and opportunities. *Advanced Industrial and Engineering Polymer Research*, 8–18. https://doi.org/10.1016/j.aiepr.2019.11.003

Tsuiki, K., Fujisawa, H., Itoh, A., Sato, M., & Fujita, N. (2016). Alterations of starch structure lead to increased resistant starch of steamed rice: identification of high resistant starch rice lines. *Journal of Cereal Science, 68*, 88–92. https://doi.org/10.1016/j.jcs.2016.01.002

van Soest, J. J. G., Tournois, H., de Wit, D., & Vliegenthart, J. F. G. (1995). Short-range structure in (partially) crystalline potato starch determined with attenuated total reflectance Fourier-transform IR spectroscopy. *Carbohydrate Research, 279*(C), 201–214. https://doi.org/10.1016/0008-6215(95)00270-7

Vanier, N. L., El Halal, S. L. M., Dias, A. R. G., & da Rosa Zavareze, E. (2017). Molecular structure, functionality and applications of oxidized starches: a review. *Food Chemistry, 221*, 1546–1559. https://doi.org/10.1016/j.foodchem.2016.10.138

Villar, M. A., Barbosa, S. E., García, M. A., Castillo, L. A., & López, O. V. (2017). Starch-based materials in food packaging: processing, characterization and applications. In *Starch-based materials in food packaging: processing, characterization and applications* (pp. 1–321). Elsevier Inc. http://www.sciencedirect.com/science/book/9780128094396.

Wang, H., Yang, Q., Gao, L., Gong, X., Qu, Y., & Feng, B. (2020). Functional and physicochemical properties of flours and starches from different tuber crops. *International Journal of Biological Macromolecules, 148*, 324–332. https://doi.org/10.1016/j.ijbiomac.2020.01.146

Wang, S., Li, C., Copeland, L., Niu, Q., & Wang, S. (2015). Starch retrogradation: a comprehensive review. *Comprehensive Reviews in Food Science and Food Safety, 14*(5), 568–585. https://doi.org/10.1111/1541-4337.12143

Wang, S., Sun, Y., Wang, J., Wang, S., & Copeland, L. (2016). Molecular disassembly of rice and lotus starches during thermal processing and its effect on starch digestibility. *Food and Function, 7*(2), 1188–1195. https://doi.org/10.1039/c6fo00067c

Wang, S., Yu, J., & Yu, J. (2008). The semi-crystalline growth rings of C-type pea starch granule revealed by SEM and HR-TEM during acid hydrolysis. *Carbohydrate Polymers, 74*(3), 731–739. https://doi.org/10.1016/j.carbpol.2008.03.001

Xiao, H., Yang, F., Lin, Q., Zhang, Q., Zhang, L., Sun, S., Han, W., & Liu, G. Q. (2020). Preparation and characterization of broken-rice starch nanoparticles with different sizes. *International Journal of Biological Macromolecules, 160*, 437–445. https://doi.org/10.1016/j.ijbiomac.2020.05.182

Ya Bogracheva, T., Wang, Y. L., & Hedley, C. L. (2001). The effect of water content on the ordered/disordered structures in starches. *Biopolymers, 58*(3), 247–259. https://doi.org/10.1002/1097-0282(200103)58:3<247::AID-BIP1002>3.0.CO;2-L

Yu, L., & Christie, G. (2001). Measurement of starch thermal transitions using differential scanning calorimetry. *Carbohydrate Polymers, 46*(2), 179–184. https://doi.org/10.1016/S0144-8617(00)00301-5

Zabot, G. L., Silva, E. K., Emerick, L. B., Felisberto, M. H. F., Clerici, M. T. P. S., & Meireles, M. A. A. (2019). Physicochemical, morphological, thermal and pasting properties of a novel native starch obtained from annatto seeds. *Food Hydrocolloids, 89*, 321–329. https://doi.org/10.1016/j.foodhyd.2018.10.041

Zhang, X., Golding, J., & Burgar, I. (2002). Thermal decomposition chemistry of starch studied by 13C high-resolution solid-state NMR spectroscopy. *Polymer, 43*(22), 5791–5796. https://doi.org/10.1016/S0032-3861(02)00546-3

Zhu, F. (2017). NMR spectroscopy of starch systems. *Food Hydrocolloids, 63*, 611–624. https://doi.org/10.1016/j.foodhyd.2016.10.015

Zhu, F. (2018). Relationships between amylopectin internal molecular structure and physicochemical properties of starch. *Trends in Food Science & Technology, 78*, 234–242. https://doi.org/10.1016/j.tifs.2018.05.024

Zou, J., Xu, M., Wen, L., & Yang, B. (2020). Structure and physicochemical properties of native starch and resistant starch in Chinese yam (*Dioscorea opposita* Thunb.). *Carbohydrate Polymers, 237*. https://doi.org/10.1016/j.carbpol.2020.116188

Biocomposites potential for nanotechnology

Sergiana dos Passos Ramos[1], *Michele Giaconia*[1], *Monize Burck*[1], *Daniella Carisa Murador*[1], *Maria Carolina Bezerra Di-Medeiros Leal*[2], *Ailton Cesar Lemes*[3], *Mariana Buranelo Egea*[4] *and* *Anna Rafaela Cavalcante Braga*[5]
[1]Department of Biosciences, Universidade Federal de São Paulo (UNIFESP), Santos, São Paulo, Brazil; [2]Federal University of São Carlos (UFSCar), Nuclear Magnetic Resonance Laboratory, São Carlos, Brazil; [3]Department of Biochemical Engineering, Universidade Federal do Rio de Janeiro (UFRJ), Technological Center, School of Chemistry, Ilha do Fundão, Rio de Janeiro, Brazil; [4]Goiano Federal Institute of Education, Science and Technology, Rio Verde, Goiás, Brazil; [5]Department of Chemical Engineering, Universidade Federal de São Paulo (UNIFESP), Diadema, Brazil

1. Highlights

The main areas covered in this chapter include:

- Replacing standard composites with biodegradable composites is more advantageous.
- Nanotechnology is a profitable investment for several types of industry.
- The main advantages of nanobiocomposites for food packaging are the extension of shelf-life and better quality of food.
- The safer and prolonged sensory quality of the food product are characteristics promoted by utilizing nanobiocomposites.
- An application is a sensor that can detect contaminants, gases, or microorganisms in packaged foods.
- Nanobiocomposites have versatility for the cosmetic area due to their lipophilic and hydrophilic delivery systems.
- Nanobiocomposites can improve the production of hair and skin products.
- It is possible to create microenvironments capable of inducing cell growth and differentiation.
- Nanocomposites can be synthesized with the ability to self-regenerate and/or self-repair.

2. Biocomposites and nanotechnology

Composites can be defined as any multiphase material which results in the improvement of the quality of isolated features. Two main classes can be highlighted, being

Biodegradable Polymers, Blends and Composites. https://doi.org/10.1016/B978-0-12-823791-5.00001-6

plastic composites and biodegradable or green composites. Most plastic composites possess several properties that have promoted their vast use in recent years, such as light-impermeable, high-strength materials. However, only small percentage of the total produced plastic polymers is recycled (Faruk et al., 2012).

The demand for biodegradable composites has grown worldwide in recent years, mainly in order to reduce the environmental impact caused by structural materials produced using petroleum industry products. Considering this information, it is critical to develop and apply renewable polymers, also known as biopolymers, to substitute nondegradable composites based mainly on petroleum-based polymers (Braga et al., 2021; de Oliveira-Filho et al., 2020).

Several types of biocomposites have been developed in recent years with technological advances, generating some classifications concerning their structure, production forms, and materials, among others. The combination of organic natural materials, inorganic or organic polymers, and nanostructures has presented elevated potential in the format of nanobiocomposites, thus refining the mechanical performances of these structures and, consequently, escalating the industrial and academic areas of application (Hasan et al., 2020).

Recently, various inorganic nanoparticles have been studied for incorporating them with biofibers in the matrix to form nanobiocomposites, especially for their biodegradability. This results in developing an interfacial bond between the biofibers and polymers in a composite system, whereas the organic phase helps to form an inorganic matrix. Biocomposites have been mainly used in the construction, textile, food, biotechnology, and biomedical sectors, especially for the development of building materials with superior flexibility and improved physical, mechanical, and functional properties (Bordes et al., 2009; Elshaarawy et al., 2019; Hasan et al., 2020).

The nanodimensional phase has a remarkable effect on nanobiocomposite properties, especially the thermal, electrical, optical, mechanical, catalytic, and electrochemical properties. The outstanding performance characteristics of biocomposites and nanobiocomposites have made them superior to conventional composites.

Fig. 17.1 shows the potential formation mechanism of nanobiocomposites. Therefore, the need to understand and study various nanobiocomposites and their formation and functional perspectives are essential.

3. Nanobiocomposite applications

Biodegradable polymers are (1) directly extracted from marine or agricultural resources (e.g., natural rubber, cellulose, starch, chitin, and collagen), (2) obtained

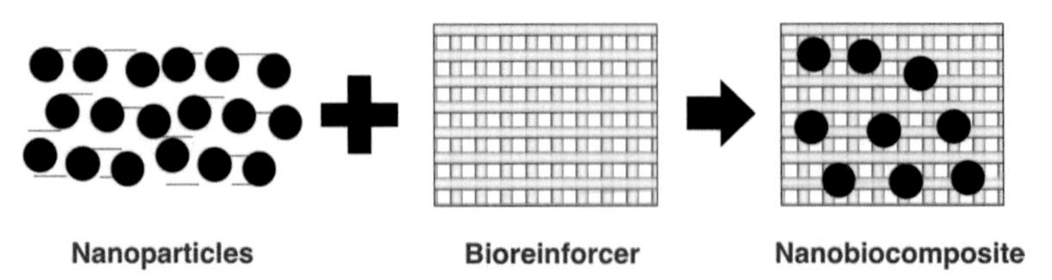

Figure 17.1 Production scheme for nanobiocomposites.

through microorganisms, such as genetically modified bacteria [e.g., bacterial cellulose (BC), poly(hydroxyalkonoates) (PHAs), and poly(hydroxybutyrate) (PHB)], and (3) achieved by a synthetic process involving biobased resources [e.g., polycaprolactone (PCL), polylactic acid (PLA)] (Jamróz et al., 2019; Ramos et al., 2018; Valdés et al., 2017). Nanobiocomposites are a coupling of a polymer matrix and nanometric neutral particles that reveal valuable properties frequently different from those of conventional polymers, and have gained notoriety over the past few years (Giaconia et al., 2020) since upcoming innovative technologies request materials with specific properties, such as improved thermogravimetric stability, superior tensile strength, degree of miscibility, and so forth.

Unquestionably, the ecofriendly approach associated with the feasibility of scaling up and promising applicability is fundamental to meeting the dissemination of nanobiocomposites in the market. Table 17.1 show examples of nanobiocomposite applications found in Scopus-indexed journals from 2010 to 2020 that were outlines, including tissue engineering and biomedical fields (Belaid et al., 2020; Saneinezhad et al., 2020), food packaging (Sánchez-García et al., 2010), and water remediation waste (Baruah et al., 2020).

These studies underline that nanobiocomposite utilization comprises multifaceted sectors. Understanding the interplay between nanoparticles and polymers is essential to developing new nanocomposites for efficient application in the production of drugs, biosensors, biotechnological devices, car elements, and textile products, etc. Along these lines, nanobiocomposites are mostly health-related, followed by packaging applications, which are extensively discussed in the following sections. However, it is worth mentioning that great biocompatibility, low toxicity, safety as a fundamental issue in biomedical perspectives and relative to the packaging industry, high gas barrier for providing antimicrobial protection, thermal resistance, and biodegradation are essential properties.

Hydroxyapatite (HA) is a widely used biomaterial for bone applications (Silva et al., 2005). Carbonated hydroxyapatite/SiO_2 nanobiocomposites demonstrated compressive strength analogous to human bone and improved physical and mechanical properties, and intensified in vitro bioactivity behavior (Taha et al., 2020).

Graphene biohybrids have been recognized as an excellent material for innovating multifunctional tissue scaffolds, mainly focused on bone and neural regeneration, due to their higher biocompatibility. However, the side effects of in vivo application of graphene biohybrids still need to be carefully studied in the future (Ma et al., 2019). In addition, graphene can improve conventional matrices' physical properties, such as polymeric films, hydrogels, and electro-spun fibers (Ma et al., 2019). Regarding its nanobiocomposite conformation, a hydrogel is underpinned as a dressing material for healing purposes as a convenient choice for skin drug delivery (Nouri, 2019). κ-Carrageen colloids are interesting for drug delivery as well; understanding the interaction between κ-carrageen and Au is vital for reaching controllable nanobiocomposite colloids (Gasilova et al., 2019). Stable colloidal suspensions of surface active maghemite nanoparticle (SAMN)-based nanobiocomposites have potential for the preparation of self-assembled opsonized nanoparticles as promising candidates for

Table 17.1 Studies regarding nanobiocomposite applications and potential applications.

Application	Natural source	Nanobiocomposite	Author	Year
Biomedical	–	Hydroxyapatite/SiO_2	Taha et al.	2020
	Anaerobic bacteria	Bacterial cellulose/polypyrrole	Mashkour	2017
	κ-Carrageenan polysaccharide	Au/κCAR	Gasilova et al.	2019
Cancer therapeutics	Seaweed polysaccharide	Bioaldehyde/Ag	Kholiya et al.	2020
	Catharanthus roseus flower extract	Cu/*C. roseus* extract	Baskar, Sathivel, George	2016
Tissue engineering	–	Polylactic acid/exfoliated boron nitride	Belaid et al.	2020
	–	Zirconia/hydroxyapatite	Youness; Taha; Ibrahim	2020
Pressure sensor	Sea sponge (*Phylum porifera*)	Polydopamine (pDA)/graphene oxide/Ag/sponges (DGA-S)	Dong et al.	2018
Biosensor	–	Glucose oxygenase/poly(1,2-diaminobenzene) (Prussian Blue)/Au/TiO_2	Gao et al.	2014
Water remediation	Sugarcane bagasse, rice husk	Nanocellulose/iron oxide	Baruah et al.	2020
	–	Zn–Ag/montmorillonite/Polylactic acid	Sahithya; Das	2016
	Gum acacia, chitosan	Plant gum-based/ZnO/chitosan	Varghese; Das; Das	2016
	–	Functionalized cellulose/Preyssler/Pb	Saneinezhad et al.	2020
Packaging	–	Poly(ethymethacrylate)/graphene oxide/Ag	Mohanty and Swain	2019
	Alginate, shaddock peel	Sodium alginate/ε-polylysine/cellulose nanocrystals (CNCs)	Tang et al.	2017
	–	poly(ε-caprolactone)/hydroxytyrosol/montmorillonite	Beltrán et al.	2014
	Cinnamon pectin	Cinnamon nanoemulsion/pectin	Moura et al.	2014
Gas barrier	Gelatin industrial by-product (hides and carcasses)	Graphene-like/boron nitrite/gelatin	Biscarat et al.	2015

diagnostic and therapeutic applications in biomedicine because, when introduced into the cell culture medium, they spontaneously bind covalently to specific proteins (Venerando et al., 2013).

Continuing to the health segment, nanomedicine studies are revealing the potential for cancer cell imaging through jacalin enzyme with gold nanoparticles (Marangoni et al., 2013); copper nanoparticles for an anticancer drug (Baskar et al., 2017); antibacterial activity of collagen titanium dioxide/Ag (Spoiala et al., 2015), etc. A pressure sensor for monitoring health was developed by Choo et al. (2019) through a dip-coated process with the hierarchical structure of sea sponges and composites conductive networks of polydopamine-reduced graphene oxides and silver nanowires. The pressure sensor presented excellent elasticity, and durable strain−stress properties coupled with significant advantages of biomass, reduced cost, and easy fabrication, increasing the attractiveness for future electronic skins. Enzymatic glucose biosensors with remarkably high sensitivity have been achieved (Gao et al., 2015) through glucose oxidase and Prussian Blue (PB) with AuNP and TiO_2 nanotube arrays as a supporting electrode. A synergistic effect was observed through the high electrocatalytic activity of PB and the biocompatibility of AuNPs that enhanced the analytical performance of electrodes.

In the packaging industry, improving gas barriers, biodegradability, and antimicrobial behavior are essential attributes. Starch hybrid poly(ethymethacrylate)/graphene oxide/Ag nanobiocomposite showed the characteristics mentioned above and chemical and thermal improvements, resulting in a suitable material for packaging (Mohanty & Swain, 2018). Beltrán et al. (2014) achieved potential for active food packaging using poly(ε-caprolactone), hydroxytyrosol, and commercial montmorillonite. In addition, hydroxytyrosol(3,4-dihydroxyphenylethanol) (HT) is an antioxidant from olives that could be obtained from agricultural resources and turned into higher added-value active food-packaging additives. Reductions in solubility and increased mechanical properties were observed in nanobiocomposites containing cinnamon nanoemulsions and a pectin-formed film with great potential for food packaging (Moura et al., 2014). Underlining the gas barrier and its possible that the water content permeate the food packaging, the synthesized graphene-like/boron nitrite/gelatin nanocomposite presented good dispersion in the matrix and outstanding capability to improve the crystallinity and barrier properties of gelatin (Biscarat et al., 2015).

Unequivocally, water remediation is an immediate concern. High adsorptive performance, magnetic recoverability, and recyclability are important factors for water remediation; therefore, nanocellulose iron oxide nanobiocomposites were produced from rice husk- and sugarcane bagasse-derived nanocelluloses with a view to removing arsenic and associated contaminants from groundwater samples. The nanobiocomposite showed an efficiency of approximately 99% under the evaluated conditions (Baruah et al., 2020). Sahithya et al. (2016) analyzed the effectiveness of removing organophosphate insecticide from an aqueous solution and found that bimetallic Zn−Ag nanoparticles with montmorillonite and PLA demonstrated the best insecticide removal as compared with chitosan and gum ghatti blends.

As a reinforcing agent, it was observed that modified starch nanoparticles loaded with natural rubber demonstrated superior strength and elongation to conventional composites over a wide range of temperatures. Thus, the nanobiocomposite is promising for vibration-damping applications and replacing carbon black on rubber (Valodkar & Thakore, 2011). Referring to the automotive sector, the use of a natural fiber composite for interior and exterior applications is attractive because of their lightweight, impact-resistant, nontoxic, and no-sharp edged fracture in the case of a car accident properties (Gholampour & Ozbakkaloglu, 2020). However, to the best of our knowledge, studies involving nanobiocomposites and car applications remain scarce, given that previous searched for composites related to the automobile industry (Kausar, 2021). Notwithstanding the above, Abdelwahab et al. (2020) identified great mechanical properties and dimensional stability in Polyamide 6 (PA 6), biocarbon (*Miscanthus* fiber), and nanoclay nanobiocomposites. Idumah and Hassan (2017) described improved mechanical, morphological, and thermal properties through graphene nanoparticles with biofibers (*Hibiscus cannabinus*). Both studies suggested that these lightweight nanobiocomposites have potential to be applied in the automotive sector.

It is possible to obtain HA using an environment-friendly process from fish skeletons, with cost reduction and biocompatibility (Granito et al., 2028). Similarly, collagen from leather biowaste (Mandal et al., 2016; Meiyazhagan et al., 2015) or gelatin from hides and carcass by-product of an industrial process (Biscarat et al., 2015) are green, simple, and cost-practical approaches to biopolymers. Collagen blended with Al_2O_3 nanoparticles (NPs) performs as a potential load-bearing application, such as in hip prostheses, due to increased thermal stability and lower water uptake resulting in superior tensile, tear, and flexural strength characteristics. Further, in vitro and in vivo biocompatibility and efficacy must be assessed (Mandal et al., 2016). Thin electrically conducting hybrid films obtained from collagen from leather industry waste, with guar gum, and two carbon nanotubes resulted in flexible and electrically conducting material for wide applications such as biosensors and electronic devices (Meiyazhagan et al., 2015). Therefore, using natural resources, especially biowaste, is an outstanding method for an environmentally appropriate process accompanied by lower expenditures and greater efficiency.

Another nanobiocomposite-inherent aspect that should be undertaken is overseeing the use and manufacturing of nanomaterials, through regulation by governmental or regulatory entities, in order to guarantee minimum safety requirements and satisfactory production criteria to control the migration of components in food and the human body, and also the environment (Braga et al., 2021; de Oliveira-Filho et al., 2020; Giaconia et al., 2020).

4. Nanobiocomposites and the food industry

Nanotechnology has been rapidly growing in recent years and is being widely applied in several science and technology areas for developing new materials at the nanoscale level. The food sector is included in this advancement, in which nanotechnology allows improvements in the safety and quality of food production, processing, and storage, following sustainable principles. Nanotechnology has been used in food processing, development of innovative products, nanoencapsulation for the delivery of nutrients, bioactive compounds, flavors, and aromas, in the food safety area, such as nanosensors and nanosieves, and in food packaging, which involves nanocomposites, among others (Berekaa, 2015). In practice, these approaches could positively affect the durability and integrity potential of foods, contributing to the maintenance of their nutritional, sensory, and microbiological characteristics.

Some of the most relevant advantages of nanostructures are related to their higher surface area compared to non-nano ones, which means a higher surface area of contact and, consequently, greater possibility of intense interaction between the matrix in which they are inserted and the nanoparticles (Assis et al., 2012), for example, more efficient absorption of compounds during the digestion process. In addition, the singular chemical and physical properties of nanoparticles have led to an emerging exploration of the antimicrobial potential of nanoscale materials, which makes them of great interest to be incorporated into food-packaging materials, contributing to the increased shelf-life and safety of foods for human consumption (Berekaa, 2015).

In general, polymers are the most used material in the production of nanostructures; in addition to the classic polymeric materials, a class of nanoreinforced biopolymers, known as nanobiocomposites, has gained interest in the past few years (Giaconia et al., 2020) as an emerging trend for developing new biodegradable materials has been realized, including their application in the food industry. Basically, the biodegradable polymers can be classified into three categories depending on their origin: (1) includes polymers directly extracted from biomass from marine or agro-resources, such as polysaccharides (e.g., starch, cellulose, chitosan/chitin), proteins (e.g., casein, collagen, whey, zein, soya), and lipids (e.g., cross-linked triglyceride), alone or combined; (2) involves synthetic polymers from renewable biobased monomers, such as polylactic acid (PLA, a biopolyester polymerized from lactic acid monomers); and (3) comprises polymers produced by microorganisms or genetically modified bacteria, such as poly(hydroxyalkonoates) (PHAs) and poly(hydroxybutyrate) (PHB), and bacterial cellulose (BC) (Jamróz et al., 2019; Ramos et al., 2018; Valdés et al., 2017).

Considering the food industry context, one of the most exploited fields of nanotechnology for biocomposites is the development of food packaging, representing ~50% of the market value for all nanotechnology-enabled products (Vasile, 2018). Nanomaterials can be applied in food packaging as polymer nanocomposites with high barrier properties, intelligent packaging, nanocoatings, surface biocides, active packaging, silver nanoparticles as antimicrobial agents, nutrition and nutraceuticals, nanosensors, and assays for the detection of food-relevant analytes (gases, small organic molecules, and foodborne pathogens), and bioplastics (Vasile, 2018). The main advantages related to the nanocomposites for food packing are extended shelf-life, better food quality, which involves safer and longer sensorial quality of the product, and enhanced mechanical, thermal, flexible, and barrier properties, besides the antimicrobial potential with greater activity and sensors that can detect trace contaminants, gases, or microorganisms in packaged foods (Luna & Vílchez, 2017; Ramos et al., 2018).

Several studies have reported the successful application of nanobiocomposites in the development of novel alternatives for food packaging. Starch, cellulose, zein, and chitin are some of the most common nanoreinforced biopolymers used in food packaging. Starch is a polymer obtained from agro-resources (e.g., potato, maize, pea, rice, wheat, cassava, waxy maize, amylomaize) that has received considerable attention due to its biodegradability, wide availability, and low cost, and although the structural units are identical among the various starches, the different amylose and amylopectin contents influence its physicochemical properties and, consequently, the size of native granules (Ramos et al., 2018). The nanobiocomposite films produced using starch are nontoxic and environment friendly. However, further investigation regarding some limitations of this biopolymer still deserves attention, such as the moisture sensitivity and weak barrier property, which can hinder its application as a commercial packaging material (Jamróz et al., 2019).

Cellulose is the most abundant polymer in nature, presenting interesting mechanical properties, chemical structures, and reinforcing effects beyond its cost-effectiveness, as agricultural wastes can be a rich source of this material, such as banana rachis, or by fermentation through *Gluconacetobacter xylinus* bacteria (Alila et al., 2013). Thus, considering these characteristics, cellulose can be turned into a very attractive nanoreinforced material used for the preparation of nanobiocomposites that are useful for food packaging (Ramos et al., 2018). There are three types of cellulose used on the nanoscale that can be applied as biopolymers in food packaging: nanocrystalline cellulose, nanofibrillated cellulose, and bacterial nanocellulose; depending on the source of origin and the method of extraction, they can differ in morphology, degree of crystallinity, particle size, and some other properties (Jamróz et al., 2019). Bacterial nanocellulose, in particular, presents practical biotechnological conditions for food production and unique thermomechanical characteristics and biocompatibility, making it promising for an extensive range of uses, besides being renewable and quickly produced (Giaconia et al., 2020). Nanobiocomposites produced with bacterial cellulose have been investigated in various studies regarding food and bioprocess industries, such as texture modifiers, bioactive compounds, antimicrobial carriers, and for stabilization of olive oil pickering emulsion, among others (Giaconia et al., 2020; Yan et al., 2017).

Chitosan is the second most abundant polysaccharide polymer in nature after cellulose, and is a deacetylated derivative of chitin, which is also a polysaccharide polymer (Miteluţ et al., 2015). Chitosan as biopolymer (that can also be used in the nanoscale) is a new material for food packaging due to some favorable characteristics: it can be obtained from marine biomass resource, being biodegradable and nontoxic; it confers strength and elongation properties; it has selective permeability to gases (CO_2 and O_2) and good mechanical properties; but their applications can be limited due to its hydrophilic nature, conferring a poor barrier to moisture (Miteluţ et al., 2015). On the other hand, chitosan also has inherent antimicrobial activity against a wide range of bacteria, yeasts, and fungi, so, along with the nanotechnology support, chitosan nanocomposite-based films can present expressive antimicrobial activity, enhancing the shelf-life and food safety (Novak et al., 2019).

Along with starch, the biobased polymer polylactic acid (PLA) is one of the most exciting polymers for food packaging because of its compelling balance of properties and it is produced on an industrial scale and is commercially available

(Valdés et al., 2014). PLA is a thermoplastic polyester obtained by fermentation of lactic acid, and so 100% bio-based, being produced from renewable material (Giaconia et al., 2020). PLA presents excellent transparency and relatively good water resistance, showing widespread applications as packaging barrier transparent film bottles, thermoformed products, trays, and blisters, among others (Ramos et al., 2018; Valdés et al., 2014). Furthermore, PLA is appropriate for packaging short-shelf-life products, like fruits and vegetables, that do not require any specific gas barrier and can be packaged in atmospheric conditions (Ramos et al., 2018).

The use of waste residues as biomass resources of biopolymers has also been a focus of study since, in the last few years, processes involving green approaches have become primordial. In this context, pectin represents a natural biopolymer in the cell walls of many plants that can be extracted from them and used in different applications, including food packaging. Mellinas et al. (2020) reported that there is an emerging research field with a multitude of food and beverage industry applications related to pectin and combinations with other biopolymers to improve some important properties, such as antioxidant and antimicrobial performance, and flexibility to obtain the films. As initially described, the biopolymers can be used alone or combined, and as each of them exhibits some limitations, their blending can be a promising alternative.

Polymer food packaging usually results in modified physical and mechanical properties compared to materials made of individual components, so allowing the achievement of the desired properties (Miteluţ et al., 2015). For example, Yu et al. (2017) developed biopolymer-based edible nanocomposite films using cellulose nanofibrils, corn starch, and chitosan, and concluded that the cellulose nanofibrils effectively enhanced the antimicrobial effect and barrier properties of the nanocomposite films, showing great potential in applications of active packaging for food products. Ramos et al. (2018) indicated that the chitosan–cellulose blends are of great interest because of these two biopolymers' structural similarity, allowing the combination of chitosan physicochemical properties and the mechanical ones of cellulose, resulting in a final material with superior mechanical properties and transparency. Yang et al. (2016) showed that cellulose nanocrystals and lignin nanoparticles could be successfully applied as reinforcement agents in the production of PLA nanocomposite films, improving their tensile strength and elastic modulus, as well as displaying antibacterial activity toward harmful pathogens (e.g., *Pseudomonas syringae*).

In this context, Sundramoorthy et al. (2018) reported graphene-based nanomaterials as ideally suited for use in biosensors and other devices and techniques used in food quality analysis because of their high conductivity, mechanical flexibility, amenability for versatile surface functionalization, ultrahigh surface area, and biocompatible characteristics. The authors also discussed the importance of developing novel smart food packaging systems, such as active and intelligent packaging (e.g., time–temperature indicators, integrity indicators, freshness indicators, radio frequency identification), which contribute significantly to improved food safety and quality. Another interesting use of nanobiocomposites in the food field is related to bioactive compounds' nanoencapsulation since they are highly unstable, have low solubility, and are poorly bioavailable. Therefore, in this case, the nanoencapsulation could preserve the bioactive compounds from environmental stresses and improve their physicochemical properties and health-promoting and antidisease activities (Shishir et al., 2018), solving their restrictions regarding their industrial applications.

5. Cosmetics applications of nanobiocomposites

The importance of the cosmetics industry has been increasing constantly over the years, as well as the concern to meet consumer expectations (Santos et al., 2019). Furthermore, scientists are continually looking for more natural products, that is why active biologic ingredients are enhancing the cosmetic area, known as cosmeceuticals. Another particular reason to incorporate bioactive ingredients in beauty products is related to their therapeutic benefits (Aziz et al., 2019). Looking for enhanced cosmetics benefits, nanobiocomposites may be broadly used in the formulation of skin products, such as make-up, sunscreen, antiwrinkle and skin whitening creams, deodorants, and also for hair products, such as shampoos, conditioners, and serums, considering their versatility for lipophilic and hydrophilic delivery systems (Kalouta et al., 2020; Santos et al., 2019).

Nanotechnology has been actively applied and studied in cosmetics and cosmeceutical areas as this innovative method can improve the solubility and stability of active ingredients and promote their skin penetration efficiently and control the rheological properties, such as sensory characteristics. These aspects may be developed in different products, providing several advantages, for example, long-lasting perfume, UV protection, and antiaging effect, which have been achieved through the use of a large variety of nanobiocomposites, such as polymeric nanoparticles, nanoemulsions, solid lipid nanoparticles (SLNs), liposomes, micelles, dendrimers, and metal nanoparticles (Aziz et al., 2019; Kalouta et al., 2020; Santos et al., 2019).

Regarding the variety of make-up products, Santos et al. (2019) performed a compilation of some patents regarding beauty cosmetics to catalog the products available on the market. Among the found products, a gel formulation was created capable of disguising wrinkles through the use of biocomposites produced with TiO_2 NPs and color pigments. A global cosmetic industry developed the product, with the gel forming a film that results in space-filling and changes the light's refraction and diffusion when applied to the skin.

Although there are many high-quality make-up products in the market currently, they do not exhibit the expected effect when the skin produces sebum excessively. Therefore, some solutions have been created to minimize this production by utilizing skin foundations with nanobiocomposites containing glycyrrhizin acid encapsulated into hydrophobic silicone-coated chitosan NPs, promoting prolonged inhibition of sebum production. Another product compiled by those authors was directed at the eyes, being an eye mascara produced with Fe_2O_3 NPs, which achieved long-lasting water- and sweat-proof effects, explained by its lipophilic and hydrophobic characteristics. In addition, the last product described was for lip care and was created with the substitution of toxic pigments by Au and AgNPs pigments which exhibit yellow and red colors, respectively. Thus, it is possible to mix them and create a variety of colored lipsticks. In the nail care field, Lau et al. (2017) developed NPs using a laser, which allows the incorporation of metal NPs into nail varnish and other viscous environments. These metallic NPs present some desired properties, such as optical effects, while the increased durability of color provides hardness and resistance to harm in addition to antibacterial activity (Lau et al., 2017).

Sunscreens are one of the first products to use nanotechnology in their formulations, and their efficacy has been advanced in recent years. In order to improve sun protection, several nanobiocomposites have been tested and produced. Morin, a flavonoid, can be incorporated by PLGA NPs (Mutalik et al., 2015), AuNPs with phytolatex can also be used (Borase et al., 2014), as well as the encapsulation of padimate-O in bioadhesive NPs, that not only enables sun protection but also reduces double-strand DNA breaks, preventing sun-induced damage (Deng et al., 2015).

The aging process sometimes is widely considered to be unwanted, and so novel investigations are seeking more effective carriers that can enhance the skin permeation of vitamins, antioxidants, and other actives, and their desired stability into products with better antiaging activity. In this sense, nanobiocomposites appear to be a revolutionary method to do so, by applying SLNs and NLCs to deliver vitamin E, resveratrol, epigallocatechin gallate (Chen et al., 2017), or nanoemulsion for coenzyme Q10 (El-Leithy et al., 2018), and also transfersomes for hyaluronic acid and epigallocatechin gallate (Avadhani et al., 2017).

Another recurring product that exploits the advantages of nanotechnology is deodorant. Santos et al. (2019) described a patent in which compositions utilizing carbon-based NPS could be used to substitute traditional odor removers and thinking about enhancing the fragrance durability. The perfume molecules can be nanoencapsulated into the polymer poly-L-lactic acid (PLA), as an alternative to protect and deliver fragrances (Hosseinkhani et al., 2015). In the cosmeceutical arena, the use of tea tree extract and pomegranate peel extract has taken place, with antiwrinkle action, acne treatment, and antioxidant and antiinflammatory activity, respectively, to formulate facial creams through an electrodynamic technique (Kalouta et al., 2020).

Whitening skin cosmetics have also improved their functionality through nanotechnology. Some researchers have used SLN to incorporate a tyrosinase inhibitor (So et al., 2010), gold nanoparticles to include *Panax ginseng* extract, a plant with antioxidant and antiaging properties (Jiménez et al., 2018), chitosan nanoparticles as a carrier for arbutin, which can inhibit excessive melanin production (Ayumi et al., 2019), and also linoleic acid delivered by ethosomes and transfersomes, combating skin hyperpigmentation disorders (Celia et al., 2012), in topical formulations. These applications have shown meaningfully reduced levels of tyrosinase activity and melanin synthesis, with excellent stability.

Nanosystems for hair problems have also been studied to avoid hair loss and/or to improve its growth. Nanobiocomposites produced by roxithromycin loaded into polymeric NPs can be used as a hair-restoring agent as this structure can act directly on the hair bulb (Główka et al., 2014). In addition, Santos et al. (2019) also related some patents in this area, including formulation with AuNPs because of their capacity to promote blood circulation, with stimulation of the scalp, hormone secretion, and cell division; all these actions stimulate hair growth and, consequently, help prevent its loss. Other NPs able to promote hair growth are composed of fermented extract of ginger, vinegar, ginger root, and coconut oil in a calcium alginate hydrogel. Nanotechnology can also increase the absorption of silicone oil into hair when involved by nanoemulsions, which may be used in shampoo formulations (Hu et al., 2012).

6. Nanobiocomposites in biomedical areas

Nanobiocomposites have been widely used and studied in biomedical applications due to their excellent mechanical and biological capacities, in addition to not undergoing noticeable degradation due to the action of the immune system. However, the advantages of these materials depend on their application, so it is necessary first to define the use for which they are being developed. Since many drugs in their macro- or micromolecular formulation have low bioavailability and pharmacokinetics, polymers on a nanoscale associated with nanoparticles have gained notoriety to improve this aspect.

Nanobiocomposite applications can be interesting due to the compelling improvements related to their biocompatibility, bioavailability, safety, increased permeability, and low toxicity. This association can create microenvironments capable of inducing cell growth and differentiation, in addition to carrying drugs to a specific target without exhibiting cytotoxicity. Tissue engineering becomes an ally for the development of these nanobiocomposites since these materials must possess an extracellular matrix with a three-dimensional structure very similar to human tissues (George et al., 2019; Mozumder et al., 2017).

According to Mozumder et al. (2017), the concept of using polymeric composites in biomedical applications emerged in the 1980s with the use of polyethylene associated with hydroxyapatite to replace bone tissues. From that, many studies were directed to the improvement of these biomaterials regarding mechanical, thermal, and biological aspects. The particle size used in polymeric composites directly interfered with the properties of these materials since the characteristics were enhanced as the size decreased. The increases in the contact surface of these materials imply a significant rise regarding compatibility and distribution, crucial factors in biomedical applications (Mozumder et al., 2017).

Currently, nanobiocomposites may present two-dimensional or three-dimensional structures, and their topography and architecture are two of the most important characteristics required for applications in tissue engineering. However, associating these two characteristics has been a challenge, and therefore, several techniques have been applied in the production of these materials in a successful way for cell growth and tissue regeneration (Moonesi Rad et al., 2019).

Besides the architecture of the nanobiocomposites, another relevant factor is the nature of the polymer, as already mentioned. Natural polymers such as proteins and carbohydrates are widely used mainly due to the ease with which the human body recognizes and degrades these compounds, presenting a higher interaction with human tissue. However, as it is usually obtained from animal raw material, natural polymers can have a high degree of impurities for use as scaffolds (Mishra, 2019). Synthetic polymers, on the other hand, have greater structural uniformity, have a high degree of purity, and are easily reproduced. In addition, many of these polymers do not have satisfactory mechanical properties for this application. Therefore, combining two or more polymers for the production of scaffolds becomes essential as it allows improvements in several features, adapting the material for biomedical applications (Mishra, 2019; Mozumder et al., 2017).

Many polymers are applied in medicine, but the association with other polymers and/or specific particles, in the form of biocomposites, promotes successful applications. The polymeric matrix for the formation of these nanobiocomposites must be chosen according to the intended function. Because of this, several kinds of research have been directed toward the development of smart polymers. The hydrogel polymer composites are an example of intelligent polymers that have high porosity and molecular structure with great hydration capacity, and are very similar to the tissue microenvironment (Singla et al., 2017). Although hydrogels are mostly formed by insoluble polymers (such as chitosan), the formation of composites guarantees high hydrophilic property and can be developed from both natural and synthetic polymers (Tashakkorian et al., 2020). Due to their high hydration, these compounds can be molded into the desired sizes and shapes, which are generally associated with nanomaterials improving their mechanical and biological properties. The application of these nanocomposites is more closely linked to the recovery of human skin wounds and the regeneration of bone tissues (Amaral et al., 2020; Belluzo et al., 2020; Singla et al., 2017; Tashakkorian et al., 2020).

Still, in regard to smart polymers, nanocomposites are being synthesized with the ability to self-regenerate or self-repair, mimicking the self-healing observed by various tissues of living organisms. Self-healing polymers have the ability to repair mainly material damage caused by the environment due to the exposure time. This mechanism can be activated automatically or driven by an external agent. Nanoparticles have been added to these materials mainly, forming nanocomposites that can be used for long periods of time without significant deterioration or loss of function (Thakur & Kessler, 2015; Wang et al., 2020; Wu et al., 2019; Yue et al., 2018).

Another category of smart materials is the memory-shape polymers. These are polymers capable of changing from a temporary to a permanent form through a thermal, luminous, magnetic, or electrical stimulus, with the induction being dependent on the application. These nanocomposites can improve the mechanical properties and cellular compatibility (Mozumder et al., 2017). One of the most significant advances achieved by these materials is their use in minimally invasive surgeries, as they can be designed to fit inside small incisions. After being implanted in the body, they assume their functional forms (Cheng et al., 2019). Hydroxyapatite nanoparticles associated with poly(ε-caprolactone) (c-PCL) were used to form scaffolds, and their volume was decreased by thermal compression. This scaffold is able to return to its original shape when subjected to body temperature. These nanocomposites have not only excellent mechanical properties but also the ability to induce bone regeneration (Cheng et al., 2019; Liu et al., 2014; Mozumder et al., 2017).

In general, the applications of nanobiocomposites in the biomedical area are mainly related to tissue bone regeneration, cancer therapy, and biosensors for diagnosis. Regardless of the type of application, a determining factor for the use of these nanobiocomposites is related to the cytotoxicity and biocompatibility of the material developed.

The scientific literature presents several studies concerning biomedical applications of nanobiocomposites, and some of them are presented as follows. In order to verify the effectiveness and ensure the delivery of two anticancer drugs (camptothecin and 3,3′- diindolylmethane), a nanobiocomposite was produced by functionalizing chitosan with graphene oxide nanoparticles and decorated with folic acid. In vitro analyses showed that the association resulted in delayed and controlled drug release, and in vivo trials showed the antitumor action of drugs (Deb et al., 2018). Chitosan was also used as a polymeric matrix to produce nanocomposites containing iron oxide and selenium in order to inhibit the growth of breast cancer cells. Nanoparticles with 5−9 nm diameter were obtained and induced cell death in tumor cells after 24 h of incubation, showing the effectiveness of the drug-delivery system (Hauksdóttir & Webster, 2018). In addition, selenium nanoparticles were incorporated in poly L-lactic acid, and the developed nanocomposite was able to inhibit the growth of recurrent bone cancer cells in addition to promoting osteoblast functions (Stolzoff & Webster, 2016). A biomimetic nanocarrier was developed to enhance the chemotherapeutic efficacy of doxorubicin and icotinib in chemoresistant lung cancer cells. The nanoparticles of the membrane biomimetic were prepared with poly(amidoamine) dendrimers (PAMAN-PC) loaded by the drugs inside. The system inhibited tumor growth in vivo without any apparent side effects (Wu et al., 2019).

To repair a mandibular bone defect, cross-linked poly(ε-caprolactone) (c-PCL) and hydroxyapatite nanoparticles were used to synthesize a scaffold. In vivo, analyses showed good cytocompatibility, and it was possible to prove bone regeneration (Liu et al., 2014). Nanobiocomposite scaffolds of chitosan, carboxymethyl cellulose, and silver nanoparticles were developed to ensure strong mechanical and antimicrobial activity in bone tissue engineering applications. The addition of the nanocomposite allowed sufficient porosity to the material, and the mechanical resistance was significantly improved, in addition to exhibiting excellent antimicrobial activity (Hasan et al., 2018). For the regeneration of bone tissues, the nanocomposites must present mechanical strength and porosity and an environment capable of inducing cell proliferation to promote osteoinduction. A nanobiocomposite scaffold was developed, associating gelatin, alginate, and graphene oxide for bone tissue engineering. The incorporation of graphene oxide in the polymeric matrix resulted in chemical and physical changes in the scaffold, resulting in good fixation and proliferation of mesenchymal cells. In addition, the composition of the nanobiocomposite expresses osteoinduction factors, showing the potential use of this material for bone regeneration (Dutt et al., 2019).

For the purpose of diagnosing diseases, biosensors have been developed from nanocomposites with multiple functions of magnetism, fluorescence, and bioaffinity. For this, Fe_3O_4 nanoparticles, paramagnetic ions, were incorporated into nanocomposites for tumor detection. Marking agents for detection, such as fluorescein (Li et al., 2016), or substrates, such as gold or silver particles (Zhao et al., 2019), were incorporated into the polymeric matrices and directed to the tumor target due to the electromagnetic properties. These nanocomposites have been identified as promising for the detection of tumors for early diagnosis (Li et al., 2016; Yan et al., 2018; Zhao et al., 2019).

In summary, the success in the development of scaffolds similar to the extracellular matrix depends mainly on cell adhesion, proliferation, and differentiation, and the incorporation and adherence of stem cells are closely linked to the pore size and diameter of the polymeric matrix. In addition, infections that occur due to the insertion of scaffolds are a severe limitation that must be overcome. A possible solution to overcome this situation would be the insertion of antimicrobial or antifungal agents into the scaffolds or the joint use of nanoparticles during the development of these materials.

7. Trends and gaps in knowledge

An inherent aspect of nanobiocomposites that should be undergone is the overseeing of the use and manufacturing of nanomaterials, through regulation by governmental or regulatory entities, in order to guarantee minimum safety requirements and satisfactory production criteria to control the migration of components into food and the human body, and also the environment (Giaconia et al., 2020).

In summary, as briefly discussed in this section, most investigations into the application of nanobiocomposites in the food industry involve, by far, food packaging. However, it is worth highlighting that nanobiocomposites have also been indicated for other important applications, such as biosensors capable of detecting the presence of several deleterious compounds in foods. It is also essential to consider the development of novel smart food-packaging systems, such as active and intelligent packaging (e.g., time–temperature indicators, integrity indicators, freshness indicators, radio frequency identification), which contributes significantly to increased food safety and quality.

Nanotechnology has been actively applied and studied in cosmetics and cosmeceutical areas as this innovative method can improve the solubility and stability of active ingredients and promote their skin penetration efficiently while also controlling the rheological properties, such as sensory characteristics. in this way, the cosmetic industry is a promising field to be better explored for nanotechnology applications.

References

Abdelwahab, M., Codou, A., Anstey, A., Mohanty, A. K., & Misra, M. (2020). Studies on the dimensional stability and mechanical properties of nanobiocomposites from polyamide 6-filled with biocarbon and nanoclay hybrid systems. *Composites Part A: Applied Science and Manufacturing, 129*, 105695. https://doi.org/10.1016/j.compositesa.2019.105695

Alila, S., Besbes, I., Vilar, M. R., Mutjé, P., & Boufi, S. (2013). Non-woody plants as raw materials for production of microfibrillated cellulose (MFC): a comparative study. *Industrial Crops and Products, 41*(1), 250–259. https://doi.org/10.1016/j.indcrop.2012.04.028

Amaral, D. L. A. S., Zanette, R. de S. S., Souza, G. T., de Silva, S. A., da Aguiar, J. A. K., Marcomini, R. F., Carmo, A. M. R., Nogueira, B. V., Barros, R. J., da, S., Silva, F. de S., Santos, M. de O., Munk, M., Brandão, H. de M., & Maranduba, C. M. da C. (2020). Induction of osteogenic differentiation by demineralized and decellularized bovine extracellular matrix derived hydrogels associated with barium titanate. *Biologicals: Journal of the International Association of Biological Standardization, 66*(September 2019), 9–16. https://doi.org/10.1016/j.biologicals.2020.06.003

Assis, L. M. de, Zavareze, E. da R., Prentice-Hernández, C., & Souza-Soares, L. A. de (2012). Revisão: características de nanopartículas e potenciais aplicações em alimentos. *Brazilian Journal of Food Technology, 15*(2), 99–109. https://doi.org/10.1590/s1981-67232012005000004

Avadhani, K. S., Manikkath, J., Tiwari, M., Chandrasekhar, M., Godavarthi, A., Vidya, S. M., Hariharapura, R. C., Kalthur, G., Udupa, N., & Mutalik, S. (2017). Skin delivery of epigallocatechin-3-gallate (EGCG) and hyaluronic acid loaded nano-transfersomes for antioxidant and anti-aging effects in UV radiation induced skin damage. *Drug Delivery, 24*(1), 61–74. https://doi.org/10.1080/10717544.2016.1228718

Ayumi, N. S., Sahudin, S., Hussain, Z., Hussain, M., & Samah, N. H. A. (2019). Polymeric nanoparticles for topical delivery of alpha and beta arbutin: preparation and characterization. *Drug Delivery and Translational Research, 9*(2), 482–496. https://doi.org/10.1007/s13346-018-0508-6

Aziz, Z. A. A., Mohd-Nasir, H., Ahmad, A., Siti, S. H., Peng, W. L., Chuo, S. C., Khatoon, A., Umar, K., Yaqoob, A. A., & Mohamad Ibrahim, M. N. (2019). Role of nanotechnology for design and development of cosmeceutical: application in makeup and skin care. *Frontiers in Chemistry, 7*(November), 1–15. https://doi.org/10.3389/fchem.2019.00739

Baruah, J., Chaliha, C., Kalita, E., Nath, B. K., Field, R. A., & Deb, P. (2020). Modelling and optimization of factors influencing adsorptive performance of agrowaste-derived Nanocellulose Iron Oxide Nanobiocomposites during remediation of Arsenic contaminated groundwater. *International Journal of Biological Macromolecules, 164*, 53–65. https://doi.org/10.1016/j.ijbiomac.2020.07.113

Baskar, K., Anusuya, T., & Venkatasubbu, G. D. (2017). Mechanistic investigation on microbial toxicity of nano hydroxyapatite on implant associated pathogens. *Materials Science and Engineering: C, 73*, 8–14. https://doi.org/10.1016/j.msec.2016.12.060

Belaid, H., Nagarajan, S., Barou, C., Huon, V., Bares, J., Balme, S., Miele, P., Cornu, D., Cavaillès, V., Teyssier, C., & Bechelany, M. (2020). Boron nitride based nanobiocomposites: design by 3D printing for bone tissue engineering. *ACS Applied Bio Materials, 3*(4), 1865–1874. https://doi.org/10.1021/acsabm.9b00965

Belluzo, M. S., Medina, L. F., Molinuevo, M. S., Cortizo, M. S., & Cortizo, A. M. (2020). Nanobiocomposite based on natural polyelectrolytes for bone regeneration. *Journal of Biomedical Materials Research Part A, 108*(7), 1467–1478. https://doi.org/10.1002/jbm.a.36917

Beltrán, A., Valente, A. J. M., Jiménez, A., & Garrigós, M. C. (2014). Characterization of poly(ε-caprolactone)-based nanocomposites containing hydroxytyrosol for active food packaging. *Journal of Agricultural and Food Chemistry, 62*(10), 2244–2252. https://doi.org/10.1021/jf405111a

Berekaa, M. M. (2015). Nanotechnology in food Industry; advances in food processing, packaging and food safety review article. *International Journal of Current Microbiology and Applied Sciences, 4*(5), 345–357. https://www.researchgate.net/publication/306017224.

Biscarat, J., Bechelany, M., Pochat-Bohatier, C., & Miele, P. (2015). Graphene-like BN/gelatin nanobiocomposites for gas barrier applications. *Nanoscale, 7*(2), 613–618. https://doi.org/10.1039/C4NR05268D

Borase, H. P., Patil, C. D., Salunkhe, R. B., Suryawanshi, R. K., Salunke, B. K., & Patil, S.v. (2014). Phytolatex synthesized gold nanoparticles as novel agent to enhance sun protection factor of commercial sunscreens. *International Journal of Cosmetic Science, 36*(6), 571–578. https://doi.org/10.1111/ics.12158

Bordes, P., E, P., & Averous, L. (2009). Nano-biocomposites: biodegradable polyester/nanoclay systems. *Progress in Polymer Science, 34*(2), 125–155. https://doi.org/10.1016/j.progpolymsci.2008.10.002

Braga, A. R. C., Lemes, A. C., & de Rosso, V. V. (2021). Polymer nanocomposite's applications in food and bioprocessing industry. In *Handbook of polymer nanocomposites for industrial applications* (pp. 237–250). Elsevier. https://doi.org/10.1016/B978-0-12-821497-8.00007-1

Celia, C., Cilurzo, F., Trapasso, E., Cosco, D., Fresta, M., & Paolino, D. (2012). Ethosomes® and transfersomes® containing linoleic acid: physicochemical and technological features of topical drug delivery carriers for the potential treatment of melasma disorders. *Biomedical Microdevices, 14*(1), 119–130. https://doi.org/10.1007/s10544-011-9590-y

Chen, J., Wei, N., Lopez-Garcia, M., Ambrose, D., Lee, J., Annelin, C., & Peterson, T. (2017). Development and evaluation of resveratrol, Vitamin E, and epigallocatechin gallate loaded lipid nanoparticles for skin care applications. *European Journal of Pharmaceutics and Biopharmaceutics, 117*, 286–291. https://doi.org/10.1016/j.ejpb.2017.04.008

Cheng, X., Fei, J., Kondyurin, A., Fu, K., Ye, L., Bilek, M. M. M., & Bao, S. (2019). Enhanced biocompatibility of polyurethane-type shape memory polymers modified by plasma immersion ion implantation treatment and collagen coating: an in vivo study. *Materials Science and Engineering: C, 99*, 863–874. https://doi.org/10.1016/j.msec.2019.02.032

Choo, D. C., Bae, S. K., & Kim, T. W. (2019). Flexible, transparent patterned electrodes based on graphene oxide/silver nanowire nanocomposites fabricated utilizing an accelerated ultraviolet/ozone process to control silver nanowire degradation. *Scientific Reports, 9*(1), 5527. https://doi.org/10.1038/s41598-019-41909-4

de Oliveira-Filho, J. G., Lemes, A. C., Braga, A. R. C., & Egea, M. B. (2020). Biodegradable eco-friendly packaging and coatings incorporated of natural active compounds. In *Food packaging* (pp. 171–206). CRC Press.

Deb, A., Andrews, N. G., & Raghavan, V. (2018). Natural polymer functionalized graphene oxide for co-delivery of anticancer drugs: in-vitro and in-vivo. *International Journal of Biological Macromolecules, 113*, 515–525. https://doi.org/10.1016/j.ijbiomac.2018.02.153

Deng, Y., Ediriwickrema, A., Yang, F., Lewis, J., Girardi, M., & Saltzman, W. M. (2015). A sunblock based on bioadhesive nanoparticles. *Nature Materials, 14*(12), 1278–1285. https://doi.org/10.1038/nmat4422

Dutt, S., Bhaskar, R., Singh, H., & Yadav, I. (2019). Development of a nanocomposite scaffold of gelatin – alginate – graphene oxide for bone tissue engineering. *International Journal of Biological Macromolecules, 133*, 592–602. https://doi.org/10.1016/j.ijbiomac.2019.04.113

El-Leithy, E. S., Makky, A. M., Khattab, A. M., & Hussein, D. G. (2018). Optimization of nutraceutical coenzyme Q10 nanoemulsion with improved skin permeability and anti-wrinkle efficiency. *Drug Development and Industrial Pharmacy, 44*(2), 316–328. https://doi.org/10.1080/03639045.2017.1391836

Elshaarawy, R. F. M., Seif, G. A., El-Naggar, M. E., Mostafa, T. B., & El-Sawi, E. A. (2019). In-situ and ex-situ synthesis of poly-(imidazolium vanillyl)-grafted chitosan/silver nanobiocomposites for safe antibacterial finishing of cotton fabrics. *European Polymer Journal, 116*(January), 210–221. https://doi.org/10.1016/j.eurpolymj.2019.04.013

Faruk, O., Bledzki, A. K., Fink, H. P., & Sain, M. (2012). Biocomposites reinforced with natural fibers: 2000–2010. *Progress in Polymer Science, 37*(11), 1552–1596. https://doi.org/10.1016/j.progpolymsci.2012.04.003

Gao, Z.-D., Qu, Y., Li, T., Shrestha, N. K., & Song, Y.-Y. (2015). Development of amperometric glucose biosensor based on Prussian Blue functionlized TiO_2 nanotube arrays. *Scientific Reports, 4*(1), 6891. https://doi.org/10.1038/srep06891

Gasilova, E. R., Aleksandrova, G. P., Volchek, B. Z., Vlasova, E. N., & Baigildin, V. A. (2019). Smart colloids containing ensembles of gold nanoparticles conjugated with κ-carrageenan. *International Journal of Biological Macromolecules, 137*, 358–365. https://doi.org/10.1016/j.ijbiomac.2019.06.215

George, A., Shah, P. A., & Shrivastav, P. S. (2019). Natural biodegradable polymers based nano-formulations for drug delivery: a review. *International Journal of Pharmaceutics, 561*(December 2018), 244–264. https://doi.org/10.1016/j.ijpharm.2019.03.011

Gholampour, A., & Ozbakkaloglu, T. (2020). A review of natural fiber composites: properties, modification and processing techniques, characterization, applications. *Journal of Materials Science, 55*(3), 829–892. https://doi.org/10.1007/s10853-019-03990-y

Giaconia, M. A., Ramos, S., dos, P., Pereira, C. F., Lemes, A. C., de Rosso, V. V., & Braga, A. R. C. (2020). Overcoming restrictions of bioactive compounds biological effects in food using nanometer-sized structures. *Food Hydrocolloids, 107*(February), 105939. https://doi.org/10.1016/j.foodhyd.2020.105939

Główka, E., Wosicka-Frąckowiak, H., Hyla, K., Stefanowska, J., Jastrzebska, K., Klapiszewski, Ł., Jesionowski, T., & Cal, K. (2014). Polymeric nanoparticles-embedded organogel for roxithromycin delivery to hair follicles. *European Journal of Pharmaceutics and Biopharmaceutics, 88*(1), 75–84. https://doi.org/10.1016/j.ejpb.2014.06.019

Granito, R. N., Renno, A. C. M., Yamamura, H., Almeida, M. C., Ruiz, P. L. M., & Ribeiro, D. A. (2020). Hydroxyapatite from fish for bone tissue engineering: a promising approach. *International Journal of Molecular and Cellular Medicine Spring, 7*(2), 80–90. https://doi.org/10.22088/IJMCM.BUMS.7.2.80

Hasan, A., Waibhaw, G., Saxena, V., & Pandey, L. M. (2018). Nano-biocomposite scaffolds of chitosan, carboxymethyl cellulose and silver nanoparticle modified cellulose nanowhiskers for bone tissue engineering applications. *International Journal of Biological Macromolecules, 111*, 923–934. https://doi.org/10.1016/j.ijbiomac.2018.01.089

Hasan, K. M. F., Horváth, P. G., & Alpár, T. (2020). Potential natural fiber polymeric nanobiocomposites: a review. *Polymers, 12*(5), 1072. https://doi.org/10.3390/polym12051072

Hauksdóttir, H. L., & Webster, T. J. (2018). Selenium and iron oxide nanocomposites for magnetically-targeted anti-cancer applications. *Journal of Biomedical Nanotechnology, 14*(3), 510–525. https://doi.org/10.1166/jbn.2018.2521

Hosseinkhani, B., Callewaert, C., Vanbeveren, N., & Boon, N. (2015). Novel biocompatible nanocapsules for slow release of fragrances on the human skin. *New Biotechnology, 32*(1), 40–46. https://doi.org/10.1016/j.nbt.2014.09.001

Hu, Z., Liao, M., Chen, Y., Cai, Y., Meng, L., Liu, Y., Lv, N., Liu, Z., & Yuan, W. (2012). A novel preparation method for silicone oil nanoemulsions and its application for coating hair with silicone. *International Journal of Nanomedicine, 7*, 5719–5724. https://doi.org/10.2147/IJN.S37277

Idumah, C. I., & Hassan, A. (2017). *Hibiscus Cannabinus* fiber/PP based nano-biocomposites reinforced with graphene nanoplatelets. *Journal of Natural Fibers, 14*(5), 691–706. https://doi.org/10.1080/15440478.2016.1277817

Jamróz, E., Kulawik, P., & Kopel, P. (2019). The effect of nanofillers on the functional properties of biopolymer-based films: a review. *Polymers, 11*(4), 675. https://doi.org/10.3390/polym11040675

Jiménez, Z., Kim, Y. J., Mathiyalagan, R., Seo, K. H., Mohanan, P., Ahn, J. C., Kim, Y. J., & Yang, D. C. (2018). Assessment of radical scavenging, whitening and moisture retention activities of Panax ginseng berry mediated gold nanoparticles as safe and efficient novel cosmetic material. *Artificial Cells, Nanomedicine and Biotechnology, 46*(2), 333–340. https://doi.org/10.1080/21691401.2017.1307216

Kalouta, K., Eleni, P., Boukouvalas, C., Vassilatou, K., & Krokida, M. (2020). Dynamic mechanical analysis of novel cosmeceutical facial creams containing nano-encapsulated natural plant and fruit extracts. *Journal of Cosmetic Dermatology, 19*(5), 1146–1154. https://doi.org/10.1111/jocd.13133

Kausar, A. (2021). Self-healing polymer/carbon nanotube nanocomposite: a review. *Journal of Plastic Film and Sheeting, 37*(2), 160–181. https://doi.org/10.1177/8756087920960195

Lau, M., Waag, F., & Barcikowski, S. (2017). Direct integration of laser-generated nanoparticles into transparent nail polish: the plasmonic "goldfinger". *Industrial and Engineering Chemistry Research, 56*(12), 3291–3296. https://doi.org/10.1021/acs.iecr.7b00039

Li, Z., Li, S., Zhou, X., Sun, L., Zhang, Q., Pan, Y., & Zhao, Q. (2016). Synthesis of multifunctional nanocomposites and their application in imaging and targeting tumor cells in vitro. *Artificial Cells, Nanomedicine and Biotechnology, 44*(5), 1236–1246. https://doi.org/10.3109/21691401.2015.1019667

Liu, R., Zhu, Y., Chen, J., Wu, H., Shi, L., Wang, X., & Wang, L. (2014). Characterization of ACE inhibitory peptides from mactra veneriformis hydrolysate by nano-liquid chromatography electrospray ionization mass spectrometry (Nano-LC-ESI-MS) and molecular docking. *Marine Drugs, 12*(7), 3917–3928. https://doi.org/10.3390/md12073917

Luna, J., & Vílchez, A. (2017). Polymer nanocomposites for food packaging. In *Emerging nanotechnologies in food science* (pp. 119–147). Elsevier. https://doi.org/10.1016/B978-0-323-42980-1.00007-8

Ma, L., Zhou, M., He, C., Li, S., Fan, X., Nie, C., Luo, H., Qiu, L., & Cheng, C. (2019). Graphene-based advanced nanoplatforms and biocomposites from environmentally friendly and biomimetic approaches. *Green Chemistry, 21*(18), 4887–4918. https://doi.org/10.1039/C9GC02266J

Mandal, A., Katheem Farhan, M., & Sastry, T. P. (2016). Effect of reinforced Al_2O_3 nanoparticles on collagen nanobiocomposite from chrome-containing leather waste for biomedical applications. *Clean Technologies and Environmental Policy, 18*(3), 765–773. https://doi.org/10.1007/s10098-015-1045-3

Marangoni, V. S., Paino, I. M., & Zucolotto, V. (2013). Synthesis and characterization of jacalin-gold nanoparticles conjugates as specific markers for cancer cells. *Colloids and Surfaces B: Biointerfaces, 112*, 380–386. https://doi.org/10.1016/j.colsurfb.2013.07.070

Meiyazhagan, A., Thangavel, S., Daniel, P.,H., Pulickel, M.,A., & Palanisamy, T. (2015). Electrically conducting nanobiocomposites using carbon nanotubes and collagen waste fibers. *Materials Chemistry and Physics, 157*, 8–15. https://doi.org/10.1016/j.matchemphys.2015.03.005

Mellinas, C., Ramos, M., Jiménez, A., & Garrigós, M. C. (2020). Recent trends in the use of pectin from agro-waste residues as a natural-based biopolymer for food packaging applications. *Materials, 13*(3), 673. https://doi.org/10.3390/ma13030673

Mishra, P. K. (2019). Nano-engineered flavonoids for cancer protection. *Frontiers in Bioscience, 24*(6), 4771. https://doi.org/10.2741/4771

Miteluţ, A. C., Tănase, E. E., Popa, V. I., & Popa, M. E. (2015). Sustainable alternative for food packaging: chitosan biopolymer – a review. *AgroLife Scientific Journal, 4*(December), 52–61.

Mohanty, F., & Swain, S. K. (2018). Effect of graphene platelets on the thermal and conducting properties of poly(ethyl methacrylate). *Advances in Polymer Technology, 37*(5), 1316–1322. https://doi.org/10.1002/adv.21790

Moonesi Rad, R., Atila, D., Akgün, E. E., Evis, Z., Keskin, D., & Tezcaner, A. (2019). Evaluation of human dental pulp stem cells behavior on a novel nanobiocomposite scaffold prepared for regenerative endodontics. *Materials Science and Engineering: C, 100*(February), 928–948. https://doi.org/10.1016/j.msec.2019.03.022

Moura, M. R. de, Aouada, F. A., Souza, J. R., & Mattoso, L. H. C. (2014). Preparation of new active edible nanobiocomposite containing cinnamon nanoemulsion and pectin. *Polímeros, 24*(4), 486–490. https://doi.org/10.1590/0104-1428.1508

Mozumder, M. S., Mairpady, A., & Mourad, A. H. I. (2017). Polymeric nanobiocomposites for biomedical applications. *Journal of Biomedical Materials Research Part B: Applied Biomaterials, 105*(5), 1241–1259. https://doi.org/10.1002/jbm.b.33633

Mutalik, S., Shetty, P. K., Venuvanka, V., Jagani, H. V., Gejjalagere, C. H., Nayak, U. Y., Musmade, P. B., Reddy, M. S., Kalthur, G., Udupa, N., Ligade, V. S., & Rao, C. M. (2015). Development and evaluation of sunscreen creams containing morin-encapsulated nanoparticles for enhanced UV radiation protection and antioxidant activity. *International Journal of Nanomedicine, 10*, 6477. https://doi.org/10.2147/IJN.S90964

Nouri, A. (2019). Chitosan nano-encapsulation improves the effects of mint, thyme, and cinnamon essential oils in broiler chickens. *British Poultry Science*. https://doi.org/10.1080/00071668.2019.1622078

Novak, U., Bajić, M., Kõrge, K., Oberlintner, A., Murn, J., Lokar, K., Triler, K. V., & Likozar, B. (2019). From waste/residual marine biomass to active biopolymer-based packaging film materials for food industry applications – a review. *Physical Sciences Reviews*, 1–24. https://doi.org/10.1515/psr-2019-0099

Ramos, Ó. L., Pereira, R. N., Cerqueira, M. A., Martins, J. R., Teixeira, J. A., Malcata, F. X., & Vicente, A. A. (2018). Packaging and their effect in food quality and safety. In *Food packaging and preservation*. Elsevier Inc. https://doi.org/10.1016/B978-0-12-811516-9/00008-7

Sahithya, K., Das, D., & Das, N. (2016). Adsorptive removal of monocrotophos from aqueous solution using biopolymer modified montmorillonite–CuO composites: equilibrium, kinetic and thermodynamic studies. *Process Safety and Environmental Protection, 99*, 43–54. https://doi.org/10.1016/j.psep.2015.10.009

Sánchez-García, M. D., Hilliou, L., & Lagarón, J. M. (2010). Morphology and water barrier properties of nanobiocomposites of κ/ι-Hybrid carrageenan and cellulose nanowhiskers. *Journal of Agricultural and Food Chemistry, 58*(24), 12847–12857. https://doi.org/10.1021/jf102764e

Saneinezhad, S., Mohammadi, L., Zadsirjan, V., Bamoharram, F. F., & Heravi, M. M. (2020). Silver nanoparticles-decorated Preyssler functionalized cellulose biocomposite as a novel and efficient catalyst for the synthesis of 2-amino-4H-pyrans and spirochromenes. *Scientific Reports, 10*(1), 14540. https://doi.org/10.1038/s41598-020-70738-z

Santos, A. C., Morais, F., Simões, A., Pereira, I., Sequeira, J. A. D., Pereira-Silva, M., Veiga, F., & Ribeiro, A. (2019). Nanotechnology for the development of new cosmetic formulations. *Expert Opinion on Drug Delivery, 16*(4), 313–330. https://doi.org/10.1080/17425247.2019.1585426

Shishir, M. R. I., Xie, L., Sun, C., Zheng, X., & Chen, W. (2018). Advances in micro and nano-encapsulation of bioactive compounds using biopolymer and lipid-based transporters. *Trends in Food Science and Technology, 78*(May), 34–60. https://doi.org/10.1016/j.tifs.2018.05.018

Silva, R. V., Camilli, J. A., Bertran, C. A., & Moreira, N. H. (2005). The use of hydroxyapatite and autogenous cancellous bone grafts to repair bone defects in rats. *International Journal of Oral and Maxillofacial Surgery, 34*(2), 178–184. https://doi.org/10.1016/j.ijom.2004.06.005

Singla, R., Soni, S., Patial, V., Kulurkar, P. M., Kumari, A., Magesh, S., Padwad, Y. S., & Yadav, S. K. (2017). In vivo diabetic wound healing potential of nanobiocomposites containing bamboo cellulose nanocrystals impregnated with silver nanoparticles. *International Journal of Biological Macromolecules, 105*, 45–55. https://doi.org/10.1016/j.ijbiomac.2017.06.109

So, J.-W., Kim, S., Park, J.-S., Kim, B.-H., Jung, S.-H., Shin, S.-C., & Cho, C.-W. (2010). Preparation and evaluation of solid lipid nanoparticles with JSH18 for skin-whitening efficacy. *Pharmaceutical Development and Technology, 15*(4), 415–420. https://doi.org/10.3109/10837450903262066

Spoiala, A., Voicu, G., Ficai, D., Ungureanu, C., Albu, M. G., Vasile, B. S., Ficai, A., & Andronescu, E. (2015). Collagen/TiO_2-ag composite nanomaterials for antimicrobial applications. *Scientific Bulletin-University Politehnica of Bucharest, 77*(4), 275.

Stolzoff, M., & Webster, T. J. (2016). Reducing bone cancer cell functions using selenium nanocomposites. *Journal of Biomedical Materials Research Part A, 104*(2), 476–482. https://doi.org/10.1002/jbm.a.35583

Sundramoorthy, A. K., Kumar, T. H. V., & Gunasekaran, S. (2018). Graphene-based nanosensors and smart food packaging systems for food safety and quality monitoring. In *Graphene bioelectronics* (pp. 267–306). Elsevier.

Taha, M. A., Youness, R. A., & Ibrahim, M. (2020). Biocompatibility, physico-chemical and mechanical properties of hydroxyapatite-based silicon dioxide nanocomposites for biomedical applications. *Ceramics International, 46*(15), 23599–23610. https://doi.org/10.1016/j.ceramint.2020.06.132

Tashakkorian, H., Hasantabar, V., Mostafazadeh, A., & Golpour, M. (2020). Transparent chitosan based nanobiocomposite hydrogel: synthesis, thermophysical characterization, cell adhesion and viability assay. *International Journal of Biological Macromolecules, 144*, 715–724. https://doi.org/10.1016/j.ijbiomac.2019.10.157

Thakur, V. K., & Kessler, M. R. (2015). Self-healing polymer nanocomposite materials: a review. *Polymer, 69*, 369–383. https://doi.org/10.1016/j.polymer.2015.04.086

Valdés, A., Mellinas, A. C., Ramos, M., Garrigós, M. C., & Jiménez, A. (2014). Natural additives and agricultural wastes in biopolymer formulations for food packaging. *Frontiers in Chemistry, 2*(FEB), 1–10. https://doi.org/10.3389/fchem.2014.00006

Valdés, A., Ramos, M., Beltrán, A., Jiménez, A., & Garrigós, M. C. (2017). State of the art of antimicrobial edible coatings for food packaging applications. *Coatings, 7*(4), 56. https://doi.org/10.3390/coatings7040056

Valodkar, M., & Thakore, S. (2011). Organically modified nanosized starch derivatives as excellent reinforcing agents for bionanocomposites. *Carbohydrate Polymers, 86*(3), 1244–1251. https://doi.org/10.1016/j.carbpol.2011.06.020

Vasile, C. (2018). Polymeric nanocomposites and nanocoatings for food packaging: a review. *Materials, 11*(10), 1834. https://doi.org/10.3390/ma11101834

Venerando, R., Miotto, G., Magro, M., Dallan, M., Baratella, D., Bonaiuto, E., Zboril, R., & Vianello, F. (2013). Magnetic nanoparticles with covalently bound self-assembled protein corona for advanced biomedical applications. *Journal of Physical Chemistry C, 117*(39), 20320–20331. https://doi.org/10.1021/jp4068137

Wang, C., Liang, C., Wang, R., Yao, X., Guo, P., Yuan, W., Liu, Y., Song, Y., Li, Z., & Xie, X. (2020). The fabrication of a highly efficient self-healing hydrogel from natural biopolymers loaded with exosomes for the synergistic promotion of severe wound healing. *Biomaterials Science, 8*(1), 313–324. https://doi.org/10.1039/c9bm01207a

Wu, P., Yin, D., Liu, J., Zhou, H., Guo, M., Liu, J., Liu, Y., Wang, X., Liu, Y., & Chen, C. (2019). Cell membrane based biomimetic nanocomposites for targeted therapy of drug resistant EGFR-mutated lung cancer. *Nanoscale, 11*(41), 19520–19528. https://doi.org/10.1039/c9nr05791a

Yan, D., Liu, X., Deng, G., Yuan, H., Wang, Q., Zhang, L., & Lu, J. (2018). Facile assembling of novel polypyrrole nanocomposites theranostic agent for magnetic resonance and computed tomography imaging guided efficient photothermal ablation of tumors. *Journal of Colloid and Interface Science, 530*, 547–555. https://doi.org/10.1016/j.jcis.2018.07.001

Yan, H., Chen, X., Song, H., Li, J., Feng, Y., Shi, Z., Wang, X., & Lin, Q. (2017). Synthesis of bacterial cellulose and bacterial cellulose nanocrystals for their applications in the stabilization of olive oil pickering emulsion. *Food Hydrocolloids, 72*, 127–135. https://doi.org/10.1016/j.foodhyd.2017.05.044

Yang, W., Fortunati, E., Dominici, F., Giovanale, G., Mazzaglia, A., Balestra, G. M., Kenny, J. M., & Puglia, D. (2016). Synergic effect of cellulose and lignin nanostructures in PLA based systems for food antibacterial packaging. *European Polymer Journal, 79*, 1–12. https://doi.org/10.1016/j.eurpolymj.2016.04.003

Yu, Z., Alsammarraie, F. K., Nayigiziki, F. X., Wang, W., Vardhanabhuti, B., Mustapha, A., & Lin, M. (2017). Effect and mechanism of cellulose nanofibrils on the active functions of biopolymer-based nanocomposite films. *Food Research International, 99*, 166–172. https://doi.org/10.1016/j.foodres.2017.05.009

Yue, S., Wu, J., Zhang, Q., Zhang, K., Weir, M. D., Imazato, S., Bai, Y., & Xu, H. H. K. (2018). Novel dental adhesive resin with crack self-healing, antimicrobial and remineralization properties. *Journal of Dentistry, 75*(May), 48–57. https://doi.org/10.1016/j.jdent.2018.05.009

Zhao, X., Zeng, L., Hosmane, N., Gong, Y., & Wu, A. (2019). Cancer cell detection and imaging: MRI-SERS bimodal splat-shaped Fe_3O_4/Au nanocomposites. *Chinese Chemical Letters, 30*(1), 87–89. https://doi.org/10.1016/j.cclet.2018.01.028

Zein-based blends and composites

18

Mariana Buranelo Egea [1], *Josemar Gonçalves de Oliveira Filho* [2], *Anna Rafaela Cavalcante Braga* [3], *Maria Carolina Bezerra Di-Medeiros Leal* [4], *Jesús María Frías Celayeta* [5] *and Ailton Cesar Lemes* [6]
[1]Goiano Federal Institute of Education, Science and Technology, Rio Verde, Goiás, Brazil; [2]São Paulo State University (UNESP), School of Pharmaceutical Sciences, Araraquara, Brazil; [3]Department of Chemical Engineering, Universidade Federal de São Paulo (UNIFESP), Diadema, Brazil; [4]Federal University of São Carlos (UFSCar), Nuclear Magnetic Resonance Laboratory, São Carlos, Brazil; [5]Dublin Institute of Technology, Environmental Sustainability and Health Institute (EHSI), College of Sciences and Health, Dublin, Ireland; [6]Federal University of Rio de Janeiro (UFRJ), School of Chemistry, Department of Biochemical Engineering, Rio de Janeiro, Brazil

1. Introduction: chemistry and properties of zein molecules

Maize (*Zea mays*), a grass of the Poaceae family, has been widely used as a food staple because it contains more than 50% proteins (mainly prolamins, also known as zeins) in its kernel. However, zein, which is rich in glutamine and proline, does not contain lysine and tryptophan, and is therefore considered of low protein quality for human and animal food (Li & Song, 2020). Hybrid varieties have been developed in an attempt to increase the content of lysine and tryptophan (such as opaque2, o2) with the advantage of not altering the opacity of the kernel (opaco16, o16) (Hossain et al., 2019).

The large family of zeins is classified as α-type (most abundant and divided into subclasses of 19 and 22 kDa based on their apparent molecular weights) (Li & Song, 2020), β-type (one of 15 kDa) (Pedersen et al., 1986), γ-type (three of 16, 27, and 50 kDa) (Prat et al., 1987; Woo et al., 2001), and δ-type (two of 10 and 18 kDa) (Chui & Falco, 1995; Kirihara et al., 1988). The formation of zein in the rough endoplasmic reticulum begins with the deposition of γ- and β-zeins forming the protein bodies and then the migration of α- and δ-zeins to the center of the molecule which at the end is protected for a layer of γ-zein. In α-zein, the 19 kDa molecules are concentrated in the center, while the 22 kDa molecules are located in the outermost parts (Lending & Larkins, 1989; Li & Song, 2020),

Corn proteins are important for the food industry as an ingredient in the development of new products. Zein—a relatively hydrophobic molecule—has demonstrated excellent solubility and therefore has excellent foaming and emulsion properties (Teklehaimanot & Emmambux, 2019). The amino acid composition in the zein

Biodegradable Polymers, Blends and Composites. https://doi.org/10.1016/B978-0-12-823791-5.00009-0

molecule is responsible for its properties when this protein is applied in the food industry. For example, γ- and β-zein subclasses are rich in cysteine that can form disulfide bonds and stabilize forms (King et al., 2016).

The quality of zein-like proteins in corn is an important factor because, in addition to impacting the nutritional quality, it also impacts the quality of starch extracted from the endosperm, which is the main product of the corn industry. Changes in the zein profile, either for a selection of genotypes and fertilizing the soil with nitrogen or for environmental reasons, affect the hardness of the endosperm, which in turn decreases enzyme accessibility to starch granules, which can decrease the yield in fermentation processes or change properties such as the level of gelatinization or vitreousness of starch that interferes with its use as an ingredient (Gerde et al., 2017; Kljak et al., 2018; Tamagno et al., 2016).

These chemical characteristics of zein determine its properties and impact the application of this molecule as an important ingredient for films, coatings, and nanotechnology applications in the food industry. In this chapter, the properties of zein as a molecule are reviewed, in addition to discussing the applications of this compound in edible films and coatings for food as well as its application in nanotechnology as an encapsulating material.

2. Zein as an ingredient for ecofriendly films

The food industry is increasingly concerned with offering products that fully meet the needs and desires of consumers. In this sense, it pursues the elaboration of diversified, nutritious and, above all, safer products with a long useful life. In the food packaging sector, the search is for more attractive packaging, protection against external factors, and less environmental impact, using natural and renewable components (de et al., 2020; Lemes et al., 2016; Santeramo et al., 2018).

To meet this market requirement, biopolymer films, formed separately from food and applied later, are applied for protection and containment or act as a vehicle for active compounds. Film production consists of the formation of a dispersion (solid in water) using some polymeric solution (carbohydrates, proteins, lipids, or their combination) that varies according to the purpose of application of the film. Then the homogenized solution was applied in molds that give a specific shape (casting) before submitted to drying. Subsequently, the dry film is removed from the mold (detaching) and applied to food (Fig. 18.1) (da Costa et al., 2019; Dhumal & Sarkar, 2018; Pinheiro et al., 2010; Spasojević et al., 2019).

Films can be produced with different functionalities and characteristics (color, homogeneity, thickness, and mechanical properties, among others), which are dependent on their composition, the form of production, and function (Drobny, 2020, 2020; Feldman, 2001; Pinheiro et al., 2010, 2010; Zhang et al., 2019, 2020). Among their functions, they can act as protective barriers against the transport of gases, vapors, liquids, and light (Drobny, 2020; Feldman, 2001; Kashiri et al., 2017; Pinheiro et al., 2010; Zhang et al., 2019), as well being applied as containers

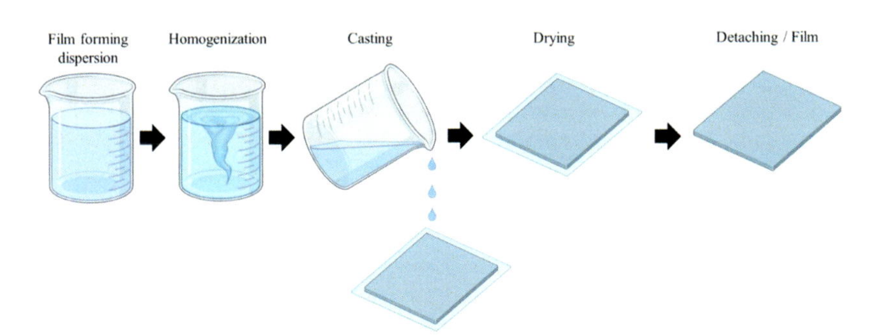

Figure 18.1 Generic film production scheme (casting) for food applications.

(Pascall & Lin, 2013), which in turn helps controlling microbial growth (G. et al., 2011) and decreasing oxidation (Pascall & Lin, 2013) and respiration rates (Dhumal & Sarkar, 2018), among others.

Zein, a prolamine found in corn (Shukla & Cheryan, 2001), has been widely used in film production due to its thermoplastic properties (Ghanbarzadeh et al., 2007), absence of toxicity (Zhou & Wang, 2021), high degree of polymerization (De Almeida et al., 2018), and high proportion of nonpolar amino acid residues that result in high hydrophobicity (Zhang et al., 2020). The occurrence of a hydrophobic interaction between zein and fatty acids promotes an increase in water resistance in films (Ghanbarzadeh & Oromiehi, 2008) and improves the mechanical properties, including better elasticity and greater tensile strength, as well as lower vapor permeability of water, oxygen, and carbon dioxide (Zhang et al., 2019, 2020). These characteristics may vary according to the purpose of the application of the film, giving different properties of barriers, thicknesses, mechanical properties, bioactivities and, also, characteristics of attractiveness by the consumer, such as transparency or modification of its color. Several studies have reported the characteristics related to the application of zein in the preparation of polymeric films for the most diverse applications, either as a single component or associated with other components forming composite films (Table 18.1).

Zein film has nonhomogeneous, discontinuous, and brittle characteristics and therefore, Scramin et al. (2011) evaluated the addition of oleic acid as a plasticizer. These authors demonstrated that the zein-based film with oleic acid promoted an improvement in elasticity with an increase in the deformation at the break and greater adhesion strength of the film. This behavior occurs because oleic acid is efficient in reducing the hydrophobic characteristics of the film surface and the roughness. Meanwhile, De Almeida et al. (2018) compared the addition of pure oleic acid and edible vegetable oils (macadamia, buriti, and olive oil) in zein-based film. The addition of oils, especially macadamia and buriti, provided similar structures with the homogeneous presence of rounded components that resemble pores on the surface of zein-based films, which was greater with the addition of oleic acid and olive oil. The film showed thermal degradation in the range of 270–415°C and higher percentages of elongation at break with the addition of oleic acid in the matrix.

Table 18.1 Film characteristics in relation to the addition of zein and the combination with plasticizers and other components.

Compounds	Application	Characteristics conferred by the addition of zein	References
Zein + oleic acid	Film production	↑ elasticity, deformation at break, and adhesion strength; and ↓ hydrophobicity and roughness	Scramin et al. (2011)
Zeína + pure oleic acid and edible vegetable oils (macadamia, buriti, and olive oil)	Film	↑ homogeneity (oleic acid and olive oil) and stretching (oleic acid) = thermal degradation; and ↓ elongation (absence of fatty acids in the matrix)	De Almeida et al. (2018)
Zein + essential oil of *Zataria multiflora* Boiss	Film and packaging for pasteurized milk	↑ yellowish color (increased [zein]), protection against UV light, and disintegration in contact with water; and ↓ transparency (increase [zein]). The inhibition against *L. monocytogenes* or *E. coli* and transparent and flexible films (low zein concentrations)	Kashiri et al. (2017)
Zeine + methylcellulose + plasticizers (oleic acid and polyethylene glycol)	Film production	↑ stretching and solubility (methylcellulose and PEG); and ↓ water vapor permeability (zein and oleic acid). Smooth, homogeneous, and flexible film, inhibition against *E. coli,* and absence of toxicity.	Zhou and Wang (2021)
Zein + gelatin	Film production	↓ moisture, solubility (increased [zein]), water vapor and oxygen barrier properties, and luminosity; and ↑ hydrophobicity and tensile strength (low concentrations of zein);	Ahammed et al. (2020)
Zein + chitosan	Mushroom packaging	↑ thermal stability, elasticity, tensile strength, and yellowish color; and ↓ oxygen permeability, permeability to carbon dioxide, weight loss, respiration rate, and transparency	Zhang et al. (2019)
Zein + chitosan + α-tocopherol	Mushroom packaging (*Agaricus bisporus*)	↑ shelf-life, antioxidant activity, and firmness of mushrooms; and ↓ weight loss, browning index, respiration rate, permeability to O_2, and CO_2 production	Zhang et al. (2020)
Zein + chitosan + plasticizers (glycerol, PEG-400 and sorbitol)	Film production	↑ permeability (PEG-400), opacity, tensile strength (addition of plasticizer), and water vapor permeability; and ↓ crystallinity	Sun et al. (2020)

Zhou and Wang (2021) also combined zein with methylcellulose and plasticizers (oleic acid and polyethylene glycol—PEG) to obtain the film. This combination resulted in smooth, homogeneous, and flexible films, as well as decreasing water vapor permeability. The low solubility of this film can be altered with the addition of methylcellulose and PEG, allowing the production of films with different properties. Regarding the antimicrobial activity of the film, it was found that the addition of thymol (0.1−0.20 g/g) resulted in protection against *E. coli*. Finally, it was verified that the films containing zein did not present toxicity to human epidermal cells, indicating no risk of toxicity.

Zein also proved to be a good enhancer of the properties of gelatin-based film for application in wet foods (Ahammed et al., 2020). The addition of zein promoted the formation of films with lower moisture values and higher hydrophobicity, resulting in a reduction in the solubility of the films in water proportional to the increase in the concentration of zein in the matrix. Regarding the tensile strength, it was found that zein in small concentrations improves the tensile strength but seems to affect the structure and make the film brittle in large concentrations. The increase in the concentration of zein in the film also affected the luminosity, and its association with the other components seems to have a better effect on light transmission, causing better protection against external light.

The zein added to the chitosan-based film was able to promote better barrier characteristics, decreasing the permeability to oxygen and carbon dioxide by the presence of polar interactions that reduces the solubility of these molecules, which in turn prevented their release. The addition of zein increased the thermal stability of the films, resulting in lower values of weight loss and lower respiration rate, and promoted better elasticity and greater tensile strength. On the other hand, the addition of zein promoted a color change from transparent to yellowish hue (b* values increased), which can compromise consumer attractiveness (Zhang et al., 2019).

Sun et al. (2020) verified the alteration of the properties of zein-based films with chitosan, from the addition of different plasticizers (glycerol, PEG-400, and sorbitol). The increase in the concentration of plasticizer, especially PEG-400, increased the permeability of the films and opacity and decreased their crystallinity and tensile strength due to reduced interaction of the biopolymer chains. The water vapor permeability intensified with the addition of plasticizers, especially with the addition of glycerol due to its lower molecular weight, and also an increase in permeation of O_2 and CO_2 was observed (from the increase in the concentration of PEG-400, due to better compatibility between these components).

Zein films also have been used in controlled-release systems for bioactive components. Kashiri et al. (2017) applied *Zataria multiflora* Boiss essential oil−rich in phenolic monoterpene carvacrol and its isomer, thymol−in order to act as an antimicrobial, especially against *Listeria monocytogenes* and *Escherichia coli*. The use of zein provided transparent and flexible films; however, a loss of transparency and an increase in yellowish tones were observed with an increase in the concentration of zein. Despite the lower transparency, the films presented an excellent barrier against UV-C light, which can be useful in preventing oxidation in foods packed in zein films, especially fat-rich foods. In addition, they presented resistance to disintegration in contact with water—important for application in foods with a high water content—and, even allowed the release of bioactive components, which was more intense at higher temperatures and in the first hour of contact with water. Regarding microbial inhibition, minimum inhibitory concentration values for *L. monocytogenes* and *E. coli* were 50 and 100 ppm, respectively, after the addition of essential oils to zein film. The zein + essential oil films were also applied in the coating of polypropylene bags for pasteurized milk inoculated with *L. monocytogenes* and *E. coli* and the films were able to reduce the microbial load over the 6 days of product storage. The data demonstrated that zein was as an excellent matrix to carry bioactive components with effects on food preservation.

In another application, zein was combined with chitosan and α-tocopherol and also worked properly in the production of films for packaging mushrooms (*Agaricus bisporus*), promoting an increase in the shelf-life. Due to its properties, the film resulted in lower weight loss values due to lower water permeability and darkening index (↓ 1.6 times after 12 days of storage) due to high antioxidant activity. Mushrooms with films resulted in a lower respiration rate—the result of lower permeability to O_2 and also lower CO_2 production—and higher firmness with less damage to the characteristics of the product (Zhang et al., 2020).

The different applications of zein-based films or its combination with other polymeric materials demonstrate the feasibility of its use in films. In addition, it is evident that its properties, including mechanical, thermal, and barrier properties, can be adjusted based on the optimization of the matrix components and, also, by the addition of plasticizing and bioactive components, in order to promote the desirable characteristics for the preservation of food or any other product coated by film.

3. Zein as an ingredient for coatings

Edible coatings have been used as a strategy to increase shelf-life and maintain food stability during storage. Edible coatings are defined as a thin film, imperceptible to the naked eye, formed on the food surface, acting as a protective layer against gas diffusion, water migration, aroma changes, and different solute changes (Fig. 18.2) (Riva et al., 2020). The application of this layer is carried out mainly by immersion or spraying on the food surface (Thakur et al., 2019; Yousuf et al., 2018).

The coating materials are formulated from various types of organic macromolecules such as proteins, polysaccharides, and lipids, as well as the mixture of more than one of these materials forming composite materials. As the covering materials become part of the food and are consumed as such, the raw materials used for their production must be of food origin and considered GRAS (generally recognized as safe), that is, they are nontoxic and safe for the environment and use in food and for human consumption (FDA, 2013).

Proteins such as collagen gelatin, gluten, mung bean protein, corn zein, soy protein, and casein have good gas barrier properties, do not maintain anaerobic conditions, and therefore have been used to develop edible coatings, increasing the shelf-life of foods and improving their quality (Chen et al., 2019). Among these proteins, zein presents

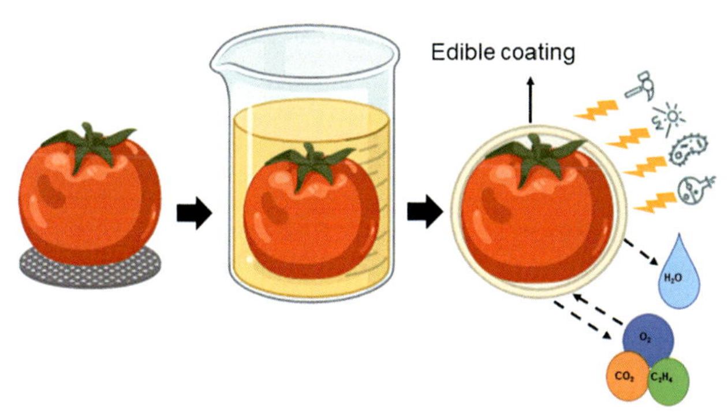

Figure 18.2 Application of edible coatings using an immersion technique and their main functions for food preservation.

itself as an essential coating material, with biodegradability, low cost, bioavailability, long history of use, hydrophobicity, adhesion, high film-forming capacity, good binding capacity with drugs, elasticity, the potential for cellular absorption, and with the characteristics of a natural and renewable source (Luo et al., 2013; Quispe-Condori et al., 2011).

The hydrophobicity of zein occurs due to its large amount of hydrophobic amino acid residues, which provides zein coatings with low water vapor permeability compared to most other proteins (Moradi et al., 2016), and also with good gas (Arcan & Yemenicioğlu, 2011), making it effective in decreasing the presence of gases between the products and their surrounding environment, and the respiration rate of the products. Table 18.2 shows edible coatings with the incorporation of zein, their applications, and the main results obtained.

Zein has been successfully applied as a coating material for fruits, vegetables, and seeds (Assis & Leoni, 2009; Bai et al., 2003; Zapata et al., 2008). Coatings of zein from the concentration of 4% decrease the respiration rate of the fruits, as well as the coatings, and were able to provide adequate brightness to "Gala" apples and to maintain the overall quality of the fruit comparable with a commercial shellac coating (Bai et al., 2003). In tomatoes, zein-based coatings were effective in maintaining the postharvest quality of tomatoes by reducing the respiration rate, ethylene production, softening, color evolution, and weight loss compared to the control group, as well as increasing the sugar and organic acid contents, and the results of sensory analysis compared to the control group (Zapata et al., 2008). In beet and broccoli seeds, the coatings were able to delay the sprouting and germination of undesirable seeds during storage by at least 11 days. This action was attributed to the physical barrier property of the film that reduces the permeability of water into the seeds (Assis & Leoni, 2009).

Despite improving the water barrier property characteristics, which in turn improves the shelf-life of food products and seeds, zein coatings have been shown to have a brittle nature. To overcome this limitation, a strategy is to develop composite coatings, that is, a combination of zein and other biomolecules, mainly lipids, carbohydrates, and plasticizing, and cross-linking compounds (Scramin et al., 2011).

Table 18.2 Edible coatings with incorporation of zein, their application, and the main results obtained.

Coating composition	Food	Results	References
Zein	Apple	↑ brightness; and ↓ respiration rate	Bai et al. (2003)
Zein	Tomato (*Solanum lycopersicon* Mill)	↑ sugars, organic acids (mainly ascorbic acid), and sensory acceptability; and ↓ respiration rate, ethylene production, softening, color evolution, and loss of mass	Zapata et al. (2008)
Zein	Beet seeds (*Beta vulgaris* L.) and broccoli (*Brassica oleraceae* var. italic L.)	↓ sprouting and germination	Assis and Leoni (2009)
Zein/oleic acid	Pears (*Pyrus communis* L.)	↓ loss of fresh mass	Scramin et al. (2011)
Zein/carnauba wax	Date palm	↑ shelf-life and sugar degradation; and ↓ fruit ripening and loss of fresh weight	Mehyar et al. (2014)
Zein/oleic acid/xanthan gum	Cheese	↓ mass loss and microbiological contamination	Pena-Serna et al. (2016)
Zein/curcumin	Apples	↑ inhibition of the growth of *Botrytis cinerea* and *Penicillium expansum*	Yilmaz et al. (2016)
Zein/tannic acid	Guava	↓ visual appearance, chlorophyll content, color, loss of mass, softening, soluble solids, peak breathing, H_2O_2, peaks of superoxide dismutase activity, and ethylene production	Santos et al. (2018)
Zein/essential and eugenol, carvacrol, and thymol	Melon	↓ loss of fresh mass	Boyacı et al. (2019)
Chitosan/zein-cinnamaldehyde nanocellulose	Mango	↓ yellowing, breathing rate, loss of mass, vitamin C content, and accumulation of malondialdehyde	Xiao et al. (2021)

Scramin et al. (2011) evaluated the effect of combining zein with oleic acid in order to overcome the characteristic fragility of zein coatings for application as a coating on pears. The addition of oleic acid improved the topography of the coating, making it smoother, which was directly related to the increased concentration of oleic acid. The addition of oleic acid resulted in changes in the mechanical properties (reduction in the elasticity module, increase in elongation at break, and less influence on tensile strength). When the coating was applied to the pear surface, the formulation with 0.25% oleic acid preserved the pear mass for 12 days at room temperature.

In another study, Pena-Serna et al. (2016) developed two zein-based coatings, one based on oleic acid and the other with oleic acid and xanthan gum, and evaluated their effects on the preservation of a "Minas Padrão" type cheese during 56 days of storage. The coatings were efficient in reducing the mass loss of the cheeses by ~30% and avoiding microbiological contamination for more than 50 days in comparison with the uncoated cheeses that showed contamination after 21 days of storage.

Yilmaz et al. (2016) developed a coating combining electrified zein and curcumin as a strategy to prevent fungal deterioration on the surface of apples inoculated with *Botrytis cinerea* and *Penicillium expansum*. The combination of zein and curcumin at concentrations of 2.5% and 5% was subjected to electrospinning to produce cylindrical and ultrafine polymeric nanofibers (<350 nm in diameter). The coatings were effective in inhibiting the in vitro growth of fungal phytopathogens, exhibiting ~40–50% mycelial growth inhibition. In vivo studies demonstrated ~50% reduction in the diameter of the lesion measured in coated apples infected with *Penicillium expansum*.

Santos et al. (2018) compared the effect of zein coating and cross-linked zein with tannic acid on maintaining the postharvest quality of guava fruits. The coating with a combination of zein and tannic acid was more effective in reducing mass loss, softening, color changes, ethylene production, and oxidative stress in guava fruits compared with a zein coating. The greater efficiency of the zein and tannic acid coating was attributed to the cross-linking of the zein, which probably resulted in decreased gas permeability, promoting lower respiration rate, and lower ROS production, delaying the ripening process, and extending the guava stability.

Zein seems to be an interesting material for the production of composite coatings. A chitosan/zein-cinnamaldehyde nanocomposite coating material that showed improved barrier properties such as crack-free surface and a more flexible and dense structure with strong cross-linking was used to preserve mangoes. The coating formulation containing chitosan (35.0 g/L), zein (3.0 g/L), nanocellulose (25.0 g/L), acetic acid (0.26 M/L), and cinnamaldehyde (3.0 g/L) was applied as a mango coating material and improved the postharvest characteristics of the fruit, reducing yellowing, respiration rate, loss of mass, and accumulation of malondialdehyde. The nanocomposite coating triggered the appropriate oxidative burst and induced the stress response in mangoes, which were characterized by the accumulation of H_2O_2, and the activation of the antioxidant system, including superoxide dismutase (SOD) and vitamin C. The chitosan/zein-cinnamaldehyde nanocomposite coating is a promising material for application as a coating to maintain the postharvest quality of mangoes and other fruits (Xiao et al., 2021).

In another study, Boyacı et al. (2019) developed a zein-based coating with eugenol (EUG), carvacrol (CAR), and thymol (THY) essential oils added. The essential oils acted to reduce the fragility typical of zein due to their plasticizing behavior. The films showed antibacterial activity against *L. innocua* and *E. coli*. The zein-based coating containing 2% EUG reduced *L. innocua* and *E. coli* counts in inoculated melons by 2–3 decimals. The best oxygen barrier properties were observed for the formulation with 3% EUG, which was attributed to an essential oil homogenization effect that ended up leading to a denser network without holes. This work suggested that flexible zein coatings containing essential oil could inhibit pathogens embedded in the rough surface of the melon skin.

As already noted, zein, due to its hydrophobic properties, represents an important biopolymer for application, alone or in combination with other biopolymers, in the development of functional and edible coating materials, with potential for applications in maintaining food quality and conservation.

4. Nanotechnology applications with zein

In the last decade, efforts have been made to evaluate the applications of zein/zein-based nanomaterials considering the different areas such as food packaging and nutrition sectors, including the cosmetics, pharmaceuticals, and biomedical industries, among others. The tendency to evaluate the potential uses of this corn by-product with renewable resources for the nanoencapsulation of bioproducts has been gaining momentum, especially for food ingredients and nutrients, since zein possesses several characteristics encouraging its use, particularly its hydrophobic/hydrophilic character, and film/fiber forming and antioxidant properties (Tuan Zainazor et al., 2020).

Understanding this potential, it was possible to create, visualize, and explore the bibliometric map of 69 articles from a search of the Elsevier Scopus database. The keywords inserted were: "(Zein AND nanotechnology)," and the search included articles from 2008 to 2021, generating the data presented in Fig. 18.3. The information provided in this section should be useful for the arrangement and explanation of prospective studies investigating the potential zein in nanotechnology application.

As is well known, biopolymers have been highlighted in recent years as substitutes for synthetic polymers, nevertheless they have some drawbacks, including low degradation temperatures and high-water vapor permeabilities. Considering these limitations, the combination of the utilization of additives and applying nanotechnology to enhance their properties have been done in order to overcome them (Pérez-Guzmán & Castro-Muñoz, 2020).

According to the literature, the usual size of nanostructures obtained using zein is 50–200 nm and the functions of zein nanomaterials were multiple, from the role of carrier of delivery systems to the shell core for encapsulation systems. In addition, zein nanomaterials have been used for food and nutrient components encapsulation and the evaluation of the bioavailability of several of them was enhanced due to their incorporation into zein nanostructures (Li & Song, 2020), in addition, bioactive compounds with potential applications for food and nutrition sectors were stabilized using zein/zein-based nanomaterials (Kasaai, 2018).

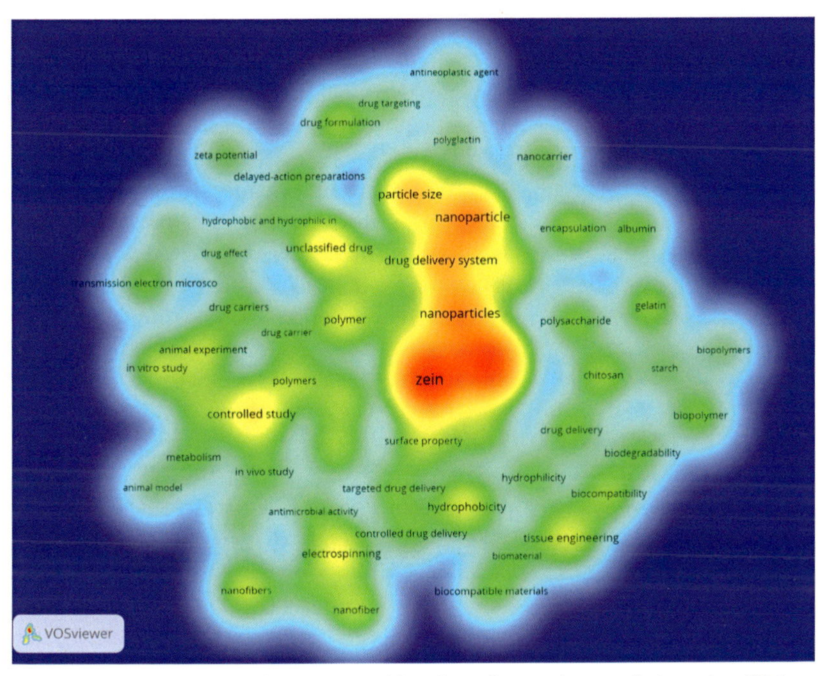

Figure 18.3 Density visualization map resulting from Scopus keywords insertion "Zein AND nanotechnology" (2008–21) using VosViewer software version 1.6.13.

Zein, a molecule that presents a positive charge, is appropriate for the release of negatively charged targets such as drugs, foodstuffs, and nutrients. In this way, a wide range of isoelectric points for zein is an appropriate condition for delivering different nutrients into the body. Because of the numerous advantages of zein compared with other proteins, several studies have explored the potential of zein nanostructure applications for food and nutrition indications, especially as zein is a green polymer that is recognized as safe for food applications (Weissmueller et al., 2016).

Horuz and Belibağlı (2018) studied the nanoencapsulation of carotenoids obtained from tomato peels into zein nanofibers using electrospinning. These authors produced zein nanofibers containing carotenoids extracted from tomato peel and compared them with nonencapsulated extract, with nanofibers presenting better retention of lycopene and antioxidant activity during 14-days' storage.

Besides the applications related to food, the use of zein as a nanomaterial was applied for several other uses, including biomedical products. Tissue engineering is one of the most challenging fields of research as it provides current alternative protocols and materials for the regeneration of damaged tissue. Studies have highlighted zein as a prospective biopolymer for tissue engineering uses considering nanotechnology processes (Pérez-Guzmán & Castro-Muñoz, 2020). Brahatheeswaran et al. (2012) produced fluorescent curcumin-zein-nanofiber scaffolds also for biomedical uses and the produced structures showed uniform and smooth round form with a narrow distribution of diameters. The in vitro drug release of the zein nanofibers revealed that a marked burst liberation phase was first detected, followed by gradual release throughout the remaining time. The authors concluded that the study indicated that the zein curcumin nanofiber scaffolds possess great potential as new biopolymers.

A new way to use zein was approached by Gott et al. (2014). These authors produced nanoparticles with zein and chitosan containing the test chemical rhodamine B, utilized as films to coat food, and after that the test organism (freshwater amphipod *Hyalella azteca*) was fed. Both zein and chitosan nanostructures presented a significantly lower release rate of rhodamine B into water than food dyed with rhodamine B without biopolymer nanoparticles. Thus, the authors concluded that zein and chitosan can be considered successful in nanoencapsulating and releasing compounds to aquatic invertebrates.

5. Final considerations

Zein has demonstrated potential for being ecofriendly, biodegradable, atoxic, low cost, and having bioavailable properties. Zein has hydrophobic properties that are important for a number of applications, alone or in combination with other biopolymers, in the development of functional and edible film and coating materials, for maintaining food quality. For its application, zein demonstrated film-forming capacity resulting in films with good mechanical properties to be applied to food; good adhesion for use as edible coatings, resulting in good gas and water barrier properties; and adequate binding capacity with drugs or biocompounds with the possibility of use in encapsulated form because it presents good binding and releases with bioactive compounds. Thus, zein is a green polymer recognized as being safe for food, medical, and engineering applications.

References

Ahammed, S., Liu, F., Khin, M. N., Yokoyama, W. H., & Zhong, F. (2020). Improvement of the water resistance and ductility of gelatin film by zein. *Food Hydrocolloids, 105*. https://doi.org/10.1016/j.foodhyd.2020.105804

Arcan, I., & Yemenicioğlu, A. (2011). Incorporating phenolic compounds opens a new perspective to use zein films as flexible bioactive packaging materials. *Food Research International, 44*(2), 550–556. https://doi.org/10.1016/j.foodres.2010.11.034

Assis, O. B. G., & Leoni, A. M. (2009). Protein hydrophobic dressing on seeds aiming at the delay of undesirable germination. *Scientia Agricola, 66*(1), 123–126. https://doi.org/10.1590/s0103-90162009000100017

Bai, J., Alleyne, V., Hagenmaier, R. D., Mattheis, J. P., & Baldwin, E. A. (2003). Formulation of zein coatings for apples (*Malus domestica* Borkh). *Postharvest Biology and Technology, 28*(2), 259–268. https://doi.org/10.1016/S0925-5214(02)00182-5

Boyacı, D., Iorio, G., Sozbilen, G. S., Alkan, D., Trabattoni, S., Pucillo, F., Farris, S., & Yemenicioğlu, A. (2019). Development of flexible antimicrobial zein coatings with essential oils for the inhibition of critical pathogens on the surface of whole fruits: test of coatings on inoculated melons. *Food Packaging and Shelf Life, 20*. https://doi.org/10.1016/j.fpsl.2019.100316

Brahatheeswaran, D., Mathew, A., Aswathy, R. G., Nagaoka, Y., Venugopal, K., Yoshida, Y., Maekawa, T., & Sakthikumar, D. (2012). Hybrid fluorescent curcumin loaded zein electrospun nanofibrous scaffold for biomedical applications. *Biomedical Materials, 7*(4), 045001.

Chen, H., Wang, J., Cheng, Y., Wang, C., Liu, H., Bian, H., Pan, Y., Sun, J., & Han, W. (2019). Application of protein-based films and coatings for food packaging: a review. *Polymers, 11*(12), 2039.

Chui, C. F., & Falco, S. C. (1995). A new methionine-rich seed storage protein from maize. *Plant Physiology, 107*(1). https://doi.org/10.1104/pp.107.1.291, 291–291.

da Costa, R. D. S., da Cruz Rodrigues, A. M., Borges Laurindo, J., & da Silva, L. H. M. (2019). Development of dehydrated products from peach palm–tucupi blends with edible film characteristics using refractive window. *Journal of Food Science and Technology, 56*(2), 560–570. https://doi.org/10.1007/s13197-018-3454-x

De Almeida, C. B., Corradini, E., Forato, L. A., Fujihara, R., & Filho, J. F. L. (2018). Microstructure and thermal and functional properties of biodegradable films produced using zein. *Polimeros, 28*(1), 30–37. https://doi.org/10.1590/0104-1428.11516

de, O.-F. J. G., Cesar, L. A., Cavalcante, B. A. R., & Buranelo, E. M. (2020). *Biodegradable eco-friendly packaging and coatings incorporated of natural active compounds* (pp. 171–206). Informa UK Limited. https://doi.org/10.1201/9780429322129-6

Dhumal, C. V., & Sarkar, P. (2018). Composite edible films and coatings from food-grade biopolymers. *Journal of Food Science and Technology, 55*(11), 4369–4383. https://doi.org/10.1007/s13197-018-3402-9

1 - introduction. In Drobny, J. G. (Ed.), *Applications of fluoropolymer films*, (pp. 3–38). (2020) (pp. 3–38). William Andrew Publishing.

FDA. US Food and Drug Administration. (2013). *GRAS substances (SCOGS) database.* Retrieved from https://www.ecfr.gov/current/title-21/chapter-I/subchapter-B/part-184/subpart-B/section-184.1984. (Accessed 21 September 2021).

Feldman, D. (2001). Polymer barrier films. *Journal of Polymers and the Environment, 9*(2), 49–55.

G., Z. K., P., K. K., & G., B. C. (2011). Biopolymer-based films as carriers of antimicrobial agents. *Procedia Food Science*, 190–196. https://doi.org/10.1016/j.profoo.2011.09.030

Gerde, J. A., Spinozzi, J. I., & Borrás, L. (2017). Maize kernel hardness, endosperm zein profiles, and ethanol production. *Bioenergy Research, 10*(3), 760–771. https://doi.org/10.1007/s12155-017-9837-4

Ghanbarzadeh, B., Musavi, M., Oromiehie, A. R., Rezayi, K., Razmi Rad, E., & Milani, J. (2007). Effect of plasticizing sugars on water vapor permeability, surface energy and microstructure properties of zein films. *Lebensmittel-Wissenschaft und -Technologie- Food Science and Technology, 40*(7), 1191–1197. https://doi.org/10.1016/j.lwt.2006.07.008

Ghanbarzadeh, B., & Oromiehi, A. R. (2008). Biodegradable biocomposite films based on whey protein and zein: barrier, mechanical properties and AFM analysis. *International Journal of Biological Macromolecules, 43*(2), 209–215. https://doi.org/10.1016/j.ijbiomac.2008.05.006

Gott, R., Luo, Y., Wang, Q., & Lamp, W. (2014). Development of a biopolymer nanoparticle-based method of oral toxicity testing in aquatic invertebrates. *Ecotoxicology and Environmental Safety, 104*, 226–230.

Horuz, T., & Belibağlı, K. (2018). Nanoencapsulation by electrospinning to improve stability and water solubility of carotenoids extracted from tomato peels. *Food chemistry, 268*, 86–93.

Hossain, F., Sarika, K., Muthusamy, V., Zunjare, R. U., & Gupta, H. S. (2019). Quality protein maize for nutritional security. In *Quality breeding in field crops* (pp. 217–237). Springer International Publishing. https://doi.org/10.1007/978-3-030-04609-5_11

Kasaai, M. R. (2018). Zein and zein -based nano-materials for food and nutrition applications: a review. *Trends in Food Science and Technology, 79*, 184–197. https://doi.org/10.1016/j.tifs.2018.07.015

Kashiri, M., Cerisuelo, J. P., Domínguez, I., López-Carballo, G., Muriel-Gallet, V., Gavara, R., & Hernández-Muñoz, P. (2017). Zein films and coatings as carriers and release systems of Zataria multiflora Boiss. essential oil for antimicrobial food packaging. *Food Hydrocolloids, 70*, 260–268. https://doi.org/10.1016/j.foodhyd.2017.02.021

King, B. L., Taylor, J., & Taylor, J. R. N. (2016). Formation of a viscoelastic dough from isolated total zein (α-, β- and γ-zein) using a glacial acetic acid treatment. *Journal of Cereal Science, 71*, 250–257. https://doi.org/10.1016/j.jcs.2016.09.005

Kirihara, J. A., Petri, J. B., & Messing, J. (1988). Isolation and sequence of a gene encoding a methionine-rich 10-kDa zein protein from maize. *Gene, 71*(2), 359–370. https://doi.org/10.1016/0378-1119(88)90053-4

Kljak, K., Duvnjak, M., & Grbeša, D. (2018). Contribution of zein content and starch characteristics to vitreousness of commercial maize hybrids. *Journal of Cereal Science, 80*, 57–62. https://doi.org/10.1016/j.jcs.2018.01.010

Lemes, A. C., Sala, L., Ores, J. D. C., Braga, A. R. C., Egea, M. B., & Fernandes, K. F. (2016). A review of the latest advances in encrypted bioactive peptides from protein-richwaste. *International Journal of Molecular Sciences, 17*(6). https://doi.org/10.3390/ijms17060950

Lending, C. R., & Larkins, B. A. (1989). Changes in the zein composition of protein bodies during maize endosperm development. *The Plant Cell Online, 1*(10), 1011–1023. https://doi.org/10.1105/tpc.1.10.1011

Li, C., & Song, R. (2020). The regulation of zein biosynthesis in maize endosperm. *Theoretical and Applied Genetics, 133*(5), 1443–1453. https://doi.org/10.1007/s00122-019-03520-z

Luo, Y., Teng, Z., Wang, T. T. Y., & Wang, Q. (2013). Cellular uptake and transport of zein nanoparticles: effects of sodium caseinate. *Journal of Agricultural and Food Chemistry, 61*(31), 7621–7629. https://doi.org/10.1021/jf402198r

Mehyar, G., El Assi, N., Alsmairat, N., & Holley, R. (2014). Effect of edible coatings on fruit maturity and fungal growth on Berhi dates. *International journal of food science & technology, 49*(11), 2409–2417.

Moradi, M., Tajik, H., Razavi Rohani, S. M., & Mahmoudian, A. (2016). Antioxidant and antimicrobial effects of zein edible film impregnated with Zataria multiflora Boiss. essential oil and monolaurin. *Lebensmittel-Wissenschaft und -Technologie- Food Science and Technology, 72*, 37–43. https://doi.org/10.1016/j.lwt.2016.04.026

Pascall, M., & Lin, S. (2013). The application of edible polymeric films and coatings in the food industry. *Journal of Food Processing & Technology, 4*(2), 1–2. https://doi.org/10.4172/2157-7110.1000e116

Pedersen, K., Argos, P., Naravana, S. V. L., & Larkins, B. A. (1986). Sequence analysis and characterization of a maize gene encoding a high-sulfur zein protein of M(r) 15,000. *Journal of Biological Chemistry, 261*(14), 6279–6284.

Pena-Serna, C., Penna, A., & Lopes Filho, J. (2016). Zein-based blend coatings: Impact on the quality of a model cheese of short ripening period. *Journal of Food Engineering, 171*, 208–213.

Pérez-Guzmán, C., & Castro-Muñoz, R. (2020). A review of zein as a potential biopolymer for tissue engineering and nanotechnological applications. *Processes, 8*(11).

Pinheiro, A. C., Cerqueira, M. A., Souza, B. W. S., Martins, J. T., Teixeira, J. A., & Vicente, A. A. (2010). Utilização de revestimentos/filmes edíveis para aplicações alimentares. *Boletim de Biotecnologia, 1*, 18–28.

Prat, S., Pérez-Grau, L., & Puigdomènech, P. (1987). Multiple variability in the sequence of a family of maize endosperm proteins. *Gene, 52*(1), 41–49. https://doi.org/10.1016/0378-1119(87)90393-3

Quispe-Condori, S., Saldaña, M. D. A., & Temelli, F. (2011). Microencapsulation of flax oil with zein using spray and freeze drying. *Lebensmittel-Wissenschaft und -Technologie-Food Science and Technology, 44*(9), 1880–1887. https://doi.org/10.1016/j.lwt.2011.01.005

Riva, S. C., Opara, U. O., & Fawole, O. A. (2020). Recent developments on postharvest application of edible coatings on stone fruit: a review. *Scientia Horticulturae, 262*. https://doi.org/10.1016/j.scienta.2019.109074

Santeramo, F. G., Carlucci, D., De Devitiis, B., Seccia, A., Stasi, A., Viscecchia, R., & Nardone, G. (2018). Emerging trends in European food, diets and food industry. *Food Research International, 104*, 39–47. https://doi.org/10.1016/j.foodres.2017.10.039

Santos, T. M., Souza Filho, M.d. S. M., Silva, E.d. O., Silveira, M. R. S.d., Miranda, M. R. A.d., Lopes, M. M. A., & Azeredo, H. M. C. (2018). Enhancing storage stability of guava with tannic acid-crosslinked zein coatings. *Food Chemistry, 257*, 252–258. https://doi.org/10.1016/j.foodchem.2018.03.021

Scramin, J. A., de Britto, D., Forato, L. A., Bernardes-Filho, R., Colnago, L. A., & Assis, O. B. G. (2011). Characterisation of zein-oleic acid films and applications in fruit coating. *International Journal of Food Science and Technology, 46*(10), 2145–2152. https://doi.org/10.1111/j.1365-2621.2011.02729.x

Shukla, R., & Cheryan, M. (2001). Zein: the industrial protein from corn. *Industrial Crops and Products*, 13(3), 171–192. https://doi.org/10.1016/S0926-6690(00)00064-9.

Spasojević, L., Katona, J., Bučko, S., Savić, S. M., Petrović, L., Milinković Budinčić, J., Tasić, N., Aidarova, S., & Sharipova, A. (2019). Edible water barrier films prepared from aqueous dispersions of zein nanoparticles. *Lebensmittel-Wissenschaft und -Technologie, 109*, 350–358. https://doi.org/10.1016/j.lwt.2019.04.038

Sun, Y., Liu, Z., Zhang, L., Wang, X., & Li, L. (2020). Effects of plasticizer type and concentration on rheological, physico-mechanical and structural properties of chitosan/zein film. *International Journal of Biological Macromolecules*, 143, 334–340. https://doi.org/10.1016/j.ijbiomac.2019.12.035.

Tamagno, S., Greco, I. A., Almeida, H., Di Paola, J. C., Martí Ribes, F., & Borrás, L. (2016). Crop management options for maximizing Maize Kernel hardness. *Agronomy Journal, 108*(4), 1561–1570. https://doi.org/10.2134/agronj2015.0590

Teklehaimanot, W. H., & Emmambux, M. N. (2019). Foaming properties of total zein, total kafirin and pre-gelatinized maize starch blends at alkaline pH. *Food Hydrocolloids, 97*. https://doi.org/10.1016/j.foodhyd.2019.105221

Thakur, R., Pristijono, P., Scarlett, C. J., Bowyer, M., Singh, S. P., & Vuong, Q. V. (2019). Starch-based films: major factors affecting their properties. *International Journal of Biological Macromolecules, 132*, 1079–1089. https://doi.org/10.1016/j.ijbiomac.2019.03.190

Tuan Zainazor, T. C., Fisal, A., Goh, E. G., Che Sulaiman, N. F., & Sarbon, N. M. (2020). Emerging of bio-nano composite gelatine-based film as bio-degradable food packaging: a review. *Food Research, 4*(4), 944–956. https://doi.org/10.26656/FR.2017.4(4).365

Weissmueller, N. T., Lu, H. D., Hurley, A., & Prud'Homme, R. K. (2016). Nanocarriers from GRAS zein proteins to encapsulate hydrophobic actives. *Biomacromolecules, 17*(11), 3828–3837. https://doi.org/10.1021/acs.biomac.6b01440

Woo, Y. M., Hu, D. W. N., Larkins, B. A., & Jung, R. (2001). Genomics analysis of genes expressed in maize endosperm identifies novel seed proteins and clarifies patterns of zein gene expression. *The Plant Cell Online, 13*(10), 2297–2317. https://doi.org/10.1105/tpc.13.10.2297

Xiao, J., Gu, C., Zhu, D., Huang, Y., Luo, Y., & Zhou, Q. (2021). Development and characterization of an edible chitosan/zein-cinnamaldehyde nano-cellulose composite film and its effects on mango quality during storage. *Lebensmittel-Wissenschaft und -Technologie, 140*. https://doi.org/10.1016/j.lwt.2020.110809

Yilmaz, A., Bozkurt, F., Cicek, P. K., Dertli, E., Durak, M. Z., & Yilmaz, M. T. (2016). A novel antifungal surface-coating application to limit postharvest decay on coated apples: molecular, thermal and morphological properties of electrospun zein–nanofiber mats loaded with curcumin. *Innovative Food Science & Emerging Technologies, 37*, 74–83. https://doi.org/10.1016/j.ifset.2016.08.008

Yousuf, B., Qadri, O. S., & Srivastava, A. K. (2018). Recent developments in shelf-life extension of fresh-cut fruits and vegetables by application of different edible coatings: a review. *Lebensmittel-Wissenschaft und -Technologie- Food Science and Technology, 89*, 198–209. https://doi.org/10.1016/j.lwt.2017.10.051

Zapata, P. J., Guillén, F., Martínez-Romero, D., Castillo, S., Valero, D., & Serrano, M. (2008). Use of alginate or zein as edible coatings to delay postharvest ripening process and to maintain tomato (*Solanum lycopersicon* Mill) quality. *Journal of the Science of Food and Agriculture, 88*(7), 1287–1293. https://doi.org/10.1002/jsfa.3220

Zhang, L., Liu, Z., Sun, Y., Wang, X., & Li, L. (2020). Combined antioxidant and sensory effects of active chitosan/zein film containing α-tocopherol on *Agaricus bisporus*. *Food Packaging and Shelf Life, 24*. https://doi.org/10.1016/j.fpsl.2020.100470

Zhang, L., Liu, Z., Wang, X., Dong, S., Sun, Y., & Zhao, Z. (2019). The properties of chitosan/zein blend film and effect of film on quality of mushroom (*Agaricus bisporus*). *Postharvest Biology and Technology, 155*, 47–56. https://doi.org/10.1016/j.postharvbio.2019.05.013

Zhou, L., & Wang, Y. (2021). Physical and antimicrobial properties of zein and methyl cellulose composite films with plasticizers of oleic acid and polyethylene glycol. *Lebensmittel-Wissenschaft und -Technologie, 140*. https://doi.org/10.1016/j.lwt.2020.110811

Biodegradable polymer blends and composites from renewable resources

19

L. Rajeshkumar
Department of Mechanical Engineering, KPR Institute of Engineering and Technology, Coimbatore, Tamilnadu, India

1. Introduction

Plastics and polymers have become inseparable parts of human life. Nevertheless, concerns over the increasing amounts of plastic wastes and their adverse effects on the environment have attracted various organizations and other groups to make this a significant research and news topic. The history of the use of naturally available polymeric materials dates back to ancient times, when these materials initially were used by mankind for many applications. In the early 19th century, polymeric materials were obtained from the modification of naturally available materials like cellulose, natural rubber, and casein. Biodegradable materials derived from synthetic oil-based or other renewable materials have been characterized by advantageous features such as a high rate of biodegradability, high compostability, high recycling ability, and low carbon footprint (Haider et al., 2019). Hence, researchers currently are investigating the possibilities of developing bioplastics through sustainable design principles to meet the needs of reducing the overwhelming pollution caused by conventional plastics and enhance the end-of-life options of biodegradable materials, which in turn will benefit the economy. Current research focuses primarily on developing such biodegradable materials from waste materials like biomass, algal cells, microbes, and waste streams, which in turn reduces the amount of carbon in all those structures. This also reduces the presence of carbon from the production process of biodegradable polymers. Additionally, the method of sustainable production and recycling of these biodegradable polymers is completely in line with the strategies and policies laid out in the European circular economy strategy and the UN sustainable development goals (Ramesh et al., 2020; Zhao et al., 2020). Production of bioplastics at the global level requires complex estimation for its production quantity as bioplastics has seen a great increase in interest from various industrial sectors. Hence the production quantity of biopolymers can be calculated only through forecasting techniques (Kargarzadeh et al., 2018).

It was reported in a few articles that the production of biopolymers will witness a growth from 2.12 million tons in 2018 to 2.62 million tons in 2023 based on the current rate of production. It was also stated that Europe and Asia would hold the top two positions in the production of biodegradable polymers and that these continents are

Biodegradable Polymers, Blends and Composites. https://doi.org/10.1016/B978-0-12-823791-5.00015-6

acting as warehouses for the production and consumption of biopolymers. Industrial applications have seen a vast usage of biobased polymers and this is expected to multiply in the coming years due to the innovations taking place with respect to the production and utilization of bioplastics. Different production strategies are being adopted by industries for producing biopolymers due to the circular economy and sustainability relevance, which in turn result in positive growth and technological advancements in these industries (Nakajima et al., 2017; Papageorgiou, 2018; Ramesh & Rajeshkumar, 2018). The maximum production of biopolymers is currently that of natural polymeric materials like thermoplastic starch (TPS), cellulose, and their blends due to their potentiality to replace conventionally used polymers in flexible film packaging-related applications. Among the numerous biopolymers, PLA has received much attention lately due to its versatility and adoptability toward processing and applications in the academic and industrial sectors due to its recent technological advancements. PLA can be a substitute for polypropylene and polystyrene in packaging and other complex structural and nonstructural applications as it possesses better load-bearing properties that are on par with those of nonbiodegradable polymers (Balart et al., 2020; Uyama, 2018). Besides PLA, polyhydroxyalkanoates (PHA) are being employed widely in many applications and this material holds the highest number of patents filed at the international level. However, polybutylene succinate (PBS) and polybutylene adipate coterephthalate (PBAT) rank high in global production capacities when compared with PHA polymers. It was stated that PBS and PBAT contribute about 4.4% and 13.5% of global production of biodegradable polymers, while PHA contributes about 1.3%, which is approximately 26,000 tons. It was also found by some of the earlier researchers that the biodegradability of a polymeric material is not governed by the polymeric molecular structure or its source of origin. This can be clearly seen in PCL and PBAT polymers whose origin is from fossil fuels although they have high biodegradability. Fig. 19.1 depicts the various classes of biopolymers based on their origin and biodegradability, along with the chemical structures of some elements (Laycock et al., 2017; Prajapati et al., 2019).

Biobased polymeric materials have their own importance in many fields such as science, technology, and engineering, and these materials have created much interest for research and development in various areas. Biopolymers are classified into three types, as shown in Fig. 19.1. First is the natural biobased polymers arising directly from the natural sources or biomass, which are modified chemically, examples include modified starch, cellulose, chitin, starch, and cellulose acetate. Second is designated as biopolymers which are synthesized biologically from microorganisms and plants, examples include polyglutamic acid and PHA. The third type is artificial or synthetic polymers which are prepared in the laboratory, examples of which include synthetic polymers such as polylactide, PBS, polyethylene terephthalate (PET), and polyolefins (RameshKumar et al., 2020; Saravana Kumar et al., 2017). Waste generated from packaging industries and specifically wastes from nonbiodegradable polymers are a major part of solid wastes which create environmental hazards. The management of packaging material waste that is discarded after use is also a problem, and the littering of such wastes has significantly increased lately. Polyethylene (PE) is the most widely used petroleum-based polymeric material in the packaging industry. It is a well-known fact that the disposal of any petroleum-based polymer is very difficult either in coastal

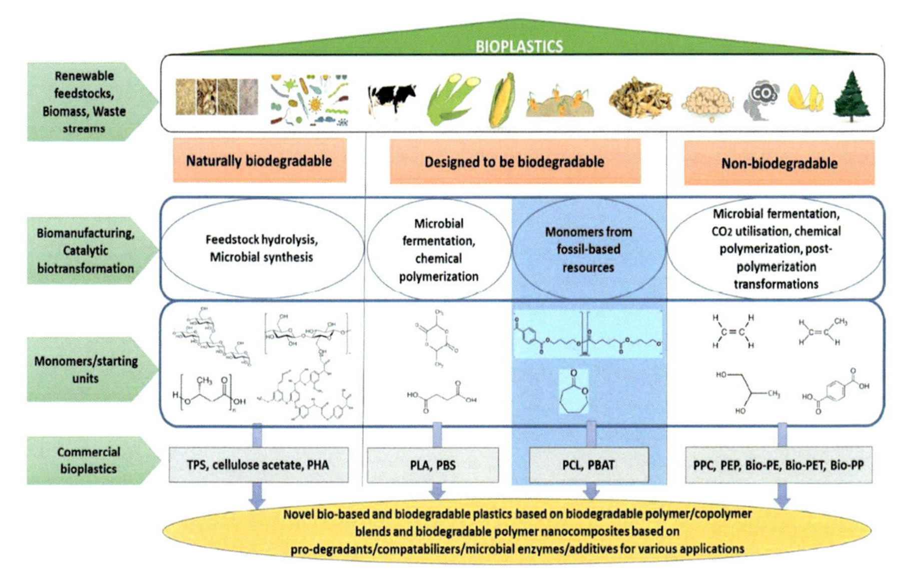

Figure 19.1 Approaches to the production of biobased polymers.
Adapted from (n.d.) https://doi.org/10.1016/j.cogsc.2019.12.005

areas or in land and, even if they are disposed of, they pose a major threat of contamination at different levels. Hence, in order to overcome all the above problems with respect to packaging materials, the manufacturing of materials using biodegradable polymers and specifically antimicrobial materials might enhance the suitability for customers and render eco-friendly and hygienic products. Such manufacturing of packaging material from antimicrobial elements may increase the ability of the material to be stored and reduce the formation of fungi or bacteria on those materials, which may cause a hazard to human health. It is also a fact that the biodegradability and antimicrobial activity are two different aspects which have to be properly addressed while manufacturing a material for packaging applications, although both fall under the category of green materials. Starch and cellulose are the most commonly used biopolymers and are characterized by some advantageous features such as biocompatibility, inexpensiveness, eco-friendly nature, renewability, and biodegradability. Hence the use of such biopolymers has been witnessing steady growth in research and industrial applications during recent times (La et al., 2021; Saravana Kumar et al., 2017; Sathish et al., 2014; Wróblewska-Krepsztul et al., 2018). Fig. 19.2 depicts the biological cycle of biopolymeric materials starting from their origin and ending with recycling in an eco-friendly manner.

2. Biodegradable polymers and blends

Biodegradable polymers are the potential materials to overcome all the problems of waste pollution, litter, etc. caused by nonbiodegradable polymeric materials. However, the utilization of such advantageous biodegradable plastics is very limited due to their

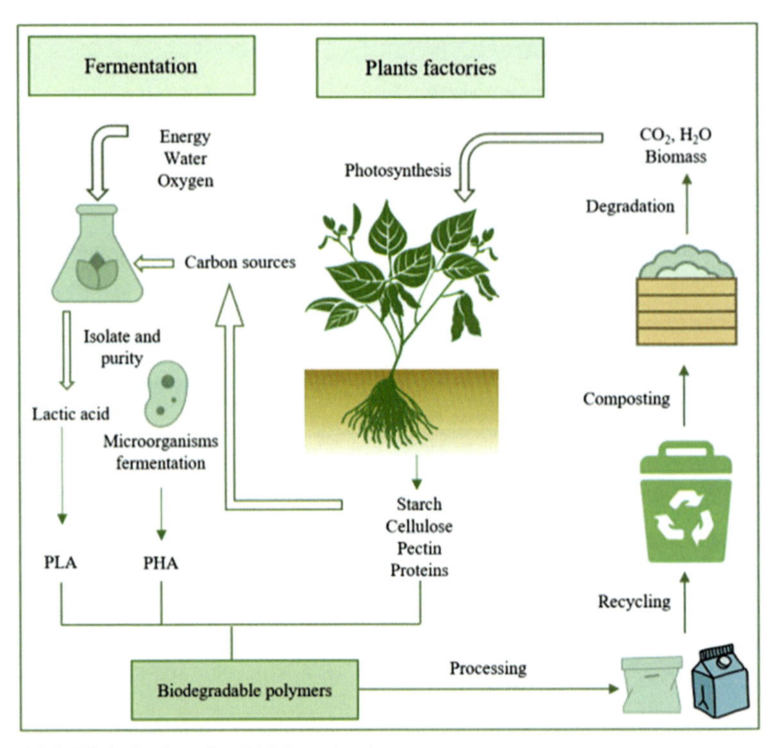

Figure 19.2 Biological cycle of biobased polymers.
Adapted from (n.d.) https://doi.org/10.1016/j.aiepr.2019.11.002

expense and their selection range, causing many limitations for various end applications (Mangaraj et al., 2019). Hence, many researchers have stated that the expensive nature of biodegradable polymers could be eradicated by preparing a blend with other renewable materials, which in turn would broaden their industrial application prospects. A biopolymer material is considered to be completely biodegradable when it has originated from a source which is purely natural and it undergoes complete decomposition without affecting the environment after its useful lifetime. Some elements derived from bioresources like natural oils and polysaccharides such as cellulose, starch, proteins, lignin, and chitosan are ideal biodegradable polymers, while materials like PCL and PLA have a synthetic origin but fall under the category of biodegradable polymers. Among the aforesaid sources, natural oils are considered to be a potential substitute for chemical raw materials and the oils are usually derived from both animal and plant sources extensively. Many studies have indicated that triglycerides find their major applications in agricultural chemicals, inks, and coatings. Various novel methods to convert triglyceride oils into polymers and monomers have been adopted in most of the above applications. Apart from these materials, nanocellulose is a very commonly used newer type of nanomaterial owing to its relatively better physical and chemical characteristics. Such nanomaterials are characterized by some advantageous features such as high flexibility, better strength, low density, and so lower final weight, and the potentiality of the nanocellulose to alter the surface chemistry of the incorporated material (Iqbal et al., 2019; Ramesh et al., 2020).

2.1 Starch

Starch is the most widely and commonly used biopolymeric material synthesized purely from natural and renewable resources. Starch-based polymers are in high demand currently due to starch being relatively cheap, completely biodegradable, and widely available. In order to reduce the manufacturing cost of biopolymers, blending of any biodegradable polymer with thermoplastic starch (TPS) has been proven to be an effective method. Starch is also the most plentiful, widely available, and renewable class of polysaccharide material obtained from plants. Starch is comprised of two major glucose-type polymers, amylopectin and amylose. Amylopectin is made from a short α-1,4-connected D-glucose chain of atoms with α-1,6-branched bonds and amylose is a polymeric chain of D-glucose atomic chain linked together by α-1,4-branched bonds. Although starch is considered to be completely biodegradable, low cost, and able to produce film form components with permeability of oxygen at lower levels and can be handled easily, it faces a major constraint in its usage due to its hydrophilicity and brittleness. Hence, starch-based polymers face limitations in such a way that they cannot be applied in common applications such as packaging of foods and substitutes for plastic bags. In order to overcome this limitation and in order to improve the processing capability and flexibility, starch was converted into TPS by the use of different plasticizers such as sorbitol, glycerol, and glycol by various researchers under the action of shear stress and heat during the extrusion process (Bastioli, 2020).

Recently, starch has been widely used in the packaging field due to its cheap, biodegradable, and available nature, but the application has not been as widely used as was expected. In addition, modified and converted starch has been proven to be a substandard alternative for conventional polymeric materials owing to the hydrophilicity and lower mechanical characteristics despite being modified by the use of plasticizers. Many researchers have explored the properties, structure, and processing of TPS/plasticized starch in biodegradable polymer blends for applications such as packaging and many other applications with a shorter lifetime. Hence, novel sustainable starch-based biodegradable polymer blends have been developed by blending starch with other biodegradable polymers for packaging-related applications. It has been also stated in previous research that the rate of biodegradation of the materials developed by blending starch and other starch-based material with biodegradable polymers was higher than for nonblended starch-based materials (Bulatović et al., 2021; Muthuraj et al., 2018).

2.2 Poly-lactic acid

PLA is one of the most commonly used biodegradable polymers—equal as starch—and is a major variant of aliphatic polyester that is obtained from lactic acid, which in turn results from the fermentation of naturally available plants like corn and sugar beet. Similar to starch, this biodegradable polymer is also cheap and plentiful, and so it has been an area of great interest from researchers and industrialists. Additionally, PLA is characterized by advantages such as biocompatibility, commercially available, completely biodegradable, ease of processing, and a high degree of transparency, which are the prime reasons for its wider utilization (Ramesh & Rajeshkumar, 2021). Normally, fermentation of petrochemicals or fermentation using bacteria results in the formation of lactic acid. As a result of ring opening polymerization reaction of lactide monomer contained in lactic acid or by making L- or D-lactic acid to undergo condensation polymerization, PLA is manufactured. If PLA is needed with a lower molecular weight, then lactic acid is made to undergo polycondensation directly. On the other hand, PLA with higher molecular weight exhibits better mechanical characteristics, which are prepared by two different methods, ring-opening polymerization reaction and azeotropic condensation polymerization of lactic acid (Ramesh & Rajesh Kumar, 2020).

During the previous decade, reinforcing PLA matrix with nanocellulosic materials to obtain PLA-based biocomposites was attempted by various researchers. It can also be seen from the publication records that almost 200 publications were made pertaining to PLA biocomposites. It was reported in various studies that when the original nanocellulose was reinforced in a PLA matrix to obtain biocomposites based on PLA, the mechanical strength and stiffness of the resultant composites were enhanced, along with the elastic modulus of the composites, but the elastic nature of the composites decreased beyond a certain limit (Guo et al., 2020). Low load-bearing capacity PLA-based biocomposites resulted from the direct reinforcement of nanocellulose fibers into the PLA matrix owing to the poor interfacial adhesion between the hydrophobic matrix and hydrophilic filler materials. Hence, the studies stated that the presence of 0.5–2 wt.% of nanocellulose within the PLA matrix would render optimum results. Techniques of chemical surface modification such as polymer grafting or derivatization and physical modification methods like macromolecule or surfactant coating were used to enhance the compatibility and dispersion of nanocellulose fillers within hydrophobic and nonpolar PLA matrix, which in turn increased the interfacial characteristics between the filler and matrix. The above methods and techniques not only rendered better interfacial adhesion characteristics but also enhanced the performance of nanocellulose PLA biocomposites substantially, and they specifically exhibited better mechanical characteristics for the resulting surface-modified composites (Silva et al., 2020; Ramesh & Rajeshkumar, 2021; Zuo et al., 2020).

As stated earlier, PLA is normally synthesized from lactide, which is a cyclic dimer, through a process called ring-opening polymerization (ROP). This is the usual process carried out in almost all industries for the synthesis of PLA, but in some cases polycondensation of lactic acid directly has also been carried out. Due to the higher potentiality of biological production of L-lactic acid when compared with D-lactic acid, poly-L-lactic acid (PLLA) experiences a higher production rate and thus has more commercial value. In some experiments, an insoluble solution precipitant, namely stereocomplex PLA (sc-PLA), has been reported which was considered to be the complex form of poly-D-lactide acid (PDLA) and PLLA. The original solubility

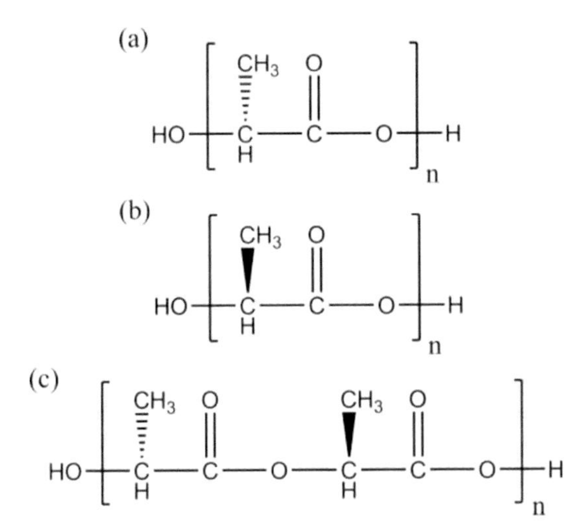

Figure 19.3 Chemical structure of polylactic acid.
Adapted from (n.d.) https://doi.org/10.1080/15583724.2017.1380039

nature of homo-chiral-shaped PLAs was reduced due to the occurrence of changes in the chemical characteristics of PLA at the time of development of complex sc-PLA elements. This forced experimenters to precipitate the sc-PLA granules from the PLA crystallite form selectively. Fig. 19.3 depicts the normal chemical structure and existence of chirality in PLA (Aziz et al., 2020).

2.3 Poly-caprolactone

PCL is another thermoplastic polymer with better biodegradable characteristics, less viscosity, better thermal processing ability, and a lower melting point, which is normally manufactured from ε-caprolactone through the polymerization reaction. The application of PCL in packaging and other industrial applications demanding biodegradability is limited due to their lower melting point, which in turn could be due to their reduced mechanical characteristics and poor barrier characteristics. Hence, in order to widen the application spectrum of PCL-based biopolymers, they are normally blended with other types of biodegradable polymers such as cellulose acetate butyrate, PLA, and cellulose propionate, which enhances the adhesion, dyeability, and resistance toward stress cracking of PCL biopolymers (Chan et al., 2018).

PCL is a synthetic biodegradable polymer derived from an aliphatic polyester which has a hydrophobic nature and was characterized by a melting point as low as 55−60°C and −60°C glass transition temperature. Biodegradable composites can be prepared from PCL by combining them with polymeric matrix materials which has attracted a great deal of attention due to its commercial availability, physical characteristics, and biodegradability. Studies have stated that nanocellulose could be reinforced only in smaller proportions with PCL matrix due to the fact that the interfacial bond existing with nanocellulose-reinforced PCL biocomposites might have developed due to the direct mixing of chemically incompatible hydrophobic PCL and hydrophilic nanocellulose fibers. In many of these studies, it has been concluded that such incompatibility resulting in poor interfacial characteristics could be improved by the surface modification of nanocellulose fibers, which enhances the compatibility of the reinforcement with other elements in the biocomposites. The solubility of the nanocellulose fibers within the PCL polymer matrix could be effectively enhanced through polymer chain surface grafting directly onto the fiber surface. The surface polarity of the nanocellulose particles could be readily regulated by grafting the PCL polymer chains onto the surfaces of the nanocellulose particles adopting "grafting from" and "grafting onto" strategies. This makes the modified nanocellulose a suitable candidate for reinforcement in PCL-based biobased composites (Lee et al., 2020; Ramesh et al., 2021a; Ramesh et al., 2020).

2.4 Poly-hydroxyalkanoate/poly-hydroxybutyrates

PHA is a biodegradable polyester derived from different hydroxyalkanoates through microbial fermentation that could be used as another potential material in various areas such as agricultural, medical, and packaging applications. Hydroxyalkanoate monomers are the starting materials for PHA and are members of the polyester family. These materials are characterized by nontoxicity, a low melting point, high degree of crystallinity with thermoplastic elastomer molecules, better chemical and physical properties, good biocompatibility, and better resistance toward UV light, with all the characteristics of PHA being governed by the configuration of monomers. Despite their better properties, their applications are limited due to their high thermal degradation, and their incompatibility toward thermal processing methods and lower mechanical properties. PHB is another commonly used biodegradable polymer which is considered to be another representative element of the alkanoates family which is inherently characterized by a high degree of crystallinity (Ramesh et al., 2020; Omaníková et al., 2020). PHA and PHB are the most commonly considered candidates for food-packaging applications with a short life span. Due to the formation of R- and S-hydroxybutyrates along with some other nontoxic compounds under anaerobic and aerobic conditions from PHA depolymerases and PHA hydroxylases, these materials are characterized by a high degree of biodegradability. Naturally, PHAs exist as poly-3-hydroxybutyrate-co-3-hydroxyvalerate (P3HB-co-3HV) and poly-3-hydroxybutyrate (P3HB) homogeneous polymers. In bacteria, PHAs exist as pure polymeric granules which are naturally used as an energy-storage medium and this presence is similar to the starch present in plants and fat present in animals (Thirmizir et al., 2020).

The commercial production of PHAs involves fatty acid transformation from a high-energy feedstock and allowing the bacteria to feed on the same. At the same time, when PHAs are produced in industrial mass production, isolation of cells and their lyse were carried out, followed by some feast–famine cycles. Granules of PHA polymers were then extracted from the lysed cells, purified, and converted into powder or pellets. The use of wastewater as an energy-rich raw material for the production of PHAs was tried, and trials were carried out for the production of PHAs from energy-rich pure feedstock. Meanwhile, PHA production could be

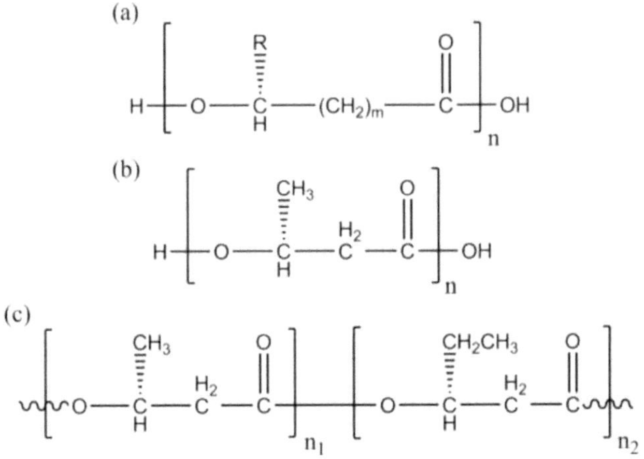

Figure 19.4 Chemical structure of PHA/PHB.
Adapted from (n.d.). https://doi.org/10.1080/15583724.2017.1380039

enhanced through genetic alteration, where the production was improved by the bacteria producing PHA and also the plants were able to synthesize PHA by themselves (Abioye & Obuekwe, 2020; Ramesh et al., 2021). Fig. 19.4 illustrates the isotopes and normal chemical structures of PHA and PHB which were present in a normal element.

Mechanical characteristics of PHB with 70% crystallinity were comparatively high and this behavior is similar to that of other nonbiodegradable polymers like polyethylene. The characteristics of PHB such as water permeability, barrier properties, and excellent aromatic behavior were due to its lamellar arrangement, which in turn contributes to the employment of PHB in food-packaging applications. The blending of PHB with other types of polymers in a molten state would be highly feasible due to its low melting point when compared with PLA, which has a relatively higher melting point. Nevertheless, the applications of PHB face many constraints due to its lower mechanical properties, complex processing, low barrier properties, low thermal stability, and highly brittle nature. Various researchers have tried to overcome the aforesaid limitations through experimental studies and attempts have been made to successfully employ PHA/PHB in various applications. In one such trial, PHB was blended with plasticized PLA along with the incorporation of catechin through a melt processing method and their mechanical properties were analyzed. The results showcased that the mechanical properties of plasticized PLA were improved by the addition of PHB–catechin blends and they displayed a high degree of potentiality for the packaging of fatty foods through a pure biodegradable active material (Zarrintaj et al., 2020).

2.5 Biopolymer blends

In order to enhance the biodegradability of existing materials or to develop novel biodegradable materials, blending of one or more biodegradable materials with other materials has been used as a most common strategy. Formerly, before completely biodegradable materials were developed, blending of starch with low-density polyethylene (LDPE), polyolefin, and linear LDPE was carried out to manufacture eco-friendly and compostable materials for packaging applications specifically. However, these materials were not fully biodegradable. Usually, compounding and blending of biopolymers with one another would be carried out through a film-blowing or cast-filming technique, preceded by extrusion via a twin-screw extruder machine. While compounding in the above-mentioned technique, a third element apart from the blended elements, known as the compatibilizer, would be employed in order to enhance the compatibility between the precursors. Maleic anhydride grafted within PE is the most commonly used compatibilizer and the content of the compatibilizer has to be comparatively higher than the content of the lower proportioned precursor (Ramesh et al., 2021). Experiments have been performed by blending a TPS/PE biodegradable polymer blend, and the thermal characteristics such as differential scanning calorimetry. The results indicated that the dispersion of TP within PE was found to be better when a compatibilizer was added and the results of DSC indicated that the crystallization of PE was enhanced due to the presence of starch, which had a strong influence on crystallization of the blend. It was also stated in the study that the mechanical characteristics necessary for the packaging applications were as low as 15% by weight of the entire blend and the amount of starch required to render a completely compostable material was as high as 60% by weight of the entire blend. However, the equilibrium between the better mechanical characteristics and biodegradability has not been established until now (Maraveas, 2020; Maghsoudi et al., 2020).

Some authors analyzed the HDPE blend with TPS with the aid of modifiers, such as deep eutectic solvents, and found that the ductility of the blend was enhanced due to the incorporation of modifiers. Pure starch material possesses some weaker properties such as high shrinkage, low resilience, and high sensitivity toward moisture, however these properties can be improved by blending PCL with starch, at least at a lower concentration which has been proven to be a potential strategy by many researchers. It was also stated that the impact strength and dimensional stability of natural starch were improved notably when it was blended with PCL. The glass transition temperature and mechanical characteristics of the TPS/PCL blend were greatly influenced by the blend composition and weight fraction of the plasticizer used. It was noted by the experimenters that TPS/PCL and TPS/PLA blends were mostly alike in terms of end applications and rendered compatibility. Reinforcing starch/PCL blends with nanoclay and fibers resulted in the formation of blends which have been experimentally analyzed by many researchers. Some other properties of TPS/PCL blends such as compatibilization, hydrolytic stability, and rate of degradation were also enhanced when fibers and nanoclay were reinforced with PCL blend. When starch is blended with PLA in sufficient proportions with a compatibilizer as catalyst it renders a biodegradable polymer blend in the form of thin films with high performance when utilized for packaging applications. This became possible due to the development of their individual characteristics and the properties of the multilayer films developed as a result of their blending, specifically for food packaging-related applications. Many previous studies have focused on sustaining the biodegradability of starch/PLA blends but minimizing the cost of their manufacture. In the current scenario, research focus has been given to enhancing the strength characteristics of starch/PLA blends with lower production costs. During manufacturing of the starch/PLA blend, the use of a compatibilizer enhances the compatibility between the individual elements, and this in turn improves the toughness of the blends. Some experimental trials have been carried out by blending starch with amorphous bilayer PLA through a compression molding technique and conventional casting with blends with and without cinnamaldehyde as the compatibilizer. The results indicated that the water vapor barrier and tensile characteristics of the bilayer film blends were higher as compared with the individual starch films. It was also noted that the transparency of film blends was relatively higher and the oxygen barrier was almost the same for film blends as compared with individual starch films (Bhuvaneswari et al., 2021; Jiang et al., 2020).

Some of the nonbiodegradable polymers were also converted into biodegradable polymers by polymer blends. In order to enhance the biodegradability of PET, bio-based terephthalic acid (TPA) was manufactured and converted into bio-TPA using the biomass feedstock derived from natural resources. Biomass feedstock derived from bio-PET could be the result of a combination of bioethylene glycol (EG) and bio-TPA theory. Some experiments have focused on the manufacture of functional plant oil blends such as epoxidized linseed oil (ELO) and epoxidized soybean oil (ESO) from biodegradable epoxy with tetraethyl ammonium bromide as catalyst and methyl tetrahydrophthalic anhydride (MHPA) as the curing agent the synthesis. Soft plant oil-based epoxidized blends were produced after a curing reaction which took place at about 130°C. The feed molar ratio of anhydride/oxirane influenced the mechanical and thermal characteristics of ESO/MHPA polymer blends. From the dynamic mechanical analysis results, it was found by the authors that the glass transition temperatures for ELO/MHPA and ESO/MHPA blends were 110 and 35°C, respectively. Similarly, the tensile strength and storage modulus in the rubbery region for an ELO/MHPA polymer blend were found to be higher when compared with ESO/MHPA polymer blends since ELO blends were characterized by maximum cross-linking density. The shape memory capacities of ELO/MHPA polymer blends was found to be exceptionally good, while for ESO/MHPA they were not better as they were unstable in any of the formed shapes (Sazali et al., 2020).

Some experiments have focused on preparing branched polylactic acid (BP) through a lactide ring polymerization process with castor oil as the catalyst. It was stated by the authors that by governing the lactide and castor oil feed ratio, the molecular weight of the BP was varied as per the requirement. When BP was used as a polymer plasticizer for the production of PLLA, the polymeric blend of PLLA/BP film exhibited an appreciable plasticization effect. This effect became possible even with a low BP content, which could completely plasticize PLLA. However, some parameters like crystallization and the glass transition temperature of PLLA reduced to a certain extent when BP was incorporated as a plasticizer. It was noticed that as the amount of BP was increased, the aforementioned characteristics gradually decreased, and at the same time, when BP was further added into PLLA, its crystal size was reduced by a significant amount. It was seen from previous research that some characteristic behaviors of BP/PLLA blends were not noted in the earlier experimental works which used low-molecular-weight plasticizers and they exhibited many signature characteristics retained by BP/PLLA polymeric blends. The stability of the BP/PLLA polymer blend was enhanced due to the incorporation of BP, which is a polymer plasticizer that retards the volatilization or migration of plasticizer molecules. Some experiments have been performed to develop a synthesis method for ESO through a plant oil-based green composite with a porous PHB monolith. Fig. 19.5 depicts the process involved in the synthesis of ESO through the PHB monolith.

A monolith is a specific class of functional material that is comprised of a 3D porous structure with a continuous chain of molecules. Thermally induced phase separation (TIPS) was the technique used for the manufacturing of PHB monoliths with DMSO as the solvent material. Initially, the PHB monolith was submerged within an ESO solution in the presence of thermally latent catalyst, and curing of the above blend renders PHB/ESO composites. It was observed that the PHB/ESO composites were highly transparent in nature owing to the presence of a fibrous nanoscale structure in the PHB monolith. It was noted from the DMA and DSC tests of PHB/ESO composites that the curves exhibited two peaks pertaining to two different melting behaviors and two different glass transition temperatures. It was also noted that the tensile strength and Young's modulus of the PHB/ESO composites were higher than for the individual ESO polymeric materials. The results also showed that some other mechanical characteristics like toughness and strain at break improved for the resulting ESO/PHB composite solely due to the incorporation of the PHB monolith. From all the above discussions, it can be concluded that the production of biodegradable blends and polymers is the result of current research trends, but the application necessitates properties of the blends to undergo further development (Ramesh et al., 2021; Plavec et al., 2020; Ramesh et al., 2021).

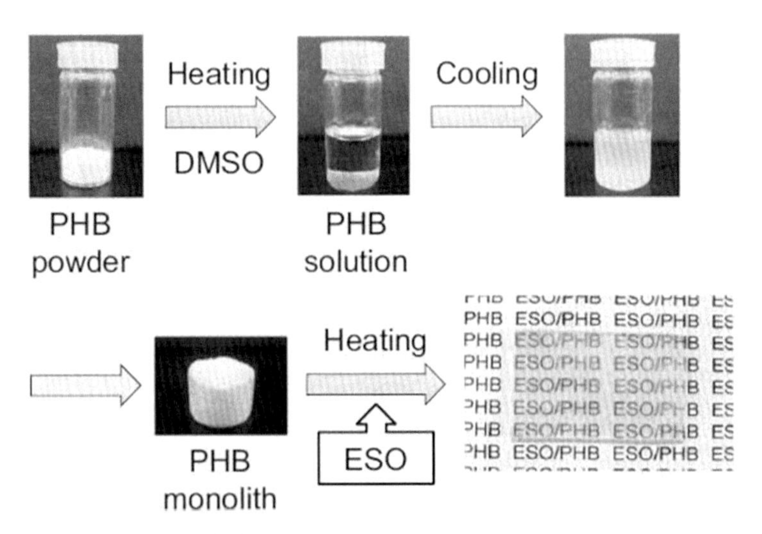

Figure 19.5 ESO/PHB synthesis process.
Adapted from (n.d.). https://doi.org/10.1038/s41428-018-0097-8

3. Manufacturing of biopolymer blends

Composites, which include biodegradable polymers, fall into this exclusive sector of polymers that are environmentally friendly (biocompatible and decomposable) as shown in Fig. 19.6. These biopolymers can be produced according to four dissimilar groupings based on the types of anticipated products and the availability of the necessary ingredients or starting materials. The classifications are the four biochemical amalgamation approaches, the four microbial amalgamation approaches, the four biopolymer mixtures, and the nonconsumable resources. A commercially relevant application for biodegradable polymers will select biodegradable polymers associated with the physical characteristics of the polymers. The utmost importance of ductile strength and yield strength is witnessed in construction-related situations (Ojogbo et al., 2020; Ramesh et al., 2021).

3.1 Production from agricultural wastes

When looking at whether a plant-based polymeric starting material can be produced from agricultural wastes, two variables should be taken into consideration: the availability of the material in sufficient amounts, and the cost of the material. Since sewage sludge is a starting point for raw material, that is used in the development of nitrogen adhesives, biocides, and oxidative additives, agriculture has a positive impact on the environment. As a result, potato vegetable oils are key sources of cellulose, which are the appropriate replacements of synthetic plasticizers. The most important task of the additives is to assist in the development of permeability and corrosion resistance in biobased polymers. There has been little scientific investigation done into the effectiveness of potato phospholipid fillers compared to phenol and other artificial plasticizers, and this has limited their commercial applications. The utilization of agricultural waste products like obtains of tamarind seeds, fruit juice leaf extract,

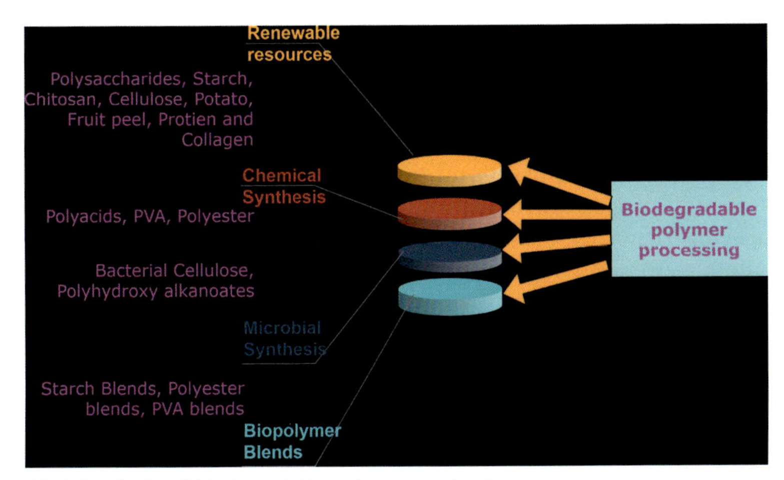

Figure 19.6 Methods of biodegradable polymer production.

proto-catechuic acidic, seed pulled out of grape fruit, and derivatives of turmeric seeds are used in the expansion of antioxidant essences. Isothymalonitrile (IT), tansy herb root extracts (TE), basil leaf extract (BE), ginger root extract (GE), burdock root extract (BT), lettuce leaf extract (LE), red raspberry leaf extract (RR), yellow dock root extract (DD), grape seed extract (GE), olive leaf extract (LE), arugula leaf extract (AR), oregano leaf extract (OR), thistle extract (ST), jalapeno peppers (also known as cayenne peppers), noni fruit, berries, purple carrots, mangos, guavas, and dragon fruit (ME) are some of the most commonly used plant based biopolymer materials for the preparation of biopolymer composites. To synthesize polymers from lignocellulose, a combination of acid and alcohol products of cellulose, polylactic acid, and polyhydroxyalkanoates, for the treatment of polymers are basically derived from bioadditive precursors containing lignocellulose fibers, cellulose esters, polylactic acid, and polyhydroxyalkanoates (PHA). Plant fibers like those found in cotton, flax, jute, kapok, and sugarcane are commonly used in the manufacture of plastic substitutes. The physical qualities of the final good are primarily enabled by the method of extraction. Due to their ability to increase the mechanical strength, sources of energy of polysaccharides, starch, chitosan, cellulose, potato and cassava peel, fruit peels, proteins, and collagen are preferred. The polymer blends described above are based on polyethylene glycol (PEG), polyvinyl alcohol (PVA), polyvinyl acetate (PVAc), or poly-acrylamide (PAA). Conversely, even if the fiber is a biobased substance, the originator, derived from biomass, is nonbiodegradable because of the superior level of replacement. This energy source differs from other energy sources, which are derived from plants, as waste materials contains of garbages, mostly by nourishment dispensation such as coconut shells, potato peels, fruit peels, and fruit seeds, which have traditionally been discarded as waste in farms and food processing facilities (Priya et al., 2020; Ramesh et al., 2021a).

3.2 Production of biopolymers from vegetable extracts and winery wastes

Winery agro-waste in the form of pomace (grape pomace) is the primary product of Merlot grapes. An alternative to allowing them to decay, vineyard agricultural wastes serve as a proper foundation of amalgams that are processed using coagulation methods, and pressurized liquid extraction (PLE). These polyhydroxyalkanoates (PHA) excerpts are blended with marketing polyhydroxyalkanoate (PHA) to form the polyhydroxyalkanoate (PHA) matrix. Finally, during the final stage of production, the biopolymer is mixed with PHBV, a copolyester that consists of hydroxyvaleric acid to form active biocomposites. It should be noted that dried sugar beet pulp and grape pomace are excellent foundations for biocomposites related to the addition of carbocal in the dried mash (Gul et al., 2020). Carbocal properties are enhanced with the addition of an LLDPE-carbocal biopolymer that is created through blending, grinding, and washing of leaf mulch. An assessment of the ductile characteristics revealed that a maximum carbocal content enhanced the Young's modulus, but at the expense of the elongation at break. Very little necking and a small amount of plastic deformation were found. The types of bacteria that affect the growth of lactic acid and poly-lactic acid for fermentation and hydrolysis are influenced by specific strains and the affordability of agro-wastes used as the raw material (Ramesh et al., 2021a). An important strain in the commercial applications of fungi and bacteria is *Rhizopus*, and also *Pediococcus* and *Streptococcus* (Ramesh et al., 2021a). As the number of different bacteria and fungi species that are available increases, so does the influence on the final product's material properties (including biological properties, morphological characteristics, and psychological characteristics) by virtue of the various fermentation methods that are used in making fermentable sugars, such as starch and fiber.

Pineapple peel is utilized as a source of biopolymers for the manufacture of biobased polymers, which involves a standard process that consists of the removal of biopolymers from farming waste. To start the chemical proportion is analyzed, specifically the C/N and C/P ratios, which indicates what the yields of polymer will be. This approach, using tomato pomace for poly-condensation and pineapple peels for poly-esterification, resembles the method used to produce polymers from pineapple skin (Ramesh et al., 2021). The entire process is common when excluding the melting poly-condensation step. Excluding the processing of biopolymer skin of fruits has proven to be useful in improving the mechanical characteristics of the produced polymers. Based on the above, some authors have hypothesized that the inclusion of lemon skin powder at from 10 to 30 percent and sweet lime skin powder works to enhance the mechanical durability of the actual fibers and epoxy gums. Also, there was a good distribution of particles and good adhesion of particles to the matrix. Although the lemon and its skin exhibited good characteristics in the encouragement of the constructions, the durability of the lemon and sweet lime fruits remains questionable; as lemon and sweet lime fruits are edible, the increased usage of skin of these fruits is uncertain. In high-value marketplaces, the fruit skins are used to add significant value-added foodstuffs, such as bioactive polyphenols (Boyacioglu et al., 2020). Antioxidant capabilities have been shown in the scientific literature to be present in phenol-containing compounds (Muthuraj et al., 2018). Another option is that the skins are used in the formation of cosmetic products that are manufactured at home.

3.3 Microbial synthesis and chemical extraction

Biodegradable polymers are also produced by Gram-negative and Gram-positive bacteria when given the proper conditions, which include nutrients and carbon-rich waste. Bacterial poly(chain) production is activated by a variation in pH values, minimal availability of key sources such as phosphorus and nitrogen, and the specific conditions in which the culture is kept (Zhong et al., 2020). Other factors that influence the rate of bacterial poly(chain) production include the bacteria's nutritional requirements, how they are maintained, and what type of media they are in (Rogovina et al., 2020). Microalgae are essential to the biological carbon fixation procedures that result in the growth of biopolymers, which in turn allows for the development of carbon-based biopolymers. The last step in the polysaccharide formation process is the production of branched polysaccharides. Currently, the most widely used biobased biopolymer is PHA, which is synthesized from microbes. Rice bran is naturally a mixture of amino acids, carbohydrates, proteins, and other biobased polymers.

Sinorhizobium meliloti MTCC 100 bacteria catalyzes the formation of these polymers. Bacteria used in the study were selected due to their potential influence on the atmosphere being less significant compared to other species, and also due to their production in a significant number of agricultural wastes. Bacteriological combination techniques have also shown effectiveness in the treating of poly-β-hydroxybutyric acid (PHB)—a biodegradable and elevated potent PHA biopolymer (Ramesh, Rajeshkumar, Bhoopathi, Vigneshkumar, & Elango, 2021). The amalgamation of the biopolymer is reliant on the accessibility of a carbon-rich precursor, in which the bacteria is used as a foundation of foodstuffs and energy. Unlike other microbial-synthesized biopolymers, PHB has good mechanical properties, making it appropriate for maximum-strength sectors. In particular, it has comparable mechanical characteristics to petroleum-based biopolymers such as PP.

In order to obtain the lignin and cellulose resources from agro-wastes/food wastes, the compounds involved in chemical synthesis methods involve the use of acids and alkalis. The biopolymer mechanical properties are determined by the characteristics of the treatment processes [formation of practical collections (C=O, and C—O—C, among others), which impact the mechanical characteristics of the biopolymer with the help of the chemical additives]. There are, however, some negatives associated with the chemical process. One is that it produces the by-products of furan, carboxylic acids, and lignin-derived phenols that have a negative effect on enzymatic activity, which is a crucial part of the fermentation phase (Ramesh et al., 2021). Inhibiting the growth of acid- and alkali-tolerant yeast or molds through the addition of a variety of white-rot fungus called *Ceriporiopsis subvermispora* prior to the fermentation process helps to eliminate the negative impact on the fermentation process (ÇELİK & Eylem, 2020). While using microorganisms in chemical amalgamation does not invalidate the claim that artificial chemicals are hazardous to the atmosphere and lessen the derivatives of using agricultural precursors in place of PE, PET, and PP, the use of these microbes in chemical synthesis has not been proven to be a viable method for solving the world's increasing needs for industrial chemicals.

4. Biodegradable composites

4.1 Use of natural fibers as reinforcements

Some experimenters found that the consequence of flax fibers on the mechanical characteristics of biodegradable polymers (polyesters, polysaccharides, and starch blends) had been conveyed by a very few researchers previously. The extrusion process was used to prepare the composites. The fiber content varied from 10% to 35% with respect to the total weight of the composite. The ductile strength of the strengthened polymer was improved due to a chemical interface between polysaccharides and plant fibers. Thermoplastic wheat starch was reinforced with a fourfold increment (37 N/mm^2) with additional material (in this case, additional plastic). Filling reinforced with a mixture of cellulose diacetate and starch blend causes stress to increase from 52% to 55 N/mm^2 and from 25% to 64 N/mm^2. The thickness (between 40 and

2000 μm) has a significant impact on the weight loss, biodegradability, and degrees of degradation in pure PEA films and PEA/flax fiber composites, with numerous experiments that describe these outcomes (Di & Laura, 2020). The combination of improved transportation, circulation of water, and incorporating original fibers into the polymer matrix allowed biodegradation of the PEA matrix. The weight loss percentage of PEA was 35% in this case, due to the degradation of the PEA films of 500 μm or less within 6 weeks, and the thickness of PEA films in this case was approximately 2000 μm. While pure PEA films degraded at a slower rate, PEA/flax composites deteriorated at a faster rate. The samples were examined with microscopic techniques to learn more about their characteristics.

Only a small number of authors have investigated the mechanical nature of jute fiber-strengthened biodegradable polyester composites. A comparison of the properties of the various natural fiber-based thermoplastic composites using experimental research of the technical properties has been carried out. The jute fibers were imbedded within the polyester matrix and exhibited distinct differences in their chemical characteristics, which can be determined through morphological analysis. It has been previously reported that biodegradable PEA matrix composites were strengthened with jute fibers which were superficially altered. The different processes employed in treating jute fabrics, such as bleaching, dewaxing, alkali treatment, cyanothylation, and vinyl grafting, are referred to as surface modifications. To supplement their research, the researchers looked into different surface treatments and fiber compositions also. Images obtained with an scanning electron microscope showed that the great fiber–matrix interaction was apparent to the naked eye. This additional strength is due to surface modifications, and it allows for more stress and strain to be placed on the product. When alkali treatment and cyanoethylation were performed, the composites improved in terms of physical properties when related to another changes. The tensile strength of alkali-treated composites increased by 40% when related to a PEA matrix that had not been treated with alkali. While studies of weight loss show that it is directly proportional to the stability of the prepared composites, degradation studies show that it is linearly proportional to the strength of the composites. For developing biocomposites based on blends of PLA, PBS, and bamboo fiber (BF), where lysine-based diisocyanate (LDI) was used as a coupling agent, it was found that lysine was a component of bamboo fiber. It appears that the degradation rate of LDI-added biocomposites was slower than that of the enzymatic decay of the same. This integration of LDI into the mixtures shows improved tensile properties, water resistance, interface interaction, crystallization temperature, enthalpy, thermal degradation temperature, and a reduction in the fusion heat. The cross-linking among the matrix and the fibers works to quench the thermal flow, and no variation was observed at an elevated temperature (Balaji et al., 2020; de Lima Barizão et al., 2020; Ramesh et al., 2021; Rahman et al., 2020).

Various experimenters have reported on compositions that use surplus timber and phthalic anhydride (PA)-modified PVA with a plasticizer. Suji flour (also called semolina, which can be abrasive, with extracted fibers of hard wheat, other variants of wheat, and other foods, such as rice and maize) was also added to the PA/PVA matrix. The PA/PVA ratio affected all of the aforementioned changes in composite structure,

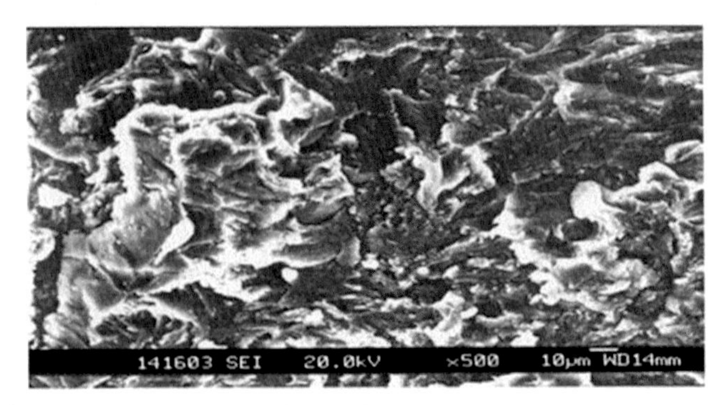

Figure 19.7 SEM image of fractured fir saw dust.
Adapted from (n.d.). https://doi.org/10.1007/s10924-017-0985-6

as well as all of the following: elasticity, rupture, elastic modulus, mechanical properties, and weight loss. Soil burial tests were used to measure the rate of degradation and then an FTIR and SEM analysis of the resulting composite was carried out. The composites prepared beforehand demonstrated a reduction in their mechanical properties (Jayanth et al., 2018). Since it is possible to add mechanical and decaying characteristics to alter the wood powder thermoplastics, there are numerous practical applications for these thermoplastics. It was discovered that when the response heat, the proportion of the aqueous corrosive solution, and the phase change catalyst were modified, there were considerable variations in the thermal and mechanical properties. The SEM micrograph in Fig. 19.7 shows raw fir saw dirt, benzylated fir saw dirt, and surfaces of molded pieces that have been fractured.Petroleum-based constructional applications, with reasonable mechanical properties and maximum biodegradability, are considered to be an alternative to the use of composites in structural applications (Chen et al., 2021; Gonçalves et al., 2020; Ramesh et al., 2021; Tessanan et al., 2020).

5. Conclusion and future perspectives

The removal of polymer products that are not degradable has an environmental impact including the burning of fossil fuels. Biopolymer companies producing financially significant amounts of agro-waste, where the waste occurs at grocery and farm level, and that there is no waste filtering as well as disposal mechanism. Furthermore, the accessibility of agricultural wastes varies globally which is a key component that impacted the final characteristics of the composites developed. The above-mentioned problems have seen great progress in the area of recyclable plastic. Bioavailability and sensory attributes are among the characteristics that can be found in the latter products. Nanoparticles are used to improve the quality of compostable carbon fibers, and these compostable nanoparticles have a wider range of applications. Researchers have developed biodegradable materials using renewable resources and crops, and several studies have produced reports illustrating the end applications of these materials in the automotive industry, medical field (tissue engineering, orthopedic), ceiling tiles, mailing lists, food packaging, biocoatings, construction, and other fields. The use of biopolymers in construction and agriculture is dependent on the accessibility of chemical polymers that strike a balance between the tensile properties, biodegradation, and environmental impact. Environmentally friendly nanostructures have primarily been developed for device applications, packaging, and biomedicine. However, due to the constraints in areas such as flexural strength, processability, and high strength, environmentally friendly equivalents remain limited when it comes to the implementation of standard polymer composites. Price and industrial considerations need to be given more consideration. Researchers will probably need to rely on nanofillers to improve composite processes in the future, and effective processes that are tricky for this genetic engineering to study. In the future, overcoming these constraints will create many opportunities for biodegradable products to become particularly suitable for various manufacturers and thus reduce (or even eliminate) the use of nonbiodegradable equipment. Compostable polymer composites could replace most existing materials in the future, which is critical for the sustainability of mankind; it is therefore an important challenge to analyze nanofillers in order to improve biodegradable lightweight structures.

References

Abioye, A. A., & Obuekwe, C. C. (2020). Investigation of the biodegradation of low-density polyethylene-starch Bi-polymer blends. *Results in Engineering, 5*, 100090.

Aziz, S. B., Hamsan, M. H., Kadir, M. F. Z., & Woo, H. J. (2020). Design of polymer blends based on chitosan: POZ with improved dielectric constant for application in polymer electrolytes and flexible electronics. *Advances in Polymer Technology*, 2020.

Balaji, D., Ramesh, M., Kannan, T., Deepan, S., Bhuvaneswari, V., & Rajeshkumar, L. (2020). Experimental investigation on mechanical properties of banana/snake grass fiber reinforced hybrid composites. *Materials Today: Proceedings, 42*, 350–355. https://doi.org/10.1016/j.matpr.2020.09.548

Balart, R., Montanes, N., Franco, D., Boronat, T., & Torres-Giner, S. (2020). *Environmentally friendly polymers and polymer composites* (p. 4892).

Bastioli, C. (2020). In *Handbook of biodegradable polymers*. Walter de Gruyter GmbH & Co KG.

Bhuvaneswari, V., Priyadharshini, M., Deepa, C., Balaji, D., Rajeshkumar, L., & Ramesh, M. (2021). Deep learning for material synthesis and manufacturing systems: a review. *Materials Today: Proceedings, 46*(9), 3263–3269. https://doi.org/10.1016/j.matpr.2020.11.351

Boyacioglu, S., Kodal, M., & Ozkoc, G. (2020). A comprehensive study on shape memory behavior of PEG plasticized PLA/TPU bio-blends. *European Polymer Journal, 122*, 109372.

Bulatović, V. O., Mandić, V., Kučić Grgić, D., & Ivančić, A. (2021). Biodegradable polymer blends based on thermoplastic starch. *Journal of Polymers and the Environment, 29*(2), 492–508.

ÇELİK, M., & Eylem, K. (2020). Usage of plant-based biopolyethylene in biocomposite production and polymer blends. *Journal of Textiles and Engineer, 27*, 119 (2020).

Chan, C. M., Vandi, L.-J., Pratt, S., Peter, H., Richardson, D., Werker, A., & Laycock, B. (2018). Composites of wood and biodegradable thermoplastics: a review. *Polymer Reviews, 58*(3), 444−494.

Chen, W., Qi, C., Yao, L., & Tao, H. (2021). The degradation investigation of biodegradable PLA/PBAT blend: thermal stability, mechanical properties and PALS analysis. *Radiation Physics and Chemistry, 180*, 109239.

de Lima Barizão, C., Crepaldi, M. I., de Oliveira, S. O., de Oliveira, A. C., Martins, A. F., Garcia, P. S., & Bonafé, E. G. (2020). Biodegradable films based on commercial κ-carrageenan and cassava starch to achieve low production costs. *International Journal of Biological Macromolecules, 165*, 582−590.

Di, L., & Laura, M. (2020). Poly (l-Lactic acid)/poly (butylene succinate) biobased biodegradable blends. *Polymer Reviews*, 1−37.

Gonçalves, L. M. G., Rocha Rigolin, T., Maia Frenhe, B., & Helena Prado Bettini, S. (2020). On the recycling of a biodegradable polymer: multiple extrusion of poly (lactic acid). *Materials Research, 23*(5).

Gul, S., Awais, M., Jabeen, S., & Farooq, M. (2020). Recent trends in preparation and applications of biodegradable polymer composites. *Journal of Renewable Materials, 8*(10), 1305−1326.

Guo, Y., Zuo, X., Yuan, X., Tang, J., Gouzman, M., Fang, Y., Zhou, Y., Wang, L., Yu, Y., & Rafailovich, M. H. (2020). Engineering thermally and electrically conductive biodegradable polymer nanocomposites. *Composites Part B: Engineering, 189*, 107905.

Haider, T. P., Völker, C., Kramm, J., Landfester, K., & Wurm, F. R. (2019). Plastics of the future? The impact of biodegradable polymers on the environment and on society. *Angewandte Chemie International Edition, 58*(1), 50−62.

Iqbal, N., Samad Khan, A., Asif, A., Yar, M., Haycock, J. W., & Ur Rehman, I. (2019). Recent concepts in biodegradable polymers for tissue engineering paradigms: a critical review. *International Materials Reviews, 64*(2), 91−126.

Jayanth, D., Sathish Kumar, P., Chandra Nayak, G., Kumar, J. S., Pal, S. K., & Rajasekar, R. (2018). A review on biodegradable polymeric materials striving towards the attainment of green environment. *Journal of Polymers and the Environment, 26*(2), 838−865.

Jiang, T., Duan, Q., Zhu, J., Liu, H., & Long, Y. (2020). Starch-based biodegradable materials: challenges and opportunities. *Advanced Industrial and Engineering Polymer Research, 3*(1), 8−18.

Kargarzadeh, H., Huang, J., Lin, N., Ahmad, I., Mariano, M., Dufresne, A., Thomas, S., & Gałeski, A. (2018). Recent developments in nanocellulose-based biodegradable polymers, thermoplastic polymers, and porous nanocomposites. *Progress in Polymer Science, 87*, 197−227.

La, M., Paolo, F., Ceraulo, M., Testa, P., & Morreale, M. (2021). Biodegradable polymers for the production of nets for agricultural product packaging. *Materials, 14*(2), 323.

Laycock, B., Nikolić, M., Colwell, J. M., Gauthier, E., Peter, H., Bottle, S., & George, G. (2017). Lifetime prediction of biodegradable polymers. *Progress in Polymer Science, 71*, 144−189.

Lee, L.-T., He, S.-P., & Huang, C.-F. (2020). Enhancement of crystallization behaviors in quaternary composites containing biodegradable polymer by supramolecular inclusion complex. *Crystals, 10*(12), 1137.

Maghsoudi, S., Taghavi Shahraki, B., Rabiee, N., Fatahi, Y., Dinarvand, R., Tavakolizadeh, M., Ahmadi, S., et al. (2020). Burgeoning polymer nano blends for improved controlled drug release: a review. *International Journal of Nanomedicine, 15*, 4363.

Mangaraj, S., Yadav, A., Bal, L. M., Dash, S. K., & Mahanti, N. K. (2019). Application of biodegradable polymers in food packaging industry: a comprehensive review. *Journal of Packaging Technology and Research, 3*(1), 77–96.

Maraveas, C. (2020). Production of sustainable and biodegradable polymers from agricultural waste. *Polymers, 12*(5), 1127.

Muthuraj, R., Misra, M., & Kumar Mohanty, A. (2018). Biodegradable compatibilized polymer blends for packaging applications: a literature review. *Journal of Applied Polymer Science, 135*(24), 45726.

Nakajima, H., Dijkstra, P., & Loos, K. (2017). The recent developments in biobased polymers toward general and engineering applications: polymers that are upgraded from biodegradable polymers, analogous to petroleum-derived polymers, and newly developed. *Polymers, 9*, 523–610.

Ojogbo, E., Ogunsona, E. O., & Mekonnen, T. H. (2020). Chemical and physical modifications of starch for renewable polymeric materials. *Materials Today Sustainability, 7*, 100028.

Omaníková, L., Bočkaj, J., Černák, M., Plavec, R., Feranc, J., & Jurkovič, P. (2020). Influence of composition and plasma power on properties of film from biodegradable polymer blends. *Polymers, 12*(7), 1592.

Papageorgiou, G. Z. (2018). *Thinking green: sustainable polymers from renewable resources* (p. 952).

Plavec, R., Hlaváčiková, S., Omaníková, L., Feranc, J., Vanovčanová, Z., Tomanová, K., Bočkaj, J., et al. (2020). Recycling possibilities of bioplastics based on PLA/PHB blends. *Polymer Testing, 92*, 106880.

Prajapati, S. K., Jain, A., Jain, A., & Jain, S. (2019). Biodegradable polymers and constructs: a novel approach in drug delivery. *European Polymer Journal, 120*, 109191.

Priya, C. N., Muruganandham, R., Muthuvinayagam, M., & Vahini, M. (2020). Studies on pectin-polyvinyl alcohol–based biodegradable polymer blend electrolytes. *Materials Performance and Characterization, 9*(1), 692–700.

Rahman, M., Paul, S., & Mahbub, F. (2020). Biodegradability and composite coatings: past, present and future prospects. In *Composite materials: applications in engineering, biomedicine and food science* (pp. 399–415). Cham: Springer.

Ramesh, M., & Rajeshkumar, L. (2018). Wood flour filled thermoset composites. Thermoset composites: preparation, properties and applications. *Materials Research Foundations, 38*, 33–65.

Ramesh, M., Deepa, C., Niranjana, K., Rajeshkumar, L., Bhoopathi, R., et al. (2021). Influence of Haritaki (Terminalia chebula) nano-powder on thermo-mechanical, water absorption and morphological properties of Tindora (Coccinia grandis) tendrils fiber reinforced epoxy composites. *Journal of Natural Fibers*. https://doi.org/10.1080/15440478.2021.1921660. In press.

Ramesh, M., Deepa, C., Rajesh Kumar, L., Sanjay, M. R., & Siengchin, S. (2020). Life-cycle and environmental impact assessments on processing of plant fibres and its biocomposites: a critical review. *Journal of Industrial Textiles*. https://doi.org/10.1177/1528083720924730, 1528083720924730.

Ramesh, M., Deepa, C., Rajeshkumar, L., et al. (2021). Influence of fiber surface treatment on the tribological properties of Calotropis gigantea plant fiber reinforced polymer composites. *Polymer Composites, 42*(9), 4308–4317. https://doi.org/10.1002/pc.26149

Ramesh, M., Deepa, C., Tamil Selvan, M., Rajeshkumar, L., Balaji, D., & Bhuvaneswari, V. (2020). Mechanical and water absorption properties of Calotropis gigantea plant fibers reinforced polymer composites. *Materials Today: Proceedings, 46*(9), 3367–3372. https://doi.org/10.1016/j.matpr.2020.11.480

Ramesh, M., & Rajesh Kumar, L. (2020). Bioadhesives. In Inamuddin, et al. (Eds.), *Green adhesives: preparation, properties and applications* (pp. 145—164). Scrivener Publishing.

Ramesh, M., Rajesh Kumar, L., Khan, A., & Mohamed Asiri, A. (2020). Self-healing polymer composites and its chemistry. In *Self-healing composite materials* (pp. 415—427). Woodhead Publishing.

Ramesh, M., Maniraj, J., & Rajesh Kumar, L. (2021). Biocomposites for energy storage. *Bio-based Composites: Processing, Characterization, Properties, and Applications*, 123—142.

Ramesh, M, & Rajeshkumar, L (2021). Case-Studies on Green Corrosion Inhibitors. In Inamuddin, et al. (Eds.)*, 107. Sustainable Corrosion Inhibitors* (pp. 204—221). Materials Research Foundations.

Ramesh, M, & Rajeshkumar, L (2021). Technological Advances in Analyzing of Soil Chemistry. In *Applied Soil Chemistry: 1. Applied Soil Chemistry* (pp. 61—78). John Wiley & Sons.

Ramesh, M., Rajeshkumar, L., Balaji, D., et al. (2021). Self-Healable Conductive Materials. In Inamuddin, et al. (Eds.), *Self-Healing Smart Materials and Allied Applications* (pp. 297—319). John Wiley & Sons.

Ramesh, M, Rajeshkumar, L, & Balaji, D (2021). Influence of Process Parameters on the Properties of Additively Manufactured Fiber-Reinforced Polymer Composite Materials: A Review. *Journal of Materials Engineering and Performance, 30*(7), 4792—4807. https://doi.org/10.1007/s11665-021-05832-y

Ramesh, M., Rajeshkumar, L., & Balaji, D. (2021). Aerogels for insulation applications. *Aerogels II: Preparation, Properties and Applications, 98*, 57—76.

Ramesh, M., Rajeshkumar, L., Bhoopathi, R., Vigneshkumar, N., & Elango, K. S. (2021). Carbon substrates: a review on fabrication, properties and applications. *Carbon Letters, 31*, 557—580. https://doi.org/10.1007/s42823-021-00264-z

Ramesh, M., RajeshKumar, L., & Bhuvaneshwari, V. (2021). Bamboo fiber reinforced composites. In *Bamboo fiber composites* (pp. 1—13). Singapore: Springer.

Ramesh, M., Rajeshkumar, L, & Balaji, D (2021). Mechanical and Dynamic Properties of Ramie Fiber-Reinforced Composites. In Rajini Nagarajan, et al. (Eds.), *Mechanical and Dynamic Properties of Biocomposites* (pp. 275—322). John Wiley & Sons.

Ramesh, M., Rajeshkumar, L., Balaji, D., & Bhuvaneswari, V. (2021). Green composite using agricultural waste reinforcement. In *Green composites* (pp. 21—34). Singapore: Springer.

Ramesh, M, Rajeshkumar, L, Deepa, C, Tamilselvan, M, et al. (2020). Impact of Silane Treatment on Characterization of *Ipomoea Staphylina* Plant Fiber Reinforced Epoxy Composites. *Journal of Natural Fibers*. https://doi.org/10.1080/15440478.2021.1902896. In press.

RameshKumar, S., Shaiju, P., & O'Connor, K. E. (2020). Bio-based and biodegradable polymers-State-of-the-art, challenges and emerging trends. *Current Opinion in Green and Sustainable Chemistry, 21*, 75—81.

Rogovina, S., Zhorina, L., Gatin, A., Prut, E., Kuznetsova, O., Yakhina, A., Olkhov, A., et al. (2020). Biodegradable polylactide—poly (3-hydroxybutyrate) compositions obtained via blending under shear deformations and electrospinning: characterization and environmental application. *Polymers, 12*(5), 1088.

Saravana Kumar, A., Maivizhi Selvi, P., & Rajeshkumar, L. (2017). Delamination in drilling of sisal/banana reinforced composites produced by hand lay-up process. In *Applied mechanics and materials* (vol. 867, pp. 29—33).

Sathish, S., Kumaresan, M., Karthi, N., & Dhilip Kumar, T. (2014). Tensile and impact properties of natural fiber hybrid composite materials. *International Journal of Modern Engineering Research, 4*, 9—12.

Sazali, N., Ibrahim, H., Jamaludin, A. S., Mohamed, M. A., Salleh, W. N. W., & Abidin, M. N. Z. (2020). Degradation and stability of polymer: a mini review. In *IOP conference series: materials science and engineering* (vol. 788, p. 012048). IOP Publishing, 1.

Silva, M., Ferreira, F. N., Alves, N. M., & Paiva, M. C. (2020). Biodegradable polymer nanocomposites for ligament/tendon tissue engineering. *Journal of Nanobiotechnology, 18*(1), 23.

Tessanan, W., Chanthateyanonth, R., Yamaguchi, M., & Phinyocheep, P. (2020). Improvement of mechanical and impact performance of poly (lactic acid) by renewable modified natural rubber. *Journal of Cleaner Production, 276*, 123800.

Thirmizir, M. Z. A., Mohd Ishak, Z. A., & Salim, M. S. (2020). Compatibilization and cross-linking in biodegradable thermoplastic polyester blends. In *Reactive and functional polymers volume two* (pp. 23–89). Cham: Springer.

Uyama, H. (2018). Functional polymers from renewable plant oils. *Polymer Journal, 50*(11), 1003–1011.

Wróblewska-Krepsztul, J., Rydzkowski, T., Gabriel, B., Szczypiński, M., Klepka, T., & Thakur, V. K. (2018). Recent progress in biodegradable polymers and nanocomposite-based packaging materials for sustainable environment. *International Journal of Polymer Analysis and Characterization, 23*(4), 383–395.

Zarrintaj, P., Reza Saeb, M., Jafari, S. H., & Mozafari, M. (2020). Application of compatibilized polymer blends in biomedical fields. In *Compatibilization of polymer blends* (pp. 511–537). Elsevier.

Zhao, L., Rong, L., Zhao, L., Yang, J., Wang, L., & Sun, H. (2020). Plastics of the future? The impact of biodegradable polymers on the environment. *Microplastics in Terrestrial Environments: Emerging Contaminants and Major Challenges*, 423–445.

Zhong, Y., Godwin, P., Jin, Y., & Xiao, H. (2020). Biodegradable polymers and green-based antimicrobial packaging materials: a mini-review. *Advanced Industrial and Engineering Polymer Research, 3*(1), 27–35.

Zuo, X., Yuan, X., Zhou, Y., Yin, Y., Wang, L., Chuang, Y.-C., Chang, C.-C., Rafailovich, M. H., & Guo, Y. (2020). The use of low cost, abundant, homopolymers for engineering degradable polymer blends: compatibilization of poly (lactic acid)/styrenics using poly (methyl methacrylate). *Polymer, 186*, 122010.

Electrically conductive biodegradable polymer blends and composites

Ravi Prakash Magisetty [1,2,3], *Aarsha Surendren* [1,2,3], *Naga Srilatha Cheekuramelli* [1,2,3] and *Radhamanohar Aepuru* [1,2,3]
[1]Materials Engineering Division, Defence Institute of Advanced Technology, Pune, Maharashtra, India; [2]Central Institute of Petrochemicals Engineering and Technology (CIPET): Institute of Plastics Technology (IPT), Kochi, Kerala, India; [3]Departamento de Ingeniería Mecánica, Facultad de Ingeniería, Universidad Tecnológica Metropolitana, Santiago, Chile

1. Introduction

The utilization of polymeric materials is increasing rapidly due to their functional characteristics in rendered potential applications and has become a part of everyday life. Recently, electrically conducting polymers have been successfully employed in various electronic applications, including photovoltaics (PVs) (Sims et al., 2012), light-emitting devices (LEDs) (Mathai et al., 2002), and field-effect transistors (FETs) (Horowitz, 1998). However, the environment and the ecosystem's sustainable developments require electrical conducting biodegradable polymeric systems. These biodegradable materials are useful also in the field of biomedical engineering. The additional functionality of biocompatibility facilitates possible biomedication with a compatible biological system (Grainger, 1999). Therefore, researchers are looking for alternatives as electrically conductive biodegradable polymers, blends, and composites (Bogoeva-Gaceva & Dimko, 2013).

These advanced functional materials come under the heading of biomaterials, and it is postulated that the biological material-induced interactions facilitate the medication or ecosystem to degrade it (Biomaterials, 2009). Thus, researchers have described various kinds of materials either experimentally and/or theoretically. The quantified responses interact with the biological systems and/or ecosystem and the materials meet the eco-friendly and bio needs.[1] Further, they suggested that polymers possess significant properties that can be altered by flexibly controlling their molecular chemistry and chemical structure, leading to great diversity and versatility of chemical and physical functionalities. Since then, various synthetic methods have been developed to meet the demand for a clean and green, sustainable pollution-free environment with the reduction of fossil fuel utilization. Also, investigators are discovering new avenues to

[1]Bio needs means the biomaterial exibits biologically compatible functional properties.

Biodegradable Polymers, Blends and Composites. https://doi.org/10.1016/B978-0-12-823791-5.00021-1

replace plastics with biodegradable materials via renewable energy sources. The developed biomaterials suffer from poor mechanical stability, deprived long-term durability, low chemical resistance, and other functional characteristics. To overcome these drawbacks, newly available materials that can tolerate and withstand a wide range of functional applications are essential. Recently, researchers have demonstrated biopolymeric material blends or other kinds of filler systems such as micro/nano-sized fillers incorporated or reinforced to produce biocomposites or nanobiocomposites. Significantly, nano-level morphologies exhibit greater precise surface area, lower density, and optimum surface energy than the conformist micromorphologies rendered with advanced functional properties via well-established materialized interactions that are required for some applications. Moreover, composites have attractive characteristics with low cost, and are replacements for metals and alloys in diversified applications. Recently, carbon filler-based electrically conducting biodegradable polymeric composite systems have been demanding functional applications (Ogunsona et al., 2017). Thus, these bionanocomposites are advantageous to the applications, include packaging, nanomedicine, energy, agriculture electronics, biomedical electronics, etc.

In this context, current advancements in electrically conductive biodegradable polymeric systems, particularly cyclopolymerized 1,6-heptadienes and their derivatives, have been described. Further, the biodegradable 1,6-heptadienes produced via various synthetic methods are described. In addition, for biodegradable polymeric systems, the need for electrically conducting biodegradable and biocompatible polymeric systems, and their applications are highlighted. This chapter describes the biodegradable polymeric systems with extensive functionalities suitable for eco-friendly and sustainable development.

2. Biodegradable polymeric systems, biodegradability, and compatibility

Biomaterials can be either biocompatible, that is, durable, or biodegradable, or both, that is, biocompatible and biodegradable. For example, biopolyethylene terephthalate is known as bio-PET. It is a nonbiodegradable but biocompatible material that has been available for some years. Then, based on extensive research and experimental investigations, researchers developed fissile fuel-based PET. For example, the best plant-based example is a plant bottle. This is made from plants and is fully recyclable and an alternative to the traditional bio-PET. Novel hybrid composites based on bio-PET with functional electronic conductivity could be a prominent composite material for electronic applications (Kuciel & Mazur, 2019). Similarly, polylactic acid (PLA) is a biodegradable polymer, and PLA-derived functionalized compounds and composites have been developed widely, with these materials facilitating the production of many commercially available electronic products, even microscale or nanoscale PLA-derived electronic products with different shapes can be realized with the aid of novel 3-dimensional additive printing technique, known as 3D printing (Inoue et al., 2007; Magisetty & Cheekuramelli, 2019; Raquez et al., 2013). This is a widely used polymer for many commercial and industrial products and protective packaging systems (Magisetty & Cheekuramelli, 2019). The World Food Development Authority has also recognized and approved PLA as a biodegradable polymer, that is, it suggested that this polymer can be useful as a biodegradable and compatible material for eco-friendly sustainable development. This instigated a path for developing a wide range of PLA-based biomedical electronic systems and wearable electronic devices (Geczy et al., 2011; Schramm et al., 2012). Also, PHA is a polyester-based polymer produced in nature by numerous microorganisms, including lipids or bacterial fermentation. This PHA is also biodegradable and offers to produce different PHA-derived plastic materials with novel properties, such as electronic properties (Sun et al., 2018; Philp et al., 2013). The above-delineated polymeric systems are approved as either biologically compatible, biodegradable, or both. However, although approved by the FDA, it does not mean that these materials degrade biologically when exposed to the environment. Biodegradation or biodegradability does not describe that the material or product degrades automatically under any environmental exposure. According to the FDA rules and regulations, biodegradation or biodegradability means that the right treatment is necessary at the product's end of life. In most cases, industrial composting with specific conditions is needed to ensure biodegradation within a reasonable period of time. Otherwise, it takes a longer time frame to degrade in open-environment conditions. Industrial compost needs to biodegrade within 6 months as per the European standards EN14995 and EN13432. According to these standards, compostable material needs to be approved before compositing in an industrial plant. Material compositing in industrial plants means that the material undergoes predefined conditions, which are a part of the biodegradation process according to the standards. Predefined conditions include time, temperature, humidity, fungi, bacteria, etc. In addition to a plastic material's biodegradability, material or product recycling or reusability are also important considerations. This is because composting of materials only makes sense for specific applications where recycling is too difficult. According to a recent UN report, the United States produces 6.3 million tons or 14% of the world's electronic waste. Worldwide, almost 45 million tons of electronic waste was discarded in 2016. Out of that staggering amount of electronic waste, only 20% was used with the remaining 80% making its way to industrial or landfill composting stations. This suggests that electronic waste composting is a better option than recycling because recycling is not economically feasible and is difficult to perform. However, among the products, those with different kinds of fossil and biobased plastics can be separated through physical and mechanical processes in recycling plants during the recycling process. The separated plastic material can be reused for new products after processing in the form of granulizing or remelting. During recycling, purity is an important consideration in order to reuse those materials for the production of products or devices. The purity of a material deals with its functionality for a particular application; if a material does not have 100% yield that material may reduce its application efficiency and reliability, therefore, the yield is most important when we consider functional properties such as conductivity in electronic applications. Some polymers, for example bio-PE or bio-PET, are chemically identical to that of fossil fuel-based versions of PE and PET. They can be perfectly integrated during an established recycling stream in order to maintain purity. However, in other cases, for example, biobased plastics which are not chemically identical and those that need to be recycled in separate streams for each material type, the purity of the recycle stream is important. In comparison to the fossil-based plastics, the CO_2 from increased biobased plastics was recently captured, in order to avoid greenhouse gas release into atmosphere, recycling of some plastics needs to be followed. Some countries have more stringent regulations involving product environmental claims when it comes considerations to biodegradable claims, with regulations to limit the use of certain items. This is due to warning labels stating that items have been found to cause cancer defects in newborns or in the reproductive system.

2.1 Biodegradability

The environmentally friendly or ecofriendly environment is more important to live and produce sustainable and eco-friendly communities. Sustainable environmental processes refer to goods and products that have minimal impact or harm upon the ecosystem. Industrial composting is one of the major processes to minimize harm to the ecosystem related to plastic and harmful material-based products. In industrial composting, the material heat is at a high enough temperature that allows microbes to break it down. Without that intense heat and critical conditions, the material will not degrade on its own in an acceptable time frame. Without the predefined industrial compositing environmental conditions, the probability of material degradation takes a long time. It starts after 100 years or more, with either in-home compost heaps or even landfills. If the long time frame degradation process starts or ends up in marine environments, they may degrade by breaking down into microsized pieces. These can last for decades, leading to unpredicted harmful conditions for marine life. This is dangerous for marine animals and the ecosystem. For example, PLA is not really biodegradable on its own or in the ocean, however it can be composted in an industrial facility. The compostable materials which are suitable for industrial composting have to meet the following specific requirements (Fig. 20.1).

Above all, it is suggested that biodegradable polymers are often not as biodegradable as they claim to be. It is always reliant on the surrounding environment. Therefore, controllable environmental conditions in industrial composting stations are required for biodegradation within a reasonable time frame. The biodegradation process is illustrated in Figs. 20.1 and 20.2. The stages of the biodegradation process and its end-product realization are described in Table 20.1.

2.2 Biocompatibility

Electrically conductive polymer blends and composites, particularly useful in biomedical electronics, are needed with a particular characteristic, which is biocompatibility. However, the understanding of biocompatibility has long been an experimental investigation. It is complicated because of problematic discretization, which exists between investigators thoughts and their experimental evaluations. According to the benchtop animal empirical conclusions and clinical investigation, researchers have pointed out the following findings (Fig. 20.3).

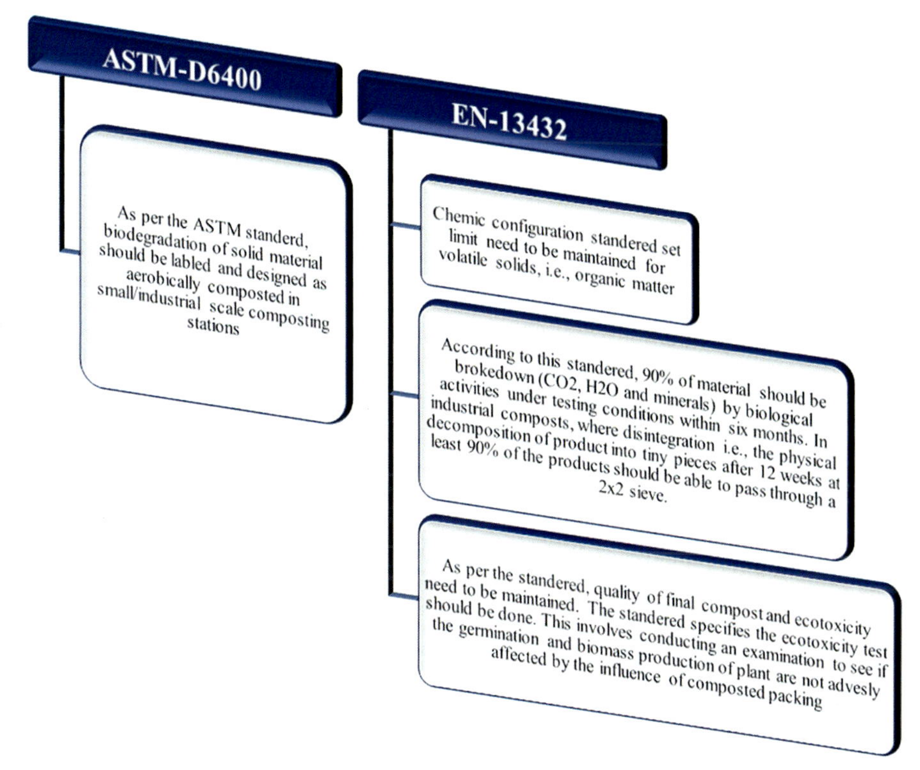

Figure 20.1 Biodegradability standards.

3. Necessity for electrically conducting biodegradable and biocompatible polymeric systems

More than 44 million tons of electronic and electrical waste were produced globally in 2017. A global eco-friendly and sustainable survey report stated that millions of tons of plastic waste are buried in landfill sites every year. This tremendously increasing e-waste leads to unrecoverable damage to the Earth's ecosystem and marine system. In 2019, one report stated that, globally, e-waste production is on track to reach 120 million tons per year by 2050 if the current trend is to continue. Of the total e-waste, less than 20% is formally recycled. The remaining 80% either ends up in landfill or is informally recycled. Exposure to these e-waste substances such as electrical polymers, blends, and composites with lead and cadmium substances to workers is hazardous and carcinogenic. Also, e-waste in landfills contaminates soil and/or groundwater, putting food supply systems and water sources at risk. If the long frame degradation process starts in marine environments, it can lead to abnormal marine life conditions, which are dangerous for aquatic life and the ecosystem. To avoid this e-waste and to protect our ecosystem and marine system, researchers are seeking permanent solutions by expanding the research and development into biodegradable

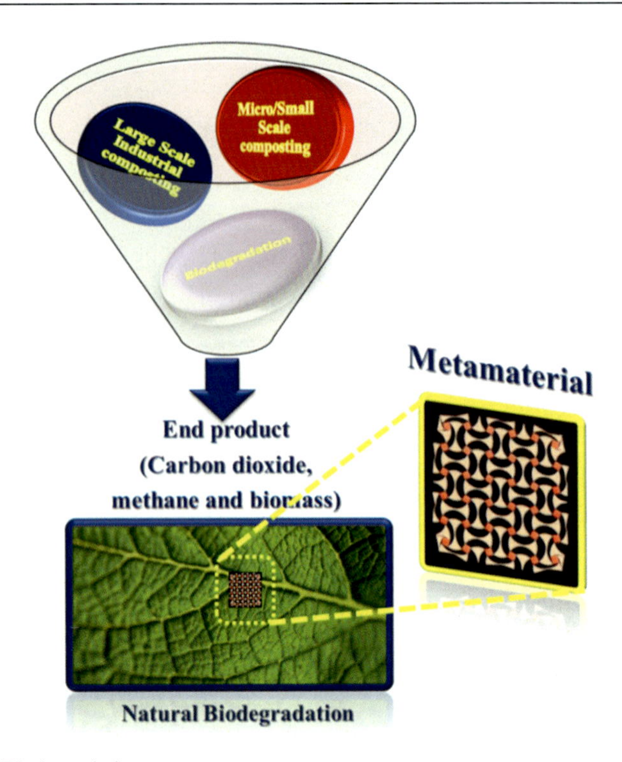

Figure 20.2 Biodegradation process.

materials (Waste, 2016). According to the ASTM and ISO standards, degradable polymers undergo chemical changes in their chemical structure by various conditional environments (Kolybaba et al., 2003). There are many sources of biodegradable plastics, from natural to synthetic polymers. Natural biopolymers are available in large quantities from renewable sources. At the same time, synthetic polymers are produced from nonrenewable sources such as petroleum resources, etc. Such biodegradable system's functional properties can be tailored by various synthetic routes. These have been widely explored and suggested for electrical and electronic applications. The blends and composites of different biopolymers are also expanding toward electrical and electronic applications due to their extraordinary multifunctionalities. Moreover, in the case of biomedical applications, biocompatibility is also a predominant parametric feature for the treatment. Biocompatibility is a quality of being compatible with living systems, and not toxic or injurious or causing immunological diseases. This kind of biocompatible feature, besides biodegradability, can be attained with grafting technology of compatible functional groups via various synthetic routes. The electrically conducting biodegradable and biocompatible polymeric system applications are shown in Fig. 20.4.

Table 20.1 Stages of the biodegradation process and its end-product realization.

Physicomechanical activities for degradation		
Name	**Process**	**Outcome**
Micro-, small-, medium-scale industrial composting (initial stage)	Biodegradation, i.e., breakdown/disintegration, occurs at the molecular scale of material by multiple factors, for example, water, wind, ultraviolet rays, and other weather	Carbon dioxide, methane, and biomass (takes much longer time to degrade)
Large-scale industrial composting (intermediate stage)	Biodegradation, i.e., breakdown/disintegration, takes place in an industrial facility to produce compost: Controlled environmental conditions, high temperatures	Carbon dioxide, methane, and biomass (the outcome of this process can be obtained in less time due to controlled environmental conditions)
Biodegradation (final stage complete degradation)	Chemical, thermal, and mechanical processes break down the material to carbon dioxide, methane, and biomass. The methods include UV degradation, thermo-oxidative degradation, microbiological degradation, chemical solubility-induced degradation, electrochemical degradation, and mechanical degradation	The end products are carbon dioxide, methane, and biomass

4. Electrically conducting biodegradable polymeric systems

Organic materials' conventional designs enable various tailorable functional characteristics in a controllable way and continue to be one of the most extensive domains in extant polymer exploration (Li et al., 2020; Probst et al., 2019; Podiyanachari et al., 2020). Since the initial realization of these organic polymers in the late 1970s, a wide range of organic conducting polymers for the ascribed applications have flourished due to their superlative functional electronic and photonic characteristics (Li et al., 2020; Probst et al., 2019; Podiyanachari et al., 2020; Liu et al., 2020). In recent years, miniaturization of electronic appliances has been at the forefront of research in the electronic semiconductor industry (Li et al., 2020; Probst et al., 2019). The nano/microscale development accumulation of electronic components for various

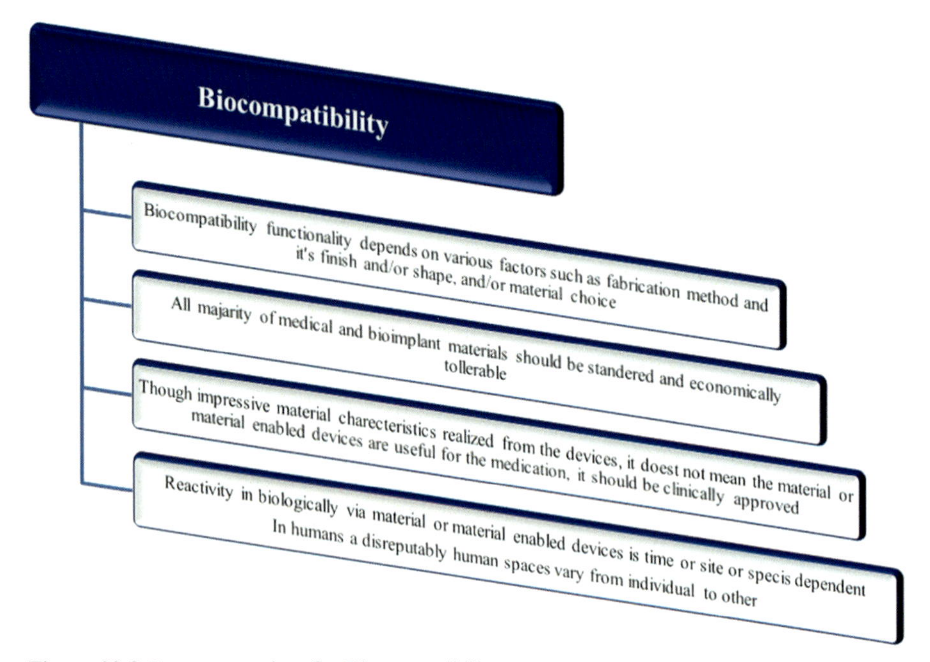

Figure 20.3 Important points for Biocompatibility.

diverse digitalized logical functions is crucial to the next-generation microelectronics. Owing to their flexibility, production cost, electrochemical characteristics, chemical stability, and other functional properties, conducting polymers have emerged as an alternative functional material to build these miniaturized electronic components and devices (Li et al., 2020; Probst et al., 2019; Podiyanachari et al., 2020). For example, conducting polymers have been successfully employed in photovoltaics (Sims et al., 2012), light-emitting devices (Mathai et al., 2002), and field-effect transistors (Horowitz, 1998). Many gas sensors are also made using polymers because of their design flexibility and the formation of selective layers in which the analyte gas and matrix interactions preeminently initiate the mechanism of physical parameter transduction (Gerard, 2002; McQuade et al., 2000). Recently, the discovery has been made that it is possible to control the conducting characteristics ranging from insulating polymers to metals by various synthetic methods. This has led to substantial efforts to produce conducting polymer materials for the electronic industry (Guimard et al., 2007; Probst et al., 2019; Pasini & Nitti, 2020). Among them, ring-closing metathesis (RCM) is a kind of cyclopolymerization technique. It is a more attractive and efficient method to synthesize functionalized macrocycle-instigated conducting polymers but has been less explored (Pasini & Nitti, 2020; Kang et al., 2011). 1,6-Heptadienes-based ring-closing polymer (RCP) synthesized via cyclopolymerization technique has been recently reported by Tae-Lim Choi and colleagues (Kang et al., 2011) Its applications need to be explored due to the probability of polyene cyclopolymerization providing a highly π-conjugated system in the polymer backbone

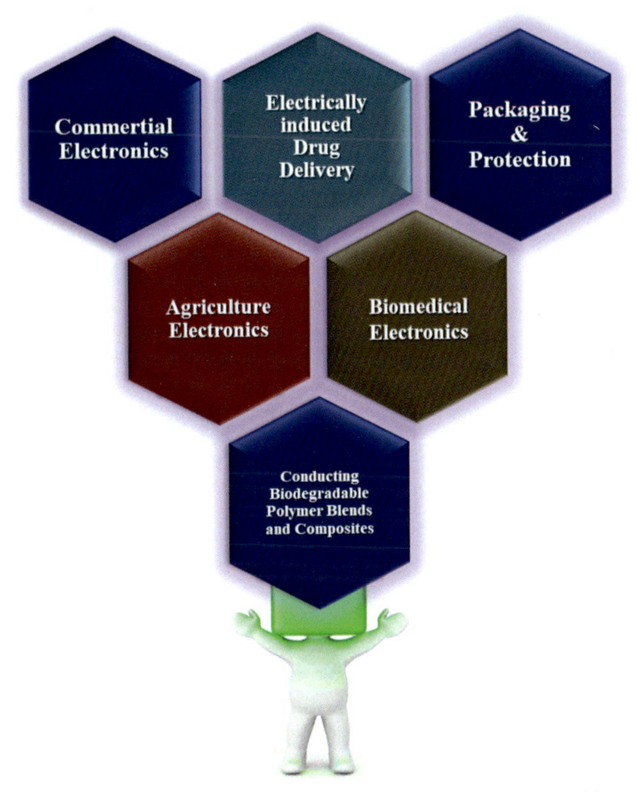

Figure 20.4 Applications of electrically conducting biodegradable and/or biocompatible polymer systems.

(Kang et al., 2011; Liu et al., 2020; Pasini & Nitti, 2020; Podiyanachari et al., 2020). Possible backbone chemical structures are illustrated in Fig. 20.5, which additionally provides a red-orange color to the solution (Kang et al., 2011; Liu et al., 2020; Pasini & Nitti, 2020; Podiyanachari et al., 2020). That is due to conjugation in the backbone of the polymer RCP giving a nice red-orange color to the polymer. To confirm this UV-vis absorption a wavelength measurement-enabled absorption resonance peak is

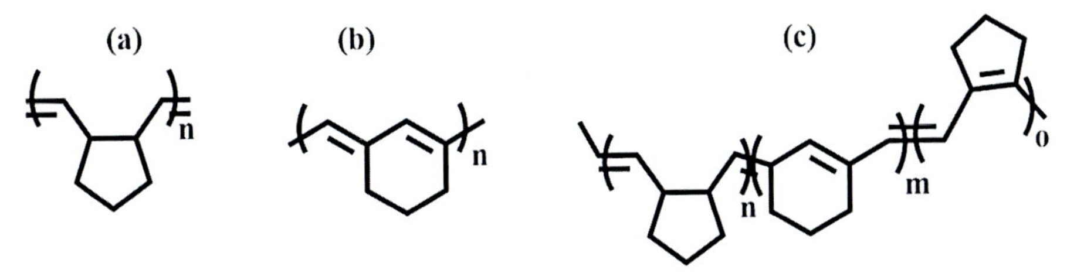

Figure 20.5 1-6-Heptadiene (a) five-membered and (b) six-membered ring structures.

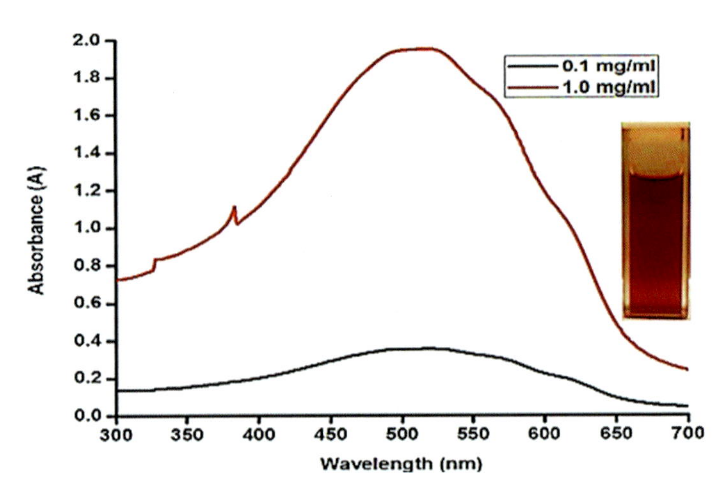

Figure 20.6 UV-vis absorption wavelength study of RCP.

illustrated in Fig. 20.6, where, the 550–600 nm wavelength range is in the absorption of orange color. Thus the study investigation reveals 1-6-heptadienes-based conducting RCP as a UV-active polymer. Such cyclopolymers are highly active in the UV region and help in studying its nature without attaching any fluorescent moiety (Li et al., 2020; Mathai et al., 2002; Probst et al., 2019).

However, electrically conducting 1,6-heptadiene facilitates different morphologies via the controlled cyclopolymerization technique. For example, FESEM measurement of synthesized poly(1,6-heptadiene)s. This sample was prepared for FESEM measurement by dissolving 1 mg of poly(1,6-heptadiene)s organic compound in 10 mL of HPLC tetrahydrofuran-THF. It was subsequently equilibrated for 6 h, and then a solution droplet deposited onto the copper grids coated with carbon followed by air drying under ambient conditions. Interestingly, the micrograph shows a highly extended nanotube structure with a uniform diameter of ~55 nm for poly(1,6-heptadiene)s. Further, the result from Materials Studio shown in Fig. 20.7, suggest the poly(1,6-heptadiene)s are in the form of a helical nanotube structure and is consistent with the structural investigations reported by Choi et al. (Choi et al., 2000). Moreover, the method of experimental and theoretical analysis and confirmation of coil-to-rod transition suggests the cyclopolymerization-assisted heptadiene coil-to-rod formation activity, which is attributable to the experimental and theoretical distress factors.

The plotted graph of SAXS data of poly(1,6-heptadiene)s, where the lower "q" region of poly(1,6-heptadiene)s peak intensity is associated with the lamellar periodicity (Winokur et al., 1989). The intensified peak of poly(1,6-heptadiene)s is either due to the intermolecular atomic distances or intramolecular atomic distances (Ciprelli et al., 1995; Liu et al., 2013; Tashiro et al., 1991; Yang et al., 1990). Yang et al., 1990 demonstrated that the crystalline region is due to the rod-type structure, where the crystalline regions are in the polymer readily assemble to form a rod-type structure after being dissolved or hydrolyzed. This could suggest that the FESEM-described rod

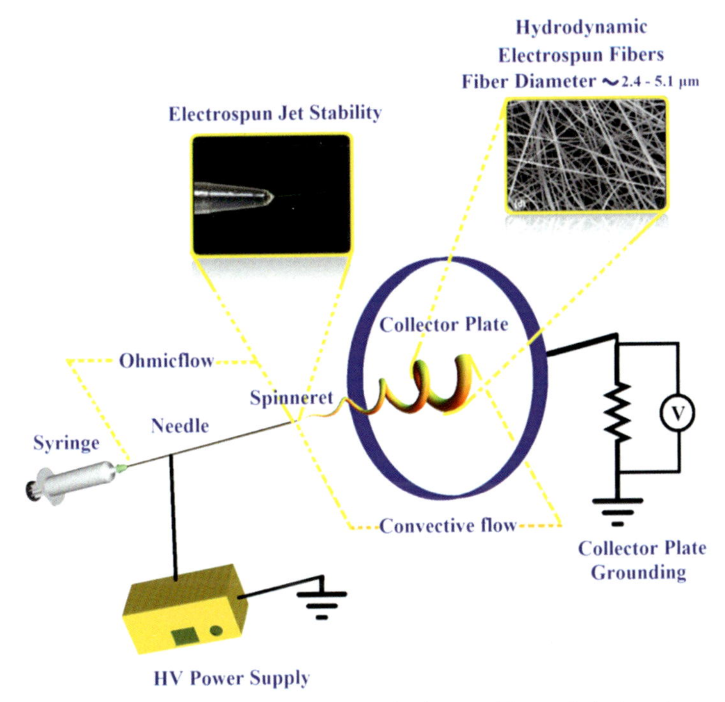

Figure 20.7 Schematic illustration of an electrospinning machine and electrospinning of RCP/ABS microfibers.

formation is in good agreement with the SAXS data. However, the small percentage of the Bragg diffraction peak renders a wide diffraction peak, which is probably attributable to the disordered chain segments and could suggest the helical nature of poly(1,6-heptadiene)s.

Moreover, Song et al. developed bridge-like structures via the metathesis-cyclopolymerization technique. High-resolution transmission electron microscopy (HRTEM) analysis suggests that the parallel stripes like a ladder in the same direction, which is attributed to flat aromatic perylene bisimide (PBI) core close edge on orientation concerning the substrate surface. There is a strong interaction between the molecules. Presumably, $\pi-\pi$ interactions between the polymeric backbone in the horizontal dimensions may account for this highly oriented ladder-like morphology with a ladder length of around 4.1 nm. Each ladder consists of 12 monomeric units (Song et al., 2015). Such a kind of ladder network form of polymers and their binary blends has been widely explored for molecular electronic devices. Their ladder structure endows a smaller optical band-gap, greater electronic delocalization, and substantially enhanced electron transport (Kim et al., 2019). Above all, delineations elucidating that the electrically conducting 1,6-heptanes with different morphologies (rod-, helical-, or ladder-like structures, etc.) obtained via controlled cyclopolymerization technique could probably lead to the realization of electronics and even molecular-level digital bit control in electronic gates.

The weak mechanical property is the most serious problem of polymers obtained using the RCM (1,6-heptadienes based ring-closing polymer) technique (Cao et al., 2014; Mao et al., 2015). Additionally, induced poor hydrophobicity, cell adhesion ability, and thermoplasticity limit the efforts to make a high impact on the electronic industry (Cao et al., 2014; Mao et al., 2015). To overcome these drawbacks, the modern miniaturized electronic industry has recommended unique superlative engineered materials with functional electrical characteristics via utilization of a novel hydrodynamic-electrospun fiber formation technique (the electrospun fiber formation technique is illustrated in Fig. 20.8) (Ali et al., 2019). Recently, researchers have performed contemporary studies on these engineered materials to accomplish significant functional characteristics such as conductivity and permittivity with their improved mechanical and structural properties using nonconductive or conductive organic polymers or their blends decorated with synthetic conducting polymers (Afzal et al., 2009; Sarma et al., 2002; Yakuphanoglu et al., 2006). For example, the research group of Huang et al. (2017) demonstrated the PVA/PVDF hyperstretchable self-powered piezoelectric microfiber sensor with enhanced mechanical characteristics. Dan Li et al. demonstrated electrospun fibrous uniaxial conductive arrays for electronics (Li et al., 2003). Dan Li et al suggested that with the incorporation of various filler types such as graphite carbon, metal oxides, and conventional organic polymers, and by controlling of electrospinning parameters, new shapes and electronic components can be realized. Bin Sun et al. suggested flexible and stretchable electrospun fibrous nonwoven mats for electronic devices, including strain sensors and pressure sensors, supercapacitors, electrodes, and field-effect transistors (Sun et al., 2014). Moreover, synthesized electrospun ABS terpolymer/poly(1,6-heptadiene) microfibers with the aid of an electrospun processing technique have been illustrated in Fig. 20.8. It has been suggested that the fibers enabled electronic device performance to be improved

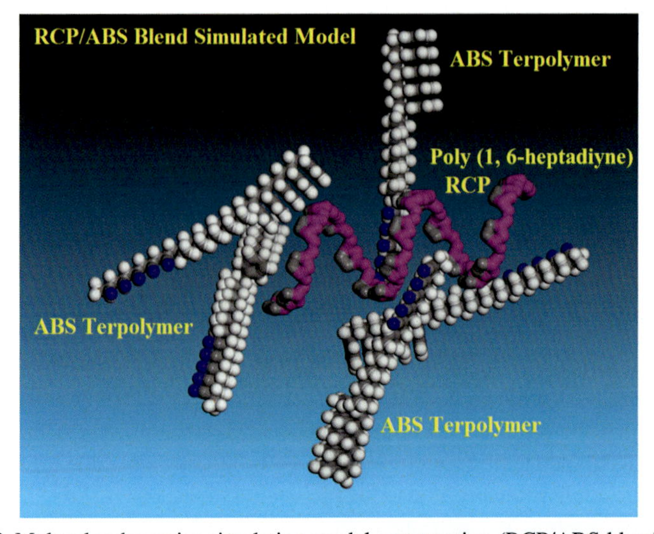

Figure 20.8 Molecular dynamics simulation model construction (RCP/ABS blend).

by accurately controlling large scale and precise deposition and flexibility during the spin process. However, in electrospun polymeric blend materials (such as conductive and insulating polymeric blend or conductive and conductive polymeric blend) miscibility or immiscibility functionality depends on its individual material properties (Duan et al., 2017; Yakuphanoglu et al., 2006).

For example, recently, immiscible polystyrene and polypropylene blends were developed using percolated conducting polyaniline/polyamide (PANI/PA) filler combined with polystyrene-block-poly(ethylene-ran-butylene)-block-polystyrene-graft-maleic anhydride compatibilizer. Here the compatibilizer was used to tailor the morphology and dispersion of the PANI/PA filler. Selective localization of PANI/PA in the PS phase with improved filler dispersion is achieved with the optimum masterbatch preparation technique followed by its optimized sequence being added to the blend components. The results suggested an increase in the DC conductivity by 6 times compared with that of the neat blend at an effective 4–8 wt.% of PANI concentration. Finally, they suggested that this significant effect in conductivity is attributable to the interactions between polymer functionality and concentration of the filler and the compatibilizer on the filler connectivity in the blend (Bharati et al., 2020). The phase separation effect on electronic properties is a fundamental investigation attributable to statically valued conclusions obtained from the molecular dynamics simulation techniques and further leads to an efficient electrically conductive/insulative blend or composite design. For example, recently, Saha and Bhowmick reported molecular dynamics simulation results for an immiscible polymer blend system based on poly(vinylidene fluoride) and hydrogenated nitrile rubber (PVDF/HNBR) via atomistic modeling simulation. They developed a thermoplastic elastomer based on PVDF/HNBRs, which had loadings of 30/70 and 40/60 for PVDF/HNBR, respectively. The reported blend is immiscible and possesses a well-dispersed but phase-separated biphasic morphology, as confirmed by the FESEM results. The immiscibility of the PVDF/HNBR blend system is also confirmed by comparing the simulated glass transition temperatures (T_g) with experimentally analyzed T_g via dynamic mechanical analysis (DMA). For the PVDF/HNBR blend of 30/70 loading, the simulated T_g was found to be $-54°C$ and $-15°C$, for the plastic phase and rubber phase, respectively. In contrast, the experimental T_g was $-37°C$ and $-20°C$. For PVDF/HNBR blend of 40/60 loading, the simulated T_g was found to be $-55°C$ and $-8°C$, for the plastic phase and rubber phase, respectively. In contrast, the experimental T_g was $-38°C$ and $-19°C$. Dual T_g for the polymer blend system indicates that the polymers are physically dispersed but not molecularly mixed, confirming the polymer blend's immiscible nature (Huang et al., 2012). Huang et al. reported phase separation behavior in immiscible polymer blends using dissipative particle dynamics simulations. They investigated the effect of Janus nanospheres on the blend system and observed that the Janus nanospheres influence the domain growth in the immiscible polymer blends, which is attributable to the intrinsic equatorial adsorption and the low desorption probability at the blend interface, thereby lowering the interfacial tension and causing the phase separation polymer blend system (Saha & Bhowmick, 2017). Analogously in the case of an ABS terpolymer/poly(1,6-heptadiene) blend system, phase separation dynamics makes a more immiscible nature at the end blend system, and a molecular dynamics simulation model construction of an RCP/ABS blend system has been illustrated in Fig. 20.7. However, the proper dispersion and concentration of this blend with a controlled spin process could lead to an efficient electrically conducting mat for novel electronic applications. The above simulation dynamics instigated delineations to reveal that the separation effect/different phases toward impedance mismatch/phase interference are a predominant parametric feature to accomplish insulating characteristics. Meanwhile, in the case of the electrically conductive blend, a proper compatibilizer/mediator/interlink between the two different phases provides an impedance match or establishes an interference-free path for the electron motion either intra- or intercharge carrier motion in a conducting polymer blend system leading to the realization of an efficient electrically conductive polymer. Finally, the authors envisioned that such a kind of electrospun fibrous nonwoven mat with enhanced mechanical and electric properties could be potentially utilized in next-generation electronics, such as biomedical sensors, human interface machines, artificial skins, and humanoid robots (Duan et al., 2017; Yakuphanoglu et al., 2006).

Moreover, researchers have successively demonstrated poly(1,6-heptadienes) functional derivatives (such as blends, functionalized polymeric systems, and nanocomposites, etc.) for advanced, diverse field functional applications ranging from optical fibers, flat panel displays, waveguides, dielectric encapsulants for electronic packaging, oil spill management and sacrificial materials for microelectromechanical systems (MEMS), sensors, and electrical circuit boards (Bhattacharya et al., 2013; Fujiwara et al., 2007; Mane et al., 2015; Mukherjee et al., 2016; Rao et al., 2012; Sarkar & Shunmugam, 2014). Furthermore, Rao et al. successfully demonstrated the synthesis of norbornene-derived doxorubicin block-copolymers via the ring-opening metathesis (RCM) polymerization technique (Rao et al., 2012). Such synthetic polymers and their derivatives exhibiting patterned nanostructures have been well exploited in microelectronics, magnetoelectronic and magnetic memory storage devices, and dielectrics (Liu et al., 2020; Pasini & Nitti, 2020). As well as well-shielded drug moieties, the same synthesized block copolymers have shown a remarkable tendency of rendering acid-triggered drug release at a mildly acidic pH of 5.5−6 demonstrating them as promising carriers for pH-triggered intracellular administration of hydrophobic anticancer drugs. These concomitants all quantitatively divulge that these conjugated synthetic poly(1-6-heptadiene)s and their derivatives can be successfully utilized in potential next-generation electronic applications such as biomedical sensors, batteries, supercapacitors, and electrical circuit boards, etc. (Bhattacharya et al., 2013; Fujiwara et al., 2007; Mane et al., 2015; Mukherjee et al., 2016; Rao et al., 2012; Sarkar & Shunmugam, 2014).

In the face of growing environmental concerns, biodegradable and biosource plastic nanocomposites are emerging as a new class of materials. Though researchers have explored various kinds of polymer composites and blends for diversified electronics, biodegradable polymeric blends or composites are dominantly required to maintain an eco-friendly ecological system, water purity, food-based customized goods, and electrical and electronic applications, along with the biodegradability. Compatible resin materials are also required in the case of biomedical devices and bioimplants. Though the applications of cyclopolymerized poly(1,6-heptadiene)s have not been explored fully, its chemistry and the chemical structure of cyclopolymerization-derived polymers have been investigated and understood in depth. In a recent study, researchers developed pH-dependent degradation of PMAH and P(MAH-co-MAA) via a cyclopolymerization technique. During polymerization, the two methacrylate bonds either undergo polymerization that creates an intermolecular cross-linked network or intercross-linking cyclopolymerization that creates six-membered rings. The hydrolysis of anhydride groups in the MAH moiety as either the cross-linking structure or side-chain rings, both result in two carboxylic groups, and the polymer transforms into a homopolymer of PMMA. Owing to the excellent solubility of PMMA in an aqueous solution, the insoluble PMAH and P(MAH-co-MAA) become soluble upon hydrolysis of the anhydride group, leading to degradation of the polymeric system (Su et al., 2020). Recently, polypropylene/polylactide/nanoclay blend/composite films with/without prooxidant compatibilizers were prepared and aerobically degraded to measure the CO_2 evaluation under controlled composting conditions as per ASTM S5338. The first-order Komilis model in series with a flat lag phase was postulated involving two stages: hydrolysis of stable carbon and rapid mineralization. The first rate-limiting stage comprises three possible parallel paths with the solid hydrolysis of readily, moderately, and slowly hydrolyzable carbon fractions. The model parameters were computed after correlating with the experimental data using nonlinear regression analysis. The results of the model characteristic parameters, nonexperimental data using nonlinear regression analysis suggest that the characteristic model parameters, ungraded/hydrolyzable/mineralizable intermediate carbon kinetics and degradation curves, exhibit two distinct kinetic regimes. The first regime is comprised of slowly and moderately hydrolyzed carbon and is shown by the first four films without prooxidants. This causes low degradability and degradation rate. The second regime is comprised of the readily and moderately hydrolyzable carbon and is shown by another film containing prooxidants. They exhibit relatively high degradability and degradation rate, which peaks at around the 11−14th day in the range of 0.219%−0.268% per day. The values of their moderately hydrolyzable carbon fractions and the corresponding hydrolysis rates are significantly higher than those of the first regime. For the first regime, the degradability and degradation rate decreases with an increase in the slowly hydrolyzable carbon impervious to microbial attack. Their degradation rate profiles show the absence of a growth phase due to the absence of readily hydrolyzable carbon. The rate decreases monotonically starting from the minimum value ranging from 0.043% to 0.180% per day. The approach presented can also be implemented to model and design equipment for other waste biodegradation systems (Sable et al., 2019). Based on the above hydrolysis delineations, cyclopolymerized poly(1,6-heptadiene)s is also possible to made into biodegradable poly(1,6-heptadiene)s, which could lead to a potential electrically conducting polymer material for various electronic components. Further, its derivatives provide possible solutions to the unavoidable drawbacks and disadvantages raised by traditional electronics materials.

Moreover, researchers have explored various kinds of composites that electronically stimulate the biological parts, for example, in the human body such as heartbeat monitoring systems, nanogenerators for heart rhythms and fabrication, biobased electronic chips to stimulate brain frequencies, electronic circuit-enabled eye lenses for stimulation, or even in agriculture for growth-monitoring systems. Recently, composites based on synthetic matrices and nanostructures with novel properties have attracted attention due to their potential applications in biomedicine, wastewater treatment, and electronics. Therefore researchers are extensively investigating multifunctional composites. Aguilar et al. investigated a smart bioplastic based on cellulose films with adsorbed magnetic particles. A delignification of cotton and sugarcane base was used to obtain biopolymers. These composites were obtained by immersion of the cellulose-based biopolymers into a nanometer-sized $MnFe_2O_4$ ferrofluid. The magnetic measurements were conducted, revealing that the superparamagnetic hysteresis shape is maintained in the bioplastics. However, the optical transparency did not show a drastic effect after the adsorption of magnetic particles. In addition, the thermal stability is described as being independent of the source of extraction and deposition of nanoparticles (Aguilar et al., 2019). This kind of biodegradable and biocompatible composite with functional characteristics, either magnetic or electric characteristics such as dielectric permittivity, dielectric loss, electronic conductivity, and its radiation phenomena and other functionalities such as hydrophobicity, in a molecular scale, enabled electronic devices to be useful as potential candidates for bioimplanted devices to stimulate biological activities (Magisetty, Hemnath, Shukla et al., 2020; Malik et al., 2020; Magisetty & Naga Srilatha, 2021; Magisetty, Shukla et al., 2020; Magisetty, Prajapati et al., 2019; Magisetty, Kumar et al., 2018; Magisetty et al., 2018a; Magisetty et al., 2018b; Magisetty, Hemnath, Shukla et al., 2020; Magisetty & Marutheswar, 2011; Magisetty & Cheekuramelli, 2019; Magisetty, Raj et al., 2020; Magisetty, Shukla et al., 2019).

5. Conclusion

In this chapter, the advancements in electrically conductive biodegradable materials have been described in regard to their functional probabilities. Especially, this chapter focuses on the cyclopolymerization-derived electrically conducting 1,6-heptadienes and their derivatives. Further, the hydrolyzed techniques suggested that the 1,6-heptadienes could be made into biodegradable 1,6-heptadienes. Moreover, the most exciting facts about biodegradable polymeric systems are the biodegradability and compostability. Moreover, the necessity for electrically conducting biodegradable and biocompatible polymeric systems and their electrical and electronic applications are highlighted. Further, this chapter illustrates electrically conductive biodegradable materials with widespread functionalities that could be used as excellent candidate materials to deliver efficient and low-cost ecofriendly and sustainable development of electrical and electronic applications.

Acknowledgment

The authors gratefully acknowledges the FONDECYT Postdoctoral Project No.: 3180172, Government of Chile, Santiago.

References

Afzal, A. B., Akhtar, M. J., Nadeem, M., & Hassan, M. M. (2009). Investigation of structural and electrical properties of polyaniline/gold nanocomposites. *Journal of Physical Chemistry C, 113*(40), 17560–17565.

Aguilar, N. M., Arteaga-Cardona, F., de Anda Reyes, M. E., Gervacio-Arciniega, J. J., & Salazar-Kuri, U. (2019). Magnetic bioplastics based on isolated cellulose from cotton and sugarcane bagasse. *Materials Chemistry and Physics, 238*, 121921.

Ali, N. N., Atassi, Y., Salloum, A., Malki, A., Jafarian, M., & Almarjeh, R. K. B. (2019). Lightweight broadband microwave absorbers of core–shell (polypyrrole/NiZn ferrite) nanocomposites in the X-band: insights on interfacial polarization. *Journal of Materials Science: Materials in Electronics, 30*(7), 6876–6887.

Bharati, A., Hejmady, P., Van der Donck, T., Seo, J. W., Cardinaels, R., & Moldenaers, P. (2020). Developing conductive immiscible polystyrene/polypropylene blends with a percolated conducting polyaniline/polyamide filler by tuning its specific interactions with the styrene-based triblock compatibilizer grafted with maleic anhydride. *Journal of Applied Polymer Science, 137*(9), 48433.

Bhattacharya, S., Sarkar, S., & Shunmugam, R. (2013). Unique norbornene polymer based "in-field" sensor for As(iii). *Journal of Materials Chemistry, 1*(29), 8398.

Biomaterials, D. W. (2009). *Undefined on the nature of biomaterials*. Elsevier. and.

Bogoeva-Gaceva, G., & Dimko, D. (2013). Biocomposites based on poly (lactic acid) and kenaf fibers: effect of micro-fibrillated cellulose. *Macedonian Journal of Chemistry and Chemical Engineering, 32*(2), 331–335. mjcce.org.mk.

Cao, Z., Dou, C., & Dong, S. (2014). Scaffolding biomaterials for cartilage regeneration. *Journal of Nanomaterials*, 1–8, 2014.

Choi, S.-K., Gal, Y.-S., Jin, S.-H., & Kim, H. K. (2000). Poly(1,6-heptadiyne)-Based materials by metathesis polymerization. *Chemical Reviews, 100*(4), 1645–1682.

Ciprelli, J.-L., Clarisse, C., & Delabouglise, D. (1995). Enhanced stability of conducting poly(3-octylthiophene) thin films using organic nitrosyl compounds. *Synthetic Metals, 74*(3), 217–222.

Duan, Y., Ding, Y., Xu, Z., Huang, Y., & Yin, Z. (2017). Helix electrohydrodynamic printing of highly aligned serpentine micro/nanofibers. *Polymers, 9*(12), 434.

Fujiwara, M., Shirato, Y., Owari, H., Watanabe, K., Matsuyama, M., Takahama, K., Mori, T., Miyao, K., Choki, K., Fukushima, T., Tanaka, T., & Koyanagi, M. (2007). Novel optical/electrical printed circuit board with polynorbornene optical waveguide. *Japanese Journal of Applied Physics, 46*(4B), 2395–2400.

Geczy, A., Lener, V., Hajdu, I., & Illyefalvi-Vitez, Z. (2011). Low temperature soldering on biopolymer (PLA) Printed Wiring Board substrate. In *Proceedings of the 2011 34th International Spring Seminar on Electronics Technology* (pp. 57–62).

Gerard, M. (2002). Application of conducting polymers to biosensors. *Biosensors and Bioelectronics, 17*(5), 345–359.

Grainger, D. W. (1999). The Williams dictionary of biomaterials. *Materials Today, 2*(3), 29.

Guimard, N. K., Gomez, N., & Schmidt, C. E. (2007). Conducting polymers in biomedical engineering. *Progress in Polymer Science, 32*(8–9), 876–921.

Horowitz, B. G. (1998). *Organic field-effect transistors* (vol. 5, pp. 365–377).

Huang, M., Li, Z., & Guo, H. (2012). The effect of Janus nanospheres on the phase separation of immiscible polymer blends via dissipative particle dynamics simulations. *Soft Matter, 8*(25), 6834.

Huang, Y., Ding, Y., Bian, J., Su, Y., Zhou, J., Duan, Y., & Yin, Z. (2017). Hyper-stretchable self-powered sensors based on electrohydrodynamically printed, self-similar piezoelectric nano/microfibers. *Nano Energy, 40*, 432–439.

Inoue, K., Serizawa, S., Yamashiro, M., & Iji, M. (2007). Highly functional bioplastics (PLA compounds) used for electronic products. In *Polytronic 2007 - 6th Conference on Polymers and Adhesives in Microelectronic and Photonics* (pp. 73–76).

Kang, E.-H., Lee, I. S., & Choi, T.-L. (2011). Ultrafast cyclopolymerization for polyene synthesis: living polymerization to dendronized polymers. *Journal of the American Chemical Society, 133*(31), 11904–11907.

Kim, F. S., Park, C. H., Na, Y., & Jenekhe, S. A. (2019). Effects of ladder structure on the electronic properties and field-effect transistor performance of Poly(-benzobisimidazobenzophenanthroline). *Organic Electronics, 69*, 301–307.

Kolybaba, M., Tabil, L. G., Panigrahi, S., Crerar, W. J., Powell, T., & Wang, B. (2003). *Biodegradable polymers: past, present, and future. ASABE/CSBE North cent. Intersect. Meet.*

Kuciel, S., & Mazur, K. (2019). Novel hybrid composite based on bio-PET with basalt/carbon fibre. *IOP Conference Series: Materials Science and Engineering, 634*, 012009.

Li, D., Wang, Y., & Xia, Y. (2003). Electrospinning of polymeric and ceramic nanofibers as uniaxially aligned arrays. *Nano Letters, 3*(8), 1167–1171.

Li, H., Zhang, H., Liao, X., Sun, R., & Xie, M. (2020). Incorporating trifunctional 1,6-heptadiyne moiety into polyacetylene ionomer for improving its physical and conductive properties. *Polymer Chemistry, 11*(19), 3322–3331.

Liu, X., Dai, K., Hao, X., Zheng, G., Liu, C., Schubert, D. W., & Shen, C. (2013). Crystalline structure of injection molded β-isotactic polypropylene: analysis of the oriented shear zone. *Industrial and Engineering Chemistry Research, 52*(34), 11996–12002.

Liu, X., Liu, F., Liu, W., & Gu, H. (2020). ROMP and MCP as versatile and forceful tools to fabricate dendronized polymers for functional applications. *Polymer Reviews*, 1–53.

Magisetty, R., & Naga Srilatha, C. H. (2021). *Bioepoxy polymer, blends and composites derived utilitarian electrical, magnetic and optical properties. Bio-based epoxy polymers, blends and composites: synthesis, properties, characterization and applications* (pp. 249–266).

Magisetty, R., & Cheekuramelli, N. S. (2019). Additive manufacturing technology empowered complex electromechanical energy conversion devices and transformers. *Applied Materials Today, 14*, 35–50.

Magisetty, R. P., & Marutheswar, G. V. (2011). The future generation of low energy loss HVDC power system using HTS technology. *International Journal on Recent Trends in Engineering and Technology, 6*(2).

Magisetty, R., Hemnath, N. R., Shukla, A., Shunmugam, R., & Kandasubramanian, B. (2020). Poly (1, 6-heptadiyne)/NiFe2O4 composite as capacitor for miniaturized electronics. *Polymer-Plastics Technology and Materials, 59*(18), 2018–2026.

Magisetty, R., Kumar, P., Kumar, V., Shukla, A., Kandasubramanian, B., & Shunmugam, R. (2018). NiFe2O4/poly (1, 6-heptadiyne) Nanocomposite energy-storage device for electrical and electronic applications. *ACS Omega, 3*(11), 15256–15266.

Magisetty, R., Shukla, A., & Kandasubramanian, B. (2018). Magnetodielectric microwave radiation absorbent materials and their polymer composites. *Journal of Electronic Materials, 47*(11), 6335–6365.

Magisetty, R., Shukla, A., & Kandasubramanian, B. (2018). Dielectric, hydrophobic investigation of ABS/NiFe 2O4 nanocomposites fabricated by atomized spray assisted and solution casted techniques for miniaturized electronic applications. *Journal of Electronic Materials, 47*(9), 5640–5656.

Magisetty, R., Prajapati, D., Ambekar, R., Shukla, A., & Kandasubramanian, B. (2019). β-Phase Cu-phthalocyanine/acrylonitrile butadiene styrene terpolymer nanocomposite film technology for organoelectronic applications. *Journal of Physical Chemistry C, 123*(46), 28081–28092.

Magisetty, R., Shukla, A., & Kandasubramanian, B. (2019). Terpolymer (ABS) cermet (Ni-NiFe2O4) hybrid nanocomposite engineered 3D-carbon fabric mat as a X-band electromagnetic interference shielding material. *Materials Letters, 238*, 214–217.

Magisetty, R., Hemanth, N. R., Kumar, P., Shukla, A., Shunmugam, R., & Kandasubramanian, B. (2020). Multifunctional conjugated 1, 6-heptadiynes and its derivatives stimulated molecular electronics: future moletronics. *European Polymer Journal, 124*, 109467.

Magisetty, R., Raj, A. B., Datar, S., Shukla, A., & Kandasubramanian, B. (2020). Nanocomposite engineered carbon fabric-mat as a passive metamaterial for stealth application. *Journal of Alloys and Compounds, 848*, 155771.

Magisetty, R., Shukla, A., & Kandasubramanian, B. (2020). Molecular dynamic simulation constructed interaction parameter investigation between poly (1, 6-heptadiyne) and NiFe2O4 in nanocomposite. *Materials Today: Proceedings, 24*, 1720–1728.

Malik, A., Magisetty, R., Kumar, V., Shukla, A., & Kandasubramanian, B. (2020). Dielectric and conductivity investigation of polycarbonate-copper phthalocyanine electrospun nonwoven fibres for electrical and electronic application. *Polymer-Plastics Technology and Materials, 59*(2), 154–168.

Mane, S. R., Sarkar, S., N., V. R., Sathyan, A., & Shunmugam, R. (2015). An efficient method to prepare a new class of regioregular graft copolymer via a click chemistry approach. *RSC Advances, 5*(90), 74159–74161.

Mao, X.-M., Zhan, Z.-J., Grayson, M. N., Tang, M.-C., Xu, W., Li, Y.-Q., Yin, W.-B., Lin, H.-C., Chooi, Y.-H., Houk, K. N., & Tang, Y. (2015). Efficient biosynthesis of fungal polyketides containing the dioxabicyclo-octane ring system. *Journal of the American Chemical Society, 137*(37), 11904–11907.

Mathai, C. J., Saravanan, S., Anantharaman, M. R., Venkitachalam, S., & Jayalekshmi, S. (2002). Effect of iodine doping on the band-gap of plasma polymerized aniline thin films. *Journal of Physics D Applied Physics, 35*(17), 318.

McQuade, D. T., Pullen, A. E., & Swager, T. M. (2000). Conjugated polymer-based chemical sensors. *Chemical Reviews, 100*(7), 2537–2574.

Mukherjee, M., Ganivada, M. N., Venu, P., Kanjilal, P., & Shunmugam, R. (2016). Unique nanotubes from polynorbornene derived graphene sheets. *RSC Advances, 6*(47), 40691–40697.

Ogunsona, E. O., Misra, M., & Mohanty, A. K. (2017). Impact of interfacial adhesion on the microstructure and property variations of biocarbons reinforced nylon 6 biocomposites. *Composites Part A: Applied Science and Manufacturing, 98*, 32–44.

Pasini, D., & Nitti, A. (2020). Free radical cyclopolymerization: a tool towards sequence control in functional polymers. *European Polymer Journal, 122*, 109378.

Philp, J. C., Bartsev, A., Ritchie, R. J., Baucher, M.-A., & Guy, K. (2013). Bioplastics science from a policy vantage point. *New Biotechnology, 30*(6), 635–646.

Podiyanachari, S. K., Moncho, S., Brothers, E. N., Al-Meer, S., Al-Hashimi, M., & Bazzi, H. S. (2020). One-pot tandem ring-opening and ring-closing metathesis polymerization of disubstituted cyclopentenes featuring a terminal alkyne functionality. *Macromolecules*, 0c00462.

Probst, P., Elser, I., Schowner, R., Benedikter, M. J., & Buchmeiser, M. R. (2019). Regio and stereospecific cyclopolymerization of α,ω diynes by cationic molybdenum imido alkylidene N heterocyclic carbene complexes. *Macromolecular Rapid Communications*, 1900398.

Rao, N.,V., Mane, S., Kishore, A., Das Sarma, J., & Shunmugam, R. (2012). Norbornene derived doxorubicin copolymers as drug carriers with pH responsive hydrazone linker. *Biomacromolecules, 13*(1), 221–230.

Raquez, J.-M., Habibi, Y., Murariu, M., & Dubois, P. (2013). Polylactide (PLA)-based nanocomposites. *Progress in Polymer Science, 38*(10–11), 1504–1542.

Sable, S., Mandal, D. K., Ahuja, S., & Bhunia, H. (2019). Biodegradation kinetic modeling of oxo-biodegradable polypropylene/polylactide/nanoclay blends and composites under controlled composting conditions. *Journal of Environmental Management, 249*, 109186.

Saha, S., & Bhowmick, A. K. (2017). Computer aided simulation of thermoplastic elastomer from poly (vinylidene fluoride)/hydrogenated nitrile rubber blend and its experimental verification. *Polymer, 112*, 402–413.

Sarkar, S., & Shunmugam, R. (2014). Polynorbornene derived 8-hydroxyquinoline paper strips for ultrasensitive chemical nerve agent surrogate sensing. *Chemical Communications, 50*(62), 8511–8513.

Sarma, T. K., Chowdhury, D., Paul, A., & Chattopadhyay, A. (2002). Synthesis of Au nanoparticle–conductive polyaniline composite using H2O2 as oxidising as well as reducing agent. *Chemical Communications*, (10), 1048–1049.

Schramm, R., Reinhardt, A., & Franke, J. (2012). Capability of biopolymers in electronics manufacturing. In *2012 35th International Spring Seminar on Electronics Technology* (pp. 345–349).

Sims, L., Egelhaaf, H. J., Hauch, J. A., Kogler, F. R., & Steim, R. (2012). Plastic solar cells. *Comprehensive Renewable Energy, 1*(1), 439–480.

Song, W., Han, H., Wu, J., & Xie, M. (2015). A bridge-like polymer synthesized by tandem metathesis cyclopolymerization and acyclic diene metathesis polymerization. *Polymer Chemistry, 6*(7), 1118–1126.

Su, C., Hu, Y., Song, Q., Ye, Y., Gao, L., Li, P., & Ye, T. (2020). Initiated chemical vapor deposition of graded polymer coatings enabling antibacterial, antifouling, and biocompatible surfaces. *ACS Applied Materials and Interfaces, 12*(16), 18978–18986.

Sun, B., Long, Y.-Z., Chen, Z.-J., Liu, S.-L., Zhang, H.-D., Zhang, J.-C., & Han, W.-P. (2014). Recent advances in flexible and stretchable electronic devices via electrospinning. *Journal of Materials Chemistry C, 2*(7), 1209–1219.

Sun, J., Shen, J., Chen, S., Cooper, M., Fu, H., Wu, D., & Yang, Z. (2018). Nanofiller reinforced biodegradable PLA/PHA composites: current status and future trends. *Polymers, 10*(5), 505.

Tashiro, K., Ono, K., Minagawa, Y., Kobayashi, M., Kawai, T., & Yoshino, K. (1991). Structure and thermochromic solid-state phase transition of poly (3-alkylthiophene). *Journal of Polymer Science Part B: Polymer Physics, 29*(10), 1223–1233.

Waste. (2016). *Waste & packaging*. Unilever.

Winokur, M. J., Spiegel, D., Kim, Y., Hotta, S., & Heeger, A. J. (1989). Structural and absorption studies of the thermochromic transition in poly(3-hexylthiophene). *Synthetic Metals, 28*(1–2), 419–426.

Yakuphanoglu, F., Basaran, E., Şenkal, B. F., & Sezer, E. (2006). Electrical and optical properties of an organic semiconductor based on polyaniline prepared by emulsion polymerization and fabrication of Ag/Polyaniline/n-Si Schottky diode. *The Journal of Physical Chemistry B, 110*(34), 16908–16913.
Yang, R., Evans, D. F., Christensen, L., & Hendrickson, W. A. (1990). Scanning tunneling microscopy (STM) evidence of semicrystalline and helical conducting polymer structures. *The Journal of Physical Chemistry, 94*(15), 6117–6122.

Biodegradable polymer blends and composites for biomedical applications

21

Noor Izyan Syazana Mohd Yusoff[1,2,3,4], *Mat Uzir Wahit*[1,2], *Weng Hong Tham*[1], *Tuck-Whye Wong*[3,4], *Xiau Yeen Lee*[5] *and Farah Hidayah Jamaludin*[3]
[1]School of Chemical and Energy Engineering, Faculty of Engineering, Universiti Teknologi Malaysia, Skudai, Johor, Malaysia; [2]Centre for Advanced Composite Materials (CACM), Universiti Teknologi Malaysia, Skudai, Johor, Malaysia; [3]School of Biomedical Engineering and Health Sciences, Faculty of Engineering, Universiti Teknologi Malaysia, Skudai, Johor, Malaysia; [4]Advance Membrane Technology Research Centre (AMTEC), Universiti Teknologi Malaysia, Skudai, Johor, Malaysia; [5]Centre for Advanced Materials, Tunku Abdul Rahman University College (TARUC), Setapak, Kuala Lumpur, Malaysia

1. Introduction

Polymers have emerged as important materials to replace traditionally used natural materials such as wood, natural fiber, and natural rubber. A wide variety of synthetic polymers have been developed for use in agriculture, construction, packaging, transportation, and medical applications. In the past few decades, the emergence of biomedical engineering had spurred the development of novel biomaterials. In particular, polymeric biomaterials, both natural and synthetic biodegradable polymers, have been extensively studied and have been used in various biomedical fields (Aravamudhan et al., 2014; Chen et al., 2013; Kucinska-Lipka et al., 2015; Lee et al., 2014; Zhang et al., 2013).

Natural biopolymers commonly used in biomaterials such as collagen (Duan et al., 2007), elastin (Daamen et al., 2007), silk (Harkin et al., 2011), and polysaccharides (Aravamudhan et al., 2014) offer several attractive advantages such as excellent biocompatibility, unique mechanical properties, and hydrolytic and enzymatic degradability. However, the drawbacks of natural biopolymers are also very significant, including the potential to induce an immunogenic response, complex purification techniques, processing variability, and unstable material supply (Nair & Laurencin, 2006, 2007). In contrast, synthetic biodegradable polymers offer more advantages due to their flexibility in synthesis and excellent reproducibility.

These advantages have attracted the interest of researchers in choosing synthetic polymers as the material of choice for biomedical applications. Synthetic polymers have become attractive alternatives for several reasons: (1) the polymers can be designed to be biodegradable, which can overcome the issues associated with revision surgeries (Nair & Laurencin, 2006); (2) the polymers can be tailored to closely mimic

Biodegradable Polymers, Blends and Composites. https://doi.org/10.1016/B978-0-12-823791-5.00016-8

the mechanical profiles of natural tissues by combining various types of monomers; (3) using monomers that usually take part in the human metabolic process could improve the biocompatibility of the polymers and, simultaneously, the degradation products could be absorbed and removed from the body (Rai et al., 2012); and (4) the properties of synthetic polymers can be further enhanced by the addition of bioceramics such as calcium phosphate, bioactive glass, hydroxyapatite, and others.

Polymeric biomaterials are specifically synthetic biodegradable polymers that have been extensively investigated for biomedical applications due to their specific properties such as biocompatibility, biodegradability, and bioactivity (Nair & Laurencin, 2007; Piskin, 1995). Biocompatibility is the first criterion for a material to be considered as a biomaterial. Since the material is designed to be used in direct contact with living tissues, it is considered biocompatible if it does not cause unacceptable harmful effects such as unresolved irritation, immunogenicity, or cytotoxicity upon implantation in the body (Williams, 1999). Polymeric biomaterials offered better biocompatibility, which can eliminate adverse events such as thrombosis, inflammation, toxicity, and allergic reactions (Liu et al., 2012) when compared to existing permanent implants.

Polymeric biomaterials must be biocompatible, which means that the materials should not elicit a sustained inflammatory or toxic response after implantation in the human body. The materials should be able to degrade hydrolytically or enzymatically, and the degradation products should be nontoxic, and able to be metabolized and cleared from the body. At the same time, the rate of degradation should match the regeneration process. Next, the materials should have sufficient mechanical properties to provide mechanical support during the healing process and not collapse during the patient's normal activities. Also, the materials should have good processability so that they can be shaped into complicated geometries required for specific applications (Gunatillake et al., 2003; O'Brien, 2011). One class of materials that has this versatility is synthetic biodegradable polymers.

Biodegradable polymers can be separated into two major categories depending on the mode of degradation (Hayashi, 1994; Nair & Laurencin, 2007): hydrolytically degradable polymers and enzymatically degradable polymers. Hydrolytically degradable polymers refer to the polymers that can be degraded by reaction with water, while enzymatically degradable polymers are polymers that can be degraded via enzyme-specific reactions. The mechanical properties of biodegradable polymers that will vary with time due to biodegradation are suitable for tissue-engineering applications. The degrading matrix will gradually transfer stress to the newly regenerated tissues and, hence, it can improve tissue formation (Serrano et al., 2010). The bioactivity of biodegradable polymers can be improved by the introduction of some biologically and pharmacologically active substances and surface modification (Emadi et al., 2010; Jiao & Cui, 2007).

Synthetic biodegradable polyesters are commonly chosen as polymeric materials, thus they meet the criteria for biomedical applications. Biodegradable thermoplastic polyesters are prepared by ring-opening polymerization, such as poly(glycolic acid) (PGA), poly(lactic acid) (PLA), poly(caprolactone) (PCL), and their copolymers have been extensively investigated for biomedical applications due to their biocompatibility, biodegradability, bioresorbability, and versatility in the synthesis process together with their tunable mechanical and degradation properties (Mei et al., 2004; Nair & Laurencin, 2007). These thermoplastic polyesters have been explored for potential uses as materials for sutures, orthopedic fixation devices, and drug-delivery vehicles (Ashammakhi & Rokkanen, 1997; Athanasiou et al., 1996; Nair & Laurencin, 2006; Seyednejad et al., 2011; Sokolsky-Papkov et al., 2007).

The main focus of this chapter is to discuss the synthesis of a novel biodegradable polyester and its biocomposites which are suitable to be used as biomaterials for biomedical applications (Tham, Wahit, Kadir, & Wong, 2012; Lee et al., 2016; Wong et al., 2014, 2020). Biomaterials developed from metals are strong, tough, and ductile; however, being stiff, permanent, or nondegradable in nature, with high density, and the possibility of corrosion and require future surgery to remove the implant are the major drawbacks. Bioactive ceramics offer excellent biocompatibility and good mechanical properties; however, they are brittle, not resilient, and difficult to fabricate. Polymeric biomaterials with mechanical versatility close to that of human tissue are the most popular materials studied and applied for tissue engineering applications, except in orthopedics (Williams, 2003; Zhang et al., 2014).

2. Biodegradable synthetic polymers

Natural biodegradable polymers have excellent biocompatibility and bioactivity, their structure closely mimics components in host tissues, they possess unique mechanical characteristics, and are enzymatic or hydrolytic degradable. However, the use of natural polymers in biomedical applications is limited due to several significant disadvantages which include (1) strong antigenic responses, (2) possibility of disease transmission, (3) insufficient mechanical strength, (4) in vivo rates of degradation are difficult to evaluate due to differences in enzyme concentrations in different parts of living organs and tissues, (5) complex purification steps, and (6) batch-to-batch variation in properties (Nair & Laurencin, 2006; Hayashi, 1994; Nair & Laurencin, 2007).

Compared to natural polymers, synthetic polymers are more biologically inert, by controlling the synthesis conditions, the mechanical properties and degradation rates can be manipulated; synthetic polymers can be designed into various shapes, have predictable properties, and material impurities can be controlled better than with natural polymers. Since the synthesis conditions are well known and controlled, synthetic polymers offer batch-to-batch uniformity, with reproducible mechanical and physical properties. Due to the simple structure and well-known monomeric unit of synthetic polymers, the possible risks such as toxicity, immunogenicity, and favoring of infections upon implantation are reduced.

Huge efforts have been made to develop synthetic biodegradable polymers that mimic natural tissues (Ma, 2008). They are suitable for use in the recently growing biomedical fields including tissue engineering, bionanotechnology, gene therapy, regenerative medicine, and drug delivery (Martina & Hutmacher, 2007; Woodruff & Hutmacher, 2010). Polymeric biomaterials are currently dominated by thermoplastic polyesters such as PLA, PGA, PCL, and their copolymers (Huang et al., 2014;

Pintado-Sierra et al., 2014; Seal et al., 2001). Some other examples of synthetic biodegradable polymers developed for medical applications include polyurethane (Cherng et al., 2013), polyhydroxyalkanoates (Chen and Wu, 2005), polyanhydrides (Li, Deng, & Stephens, 2002), and poly(ester amide) (Hemmrich et al., 2008).

However, currently available biodegradable polymers are still insufficient due to their inability to cover the mechanical and degradation properties required for each specific and unique medical application. For example, natural tissues such as the bladder, heart valves, cartilage, and ligament exhibit different magnitude of mechanical properties. Hence, the current exploration of novel biodegradable polymers has been focused on designing polymers in which their properties can be tailored to extend over a wide range for biomedical applications.

Recently, there has been increasing research activity toward the development of thermoset biodegradable polyester synthesized through catalyst-free polycondensation as biomaterials for soft-tissue engineering. Poly(glycerol sebacate) (PGS) is the thermoset biodegradable polyester developed from this method that is most studied (Wang et al., 2002). PGS polyesters synthesized from glycerol and sebacic acid demonstrated excellent biocompatibility together with tunable mechanical properties and degradation rates by simply adjusting the feed monomers and curing conditions (Li et al., 2013; Rai et al., 2012). After that, Bruggeman et al. (2008) used this route to develop biodegradable polyester from sebacic acid with different polyol monomers. The resulting polyesters, poly(polyol sebacate) (PPS), exhibited a range of mechanical properties and degradation rates and may potentially be developed for biomedical applications (Bruggeman et al., 2008).

The recent developments in synthetic biodegradable polymers also focused on biodegradable elastomers, as tabulated in Table 21.1. This is due to their potential in the emerging field of soft-tissue engineering applications where the mechanical and elastomeric properties of the polymer scaffolds can mimic those of human organs and tissues (Amsden, 2007; Barrett & Yousaf, 2009; Gunatillake et al., 2003; Shi et al., 2009). They are capable of maintaining their geometries and recovering from multiple deformations while not causing inflammation to the surrounding tissue in a mechanically demanding environment (Shi et al., 2009; Tran et al., 2009). Scaffolds developed from biodegradable elastomers have shown excellent biocompatibility, mechanical properties that mimic those natural tissues, and enhanced cellular behavior and tissue viability (Barrett & Yousaf, 2009; Gunatillake et al., 2003; Tran et al., 2009).

As mentioned above, biocompatibility is an important characteristic for a biomaterial because it is designed to be used in intimate contact with living tissue. However, the requirements for this biocompatibility are complex, varying with specific medical applications. For example, a material used satisfactorily in orthopedic surgery may be inappropriate for cardiovascular applications because of its thrombogenic properties (Macocinschi, Filip, & Vlad, 2011). For biodegradable polymers, their biocompatibility can be influenced by the chemical, physical, and biological properties. Specifically, the material composition, micro- or nanostructure, morphology, crystallinity, mechanical properties, hydrophilicity, porosity, surface chemical composition, degradation profile and product, toxicity, additive, and catalyst are among the possible factors that should be first considered (Williams, 2008).

Table 21.1 Mechanical properties of biodegradable synthetic polymer. (a) PGA: polyglycolide; (b) PLLA: poly(L-lactide); (c) PDLLA: poly(D,L-lactide); (d) PCL: polycaprolactone; (e) PTMC: poly(trimethylene carbonate); (f) PLGA: poly(lactide-co-glycolide); (g) PLCL: poly(lactide-co-caprolactone); (h) PHB: poly(3-hydroxybutyrate); (i) PH4B: poly(4-hydroxybutyrate); (j) PGS: poly(glycerol-sebacate), ratio of 1:1, postpolymerized (120°C, 77 h); (k) PGSA: poly(glycerol sebacate) acrylate, mole ratio of 1:1, degree of acrylation of 0.17–0.54, polymerization 120°C for 77 h + photopolymerization of 10 min; (L) POC: poly(1,8-octanediol-citrate), ratio of 1:1, postpolymerization (80–120°C, vacuum or no, for 1–6 days).

Polymer	Young's modulus (MPa)	Tensile strength (MPa)	Elongation at break (%)	Degradation	References
PGA (a)	6900	70	<3	1.5 months	(Martin & Williams, 2003)
PLLA (b)	1200–2700	28–50	6	16–60 months	(Martin & Williams, 2003)
PDLLA (c)	1900–2400	29–35	6	–	(Yang, Leong, Du, & Chua, 2001; Barrett & Yousaf, 2009)
PCL (d)	0.21–0.34	20.7–34.5	300–500	>24 months	(Barrett & Yousaf, 2009)
PTMC (e)	6.3–6.8	12–24	820–831	>8 weeks	(Pêgo, Poot, Grijpma, & Feijin, 2003; Zhang, Kujier, Bulstra, Grijpma, & Feijen, 2006)
PLGA (f)	1.4–2.8	41.4–55.2	3–10	1–12 months	(Rezwan, Chen, Blaker, & Boccaccini, 2006; Seal, Otero, & Panitch, 2001)
PLCL (g)	0.192–68.57	0.57–8.55	175–854.4		(Barrett & Yousaf, 2009)
P3HB (h)	2.5	36	3	24 months	(Martin & Williams, 2003)
P4HB (i)	0.07	50	1000	2–12 months	(Martin & Williams, 2003)
PGS (j)	0.282	>0.5	>267	1 month	(Wang et al., 2002)
PGSA (k)	0.048–1.375	0.054–0.498	47.4–170	>48 h	(Nijst et al., 2007)
POC (l)	1.85–6.44	2.93–5.80	117–367	>26 weeks	(Yang, Webb, & Ameer, 2004; Yang, Webb, Pickerill, Hageman, & Ameer, 2006)

3. Preparation of biodegradable synthetic polyesters

The majority of the currently used and under-investigated biodegradable polymers are members of the polyester family. Polyesters were historically the first family of synthetic polymers developed from condensation polymerization (Carothers, 1929; Carothers & Arvin, 1929). Polyesters can be designed into biocompatible biomaterials that extend over a wide range of rigidities, flexibilities, and strengths. These can be done by incorporating different choices of multifunctional monomers, changing the synthesizing parameters (i.e., temperature, pressure, and duration) and for certain polyesters, tuning the postcuring conditions. In addition, the degradation rates of polyesters can be easily tuned to meet the specifications of a particular application.

Okada and friends reviewed different synthetic routes for developing biodegradable polyesters (Masahiko, 2002). Polyesters can be synthesized by polycondensation of combinations of diols and dicarboxylic acids, self-polycondensation of hydroxy acids, or by ring-opening polymerization. For example, PLA can be prepared by either condensation polymerization of hydroxy acids or from ring-opening polymerization of lactones and lactides. Biodegradable polyesters such as PGS and POC are among the examples of polyesters synthesized from polycondensation of a mixture of diols and dicarboxylic acids. Recently, enzymatic polyesterification also has been investigated to synthesize biodegradable polyesters.

The polycondensation reaction, which is a simple, convenient, and cost-effective method to develop polyesters, is involved in the esterification reaction between alcohol and acid monomers with two or more functional groups. For example, diacids and diols, diacids chlorides and diols, and diesters and diols can be used to synthesize polyester. Network polyesters are formed if the monomers used have a functional group greater than two. However, the main limitation of this method is that harsh synthesizing conditions (vacuum, high temperatures, and long reaction times) are required in order to achieve a high-molecular-weight product. This reaction is also limited to an equilibrium where the byproducts, for example, water, must be removed during the reaction to obtain high-molecular-weight polymers and greater conversion (Edlund & Albertsson, 2003; Pang et al., 2006).

An alternative method to develop high-molecular-weight polyester is from ring-opening polymerization of cyclic monomers. Compared to polycondensation, this route offered shorter reaction times, the ability to work under mild reaction conditions, and reaction conversions are not affected by the byproducts and high-molecular-weight polymers (Madhavan Nampoothiri, Nair, & John, 2010; Williams, 2007). Taking cyclic lactones as an example, aliphatic polyesters can be synthesized through ring-opening polymerization from lactones with various ring sizes (Williams, 2007). A wide range of molecular weights and rates of polymerization can be achieved by using different initiators and organometallic catalysts and has been extensively investigated (Gupta & Kumar, 2007; Jérôme & Lecomte, 2008). However, while organometallic compounds are characterized by high toxicity, it is practically impossible to completely remove the catalysts from the polymers, which is one of the major limitations of this route.

Enzymatic polyesterification is an alternative approach that uses a biocatalyst, such as lipase enzymes, to catalyze the condensation of diols and diacids or diesters monomers. Biocatalysts are very suitable to be used with polyesters of bacterial origin and enzymatically degradable polyesters such as polyhydroxyalkanoates (PHA), poly(lactic acid), and poly(carbonate). Enzymatic polyesterification has been extensively investigated and reviewed by Kobayashi and colleagues (Kobayashi, 1999; Kobayashi et al., 2001; Kobayashi & Makino, 2009) and other researchers (Albertsson & Srivastava, 2008; Kadokawa & Kobayashi, 2010; Uyama, 2007). The advantages of enzymatic polyesterification include (Albertsson & Srivastava, 2008; Varma, Ann-Christine, Ritimoni, & Srivastava, 2005):

- Mild reaction conditions with high selectivity;
- Enzymes can be easily separated after the reaction and are recyclable;
- Enzymes can be used in bulk media, organic media, and various interfaces;
- The synthesized polymers have well defined structures;
- When lipases are used as the catalysts for polyester synthesis, they do not require the removal of water and air.

Under normal polymerization conditions, enzymes have shown the ability to polymerize macrolides compared to organometallic catalysts that can only polymerize small (four to seven members) cyclic lactones.

Theoretically, polyesters are all degradable polymers because esterification is thermodynamically a reversible reaction. In practice, the backbones of aromatic polyesters are hydrophobic, hence the labile bonds are protected. Therefore, aliphatic polyesters are degradable only if the polymers have reasonable short methylene chains between the ester bonds. Aliphatic polyesters undergo chemical degradation by hydrolytic cleavage of the ester bonds in the polymer's backbone, or by enzymatic degradation or a combination of these two processes (Göpferich, 1996, 1997; Vert, 2005).

4. Biodegradable synthetic thermally cross-linked polyester blends

Many new polymerization strategies, including thermal polycondensation, are being explored to develop biodegradable polymers to cover various biomedical applications. A catalyst-free polycondensation technique to prepare polyol-based biodegradable polyesters is attractive in terms of its feasibility. The chances of the synthesized polyesters inducing a toxic response are minimized because no toxic catalyst or solvent was involved in the polymerization process. The final properties of the polymers will vary and depend on the choice of monomer, reaction, and curing conditions. The following sections discuss polyol-based biodegradable polyesters synthesized from this method.

4.1 Poly(sorbitol sebacate malate) blend

Poly(sorbitol sebacate malate) (PSSM) thermoset polyesters were prepared by a catalyst-free polycondensation reaction based on the schematic represented in Fig. 21.1 (Tham, 2015). The synthesis was carried out at different conditions by altering either the malic acid feed ratio or the postpolymerization duration to prepare PSSM polyesters with a range of properties. In brief, PSSM prepolymer was synthesized by polycondensation of sorbitol, sebacic acid, and malic acid, with an excess of the stoichiometric alcohol groups. The reaction was carried out at 150°C for 1 h under a high flow rate of nitrogen gas without using catalyst or solvent to aid the polyesterification process.

4.2 Poly(xylitol-co-dodecanedioate) blend

Wong (2015) had successfully synthesized xylitol-based thermally cross-linked polyesters through catalyst-free polymerization. Spectroscopic analysis by FTIR proved the formation of ester structures, while quantification of polymer cross-linking was done by sol−gel content analysis. WAXD pattern and DSC thermograms of poly(xylitol-co-1,12-dodecanedioate) (PXDD) proved that PXDD is a semi-crystalline polymer.

Interestingly, the formation of a polymer crystalline structure within the PXDD matrix could be functionalized to enable shape-memory properties. At high temperatures ($T_{high_} > T_m$), all crystalline structures melted so PXDD is rubber-like, but the whole matrix still exists in the solid state at this stage because the polymer chains were cross-linked. However, PXDD becomes tough as soon as it is cooled to a low temperature ($T_{low_} < T_m$). The formation of polymer crystalline upon cooling restricts the mobility of polymer chains. To study the shape-memory properties of PXDD requires a combination of understandings of the polymer molecular architecture with a shape-programming technique (Lendlein, 2010).

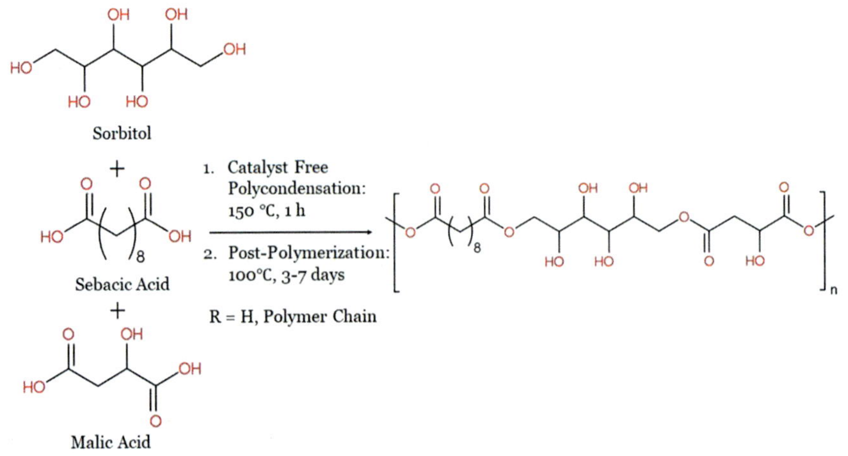

Figure 21.1 A synthesis scheme of poly(sorbitol sebacate malate).

4.3 Poly(1,8-ocatnediol-glycerol-dodecanedioate) blend

Poly(1,8-ocatnediol-glycerol-dodecanedioate) (POGDA) biodegradable polyesters composed of monomers from renewable resources were synthesized in this study with a solvent-free method (Lee, & Wahit, 2015). The results for gel content, swelling, UV-vis spectroscopy, DSC, XRD patterns, and biodegradation show that the properties of the POGDAs could be fine-tuned through variations in the molar ratio of the reactants. Lee et al. found that the shape-memory effect for POGDA was achieved by alterations in the cross-linking density and crystallinity to obtain hard and soft segments that remember permanent and temporary shapes. A polymer with tunable material properties by tailoring the monomer molar ratio is expected to have broad applications in medical fields such as drug-delivery systems and tissue engineering.

5. Biodegradable synthetic thermally cross-linked polyester composites

5.1 Poly(sorbitol sebacate malate)/hydroxyapatite composites

Novel hydroxyapatite (HAp)/poly(sorbitol sebacate malate) (PSSM) composites are a potential application in soft-tissue engineering developed by Tham and coworkers. The composites consist of a biodegradable polyester prepared from sorbitol, sebacic acid, and malic acid together with various amounts of HA (5, 10, and 15 wt.%) (Tham, Wahit, Kadir, & Wong, 2012). The effects of different weight percent of HA on the properties of composites were studied. Tensile, thermal, and chemical characterization were performed on the composites. Fourier transform infrared (FT-IR) spectroscopy was performed to analyze the interactions between the HA and PSSM matrix. Tensile tests were conducted to evaluate the mechanical properties of the composites and differential scanning calorimetry (DSC) was carried out to study the thermal properties of HA/PSSM composites. The Young's modulus and tensile strength of composites were increased with a higher concentration of HA. However, the elongation at break decreased with a higher weight percent of HA, as tabulated in Table 21.2. DSC analysis found that the glass transition temperature (T_g) of all the samples was slightly higher than at room temperature.

Fourier transform infrared (FT-IR) spectroscopy was performed to analyze the interactions between the HA and PSSM matrix. The intense peak between 1690–1750 cm^{-1} in the spectrum was assigned to carbonyl (C=O) groups, which proves the formation of ester bonds. Compared to sample C0, the carbonyl peak for sample C5 showed a small shift to the right (1730 cm^{-1}); this change could be attributed to the formation of hydrogen bonding between C=O groups of PSSM and the surface hydroxyl (–OH) group of HA. This phenomenon demonstrated that most of the terminal hydroxyl groups had reacted to form ester bonds. For sample C5, the band assigned for hydroxyl groups was still available after curing. This may be related to the incorporation of HA which hindered the esterification, causing many unreacted terminal hydroxyl groups to remain in the matrix.

Table 21.2 Mechanical and thermal properties of PSSM/HA composites.

Sample	Tensile strength at yield (MPa)	Tensile modulus (MPa)	Elongation at break (%)	Glass transition temperature, T_g (°C)
C0	16.20 ± 1.73	626.96 ± 81.04	49.04 ± 4.30	39.83
C5	19.74 ± 1.71	710.56 ± 100.34	36.06 ± 6.70	40.71
C10	22.13 ± 4.73	875.42 ± 157.58	27.04 ± 3.50	41.47
C15	23.96 ± 2.56	1026 ± 105.12	10.10 ± 2.10	42.61

From this study, the composites of HA and PSSM were successfully prepared. All the monomers, sorbitol, sebacic acid, malic acid, and hydroxyapatite are biocompatible with the human body. The preparation of the composites is relatively simple and did not involve the use of harsh solvent or catalyst.

5.2 Poly(xylitol-co-dodecanediote)/hydroxyapatite composites

Wong et al. (2014) developed a novel biobased composite shape-memory polymer (SMP) assigned as PXDD/HA (SMP) from poly(xylitol-co-dodecanedioate) (PXDD) and hydroxyapatite (HA). The polymer matrix, PXDD, was synthesized from polycondensation of xylitol, a biobased sugar polyol, and 1,12-dodecanedioic acid, a biobased dicarboxylic acid commercially produced from *Vernonia galamensis* oil. On the other hand, HA is an inorganic compound commonly applied for bone fillers and implant materials owing to its close resemblance to the mineral phase of natural bone (Fidancevska, Ruseska, Bossert, Lin, & Boccaccini, 2007; Hatzistavrou et al., 2010).

Thus, reinforcing PXDD with HA produces an SMP suitable for bone implant application, PXDD/HA. The biocomposite was synthesized through melt-blending of 1,12-dodecanedioic acid and xylitol in equimolar amounts at 165°C with vigorous stirring at 250−300 rpm, which allows polycondensation to take place. Then, HA was added to the molten mixture followed by a gradual decrease in oil bath temperature to 155°C. After 5 h, the prepolymer/HA composites were collected and cast directly on a preheated Teflon mold and left to cure at 140°C for 48 h. To study the effect of HA on chemical, structural, thermal, and shape-memory properties of the biocomposite, the amount of HA filler applied was varied to 10, 15, and 20 wt.% with respect to the total weight of the composite.

Through FTIR analysis, it was discovered that the incorporation of HA did not affect the chemical structure of PXDD. DSC results further confirm this hypothesis as melting temperature, T_m, remains at 55°C despite the increase in HA content. However, the crystallization temperature, T_c, increases with the increase in HA content. This indicates that during the cooling down of the PXDD/HA from its amorphous state, the HA particles may have provided nucleation sites for the PXDD chain segments to crystallize at a higher temperature (Wang, Kempen, Yaszemski, & Lu, 2009). Similarly, the melting enthalpies (H_m) and crystallinity, X_c, obtained from XRD analysis of PXDD/HA composites increased with increasing HA content. This confirms that the crystallinity of PXDD/HA increases with increased HA content (Sonseca, Peponi, Sahuquillo, Kenny, & Giménez, 2012).

Therefore, by increasing the HA content in the composite, the shape fixity, R_f (%), of the PXDD/HA composite was also improved significantly as higher X_c elevates dimension stability of temporary shape (Guo et al., 2011; Xue, Dai, & Li, 2010). However, the shape recovery ratio, R_r, remains unchanged with a value of approximately 100%. Interestingly, PXDD/HA with 20 wt.% of HA exhibited R_r of approximately 100% at a recovery temperature, T_{rec}, of 48.1°C within a narrow temperature range, with a T_{rec} that is higher than human body temperature, 36.9°C, as shown in Fig. 21.2.

Recently, (Wong et al., 2020) investigated two counteracting effects caused by HA addition within the poly[xylitol-(1,12-dodecanedioate)] (PXD) matrix on the structural and thermal properties. The blank PXD and all the biobased PXDHCy showed a good shape fixity ratio (R_f) higher than 93% and shape-recovery ratio (R_r) higher than 99%. This study also showed that polyester degraded hydrolytically. The tensile results show that at ambient temperature, PXD and PXDHCy composites act like a thermoplastic polymer and undergo permanent yielding as well as stress whitening because of the tensile force. Conversely, all samples show elastomeric properties when heated above T_m. From DSC testing, it can be seen that the crystallization temperatures (T_c) of all the composites were higher than neat polyester. This is due to the dominance of the physical interaction between HA and PXD matrix and as a result the PXD matrix is able to crystallize on the surface of HA particles at higher temperatures compared to

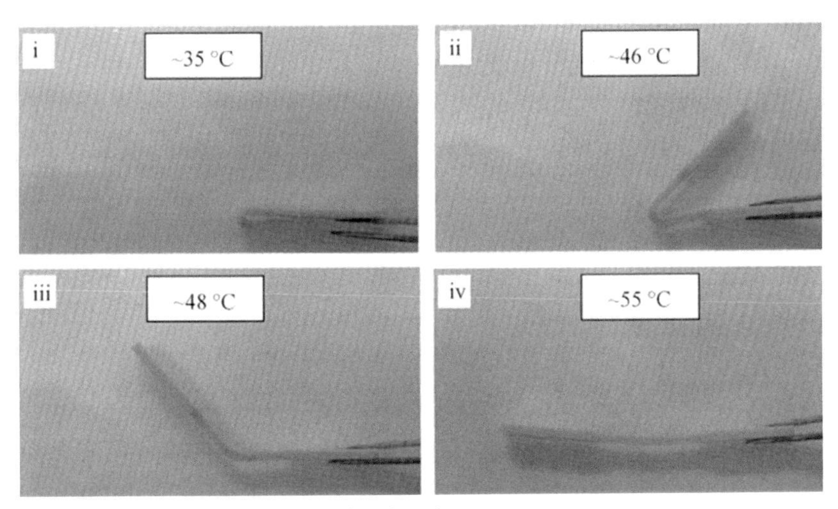

Figure 21.2 Images of PXDD/HA performing shape recovery.
From Wong, T. W., Wahit, M. U., Abdul Kadir, M. R., Soheilmoghaddam, M., & Balakrishnan, H. (2014). A novel poly(xylitol-co-dodecanedioate)/hydroxyapatite composite with shape-memory behavior. Materials Letters, *126*, 105–108. https://doi.org/10.1016/j.matlet.2014.04.020.

neat PXD. Nevertheless, the existence of the physical interaction between polymer matrix and HA particles will also provide a higher thermal transition temperature (Lei et al., 2009). The PXD polyester biobased composites were prepared with biobased potential to be used in biomedical applications.

5.3 Poly(1,8-ocatnediol-glycerol-dodecanedioate)/hydroxyapatite composites

Incorporation of HA particles into POGDA was done using the same solvent- and catalyst-free polycondensation polymerization method to improve the mechanical properties and biocompatibility of POGDA polyester (Lee et al., 2016). Loading of HA particles and their concentration affected the material properties in various ways. The T_c and T_m of POGDA/HA composites were increased by the loading of HA in the polyester. HA acted as a nucleating agent in the POGDA matrix and facilitated the crystallization process. However, HA particles restricted the movement of molecular chains at higher concentrations and caused a decrease in the crystallinity of the composites. Loading of HA particles significantly improved the mechanical properties of the composites but overloading of HA resulted in a decrease in elongation at break.

In vitro biodegradation tests revealed the ability of HA particles to slow down the degradation rate of the composites compared to POGDA (0.5 Gly) by acting as a buffer during degradation. The cell morphology cultured with POGDA/HA composites appeared healthier compared to POGDA (0.5 Gly) polyester, which indicated the improved biocompatibility of HA particles. The scaffolds made from HA 1% composites were fabricated in the present study using the salt-leaching method without using any solvent. TGA tests confirmed that the extraction of salt from the scaffolds was completed. With a mixing ratio of 1 HA 1%:3 sieved salt, porosity of the S1:3 scaffolds were approximately 68% with a pore size in the range 150−355 μm. The size of pores in scaffolds can be manipulated with the size of the sieved salt, while the porosity of the scaffolds can be easily altered with the different mixing ratios of prepolymer and sieved salt. The shape-memory effect showed by the S1:3 scaffolds indicates the loading of HA and scaffold fabrication method did not affect the temperature-responsive behavior of POGDA (0.5 Gly). Good biocompatibility revealed by S1:3 scaffolds in the Alamar Blue assay as the fibroblasts showed the ability to proliferate in the scaffolds.

6. Remarks and future directions

This chapter discusses the properties of some of our research in biodegradable polymeric materials for potential biomedical applications. Previous scientific research shows the biocompatibility of biodegradable polyester. Current research synthesized biodegradable and thermoresponsive polyesters and composites using monomers from renewable resources and using HA as a filler. Several tests have been conducted to understand the properties of the materials and their potential in the biomedical field. Researchers are paying full attention in order to produce a multifunctional polymeric system due to the increasing importance of biomedical applications. Therefore, many of them have put a lot of effort into obtaining a whole new series of biocompatible polymers with controllable biodegradation rate and mechanical properties.

Acknowledgment

This work was supported by Universiti Teknologi Malaysia (UTM). The research was funded by the Ministry of Science, Technology and Innovation for E-Science grant (Vote No: 79400) and Ministry of Higher Education (MOHE) under Prototype Research Grant Scheme (PRGS) No: 4L608. The authors would also like to acknowledge UTM for the PDRU grant (Q.J130000.21A2.05E46).

References

Albertsson, A. C., & Srivastava, R. K. (2008). Recent developments in enzyme-catalyzed ring-opening polymerization. *Advanced Drug Delivery Reviews, 60*(9), 1077–1093. https://doi.org/10.1016/j.addr.2008.02.007

Amsden, B. (2007). Curable, biodegradable elastomers: emerging biomaterials for drug delivery and tissue engineering. *Soft Matter, 3*(11), 1335–1348. https://doi.org/10.1039/b707472g

Aravamudhan, A., Ramos, D. M., Nada, A. A., & Kumbar, S. G. (2014). Natural polymers: polysaccharides and their derivatives for biomedical applications. In *Natural and synthetic biomedical polymers* (pp. 67–89). Elsevier Inc. https://doi.org/10.1016/B978-0-12-396983-5.00004-1

Ashammakhi, N., & Rokkanen, P. (1997). Absorbable polyglycolide devices in trauma and bone surgery. *Biomaterials, 18*(1), 3–9. https://doi.org/10.1016/S0142-9612(96)00107-X

Athanasiou, K. A., Niederauer, G. G., & Agrawal, C. M. (1996). Sterilization, toxicity, biocompatibility and clinical applications of polylactic acid/polyglycolic acid copolymers. *Biomaterials, 17*(2), 93–102. https://doi.org/10.1016/0142-9612(96)85754-1

Barrett, D. G., & Yousaf, M. N. (2009). Design and applications of biodegradable polyester tissue scaffolds based on endogenous monomers found in human metabolism. *Molecules, 14*(10), 4022–4050. https://doi.org/10.3390/molecules14104022

Bruggeman, J. P., de Bruin, B. J., Bettinger, C. J., & Langer, R. (2008). Biodegradable poly(-polyol sebacate) polymers. *Biomaterials, 29*(36), 4726–4735. https://doi.org/10.1016/j.biomaterials.2008.08.037

Carothers, W. H. (1929). Studies on polymerization and ring formation. I. An introduction to the general theory of condensation polymers. *Journal of the American Chemical Society, 51*(8), 2548–2559. https://doi.org/10.1021/ja01383a041

Carothers, W. H., & Arvin, J. A. (1929). Studies on polymerization and ring formation. II. Polyesters. *Journal of the American Chemical Society, 51*(8), 2560–2570. https://doi.org/10.1021/ja01383a042

Chen, G. Q., & Wu, Q. (2005). The application of polyhydroxyalkanoates as tissue engineering materials. *Biomaterials, 26*(33), 6565–6578. https://doi.org/10.1016/j.biomaterials.2005.04.036

Chen, Q., Liang, S., & Thouas, G. A. (2013). Elastomeric biomaterials for tissue engineering. *Progress in Polymer Science, 38*(3–4), 584–671. https://doi.org/10.1016/j.progpolymsci.2012.05.003

Cherng, J. Y., Hou, T. Y., Shih, M. F., Talsma, H., & Hennink, W. E. (2013). Polyurethane-based drug delivery systems. *International Journal of Pharmaceutics, 450*(1–2), 145–162. https://doi.org/10.1016/j.ijpharm.2013.04.063

Daamen, W. F., Veerkamp, J. H., van Hest, J. C. M., & van Kuppevelt, T. H. (2007). Elastin as a biomaterial for tissue engineering. *Biomaterials, 28*(30), 4378–4398. https://doi.org/10.1016/j.biomaterials.2007.06.025

Duan, X., McLaughlin, C., Griffith, M., & Sheardown, H. (2007). Biofunctionalization of collagen for improved biological response: scaffolds for corneal tissue engineering. *Biomaterials, 28*(1), 78–88. https://doi.org/10.1016/j.biomaterials.2006.08.034

Edlund, U., & Albertsson, A. C. (2003). Polyesters based on diacid monomers. *Advanced Drug Delivery Reviews, 55*(4), 585–609. https://doi.org/10.1016/S0169-409X(03)00036-X

Emadi, R., Tavangarian, F., & Esfahani, S. I. R. (2010). Biodegradable and bioactive properties of a novel bone scaffold coated with nanocrystalline bioactive glass for bone tissue engineering. *Materials Letters, 64*(13), 1528–1531. https://doi.org/10.1016/j.matlet.2010.04.011

Fidancevska, E., Ruseska, G., Bossert, J., Lin, Y. M., & Boccaccini, A. R. (2007). Fabrication and characterization of porous bioceramic composites based on hydroxyapatite and titania. *Materials Chemistry and Physics, 103*(1), 95–100. https://doi.org/10.1016/j.matchemphys.2007.01.015

Göpferich, A. (1996). Mechanisms of polymer degradation and erosion. *Biomaterials, 17*(2), 103–114. https://doi.org/10.1016/0142-9612(96)85755-3

Göpferich, A. (1997). Polymer bulk erosion. *Macromolecules, 30*(9), 2598–2604. https://doi.org/10.1021/ma961627y

Gunatillake, P. A., Adhikari, R., & Gadegaard, N. (2003). Biodegradable synthetic polymers for tissue engineering. *European Cells and Materials, 5*, 1–16. https://doi.org/10.22203/eCM.v005a01

Guo, B., Chen, Y., Lei, Y., Zhang, L., Zhou, W. Y., Rabie, A. B. M., & Zhao, J. (2011). Biobased Poly(propylene sebacate) as shape memory polymer with tunable switching temperature for potential biomedical applications. *Biomacromolecules, 12*(4), 1312–1321. https://doi.org/10.1021/bm2000378

Gupta, A. P., & Kumar, V. (2007). New emerging trends in synthetic biodegradable polymers - polylactide: A critique. *European Polymer Journal, 43*(10), 4053–4074. https://doi.org/10.1016/j.eurpolymj.2007.06.045

Harkin, D. G., George, K. A., Madden, P. W., Schwab, I. R., Hutmacher, D. W., & Chirila, T. V. (2011). Silk fibroin in ocular tissue reconstruction. *Biomaterials, 32*(10), 2445–2458. https://doi.org/10.1016/j.biomaterials.2010.12.041

Hatzistavrou, E., Chatzistavrou, X., Papadopoulou, L., Kantiranis, N., Kontonasaki, E., Boccaccini, A. R., & Paraskevopoulos, K. M. (2010). Characterisation of the bioactive behaviour of sol–gel hydroxyapatite–CaO and hydroxyapatite–CaO–bioactive glass composites. *Materials Science and Engineering: C, 30*(3), 497–502. https://doi.org/10.1016/j.msec.2010.01.009

Hayashi, T. (1994). Biodegradable polymers for biomedical uses. *Progress in Polymer Science, 19*(4), 663–702. https://doi.org/10.1016/0079-6700(94)90030-2

Hemmrich, K., Salber, J., Meersch, M., Wiesemann, U., Gries, T., Pallua, N., & Klee, D. (2008). Three-dimensional nonwoven scaffolds from a novel biodegradable poly(ester amide) for tissue engineering applications. *Journal of Materials Science: Materials in Medicine, 19*(1), 257–267. https://doi.org/10.1007/s10856-006-0048-3

Huang, R., Zhu, X., Tu, H., & Wan, A. (2014). The crystallization behavior of porous poly(lactic acid) prepared by modified solvent casting/particulate leaching technique for potential use of tissue engineering scaffold. *Materials Letters, 136*, 126–129. https://doi.org/10.1016/j.matlet.2014.08.044

Jérôme, C., & Lecomte, P. (2008). Recent advances in the synthesis of aliphatic polyesters by ring-opening polymerization. *Design and Development Strategies of Polymer Materials for Drug and Gene Delivery Applications, 60*(9), 1056–1076. https://doi.org/10.1016/j.addr.2008.02.008

Jiao, Y. P., & Cui, F. Z. (2007). Surface modification of polyester biomaterials for tissue engineering. *Biomedical Materials, 2*(4), R24–R37. https://doi.org/10.1088/1748-6041/2/4/R02

Pang, K., Kotak, R., & Tonelli, A. (2006). Review of conventional and novel polymerization processes for polyesters. *Progress in Polymer Science*, 1009–1037. https://doi.org/10.1016/j.progpolymsci.2006.08.008

Kadokawa, J.i., & Kobayashi, S. (2010). Polymer synthesis by enzymatic catalysis. *Current Opinion in Chemical Biology, 14*(2), 145–153. https://doi.org/10.1016/j.cbpa.2009.11.020

Kobayashi, S. (1999). Enzymatic polymerization: a new method of polymer synthesis. *Journal of Polymer Science Part A: Polymer Chemistry, 37*(16), 3041–3056. https://doi.org/10.1002/(SICI)1099-0518(19990815)37:16<3041::AID-POLA1>3.0.CO;2-V

Kobayashi, S., & Makino, A. (2009). Enzymatic polymer synthesis: an opportunity for green polymer chemistry. *Chemical Reviews, 109*(11), 5288–5353. https://doi.org/10.1021/cr900165z

Kobayashi, S., Uyama, H., & Kimura, S. (2001). Enzymatic polymerization. *Chemical Reviews, 101*(12), 3793–3818. https://doi.org/10.1021/cr990121l

Kucinska-Lipka, J., Gubanska, I., Janik, H., & Sienkiewicz, M. (2015). Fabrication of polyurethane and polyurethane based composite fibres by the electrospinning technique for soft tissue engineering of cardiovascular system. *Materials Science and Engineering: C, 46*, 166–176. https://doi.org/10.1016/j.msec.2014.10.027

Lee, X. Y., & Wahit, M. U. (2015). Synthesis of biodegradable polyester by polycondensation with tunable properties. *Advanced Materials Research, 1119*, 418–422. https://doi.org/10.4028/www.scientific.net/AMR.1119.418

Lee, X. Y., Wahit, M. U., & Adrus, N. (2016). Biodegradable and temperature-responsive thermoset polyesters with renewable monomers. *Journal of Applied Polymer Science, 133*(40). https://doi.org/10.1002/app.44007

Lee, A. Y., Mahler, N., Best, C., Lee, Y. U., & Breuer, C. K. (2014). Regenerative implants for cardiovascular tissue engineering. *Translational Research, 163*(4), 321–341. https://doi.org/10.1016/j.trsl.2014.01.014

Lei, L., Li, L., Zhang, L., Chen, D., & Tian, W. (2009). Structure and performance of nano-hydroxyapatite filled biodegradable poly((1,2-propanediol-sebacate)-citrate) elastomers. *Polymer Degradation and Stability, 94*(9), 1494–1502. https://doi.org/10.1016/j.polymdegradstab.2009.04.034

Lendlein, A. (2010). Progress in actively moving polymers. *Journal of Materials Chemistry, 20*(17), 3332–3334. https://doi.org/10.1039/c004361n

Li, L. C., Deng, J., & Stephens, D. (2002). Polyanhydride implant for antibiotic delivery - from the bench to the clinic. *Advanced Drug Delivery Reviews, 54*(7), 963–986. https://doi.org/10.1016/S0169-409X(02)00053-4

Li, Y., Huang, W., Cook, W. D., & Chen, Q. (2013). A comparative study on poly(xylitol sebacate) and poly(glycerol sebacate): mechanical properties, biodegradation and cytocompatibility. *Biomedical Materials, 8*(3), 035006. https://doi.org/10.1088/1748-6041/8/3/035006

Liu, Q., Jiang, L., Shi, R., & Zhang, L. (2012). Synthesis, preparation, in vitro degradation, and application of novel degradable bioelastomers - a review. *Progress in Polymer Science, 37*(5), 715–765. https://doi.org/10.1016/j.progpolymsci.2011.11.001

Ma, P. X. (2008). Biomimetic materials for tissue engineering. *Advanced Drug Delivery Reviews, 60*(2), 184–198. https://doi.org/10.1016/j.addr.2007.08.041

Macocinschi, D., Filip, D., & Vlad, S. (2011). Natural-based polyurethane biomaterials for medical applications. *Biomaterials applications for nanomedicine*. IntechOpen.

Madhavan Nampoothiri, K., Nair, N. R., & John, R. P. (2010). An overview of the recent developments in polylactide (PLA) research. *Bioresource Technology, 101*(22), 8493–8501. https://doi.org/10.1016/j.biortech.2010.05.092

Martin, D. P., & Williams, S. F. (2003). Medical applications of poly-4-hydroxybutyrate: a strong flexible absorbable biomaterial. *Biochemical Engineering Journal, 16*(2), 97–105. https://doi.org/10.1016/S1369-703X(03)00040-8

Martina, M., & Hutmacher, D. W. (2007). Biodegradable polymers applied in tissue engineering research: a review. *Polymer International, 56*(2), 145–157. https://doi.org/10.1002/pi.2108

Masahiko, O. (2002). Chemical syntheses of biodegradable polymers. *Progress in Polymer Science*, 87–133. https://doi.org/10.1016/s0079-6700(01)00039-9

Mei, Y., Kumar, A., Gao, W., Gross, R., Kennedy, S. B., Washburn, N. R., Amis, E. J., & Elliott, J. T. (2004). Biocompatibility of sorbitol-containing polyesters. Part I: synthesis, surface analysis and cell response in vitro. *Biomaterials, 25*(18), 4195–4201. https://doi.org/10.1016/j.biomaterials.2003.10.087

Nair, L. S., & Laurencin, C. T. (2006). Polymers as biomaterials for tissue engineering and controlled drug delivery. *Advances in Biochemical Engineering/Biotechnology, 102*, 47–90. https://doi.org/10.1007/b137240

Nair, L. S., & Laurencin, C. T. (2007). Biodegradable polymers as biomaterials. *Polymers in Biomedical Applications, 32*(8), 762–798. https://doi.org/10.1016/j.progpolymsci.2007.05.017

Nijst, C. L. E., Bruggeman, J. P., Karp, J. M., Ferreira, L., Zumbuehl, A., Bettinger, C. J., & Langer, R. (2007). Synthesis and characterization of photocurable elastomers from poly(glycerol-*co*-sebacate). *Biomacromolecules, 8*(10), 3067–3073. https://doi.org/10.1021/bm070423u

O'Brien, F. J. (2011). Biomaterials & scaffolds for tissue engineering. *Materials Today, 14*(3), 88–95. https://doi.org/10.1016/S1369-7021(11)70058-X

Pêgo, A. P., Poot, A. A., Grijpma, D. W., & Feijin, J. (2003). Biodegradable elastomeric scaffolds for soft tissue engineering. *Journal of Controlled Release, 87*(1–3), 69–79. https://doi.org/10.1016/s0168-3659(02)00351-6

Pintado-Sierra, M., Delgado, L., Aranaz, I., Marcos-Fernández, Á., Reinecke, H., Gallardo, A., Zeugolis, D., & Elvira, C. (2014). Surface hierarchical porosity in poly (ε-caprolactone) membranes with potential applications in tissue engineering prepared by foaming in supercritical carbon dioxide. *The Journal of Supercritical Fluids, 95*, 273–284. https://doi.org/10.1016/j.supflu.2014.09.019

Piskin, E. (1995). Biodegradable polymers as biomaterials. *Null, 6*(9), 775–795. https://doi.org/10.1163/156856295X00175

Rai, R., Tallawi, M., Grigore, A., & Boccaccini, A. R. (2012). Synthesis, properties and biomedical applications of poly(glycerol sebacate) (PGS): a review. *Progress in Polymer Science, 37*(8), 1051–1078. https://doi.org/10.1016/j.progpolymsci.2012.02.001

Rezwan, K., Chen, Q. Z., Blaker, J. J., & Boccaccini, A. R. (2006). Biodegradable and bioactive porous polymer/inorganic composite scaffolds for bone tissue engineering. *Biomaterials, 27*(18), 3413–3431. https://doi.org/10.1016/j.biomaterials.2006.01.039

Seal, B. L., Otero, T. C., & Panitch, A. (2001). Polymeric biomaterials for tissue and organ regeneration. *Materials Science and Engineering: R: Reports, 34*(4), 147–230. https://doi.org/10.1016/S0927-796X(01)00035-3

Serrano, M. C., Chung, E. J., & Ameer, G. A. (2010). Advances and applications of biodegradable elastomers in regenerative medicine. *Advanced Functional Materials, 20*(2), 192–208. https://doi.org/10.1002/adfm.200901040

Seyednejad, H., Ghassemi, A. H., Van Nostrum, C. F., Vermonden, T., & Hennink, W. E. (2011). Functional aliphatic polyesters for biomedical and pharmaceutical applications. *Journal of Controlled Release, 152*(1), 168–176. https://doi.org/10.1016/j.jconrel.2010.12.016

Shi, R., Chen, D., Liu, Q., Wu, Y., Xu, X., Zhang, L., & Tian, W. (2009). Recent advances in synthetic bioelastomers. *International Journal of Molecular Sciences, 10*(10), 4223–4256. https://doi.org/10.3390/ijms10104223

Sokolsky-Papkov, M., Agashi, K., Olaye, A., Shakesheff, K., & Domb, A. J. (2007). Polymer carriers for drug delivery in tissue engineering. *Advanced Drug Delivery Reviews, 59*(4–5), 187–206. https://doi.org/10.1016/j.addr.2007.04.001

Sonseca, A., Peponi, L., Sahuquillo, O., Kenny, J. M., & Giménez, E. (2012). Electrospinning of biodegradable polylactide/hydroxyapatite nanofibers: Study on the morphology, crystallinity structure and thermal stability. *Polymer Degradation and Stability, 97*(10), 2052–2059. https://doi.org/10.1016/j.polymdegradstab.2012.05.009

Tham, W. H (2015). *Synthesis and characterization of novel citric acid-based polyester elastomer*. Universiti Teknologi Malaysia.

Tham, W. H., Wahit, M. U., Kadir, M. R. A, & Wong, T. (2012). Mechanical and thermal properties of biodegradable hydroxyapatite/poly (sorbitol sebacate malate) composites. *Songklanakarin Journal of Science and Technology, 35*, 57–67.

Tran, R. T., Zhang, Y., Gyawali, D., & Yang, J. (2009). Recent developments on citric acid derived biodegradable elastomers. *Recent Patents on Biomedical Engineering, 2*(3), 216–227. https://doi.org/10.2174/1874764710902030216

Uyama, H. (2007). Enzymatic polymerization. In *Future directions in biocatalysis* (pp. 205–251). Elsevier. https://doi.org/10.1016/B978-044453059-2/50010-8

Varma, I. K., Ann-Christine, A., Ritimoni, R., & Srivastava, R. (2005). Enzyme catalyzed synthesis of polyesters. *Progress in Polymer Science*, 949–981. https://doi.org/10.1016/j.progpolymsci.2005.06.010

Vert, M. (2005). Aliphatic polyesters: great degradable polymers that cannot do everything. *Biomacromolecules, 6*(2), 538–546. https://doi.org/10.1021/bm0494702

Wang, S., Kempen, D. H. R., Yaszemski, M. J., & Lu, L. (2009). The Roles of matrix polymer crystallinity and hydroxyapatite nanoparticles in modulating material properties of photo-crosslinked composites and bone marrow stromal cell responses. *Biomaterials, 30*(20), 3359–3370. https://doi.org/10.1016/j.biomaterials.2009.03.015

Wang, Y., Ameer, G. A., Sheppard, B. J., & Langer, R. (2002). A tough biodegradable elastomer. *Nature Biotechnology, 20*(6), 602–606. https://doi.org/10.1038/nbt0602-602

Williams, C. K. (2007). Synthesis of functionalized biodegradable polyesters. *Chemical Society Reviews, 36*(10), 1573–1580. https://doi.org/10.1039/b614342n

Williams, D. F. (2003). Biomaterials and tissue engineering in reconstructive surgery. *Sadhana, 28*(3), 563–574. https://doi.org/10.1007/BF02706447

Williams, D. F. (1999). *The Williams dictionary of biomaterials.*

Williams, D. F. (2008). On the mechanisms of biocompatibility. *Biomaterials, 29*(20), 2941–2953. https://doi.org/10.1016/j.biomaterials.2008.04.023

Wong, T. (2015). *Synthesis and characterization of novel biodegradableand biocompatible xylitol-based shape-memory polymer/hydroxyapatite composites.* Universiti Teknologi Malaysia.

Wong, T. W., Behl, M., Syazana Mohd Yusoff, N. I., Li, T., Wahit, M. U., Ismail, A. F., Zhao, Q., & Lendlein, A. (2020). Bio-based composites from plant based precursors and hydroxyapatite with shape-memory capability. *Composites Science and Technology, 194.* https://doi.org/10.1016/j.compscitech.2020.108138

Wong, T. W., Wahit, M. U., Abdul Kadir, M. R., Soheilmoghaddam, M., & Balakrishnan, H. (2014). A novel poly(xylitol-co-dodecanedioate)/hydroxyapatite composite with shape-memory behaviour. *Materials Letters, 126*, 105–108. https://doi.org/10.1016/j.matlet.2014.04.020

Woodruff, M. A., & Hutmacher, D. W. (2010). The return of a forgotten polymer - polycaprolactone in the 21st century. *Progress in Polymer Science, 35*(10), 1217–1256. https://doi.org/10.1016/j.progpolymsci.2010.04.002

Xue, L., Dai, S., & Li, Z. (2010). Biodegradable shape-memory block co-polymers for fast self-expandable stents. *Biomaterials, 31*(32), 8132–8140. https://doi.org/10.1016/j.biomaterials.2010.07.043

Yang, J, Webb, A. R., & Ameer, G. A. (2004). Novel Citric Acid-Based Biodegradable Elastomers for Tissue Engineering. *Advanced Materials, 16*(6), 511–516. https://doi.org/10.1002/adma.200306264

Yang, J., Webb, A. R., Pickerill, S. J., Hageman, G., & Ameer, G. A. (2006). Synthesis and evaluation of poly(diol citrate) biodegradable elastomers. *Biomaterials, 27*(9), 1889–1898. https://doi.org/10.1016/j.biomaterials.2005.05.106

Yang, S., Leong, K., Du, Z., & Chua, C. (2001). The design of scaffolds for use in Tissue engineering. Part I. traditional factors. *Tissue Engineering, 7*(6), 679–689. https://doi.org/10.1089/107632701753337645

Zhang, Y., Chan, H. F., & Leong, K. W. (2013). Advanced materials and processing for drug delivery: the past and the future. *Advanced Drug Delivery Reviews, 65*(1), 104–120. https://doi.org/10.1016/j.addr.2012.10.003

Zhang, Z., Kujier, R., Bulstra, S. K., Grijpma, D. K., & Feijen, J. (2006). The in vivo and in vitro degradation behavior of poly(trimethylene carbonate). *Biomaterials, 27*(9), 1741–1748. https://doi.org/10.1016/j.biomaterials.2005.09.017

Zhang, Z., Ortiz, O., Goyal, R., Kohn, J., Lanza, R., Langer, R., & Vacanti, J. (2014). *Chapter 23 - biodegradable polymers* (pp. 441–473). Academic Press. https://doi.org/10.1016/B978-0-12-398358-9.00023-9

Biodegradable polymer blends for tissue engineering

Aarsha Surendren[1,2], *Naga Srilatha Cheekuramelli*[1,2] *and Ravi Prakash Magisetty*[1,2]
[1]Central Institute of Petrochemicals Engineering and Technology (CIPET): Institute of Plastics Technology (IPT), Kochi, Kerala, India; [2]Materials Engineering Division, Defence Institute of Advanced Technology, Pune, Maharashtra, India

1. Introduction

Tissue engineering is an amenable approach to the regeneration of biological substitutes that restore, maintain, or recuperate tissue function (Langer and Vacanti, 1993). The ultimate goal of tissue engineering is to repair or replace organs or tissues that have failed through congenital abnormalities, genetic errors, and traumatic injuries (Langer and Vacanti, 1993). The selection of the correct cells, the correct environment, the correct biomolecules that impart cells to be healthy and productive, and the presence of physical and mechanical forces for the development of cells are four major factors that affect tissue engineering (Christy et al., 2020). Tissue engineering is classified into two types, soft-tissue engineering and hard-tissue engineering, where soft-tissue engineering deals with skin, ligaments, blood vessels, cardiac patches, skeletal muscles, and nerves. In contrast, hard-tissue engineering deals with bone (Abedalwafa et al., 2013). Currently, the need for artificial tissues and organs is rapidly increasing, and donor organ availability is decreasing (Kramschuster & Turng, 2013). The chronic shortage of organs, especially kidneys and livers, has prompted scientists to search for an alternative way to develop artificial biological organs. The development of the tissue engineering process had paved an excellent way for millions of patients to regain their health with a long life span. Biologically active artificial organs are usually developed using different cells and biomaterials, which replicate the shape and function of mammalian organs.

Biomaterials are any natural or synthetic materials used to manufacture devices for a particular part or function of a body in a reliable and physiologically acceptable way (Park & Lakes, 2007). They also act as a three-dimensional microstructure for cellular interaction and a reservoir for cell-signaling molecules. According to the origin of biomaterials, biomaterials used for scaffold engineering have been classified into four types: protein-based biomaterials, polysaccharide-based biomaterials, ceramic-based biomaterials, and polymer-based biomaterials.

In the current scenario, a wide variety of biomaterial scaffolds have been used for tissue engineering, including metals, ceramics, polymers, and their blends and composites. However, among these, polymeric biomaterials and their blends and

Biodegradable Polymers, Blends and Composites. https://doi.org/10.1016/B978-0-12-823791-5.00022-3

composites are gaining increasing attention in the field of tissue engineering owing to their biodegradability and lower immune rejection rate than conventional metals and ceramics. This chapter discusses the current trends in biodegradable polymer blends as tissue engineering scaffold materials and also highlights the different polymeric biomaterials and polymeric blends along with their property evaluation and applications in the field of tissue engineering.

2. Tissue engineering

Tissue engineering is an interdisciplinary field that deals with developing three-dimensional structurally, functionally, and mechanically stable biological substitutes that are fabricated via living cells and signaling molecules embedded in a natural or synthetic scaffold material, which thus replace or heal damaged tissue or organs (Murphy & Mikos, 2007). There are three major components of tissue engineering, including cell−cell signaling and scaffolds (Fig. 22.1).

Cells are building blocks of tissue that produce protein, carry the function of cells, and provide tissues with repairing properties. According to the source of cell extraction, they have been categorized as autologous cells, allogeneic cells, xenogeneic cells, isogenic cells, and stem cells (Fig. 22.2).

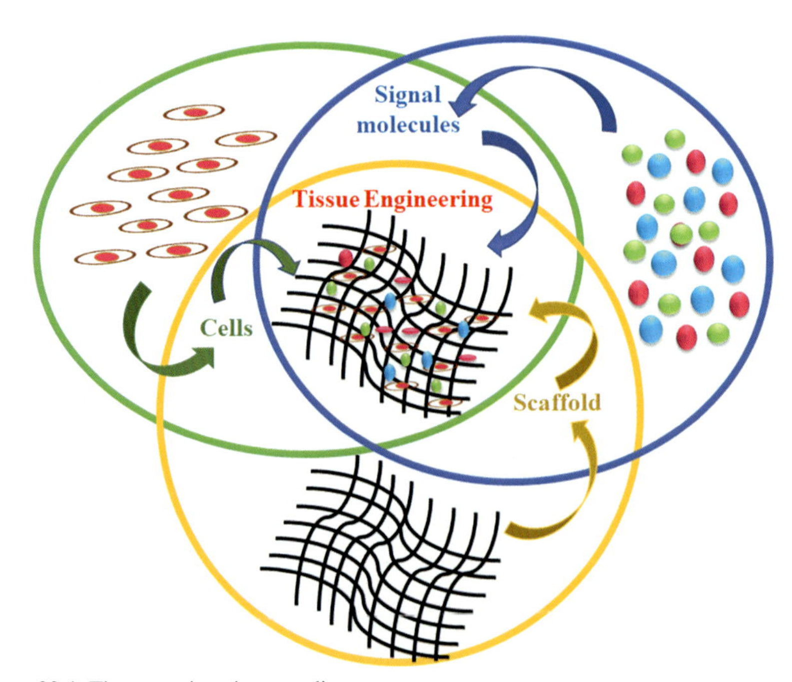

Figure 22.1 Tissue engineering paradigm.

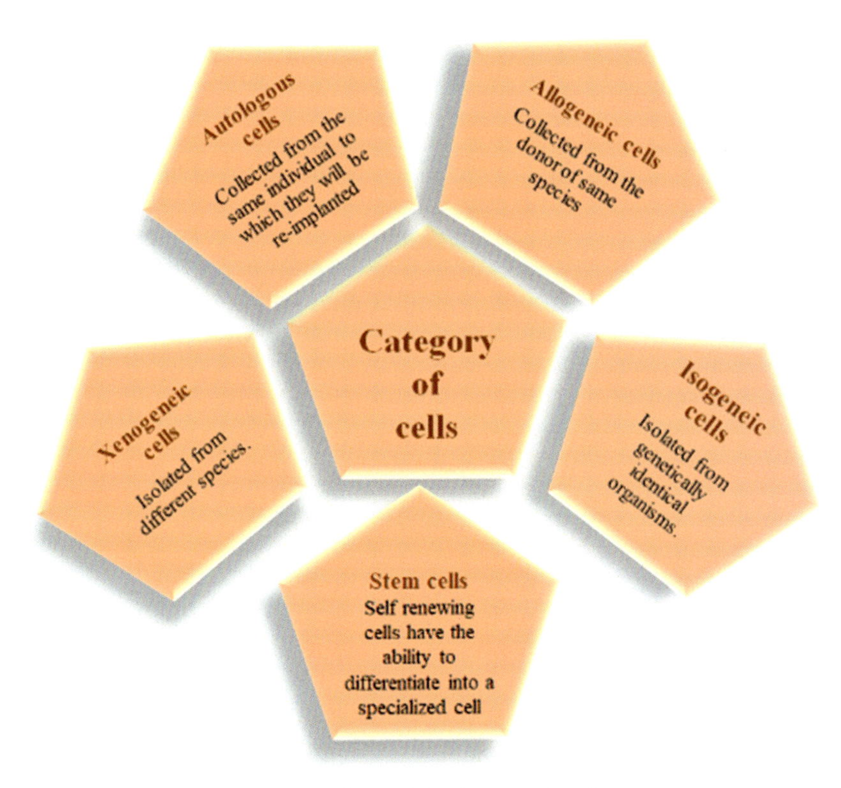

Figure 22.2 Different categories of cells used for tissue engineering.

The stem cell is a cell with the unique ability to develop into a specialized cell type in the body. In the future, they may be used for replacing damaged cells, ex., lost due to disease. Some of the types of stem cells has been presented in Table 22.1.

Signaling molecules, also referred to as biological modifiers, are proteins or mechanical stimulations that are different for different tissues, which regulate tissue growth functions. These are biomolecules released from scaffolds, incorporated into

Table 22.1 Stem cell types and examples.

S. no.	Stem cells	Description	Examples
1	Totipotent	Produce all cell types	Spores and zygotes
2	Pluripotent	Produce most cell types	Embryonic stem cells
3	Multipotent	Produce more than one cell type	Fetal tissue cord blood
4	Oligopotent	Produce only a few cell types	Lymphoid or myeloid stem cells
5	Unipotent	Produce only one cell type	Adult muscle stem cells

scaffolds, or added to the cell culture media to lead to a controlled and specific functional tissue growth. Signaling molecules may be small molecules such as hormones or corticosteroids, proteins and peptides such as cytokines, morphogens, or growth factors, and oligonucleotides such as DNA or RNA.

Scaffolds are porous three-dimensional materials which provide mechanical strength and structural integrity to tissues or organs and also allow cells to grow, adhere, proliferate, and secrete extracellular matrix (ECM) (Murphy & Mikos, 2007). Scaffolds also provide a suitable environment for new tissue regeneration and nutrient transport. An ideal tissue engineering scaffold should possess high porosity, pore size, commercially feasible fabrication technology, biocompatibility, and controllable biodegradation kinetics to promote natural healing (Krithica et al., 2012). The high porosity and pore size of scaffolds provides proper cell seeding and diffusion of nutrients through the whole structure. Potential materials with these features include bioactive ceramics, metals, natural or synthetic polymers, and combinations of these (Ratner, 1996).

Scaffolds prepared from biomaterials are widely gaining attention in the reconstruction of soft tissues such as skin or adipose tissue or for hard tissues such as bones. The ideal biomaterial should possess key characteristics such as biocompatibility, biodegradability, porosity, mechanical strengths, and no immunogenicity (Kramschuster & Turng, 2013).

The biocompatibility of scaffolds refers to the bioinert, bioactive, and bioresorbable nature of scaffold material according to the type of tissue growth. The implanted biomaterial scaffolds should exhibit suitable biocompatibility with the cells without immune rejection and controllable degradation rate. A controllable biodegradability rate of scaffold material is needed for attaining native tissue growth and structural support for regaining the tissue architecture of damaged tissue. It also provides mechanical stability to the regenerated cellular tissues (Sundar et al., 2020).

The porous architecture of the scaffold is important for implant performance. Homogenously interconnected three-dimensional porous scaffolds developed from polymers or ceramics materials with a higher surface area to volume ratio enable the cells to interact and adhere effectively into the voids and encourage new tissue growth (Dhandayuthapani et al., 2011). The different architectures of scaffolds have been developed for tissue regeneration, such as fiber mesh type, acceptable filament type, sponge type, 3D-printed, and/or injectable hydrogel model. Different fabrication techniques have been developed for engineering scaffolds for tissue engineering with external geometry, surface property, good pore size and porosity, mechanical properties, and interconnectivity (Table 22.2).

3. Biomaterials for tissue engineering

Natural and synthetic biomaterials for tissue engineering applications are presented in Fig. 22.3. Natural biomaterials are renewable materials derived from animals, plants, and microorganisms. In the early stage of tissue engineering, natural polymeric biomaterials were used as scaffolding material due to the presence of extracellular matrix

Table 22.2 Different fabrication techniques used for engineering scaffolds (Subia et al., 2010; Thavornyutikarn et al., 2014).

S. no.	Techniques	Advantages	Disadvantages
1	Solvent casting	Simple and inexpensive method, controllable porosity and pore dimension, control over crystallinity	Poor mechanical strength, poor degree of pore-interconnectivity, toxic solvent residues
2	Particulate leaching	Easy controllable pore size, desirable crystallinity, lower polymer usage	Limited mechanical property, residual solvents and porogens, difficult to control pore shape and interpore openings
3	Gas forming	Free of toxic solvents, low-temperature process, controllable porosity and pore size, suitable for heat-sensitive biological agents	Limited pore size, limited mechanical integrity, inadequate pore interconnectivity, possibility of closed pore structure, outer skin formation
4	Phase separation	Pore structure with higher interconnectivity, easily combinable with other fabrication techniques, allows incorporation of bioactive agents	Limited range of pore size, shrinkage issue, long time to sublime solvent, residual solvents
5	Electrospinning/textile technology	High surface area to volume ratio, simple process, high interconnected porosity, control over porosity	Insufficient mechanical strength for use in load-bearing applications, pore size decreases with fiber thickness, residual solvents
6	Rapid prototyping	Excellent control over geometry and porosity, no need for supporting material	Limited polymer type, expensive process
7	Freeze drying	Leaching steps are not required, Low-temperature process	Long processing hours, small pore size, limited mechanical strength
8	Fiber bonding	High surface to volume ratio, high porosity	Poor mechanical property
9	Fiber mesh	Large surface area for cell attachment, rapid nutrient diffusion	Poor structural stability

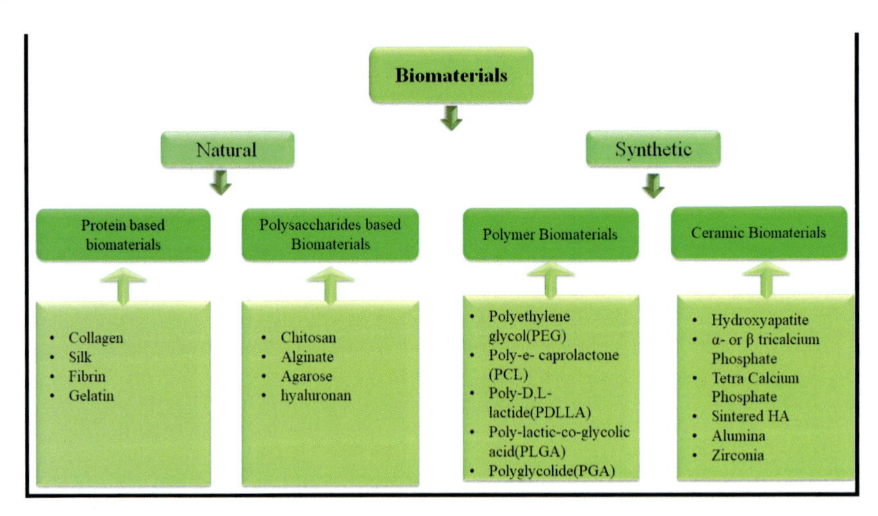

Figure 22.3 Natural and synthetic biomaterials for tissue engineering applications.

components or the similarity of natural polymers with ECM. These polymeric materials exhibit wide applications in regenerative medicine and tissue engineering because of their complex molecular structure, pseudoplastic nature, water-binding aptitude, biocompatibility, gelation behavior, and biodegradability. In addition, these materials possess different amino, carboxylic, and hydroxyl groups that are proficient at modifying by chemical or enzymatic means (Broderick et al., 2005; Chen et al., 2003; Cui, 2005; Kurita, 2001).

Collagen is an abundant fibrous protein present in the extracellular matrix of mammalian connective tissues. It is also found in articular cartilage and fibrocartilage connective tissues. The versatility, biocompatibility, biodegradability, and low immunogenicity of collagen with human tissue have made it an attractive biomaterial for tissue engineering applications (Castile et al., 2001). The collagen-based scaffolds are ideal for tissue regeneration due to their characteristic properties such as permeability, stability, hydrophilicity, and porous structure (Oliveira et al., 2010). The presence of the RGD sequence in collagen protein provides good adhesion, proliferation, and growth. The presence of serine and collagenase proteases allows natural degradation of biomaterial by the cells of specific tissues (Castile et al., 2001; Park & Lakes, 1992). Collagen-based biomaterials are frequently used in the repair and replacement of skin, tendons, vascular grafts, dental implants, heart valves, and bone regeneration. It can be easily isolated and extracted from human or animal tissues. Its mechanical and biological properties can be altered by treating with chemical or physical cross-linkers or functionalizing with inorganic or organic polymers (Brandon et al., 2009; Park & Lakes, 1992; Weadock et al., 1995).

Collagen scaffolds have been widely used in artificial skin regeneration or construction for severe skin burns and wounds (Singh et al., 2011; Westgate et al., 2012). Inayat et al. studied collagen-based nanogel from curcumin loaded fish scale collagen-HPMC nanogel for dermatological applications (Pathan et al., 2019).

In vivo and ex vivo studies in a rat model revealed that the effect of curcumin nanoemulsion and extracted collagen showed higher wound contraction value and prolonged release with better stability in a wound-healing application (Pathan et al., 2019).

Gelatin is a protein-based biomaterial derived from collagen by acidic or basic hydrolysis of different mammalian tissues. It is a promising natural polymeric scaffold material for 3D cell cultures with abundant availability, and is low toxic, inexpensive, biocompatible, and has ease of chemical modifications with biomolecules (Naahidi et al., 2017). The limited mechanical property, high viscosity, and rapid enzymatic degradation rate of gelatin have limited its applications, and thus encouraged researchers to seek different modification methodologies such as chemical and physical cross-linking or bending or composite preparations. Chemical and physical cross-linking were performed using glutaraldehyde or methacrylic anhydride or with carbo-imide, but its applications were constrained. It increases the cytotoxic nature and immune rejection with the host body. Han et al. (2014) blended gelatin with chitosan to prepare a sponge scaffold with good biocompatibility and mechanical strength for skin tissue regeneration applications.

Other protein-based biomaterials for tissue engineering include elastin, soybean, silk fibroin, actin, keratin, and myosin. Elastin is a structural protein present in the ECM of connective tissues such as skin, esophagus, and blood vessels (Joddar & Ramamurthi, 2006; Li et al., 2005). The microfibrillar components present in the elastin have the tendency to calcify upon tissue implantation. The difficulty in purification and the insoluble nature have limited the use of elastin as a biomaterial for tissue engineering (Daamen et al., 2005). Soy protein extracted from soybean is a renewable resource material that is economically feasible, water-resistant, and has good storage ability (Hinds et al., 2006). Along with these features, its similarity with tissue components and lower susceptibility to thermal degradation make it an ideal candidate for biomedical engineering, and especially membranes and thermoplastic soy material-based scaffolds were used for tissue engineering (Mano et al., 1999). Silva et al. prepared a 3D porous structure from a chitosan-soy protein blend in the presence of tetraethyl orthosilicate by combining a sol−gel methodology with freeze-drying technology, resulting in the development of a novel structure with good physiochemical properties, mechanical stability, and degradation for cartilage tissue engineering (Silva et al., 2006). Silk fibroin is a fibrous protein produced by silkworms, where fibroin has promising properties like water and oxygen permeability, cell adhesion, cell growth factors, slow degradation, good mechanical properties (tensile strength with flexibility), and a low inflammatory response for bone and cartilage tissue engineering scaffold applications. The blending of fibroin with poly(ethylene glycol oxide), chitosan, or other synthetic polymers has improved the mechanical properties and biocompatibility of scaffold (Jin et al., 2002; Gobin et al., 2005).

For more than over two decade these carbohydrate-based versatile biomaterials have increased their importance in biological studies with growing interest in the field of polysaccharides as a scaffold material for tissue engineering. Polysaccharide-based natural biomaterials include chitosan alginate, agarose, hyaluronan, dextran, cellulose, and pullulan. They possess attractive features such as being renewable, low cost, biologically acceptable, biologically degradable, and having ease of derivatization.

Further, cell interactions can be improved by different purification techniques and backbone modifications (Tiwari et al., 2019). Along with these advantages, polysaccharides have certain limitations including rapid hydrolysis, low mechanical strength, and inconsistency in molecular weight distribution and branching. Hence, rather than using pure polymer scaffold systems, polysaccharides are mostly used along with other natural polymers or modified or cross-linked with other functional materials to meet tissue engineering application requirements.

In Fig. 22.4, chitosan (copolymer of D-glucose amine and N-acetyl-D-glucose amine) is seen as a linear semicrystalline polysaccharide biomaterial derived from chitin, which is the second most abundant biopolymer after cellulose, commonly found in crustacean shells and mollusks. It is prepared by deacetylation of chitin, where the degree of deacetylation determines the properties of chitosan. The polycationic nature of chitosan supports cell attachment, proliferation, and migration. The structural resemblance of chitosan with extracellular proteoglycans, hydrophilic nature, mucoadhesion, and hemostatic action enables its applications in regenerative medicine (Tiwari et al., 2019).

The minimum immune rejection of chitosan with a low or null amount of fibrous encapsulation has appeared to be typical for skin, bone, and cartilage tissue scaffold applications (Li et al., 2020). The ease of chitosan processability has enabled different molded shapes like porous scaffolds, membranes, gels, tubes, and particles. Chitosan is thus considered to be the most promising natural biomaterial, with chemical versatility and a reasonable degradation rate for tissue engineering applications.

In Fig. 22.4, alginate [block copolymer of (1−4) linked β-D mannuronic acid and α-L-guluronic acid monomers] is shown an anionic heteropolysaccharide derived from the cell walls of brown seaweed. It is a natural polysaccharide with excellent biocompatibility, biodegradability, and chelating property for the application of supporting

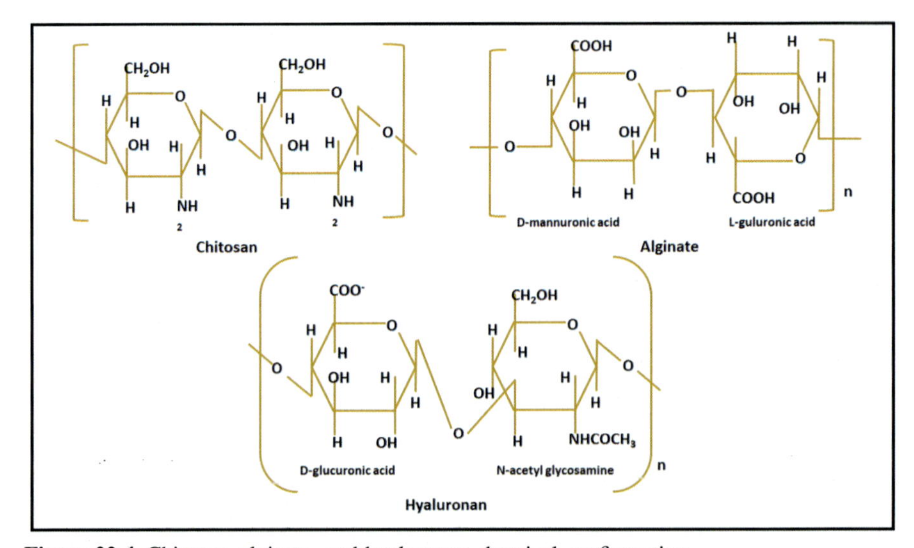

Figure 22.4 Chitosan, alginate, and hyaluronan chemical configuration.

matrix for tissue repair. The properties such as mechanical strength and degradation rate of alginate-based biomaterials usually depend on the molecular weight of the alginate. During in vivo and in vitro studies, alginate-based biomaterials were shown to have histocompatibility and low cytotoxicity (Sun & Tan, 2013; Vanacker et al., 2012). Also, alginate-based matrix delivery systems had greater significance in fabricating bioactive biomaterials for wound healing, cartilage repair, and bone regeneration applications.

In Fig. 22.4, hyaluronan, also known as hyaluronic acid [composed of (β-1,4-D-glucuronic acid and β-1,3-N-acetyl-D-glucosamine)] is shown as a versatile hydrophilic linear polysaccharide found in extracellular matrix or tissues. It is a compactable tissue biomaterial that promotes cell proliferation, differentiation, and is degradable with nontoxic products (Zhu et al., 2017). A large number of hydroxyl and carboxyl functional groups present in hyaluronic acid can be modified by chemical methods or by cross-linking. The degree of cross-linking or chemical modification determines the physical properties, chemical properties, and mechanical strength. The most extensively used tissue repair application of hyaluronic acid is in cartilage tissue engineering as it acts as a stable scaffold material for chondrocyte growth and also regulates the function of chondrocytes (Zhu et al., 2017). It is also ideal for bone regeneration and osteochondral tissue regeneration (Park et al., 2014).

For the application of scaffolding materials, biomaterials should possess properties such as surface chemistry, high molecular weight, structural stability, hydrophilicity or hydrophobicity, surface energy, lubrication, degradation, and erosion mechanism (Dhandayuthapani et al., 2011). Synthetic polymeric biomaterials are another class of materials that have drawn attention in tissue regeneration. The high surface to volume ratio, high porosity with limited pore size, tunable physicochemical properties, biocompatibility, biodegradable quality, and reproducible mechanical properties are particular features of synthetic polymers (Asghari et al., 2017). Synthetic biomaterials have processing flexibility and no immunogenicity compared with natural extracellular matrix protein (Liu et al., 2012). Polymers such as polyglycolic acid, polycaprolactone, polylactic acid, and their copolymers are mainly used in regenerative medicine.

Polyglycolic acid is an aliphatic polyester exploited for its application in the biomedical field. It is prepared by ring-opening polymerization of glycolic acid. It possesses optimum crystallinity, high melt strength, and is less soluble in organic solvents. In addition to these features, it has good tissue compatibility, biodegradability, and reproducible mechanical properties like elongation, tensile strength, and knot retention (Boland et al., 2001). The hydrophilic nature of polyglycolic acid provides gradual degradation within 2–4 weeks in the case of monofilaments (Wong & Mooney, 1997).

Polycaprolactone is a semicrystalline polymer, prepared by ring-opening polymerization of ε-caprolactone with a glass transition temperature of −60°C and a melting point of approximately 59–64°C. Imperceptible degradation and high permeability of polycaprolactone define its role in long-term implant applications (Gomes et al., 2017). The properties, such as exceptional viscoelasticity and admirable blend-compatibility, make it an attractive material for 3D-printing (Cheng & Chen, 2017). Hydrophobicity and the absence of a bioactive surface limit its applications in regenerative medicine. To improve cell interaction and proliferation, several modification techniques like plasma treatment, physical absorption, and surface graft polymerization have been employed (Li et al., 2017).

Polylactic acid is an aliphatic linear biodegradable polymer produced from renewable resources like sugar cane or corn. PLA is widely used in the biomedical field due to its biological compatibility and controlled degradation. The comparable elastic modulus of PLA with human bone makes it ideal for scaffolding bone tissue regeneration (Liu et al., 2020). The hydrophobicity and bio-inertness of PLA decrease the cell adhesion and degradation rate, limiting its applications in tissue engineering. To overcome these limitations, PLA is chemically modified by diverse mechanisms such as plasma treatments, chemical etching, grafting, and copolymerization. It is also enhanced by composite preparation by blending with other compactable polymers, which improves the mechanical, barrier, and biological properties (Bhushan & Kumar, 2019; Goreninskii et al., 2019).

4. Biodegradable polymer blends for tissue engineering

Over the past few decades, the development of biological structures that resemble the architecture and properties of body tissues or organs has been challenging for scientists and researchers. Scaffolds are an essential structural device that support the growth of cells to form new tissues and also provide mechanical stability to attain the desired tissue formation (Table 22.3). The primary need for the successful functioning of scaffold material involves numerous features such as biocompatibility, biodegradability, mechanical strength, cell attachment, 3D architecture, and incorporation of ECM components (Boccaccini & Blaker, 2005). The incorporation of extracellular matrix components supports cell growth and proliferation, along with survival and migration of cells, thus enhancing tissue formation and growth.

Generally, scaffolds are prepared from biodegradable polymers that have the ability to degrade and disappear during cell growth and tissue formation without harming the host body. Initially, protein-based or polysaccharide-based polymeric biomaterials were used as scaffolding materials. Further, aliphatic polyester-based polymers such as polylactic acid (PLA), poly-ε-caprolactone (PCL), polyglycolic acid (PGA), polylactic-co-glycolic acid (PLGA), and their copolymers have attained significant attention in scaffold preparation due to their unique ability in hydrolytic degradation from ester linkage (Aguilar et al., 2005; Boccaccini & Blaker, 2005). In addition, biodegradable polyurethane and polyethylene glycols were also used. These synthetic biomaterials lack cell adhesion properties, but they possess good mechanical properties, tailorable porosity, and have a longer shelf-life than natural biomaterials (Paun et al., 2015). The versatility and ease of chemical modification and degradation properties have attracted their role in scaffold designing. The greatest challenge faced in tissue regeneration is to design a biomaterial scaffold mimicking native extracellular matrix with physicomechanical physiognomies and biological properties that regulate tissue function and growth (Mao et al., 2017).

Table 22.3 Different biodegradable polymer blends used in tissue engineering applications.

S. no.	Composite blend	Method of fabrication	Properties achieved	Application	References
1	Alginate/collagen	Freeze-drying/ lyophilization	Mechanical stability, shape-memory property, support cell proliferation, ECM deposition	Annulus fibrous tissue regeneration	Guillaume et al. (2015)
2	Silk fibroin/collagen	Blending	Biocompatibility, excellent mechanical strength, rapid degradation	Peripheral nerve regeneration	Li et al. (2015)
3	Collagen/elastin	Blending	Promising mechanical performance, efficient neovascularization, enhancing cell infiltration	Ventral hernia repair	Minardi et al. (2017)
4	Fibrin/collagen	Lyophilization	Enhanced cell attachment, proliferation, and osteoblast differentiation	Bone regeneration	Kim et al. (2013)
5	3-D collagen/ hyaluronic acid scaffolds	Freeze-drying technique	Improved the thermal stability, compressive modulus, and elastic collapse stress of the composite, support the growth and differentiation of mammary stromal tissue	Adipose tissue engineering	Davidenko et al. (2010)
6	Chitosan-γ-poly(caprolactone)/ poly(caprolactone)	Electrospinning	Increased initial cellular attachment and proliferation, cell vitality, adequate mechanical properties, fiber diameter, and interstitial volume	Bioactive implants	de Cassan et al. (2020)
7	Collagen/PEG	In situ gelation	Tunable mechanical property, biodegradation, assisting cell adhesion and proliferation	Injectable tissue scaffold	Lalwani et al. (2013)
8	Glycol/chitosan/ hyaluronic acid blend hydrogel	Sol−gel method	Enhanced physical stability, mechanical properties, cell binding affinity, tissue compatibility	Injectable tissue scaffold	Lee et al. (2020)
9	PCL/collagen	Bioextruder	Better biological properties such as adhesion, growth, and proliferation of cells	Tissue engineering	Sousa et al. (2013)

Continued

Table 22.3 Continued

S. no.	Composite blend	Method of fabrication	Properties achieved	Application	References
10	Collagen/PVA	UV combined electrospinning	Greater surface area for cell attachment and spreading, better noncytotoxic behavior	Tissue engineering	Oktay et al. (2015)
11	PLGA/PCL/TEC	3D printing	Good biocompatibility, improve the success rate of the operation process, reduce the postoperative risk	Urethra tissue engineering	Xu et al. (2020)
12	Poly(glycerol sebacate) (PGS)/poly(ε-caprolactone) (PCL)	Electrospinning	Oxygen-releasing electro-spun composite scaffolds, good antibacterial performance, improved degradation behavior	Tissue engineering	Abudula et al. (2020)
13	Chitosan/poly(lactide-co-glycolide) (PLAGA)	Lyophilization	Load-bearing enhanced mechanical properties	Bone regeneration	Jiang et al. (2010)
14	Polycaprolactone, PCL/chitosan (CS)/gelatin blend	Electrospinning	Better physical properties, cell adhesion ratio	Skin tissue engineering	Gomes et al. (2017)
15	Chitosan/agarose	Freeze-drying	Improved cellular adhesion and proliferation, boosting cell–matrix interactions, biocompatible, and support cell attachment and in vitro proliferation	Nasal cartilage tissue engineering	Garakani et al. (2020)

It is difficult to achieve or satisfy all the scaffold requirements with a single polymer; thus, composite material preparations, such as polymer blends, lead the way in tissue engineering. A blend prepared by the combination of natural and synthetic biomaterials exhibits combinations of properties such as physicochemical properties, processability, mechanical properties, biological properties, tissue compatibility, and biodegradability.

For example, He et al. developed an ECM mimicked tissue-engineered nanofiber scaffold by blending electrospun collagen with poly(L-lactic acid)-co-poly-ε-caprolactone) with good endothelialization for a vascular graft for a blood vessel tissue regeneration application. The prepared blend achieved better cell spreading morphology and attachment compared with prepolymer nanofibers (He et al., 2005).

One of the major hurdles faced in tissue engineering is the inability to monitor and control the function of an engineered tissue following transplantation. Recent years have seen major developments in the field by integrating electronics within engineered tissues. However, material functional electronic properties such as dielectric permittivity, dielectric loss, electron conductivity, electromagnetic properties and their radiation phenomena, and material interactions with the electromagnetic fields, at a molecular level, need to be explored (Magisetty, Hemnath, Shukla et al., 2020; Malik et al., 2020; Magisetty & Naga Srilatha, 2021; Magisetty, Shukla et al., 2020; Magisetty, Prajapati et al., 2019; Magisetty, Kumar et al., 2018; Magisetty et al., 2018a; Magisetty et al., 2018b; Magisetty, Hemnath, Kumar et al., 2020; Magisetty & Marutheswar, 2011; Magisetty & Cheekuramelli, 2019; Magisetty, Raj et al., 2020; Magisetty, Shukla et al., 2019). This may facilitate efficient diagnosis of tissues for better reliability and also provide a path for understanding neural dynamics by enabling neural interfaces and communications.

5. Conclusion

This chapter mainly focuses on the biodegradable polymeric materials and their blends as scaffolding materials for tissue engineering applications. Natural biomaterials have admirable biological properties such as cell adhesion, proliferation, and properties similar to native ECM components. They have poor mechanical properties, whereas synthetic polymeric biomaterials have low biological properties but are efficient in structural stability and mechanical strength. Thus, both natural and synthetic biomaterials have their specific advantages and disadvantages. To overcome these limitations and to fabricate scaffolds with appropriate mechanical and biological features, composite blend preparation is an ideal solution. In the current scenarios, a great deal of research is on-going to achieve ideal scaffolding materials for specific regenerative tissue applications.

References

Abedalwafa, M., Wang, F., Wang, L., & Li, C. (2013). Biodegradable poly-epsilon-caprolactone (PCL) for tissue engineering applications: a review. *Reviews on Advanced Materials Science, 34*(2), 123–140.

Abudula, T., Gauthaman, K., Hammad, A. H., Joshi Navare, K., Alshahrie, A. A., Bencherif, S. A., Tamayol, A., & Memic, A. (2020). Oxygen-releasing antibacterial nanofibrous scaffolds for tissue engineering applications. *Polymers, 12*(6), 1233.

Aguilar, C. A., Lu, Y., Mao, S., & Chen, S. (2005). Direct micro-patterning of biodegradable polymers using ultraviolet and femtosecond lasers. *Biomaterials, 26*(36), 7642–7649.

Asghari, F., Samiei, M., Adibkia, K., Akbarzadeh, A., & Davaran, S. (2017). Biodegradable and biocompatible polymers for tissue engineering application: a review. *Artificial cells, Nanomedicine, and Biotechnology, 45*(2), 185–192.

Bhushan, B., & Kumar, R. (2019). Plasma treated and untreated thermoplastic biopolymers/biocomposites in tissue engineering and biodegradable implants. In *Materials for biomedical engineering* (pp. 339–369). Elsevier.

Boccaccini, A. R., & Blaker, J. J. (2005). Bioactive composite materials for tissue engineering scaffolds. *Expert Review of Medical Devices, 2*(3), 303–317.

Boland, E. D., Wnek, G. E., Simpson, D. G., Pawlowski, K. J., & Bowlin, G. L. (2001). Tailoring tissue engineering scaffolds using electrostatic processing techniques: a study of poly (glycolic acid) electrospinning. *Journal of Macromolecular Science, Part A, 38*(12), 1231–1243.

Brandon, V. S., Shahana, S. K., Omar, Z. F., Ali, K., & Nicholas, A. P. (2009). Hydrogels in regenerative medicine. *Advanced Materials, 21*(32–33), 3307–3329.

Broderick, E. P., O'Halloran, D. M., Rochev, Y. A., Griffin, M., Collighan, R. J., & Pandit, A. S. (2005). Enzymatic stabilization of gelatin-based scaffolds. *Journal of Biomedical Materials Research Part B: Applied Biomaterials, 72*(1), 37–42. The Japanese Society for Biomaterials, and The Australian Society for Biomaterials and the Korean Society for Biomaterials.

Castile, J. D., Taylor, K. M., & Buckton, G. (2001). The influence of incubation temperature and surfactant concentration on the interaction between dimyristoylphosphatidylcholine liposomes and poloxamer surfactants. *International Journal of Pharmaceutics, 221*(1–2), 197–209.

Chen, T., Embree, H. D., Brown, E. M., Taylor, M. M., & Payne, G. F. (2003). Enzyme-catalyzed gel formation of gelatin and chitosan: potential for in situ applications. *Biomaterials, 24*(17), 2831–2841.

Cheng, Y. L., & Chen, F. (2017). Preparation and characterization of photocured poly (ε-caprolactone) diacrylate/poly (ethylene glycol) diacrylate/chitosan for photopolymerization-type 3D printing tissue engineering scaffold application. *Materials Science and Engineering: C, 81*, 66–73.

Christy, P. N., Basha, S. K., Kumari, V. S., Bashir, A. K. H., Maaza, M., Kaviyarasu, K., Arasu, M. V., Al-Dhabi, N. A., & Ignacimuthu, S. (2020). Biopolymeric nanocomposite scaffolds for bone tissue engineering applications–A review. *Journal of Drug Delivery Science and Technology, 55*, 101452.

Cui, S. W. (Ed.). (2005). *Food carbohydrates: chemistry, physical properties, and applications* (vol. 2005, pp. 357–405). Boca Raton: CRC press. Taylor & Francis Group.

Daamen, W. F., Nillesen, S. T. M., Hafmans, T., Veerkamp, J. H., Van Luyn, M. J. A., & Van Kuppevelt, T. H. (2005). Tissue response of defined collagen–elastin scaffolds in young and adult rats with special attention to calcification. *Biomaterials, 26*(1), 81–92.

Davidenko, N., Campbell, J. J., Thian, E. S., Watson, C. J., & Cameron, R. E. (2010). Collagen–hyaluronic acid scaffolds for adipose tissue engineering. *Acta Biomaterialia, 6*(10), 3957–3968.

de Cassan, D., Becker, A., Glasmacher, B., Roger, Y., Hoffmann, A., Gengenbach, T. R., Easton, C. D., Hänsch, R., & Menzel, H. (2020). Blending chitosan-g-poly (caprolactone) with poly (caprolactone) by electrospinning to produce functional fiber mats for tissue engineering applications. *Journal of Applied Polymer Science, 137*(18), 48650.

Dhandayuthapani, B., Yoshida, Y., Maekawa, T., & Kumar, D. S. (2011). Polymeric scaffolds in tissue engineering application: a review. *International Journal of Polymer Science, 2011.*

Garakani, S. S., Khanmohammadi, M., Atoufi, Z., Kamrava, S. K., Setayeshmehr, M., Alizadeh, R., Faghihi, F., Bagher, Z., Davachi, S. M., & Abbaspourrad, A. (2020). Fabrication of chitosan/agarose scaffolds containing extracellular matrix for tissue engineering applications. *International Journal of Biological Macromolecules, 143*, 533–545, 0.

Gobin, A. S., Froude, V. E., & Mathur, A. B. (2005). Structural and mechanical characteristics of silk fibroin and chitosan blend scaffolds for tissue regeneration. *Journal of Biomedical Materials Research Part A, 74*(3), 465–473.

Gomes, S., Rodrigues, G., Martins, G., Henriques, C., & Silva, J. C. (2017). Evaluation of nanofibrous scaffolds obtained from blends of chitosan, gelatin and polycaprolactone for skin tissue engineering. *International Journal of Biological Macromolecules, 102*, 1174–1185.

Goreninskii, S. I., Guliaev, R. O., Stankevich, K. S., Danilenko, N. V., Bolbasov, E. N., Golovkin, A. S., Mishanin, A. I., Filimonov, V. D., & Tverdokhlebov, S. I. (2019). "Solvent/non-solvent" treatment as a method for non-covalent immobilization of gelatin on the surface of poly (l-lactic acid) electrospun scaffolds. *Colloids and Surfaces B: Biointerfaces, 177*, 137–140.

Guillaume, O., Naqvi, S. M., Lennon, K., & Buckley, C. T. (2015). Enhancing cell migration in shape-memory alginate–collagen composite scaffolds: in vitro and ex vivo assessment for intervertebral disc repair. *Journal of Biomaterials Applications, 29*(9), 1230–1246.

Han, F., Dong, Y., Su, Z., Yin, R., Song, A., & Li, S. (2014). Preparation, characteristics and assessment of a novel gelatin–chitosan sponge scaffold as skin tissue engineering material. *International Journal of Pharmaceutics, 476*(1–2), 124–133.

He, W., Yong, T., Teo, W. E., Ma, Z., & Ramakrishna, S. (2005). Fabrication and endothelialization of collagen-blended biodegradable polymer nanofibers: potential vascular graft for blood vessel tissue engineering. *Tissue Engineering, 11*(9–10), 1574–1588.

Hinds, M. T., Rowe, R. C., Ren, Z., Teach, J., Wu, P. C., Kirkpatrick, S. J., Breneman, K. D., Gregory, K. W., & Courtman, D. W. (2006). Development of a reinforced porcine elastin composite vascular scaffold. *Journal of Biomedical Materials Research Part A, 77*(3), 458–469.

Jiang, T., Nukavarapu, S. P., Deng, M., Jabbarzadeh, E., Kofron, M. D., Doty, S. B., Abdel-Fattah, W. I., & Laurencin, C. T. (2010). Chitosan–poly (lactide-co-glycolide) microsphere-based scaffolds for bone tissue engineering: in vitro degradation and in vivo bone regeneration studies. *Acta Biomaterialia, 6*(9), 3457–3470.

Jin, H. J., Fridrikh, S. V., Rutledge, G. C., & Kaplan, D. L. (2002). Electrospinning *Bombyx mori* silk with poly (ethylene oxide). *Biomacromolecules, 3*(6), 1233–1239.

Joddar, B., & Ramamurthi, A. (2006). Fragment size-and dose-specific effects of hyaluronan on matrix synthesis by vascular smooth muscle cells. *Biomaterials, 27*(15), 2994–3004.

Kim, B. S., Kim, J. S., & Lee, J. (2013). Improvements of osteoblast adhesion, proliferation, and differentiation in vitro via fibrin network formation in collagen sponge scaffold. *Journal of Biomedical Materials Research Part A, 101*(9), 2661–2666.

Kramschuster, A., & Turng, L. S. (2013). 17-Fabrication of tissue engineering scaffolds A2-ebnesajjad, sina. *Handbook of Biopolymers and Biodegradable Plastics* (pp. 1–472). Hardcover. ISBN: 9781455728343, eBook ISBN: 9781455730032.

Krithica, N., Natarajan, V., Madhan, B., Sehgal, P. K., & Mandal, A. B. (2012). Type I collagen immobilized poly (caprolactone) nanofibers: characterization of surface modification and growth of fibroblasts. *Advanced Engineering Materials, 14*(4), B149–B154.

Kurita, K. (2001). Controlled functionalization of the polysaccharide chitin. *Progress in Polymer Science, 26*(9), 1921–1971.

Lalwani, G., Henslee, A. M., Farshid, B., Parmar, P., Lin, L., Qin, Y. X., Kasper, F. K., Mikos, A. G., & Sitharaman, B. (2013). Tungsten disulfide nanotubes reinforced biodegradable polymers for bone tissue engineering. *Acta Biomaterialia, 9*(9), 8365–8373.

Langer, R., & Vacanti, J. P. (1993). Tissue engineering. *Science, 260*, 920–926.

Lee, E. J., Kang, E., Kang, S. W., & Huh, K. M. (2020). Thermo-irreversible glycol chitosan/hyaluronic acid blend hydrogel for injectable tissue engineering. *Carbohydrate Polymers*, 116432.

Li, M., Mondrinos, M. J., Gandhi, M. R., Ko, F. K., Weiss, A. S., & Lelkes, P. I. (2005). Electrospun protein fibers as matrices for tissue engineering. *Biomaterials, 26*(30), 5999–6008.

Li, P. J., Jin, T., Luo, D. H., Shen, T., Mai, D. M., Hu, W. H., & Mo, H. Y. (2015). Effect of prolonged radiotherapy treatment time on survival outcomes after intensity-modulated radiation therapy in nasopharyngeal carcinoma. *PloS One, 10*(10), e0141332.

Li, Y., Li, X., Zhao, R., Wang, C., Qiu, F., Sun, B., Ji, H., Qiu, J., & Wang, C. (2017). Enhanced adhesion and proliferation of human umbilical vein endothelial cells on conductive PANI-PCL fiber scaffold by electrical stimulation. *Materials Science and Engineering: C, 72*, 106–112.

Li, Y., Zhang, K., Nie, M., & Wang, Q. (2020). Application of compatibilized polymer blends in packaging. In *Compatibilization of polymer blends* (pp. 539–561). Elsevier.

Liu, X., Holzwarth, J. M., & Ma, P. X. (2012). Functionalized synthetic biodegradable polymer scaffolds for tissue engineering. *Macromolecular Bioscience, 12*(7), 911–919.

Liu, S., Qin, S., He, M., Zhou, D., Qin, Q., & Wang, H. (2020). Current applications of poly (lactic acid) composites in tissue engineering and drug delivery. *Composites Part B: Engineering*, 108238.

Magisetty, R., & Naga Srilatha, C. H. (2021). *Bio epoxy polymer, blends and composites derived utilitarian electrical, magnetic and optical properties. Bio-based epoxy polymers, blends and composites: synthesis, properties, characterization and applications* (pp. 249–266).

Magisetty, R., & Cheekuramelli, N. S. (2019). Additive manufacturing technology empowered complex electromechanical energy conversion devices and transformers. *Applied Materials Today, 14*, 35–50.

Magisetty, R. P., & Marutheswar, G. V. (2011). The future generation of low energy loss HVDC power system using HTS technology. *International Journal on Recent Trends in Engineering and Technology, 6*(2).

Magisetty, R., Kumar, P., Kumar, V., Shukla, A., Kandasubramanian, B., & Shunmugam, R. (2018). NiFe2O4/poly (1, 6-heptadiyne) Nanocomposite energy-storage device for electrical and electronic applications. *ACS Omega, 3*(11), 15256–15266.

Magisetty, R., Shukla, A., & Kandasubramanian, B. (2018). Magnetodielectric microwave radiation absorbent materials and their polymer composites. *Journal of Electronic Materials, 47*(11), 6335–6365.

Magisetty, R., Shukla, A., & Kandasubramanian, B. (2018). Dielectric, hydrophobic investigation of ABS/NiFe 2O4 nanocomposites fabricated by atomized spray assisted and solution casted techniques for miniaturized electronic applications. *Journal of Electronic Materials, 47*(9), 5640–5656.

Magisetty, R., Prajapati, D., Ambekar, R., Shukla, A., & Kandasubramanian, B. (2019). β-Phase Cu-phthalocyanine/acrylonitrile butadiene styrene terpolymer nanocomposite film technology for organoelectronic applications. *Journal of Physical Chemistry C, 123*(46), 28081–28092.

Magisetty, R., Shukla, A., & Kandasubramanian, B. (2019). Terpolymer (ABS) cermet (Ni-NiFe2O4) hybrid nanocomposite engineered 3D-carbon fabric mat as a X-band electromagnetic interference shielding material. *Materials Letters, 238*, 214–217.

Magisetty, R., Hemanth, N. R., Shukla, A., Shunmugam, R., & Kandasubramanian, B. (2020). Poly (1, 6-heptadiyne)/NiFe2O4 composite as capacitor for miniaturized electronics. *Polymer-Plastics Technology and Materials, 59*(18), 2018–2026.

Magisetty, R., Shukla, A., & Kandasubramanian, B. (2020). Molecular dynamic simulation constructed interaction parameter investigation between poly (1, 6-heptadiyne) and NiFe2O4 in nanocomposite. *Materials Today: Proceedings, 24*, 1720–1728.

Magisetty, R., Hemanth, N. R., Kumar, P., Shukla, A., Shunmugam, R., & Kandasubramanian, B. (2020). Multifunctional conjugated 1, 6-heptadiynes and its derivatives stimulated molecular electronics: future moletronics. *European Polymer Journal, 124*, 109467.

Magisetty, R., Raj, A. B., Datar, S., Shukla, A., & Kandasubramanian, B. (2020). Nanocomposite engineered carbon fabric-mat as a passive metamaterial for stealth application. *Journal of Alloys and Compounds, 848*, 155771.

Malik, A., Magisetty, R., Kumar, V., Shukla, A., & Kandasubramanian, B. (2020). Dielectric and conductivity investigation of polycarbonate-copper phthalocyanine electrospun nonwoven fibres for electrical and electronic application. *Polymer-Plastics Technology and Materials, 59*(2), 154–168.

Mano, J. F., Vaz, C. M., Mendes, S. C., Reis, R. L., & Cunha, A. M. (1999). Dynamic mechanical properties of hydroxyapatite-reinforced and porous starch-based degradable biomaterials. *Journal of Materials Science: Materials in Medicine, 10*(12), 857–862.

Mao, Y., Hoffman, T., Wu, A., Goyal, R., & Kohn, J. (2017). Cell type–specific extracellular matrix guided the differentiation of human mesenchymal stem cells in 3D polymeric scaffolds. *Journal of Materials Science: Materials in Medicine, 28*(7), 100.

Minardi, S., Taraballi, F., Wang, X., Cabrera, F. J., Van Eps, J. L., Robbins, A. B., Sandri, M., Moreno, M. R., Weiner, B. K., & Tasciotti, E. (2017). Biomimetic collagen/elastin meshes for ventral hernia repair in a rat model. *Acta Biomaterialia, 50*, 165–177.

Murphy, M. B., & Mikos, A. G. (2007). Chapter 22: polymer scaffold fabrication. *Principles of Tissue Engineering, 3*, 309–319.

Naahidi, S., Jafari, M., Logan, M., Wang, Y., Yuan, Y., Bae, H., Dixon, B., & Chen, P. (2017). Biotechnol. Biocompatibility of hydrogel-based scaffolds for tissue engineering applications. *Biotechnology Advances, 35*(5), 530–544. https://doi.org/10.1016/j.biotechadv.2017.05.006

Oktay, B., Kayaman-Apohan, N., Erdem-Kuruca, S., & Süleymanoğlu, M. (2015). Fabrication of collagen immobilized electrospun poly (vinyl alcohol) scaffolds. *Polymers for Advanced Technologies, 26*(8), 978–987.

Oliveira, S. M., Ringshia, R. A., Legeros, R. Z., Clark, E., Yost, M. J., Terracio, L., & Teixeira, C. C. (2010). An improved collagen scaffold for skeletal regeneration. *Journal of Biomedical Materials Research Part A, 94*(2), 371–379.

Park, J., & Lakes, R. S. (2007). *Biomaterials: an introduction*. Springer Science & Business Media.

Park, J. B., & Lakes, R. S. (1992). Structure-property relationships of biological materials, biomaterials. an introduction. *Biomaterials*. Springer, Boston, MA: Springer. https://doi.org/10.1007/978-1-4757-2156-0_9.

Park, J. Y., Choi, J. C., Shim, J. H., Lee, J. S., Park, H., Kim, S. W., Doh, J., & Cho, D. W. (2014). A comparative study on collagen type I and hyaluronic acid dependent cell behavior for osteochondral tissue bioprinting. *Biofabrication, 6*(3), 035004.

Pathan, I. B., Munde, S. J., Shelke, S., Ambekar, W., & Mallikarjuna Setty, C. (2019). Curcumin loaded fish scale collagen-HPMC nanogel for wound healing application: ex-vivo and In-vivo evaluation. *International Journal of Polymeric Materials and Polymeric Biomaterials, 68*(4), 165–174.

Paun, I. A., Zamfirescu, M., Mihailescu, M., Luculescu, C. R., Mustaciosu, C. C., Dorobantu, I., Calenic, B., & Dinescu, M. (2015). Laser micro-patterning of biodegradable polymer blends for tissue engineering. *Journal of Materials Science, 50*(2), 923–936.

Ratner, B. D. (1996). *Biomaterials science. An introduction to materials in medicine*. In A. S. Hoffman, F. J. Shoen, & J. E. Lemons (Eds.) (p. 3750). New York: Academic Press, 1996.

Silva, S. S., Oliveira, J. M., Mano, J. F., & Reis, R. L. (2006). Physicochemical characterization of novel chitosan-soy protein/TEOS porous hybrids for tissue engineering applications. In *Materials science forum* (vol. 514, pp. 1000–1004). Trans Tech Publications Ltd.

Singh, O., Gupta, S. S., Soni, M., Moses, S., Shukla, S., & Mathur, R. K. (2011). Collagen dressing versus conventional dressings in burn and chronic wounds: a retrospective study. *Journal of Cutaneous and Aesthetic Surgery, 4*(1), 12.

Sousa, I., Mendes, A., & Bártolo, P. J. (2013). PCL scaffolds with collagen bioactivator for applications in tissue engineering. *Procedia Engineering, 59*, 279–284.

Subia, B., Kundu, J., & Kundu, S. C. (2010). Biomaterial scaffold fabrication techniques for potential tissue engineering applications. *Tissue Engineering*, 141.

Sun, J., & Tan, H. (2013). Alginate-based biomaterials for regenerative medicine applications. *Materials, 6*(4), 1285–1309.

Sundar, G., Joseph, J., John, A., & Abraham, A. (2020). Natural collagen bioscaffolds for skin tissue engineering strategies in burns: a critical review. *International Journal of Polymeric Materials and Polymeric Biomaterials*, 1–12.

Thavornyutikarn, B., Chantarapanich, N., Sitthiseripratip, K., Thouas, G. A., & Chen, Q. (2014). Bone tissue engineering scaffolding: computer-aided scaffolding techniques. *Progress in Biomaterials, 3*(2), 61–102.

Tiwari, S., Patil, R., & Bahadur, P. (2019). Polysaccharide based scaffolds for soft tissue engineering applications. *Polymers, 11*(1), 1.

Vanacker, J., Luyckx, V., Dolmans, M. M., Des Rieux, A., Jaeger, J., Van Langendonckt, A., Donnez, J., & Amorim, C. A. (2012). Transplantation of an alginate–matrigel matrix containing isolated ovarian cells: first step in developing a biodegradable scaffold to transplant isolated preantral follicles and ovarian cells. *Biomaterials, 33*(26), 6079–6085.

Weadock, K. S., Miller, E. J., Bellincampi, L. D., Zawadsky, J. P., & Dunn, M. G. (1995). Physical crosslinking of collagen fibers: comparison of ultraviolet irradiation and dehydrothermal treatment. *Journal of Biomedical Materials Research, 29*(11), 1373–1379.

Westgate, S., Cutting, K. F., DeLuca, G., & Asaad, K. (2012). Collagen dressings made easy. *Wounds UK, 8*(1), 1–4.

Wong, W. H., & Mooney, D. J. (1997). Synthesis and properties of biodegradable polymers used as synthetic matrices for tissue engineering. In *Synthetic biodegradable polymer scaffolds* (pp. 51–82). Birkhäuser Boston.

Xu, Y., Meng, Q., Jin, X., Liu, F., & Yu, J. (2020). Biodegradable scaffolds for urethra tissue engineering based on 3D printing. *ACS Applied Bio Materials, 3*(4), 2007–2016.

Zhu, Z., Wang, Y. M., Yang, J., & Luo, X. S. (2017). Hyaluronic acid: a versatile biomaterial in tissue engineering. *Plastic and Aesthetic Research, 4*, 219–227.

Additive manufacturing with biodegradable polymers

Daniele Rigotti and Alessandro Pegoretti
University of Trento, Department of Industrial Engineering, Trento, Italy

1. Introduction

Since human behavior in the last century has been found to affect the global Earth's climate, extraction of raw materials and manufacturing processes are gaining a fundamental role in preserving our environment (DellaSala, 2013a; 2013b; Ebi, 2011). The widespread new technologies like additive manufacturing (AM) could contribute to reducing the human footprint (Gebler et al., 2014). In fact, the intrinsic nature of this process leads to material saving, no production of scrap like in subtractive processes, where the excess material is removed to form the final object. Waste material can be turned into final products with a lower amount of energy and chemical treatment (Ford & Despeisse, 2016). Additive manufacturing gives us the possibility of producing an object in any place, in our home, inside the International Space Station, or on the surface of another planet (Cesaretti et al., 2014; Kading & Straub, 2015; Wong & Pfahnl, 2014), where this object is really required. This will lead to a reduction in CO_2 emissions due to the lower transportation needs. The development of biopolymers either from renewable or biodegradable resources represents the next step to improve the "green" benefits of 3D printing. In fact, biopolymers provide several additional advantages if compared to conventional plastics, such as a reduced carbon footprint and additional waste management options. Biopolymers are gaining interest in an increasing number of applications from packaging, agriculture, consumer electronics, textiles, to the automotive sector (George et al., 2020). Moreover, AM enables on-demand solutions for a wide spectrum of needs ranging from personal protection equipment to medical devices in the case of supply–demand deficiency triggered by socioeconomic crises and disruptions of the supply chains (Choong et al., 2020).

2. Additive manufacturing

Additive manufacturing is an increasingly popular manufacturing technique that allows the production of solid objects in a new and disruptive way. It can be defined as one of the most important tools in a new industrial revolution because it changes completely the way objects are formed. It allows people to make customized products without using expensive tools or molds, so drastically decreasing costs and increasing accessibility to manufactured items (Berman, 2012). This technology consists of the

Biodegradable Polymers, Blends and Composites. https://doi.org/10.1016/B978-0-12-823791-5.00026-0

production of 3D objects starting from a digital model. It has been gradually introduced in many fields, from the aerospace to the automotive and bioengineering industries, with traditional materials such as polymers, metals, and ceramics, but also food (Mantihal et al., 2020), biotissues (Ebnesajjad & Drobny, 2007), and nanocomposites (Dorigato et al., 2017; Dul et al., 2018, 2016). It can be used to produce many objects, even those with complex shapes, and in a relative short time, and as a direct consequence of this ability it is useful for the construction of concept models and functional prototypes.

The first 3D printer was developed by Chuck Hull in 1984 (Dastbaz et al., 2015) when, working for the company 3D Systems Corp., he developed the stereolithography (STL) that is able to cure selective parts of a liquid photopolymer layer using a laser. With this apparatus, Hull was able also to develop the STL file format that is now the most used format for the exchange of 3D models and can easily be processed by software for filling, slicing, and cutting—Slic3r. In the following years, many other AM devices were created and also the media interest increased, in particular when the first open-source extrusion-based plastic printer was developed at the University of Bath. The Rep-Rap (**rep**licating **rap**id prototyper) opened the door for a market of low-cost 3D printers that were the perfect tool for a widespread design (Berman, 2012).

To prepare a digital file for printing, the 3D modeling software "slices" the final model into hundreds or thousands of horizontal layers. When the sliced file is uploaded in a 3D printer, the object can be created layer by layer. The 3D printer reads every slice (or 2D image) and creates the object, layer by layer, resulting in a three-dimensional object.

Additive manufacturing has the potential to greatly accelerate innovation, compress supply chains, minimize materials and energy usage, and reduce waste (Dastbaz et al., 2015). Below, a list of some benefits of AM technology (Christian Weller, 2015):

Lower energy consumption: AM can contribute to saving energy by eliminating production steps, using substantially less material, enabling reuse of by-products, and producing lighter objects (Dastbaz et al., 2015). There is a large reduction in energy consumption compared with traditional processes such as casting, that require large-scale heating and cooling cycles. It has been calculated that AM has the potential to reduce the global energy demand by 2.54−9.30 EJ and CO_2 emissions for 130.5−525.5 Mt of industrial manufacturing by 2025 (Gebler et al., 2014).

Less waste: due to its additive characteristic, AM can reduce material needs and costs by up to 90% compared with conventional subtracting techniques. AM can also reduce the costs and disposal at the end of life of machine components through avoidance of the tools, dies, and materials scrap associated with classic manufacturing processes (Dastbaz et al., 2015).

Reduced time to market: items can be fabricated as soon as the 3D digital model of the part has been created, eliminating the need for expensive and time-consuming part tooling and prototype fabrication.

Innovation: AM enables designs with novel geometries that would be difficult or impossible to achieve using conventional processes, which can improve a component's engineering performance. Novel geometries enabled by AM technologies can also lead to performance and environmental benefits in a component's product application.

Part consolidation: AM permits to design products with fewer, more complex parts, so reducing the number of components. Fewer parts means less time and labor are required for assembling the product, again contributing to a reduction in overall manufacturing costs.

Lightweighting: with the elimination of tooling and the ability to create complex shapes, AM enables the design of parts that can often be made to the same functional specifications as conventional parts, but with less material.
Agility to manufacturing operations: AM enables a quick response to markets. Spare parts can be produced on demand, reducing or eliminating the need for stockpiles and complex supply chains (Ford & Despeisse, 2016).

Polymers, due to their low costs and easier manufacturability with respect to most traditional building materials, have the potentialities for a large market penetration and wider user accessibility (Huang et al., 2015). Although plastic materials are not involved in the development of structural mechanical components like metals, they still represent the majority of the market. Polymers in AM will represent a greater choice over metals in many fields from consumer products, sustainable applications, advanced manufacturing, to biomedical devices (Baumann & Roller, 2017). Liquid monomers or polymers in the form of solid materials, such as filaments in fused filament fabrication (FFF), represent the largest part of the market over powders due to difficulties in handling of the latter by most nonmanufacturing consumers (Forster, 2015).

There are several polymeric materials available for additive manufacturing and the selection depends on the method used and the mechanical properties to be achieved. Some of the most important polymers for AM are acrylonitrile-styrene-butadiene (ABS), polycarbonate (PC), polylactide (PLA), polystyrene (PS), polyamide (PA), and polyurethane (PU). These materials are mainly used for low-performance components or for prototyping designs, but the demand for new higher performance polymer materials and composites is growing. High temperatures and chemically resistant polymers such as polyphenylene sulfide (PPS), polyetherimide (PEI), polyphenylsulfone (PPSU), and polyether ether ketone (PEEK) have been studied in the last few years for application in AM due to their high performances (Haleem & Javaid, 2019; Hoskins et al., 2018). Wide interest is emerging in the use of composite and nanocomposite materials in additive manufacturing to improve the mechanical properties or introduce new functions to commercially available polymers, such as thermal and -electrical conductivity (De Leon et al., 2016).

3. Additive manufacturing techniques

Three-dimensional (3D) printing technologies have seen significant developments in the past decade, in part due to many different varieties of 3D printers becoming available. Depending on the type of 3D printer used, materials with different properties are required for a proper 3D processing. For example, direct write (DW) 3D printing, a form of material extrusion, requires high-viscosity ink with appreciable shear thinning behavior for material deposition and shape holding during 3D printing. In contrast, inkjet-based 3D printing requires low-viscosity inks with strict control of the physical properties, such as stable surface tension. As a result, various materials have their own unique processing requirement needs, while also providing different properties and functionalities to the final printed part. The American Society for Testing and Materials (ASTM) developed a set of standards that classify the AM processes into seven categories according to "ASTM F42 – Additive Manufacturing":

Material extrusion, additive manufacturing process in which material is selectively dispensed through a nozzle or orifice;
Material jetting, additive manufacturing process in which droplets of build material are selectively deposited;
Vat photopolymerization, additive manufacturing process in which liquid photopolymer is placed in a vat;
Powder bed fusion, additive manufacturing process in which thermal energy selectively fuses regions of a powder bed;
Binder jetting, additive manufacturing process in which a liquid bonding agent is selectively deposited to join powder materials;
Directed energy deposition, additive manufacturing process in which focused thermal energy is used to fuse materials by melting as they are being deposited;
Sheet lamination, additive manufacturing process in which sheets of material are bonded to form a part.

In Table 23.1, the principal AM processes are reported.

Table 23.1 Classification of the principal AM processes.

Process type	Brief description	Related technologies
Material extrusion	Material is selectively dispensed through a nozzle	FFF, DW
Material jetting	Droplets of build material are selectively deposited	Multi-jet modeling (MJM)
Vat photopolymerization	Liquid photopolymer in a vat is selectively cross-linked by light-activated polymerization	Stereolithography (SLA), digital light processing (DLP)
Powder bed fusion	Thermal energy selectively fuses regions of a powder bed	Electron beam melting (EBM), selective laser sintering (SLS)
Binder jetting	A liquid bonding agent is selectively deposited to join powder materials	Powder bed and inkjet head (PBIH), plaster-based 3D printing (PP)
Directed energy deposition	Focused thermal energy is used to fuse materials by melting as the material is being deposited	Laser metal deposition (LMD)
Sheet lamination	Sheets of material are bonded to form an object	Laminated object manufacturing (LOM), ultrasonic consolidation (UC)

3.1 Material extrusion

The most commonly used technology in this process is FFF. In this technology, a plastic filament or metal wire is unwound from a coil and supplied to an extrusion nozzle which can turn the flow on and off. The nozzle is heated to melt the material and can be moved in both horizontal and vertical directions by a numerically controlled mechanism, directly controlled by a computer-aided manufacturing (CAM) software package. The object is produced by extruding layers of melted material that solidify upon cooling immediately after extrusion from the nozzle (Fig. 23.1).

This technology is most widely used with two types of plastic filament material, that is, acrylonitrile butadiene styrene (ABS) and polylactic acid (PLA), but many other materials are available ranging in properties from wood filled, conductive, flexible, etc. FFF was invented by Scott Crump in the late 1980s. After patenting this technology, he started the company Stratasys in 1988. The software that comes with this technology automatically generates support structures if required. The machine dispenses two materials, one for the model and one for a disposable support structure. The most common mode of failure in the FFF process is the filament buckling above the extruder. For a particular filament, a careful balance of the compressive modulus (E) and the apparent viscosity (η_a) is essential for a successful FFF printing process (Venkataraman, 2000). If the ratio E/η_a of a material exceeds a critical value filament will buckle. E/η_a has been widely used to determine the printability of a filament during the FDM process and can be calculated according to Eq. (23.1).

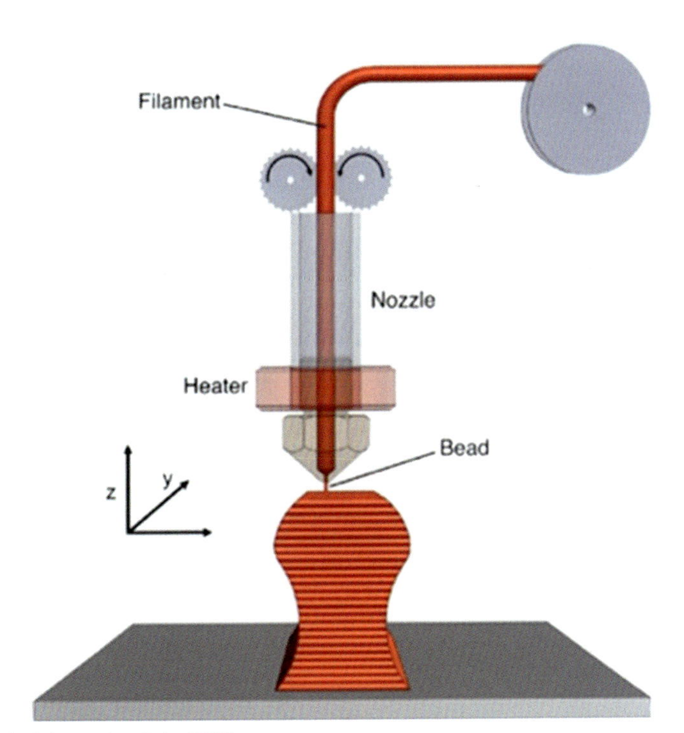

Figure 23.1 Schematic of the FFF process.
From Osswald, T. A., Puentes, J., & Kattinger, J. (2018). Fused filament fabrication melting model. *Additive Manufacturing, 22*, 51–59. https://doi.org/10.1016/j.addma.2018.04.030.

$$\frac{E}{\eta_a} = \frac{8\,Q\,l\,(L/R)^2}{\pi^3\,r^4\,k} \tag{23.1}$$

where Q is the volumetric flow rate of the melt, l is the length, r is the radius of the heating liquefier, L/R is the aspect ratio of the filament above the liquefier with a scaling factor k.

DW is another extrusion-based AM process where a viscous ink is used as raw material and is extruded from a syringe by pneumatic pressure overcoming the problem of buckling in FFF. Unlike with FDM, DW enables the printing of various polymers, also in the gelled form. The rigidity of the material should increase significantly after printing, enabling the printed part to maintain its geometry (Bekas et al., 2019).

3.2 Material jetting

In this process, material is ejected in droplets through a small-diameter nozzle, similar to the way a common inkjet paper printer works. The material is applied layer-by-layer to a build platform to form a 3D object and then hardened by UV light or heat (Fig. 23.2).

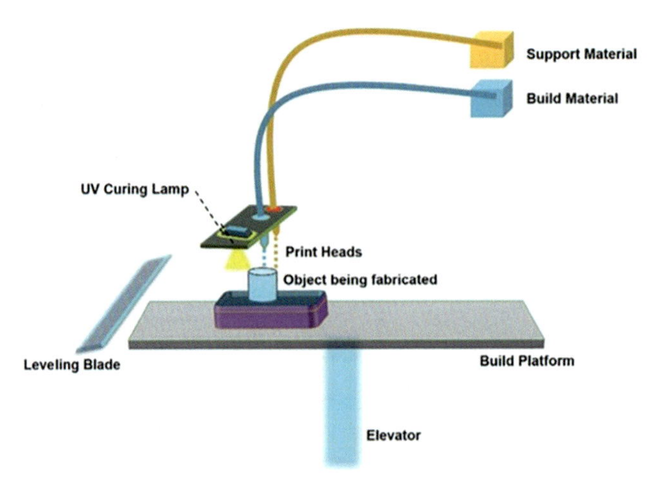

Figure 23.2 Schematic representation of the material jetting process.
Data from Sireesha, M., Lee, J., Kranthi Kiran, A. S., Babu, V. J., Kee, B. B. T., & Ramakrishna, S. (2018). A review on additive manufacturing and its way into the oil and gas industry. *RSC Advances, 8*(40), 22460–22468. https://doi.org/10.1039/c8ra03194k.

3.3 Vat photopolymerization

In a 3D printer based on the vat photopolymerization method, a photopolymerizable resin in a container is selectively hardened with a UV light source. The most commonly used technologies in these processes are stereolithography (SLA) and digital light processing (DLP) (Fig. 23.3). SLA employs a vat of liquid ultraviolet-curable

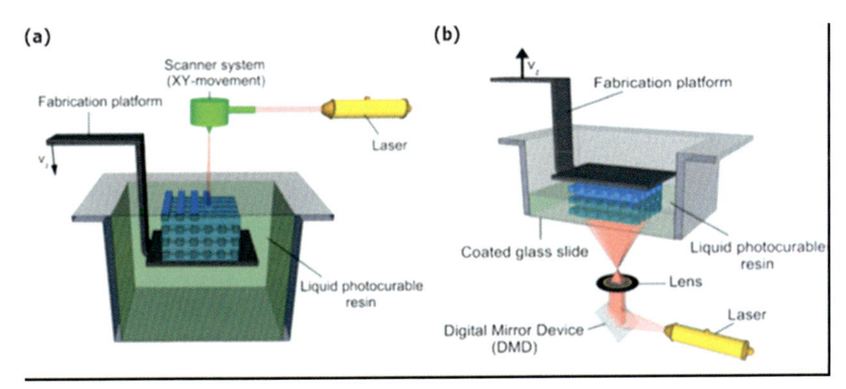

Figure 23.3 (a) SLA type 3D printer and (b) DLP type 3D printer.
Data from Ko, D.-H., Gyak, K.-W., & Kim, D. (n.d.). Emerging microreaction systems based on 3D printing techniques and separation technologies. *Journal of Flow Chemistry, 7*, 1–10. https://doi.org/10.1556/1846.2017.00013.

photopolymer resin and an ultraviolet laser to build the object's layers one at a time. For each layer, the laser beam traces a cross-section of the part pattern on the surface of the liquid resin. Exposure to the ultraviolet laser light cures and solidifies the pattern traced on the resin and joins it to the layer below. After the pattern has been traced, the SLA's elevator platform descends by a distance equal to the thickness of a single layer, typically 0.05–0.15 mm (0.002–0.006″). Then, a resin-filled blade sweeps across the cross-section of the part, recoating it with fresh material. On this new liquid surface, the subsequent layer pattern is traced, joining the previous layer. The complete three-dimensional objects are formed by this project. Stereolithography requires the use of supporting structures which serve to attach the part to the elevator platform and to hold the object that otherwise would float in the basin filled with liquid resin. These are manually removed after the object is finished. In contrast, DLP employs a digital mask projection to trigger localized photopolymerization, and enables high-efficiency fabrication of 3D hydrogel structures with high precision ranging from 1 to 100 μm (Shen et al., 2020).

3.4 Powder bed fusion

The most commonly used technology for powder fusion is represented by selective laser sintering (SLS). This technology uses a high-power laser beam to melt small particles of polymeric, metallic, ceramic, or glassy powders into a mass that has the desired three-dimensional shape. The laser selectively fuses the powdered material by scanning the cross-sections (or layers) generated by the 3D modeling program on the surface of a powder bed (Fig. 23.4). After each cross-section is scanned, the powder bed is lowered by one-layer thickness. Then a new layer of material is applied on top and the process is repeated until the object is completed. Powder grains not scanned by the laser beam constitute a support structure for the growing object. Therefore there is no need for a support structure (Lipson & Kurman, 2013). Laser sintering

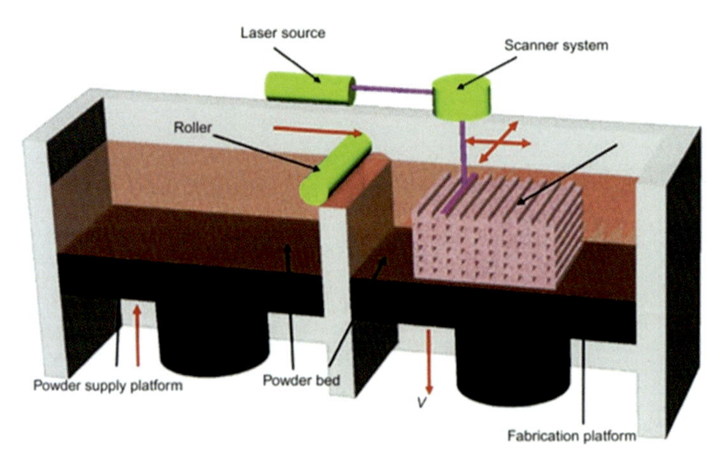

Figure 23.4 SLS system schematic.
From Wang, X., Jiang, M., Zhou, Z., Gou, J., & Hui, D. (2017). 3D printing of polymer matrix composites: A review and prospective. *Composites Part B: Engineering, 110*, 442–458. https://doi.org/10.1016/j.compositesb.2016.11.034.

can be used to create metallic, polymeric, and ceramic objects. The degree of detail is limited only by the precision of the laser and the fineness of the powder, therefore, with this type of printer, it is possible to create especially detailed and delicate structures (Hoy, 2013).

3.5 Binder jetting

For binder jetting two materials are used: a powder base material and a liquid binder. In the build chamber, powder is spread in equal layers and the binder is applied through jet nozzles that "glue" the powder particles in the shape of a 3D object (Fig. 23.5). After the printing is finished, the remaining powder is cleaned off and used for the next object. This technology was first developed at the Massachusetts Institute of Technology in 1993 and in 1995 Z Corporation obtained an exclusive license.

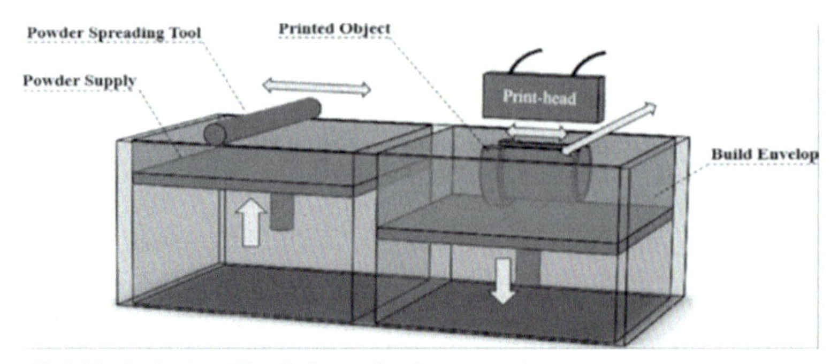

Figure 23.5 Binder jetting 3D printing technology overview.
From Ziaee, M., & Crane, N. B. (2019). Binder jetting: A review of process, materials, and methods. *Additive Manufacturing, 28*, 781–801. https://doi.org/10.1016/j.addma.2019.05.031.

The technology of binder jetting uses plastic powders or sand and was configured to create models for precision casting or casting cores as well as for creating illustration models for architecture. Poor stability and high porosity of the parts are significant disadvantages of this technology. In the case of poly(methyl methacrylate) (PMMA), the binder reacts with the powder and solidifies the printed areas. Afterward, the parts have to remain in the powder-bed for 24 h before being removed. The tensile strength of these parts was measured to be about 4.5 MPa; a posttreatment by infiltrating the part with wax or epoxide resin raises the tensile strength up to 28 MPa. Due to the solidification mechanism, the accuracy of this technology is not as high as that of a laser beam. The binder jetting technology has high potential for the efficient production of parts. One disadvantage is the low density of the parts. To enlarge the field of application for this technology, the manufacturing system was extended by integrating different functions (Glasschroeder et al., 2015).

3.6 Directed energy deposition

This process is mostly used in the high-tech metal industry and in rapid manufacturing applications. The 3D printing apparatus is usually attached to a multi-axis robotic arm and consists of a nozzle that deposits metal powder or wire on a surface and an energy source (laser, electron beam, or plasma arc) that melts it, forming a solid object as shown in Fig. 23.6. Applications include design visualization, prototyping/CAD, metal casting, architecture, education, geospatial, health care and entertainment/retail.

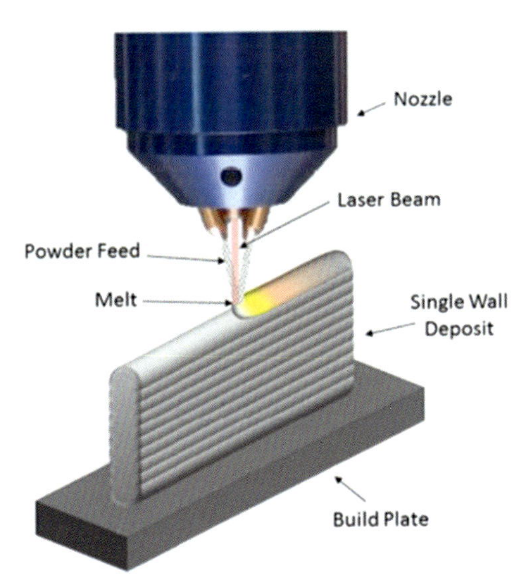

Figure 23.6 Schematic of a directed energy deposition system.
From Reichardt, A., Dillon, R. P., Borgonia, J. P., Shapiro, A. A., McEnerney, B. W., Momose, T., & Hosemann, P. (2016). Development and characterization of Ti–6Al–4V to 304L stainless steel gradient components fabricated with laser deposition additive manufacturing. *Materials & Design, 104*, 404–413. https://doi.org/10.1016/j.matdes.2016.05.016.

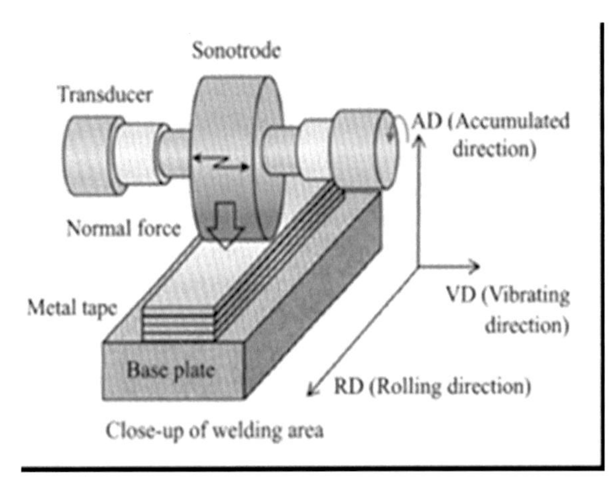

Figure 23.7 Simplified model of ultrasonic sheet metal 3D printing.
From Neikov, O. D., Naboychenko, S. S., Yefimov, N. A., & Neikov, O. D. (2019). Chapter 13: Powders for additive manufacturing processing. In *Handbook of non-ferrous metal powders* (Second Edition) (pp. 373–399). Elsevier. https://doi.org/10.1016/B978-0-08-100543-9.00013-0.

3.7 Sheet lamination

Sheet lamination involves materials in sheets which are bound together with the application of external pressure. The sheets can be metallic, polymeric, or paper. Metal sheets are welded together by ultrasonic welding in layers and then milled into the correct shape by a computerized numerical control (CNC) machine (Fig. 23.7). Paper sheets can also be used with adhesive glue and cut into shape by precise blades.

4. Biodegradable polymers in additive manufacturing

4.1 Polylactic acid (PLA)-based material

Polylactic acid (PLA) includes a relatively large category of polymers from renewable resources. They are not only compostable and biocompatible, but also processable with most standard processing equipment (Rafael, 2010; Ren, 2010). PLA is a compostable polymer derived from renewable sources; it has been viewed as a potential material to reduce the societal solid waste disposal problem. Its low toxicity, along with its environmentally benign characteristics, has made PLA an ideal material for food packaging and other consumer products. PLA is the front runner in the emerging bioplastics market with the best availability and most attractive performance/cost ratio (Lunt, 1998). The production of the aliphatic polyester from lactic acid, a naturally occurring acid and bulk-produced food additive, is relatively straightforward. PLA is a thermoplastic material with stiffness and light transmissibility similar to polystyrene or poly(ethylene terephthalate). PLA is used for many different applications, from packaging to agricultural products and disposable materials, as well as in the medicine, surgery, and pharmaceutical fields (Rafael, 2010; Ren, 2010).

PLA filament has gained wide acceptance within additive manufacturing partly because it is made from renewable resources and also because of its mechanical properties. It is often the preferred choice for beginners in 3D printing as it is an easily processable material. PLA is a semicrystalline polymer with a melting temperature of 180°C. This means that when printing with PLA, the use of a heated printing bed is not necessary, nor is a closed chamber a necessity (Jerez-Mesa et al., 2017). PLA is becoming increasingly popular as a biodegradable engineering plastic for 3D printing applications due to its easy processability compared to other biopolymers (Auras et al., 2004). Within academic research, 3D-printed PLA objects have demonstrated great potential in drug-delivery systems and tissue engineering. Boetker et al. (2016) focused their attention on the design of a device for the release of an active drug (nitrofurantoin) based on a 3D-printed PLA geometry intended for flexible dosing and precision medication. Nitrofurantoin, hydroxypropyl methylcellulose (HPMC), and polylactic acid were successfully coextruded with up to 40% HPMC content, and subsequently 3D printed into disk geometries. HPMC worked as a water-soluble excipient that facilitated the formation of a porous network in the printed geometries. A printing temperature of 190–200°C was selected for all the mixtures to avoid the thermal degradation of the drug and assuring sufficient flow viscosity for the deposition of the material. Nitrofurantoin remained in its original solid form during both hot-melt extrusion and subsequent 3D printing, as evidenced by DSC and Raman spectroscopy. In vitro release experiments conducted in phosphate-buffered saline showed that the initial rate and the overall total release of drug were correlated to the amount of water-soluble excipient inserted in the delivered 3D-printed device.

In tissue engineering, hydroxyapatite (HA) demonstrates high biological compatibility and bioactivity, ability to stimulate osteogenesis, coalesce with bone, and it serves as material for bone tissue growth due to the similar chemical composition with human bone (Wei & Ma, 2004). According to Dubinenko et al. (2020), FFF 3D-printed PLA/HA scaffolds could be prepared starting from solution mixing of the two components followed by filament extrusion. This path allowed a good dispersion of the filler inside the PLA matrix to be obtained. Cylindrical-shaped scaffolds were prepared by FFF at 220°C and a significant increase in the stiffness was detected, reaching 8111 ± 714 MPa at 50 wt.% of HA. Morphological properties of 3D-printed porous PLA and PLA/HA scaffolds and the dependence for the dimensional accuracy on the composition of the filaments were investigated instead by Gendviliene et al. (2020). PLA and 10 wt.% HA were mixed during filament extrusion at 145°C through a 1.75 mm die. Scaffolds were printed at 210°C with a layer height of 0.4 mm and a hexagonal internal geometry. Although the calculated porosity of the scaffolds deviated from the STL file, there were no statistically significant differences found between scaffold groups ($P > 0.05$) or between the porosity of scaffolds and the porosity of the original STL file. High fidelity of the printed scaffold to the designed model is a crucial condition since the cell response to an implant is driven by the morphology of the implant itself (Zafar et al., 2020).

FFF has been used by Wu et al. (2020) to print bone trabecular models with PLA/HA composites and to evaluate the morphology and mechanical properties of the printed models. PLA and HA powder from 5 wt.% to 15 wt.% were mixed in a solvent solution and then extruded to obtain a filament with a diameter of about 2.85 mm Three-dimensional models to be printed were obtained from microcomputed tomography of trabecular bone isolated from a femoral head and were utilized to generate bone models for additive manufacturing. A desktop FFF 3D printer was used to deposit material in the designed path at a temperature of 210°C in 60 μm layer thickness. A screw pull-out test on trabecular 3D-printed bones with a titanium orthopedic cancellous screw showed an increase of 100 and 200 N in mean pullout load, with the incorporation of 5 and 15 wt.% of HA, respectively, compared to neat PLA, that sustained a force of about 1250 N. The surface morphology of trabecular structure was determined by the nozzle size of the FFF machine that limited the final resolution of the object. Smaller nozzle sizes, with a diameter lower than 0.6 mm, tended to be clogged by HA agglomerates.

FFF allows not only the preparation of porous scaffolds able to mimic the targeted tissue but also complex, customizable, and bioresponsive orthopedic implants as shown by Dhandapani et al. (2020). PLA cortical screws with tunable porosity were developed by FFF, which facilitates the fabrication of complex structures with high accuracy, low cost, rapid product development, and design freedom (Fig. 23.8). Layer-by-layer deposition of biodegradable polymer employed in the development of porous orthopedic screws ensured a gradual dissolution and complete metabolic resorption, thereby overcoming the limitations of conventional metallic screws. Micro-CT analysis and scanning electron micrographs of screws with 45% fill density proved the presence of interconnected pores similar to natural bone tissue (~300 μm) without affecting the mechanical strength of the screw, thus avoiding the stress-shielding effect. Human bone marrow-derived mesenchymal stem cells cultured on the screws confirmed differentiation toward osteoblast lineage and similarly osteoblast-like cells cultured over 21 days showed mineralization with largely infiltrated cells in porous screws. These porous screws showed significantly increased vascularization in a rat subcutaneous implantation as compared to control screws.

PLA has been also intensively used to prepare biobased and biodegradable composites filled with cellulosic materials to reduce the overall costs and improve the mechanical properties (Fortunati et al., 2014). The effect of wood flour content from 0% to 50% by weight in PLA-based 3D-printed parts was studied by Kariz et al. (2018). Wood particles with a mesh size lower than 0.237 mm were compounded with PLA and pelletized. Specimens were printed at 230°C through a nozzle of 0.4 mm and a layer thickness of 0.19 mm. The surface of 3D-printed objects became increasingly rougher as the wood content increased, along with the presence of voids and clusters of wood particles. Ayrilmis (2018) found that not only the wood content affected the roughness of 3D-printed parts, but also the layer thickness. When the layer thickness was reduced, the accuracy of the printed sample was improved and so the surface roughness decreased. Mechanical properties of wood-reinforced PLA 3D-printed objects were also studied by Bhagia et al. (2020), in particular, the variations in wood structure were examined in depth. For this purpose, 70 *Populus trichocarpa* trees (poplar), that were grown under comparable environment conditions, were acquired and wood samples underwent the same material processing, printing, and tensile testing. Wood flour obtained through ball milling was mixed with PLA, the

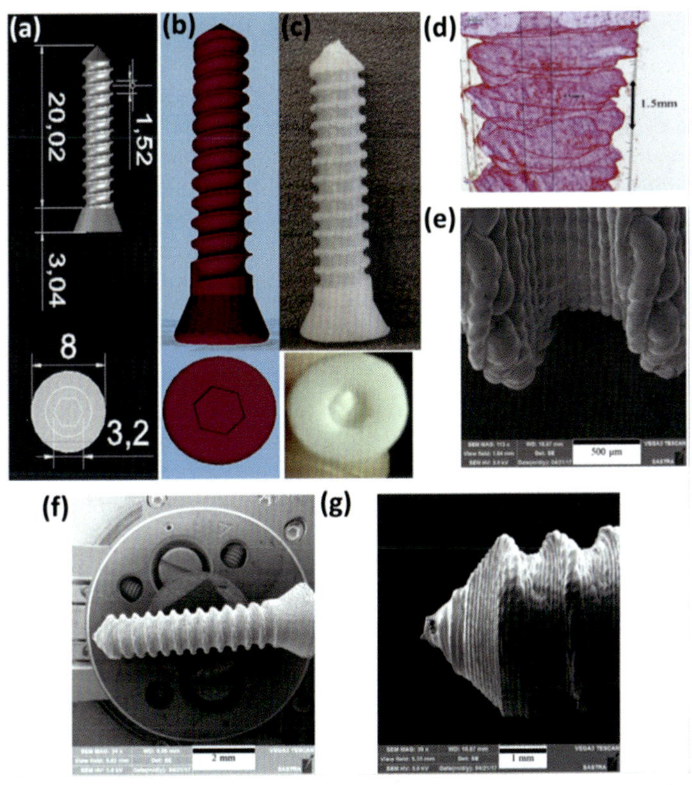

Figure 23.8 (a) AUTOCAD designed screw in mm; (b) STL file shows the 3D printable screw model; (c) 3D-printed cortical screw; (d, e) pitch length of printed screw measured using micro-CT and SEM; (f, g) uniform layers were observed forming a rough surface from the head to the tapered end.

Data from Dhandapani, R., Krishnan, P. D., Zennifer, A., Kannan, V., Manigandan, A., Arul, M. R., Jaiswal, D., Subramanian, A., Kumbar, S. G., & Sethuraman, S. (2020). Additive manufacturing of biodegradable porous orthopedic screw. *Bioactive Materials, 5*(3), 458–467. https://doi.org/10.1016/j.bioactmat.2020.03.009.

composite was printed using an FFF machine at 230°C at 20 mm s^{-1}. The mechanical properties were found to be significantly affected by the natural diversity among poplar structural characteristics. The median tensile stresses at yield and at break were ~50 and ~48 MPa, respectively, while the elastic modulus was 3.65 GPa and it was found to be correlated to the concentration of β-glycosidic of polysaccharides bond of poplar. Layer height, infill level, and nozzle diameter are the most significant parameters affecting the fatigue life of PLA-based composite reinforced with wood processed through FFF as evaluated through rotating bending fatigue tests by Travieso-Rodriguez et al. (2020). Specimens were printed at a nozzle temperature ranging from 170 to 185°C from a commercial wood–PLA-based filament. The specimen geometry was reproduced according to the ASTM D7774 standard. The effects of layer height, nozzle diameter, infill density, and extrusion velocity were studied through a Taguchi experimental design. Also in this work, it was underlined that the incorporation of wood fibers has a negative effect on the PLA matrix in terms of mechanical behavior, as they reduce the adhesion between fibers, and increase the voids among them. The authors concluded that the usage of wood-filled PLA filaments is preferable for objects where the aspect is more important than function. Cork, a renewable biodegradable material obtained from the outer bark of an oak tree, was investigated by Daver et al. (2018) as a filler for 3D-printable PLA-based composites. Cork powder with a mean size of 446 μm was melt-mixed with PLA at 200°C. Tributyl citrate was used as a plasticizer to improve the printability of the material. Specimens were obtained with an FFF 3D-printing machine and material was deposited at 230°C through a 0.8 mm diameter nozzle in 0.4 mm layers. The mechanical properties were negatively affected by the presence of cork and the addition of a plasticizer mitigated only the reduction of the elongation at break to a level similar to that of neat PLA. Since cellulose possesses a hierarchical structure, different sizes of cellulosic material can be used as filler in polymeric matrices. In fact, from wood pulp, through different techniques and treatments, it is possible to obtain both micro- and nano-sized particles. Three-dimensional printable filaments made of mechanically processed wood pulp PLA composite with up to 30 wt.% of micro/nanocellulose (MNC) constitution were developed by Wang et al. (2017). Wood pulp was swelled in PEG400 and then agitated by mechanical disintegration with a colloidal mill at room temperature with a rotor speed of 6000 rpm for 10 min, PEG400 was removed by washing with dichloromethane. The obtained MNC paste was modified with a silane coupling agent to improve the interfacial adhesion. MNC, PEG6000, and PLA were mixed in DCM and upon drying extruded at 190°C in 1.75 mm in diameter filaments. Specimens were printed with an FFF machine at 190°C with a 0.4 mm nozzle at 50 mm/s in 0.2 mm layers. An optimal composition for printing was found for 30 wt.% cellulose, 65% PLA, and 5% PEG6000 with a good surface quality, as seen in Fig. 23.9.

Figure 23.9 Samples of 3D-printed objects, including double-balls standing on the shelf, buckets, half-baskets, and sticks in elongated and dumbbell shape that were used for the testing of mechanical properties.
From Wang, Z., Xu, J., Lu, Y., Hu, L., Fan, Y., Ma, J., & Zhou, X. (2017). Preparation of 3D printable micro/nanocellulose-polylactic acid (MNC/PLA) composite wire rods with high MNC constitution. *Industrial Crops and Products, 109*, 889–896. https://doi.org/10.1016/j.indcrop.2017.09.061.

At the optimal conditions, the mechanical properties of 3D products from MNC/PLA composites were maintained at a comparable level to that of neat PLA with elongation at break of 12%, tensile strength of 59.7 MPa, and flexural strength of 50.7 MPa.

Nanocrystalline cellulose (NCC) has the potentiality for large-scale manufacture of green composites for various applications, including packaging and biomedical fields. PLA/NCC composites were prepared with 1, 2, 5, and 10 wt.% NCC and printed through FFF by Dinesh Kumar et al. (2020). The results showed that the crystallinity of the polymer matrix increased with the addition of just 1 wt.% cellulose with the highest cold crystallization peak when compared to the neat PLA. A PLA/NCC filament of 1.75 mm in diameter was obtained through a single-screw extruder. Specimens were printed through a 0.35 mm nozzle diameter, 0.20 mm layer height, 230°C nozzle temperature, and 60°C bed temperature. The results of the characterization of 3D-printed samples showed that an increase in tensile modulus of 50% and yield strength of 18% was reached for 1 wt.% NCC composites, when compared with neat PLA. However, a strong decrease in the strain at break was detected for all the investigated NCC concentrations. The dispersion of NCC in the PLA matrix was poor, due to the different chemical natures of the two materials, which made it difficult for both materials to interact with each other. This problem could be solved or at least mitigated by employing a surface treatment process on cellulose particles (Checchetto et al., 2020; Dickmann et al., 2020; Rigotti et al., 2019, 2020). A cellulose nanofiber PLA mixture was extruded into filaments, and subsequently 3D printed with an FFF technique into composites by Dong et al. (2019). Moreover, the effect of post treatment annealing was studied. PLA monomers were grafted on CNFs via ring-opening polymerization and subsequently dissolved in PLA to prepare a NCC-PLA masterbatch. Three-dimensional printed composites containing CNFs with a loading rate from 1 to 3 wt.% were deposited at 210°C from a 0.6 mm diameter nozzle at a speed of 1800 mm s^{-1} in 0.2 mm layer thickness. XRD patterns showed that annealing could induce the formation of PLA crystalline regions, as seen in XRD patterns, that could hamper the free motion of PLA molecular chains, upshifting T_g, and improving composite resistance at both low and high temperatures. Dynamic storage modulus and static flexural modulus of 3 wt.% annealed composites increased 0.72 times and 1.56 times at 35°C, and 2 times and 1.52 times at 70°C in comparison with a neat unannealed 3D-printed sample. L. Li et al., 2019; V.C.-F. Li et al., 2019 (Li et al., 2019) found that the addition of a small amount of TEMPO-oxidized bacterial cellulose (BC) could effectively improve the mechanical properties and crystal properties of 3D-printed PLA, acting as a heterogeneous nucleating agent and enhancing PLA due to its high aspect ratio and excellent mechanical properties. Compared to wood cellulose, bacterial cellulose has the advantage that it does not require the removal of impurities such as lignin and hemicellulose. Uniformly dispersed PLA/BC nanocomposites were prepared based on the Pickering emulsion approach. Composite filaments were extruded with a single screw, followed by 3D printing with FFF technique. Tensile strength, elongation at break, maximum bending strength, and elastic modulus of the PLA increased by 9.2% (from 29.11 to 31.79 MPa), 202% (from 6.42% to 19.45%), 45% (from 50.99 to 73.93 MPa) and 49% (from 2.16 to 3.22 GPa), respectively, when 1.5 wt.% of BC was added to PLA.

SLS permits the production of nearly any designed 3D geometry with higher accuracy than FFF and allows the processing of a wider range of biocompatible and biodegradable materials. The possibility of processing PLA and a PLA/nanoclay composite by SLS was accurately investigated by Bai et al. (2017). The PLA/nanoclay composite was prepared by melt mixing and then cryogenically ground to obtain a powder with an average particle size of 30 μm. The effect of laser power, bed temperature, and scan count were related to the flexural modulus of printed specimen. Increasing the bed temperature from 60 to 80°C caused an increase in the flexural modulus in neat PLA powders. Increasing the laser power, from 15 to 17 W, had the same stiffening effect due to an improved sintering effect between particles, reaching an improvement of 41.5% in flexural modulus for PLA/nanoclay printed at the highest power. An impressive effect was found in double scan, that is, when each layer was scanned twice with the laser. Compared to a single-scanned sample, the flexural modulus for parts built with a double scan experienced a large increase, in particular, 196.6% for PLA and 158.3% for PLA/nanoclay, respectively. This result was related to the fact that a double scan can deliver the laser energy to each layer in a more gradual manner, resulting in the release of structural stresses in sintered parts, which results in less shrinkage and curling. When a double scan was applied, there was no visible increase in curling of the sintered components with 17 W laser power, however, when this power was exceeded, curling started to appear. Nevertheless, a double scan required doubled amount of time for the whole printing process. Since large bone defects cannot be healed by the body itself, continuous efforts are being put into the development of 3D scaffolds for bone tissue engineering. Anatomical structures with high bulk density based on individual 3D data from computer tomography models made of PLA and β-tricalcium phosphate can be easily processed by SLS to produce individually tailored bioresorbable implants. A key step is the identification of process parameters such as laser power, scanning velocity, and laser beam diameter to avoid degradation of the polymer during melting and to prevent a phase transition of β-TCP (Verdonck et al., 2009). Manufacturing of bone substitute implants able to dissolve inside the body and to be replaced by new bone represents an innovative approach in regenerative medicine. For this purpose, the pore structure of the implant is essential for bone ingrowth, especially for critical-size defects and SLS allows the manufacture of implants with a defined internal porous structure with customized degradation time and optimized mechanical properties. Lindner et al. (2011) developed biodegradable implants made of 50% PDLLA and 50% β-TCP using SLS. The idea was to melt PLA in order to glue together the β-TCP particles homogeneously distributed in the building plate powder. A CO_2 laser beam with a wavelength of 10.6 μm was scanned over the powder bed and suitable process parameters were selected by tuning the laser power, the scanning speed, the hatching distance, the layer thickness, and the scanning path. No phase transition of β-TCP or new phases were detected by X-ray diffraction analysis upon SLS treatment. Stress at break of the printed specimen measured in a four-points bending test was found to increase as the laser power increased, notably from 13 ± 1 MPa to 23 ± 1 MPa for 0.3 and 0.5 W, respectively. Printed material showed a linear elastic behavior up to the point of catastrophic failure, making this material only suitable for a nonload-bearing or even low-load-bearing implantation site. In a subsequent study, the influences of particle size and melt viscosity of the PLA/β-TCP powder on the processability and on the mechanical strength of 3D-printed parts were investigated (Gayer et al., 2018). Powder with a smaller particle size and lower molecular weight showed improved melting behavior due to lower melt viscosity. The formation of a homogeneous melt film in a low-molecular-weight sample resulted in higher thermal conductivity that improved the resistance to thermal degradation, despite its lower molecular weight (MW). The ability to form a homogeneous melt film was found to be related to the loss factor, since a melt with a higher loss factor is more viscous than elastic. Instead, smaller β-TCP particles were observed to increase the melt viscosity due to the higher friction that is caused by their larger surface/volume ratio decreasing the processability through SLS. Lower melt viscosity led to higher densification due to easier coalescence of the powder particles that resulted in lower porosity. This behavior produced specimens with higher bending strength with respect to those printed with higher MW PLA, in fact, the higher relative density compensated for the lower MW. As previously reported, in order to obtain parts with high mechanical strength, complete coalescence of the powder particles should be reached with a processing time as short as possible to ensure high productivity. Since no external shear forces are applied to the melt during SLS processing, it is assumed that the melting behavior is mainly determined by the zero-shear viscosity. Therefore, coalescence occurring by viscous flow for two identical spherical polymer particles can be described by the Frenkel–Eshelby model (Shaler, 1949). High surface tension, low melt viscosity, and small particle radius are the characteristics sought in materials for SLS application. With these considerations in mind, it was possible to print 3D scaffolds with interconnected pore structure with great potential to be used as patient-specific bone replacement implants (Fig. 23.10) (Gayer et al., 2019).

While rapid prototyping methods such as FFF and SLS are restricted to simple architectures, stereolithography allows the preparation of structures with more complex design. This allows the preparation of scaffolds for tissue engineering with architectures optimized for cell seeding and culturing. However, to apply PLA macromers in SLA, the macromer must be in the liquid state by heating or diluting and network formation by photocross-linking is required. Jansen et al. (2009) produced 3D-printed scaffolds via SLA by photocross-linking functionalized fumaric acid monoethyl ester (FAME), three-armed poly(D,L-lactide) oligomers using N-vinyl-2-pyrrolidone (NVP) as diluent and comonomer. Fumaric acid is naturally found in the body, in particular when human skin is exposed to sunlight, and so it gains attention for end-functionalization (Products, 2008). Solutions containing macromer, NVP, and Irgacure 2959 photoinitiator were mixed and exposed to 365 nm UV light for 15 min in a nitrogen atmosphere. After cross-linking, the samples were rinsed with acetone and dried under a nitrogen flow at 90°C. Scaffolds of approximately 4.5 mm × 4.5 × 1.8 mm were obtained with a porosity of 80% and an average pore size of 256 μm. The small deviations from the original virtual model (less than 5%) were attributed to inaccuracies in the digitalization procedure, in the imaging, in the 3D reconstruction of the scanning data, or in the method of assigning sizes to pores. The open pore volume was measured using computed tomography (CT) scan,

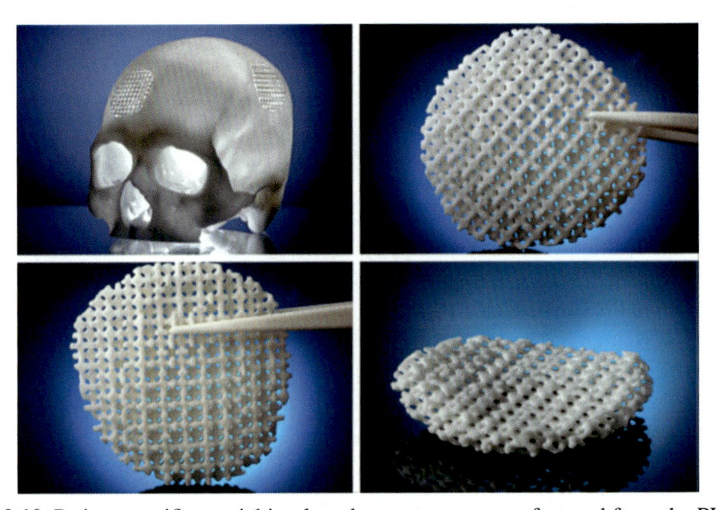

Figure 23.10 Patient-specific cranial implant demonstrators manufactured from the PLLA-1.0/CC (77/23) composite using a modified Formiga P 110 laser sintering machine. The pore structures had a designed porosity of about 72% and a strut thickness of about 1 mm. The skull was also manufactured by SLS but from polyamide 12.
Data from Gayer, C., Ritter, J., Bullemer, M., Grom, S., Jauer, L., Meiners, W., Pfister, A., Reinauer, F., Vucak, M., Wissenbach, K., Fischer, H., Poprawe, R., & Schleifenbaum, J. H. (2019). Development of a solvent-free polylactide/calcium carbonate composite for selective laser sintering of bone tissue engineering scaffolds. *Materials Science and Engineering C: Materials for Biological Applilications, 101*, 660–673. https://doi.org/10.1016/j.msec.2019.03.101.

illustrating that it represents more than 90% of the total pore volume. This will allow a high permeability of the scaffold for nutrients and metabolites in the cell-seeding procedure and during cell culture or implantation. Diluents are essential components in the formulation of cross-linkable solutions for SLA as viscosity modifier, however, N-vinyl-2-pyrrolidone, due to its oil-derived origin, should be replaced with a biobased diluent, as proposed by Melchels et al. (2009). Ethyl lactate was used as a biobased nonreactive diluent in PLA-based resin for SLA. Suitable resin compositions were prepared to print porous structures with predesigned architectures at high resolution. Therefore, methacrylate end-functionalized poly(D,L-lactide) oligomers of varying molecular architectures were synthesized and photocross-linked in the presence of ethyl lactate, Lucirin TPO-L visible light photoinitiator, hydroquinone inhibitor, and Orasol Orange G dye. Scaffolds with a gyroid architecture were printed projecting a blue light pattern with a wavelength peak at 440 nm of 1280 × 1024 pixels, each measuring 32 × 32 μm^2 in layers of 25 μm (Fig. 23.11). Porous scaffolds with a gyroid architecture were obtained with a very high precision determined by the overcure distance of only 7 μm. This open and regular structure facilitated the penetration of water into PLA scaffolds and so the cell seeding could be performed effortlessly by simply pipetting the cell suspension onto the scaffold. Osteoblasts demonstrated good adhesion to these 3D-printed scaffolds and the proliferation rate was comparable to that on high-molecular-weight PLA and tissue culture polystyrene.

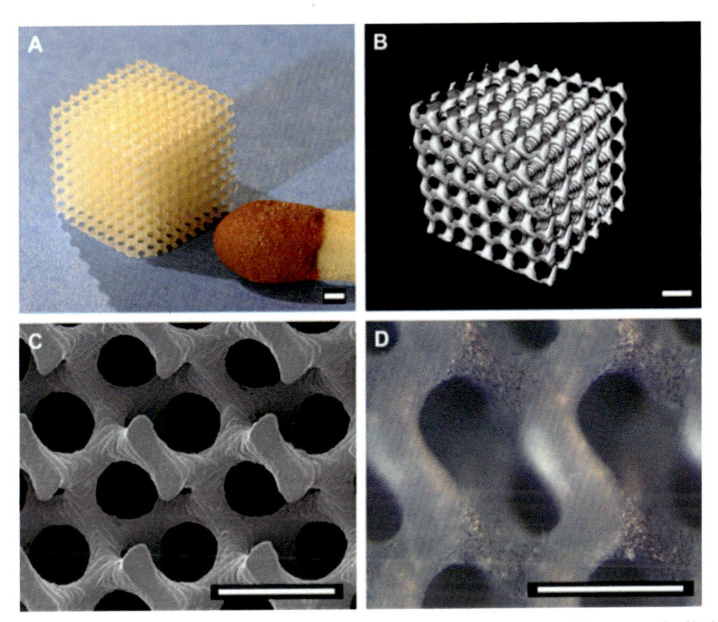

Figure 23.11 Images of PDLLA network scaffolds with a gyroid architecture, built by stereolithography: (a) photograph, (b) mCT visualization, and (c) SEM image. In (d) a light microscopy image is shown for a scaffold seeded with mouse preosteoblasts after 1 d of culturing. Scale bars represent 500 mm.
From Melchels, F. P., Feijen, J., & Grijpma, D. W. (2009). A poly(D,l-lactide) resin for the preparation of tissue engineering scaffolds by stereolithography. *Biomaterials, 30*(23–24), 3801–3809. https://doi.org/10.1016/j.biomaterials.2009.03.055.

4.2 Polyhydroxyalkanoates (PHA) and their copolymers

Polyhydroxyalkanoates (PHA) are a family of aliphatic polyesters extensively considered for biomedical applications due to their well-known biodegradability and biocompatibility. The broad variety of macromolecular structures provided by the various PHA homo- and copolymers offers good flexibility in terms of processing and mechanical properties, similar to that of oil-based aliphatic polyesters (George & Ipsita Roy, 2014).

Poly[3-hydroybutyrate] (PHB) and poly[3-hydroxybutyrate-co3-hydroxyvalerate] (PHBV) copolymers have been investigated in depth for bone tissue engineering applications where SLS is one of the most promising techniques for manufacturing polymer scaffolds with suitable mechanical properties. PHBV powder was used as a raw material to fabricate porous cylindrical scaffolds using an SLS system equipped with a CO_2 laser with a power that varied from 3 to 7 W, an inherent wavelength of 10.6 μm, and a spot size of 420 μm (Diermann et al., 2018). The laser scan speed was fixed at 5000 mm s^{-1}, while an inert N_2 environment with an oxygen content below 5% at 125°C was kept during the entire process. The scaffold microstructure was largely determined by the laser energy density used during the SLS process. An increase in the energy resulted in higher relative densities, stronger interlayer connections, and a reduced quantity of residual powder trapped inside the pores. An increase in relative density from 20.3% to 41.1% resulted in a higher maximum compressive modulus and strength of 36.4 and 6.7 MPa, respectively. The fabricated scaffolds reached a maximum porosity of 59%, which corresponded to ttoo a compressive elastic modulus of 36.4 and 31.8 MPa when tested in normal and lateral directions, respectively, and a compressive strength of 6.7 and 4.1 MPa, respectively. SLS was employed to produce 3D scaffolds based on nanosized calcium phosphate (Ca–P) and PHBV microspheres in order to develop a biomimetic environment for osteoblastic cell attachment, proliferation, and differentiation for bone tissue engineering applications (Duan et al., 2010). The Ca–P/PHBV microspheres were prepared using a solid-in-oil-in-water emulsion/solvent evaporation method. A scaffold model with a 3D orthogonal periodic porous architecture was printed with a laser characterized by a power intensity of 15 W and spot size of 457 μm. The scan speed was set at 1257 mm s^{-1}, bed temperature at 35°C, and layers of 0.1 mm were subsequently sintered (Fig. 23.12). Printing parameters were optimized with a three-factors three-levels factorial design in which laser power, scan spacing, and layer thickness were related to the structure and handling stability of the scaffold, the dimensional accuracy, and the mechanical properties in the compression test (Duan et al., 2011). The sintered scaffolds possessed controlled material distribution and a hierarchical interconnected

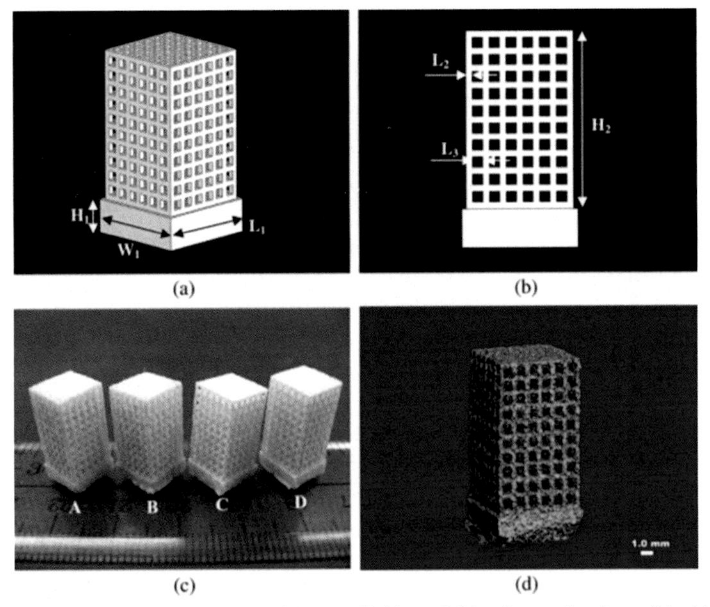

Figure 23.12 (a) Schematic diagram of the scaffold model in diametric view; (b) side view of the scaffold model; (c) scaffolds produced by SLS: (A) PHBV; (B) Ca–P/PHBV; (C) PLLA; (D) CHAp/PLLA. (d) MicroCT image of a Ca–P/PHBV scaffold.
From Duan, B., Wang, M., Zhou, W. Y., Cheung, W. L., Li, Z. Y., & Lu, W. W. (2010). Three-dimensional nanocomposite scaffolds fabricated via selective laser sintering for bone tissue engineering. *Acta Biomaterialia, 6*(12), 4495–4505. https://doi.org/10.1016/j.actbio.2010.06.024.

porous structure. In vitro biological evaluation revealed that human osteoblasts showed high cell viability, normal morphology, and proper phenotype expression after 3 days culture. The incorporation of Ca−P nanoparticles significantly improved cell proliferation if related to the neat PHBV scaffold. After 7 days culture, cells were well attached and spread over the strut surface and interacted favorably with all scaffolds. These microspheres were found to be suitable to be encapsulated with biomolecules, enhancing them with more functions for bone tissue engineering applications or making them suitable for localized delivery of therapeutics while retaining a good dimensional accuracy during the SLS process (Duan et al., 2010).

PHB scaffolds for bone tissue applications were produced by SLS and subsequently functionalized with osteogenic growth peptide by Saska et al. (2018). PHB powder was selectively sintered using a CO_2 laser with a power of 8 W and beam spot size of 450 μm. Printing was performed at a speed of 2000 mm s^{-1}, temperature of 100°C, with a scan spacing and layer thickness of 0.15 and 0.18 mm, respectively. Peptides were incorporated inside the scaffold by swelling and adsorption. The obtained scaffolds showed a hierarchical structure with an intrinsic porosity of 56% and pore size in the 500−700 μm range. The compressive strength and Young's modulus of the PHB scaffolds were 0.72 and 4.9 MPa, respectively. In vitro assays were performed using bone marrow stem cells, scaffolds showed good cell viability and proliferation plus improved morphological differentiation in the samples containing osteogenic growth peptide.

The processability of PHAs using FFF is a challenge due to the low thermal stability during melt processing. The drop in the molecular weight can be so drastic that the final mechanical properties often do not fulfill the theoretical expectations. It was found (Fig. 23.13) that the low thermal stability and the massive decrease in viscosity

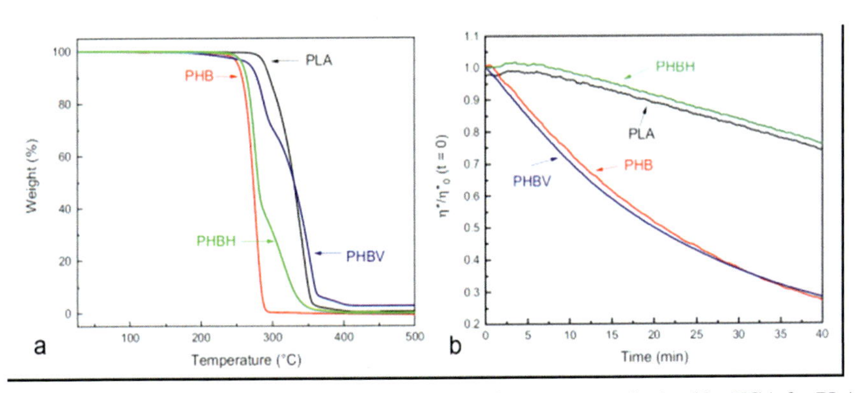

Figure 23.13 (a) Weight of the samples as a function of temperature obtained by TGA for PLA, PHB, PHBV and PHBH and (b) evolution of complex viscosity, $\eta^*(t)/\eta^*0$ ($t = 0$) versus time of PLA, PHB, PBHV, and PHBH.
From Kovalcik, A., Sangroniz, L., Kalina, M., Skopalova, K., Humpolicek, P., Omastova, M., Mundigler, N., & Muller, A. J. (2020). Properties of scaffolds prepared by fused deposition modeling of poly(hydroxyalkanoates). *International Journal of Biological Macromolecules, 161*, 364−376. https://doi.org/10.1016/j.ijbiomac.2020.06.022.

of PHB and PHBV precluded their use for FFF (Kovalcik et al., 2020). On the other hand, the longer alkyl side chain confers to poly[3-hydroxybutyrate-co-3-hydroxyhexanoate) (PHBH) lower crystallinity and a broader thermal processing window in comparison to PHB and PHBV (comparable to that of PLA).

PHBH was successfully applied in FFF by Valentini et al. (2019) for the fabrication of a dumbbell specimen for further mechanical investigation. The thermal behavior of PHBH was improved by the addition of various amounts of nano-fibrillated cellulose to the polymer matrix. The onset temperature was raised from 253°C with neat polymer to 266°C when 3 wt.% of nanocellulose was added. Composite materials were prepared via solution mixing and then extruded in filaments with 1.75 mm diameter via a single-screw extruder at 150°C. Special attention was paid to the selection of the temperature profile during 3D printing. The bed temperature was set at 75°C to avoid crystallization during the process and a nozzle profile from 200 to 180°C was selected in order to avoid degradation of the material and to minimize the risk of warping and detachment. In this way, ISO 527 type 1BA specimens were 3D printed. The highest mechanical properties were detected for a concentration of nanocellulose as low as 0.5 wt.%, in particular, an elastic modulus of 897 and 1259 MPa was recorded for neat PHBH and 0.5 wt.%, respectively, while a stress at break of 21.6 and 23.1 MPa was found for neat PHBH and 0.5 wt.%, respectively. Strain at break was not impaired and remained around 10%. PHB and PHBV can be applied to FFF if compounded to other materials with a higher thermal stability, such as PLA (Findrik Balogova et al., 2018) or PCL (Kosorn et al., 2017). In the latter case, PHBV can also enhance the bioresponse if FFF scaffolds were applied to tissue engineering.

However, the application of melt-based AM for fabricating PHA scaffolds is limited by the relatively small temperature processing window. To overcome this problem, Puppi et al. proposed an innovative phase inversion-based AM approach to obtain poly(3-hydroxybutyrate-co-3-hydroxyexanoate) (PHBHH) scaffolds via processing of polymer/solvent/nonsolvent ternary mixtures in a nonsolvent bath (Puppi et al., 2020). Scaffolds were fabricated by the so-called computer-aided wet spinning (CAWS) technique, enabling simultaneous control of the feed rate of a polymeric solution and its laydown pattern. A PHBHH/chloroform/ethanol mixture was placed into a glass syringe fitted with a metallic needle with a diameter of 0.4 mm and then injected at a controlled feeding rate directly into an ethanol bath. In this way, scaffolds characterized by a dual-scale porosity were fabricated with an interconnected network of macropores plus a microporosity, developed as a result of the phase inversion process during polymer solidification (Fig. 23.14). Mechanical properties and also the response of cells during in vitro tests were considerably affected by this peculiar structural morphology.

4.3 Polycaprolactone (PCL)

Polycaprolactone (PCL) is a synthetic biodegradable semicrystalline polyester with a melting point as low as 60°C. It also has a very low glass-transition temperature ($T_g = -72/-60$°C) than other biodegradable polymers, which improves its biodegradability despite its high degree of crystallinity, typically at 50% (Gao et al., 2012;

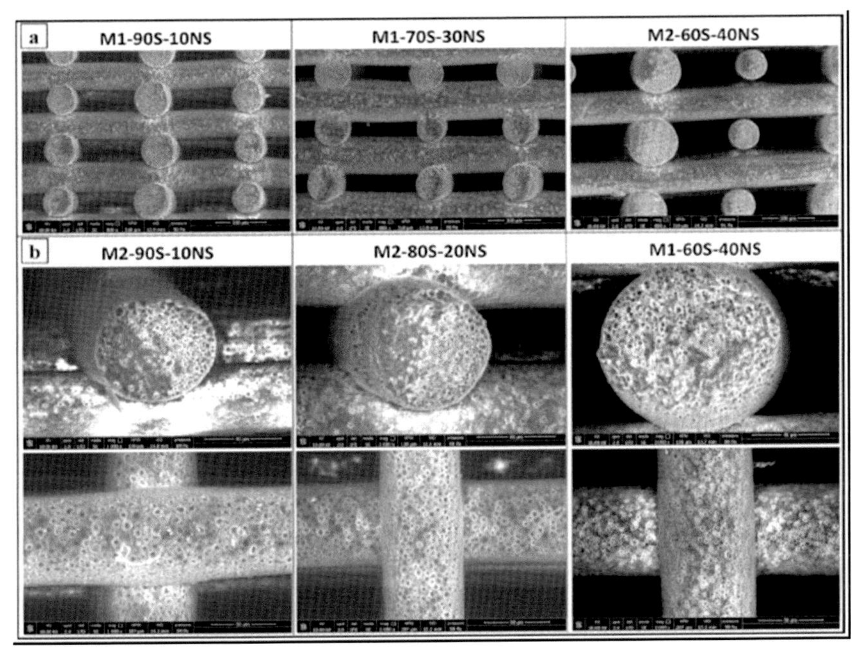

Figure 23.14 High-magnification morphological analysis of PHBHHx scaffolds: representative SEM micrographs of (a) perpendicular cross-section of scaffolds with different Z axis inter-fiber distance (800×; scale bar: 100 μm); (b) fiber cross-section (3000×; scale bar: 40 μm) and fiber external surface (2000×; scale bar: 50 μm) of scaffolds with different micropore concentrations.
From Puppi, D., Braccini, S., Ranaudo, A., & Chiellini, F. (2020). Poly(3-hydroxybutyrate-co-3-hydroxyexanoate) scaffolds with tunable macro- and microstructural features by additive manufacturing. *Journal of Biotechnology, 308*, 96–107. https://doi.org/10.1016/j.jbiotec.2019.12.005.

Wypych & Wypych, 2016). PCL is produced by ring-opening polymerization of ε-caprolactone using a catalyst such as stannous octanoate (Ebnesajjad & McKeen, 2013). PCL is degraded by hydrolysis of its ester linkages in physiological conditions and has therefore received a great deal of attention for use as an implantable biomaterial. In particular, it is especially interesting for the preparation of long-term implantable devices. PCL has been widely used as a raw material due to its high thermal stability, FDA approval, cost-effectiveness, and wide availability (Woodruff & Hutmacher, 2010). However, its intrinsic hydrophobic nature and the absence of biological recognition sites result in poor cell attachment and proliferation when used as part of a construct for tissue engineering (Zhang et al., 2017).

Porous 3D scaffolds made of PCL can be produced directly from PCL monomers, Elomaa et al. prepared photocross-linkable PCL resin to be applied in SLA (Elomaa et al., 2011). Three-armed hydroxyl-terminated PCL oligomers were synthesized by ring-opening polymerization of Ɛ-caprolactone monomers. From the methacrylation of the hydroxyl-terminated oligomers using methacrylic anhydride, PCL macromers were obtained and Irgacure 369 initiator added to the molten macromers was used to begin the cross-linking reaction upon light exposure. Three-dimensional scaffolds with an interconnected pore structure (Fig. 23.15) were then printed using an SLA apparatus equipped with a digital micromirror device that projects 1280 × 1024 pixels of blue light, with an intensity of 1600 mW dm^{-2}, onto the transparent nonadhesive bottom surface of a resin reservoir surface with an exposure time of 10 s to cross-link a resin layer of 25 μm in thickness. To avoid the use of solvents, the resin reservoir was heated at about 45°C to lower the viscosity just enough to facilitate the SLA processing. The scaffolds accurately matched the computer-aided designs, with no visible material shrinkage and an average pore size of 465 μm. The interconnectivity of the pores was high, suggesting a good possibility for these structures in cell seeding and implanting. Gelatin alone has not been widely used in SLA-based 3D printing of tissue scaffolds due to its difficult viscosity control. However, to improve the biomimetic of PCL, gelatin from different sources can be blended together with PCL-based ink and printed with SLA technology (Elomaa et al., 2011). The obtained resins can promote the printing fidelity and cell viability of tissue scaffolds produced via SLA. PCL/gelatin scaffolds were also successfully printed through DW with bone replacement applications in mind (Duymaz et al., 2019). PCL and gelatin in a ratio of 10:0.25 and various concentrations of low-molecular-weight *Halomonas levan*, a polysaccharide produced by several microorganisms and plants, were dissolved in a dichloromethane (DCM)/dimethyl sulfoxide (DMSO) mixture. The obtained ink was printed through a 0.4 mm nozzle with a speed between 100 and 150 mm s^{-1} at a temperature bed of 50°C. The shape uniformity improved as dense appearance and regular pores were obtained by increasing the amount of *Halomonas levan* due to increased viscosity of the ink. However, this was accompanied by a decrease in the mechanical properties of the scaffolds.

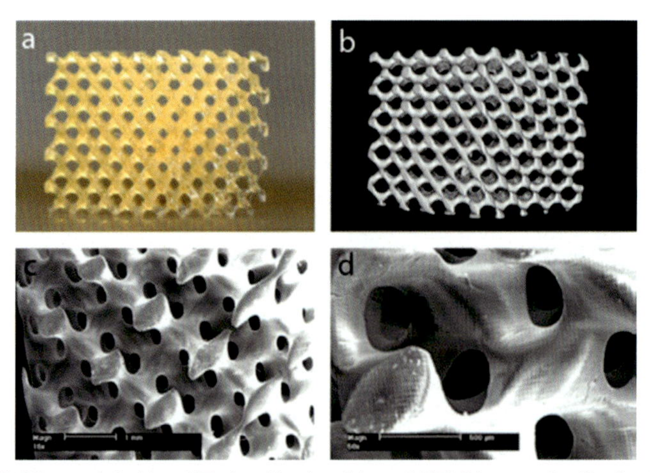

Figure 23.15 Photograph (a), μCT visualization (b), and SEM images (c, d) of a scaffold built by stereolithography.
From Elomaa, L., Teixeira, S., Hakala, R., Korhonen, H., Grijpma, D. W., & Seppala, J. V. (2011). Preparation of poly(epsilon-caprolactone)-based tissue engineering scaffolds by stereolithography. *Acta Biomaterialia, 7*(11), 3850–3856. https://doi.org/10.1016/j.actbio.2011.06.039.

PCL scaffolds fabricated using SLS have mechanical properties within an appropriate physiological range, offer the chance to be loaded with therapeutical drugs, and support the in-growth of bone tissue to be applied as bone tissue replacement (Salmoria et al., 2016). Williams et al. (2005) started from a PCL powder with a particle size distribution in the 10−100 mm range to produce bone scaffold via SLS. The powder was preheated at 49.5°C, a laser with a beam spot size of 450 μm and a power of 4.5 W was scanned above the building bed at a speed of 1.257 m s^{-1}. Cylindrical porous scaffolds were printed layer-by-layer with a layer thickness of 100 μm and with a three-dimensional orthogonal periodic structure. Compressive modulus and yield strength of 3D-printed scaffold ranged from 52 to 67 MPa and 2.0−3.2 MPa, respectively. These values agreed with those predicted from a finite element analysis performed on microcomputed tomography images of the samples. The obtained PCL scaffolds were able to promote the tissue in-growth when human gingival fibroblasts were seeded, thus confirming the applicability as replacement tissue in bone defects. The addition of ceramic fillers to SLS 3D-printed PCL was relatively easy and various ceramics such as TiO_2, Al_2O_3, ZrO_2, and HA were successfully incorporated in the polymer matrix by mechanical mixing (Shishkovsky & Scherbakov, 2012). In bone scaffold applications, the combination between HA, which acted as the ceramic substrate for regeneration of bone tissues, and PCL, which acted as the polymeric binder, resulted in a scaffold with good mechanical properties and an improved cell differentiation and tissue replacement (Wiria et al., 2007). Scaffolds were produced with PCL and 30 wt.% HA via SLS and the main process parameters were investigated by a design of experiment (DOE) approach by Eosoly et al. (2010). A PCL powder with an average size of 125 μm was mixed with HA powder with average size of 38 μm and a continuous-wave CO_2 laser focused to a 410 μm spot was used to fabricate the scaffolds in Fig. 23.16. According to the design plan, laser power and scan spacing were varied while laser speed, bed temperature, and layer thickness were kept constant. Mechanical characterization revealed a dominant effect of scan spacing above laser power in all the building directions. The stiffness of the scaffolds was dependent on the building direction with lowest values in the x direction, where scan lines were parallel to the loading direction with a maximum along the y direction, where scan lines were perpendicular to loading. Building directions had an effect also on the failure mechanism under compression, in the x direction, the scaffold exhibited an elastic recoverable buckling, while in the other two directions hinge formation was prominent. The incorporation of a ceramic filler resulted in an increase in stiffness compared to neat PCL, followed by a significant reduction in both strength and strain at break (Doyle et al., 2015). In vitro tests with osteoblast cells showed that they were able to grow on the scaffolds with a preferential distribution around the macropores. Moreover, it was highlighted that differences in powder composition that change the sintering behavior and the mechanical properties of the scaffold particularly under cell culture conditions should be taken into account for SLS processing of scaffolds (Eosoly et al., 2010). Innovative methods, as an alternative to the mechanical mixing of PCL and HA, were investigated and a microsphere-based strategy could

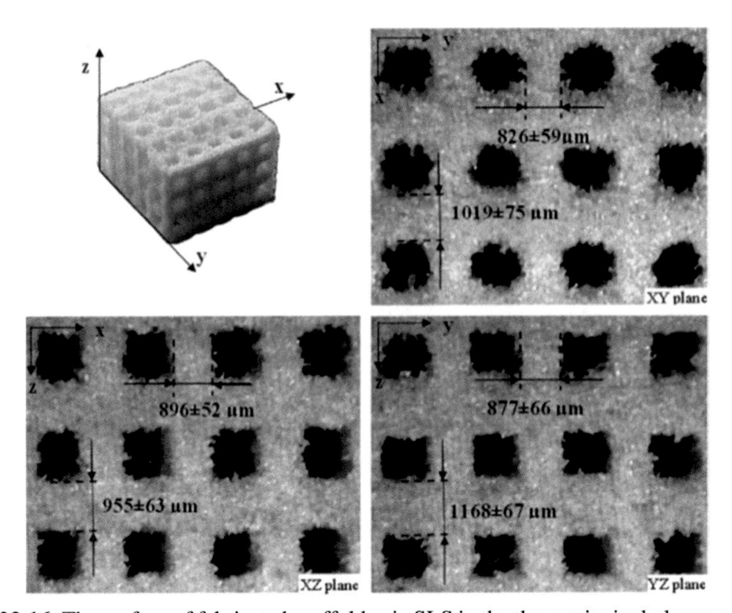

Figure 23.16 The surface of fabricated scaffolds via SLS in the three principal planes, with strut sizes (average ± standard deviation) measured in the different directions.
From Eosoly, S., Brabazon, D., Lohfeld, S., & Looney, L. (2010). Selective laser sintering of hydroxyapatite/poly-epsilon-caprolactone scaffolds. *Acta Biomaterialia, 6*(7), 2511−17. https://doi.org/10.1016/j.actbio.2009.07.018.

significantly enhance the molding precision by the regular shape and good mobility of microspheres (Du et al., 2015). PCL/HA microspheres were obtained by emulsification and solvent evaporation method and used in the SLS apparatus for the fabrication of bone scaffold. HA was able to appreciably stimulate the spread and osteogenic differentiation of mesenchymal stem cells compared with neat PCL.

Since cellular activity is considerably affected by the presence of biologically active sites on the surface of the material, that are almost completely absent on synthetic polymers, incorporation of bio-active ceramic fillers or surface modification must be adopted to improve the cell viability and growth. Chen et al. (2014) reported a positive effect on chondrocyte growth when PCL 3D-printed scaffolds were modified with the deposition of gelatin and collagen. Cylindrical specimens were printed from PCL powder with a laser power of 2 W, particle bed temperature of 40°C and a scanning speed of 500 mm/s. Surface modification was done by immersion of the scaffold in gelatin or collagen for 2 days. Collagen was found to be superior with respect to gelatin in biological response and an in vivo test showed evident tissue infiltration after 8 weeks and no inflammatory activity. If collagen had a lower efficiency on stimulating cell attachment, proliferation, and differentiation than collagen, a worse effect was found for O_2 plasma modification of PCL scaffold (Van Bael et al., 2013). Porosity and pore architecture are critical to encourage cell migration and to ensure adequate flow of medium and waste through the scaffold. This is particularly critical when tissues with a high number of cells must be seeded, such as cardiac ones. Yeong et al. (2010)

produced a 3D-printed PCL scaffold via SLS relying on a prototype system, called "computer-aided system for tissue scaffolds," that it was able to design scaffolds through parametric algorithms. PCL powder with average size of 100 μm was selectively sintered according to the designed pattern with a 3 W laser at a scanning speed of 3.8 m s^{-1} and a bed temperature of 40°C. Tensile stiffness and tensile yield strength of a single sintered PCL strut were measured and correlated to the mechanical properties of the printed scaffold as a function of its porosity. A logarithm relation between the mechanical properties of a single strut and those of the entire scaffold was found with a high R^2 value. Fluorescence images showed myoblast cells located throughout the scaffold and a regular population of cells was preserved throughout 21 days of culturing, proving the possibility of using PCL SLS scaffolds for cardiac tissue engineering.

The low melting temperature of PCL makes it attractive to be processed with 3D printing material extrusion techniques such as hot-extrusion (HE)-DW and FFF upon the production of a suitable filament. HE-DW of PCL was performed with a bioplotter equipped with a pneumatic dispensing system through a 27-gauge needle at 120°C at a pressure of 5 bar and printing speed of 240 mm/min (Tardajos et al., 2018). A lower temperature, as low as 90°C, can be used in pellet extrusion deposition where the force to extrude the molten polymer is much higher due to pressure created by a turning screw (Shor et al., 2007). PCL bioresorbable 3D scaffolds with a fully interconnected pore network with various designs were produced with FFF by Hutmacher (2000). Filaments were extruded with a fiber-spinning machine at 140°C through spinnerets with a die exit diameter of 1.63 mm. The piston speed was set at 10 mm/min and the combination of temperature, piston speed, and height drop to water quenching settings produced a filament diameter of 1.70 ± 0.10 mm. Scaffold specimens were printed with PCL filaments with an FFF 3D Modeler RP system from Stratasys Inc. with a nozzle temperature of 120°C. Two different lay-down patterns were investigated and scaffolds were built: 0/60/120 degree patterns showed compressive modulus and yield strength of 41.9 ± 3.5 MPa and 3.1 ± 0.1 MPa, respectively, while the 0/72/144/36/108 degree pattern exhibited values of 20.2 ± 1.7 and 2.4 ± 0.1 MPa, respectively. PCL scaffolds showed excellent biocompatibility with human fibroblast and periosteal cell culture systems. Cells were able to colonize the struts and bars of the scaffold and created a cell-to-cell and cell-to-extracellular-matrix interconnective network throughout the entire 3D structure. According to the mechanical tests, porous PCL scaffolds behaved similarly to a porous material undergoing compression. Mechanical properties depended on porosity, regardless of the lay-down pattern and channel size (Iwan Zein, 2002). A power law relationship between modulus, yield strength, yield strain, and porosity was found as predicted by models of porous structures composed of beams with square cross-section. In situ impregnation with reinforcing fibers can be an interesting approach for enhancing the mechanical properties of FFF 3D-printed parts. In a recent work by Hedayati et al. (2020), a biodegradable wire of poly-glycolic acid (PGA) was used to improve the tensile properties of the PCL matrix through modified FFF technology. A continuous PGA yarn was guided into the nozzle tip through a specially made orifice on the side of a particularly designed nozzle for the simultaneous in-melt feeding of continuous fibers. Material was extruded at 160°C with a printing speed of 5 mm s^{-1}. The tensile strength and elastic modulus of the reinforced PCL reached values of 79.7 MPa and 3.5 GPa, respectively, with increases of 374% and 775%, respectively, compared to the nonreinforced PCL. Due to the low printing temperature of PCL, bioactive components can be added to the polymer matrix and realize 3D-printed objects with various functions. Antibacterial FFF materials can be prepared by direct addition of organic antibacterial agents into the PCL filament such as quaternary ammonium or polyhexamethylene biguanidine (Zhao et al., 2020). The antibacterial efficiency of 3D-printed samples was higher than 99.99% against both *S. aureus* and *E. coli* when 1 phr of antibacterial components was added. PCL processed via FFF was suitable to produce drug-loaded devices. Kempin et al. (2017) used fluorescent quinine as a model drug to visualize drug distribution in filaments and implants. Quinine-loaded filaments were produced by solvent casting and subsequent extrusion. Hollow cylinders were fabricated with a standard FFF machine with a 0.35 mm nozzle at a temperature of 53°C and printing speed of 24 mm s^{-1}. The drug release was investigated in under-sink conditions at 7.4 pH. Quinine release from PCL 3D-printed objects was 76.4 ± 1.8% after 58 days. Drug release kinetics followed two clear steps, the first one where the quinine that is located at the implant surface dissolves rapidly and the second one described by a slower release rate due to slow drug diffusion through the polymeric matrix.

Another advantage of thermoplastic PCL is the possibility to use it as a healing agent in epoxy systems due to the low melting temperature. Healing will occur via chain entanglement of the PCL molecules above the melting temperature. A PCL-rich interphase should work also for toughness enhancement in a fiber-reinforced composite material. PCL was directly deposited through FFF technique on the surface of reinforcing fabrics in different patterns by Szebenyi et al. (2017). A rectangular 2D pattern was printed over a unidirectional carbon weave by FFF with a nozzle temperature of 180°C and bed temperature of 40°C. Six of these so-produced carbon layers were stacked up and infiltrated with epoxy by hand layup followed by vacuum pressing. PCL, previously deposited on fabrics, may be partially or completely dissolved during infiltration and following curing. However, even in the latter case PCL, being under constraints, remains in the interphase, therefore a PCL-rich interphase develops. A noticeable improvement in ductility was found under static flexure after the incorporation of a PCL-rich interphase. A certain restoration of the mechanical properties (around 15% of the pristine value of the maximum bending stress) was detected in static flexure after heat treatment of the previously fractured specimen. A three-dimensional approach to improve the toughness and self-healing ability of epoxy systems was investigated by Dorigato et al., 2017, 2020; Dul et al., 2016 with the fabrication of an interconnected cocontinuous 3D scaffold of PCL through FFF and resin impregnation (Fig. 23.17). In this way, the final morphology of the epoxy-impregnated PCL blend can be controlled not only by the thermodynamics of mixing of the two phases, but it also can be accurately designed and tuned. It was seen that during fracture toughness tests the crack mainly propagated within the epoxy phase, while the progressive yielding of the PCL phase contributed to energy absorption through plastic deformation. Under impact conditions, the particular 3D structure obtained through

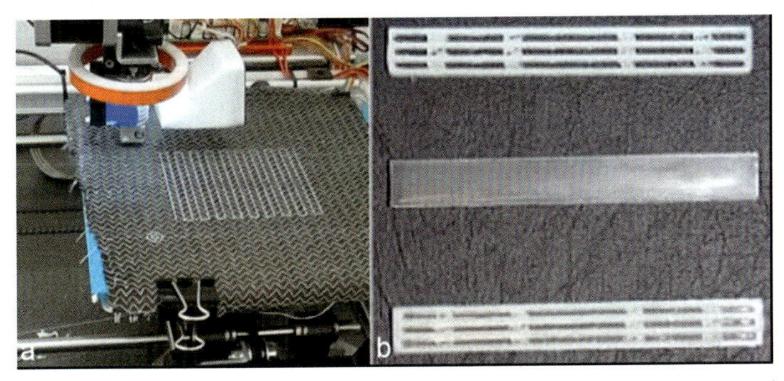

Figure 23.17 Three-dimensional-printed PCL scaffold, epoxy specimen, and a combination of both.
Data from Dorigato, A., Rigotti, D., & Pegoretti, A. (2020). Novel poly(caprolactone)/epoxy blends by additive manufacturing. *Materials (Basel, Switzerland), 13*(4), 819. https://doi.org/10.3390/ma13040819.

FFF was able to fully express the potential of PCL as an energy absorber. The healing efficiency according to the maximum force reached by the samples after and before healing was found to be around 13% in the case of a steady fracture propagation test and 40% under impact conditions.

4.4 Cellulose-derived materials

Cellulose is the most abundant natural polymer, the repeat unit of which is derived from glucose. It is a crucial component of the cell wall of different plants. Along with plants, cellulose is also present in a large variety of living species, including algae, fungi, bacteria, and even sea animals, such as tunicates. Although widely found in nature, at an industrial scale cellulose is derived almost entirely from cotton and wood. Wood cellulose nanofibrils have a typical width of 3–5 nm and length of several micrometers, and wrap around each other with strong hydrogen bonds, forming aggregates of larger microfibrils. Microfibrils have a typical width of 5–50 nm, which in turn aggregate to larger cellulose fibrils. The microfibrils reinforce the wood structure with an intrachain and interchain hydrogen bond network (Poletto et al., 2015). The plant fibers can be considered as naturally occurring composites due to their unique structure, in which the semicrystalline cellulose microfibrils act as reinforcing elements in an amorphous matrix mostly made up of lignin and hemicellulose. While the matrix performs as a natural barrier to microbial degradation and serves as mechanical protection, the cellulose microfibrils have the function of providing rigidity and structural stability to the cell walls of the fibers (Alberts, 2002). Cellulose fibers are the main load-bearing component in trees and plants due to the high modulus of their crystalline part, which can reach 140 GPa (Page & El-Hosseiny, 1983). The most important variables that determine the overall properties of wood material are the structure, microfibrillar angle, cell dimensions, defects, and chemical composition of fibers (Fan et al., 2017). However, isolation of cellulose in the nano-size range is required in order to take full advantage of its inherent properties. When the crystalline regions of cellulose nanofibrils are isolated through an acid treatment, a nanocellulosic material referred to as nanocrystalline cellulose (NCC) is produced while that obtained mainly by mechanical disintegration is called cellulose nanofibrils (CNFs).

An aqueous NCC solution can be used as an ink for DW 3D printing of textured cellular architectures as studied by Siqueira et al. (2017). Freeze-dried NCC, prepared via sulfuric acid hydrolysis of eucalyptus pulp, were added to deionized water with a concentration of 20 wt.% and loaded in a 3D bioprinter. The ink was driven pneumatically through a micronozzle under pressure ranging from 2 to 4 bar at a speed of 10–20 mm s^{-1}. The printed structures showed a high degree of shear-induced alignment of NCC particles along the printing direction, similar to the microreinforcing effect in plant cell walls. This method offers new opportunities for designing lightweight, sustainable composites with tailored architectures and mechanical properties. N-methylmorpholine-N-oxide (NMMO) is an environmentally friendly, nonvolatile, nontoxic solvent suitable to physically dissolve cellulose without any pretreatment, for this reason, solutions of cellulose and NMMO are promising inks for advanced applications with precisely controlled structures, as suggested by Li et al. (2018). Cellulose was dissolved in a water solution of NMMO through stirring at 115°C. Three-dimensional printing was performed through a nozzle at 70°C. The extruded gel fiber solidified quickly when it was exposed to the air at room temperature and could pile up freely. When performed at a higher temperature, such as 80°C, the extruded fiber could not solidify and a gel object was printed (Fig. 23.18). In this way, cylindrical specimens were obtained with an interconnected porous structure.

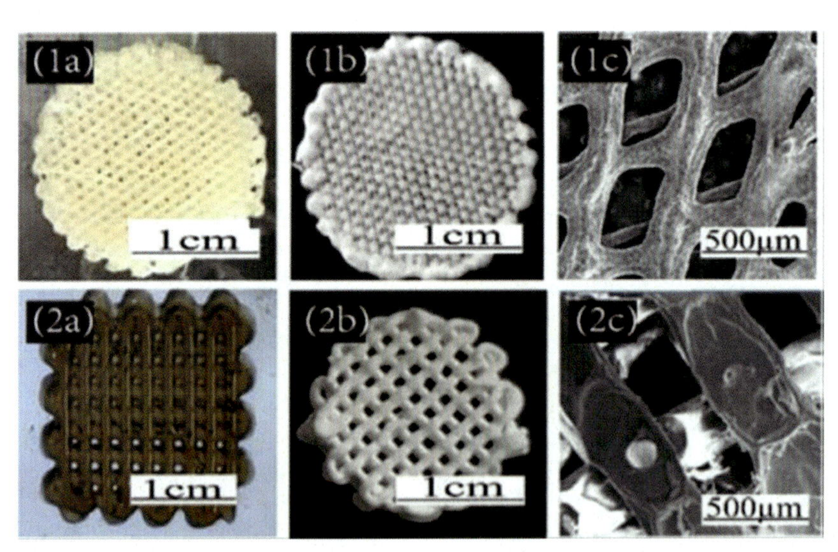

Figure 23.18 Photographs of 3D-bioprinted cellulose/NMMO intermediate objects in the solid form (1a) and in the gel form (2a); Photographs and SEM images of the final cellulose objects from the solid form (1b, 1c) and from the gel form (2b, 2c) after freeze-drying.
From Li, L., Zhu, Y., & Yang, J. (2018). 3D bioprinting of cellulose with controlled porous structures from NMMO. *Materials Letters, 210*, 136–138. https://doi.org/10.1016/j.matlet.2017.09.015.

The elastic moduli in compression were 12.9 and 7.6 MPa for cellulose printed in the solid form and gel form, respectively. Compressive stresses were 5.7 and 4.1 MPa, respectively, thus suggesting that printed cellulose products from NMMO solution were particularly resistant to compressive stress.

Various modifications of the nanocellulose surface can be performed to improve the printability. A combination of carboxymethylation and periodate oxidation can produce a homogeneous material with higher consistency. In fact, as demonstrated by Rees et al. (2015), a stabilized structure can be formed by ionic links between the carboxyl groups (COO-) and the divalent cations (Ca^{2+}) upon cross-linking with $CaCl_2$, and appropriate rheology. TEMPO-oxidized nanocellulose gel, instead, resulted in 3D structures that collapsed, and no defined deposited tracks could be detected due to the poor viscosity. Shear thinning properties of nanofibrillated cellulose aqueous solution were combined with alginate by Markstedt et al. (2015) to obtain a bioink for printing a living soft tissue with cells. The bioink was printed using a 3D DW bioprinter consisting of a microvalve, based on electromagnetic jet technology, with a 300 μm diameter nozzle. The flow rate was controlled by adjusting the dispensing pressure (20–60 kPa), the valve opening time (400–1200 μs), and the dosing distance (0.05–0.07 mm). Human chondrocytes were incorporated in the NFC/alginate ink to evaluate the cell viability after printing. After the 3D printing process, the obtained structures were cross-linked by soaking in a $CaCL_2$ solution. To demonstrate the potential use of nanocellulose for 3D bioprinting of living tissues and organs, different shapes were 3D printed, such as a human ear and a sheep meniscus (Fig. 23.19). Rheological properties of the prepared inks were dominated by the concentration of cellulose, while shear thinning guaranteed a high shape fidelity and increased printing resolution. Upon cross-linking, mechanical properties were related to the amount of alginate highlighting the fact that tailored mechanical properties could be obtained changing the composition of the ink and the printing pattern. A bioink composed of nanofibrillated cellulose and alginate is a suitable hydrogel for 3D bioprinting with living cells for growth of cartilage tissue as emerged from the positive results of a cytotoxicity test and cell viability analysis of the 3D-printed product. Leppiniemi et al. improved the printing quality of NFC/alginate inks replacing part of the water with glycerin (Leppiniemi et al., 2017). The increase of nonvolatile components in the hydrogel led to a minimization of the shrinkage of the printed specimen so the capacity to retain its shape was improved also before being cured. The objects enriched with glycerin were characterized by a higher elasticity upon curing compared to the hard and fragile ones without glycerin. Markstedt et al. (2017) suggested replacing cross-linkable alginate with hemicellulose to obtain an all-wood-based 3D printable ink. In particular, xylan, a hemicellulose extracted from spruce, was functionalized with tyramine to make it cross-linkable. Cross-linking was done in H_2O_2 upon printing to form freestanding structures. Because the printed ink formed a gel with a high water content, the best mechanical properties and printability were found for the highest amount of xylan, that is, around 10 wt.% of the final concentration.

A screw extrusion-based DW system is suitable to deposit viscous inks with a higher content of short natural cellulose fibers with carboxymethyl cellulose as demonstrated by Thibaut et al. (2019). The 3D printing apparatus consisted of a screw-driven

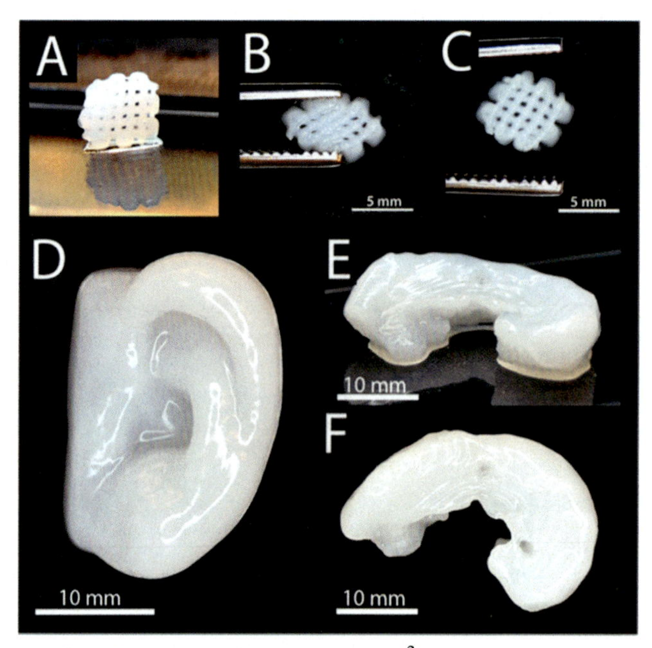

Figure 23.19 (a) 3D printed small grids (7.2 × 7.2 mm^2) with Ink8020 after cross-linking. (b) The shape of the grid deforms while squeezing, and (c) it is restored after squeezing. (d) 3D printed human ear and (e and f) sheep meniscus with nanocellulose/alginate ink. Side view (e) and top view (f) of meniscus.
Data from Markstedt, K., Mantas, A., Tournier, I., Martinez Avila, H., Hagg, D., & Gatenholm, P. (2015). 3D bioprinting human chondrocytes with nanocellulose-alginate bioink for cartilage tissue engineering applications. *Biomacromolecules, 16*(5), 1489–1496. https://doi.org/10.1021/acs.biomac.5b00188.

extruder with a steel nozzle having an outlet diameter d varying from 0.5 to 0.9 mm mounted on a traditional FFF machine. This device allowed the use of highly viscous pastes at printing speeds of the same order of magnitude of those typically reached in the FFF process, that is, ranging from 10 to 50 mm s^{-1}. The rotation velocity of the screw was approximately 10 rpm during printing. A paste with a cellulose fiber content of 30 wt.%, a carboxymethyl cellulose content of 12.5 and 57.5 wt.% of distilled water was applied as a biobased ink for 3D printing. This formulation exhibited a pronounced shear thinning behavior and yield stress after deposition that made it possible to print complex parts with limited and anisotropic deformation that maintain fidelity with the initial 3D digital model.

In additive manufacturing, a sacrificial support material is usually needed to fabricate highly complex structures. However, the removal of support structures in AM may damage printed parts and generate chemical hazards. Cellulose-based hydrogels can be a promising 3D printing support material due to their sustainability, renewability, and potential recyclability. Different kinds of inks have been developed based on NCC (Li et al., 2019) and hydroxypropyl methylcellulose and methylcellulose (Polamaplly et al., 2019) to be applied in SLA and FFF, respectively. The proposed inks had the advantage of being easily dissolvable in water, they possessed lower bondability with other polymer matrices, and offered a reduced carbon footprint. Nevertheless, the application of these inks required significant modifications of the original printing devices which makes it difficult for the application of these systems to machines already available on the market.

A NCC nanocomposite hydrogel precursor suitable for 3D printing via stereolithography was investigated by Palaganas et al. (2017) to form complex architectures exhibiting enhanced properties intended for tissue engineering constructs. NCC extracted from abaca pulp fibers through acid hydrolysis were incorporated into a poly(ethylene glycol) diacrylate (PEGDA) matrix. Lithium phenyl(2,4,6-trimethylbenzoyl) phosphinate (LAP) was the initiator of the radical polymerization at ambient temperature upon exposure to UV light. Specimens were printed using a laser characterized by a wavelength of 405 nm, power of 250 mW, and spot size of 140 μm. SLA offered the opportunity to fabricate complex architectures, such as a human ear (Fig. 23.20), that was potentially suitable for reconstructive surgery since the main components of the resin, PEGDA, NCC, and LAP are all known for their biocompatibility.

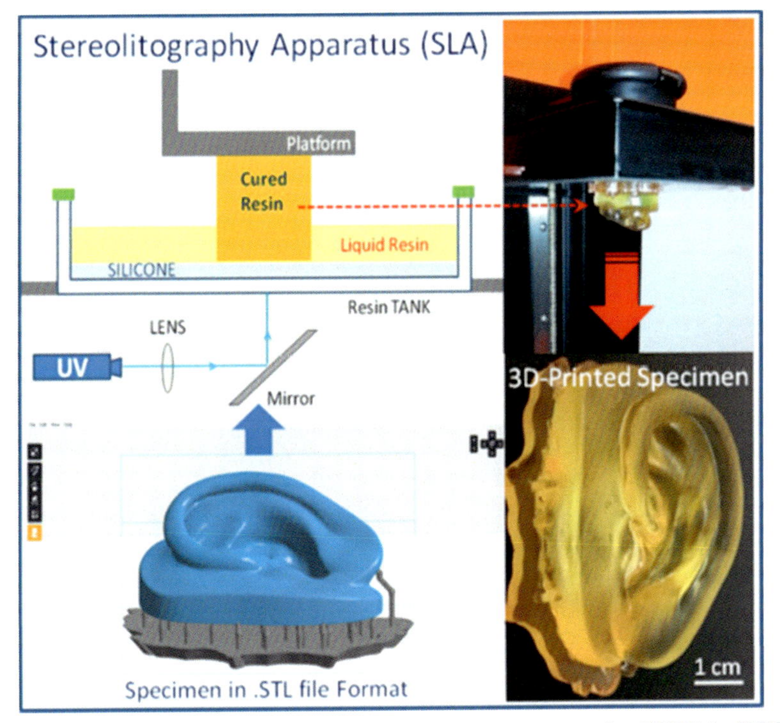

Figure 23.20 Three-dimensional printing of a human ear construct using PEGDA–NCC hydrogel via SLA potentially suitable for tissue engineering applications.
Data from Palaganas, N. B., Mangadlao, J. D., de Leon, A. C. C., Palaganas, J. O., Pangilinan, K. D., Lee, Y. J., & Advincula, R. C. (2017). 3D printing of photocurable cellulose nanocrystal composite for fabrication of complex architectures via stereolithography. *ACS Applied Materials and Interfaces, 9*(39), 34314–34324. https://doi.org/10.1021/acs,ami.7b0,9223.

Cellulose acetate scaffolds were made by selective laser sintering by Salmoria et al. (2016) and the effects of laser power, laser scan speed, and particle size on the scaffold properties were evaluated. The adjustment of process parameters was essential to guarantee the processability of biodegradable cellulose-based polymers by SLS. Specimens with 250 μm layer thickness were produced using a laser with 0.171 J/mm^2 energy density, scan speed of 39.8 mm/s, and the powder bed was maintained at 145°C. Smaller particle sizes resulted in a higher degree of sintering and closed porosity, which resulted in superior mechanical properties with respect to larger particle size. Laser energy and scanning speed were finely tuned to avoid the degradation of cellulose acetate at 290°C. Cellulosic polymers such as hydroxypropyl cellulose (HPC) are gaining interest in the pharmaceutical and food industries due to their high safety profile as accelerated drug-release materials for orally disintegrable tablets. With this purpose in mind, Fina et al. (2018) studied the application of HPC in SLS. This manufacturing method was selected because it has the advantage when applied in food or pharmaceutical applications that it is a solvent-free process compared to other techniques, for example, DW. HPC was sintered with the aid of a 2.3 W blue diode laser (445 nm) and maintained at a temperature of 135°C during the process. The surface was scanned at a speed ranging from 100 to 300 mm s^{-1} in a pattern based on the .STL file. Printing speed, that can be translated into energy deposited on the substrate, affected not only the porosity and mechanical properties of the fabricated objects but also the drug-release profile. Faster speed produced items with lower density and lower mechanical properties but a faster release rate for the investigated drug. Tablets fabricated with a scanning speed of 100 mm/s showed a paracetamol release that was complete after approximately 4 h, while tablets fabricated using a laser scanning speed of 300 mm/s exhibited the shortest dissolution profile with complete drug release at approximately 2 h. FFF 3D-printed disintegrable tablets made of HPC for drug release were also studied by Arafat et al. (2018). Theophylline drug and HPC were mixed together in a twin-screw extruder and a filament for the 3D printer was produced. Material was extruded at 120°C through a 1.7 mm die. Tablets were printed at 150 mm s^{-1} through a 0.4 mm nozzle in layers of 300 μm thickness. The tablets were designed as a jointed system of different blocks divided by bridges (Fig. 23.21); this layout should improve the efficiency of drug release over time. This innovative approach was capable of achieving faster disintegration and dissolution with respect to conventional formulation methods. Computer-aided designed tablets with their complex geometries showed the potential to engineer the mechanical and functional aspects of a traditional pharmaceutical product.

4.5 Silk

Silk is a natural polymer sourced mainly from silkworms and spiders. It is composed of two proteins, fibroin and sericin. Fibroin is the structural protein that provides strength and stability, whereas the water-soluble sericin acts as a glue and is responsible for holding the fibroin filaments together (Series & Rama, 2020). Silk fibroin (SF) has been intensively investigated for biomedical applications due to its biocompatibility, biodegradability, high tensile strength, and excellent biological characteristics such

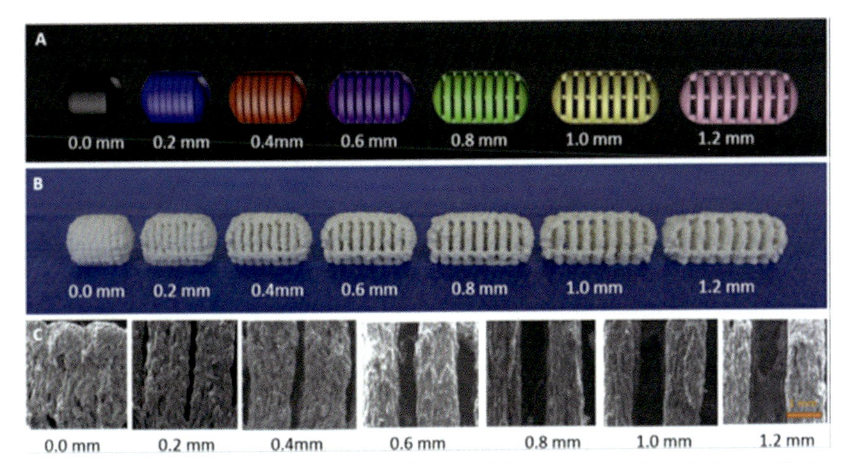

Figure 23.21 (a) Rendered images and (b) photographic images of tablet designs with 1 mm block and increasing interblock spacing: 0, 0.2, 0.4, 0.6, 0.8, 1.0 and 1.2 mm. (c) SEM images of interchannel spaces for these tablets.
From Arafat, B., Wojsz, M., Isreb, A., Forbes, R. T., Isreb, M., Ahmed, W., Arafat, T., & Alhnan, M. A. (2018). Tablet fragmentation without a disintegrant: A novel design approach for accelerating disintegration and drug release from 3D printed cellulosic tablets. E*uropean Journal of Pharmaceutical Sciences, 118*, 191–199. https://doi.org/10.1016/j.ejps.2018.03.019.

as proliferation and adherence of various cells, and low inflammation. It can be processed into several formats, such as scaffolds, films, gels, membranes, powders, and porous sponges under all aqueous processing conditions and light polymerization. By using the common processing methods only constructs with limited complexity can be generated. On the other hand, 3D printing allows silk to be printed into more intricate designs, increasing its potential applications (Wani et al., 2020).

DW 3D printing represents an agile and inexpensive way to obtain complex 3D morphologies with tailored mechanical properties and chemistry. Inks composed of silk fibroin from *Bombyx mori* were prepared by Ghosh et al. (2008). SF solution was obtained using a well-known method described in the literature. After a first stage of degumming in hot water and Na_2CO_3 to remove sericin, SF was extracted with a treatment in LiBr then dialyzed and centrifugated. The silk fibroin ink was extruded through a computer-controlled microcapillary nozzle under an applied pressure of 20–70 kPa at a constant deposition speed of 2 mm s^{-1} into a methanol/water bath to obtain rapid coagulation and good shape stability. The scaffolds with a size of 2 mm × 2 mm shown in Fig. 23.22 were built layer-by-layer with a fiber diameter of around 5 μm, pore size of 100 μm, and elastic modulus of 5.64 ± 1.36 GPa calculated from nanoindentation measurement. These 3D-printed scaffolds can find potential applications in tissue engineering as they are able to promote cell adhesion and growth of human bone marrow-derived mesenchymal stem cells due to the precise geometrical control of their architecture and their biocompatibility. The same approach offers a new route to produce optical waveguides, both biodegradable and biocompatible, for biophotonic elements such as sensors, imaging, biomicroelectromechanical

Figure 23.22 (a) Schematic illustration of 3D direct ink writing of silk fibroin in liquid reservoir. Representative 3D structures of (b) square lattice and (c) circular web. (d) Magnified image of direct writing silk fiber.
Data from Ghosh, S., Parker, S. T., Wang, X., Kaplan, D. L., & Lewis, J. A. (2008). Direct-write assembly of microperiodic silk fibroin scaffolds for tissue engineering applications. *Advanced Functional Materials, 18*(13), 1883–89. https://doi.org/10.1002/adfm.200800040.

systems devices, and therapeutics (Parker et al., 2009). The optical loss measurements demonstrated that printed silk waveguides may find potential applications in a variety of optical devices.

HA-loaded silk scaffolds were 3D printed using a DW technique with the gradient pore size ranging from 200 to 750 μm (Sun et al., 2012). Pore dimensions hardly affected the tissue outcome and the speed of migration of different cell types. Moreover, by combining two different materials in the same scaffold, silk and HA, bone tissue formation can be promoted with the presence of vascular structures. In fact, HA promoted an osteogenesis mechanism, while silk favored endothelial growth and migration inside the scaffold. To improve the printability of SF/HA inks, sodium alginate was used to bind mineralized SF particles together (Huang et al., 2019). Mineralized SF powder was obtained precipitating HA directly on SF from a Ca/P aqueous solution and then lyophilized. SF/HA powder was mixed together with a solution of sodium alginate and extruded with a DW 3D printer from a 0.4 mm nozzle. Cylindrical scaffolds were fabricated with an angle between subsequent layers of 60 degrees to obtain a more complex pore structure to avoid the leakage of the seeded cells and a circular porosity of around 300 μm in diameter. The increasing amount of sodium alginate in the scaffolds led to an improvement in the compressive strength, with values of 17.5 ± 0.9 MPa, 14.7 ± 2.1 MPa, 10.3 ± 1.6 MPa, and 6.1 ± 1.3 MPa moving from neat SF to 3%, 5%, and 10% of SF/HA, respectively. The same trend was observed for compressive modulus with values of 1033 ± 199 MPa, 873 ± 112 MPa, 650 ± 172 MPa, and 674 ± 79 MPa, respectively. In vitro analysis highlighted that scaffolds produced with SF/HA particles promoted the proliferation and osteogenic differentiation of osteoblastic phenotype cells. Due to their low mechanical properties, it is difficult to apply 3D-printed scaffolds of collagen in bone tissue engineering, despite their outstanding biocompatibility. Lee et al. (2018) sidestepped the problem of reinforcing the collagen scaffolds with SF. A solution of collagen and SF was extruded from a nozzle with diameter 0.3 mm at a speed of 10 mm s^{-1} on a refrigerated plate at −40°C. Scaffolds were freeze-dried and cross-linked in an acidic solution. Afterward, the samples were treated in methanol to induce β-sheet formation in the SF. Compressive modulus increased 10-fold with the addition of SF in comparison to the control scaffold made of pure collagen, that is, from 30 ± 2 kPa to 300 ± 40 kPa. Gelation in SF solution corresponds to the formation of an insoluble β-sheet structure from a random coil one and is a crucial step in DW 3D printing to obtain self-standing structures. This step usually requires a postprocessing chemical treatment; however, it is possible to skip this stage with an appropriate printing technique. Low-molecular- weight polyethylene-glycols (PEG) can be used to induce silk gelation, so, DW in PEG with the addition of nanoclay, to obtain a high recovery and thixotropic behavior in the medium bath, would allow for silk to be printed as a fluid and be physically cross-linked, while suspended in the colloidal system (Rodriguez et al., 2018). PEG can be also mixed directly with SF solution just before printing to obtain self-standing bioinks. This allows the induction of rapid gelation of the silk, maintaining a shear thinning behavior of the ink and therefore good printability (Zheng et al., 2018).

DW 3D printing offers the possibility of building scaffolds with porosity ranging from 100 μm to mm, which is suitable to guarantee nutrition supply and growth of the cells, however, nano-dimensional features on the walls of these pores can improve the growth of the tissue and the colonization of the scaffold. To exploit this idea, the combination of a bottom-up template assembly and a top-down approach was studied by Sommer et al. (2016). Silk-based ink was deposited with a DW 3D printing apparatus upon the incorporation of sacrificial particles, in particular wax and PCL microparticles obtained from microfluidics, to design the porosity at different size levels after a dissolution stage. This approach offered a path to replicate the complex hierarchical structure present in the natural world, controlling the morphology of the material over different length scales. Moreover, the possibility to coat these microparticles with nanostructures to further control the morphology at a smaller length scale was suggested. Hierarchical pore structure is a promising approach in tissue engineering, offering a biomimetic morphology for cell adhesion and growth. Bacterial cellulose nanofibers (BCNFs) were incorporated in silk fibroin (SF)/gelatin hydrogel 3D-printed scaffolds to enhance their structural resolution and improve their mechanical properties with the aim of matching those of natural tissues (Huang et al., 2019).

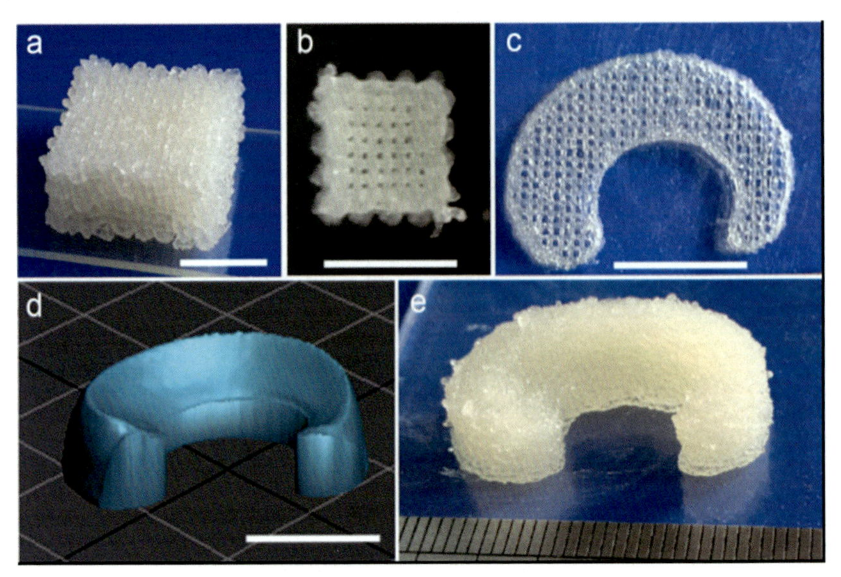

Figure 23.23 (a) 3D view of printed macrostructure scaffolds (BCNFs-0.70 wt.%, similar to b–e). (b) Top view of 3D.printed square scaffolds. (c) Photographs of scaffolds based on the 3D model of lateral meniscus with the internal architecture of 90 degrees/0 degree strand structure. (d) The smoothed 3D model of human meniscus. (e) 3D.printed human meniscus. Scale bar is 10 mm.
From Huang, L., Du, X., Fan, S., Yang, G., Shao, H., Li, D., Cao, C., Zhu, Y., Zhu, M., & Zhang, Y. (2019). Bacterial cellulose nanofibers promote stress and fidelity of 3D-printed silk based hydrogel scaffold with hierarchical pores. *Carbohydrate Polymers, 221*, 146–156. https://doi.org/10.1016/j.carbpol.2019.05.080.

SK, gelatin, and BCNFs were mixed to form a printable ink. This paste was deposited through a 0.4 mm nozzle with a pressure of 1.0–2.0 bar at a speed of 3.0–5.2 mm s^{-1}. Different scaffold geometries were printed, in particular, a human meniscus, shown in Fig. 23.23, was fabricated to demonstrate the feasibility of this system. The hydrogel scaffolds were freeze-dried and then treated with ethanol and genipin to favor the gelation of SF and cross-linking of gelatin, respectively. The addition of BCNFs increased the mechanical properties of the printed scaffolds. Compressive stress increased by about twofold, reaching a value of 65 kPa at 30% strain upon the addition of 0.7 wt.% of BCNFs, matching the mechanical requirements of the targeted soft tissue. Hierarchical prestructure was achieved by designing printing pattern and lyophilization after extrusion. Large pores with diameters from 300 to 600 μm, that ensured sufficient nutrient supply, were designed during DW printing while microporosity ranging from 10 to 20 μm on the walls of strands served as a host for cellular infiltration. In vitro and in vivo tests demonstrated good biocompatibility of the fabricated scaffolds supporting cell infiltration and tissue ingrowth, matching the mechanical properties of the native tissue. Three-dimensional scaffolds printed with SF and cellulose, with tailored hierarchical porosity structure, provided by freeze-drying treatment, may possess the potential to facilitate tracheal epithelium proliferation (Zhong et al., 2019).

A transparent silk sericin-based hydrogel scaffold for wound-dressing application was prepared by Chen et al. through DW 3D printing and ultraviolet cross-linking (Chen et al., 2014). Silk sericin was extracted from *Bombyx mori* cocoons through dissolution in water and recovered via centrifugation and filtration. Silk sericin, gelatin, and Irgacure 2959 were mixed with phosphate buffer solution and the solution was loaded in a syringe with a nozzle diameter of 160 μm. Porous scaffolds were 3D printed and immediately solidified by ultraviolet radiation for 1 min, therefore, the free radical polymerization was able to sustain the newly produced hydrogel structures. The fabricated scaffolds were suitable for the visualization of wound care, demonstrating a homogeneous structure, high transparency, and controllable degradability tuning the degree of cross-linking. Moreover, the 3D-printed hydrogel scaffold supported the proliferation and stratification of keratinocytes, showing great potential for advanced wound care.

DW printers are compatible with hydrogels of various viscosity, however, in the case of materials loaded with cells, the large mechanical stresses in the nozzle and a relatively long printing time could reduce cell viability. In this scenario, digital light processing (DLP) technology offers a suitable alternative. The short printing time and nozzle-free technique ensure a higher cell viability in comparison to other techniques. A bioink from silk fibroin (SF) for DLP 3D bioprinting was developed by Kim et al. (2018) through a methacrylation process using glycidyl methacrylate (GMA) during the fabrication of silk fibroin solution. Degummed silk from *Bombyx mori* was treated with LiBr to favor the reaction with GMA and freeze-dried. The obtained powder was dissolved in water and the photoinitiator LAP was added. A DLP printer working with a wavelength of 365 nm and a resolution of 30 μm was used to print the designed 3D morphology. A spatial accuracy between 66 and 142 μm was found. Scaffolds showed a uniform and interconnected porosity. The mechanical properties of 3D-printed objects increased with the concentration of fibroin in the hydrogel. Compressive and tensile stress at break values of 0.91 and 0.75 MPa were reached at a concentration of 30 wt.%. When this ink was loaded with cells, the rapid printing time did not damage the cells that remained well distributed inside the scaffold. In another study by Hong et al. (2020), the previously described bioink was used for the construction of an artificial trachea through DLP 3D printing and its in vivo performances were studied. Ink was loaded with chondrocytes extracted from the designed animal system, in this case a rabbit, before DLP printing. Six weeks after transplantation, the 3D-printed silk hydrogel showed great efficiency in replacing defective cartilage tissue, guiding the proliferation and growth of the cell toward a physiological tissue.

One of the main challenges in 3D printing of cell-loaded scaffolds with DLP technology is to maintain a stable cell distribution inside a low viscous ink, avoiding sedimentation (Fig. 23.24). Na et al. (2018) suggested using silk fibroin as a viscosity modifier to retard the cell sedimentation in photocross-linkable DLP bioinks. NIH 3T3 cells were stained with a fluorescence dye and mixed with gelatin and silk fibroin to investigate the cell dispersion with the aid of a fluorescence microscope. Bioink was printed with a DLP 3D printer using eosin Y as a photoinitiator and NVP and TEA as

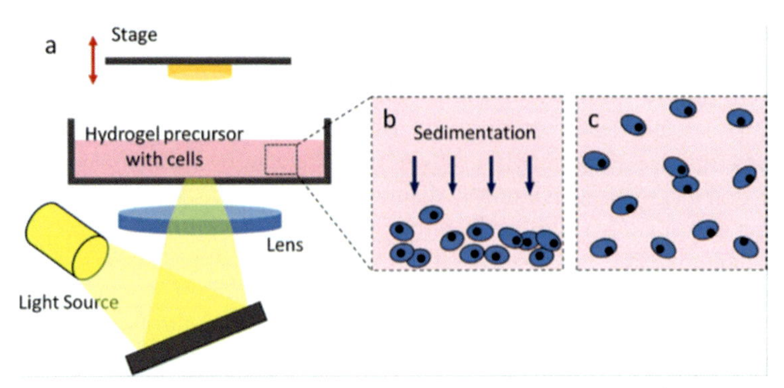

Figure 23.24 DLP printing process with bioink. (a) Schematic diagram of DLP 3D printing, (b) cell sedimentation with the low-viscosity hydrogel precursor solution and (c) stable dispersion of the cells in the viscous solution.
From Na, K., Shin, S., Lee, H., Shin, D., Baek, J., Kwak, H., Park, M., Shin, J., & Hyun, J. (2018). Effect of solution viscosity on retardation of cell sedimentation in DLP 3D printing of gelatin methacrylate/silk fibroin bioink. *Journal of Industrial and Engineering Chemistry, 61*, 340–347. https://doi.org/10.1016/j.jiec.2017.12.032.

coinitiators. The shear viscosity and the stiffness of the silk fibroin-loaded solution increased as the amount of incorporated SF particles increased, improving also the dispersion of the seeded cells.

4.6 *Starch derivative*

Starches are biodegradable biopolymers obtained naturally from different sources, and are widely used in different applicative fields, such as foodstuffs, animal feed, chemicals, petrochemicals, textiles, paper, and tissue engineering. Starch-based inks strongly behave as non-Newtonian fluids and show a shear-thinning effect. At a given shear rate, the viscosity of a starch water solution increases as the starch concentration increases (Chen et al., 2019).

Different sources of starch are under study to produce inks for direct write 3D printing. Chen et al. (2019) investigated the possibility of using rice starch, potato starch, and corn starch. The rheological properties were evaluated mimicking a 3D printing process and these characteristics were related to the printability of the investigated systems. It was found that rice starch at 15%–25% (w/w) was the best among the proposed inks in terms of printability. This was due to the suitable storage modulus and loss tangent at relatively low concentration that resulted in a better shape retention capability and high resolution. Increasing the concentration of rice starch (>30%) did not lead to better results due to the inconsistency of the extrudate thread that caused poor printability, as reported in Fig. 23.25.

With the aim of producing resorbable bone scaffolds with enhanced bioactivity and mechanical strength, a gelatinized corn starch slurry was used to produce a ceramic bioink by Koski et al. (2018). Corn starch was dissolved in cold water to create gelatinized starch and HA powder was then added. Different compositions were printed with the aid of a syringe mounted on a 3D printer in the form of cylinders (Fig. 23.26).

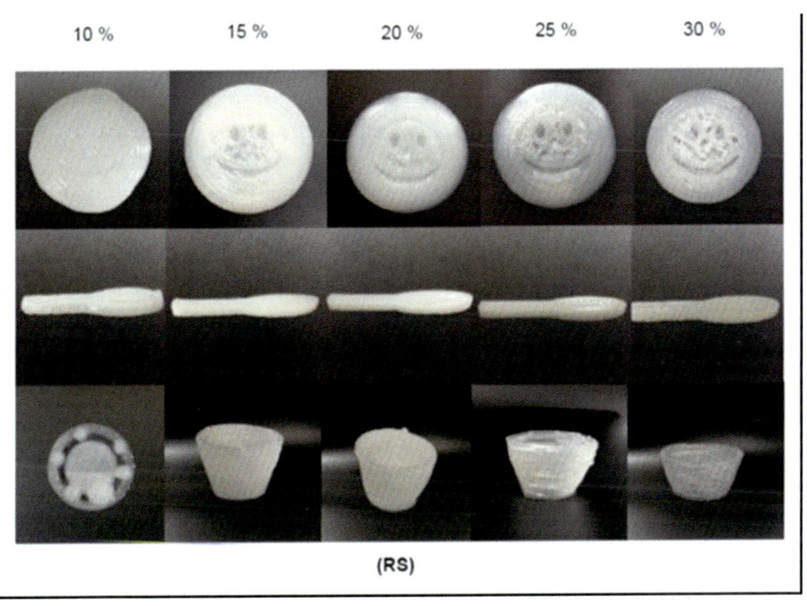

Figure 23.25 HE-3D printed objects using rice starch at 80°C at different concentrations.
From Chen, H., Xie, F., Chen, L., & Zheng, B. (2019). Effect of rheological properties of potato, rice and corn starches on their hot-extrusion 3D printing behaviors. *Journal of Food Engineering, 244*, 150–158. https://doi.org/10.1016/j.jfoodeng.2018.09.011.

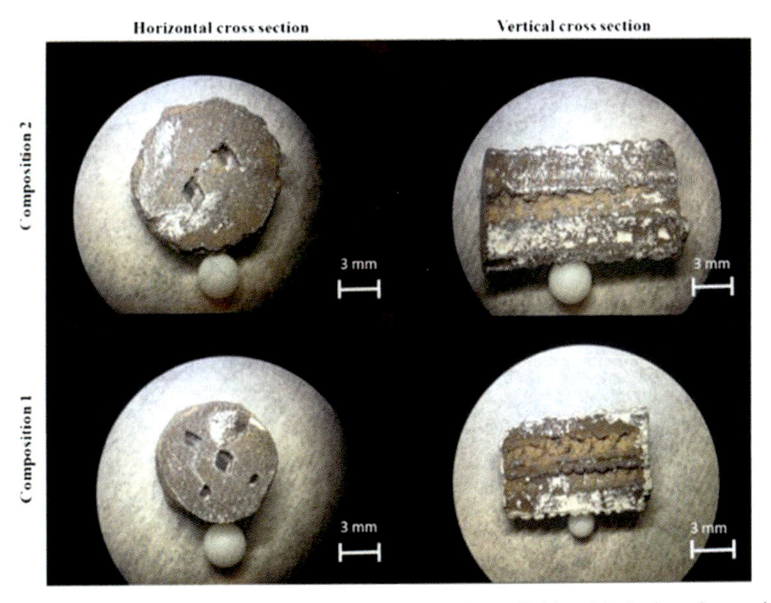

Figure 23.26 (a) Stereoscope images of HA, corn starch scaffolds with designed porosity, showing open pore and interconnected pore architecture. (b) Representation of fabrication of starch-HA composite scaffold.
From Koski, C., Onuike, B., Bandyopadhyay, A., & Bose, S. (2018). Starch-hydroxyapatite composite bone scaffold fabrication utilizing a slurry extrusion-based solid freeform fabricator. *Additive Manufacturing, 24*, 47–59. https://doi.org/10.1016/j.addma.2018.08.030.

Printing parameters were optimized based on the ratio between starch and HA powder and nozzle diameter. The compression strength of the printed object was evaluated. It was shown that the increase in starch concentration from 3.8 to 5.1 wt.% led to a significant increase in compressive strength from 4.07 ± 0.66 MPa to 10.35 ± 1.10 MPa. This behavior was associated with two possible mechanisms. The first is the conversion of the gelled starch into a strong reinforcing phase during the hardening of calcium phosphate-based slurries, while the second one refers to the interaction between amylose and hydroxyapatite via hydrogen bonding that acts as an interlocking phase for apatite crystals reaching higher mechanical properties as highlighted in a subsequent work by Koski et al. (2018).

Dimensional stability is a key parameter to evaluate the printability of a material in DW techniques. Material in the form of paste must be able to support its own weight during deposition without spreading on the build plate. Evaluation of the printability of cassava starch with DW technology was investigated by Maniglia et al. focusing in particular on the effect of ozone processing and gelatinization conditions (Maniglia et al., 2019). Starch concentration was found to be a critical parameter that influenced the apparent viscosity of the ink and therefore the printability of the object. At the end of the work, a water-based solution with a starch concentration of 10.7 wt.% was selected as optimum. While the gelatinization temperature has been found to play a minimal role on the rheological behavior of the slurries, ozone treatment for at least 30 min brought a significative improvement in printability, promoting starch oxidation and reducing the molecular weight, as is clearly visible in Fig. 23.27. Geometrical

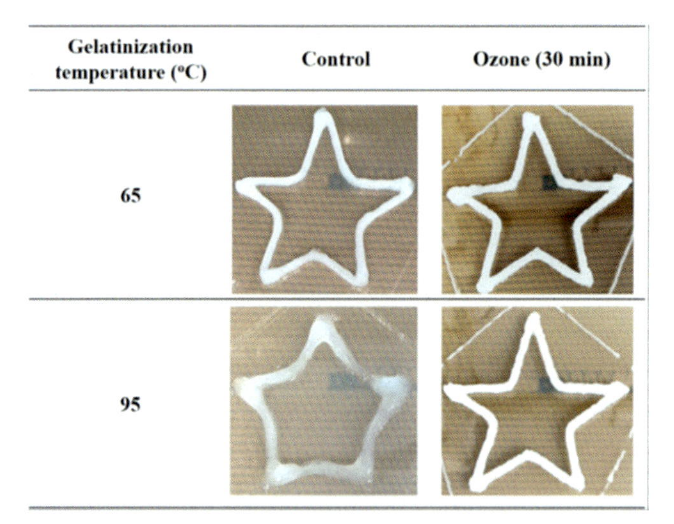

Figure 23.27 Images of the stars obtained by a 3D printer of cassava starch gels: control (native starch) and starch ozonated for 30 min, produced at gelatinization temperatures of 65 and 95°C, and stored for 7 days at 5°C before printing.
From Maniglia, B. C., Lima, D. C., Matta Junior, M. D., Le-Bail, P., Le-Bail, A., & Augusto, P. E. D. (2019). Hydrogels based on ozonated cassava starch: Effect of ozone processing and gelatinization conditions on enhancing 3D-printing applications. *International Journal of Biological Macromolecules, 138*, 1087–1097. https://doi.org/10.1016/j.ijbiomac.2019.07.124.

features, in terms of dimension stability and printing defects, can be easily evaluated for 3D-printed starch with a camera-based method, as proposed by Fahmy et al. (2020). The behavior of printing ink and the effect of viscoelastic properties during material extrusion were studied using two cameras to capture the side and top views of the build and related to the water content of the paste. It was found that as the water content in the printing ink was reduced an increase in the static yield strength (i.e., the cross-over point between the elastic and viscous behaviors of the suspension) can be observed. This effect led also to a time delay at the beginning of the deposition process caused by a weak adhesive force between the ink and the building plate. Furthermore, the increased hydration decreased the complex modulus of the material which indicated a softening of the extruded paste affecting the geometrical stability of the layered structure. However, this softening effect was counterbalanced by the formation of a gluten network in the hydrated starch.

It has been also proven that starch can be 3D-printed also in the form of powder. In fact, Lam (2002) used a binder jetting technology to realize complex scaffold geometries characterized by a fully interconnected pore network using water as a binder. The mechanical properties of the green parts were improved by a thermal treatment to remove water and a subsequent process in which the porous structure was infiltrated by a solution of PLA and PCL. Compressive properties were highly affected by the amount of designed porosity inside the specimen. Elastic modulus decreased as the porosity increased. At the same time, a higher degree of porosity allowed a higher uptake of polymer solution and so an increase in yield strength.

Hot-extrusion 3D printing can help with the destruction of intermolecular hydrogen bonds in starch and ensure full gelatinization of the paste with the aid of a higher extrusion temperature in comparison to traditional DW. Liu et al. investigated the 3D printability of potato starch processed through HE-3DP (Liu et al., 2020). An optimal concentration of potato starch was found to be between 15% and 20%, that guaranteed a good printability and a great dimension stability. Concentrations equal to or greater than 25% tended to block the nozzle, characterized by a diameter of 0.8 mm. The temperature at which the best results were reached was 70°C, which corresponds to the maximum of G′ and to the full gelatinization of the starch. A higher extrusion temperature led to a worsening of the mechanical properties of the obtained specimen. Similar results in terms of starch concentration and printing temperature were found by Zeng et al. (2020). The optimal parameters were 20% and 70–75°C for corn starch and 15%–20% and 75–80°C for rice starch. The authors suggested that the printing temperature should be higher than the gelatinization temperature to ensure complete gelatinization of the system. However, higher temperatures could damage the 3D structure of the gel due to the higher mobility of molecular chains.

4.7 Chitosan

Chitosan (CS), a deacetylated form of chitin, is the second most abundant natural biopolymer after cellulose. It is extracted from the exoskeleton of crustaceans like crabs and shrimps (Gopi et al., 2020). Chitosan is a biocompatible polymer that possesses a series of properties such as nontoxicity, excellent biodegradability, and antibacterial activity (Gopi, Thomas, Pius, Amalraj, et al., 2020). Chitosan has attracted a great deal of attention because it is a multifunctional material that can be applied in a range of biomedical applications such as tissue engineering, drug-delivery media, and enzyme immobilization for biosensing. Chitosan can be applied to fabricate nanoparticles, wire, fiber, fabric, film, hydrogel, bandages, and devices for drug release, suture lines, wound dressings, antibacterial coatings, cell cultures, and scaffolds for tissue engineering (Jennings et al., 2017).

Three-dimensional printed films based on chitosan were prepared by Hafezi et al. (2019) for tissue engineering and in particular for wound-healing purposes. The 3D-structured films were made by jet deposition technology in which air pressure was used to push the polymer solution through a nozzle of 400 μm in diameter. The bioink was composed of chitosan cross-linked with genipin that acts not only as a cross-linker but also provides antiinflammatory and antibacterial functions to the film. Plasticizers such as glycerol and polyethylene glycol were added to the ink to improve the flexibility of the film and so the adhesion to the hypothetically corrupted skin. A 3D-printed cross-linked film of chitosan was able to release a therapeutic drug in a tailored time frame, thus representing a promising dressing for chronic wound application. Extrusion-based 3D-printing technologies could also be employed to produce porous nanocomposite films made of chitosan thanks to the precise spatial control providing a new idea for the preparation of food-packaging materials (Wang et al., 2017).

Chitosan with its ability to enhance biological responses can be efficiently blended with PLA properties in biomedical applications where biochemical activity, 3D architecture, and mechanical properties need to be specifically tailored to the tissue where it will be required to perform. PLA/chitosan scaffolds can be printed with FFF technology obtaining a tailored porosity to mimic rigid human tissues and the PLA/chitosan system represents an important alternative in tissue engineering possibly extending to bone regeneration (Rojas-Martínez et al., 2020). Singh et al. (2020) prepared a PLA/chitosan scaffold through FFF 3D-printing technology and a design of experiment approach was employed to correlate the effects of chitosan loading, infill density, and annealing temperature to the mechanical behavior of composite scaffolds. Chitosan was melt-blended with PLA at concentrations ranging from 1 to 2 wt.%. Higher levels of chitosan were reported to cause difficulties in the extrusion of the filaments. The 3D-printed specimens were printed with a nozzle temperature of 195°C, bed temperature 80°C, and a nozzle diameter of 0.2 mm. Chitosan particles caused discontinuities inside the material and promoted the slippage of the polymeric chains due to the lack of physical and chemical attraction. This effect was correlated with a decrease in the tensile strength in the 3D-printed specimen. Infill density played an important role in increasing the compressive strength and determined the failure mode of the object from buckling at higher levels of infill to crushing at lower levels.

Three-dimensional microfabrication via stereolithography can be employed to produce chitosan-based scaffolds for tissue engineering applications. Akopova et al. (2015) produced 3D scaffolds via SL technologies starting from chitosan obtained through a solvent-free method that used shear deformation. Chitosan was mixed with a photosensitive curing agent (Irgacure 2959) and printed with a laser wavelength

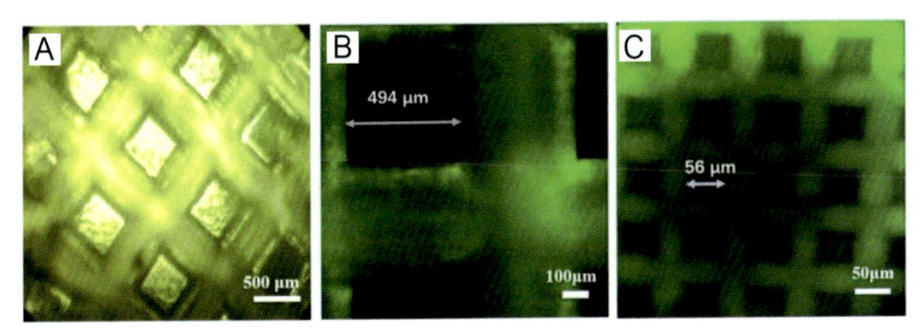

Figure 23.28 (A−C) Micromorphology of lattice structures of CHI-MA hydrogels with different mesh sizes fabricated by DLP printing.
From Shen, Y., Tang, H., Huang, X., Hang, R., Zhang, X., Wang, Y., & Yao, X. (2020). DLP printing photocurable chitosan to build bio-constructs for tissue engineering. *Carbohydrate Polymers, 235*. https://doi.org/10.1016/j.carbpol.2020.115970.

of 1050 and 325 nm. The produced scaffolds, presented in Fig. 23.28, were able to support the adhesion and growth of the seeded cells with no cytotoxic effects. A photocurable chitosan-based bioink was developed by Shen et al. (2020) which can be applied for photolithography, in particular it was applied to a 3D-printing DLP technology. A solution of chitosan and methacrylic anhydride was printed with the aid of a blue light with a wavelength of 405 nm upon the addition of lithium phenyl-2,4,6-trimethyl benzoyl phosphinate (LAP) as a photoinitiator. Mammalian cells were added to the ink to demonstrate the printability and cytocompatibility of the photocured hydrogel processes via DLP. The wavelength of blue light was not lethal for the cell if the exposure was relatively short, at less than 30 s. This range of time was sufficient for photopolymerization of the chitosan scaffold leading to a scaffold with a high resolution of 50 μm (Fig. 23.29). Cytotoxicity was investigated looking at the scaffold with a confocal microscope after a proper staining stage. It was demonstrated that the cells were uniformly distributed inside the printed structure and that the hydrogel was able to sustain the viability of mammalian cell lines after 14 days under

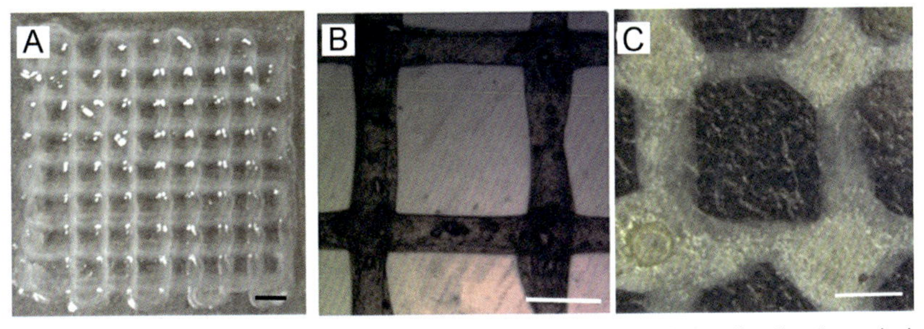

Figure 23.29 Representative images of four-layer bioprinted grid immediately after printing (a,b) or after freeze-drying (c). Scale bars: 1 mm.
From Tonda-Turo, C., Carmagnola, I., Chiappone, A., Feng, Z., Ciardelli, G., Hakkarainen, M., & Sangermano, M. (2020). Photocurable chitosan as bioink for cellularized therapies toward personalized scaffold architecture. *Bioprinting, 18*. https://doi.org/10.1016/j.bprint.2020.e00082.

standard culture conditions. Photocurable inks can be combined with DW 3D printing to merge the easiness of printing with a liquid ink that gels after deposition and the opportunity to photocross-link the obtained structure to improve its geometrical stability over time. Nevertheless, the resolution obtainable with the combination of these two techniques cannot match that obtained via stereolithography (Tonda-Turo et al., 2020).

In DW 3D printing, hydrogels are commonly the main feeding material deposited through a syringe to fabricate the desired object. These viscous solutions can be printed with cells in suspension or without to form scaffolds to support the cell proliferation in a second moment. Chitosan solution was successfully printed through DW technology and interesting mechanical properties can be achieved with a subsequent a cross-linking step. The pathway from hydrogel to a solid structure is given by intermolecular forces among the polymer chains. For chitosan, pH neutralization is enough to trigger the gelation process (Heidenreich et al., 2020). However, the acidic environment of the chitosan hydrogel can cause necrosis of the cells embedded in the ink. Possible remedies are the use of cross-linking agents such as genipin immediately after the deposition or the use of N,O-carboxylmethyl chitosan (NOCC), a derivative of pure chitosan. Chitosan-based inks could have the potential to be used for a wide range of tissue-engineering applications using nontoxic cross-linking agents and by gelling under neutral conditions. Applications of chitosan-based inks for DW 3D printing, with the ability to deliver cells, biomaterials, and growth factors within a scaffold, can range from neural tissue growth (Butler et al., 2020) to cartilage regeneration (Sadeghianmaryan et al., 2020). Zhao et al. (2020) investigated the DW 3D printing processes, mechanical properties, degradability, and preliminary cytotoxicity of chitosan ducts for soft-tissue restorations. In particular, a glycolic acid solution was used for the preparation of the chitosan ink. Three methods based on extrusion and 3D printing were applied for manufacturing chitosan ducts and fulfilling different soft-tissue requirements (Fig. 23.30). In particular, one method involved 3D printing with the aid of a rod support to fabricate the duct sample shape. Different wall thicknesses of the ducts were easily achieved by the combination of 3D printing and a rotary axis. A double-nozzle 3D printer was used to manufacture the complicated structure of a CS duct using PLA as support in the last method investigated. The produced ducts have been proven to be suitable for soft-tissue restoration, showing excellent matching with the mechanical properties of natural tissues.

The 3D printing of porous scaffolds offers the possibility of modulating the shape and size of the pores inside the object, thus obtaining a very high area/volume ratio. Bergamonti et al. (2019) used 3D-printed chitosan as substrate to embed TiO_2 particles with a size of 20 nm for the removal of pharmaceutical compounds in wastewater through photolytic degradation. An acetic acid solution was used to dissolve chitosan at a concentration of 6 w/v% and nanometric TiO_2 was homogeneously dispersed through mechanical stirring. The prepared ink was extruded from a nozzle with a diameter of 192 μm and deposited over a Peltier cooled bed. The prepared scaffold was exposed to ammonia vapors in order to activate the gelation of chitosan with exposure to an alkaline environment. The mechanical properties of the 3D-printed scaffolds were suitable for handling and manipulation in industrial environments. The chemical

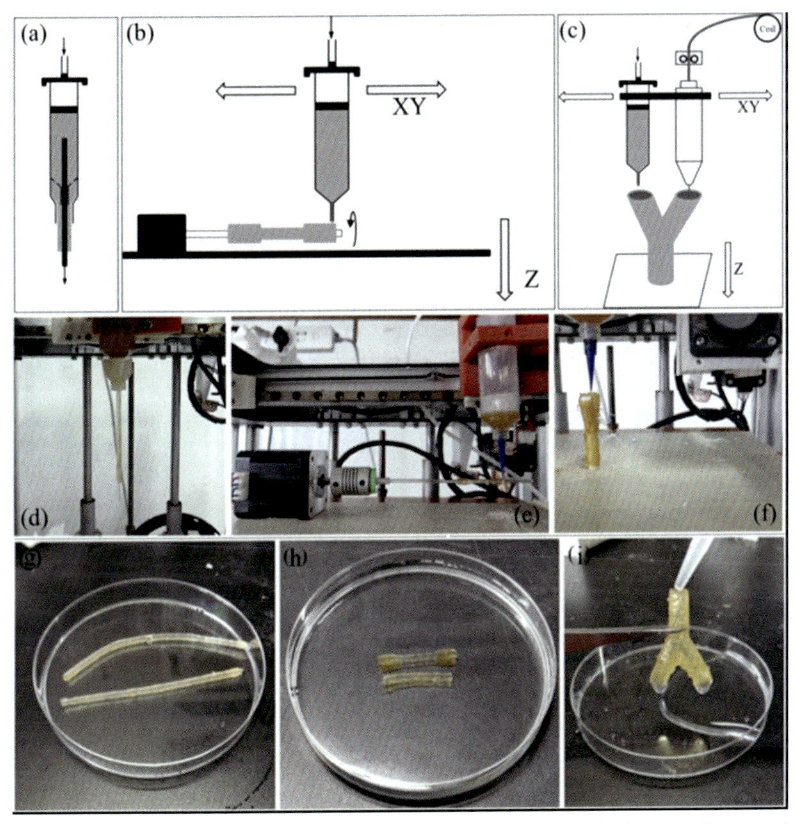

Figure 23.30 Preparation of ducts: (a) and (d) direct extrusion of CS slurry and supporting; (b) and (e) combination of rotary axis and extrusion-based 3D printing; (c) and (f) combination of fused deposition modeling and extrusion-based 3D printing, (g) CS specimens fabricated by method (a), (h) CS specimens fabricated by method (b), and (i) CS specimens fabricated by method (c).
From Zhao, C. Q., Liu, W. G., Xu, Z. Y., Li, J. G., Huang, T. T., Lu, Y. J., Huang, H. G., & Lin, J. X. (2020). Chitosan ducts fabricated by extrusion-based 3D printing for soft-tissue engineering. *Carbohydrate Polymers, 236*, 116058. https://doi.org/10.1016/j.carbpol.2020.116058.

composition and architecture of 3D scaffolds made with PLA and chitosan printed through the DW technique were studied by Almeida et al. (2014). In particular, the polarization of human macrophages toward an antiinflammatory phenotype affecting the curvature of the cytoskeleton actin fibers was correlated to the surface topography/chemistry via study of the cytokine profile. Scaffolds were printed from a Teflon tip with diameter of 150 μm at 40°C with a speed of 7 mm s^{-1}, moreover, NaOH in ethanol was dispensed drop by drop on top of the printed struts to cross-link the polymer structure during the fabrication process. Different geometries were obtained and it was proved that macrophages were mainly influenced by material properties and to a lower extent by scaffold features, such as pore geometry. Barrier function and the ability to guide tissue morphogenesis are the most important functions of bio-engineered skin in wound-healing applications. Chitosan with its antimicrobial properties and the capability to trigger hemostasis has attracted wide interest in the field of tissue engineering. Ng et al. (2016) developed a polyelectrolyte gelatin-chitosan hydrogel optimized for 3D bioprinting to achieve high shape fidelity of the printed 3D constructs and good biocompatibility with fibroblast skin cells. The hydrogel was deposited via DW printing at room temperature to obtain a good spatial control over deposition of selected biomaterials at specific regions to fabricate customizable tissue-engineered constructs. The printability of various chitosan-based hydrogels was evaluated using a combination of different printing pressures from 1 to 3.5 bars and feed rates from 600 to 1000 mm/min using a constant nozzle diameter of 210 μm. This kind of ink exhibited a sufficiently high viscosity, resulting in good shape fidelity of the printed constructs and reaching a resolution for a grid-like pattern of 300 μm.

Three-dimensional printed scaffolds of chitosan were studied in vivo by Intini et al. (2018) to treat wounds on rat models. These scaffolds exhibited exceptional properties in terms of biocompatibility toward various skin-associated human cell lines. Cells were seeded into the scaffolds prior to the implantation and it was seen that these scaffolds were able to promote faster regeneration of skin tissue with respect to commercial products, thus suggesting the potential of chitosan scaffolds for the treatment of dermal injuries. Due to the lower temperatures involved in DW 3D printing, an anesthetic drug could be also successfully incorporated into the chitosan hydrogel wound dressing (Long et al., 2019). Drug release from the scaffold could start together with biodegradation of the material that also offered a benefit for wound application as the adhesion strength would reduce over time, minimizing tissue damage or scar formation. Chitosan-based 3D-printed scaffolds showed promising features not only in skin tissue applications but also in hard-tissue engineering. Ergul et al. (2019) produced a scaffold suitable for human mesenchymal stem cells via DW technology, extruding a hydrogel composed of a mix of chitosan and polyvinyl alcohol with the addition of hydroxyapatite and a bone morphogenetic protein. Chitosan at a concentration of 2.5 wt.% was dissolved in 2% acetic acid solution and mixed with a 1 wt.% PVA solution in distilled water at a temperature of 50°C. After mixing, a solution of synthesized HA powder was added. When these scaffolds were printed with a content of hydroxyapatite of around 15 wt.% (Fig. 23.31), they reached elastic modulus values, suitable for bone substitute materials, of around 91 MPa. In vitro experiments demonstrated that the proliferation rate of human mesenchymal stem cells increased significantly in 3D-printed CH/PVA/HA scaffolds compared to the control material.

In contrast to permanent ones, biodegradable implants offer permanent treatment of tissue injury and not just a temporary solution. Zafeiris et al. (2020) 3D-printed hydrogels consisting of hydroxyapatite nanocrystals with chitosan and L-arginine with DW technology and, to improve their mechanical properties and dimension stability, genipin, a natural cross-linking agent was added. Scaffolds were printed through a 0.4 mm nozzle and printing parameters were optimized to reach optimal resolution, in particular 1.00 mm/s flow speed and 5.00 mm/s speed. Biocompatibility was evaluated using MG63 human osteosarcoma cells for 7 days of culturing. SEM and

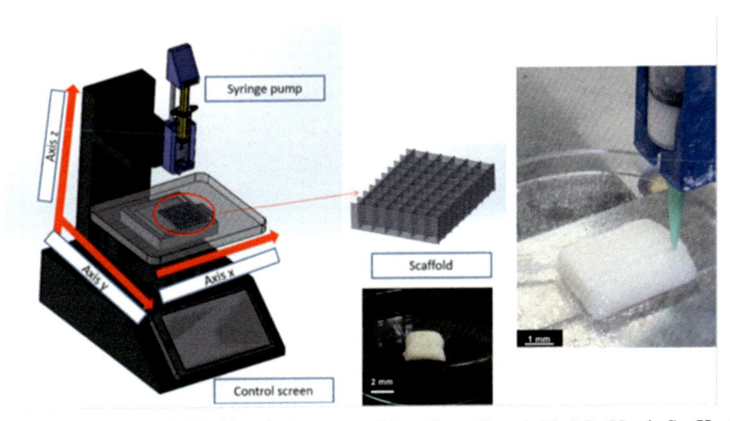

Figure 23.31 Experimental 3D-bioprinting setup.Data from Ergul, N. M., Unal, S., Kartal, I., Kalkandelen, C., Ekren, N., Kilic, O., Chi-Chang, L., & Gunduz, O. (2019). 3D printing of chitosan/poly(vinyl alcohol) hydrogel containing synthesized hydroxyapatite scaffolds for hard-tissue engineering. *Polymer Testing, 79*, 106006. https://doi.org/10.1016/j.polymertesting.2019.106006.

confocal observations revealed that chitosan scaffolds were a friendly environment for the seeded cells, allowing them to adhere on the surface, increase their population, and maintain high levels of viability. It was found that the printing fidelity of chitosan hydrogel can be improved without affecting printing flow by the addition of silk fibroin (Zhang et al., 2017). Silk material in different forms was combined with chitosan and used as inks for 3D printing by Zhang et al. (2018). Silk nanofibers with a diameter of less than 1 μm and length of a few hundred microns provided the highest cell proliferation and differentiation due to the preference of the cells for the nanotopography and higher surface roughness that favors cell attachment with respect to other types of SF with higher dimensions.

It is well known that pore structure regulates the degradation of scaffold and drives the formation of new bone tissue. Scaffolds that simulate natural bone components were prepared by Chen et al. (2020) using chitosan and hydroxyapatite by DW 3D printing. To efficiently treat bone defects, polydopamine, a strong adhesive produced by mussels, was added to the formulation of the ink. The scaffolds were printed from a 0.4 mm nozzle at room temperature, the extruder air pressure was 0.08−0.5 MPa, and the speed of printing was 12−18 mm/s. After the preparation of the scaffolds, genipin was used to perform the cross-link of the chitosan hydrogel. Scaffolds were freeze-dried and irradiated before in vitro and in vivo tests. The elastic modulus in compression of the scaffolds had a remarkable potential as material for repairing bone defects, and was measured at 8.9 ± 2.0 MPa. To evaluate the bone-repairing ability of the produced scaffolds an in vivo evaluation was performed on bone defects in rabbit femurs. In Fig. 23.32(a and b), chitosan scaffolds and commercial control materials based on collagen and hydroxyapatite are shown, meanwhile, in Fig. 23.32(c and d) the two materials implanted in the femur are shown 4 and 12 weeks after surgery. Fig. 23.32(e) shows a micro-CT of the treated bone to evaluate the repair quality of the 3D-printed scaffolds. This evaluation showed that the produced chitosan-based

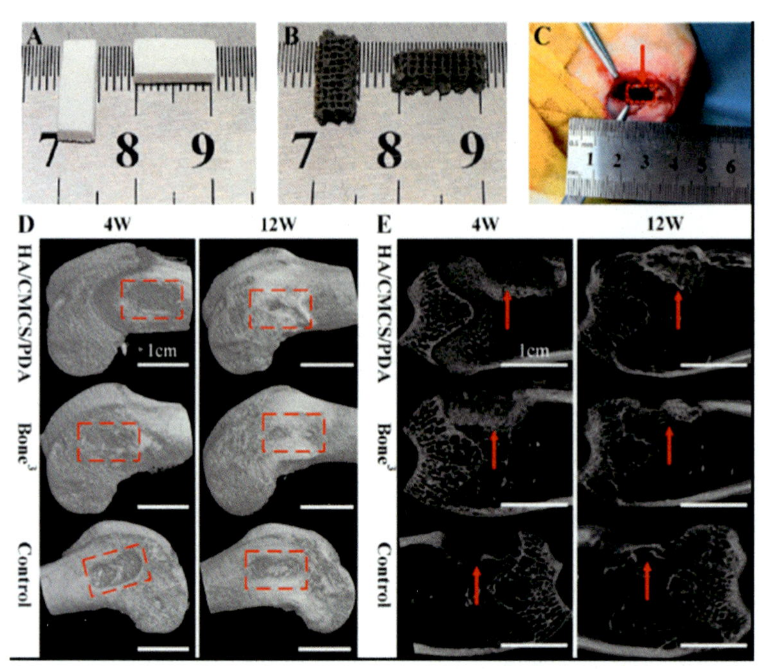

Figure 23.32 Bone defect repair in vivo. (a) Commercial bone repair materials: Bone3. (b) 3D-printed porous scaffolds: HA/CMCS/PDA scaffolds. (c) HA/CMCS/PDA scaffold implanted into the femoral defect of rabbit. (d) Representative three-dimensional images of repair after implantation at 4 and 12 weeks. (e) Representative two-dimensional micro-CT pictures corresponding to (d) after implantation at 4 and 12 weeks.
From Chen, T., Zou, Q., Du, C., Wang, C., Li, Y., & Fu, B. (2020). Biodegradable 3D printed HA/CMCS/PDA scaffold for repairing lacunar bone defect. *Materials Science and Engineering: C, 116*. https://doi.org/10.1016/j.msec.2020.111148.

scaffolds were able to promote bone regeneration in rabbit femoral defects. In fact, implanted scaffold was gradually replaced by newly formed bone tissue driven by 3D structure upon degradation.

4.8 Poly(vinyl alcohol)

Poly(vinyl alcohol) (PVA) is the most commercially used water-soluble polymer. It is tasteless, odorless, biodegradable, and biocompatible. Due to its water solubility and biodegradability, PVA is used in a wide range of applications such as water-soluble films, bindings for pigments and fibers, dip-coated articles, protective strippable coatings, manufacturing of detergents and cleansing agents, adhesives, emulsion paints, and solution cast films. PVA has excellent film-forming, emulsifying, and adhesive properties. It is characterized by a high tensile strength, good flexibility, good oxygen barrier properties, and good solvent resistance properties (Goodship & Jacobs, 2005).

Hydroxyapatite-based bone scaffolds can be produced by using various 3D printing technologies. However, the binder solution remains a challenge to solve for the production of bioceramic scaffold through binder jetting additive manufacturing. The green mechanical strength and resolution of the printed objects are mainly determined by the binder and its interaction with the HA particles. PVA, with its exclusive properties of good solubility in water and the ability to act as a gluing agent, is an optimal candidate for the preparation of water-based polymeric binders for binder jetting technology.

Chai et al. (2020) found that only the concentrations of PVA in water from 0.8 to 1.5 wt.% can be used for printing. Fabrication of a porous scaffold was not permitted for concentrations below 0.5 wt.%, while printing head clogging was observed for concentrations above 2.0 wt.%. The size and number of distinguishable pores on the surface of the produced scaffold were the parameters to estimate the printability of PVA binders, in particular, smaller diameters are associated with better printing resolution. HA-based porous scaffolds were produced with powder-based 3D printer spraying binders with various PVA concentrations in a ratio of 0.48 binder/powder above the bed surface. It was found that the best printing resolution was achieved with a concentration of 1.0 wt.% of PVA. Scaffolds printed by 1.0 wt.% PVA showed also the best compressive strength of about 4 MPa. The clogging of the printing head can be avoided by blending the PVA adhesive to the HA powder before the printing process. PVAs with low and high molecular weight (MW) were blended with HA from 10 to 30 wt.% by Zhou, Lennon, Buchanan, McCarthy, & Dunne, (2020). Different ratios between PVA and HA powder were studied, 10:90, 20:80, or 30:70 wt.%, with HA being the main component. Various molecular weights for the PVA were taken into consideration, ranging from 70 up to 228 kg mol^{-1}. The 3D printed scaffolds were analyzed by μ-CT scanning to visualize the internal pore structure and the obtained images were overlapped with the original CAD design to estimate the geometrical accuracy. The best results were obtained with the highest MW PVA at 30:70 concentration, reaching an accuracy of 85% and a green compression strength of 5.63 ± 0.27 MPa. PVA in powder form can be also applied to form highly flexible composites through binder jetting. A properly designed ink can be sprayed on the PVA powder bed to fabricate a multimaterial and functionally graded object without the need for assembly or bonding (Shen et al., 2020).

PVA is commonly used in textile and paper applications as an emulsifier, colloidal particle stabilizer, adhesive, and coating agent (Wiśniewska et al., 2015). This makes PVA a good starting point to produce nanocomposite inks for DW 3D printing. Angjellari et al. (2017) produced novel inks dispersing detonation nanodiamond (DND) in PVA suitable to be extruded with a DW apparatus. A Mendel 3-RepRapPro© 3D printer custom-assembled for layering slurries, shown in Fig. 23.33, with different physical properties, was used to print PVA inks (10 wt.% in water) with various amounts of DND. A nano-indentation technique evidenced that even small additions of DND led to significant enhancement in the hardness and indentation modulus of fabricated structures in comparison to neat material. The indentation modulus was found to increase up to 200% at 5 wt.% loading of DND.

Bioactive ceramics can be extruded with this technique to obtain functionalized scaffolds for tissue-engineering applications. Alonso et al. (2019) studied the possibility of using PVA solution to homogeneously disperse Ag, TiO_2, sodium hexatitanate

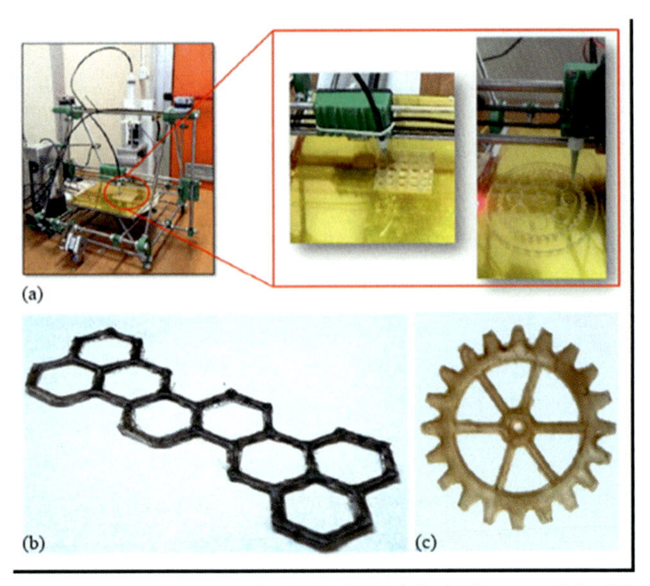

Figure 23.33 Objects printed by AM using PVA-DND inks (a) Images of the 3D printer during printing sessions. Objects printed by AM using PVA-DND inks: (b) a honeycomb structure (2 mm thick); (c) a wheel-like structure (3 mm thick).
From Angjellari, M., Tamburri, E., Montaina, L., Natali, M., Passeri, D., Rossi, M., & Terranova, M. L. (2017). Beyond the concepts of nanocomposite and 3D printing: PVA and nanodiamonds for layer-by-layer additive manufacturing. *Materials & Design, 119*, 12–21. https://doi.org/10.1016/j.matdes.2017.01.051.

($Na_2Ti_6O_{13}$), and tricalcium phosphate [$Ca_3(PO_4)_2$] particles to produce scaffolds promoting living tissue regeneration through DW 3D printing. A 10% aqueous solution of PVA was used as the starting point for the ink preparation and different slurries were printed with a Prusa i3 through a 31G needle. A similar approach was used also by Aki et al. (2020) to fabricate solid scaffolds for bone tissue replacement incorporating boron nitride and bacterial cellulose in a PVA solution. Cellulose due to its hydrophilic behavior can be easily dispersed in PVA water solution. Cataldi et al. (2018) prepared filaments reinforced with nanocellulose suitable to be printed with FFF 3D printing technologies. In particular, cellulose nanocrystals obtained from acid treatment of microcellulose were mixed together with PVA in a water solution. Upon drying, the material was granulated and extruded at 175°C to form filaments with a diameter of 1.7 mm and concentration of nanocellulose from 2 to 20 wt.%. The mechanical characterization on a 3D-printed dumbbell specimen, printed at 230°C, highlighted the positive effect of cellulose on the rubbery PVA, improving not only the stiffness of the material but also its toughness, reaching an increase of 73% for the stress at break when 5 wt.% nanocellulose was added in respect of the neat polymer.

PVA in FFF technology is mostly used as a support material for overhanging parts (Rigotti et al., 2018) or in the fabrication of temporary molds (Hassanajili et al., 2019), because it can be removed easily without any machining process by simply dissolving it in hot water. The easy extrudability of PVA makes it suitable to produce ceramic composite 3D-printed structures through FFF technology. Janek et al. (2020) studied the mechanical properties of HA-PVA filaments for FFF 3D printing. They focused their attention on filament buckling, which may occur during filament feeding into the liquefier, especially in the case of rubbery materials. Filaments with a diameter of 1.75 mm and an inorganic filler concentration of 50 wt.% were extruded at 180°C and tested to measure the elastic modulus, the compression strength, and the viscosity values. The critical buckling pressure calculated from Euler analysis using measured elastic modulus revealed underestimated critical pressure values about 2.5–5.0 times if compared to values of maximal filament compressive pressure loads simulating buckling. The thermal behavior, printability, microstructure, and mechanical properties of composite scaffolds made of PVA/β-tricalcium phosphate (β-TCP) fabricated using FFF were investigated by Chen et al. (2019) for potential bone repair and regeneration scaffolds in tissue-engineering applications. A solid-state shear milling was used to homogeneously disperse β-TCP particles into the PVA resulting in the formation of hydrogen bonding between β-TCP and PVA, which helped to improve the interface strength of the material. Filaments were extruded with a single-screw extruder and the temperature zones were set at 0, 180, 180, and 120°C from the feeding section to the die. The printability window (see Fig. 23.34) of the material was established based on the ratio of its compressive modulus to the apparent viscosity. The addition of β-TCP improves the printability of the materials to some degree and an optimized printing speed of 200 mm/s with a shear rate of 110 s^{-1} was selected to avoid buckling of the filament during the FFF process.

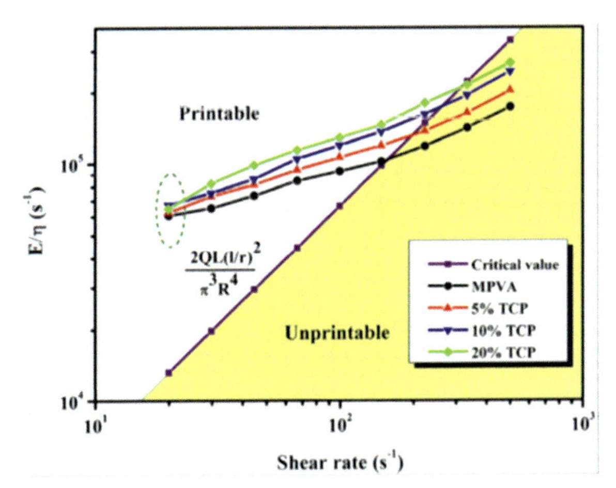

Figure 23.34 The ratio of E/η plotted as a function of shear rate for the MPVA and PVA/β-TCP composite filaments.
From Chen, G., Chen, N., & Wang, Q. (2019). Fabrication and properties of poly(vinyl alcohol)/β-tricalcium phosphate composite scaffolds via fused deposition modeling for bone tissue engineering. *Composites Science and Technology, 172*, 17–28. https://doi.org/10.1016/j.compscitech.2019.01.004.

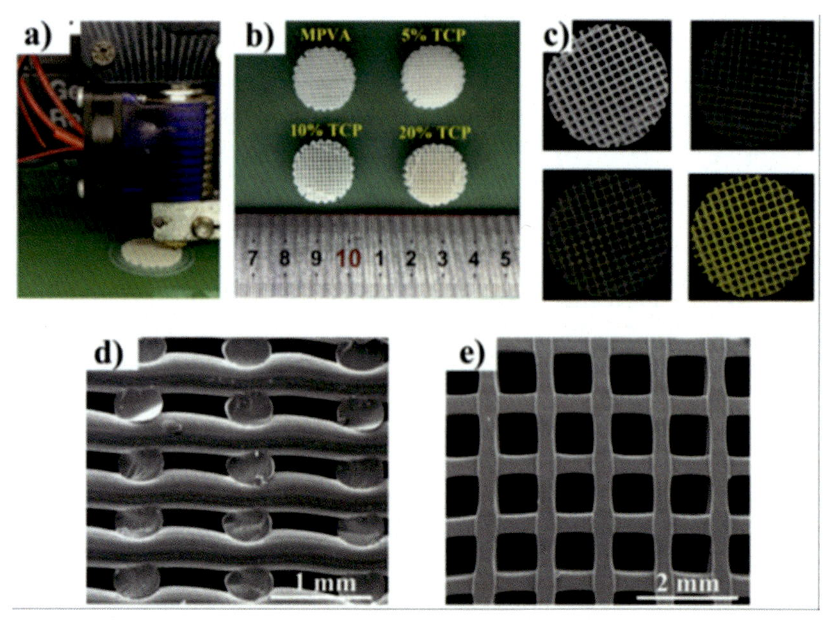

Figure 23.35 (a) Optical image of the scaffold fabrication process; (b–c) optical and corresponding μCT images of MPVA and PVA/β-TCP composite scaffolds; (d–e) SEM images of the cross-section and top view of a representative composite scaffold, respectively. From Chen, G., Chen, N., & Wang, Q. (2019). Fabrication and properties of poly(vinyl alcohol)/β-tricalcium phosphate composite scaffolds via fused deposition modeling for bone tissue engineering. *Composites Science and Technology, 172*, 17–28. https://doi.org/10.1016/j.compscitech.2019.01.004.

Scaffolds were printed at 175°C from a 0.4 mm nozzle at a printing speed of 200 mm/s (Fig. 23.35). The obtained composite scaffolds were well-structured with an interconnected porosity. Compression strength increased from 8.3 to 10.7 kPa with an increase in the β-TCP content up to 20 wt.%. Moreover, enhanced biocompatibility, which is positive for cell adhesion and proliferation, was observed in an in vitro cell culture.

Additive manufacturing plays an important role in the development of personalized medicine due to its lower cost. PVA, due to its biodegradability, finds a great number of applications in the pharmaceutical industry and drug-delivery systems. Different approaches have been followed to obtain 3D-printed tablets to dispense selected drugs with a tailored time profile. The benefit of capsules with tailored degradation includes the removal of excipients needed for formulation of controlled-release dosage forms. Commercial PVA filaments can be loaded with the selected drug through swelling. Goyanes et al. (2014) prepared 3D-printed tablets with a different infill percentage and geometries resulting in various drug-release profiles. The poor printability of pharmaceutical-grade polymers loaded with drugs during FFF 3D printing can be enhanced after blending PVA with others polymers such as PEG or PEO and plasticizers such as polysorbate (Alhijjaj et al., 2016). However, the thermal degradation of the drug during the 3D printing process may not be appropriate if the printing temperature is higher than that of the drug degradation temperature (Goyanes et al., 2015). To overcome this problem a solution can be to produce PVA capsules that are filled upon printing. The dissolution of the material and the drug release profile can be modulated by using different thicknesses. Matijašić et al. produced modular PVA capsules which were concentrically compartmental with different thicknesses through FFF. In this way, the active agent can be simply loaded into the capsules, avoiding exposure of the drug to high temperatures. Therefore, the release profiles can be modified just by changing the geometry or combining different parts of 3D-printed soluble capsules (Matijašić et al., 2019).

5. Conclusions

Since its introduction in the 1980s, AM technology has made considerable progress in terms of direct transformation of CAD drawings into real products, making complex and intricate objects with mass customization capabilities. Its main advantages are a reduction of the overall product development time and time to market, the combination of different production stages into one, and production from a wide variety of materials with minimal wastage and maximum use of materials. AM has been demonstrated to have huge potential in a wide area of applications and can offer new opportunities for the use of biodegradable materials. In the last few decades, industrial and academic researchers have focused their attention on these materials since they could contribute to a more sustainable development as they offer an advantage from a cost-effectiveness point of view, with a wide range of disposal options and a lower environmental impact. The demand for more sustainable and biodegradable polymeric materials is increasing in the pursuit of a "green" life, which highlights the significant importance of the right materials to achieve this goal. The latest innovations in 3D printing of biomaterials open numerous applications in the aerospace, automotive, textile, food, and biomedical industries, and are considered to represent the future of sustainable production. A large advantage is that an object produced by 3D printing of organic biodegradable materials can be fully recycled at the end of its life cycle to create a new part or disposed of as a biodegradable waste (compost) without damaging the natural environment. Natural-derived biopolymers, including lignocellulosic materials and materials from crustacean exoskeletons, represent the most abundant biobased and renewable feedstocks for different 3D printing technologies. Blending or modification with other components to add or improve the material's processability and functionality will definitely expand the application areas of the natural-derived biopolymers. Technical advances and the associated feasibility in the development of natural-derived biopolymers as feedstock formulations for a variety of AM technologies have been confirmed and demonstrated in recent years, at least at the scale of proof-of-concept level. Recent progress in the design and development of biodegradable polymers and relevant processing approaches is providing a unique means for AM development of biomedical devices. Particularly, 3D printing provides a promising pathway to manufacture functional biomaterials with tremendous technological potential to customize biodegradable systems composition, micro/nanostructure, macroshape, mechanical behavior, and functional properties. Due to these exciting features, AM is revolutionizing the approach to a wide range of biomedical applications, including, among others, prostheses, surgical implants, tissue engineering, in vitro tissue modeling, and controlled drug release. The 3D printed scaffolds represent a rapidly growing field with substantial potential to have a great impact on health care and medicine, including personalized devices, over the coming decades. The continuous growth of new printable biomaterials, together with the development of advanced-materials engineering tools, the constantly increasing automation and digital control of the design and fabrication processes, in relationship to the resulting properties of the fabricated items, are expected in the near future to tremendously impact the biomedical industry.

However, there remains a large number of weak points to be improved, such as in the material processability, degradation, chemical, biological, and mechanical properties before achieving the full potential of biopolymers. Once these challenges are overcome, 3D printing technology will no doubt accelerate with rapid growth for biopolymer applications in a larger variety of areas. Therefore, at the end of this chapter, it can be forecast that the future materials development in AM will be progressively increase the utilization of those biodegradable feedstocks.

References

Aki, D., Ulag, S., Unal, S., Sengor, M., Ekren, N., Lin, C.-C., Yılmazer, H., Ustundag, C. B., Kalaskar, D. M., & Gunduz, O. (2020). 3D printing of PVA/hexagonal boron nitride/bacterial cellulose composite scaffolds for bone tissue engineering. *Materials and Design, 196*. https://doi.org/10.1016/j.matdes.2020.109094

Akopova, T. A., Demina, T. S., Bagratashvili, V. N., Bardakova, K. N., Novikov, M. M., Selezneva, I. I., Istomin, A. V., Svidchenko, E. A., Cherkaev, G. V., Surin, N. M., & Timashev, P. S. (2015). Solid state synthesis of chitosan and its unsaturated derivatives for laser microfabrication of 3D scaffolds. *IOP Conference Series: Materials Science and Engineering, 87*. https://doi.org/10.1088/1757-899x/87/1/012079

Alberts, B., Johnson., A., & Lewis, J. (2002). Molecular biology of the cell (4th edition.). *Garland Science*. https://www.ncbi.nlm.nih.gov/books/NBK26928/.

Alhijjaj, M., Belton, P., & Qi, S. (2016). An investigation into the use of polymer blends to improve the printability of and regulate drug release from pharmaceutical solid dispersions prepared via fused deposition modeling (FDM) 3D printing. *European Journal of Pharmaceutics and Biopharmaceutics, 108*, 111–125. https://doi.org/10.1016/j.ejpb.2016.08.016

Almeida, C. R., Serra, T., Oliveira, M. I., Planell, J. A., Barbosa, M. A., & Navarro, M. (2014). Impact of 3-D printed PLA- and chitosan-based scaffolds on human monocyte/macrophage responses: unraveling the effect of 3-D structures on inflammation. *Acta Biomaterialia, 10*(2), 613–622. https://doi.org/10.1016/j.actbio.2013.10.035

Alonso, M., Enrique Guerrero-Beltrán, C., & Ortega-Lara, W. (2019). Design and characterization of Gelatin/PVA hydrogels reinforced with ceramics for 3D printed prosthesis. *Materials Today: Proceedings, 13*, 324–331. https://doi.org/10.1016/j.matpr.2019.03.161

Angjellari, M., Tamburri, E., Montaina, L., Natali, M., Passeri, D., Rossi, M., & Terranova, M. L. (2017). Beyond the concepts of nanocomposite and 3D printing: PVA and nanodiamonds for layer-by-layer additive manufacturing. *Materials and Design, 119*, 12–21. https://doi.org/10.1016/j.matdes.2017.01.051

Arafat, B., Wojsz, M., Isreb, A., Forbes, R. T., Isreb, M., Ahmed, W., Arafat, T., & Alhnan, M. A. (2018). Tablet fragmentation without a disintegrant: a novel design approach for accelerating disintegration and drug release from 3D printed cellulosic tablets. *European Journal of Pharmaceutical Sciences, 118*, 191–199. https://doi.org/10.1016/j.ejps.2018.03.019

Auras, R., Harte, B., & Selke, S. (2004). An overview of polylactides as packaging materials. *Macromolecular Bioscience, 4*(9), 835–864. https://doi.org/10.1002/mabi.200400043

Ayrilmis, N. (2018). Effect of layer thickness on surface properties of 3D printed materials produced from wood flour/PLA filament. *Polymer Testing, 71*, 163–166. https://doi.org/10.1016/j.polymertesting.2018.09.009

Bai, J., Goodridge, R. D., Hague, R. J. M., & Okamoto, M. (2017). Processing and characterization of a polylactic acid/nanoclay composite for laser sintering. *Polymer Composites, 38*(11), 2570–2576. https://doi.org/10.1002/pc.23848

Baumann, F., & Roller, D. (2017). *Survey on additive manufacturing, cloud 3D printing and services.*

Bekas, D. G., Hou, Y., Liu, Y., & Panesar, A. (2019). 3D printing to enable multifunctionality in polymer-based composites: a review. *Composites Part B: Engineering, 179*, 107540. https://doi.org/10.1016/j.compositesb.2019.107540

Bergamonti, L., Bergonzi, C., Graiff, C., Lottici, P. P., Bettini, R., & Elviri, L. (2019). 3D printed chitosan scaffolds: a new TiO2 support for the photocatalytic degradation of amoxicillin in water. *Water Research, 163*, 114841. https://doi.org/10.1016/j.watres.2019.07.008

Berman, B. (2012). 3-D printing: the new industrial revolution. *Business Horizons, 55*(2), 155–162. https://doi.org/10.1016/j.bushor.2011.11.003

Bhagia, S., Lowden, R. R., Erdman, D., Rodriguez, M., Haga, B. A., Solano, I. R. M., Gallego, N. C., Pu, Y., Muchero, W., Kunc, V., & Ragauskas, A. J. (2020). Tensile properties of 3D-printed wood-filled PLA materials using poplar trees. *Applied Materials Today, 21*. https://doi.org/10.1016/j.apmt.2020.100832

Boetker, J., Water, J. J., Aho, J., Arnfast, L., Bohr, A., & Rantanen, J. (2016). Modifying release characteristics from 3D printed drug-eluting products. *European Journal of Pharmaceutical Sciences, 90*, 47–52. https://doi.org/10.1016/j.ejps.2016.03.013

Butler, H. M., Naseri, E., MacDonald, D. S., Andrew Tasker, R., & Ahmadi, A. (2020). Optimization of starch- and chitosan-based bio-inks for 3D bioprinting of scaffolds for neural cell growth. *Materialia, 12*. https://doi.org/10.1016/j.mtla.2020.100737

Lam, C. X. F. (2002). Scaffold development using 3D printing with a starch-based polymer. *Materials Science and Engineering C: Materials for Biological Applications, 20*(1–2), 49–56. https://doi.org/10.1016/S0928-4931(02)00012-7

Cataldi, A., Rigotti, D., Nguyen, V. D. H., & Pegoretti, A. (2018). Polyvinyl alcohol reinforced with crystalline nanocellulose for 3D printing application. *Materials Today Communications, 15*, 236–244. https://doi.org/10.1016/j.mtcomm.2018.02.007

Cesaretti, G., Dini, E., De Kestelier, X., Colla, V., & Pambaguian, L. (2014). Building components for an outpost on the Lunar soil by means of a novel 3D printing technology. *Acta Astronautica, 93*, 430–450. https://doi.org/10.1016/j.actaastro.2013.07.034

Chai, W., Wei, Q., Yang, M., Ji, K., Guo, Y., Wei, S., & Wang, Y. (2020). The printability of three water based polymeric binders and their effects on the properties of 3D printed hydroxyapatite bone scaffold. *Ceramics International, 46*(5), 6663–6671. https://doi.org/10.1016/j.ceramint.2019.11.154

Checchetto, R., Rigotti, D., Pegoretti, A., & Miotello, A. (2020). Chloroform desorption from poly(lactic acid) nanocomposites: a thermal desorption spectroscopy study. *Pure and Applied Chemistry, 92*(3), 391–398. https://doi.org/10.1515/pac-2018-1216

Chen, C. H., Lee, M. Y., Shyu, V. B., Chen, Y. C., Chen, C. T., & Chen, J. P. (2014). Surface modification of polycaprolactone scaffolds fabricated via selective laser sintering for cartilage tissue engineering. *Materials Science and Engineering :C Materials for Biological Applications, 40*, 389–397. https://doi.org/10.1016/j.msec.2014.04.029

Chen, G., Chen, N., & Wang, Q. (2019). Fabrication and properties of poly(vinyl alcohol)/β-tricalcium phosphate composite scaffolds via fused deposition modeling for bone tissue engineering. *Composites Science and Technology, 172*, 17–28. https://doi.org/10.1016/j.compscitech.2019.01.004

Chen, H., Xie, F., Chen, L., & Zheng, B. (2019). Effect of rheological properties of potato, rice and corn starches on their hot-extrusion 3D printing behaviors. *Journal of Food Engineering, 244*, 150–158. https://doi.org/10.1016/j.jfoodeng.2018.09.011

Chen, T., Zou, Q., Du, C., Wang, C., Li, Y., & Fu, B. (2020). Biodegradable 3D printed HA/CMCS/PDA scaffold for repairing lacunar bone defect. *Materials Science and Engineering: C, 116*. https://doi.org/10.1016/j.msec.2020.111148

Choong, Y. Y. C., Tan, H. W., Patel, D. C., Choong, W. T. N., Chen, C.-H., Low, H. Y., Tan, M. J., Patel, C. D., & Chua, C. K. (2020). The global rise of 3D printing during the COVID-19 pandemic. *Nature Reviews Materials, 5*(9), 637–639. https://doi.org/10.1038/s41578-020-00234-3

Christian Weller, R. K., & Piller, F. T. (2015). Economic implications of 3D printing: market structure models in light of additive manufacturing revisited. *International Journal of Production Economics, 164*, 43–56.

Dastbaz, M., Pattinson, C., Akhgar, B., Wilkinson, S., & Cope, N. (2015). 3D printing and sustainable product development. In *Green Information technology* (pp. 161–183). https://doi.org/10.1016/b978-0-12-801379-3.00010-3

Daver, F., Lee, K. P. M., Brandt, M., & Shanks, R. (2018). Cork–PLA composite filaments for fused deposition modelling. *Composites Science and Technology, 168*, 230–237. https://doi.org/10.1016/j.compscitech.2018.10.008

De Leon, A. C., Chen, Q., Palaganas, N. B., Palaganas, J. O., Manapat, J., & Advincula, R. C. (2016). High performance polymer nanocomposites for additive manufacturing applications. *Reactive and Functional Polymers, 103*, 141–155. https://doi.org/10.1016/j.reactfunctpolym.2016.04.010

DellaSala, D. A. (2013a). Global change. In *Reference module in earth systems and environmental sciences*. Elsevier. https://doi.org/10.1016/B978-0-12-409548-9.05355-0

DellaSala, D. A. (2013b). The carbon cycle and global change: Too much of a good thing. In *Reference module in Earth systems and environmental sciences*. Elsevier. https://doi.org/10.1016/B978-0-12-409548-9.05874-7

Dhandapani, R., Krishnan, P. D., Zennifer, A., Kannan, V., Manigandan, A., Arul, M. R., Jaiswal, D., Subramanian, A., Kumbar, S. G., & Sethuraman, S. (2020). Additive manufacturing of biodegradable porous orthopaedic screw. *Bioactive Materials, 5*(3), 458–467. https://doi.org/10.1016/j.bioactmat.2020.03.009

Dickmann, M., Tarter, S., Egger, W., Pegoretti, A., Rigotti, D., Brusa, R. S., & Checchetto, R. (2020). Interface nanocavities in poly (lactic acid) membranes with dispersed cellulose nanofibrils: their role in the gas barrier performances. *Polymer, 202*. https://doi.org/10.1016/j.polymer.2020.122729

Diermann, S. H., Lu, M., Zhao, Y., Vandi, L. J., Dargusch, M., & Huang, H. (2018). Synthesis, microstructure, and mechanical behaviour of a unique porous PHBV scaffold manufactured using selective laser sintering. *Journal of the Mechanical Behavior of Biomedical Materials, 84*, 151–160. https://doi.org/10.1016/j.jmbbm.2018.05.007

Hutmacher, D. W. (2000). *Mechanical properties and cell cultural response of polycaprolactone scaffolds designed and fabricated via fused deposition modeling*.

Dinesh Kumar, S., Venkadeshwaran, K., & Aravindan, M. K. (2020). Fused deposition modelling of PLA reinforced with cellulose nano-crystals. *Materials Today: Proceedings, 33*, 868–875. https://doi.org/10.1016/j.matpr.2020.06.404

Dong, J., Mei, C., Han, J., Lee, S., & Wu, Q. (2019). 3D printed poly(lactic acid) composites with grafted cellulose nanofibers: effect of nanofiber and post-fabrication annealing treatment on composite flexural properties. *Additive Manufacturing, 28*, 621–628. https://doi.org/10.1016/j.addma.2019.06.004

Dorigato, A., Moretti, V., Dul, S., Unterberger, S., & Pegoretti, A. (2017). Electrically conductive nanocomposites for fused deposition modelling. *Synthetic Metals, 226*, 7–14. https://doi.org/10.1016/j.synthmet.2017.01.009

Dorigato, A., Rigotti, D., & Pegoretti, A. (2020). Novel poly(caprolactone)/epoxy blends by additive manufacturing. *Materials, 13*(4), 819. https://doi.org/10.3390/ma13040819

Doyle, H., Lohfeld, S., & McHugh, P. (2015). Evaluating the effect of increasing ceramic content on the mechanical properties, material microstructure and degradation of selective laser sintered polycaprolactone/beta-tricalcium phosphate materials. *Medical Engineering and Physics, 37*(8), 767–776. https://doi.org/10.1016/j.medengphy.2015.05.009

Du, Y., Liu, H., Shuang, J., Wang, J., Ma, J., & Zhang, S. (2015). Microsphere-based selective laser sintering for building macroporous bone scaffolds with controlled microstructure and excellent biocompatibility. *Colloids and Surfaces B: Biointerfaces, 135*, 81–89. https://doi.org/10.1016/j.colsurfb.2015.06.074

Duan, B., Cheung, W. L., & Wang, M. (2011). Optimized fabrication of Ca-P/PHBV nanocomposite scaffolds via selective laser sintering for bone tissue engineering. *Biofabrication, 3*(1), 015001. https://doi.org/10.1088/1758-5082/3/1/015001

Duan, B., Wang, M., Zhou, W. Y., Cheung, W. L., Li, Z. Y., & Lu, W. W. (2010). Three-dimensional nanocomposite scaffolds fabricated via selective laser sintering for bone tissue engineering. *Acta Biomaterialia, 6*(12), 4495–4505. https://doi.org/10.1016/j.actbio.2010.06.024

Dubinenko, G. E., Zinoviev, A. L., Bolbasov, E. N., Novikov, V. T., & Tverdokhlebov, S. I. (2020). Preparation of Poly(L-lactic acid)/Hydroxyapatite composite scaffolds by fused deposit modeling 3D printing. *Materials Today: Proceedings, 22*, 228–234. https://doi.org/10.1016/j.matpr.2019.08.092

Dul, S., Fambri, L., & Pegoretti, A. (2018). Filaments production and fused deposition modelling of ABS/carbon nanotubes composites. *Nanomaterials, 8*(1). https://doi.org/10.3390/nano8010049

Dul, S., Fambri, L., & Pegoretti, A. (2016). Fused deposition modeling with ABS-graphene nanocomposites. *Composites Part A: Applied Science and Manufacturing, 85*, 181–191. https://doi.org/10.1016/j.compositesa.2016.03.013

Duymaz, B. T., Erdiler, F. B., Alan, T., Aydogdu, M. O., Inan, A. T., Ekren, N., Uzun, M., Sahin, Y. M., Bulus, E., Oktar, F. N., Selvi, S. S., ToksoyOner, E., Kilic, O., Bostan, M. S., Eroglu, M. S., & Gunduz, O. (2019). 3D bio-printing of levan/polycaprolactone/gelatin blends for bone tissue engineering: characterization of the cellular behavior. *European Polymer Journal, 119*, 426–437. https://doi.org/10.1016/j.eurpolymj.2019.08.015

Ebi, K. L. (2011). Climate change and gealth A2 - Nriagu, J.O. In *Encyclopedia of environmental health* (pp. 680–689). Elsevier. https://doi.org/10.1016/B978-0-444-52272-6.00165-3

Ebnesajjad, S., & Drobny, J. G. (2007). Handbook of thermoplastic elastomers. In *Plastics design library* (p. 464). Matthew Deans.

Ebnesajjad, S., & McKeen, L. W. (2013). 1 - introduction to use of plastics in food packaging. In *Plastic films in food packaging* (pp. 1–15). William Andrew Publishing. https://doi.org/10.1016/B978-1-4557-3112-1.00001-6

Elomaa, L., Teixeira, S., Hakala, R., Korhonen, H., Grijpma, D. W., & Seppala, J. V. (2011). Preparation of poly(epsilon-caprolactone)-based tissue engineering scaffolds by stereolithography. *Acta Biomaterialia, 7*(11), 3850–3856. https://doi.org/10.1016/j.actbio.2011.06.039

Eosoly, S., Brabazon, D., Lohfeld, S., & Looney, L. (2010). Selective laser sintering of hydroxyapatite/poly-epsilon-caprolactone scaffolds. *Acta Biomaterialia, 6*(7), 2511–2517. https://doi.org/10.1016/j.actbio.2009.07.018

Ergul, N. M., Unal, S., Kartal, I., Kalkandelen, C., Ekren, N., Kilic, O., Chi-Chang, L., & Gunduz, O. (2019). 3D printing of chitosan/poly(vinyl alcohol) hydrogel containing synthesized hydroxyapatite scaffolds for hard-tissue engineering. *Polymer Testing, 79*, 106006. https://doi.org/10.1016/j.polymertesting.2019.106006

Fahmy, A. R., Becker, T., & Jekle, M. (2020). 3D printing and additive manufacturing of cereal-based materials: quality analysis of starch-based systems using a camera-based morphological approach. *Innovative Food Science and Emerging Technologies, 63*. https://doi.org/10.1016/j.ifset.2020.102384

Fan, M., Fu, F., & Djafari Petroudy, S. R. (2017). 3 - physical and mechanical properties of natural fibers. In *Advanced high strength natural fibre composites in construction* (pp. 59–83). Woodhead Publishing. https://doi.org/10.1016/B978-0-08-100411-1.00003-0

Fina, F., Madla, C. M., Goyanes, A., Zhang, J., Gaisford, S., & Basit, A. W. (2018). Fabricating 3D printed orally disintegrating printlets using selective laser sintering. *International Journal of Pharmaceutics, 541*(1–2), 101–107. https://doi.org/10.1016/j.ijpharm.2018.02.015

Findrik Balogova, A., Hudak, R., Toth, T., Schnitzer, M., Feranc, J., Bakos, D., & Zivcak, J. (2018). Determination of geometrical and viscoelastic properties of PLA/PHB samples made by additive manufacturing for urethral substitution. *Journal of Biotechnology, 284*, 123–130. https://doi.org/10.1016/j.jbiotec.2018.08.019

Ford, S., & Despeisse, M. (2016). Additive manufacturing and sustainability: an exploratory study of the advantages and challenges. *Journal of Cleaner Production, 137*, 1573–1587. https://doi.org/10.1016/j.jclepro.2016.04.150

Forster, A. M. (2015). Materials testing standards for additive manufacturing of polymer materials: state of the art and standards applicability. *NIST Interagency/Internal Report (NISTIR), 8059*, 67–123.

Fortunati, E., Luzi, F., Puglia, D., Dominici, F., Santulli, C., Kenny, J. M., & Torre, L. (2014). Investigation of thermo-mechanical, chemical and degradative properties of PLA-limonene films reinforced with cellulose nanocrystals extracted from *Phormium tenax* leaves. *European Polymer Journal, 56*, 77–91. https://doi.org/10.1016/j.eurpolymj.2014.03.030

Gao, F., McLauchlin, A. R., & Thomas, N. L. (2012). 13 - biodegradable polymer nanocomposites. In *Advances in polymer nanocomposites* (pp. 398–430). Woodhead Publishing. https://doi.org/10.1533/9780857096241.2.398

Gayer, C., Abert, J., Bullemer, M., Grom, S., Jauer, L., Meiners, W., Reinauer, F., Vucak, M., Wissenbach, K., Poprawe, R., Schleifenbaum, J. H., & Fischer, H. (2018). Influence of the material properties of a poly(D,L-lactide)/beta-tricalcium phosphate composite on the processability by selective laser sintering. *Journal of the Mechanical Behavior of Biomedical Materials, 87*, 267–278. https://doi.org/10.1016/j.jmbbm.2018.07.021

Gayer, C., Ritter, J., Bullemer, M., Grom, S., Jauer, L., Meiners, W., Pfister, A., Reinauer, F., Vucak, M., Wissenbach, K., Fischer, H., Poprawe, R., & Schleifenbaum, J. H. (2019). Development of a solvent-free polylactide/calcium carbonate composite for selective laser sintering of bone tissue engineering scaffolds. *Materials Science Engineering C: Materials for Biological Applications, 101*, 660–673. https://doi.org/10.1016/j.msec.2019.03.101

Gebler, M., Schoot Uiterkamp, A. J. M., & Visser, C. (2014). A global sustainability perspective on 3D printing technologies. *Energy Policy, 74*, 158–167. https://doi.org/10.1016/j.enpol.2014.08.033

Gendviliene, I., Simoliunas, E., Rekstyte, S., Malinauskas, M., Zaleckas, L., Jegelevicius, D., Bukelskiene, V., & Rutkunas, V. (2020). Assessment of the morphology and dimensional accuracy of 3D printed PLA and PLA/HAp scaffolds. *Journal of the Mechanical Behavior of Biomedical Materials, 104*, 103616. https://doi.org/10.1016/j.jmbbm.2020.103616

George, A., Sanjay, M. R., Srisuk, R., Parameswaranpillai, J., & Siengchin, S. (2020). A comprehensive review on chemical properties and applications of biopolymers and their composites. *International Journal of Biological Macromolecules, 154*, 329–338. https://doi.org/10.1016/j.ijbiomac.2020.03.120

George, A. K., & Ipsita Roy, V. P. M. (2014). Polyhydroxyalkanoate (PHA) based blends, composites and nanocomposites. In *Green chemistry series* (pp. P001–P006). The Royal Society of Chemistry. https://doi.org/10.1039/9781782622314-fp001

Ghosh, S., Parker, S. T., Wang, X., Kaplan, D. L., & Lewis, J. A. (2008). Direct-write assembly of microperiodic silk fibroin scaffolds for tissue engineering applications. *Advanced Functional Materials, 18*(13), 1883–1889. https://doi.org/10.1002/adfm.200800040

Glasschroeder, J., Prager, E., & Zaeh, M. F. (2015). Powder-bed-based 3D-printing of function integrated parts. *Rapid Prototyping Journal, 21*(2), 207–215. https://doi.org/10.1108/rpj-12-2014-0172

Goodship, V., & Jacobs, D. (2005). *Polyvinyl alcohol materials, processing and applications*. Smithers Rapra Technology.

Gopi, S., Thomas, S., Pius, A., Amalraj, A., Jude, S., & Gopi, S. (2020). Chapter 1 - polymer blends, composites and nanocomposites from chitin and chitosan; manufacturing, characterization and applications. In *Handbook of chitin and chitosan* (pp. 1–46). Elsevier. https://doi.org/10.1016/B978-0-12-817968-0.00001-9

Gopi, S., Thomas, S., Pius, A., Hossain, M. R., Mallik, A. K., & Rahman, M. M. (2020). Chapter 7: Fundamentals of chitosan for biomedical applications. In *Handbook of chitin and chitosan* (pp. 199–230). Elsevier. https://doi.org/10.1016/B978-0-12-817966-6.00007-8

Goyanes, A., Buanz, A. B., Basit, A. W., & Gaisford, S. (2014). Fused-filament 3D printing (3DP) for fabrication of tablets. *International Journal of Pharmaceutics, 476*(1–2), 88–92. https://doi.org/10.1016/j.ijpharm.2014.09.044

Goyanes, A., Buanz, A. B., Hatton, G. B., Gaisford, S., & Basit, A. W. (2015). 3D printing of modified-release aminosalicylate (4-ASA and 5-ASA) tablets. *European Journal of Pharmaceutics and Biopharmaceutics, 89*, 157–162. https://doi.org/10.1016/j.ejpb.2014.12.003

Hafezi, F., Scoutaris, N., Douroumis, D., & Boateng, J. (2019). 3D printed chitosan dressing crosslinked with genipin for potential healing of chronic wounds. *International Journal of Pharmaceutics, 560*, 406–415. https://doi.org/10.1016/j.ijpharm.2019.02.020

Haleem, A., & Javaid, M. (2019). Polyether ether ketone (PEEK) and its manufacturing of customised 3D printed dentistry parts using additive manufacturing. *Clinical Epidemiology and Global Health, 7*(4), 654–660. https://doi.org/10.1016/j.cegh.2019.03.001

Hassanajili, S., Karami-Pour, A., Oryan, A., & Talaei-Khozani, T. (2019). Preparation and characterization of PLA/PCL/HA composite scaffolds using indirect 3D printing for bone tissue engineering. *Materials Science and Engineering C: Materials for Biological Applications, 104*, 109960. https://doi.org/10.1016/j.msec.2019.109960

Hedayati, S. K., Behravesh, A. H., Hasannia, S., Bagheri Saed, A., & Akhoundi, B. (2020). 3D printed PCL scaffold reinforced with continuous biodegradable fiber yarn: a study on mechanical and cell viability properties. *Polymer Testing, 83*, 106347. https://doi.org/10.1016/j.polymertesting.2020.106347

Heidenreich, A. C., Pérez-Recalde, M., González Wusener, A., & Hermida, É. B. (2020). Collagen and chitosan blends for 3D bioprinting: a rheological and printability approach. *Polymer Testing, 82*. https://doi.org/10.1016/j.polymertesting.2019.106297

Hong, H., Seo, Y. B., Kim, D. Y., Lee, J. S., Lee, Y. J., Lee, H., Ajiteru, O., Sultan, M. T., Lee, O. J., Kim, S. H., & Park, C. H. (2020). Digital light processing 3D printed silk fibroin hydrogel for cartilage tissue engineering. *Biomaterials, 232*, 119679. https://doi.org/10.1016/j.biomaterials.2019.119679

Hoskins, T. J., Dearn, K. D., & Kukureka, S. N. (2018). Mechanical performance of PEEK produced by additive manufacturing. *Polymer Testing, 70*, 511–519. https://doi.org/10.1016/j.polymertesting.2018.08.008

Hoy, M. B. (2013). 3D printing: making things at the Library. *Medical Reference Services Quarterly, 32*(1), 94–99. https://doi.org/10.1080/02763869.2013.749139

Huang, L., Du, X., Fan, S., Yang, G., Shao, H., Li, D., Cao, C., Zhu, Y., Zhu, M., & Zhang, Y. (2019). Bacterial cellulose nanofibers promote stress and fidelity of 3D-printed silk based hydrogel scaffold with hierarchical pores. *Carbohydrate Polymers, 221*, 146–156. https://doi.org/10.1016/j.carbpol.2019.05.080

Huang, T., Fan, C., Zhu, M., Zhu, Y., Zhang, W., & Li, L. (2019). 3D-printed scaffolds of biomineralized hydroxyapatite nanocomposite on silk fibroin for improving bone regeneration. *Applied Surface Science, 467*(468), 345–353. https://doi.org/10.1016/j.apsusc.2018.10.166

Huang, Y., Leu, M. C., Mazumder, J., & Donmez, A. (2015). Additive manufacturing: Current state, future potential, gaps and needs, and recommendations. *Journal of Manufacturing Science and Engineering, 137*(1), 1–10. https://doi.org/10.1115/1.4028725

Intini, C., Elviri, L., Cabral, J., Mros, S., Bergonzi, C., Bianchera, A., Flammini, L., Govoni, P., Barocelli, E., Bettini, R., & McConnell, M. (2018). 3D-printed chitosan-based scaffolds: an in vitro study of human skin cell growth and an in-vivo wound healing evaluation in experimental diabetes in rats. *Carbohydrate Polymers, 199*, 593–602. https://doi.org/10.1016/j.carbpol.2018.07.057

Iwan, Z. (2002). Fused deposition modeling of novel scaffold architectures for tissue engineering applications. *Biomaterials, 23*, 1169–1185.

Janek, M., Žilinská, V., Kovár, V., Hajdúchová, Z., Tomanová, K., Peciar, P., … Bača, Ľ. (2020). Mechanical testing of hydroxyapatite filaments for tissue scaffolds preparation by fused deposition of ceramics. *Journal of the European Ceramic Society, 40*(14), 4932–4938. https://doi.org/10.1016/j.jeurceramsoc.2020.01.061

Jansen, J., Melchels, F. P. W., Grijpma, D. W., & Feijen, J. (2009). Fumaric acid monoethyl ester-functionalized poly(d,l-lactide)/N-vinyl-2-pyrrolidone resins for the preparation of tissue engineering scaffolds by stereolithography. *Biomacromolecules, 10*(2), 214–220. https://doi.org/10.1021/bm801001r

Jennings, J. A., Bumgardner, J. D., Vunain, E., Mishra, A. K., & Mamba, B. B. (2017). 1: Fundamentals of chitosan for biomedical applications. In *Chitosan Based Biomaterials* (vol. 1, pp. 3–30). Woodhead Publishing. https://doi.org/10.1016/B978-0-08-100230-8.00001-7

Jerez-Mesa, R., Travieso-Rodriguez, J. A., Llumà-Fuentes, J., Gomez-Gras, G., & Puig, D. (2017). Fatigue lifespan study of PLA parts obtained by additive manufacturing. *Procedia Manufacturing, 13*, 872–879. https://doi.org/10.1016/j.promfg.2017.09.146

Kading, B., & Straub, J. (2015). Utilizing in-situ resources and 3D printing structures for a manned Mars mission. *Acta Astronautica, 107*, 317–326. https://doi.org/10.1016/j.actaastro.2014.11.036

Kariz, M., Sernek, M., Obućina, M., & Kuzman, M. K. (2018). Effect of wood content in FDM filament on properties of 3D printed parts. *Materials Today Communications, 14*, 135–140. https://doi.org/10.1016/j.mtcomm.2017.12.016

Kempin, W., Franz, C., Koster, L. C., Schneider, F., Bogdahn, M., Weitschies, W., & Seidlitz, A. (2017). Assessment of different polymers and drug loads for fused deposition modeling of drug loaded implants. *European Journal of Pharmaceutics and Biopharmaceutics, 115*, 84–93. https://doi.org/10.1016/j.ejpb.2017.02.014

Kim, S. H., Yeon, Y. K., Lee, J. M., Chao, J. R., Lee, Y. J., Seo, Y. B., Sultan, M. T., Lee, O. J., Lee, J. S., Yoon, S. I., Hong, I. S., Khang, G., Lee, S. J., Yoo, J. J., & Park, C. H. (2018). Precisely printable and biocompatible silk fibroin bioink for digital light processing 3D printing. *Nature Communications, 9*(1), 1620. https://doi.org/10.1038/s41467-018-03759-y

Koski, C., Onuike, B., Bandyopadhyay, A., & Bose, S. (2018). Starch-hydroxyapatite composite bone scaffold fabrication utilizing a slurry extrusion-based solid freeform fabricator. *Additive Manufacturing, 24*, 47–59. https://doi.org/10.1016/j.addma.2018.08.030

Kosorn, W., Sakulsumbat, M., Uppanan, P., Kaewkong, P., Chantaweroad, S., Jitsaard, J., Sitthiseripratip, K., & Janvikul, W. (2017). PCL/PHBV blended three dimensional scaffolds fabricated by fused deposition modeling and responses of chondrocytes to the scaffolds. *Journal of Biomedical Materials Research Part B: Applied Biomaterials, 105*(5), 1141–1150. https://doi.org/10.1002/jbm.b.33658

Kovalcik, A., Sangroniz, L., Kalina, M., Skopalova, K., Humpolicek, P., Omastova, M., Mundigler, N., & Muller, A. J. (2020). Properties of scaffolds prepared by fused deposition modeling of poly(hydroxyalkanoates). *International Journal of Biological Macromolecules, 161*, 364–376. https://doi.org/10.1016/j.ijbiomac.2020.06.022

Lee, H., Yang, G. H., Kim, M., Lee, J., Huh, J., & Kim, G. (2018). Fabrication of micro/nanoporous collagen/dECM/silk-fibroin biocomposite scaffolds using a low temperature 3D printing process for bone tissue regeneration. *Materials Science and Engineering C: Materials for Biological Applications, 84*, 140–147. https://doi.org/10.1016/j.msec.2017.11.013

Leppiniemi, J., Lahtinen, P., Paajanen, A., Mahlberg, R., Metsa-Kortelainen, S., Pinomaa, T., Pajari, H., Vikholm-Lundin, I., Pursula, P., & Hytonen, V. P. (2017). 3D-printable bioactivated nanocellulose-alginate hydrogels. *ACS Applied Materials and Interfaces, 9*(26), 21959–21970. https://doi.org/10.1021/acsami.7b02756

Li, L., Zhu, Y., & Yang, J. (2018). 3D bioprinting of cellulose with controlled porous structures from NMMO. *Materials Letters, 210*, 136–138. https://doi.org/10.1016/j.matlet.2017.09.015

Li, L., Chen, Y., Yu, T., Wang, N., Wang, C., & Wang, H. (2019). Preparation of polylactic acid/TEMPO-oxidized bacterial cellulose nanocomposites for 3D printing via Pickering emulsion approach. *Composites Communications, 16*, 162–167. https://doi.org/10.1016/j.coco.2019.10.004

Li, V. C.-F., Kuang, X., Hamel, C. M., Roach, D., Deng, Y., & Qi, H. J. (2019). Cellulose nanocrystals support material for 3D printing complexly shaped structures via multi-materials-multi-methods printing. *Additive Manufacturing, 28*, 14–22. https://doi.org/10.1016/j.addma.2019.04.013

Lindner, M., Hoeges, S., Meiners, W., Wissenbach, K., Smeets, R., Telle, R., Poprawe, R., & Fischer, H. (2011). Manufacturing of individual biodegradable bone substitute implants using selective laser melting technique. *Journal of Biomedical Materials Research Part A, 97*(4), 466–471. https://doi.org/10.1002/jbm.a.33058

Lipson, H., & Kurman, M. (2013). *Fabricated: the new world of 3D printing*. John Wiley and Sons, Inc.

Liu, Z., Chen, H., Zheng, B., Xie, F., & Chen, L. (2020). Understanding the structure and rheological properties of potato starch induced by hot-extrusion 3D printing. *Food Hydrocolloids, 105*, 105812. https://doi.org/10.1016/j.foodhyd.2020.105812

Long, J., Etxeberria, A. E., Nand, A. V., Bunt, C. R., Ray, S., & Seyfoddin, A. (2019). A 3D printed chitosan-pectin hydrogel wound dressing for lidocaine hydrochloride delivery. *Materials Science and Engineering C: Materials for Biological Applications, 104*, 109873. https://doi.org/10.1016/j.msec.2019.109873

Lunt, J. (1998). Large-scale production, properties and commercial applications of polylactic acid polymers. *Polymer Degradation and Stability, 59*(1), 145–152. https://doi.org/10.1016/S0141-3910(97)00148-1

Maniglia, B. C., Lima, D. C., Matta Junior, M. D., Le-Bail, P., Le-Bail, A., & Augusto, P. E. D. (2019). Hydrogels based on ozonated cassava starch: effect of ozone processing and gelatinization conditions on enhancing 3D-printing applications. *International Journal of Biological Macromolecules, 138*, 1087–1097. https://doi.org/10.1016/j.ijbiomac.2019.07.124

Mantihal, S., Kobun, R., & Lee, B.-B. (2020). 3D food printing of as the new way of preparing food: a review. *International Journal of Gastronomy and Food Science, 22*, 100260. https://doi.org/10.1016/j.ijgfs.2020.100260

Markstedt, K., Escalante, A., Toriz, G., & Gatenholm, P. (2017). Biomimetic inks based on cellulose nanofibrils and cross-linkable xylans for 3D printing. *ACS Applied Materials and Interfaces, 9*(46), 40878–40886. https://doi.org/10.1021/acsami.7b13400

Markstedt, K., Mantas, A., Tournier, I., Martinez Avila, H., Hagg, D., & Gatenholm, P. (2015). 3D bioprinting human chondrocytes with nanocellulose-alginate bioink for cartilage tissue engineering applications. *Biomacromolecules, 16*(5), 1489–1496. https://doi.org/10.1021/acs.biomac.5b00188

Matijašić, G., Gretić, M., Vinčić, J., Poropat, A., Cuculić, L., & Rahelić, T. (2019). Design and 3D printing of multi-compartmental PVA capsules for drug delivery. *Journal of Drug Delivery Science and Technology, 52*, 677–686. https://doi.org/10.1016/j.jddst.2019.05.037

Melchels, F. P., Feijen, J., & Grijpma, D. W. (2009). A poly(D,L-lactide) resin for the preparation of tissue engineering scaffolds by stereolithography. *Biomaterials, 30*(23–24), 3801–3809. https://doi.org/10.1016/j.biomaterials.2009.03.055

Na, K., Shin, S., Lee, H., Shin, D., Baek, J., Kwak, H., Park, M., Shin, J., & Hyun, J. (2018). Effect of solution viscosity on retardation of cell sedimentation in DLP 3D printing of gelatin methacrylate/silk fibroin bioink. *Journal of Industrial and Engineering Chemistry, 61*, 340–347. https://doi.org/10.1016/j.jiec.2017.12.032

Ng, W. L., Yeong, W. Y., & Naing, M. W. (2016). Polyelectrolyte gelatin-chitosan hydrogel optimized for 3D bioprinting in skin tissue engineering. *International Journal of Bioprinting, 2*(0). https://doi.org/10.18063/ijb.2016.01.009

Page, D. H., & El-Hosseiny, F. (1983). Mechanical properties of single wood pulp fibres. *Journal of Pulp and Paper Science, 9*(4), 99–100.

Palaganas, N. B., Mangadlao, J. D., de Leon, A. C. C., Palaganas, J. O., Pangilinan, K. D., Lee, Y. J., & Advincula, R. C. (2017). 3D printing of photocurable cellulose nanocrystal composite for fabrication of complex architectures via stereolithography. *ACS Applied Materials and Interfaces, 9*(39), 34314–34324. https://doi.org/10.1021/acsami.7b09223

Parker, S. T., Domachuk, P., Amsden, J., Bressner, J., Lewis, J. A., Kaplan, D. L., & Omenetto, F. G. (2009). Biocompatible silk printed optical waveguides. *Advanced Materials, 21*(23), 2411–2415. https://doi.org/10.1002/adma.200801580

Polamaplly, P., Cheng, Y., Shi, X., Manikandan, K., Zhang, X., Kremer, G. E., & Qin, H. (2019). 3D printing and characterization of hydroxypropyl methylcellulose and methylcellulose for biodegradable support structures. *Polymer, 173*, 119–126. https://doi.org/10.1016/j.polymer.2019.04.013

Poletto, M., Börjesson, M., & Westman, G. (2015). Crystalline nanocellulose — preparation, modification, and properties. In *Cellulose, fundamental aspects and current trends*. IntechOpen. https://doi.org/10.5772/61899

Products, C. E. C. E. C. (2008). *Active ingredients used in cosmetics: safety survey*. Council of Europe Pub. https://books.google.it/books?id=tWJLUWT06BEC.

Puppi, D., Braccini, S., Ranaudo, A., & Chiellini, F. (2020). Poly(3-hydroxybutyrate-co-3-hydroxyexanoate) scaffolds with tunable macro- and microstructural features by additive manufacturing. *Journal of Biotechnology, 308*, 96–107. https://doi.org/10.1016/j.jbiotec.2019.12.005

Rafael, A. (2010). Poly(lactic acid): synthesis, structures, properties, processing, and applications. In *Wiley series on polymer engineering and technology*. John Wiley & Sons.

Rees, A., Powell, L. C., Chinga-Carrasco, G., Gethin, D. T., Syverud, K., Hill, K. E., & Thomas, D. W. (2015). 3D bioprinting of carboxymethylated-periodate oxidized nanocellulose constructs for wound dressing applications. *BioMed Research International, 2015*, 925757. https://doi.org/10.1155/2015/925757

Ren, J. (2010). *Biodegradable poly(lactic acid): Synthesis, modification, processing and applications*. Springer.

Rigotti, D., Checchetto, R., Tarter, S., Caretti, D., Rizzuto, M., Fambri, L., & Pegoretti, A. (2019). Polylactic acid-lauryl functionalized nanocellulose nanocomposites: microstructural, thermo-mechanical and gas transport properties. *Express Polymer Letters, 13*(10), 858–876. https://doi.org/10.3144/expresspolymlett.2019.75

Rigotti, D., Fambri, L., & Pegoretti, A. (2018). Polyvinyl alcohol reinforced with carbon nanotubes for fused deposition modeling. *Journal of Reinforced Plastics and Composites, 37*(10), 716–727. https://doi.org/10.1177/0731684418761224

Rigotti, D., Pegoretti, A., Miotello, A., & Checchetto, R. (2020). Interfaces in biopolymer nanocomposites: their role in the gas barrier properties and kinetics of residual solvent desorption. *Applied Surface Science, 507*. https://doi.org/10.1016/j.apsusc.2019.145066

Rodriguez, M. J., Dixon, T. A., Cohen, E., Huang, W., Omenetto, F. G., & Kaplan, D. L. (2018). 3D freeform printing of silk fibroin. *Acta Biomaterialia, 71*, 379–387. https://doi.org/10.1016/j.actbio.2018.02.035

Rojas-Martínez, L. E., Flores-Hernandez, C. G., López-Marín, L. M., Martinez-Hernandez, A. L., Thorat, S. B., Reyes Vasquez, C. D., Del Rio-Castillo, A. E., & Velasco-Santos, C. (2020). 3D printing of PLA composites scaffolds reinforced with keratin and chitosan: effect of geometry and structure. *European Polymer Journal, 141*. https://doi.org/10.1016/j.eurpolymj.2020.110088

Sadeghianmaryan, A., Naghieh, S., Sardroud, H. A., Yazdanpanah, Z., Soltani, Y. A., Sernaglia, J., & Chen, X. (2020). Extrusion-based printing of chitosan scaffolds and their in vitro characterization for cartilage tissue engineering. *International Journal of Biological Macromolecules, 164*, 3179–3192. https://doi.org/10.1016/j.ijbiomac.2020.08.180

Salmoria, G. V., Cardenuto, M. R., Roesler, C. R. M., Zepon, K. M., & Kanis, L. A. (2016). PCL/Ibuprofen implants fabricated by selective laser sintering for orbital repair. *Procedia CIRP, 49*, 188−192. https://doi.org/10.1016/j.procir.2015.11.013

Saska, S., Pires, L. C., Cominotte, M. A., Mendes, L. S., de Oliveira, M. F., Maia, I. A., da Silva, J. V. L., Ribeiro, S. J. L., & Cirelli, J. A. (2018). Three-dimensional printing and in vitro evaluation of poly(3-hydroxybutyrate) scaffolds functionalized with osteogenic growth peptide for tissue engineering. *Materials Science and Engineering C: Materials for Biological Applications, 89*, 265−273. https://doi.org/10.1016/j.msec.2018.04.016

Series, T. T. I. B., Rama, N. (2020). *Silk: materials, processes, and applications*. Woodhead Publishing.

Shaler, A. J. (1949). Seminar on the kinetics of sintering. *JOM, 1*(11), 796−813. https://doi.org/10.1007/BF03398399

Shen, X., Chu, M., Hariri, F., Vedula, G., & Naguib, H. E. (2020). Binder jetting fabrication of highly flexible and electrically conductive graphene/PVOH composites. *Additive Manufacturing, 36*. https://doi.org/10.1016/j.addma.2020.101565

Shen, Y., Tang, H., Huang, X., Hang, R., Zhang, X., Wang, Y., & Yao, X. (2020). DLP printing photocurable chitosan to build bio-constructs for tissue engineering. *Carbohydrate Polymers, 235*. https://doi.org/10.1016/j.carbpol.2020.115970

Shishkovsky, I., & Scherbakov, V. (2012). Selective laser sintering of biopolymers with micro and nano ceramic additives for medicine. *Physics Procedia, 39*, 491−499. https://doi.org/10.1016/j.phpro.2012.10.065

Shor, L., Guceri, S., Wen, X., Gandhi, M., & Sun, W. (2007). Fabrication of three-dimensional polycaprolactone/hydroxyapatite tissue scaffolds and osteoblast-scaffold interactions in vitro. *Biomaterials, 28*(35), 5291−5297. https://doi.org/10.1016/j.biomaterials.2007.08.018

Singh, S., Singh, G., Prakash, C., Ramakrishna, S., Lamberti, L., & Pruncu, C. I. (2020). 3D printed biodegradable composites: an insight into mechanical properties of PLA/chitosan scaffold. *Polymer Testing, 89*. https://doi.org/10.1016/j.polymertesting.2020.106722

Siqueira, G., Kokkinis, D., Libanori, R., Hausmann, M. K., Gladman, A. S., Neels, A., Tingaut, P., Zimmermann, T., Lewis, J. A., & Studart, A. R. (2017). Cellulose nanocrystal inks for 3D printing of textured cellular architectures. *Advanced Functional Materials, 27*(12), 1604619. https://doi.org/10.1002/adfm.201604619

Sommer, M. R., Schaffner, M., Carnelli, D., & Studart, A. R. (2016). 3D printing of hierarchical silk fibroin structures. *ACS Applied Materials and Interfaces, 8*(50), 34677−34685. https://doi.org/10.1021/acsami.6b11440

Sun, L., Parker, S. T., Syoji, D., Wang, X., Lewis, J. A., & Kaplan, D. L. (2012). Direct-write assembly of 3D silk/hydroxyapatite scaffolds for bone co-cultures. *Advanced Healthcare Materials, 1*(6), 729−735. https://doi.org/10.1002/adhm.201200057

Szebenyi, G., Czigany, T., Magyar, B., & Karger-Kocsis, J. (2017). 3D printing-assisted interphase engineering of polymer composites: concept and feasibility. *Express Polymer Letters, 11*(7), 525−530. https://doi.org/10.3144/expresspolymlett.2017.50

Tardajos, M. G., Cama, G., Dash, M., Misseeuw, L., Gheysens, T., Gorzelanny, C., Coenye, T., & Dubruel, P. (2018). Chitosan functionalized poly-ε-caprolactone electrospun fibers and 3D printed scaffolds as antibacterial materials for tissue engineering applications. *Carbohydrate Polymers, 191*, 127−135. https://doi.org/10.1016/j.carbpol.2018.02.060

Thibaut, C., Denneulin, A., Rolland du Roscoat, S., Beneventi, D., Orgeas, L., & Chaussy, D. (2019). A fibrous cellulose paste formulation to manufacture structural parts using 3D printing by extrusion. *Carbohydrate Polymers, 212*, 119−128. https://doi.org/10.1016/j.carbpol.2019.01.076

Tonda-Turo, C., Carmagnola, I., Chiappone, A., Feng, Z., Ciardelli, G., Hakkarainen, M., & Sangermano, M. (2020). Photocurable chitosan as bioink for cellularized therapies towards personalized scaffold architecture. *Bioprinting, 18*. https://doi.org/10.1016/j.bprint.2020.e00082

Travieso-Rodriguez, J. A., Zandi, M. D., Jerez-Mesa, R., & Lluma-Fuentes, J. (2020). Fatigue behavior of PLA-wood composite manufactured by fused filament fabrication. *Journal of Materials Research and Technology, 9*(4), 8507–8516. https://doi.org/10.1016/j.jmrt.2020.06.003

Valentini, F., Dorigato, A., Rigotti, D., & Pegoretti, A. (2019). Polyhydroxyalkanoates/fibrillated nanocellulose composites for additive manufacturing. *Journal of Polymers and the Environment, 27*(6), 1333–1341. https://doi.org/10.1007/s10924-019-01429-8

Van Bael, S., Desmet, T., Chai, Y. C., Pyka, G., Dubruel, P., Kruth, J. P., & Schrooten, J. (2013). In vitro cell-biological performance and structural characterization of selective laser sintered and plasma surface functionalized polycaprolactone scaffolds for bone regeneration. *Materials Science and Engineering C: Materials for Biological Applications, 33*(6), 3404–3412. https://doi.org/10.1016/j.msec.2013.04.024

Venkataraman, N. (2000). Feedstock material property – process relationships in fused deposition of ceramics (FDC). *Rapid Prototyping Journal, 6*(4), 244–253. https://doi.org/10.1108/13552540010373344

Verdonck, P., Nyssen, M., Haueisen, J., Hoeges, S., Lindner, M., Fischer, H., Meiners, W., & Wissenbach, K. (2009). *Manufacturing of bone substitute implants using Selective Laser Melting* (pp. 2230–2234). Springer Berlin Heidelberg.

Wang, Z., Xu, J., Lu, Y., Hu, L., Fan, Y., Ma, J., & Zhou, X. (2017). Preparation of 3D printable micro/nanocellulose-polylactic acid (MNC/PLA) composite wire rods with high MNC constitution. *Industrial Crops and Products, 109*, 889–896. https://doi.org/10.1016/j.indcrop.2017.09.061

Wani, S. U. D., Gautam, S. P., Qadrie, Z. L., & Gangadharappa, H. V. (2020). Silk fibroin as a natural polymeric based bio-material for tissue engineering and drug delivery systems-A review. *International Journal of Biological Macromolecules, 163*, 2145–2161. https://doi.org/10.1016/j.ijbiomac.2020.09.057

Wei, G., & Ma, P. X. (2004). Structure and properties of nano-hydroxyapatite/polymer composite scaffolds for bone tissue engineering. *Biomaterials, 25*(19), 4749–4757. https://doi.org/10.1016/j.biomaterials.2003.12.005

Williams, J. M., Adewunmi, A., Schek, R. M., Flanagan, C. L., Krebsbach, P. H., Feinberg, S. E., Hollister, S. J., & Das, S. (2005). Bone tissue engineering using polycaprolactone scaffolds fabricated via selective laser sintering. *Biomaterials, 26*(23), 4817–4827. https://doi.org/10.1016/j.biomaterials.2004.11.057

Wiria, F. E., Leong, K. F., Chua, C. K., & Liu, Y. (2007). Poly-epsilon-caprolactone/hydroxyapatite for tissue engineering scaffold fabrication via selective laser sintering. *Acta Biomaterialia, 3*(1), 1–12. https://doi.org/10.1016/j.actbio.2006.07.008

Wiśniewska, M., Ostolska, I., Szewczuk-Karpisz, K., Chibowski, S., Terpiłowski, K., Gun'ko, V. M., & Zarko, V. I. (2015). Investigation of the polyvinyl alcohol stabilization mechanism and adsorption properties on the surface of ternary mixed nanooxide AST 50 (Al_2O_3–SiO_2–TiO_2). *Journal of Nanoparticle Research, 17*(1), 12. https://doi.org/10.1007/s11051-014-2831-2

Wong, J. Y., & Pfahnl, A. C. (2014). 3D printing of surgical instruments for long-duration space missions. *Aviation Space and Environmental Medicine, 85*(7), 758–763. https://doi.org/10.3357/ASEM.3898.2014

Woodruff, M. A., & Hutmacher, D. W. (2010). The return of a forgotten polymer—polycaprolactone in the 21st century. *Progress in Polymer Science, 35*(10), 1217–1256. https://doi.org/10.1016/j.progpolymsci.2010.04.002

Wu, D., Spanou, A., Diez-Escudero, A., & Persson, C. (2020). 3D-printed PLA/HA composite structures as synthetic trabecular bone: a feasibility study using fused deposition modeling. *Journal of Mechanical Behaviors and Biomedical Materials, 103*, 103608. https://doi.org/10.1016/j.jmbbm.2019.103608

Wypych, G., & Wypych, G. (2016). PCL poly(ϵ-caprolactone). In *Handbook of polymers* (Second Edition, pp. 323–325). ChemTec Publishing. https://doi.org/10.1016/B978-1-895198-92-8.50099-9

Yeong, W. Y., Sudarmadji, N., Yu, H. Y., Chua, C. K., Leong, K. F., Venkatraman, S. S., Boey, Y. C., & Tan, L. P. (2010). Porous polycaprolactone scaffold for cardiac tissue engineering fabricated by selective laser sintering. *Acta Biomaterialia, 6*(6), 2028–2034. https://doi.org/10.1016/j.actbio.2009.12.033

Zafar, M. S., Khurshid, Z., Khan, A. S., Najeeb, S., Sefat, F., Kumar, N., Ali, S., Kumar, B., Zafar, M. S., & Khurshid, Z. (2020). 5 - hydroxyapatite and nanocomposite implant coatings. In *Dental implants* (pp. 69–92). Woodhead Publishing. https://doi.org/10.1016/B978-0-12-819586-4.00005-6

Zafeiris, K., Brasinika, D., Karatza, A., Koumoulos, E. P., Karoussis, I. K., Kyriakidou, K., & Charitidis, C. A. (2020). Additive manufacturing of hydroxyapatite – chitosan – Genipin composite scaffolds for bone tissue engineering applications. *Materials Science and Engineering: C*. https://doi.org/10.1016/j.msec.2020.111639

Zeng, X., Chen, H., Chen, L., & Zheng, B. (2020). Insights into the relationship between structure and rheological properties of starch gels in hot-extrusion 3D printing. *Food Chemistry*, 128362. https://doi.org/10.1016/j.foodchem.2020.128362

Zhang, X., Cama, G., Mogosanu, D. E., Houben, A., & Dubruel, P. (2017). 3: synthetic biodegradable medical polyesters: poly-ε-caprolactone. In *Science and principles of biodegradable and bioresorbable medical polymers* (pp. 79–105). Woodhead Publishing. https://doi.org/10.1016/B978-0-08-100372-5.00003-9

Zhao, C. Q., Liu, W. G., Xu, Z. Y., Li, J. G., Huang, T. T., Lu, Y. J., Huang, H. G., & Lin, J. X. (2020). Chitosan ducts fabricated by extrusion-based 3D printing for soft-tissue engineering. *Carbohydrate Polymers, 236*, 116058. https://doi.org/10.1016/j.carbpol.2020.116058

Zhao, Y. Q., Yang, J. H., Ding, X., Ding, X., Duan, S., & Xu, F. J. (2020). Polycaprolactone/polysaccharide functional composites for low-temperature fused deposition modelling. *Bioactive Materials, 5*(2), 185–191. https://doi.org/10.1016/j.bioactmat.2020.02.006

Zheng, Z., Wu, J., Liu, M., Wang, H., Li, C., Rodriguez, M. J., Li, G., Wang, X., & Kaplan, D. L. (2018). 3D bioprinting of self-standing silk-based bioink. *Advanced Healthcare Materials, 7*(6), e1701026. https://doi.org/10.1002/adhm.201701026

Zhong, N., Dong, T., Chen, Z., Guo, Y., Shao, Z., & Zhao, X. (2019). A novel 3D-printed silk fibroin-based scaffold facilitates tracheal epithelium proliferation in vitro. *Journal of Biomaterials Applications, 34*(1), 3–11. https://doi.org/10.1177/0885328219845092

Zhou, Z., Lennon, A., Buchanan, F., McCarthy, H. O., & Dunne, N. (2020). Binder jetting additive manufacturing of hydroxyapatite powders: effects of adhesives on geometrical accuracy and green compressive strength. *Additive Manufacturing, 36*, 101645. https://doi.org/10.1016/j.addma.2020.101645

Further readings

Alfonso Jimenez, M. P., & Ruseckaite, R. (2013). Poly(lactic acid): Science and technology processing, properties, additives and applications. In *Polymer chemistry series No. 12*. RSC Publishing.

Chen, C. S., Zeng, F., Xiao, X., Wang, Z., Li, X. L., Tan, R. W., Liu, W. Q., Zhang, Y. S., She, Z. D., & Li, S. J. (2018). Three-dimensionally printed silk-sericin-based hydrogel scaffold: a promising visualized dressing material for real-time monitoring of wounds. *ACS Applied Materials and Interfaces, 10*(40), 33879–33890. https://doi.org/10.1021/acsami.8b10072

Duan, B., & Wang, M. (2010). Encapsulation and release of biomolecules from Ca–P/PHBV nanocomposite microspheres and three-dimensional scaffolds fabricated by selective laser sintering. *Polymer Degradation and Stability, 95*(9), 1655–1664. https://doi.org/10.1016/j.polymdegradstab.2010.05.022

Elomaa, L., Keshi, E., Sauer, I. M., & Weinhart, M. (2020). Development of GelMA/PCL and dECM/PCL resins for 3D printing of acellular in vitro tissue scaffolds by stereolithography. *Materials Science and Engineering: C*, 110958. https://doi.org/10.1016/j.msec.2020.110958

Eosoly, S., Vrana, N. E., Lohfeld, S., Hindie, M., & Looney, L. (2012). Interaction of cell culture with composition effects on the mechanical properties of polycaprolactone-hydroxyapatite scaffolds fabricated via selective laser sintering (SLS). *Materials Science and Engineering: C, 32*(8), 2250–2257. https://doi.org/10.1016/j.msec.2012.06.011

Koski, C., & Bose, S. (2019). Effects of amylose content on the mechanical properties of starch-hydroxyapatite 3D printed bone scaffolds. *Additive Manufacturing, 30*, 100817. https://doi.org/10.1016/j.addma.2019.100817

Salmoria, G. V., Klauss, P., Paggi, R. A., Kanis, L. A., & Lago, A. (2009). Structure and mechanical properties of cellulose based scaffolds fabricated by selective laser sintering. *Polymer Testing, 28*(6), 648–652. https://doi.org/10.1016/j.polymertesting.2009.05.008

Wang, Y., Yi, S., Lu, R., Sameen, D. E., Ahmed, S., Dai, J., ... Liu, Y. (2020). Preparation, characterization, and 3D printing verification of chitosan/halloysite nanotubes/tea polyphenol nanocomposite films. *International Journal of Biological Macromolecules, 166*, 32–44. https://doi.org/10.1016/j.ijbiomac.2020.09.253

Zhang, J., Allardyce, B. J., Rajkhowa, R., Kalita, S., Dilley, R. J., Wang, X., & Liu, X. (2019). Silk particles, microfibres and nanofibres: a comparative study of their functions in 3D printing hydrogel scaffolds. *Materials Science and Engineering C: Materials for Biological Applications, 103*, 109784. https://doi.org/10.1016/j.msec.2019.109784

Zhang, J., Allardyce, B. J., Rajkhowa, R., Zhao, Y., Dilley, R. J., Redmond, S. L., Wang, X., & Liu, X. (2018). 3D printing of silk particle-reinforced chitosan hydrogel structures and their properties. *ACS Biomaterials Science and Engineering, 4*(8), 3036–3046. https://doi.org/10.1021/acsbiomaterials.8b00804

Production of biodegradable films and blends from proteins

G. Rajeshkumar, S. Arvindh Seshadri, R. Ronia Richelle, K. Madhu Mitha and V. Abinaya
Department of Mechanical Engineering, PSG Institute of Technology and Applied Research, Coimbatore, Tamil Nadu, India

1. Introduction

Materials that are derived from petroleum resources are used in a large number of applications, and in different forms. Among these, an important industry that uses petroleum-based products is the packaging industry, which uses plastics in the form of films. Petroleum-based plastics find applications in the packaging industry because of their meritorious properties such as easy formability, easy manufacturability, lightweight, superior chemical resistance, and good mechanical and barrier properties like tensile strength, aroma transmission, and oxygen permeability (Abdul Khalil et al., 2018; Pirsa & Sharifi, 2020). However, there are some serious issues faced when petroleum-based polymer films are used. These include: an increase in carbon footprint, nonbiodegradable and nonrecyclable nature, toxic when burnt, difficult to dispose of etc. (Rajeshkumar et al., 2021; Saurabh et al., 2013). Therefore, researchers have turned to biodegradable polymers in order to manufacture films and blends for packaging applications. In addition to packaging applications, protein-based films are also conventional materials in applications such as antimicrobial coatings, tissue engineering and wound repair scaffolds, sensing and diagnostic devices, and optical applications (Gopalakrishnan et al., 2021; Krekic et al., 2019; Nagarjun et al., 2021; Rajeshkumar et al., 2020). These polymers are safe as they decompose into simple chemical components such as CO_2, water, methane, and biomass, under suitable conditions of temperature, humidity, and presence of oxygen. Additionally, they do not leave behind any toxic by-products (Asadi & Pirsa, 2020; Brodhagen et al., 2015; Sintim et al., 2019).

Biodegradable films and blends can be produced from various sources, namely: polysaccharides, proteins, bioplastics, lipids, etc. Of these, plant and animal proteins are used predominantly. This is due to their widespread availability, biodegradability, nutritive value, and good film-formation ability. Additionally, the mechanical and barrier properties of protein-based films are better than those of polysaccharides or lipid-based films (Kashiri et al., 2016; Nuanmano et al., 2015; Saremnezhad et al., 2011). A chemical potential is generated due to the distribution of charge, polar and nonpolar amino acids. Thus, a well-adhered film is produced by the resulting interactions between the molecules. Different interactions such as hydrogen bonding, covalent

Biodegradable Polymers, Blends and Composites. https://doi.org/10.1016/B978-0-12-823791-5.00003-X

bonding, electrostatic interactions, and van der Waals forces help in the formation of protein films and blends (Baldwin et al., 2011). In addition, there is another advantage of using proteins to manufacture biodegradable films and blends. Due to the presence of different amino acid functional groups, they have several sites for chemical interaction. This enables researchers to tailor and enhance the properties of protein-based films and blends (Barone & Schmidt, 2006).

Two major methods can be used to produce biodegradable films and blends: wet and dry processes. In the wet process, which is also known as solvent casting, the biopolymers are dissolved in a suitable film-forming solution or dispersion before being cast over a base. In the dry process, such as thermoplastic extrusion, the biopolymer is plasticized and heated above its glass transition temperature under low moisture (Cuq et al., 1998; Khwaldia et al., 2004). The properties of various biodegradable blends and films produced using these methods are reviewed in this chapter.

2. Protein sources to produce biodegradable films and blends

Proteins are a group of molecules (and also a polymer), whose fundamental constituents are units known as amino acids. The amino acids are connected in such a way that two monomers, one that provides H^+ and other OH^-, are adjacent to each other. The difference lies in the functional group R. Proteins have nearly 20 different types of monomers and their order and number can be altered to obtain a wide range of proteins (Pirsa & Sharifi, 2020).

Proteins are biopolymers that exhibit multifunctional properties due to their complex structure and high intermolecular binding ability that is possible through various bond types. Proteins form a strong, cohesive film with viscoelastic properties when they interact with other molecules at the interface. The film-forming ability of proteins relies on various factors such as charge, thermal stability, flexibility, molecular weight, etc. (Danijela et al., 2015; Koshy et al., 2015).

A large number of proteins have been utilized to produce biodegradable films and blends. These can be classified under three categories: animal proteins, plant proteins, and legume seed proteins (Fig. 24.1). Some of the animal proteins used include gelatin, collagen, myofibrillar proteins, whey proteins, and casein. Gelatin and collagen are obtained from waste by-products of the meat, fish, and poultry industries. Gelatin possesses superior barrier properties and good mechanical properties; however it is highly sensitive to humidity (Liu et al., 2017; Maryam Adilah & Nur Hanani, 2016). Casein, which is derived from milk, can be utilized to produce films which are water insoluble, opaque, and have low permeability toward oxygen (Grujic et al., 2017). Whey protein (a by-product of cheese production), can be used to prepare films with superior barrier properties. However, it has low strength and is highly permeable toward water (Galus & Kadzińska, 2016). Myofibrillar proteins are derived from meat; and biodegradable films and blends with good tensile properties can be produced from them (Lee et al., 2015).

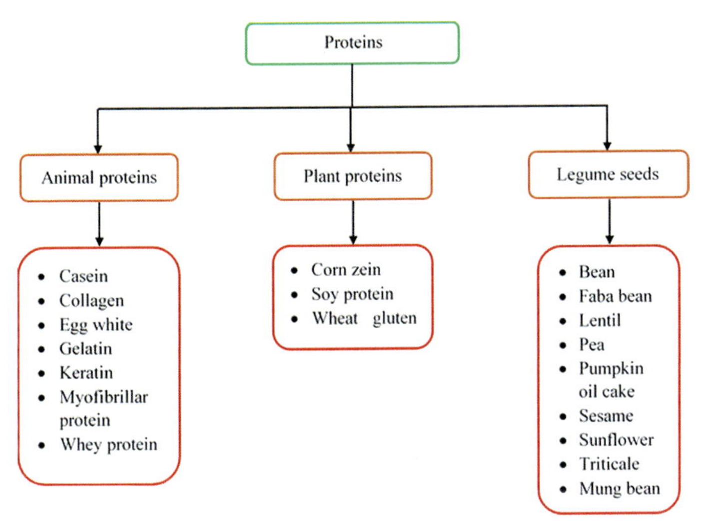

Figure 24.1 Different types of proteins based on their sources.

Proteins derived from plants include corn zein, wheat gluten, and soy protein. Wheat gluten can be utilized to prepare films with good barrier properties, but they are brittle in nature and highly sensitive to humidity (Rocca-Smith et al., 2016). Corn zein is a by-product of corn starch production and exhibits favorable properties such as good film-forming capacity, mechanical, and barrier properties (Kashiri et al., 2016). Soy protein isolate is obtained from soybean oil production and is used to make films due to its wide availability, low cost, and high protein content. Films made from soy protein exhibit good mechanical and superior barrier properties (Wu et al., 2021). In addition, proteins that are derived from legume seeds such as pea, lentil, faba bean, sunflower, mung bean etc., are also used to produce biodegradable films and blends as they possess good film-forming ability (Calva-Estrada, Jiménez-Fernández, & Lugo-Cervantes, 2019; Moghadam et al., 2020). The properties, advantages, and disadvantages of protein-based films derived from different sources are detailed in Table 24.1.

3. Production of biodegradable films and blends

Biodegradable films and blends can be prepared using two different techniques: dry and wet processes (Fig. 24.2). In the dry process, the proteins are either extruded or compression molded, while in the wet process, which is also known as solution/solvent casting, the proteins are dissolved in a suitable solvent and then the solvent is evaporated to obtain the film (Hernandez-Lzquierdo & Krochta, 2009). The proteins must be denatured initially with the addition of suitable chemicals or using heat in order to enable the better extension of their structure. This will enhance an interaction between the protein chains, which will help in the formation of strong, cohesive protein films (Dhall, 2013).

Table 24.1 Properties, advantages, and disadvantages of protein-based films from different sources (Calva-Estrada, Jiménez-Fernández, & Lugo-Cervantes, 2019).

Type of protein	Protein	Tensile strength (MPa)	Elongation at break (%)	Water vapor permeability (10^{-10} g m/Pa s m^2)	Advantages	Disadvantages
Animal protein	Albumen	6.5–9.5	85–155	–	Good transparency and elasticity; can be produced at low temperatures	Highly permeable to water vapor
	Casein	2–77	2–130	1.6–11	Good mechanical properties; low oxygen permeability	Opaque; highly permeable to water vapor
	Gelatin	10–80	20–140	0.3–36	Superior mechanical and barrier properties; highly elastic in nature with good film-forming ability	Highly permeable to water vapor
	Myofibrillar protein	5–12	35–175	0.6–35	Good tensile properties; good film-forming ability	Highly permeable to water vapor; brittle in nature
	Whey protein	0.8–16	13–196	2–36	Superior barrier properties; good film-forming ability	Highly permeable to water vapor; poor tensile properties
Plant proteins	Corn zein	18–33	2.6–3.6	0.3–1.4	Good film-forming ability; superior tensile properties; low permeability toward water and oxygen	Brittle in nature
	Soy protein	2–7	14–170	8–19.5	Good film-forming ability	Highly permeable to water vapor; poor mechanical properties
	Wheat gluten	5–18	2–90	7.33	Good film-forming ability; good barrier properties	Highly permeable to water vapor; brittle in nature
Legume seed proteins	Faba bean protein	0.5–4	40–200	1.6–2.9	Color and semitransparency; good film-forming ability	Brittle in nature
	Lentil protein	4.24	58.22	3.09		
	Pea protein	6.3	37	11–20		
	Mung bean	3.33	81.18	2.99		

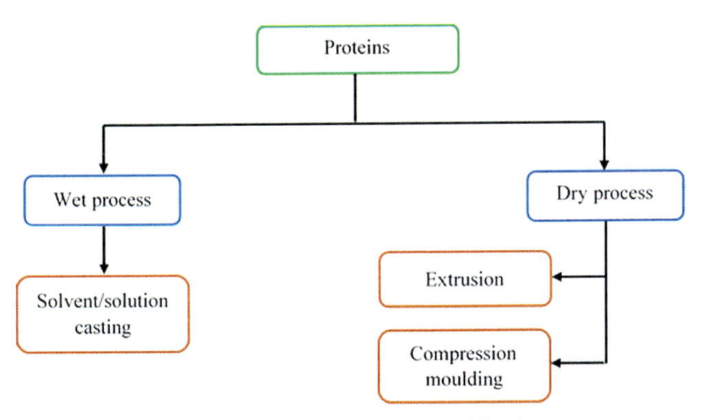

Figure 24.2 Methods to produce biodegradable films and blends.

There are three major steps that take place during the production of biodegradable films and blends from proteins: (1) the breakage of low-energy intermolecular bonds that stabilize the proteins, (2) the arrangement and orientation of polymer chains, and (3) the formation of a 3-D network stabilized by new bonds and interactions (Jerez et al., 2007; Pommet et al., 2005).

4. Solution/solvent casting

Solvent casting is a method that is predominantly used to prepare biodegradable films and blends, especially in laboratories by researchers. This method is preferred as it requires simple equipment like Petri dishes, and is efficient and cost-effective. In solvent casting, raw materials including proteins and additives are dispersed and dissolved in a suitable solvent such as water. In addition to the raw materials, other substances such as plasticizers, cross-linking agents, nanoparticles, antimicrobial agents, etc., can be incorporated to improve certain characteristics of films and blends. Some parameters such as pH, time, and temperature are controlled to obtain films and blends with optimum properties (Chonhenchob et al., 2007). This is followed by removal of the solvent that was used to dissolve the proteins and other substances. The films are dried at suitable temperature and humidity to obtain free-standing films with the required properties. The rate and temperature of drying have a major influence on the mechanical properties of the film. It is better to use infrared rays for drying rather than conventional heat sources (Srinivasa et al., 2004). In the last step, the protein-based films are conditioned, by storing them under specific humidity conditions in order to have a uniform moisture content for further analysis, as they are permeable to water vapor (Saurabh et al., 2013). The stages involved in the solvent casting method are depicted in Fig. 24.3.

Nogueira and Martins (2019) prepared biodegradable films and blends from different proteins using a solvent casting method. Individual films were produced using hake protein, wheat gluten, and corn zein, while blends were prepared

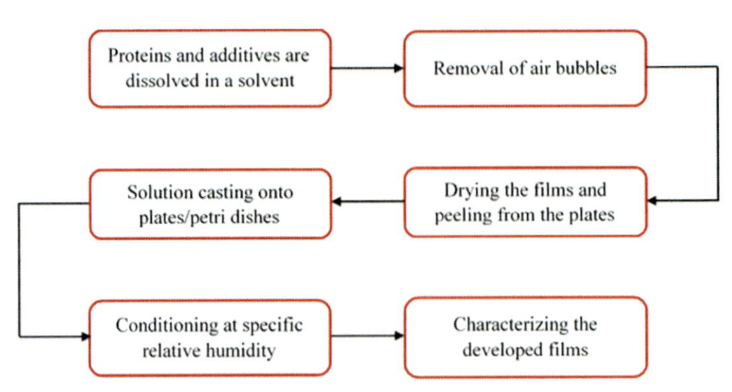

Figure 24.3 Stages involved in the solvent casting method for producing biodegradable films/blends from proteins.

by combining hake protein and wheat gluten (H/G), and corn zein and wheat gluten (C/G). Appropriate concentrations of each of the three proteins were dissolved in a suitable solvent. Further, the solutions were homogenized and mechanically stirred continuously. Once the proteins were completely dissolved, the solutions were spread on Petri dishes and dried using a hot-air oven. The drying process was followed by storing the films at 55% relative humidity and 25°C for subsequent characterization. A similar process was followed for the preparation of blends, where the blends were prepared in equal proportions (1:1). The corn zein films exhibited poor mechanical properties and high solubility, while hake protein films were resistant to water. Films made from wheat gluten had higher elongation at break. The authors noted that the properties of blends were intermediate when compared to the properties of their individual constituent films. The blend prepared using hake protein and wheat gluten was found to possess homogeneous structure, good mechanical and thermal properties, and transparency, due to which, the authors suggested that it could be used for packaging applications.

Khanzadi et al. (2015) produced biodegradable films blends using whey protein (WP) and pullulan (PL), by varying their respective ratios (30:70, 50:50, and 70: 30% w/w). The films were produced with the help of a solvent casting method. Whey protein was dissolved in deionized water and heated for half an hour at 90°C. Similarly, pullulan was dissolved in deionized water separately. Further, the solutions were cooled and mixed at the required ratios. Glycerol, which acted as a plasticizer, was mixed into the solution and the resultant solution was homogenized. In order to enhance the properties of the blends, different percentages of beeswax (BW) (0%–30% $w/w_{glycerol}$) were added and a rotor stator homogenizer was used to homogenize the solution. This was followed by transferring the solution to Teflon plates. The films were dried at 25°C for 2 days in a hot-air oven and the films were conditioned at 50% RH and 25°C for further analysis. Better physical and mechanical properties, that is, low water content, water permeability and water solubility, and high elongations, were obtained when higher ratios of whey protein and pullulan were used. The addition of beeswax led to an increase in tensile strength by 7 MPa, reduced water solubility by 12%, and enhanced color indices of the films. The ideal properties were obtained in films: WP70:PL30, containing 30% BW, which could be used as food-packaging materials.

Deng et al. (2018) produced biodegradable films of gelatin, corn zein, and a combination of the two (gelatin/corn zein), using a solvent casting method. To prepare the film-forming solution, 30% (w/v) of the protein was dissolved in 80% (v/v) aqueous acetic acid. For the blend, gelatin and corn zein were mixed in the ratio of 1:1. In the solvent casting process, the film-forming solutions were spread over aluminum foil, followed by air drying for 1 day at 25°C and 50% RH. The obtained films were subjected to various characterization tests. The melting points of corn zein, gelatin, and gelatin/corn zein films were 96.5, 94.16, and 90.59°C, respectively. All three films possessed a hydrophilic surface, with the contact angle of gelatin/corn zein film (53.5°C) between those of gelatin and corn zein films. The gelatin and corn zein were mixed heterogeneously in the blended film as corn zein particles prevented intramolecular aggregation of gelatin chains, as a result of which, it exhibited poor solvent resistance.

da Silva Pereira and colleagues (Vasconcelos da Silva Pereira et al., 2019) manufactured protein-based films from myofibrillar proteins under different processing conditions, in order to obtain films with ideal properties. The films were produced using a solvent casting method. Different amounts of protein and plasticizer (glycerol) were utilized during the preparation of the films. The films produced exhibited optimum properties under the following conditions: 1.13% (w/v) proteins, 35.96% (w/w) plasticizer, and 25.96°C drying temperature. The tensile strength, elongation, and water vapor permeability of the films under these conditions were 6.31 MPa, 160.38%, and 5.8×10^{-11} g m/Pa s m^2, respectively.

5. Dry processes (thermomechanical processing)

A major limitation of the solvent casting method is that it cannot be scaled up to industrial levels. Therefore, a dry process can be utilized to manufacture protein-based films by forming them under the application of heat and/or pressure (Balaguer et al., 2014). Extrusion and compression molding are two methods which have high efficacy and excellent potential for mass production of protein-based films, due to their shorter operation times. However, there is a difficulty faced when carrying out these processes. Proteins have complex and heterogeneous intermolecular interactions, which cause a lack of flow region during manufacturing (Hernandez-Izquierdo & Krochta, 2008). Hence, there is a need to use certain chemical agents during manufacturing, which will break these intermolecular bonds. Although extrusion and compression molding are well defined methods, the process parameters need to be optimized to produce protein-based films with optimum properties.

Guerrero et al. (2010) produced biodegradable films from soy protein using three different manufacturing methods: solvent casting, compression molding, and freeze drying followed by compression molding (Fig. 24.4). Glycerol, which is a plasticizer,

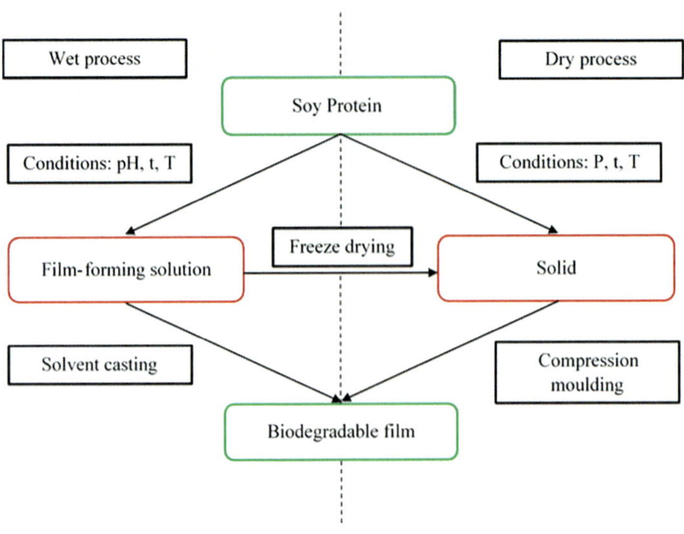

Figure 24.4 Production of soy protein-based films using different methods (Guerrero et al., 2010).

was also added to the films. In compression molding, soy protein and glycerol with different ratios [50/50, 60/40, and 70/30 (w/w)] were mixed. The mixtures were blended for 5 min and then hot pressed. The platens of the press were preheated to 150°C and subjected to a pressure of 12 MPa for 2 min. This was followed by the cooling of plates. The films were removed after the plates were cooled.

In solvent casting, the soy protein was dissolved in distilled water and stirred at 150 rpm and 80°C for 30 min. This was followed by the addition of glycerol and the solution was further stirred at the same conditions for another 30 min. The resulting solution was divided into two portions. The first one was transferred onto a Petri dish, while the second portion was freeze dried. The powder obtained after freeze drying was compression molded.

The protein films exhibited their best properties when the glycerol content was between 30% and 40%. Below 30% the films were brittle, while above 40% glycerol, the films were sticky. A higher amount of glycerol leads to lower tensile strength and higher elongation as glycerol reduces the interactions between protein chains and increases their mobility. Soy protein-based films produced using compression molding exhibited the best mechanical properties. This was followed by films obtained from freeze drying. Films that were produced using solvent casting had low tensile strength and elongation at break. The authors concluded that compression molding is better than the solvent casting method, as it produces films with optimum properties and it can also be used for mass production (Guerrero et al., 2010).

Nur Hanani et al. (2014) investigated the influence of processing temperature and pH on the mechanical and barrier properties of gelatin and corn oil composite films, produced through extrusion. The extrusion temperature was varied from 90 to 130°C, whereas pH was varied from 5.7 to 8.7. The films were produced using a twin-screw corotating extruder, with an aspect ratio (L/D) of 25:1 and screw diameter of 19 mm. The screw speed during the extrusion process was set at 300 rpm. The tensile strength of the obtained films increased and elongation decreased, with the increase in processing temperature. At 130°C, the highest tensile strength of 5.37 MPa was obtained. On the other hand, the elongation was maximum at 90°C (2.1%). It was also seen that the water vapor permeability of the films declined as temperature was increased from 90 to 130°C. The oxygen permeability of the films increased with an increase in the processing temperature. The gelatin films produced at a temperature of 120°C had the highest oxygen permeability and lowest water vapor permeability. The authors concluded that extrusion is a promising process for the manufacture gelatin films for food-packaging applications, however additional research is required to optimize the process.

6. Conclusion and future perspectives

The use of proteins derived from plant, animal, and legume seed sources has been shown to be promising in the development of biodegradable packaging films and blends for packaging applications. The wet method is more widely used to manufacture films than the dry method as these materials lack defined melting points and decompose when heated. Furthermore, in the wet process the parameters such as pH, time, and temperature are controlled to obtain films and blends with optimum properties. Because of the synergistic effect, biodegradable films provide a wide range of properties that are required by the packaging industry. A large number of protein sources are available all over the world, many of which have yet to be discovered. In this field, there is a huge research opportunity for researchers to develop novel biodegradable films and blends with favorable mechanical and barrier properties, which will assist the packaging industry. Undoubtedly, biodegradation offers an attractive route to environmental waste management.

References

Abdul Khalil, H. P. S., Banerjee, A., Saurabh, C. K., Tye, Y. Y., Suriani, A. B., Mohamed, A., Karim, A. A., Rizal, S., & Paridah, M. T. (2018). Biodegradable films for fruits and vegetables packaging application: preparation and properties. *Food Engineering Reviews, 10*(3), 139–153. https://doi.org/10.1007/s12393-018-9180-3

Asadi, S., & Pirsa, S. (2020). Production of biodegradable film based on polylactic acid, modified with lycopene pigment and TiO_2 and studying its physicochemical properties. *Journal of Polymers and the Environment, 28*(2), 433–444. https://doi.org/10.1007/s10924-019-01618-5

Balaguer, M. P., Gomez-Estaca, J., Cerisuelo, J. P., Gavara, R., & Hernandez-Munoz, P. (2014). Effect of thermo-pressing temperature on the functional properties of bioplastics made from a renewable wheat gliadin resin. *Lebensmittel-Wissenschaft und -Technologie- Food Science and Technology, 56*(1), 161–167. https://doi.org/10.1016/j.lwt.2013.10.035

Baldwin, E. A., Hagenmaier, R. D., & Bai, J. (2011). Edible coatings and films to improve food quality. In *Edible coatings and films to improve food quality, second edition* (second edition, pp. 1−417). CRC Press. https://www.taylorfrancis.com/books/e/9781420059663.

Barone, J. R., & Schmidt, W. F. (2006). Nonfood applications of proteinaceous renewable materials. *Journal of Chemical Education, 83*(7), 1003−1009. https://doi.org/10.1021/ed083p1003

Brodhagen, M., Peyron, M., Miles, C., & Inglis, D. A. (2015). Biodegradable plastic agricultural mulches and key features of microbial degradation. *Applied Microbiology and Biotechnology, 99*(3), 1039−1056. https://doi.org/10.1007/s00253-014-6267-5

Calva-Estrada, S. J., Jiménez-Fernández, M., & Lugo-Cervantes, E. (2019). Protein-based films: advances in the development of biomaterials applicable to food packaging. *Food Engineering Reviews, 11*, 78−92. https://doi.org/10.1007/s12393-019-09189-w

Chonhenchob, V., Chantarasomboon, Y., & Singh, S. P. (2007). Quality changes of treated fresh-cut tropical fruits in rigid modified atmosphere packaging containers. *Packaging Technology and Science, 20*(1), 27−37. https://doi.org/10.1002/pts.740

Cuq, B., Gontard, N., & Guilbert, S. (1998). Proteins as agricultural polymers for packaging production. *Cereal Chemistry, 75*(1), 1−9. https://doi.org/10.1094/CCHEM.1998.75.1.1

Danijela, S., Vera, L., Senka, P., & Nevena, H. (2015). Edible films and coatings: sources, properties and application. *Food and Feed Research*, 11−22. https://doi.org/10.5937/ffr1501011s

Deng, L., Kang, X., Liu, Y., Feng, F., & Zhang, H. (2018). Characterization of gelatin/zein films fabricated by electrospinning vs solvent casting. *Food Hydrocolloids, 74*, 324−332. https://doi.org/10.1016/j.foodhyd.2017.08.023

Dhall, R. K. (2013). Advances in edible coatings for fresh fruits and vegetables: a review. *Critical Reviews in Food Science and Nutrition, 53*(5), 435−450. https://doi.org/10.1080/10408398.2010.541568

Galus, S., & Kadzińska, J. (2016). Moisture sensitivity, optical, mechanical and structural properties of whey protein-based edible films incorporated with rapeseed oil. *Food Technology and Biotechnology, 54*(1), 78−89. https://doi.org/10.17113/ftb.54.01.16.3889

Gopalakrishnan, S., Xu, J., Zhong, F., & Rotello, V. M. (2021). Strategies for fabricating protein films for biomaterial applications. *Advanced Sustainable Systems, 5*(1). https://doi.org/10.1002/adsu.202000167

Grujic, R., Vukic, M., & Gojkovic, V. (2017). Application of biopolymers in the food industry. In *Advances in applications of industrial biomaterials* (pp. 103−119). Springer International Publishing. https://doi.org/10.1007/978-3-319-62767-0_6

Guerrero, P., Retegi, A., Gabilondo, N., & De La Caba, K. (2010). Mechanical and thermal properties of soy protein films processed by casting and compression. *Journal of Food Engineering, 100*(1), 145−151. https://doi.org/10.1016/j.jfoodeng.2010.03.039

Hernandez-Izquierdo, V. M., & Krochta, J. M. (2008). Thermoplastic processing of proteins for film formation - a review. *Journal of Food Science, 73*(2), R30−R39. https://doi.org/10.1111/j.1750-3841.2007.00636.x

Hernandez-Lzquierdo, V. M., & Krochta, J. M. (2009). Thermal transitions and heat-sealing of glycerol-plasticized whey protein films. *Packaging Technology and Science, 22*(5), 255−260. https://doi.org/10.1002/pts.847

Jerez, A., Partal, P., Martínez, I., Gallegos, C., & Guerrero, A. (2007). Protein-based bioplastics: effect of thermo-mechanical processing. *Rheologica Acta, 46*(5), 711−720. https://doi.org/10.1007/s00397-007-0165-z

Kashiri, M., Cerisuelo, J. P., Domínguez, I., López-Carballo, G., Hernández-Muñoz, P., & Gavara, R. (2016). Novel antimicrobial zein film for controlled release of lauroyl arginate (LAE). *Food Hydrocolloids, 61*, 547–554. https://doi.org/10.1016/j.foodhyd.2016.06.012

Khanzadi, M., Jafari, S. M., Mirzaei, H., Chegini, F. K., Maghsoudlou, Y., & Dehnad, D. (2015). Physical and mechanical properties in biodegradable films of whey protein concentrate-pullulan by application of beeswax. *Carbohydrate Polymers, 118*, 24–29. https://doi.org/10.1016/j.carbpol.2014.11.015

Khwaldia, K., Ferez, C., Banon, S., Desobry, S., & Hardy, J. (2004). Milk proteins for edible films and coatings. *Critical Reviews in Food Science and Nutrition, 44*(4), 239–251. https://doi.org/10.1080/10408690490464906

Koshy, R. R., Mary, S. K., Thomas, S., & Pothan, L. A. (2015). Environment friendly green composites based on soy protein isolate - a review. *Food Hydrocolloids, 50*, 174–192. https://doi.org/10.1016/j.foodhyd.2015.04.023

Krekic, S., Nagy, D., Taneva, S. G., Fábián, L., Zimányi, L., & Dér, A. (2019). Spectrokinetic characterization of photoactive yellow protein films for integrated optical applications. *European Biophysics Journal, 48*(5), 465–473. https://doi.org/10.1007/s00249-019-01353-8

Lee, J. H., Lee, J., & Song, K. B. (2015). Development of a chicken feet protein film containing essential oils. *Food Hydrocolloids, 46*, 208–215. https://doi.org/10.1016/j.foodhyd.2014.12.020

Liu, F., Chiou, B. S., Avena-Bustillos, R. J., Zhang, Y., Li, Y., McHugh, T. H., & Zhong, F. (2017). Study of combined effects of glycerol and transglutaminase on properties of gelatin films. *Food Hydrocolloids, 65*, 1–9. https://doi.org/10.1016/j.foodhyd.2016.10.004

Maryam Adilah, Z. A., & Nur Hanani, Z. A. (2016). Active packaging of fish gelatin films with *Morinda citrifolia* oil. *Food Bioscience, 16*, 66–71. https://doi.org/10.1016/j.fbio.2016.10.002

Moghadam, M., Salami, M., Mohammadian, M., Khodadadi, M., & Emam-Djomeh, Z. (2020). Development of antioxidant edible films based on mung bean protein enriched with pomegranate peel. *Food Hydrocolloids, 104*. https://doi.org/10.1016/j.foodhyd.2020.105735

Nagarjun, J., Kanchana, J., RajeshKumar, G., Manimaran, S., & Krishnaprakash, M. (2021). Enhancement of mechanical behavior of PLA matrix using tamarind and date seed micro fillers. *Journal of Natural Fibers*, 1–13. https://doi.org/10.1080/15440478.2020.1870616

Nogueira, D., & Martins, V. G. (2019). Use of different proteins to produce biodegradable films and blends. *Journal of Polymers and the Environment, 27*(9), 2027–2039. https://doi.org/10.1007/s10924-019-01494-z

Nuanmano, S., Prodpran, T., & Benjakul, S. (2015). Potential use of gelatin hydrolysate as plasticizer in fish myofibrillar protein film. *Food Hydrocolloids, 47*, 61–68. https://doi.org/10.1016/j.foodhyd.2015.01.005

Nur Hanani, Z. A., O'Mahony, J. A., Roos, Y. H., Oliveira, P. M., & Kerry, J. P. (2014). Extrusion of gelatin-based composite films: effects of processing temperature and pH of film forming solution on mechanical and barrier properties of manufactured films. *Food Packaging and Shelf Life, 2*(2), 91–101. https://doi.org/10.1016/j.fpsl.2014.09.001

Pirsa, S., & Sharifi, K. A. (2020). A review of the applications of bioproteins in the preparation of biodegradable films and polymers. *The Journal of Physical Chemistry Letters, 1*, 47–58.

Pommet, M., Redl, A., Guilbert, S., & Morel, M. H. (2005). Intrinsic influence of various plasticizers on functional properties and reactivity of wheat gluten thermoplastic materials. *Journal of Cereal Science, 42*(1), 81–91. https://doi.org/10.1016/j.jcs.2005.02.005

Rajeshkumar, G., Naveen Kumar, K., Aravind, M., Santhosh, S., Gowtham Keerthi, T. K., & Arvindh Seshadri, S. (2021). A comprehensive review on mechanical properties of natural cellulosic fiber reinforced PLA composites. In *Lecture notes in mechanical engineering* (pp. 227–237). Springer Science and Business Media Deutschland GmbH. https://doi.org/10.1007/978-981-15-9809-8_19

Rajeshkumar, G., Vishnupriyan, R., & Selvadeepak, S. (2020). Tissue mimicking material an idealized tissue model for clinical applications: a review. *Materials Today: Proceedings, 22*, 2696–2703. https://doi.org/10.1016/j.matpr.2020.03.400

Rocca-Smith, J. R., Marcuzzo, E., Karbowiak, T., Centa, J., Giacometti, M., Scapin, F., Venir, E., Sensidoni, A., & Debeaufort, F. (2016). Effect of lipid incorporation on functional properties of wheat gluten based edible films. *Journal of Cereal Science, 69*, 275–282. https://doi.org/10.1016/j.jcs.2016.04.001

Saremnezhad, S., Azizi, M. H., Barzegar, M., Abbasi, S., & Ahmadi, E. (2011). Properties of a new edible film made of faba bean protein isolate. *Journal of Agricultural Science and Technology A, 13*(2), 181–192. http://jast.journals.modares.ac.ir/jufile?c2hvd1BERj0zOTc=.

Saurabh, C. K., Gupta, S., Bahadur, J., Mazumder, S., Variyar, P. S., & Sharma, A. (2013). Radiation dose dependent change in physiochemical, mechanical and barrier properties of guar gum based films. *Carbohydrate Polymers, 98*(2), 1610–1617. https://doi.org/10.1016/j.carbpol.2013.07.041

Sintim, H. Y., Bandopadhyay, S., English, M. E., Bary, A. I., DeBruyn, J. M., Schaeffer, S. M., Miles, C. A., Reganold, J. P., & Flury, M. (2019). Impacts of biodegradable plastic mulches on soil health. *Agriculture, Ecosystems and Environment, 273*, 36–49. https://doi.org/10.1016/j.agee.2018.12.002

Srinivasa, P. C., Ramesh, M. N., Kumar, K. R., & Tharanathan, R. N. (2004). Properties of chitosan films prepared under different drying conditions. *Journal of Food Engineering, 63*(1), 79–85. https://doi.org/10.1016/S0260-8774(03)00285-1

Vasconcelos da Silva Pereira, G., Vasconcelos da Silva Pereira, G., Furtado de Araujo, E., Maria Paixão Xavier, E., Regina Sarkis Peixoto Joele, M., & de Fátima Henriques Lourenço, L. (2019). Optimized process to produce biodegradable films with myofibrillar proteins from fish byproducts. *Food Packaging and Shelf Life, 21*. https://doi.org/10.1016/j.fpsl.2019.100364

Wu, H., Lu, J., Xiao, D., Yan, Z., Li, S., Li, T., Wan, X., Zhang, Z., Liu, Y., Shen, G., Li, S., & Luo, Q. (2021). Development and characterization of antimicrobial protein films based on soybean protein isolate incorporating diatomite/thymol complex. *Food Hydrocolloids, 110*, 106138. https://doi.org/10.1016/j.foodhyd.2020.106138

Biodegradable polymer blends and composites for food-packaging applications

M. Ramesh and M. Muthukrishnan
Department of Mechanical Engineering, KIT—Kalaignarkarunanidhi Institute of Technology, Coimbatore, Tamil Nadu, India

1. Introduction

The global food-packaging market was around USD 303 billion in 2020, and this is expected to increase further due to the quickening pace of life which results in changes to eating habits. This has paved way for many online food-delivery portals which deliver ready-to-serve food. As a result, more care has been taken for food packaging, which protects food from various external factors such as odors, dust, temperature, light, germs and bacteria, UV protection, transparency, physical damage, etc. (Carocho et al., 2015). Thus, food industries are giving great priority to food packaging which improves the shelf-life and quality of food. The challenge to manufacturers is to constantly come up with new cost-effective innovative packaging techniques to ensure quality and freshness of the food at different levels of storage. Thus, manufacturers are paying increased attention to the types of materials used for packing and the design of these packages. Generally, food packages come in different forms, like polymer-based film carry bags, wrappers, boxes, bottles, trays, edible packing materials, etc. The functions of these packages are to: (1) contain the food packaged material with a proper fit; (2) preserve the food material from the surrounding environment; (3) maintain the quality of the food without a loss of taste or odor; (4) suitability for transport without any physical damage between leaving the warehouse and reaching the customer; and (5) provide proper information to customers about the content of the packaged material.

1.1 Types of food packages

Packages are generally classified into three categories based on their primary functions. Primary packaging comes into direct contact with the food material by being wrapped around it. Materials like metals, plastic containers, paper, and glass are generally used as primary packaging materials. These packaging materials should be nonreactive and should retain the properties of the food material. Secondary packaging includes materials like biodegradable jute, paperboard cartons and plastics containers, tray, boxes, etc. Secondary packaging is comprised of several primary packages.

Biodegradable Polymers, Blends and Composites. https://doi.org/10.1016/B978-0-12-823791-5.00004-1

Tertiary packaging ensures safe handling of the primary and secondary packages avoiding any physical damage, while being transported using large wooden or metal boxes.

1.2 Materials used for food packaging

1. Glass: Glass comes in the form of containers, bottles, bowls, trays, cups, etc., and is the oldest form of packing material owing to its inertness toward reactive food materials like vinegar, salt, medicines, etc. and at the same time retaining the aromatic nature of the food material. Glass has the advantage of being transparent, chemically inert, with the ability to be recycled multiple times without losing its indigenous properties. The drawbacks are that it is heavy and brittle in nature, which demands careful secondary and tertiary packaging during transport. It is used in other materials such as rubber, metals, and plastics for packaging food materials.
2. Metals: Aluminum cans are the most preferred materials for primary packaging. Aluminum is widely used for its antimicrobial property. For food items that need to be packed in a hot state or under high pressure, metals are the preferred materials.
3. Paperboards: These can be made from recycled paper pulp and are widely used to store liquid edibles, dry fruits, fast food, and frozen food. The drawback is that they are not inert to external contaminants and low volatile additives that can be transferred to the food materials through them.
4. Polymers: Polymers are the most widely used food-packaging material in the industry and offer flexibility in design. Some of the inherent characteristics of polymers are light weight, less expensive, easy processability, and flexibility in molding to any shape and size. Polymers are the most suitable candidates for production lines due to their inherent properties including better thermal, mechanical properties, heat sealability, and printability, which satisfy various norms of the food-packaging industry. Polymers are classified as petroleum-based polymers and fossil-based polymers. Petroleum-based polymers are known for their lightweight, transparent, stable, high-strength, cost-effective, chemical-resistant characteristics, and ease of processability and recyclability (Rubio et al., 2004). Some of the most prominent polymer materials widely in use are polyolefins (PO) and polyesters, polystyrene (PS), polyvinyl chloride (PVC), ethylene vinyl alcohol (EVOH), polyamide (PA), etc. (Lau & Wong, 2000) (Fig. 25.1).

One of the prominent materials used in domestic applications are polyethylene terephthalate (PET) used in beverages industries. Polycarbonate (PC) is a clear, durable, and heat-resistant polymer used for water bottles and sterilizable baby bottles. PVC, owing to its better thermoforming characteristics, and is used in meat-packaging industries; for flexible single-layer film packaging, polyvinylidene chloride (PVdC) is used. However, many research studies have proven that petroleum-based plastics pose serious environmental and health issues. Their degradability issues lead to huge accumulation of plastic wastage in all corners of the world, and there are also issues such as monomer residue, stabilizers, and plasticizers in plastic lead, and the release of bisphenol through the use of cleaning detergents leads to severe health concern. Synthetic plastic wastes contribute around 140 million tons annually, causing serious environmental and health concerns (Shimao, 2001). These wastes prove to be a challenge for wastewater treatment plants and contaminate

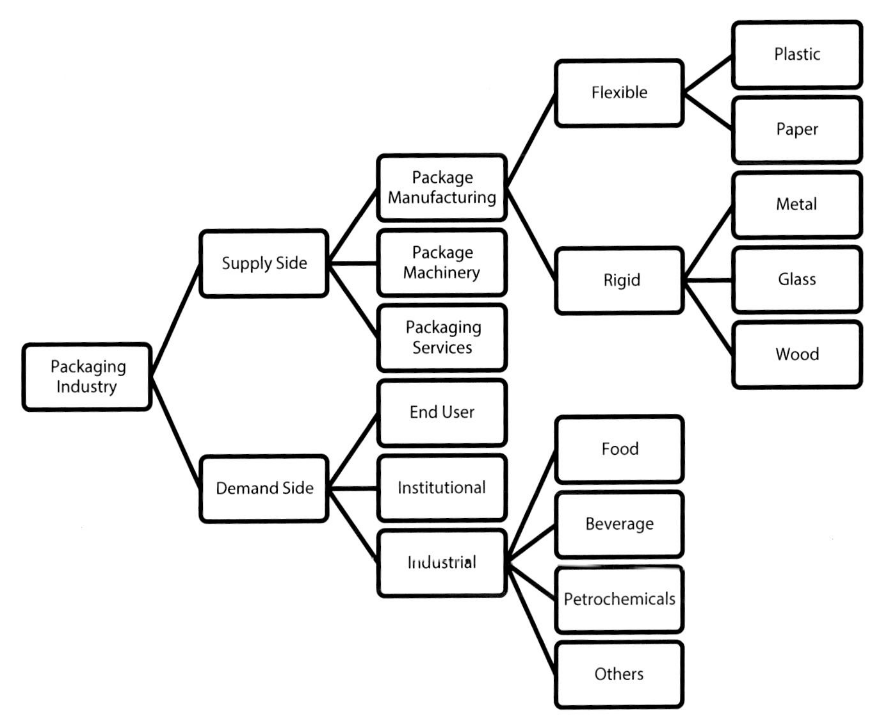

Figure 25.1 Packaging industry.

ground and surface water. Also, they contribute a major portion of the municipal solid waste in landfills because they are resistant to physical and chemical degradation. Thus, many industries have taken the initiative in increasing their research efforts on biodegradable polymer blends. Currently, food-packaging industries are facing high demands due to customer awareness and high expectations of various quality, packed, lifestyle foods.

1.3 Biodegradable plastics for food packaging

The food-packaging industries have focused their research into developing biodegradable polymers that are environment friendly without violating the norms of food packaging, that is, reduce, recycle, and reuse. Thus the sustainable development of biodegradable new materials should balance the key factors of economic, social, and environmental (Miller, 2014). Some of the advantages of biodegradable polymers are: (1) they play a major role in reducing dependency on petroleum materials; (2) lower CO_2 emissions by 30%–70%, which leads to a healthier environment; (3) rely on biobased renewable resources; (4) less energy consumption; (5) less landfill as they are degradable; and (6) safe to handle. However, biodegradation also possesses certain disadvantages, such as: (1) involves complex processes for the treatment of biomaterials and thus requires costly processing equipment; (2) more diversion of agricultural land to produce biomaterials than producing food; (3) there is some serious research on the toxicity levels emitted by biomaterials on degradation; and (4) another issue is that these polymers tend to produce methane gas in landfills.

1.3.1 Polymer blends

A polymer blend is a group of dissimilar polymers that are mixed together to achieve the desired characteristics or property by forming a new material with or without a chemical interaction between them. Polymer blends are frequently used in the food-packaging industry to (1) improve the processability of a natural polymer to suit the industrial applications; (2) improve the physical, chemical, and mechanical properties of the biodegradable material; and (3) adapt to frequent changes in demand for new materials for the food-packaging industry. In comparison to copolymerization, polymer blends offer a low-cost process and instead of complicated chemical processes, simple physical processes are involved to achieve the desired characteristics. The factors that determine the performance of polymer blends are morphology, miscibility, and compatibility. The morphology of the immiscible polymer blends is determined by several parameters such as the concentration of the blended materials, viscosity ratio, compatibility between the blend polymers, and interfacial tension between the polymers.

Miscibility is a thermodynamic term that describes a number of phases within the polymer blends. Polymers are based on miscibility, which are again subgrouped into: (1) completely miscible polymers; (2) partially miscible polymers; and (3) completely immiscible polymers. The glass transition temperature (T_g) and morphology are used to determine the miscibility of polymer blends. For miscible polymers, the T_g value will be a single entity, whereas for heterogeneous partially immiscible polymer blends, the T_g value will be shifted toward one of the polymers and, for completely immiscible polymer blends, there will be two values of T_g for each material. Compatibility is a method of improving the inherent properties of the miscible and partially miscible polymer blends. Also, the morphology of the polymer blends turns from coarse morphology to fine morphology which improves the adhesion and surface tension by allowing stress transfer between the phases that improves the performance of the polymer blends.

1.3.2 Classification of biodegradable polymers

Biodegradable polymers are briefly classified into various groups based on their origin, synthesis method, chemical composition, applications, etc. Generally, a broader classification of biopolymers is based on their origin only. Accordingly, they are widely classified as (1) natural polymers that are developed from natural or renewable resources and (2) biodegradable polymers that are developed from petrochemical products. Again, natural polymers are further divided into polymers that are developed based on (1) naturally occurring biodegradable polymers; (2) biodegradable polymers derived from renewable resources; and (3) biopolymers derived from plant oils.

Natural polymers are obtained in nature in the form of biomass and are generally derived either from forestry or agricultural resources. Forestry biomass includes forest wastes and shrubs, and from agricultural feedstocks polysaccharides, lipids, and proteins can be derived. Polysaccharides include starch, cellulose, lignin, pectin, alginates, chitin, agar, carrageenan, etc., while proteins derived from animal feedstocks include whey, soy, casein, and gluten. The second type of natural polymers is microbial polymers that are developed by microorganisms through carbon substrate fermentation. Some examples of microbial polymers are poly(3-hydroxybutyrate) and poly(3-hydroxybutyrate-*co*-3-hydroxyvalerate) (PHBV) that come under polyhydroxyalkanoates (PHA).

The third type is bioderived polymers that are made partially from renewable resources. For example, starch can be used to produce lactic acid and polylactic acid. Similarly, sugarcane can be used to produce ethylene, which is in turn used in the manufacture of ethylene. Fig. 25.2 gives the classification of biodegradable polymers that are used in good packaging applications. In the food-packaging industry, plastics contribute more than 50% of the total materials used. The global initiative toward biodegradable polymers has increased its demand and it is predicted that there will be a more than 70% rise in the consumption of these biodegradable polymers. Among the various biodegradable polymers that are used, starch-based thermoplastics, cellulose, PLA, PBS, P(H3B), and CA are widely used in areas like food packaging, thermo-formed food catering, trash bags, and short-life products made by injection molding, etc.

2. Biobased polysaccharides

Polysaccharides are classified as biological polymers that include starch, cellulose, and chitosan and are abundantly derived directly from plants (cellulose), smaller

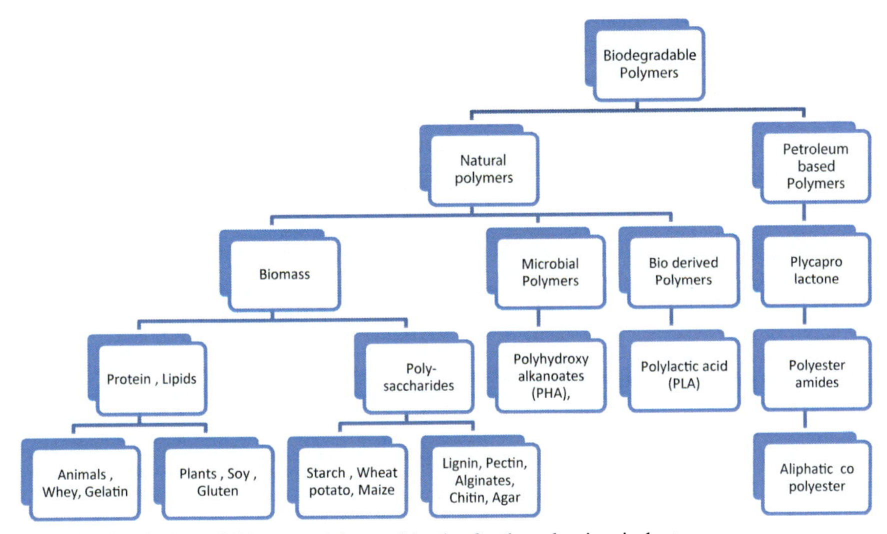

Figure 25.2 Biodegradable materials used in the food-packaging industry.

microorganisms, and marine organisms (chitin and chitosan) (Ramesh & Tharanathan, 2003). Polysaccharides are widely used in the food industry as edible coatings and in the packaging industry owing to their ability to form membranes and films. These membranes and films at moderate temperature and humidity provide good protection for food materials against harmful gases and improve the shelf-life. Thus they are widely used as primary packaging materials in the food, medical, and pharmaceutical industries, etc. and are slowly replacing synthetic polymers in the packaging industry (Han, 2014; Falguera et al., 2011). Apart from packaging, they are also used as edible coatings in the food industry. However, polysaccharides are hydrophilic in nature and are stiffer than conventional synthetic polymer packaging films, which prevents their use in commercial packaging applications (Debeaufort et al., 1998).

2.1 Polysaccharides from animals

Chitin is the one of the biopolymers that is most abundant in nature. It is primarily derived from the exoskeletons of arthropods. Chitin is composed of N-acetyl-D-glucosamine and is basically an acetylated polysaccharide. It is derived by a chemical extraction process (Thakur & Thakur, 2016; Freitas et al., 2014). On the other hand, chitosan, a polysaccharide, is derived from chitin under varying processing parameters such as temperature, chitin chemical properties, incubation time, alkali concentration, etc. Chitosan can be derived from chitin under high-temperature and alkali conditions. Chitosan is composed of *N*-acetylglucosamine and glucosamine that are linked by β-(1−4)-glycosidic bonds (Fig. 25.3) (Mati-Baouche et al., 2014). Chitosan possess antimicrobial properties and is insoluble in water (Rinaudo, 2006). Further, the addition of glycerol and mechanical kneading results in thermoplastic materials with improved mechanical properties (Epure et al., 2011). Good film-forming properties allow membranes and coatings to be developed to act as food preservatives. There are biodegradable, biocompatible, nontoxic, bioadhesive, renewable, and commercially useable shitosan membranes. In addition, chitosan membranes are stated to be semipermeable to gases with low oxygen permeability, which is necessary for the preservation of certain food items, and provide moderate water vapor barriers (Pereda et al., 2009).

The barrier properties of chitosan are reduced by its high sensitivity toward water, and the water resistance can be improved without affecting the biodegradability by (1) blending chitosan with polymers like polyvinyl alcohol, gelatine, cellulose, etc.

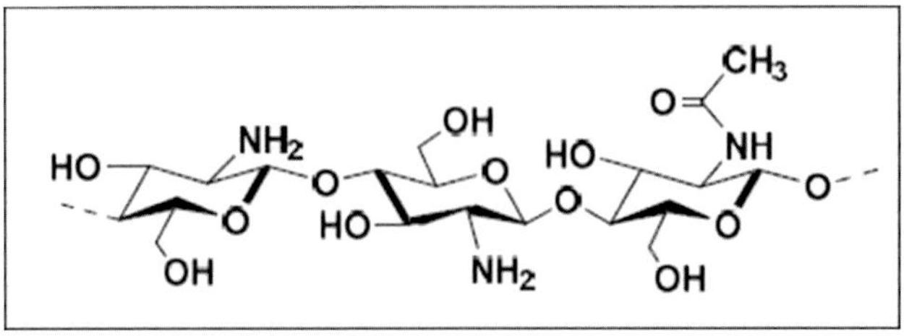

Figure 25.3 Chemical structure of chitosan (Mati-Baouche et al., 2014).

and (2) using glycerol, citric acid, and genipin as plasticizers and cross-linkers (Zemljič et al., 2013). Also, in the meat-packaging industry, chitosan's antimicrobial properties are used to combat meat pathogens by being reverse-coated onto polyethylene terephthalate (PET) (Zemljič et al., 2013). Similarly, chitosan is coupled with lysozyme to enhance the antimicrobial properties of mozzarella cheese against *Escherichia coli* and *Listeria monocytogenes* (Duan et al., 2007). Mozzarella cheese coated with chitosan−lysosome films resulted in 0.32−1.40 log reductions against the pathogens. Chitosan combined with rosemary oil was proved to prolong the shelf-life of turkey and chicken meat and prevent lipid oxidation, microbial attacks, and discoloration (Vasilatos & Savvaidis, 2013). Similarly chitosan combined with thyme oil (El-Obeid et al., 2018), cinnamon oil (Ojagh et al., 2010), and clove oil (Vieira et al., 2019) improves the shelf-life of seafood.

Recent innovations like chitosan films embedded with rice bran anthocyanins enable intelligent food packaging in the fish industry, allowing the quality of seafood to be monitored by color change. Here, chitosan is used as an edible film that has better antioxidant properties, possessing a UV barrier effect and pH sensitivity nature (Wu et al., 2019). Chitosan aerogels developed by combining gelation cross-linked with formaldehyde followed by CO_2 supercritical drying result in a transparent and low thermal conductive polymer film that is used as an insulator for maintaining the temperature of temperature-dependent food packs (Takeshita & Yoda, 2015). These chitosan aerogels possess excellent mechanical toughness, are biodegradable, and have higher insulating capacity than conventional aerogels.

2.2 Polysaccharides from plants

2.2.1 Starch

Starch is one of the polysaccharides produced by agrofeeds like corn, wheat, tapioca, rice, cassava, and potato for energy storage (Dai et al., 2019). It is also called amylum, and is a polymeric carbohydrate comprised of amylose and amylopectin. The chemical structure of starch, as shown in Fig. 25.4, shows α-1,4-linked D-glucose units (amylose) that are connected by glycosidic bonds (amylopectin) (Nešić et al., 2020). Starch is widely used in the food-packaging industry for its biodegradability, custom modifications, and very low cost.

One of the commercially available starches is thermoplastic starch (TPS), which is prepared by exposing starch to shear and high pressure in the presence of plasticizers like water, glycol, and oils (Meng et al., 2019; Volpe et al., 2018). TPS is further processed into thin films, bags, and food container under various industrial processes such as compression molding, extrusion, and blowing (Blohm & Heinze, 2019). Generally, parameters that determine the suitability of starch as food-packaging materials depend on the biosource, genotype, morphology and size distribution of granules, amylose/amylopectin ratio, composition, and pH (Cruz-Romero & Kerry, 2008; Babu et al., 2013). Starch films are widely used in the food industry due to their inherent biodegradability, nontoxic, colorless, and tasteless properties. They are also edible in nature and are an effective oxygen barrier. On the other hand, they have their own

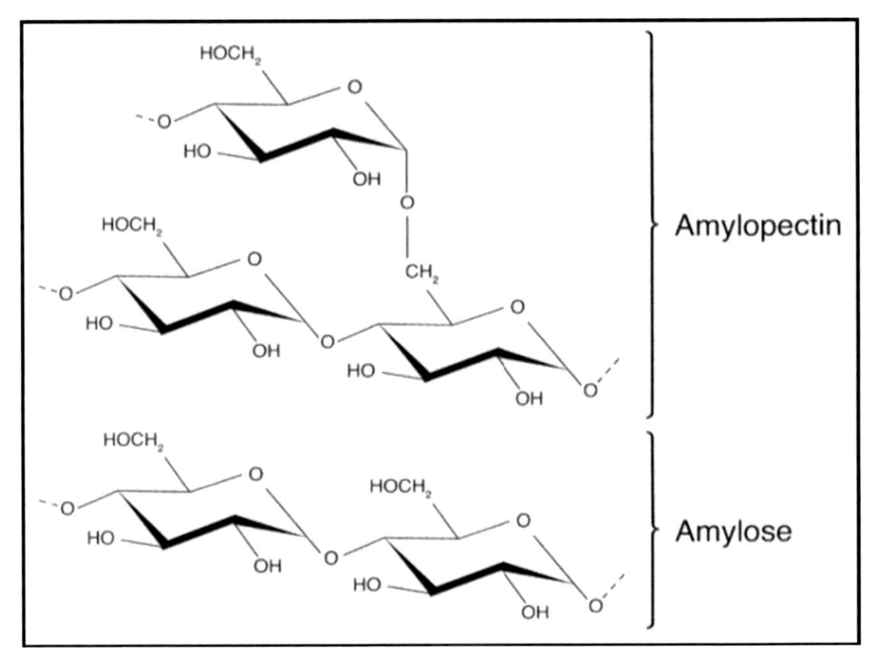

Figure 25.4 Chemical structure of starch (Nešić et al., 2020).

disadvantages like moisture absorption receptiveness and lower mechanical strength (Pelissari et al., 2019). The solution to overcome those disadvantages and make them suitable for the food-packaging industry has been widely studied, and several solutions have been proposed. Many researchers have proposed mixing starch with organic products and essential oils to strengthen its properties, thus improving its shelf-life in the form of the film blends, aerogels, edible foils, etc., and at the same time making them antimicrobial in nature (Arroyo et al., 2019; Mc Clements, 2018; Moreno et al., 2018; Panrong et al., 2019; Ubeyitogullari & Ciftci, 2016; Yıldırım-Yalçın et al., 2019; Zhu, 2017). Inorganics nanomaterials like Ag, ZnO, CuO, nanoclays, and carbon nanotubes are also used in the reinforcement of starch films (Chaudhary et al., 2020; Jayakumar et al., 2019; Peighambardoust et al., 2019; Shahabi-Ghahfarrokhi & Babaei-Ghazvini, 2019; Castillo et al., 2017).

Similarly, biopolymers like PVA, PVP, polysaccharides, proteins, and biodegradable polyesters like polylactic acid (PLA), polycaprolactone (PCL), PBS, etc., are used in the preparation of biodegradable starch films. Thus starch is used in areas where food-packaging materials need to be hydrophilic in nature with retrogradation in the form of membranes and edible coatings with low oxygen permeability. In order to address these disadvantages, research has been carried out, primarily using plasticizers, which increase the chain mobility and enhance durability, to manufacture starch plastics with mechanical properties comparable to those derived from polyolefin. Polyols, such as glycerol, glycol, and sorbitol, are the most widely used plasticizers (Liu et al., 2017; Mathew et al., 2019; Mlalila et al., 2018; Qin et al., 2020).

Other approaches studied include the design of mixtures and composites and the structural modification of starch to create a biodegradable material with sufficient mechanical strength, flexibility, and water barrier properties for use as a packaging material. Combining starch with more hydrophobic polymers (PCL or PLA) (Tavares et al., 2019; Ortega-Toro et al., 2015) and their composites with clay nanoparticles (Jiménez et al., 2012) has been commonly studied. Starch films prepared from rye starch extract are used in chicken wrapping. These films have been found to prevent lipid oxidation, increase film elasticity, improve antioxidant properties, and hamper the development of radical species that affect the chicken shelf-life (Go & Song, 2019). Similarly, starch film with clove oil yields better results in the packing of cheese (Yang et al., 2018). In another research, tapioca starch films blended with carvacrol or cymophenol, $C_6H_3(CH_3)(OH)C_3H_7$ are used as a bioactive film which is characterized by its inhibition of bacterial growth. With regard to food packaging, carvacrol reduces not only the tensile properties of the film but also its water-wetting properties and solubility (Homayouni et al., 2017). Food-packaging films of starch in combination with chitosan, gallic acid, and carvacrol are used for their antimicrobial property in protecting ham from foodborne pathogens and harmful microbes (Zhao et al., 2019). Also, casava starch films containing rosemary antioxidant extracts (Piñeros Hernandez et al., 2017) and cowpea starch and maqui berry extracts (Baek et al., 2019) have UV barrier properties to improve the shelf-lives of salmon, banana, etc. (Thakur et al., 2019). Similarly, sweet potato starch-coated shrimps (Alotaibi & Tahergorabi, 2018) and its combination with thyme essential oil increased the shelf-life of eggs (Eddin & Tahergorabi, 2019).

Recent studies into starch led to the development of intelligent films that identify the quality of food and indicate it as a sensor by changing the film color without compromising the biodegradability, antimicrobial, and barrier properties. For example, starch film based on polyvinyl alcohol added into betalain-rich red pitaya peel (Jiménez et al., 2012) is used to change color by sensing ammonia produced by shrimps in storage, which is an indication of shrimp spoilage. pH-sensitive intelligent starch films are made by the combination of cassava starch and *L. ruthenicum* and used to indicate pork freshness by color variation (Qin et al., 2019). Starch aerogels are produced using CO_2 (Villegas et al., 2019) and are incorporated with quercetin (Franco et al., 2018) to improve the dissolution time and lower the burst-like effect. Also, by adding natural fibers, the mechanical properties and thermal conductivity properties are improved (Engel et al., 2019). Also, a combination of starch aerogels and trans-2-hexenal possesses excellent antifungal properties (Fig. 25.5).

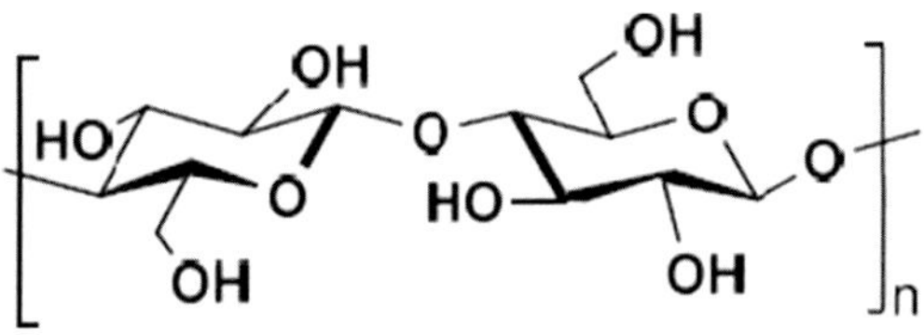

Figure 25.5 Chemical structure of cellulose (Nešić et al., 2020).

2.2.2 Cellulose

Cellulose $(C_6H_{10}O_5)_n$ is a linear homopolysaccharide and is one of the natural biopolymers abundantly available in the cell walls of all plants, and in some fungi, algae, and marine organisms (Tayeb et al., 2018). Cellulose is comprised of glucose underneath its D-glucopyranose form that is linked by glycosidic bonds (Credou & Berthelot, 2014). Cellulose has a tendency to form microfibrils and fibers owing to its regular structure with an array of hydroxyl groups, and is used in the manufacture of paper and paperboard for use in the packaging industry (Cruz-Romero & Kerry, 2008). Cellulose is characterized by its specific properties such as low density, low cost, high mechanical strength, chemical stability, resistance to strong alkali, resistance to oxidizing agents, control over viscosity properties, nontoxicity, renewability, biocompatibility, and biodegradability (Credou & Berthelot, 2014; Duan et al., 2016).

The main source of cellulose is derived from wood pulp, cotton fibers, and sugarcane bagasse that are processed with other derivatives to form paper and paperboards. However, cellulose as such cannot be used directly in the packaging industry due to its limited hydrophilic nature, increase in crystallinity over time (formation of brittle structure), and insolubility in water (Thakur & Thakur, 2016). Due to its chemical stability, it is easy to form chemical derivatives of cellulose, and some of the chemical derivatives of cellulose are ethers like carboxymethyl cellulose hydroxypropylmethyl cellulose, methylcellulose, and hydroxypropyl cellulose and esters like cellulose acetate, cellulose butyrate, and cellulose acetate-butyrate (Gelin et al., 2007).

2.2.2.1 Cellulose ester

Cellulose esterification results in combining organic acid (RCOOH) with primary and secondary hydroxyl groups of cellulose to from ester (RCOOR) and water (Kuo & Leonard, 1984). Cellulose esters are widely used in the food industry for their suitability to form coatings, biodegradable plastics, films, membranes, composites, and laminates (Edgar et al., 2001). Cellulose esters are excellent matrix binders in natural fiber-based composite materials. They are suited for industrial production processes such as injection molding, and are extruded to form sheets, and thermo-folded to form membranes and other structural components under operating temperatures of between 180 and 240°C without producing any toxins. Some of the organic derivatives of cellulose esters widely in use are cellulose acetate, cellulose triacetate, cellulose propionate, and cellulose xanthanate, and cellulose acetate butyrate (CAB) and cellulose acetate propionate (CAP) that are widely used in the formation of films, coatings, and as additives in films. These derivatives prevent defects in the surface and inhibit brittleness.

2.2.2.2 Cellulose ethers

Cellulose ethers are widely used in the food industry as gelling agents, binders, stabilizers in food mixtures in ice cream, in pie fillings as thickeners, in fruit juice and other dairy products; and are used as suspending agents. Hydroxypropyl methyl cellulose, hydroxy ethyl cellulose, and carboxymethyl cellulose are some of the cellulose ether derivatives used as additives. In order to inhibit crystallization in ice cream, juice, or icing layers, carboxymethyl cellulose is widely used. Some of the other critical properties of ethers are water-binding capacity, and being soluble, biodegradable, nontoxic, and chemically stable.

2.2.2.3 Bacterial cellulose

Bacterial cellulose (BC) is an organic compound (Chawla et al., 2009) derived from bacteria like *Komagataeibacter*, *Sarcina ventriculi*, and *Agrobacterium*, which is used in the food industry for its properties of high strength, purity, and moldability. BC exhibits enhanced water-holding ability due to the high aspect ratio of the fibrils which provides a large surface area for holding water molecules which are tightly bounded by the hydroxyl groups within the cellulose chain (Gelin et al., 2007). BC is widely used in the biomedical field, textiles, food products, coatings, and cosmetics reinforcements for optically transparent films. The addition of 10% BC in meatballs and ice creams as a fat replacer mimics similar sensory properties of hydrogenated oils and extends the shelf-life (Lin & Lin, 2004). BC is approved as GRAS (generally recognized as safe) by the FDA (Food and Drug Administration, United States). Apart from being used as a fat stabilizer, BC is used in the ice cream food industry (Fig. 25.6) as a foam stabilizer and texture modifier that helps in enhancing viscosity, increasing the meltdown time (Granger et al., 2005), and retaining ice cream for at least 60 min at room temperature (Okiyama et al., 1993).

2.2.3 Galactomannans

Galactomannans are plant-based polysaccharides consisting of a (1−4)-linked beta-D-mannopyranose mannose backbone with 1−6-linked alpha-D-galactopyranose side groups. Galactomannans are commercially extracted from the plant seeds of guar, carob, fenugreek, tara bean, locust bean, and cassia, which are distinguished primarily by the mannose to galactose ratio (Kontogiorgos, 2018). Locust bean gum and guar gum are widely available under stable prices due to which, they are preferably

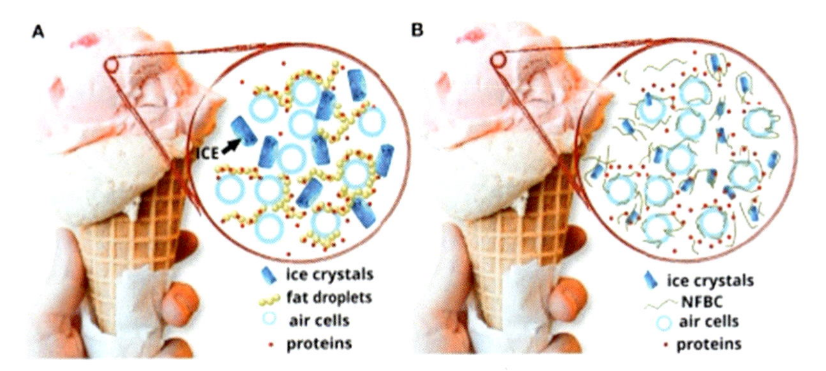

Figure 25.6 (a) Schematic representation of ice cream and (b) nanofiber BC replacing fat in ice cream (Henriette et al., 2019).

commercially used as stabilizers, thickeners, edible membranes, and coatings in the food industry, especially in the ice cream industry where it is used for improving the texture and preventing the ice cream from melting at room temperature. Galactomannans under low concentration form gelatinous solutions and exhibit resistance to pH alterations. They also exhibit characteristics of better mechanical strength, high ionic strength, improved barrier properties, and better heat-processing abilities that make it a suitable candidate for increasing the shelf-life of foods in the food-processing industry (Cerqueira et al., 2011).

2.3 Polysaccharides from algae

2.3.1 Carrageenan

This is a natural ingredient widely used in the food industry to preserve foods and drinks. It is a hydrocolloid derived from red edible seaweeds that are related to anionic sulfated linear polysaccharides made of galactose units and 3,6-anhydrogalactose (3,6-AG) that are linked together by α-1,3 and β-1,4 glycosidic linkages. The common edible seaweeds are from the Rhodophyceae family which includes the *Eucheuma*, *Chondrus*, *Gigartina*, and *Hypnea* species. However, carrageenan is used commercially in three forms of kappa, iota, and lambda. The chemical structure of these carrageenans is differentiated by the position of the ester sulfate groups as illustrated in Fig. 25.7. Kappa uses potassium ions for stiffness and rigidity thus forming a firm elastic gel, whereas iota in the presence of calcium ions forms a soft gel and in lambda, ions are not used for thickening of dairy products.

Carrageenan is a food-grade additive that is widely used in the production of edible films and coatings in the food industry. It is also an important ingredient in ice-cream, flavored milk, salad dressings, pet foods, chocolates, yogurt, and diabetic and infant foods (Necas & Bartosikova, 2013). It is also used in water applications as syrups, fruit drink powders, imitation milk, suspensions, dispersions, puddings, sauces, etc. In meat-packaging industries, it is used as a stabilizer for producing low-calorie sandwiches, sausages, and patties (Chen, Yan, Wang, Xu, & Zhang, 2010). Polymer blends like polylactide and κ-carrageenan/agar/clay nanocomposite films are used for packaging food materials by increasing their shelf-life and maintaining their quality (Rhim, 2013). Similarly, whey proteins isolate (WPI) carrageenans are used as probiotic carriers in yogurts, creams, cheeses, fermented lactic beverages, etc. (Rodríguez et al., 2014).

2.3.2 Alginates

Algins are polysaccharides derived from the cell walls of brown seaweeds. Alginates are the derivatives of algin or algin acid in association with metals like sodium and calcium to form salts. Alginate structures are comprised of nonrepeating copolymers of β-D mannuronic acid (M) and α-L-guluronic acid (G) that are bonded by glycoside. These polymers are subdivided into G blocks, M blocks, and MG blocks based on origin, maturity, and harvesting time of the aligned weeds (Fig. 25.8).

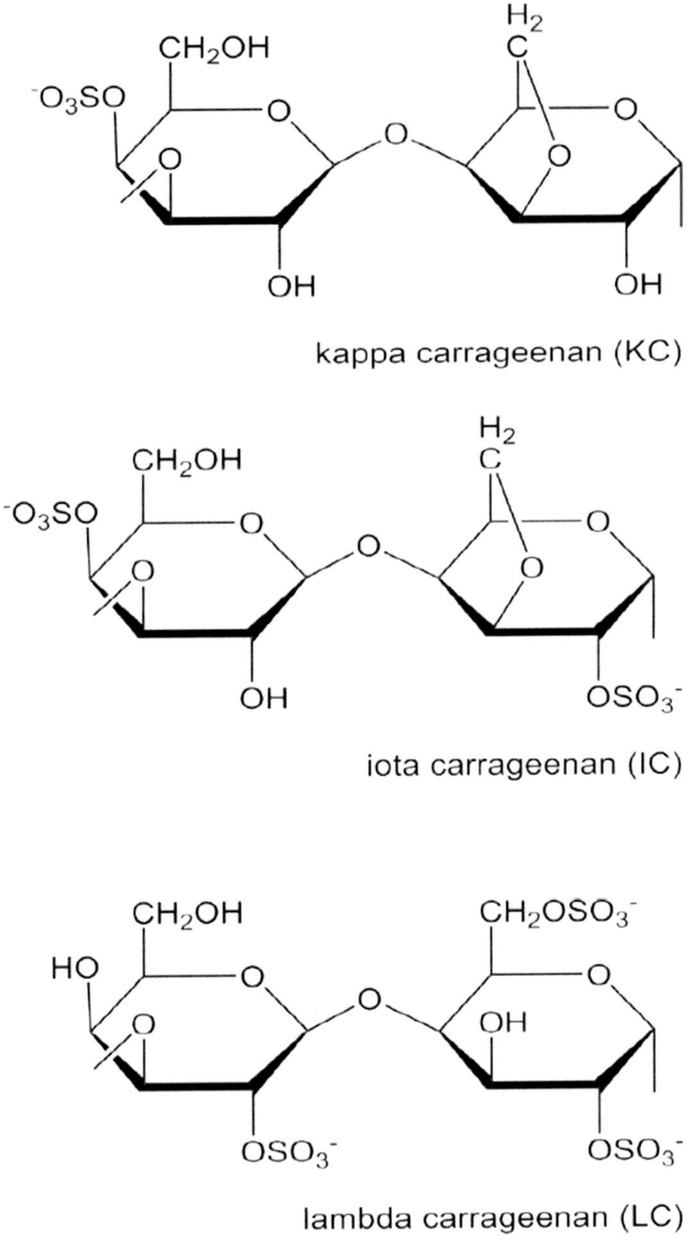

kappa carrageenan (KC)

iota carrageenan (IC)

lambda carrageenan (LC)

Figure 25.7 Chemical structure of kappa, iota, and lambda carrageenan (Dul et al., 2015).

Generally, for the food industry, brown seaweeds are cultivated and processed as sodium alginate salts. Alginates, owing to their exclusive properties such as biodegradability, water permeability, and better tensile strength are widely used in the food and food-packaging industries. Sodium alginates ($NaC_6H_7O_6$) are one of the

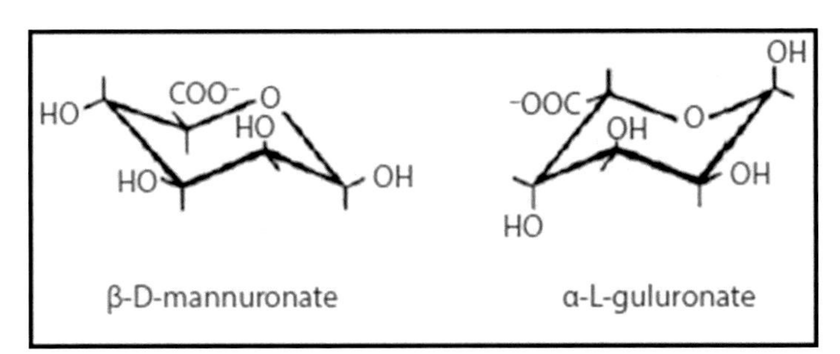

Figure 25.8 Alginates structure (Gacesa & Russell, 1990).

major ingredients of the food, fertilizer, textile printing, and pharmaceuticals industries. Sodium alginate in combination with additional compounds like natamycin increases the water vapor permeability and solubility in water (Bierhalz et al., 2012). Similarly, the addition of silver nanoparticles with sodium alginate increases the shelf-life of crops like carrots and pears (Fayaz et al., 2009). When combined with $CaCl_2$ derivatives, sodium alginates have enhanced mechanical properties and better water resistance capability (Rhim, 2004). Also, the oxygen barrier properties are improved by additional compounds like polyethyleneimine (Gu et al., 2013) and similarly potassium alginate ($KC_6H_7O_6$) and calcium alginate ($C_{12}H_{14}CaO_{12}$) are some of the other salts derived from alginates by replacing sodium ions.

Alginates are anionic and hydrophilic in nature. They are widely used in the food industry as thickening agents for ice-cream and drinks and also as a jelly agent in jellies. Calcium alginate derivatives with silver nanoparticles aid in preventing dehydration and food spoilage of vegetables like carrots (Costa et al., 2012). When calcium alginate is combined with nisin, an antibacterial peptide, it aids in suppressing bacterial growth in foods (Cutter & Siragusa, 1996). Similarly, in place of nisin, benzoic acid, propionic acid, ascorbic acid, and lactic acid are also used as antibacterial agents. Alginate cast films comprised of glycerol and sorbitol compounds are found to improve the mechanical properties of food materials (Jost et al., 2014). Alginate films mixed with cellulose exhibit high thickness due to the cross-linkage of alginate with a fibrous cellulose structure (Rhim, 2004).

2.4 Polysaccharides from microorganisms

The ability of microorganisms to produce a film-forming ability like pullulan, gellan gum, xanthan gum, and fucopal comes under this category.

2.4.1 Pullulan

Pullulans are extracellular water-soluble neutral microbial polysaccharides and are known to produce exopolysaccharides (EPS). They are synthesized by the yeast-like fungus called *A. pullulan* (*Aureobasidium pullulans*) available abundantly in

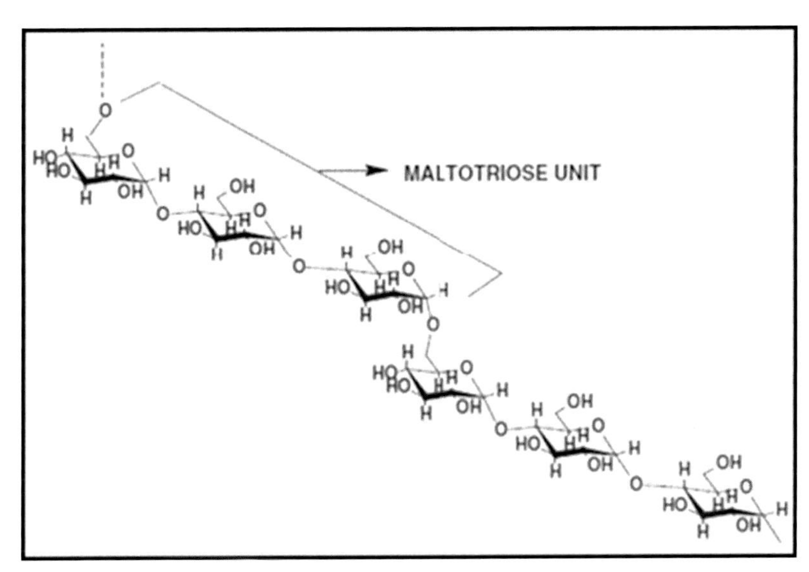

Figure 25.9 Pullulan repeating structures (Singh & Saini, 2012).

starch, sugar cultures (Cheng et al., 2010), and also from *Caesulia axillaris* flowers (Choudhury et al., 2011), forest soil, plant and animal tissues, etc. (Wu et al., 2012) (Fig. 25.9).

Pullulans are linear repeatable polymers comprising glycopyranosyl built of maltotriose subunits that are interlinked by triglucosidic linkages. Some of the parameters associated with the synthesis of pullulan are temperature, O_2 and N_2 concentration, pH, carbon source, etc. Pullulans in the food industry are used to enhance the quality of food in the form of films, additives, coatings, seasoning agents, stabilizers, intensifiers, and binders because of their inherent properties like biodegradability, inhibiting fungal growth, nontoxicity, odorless, and tasteless. As intensifiers, they are one of the major ingredients in the making of beverages like soups and sausages (Yatmaz & Turhan, 2012). They are used as low-calorie dietary fiber because of their excellent resistance to mammalian amylases (Oku et al., 1979), and they partially replace starch in baked products and cookies. Food blends of alginate, chitosan, cellulose, and starch are known for their thermal stability and better mechanical properties. They are used as stabilizers and binders for maintaining the texture and quality of mayonnaise (Singh et al., 2008) and food paste, and also to adhere nuts to cookies. Pullulans are also used as syrups that have inherent properties of moisture stability, mild sweetness, low color formulations, and better thermal stability (Singh et al., 2010). They are also used as coatings for packaged dry fruits, nuts, vegetables, noodles, and meats (Krochta & De Mulder-Johnston, 1997).

2.4.2 Gellan gum

Gellan gum is a biopolymer, specifically an anionic polysaccharide, derived from several strains of *Sphingomonas paucimobilis* bacteria. Its basic structure is comprised

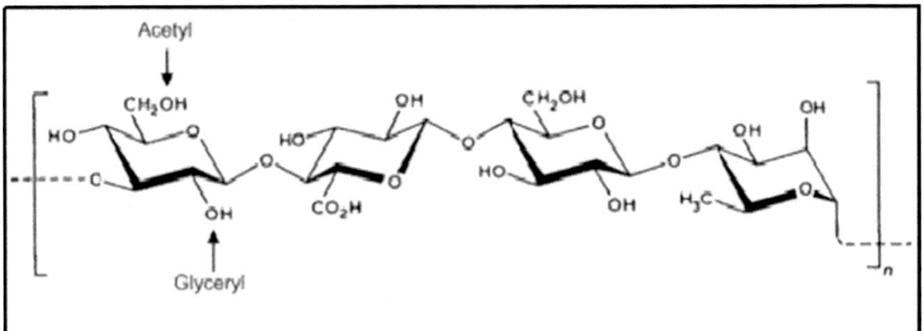

→ 3)-β-D-Glc*p*-(1 → 4)-β-D-Glc*p*A-(1 → 4)-β-D-Glc*p*-(1 → 4)-α-L-Rha*p*-(1 →

Figure 25.10 Structure of native gellan gum (Iurciuc et al., 2016).

of repeating tetrasaccharides of β-D-glucose in two units, one unit of β-D-glucuronic acid and α-L-rhamnose, and is called native gellan gum. It comes under high acetyl group gels that possess properties of softness, elasticity, nonbrittleness, and thermally reversible. On the other hand, deacetylated gellan gum is another type of gellan gum which comes under low acetyl gellan groups with the removal of acetyl groups from the native gellan gum. Low acetyl gellans are rigid, brittle, and thermally stable in nature (Fig. 25.10).

They are widely used in the food industry as stabilizers, gelling agents, film-forming agents, and thickening agents. Gellan gum is commercially sold under various brands like Kelcogel which is used as a thickening and gelling agent, while Gelrite, Phytagel, and Gel-Gro are used as solidifying agents in the food industry (Bajaj et al., 2007). FDA approved gellan gum as a food additive from 1982. Gellan gum provides a wide range of textures, from brittle to elastic. It is more effective in foods with usage in low concentrations and can be abundantly available. It is used in confectionary items like jellies, fillings, and marshmallows for structure, texture, and to reduce the set time. In instant desserts and canned puddings it is used as a starch replacer. In dietary products like gelled milk, ice-creams, and yogurts it is used to prevent moisture loss and as an agar replacer. Gellan gum is also used as a protective membrane and coating in batters for chicken and fish, and provides an effective barrier for the prevention of oil absorption and maintains crispiness (CP Kelco).

2.4.3 Xanthan gum

Xanthan gum is one of the industrial biopolymers produced from a biosource, *Xanthomonas campestris*. It is a heteropolysaccharide unit comprised of glucose, mannose, pyruvate, and acetyl units. Xanthan gums are water soluble, nontoxic, and exhibit high viscosity and high stability under varying pH and temperature (Freitas et al., 2013; Faria et al., 2011). They are expensive and used commercially in the food industry in various products such as powder beverages, fruits, pulps, vegetables, froze foods, sausages, etc. for providing texture, appearance, flavor, viscosity and enhanced oil recovery in foods.

3. Microbial polymers

3.1 Polyhydroxyalkanoates

Polyhydroxyalkanoates (PHAs) are biobased polymers chemically synthesized by natural or biobased bacterial fermentation of sugar and lipids (Lu et al., 2009). Thus, they are classified under biodegradable, biocompatible, and renewable biobased polymers and are considered as green polymers. They are used as substitutes for conventional plastics such as polypropylene and low-density polyethylene (LDPE) since they mimics their qualities in terms of physiochemical, mechanical, and thermal parameters. PHAs are conventionally polymerized under the influence of a wide variety of bacteria or microbes that accrued PHAs as carbon and energy reserves in the form of intercellular granules. Based on the monomeric unit that comprises carbon atoms, they are classified into two groups; one with three to five carbon atoms that are described as short-chain length PHAs (scl-PHAs) and the other with 6–14 carbon atoms that are considered as medium-chain length PHAs (mcl-PHAs). Commercially they are derived from municipal sludge, palm oil mill effluent, marine sediments, etc. PHAs are derived from *Alcaligenes latus*, *Cupriavidus necator*, and *Pseudomonas putida* bacterial tissue cultures, which under anerobic conditions digest these municipal sludges into volatile fatty acids like propionic, acetic, and butyric acids and further polymerizes them into PHAs. The properties of PHAs are improved by combining them with enzymes, inorganic materials, and other biopolymers. Polyhydroxybutyrate (PHB) is a type of PHA that is commercially used in various forms like fibers, films, and molds. It is crystalline in nature, water soluble, and possesses good barrier properties, and also biodegradable with good mechanical properties. PHA polymers are widely used in the food-packaging industry in the form of boxes, films, coatings, fibers, and foams. The films are used for packing perishable and short-life foods like vegetables, fruits, and dairy products. Bubble sheets are used for vegetable packaging. Molded PHA products are used as egg crates, food containers, vegetable crates, and food servicing items.

3.2 Polylactic acid

Polylactic acid is a bioderived polymer that is a thermoplastic biopolyester derived from L-lactic acid. The biosource of L-Lactic acid is corn starch, which is fermented and biodegraded by bacteria and fungi. PLAs are easily processable, biodegradable, transparent, high permeable, printable, and have low thermal resistance. However, due to their poor strength and barrier properties, their use in the food-packaging industry is limited to cups, films, and containers for short-life products. In order to improve the mechanical and barrier properties of PLA, it is blended with PHB or PHA.

4. Conclusion

Packaging is intended to avoid the destruction of food by microbial or insect attacks, in combination with the option of packaging material having sufficient air and water barrier properties. In terms of both functional and cost-effectiveness, fully biodegradable polymers have been shown to be inferior to nondegradable packaging materials. In this chapter it is concluded that biodegradable polymers such as poly(hydroxy butyrate) (PHB), PVA, chitosan, pullulan, and other modified starches (i.e., propylated hydroxy, ethylated) have potential on their own or as blended components for food-packaging applications, with several companies already manufacturing them on a large scale.

References

Alotaibi, S., & Tahergorabi, R. (2018). Development of a sweet potato starch-based coating and its effect on quality attributes of shrimp during refrigerated storage. *LWT, 88*, 203–209.

Arroyo, B. J., Santos, A. P., Melo, E. D. A. D., Campos, A., Lins, L., & Boyano-Orozco, L. C. (2019). *Bioactive compounds and their potential use as ingredients for food and its application in food packaging* (pp. 143–156). Amsterdam, Netherland: Elsevier.

Babu, R., O'Connor, K., & Seeram, R. (2013). Current progress on bio-based polymers and their future trends. *Progess in Biomaterials, 2*, 1–16.

Baek, S.-K., Kim, S., & Song, K. B. (2019). Cowpea starch films containing maqui berry extract their application in salmon packaging. *Food Package Shelf Life, 22*, 100394.

Bajaj, I. B., Survase, S. A., Saudagar, P. S., & Singhal, R. S. (2007). Gellan gum: Fermentative production, downstream processing and applications. *Food Technology and Biotechnology, 45*, 341.

Bierhalz, A. C. K., Da Silva, M. A., & Kieckbusch, T. G. (2012). Natamycin release from alginate/pectin films for food packaging applications. *Journal of Food Engineering, 110*(1), 18–25.

Blohm, S., & Heinze, T. (2019). Synthesis and properties of thermoplastic starch laurates. *Carbohydrate Research, 486*, 107833.

Carocho, M., Morales, P., & Ferreira, I. C. F. R. (2015). Natural food additives: Quo vadis. *Trends in Food Science and Technology, 45*, 284–295.

Castillo, L. A., López, O. V., García, M. A., Villar, M. A., & Barbosa, S. E. (2017). Biodegradable composites based on thermoplastic starch and Talc nanoparticles. *Handbook of Composites from Renewable Materials, 5*, 23–59.

Cerqueira, M. A., Bourbon, A. I., Pinheiro, A. C., Martins, J. T., Souza, B. W. S., Teixeira, J. A., & Vicente, A. A. (2011). Galactomannans use in the development of edible films/coatings for food applications. *Trends in Food Science and Technology, 22*, 662–671.

Chaudhary, P., Fatima, F., & Kumar, A. (2020). Relevance of nanomaterials in food packaging and its advanced future prospects. *Journal of Inorganic and Organometallic Polymers, 30*, 5180–5192.

Chawla, P. R., Bajaj, I. B., Survase, S. A., & Singhal, R. S. (2009). Microbial cellulose: Fermentative production and applications. *Food Technology and Biotechnology, 47*(2), 107–124.

Chen, H. M., Yan, X. J., Wang, F., Xu, W. F., & Zhang, L. (2010). Assessment of the oxidative cellular toxicity of carrageenan oxidative degradation product towards Caco2 cells. *Food Research International, 43*(10), 2390–2401.

Cheng, K. C., Demirci, A., Catchmark, J. M., & Puri, V. M. (2010). Modeling of pullulan fermentation by using a color variant strain of *Aureobasidium pullulans*. *Journal of Food Engineering, 98*, 353–359.

Choudhury, A. R., Saluj, P., & Prasa, G. S. (2011). Pullulan production by an osmotolerant Aureobasidium pullulans RBF-4A3 isolated from flowers of *Caesulia axillaris*. *Carbohydrate Polymers, 83*, 1547–1552.

Costa, C., Conte, A., Buonocore, G. G., Lavorgna, M., & Del Nobile, M. A. (2012). Calcium-alginate coating loaded with silver-montmorillonite nanoparticles to prolong the shelf-life of fresh-cut carrots. *Food Research International, 48*(1), 164–169.

Credou, J., & Berthelot, T. (2014). Cellulose: from biocompatible to bioactive material. *Journal of Materials Chemistry B, 2*, 4767–4788.

Cruz-Romero, M., & Kerry, J. P. (2008). Crop-based biodegradable packaging and its environmental implications. *CAB Reviews Perspectives in Agriculture Veterinary Science Nutrition and Natural Resources, 3*, 1–25.

CP Kelco. Kelcogel®gellan gum book 5th edition. Available online: http://www.appliedbioscience.com/docs/Gellan_Book_5th_Edition.pdf (accessed on 22 February 2016).

Cutter, C. N., & Siragusa, G. R. (1996). Reduction of Brochothrix thermosphacta on beef surfaces following immobilization of nisin in calcium alginate gels. *Letters in Applied Microbiology, 23*(1), 9–12.

Dai, L., Zhang, J., & Cheng, F. (2019). Effects of starches from different botanical sources and modification methods on physicochemical properties of starch-based edible films. *International Journal of Biological Macromolecules, 132*, 897–905.

Debeaufort, F., Quezada-Gallo, J.-A., & Voilley, A. (1998). Edible films and coatings: tomorrow's packagings: a review. *Critical Reviews in Food Science and Nutrition, 38*, 299–313.

Duan, J., Daeschel, M., Zhao, Y., & Park, S.-I. (2007). Antimicrobial chitosan-lysozyme (CL) films and coatings for enhancing microbial safety of mozzarella cheese. *Journal of Food Science, 72*, M355–M362.

Duan, J., Reddy, K. O., Ashok, B., Cai, J., Zhang, L., & Rajulu, A. V. (2016). Effects of spent tea leaf powder on the properties and functions of cellulose green composite films. *Journal of Environmental Chemical Engineering, 4*, 440–448.

Dul, M., Paluch, K., Kelly, H., Healy, A., Sasse, A., & Tajber, L. (2015). Self-assembled carrageenan/protamine polyelectrolyte nanoplexes—Investigation of critical parameters governing their formation and characteristics. *Carbohydrate Polymers, 123*. https://doi.org/10.1016/j.carbpol.2015.01.066

Eddin, A. S., & Tahergorabi, R. (2019). Efficacy of sweet potato starch-based coating to improve quality and safety of hen eggs during storage. *Coatings, 9*, 205.

Edgar, K. J., Buchanan, C. M., Debenham, J. S., Rundquist, P. A., Seiler, B. D., Shelton, M. C., & Tindall, D. (2001). Advances in cellulose ester performance and application. *Progress in Polymer Science, 26*(9), 1605–1688.

El-Obeid, T., Yehia, H. M., Sakkas, H., Lambrianidi, L., Tsiraki, M. I., & Savvaidis, I. N. (2018). Shelf-life of smoked eel fillets treated with chitosan or thyme oil. *International Journal of Biological Macromolecules, 114*, 578–583.

Engel, J. B., Ambrosi, A., & Tessaro, I. C. (2019). Development of biodegradable starch-based foams incorporated with grape stalks for food packaging. *Carbohydrate Polymers, 225*, 115234.

Epure, V., Griffon, M., Pollet, E., & Avérous, L. (2011). Structure and properties of glycerol-plasticized chitosan obtained by mechanical kneading. *Carbohydrate Polymers, 83*, 947–952.

Falguera, V., Quintero, J. P., Jiménez, A., Muñoz, J. A., & Ibarz, A. (2011). Edible films and coatings: structures, active functions and trends in their use. *Trends in Food Science and Technology, 22*, 292–303.

Faria, S., de Oliveira Petkowicz, C. L., de Morais, S. A. L., Terrones, M. G. H., de Resende, M. M., de França, F. P., & Cardoso, V. L. (2011). Characterization of xanthan gum produced from sugar cane broth. *Carbohydrate Polymers, 86*, 469–476.

Fayaz, A. M., Balaji, K., Girilal, M., Kalaichelvan, P., & Tvenkatesan, R. (2009). Mycobased synthesis of silver nanoparticles their incorporation into sodium alginate films for vegetable fruit preservation. *Journal of Agricultural and Food Chemistry, 57*, 6246–6625.

Franco, P., Aliakbarian, B., Perego, P., Reverchon, E., & De Marco, I. (2018). Supercritical adsorption of quercetin on aerogels for active packaging applications. *Industrial and Engineering Chemistry Research, 57*, 15105–15113.

Freitas, F., Alves, V. D., Coelhoso, I., & Reis, M. A. M. (2013). Production and food applications of microbial biopolymers. In J. A. Teixeira, & A. A. Vicente (Eds.), *Engineering Aspects of Food Biotechnology* (vol. 104). Boca Raton, FL,USA: CRC Press.

Freitas, F., Alves, V. D., Reis, M. A., Crespo, J. G., & Coelhoso, I. M. (2014). Microbial polysaccharide-based membranes: Current and future applications. *Journal of Applied Polymer Science*, 131.

Gacesa, P., & Russell, N. J. (1990). The structure properties of alginate. In *Pseudomonas Infection alginates* (pp. 29–49). Dordrecht: Springer.

Gelin, K., Bodin, A., Gatenholm, P., Mihranyan, A., Edwards, K., & Strømme, M. (2007). Characterization of water in bacterial cellulose using dielectric spectroscopy and electron microscopy. *Polymer, 48*, 7623–7631. https://doi.org/10.1016/j.polymer.2007.10.039

Go, E.-J., & Song, K. B. (2019). Antioxidant properties of rye starch films containing rosehip extract and their application in packaging of chicken breast. *Starch-Stärke, 71*.

Granger, C., Leger, A., Barey, P., Langendorff, V., & Cansell, M. (2005). Influence of formulation on the structural networks in ice cream. *International Dairy Journal, 15*, 255–262. https://doi.org/10.1016/j.idairyj.2004.07.009

Gu, C. H., Wang, J. J., Yu, Y., Sun, H., Shuai, N., & Wei, B. (2013). Biodegradable multilayer barrier films based on alginate/polyethyleneimine and biaxially oriented poly (lactic acid). *Carbohydrate Polymers, 92*(2), 1579–1585.

Han, J. H. (2014). Edible films and coatings. In *Innovations in food packaging* (pp. 213–255). Amsterdam, Netherland: Elsevier.

Henriette, M. C., Azeredo, B., Farinas, C. S., Vasconcellos, V. M., & Amanda, M. (2019). Claro3Bacterial cellulose as a raw material for food and food packaging applications. *Frontiers in Sustainable Food Systems, 3*(7), 1–14. https://doi.org/10.3389/fsufs.2019.00007

Homayouni, H., Kavoosi, G., & Nassiri, S. M. (2017). Physicochemical, antioxidant and antibacterial properties of dispersion made from tapioca and gelatinized tapioca starch incorporated with carvacrol. *LWT, 77*, 503–509.

Iurciuc, C., Savin, A., Lungu, C., Martin, P., & Popa, M. (2016). Gellan food applications. *Cellulose Chemistry and Technology, 50*(1), 1–13.

Jayakumar, A., Heera, K. V., Sumi, T. S., Joseph, M., Mathew, S., Praveen, G., Nair, I. C., & Radhakrishnan, K. R. (2019). Starch-PVA composite films with zinc-oxide nanoparticles and phytochemicals as intelligent pH sensing wraps for food packaging application. *International Journal of Biological Macromolecules, 136*, 395–403.

Jiménez, A., Fabra, M. J., Talens, P., & Chiralt, A. (2012). Effect of sodium caseinate on properties and ageing behaviour of corn starch based films. *Food Hydrocolloids, 29*, 265–271.

Jost, V., Kobsik, K., Schmid, M., & Noller, K. (2014). Influence of plasticiser on the barrier, mechanical and grease resistance properties of alginate cast films. *Carbohydrate Polymers, 110*, 309–319.

Kontogiorgos, V. (2018). *Galactomannans (guar, locust bean, fenugreek, tara). Encyclopedia of Food Chemistry* (pp. 109–113). Amsterdam, Netherlands: Elsevier. https://doi.org/10.1016/B978-0-08-100596-5.21589-8

Krochta, J. M., & De Mulder-Johnston, C. (1997). Edible and biodegradable polymer films: challenges and opportunities. *Food Technology, 51*, 61–74.

Kuo, C. M., & Leonard, A. P. (1984). *Process for esterification of cellulose using as the catalyst the combination of sulfuric acid, phosphoric acid, and a hindered aliphatic alcohol, US Patent 4480090 A*. assigned to Eastman Kodak Company.

Lau, O. W., & Wong, S. K. (2000). Contamination in food from packaging material. *Journal of Chromatography A, 882*, 255.

Lin, K. W., & Lin, H. Y. (2004). Quality characteristics of Chinese-style meatball containing bacterial cellulose (Nata). *Journal of Food Science, 69*, Q107–Q111. https://doi.org/10.1111/j.1365-2621.2004.tb13378.x

Liu, S., Li, X., Chen, L., Li, L., Li, B., & Zhu, J. (2017). Understanding physicochemical properties changes from multiscale structures of starch/CNT nanocomposite films. *International Journal of Biological Macromolecules, 104*, 1330–1337.

Lu, J., Tappel, R. C., & Nomura, C. T. (2009-08-05). Mini-Review: Biosynthesis of poly(-hydroxyalkanoates). *Polymer Reviews, 49*(3), 226–248.

Mathew, S., Jayakumar, A., Kumar, V. P., Mathew, J., & Radhakrishnan, E. (2019). One-step synthesis of eco-friendly boiled rice starch blended polyvinyl alcohol bionanocomposite films decorated with in situ generated silver nanoparticles for food packaging purpose. *International Journal of Biological Macromolecules, 139*, 475–485.

Mati-Baouche, N., Elchinger, P.-H., de Baynast, H., Pierre, G., Delattre, C., & Michaud, P. (2014). Chitosan as an adhesive. *European Polymer Journal, 60*, 198–212.

Mc Clements, D. J. (2018). Recent developments in encapsulation and release of functional food ingredients: delivery by design. *Current Opinion in Food Science, 23*, 80–84.

Meng, L., Liu, H., Yu, L., Duan, Q., Chen, L., Liu, F., Shao, Z., Shi, K., & Lin, X. (2019). How water acting as both blowing agent and plasticizer affect on starch-based foam. *Industrial Crops and Products, 134*, 43–49.

Miller, S. A. (2014). Sustainable polymers: replacing polymers derived from fossil fuels. *Polymer Chemistry, 5*, 3117.

Mlalila, N., Hilonga, A., Swai, H., Devlieghere, F., & Ragaert, P. (2018). Antimicrobial packaging based on starch, poly(3-hydroxybutyrate) and poly(lactic-co-glycolide) materials and application challenges. *Trends in Food Science and Technology, 74*, 1–11.

Moreno, O., Atarés, L., Chiralt, A., Cruz-Romero, M. C., & Kerry, J. (2018). Starch-gelatin antimicrobial packaging materials to extend the shelf life of chicken breast fillets. *LWT, 97*, 483–490.

Necas, J., & Bartosikova, L. (2013). Carrageenan: a review. *Veterinary Medicine, 58*, 187–205.

Nešić, A., Cabrera-Barjas, G., Dimitrijević-Branković, S., Davidović, S., Radovanović, N., & Delattre, C. (2020). Prospect of polysaccharide-based materials as advanced food packaging. *Molecules, 25*(1), 135. https://doi.org/10.3390/molecules25010135

Ojagh, S. M., Rezaei, M., Razavi, S. H., & Hosseini, S. M. H. (2010). Effect of chitosan coatings enriched with cinnamon oil on the quality of refrigerated rainbow trout. *Food Chemistry, 120*, 193–198.

Okiyama, A., Motoki, M., & Yamanaka, S. (1993). Bacterial cellulose IV. Application to processed foods. *Food Hydrocolloids, 6*, 503–511. https://doi.org/10.1016/S0268-005X(09)80074-X

Oku, T., Yamada, K., & Hosoya, N. (1979). *Nutrition and Food Science, 32*, 235–241.

Ortega-Toro, R., Morey, I., Talens, P., & Chiralt, A. (2015). Active bilayer films of thermoplastic starch and polycaprolactone obtained by compression molding. *Carbohydrate Polymers, 127*, 282–290.

Panrong, T., Karbowiak, T., & Harnkarnsujarit, N. (2019). Thermoplastic starch and green tea blends with LLDPE films for active packaging of meat and oil-based products. *Food Package Shelf Life, 21*, 100331.

Peighambardoust, S. J., Pournasir, N. M., & Pakdel, P. M. (2019). Properties of active starch-based films incorporating a combination of Ag, ZnO and CuO nanoparticles for potential use in food packaging applications. *Food Package Shelf Life, 22*, 1004210.

Pelissari, F. M., Ferreira, D. C., Louzada, L. B., Dos Santos, F., Corrêa, A. C., Moreira, F. K. V., & Mattoso, L. H. (2019). *Starch-based edible films and coatings* (pp. 359–420). Amsterdam, Netherland: Elsevier.

Pereda, M., Aranguren, M. I., & Marcovich, N. E. (2009). Water vapor absorption and permeability of films based on chitosan and sodium caseinate. *Journal of Applied Polymer Science, 111*, 2777–2784.

Piñeros-Hernandez, D., Medina-Jaramillo, C., López-Córdoba, A., & Goyanes, S. (2017). Edible cassava starch films carrying rosemary antioxidant extracts for potential use as active food packaging. *Food Hydrocolloids, 63*, 488–495.

Qin, Y., Liu, Y., Yong, H., Liu, J., Zhang, X., & Liu, J. (2019). Preparation and characterization of active and intelligent packaging films based on cassava starch and anthocyanins from Lycium ruthenicum Murr. *International Journal of Biological Macromolecules, 134*, 80–90.

Qin, Y., Liu, Y., Zhang, X., & Liu, J. (2020). Development of active and intelligent packaging by incorporating betalains from red pitaya (*Hylocereus polyrhizus*) peel into starch/polyvinyl alcohol films. *Food Hydrocolloids, 100*, 105410.

Ramesh, H. P., & Tharanathan, R. N. (2003). Carbohydrates - the renewable raw materials of high biotechnological value. *Critical Reviews in Biotechnology, 23*, 149–173.

Rhim, J. W. (2004). Physical and mechanical properties of water resistant sodium alginate films. *Lebensmittel-Wissenschaft und -Technologie- Food Science and Technology, 37*(3), 323–330.

Rhim, J. W. (2013). Effect of PLA lamination on performance characteristics of agar/κ-carrageenan/clay bio-nanocomposite film. *Food Research International, 51*, 714–722.

Rinaudo, M. (2006). Chitin and chitosan: properties and applications. *Progress in Polymer Science, 31*, 603–632.

Rodríguez, L. H., Calleros, C. L., González, D. J. P., & Carter, E. J. V. (2014). Lactobacillus plantarum protection by entrapment in whey protein isolate: κ-carrageenan complex coacervates. *Food Hydrocolloids, 36*, 181–188.

Rubio, A. L., Almenar, E., Munoz, P. H., Lagaro, J. M., Catala, R., & Gavara, R. (2004). Biobased plastics for food packaging applications. *Food Reviews International, 20*, 357

Shahabi-Ghahfarrokhi, I., & Babaei-Ghazvini, A. (2019). Using photo-modification to compatibilize nano-ZnO in development of starch-kefiran-ZnO green nanocomposite as food packaging material. *International Journal of Biological Macromolecules, 124*, 922–930.

Shimao, m (2001). Biodegradation of plastics. *Current Opinion in Biotechnology, 12*, 242.

Singh, R. S., & Saini, G. K. (2012). Biosynthesis of pullulan and its applications in food and pharmaceutical industry. In T. Satyanarayana, B. Johri, & A. Prakash (Eds.), *Microorganisms in sustainable agriculture and biotechnology*. Dordrecht: Springer. https://doi.org/10.1007/978-94-007-2214-9_24

Singh, R. S., Saini, G. K., & Kennedy, J. F. (2008). Pullulan: microbial sources, production and applications. *Carbohydrate Polymers, 73*, 515–531.

Singh, R. S., Saini, G. K., & Kennedy, J. F. (2010). Maltotriose syrup preparation from pullulan using pullulanase. *Carbohydrate Polymers, 80*, 401–407.

Takeshita, S., & Yoda, S. (2015). Chitosan aerogels: transparent, flexible thermal insulators. *Chemistry of Materials, 27*, 7569–7572.

Tavares, K. M., De Campos, A., Mitsuyuki, M. C., Luchesi, B. R., & Marconcini, J. M. (2019). Corn and cassava starch with carboxymethyl cellulose films and its mechanical and hydrophobic properties. *Carbohydrate Polymers, 223*, 115055.

Tayeb, A. H., Amini, E., Ghasemi, S., & Tajvidi, M. (2018). Cellulose nanomaterials-binding properties and applications: a review. *Molecules, 23*, 1–24.

Thakur, V. K., & Thakur, M. K. (2016). *Handbook of sustainable polymers: processing and applications*. Singapore: Pan Stanford Publishing.

Thakur, R., Pristijono, P., Bowyer, M., Singh, S. P., Scarlett, C. J., Stathopoulos, C. E., & Vuong, Q. V. (2019). A starch edible surface coating delays banana fruit ripening. *LWT, 100*, 341–347.

Ubeyitogullari, A., & Ciftci, O. N. (2016). Phytosterol nanoparticles with reduced crystallinity generated using nanoporous starch aerogels. *RSC Advances, 6*, 108319–108327.

Vasilatos, G., & Savvaidis, I. (2013). Chitosan or rosemary oil treatments, singly or combined to increase Turkey meat shelf-life. *International Journal of Food Microbiology, 166*, 54–58.

Vieira, B. B., Mafra, J. F., Bispo, A. S. D. R., Ferreira, M. A., Silva, F. D. L., Rodrigues, A. V. N., & Evangelista- Barreto, N. S. (2019). Combination of chitosan coating and clove essential oil reduces lipid oxidation and microbial growth in frozen stored tambaqui (*Colossoma macropomum*) fillets. *LWT, 116*, 108546.

Villegas, M., Oliveira, A. L., Bazito, R. C., & Vidinha, P. (2019). Development of an integrated one-pot process for the production and impregnation of starch aerogels in supercritical carbon dioxide. *The Journal of Supercritical Fluids, 154*, 104592.

Volpe, V., De Feo, G., De Marco, I., & Pantani, R. (2018). Use of sunflower seed fried oil as an ecofriendly plasticizer for starch and application of this thermoplastic starch as a filler for PLA. *Industrial Crops and Products, 122*, 545–552.

Wu, S., Chen, J., & Pan, S. (2012). Optimization of fermentation conditions for the production of pullulan by a new strain of Aureobasidium pullulans isolated from sea mud and its characterization. *Carbohydrate Polymers, 87*, 1696–1700.

Wu, C., Sun, J., Zheng, P., Kang, X., Chen, M., Li, Y., Ge, Y., Hu, Y., & Pang, J. (2019). Preparation of an intelligent film based on chitosan/oxidized chitin nanocrystals incorporating black rice bran anthocyanins for seafood spoilage monitoring. *Carbohydrate Polymers, 222*, 115006.

Yang, S.-Y., Cao, L., Kim, H., Beak, S.-E., & Bin Song, K. (2018). Utilization of foxtail millet starch film incorporated with clove leaf oil for the packaging of queso blanco cheese as a model food. *Starch-Stärke, 70*, 170017.

Yatmaz, E., & Turhan. (2012). Pullulan production by fermentation and usage in food industry. *Food, 37*(2), 95–102.

Yıldırım-Yalçın, M., Şeker, M., & Sadıkoğlu, H. (2019). Development and characterization of edible films based on modified corn starch and grape juice. *Food Chemistry, 292*, 6–13.

Zemljič, L. F., Tkavc, T., Vesel, A., & Šauperl, O. (2013). Chitosan coatings onto polyethylene terephthalate for the development of potential active packaging material. *Applied Surface Science, 265*, 697–703.

Zhao, Y., Teixeira, J. S., Saldaña, M. D., & Gänzle, M. G. (2019). Antimicrobial activity of bioactive starch packaging films against Listeria monocytogenes and reconstituted meat microbiota on ham. *International Journal of Food Microbiology, 305*, 108253.

Zhu, F. (2017). Encapsulation and delivery of food ingredients using starch based systems. *Food Chemistry, 229*, 542–552.

Biodegradable polymers and green-based antimicrobial packaging materials

C. Vibha [1], *Jyotishkumar Parameswaranpillai* [6], *Senthilkumar Krishnasamy* [1], *Suchart Siengchin* [5], *Aswathy Jayakumar* [1], *Sabarish Radoor* [1], *Sanjay Mavinkere Rangappa* [5], *Nisa V. Salim* [2], *Nishar Hameed* [2], *G.L. Praveen* [3] *and C.D. Midhun Dominic* [4]

[1]Department of Materials and Production Engineering, The Sirindhorn International Thai–German Graduate School of Engineering (TGGS), King Mongkut's University of Technology North Bangkok, Bangkok, Thailand; [2]Factory of the Future, Swinburne University of Technology, Hawthorn, VIC, Australia; [3]Wimpey Laboratories, Wimpey Building, Al Quoz, Dubai, United Arab Emirates; [4]Department of Chemistry, Sacred Heart College, Cochin, Kerala, India; [5]Natural Composite Research Group, Department of Materials and Production Engineering, The Sirindhorn International Thai–German Graduate School of Engineering (TGGS), King Mongkut's University of Technology North Bangkok, Bangkok, Thailand; [6]Department of Bioscience, Mar Athanasios College for Advanced Studies Thiruvalla (MACFAST), Pathanamthitta, Kerala, India

1. Introduction

The production of plastics-based materials is around 300 million tons every year, where half of these plastics are used for single-use products. The excessive use of plastics causes environmental pollution worldwide. Moreover, these plastics generate carbon emissions into the atmosphere due to burning. The nonbiodegradability of plastic waste affects both living and nonliving things. The most commonly used plastics in food-packaging industries are polyethylene terephthalate (PET), polyethylene (PE), and polypropylene (PP). These plastics are nonbiodegradable and if they are discarded or thrown away, they can remain in the environment for many years. These discarded plastics can be seen in water bodies, land, oceans, animals, birds, and fish (Abd El-Rahman et al., 2020; Ngaowthong et al., 2019; Padervand et al., 2020). For instance, over eight million tons of plastics end up in the ocean per year, and most have been found to be sedimented from the top surface to the deep ocean.

Although conventional plastics are toxic and nonbiodegradable, they are widely used in many industrial applications because of their versatility, low cost, and easy processability. However, the nonbiodegradability and damage caused by the polymer have encouraged scientists to concentrate more on the development of biodegradable plastics such as polylactic acid (PLA), poly(ε-caprolactone) (PCL), poly(butylene succinate adipate) (PBSA), poly(butylene adipate-co-terephthalate) (PBAT), and

Biodegradable Polymers, Blends and Composites. https://doi.org/10.1016/B978-0-12-823791-5.00005-3

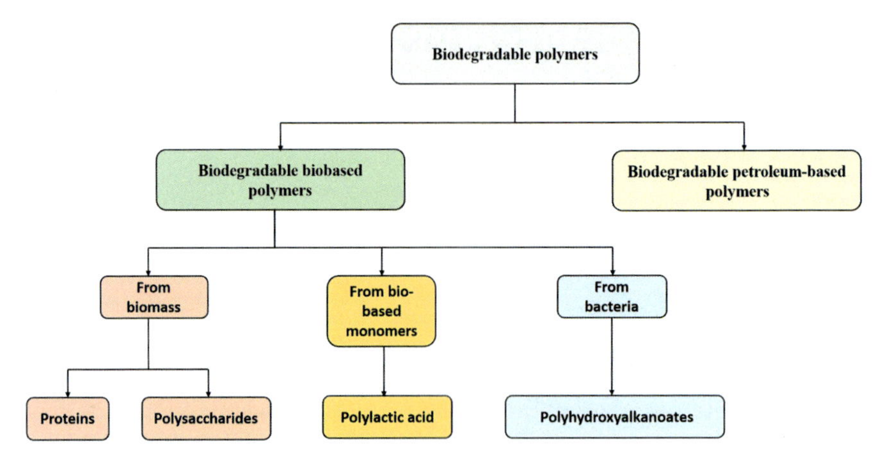

Figure 26.1 Classification of biodegradable polymers.

polyvinyl alcohol (PVOH). These biodegradable plastics offer many advantages such as being more compatible, environmentally friendly, renewable, sustainable, and nontoxic. The degradation of biodegradable plastics is induced either by (1) microorganisms through enzymatic catalysis or (2) by chemical hydrolysis resulting in H_2O, CO_2, and biomass as the end-products (Zhong et al., 2020).

The biodegradable polymers could be broadly classified as biodegradable biobased polymers and biodegradable petroleum-based polymers (Varghese et al., 2020). Fig. 26.1 shows the major and subclassifications of biodegradable polymers.

An ideal food-packaging material should be biodegradable, edible, and antimicrobial with good shelf-life without affecting the quality of the food (Chen et al., 2019). This chapter presents an overview of various biodegradable polymers and their applications in food packaging.

2. Biodegradable biobased polymers

Biodegradable biobased polymers are of biological origin. They include proteins, cellulose, starch, polyhydroxyalkanoates (PHAs), and PLA. Biodegradable biobased polymers are good replacements for synthetic nonbiodegradable polymers for many applications, especially for food-packaging applications. Edible, antimicrobial, packaging films with good shelf-life can be developed using biodegradable biobased polymers, they are also entirely degradable via hydrolytic or enzymatic cleavage. The following section deals with biodegradable biobased polymers, their properties, and their applications in food-packaging industries.

2.1 Biodegradable biobased polymers from biomass

2.1.1 Proteins

Proteins are extracted directly from biomass and are composed of amino acids units (Chen et al., 2019; Ghanbarzadeh & Almasi, 2013). Unlike lipids and carbohydrates, the different functional groups present in amino acids would help to develop protein-based polymers (Riveros et al., 2018). The protein-based biopolymers (i.e., vegetable and animal proteins) had attracted greater attention in food-packaging applications due to their film-forming ability, low price, nutritional value, ease of degradation, and higher gas barrier properties (Zubair & Ullah, 2020). A variety of vegetable proteins and animal proteins such as wheat gluten (Rovera et al., 2020), soy protein (Liang & Wang, 2018), corn zein (Vrabič Brodnjak & Tihole, 2020), casein (Picchio et al., 2018), collagen (Wu et al., 2019), whey protein (Schmid et al., 2017), and gelatin (Khan et al., 2020) are employed in the food-packaging industry. Furthermore, the properties of the protein-based biodegradable polymers can be improved with the incorporation of bionanomaterials, plasma treatments, and coatings. Table 26.1 summarizes the significant observations of recent works reported on various protein-based biopolymers (both vegetable and animal proteins) related to the packing industries.

2.1.2 *Polysaccharides*

Polysaccharides are an essential class of biopolymers made up of monosaccharides. They are biodegradable and nontoxic. Some of the vital polysaccharides are cellulose, chitosan, starch, hemicellulose, dextran, and alginate. Nowadays, many researchers are working on polysaccharide-based films in food-packaging applications, whereby the researchers intend to replace the synthetic polymer films using biodegradable polymers. It is ascribed to the biodegradability and nontoxic nature of biopolymers; moreover, these polymers offer good barrier properties against carbon dioxide (CO_2) and oxygen (O) (Ferreira et al., 2016; Nešić et al., 2020; Swain & Mohanty, 2018). However, these biodegradable polymers possess poor mechanical and low water resistance properties. Consequently, certain modifications of polysaccharide films are required to obtain satisfactory performance to use in food-packaging applications. The following section of the chapter mainly summarizes the recent advances in developing polysaccharide-based films as a food-packaging material.

2.1.2.1 Plant-based polysaccharides

2.1.2.1.1 Cellulose Cellulose is an abundant and environmentally friendly biopolymer; it is found in plant cell walls, fungi and algae, marine organisms, Gram-negative bacteria, and invertebrates (Radoor et al., 2020; Tayeb et al., 2018; Thomas et al., 2021; Vinod et al., 2020). Many researchers have reported that cellulose constitutes ~33% of all plant matter. Cotton and wood contain ~90% and ~50% of natural cellulose contents, respectively (Cellulose).

Since cellulose is a polysaccharide, it possesses many advantages such as nontoxicity, low price, lightweight, and reasonably good mechanical properties; hence it is preferred for use in the packaging industry. Furthermore, the cellulose derived from lignocellulosic biomass (i.e., plant dry waste) has attracted more attention in the food-packaging industry (Khalil et al., 2016; Qasim et al., 2020). For instance, Fig. 26.2 illustrates the production of 100% biodegradable food-packaging materials from lignocellulosic biomass.

Some of the important reported studies on cellulose-reinforced biopolymer composites for sustainable packaging are discussed in this section. Müller et al. (2009), studied the effect of the addition of cellulose fibers in starch films on mechanical

Table 26.1 Significant observations of protein-based biopolymers (both vegetable and animal proteins) related to the packing industries.

Proteins	Biopolymer film	Observation	References
Gluten	Wheat gluten/cellulose nanocrystals (CNC)/titanium dioxide (TiO_2) nanoparticles	The tensile strength, modulus, and contact angle of the wheat gluten film were increased, with the incorporation of CNC and TiO_2 nanoparticles. The water uptake and water permeability of wheat gluten were reduced with the incorporation of CNC and TiO_2 nanoparticles. The antimicrobial activity, breaking strength, and burst index of the kraft paper coated with wheat gluten/CNC/TiO_2 nanoparticles was improved	El-Wakil et al. (2015)
	Cogelation of wheat gluten with gelatin	The addition of gelatin in gluten matrix showed good reinforcement as observed from the increased storage modulus profile from the rheology study	Wesołowska-Trojanowska et al. (2020)
	Gluten + PCL	Gluten/PCL/catalyst system showed decent thermomechanical properties, and is biodegradable. This system is a good choice for shape-memory packaging material	Gutiérrez et al. (2021)
Soy proteins	Defatted soybean meal (DSM)	Cold plasma treatment (CPT) on DSM improved the tensile strength, tensile elongation, moisture barrier, and biodegradability. The study revealed that the CPT-treated DSM are promising materials for food-packaging applications	Oh et al. (2016)
	Halloysite nanotubes (HNTs)/PVOH and 1,2,3-propanetirol-diglycidyl-ether (PTGE)-incorporated soy protein isolate (SPI) films	The incorporation of modifiers synergistically improved the thermomechanical and moisture/water barrier capacity of the SPI films. The modified films are suitable for packaging applications	Liu et al. (2017)
	Pinhão seeds water extract + SPI films	The modified SPI films showed good mechanical and antioxidant properties and were recommended for packaging linseed oil	de Souza et al. (2020)

Zein	Electrospun gelatin/zein nanofiber-based films	Electrospun gelatin/zein nanofibers showed good water and ethanol resistance	Deng et al. (2018)
	Electrospun zein nanofiber coating over chitosan films	The coating of zein nanofiber over chitosan film improved the antioxidant activity. Thus, the use of film prevents the browning of freshly cut fruits	Bharathi et al. (2020)
	Edible electrospun zein fibers by incorporating thyme oil, citric acid, and nisin	The modified electrospun zein fibers showed good antibacterial efficacy and biodegradability	Aytac et al. (2020)
Casein	Casein-based film modified with *Zataraia multiflora* Bioss essential oil (EO)	The incorporation of EO in casein-based film showed good antibacterial activity against *Staphylococcus aureus*	Broumand et al. (2011)
	Casein films containing tannic acid as the cross-linking agent	Tannic acid was successfully employed as a cross-linking agent for casein film	Picchio et al. (2018)
	Extrusion of rennet casein, acid casein, and sodium caseinate	The extruded casein sheets presented reasonably good moisture resistance	Chevalier et al. (2018)
Collagen	Collagen fiber cross-linked with transglutaminase	The cross-linked collagen was modified with casein, keratin, and SPI. The casein-modified film showed the best thermal stability and mechanical properties. Meanwhile the keratin-modified film showed the best moisture resistance	Wu et al. (2019)
	Solution casting of thyme oil-modified collagen hydrolysate (CH) film	The CH was isolated from hide fleshing waste. The incorporation of thyme EO improved the flexibility of CH, however, tensile strength and flexural strength were reduced	Ocak (2020)
	Film prepared by blending collagen, methylcellulose (MC), and whey protein	Despite the different films and blends prepared, neat MC showed the best tensile strength, lowest water vapor permeability, biodegradability, and transparency	da Silva Filipini et al. (2021)

Continued

Table 26.1 Continued

Proteins	Biopolymer film	Observation	References
Whey proteins	Electrospun whey protein isolate (WPI) nanofibers containing guar gum	Bead-free electrospun nanofibers (food-grade) were produced at 9 and 0.7 wt.% of WPI and guar gum, respectively	Ramazani et al. (2019)
	Whey protein films containing lactic acid bacteria with antifungal properties (*Lactobacillus buchneri* UTAD104)	The cheese covered with whey protein incorporated with 30% (w/w) *Lactobacillus buchneri* UTAD104 presented good resistance against *P. nordicum*	Guimarães et al. (2020)
	Whey protein films containing natamycin and nanoemulsioned α-tocopherol	The modified films containing natamycin showed good antimicrobial properties. While the α-tocopherol films showed good antioxidative property. The whey protein films containing both natamycin and α-tocopherol showed good antimicrobial and antioxidative properties	Agudelo-Cuartas et al. (2020)
	Chitosan/rosemary or cinnamon extract functionalized whey protein-based laminate	The modified laminates showed improved antimicrobial and antioxidative properties	Potrč et al. (2020)
Fish proteins	Fish protein concentrate films containing sorbitol and glycerol as plasticizers	The film modified with calcium salts and glucono-δ-lactone showed good tensile properties and water barrier properties	Valdivia-López, Tecante, Granados-Navarrete, & Martínez-García, 2016
	Myofibrillar protein film containing montmorillonite nanoclay and transglutaminase	The incorporation of montmorillonite and transglutaminase improved the tensile strength and water barrier properties of the composite films	Rostamzad et al. (2016)
	Film formed via glow discharge plasma on myofibrillar proteins	The plasma treatment improved the tensile strength and melting temperature of the protein films	Romani et al. (2019)
Silk protein	Self-assembled silk fibroin on the surface of food via dip coating	The silk fibroin coating presents increased shelf-life for fruits such as strawberry and banana by reducing their respiration rate	Marelli et al. (2016)
	Silk protein coated with fluoropolymer	The coating of fluoropolymer over silk protein films improved the water barrier roperty	Numata et al. (2018)

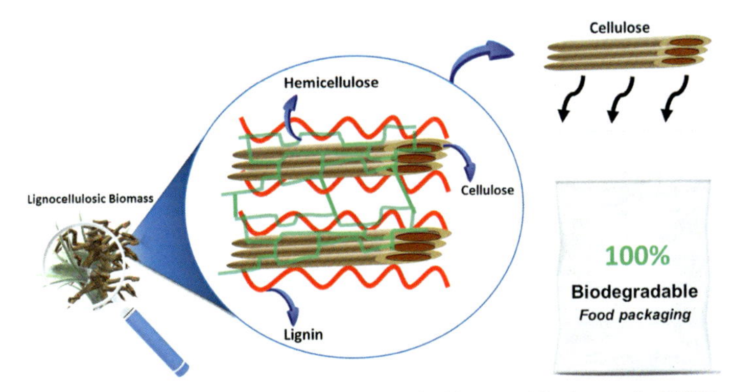

Figure 26.2 Structural constituents of lignocellulosic biomass (Qasim et al., 2020).

and water resistance properties. Analysis of the results revealed that the performance of the nanocomposites reinforced with cellulose fibers was improved due to the effective fiber reinforcement. The reinforcing effect of cellulose fibers on the mechanical and barrier properties of rice floor films was studied by Dias et al. (2011). It was reported that the mechanical and barrier properties of composites were improved. Ghaderi et al. (2014) developed all-cellulose composite films from sugarcane bagasse. The authors observed good thermomechanical properties for the fully biodegradable all cellulose composite films. Balasubramaniam et al. (2020) studied the bulk and surface modification of cellulose nanofiber films using fatty acids (lauric, palmitic, and stearic). The surface-treated films retained the tensile strength and Young's modulus, and improved the water resistance. On the other hand, the bulk modification showed the best water resistance and initial decomposition temperature; however, the tensile properties of the cellulose nanofiber films were reduced. In a recent study, Otenda et al. (2020), studied the effect of the incorporation of TEMPO-oxidized cellulose fibrils in PVOH. The results indicated that the tensile strength was increased and the water solubility was decreased by incorporating treated fibers on PVOH films. However, the transparency of PVOH was reduced.

2.1.2.1.2 Starch Starch is a renewable polysaccharide; it is predominantly available from plants. Starch-based films are biodegradable and a good alternative to thermoplastic food-packaging films (Niranjana Prabhu & Prashantha, 2018). Starches based on rice (Arvanitoyannis et al., 1998), potato (Arvanitoyannis et al., 1998), cassava (Piñeros-Hernandez et al., 2017), sweet potato (Dash et al., 2019), tapioca (Othman et al., 2019), wheat (da Silva et al., 2020), and corn (Othman et al., 2021) are commonly used for food-packaging applications. Starch could be produced in larger quantities at a lower price. The brittle and hydrophilic nature could be the among the challenges of native starch. The advantages of using starch over thermoplastic food-packaging films are its biodegradability, low cost, and that it can be converted into thermoplastic material by employing plasticizers such as sorbitol and glycerol (Khan et al., 2017). Table 26.2 summarizes some of the significant studies into starch, whereby starch was used as a food-packaging material.

Table 26.2 Reported works on food-packaging materials using starch.

Starch	Biopolymer film	Observation	References
Potato starch	Starch/clay nanocomposite films	Intercalation of nanoclay was observed in the starch film. Also, the incorporation of nanoclay in starch improved the mechanical performance of the composites	Avella et al. (2005)
Potato starch	Starch/PVA	Anthocyanins (ANT) and limonene (LIM) were incorporated in the film to identify the change in color with respect to the change in pH of food and to resist the growth of bacteria	Liu et al. (2017)
Starch (Merck, Germany)	Starch films containing Ag, ZnO, and CuO particles	The combination of Ag/ZnO/CuO nanoparticles in starch films showed good antimicrobial and mechanical performance	Peighambardoust et al. (2019)
Corn starch	Corn starch–chitosan	Corn starch/chitosan/sorbitol/grapefruit seed extract-based bionanocomposite films present good thermomechanical and antimicrobial activities	Jha (2020)
Corn starch	Corn starch film reinforced with CNC + essential oil (ho wood, cardamom, and cinnamon) by Pickering emulsion method	The performance of the film having ho wood was better compared with other essential oils	Souza et al. (2021)

2.2 Biopolymers derived from biomonomers

2.2.1 Polylactic acid

Lignocellulosic agricultural waste and starch, upon fermentation, results in biobased monomers such as bioethanol, lactide, etc. (Berezina & Maria Martelli, 2014; Mülhaupt, 2013; Saini et al., 2015). These biomonomers, on further polymerization, result in biopolymers. Lactic acid is one of the important biomonomers obtained by the anaerobic fermentation of sugarcane by the action of bacteria. Lactic acid is used for the manufacture of PLA. PLA is mainly manufactured using the ring-opening polymerization of lactide (Borkotoky et al., 2020). PLA has many advantages, such as biodegradability, easy processability, renewability, biocompatibility, and transparency. Therefore it is employed to manufacture food-packaging materials, biomedical applications, etc. (Djukić-Vuković et al., 2019; Ingrao et al., 2015). Although PLA has many advantages over synthetic polymers, it has many drawbacks such as moisture absorption, low resistance to oxygen permeation, and brittleness compared to synthetic polymers (Jem & Tan, 2020). These drawbacks limit its application in food packaging; however, they can be overcome with the modification of PLA films either by blending with poly(glycolic acid) (PGA), poly(hydroxybutyrate) PHB), etc., or with the incorporation of fillers (starch, wood flour, chitosan, and nanocellulose), essential oil, etc. (Arrieta et al., 2014; Jem & Tan, 2020; Khodayari et al., 2019; Lv et al., 2017; Nazrin et al., 2020; Râpă et al., 2016). Studies have shown that the incorporation of nanomaterials and other active agents makes PLA antimicrobial in addition to its improved performance.

2.3 Biopolymers derived from microorganisms

2.3.1 Polyhydroxyalkanoates

In recent years there has been increasing interest in developing biopolymers from bacteria and fungi. For example, cellulose, dextran, xanthan, and biopolyesters are some of the biopolymers extracted from bacteria (Moradali & Rehm, 2020; Rao et al., 2014). These biopolymers are developed in these microorganisms during their defense mechanism or as a storage material (Sukan et al., 2015). The main limitation of bacterial biopolymers is the production volume and cost. However, recently many researchers have been focusing on improving the production volume and reducing the cost by using different low-cost sustainable sources (Aljuraifani et al., 2019). Among the different bacterial polymers, PHA has great potential in food-packaging applications due to its biocompatibility, nontoxicity, and excellent moisture and oxygen barrier properties (Vijayendra & Shamala, 2014). Different types of PHAs produced by microorganisms include poly-3-hydroxybutyrate (P3HB), poly-4-hydroxybutyrate (P4HB), polyhydroxyvalerate (PHV), and poly(3-hydroxybutyrate-co-3-hydroxyvalerate) (PHBV) (Muneer et al., 2020). These PHAs are tunable solutions for traditional packaging applications. The performance and functional properties, such as the antimicrobial activity of the PHA films, can be improved by incorporating (1) long alkyl chain quaternary salt (LAQ) functionalized graphene oxide (GO-g-LAQ) (Xu et al., 2020), (2) silver nanoparticles (Castro-Mayorga et al., 2017), etc.

3. Biodegradable petroleum-based polymers

Currently, many researchers are interested in biodegradable polymers from petroleum-based resources. Common petroleum-based biodegradable polymers are PGA, PCL, poly(butylene succinate), PBSA, and PBAT. Structurally PGA is similar to PLA. It has good thermomechanical properties, high gas barrier properties, and 100% biodegradation. Owing to these properties, it has been used for food-packaging and biomedical applications (Samantaray et al., 2020). PCL is a semicrystalline polymer synthesized via ring-opening polymerization of ε-caprolactone. It has good processability, biodegradability, and biocompatibility. It is used for both food-packaging and biomedical applications (Gutiérrez & Alvarez, 2017). PBS, PBSA, and PBAT possess good thermomechanical properties, good elasticity, flexibility, processibility, biocompatibility, and have low glass transition temperature (T_g) and are widely employed in food-packaging applications (Hongsriphan & Sanga, 2018; Palai et al., 2020; Tavares et al., 2018).

4. Green packaging antimicrobial agents

In food-packaging applications, in addition to biodegradability, recently antimicrobial, antioxidant bioplastics have been preferred to enhance the shelf-life of food products. This helps to prevent the wastage of food, increases revenue, and removes the adverse effects caused by food contamination during storage. In the last few years, most of the works on food packaging have been concentrated on incorporating functional properties to the food-packaging materials, such as antimicrobial and antioxidant properties (Jafarzadeh et al., 2020). Reported works on the study of antimicrobial agents used in food-packaging materials are reported in Table 26.3.

Table 26.3 Green biodegradable food packaging with antibacterial activity.

Food packaging material	Antimicrobial agent	Bacteria studied	References
PHBV	ZnO	*E. coli* and *S. aureus*	Díez-Pascual and Diez-Vicente (2014)
PBAT	Electronspun chitosan nanofibers (CS–NF)	*S. aureus*, *B. subtilis*, *S. enteritidis*, and *E. coli*	Díez-Pascual and Díez-Vicente, (2015)
PLA	ZnO	*E. coli* and *S. aureus*	Zhang et al. (2017)
PLA	Durian skin fiber + cinnamon EO	*B. subtilis* and *E. coli*	Anuar et al. (2017)
PLA/chitosan	Rosin modified cellulose nanofiber	*E. coli* and *B. subtilis*	Niu et al. (2018)
PVA	Nanocellulose and Ag	*S. aureus* (MRSA) and *E. coli* (DH5-alpha)	Sarwar et al. (2018)
PVA/Gelatin blend	Lignin	*B. subtilis*, *S. aureus*, *E. coli*, and *P. aeruginosa*	El-Nemr et al. (2020)
Bacterial cellulose nanofibrils (BCNs)	Protein zein nanoparticles (ZNs)	*S. aureus*	Li et al. (2020)

5. Conclusion

Different types of biodegradable polymers such as (1) renewable biopolymers (i.e., from biomass, microorganisms, and biotechnology) and (2) polymers obtained from petrochemical resources have been studied as potential replacements for synthetic non-biodegradable polymers. Furthermore, this chapter has highlighted the increasing interest in using biodegradable polymers and their interest in food-packaging applications.

Nevertheless, although biodegradable polymers act as potential candidates to replace synthetic plastics, they have some serious drawbacks, including: (1) poor thermomechanical properties, (2) hydrophilicity (some bioplastics), and (3) being relatively expensive. Thus, synthetic-based polymers are used in food-packaging applications. Further research is required to use biodegradable polymers in food-packaging applications after improving their important properties such as shelf-life, antimicrobial resistance, nutritional value, and quality. The antimicrobial performance of composites can be enhanced by employing polymer blending and the incorporation of active agents. This chapter concludes that the increasing use of biodegradable polymers could reduce waste in the future; also, they could reduce the adverse effects of greenhouse gases.

References

Abd El-Rahman, K. M., Ali, S. F. A., Khalil, A. I., & Kandil, S. (2020). Influence of poly (butylene succinate) and calcium carbonate nanoparticles on the biodegradability of high density-polyethylene nanocomposites. *Journal of Polymer Research, 27*(8), 1–21.

Agudelo-Cuartas, C., Granda-Restrepo, D., Sobral, P. J., Hernandez, H., & Castro, W. (2020). Characterization of whey protein-based films incorporated with natamycin and nanoemulsion of α-tocopherol. *Heliyon, 6*(4), e03809.

Aljuraifani, A. A., Berekaa, M. M., & Ghazwani, A. A. (2019). Bacterial biopolymer (polyhydroxyalkanoate) production from low-cost sustainable sources. *MicrobiologyOpen, 8*(6), e00755.

Anuar, H., Izzati, A. N. F., Inani, S. S. N., E'zzati, M. S. N., Salimah, A. S. M., Ali, F. B., & Manshor, M. R. (2017). Impregnation of cinnamon essential oil into plasticised polylactic acid biocomposite film for active food packaging. *Journal of Packaging Technology and Research, 1*(3), 149–156.

Arrieta, M. P., López, J., Hernández, A., & Rayón, E. (2014). Ternary PLA–PHB–Limonene blends intended for biodegradable food packaging applications. *European Polymer Journal, 50*, 255–270.

Arvanitoyannis, I., Biliaderis, C. G., Ogawa, H., & Kawasaki, N. (1998). Biodegradable films made from low-density polyethylene (LDPE), rice starch and potato starch for food packaging applications: Part 1. *Carbohydrate Polymers, 36*(2–3), 89–104.

Avella, M., De Vlieger, J. J., Errico, M. E., Fischer, S., Vacca, P., & Volpe, M. G. (2005). Biodegradable starch/clay nanocomposite films for food packaging applications. *Food Chemistry, 93*(3), 467–474.

Aytac, Z., Huang, R., Vaze, N., Xu, T., Eitzer, B. D., Krol, W., MacQueen, L. A., Chang, H., Bousfield, D. W., Chan-Park, M. B., & Ng, K. W. (2020). Development of biodegradable and antimicrobial electrospun zein fibers for food packaging. *ACS Sustainable Chemistry and Engineering, 8*(40), 15354−15365.

Balasubramaniam, S. L., Patel, A. S., & Nayak, B. (2020). Surface modification of cellulose nanofiber film with fatty acids for developing renewable hydrophobic food packaging. *Food Packaging and Shelf Life, 26*, 100587.

Berezina, N., & Maria Martelli, S. (2014). CHAPTER 1:Bio-based polymers and materials. In *Renewable resources for biorefineries* (pp. 1−28). https://doi.org/10.1039/9781782620181-00001

Bharathi, S. K. V., Leena, M. M., Moses, J. A., & Anandharamakrishnan, C. (2020). Nanofibre-based bilayer biopolymer films: enhancement of antioxidant activity and potential for food packaging application. *International Journal of Food Science and Technology, 55*(4), 1477−1484.

Borkotoky, S. S., Ghosh, T., & Katiyar, V. (2020). Biodegradable nanocomposite foams: processing, structure, and properties. In V. Katiyar, A. Kumar, & N. Mulchandani (Eds.), *Advances in sustainable polymers. Materials horizons: from nature to nanomaterials.* Singapore: Springer. https://doi.org/10.1007/978-981-15-1251-3_12

Broumand, A., Emam-Djomeh, Z., Hamedi, M., & Razavi, S. H. (2011). Antimicrobial, water vapour permeability, mechanical and thermal properties of casein based Zataraia multiflora Boiss. Extract containing film. *LWT-Food Science and Technology, 44*(10), 2316−2323.

Castro-Mayorga, J. L., Fabra, M. J., Cabedo, L., & Lagaron, J. M. (2017). On the use of the electrospinning coating technique to produce antimicrobial polyhydroxyalkanoate materials containing in situ-stabilized silver nanoparticles. *Nanomaterials, 7*(1), 4.

Cellulose. In: Encyclopedia britannica. Encyclopedia britannica. Accessed April 29, 2021.

Chen, H., Wang, J., Cheng, Y., Wang, C., Liu, H., Bian, H., Pan, Y., Sun, J., & Han, W. (2019). Application of protein-based films and coatings for food packaging: a Review. *Polymers, 11*(12), 2039.

Chevalier, E., Assezat, G., Prochazka, F., & Oulahal, N. (2018). Development and characterization of a novel edible extruded sheet based on different casein sources and influence of the glycerol concentration. *Food Hydrocolloids, 75*, 182−191.

da Silva, F. T., de Oliveira, J. P., Fonseca, L. M., Bruni, G. P., da Rosa Zavareze, E., & Dias, A. R. G. (2020). Physically cross-linked aerogels based on germinated and non-germinated wheat starch and PEO for application as water absorbers for food packaging. *International Journal of Biological Macromolecules, 155*, 6−13.

da Silva Filipini, G., Romani, V. P., & Guimarães Martins, V. (2021). Blending collagen, methylcellulose, and whey protein in films as a greener alternative for food packaging: physicochemical and biodegradable properties. *Packaging Technology and Science*, 1−13, 2020.

Dash, K. K., Ali, N. A., Das, D., & Mohanta, D. (2019). Thorough evaluation of sweet potato starch and lemon-waste pectin based-edible films with nano-titania inclusions for food packaging applications. *International Journal of Biological Macromolecules, 139*, 449−458.

de Souza, K. C., Correa, L. G., da Silva, T. B. V., Moreira, T. F. M., de Oliveira, A., Sakanaka, L. S., Dias, M. I., Barros, L., Ferreira, I. C., Valderrama, P., & Leimann, F. V. (2020). Soy Protein isolate films incorporated with pinhão (*Araucaria angustifolia* (Bertol.) Kuntze) extract for potential use as edible oil active packaging. *Food and Bioprocess Technology, 13*, 998−1008.

Deng, L., Zhang, X., Li, Y., Que, F., Kang, X., Liu, Y., Feng, F., & Zhang, H. (2018). Characterization of gelatin/zein nanofibers by hybrid electrospinning. *Food Hydrocolloids, 75*, 72–80.

Dias, A. B., Müller, C. M., Larotonda, F. D., & Laurindo, J. B. (2011). Mechanical and barrier properties of composite films based on rice flour and cellulose fibers. *LWT-Food Science and Technology, 44*(2), 535–542.

Díez-Pascual, A. M., & Diez-Vicente, A. L. (2014). ZnO-reinforced poly (3-hydroxybutyrate-co-3-hydroxyvalerate) bionanocomposites with antimicrobial function for food packaging. *ACS Applied Materials and Interfaces, 6*(12), 9822–9834.

Díez-Pascual, A. M., & Díez-Vicente, A. L. (2015). Antimicrobial and sustainable food packaging based on poly (butylene adipate-co-terephthalate) and electrospun chitosan nanofibers. *RSC Advances, 5*(113), 93095–93107.

Djukić-Vuković, A., Mladenović, D., Ivanović, J., Pejin, J., & Mojović, L. (2019). Towards sustainability of lactic acid and poly-lactic acid polymers production. *Renewable and Sustainable Energy Reviews, 108*, 238–252.

El-Nemr, K. F., Mohamed, H. R., Ali, M. A., Fathy, R. M., & Dhmees, A. S. (2020). Polyvinyl alcohol/gelatin irradiated blends filled by lignin as green filler for antimicrobial packaging materials. *International Journal of Environmental Analytical Chemistry, 100*(14), 1578–1602.

El-Wakil, N. A., Hassan, E. A., Abou-Zeid, R. E., & Dufresne, A. (2015). Development of wheat gluten/nanocellulose/titanium dioxide nanocomposites for active food packaging. *Carbohydrate Polymers, 124*, 337–346.

Ferreira, A. R., Alves, V. D., & Coelhoso, I. M. (2016). Polysaccharide-based membranes in food packaging applications. *Membranes, 6*(2), 22.

Ghaderi, M., Mousavi, M., Yousefi, H., & Labbafi, M. (2014). All-cellulose nanocomposite film made from bagasse cellulose nanofibers for food packaging application. *Carbohydrate Polymers, 104*, 59–65.

Ghanbarzadeh, B., & Almasi, H. (2013). *Biodegradable polymers, biodegradation - life of science*. IntechOpen: Rolando Chamy and Francisca Rosenkranz. https://doi.org/10.5772/56230. Available from https://www.intechopen.com/books/biodegradation-life-of-science/biodegradable-polymers.

Guimarães, A. C., Ramos, Ó., Cerqueira, M., Venâncio, A., & Abrunhosa, L. (2020). Active whey protein edible films and coatings incorporating *Lactobacillus buchneri* for Penicillium nordicum control in cheese. *Food and Bioprocess Technology, 13*, 1074–1086 (2020).

Gutiérrez, T. J., & Alvarez, V. A. (2017). Films made by blending poly(ε-caprolactone) with starch and flour from sagu rhizome grown at the Venezuelan Amazons. *Journal of Polymers and the Environment, 25*, 701–716. https://doi.org/10.1007/s10924-016-0861-9

Gutiérrez, T. J., Mendieta, J. R., & Ortega-Toro, R. (2021). In-depth study from gluten/PCL-based food packaging films obtained under reactive extrusion conditions using chrome octanoate as a potential food grade catalyst. *Food Hydrocolloids, 111*, 106255.

Hongsriphan, N., & Sanga, S. (2018). Antibacterial food packaging sheets prepared by coating chitosan on corona-treated extruded poly (lactic acid)/poly (butylene succinate) blends. *Journal of Plastic Film and Sheeting, 34*(2), 160–178.

Ingrao, C., Tricase, C., Cholewa-Wójcik, A., Kawecka, A., Rana, R., & Siracusa, V. (2015). Polylactic acid trays for fresh-food packaging: a Carbon Footprint assessment. *The Science of the Total Environment, 537*, 385–398.

Jafarzadeh, S., Jafari, S. M., Salehabadi, A., Nafchi, A. M., Uthaya, U. S., & Khalil, H. A. (2020). Biodegradable green packaging with antimicrobial functions based on the bioactive compounds from tropical plants and their by-products. *Trends in Food Science & Technology, 100*, 262–277.

Jem, K. J., & Tan, B. (2020). The development and challenges of poly (lactic acid) and poly (glycolic acid). *Advanced Industrial and Engineering Polymer Research, 3*(2), 60–70.

Jha, P. (2020). Effect of plasticizer and antimicrobial agents on functional properties of bionanocomposite films based on corn starch-chitosan for food packaging applications. *International Journal of Biological Macromolecules, 160*, 571–582.

Khalil, H. A., Davoudpour, Y., Saurabh, C. K., Hossain, M. S., Adnan, A. S., Dungani, R., Paridah, M. T., Sarker, M. Z. I., Fazita, M. N., Syakir, M. I., & Haafiz, M. K. M. (2016). A review on nanocellulosic fibres as new material for sustainable packaging: process and applications. *Renewable and Sustainable Energy Reviews, 64*, 823–836.

Khan, B., Bilal Khan Niazi, M., Samin, G., & Jahan, Z. (2017). Thermoplastic starch: a possible biodegradable food packaging material—a review. *Journal of Food Process Engineering, 40*(3), e12447.

Khan, M. R., Sadiq, M. B., & Mehmood, Z. (2020). Development of edible gelatin composite films enriched with polyphenol loaded nanoemulsions as chicken meat packaging material. *CyTA - Journal of Food, 18*(1), 137–146.

Khodayari, M., Basti, A. A., Khanjari, A., Misaghi, A., Kamkar, A., Shotorbani, P. M., & Hamedi, H. (2019). Effect of poly (lactic acid) films incorporated with different concentrations of Tanacetum balsamita essential oil, propolis ethanolic extract and cellulose nanocrystals on shelf life extension of vacuum-packed cooked sausages. *Food Packaging and Shelf Life, 19*, 200–209.

Li, Q., Gao, R., Wang, L., Xu, M., Yuan, Y., Ma, L., Wan, Z., & Yang, X. (2020). Nanocomposites of bacterial cellulose nanofibrils and zein nanoparticles for food packaging. *ACS Applied Nano Materials, 3*(3), 2899–2910.

Liang, S., & Wang, L. (2018). A Natural antibacterial-antioxidant film from soy protein isolate incorporated with cortex phellodendron extract. *Polymers, 10*(1), 71.

Liu, B., Xu, H., Zhao, H., Liu, W., Zhao, L., & Li, Y. (2017). Preparation and characterization of intelligent starch/PVA films for simultaneous colorimetric indication and antimicrobial activity for food packaging applications. *Carbohydrate Polymers, 157*, 842–849.

Liu, X., Song, R., Zhang, W., Qi, C., Zhang, S., & Li, J. (2017). Development of eco-friendly soy protein isolate films with high mechanical properties through HNTS, PVA, and PTGE synergism effect. *Scientific Reports, 7*(1), 1–9.

Lv, S., Gu, J., Tan, H., & Zhang, Y. (2017). The morphology, rheological, and mechanical properties of wood flour/starch/poly (lactic acid) blends. *Journal of Applied Polymer Science, 134*(16).

Marelli, B., Brenckle, M. A., Kaplan, D. L., & Omenetto, F. G. (2016). Silk fibroin as edible coating for perishable food preservation. *Scientific Reports, 6*, 25263.

Moradali, M. F., & Rehm, B. H. (2020). Bacterial biopolymers: from pathogenesis to advanced materials. *Nature Reviews Microbiology, 18*(4), 195–210.

Mülhaupt, R. (2013). Green polymer chemistry and bio-based plastics: dreams and reality. *Macromolecular Chemistry and Physics, 214*(2), 159–174.

Müller, C. M., Laurindo, J. B., & Yamashita, F. (2009). Effect of cellulose fibers addition on the mechanical properties and water vapor barrier of starch-based films. *Food Hydrocolloids, 23*(5), 1328–1333.

Muneer, F., Rasul, I., Azeem, F., Siddique, M. H., Zubair, M., & Nadeem, H. (2020). Microbial polyhydroxyalkanoates (PHAs): efficient replacement of synthetic polymers. *Journal of Polymers and the Environment, 28*, 2301–2323.

Nazrin, A., Sapuan, S. M., Zuhri, M. Y. M., Ilyas, R. A., Syafiq, R., & Sherwani, S. F. K. (2020). Nanocellulose reinforced thermoplastic starch (TPS), polylactic acid (PLA), and polybutylene succinate (PBS) for food packaging applications. *Frontiers in chemistry, 8*.

Nešić, A., Cabrera-Barjas, G., Dimitrijević-Branković, S., Davidović, S., Radovanović, N., & Delattre, C. (2020). Prospect of polysaccharide-based materials as advanced food packaging. *Molecules, 25*(1), 135.

Ngaowthong, C., Borůvka, M., Běhálek, L., Lenfeld, P., Švec, M., Dangtungee, R., Siengchin, S., Rangappa, S. M., & Parameswaranpillai, J. (2019). Recycling of sisal fiber reinforced polypropylene and polylactic acid composites: thermo-mechanical properties, morphology, and water absorption behavior. *Waste Management, 97*, 71–81.

Niranjana Prabhu, T., & Prashantha, K. (2018). A review on present status and future challenges of starch based polymer films and their composites in food packaging applications. *Polymer Composites, 39*(7), 2499–2522.

Niu, X., Liu, Y., Song, Y., Han, J., & Pan, H. (2018). Rosin modified cellulose nanofiber as a reinforcing and co-antimicrobial agents in polylactic acid/chitosan composite film for food packaging. *Carbohydrate Polymers, 183*, 102–109.

Numata, K., Ifuku, N., & Isogai, A. (2018). Silk composite with a fluoropolymer as a water-resistant protein-based material. *Polymers, 10*(4), 459.

Ocak, B. (2020). Properties and characterization of thyme essential oil incorporated collagen hydrolysate films extracted from hide fleshing wastes for active packaging. *Environmental Science & Pollution Research, 27*, 29019–29030.

Oh, Y. A., Roh, S. H., & Min, S. C. (2016). Cold plasma treatments for improvement of the applicability of defatted soybean meal-based edible film in food packaging. *Food Hydrocolloids, 58*, 150–159.

Otenda, B. V., Kareru, P. G., Madivoli, E. S., Maina, E. G., Wanakai, S. I., & Wanyonyi, W. C. (2020). Cellulose nanofibrils from sugarcane bagasse as a reinforcing element in polyvinyl alcohol composite films for food packaging. *Journal of Natural Fibers*, 1–13.

Othman, S. H., Majid, N. A., Tawakkal, I. S. M. A., Basha, R. K., Nordin, N., & Shapi'i, R. A. (2019). Tapioca starch films reinforced with microcrystalline cellulose for potential food packaging application. *Food Science and Technology, 39*(3), 605–612.

Othman, S. H., Othman, N. F. L., Shapi'i, R. A., Ariffin, S. H., & Yunos, K. F. M. (2021). Corn starch/chitosan nanoparticles/thymol bio-nanocomposite films for potential food packaging applications. *Polymers, 13*(3), 390.

Padervand, M., Lichtfouse, E., Robert, D., & Wang, C. (2020). Removal of microplastics from the environment. A review. *Environmental Chemistry Letters*, 1–22.

Palai, B., Mohanty, S., & Nayak, S. K. (2020). Synergistic effect of polylactic acid (PLA) and Poly (butylene succinate-co-adipate)(PBSA) based sustainable, reactive, super toughened eco-composite blown films for flexible packaging applications. *Polymer Testing, 83*, 106130.

Peighambardoust, S. J., Peighambardoust, S. H., Pournasir, N., & Pakdel, P. M. (2019). Properties of active starch-based films incorporating a combination of Ag, ZnO and CuO nanoparticles for potential use in food packaging applications. *Food Packaging and Shelf Life, 22*, 100420.

Picchio, M. L., Linck, Y. G., Monti, G. A., Gugliotta, L. M., Minari, R. J., & Igarzabal, C. I. A. (2018). Casein films crosslinked by tannic acid for food packaging applications. *Food Hydrocolloids, 84*, 424–434.

Piñeros-Hernandez, D., Medina-Jaramillo, C., López-Córdoba, A., & Goyanes, S. (2017). Edible cassava starch films carrying rosemary antioxidant extracts for potential use as active food packaging. *Food Hydrocolloids, 63*, 488–495.

Potrč, S., Fras Zemljič, L., Sterniša, M., Smole Možina, S., & Plohl, O. (2020). Development of biodegradable whey-based laminate functionalised by chitosan–natural extract formulations. *International Journal of Molecular Sciences, 21*(10), 3668.

Qasim, U., Osman, A. I., Ala'a, H., Farrell, C., Al-Abri, M., Ali, M., Vo, D. V. N., Jamil, F., & Rooney, D. W. (2020). Renewable cellulosic nanocomposites for food packaging to avoid fossil fuel plastic pollution: a review. *Environmental Chemistry Letters*, 1–29.

Radoor, S., Karayil, J., Rangappa, S. M., Siengchin, S., & Parameswaranpillai, J. (2020). A review on the extraction of pineapple, sisal and abaca fibers and their use as reinforcement in polymer matrix. *Express Polymer Letters, 14*(4), 309–335.

Ramazani, S., Rostami, M., Raeisi, M., Tabibiazar, M., & Ghorbani, M. (2019). Fabrication of food-grade nanofibers of whey protein Isolate–Guar gum using the electrospinning method. *Food Hydrocolloids, 90*, 99–104.

Rao, M. G., Bharathi, P., & Akila, R. M. (2014). A comprehensive review on biopolymers. *Scientific Reviews and Chemical Communications, 4*(2), 61–68.

Râpă, M., Miteluţ, A. C., Tănase, E. E., Grosu, E., Popescu, P., Popa, M. E., Rosnes, J. T., Sivertsvik, M., Darie-Niţă, R. N., & Vasile, C. (2016). Influence of chitosan on mechanical, thermal, barrier and antimicrobial properties of PLA-biocomposites for food packaging. *Composites Part B: Engineering, 102*, 112–121.

Riveros, C. G., Martin, M. P., Aguirre, A., & Grosso, N. R. (2018). Film preparation with high protein defatted peanut flour: characterisation and potential use as food packaging. *International Journal of Food Science and Technology, 53*(4), 969–975.

Romani, V. P., Olsen, B., Collares, M. P., Oliveira, J. R. M., Prentice-Hernández, C., & Martins, V. G. (2019). Improvement of fish protein films properties for food packaging through glow discharge plasma application. *Food Hydrocolloids, 87*, 970–976.

Rostamzad, H., Paighambari, S. Y., Shabanpour, B., Ojagh, S. M., & Mousavi, S. M. (2016). Improvement of fish protein film with nanoclay and transglutaminase for food packaging. *Food Packaging and Shelf Life, 7*, 1–7.

Rovera, C., Türe, H., Hedenqvist, M. S., & Farris, S. (2020). Water vapor barrier properties of wheat gluten/silica hybrid coatings on paperboard for food packaging applications. *Food Packaging and Shelf Life, 26*, 100561.

Saini, J. K., Saini, R., & Tewari, L. (2015). Lignocellulosic agriculture wastes as biomass feedstocks for second-generation bioethanol production: concepts and recent developments. *3 Biotech, 5*(4), 337–353.

Samantaray, P. K., Little, A., Haddleton, D. M., McNally, T., Tan, B., Sun, Z., Huang, W., Ji, Y., & Wan, C. (2020). Poly (glycolic acid)(PGA): a versatile building block expanding high performance and sustainable bioplastic applications. *Green Chemistry, 22*(13), 4055–4081.

Sarwar, M. S., Niazi, M. B. K., Jahan, Z., Ahmad, T., & Hussain, A. (2018). Preparation and characterization of PVA/nanocellulose/Ag nanocomposite films for antimicrobial food packaging. *Carbohydrate Polymers, 184*, 453–464.

Schmid, M., Merzbacher, S., Brzoska, N., Müller, K., & Jesdinszki, M. (2017). Improvement of food packaging-related properties of whey protein isolate-based nanocomposite films and coatings by addition of montmorillonite nanoplatelets. *Frontiers in Materials, 4*, 35.

Souza, A. G., Ferreira, R. R., Paula, L. C., Mitra, S. K., & Rosa, D. S. (2021). Starch-based films enriched with nanocellulose-stabilized Pickering emulsions containing different essential oils for possible applications in food packaging. *Food Packaging and Shelf Life, 27*, 100615.

Sukan, A., Roy, I., & Keshavarz, T. (2015). Dual production of biopolymers from bacteria. *Carbohydrate Polymers, 126*, 47–51.

Swain, S. K., & Mohanty, F. (2018). Polysaccharides-based bionanocomposites for food packaging applications. In *Bionanocomposites for packaging applications* (pp. 191–208). Cham: Springer.

Tavares, L. B., Ito, N. M., Salvadori, M. C., Dos Santos, D. J., & Rosa, D. S. (2018). PBAT/kraft lignin blend in flexible laminated food packaging: peeling resistance and thermal degradability. *Polymer Testing, 67*, 169–176.

Tayeb, A. H., Amini, E., Ghasemi, S., & Tajvidi, M. (2018). Cellulose nanomaterials—binding properties and applications: a review. *Molecules, 23*(10), 2684.

Thomas, S. K., Begum, P. S., Midhun Dominic, C. D., Salim, N. V., Hameed, N., Rangappa, S. M., Siengchin, S., & Parameswaranpillai, J. (2021). Isolation and characterization of cellulose nanowhiskers from Acacia caesia plant. *Journal of Applied Polymer Science, 138*(15), 50213.

Valdivia-López, M. A., Tecante, A., Granados-Navarrete, S., & Martínez-García, C. (2016). Preparation of modified films with protein from grouper fish. *International Journal of Food Science, 2016*, 3926847. https://doi.org/10.1155/2016/3926847

Varghese, S. A., Siengchin, S., & Parameswaranpillai, J. (2020). Essential oils as antimicrobial agents in biopolymer-based food packaging-A comprehensive review. *Food Bioscience*, 100785.

Vijayendra, S. V. N., & Shamala, T. R. (2014). Film forming microbial biopolymers for commercial applications—a review. *Critical Reviews in Biotechnology, 34*(4), 338–357.

Vinod, A., Sanjay, M. R., Suchart, S., & Jyotishkumar, P. (2020). Renewable and sustainable biobased materials: an assessment on biofibers, biofilms, biopolymers and biocomposites. *Journal of Cleaner Production, 258*, 120978.

Vrabič Brodnjak, U., & Tihole, K. (2020). Chitosan solution containing zein and essential oil as bio based coating on packaging paper. *Coatings, 10*(5), 497.

Wesołowska-Trojanowska, M., Tomczyńska-Mleko, M., Terpiłowski, K., Muszyński, S., Nishinari, K., Nastaj, M., & Mleko, S. (2020). Co-gelation of gluten and gelatin as a novel functional material formation method. *Journal of Food Science & Technology, 57*(1), 163–172.

Wu, X., Luo, Y., Liu, Q., Jiang, S., & Mu, G. (2019). Improved structure-stability and packaging characters of crosslinked collagen fiber-based film with casein, keratin and SPI. *Journal of the Science of Food and Agriculture, 99*(11), 4942–4951.

Xu, P., Yang, W., Niu, D., Yu, M., Du, M., Dong, W., Chen, M., Lemstra, P. J., & Ma, P. (2020). Multifunctional and robust polyhydroxyalkanoate nanocomposites with superior gas barrier, heat resistant and inherent antibacterial performances. *Chemical Engineering Journal, 382*, 122864.

Zhang, H., Hortal, M., Jordá-Beneyto, M., Rosa, E., Lara-Lledo, M., & Lorente, I. (2017). ZnO-PLA nanocomposite coated paper for antimicrobial packaging application. *Lebensmittel-Wissenschaft & Technologie, 78*, 250–257.

Zhong, Y., Godwin, P., Jin, Y., & Xiao, H. (2020). Biodegradable polymers and green-based antimicrobial packaging materials: a mini-review. *Advanced Industrial and Engineering Polymer Research, 3*(1), 27–35.

Zubair, M., & Ullah, A. (2020). Recent advances in protein derived bionanocomposites for food packaging applications. *Critical Reviews in Food Science and Nutrition, 60*(3), 406–434.

Index

Note: 'Page numbers followed by "f" indicate figures and "t" indicate tables.'

Printed in the United States
by Baker & Taylor Publisher Services